67.00 100K

Capture Gamma-Ray Spectroscopy and Related Topics-1984
(International Symposium, Knoxville, Tennessee)

Fifth International Symposium on Capture Gamma-Ray Spectroscopy and Related Topics
September 10-14, 1984
Knoxville, Tennessee, USA

Sponsored by the

American Physical Society
American Nuclear Society
European Physical Society
U.S. Department of Energy
U.S. National Science Foundation
Oak Ridge National Laboratory

AIP Conference Proceedings
Series Editor: Hugh C. Wolfe
Number 125

Capture Gamma-Ray Spectroscopy and Related Topics-1984
(International Symposium, Knoxville, Tennessee)

Edited by
S. Raman
Oak Ridge National Laboratory

American Institute of Physics
New York   1985

These Proceedings were prepared for publication at the Oak Ridge National Laboratory, Oak Ridge, Tennessee 37831. This Laboratory is operated by the Martin Marietta Energy Systems, Inc., for the U.S. Department of Energy under Contract No. DE-AC05-84OR21400.

Travel funds for some invited speakers were provided by the U.S. National Science Foundation under Agreement No. PHY-8412762. Any opinion, findings, and conclusions or recommendations expressed in this publication are those of the author(s) and do not necessarily reflect the views of the National Science Foundation.

Copying fees: The code at the bottom of the first page of each article in this volume gives the fee for each copy of the article made beyond the free copying permitted under the 1978 US Copyright Law. (See also the statement following "Copyright" below). This fee can be paid to the American Institute of Physics through the Copyright Clearance Center, Inc., Box 765, Schenectady, N.Y. 12301.

Copyright © 1985 American Institute of Physics

Individual readers of this volume and non-profit libraries, acting for them, are permitted to make fair use of the material in it, such as copying an article for use in teaching or research. Permission is granted to quote from this volume in scientific work with the customary acknowldgment of the source. To reprint a figure, table or other excerpt requires the consent of one of the original authors and notification to AIP. Republication or systematic or multiple reproduction of any material in this volume is permitted only under license from AIP. Address inquiries to Series Editor, AIP Conference Proceedings, AIP, 335 E. 45th St., New York, N. Y. 10017.

L.C. Catalog Card No. 84-73303
ISBN 0-88318-324-2

DOE CONF - 840906

Preface

The Fifth International Symposium on Capture Gamma-Ray Spectroscopy and Related Topics was held at the Holiday Inn-World's Fair, Knoxville, Tennessee, September 10-14, 1984. This symposium was a sequel to meetings held at Studsvik, Sweden (1969), Petten, the Netherlands (1974), Brookhaven, USA (1978), and Grenoble, France (1981). It was sponsored by the American Physical Society, the American Nuclear Society, the European Physical Society, the U.S. Department of Energy, the U.S. National Science Foundation, and the Oak Ridge National Laboratory. This symposium was the first conference in the United States that the European Physical Society has ever sponsored.

The 5-day symposium was attended by 208 participants from 25 countries. The primary aims of the conference organizers were to maximize the interchange of ideas and discussion among the attendees and to minimize distractions. The truly international character of the meeting, its size, the high quality of the talks, the posters, the industrial exhibition, and even the weather all contributed toward making this symposium a very fruitful scientific gathering.

The main topics discussed at the symposium were nuclear models and theory ($\sim$ 11% of the contributions), photon-induced reactions ($\sim$ 11%), neutron-induced reactions ($\sim$ 38%), charged-particle-induced reactions ($\sim$ 14%), applications, including nucleosynthesis ($\sim$ 13%), and related topics ($\sim$ 13%). The Program Selection Committee, with generous input from the International Advisory Committee and the Program Advisory Committee, was responsible for the scientific program. The members of those Committees (and of the Organizing Committee) are listed below.

International Advisory Committee

G. Alaga (Yugoslavia), A. Arima (Japan), G. A. Bartholomew (Canada), G. E. Brown (USA), P. M. Endt (Holland), H. Feshbach (USA), W. A. Fowler (USA), S. S. Hanna (USA), A. Kerman (USA), A. M. Lane (UK), Li Shou-nan (PRC), C. Mahaux (Belgium), B. Mottelson (Denmark), V. G. Soloviev (USSR), A. H. Wapstra (Holland), and H. A. Weidenmüller (FRG).

Program Advisory Committee

V. L. Alexeev (USSR), B. J. Allen (Australia), C. V. K. Baba (India), F. Bečvář (Czechoslovakia), I. Bergqvist (Sweden), H. Börner (France), R. F. Casten (USA), C. Coceva (Italy), F. S. Dietrich (USA), Ding Dazhao (PRC), D. M. Drake (USA), H. Ejiri (Japan), C. D. Goodman (USA), R. W. Hoff (USA), F. Iachello (USA), H. E. Jackson, Jr. (USA), S. Joly (France), E. T. Jurney (USA), F. Käppeler (FRG), H. V. Klapdor (FRG), M. A. Lone (Canada), R. Moreh (Israel), M. V. Pasechnik (USSR), Yu. P. Popov (USSR), P. Prokofjev (USSR), D. Seeliger (GDR), O. W. B. Schult (FRG), and H. Weigmann (Belgium).

Program Selection Committee

K. Abrahams (Holland), B. Castel (Canada), R. E. Chrien (USA), T. von Egidy (FRG), J. A. Harvey (USA), J. E. Lynn (UK), N. R. Roberson (USA), C. Rolfs (FRG), K. A. Snover (USA), P. H. Stelson (USA), C. Van der Leun (Holland), and E. K. Warburton (USA).

Local Organizing Committee

S. Raman (Chairman), R. F. Carlton, S. R. Damewood, Y. A. Ellis-Akovali, N. W. Hill, B. F. McHargue, S. A. Raby, B. S. Reesor, J. K. Thacker, C. R. Vane, and R. T. Webber.

The symposium comprised 46 invited talks arranged into 9 sessions. Having decided against parallel sessions of orally presented contributed papers, the Organizing Committee arranged for two poster sessions to last a total of 7 hours; all 87 accepted contributed papers could be presented as posters. The posters were displayed in a ballroom where people congregated for refreshments. This arrangement resulted in valuable discussions among the interested parties. Attendance was high at all sessions of invited speakers. The chairpersons of the sessions (J. B. Ball, F. Iachello, G. F. Bertsch, O. W. B. Schult, C. Coceva, A. H. Snell, B. Castel, C. van der Leun, W. S. Rodney, and R. E. Chrien) deserve praise not only for keeping the sessions precisely on schedule but also for their guidance of the discussions, which were stimulating and lively.

An industrial exhibition was also held in connection with the symposium. Bicron, Canberra, EG&G Ortec, Harshaw, Nuclear Data, Nucleus, and Tennelec participated. Like plumbers and carpenters, physicists depend on the tools of their trade, and the exhibition served the purpose of bringing together industrial representatives and scientists seeking sophisticated instrumentation. The welcoming speeches at the symposium were delivered by K. C. Testerman (Mayor of Knoxville) and R. S. Wiltshire (Executive Director, ORNL).

These proceedings contain the texts of all papers presented at the conference. The material has been reproduced directly from camera-ready copy provided by the authors. Because rapid publication of the proceedings was my primary goal, I made no attempt to correct all errors, especially the typographical ones that left the meaning intact. The papers have been grouped loosely by subject, in somewhat different order from that in the conference program.

I want to thank the members of the International Advisory Committee, the Program Advisory Committee, and the Program Selection Committee. They worked hard for several months — carefully defining the topics, diligently reviewing the submitted materials, and objectively selecting the speakers. I would also like to acknowledge the important role fulfilled by the members of the Organizing Committee, who labored behind the scenes to prepare the announcements; to handle the registration, accommodations, and transportation; and to arrange the exhibition, poster session, audiovisual aids, social events, and tours. I wish to express special thanks to Bonnie Reesor for the long hours and multiple responsibilities undertaken by her as ORNL coördinator for this symposium.

Anyone (myself included) who thinks that conference fees are excessive is easily cured of this illusion by actually running a conference. I thank those whose financial support made its success possible, including the companies that participated in the exhibition and, especially, EG&G Ortec, the U.S. National Science Foundation, the U.S. Department of Energy, and the Oak Ridge National Laboratory.

My final thanks go to the speakers, the after-dinner speaker (B. B. Kinsey), the summary speaker (J. E. Lynn), and all participants for their active roles which made the symposium well worth the effort. While still maintaining the general spirit and intent of the earlier conferences by providing a forum for the discussion of neutron capture γ-ray spectroscopy, this symposium was much broader in scope. This innovation, I strongly believe, was a key to the conference's success, and I recommend it to the organizers of the Sixth Symposium (Leuven, Belgium) for their earnest consideration.

Oak Ridge, Tennessee   November 1, 1984   S. Raman

TABLE OF CONTENTS

I. NUCLEAR MODELS AND THEORY

TALKS

The Interacting Boson Model - a Review - **J. P. Elliott** ... 3

Microscopically Derived Interacting Boson Model - **T. Otsuka** ... 10

Axial Asymmetry in the IBA and an Extensive New O(6) Region Near A=130 - **R. F. Casten** ... 24

Anharmonicities in the Vibrational Spectra in Deformed Nuclei - **R. Piepenbring** ... 38

Nuclear Dynamical Supersymmetry - **Da Hsuan Feng** ... 48

New Coupling Limits, Dynamical Symmetries and Microscopic Operators of IBM/TQM - **V. Paar** ... 70

Shell-Model Predictions of Broad Trends and Fine Details of Electromagnetic Matrix Elements - **B. H. Wildenthal** ... 89

POSTERS

Local Properties of Interacting Bosons - **V. R. Manfredi** ... 103

Mixed Symmetry States in the Vibrational Limit of the IBA Model - **W. D. Hamilton, A. Irbäck, and J. P. Elliott** ... 106

Collective Magnetic Octopole Transitions - **O. Scholten** ... 109

Generalized-Seniority Mixing in Semi-Magic Nuclei - **G. Bonsignori, M. Savoia, and K. Allaart** ... 113

Configuration Dependent Pairing from the Effective Decoupling Picture - **A. K. Jain and K. Jain** ... 117

Intrinsic Excitations in Doubly Odd Nuclei - **P. C. Sood** ... 121

Shell Effects on the E1 Moments of Ra-Th Nuclei - **G. A. Leander** ... 125

II. PHOTON-INDUCED REACTIONS

TALKS

The Absorption and Scattering of Photons by the Δ Resonance - **E. Hayward** ... 131

Photon Scattering Above the Giant Dipole Resonance - **A. M. Nathan** ... 142

Nuclear Structure Studies with Photoreactions - **F. Iachello** ... 153

II. PHOTON-INDUCED REACTIONS (continued)

TALKS (continued)

Recent Results of the Göttingen-Grenoble Photon Scattering Collaboration -
M. Schumacher 166

Recent Developments in Nuclear Scattering of Capture Gamma-Rays -
O. Shahal and R. Moreh 179

Proton-Capture Gamma Rays as a Spectroscopic Tool - Ph. B. Smith 192

POSTERS

Measurement of the Photodisintegration of Deuterium - Y. Birenbaum,
R. Moreh, and S. Kahane 208

Angular Distributions for the $^2H(\gamma,n)$ Reaction Below 10 MeV - A. Wolf,
Z. Berant, Y. Birenbaum, S. Kahane, and R. Moreh 213

Resonant Absorption of Deuteron-Capture γ-Rays - F. Zijderhand,
S. W. Kikstra, S. S. Hanna, and C. van der Leun 217

Comparison of Photoabsorption by ^{16}O and ^{18}O - N. K. Sherman, W. F. Davidson,
S. Raman, W. Delbianco, and G. Kajyrs 221

Fore-Aft Asymmetry in Polarized Photoreactions - F. Saporetti and R. Guidotti ... 225

*Interacting Boson Description of the Giant Dipole Resonance in the
Lanthanide Region -* L. Zuffi, G. Maino, and A. Ventura 228

Study of E1-E2 Interference in the $^{159}Tb(\gamma,n)$ and $^{209}Bi(\gamma,n)$ Reactions -
S. Kahane, Y. Birenbaum, Z. Berant, R. Moreh, and A. Wolf 232

The $^{232}Th(\gamma,f)$ and (γ,n) Reactions Between 5 and 10 MeV - D. J. S. Findlay,
G. Edwards, N. P. Hawkes, and M. R. Sené 237

High Resolution Studies of Photofission of ^{232}Th and ^{238}U - H. X. Zhang and
H. Lancman 241

III. SLOW NEUTRONS

TALKS

Tests of Supersymmetries with the (n,γ) Reaction - D. D. Warner 247

Spectroscopy of Odd-Neutron Actinide Nuclei - H. G. Börner 262

Modeling Level Structures of Odd-Odd Deformed Nuclei - R. W. Hoff,
J. Kern, R. Piepenbring, and J. Boisson 274

III. SLOW NEUTRONS (continued)

TALKS (continued)

Investigation of E0 Transition Rates - G. G. Colvin and K. Schreckenbach 290

Structural and Statistical Aspects of Extensive Level Schemes from (n,γ) and Transfer Reactions - T. von Egidy, P. Hungerford, H. H. Schmidt, H. J. Scheerer, A. N. Behkami, G. Hlawatsch, B. Krusche, K. P. Lieb, H. G. Börner, S. A. Kerr, and K. Schreckenbach 305

Averaged Neutron Capture Data and Their Applications to Nuclear Structure and Reaction Mechanism - J. Kopecky 318

Some Open Problems in Neutron Capture γ-Rays in Rotational Nuclei - M. Stefanon 335

Nonstatistical Effects of Neutron Radiative Capture in Deformed Nuclei - F. Bečvář 345

Channel Wave Function and Channel Radiative Capture Cross Sections - Yu-Kun Ho and M. A. Lone 362

POSTERS

Study of Thermal Neutron Capture by ^{32}S - Guo Taichang, Shi Zongren, Zeng Xiantang, Li Guohua, and Ding Dazhao 376

^{74}Ge: Transitions and Levels Excited in Thermal-Neutron Capture - C. Hofmeyr, C. Franklyn, G. Barreau, H. Börner, R. Brissot, H. Faust, and K. Schreckenbach 378

Low-Lying States of ^{94}Nb - M. Bogdanović, H. Seyfarth, H. Börner, S. Kerr, F. Hoyler, K. Schreckenbach, and G. Colvin 382

Properties of Low-Lying States in ^{134}Cs - M. Bogdanović, R. Brissot, G. Barreau, K. Schreckenbach, S. Kerr, H. Börner, I. A. Kondurov, Yu. E. Loginov, V. V. Martynov, P. A. Sushkov, H. Seyfarth, T. von Egidy, P. Hungerford, H. H. Schmidt, H. J. Scheerer, A. Chalupka, and W. Kane 386

Multipolarity of Gamma Transitions in ^{140}La Produced in the $^{139}La(n,\gamma)^{140}La$ Reaction - M. P. Stojanović, J. Simić, K. Schreckenbach, and G. Colvin 390

^{144}Nd, ^{165}Dy, and ^{175}Yb Level Schemes from the (n,2γ) Reaction - V. A. Khitrov, Yu. P. Popov, A. M. Sukhovoj, and Yu. S. Yazvitsky 396

Average Intensities of Two-Quanta Cascades in ^{144}Nd, ^{165}Dy, and ^{175}Yb After the Capture of Thermal Neutrons - V. A. Khitrov, Yu. P. Popov, A. M. Sukhovoj, and Yu S. Yazvitsky 399

A Complete Identification by the Particle-Rotor Model of ^{153}Gd States up to 1 MeV - A. M. J. Spits, P. H. M. Van Assche, H. G. Börner, W. F. Davidson, D. D. Warner, K. Schreckenbach, G. Colvin, R. C. Greenwood, and C. W. Reich ... 403

III. SLOW NEUTRONS (continued)

POSTERS (continued)

Single Particle and Vibrational Bands in $^{155}Gd, ^{161}Dy,$ and ^{163}Dy -
H. H. Schmidt, P. Hungerford, T. von Egidy, H. J. Scheerer, H. G. Börner, S. A. Kerr,
K. Schreckenbach, F. Hoyler, G. Colvin, R. F. Casten, D. D. Warner, and W. Kane 406

*Investigation of Inter- and Intraband Transitions of the $0_2^+, 2_1^+, 0_3^+$ Bands
in ^{156}Gd* - F. Hoyler, K. Schreckenbach, H. G. Börner, and G. Colvin 410

Energy Levels of ^{164}Dy From the Reaction $^{163}Dy(n,\gamma)$ - T. J. Al-Janabi,
A. K. Mheemeed, S. S. Kamoon, and S. T. Ahmed 414

The Level Scheme of ^{166}Dy Obtained by Double Neutron Capture - S. A. Kerr,
F. Hoyler, K. Schreckenbach, H. G. Börner, G. Colvin, P. H. M. Van Assche,
and E. Kaerts 416

M1-E2 Mixing Ratios of Gamma-Ground Transitions in ^{168}Er - W. Gelletly,
D. D. Warner, G. C. Colvin, and K. Schreckenbach 420

*A Study of the Low-Lying States in ^{178}Hf Through the Neutron Capture
Reaction* - A. M. I. Haque, R. Richter, A. Gelberg, I. Förster, R. Rascher,
P. von Brentano, H. G. Börner, K. Schreckenbach, S. A. Kerr, G. Barreau,
R. Brissot, R. F. Casten, and D. D. Warner 423

*High Precision Neutron Capture Gamma-Ray and Conversion Electron
Measurements of ^{181}Ta* - I. Förster, H. G. Börner, P. von Brentano, G. Colvin,
A. M. I. Haque, S. A. Kerr, R. Rascher, R. Richter, and K. Schreckenbach 427

E2/M1 Mixing Ratios in ^{195}Pt from the $^{195}Pt(n,\gamma)^{196}Pt$ Reaction -
A. M. Bruce and D. D. Warner 431

Average Resonance Capture Studies of ^{102}Ru - Z. -R. Shi, R. F. Casten,
J. Stachel, and A. M. Bruce 435

On Simulation of Nonstatistical Effects in Neutron Resonance Gamma-Decay -
V. A. Knat'ko 439

IV. FAST NEUTRONS

TALKS

Radiative Capture of Nucleons in the Giant-Dipole Resonance Region -
F. S. Dietrich 445

Giant Isovector E2 Resonances Observed in (n,γ) and (γ,n) Reactions -
L. Nilsson 458

Radiative Capture in Few Nucleon Systems - H. R. Weller 470

IV. FAST NEUTRONS (continued)

TALKS (continued)

Fast Neutron Capture with a White Neutron Source - S. Wender and G. F. Auchampaugh ... 483

POSTERS

Observation of Extremely Low s-Wave Strength in the Reaction $^{136}Xe + n$ - B. Fogelberg, J. Harvey, M. Mizumoto, and S. Raman ... 493

Location of a Doorway State Using the Channel $n + \,^{207}Pb$ - L. C. Dennis and S. Raman ... 495

Neutron Strength Functions of $^{205,209}Pb$ - V. G. Soloviev, V. V. Voronov, and Ch. Stoyanov ... 499

Valence Neutron Capture in s- and p-Wave Resonances in ^{32}S - B. J. Allen and F. Z. Company ... 501

Evidence for Valence Transitions in Neutron Capture Gamma-Ray Spectra in ^{88}Sr - B. J. Allen and F. Z. Company ... 505

Gamma-Ray Strength Functions in ^{139}La and ^{141}Pr - B. J. Allen and F. Z. Company ... 509

Absolute Dipole Gamma-Ray Strength Functions for ^{176}Lu - D. G. Gardner, M. A. Gardner, and R. W. Hoff ... 513

Gamma Rays from 565-keV p-Wave Neutron Resonance and Off-Resonance Capture by ^{28}Si - M. Shimizu, M. Igashira, H. Komano, and H. Kitazawa ... 517

Excitation of Isovector M1 States in p-Wave Neutron Resonance Capture Reactions on ^{56}Fe - H. Kitazawa, H. Komano, M. Igashira, and M. Shimizu ... 520

Neutron Capture Gamma-Ray Spectra of Nuclei with $N = 82 - 118$ at the Neutron Energies of 10 - 800 keV - M. Igashira, M. Shimizu, H. Komano, H. Kitazawa, and N. Yamamuro ... 523

Investigation of Compound and Direct-Semidirect Interference Effects in the $^{89}Y(n,\gamma_0 + \gamma_1)^{90}Y$ Reaction - S. Joly ... 526

Analysis of $n + \,^{197}Au$ Cross Sections for $E_n = 0.01$-20 MeV - P. G. Young and E. D. Arthur ... 530

Level Structure of Quasi-Magic ^{96}Zr - G. Molnár, B. Fazekas, T. Belgya, and A. Veres ... 534

Decay Scheme of ^{116}Sn from $(n,n'\gamma)$ and (n,γ) Results - Z. Gácsi, J. Sa, J. L. Weil, E. T. Jurney, and S. Raman ... 539

IV. FAST NEUTRONS (continued)

POSTERS (continued)

Low-Lying Low-Spin States in ^{136}Ba Studied via $(n,n'\gamma)$ Reaction -
I. Diószegi, Cs. Maráczy, and Á. Veres 542

Angular Distribution Studies of Gamma Rays From the $^{172}Yb(n,n'\gamma)$ Reaction -
H. M. Youhana, M. A. Al-Amili, S. R. Al-Obeidi, and H. E. Abid 545

Isomer Ratio Calculations Using Modeled Discrete Levels - M. A. Gardner,
D. G. Gardner, and R. W. Hoff 547

An Estimate of the Inelastic Channel Neutron Radiative Capture Cross
Sections - J. F. Liu, Yu-Kun Ho, and M. A. Lone 551

V. CHARGED PARTICLES

TALKS

Use of Capture Reactions to Measure Short Lifetimes by the DSA Method -
J. Keinonen 557

Medium Energy Proton and Helium-3 Capture in Light Nuclei - S. L. Blatt 570

Medium Energy Gamma Rays Following Radiative Capture of Polarized Protons
on Light Nuclei - H. Ejiri, T. Shibata, Y. Nagai, T. Kishimoto, H. Ohsumi,
N. Kamikubota, and T. Satoh 582

Recent Results from Proton Capture to Giant Resonances Built on
Highly-Excited States - D. H. Dowell 597

Nuclear Spectroscopy Using TESSA Following Heavy Ion Fusion Reactions -
P. J. Nolan 612

Crystal Ball Studies of Giant Resonance Gamma Decay - J. R. Beene,
F. E. Bertrand, and M. L. Halbert 623

Observation of Radiative Capture in ^{90}Zr-Induced Fusion Reactions -
H.-G. Clerc, C. -C. Sahm, E. Tschöp, W. Schwab, K. -H. Schmidt, R. S. Simon,
J. -G. Keller, W. Reisdorf, F. Heßberger, G. Münzenberg, and B. Quint 636

The Role of Giant Resonances in Heavy-Ion Radiative Capture - A. M. Sandorfi . 647

High Energy Gamma Rays from Complex Particle Collisions - K. A. Snover 660

POSTERS

The (p,γ) Reaction on Thick Targets as a Spectroscopic Tool -
T. Paradellis 677

V. CHARGED PARTICLES (continued)

POSTERS (continued)

Analysis of the $^{11}B(p,\gamma_0)^{12}C$ Reaction in Terms of the Direct-Semidirect Model - M. Kicińska-Habior and T. Matulewicz 680

Description of Subbarrier Resonances in Radiative Proton Capture Reaction by the Effective Potential Approach - T. Matulewicz, P. Decowski, M. Kicińska-Habior, B. Sikora, and J. Töke 684

Fragmentation of the $1g_{9/2}$ Isobaric Analog Resonance in ^{53}Mn - J. Sziklai, J. A. Cameron, and I. M. Szöghy 688

Total (p,n), (p,γ), $(p,p'\gamma)$ and Differential (p,p) Cross-Section Measurements for $^{61,64}Ni$ - R. L. Hershberger, F. Gabbard, and C. E. Laird 692

The Proton-^{90}Zr Interaction at Sub-Coulomb Proton Energies - C. E. Laird, D. Flynn, R. L. Hershberger, and F. Gabbard 695

High Resolution In-Beam γ-Ray Spectroscopy - J. Kern, J.-Cl. Dousse, M. Gasser, B. Perny, and Ch. Rhême 698

Nuclear Spectroscopy with Few-Nucleon Transfer Reactions on Light Nuclei - H. J. Hauser, T. Rohwer, F. Hoyler, G. Staudt, S. Abd el-Kariem, P. Grasshoff, H. V. Klapdor, A. Körber, W. Leitner, V. Rapp, M. Walz, and D. Weinmann 701

Heavy Ion Radiative Capture of $^{12}C + ^{14}C$ into ^{26}Mg - L. Ricken, A. M. Sandorfi, D. H. Dowell, and P. Paul 705

A Study of Cerium and Neodymium Nuclei Close to the Proton Drip Line - B. J. Varley, R. Moscrop, S. Babkair, C. J. Lister, W. Gelletly, and H. G. Price 709

VI. NUCLEOSYNTHESIS

TALKS

Neutron Capture Reactions in Astrophysics - F. Käppeler 715

Capture Processes and Element Synthesis in the Universe - H. V. Klapdor 732

Status of Helium Burning of ^{12}C - H. P. Trautvetter, A. Redder, and C. Rolfs 748

Some Effects of High Temperature and Density on Neutron-Capture Nucleosynthesis - E. B. Norman and S. E. Kellogg 753

VI. NUCLEOSYNTHESIS (continued)

POSTERS

Dynamic Stellar Neutron Capture Nucleosynthesis: the Need for More Nuclear Data for the s-Process - G. J. Mathews, W. M. Howard, K. Takahashi, and R. A. Ward 766

Neutron Capture and Total Cross Sections for ^{48}Ca: Astrophysical Implications - R. F. Carlton, J. A. Harvey, N. W. Hill, and R. L. Macklin 774

The ^{163}Dy-^{163}Ho Branching: An s-Process Barometer - H. Beer, G. Walter, and R. L. Macklin 778

Target Thermalization Effect in ^{187}Os Neutron Capture - G. Reffo 782

Indirect Investigation of Proton Capture Reactions at Stellar Energies - P. Schmalbrock, T. R. Donoghue, H. J. Hausman, M. Wiescher, V. Wijekumar, C. P. Browne, A. A. Rollefson, and C. Rolfs 785

Thermonuclear Reaction Rates from (p,n) Reaction - S. Kailas 789

Neutrino-Capture and Related Problems: the Capture Cross Section in $^{69,71}Ga$ - K. Grotz, H. V. Klapdor, and J. Metzinger 793

VII. APPLICATIONS

TALKS

In Situ Neutron Capture Spectroscopy of Geological Formations - J. A. Grau, S. Antkiw, R. C. Hertzog, R. A. Manente, and J. S. Schweitzer 799

Analytical Neutron-Capture Gamma-Ray Spectroscopy: Status and Prospects - R. M. Lindstrom and D. L. Anderson 810

POSTERS

In Vivo Determination of Protein By Prompt Neutron Capture in Fibrocystic Disease - B. J. Allen, N. Blagojevic, K. Gaskin, V. Soutter, and R. Howman-Giles 820

In Situ Neutron-Induced Spectroscopy of Geological Formations with Germanium Detectors - J. S. Schweitzer and R. A. Manente 824

Physics of Recent Applications of PGNAA for On-Line Analysis of Bulk Minerals - T. Gozani 828

Nuclear Resonance-Fluorescence Analyser of Ores and Surfaces - J. C. Palathingal 847

VII. APPLICATIONS (continued)

POSTERS (continued)

Ambiguities in PIGE Caused By Different Reactions - A. Z. Kiss, E. Koltay, B. Nyakó, E. Somorjai, A. Anttila, and J. Räisänen 851

VIII. RELATED TOPICS

TALKS

Coincidence Electron Scattering (e,e'f) and Multipole Strength Functions in ^{238}U - K. A. Griffioen, P. J. Countryman, K. T. Knöpfle, K. Van Bibber, M. R. Yearian, J. G. Woodworth, D. Rowley, and J. R. Calarco 857

Sensitive Search for Neutron-Antineutron Transitions at the ILL Reactor - M. Baldo-Ceolin 871

The Beta Decay of Polarized Neutrons - P. Bopp, D. Dubbers, L. Hornig, E. Klemt, J. Last, H. Schütze, S. J. Freedman, and Otto Schärpf 881

POSTERS

Statistical Methods of Spin Assignment in Compound Nuclear Reactions - H. Mach and M. W. Johns 889

Two Photons Correlated Production at the 25 MWTh Reactor - H. Abramowicz, K. Doroba, R. Walczak, M. Górski, A. Jasiński, T. Kozlowski, W. Ratyński, M. Szeptycka, M. Szymczak, and A. Tucholski 893

An Investigation of Parity-Non-Conservation in the $^{10}B(n,\alpha)$ Reaction - F. Stecher-Rasmussen, P. J. Kok, O. N. Ermakov, J. L. Karpikhin, P. A. Krupchitsky, G. A. Lobov, and V. F. Perepelitsa 897

Parity Non-Conservation in Resonance Interaction of Polarized Nucleons with Nuclei - G. A. Lobov 900

Investigation of p-Wave Neutron Resonance Near Binding Energy with Laser Radiation - Yu-Kun Ho, F. C. Khanna, and M. A. Lone 904

Investigation of Capture Reactions Far Off Stability by β-Delayed Neutron Emission - M. Wiescher, B. Leist, W. Ziegert, H. Gabelmann, B. Steinmüller, H. Ohm, K.-L. Kratz, F.-K. Thielemann, and W. Hillebrandt 908

Direct Measurement of Natural Line Widths in Delayed-Neutron Energy Spectra - R. D. McElroy, D. D. Clark, R. L. Gill, and A. Piotrowski 912

On the Limit Resolution of a Curved-Crystal Gamma-Ray Spectrometer - V. L. Alexeev, E. K. Leushkin, L. I. Molkanov, and V. L. Rumiantsev 916

Crystal Reflectivity for Gamma Rays - E. Kaerts and P. H. M. Van Assche 918

VIII. RELATED TOPICS (continued)

POSTERS (continued)

Accurate Determination of Gamma-Ray Energies for $E \leq 2$ MeV -
E. G. Kessler, Jr., G. L. Greene, R. D. Deslattes, and H. G. Börner 921

*Study of Radiative Capture Gamma-Rays Arising from Neutron Interactions
with Iron Barriers -* **R. M. A. Maayouf and A. S. Makarious** 925

Replacement of the Vessel and Beam Tubes of the High Flux Reactor in Petten -
K. Abrahams, F. Stecher-Rasmussen, M. J. W. Weel, and the Video Team JRC 930

Upgrading the Beam Facilities of the High Flux Reactor in Petten -
F. Stecher-Rasmussen 933

IX. AFTER-DINNER SPEECH

(n,γ) Research in the Fifties - **B. B. Kinsey** 939

X. CONFERENCE SUMMARY

Concluding Remarks - **J. E. Lynn** 951

LIST OF PARTICIPANTS 965

AUTHOR INDEX 981

I. NUCLEAR MODELS AND THEORY

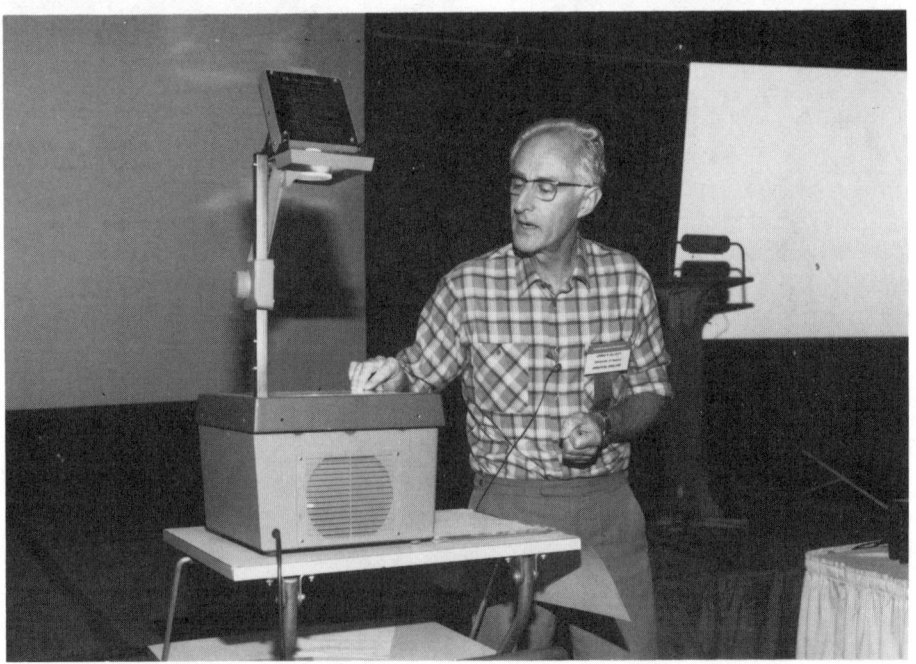

THE INTERACTING BOSON MODEL - A REVIEW

J P Elliott
School of Mathematical and Physical Sciences
University of Sussex, Brighton BN1 9QH, UK

ABSTRACT

The interacting boson model is first defined mathematically and then its two possible physical interpretations are discussed. The role of the neutron/proton degree of freedom is described. Brief mention is made of the extension to odd nuclei and finally an example of the mapping from a shell model calculation onto the IBM is given.

INTRODUCTION

It is my intention in this short talk to review the development of the IBM since it was first proposed[1] in 1975. I shall make no attempt to describe every latest version and application of the model. I hope however to give a clear definition of the model, to explain its relation to both the collective model and the shell model and to list its strengths and weaknesses. There are now a number of published reviews[2,3] which contain more detail than I can fit into 35 minutes and a more complete set of references.

THE MODEL (IBM 1)

The simplest version of the interacting boson model, referred to as IBM 1, is defined as a system of N bosons each of which may have angular momentum 0 or 2 (called s or d bosons) and which interact with one-body and two-body interactions. Extensions of the model are possible by including three-body interactions or additional types of boson. The expression Interacting Boson Appromixation (IBA) is sometimes used but it means precisely the same as IBM. As defined above, the IBM is a finite system in which the interaction Hamiltonian contains a small number (nine) of parameters. The calculation of matrix elements of the Hamiltonian in the N-boson states presents no difficulty and the matrices used in realistic applications have been of reasonable size - much smaller than those which occur in a shell-model calculation. The model does not therefore have any technical problems and a convenient computer program (PHINT) has been written by O Scholten which produces energies and transition rates for given values of N and the nine interaction parameters.

PHYSICAL INTERPRETATION

A confusing aspect of the IBM is that there are two quite different physical interpretations of the model which may or may not be related.

(i) In the first interpretation the s-boson has no physical significance (but is a mathematical device) while an s - d transition corresponds to the excitation of a quadrupole phonon in a five-dimensional collective surface model with the usual coordinates α_μ in the expression

$$R = R_o(1 + \sum_\mu \alpha_\mu Y^*_{2\mu}(\theta,\phi))$$

for the surface. It can be shown rigorously that an IBM 1 calculation with N bosons is equivalent to the solution of a collective Hamiltonian $H(\alpha_\mu, \partial/\partial\alpha_\mu)$ in an oscillator (phonon) basis with a maximum number N of quanta (phonons). The two-boson H_{IBM} corresponds to a restricted form of $H(\alpha_\mu, \partial/\partial\alpha_\mu)$ and the transformation is complicated. Nevertheless, the nine parameters in H_{IBM} are sufficient to describe situations varying from spherical equilibrium with vibrations through to non-spherical equilibrium with rotations and vibrations. In fact the algebraic structure of H_{IBM} allows analytic solutions in these two extremes and also in a limit corresponding to a stable deformation β with γ instability.

In this interpretation the IBM is simply an algebraic approximation to the simplest version of the collective model and N should be taken as large as practicable. IBM results are insensitive to N except, of course, near the truncation point at $L = 2N$. Since we know from many years experience that quadrupole collectivity is a major feature in many nuclei the IBM is guaranteed to fit such data. However in this interpretation it contains no new physics although the algebraic nature of H_{IBM} is often more convenient than the geometrical form of H_{coll}. Collective spectra are distorted by a number of effects at high energy. The truncation parameter N in the IBM clearly distorts spectra at high energy but it would seem arbitrary to try to correlate the effects of finite N with observed distortions in this interpretation.

To summarise, the first interpretation sees the IBM as an algebraic approximation to the simplest version of the collective model. It contains no new physics nor any of the microscopic aspects of the collective model such as the Nilsson potential with pairing and cranking. The IBM would be of minor interest if the first interpretation were all.

(ii) In the second interpretation each boson represents a nucleon pair and the IBM is seen as an approximation to a shell-model calculation. It is important to have such an approximation because of the great size of shell-model calculations when several single-particle levels are close together as in most heavier nuclei. It is indeed a very gross approximation since, of all possible pairs, it considers only one 0^+ and one 2^+ and furthermore it treats pairs as bosons. A typical spectrum for two neutrons (or protons) in a set of levels contains 0^+ 2^+ 4^+ ... followed by a second $0^+ 2^+$ etc. All but the first 0^+ and 2^+ are ignored. The model will therefore make sense only if states of the many-nucleon configuration can be approximately constructed from those chosen 0^+ and 2^+ pairs. But this is not an unreasonable suggestion since the BCS pairing theory, which is generally accepted, makes just this assumption for the 0^+ pairs. To regard the pairs as bosons is also not unreasonable since in the first place pair operators commute and for example if A^+ denotes the suitably normalised pair operator in a degenerate set of single particle levels then

$$[A, A^\dagger] = 1 - \hat{n}/\sum_j (j + \tfrac{1}{2})$$

where $\hat{n}$ is the number operator. Hence if the number of nucleons is small compared with the number of single particle levels the boson commutation relations are obtained.

To summarise, the second interpretation sees the IBM as an approximation to a shell-model calculation among the valence nucleons outside a system of closed shells. The closed shells are not excited so that effective charges are necessary to reproduce observed E2 effects. The collective spectra in the IBM now correspond to the coupling of the 0^+ and 2^+ pairs and the boson number N is related $N = \tfrac{1}{2}n$ to the number n of valence nucleons and is limited by the shell closure. It is a major weakness of the IBM that it does not incorporate any features of shell structure. They have to be imposed externally in some way by modification to the boson number and boson Hamiltonian. The neglect of the g-boson, and others, is another obvious weakness but experience with shell-model calculations tells us that remarkable accuracy may be achieved in a small model-space with a suitably renormalised effective Hamiltonian. At best the IBM can produce only a small sub-set of the shell-model states but it is argued that this includes the collective states which dominate the low lying spectrum of so many nuclei. A number of comparisons between shell-model and IBM calculations have been made and I shall discuss a recent one later in this talk.

THE NEUTRON-PROTON DEGREE OF FREEDOM

In the shell-model interpretation of the IBM the $J = 0$ pair may be constructed from nn, np or pp nucleons although for a heavy nucleus, in which n and p are filling different valence orbits, the np pair is absent. Hence the boson carries an extra two-fold or three-fold degree of freedom, for heavy or light nuclei respectively, in addition to the sd-space. These extensions are called IBM 2 and IBM 3 respectively[4,5]. Consequently there is a richer spectrum since the sd-space wave-function need not now be totally symmetric. The overall boson symmetry may be satisfied not only by taking total symmetry in both charge-space and sd-space but also by a product of mixed symmetries in both spaces. Hence IBM 1 can be recovered from IBM 2 or IBM 3 by choosing a Hamiltonian in which the mixed symmetry states are pushed high in energy. Most IBM 2 or IBM 3 calculations are made at, or near, this limit so that the IBM 1 success in data fitting is retained. The number of interaction parameters increases from 9 to 30 in going from IBM 1 to IBM 2 so that simplification has to be made and a very restricted Hamiltonian

$$H = E_o + \varepsilon\,(\hat{n}_s + \hat{n}_d) + \kappa\, Q_p \cdot Q_n$$

where

$$Q_p = (s^\dagger d + d^\dagger s) + \chi_p (d^\dagger d) \quad (2)$$

is often used, with four parameters. This may be too few to get a good fit unless ε is allowed to depend on N.

The lowest mixed symmetry state in the vibrational limit is a 2^+ from the configuration $s^{N-1}d$ and it is an interesting experimental question to identify it. There is recent evidence[6] that the third excited 2^+_3 state in ^{140}Ba, ^{142}Ce and ^{144}Nd at about 2 MeV has γ-decay properties $T(2^+_3 \to 0^+_1)/T(2^+_3 \to 2^+_1)$ and $\delta(2^+_3 \to 2^+_1)$ which are consistent with such a description. It is known that this 2^+_3 state does not fit into an IBM 1 analysis.

In the rotational limit a K = 1 band is the lowest with mixed symmetry and it is suggested[7,8] that the 1^+ state at 3.075 MeV in ^{156}Gd is of this kind. This state has been excited by inelastic electron scattering[9] through a strong M1 matrix element. In this limit the M1 matrix element between full symmetry and mixed symmetry is enhanced because all bosons take part whereas in the vibrational limit only the single d-boson contributes.

Note that the one-boson M1 operator in IBM 1 is just the angular momentum operator L so that M1 transitions are forbidden. In IBM 2 the operator is $g_p L_p + g_n L_n$, which allows M1 transitions when $g_n \neq g_p$. It also gives a g-factor which depends on neutron and proton number and in states of full symmetry is given by the simple formula

$$g = (g_n N_n + g_p N_p)/N$$

A reasonable fit to the g-factors of the 2^+_1 state has been obtained[7] in the region of Xe and Te with $g_n \approx 0$, $g_p \approx 1$.

Although most attention has been paid to the shell-model interpretation of IBM 2 an alternative interpretation would involve a two fluid collective model with separate shape degrees of freedom for neutron and proton densities.

ODD NUCLEI

In the collective model, odd nuclei are described by different modes of coupling of an odd nucleon to a collective core. Since IBM 1 is known to reproduce the main features of a collective core, the natural IBM picture for an odd nucleus would be a set of N bosons (s or d) with one fermion. One must then specify not only the boson hamiltonian and the single-particle energies but also the boson-fermion interaction. Many parameters are involved but some progress[10] has been made towards identifying the more important interaction terms based on the dominance of quadrupole effects and an exchange interaction between fermion and boson. It is assumed that fermion operators commute with boson operators but since this is not true for the operators of a nucleon and a nucleon pair it has to be supposed that the fermion operator is an 'undressed' nucleon operator. Both the empirical and the microscopic work on this IBFM model have been limited and it is probably wise to defer this problem until the even-even nuclei are better understood.

There have been attempts to combine the group theoretical treatment of bosons and fermions. This involves finding a common group for both types of particle, beyond the familiar rotation group in three dimensions which enables us to vector couple boson and fermion to a total angular momentum J. The two important IBM groups are O(6) for the γ-unstable limit and SU(3) for rotations. One then looks for the occurance of one of these groups in the fermion space through the shell-model group $U(2j + 1)$ or one of its sub-groups (with several orbits, $U(\Sigma (2j + 1))$). For O(6) there seem to be only accidental cases[11] such as the $j = 3/2$ shell for which SU(4) is isomorphic to O(6). More significantly, it is known in the shell-model that SU(3) is relevant for nucleons in a single oscillator shell and that 'pseudo' SU(3) extends this concept to heavier nuclei. The coupling[12] of this SU(3) to the boson SU(3) would then correspond to an alignment of the intrinsic axes of boson core and odd nucleon which corresponds to the strong coupling limit in the collective model with Nilsson orbitals. However since the IBM does not incorporate the Pauli principle it cannot predict the ordering of single-particle k-values which is the essence of the Nilsson picture.

It has been suggested[13] that the concept of supersymmetry has a role to play in the IBFM but I am not convinced of this. The multiplets of a supergroup contain both half-integer and integer spin states whereas with the usual groups all members of a multiplet are of the same type. It is mathematically possible to introduce such a group into the IBFM by introducing operators which destroy a boson and create a fermion. The multiplet would then contain, for example, (i) the ground state for N bosons, (ii) the ground state for N - 1 bosons and 1 fermion, (iii) states of (N - 2) bosons and 2 fermions ... and finally the states of N fermions. However the states (iii) are already highly excited states of the nucleus with N - 1 bosons so that data is difficult to obtain. Furthermore the energy formula from the supergroup seems to predict nothing more refined than a quadratic dependence on mass number with arbitrary coefficients which is a result common to almost any model.

MAPPING FROM A SINGLE j-SHELL ONTO IBM 3

Finally I return to the relation between the shell-model and the IBM. In a single j-shell[14] the s-boson corresponds to the $J = 0$ pair while the d-boson corresponds to the $J = 2$ pair operator modified to ensure that it always increases the seniority by two when acting on a state of full seniority. This idea may be readily extended[15] to incorporate isospin in mapping from a j-shell of neutrons and protons onto IBM 3. To get some feeling for the validity of the IBM we may then analyse the wave-functions from a standard shell-model calculation into the various boson components, including the distinction between full symmetry and mixed symmetry, and the non-boson component. Table I shows the results of such an analysis[15] for the $f_{7/2}$ nuclei, just beyond ^{40}Ca, with a shell-model interaction deduced from the observed levels of ^{42}Sc. Several conclusions may be drawn.

(i) The non-boson component is generally small, with the exception of the 4^+. Notice however that this component reduces from 61% for n = 4 to 28% for n = 6 and falls further to 5% for n = 8. The introduction of a g-boson would virtually eliminate this non-boson component.

(ii) The lowest states of each J are predominantly of full symmetry.

TABLE I Percentage analysis of shell-model wave functions into boson and non-boson components. The boson component is further analysed into symmetric (IBM 1) and mixed symmetry parts.

^{44}Ti (n = 4, T = 0)

J	s^2	sd	d^2	non-boson	Energy
2	-	6	88	6	5.0
4	-	-	39	61	2.9
2	-	89	7	4	1.2
0	83	-	17	1	0

^{46}Ti (n = 6, T = 1)

	s^3	s^2d		sd^2		d^3		non-boson	Energy
J	sym	sym	mixed	sym	mixed	sym	mixed		
2	-	1	79	10	1	7	-	2	2.5
4	-	-	-	60	11	1	0	28	2.2
2	-	86	0	3	0	5	0	6	1.2
0	82	-	-	(17)	0	-		1	0

^{48}Cr (n = 8, T = 0)

	s^4	s^3d	s^2d^2		sd^3		d^4		non-boson
J	sym	sym	sym	mixed	sym	mixed	sym	mixed	
4	-	-	84	10	-	-	(1)		5

TABLE II Excitation energies (MeV) for ^{46}Ti calculated in IBM 3 using an interaction deduced from the shell-model results for n = 4, compared with the full shell-model results.

	J = 0	2	4	2	4	6	4	2	3	1
Shell-model	0	1.2	2.2	2.5	3.0	3.2	3.5	3.5	3.6	3.7
IBM 3	0	1.0	2.0	2.5	-	3.3	3.8	3.5	3.2	3.0

(iii) The second 2^+ state in ^{46}Ti is mainly of mixed symmetry. In the shell-model calculation it has an excitation energy of 2.5 MeV and presumably corresponds to the observed level at 2.9 MeV. These results confirm that the IBM concept of symmetry in the sd-space has relevance in shell-model calculations and the group theoretical aspects of this connection are discussed elsewhere[15].

As a further test of the boson approximation we may deduce a boson Hamiltonian from the shell-model results for n = 4 (ie two bosons) and then calculate the n = 6 spectrum in the boson framework to compare with the complete shell-model spectrum for n = 6. Table II shows that the two spectra are very similar.

CONCLUSION

As an empirical model the IBM is closely related to a truncation of the quadrupole phonon version of the collective model. It contains analytic solutions in a number of physically important limits and generally it provides a convenient limited parameterisation for the analysis of collective spectra of various types.

More fundamentally, it is always possible to map nucleon pairs onto bosons and to define for any chosen shell-model system a corresponding boson problem. The difficulties are that the mapping requires more bosons than s and d and that many-boson interactions arise in the inevitable expansions, raising questions of renormalisation and convergence. Much more work needs to be done before it can be said that the IBM has a microscopic foundation as well as an empirical aspect.

Whatever the final outcome may be, the empirical use of the IBM is providing the stimulus for a more detailed experimental study of collective states, especially in the transition regions while theorists are faced with a considerable challenge to relate the empirical IBM parameters to the parameters in the effective nucleon-nucleon interaction.

REFERENCES

1. A Arima and F Iachello, Phys Rev Lett 35 1069 (1975)
2. A Arima and F Iachello, Advances in Nuclear Physics vol 13 (eds J W Negele and E Vogt) Plenum Press 1984
3. J P Elliott, Rep Prog Phys
4. A Arima, T Otsuka, F Iachello and I Talmi, Phys Lett 66B 205 (1977)
5. J P Elliott and A P White, Phys Lett 97B 169 (1980)
6. W D Hamilton, A Irback and J P Elliott (submitted to Phys Rev Lett)
7. M Sambataro and A E L Dieperink, Phys Lett 107B 249 (1981)
8. M Sambataro et al, Nucl Phys A423 333 (1984)
9. D Bohle et al, Phys Lett 137B 27 (1984)
10. R Bijker and A E L Dieperink, Nucl Phys A379 221 (1982)
11. F Iachello, Phys Rev Lett 44 772 (1980)
12. A B Balentekin et al, Phys Rev C27 1761 (1983)
13. A B Balantekin, I Bars and F Iachello, Nucl Phys A370 284 (1981)
14. T Otsuka, A Arima and F Iachello, Nucl Phys 309 1 (1978)
15. J A Evans, J P Elliott and S Szpikowski (submitted to Nucl Phys)

MICROSCOPICALLY DERIVED INTERACTING BOSON MODEL

Takaharu Otsuka
Theoretical Division, Los Alamos National Laboratory
Los Alamos, New Mexico 87545

and

Japan Atomic Energy Research Institute, Tokai, Ibaraki, Japan

ABSTRACT

The microscopic basis of the Interacting Boson Model is discussed. The IBM Hamiltonian is constructed microscopically in the following two steps. In the first step, the collective nucleon pairs of $J=0^+$ (S), 2^+ (D), etc. are mapped onto the corresponding bosons. Nucleon-nucleon interactions are also mapped onto boson-boson interactions. The mapping method for spherical nuclei is reviewed, and the mapping method for deformed nuclei, which was recently developed, is discussed in some detail. Low-lying collective states primarily consist of S and D pairs. Consequently, the corresponding boson states mainly consist of s and d bosons, while there are some admixture of g-bosons. In the second step of the derivation of the IBM Hamiltonian, effects of these g-bosons are included within the s-d boson space by a unitary transformation which transforms a combination of d and g bosons into a new d-boson. It is demonstrated that the s-d Hamiltonian thus derived indeed reproduces spectra of the original s-d-g Hamiltonian.

INTRODUCTION

I am going to talk about a microscopic derivation of the Interacting Boson Model (IBM)[1,2]. This work has been carried out in collaboration partly with Joe Ginocchio, and also partly with Naotaka Yoshinaga.

I shall begin with a brief review of the IBM from the microscopic point of view. There are two major microscopic assumptions for the IBM. Assumption I is the following[2]: There are two collective nucleon pairs of $J=0^+$ (S) and $J=2^+$ (D). The S and D pairs are coherent nucleon pairs. Since there are neutrons and protons, there are neutron S_ν and D_ν pairs, and proton S_π and D_π pairs. It is assumed that these pairs play dominant roles in low-lying collective states. In other words we truncate the gigantic shell model space to a smaller subspace constructed by S and D pairs as $|(S,D)^N\rangle$ (see Fig. 1). It is then assumed that the low-lying quadrupole collective states are in this subspace. The truncation to the S-D pair space is quite drastic, for example for the Samarium isotopes there are more than 10^{13} valence shell model states[2]. High-spin states are, on the other hand, outside this S-D pair subspace,

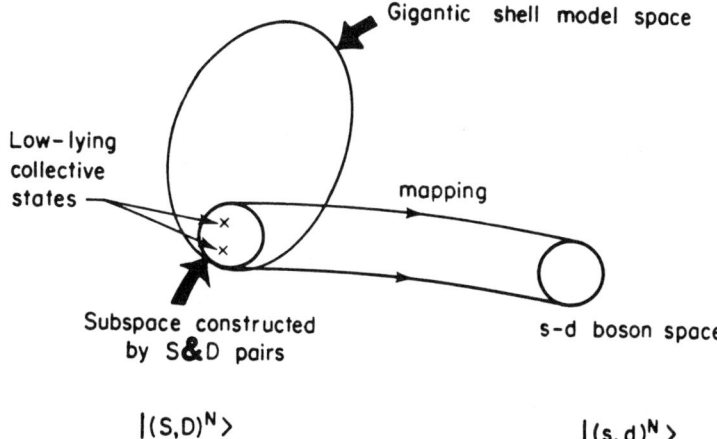

Fig. 1. Schematic representation of the microscopic assumptions for the Interacting Boson Model (see the text).

and one needs to include pairs of higher spins to describe them by extending the IBM[3].

Assumption II is that the motion of the S and D pairs can be simulated by the s and d bosons. In other words, it is assumed that energy matrix elements between S-D pair states can be reproduced by an s-d boson Hamiltonian which is obtained by mapping an S-D pair state onto the corresponding s-d boson state. The Hamiltonian of the s and d bosons are assumed further to consist of the single boson energy and the boson-boson interaction. This assumption reduces the tremendous complexity of multifermion problems. In this talk, we test the validity of these assumptions, and show how one can construct the IBM Hamiltonian from a microscopic Hamiltonian.

A BRIEF REVIEW OF SPHERICAL NUCLEI

I shall begin with a brief review on microscopic study on spherical nuclei. A prescription to derive the IBM Hamiltonian for spherical nuclei was proposed by the author, Arima and Iachello (OAI) several years ago[4]. In this method, the S-D pair states are mapped onto the s-d boson states as follows (see Fig. 1),

$$
\begin{array}{lll}
\text{fermion} & & \text{boson} \\
|S^N\rangle & \rightarrow & |s^N) \quad\quad\quad (2.1) \\
|S^{N-1}D\rangle & \rightarrow & |s^{N-1}d) \quad\quad\quad (2.2) \\
|S^{N-2}D^2(4^+)\rangle & \rightarrow & |s^{N-2}d^2(4^+)) \quad\quad\quad (2.3) \\
\vdots & & \vdots
\end{array}
$$

The boson Hamiltonian H^B is then determined by conditions[4]

$$\langle S^N|H|S^N\rangle = (s^N|H^B|s^N) \qquad (2.4)$$

$$\langle S^{N-1}D|H|S^{N-1}D\rangle = (s^{N-1}d|H^B|s^{N-1}d) \qquad (2.5)$$

$$\langle S^{N-2}D^2(L)|H|S^{N-2}D^2(L)\rangle = (s^{N-2}d^2(L)|H^B|s^{N-2}d^2(L))$$
$$(L = 0^+, 2^+, 4^+) \qquad (2.6)$$

$$\langle S^{N-2}D^2(0^+)|H|S^N\rangle = (s^{N-2}d^2(0^+)|H^B|s^N) \qquad (2.7)$$

$$\langle S^{N-2}D^2(2^+)|H|S^{N-1}D\rangle = (s^{N-2}d^2(2^+)|H^B|s^{N-1}d) . \qquad (2.8)$$

Thus, the s-d boson Hamiltonian which consists of the single boson energy and two-body boson-boson interaction is determined from matrix elements of S-D pair states with 0, 1 or 2 D-pairs[4,5].

An IBM Hamiltonian was derived[6] from a shell model Hamiltonian by this method for a fermion system with four protons and four neutrons. A surface delta interaction is assumed for the neutron-neutron channel, and also for the proton-proton channel. A quadrupole-quadrupole interaction is assumed between proton and neutron. The IBM Hamiltonian was derived from these effective nucleon-nucleon interactions. The spectrum obtained by diagonalization of this IBM Hamiltonian is shown in Fig. 2 (IBM). The spectrum exhibits the typical feature of vibrational (or spherical) nuclei.

The exact diagonalization of the shell model Hamiltonian, from which the IBM Hamiltonian was derived, was carried out. The calculated spectrum is also shown in Fig. 2 (EXACT). The agreement be-

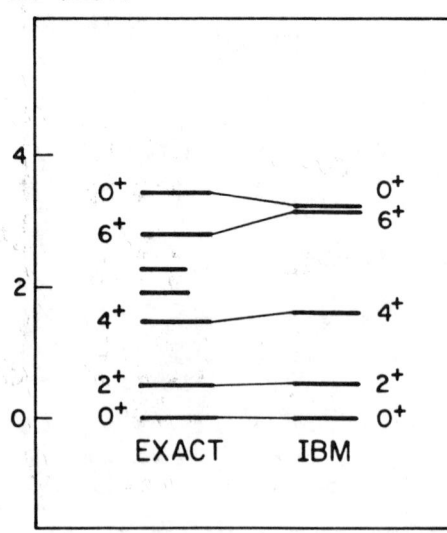

Fig. 2. Comparison between the spectrum (EXACT) calculated by diagonalizing a shell model Hamiltonian, and the spectrum (IBM) calculated by diagonalizing an IBM Hamiltonian which was derived microscopically from the above shell model Hamiltonian by the OAI method (Ref. 6). States in this figure possess the vibrational properties (see the text).

tween the two spectra is remarkable, and indicates that the present method for deriving the IBM microscopically is valid.

Using this method, many calculations have been performed[7-10]. As an example, I present the spectra of Sm isotopes calculated by Scholten[9] (See Fig. 3). Reasonable agreement can be seen between theoretical and experimental spectra for small N_ν (= the number of valence neutron pairs) which means spherical nuclei. If one looks at deformed nuclei ($N_\nu \sim 5$ in Fig. 3), however, there is more than a factor of two difference for the excitation energy of the lowest 2^+_1 state. In other words, the present prescription for deriving the IBM Hamiltonian works well for spherical nuclei, whereas it breaks down for deformed nuclei. I now turn to this problem.

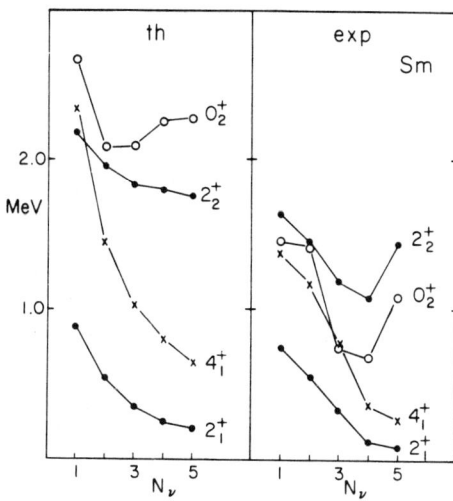

Fig. 3. Theoretical and experimental spectra of Samarium isotopes. The abscissa denotes the number of valence neutron pairs. The theoretical result was obtained by diagonalizing an IBM Hamiltonian derived microscopically in Ref. 9 by Scholten.

FERMION-BOSON MAPPING FOR DEFORMED NUCLEI

The major difference between wave functions for spherical and for deformed nuclei is the occupation probabilities of the S and the D pairs. In spherical nuclei, low-lying states mainly consist of S pairs, and contain a few D pairs. The dominant component of the ground state is actually the pure S pair state in eq. (2.7). In contrast, wave function of deformed nuclei contain S and D pairs about equally[11-14]. The OAI mapping method introduced in the previous section is based on dominance of S pairs over D pairs in low-lying states. One therefore should not expect that this method works in deformed nuclei. We shall look for another appropriate method in this section.

Before discussing this method, it is fair and useful to mention two major questions from the microscopic viewpoint concerning the IBM for deformed nuclei. The first question is the following. Since deformed nuclei occur in the middle of valence shells, there are many bosons (≥ 10) in deformed nuclei, since the number of bosons is equal to the number of valence nucleon pairs. Hence, interactions among many (≥ 3) bosons may be important, and the IBM Hamiltonian which contains only one-body and two-body interaction may not be appropriate.

The second question concerns the $J = 4^+$ pair or boson (called the g boson). The matrix element changing a d boson to a g boson is enhanced in deformed nuclei. Due to this enhancement, g bosons mix considerably, and this mixture may destroy the IBM description of deformed nuclei. I am going to answer the above questions in the subsequent two sections.

The ground state rotational band can be described by its intrinsic state ϕ, which contains 0^+, 2^+, 4^+, etc.

$$\phi^F = C_0^F |0^+\rangle + C_2^F |2^+\rangle + C_4^F |4^+\rangle + \ldots \qquad (3.1)$$

where C_I denotes an amplitude. As well know, the spin-I state in eq. (3.1) should be to a good approximation, equal to the lowest spin-I state in the exact calculation. The intrinsic state ϕ^F can be calculated by the Nilsson + BCS[15] or Hartree-Fock-Bogoljubov[16] (HFB) formalism. After particle number projection, we obtain[11]

$$\phi^F \propto (\Lambda_\pi^\dagger)^{N_\pi} (\Lambda_\nu^\dagger)^{N_\nu} |0\rangle \qquad (3.2)$$

where $N_\pi(N_\nu)$ stands for the number of valence proton (neutron) pairs, and

$$\Lambda_\tau^\dagger = x_{0(\tau)} S_\tau^\dagger + x_{2(\tau)} D_\tau^\dagger (m=0) + x_{4(\tau)} G_\tau^\dagger (m=0) + \ldots$$
$$\text{for } \tau = \pi, \nu. \qquad (3.3)$$

with $x_{I(\tau)}$ being an amplitude, and $S^\dagger$, $D^\dagger(m)$, $G^\dagger(m)$ being creation operators for $J = 0^+$, 2^+, 4^+ nucleon pairs of the magnetic quantum number m, respectively [Ref. 11]. It has been shown[11-14] that x_0^2 and x_2^2 are 30~60% and the sum $x_0^2 + x_2^2$ accounts for $\gtrsim 90\%$, while x_4^2 is $\lesssim 10\%$. Thus each Λ pair is dominated by the S and D pairs. Amplitudes $x_{L(\tau)}$ are negligibly small for $L > 6$ (Ref. 11), and will be ignored hereafter. Note that x_L in eq. (3.3) is for a pair of nucleons, while C_I in eq. (3.1) is for N-pair states. So they are completely different quantities.

We then map the Λ pair onto a λ boson;

$$\Lambda^\dagger = x_0 S^\dagger + x_2 D_0^\dagger + x_4 G_0^\dagger \rightarrow \lambda^\dagger = x_0 s^\dagger + x_2 d_0^\dagger + x_4 g_0^\dagger. \qquad (3.4)$$

where subscript π or ν is omitted. Consequently, the nucleon intrinsic state ϕ^F is mapped onto the boson intrinsic state ϕ^B as,

$$\phi^F = |\Lambda_\pi^{N_\pi} \Lambda_\nu^{N_\nu} \rangle \to \phi^B = |\lambda_\pi^{N_\pi} \lambda_\nu^{N_\nu}) \quad (3.5)$$

The boson intrinsic state ϕ^B is expanded similarly to eq. (3.1) as

$$\phi^B = c_0^B |0^+) + c_2^B |2^+) + c_4^B |4^+) + \ldots \quad (3.6)$$

We found quite recently that the fermion amplitude c_I^F and the boson amplitude c_I^B satisfy an equality,

$$c_I^F = c_I^B \quad (I = 0, 2, 4, \ldots) \quad (3.7)$$

to a very good approximation. This equality means that ϕ^F and ϕ^B in eq. (3.5) respond in the same way to the rotation of the intrinsic frame. The rotation is extremely important for deformed nuclei, because, as it well-known, many properties of the rotational band can be calculated by rotating the intrinsic frame. The probability of a spin-I member of the rotational band is, in fact, calculated by rotating ϕ^F or ϕ^B as

$$(c_I^X)^2 = \frac{2I + 1}{2} \int d(\cos\theta)\, d_{00}^I(\theta)\, \langle \phi^X | R(\theta) | \phi^X \rangle \quad (3.8)$$

where θ denotes the rotation angle with respect to the symmetry axis, $R(\theta)$ is the y-axis rotation operator, X stands for F (fermion) or B (boson), and $d_{00}^I(\theta)$ is the relevant d-function. The relation between the amplitude c_I^X and the rotation is clearly seen. I actually calculated c_I^F and c_I^B for a system with 14 protons in the Z = 50-82 shell and 16 neutrons in the N= 82-126 shell. The intrinsic wave function ϕ^F was obtained by the HFB calculation with the particle number conservation[16]. Squared values of c_I^F and c_I^B are listed in Table I. The agreement between these two are remarkable. The fermion and boson intrinsic states indeed have the same angular momentum content.

We now map the quadrupole operator Q. The following conditions are imposed:

$$\langle \Lambda^N | Q | \Lambda^N \rangle = (\lambda^N | Q^B | \lambda^N) \quad (3.9)$$

$$\langle \Lambda^N | Q\, \hat{R}_y | \Lambda^N \rangle = (\lambda^N | Q^B \hat{R}_y | \lambda^N) \quad (3.10)$$

where subscript π or ν is omitted, Q^B is the boson quadrupole operator;

Table I. Probabilities of spin-I members in the intrinsic state

I	fermion $(C_I^F)^2$	boson $(C_I^B)^2$
0^+	3.44 %	3.46 %
2^+	15.44	15.58
4^+	21.63	21.93
6^+	21.20	21.57
8^+	16.51	16.74
10^+	10.75	10.71
12^+	6.01	5.78

$$Q^B = q_1 (d^\dagger s + s^\dagger \tilde{d}) + q_2 [d^\dagger \tilde{d}]^{(2)}$$
$$+ q_3 [g^\dagger \tilde{d} + d^\dagger \tilde{g}]^{(2)} + q_4 [g^\dagger \tilde{g}]^{(2)} \qquad (3.11)$$

with q_1, q_2, etc. being coefficients to be determined, and $\hat{R}_y$ is the infinitesimal rotation operator along the y axis (i.e., $R(\theta) = \exp(i\theta \cdot \hat{R}_y)$ with $R(\theta)$ in eq. (3.8)). Eq. (3.9) means that the averaged value of Q over the states in eq. (3.1) should be reproduced by $|\lambda^N\rangle$ and Q^B. Note that ϕ^F and ϕ^B contain spin-I members in the almost same probability. Otherwise, eq. (3.9) does not make much sense.

Eq. (3.10) implies that the response of Q^B to the rotation along the y axis should be equal to that of Q in the first order. Once the conditions (3.9) and (3.10) are fulfilled, matrix elements between low-lying members of the rotational band should be reproduced by the boson calculation.

The conditions (3.9) ~ (3.10) are in practice satisfied for the quadrupole operator by

$$Q^B = (1-\varepsilon) \{ \sum_\mu \langle S|Q|D_\mu\rangle (s^\dagger d_\mu + d_\mu^\dagger s) + \sum_\mu \langle D_\mu|Q|D_{\mu'}\rangle d_\mu^\dagger d_{\mu'} + \ldots \} \qquad (3.12)$$

where ε is a blocking factor. The blocking factor ε was introduced in Ref. 17 in a different way. I point out that calculations show that the two conditions (3.9) ~ (3.10) are satisfied by a single blocking factor. The mapping in eq. (3.12) works only for the quadrupole operator. More general calculations for other operators are in progress.

In order to test the validity of the mapping $_B(3_B12)$, I calculated $\langle I | f(Q_\pi \cdot Q_\nu) | I \rangle$ and $\langle I | f(Q_\pi^B Q_\nu^B) | I \rangle$ where f is the interaction strength, and $|I\rangle$ denotes spin-I components in eq. (3.1) or in eq. (3.6). Calculated results are shown in Table II

Table II. Matrix elements of the interaction $f(Q_\pi Q_\nu)$

I	fermion	boson
0^+	14.99 MeV	15.05 MeV
2^+	14.99	15.05
4^+	14.99	15.05
6^+	14.98	15.04
8^+	14.96	14.98
10^+	14.92	14.91

for the same system as in Table I. The agreement between the fermion and the boson matrix elements is again remarkable. The mapping (3.12) is thus consistent with the angular mmentum projection, and we can indeed reproduce the fermion matrix elements of a system with 26 valence nucleons altogether.

The validity of this mapping is examined futher by looking at other matrix elements.[17] As an example, I consider matrix elements related to a state $|D_2 \Lambda^{N-1}\rangle$, where D_2 denotes the m=2 component of D_m. This state is considered to be one of the major components of the γ-band intrinsic state. Matrix elements $\langle D_2 \Lambda^{N-1}|\hat{Q}_0|D_2 \Lambda^{N-1}\rangle$ and $\langle D_2 \Lambda^{N-1}|Q_2|\Lambda^N\rangle$ are compared in Table III with the corresponding boson predictions $(d_2 \lambda^{N-1}|\hat{Q}_0^B|d_2 \lambda^{N-1})$ and $(d_2 \lambda^{N-1}|\hat{Q}_2^B|\lambda^N)$. Note that the diagonal matrix element above is related to the γ-band intrinsic quadrupole moment which is the same order of magnitude as $\langle \Lambda^N|\hat{Q}_0|\Lambda^N\rangle$. The off-diagonal one is related to the ground-gamma E2 transition which is one order of magnitude less than Q_{in}. Table III clearly demonstrates that these fermion matrix elements are reproduced very well by the mapping (3.11) [Ref. 17].

Following the above mapping procedure, the parameters of $\hat{Q}^B$ in eq. (3.11) are calculated for ^{158}Gd. (Ref. 17). Here, δ=0.25 was taken, and the pairing strength was chosen so that Δ~0.9 MeV. The Λ pair was calculated from the Nilsson+BCS.

In order to construct an s-d boson (or IBM) system, the g-boson is eliminated in the next step, and its effects are taken into account by renormalization of s-d boson terms by the method of Ref. 18. Although the renormalization method has been improved considerably since Ref. 18 as discussed in a subsequent section, essential properties in this table remain unchanged.

Parameters of the s-d boson quadrupole operator are thus calculated microscopically with the renormalization. The result of this calculation should be compared to that obtained by phenomenological IBM calculations. In Table IV, these two results are shown in a convention in which the boson quadrupole operator is defined as

Table III. Comparison between exact and boson quadrupole matrix elements (fm^2) related to the D_2 pair for (a) the 16 neutron system in the N=82-126 shell with $\delta=0.3$ and $\Delta=0.8$ MeV and for (b) the 16 proton system in the Z=50-82 shell with $\delta=0.25$ and $\Delta=0.9$ MeV. $Q_{in} = \langle\Lambda^N|\hat{Q}_0|\Lambda^N\rangle$ $(=(\lambda^N|Q_0^B|\lambda^N))$ is also shown.

case	Q_{in}	$\langle D\Lambda^{N-1}_2\|\hat{Q}_0\|D\Lambda^{N-1}_2\rangle$		$\langle D\Lambda^{N-1}_2\|\hat{Q}_2\|\Lambda^N\rangle$	
		exact	boson	exact	boson
(a)	179	150	150	12	9
(b)	108	88	91	14	13

Table IV. Parameters in the boson QQ interaction and in the boson E2 operator calculated microscopically by the present mapping method with the renormalization due to elimination of g-boson. In the column "(unr.)", the unrenormalized result is shown in parentheses. Parameters obtained by a fitting calculation and by the OAI method are also shown.

parameter	micro.	(unr.)	fitting	OAI
κ (Mev)	0.094	(0.12)	0.08 ~ 0.09	0.19
χ_π	-0.86	(-0.80)	-0.8 ~ -0.9	0.04
χ_ν	-1.18	(-1.03)	-1.1 ~ -1.2	-0.55
e^B_π (e fm)	12.0	(10.6)	12 ~ 14	14.3
e^B_ν (e fm)	10.0	(6.7)	12 ~ 14	8.4

$$T_\tau^B = d_\tau^\dagger s_\tau + s_\tau^\dagger \tilde{d}_\tau + \chi_\tau [d_\tau^\dagger \tilde{d}_\tau]^{(2)} \tag{3.13}$$

with $\tau=\pi$ (proton) or ν (neutron) and χ_τ being a parameter, and the boson proton-neutron QQ interaction is defined as $-\kappa T_\pi^B \cdot T_\nu^B$ with the strength κ. The strength of the nucleon QQ interaction is taken from Ref. 16. The boson E2 operator is written as $\hat{T}^{(E2)} = e_\pi^B T_\pi^B + e_\nu^B T_\nu^B$ with the boson effective charge e_τ^B. The nucleon effective charges are assumed to be 1.7e for protons and 0.7e for neutrons. In Table IV, a reasonable agreement is seen between the microscopically calculated parameters and those obtained by fitting calculations. In Table II, unrenormalized values of the parameters are also indicated to show changes due to the renormalization. These parameters are calculated also by the OAI method[4], where the IBM parameters are determined by states $|S^N\rangle$ and $|S^{N-1}D\rangle$ as discussed in the previous section. This method is useful in and near spherical nuclei, while it yields, in deformed regions, a too large value of κ and a too small value of $|\chi|$ compared to the results of phenomenological fitting. Large $|\chi|$ is essential to obtain rotational spectra. In fact, $|\chi|=\sqrt{7}/2$ is the SU(3) limit.[19] This discrepancy is now solved, by introducing the new mapping where the blocking from a coherent linear combination of S, D and G is included properly.

RENORMALIZATION OF g-BOSON EFFECTS

I have so far discussed on the fermion-boson mapping, in which nucleon pairs of S, D, G, etc. are mapped onto bosons s, d, g, etc. One can simply neglect effects of bosons of 6^+, 8^+, and higher. Effects of g-bosons, however, have to be included. The g-boson is mixed through the following term in the $(Q_\pi \cdot Q_\nu)$ interaction,

$$f \, q_3 (Q_\pi [g_\nu^\dagger \tilde{d}_\nu + d_\nu^\dagger \tilde{g}_\nu]^{(2)}) \tag{4.1}$$

where f is a coupling constant, and q_3 is introduced in eq. (3.11). This interaction can be rewritten as

$$f \, q_3 \, ([Q_\pi g_\nu^\dagger]^{(2)} \cdot \tilde{d}_\nu + d_\nu^\dagger \cdot [Q_\pi \tilde{g}_\nu]^{(2)}) \tag{4.2}$$

Therefore, if $[Q_\pi g_\nu^\dagger]^{(2)}$ behaves like a quadrupole boson, effects of g boson can be treated by including $[Q_\pi g_\nu^\dagger]^{(2)}$ as a part of "d" boson. For this purpose, we introduce a unitary transformation,

$$U = e^{\phi_\nu(Q_\pi \cdot [g_\nu^\dagger \tilde{d}_\nu - d_\nu^\dagger \tilde{g}_\nu]^{(2)}) + \phi_\pi(Q_\nu \cdot [g_\pi^\dagger \tilde{d}_\pi - d_\pi^\dagger \tilde{g}_\pi]^{(2)})} \tag{4.3}$$

A unitary transformation of the form $U = e^Z$ (Z is an operator) can be written in general as

$$U X U^{-1} = X + [Z, X] + \tfrac{1}{2}[Z, [Z, X]] + \cdots \quad (4.4)$$

If one has relations,

$$[Z, [Z, X]] = -\alpha^2 X \quad (4.5)$$

and

$$[Z, [Z, [Z, X]]] = -\beta^2 [Z, X] \quad (4.6)$$

one obtains,

$$U X U^{-1} = X \cos \alpha + [Z, X] \frac{\sin \beta}{\beta} \quad (4.7)$$

I shall apply eq. (4.4) to $X = d^\dagger$ with the transformation in eq. (4.3). One can obtain easily,

$$[Z, d_\nu^\dagger] = \phi_\nu [Q_\pi g_\nu^\dagger]^{(2)} \quad (4.8)$$

where Z stands for ln U for U in eq. (4.3). In deriving eq. (4.8) and also in the following, we assume that Q_π and Q_ν in U commute with any operator. In other words, Q_π and Q_ν in eq. (4.3) are treated as if they are c-numbers. Matrix elements of $[Q, X]$ (X = arbitrary operator) should be much smaller than those of Q itself for low-lying collective states, since coherent property in Q should be lost in $[Q, X]$. This assumption is clearly related to collectivity of states, and also of Q_π and Q_ν.

The double commutator is written as

$$[Z, [Z, d_\nu^\dagger]] = -\phi_\nu^2 \sum_K \sqrt{5} \sqrt{2K+1} \; W(2\,4\,K\,2 : 2\,2) \\ \times [[Q_\pi Q_\pi]^{(K)} d_\nu^\dagger]^{(2)} \quad (4.9)$$

Although K in eq. (4.9) runs from 0 to 4, the K=0 coefficient is largest, and the scalar product $[Q Q]^{(K=0)}$ is enhanced in quadrupole collective states. It is, hence, a reasonable approximation to retain only the K=0 term in eq. (4.9). Secondly, we replace the operator $[Q Q]^{(0)}$ with its expectation value $\langle [Q Q]^{(0)} \rangle$ with respect to an appropriate state, for instance, the intrinsic state of the rotational band or the ground state. Eq. (4.9) then becomes

$$-\phi_\nu^2 \langle [Q_\pi Q_\pi]^{(0)} \rangle \frac{1}{\sqrt{5}} d_\nu^\dagger \quad (4.10)$$

This equation has the same form as eq. (4.5) with $\alpha = \sqrt{\phi_\nu < [Q_\pi Q_\pi]^{(0)}/\sqrt{5}}$. We can thus evaluate all terms in eq. (4.4), and finally obtain the transformed form of the $d^\dagger$ operator as in eq. (4.7).

In practical calculations, terms of $K \neq 0$ in eq. (4.9) are also included effectively, by replacing these terms with

$$\frac{< [[Q_\pi Q_\pi]^{(K)} d^\dagger_\nu]^{(2)} >}{< [[Q_\pi Q_\pi]^{(0)} d^\dagger_\nu]^{(2)} >} \times [[Q_\pi Q_\pi]^{(0)} d^\dagger_\nu]^{(2)} \qquad (4.11)$$

where $< >$ stands for the expectation value of an appropriate state as discussed just above. In other words, effects of the $K \neq 0$ terms are included in some averaged way. Deviation or fluctuation from the approximation in eq. (4.11) can be included order by order in principle.

Based on the above prescription, all terms in the original boson Hamiltonian are transformed. The mixing angle ϕ_ν and ϕ_π in U can be determined from amplitudes in eq. (3.4) for deformed nuclei. The mixing angle will be determined in other regions by utilizing a boson analogue of the RPA approximation. It is expected that the angle stays rather constant. The major effect of the unitary transfomation is found in the single d boson energy. As already seen in eq. (4.2), the unitary transformation treats $(Qg^\dagger)^{(2)}$ as a part of the new d boson. Thus, the interaction in eq. (4.2) is transformed as a part of the single d boson energy. This actually means an enormous reduction of the single d boson energy. The reduction is expressed as,

$$Q_\pi \cdot (g^\dagger_\nu \tilde{d}_\nu + d^\dagger_\nu \tilde{g}_\nu)^{(2)} \to <(Q_\pi Q_\pi)> \hat{n}_{d_\nu} \qquad (4.12)$$

where some numerical factors are omitted. Since the ground-state expectation value of $(Q_\pi Q_\pi)$ becomes largest in the middle of the valence shell as function of proton and neutron numbers, the reduction of the single d boson energy is largest there. Note that this trend is consistent with the conclusion of phenomenological studies.

We applied the unitary transformation to an s-d-g boson Hamiltonian

$$H_{s-d-g} = \varepsilon_{d_\pi} n_{d_\pi} + \varepsilon_{g_\pi} n_{g_\pi} + \varepsilon_{d_\nu} n_{d_\nu} + \varepsilon_{g_\nu} n_{g_\nu} - f(Q_\pi Q_\nu) , \qquad (4.13)$$

where Q_π and Q_ν are obtained by eq. (3.11). The strength parameter f is adjusted so that this Hamiltonian reproduces the experimental spectrum of ^{158}Gd.

The calculated spectrum is shown in Fig. 4. The spectrum of an s-d Hamiltonian obtained by the above unitary transformation is also shown in Fig. 4. These two spectra are in an excellent agreement. The unitary transformation indeed works well. In Fig. 4, there is the spectrum of another s-d boson Hamiltonian which was obtained by dropping all g-boson terms in eq. (4.13). Although a rotational spectrum exists in this unrenormalized Hamiltonian as well, this spectrum has a scale three times larger than the other two. The comparison among these three spectra demonstrates the importance and usefulness of the unitary transformation. I note that the present method is not a perturbation, and that one can treat a rather large admixture of g bosons. After the unitary transformation, coupling between the new d boson and the new g boson becomes weak, and may be treated by perturbation, if one wished.

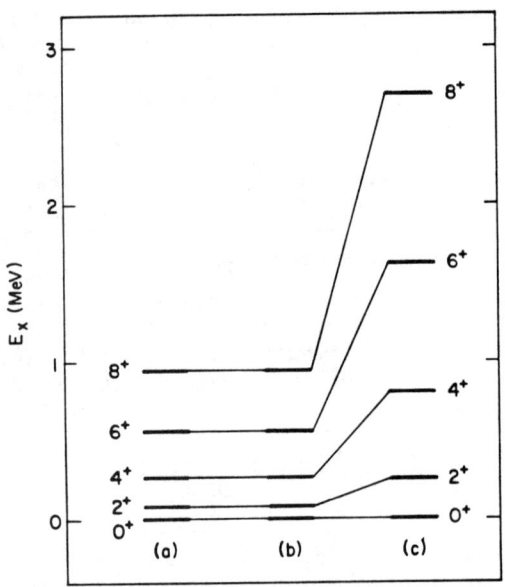

Fig. 4. Spectra obtained (a) from a Hamiltonian containing s, d, and g bosons, (b) from the renormalized s-d boson (IBM) Hamiltonian calculated from the above s-d-g boson Hamiltonian by the unitary transformation, and (c) from the s-d boson Hamiltonian obtained by dropping all g-boson terms in the above s-d-g boson Hamiltonian.

SUMMARY

I summarize this talk by mentioning the following two points.
(i) The fermion system can be mapped onto a simple boson system for low lying states in both spherical and deformed nuclei.

(ii) Most of the g-boson effects in low-lying states can be included by changing the definition of the d-boson, i.e., by introducing a unitary transformation between d and g bosons.

REFERENCES

1. A. Arima and F. Iachello, Phys. Rev. Lett. 35, 1069. (1975).
2. T. Otsuka, A. Arima, F. Iachello, and I. Talmi, Phys. Letter. 76B, 139 (1978).
3. N. Yoshida, A. Arima and T. Otsuka, Phys. Lett. 114B, 86 (1982).
4. T. Otsuka, A. Arima, and F. Iachello, Nucl. Phys. A309, 1 (1978).
5. T. Otsuka, in "Interacting Bosons in Nuclear Physics", ed. by F. Iachello (Plenum, New York, 1979).
6. T. Otsuka, Phys. Rev. Lett. 46, 710 (1981).
7. S. Pittel, P. D. Duval, and B. R. Barrett, Ann. of Phys. 144 (1982) 168.
8. Y.K. Gambhir, P. Ring and P. Schuck, Phys. Rev. C25, 2858 (1982).
9. O. Scholten, Phys. Rev. C28, 1783 (1983).
10. A. van Egmond and K. Allaart, Nucl. Phys. A425, 275 (1984).
11. T. Otsuka, A. Arima, and N. Yoshinaga, Phys. Rev. Lett. 48, 387 (1982).
12. D. R. Bes, R. A. Broglia, E. Maglione, and A. Vitturi, Phys. Rev. Lett. 48, 1001 (1982).
13. J. Dukelsky, G. G. Dussel, and H. M. Sofia, Nucl. Phys. A373, 267 (1982), and S. Pittel and J. Dukelsky, Phys. LEtter 128B, 9 (1983).
14. K. Sugawara-Tanabe and A. Arima, Phys. Lett. 100B, 87 (1982).
15. S. G. Nilsson and O. Prior, Mat.-Fys. Medd. Danske Vid. Selsk. 32, No. 16 (1961).
16. J. L. Egido and P. Ring, Nucl. Phys. A383, 189 (1982).
17. T. Otsuka, Phys. Lett. 138B, 1 (1983).
18. T. Otsuka, Nucl. Phys. A368, 244 (1981).
19. A. Arima and F. Iachello, Ann. Phys. 111, 201 (1978).

AXIAL ASYMMETRY IN THE IBA
AND AN EXTENSIVE NEW O(6) REGION NEAR A=130

R. F. Casten
Brookhaven National Laboratory
Upton, New York, 11973, USA

and

Institut fur Kernphysik
University of Koln
Koln, Germany

ABSTRACT

Although the IBA-1 contains no solutions corresponding to a rigid triaxial shape, it does contain an effective asymmetry arising from zero point motion in a γ-soft potential leading to a non-zero mean or rms γ. In the Consistent Q Formalism (CQF) of the IBA, most results of a calculation depend only on one parameter χ. A relation will be established between χ and the effective asymmetry parameter γ. The relation between the asymmetry occuring naturally in IBA-1 and the triaxiality arising from the introduction of cubic terms into the IBA Hamiltonian will be discussed. It will be shown that γ-band energy staggering is a particularly sensitive indicator of the degree of γ rigidity. Finally, an extensive new region of O(6) like Xe and Ba nuclei near A=130 will be discussed. Their remarkable similarity to Pt will be explored. Deviations from the strict O(6) limit can be described in terms of the interplay of soft and rigid axial asymmetry and calculations will be presented that interpret the Xe, Ba and Pt isotopes in this way.

1. INTRODUCTION

It is well known that the IBA-1 contains no triaxial solutions[1,2], that is, the corresponding classical potential never has an axially asymmetric minimum. Nevertheless, the O(6) limit corresponds to a γ-unstable potential with a mean γ near 30°. Moreover, calculated E2 branching ratios for deformed nuclei deviate from the Alaga rules. These facts suggest that, somehow, the IBA-1 must contain at least an effective asymmetry. It is the purpose here to show that this is, in fact, the case and to derive an effective γ that corresponds to a given IBA-1 calculation. This will be done in the context of the Consistent Q Formalism[3] (CQF) of the IBA. The origin of this asymmetry in terms of dynamical fluctuations in a γ-soft potential will be discussed as well as its relation to the more rigid asymmetric shapes that correspond to the introduction of higher order terms in the IBA-1 Hamiltonian. Lastly, a new region of O(6) nuclei near A=130 will be described and deviations from O(6) interpreted in terms of γ softness.

The discussion here is largely a report of collaborative work. Section 2 was done in collaboration with A. Aprahamian and D. D. Warner[4,5], Section 3 with K. Heyde, P. Van Isacker, and J. Jolie[6], and Section 4 with P. von Brentano[7]. The first part of this work has been reported in a recent conference at Gull Lake[8] and will therefore be abbreviated here.

2. AXIAL ASYMMETRY IN IBA-1

The question of axial asymmetry in the IBA-1 can be addressed by seeking to define an effective γ value for a geometrical model that gives the same result as an IBA calculation. It is most convenient to carry out the IBA calculations in the CQF since most results depend only on the one parameter, χ, and one simply tries to associate, to each χ value, a value for γ that gives the same result for a given observable. In the CQF, the IBA Hamiltonian for nuclei between the O(6) and SU(3) limits is[3]

$$H = -\kappa Q \cdot Q - \kappa' L \cdot L \qquad (1)$$

where

$$Q = (s^+ \tilde{d} + d^+ s) + (\chi/\sqrt{5}) (d^+ \tilde{d}) \qquad (2)$$

Since L is diagonal in the IBA basis, only the first term in H is important and thus κ is essentially an energy scale factor. The resultant wave functions and most observables therefore depend only on χ, which takes on the values $-\sqrt{35}/2 = -2.958$ in SU(3) and 0 in O(6). The transition between these limits is simply expressed in terms of a smooth variation in χ. The results of such calculations are shown in Fig. 1 for several observables and compared there with those calculated in the asymmetric rotor model[9]. A γ-χ correspondence can be extracted for a given observable if the calculated values pass through the same ranges. It is remarkable that this is precisely the case, even to the extent that the ratio $B(E2:2^+_2 \to 0^+_1)/B(E2:2^+_1 \to 0^+_1)$ maximizes at ≈ 0.07 in both calculations. The fact that the detailed behavior of each curve is different in the two calculations only means that the γ-χ correlation is non-linear. For each corresponding pair of curves in Fig. 1, a γ-χ correspondence can be extracted. The results are shown in Fig. 2 where it is apparent that the resultant γ-χ relations are nearly identical for different observables, thus suggesting that it is valid to assign a reasonably well defined γ value to any IBA calculation with eq. (1) in the CQF.

Not surprisingly, $\gamma \to 30°$ for $\chi=0$ and drops steadily as χ increases. The fact that it does not $\to 0°$ for the SU(3) limit ($\chi = -2.958$) is a direct reflection of finite boson number effects. This is easiest to see for the energy ratio shown in Fig. 1. For axial symmetry ($\gamma=0°$) this ratio $\to \infty$ in the geometrical model but it is given by

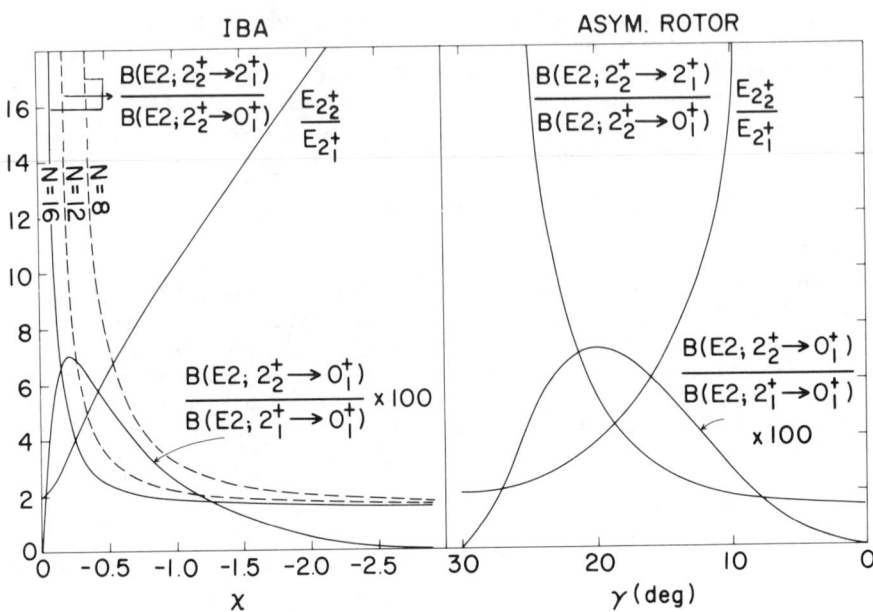

Fig. 1. Energy and B(E2) ratios in the IBA-1[4] and the asymmetric rotor model[9]. (IBA results for N=16 unless specified.)

$$E_{2_2^+}/E_{2_1^+} = [\kappa(2N-1)/(0.75\ \kappa-\kappa')] + 1 \qquad (3)$$

in the SU(3) limit of the IBA. Clearly, this can only $\to \infty$, yielding $\gamma=0°$, when $N\to\infty$. (Note: Physical values of κ and κ' have opposite sign, so the denominator in eq. (3) does not vanish.) Thus, the SU(3) limit is not a pure axial rotor, as often stated. Indeed, this is also clear from the fact that calculated branching ratios deviate from the Alaga rules, even in this limit. It is also interesting to note from the curves for N=12 and 16 in Fig. 2 that γ increases with decreasing N.

Since the IBA-1 does not contain true asymmetric minima the asymmetry just deduced must arise from dynamical fluctuations due to a softness of the potential in the γ degree of freedom. Due to this, it is clear that there must be a relation between γ-softness and mean effective γ so that large values of the latter arise from large values of the former. The above extraction of IBA-1 γ values relied on a comparison with a geometrical model with rigid asymmetry. Since asymmetry arises in the IBA precisely via γ softness, it might be questioned whether this procedure is valid. This issue has been studied by Castanos, Frank and Van Isacker[10] who devised an alternate method for extracting γ values, for the ground state, directly from the IBA wave functions. Their result is shown in

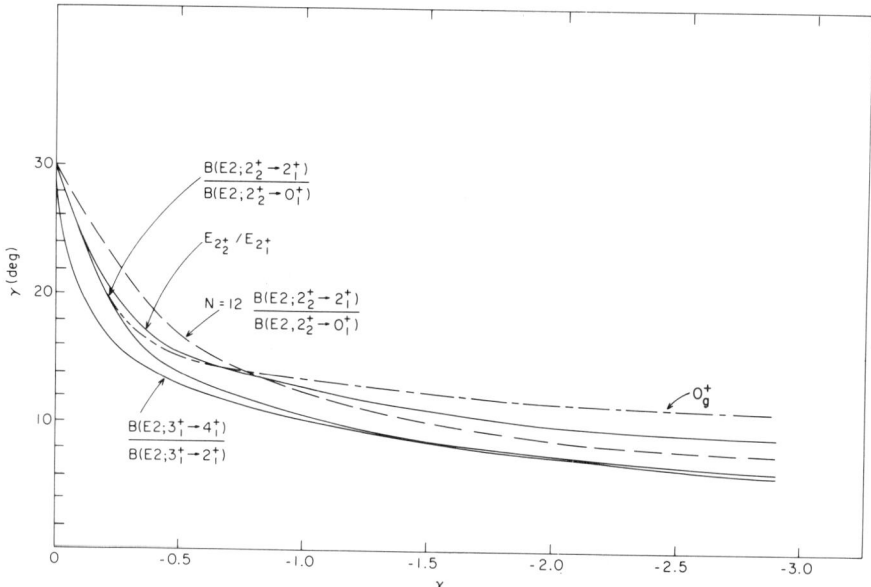

Fig. 2. Asymmetry parameter γ vs. χ for N=16 deduced for several observables. The ratio $B(E2:2^+_2 \to 2^+_1)/B(E2:2^+_2 \to 0^+_1)$ is also given for N=12. The curve labelled 0^+_g is taken from Ref. 10 and is based directly on the structure of the ground state wave function.

Fig. 2. The close agreement with the other curves suggests the approximate validity of the present procedure outlined above.

Since the one parameter characterizing the wave functions in the CQF is χ, once χ is determined for a given nucleus one can obtain an IBA prediction for the asymmetry γ by using curves similar to Fig. 2, for the appropriate N value. For the rare earth region from Gd-Os, such χ values have been extracted and are given in ref. 3. Using these values, effective γ values for the IBA have been calculated and are compared with experimental ones obtained from the known $\gamma \to g$ band relative B(E2) values, in Fig. 3. It is worth noting in passing that the cup-shape form of the IBA predictions is not due to variations in χ but is in fact an automatic outcome of the inclusion of finite boson numbers in the IBA. (Recalling Fig. 2, it is clear that lower N values imply larger γ values for a given χ.)

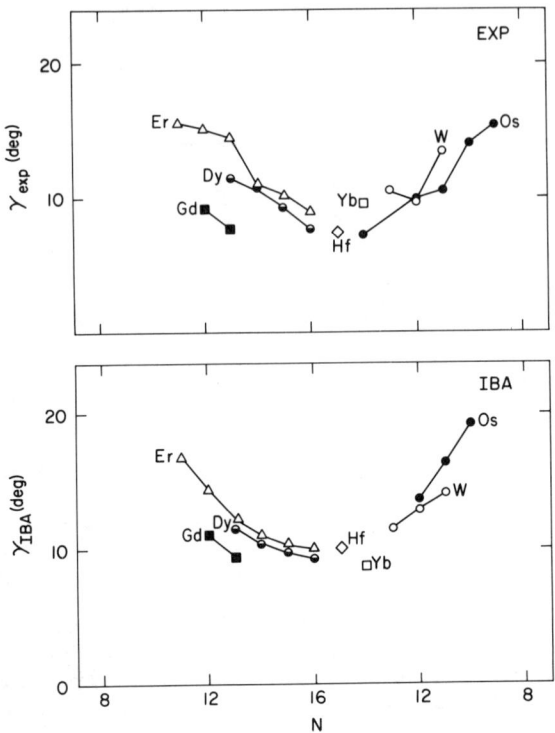

Fig. 3. Comparison of experimental and calculated γ values, plotted against the boson number N.

3. DYNAMIC VS. STATIC ASYMMETRY IN THE IBA

Axial asymmetry may be obtained in the general context of the IBA in a variety of ways. The simplest is that outlined in Section 2 where it arises naturally, in the usual IBA-1, from the associated softness of the potential in γ. Alternately, the introduction of various complications can lead to asymmetry but, as will be discussed, of a much different type. Recently, three approaches of the latter sort have been used, namely the introduction into the IBA-1 of cubic terms[11] or of g bosons[12] and, into the IBA-2 of quadrupole interactions[13] between like bosons (e.g., $Q_\pi \cdot Q_\pi$). The SU(3)* triaxial symmetry is a limiting case of the latter[13]. In all of these the minimum in the potential of the added interactions is at γ=30°, and one is then dealing with nuclei of stable asymmetric configurations relatively more rigid in γ. The question is if, and how, one can distinguish the origin and effects of asymmetry arising from these different mechanisms.

As recently discussed[8], all these methods for incorporating asymmetry in the IBA produce similar results for B(E2) values and quadrupole moments. These quantities are more sensitive to the mean γ than to dynamical fluctuations in γ. A clear difference between the various approaches, however, does exist in γ-band energies. In the rigid triaxial rotor, these levels do not follow a $J(J+1)$ law but occur in couplets, grouped as $(2_\gamma^+ 3_\gamma^+)$, $(4_\gamma^+ 5_\gamma^+)$, $(6_\gamma^+ 7_\gamma^+)$,..., and, indeed, such behavior is typical of the spectra produced by all those approaches[11-13] that correspond to potentials with minima at $\gamma=30°$.

In the simple IBA-1, however, γ deformations arise, as noted above, from γ softness and large γ_{eff} values stem from nearly γ unstable (γ independent) potentials. This is typified by the extreme case of the O(6) limit. The O(6) eigenvalue equation is

$$E(\sigma,\tau,J) = (A/4)\ \sigma(\sigma+4) + B\tau(\tau+3) + CJ(J+1) \tag{4}$$

which arises from the chain decomposition

$$U(6) \supset O(6) \supset O(5) \supset O(3). \tag{5}$$

In such a chain each successive step breaks a previous degeneracy and adds a new quantum number, a new term in the eigenvalue expression and new transition selection rules. The spirit behind a decomposition is that each step involves degeneracy breaking on a smaller energy scale so that families of levels remain together. Thus, while not rigorously required, this implies that $A >> B >> C$. Since the τ values of the γ band levels are $2^+(\tau=2)$, $3^+,4^+(\tau=3)$, $5^+,6^+(\tau=4)$, it is immediately clear that the γ-band energy staggering in O(6) is just the opposite to the triaxial case, namely (2_γ^+), $(3_\gamma^+ 4_\gamma^+)$, $(5_\gamma^+ 6_\gamma^+)$...

Since the several options[11-13] to incorporate triaxiality into the IBA are similar, the discussion below will be limited to using one of these, namely the incorporation of cubic terms. Thus, to investigate further the relation between γ softness and mean γ, one can then consider an extended CQF Hamiltonian of the schematic form

$$H = -\kappa Q \cdot Q - \kappa' L \cdot L + \theta_3 H_{cubic} \tag{6}$$

where

$$H_{cubic} \sim (d^+d^+d^+)^{(3)}(\tilde{d}\tilde{d}\tilde{d})^{(3)}$$

and study the systematic evolution of γ-band energy staggering as a function of χ and θ_3. Typical results of calculations[6] with such a Hamiltonian are shown at the top of Fig. 4 for $N=8$. A small θ_3 leaves the O(6)-like staggering intact but a more uniform spacing is achieved for larger θ_3. This change in structure results, not so much from a variation in γ (since $\gamma \approx 30°$ over the entire range of θ_3 from O(6) to pure triaxiality) but from a decrease in the

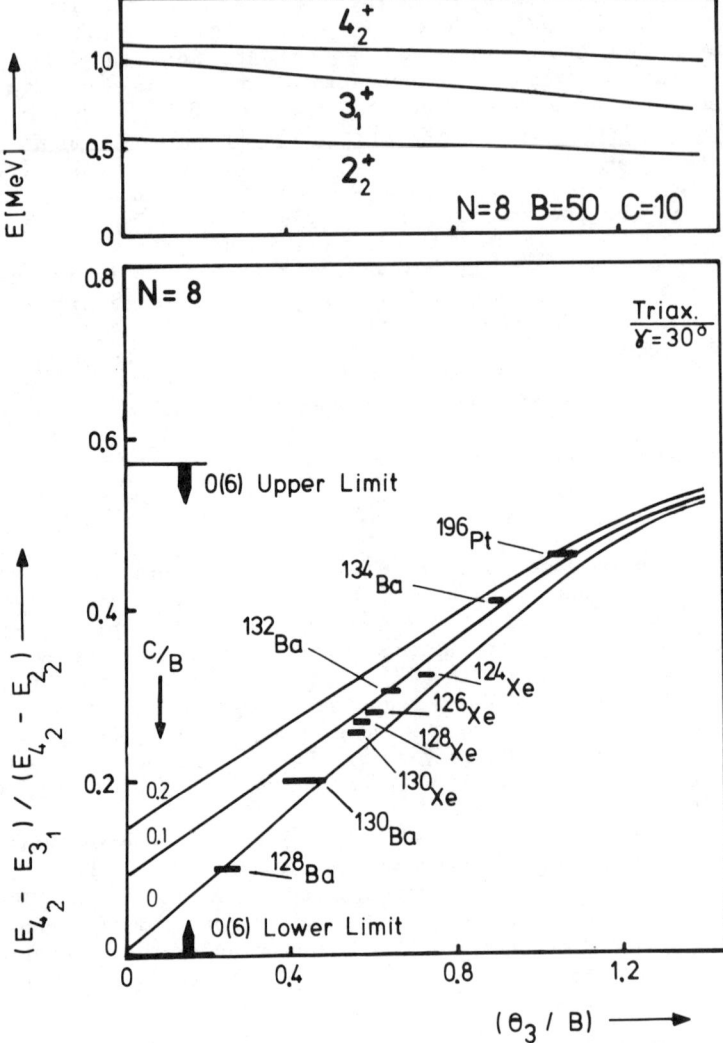

Fig. 4. Calculations of the γ band energy staggering with the Hamiltonian of eq. 6, for χ=0 and N=8. The eigenvalue equation is that of eq. 4 (the A term is irrelevant since only σ=N states are involved) with the addition of a term in θ_3. The upper part shows the change in γ band energy staggering for typical values of B and C. The lower part shows the application to the A=130 region. The energy ratio plotted depends only on θ_3/B (B is proportional to κ in eq. 4) and on C/B. Results for several C/B values are shown. Upper and lower bounds consistent with O(6) and the γ=30° triaxial ratio values are also shown. The cubic term is strongly N dependent: thus, the results for nuclei with N≠8 are only approximate. For each nucleus the energy ratio is plotted and positioned to correspond to a range of C/B values extracted from level spacings that are insensitive to the cubic term.

associated γ-softness as larger θ_3 values lead to a developing potential minimum at γ=30°.

It is apparent from these results that the introduction of cubic terms into the IBA-1 hamiltonian leads to a flexibility in the relation between γ-softness and mean γ that is absent from the usual IBA-1. Whether such flexibility is necessary in fitting actual nuclei remains an interesting question. Calculations[14] for ^{104}Ru suggest that it is needed there. Results to be presented next will show that it provides an equally useful tool in the O(6)-like region near A=130 and, indeed, probably even in ^{196}Pt.

4. AN EXTENSIVE O(6)-LIKE REGION NEAR A=130

Ever since the discovery[15] that the O(6) limit is manifested empirically in ^{196}Pt and neighboring Pt isotopes, there has been high interest in searching for new O(6) regions. There are at least two reasons for this. First, one would like to determine if Pt is an isolated peculiarity or if the O(6) coupling scheme is, like the occurrence of deformed nuclei, a periodic phenomenon. Secondly, the discovery of a symmetry region immediately provides a benchmark for the treatment of nearby regions. For example, in the case of Pt, the recognition of the O(6) symmetry immediately led to a new, simpler interpretation[15] of the entire Os-Pt transition region as undergoing an O(6)→SU(3) transition.

It was pointed out[16] rather early that nuclei such as ^{132}Ba resembled the O(6) limit but the data were sparce: states of higher τ values or those of the higher lying σ<N representations were not assigned. Many crucial B(E2) values remained unknown. Consequently, this suggestion[16] was not explored in detail. Recently, however, extensive data have been acquired, primarily by the Koln group[17-22] using the (α,nγ) and (^{13}C,xnγ) reactions. The former is particularly useful since, like (n,γ), it is highly non-selective and yet accesses as broad spin range. These data provide a wealth of new information that now permits a careful inspection of the A=130 region.

The level schemes for eight of these nuclei are collected in Fig. 5 and grouped according to their O(6) quantum numbers. For comparison a typical O(6) spectrum and that for ^{196}Pt, with energies multiplied by 1.63 (see below), are appended at the right. Tables 1 and 2 show some energy relations and relative B(E2) values.

Several points are clear upon inspection of this figure and the tables:
1) All of the nuclei closely resemble the O(6) limit. The $4^+_1/2^+_1$ and $2^+_2/2^+_1$ energy ratios are all close to 2.5. All the nuclei have, albeit to differing degrees, the typical O(6) clustering in the γ band. More generally, all exhibit clearly identifiable, closely grouped, τ multiplets. None (except ^{120}Xe which is unique in other respects as well) has a first excited (0^+_2) level near the 4^+_1 and 2^+_2 states. Rather, the first excited 0^+ state, where known, always lies near the 4^+_2 and 3^+_1 levels and, moreover, decays preferentially to the 2^+_2 state, not the 2^+_1.

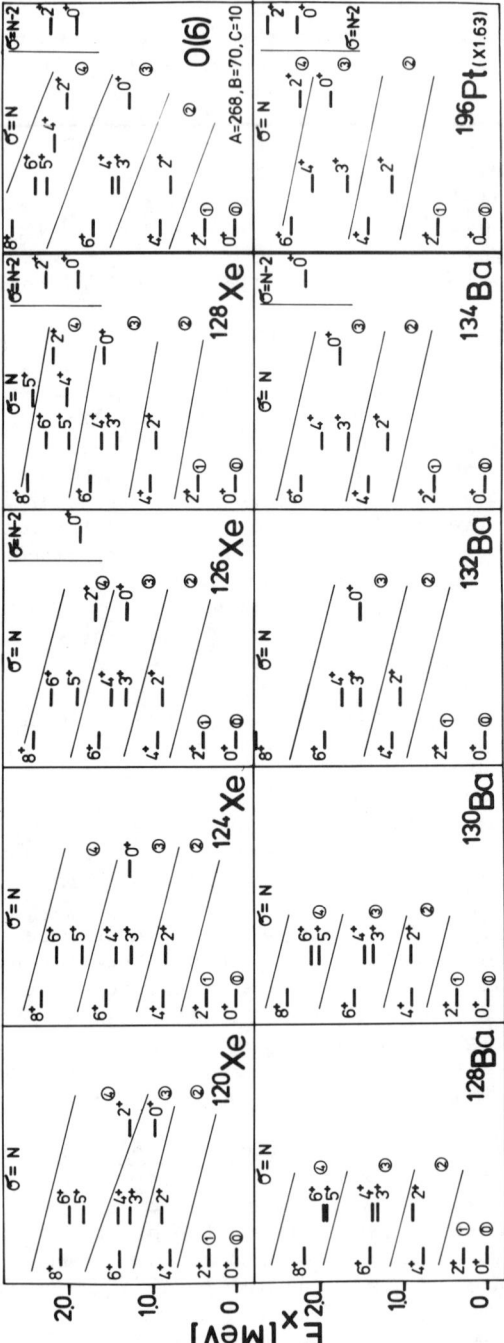

Fig. 5. Energy levels for 8 nuclei near A=130 compared to the O(6) limit and to a scaled level scheme for ^{196}Pt. The O(6) scheme is not a fit, which would differ slightly for each nucleus, but is intended to convey the general level patterns. The σ quantum numbers are given at the top and the τ quantum numbers are circled. The data are from refs. 17-25.

This is the allowed transition for a $\tau=3$ level and is one of the distinguishing features between O(6) and SU(5). In several of the nuclei, states identifiable as the heads of the $\sigma=N-2$ representation can be recognized. In ^{128}Xe, a 4^+ state with $\tau=4$ is also assigned, based on its energy and decay properties. This is the first identification of the complete $\tau=4$ multiplet since this 4^+ level has not been found in Pt to date. The B(E2) values also follow closely the O(6) predictions: allowed transitions are invariably strong or, if unobserved, the upper limits allow for strong relative B(E2) values. Forbidden transitions are always weak. Breaking of the O(6) selection rules are at the 10% or less level with the exception of two transitions at the 15% level. In some cases, such as the decay of the 3^+_γ level, the agreement with O(6) is remarkable. The only notable deviations from O(6) lie in the detailed relative sizes of the three allowed transitions from the 5^+_γ level. In addition to the results tabulated here, others, omitted for space reasons, such as the B(E2) values for the $\tau=4$ 2^+ and 4^+ states, also agree well with O(6).

2) All of the nuclei are <u>remarkably similar</u> to each other and to ^{196}Pt. The only important systematic change, across these nuclei is the $4^+_2-3^+_1$ spacing which grows with increasing mass in the Ba isotopes. The relative energies of the $\sigma=N-2$ 0^+ states are rather stable, increasing slightly with mass. Even the deviations from O(6) are similar. As will be seen below, the $4^+_2-3^+_1$ splittings are too large, and, in every case (again except ^{120}Xe), the $0^+_2(\tau=3)$ level is at or above the 3^+_1 energy, rather than below it as required by the monotonic energy ordering of states within a τ multiplet due to the J(J+1) term in eq. 4. Whatever interactions break the O(6) limit in Pt are presumably active and comparably strong in the A=130 region.

3) The factor of 1.63 multiplying the ^{196}Pt energies is just a convenient number designed to put the Pt energies on a similar scale to the other nuclei. Nevertheless, it is interesting to note that it is not far from an $A^{5/3}$ scaling which is the characteristic factor for the inertial parameter of a quadrupole ellipsoid.

4) In many respects, these Xe and Ba nuclei are better representations of O(6) than is ^{196}Pt. For example, in ^{196}Pt the $2^+_2/2^+_1$ energy ratio is lower than in Xe-Ba, the $4^+_2-3^+_1$ energy splitting is relatively larger, the $0^+(\tau=3)$ level disagrees more with the predicted sequence for the $\tau=3$ multiplet, the $0^+(\tau=3)$ level has a relatively stronger decay branch to the 2^+_1 level and the $3_\gamma \to 4_g$ transition is too strong. On the other hand, of course, more levels from the higher, $\sigma<N$, representations and more examples of the characteristic $0^+-2^+-2^+$ sequences are known in ^{196}Pt.

There are two particularly interesting aspects of these results that deserve expanded discussion. In the CQF with $\chi=0$, the Hamiltonian of eq. 1 cannot, of course, lead to three independent terms in the O(6) eigenvalue eq. 4 but rather gives a special case of O(6) in which A/4=B. This implies that the CQF implicitly

Table I: Empirical Energy Ratios near A = 130

	N	E_{4_1}/E_{2_1}	E_{2_2}/E_{2_1}	$\dfrac{E_{4_2}-E_{3_1}}{E_{2_1}}$	$\dfrac{E_{4_2}-E_{3_1}}{E_{4_2}-E_{2_2}}$	$\dfrac{E_{3_1}-E_0(\tau=3)}{E_{4_2}-E_{3_1}}$	$\dfrac{E_0(\sigma=N-2)}{E_{2_2}-E_{2_1}}$
^{120}Xe	10	2.47	2.72	0.41	0.25	+ 2.4	--
^{124}Xe	8	2.48	2.39	0.54	0.32	- 0.11	--
^{126}Xe	7	2.42	2.26	0.44	0.28	- 0.02	3.58
^{128}Xe	6	2.33	2.19	0.39	0.27	- 0.88	3.57
^{128}Ba	8	2.69	3.11	0.17	0.10	--	--
^{130}Ba	7	2.52	2.54	0.33	0.20	--	--
^{132}Ba	6	2.43	2.22	0.47	0.31	- 0.04	--
^{134}Ba	5	2.32	1.93	0.54	0.41	- 0.36	3.83
^{196}Pt	6	2.47	1.94	0.78	0.46	- 0.43	4.21
O(6)[a]	N	2.5-3.33	0-2.5	0-1.33	0-0.57	+ 1.5	$\tfrac{2}{3}$(N+1)
Triax.[a] $\gamma = 30°$	--	2.67	2.0	2.67	0.73	($-\infty$)	(∞)

[a] Ranges of values for some O(6) entries give the extrema for B>>C and C>>B in eq.4. Since one expects B to dominate C, the corresponding end of each range is favored and underlined. For the last column, the general O(6) limit of eq.5 allows any value. However, the CQF gives a special case of the O(6) limit in which A/4 = B, giving the relation indicated. In the simplest version of a triaxial rotor, 0^+ states are at infinite energy, as indicated, although of course, 3-softness can be introduced into that model. Data from refs. 17-25.

Table II: Relative B(E2) values near A = 130 compared to the O(6) limit[a]

I_i / I_f	2_γ		3_γ			4_γ				5_γ				$0_{\tau=3}$	
	$2g$	$0g$	2_γ	$4g$	$2g$	2_γ	3_γ	$4g$	$2g$	4_γ	3_γ	$6g$	$4g$	2_γ	$2g$
Nucleus															
^{120}Xe	100	5.6	100	50	2.7	100	--	62	--	--	100	--	3.1	--	--
^{124}Xe	100	3.9	100	46	1.6	100	--	91	0.4	106	100	--	3.8	100	1
^{126}Xe	100	1.4	100	47	1.1	100	--	42	1.0	127	100	--	4.9	100	9
^{128}Xe	100	1.2	100	37	2	100	--	133	1.7	88	100	204	3.7	100	14
^{128}Ba	100	9.2	--	--	--	100	--	42	1.7	≤ 44	100	≤ 56	3.0	--	--
^{130}Ba	100	5.7	100	30	1.5	100	--	89	3.9	≤ 57	100	381	6.7	--	--
^{132}Ba	100	0.2	100	73	0.2	100	--	75	2.2	--	--	--	--	100	0
^{134}Ba	100	0.6	100	40	1.0	100	14.5	77	2.5	--	--	--	--	100	4
^{196}Pt	100	0.0	100	95	0.1	100	--	109	1.7	--	--	--	--	100	16
O(6) (N = 6)	100	0	100	40	0	100	0	91	0	46	100	45	0	100	0
Rotor	100	70	(100)	(0.4	1)	(100	222)	(1	0.34)	100	100	(0.57	1)	--	--

[a] For each initial level, one transition is assigned a value of 100. The O(6) values are always for N=6: they change little with N. The rotor values are Alaga rules. In the rotor, it is arbitrarily assumed that there is a factor of 100 difference between intra- and inter-band transition B(E2) values. To emphasize this arbitrariness transitions to a given band are grouped within parentheses. Where mixing ratios are measured, the E2 strength has been used; elsewhere, pure E2 multipolarities are assumed. From refs. 17 - 25.

incorporates a relation between the τ and σ energy scales, that is, between the strength of the O(6) and O(5) degeneracy breaking steps in eq. 5. In deformed nuclei the analogous statement would concern a relation between the rotational energy scale and that of the β band intrinsic excitation. When the CQF was proposed[3], there was no a priori reason to expect this special case to exist empirically: yet, the previously deduced parameter values for ^{196}Pt, $A/4 = 46.2$ and $B = 43$ keV, were indeed nearly equal. It is all the more surprising now to observe this same special relation nearly satisfied throughout the Xe-Ba region, as evidenced by the approximate regional O(6) parameters shown in Fig. 5. This is also clear from the energy ratios $E_{0^+(\sigma=N-2)}/(E_{2^+_2}-E_{2^+_1})$ in Table 1 that hover near a value of 4 which is close to the Pt value and to the CQF prediction of 4.67 for $N=6$. The underlying significance of the linking of the τ and σ energy scales is a topic deserving further study.

The second point of special interest is the γ band energy staggering. While the $CJ(J+1)$ term in eq. 4 in principle allows for any $4^+_2-3^+_1$ spacing, one expects the τ multiplets to be closely grouped (i.e., $C \ll B$) in a good O(6) nucleus. Moreover, in any case, C can be extracted from other spacings and one can then test if the empirical $4^+_2-3^+_1$ spacing is reproduced. One finds that, in general, in this region, this is not the case: the γ band is more regularly spaced than in the O(6) limit. From Fig. 4 (upper part) one notes that this is just the effect of an increase in γ rigidity compared to the extreme γ independence of the O(6) limit, suggesting that these nuclei are γ soft but not completely γ unstable. This idea is made more quantitative in the lower part of Fig. 4 which shows calculations of γ band energies, for $N=8$ and $\chi=0(O(6))$ as a function of θ_3/B, for several C values. (See caption for a detailed explanation of the figure.) Each nucleus is plotted according to its known γ band energy values and with a range of C values extracted from the $6^+_1-4^+_2$ and $4^+_2-2^+_1$ energy spacings that are nearly independent of γ softness. If the $4^+_2-3^+_1$ spacing were consistent with O(6) all the plotted points should lie at the extreme left (i.e., at $\theta_3=0$). At the other extreme, with deviations from O(6), a wide range of values of the energy ratio plotted is possible. Thus, the figure gives, in effect, the θ_3/B value needed to reproduce consistently all the energy spacings. It is remarkable to observe then the very systematic behavior: the Xe isotopes all cluster in a narrow range and the Ba isotopes show a clear systematic increase in required triaxiality. Again, it is interesting to note that ^{196}Pt deviates from the O(6) limit more than any of these nuclei near $A=130$. It is also clear that none of them approaches very close to the pure triaxial case.

To summarize the foregoing:
1) The new data clearly point to the $A=130$ region as a good, and extensive, example of the O(6) symmetry, which is, in some respects, better and more widespread than in Pt.
2) The nuclei are all remarkably similar to each other and to the Pt isotopes, even in the characteristic deviations from O(6).

3) The relation between the σ and τ energy scales is nearly identical to Pt and both are very close to the special case of O(6) given in the Consistent Q Formalism, in which the coefficients A/4 and B are equal. It is clearly of considerable interest to explore further the deeper meaning of this relation between the energy scales of the O(6) and O(5) steps in the chain decomposition (degeneracy breaking) of the parent U(6) group.
4) Fixing the C parameter from other, more stable energy splittings, it turns out that the energy staggering (e.g., 4^+_2-3^+_1 spacing) in the γ band cannot be rigorously accounted for in O(6) but rather implies somewhat less γ softness than in the extreme limit. Calculations with a cubic term that introduces a more rigid asymmetry were able to reproduce the data with smoothly varying values of the relative strength parameter, θ_3/B. These nuclei are γ soft but not γ independent.

ACKNOWLEDGEMENTS

I am extremely grateful to A. Aprahamian, D. D. Warner, K. Heyde, P. Van Isacker, J. Jolie, and P. von Brentano who are my collaborators on this work and who have allowed me to present these results in their names as well. Many enlightening discussions with them, and with Shi Zong-ren, O. Scholten, A. Frank, P. Lipas, and J. Stachel are thankfully acknowledged. Finally, thanks are due to many of the students and staff at the Instutut fur Kernphysik in Koln (see refs. 17-22) for access to their results in the A=130 region, often prior to final publication.

Research has been performed under contract DE-AC02-76CH00016 with the United States Department of Energy and partially supported by an award from von Humboldt Foundation as well.

REFERENCES

1. J. N. Ginocchio, in *Interacting Bose-Fermi Systems in Nuclei*, ed. F. Iachello (Plenum Press, New York, 1980), p. 179, and J. N. Ginocchio and M. W. Kirson, Nucl. Phys. A350, 31 (1980); A. E. L. Dieperink, O. Scholten, and F. Iachello, Phys. Rev. Lett. 44, 1747 (1980).
2. A. Arima and F. Iachello, Phys. Rev. Lett. 35, 1069(1975) and *Advances in Nuclear Physics*, Vol. 13, ed. J. W. Negele and E. Vogt (Plenum Press, New York, 1984), Chap. 2, and references therein. See also the *Proc. of the Intern. Workshop of Collective States in Nuclei*, Suzhou, Sept. 1983, ed. L. M. Yang, X. Q. Zhou, G. O. Xu, S. S. Wu and D. H. Feng, Nucl. Phys. A421, 1c-429c (1984).
3. D. D. Warner and R. F. Casten, Phys. Rev. C28, 1798 (1983), and Phys. Rev. Lett. 48, 1385 (1982).
4. R. F. Casten, A. Aprahamian, and D. D. Warner, Phys. Rev. C29 (Rap. Comm.), 356 (1984).

5. R. F. Casten, D. D. Warner, and A. Aprahamian, Phys. Rev. C28, 894 (1983); R. F. Casten and A. Aprahamian, Phys. Rev. C29, 1919 (1984).
6. R. F. Casten et al., to be published.
7. R. F. Casten and P. von Brentano, to be published.
8. R. F. Casten, Proc. of the Intern. Workshop on Interacting Boson-Boson and Boson-Fermion Systems, Gull Lake, Michigan, May 1984, ed. O. Scholten, World Scientific Pub. Co., in press.
9. A. S. Davydov and G. F. Filippov, Nucl. Phys. 8, 237 (1958), and E. P. Grigoriev and M. P. Avotina, Nucl. Phys. 19, 248 (1960).
10. O. Castanos, A. Frank, and P. Van Isacker, Phys. Rev. Lett. 52, 263 (1984).
11. K. Heyde et al., Phys. Rev. C29, 1420 (1984).
12. K. Heyde et al., Nucl. Phys. A398, 235 (1983).
13. A. E. L. Dieperink and R. Bijker, Phys. Lett. 116B, 77 (1982).
14. J. Stachel et al., Nucl. Phys. A419, 589 (1984).
15. J. A. Cizewski et al., Phys. Rev. Lett. 40, 167 (1978); R. F. Casten and J. A. Cizewski, Nucl. Phys. A309, 477 (1978).
16. See, for example, A. Arima and F. Iachello, Ann. Phys. (N.Y.) 123, 468 (1979).
17. K. Loewenich et al., Proc. of the Intern. Symp. on Nuclear Spectroscopy and Nuclear Interactions, Osaka, 1984, to be published., and K. Loewenich, Diplomarbeit and private communication.
18. W. Gast et al., Z. Phys., to be published and private communication.
19. K. Schiffer et al., Z. Phys. A313, 245 (1983), and to be published: K. Schiffer, Diplomarbeit and private communication.
20. W. Lieberz, A. Dewald and R. Reinhardt, private communication.
21. Sun, Xianfu et al., Phys. Rev. C28, 1167 (1983) and Nucl. Phys., to be published and private communication.
22. A. Dewald et al., Z. Phys. A315, 77 (1984).
23. H. Kusakari et al., Nucl. Phys. A242, 13 (1975).
24. H. R. Hiddleston and C. P. Browne, Nucl. Data Sheets 17, 225 (1976) and 13, 133 (1974).
25. T. Tamura, K.Miyano and S. Ohya, Nucl. Data Sheets 36, 227 (1982), and Yu. V. Sergeenkov and V. M. Sigalov, Nucl. Data Sheets 34, 475 (1981).

ANHARMONICITIES IN THE VIBRATIONAL SPECTRA IN DEFORMED NUCLEI

R. PIEPENBRING

Institut des Sciences Nucléaires, 38026 Grenoble, France

ABSTRACT

Several experimental results have recently raised again the problem of the description of the anharmonicities of the vibrations observed in even-even deformed nuclei.

First, despite extensive experimental studies, one has not found any vibrational two-phonon octupole 0^+ state in $^{222-226}$Ra and $^{224-226}$Th at around double the energy of the one-phonon state. Second, the completeness of the low spin ($J<6$) energy level spectra recently obtained up to 2 MeV in the ^{168}Er exhibits clearly that the 2 γ vibrational state appears at an energy higher than twice the energy of the one phonon state.

Phenomenological explanations based on the IBA have obtained partial success but still suffer from a number of shortcomings. Different microscopic approaches have been developed : the nuclear field theory, based on perturbation theory, appeared not to be appropriate in the case of the strong anharmonicities observed in ^{168}Er. The boson expansion technique has also been used, but in this approach the Pauli principle cannot be fully taken into account. Finally, the multiphonon method has been developed to improve these two approaches in two aspects : the Pauli principle is properly treated and exact diagonalisation within collective space is used. This method is illustrated by the description of octupole vibrations in the Ra - Th region. Other possible applications are also indicated.

1. INTRODUCTION

The description of the vibrational levels in even-even deformed nuclei is a very old and challenging problem. An interpretation of the one phonon states in terms of a coherent mixture of two quasi-particle states can easily be achieved by the use of quasibosons and the harmonic assumption. For higher vibrational excitations the situation is more complicated. Several recent experimental results have clearly shown that large anharmonicities can arise as soon as one looks for the two phonon states. First, let us mention the completeness of the low spin (J < 6) energy level spectra recently obtained up to ≃ 2 MeV in ^{168}Er by use of high resolution gamma-ray studies following neutron capture [1]. In this work, anharmonic effects of the gamma motion are observed. The band head of the γ vibration has an energy of 821 keV ; the lowest candidates for the 2 γ state with K = 4 occur with band heads at 2023 keV or 2055 keV. This implies a major deviation from the harmonic vibrations since

$$\frac{E(2\gamma)}{E(\gamma)} \geq 2.5$$

Second, we would like to remind the reader that, despite very extensive experimental studies [2], it has not been possible to find any two-phonon octupole vibrational state 0^+ in $^{222-224-226}$Ra and $^{224-226}$Th at around double the energy of the one phonon state 0^-.

An interpretation of the positive-parity spectrum of ^{168}Er in terms of the interacting boson approximation (IBA) by Warner, Casten and Davidson [3] has shown partial success, but could not explain the above - mentioned anharmonicity in the γ motion. Indeed, by choosing the parameters of the IBA, so to reproduce approximatively the energy of the γ band head they found the K = 4 state at 1610 keV, i.e. more than 400 keV too low. Furthermore, it is well known that the IBA, in its actual versions cannot yet account for the negative parity states.

Different microscopic approaches have also been developed to try to understand the anharmonicities of vibrational states. The boson expansion technique [4] has extensively been used for spherical nuclei and applied to deformed nuclei [5] .

In this method the Pauli principle cannot be fully taken into account and this may lead to intruder states of no physical meaning at low energy.

Corrections to the harmonic random phase approximation (RPA) have also been attempted. Neergård an Vogel [6] have looked for the octupole case whereas Dumitrescu and Hamamoto made an analysis of the gamma motion in ^{168}Er. It was shown that a perturbative calculation based on axially symmetric RPA wave functions may not be very appropriate in the case of the strong anharmonicities observed.

A non - perturbative treatment using exact diagonalization will be more appropriate. In any case the Pauli principle has to be

properly taken into account.
The multiphonon method (MPM) has been developed to fulfill these two requirements.

In section 2 we give a sketch of this method. In section 3 it is illustrated in the case of the octupole vibrations in the Ra-Th region. Other possible applications are indicated in section 4, whereas conclusions are drawn in a last section.

2. THE MULTIPHONON METHOD

To take advantage of the collective nature of the vibrational states, the multiphonon method (MPM) introduces phonons $Q_i^\dagger$, which are defined as a superposition of two quasi-particles

$$Q_i^\dagger = \frac{1}{2} \sum_{mn} (X_i)_{mn} \alpha_m^\dagger \alpha_n^\dagger \qquad (1)$$

where
$(X_i)_{mn} = -(X_i)_{nm}$ is chosen to be an antisymmetric matrix, and $\alpha_m^\dagger$ creation operators of fermions (quasiparticles). Contrary to what is assumed in the IBA or in the quasiboson approximation the entities (1) are no longer considered as bosons since their commutation rules are now

$$[Q_1, Q_2^\dagger] = -\frac{1}{2} \mathrm{tr}(X_1 X_2) + \sum_{mn}(X_1 X_2)_{mn} \alpha_n^\dagger \alpha_m \qquad (2)$$

As an example the $\Omega_i^\dagger$ can be chosen as the Tamm-Dancoff collective phonons. In a second step, one selects the phonons $Q_i^\dagger$ which may play an important role in the problem one wants to study and builds up the multiphonon basis

$$|abc\ldots\rangle = Q_1^{\dagger a} Q_2^{\dagger b} Q_3^{\dagger c} \ldots |0\rangle \qquad (3)$$

These states (3) do generally not form an orthogonal basis. The first problem is then to calculate the overlap matrix

$$\langle a'b'c'\ldots | abc\ldots\rangle \qquad (4)$$

The next step of the MPM consists in diagonalizing the *total* Hamiltonian in the *collective* space spanned by the multiphonon basis (3). (We remind here the reader that in the harmonic TDA or RPA only a part of H is taken into account).

To calculate the matrix elements (4) and those of one and two body operators, one has two possibilities as has been shown in [8]: either an extensive use of a Wick's theorem for phonons or that of coupled recursion formulas.
Different applications [9] have shown that the second possibility is more convenient for realistic numerical calculations. The formalism gets particularly simple [10] in the case where only one type of phonons plays a role. The basis (3) can then be written:

$$|n\rangle = \frac{\Omega^{\dagger n}}{n!}|0\rangle \qquad (5)$$

Since this is just the assumption which can reasonably be made in the study of the octupole K = 0 vibrations mentioned in the introduction, we shall give here the explicit recursion formulas which are now decoupled.

For the norms $N_m = \langle n | n \rangle$ of the states (5) (where n! has been introduced to simplify as much as possible the algebraic relations) one has the recursion formula

$$n N_n = -\frac{1}{2} \sum_{\ell=0}^{n-1} N_{n-1-\ell} \, \text{tr}(\chi^{2\ell+2}) \tag{6}$$

As usual, we express the Hamiltonian of the system in terms of the quasiparticles obtained with the canonical Bogoluybov-Valatin transformation.

$$H = \mathcal{U} + H_{11} + H_{22} + H_{31} + H_{40} \tag{7}$$

with

$$H_{11} = \sum_p E_p \, \alpha_p^\dagger \alpha_p \tag{8}$$

$$H_{22} = \sum_{pqrs} S_{pqrs} \, \alpha_p^\dagger \alpha_q^\dagger \alpha_s \alpha_r \tag{9}$$

$$H_{31} = \sum_{pqrs} R_{pqrs} (\alpha_p^\dagger \alpha_q^\dagger \alpha_r^\dagger \alpha_s + \alpha_s^\dagger \alpha_r \alpha_q \alpha_p) \tag{10}$$

$$H_{40} = \sum_{pqrs} P_{pqrs} (\alpha_p^\dagger \alpha_q^\dagger \alpha_r^\dagger \alpha_s^\dagger + \alpha_s \alpha_r \alpha_q \alpha_p) \tag{11}$$

The coefficients P, R, S are assumed to be real and verify all symmetry properties of the quasiparticle operators.

With these notations, the recursion formulas for the matrix elements write

$$\langle n | H_{11} | n \rangle = -\sum_{\ell=0}^{n-1} N_{n-1-\ell} \, \mathcal{E}(\ell+1) \tag{12}$$

$$\langle n+2 | H_{40} | n \rangle = 3 \sum_{\ell=0}^{n} N_{n-\ell} \sum_{i,j=0}^{\ell} \delta_{i+j,\ell} \, \mathcal{P}(i,j) \tag{13}$$

$$\langle n+1 | H_{31} | n \rangle = -3 \sum_{\ell=0}^{n-1} N_{n-1-\ell} \sum_{i,j=0}^{\ell} \delta_{i+j,\ell} \, \mathcal{R}(i,j+1) \tag{14}$$

$$\langle n | H_{22} | n \rangle = -\sum_{\ell=0}^{n-1} N_{n-1-\ell} \sum_{i,j=0}^{\ell} \delta_{i+j,\ell} \, \mathcal{S}(i,j)$$

$$-2 \sum_{\ell=0}^{n-2} N_{n-2-\ell} \sum_{i,j=0}^{\ell} \delta_{i+j,\ell} \, \mathcal{S}'(i+1,j+1) \tag{15}$$

where the dynamical quantities $\mathcal{E}, \mathcal{P}, \mathcal{R}, \mathcal{S}$ and $\mathcal{S}'$ are :

$$\mathcal{E}(i) = \mathrm{tr}\,(E X^{2i}) \qquad (16)$$

$$\mathcal{P}(i,j) = \sum_{pqrs} P_{pqrs}\,(X^{2i+1})_{pq}\,(X^{2j+1})_{rs} \qquad (17)$$

$$\mathcal{R}(i,j) = \sum_{pqrs} R_{pqrs}\,(X^{2i+1})_{pq}\,(X^{2j})_{rs} \qquad (18)$$

$$\mathcal{S}(i,j) = \sum_{pqrs} S_{pqsr}\,(X^{2i+1})_{pq}\,(X^{2j+1})_{rs} \qquad (19)$$

$$\mathcal{S}'(i,j) = \sum_{pqrs} S_{pqsr}\,(X^{2i})_{pr}\,(X^{2j})_{qs} \qquad (20)$$

If one wants to evaluate other observables, like electromagnetic transition probabilities, one may also need the recursion relation for the operator

$$H_{20} = \sum_{pq} F_{pq}\,\alpha_p^\dagger \alpha_q^\dagger \qquad (21)$$

One simply has in this case :

$$\langle n+1|\,H_{20}\,|n\rangle = \sum_{\ell=0}^{n} N_{n-\ell}\,\mathcal{F}(\ell) \qquad (22)$$

where

$$\mathcal{F}(i) = -\,\mathrm{tr}\,(F\,X^{2i+1}) \qquad (23)$$

The eigenstates of H are obtained by the diagonalization in a basis containing all collective multiphonons (5) up to a maximum value n_{max} chosen so as to get numerical stability for the three lowest eigenvalues [e.g. In the application of section 3, $n_{max} = 12$]

These eigenstates appear then as a superposition of states (5) with different values of n. In terms of quasiparticles they would be extremly complicated.
The validity of the MPM has been checked [9] in several simple models where an exact solution is available.

3. APPLICATION OF THE MULTIPHONON METHOD TO THE ANALYSIS OF OCTUPOLE K = 0 VIBRATIONS IN THE Ra - Th REGION

In the even Ra and Th isotopes with mass number $222 \leq A \leq 228$ one observes, apart from the ground state band, negative-parity levels which can be arranged in a K = 0 rotational band.
The head of this band has spin I = 1 and appears at very low energy : less than 250 keV in some of these nuclei. One also knows the first levels of a positive-parity rotational band. The energy difference between these two band heads is especially large. As has been justified in [11], it seems reasonable to consider that the studied nuclei present a static quadrupole and a dynamic octupole deformation. The octupole vibrations are then obviously anharmonic since no 0^+ state has been found at around double the energy of the one phonon 0^- state. This latter is energically well separated from all other intrinsic excitations so that the MPM method with

only one building phonon can be used. Formulas (5-20) can therefore be applied.

The situation can still be simplified. Indeed, according to parity conservation the contribution of H_{31}, which changes the phonon number by one (see rel. 14) vanishes. The matrix of H separates into two parts, one for each parity. The vectors $|2p\rangle$ form the basis for the diagonalization of the positive parity states, whereas $|2p+1\rangle$ serve as such for the negative parity one.

We would like to emphasize that the aim of this application is not to search for a fine agreement with the experimental results. To accomplish such a task, it would be necessary to take also into account the coupling with other vibrational or non collective modes and to use a basis of type (3). According to this limited goal, we simply describe the considered nuclei as being built of neutrons and protons moving in a deformed Nilsson field and interacting through a constant monopole pairing force and a charge independent octupole - octupole force. The model Hamiltonian is

$$H = H_{sp} + H_p + H_{oct} \tag{24}$$

where

$$H_{sp} = \sum_{p,m} e_m(p) a_m^\dagger(p) a_m^\dagger(p) \tag{25}$$

$$H_p = -\sum_p G_p L_p^\dagger L_p \tag{26}$$

$$H_{oct} = -\frac{\chi}{2} \sum_{p,p'} O_p O_{p'} \tag{27}$$

with $L_p = \sum_m a_{-m}(p) a_m(p)$ (28)

$$O_p = \sum_{mn} \langle m\, p | r^3 Y_{30} | n\, p \rangle a_m^\dagger(p) a_n(p) \tag{29}$$

In these relations p refers to protons or neutrons whereas m or n labels the single particle states. The canonical Bogoluybov-Valatin transformation is used to switch from particle operators a_m to quasiparticle operators α_m (we note here that this approach of the pairing correlations introduces, as is well known, a fluctuation of the number of particles which may have some importance for the ground state energy [9] but less in the relative spectra).

The explicit form of each part of H in relations (8) to (11) are not given here ; they can be found in the literature [10]. We simply remind the reader that the P, R and S coefficients have to fulfill all the required symmetry properties.

We use the Nilsson potential with the parameters recommended by Lamm [12]. Even if this potential does not give the required level order we do not make any special single particle energy adjustments. We note that according to the limited aim of this section this may be sufficient in even - even nuclei, but certainly not in odd- A isotopes. The intrinsic matrix elements have been calculated according to the prescriptions of Boisson an Piepenbring [13]. The calculations are made for a deformation $\varepsilon = 0.15$, which is suited for the set of nuclei at the beginning of the deformed region. For the pairing strength G we use the values of Nilsson and Prior [14].

$$G_p = \frac{22}{A} \text{ MeV} \qquad G_n = \frac{16}{A} \text{ MeV}$$

Thirty active levels, equitably distributed on each side of the Fermi surface have been introduced in our calculations. This choice is a sort of compromise. If one chooses less levels one may have numerical instability, if one chooses more, one may introduce effects of states which would not be bound in a more realistic finite potential well. As usual, the $-G v^2$ single particle renormalization has been taken into account.

The X matrix of relation (1) is obtained by solving the secular equation of the Tamm - Dancoff approximation (TDA). The MPM works [10] for any choice of an antisymmetric unitary matrix X so that it is not necessary to use the same value χ in the Hamiltonian (27) as for the calculation of the TDA phonon (1). However, if one wants to restrict as much as possible the number of free parameters one may take $\chi = \chi_{TDA}$.

We adjust χ so to get the lowest calculated 0^- energy in the neighbourhood of the intrinsic band head energy deduced from experimental results.

We mention here that the observed energy of the I = 1 band head contains a rotational contribution which can roughly be estimated to ~ 10 keV.

For the retained χ value we also look if the RPA secular equation has any physical solution.

The most interesting information consists in the location of the first excited 0^+ state, which we compare to the observed one. (for this level where I = K = 0 no rotational energy has to be substracted).

In Table I we give the results for six nuclei. The following remarks can be made:

a) For the selected χ value, the TDA energies ω_{TDA} of the first excited 0^- state are quite large compared to the experimental energy $E(0^-)$, explicitly demonstrating the importance of the H_{40} term (11) neglected in the TDA.

b) The secular equation of the RPA has no real solution.

c) The calculated $E(0^+)$, energies are systematically larger than $2 E(0^-)$ showing the great anharmonicity of the octupole vibrations. They are also larger than the measured ones. The deviation

$E(0^+)$ calculated $- E(0^+)$ observed

increases with A. It is low for ^{222}Ra and ^{224}Ra but gets quite large for isobars 226 and 228.

This finding suggests that the first excited 0^+ state in ^{222}Ra and ^{224}Ra could practically be of a pure octupole nature. This is certainly no more the case in nuclei with A = 226 and 228. In these nuclei other K = 0^+ excited states have been observed [e.g. at 1042 keV in ^{228}Ra and 888 keV in ^{228}Th].

The restriction to a basis (5) with only one building phonon is than no more justified. The coupling to a quadrupole type K = 0^+ phonon may play an important role and a larger basis (3) is needed.

We have also studied the sensitivity of the results of Table I to variations of the deformation ϵ of the single particle potential and/or of the pairing strength parameters G. Qualitatively the results are not altered.

The anharmonicities are even enhanced if one introduces stronger pairing like suggested by Neergård and Vogel [6] or Chasman [15]. The application of the multiphonon method using as a building block the collective octupole K = 0^- phonon to the lightest Ra and Th even isotopes showed that it is possible to give a microscopic explanation of the existence of a K = 0^- band head at very low energies $E(0^-)$, where the random phase approximation fails. It became also clear why the two phonon states were not observed around $2E(0^-)$.

Nucleus	$E(0^-)$	$E(0^+)$	$\omega_{TDA}(0^-)$	χ
222 Ra	232. 242.13	926. 914.0	546.	3.925
224 Ra	203. 215.98	964. 916.4	511.	4.25
226 Ra	246. 253.73	997. 824.2	528.	4.35
228 Ra	464. 474.14	1209. 721.17	657.	4.20
226 Th	221. 230.4	939. 805.2	523.	3.925
228 Th	318. 327.9	1017. 831.3	566.	3.85

Table I : Calculated intrinsic energies (first line) of the first excited $K^\pi = 0^-$ and 0^+ states for octupole strength parameter χ compared to the measured band head energies (second line) where no rotational effects have been substracted and to the energy ω_{TDA} obtained in the harmonic Tamm-Dancoff approximation. All energies are in keV.

4. OTHER POSSIBLE APPLICATIONS OF THE MPM

We hereby give a non exhaustive list of other possible applications of the MPM. They mainly deal with an extension of the multiphonon basis (3).

- To extend the success of the MPM obtained in the Ra and Th with $222 \leqslant A \leqslant 226$ to the isotopes $A \geqslant 228$ we have to take into account the fact that in these isotopes a second $K = 0^+$ excited state is observed. This would need the introduction of a second building phonon with $K = 0^+$ and the use of a basis

$$|m,n\rangle = Q^{\dagger m}(K=0^+)\, Q^{\dagger n}(K=0^-)|0\rangle$$

- The anharmonicity of the γ motion in ^{168}Er would similarly need the use of a basis

$$|m,n\rangle = Q^{\dagger m}(K=2^+)\, Q^{\dagger n}(K=-2^+)|0\rangle$$

or even

$$|m,n,p\rangle = Q^{\dagger m}(K=2^+)\, Q^{\dagger n}(K=-2^+)\, Q^{\dagger p}(K=0^+)|0\rangle$$

if one wants also to introduce the effect of the β vibration. If one restricts the study to the two phonon states one needs only the basis where $m - n = 0$, 1 or 2.

- The extension of the MPM to odd A nuclei, where a quasi-particle state is introduced into the basis

$$|\mu\, a\, b\, c\ldots\rangle = \alpha_\mu^\dagger\, Q_1^{\dagger a}\, Q_2^{\dagger b}\, Q_3^{\dagger c}\ldots |0\rangle$$

is also tempting. In particular it would allow a microscopic study of the spectroscopic properties of the odd A Ra isotopes where strong couplings of the quasiparticles to the octupole mode are expected. Numerical work in this direction is in progress.

5. CONCLUSIONS

The application of the multiphonon method using as a building block the collective octupole $K = 0^-$ phonon to six nuclei with $222 \leqslant A \leqslant 228$ has clearly illustrated that the MPM appears to be a powerfull tool to study nuclei where strong anharmonicities are observed. It works even in the cases where the RPA fails, i.e. in nuclei where the ground state may get unstable with respect to the considered degree of freedom.

It is also a very efficient method to study in a microscopic way the coupling between different vibrational modes and/or between vibrations and quasiparticles.

The author is grateful to B. Silvestre-Brac for a nice cowork on several parts of the present subject.

REFERENCES

[1] W.F. Davidson, D.D. Warner, R.F. Casten, K. Schreckenbach, H.G. Börner, J. Simic, M. Stojanovic, M. Bogdanovic, S. Koicki, W. Gelletly, G.B. Orr and M.L. Stelts, J. Phys. G7 (1981) 455, G7 (1981) 843
see also

W.F. Davidson, W.R. Dixon, D.G. Burke and J.A. Cizewski, Phys. Lett. 130 B (1983) 161, E.W. Kleppinger and S.W. Yates, Phys. Rev. C 28 (1983) 943.

[2] W. Kurcewicz, N. Kaffrell, N. Trautmann, A. Plochocki, J. Zylicz, K. Stryczniewicz and I. Yutlandov, Nucl. Phys. A270 (1976) 175, W. Kurcewicz, N. Kaffrell, N. Trautmann, A. Plochocki, J. Zylicz, M. Matul and K. Stryczniewicz, Nucl. Phys. A289 (1977) 1 W. Kurcewicz, E. Ruchowska, J. Zylicz, N. Kaffrell and N. Trautmann, Nucl. Phys. A304 (1978) 77, W. Kurcewicz, E. Ruchowska, N. Kaffrell, T. Björnstad and G. Nyman, Nucl. Phys. A356 (1981) 15. E. Ruchowska, W. Kurcewicz, N. Kaffrell, T. Björnstad and G. Nyman, Nucl. Phys. A383 (1982) 1.

[3] D.D. Warner, R.F. Casten and W.F. Davidson, Phys. Rev. Lett. 45 (1980) 1761 ; Phys. Rev. C24 (1981) 1713.

[4] see for instance
B. Sørensen, Nucl. Phys. A97 (1967) 1, A119 (1968) 65, A142 (1970) 411, A217 (1973) 505, T. Kishimoto and T. Tamura, Nucl. Phys A192 (1972) 246, A270 (1976) 317.

[5] B. Silvestre-Brac and R. Piepenbring, Phys. Rev. C20 (1979) 1161

[6] K. Neergård and P. Vogel, Nucl. Phys. A149 (1970) 209, A149 (1970) 217

[7] T.S. Dumitrescu and I. Hamamoto, Nucl. Phys A383 (1982) 205, I. Hamamoto, Prog. Theor. Phys. 74 & 75 (1983) 157

[8] B. Silvestre-Brac and R. Piepenbring, Phys. Rev. C26 (1982) 2640

[9] R. Piepenbring, B. Silvestre-Brac and Z. Szymanski, Nucl. Phys. A349 (1980) 77, A378 (1982) 77. K. Jammari, R. Piepenbring and B. Silvestre-Brac, Phys. Rev. C28 (1983) 2136, R. Piepenbring, B. Silvestre-Brac and Z. Szymanski, Z. Phys. A318 (1984) to be published.

[10] B. Silvestre-Brac and R. Piepenbring, Phys. Rev. C16 (1977) 1636, C17 (1978) 364

[11] R. Piepenbring, Phys. Rev. C27 (1983) 2968

[12] I.L. Lamm, Nucl. Phys. A125 (1969) 504

[13] J.P. Boisson and R. Piepenbring, Nucl. Phys. A168 (1971) 385

[14] S.G. Nilsson and O. Prior, Mat. Fys. Medd. Dan. Vid. Selsk. 32, n°16 (1961)

[15] R.R. Chasman, Phys. Rev. Lett. 42 (1979) 630

NUCLEAR DYNAMICAL SUPERSYMMETRY

Da Hsuan Feng
National Science Foundation, Washington D.C. 20550
and Drexel University, Philadelphia, Pennsylvania 19104

1. Introduction

As is well known, the study of nuclear structure physics involves introducing models to explain the very complex data derived from various types of modern high precision experimental facilities. Several excellent talks in this conference gave a very nice account of these techniques[1]. An example of this apparent complexities is vividly demonstrated in Fig.1. In this figure, I present the "complete" level scheme for the odd-even nucleus ^{195}Pt[2] for E < 1.2MeV, measured and compiled by the Brookhaven's neutron physics group using an ingenious technique known as the Average Resonance Capture (ARC) (n,γ) reactions. Due to the fact that it would involve an exceedingly large configuration calculation, the fundamental model of nuclear structure physics, the "shell model"[3], is quite inadequate in explaining the true nature of such states. Thus, the bulk of the work done in the past was based primarily on an intuitive geometric concept, proposed first in the early 50's by Bohr and Mottelson[4]. In such a model, the nucleus is preassigned a shape and is allowed to undergo rotations and/or vibrations of that shape. The main drawback of this very appealing approach is the difficulty of a priori predicting the behavior of any nucleus. This is especially so for a sequence of neighboring nuclei.

On the other hand, considering that one has at hand a many body problem(A>100 generally), it is nonetheless surprising that below this energy, the number of states is "finite". This "simplicity" is probably

---

*Work supported by the National Science Foundation. Various aspects of this work were done in close collaboration with Y.-S. Ling(Suzhou), H.-Z. Sun and M. Zhang(Qing-Hua), Q.-Z. Han(Beijing), M. Vallieres and R. Gilmore(Drexel) and W. T. Pinkston(Vanderbilt). Strong input from R. Casten and D. D. Warner(Brookhaven) and J. A. Cizewski(Yale) are hereby acknowledged.

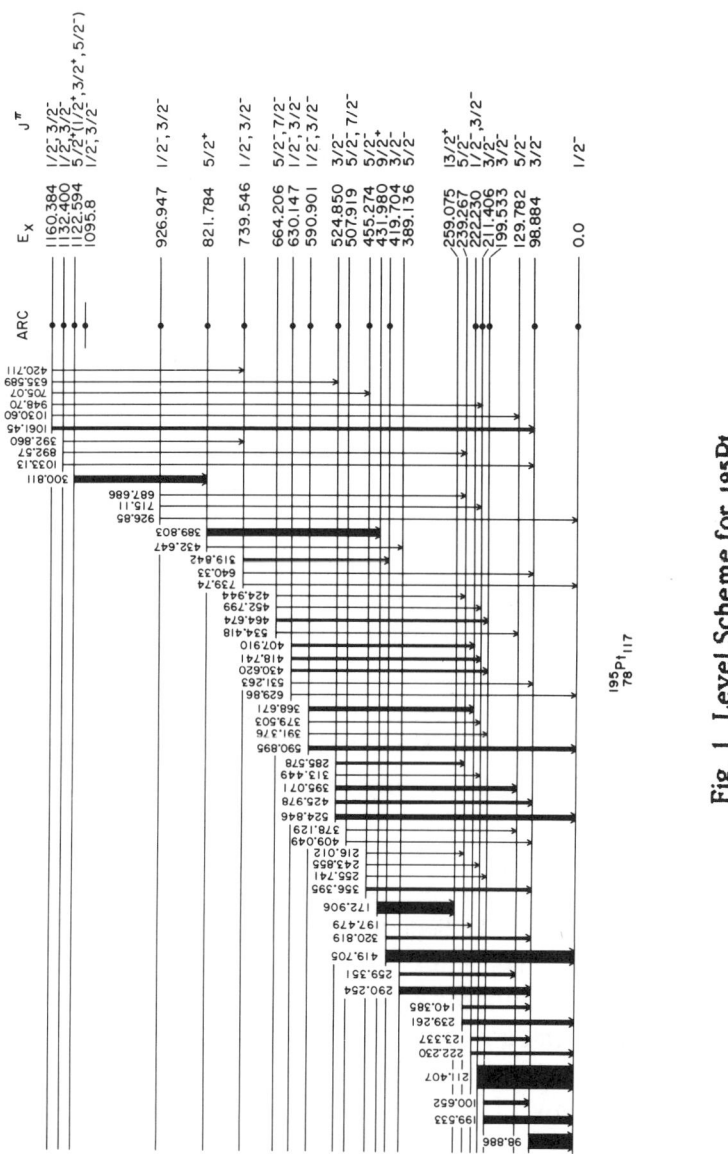

Fig. 1 Level Scheme for 195Pt

the driving force behind the early pioneers of nuclear structure physics to seek successfully a simple geometrical interpretation. The next "natural" question one may ask is whether there is a deeper understanding for such a "simplicity", derived not from an intuitive understanding, but from a more fundamental approach, namely the strong correlations of nucleons with each other. Thus enters a new and now recognized as successful model, proposed some ten years ago by Arima and Iachello[5] known as the Interacting Boson Model(IBM). The model explores what is known as dynamical symmetry and is built entirely from our knowledge of nucleon correlations. It has an inherent rich level scheme, thus giving a relatively complete picture of the collective structure of the even-even nuclei and has "predictive" power[6]. Since there are several talks about the IBM[1] in this conference, I shall not dwell on it's content here. I would merely like to make two points about the IBM. First, through the elegant use of group theory, the IBM has now taken on a geometric interpretation[7]. Second, it may have provided us with another dimension of the problem, namely we can understand the structure of the nucleus by the symmetries it possesses, eventhough from the very outset we recognize that such symmetries are an effective one. The subject matter of my talk, i.e. dynamical supersymmetry in nuclear structure physics, was infact an attempt to push the symmetry concept further and to see whether one could support such a new concept from the known data.

In this paper, I will first very briefly outline the cardinal concept of dynamical symmetry. This concept is surprizingly "trivial" in that its principle is entirely lodged in our knowledge of the first thing we learn from baby quantum mechanics: i.e. angular momentum algebra. This is discussed briefly in Section II. If you find the subsequent sections "too heavy" in group theoretic jargons, bear in mind that the concept is "just like the one in Section II". In Section III, I will succintly describe what one means when one talks about supersymmetry in nuclear structure physics. I hope I can convince you that it is only through the concept of dynamical supersymmetry can one introduce the concept of supersymmetry in nuclear physics. It is a symmetry built out of the IBM symmetry (namely U(6) from the s,d bosons). The following two sections, Section IV and V will be given to the study of two so-called multi-j dynamical symmetries, U(6/20) and U(6/12) and whether their predictions are consistent with data. In the last section, Section VI, I will outline some of the major difficulties and challenges currently facing such studies as well as the conclusions.

2. Dynamical Symmetry

One learns from quantum mechanics that the angular momentum operators L_x, L_y and L_z are closed under certain commutation relationships. Furthermore, all three operators commute with the operators L^2 and L_z (obviously for the latter). In terms of these two operators, we can construct eigenstates of the system labeled by their respective quantum numbers l and m. Thus, these eigenstates are just the familiar $|lm\rangle$. If we were to construct a hamiltonian which is a linear combination of the operators L^2 and L_z, the most general being $\alpha L^2 + \beta L_z$, then the eigenvalues are simply $\alpha l(l+1) + \beta m$. The physics is to be understood as follows: if we merely have rotational symmetry (invariance), i.e. $\beta=0$, then for each "state" (for each l), there is a 2l+1 fold degeneracy. When we "break" the rotational symmetry by the application of a preferred direction($\beta \neq 0$), the degeneracy is lifted.

In the more abstract group theoretic language, the operators L_x, L_y and L_z are the "basis" of an O(3) algebra, defined by the commutation relationships. Naturally, L_z also forms the basis of another algebra which, of course, happens to be a subalgebra of O(3). It is O(2). The operators which commute with all the basis of the algebra, L^2 and L_z are called the Casimir invariants for the respective algebras. Mathematically, we say that O(3) "contains" O(2) and write it formally as:

$$O(3) \supset O(2) \qquad (1)$$

This is called a "chain". A chain "begins" at some highest symmetry and "terminates" at a lower symmetry. Once you have a chain, you can use group theory to assign labels(or quantum numbers) for each algebra in the chain (in this case l and m) and to determine the range of values for these labels. This is followed by the construction of the irreducible representations(IR). The IRs are constructed from the eigenstates $|lm\rangle$. The procedure of obtaining quantum numbers and determine their range are known as <u>chain decomposition.</u>

Now we can define what we mean by dynamical symmetry. <u>Dynamical symmetry merely means that the Hamiltonian of a system is</u>

constructed from all the Casimir operators of a chain. A more "sophisticated" example of a chain would be what Rick Casten discussed in his talk in this conference: the O(6) limit of the IBM[1a]. This means a chain which begins at the "highest" symmetry from the model, namely U(6), with an intermediate symmetry O(6) (which is one of the three ways to break the U(6) symmetry), and terminates at the "lowest" symmetry O(3):

$$U(6) \supset O(6) \supset O(5) \supset O(3) \qquad (2)$$

Just as the O(3)⊃O(2) chain, at each intermediate algebra for eq.(2), there is certain degeneracies characterizing such algebras. Again, once you have such a chain, you will have eigenstates and eigenvalues and transitions and what have you. As long as you stay within the chain, everything remains analytic. If you leave the chain, then you will leave the "limiting symmetry" and will have to diagonalize a matrix in order to obtain eigenvalues. May I remind you that the only technical term which you may need for the next section is the "basis" of the algebra.

3. Dynamical Supersymmetry

Having defined what dynamical symmetry means, we are now in the position of defining what dynamical supersymmetry means. Very simply, it means that instead of beginning the chain at U(6) (in the IBM), we push the chain upstream to a "higher" symmetry. The reason behind this pushing is that since U(6) is an "effective" symmetry for the nucleus, there is no obvious reason to assume that it is the highest symmetry. Pushing upstream means to regard it as a symmetry broken mechanism of a still higher symmetry. Such an idea was first postulated by Iachello[8] some four years ago. The higher symmetry he proposed was dynamical supersymmetry. Now what is dynamical supersymmetry? Roughly speaking, in the context of nuclear physics, it means the symmetry, or the group structure, of a mixed boson and fermion system. The algebras of such a system, known nowadays as superalgebras, were barely introduced a decade ago.

Before discussing this new symmetry, characterized by superalgebras, let me say a few words about the "conventional" symmetries for fermion and boson systems. It is well known from Racah that for a system of fermions moving in a self consistent field, it possesses a symmetry which is characterized by a unitary algebra. The basis of this algebra is the set of bilinear creation and annihilation single particle operators $a^+_j a_{j'}$ where j and j' are the angular momentum signitures of the single particle levels. These operators are closed under commutation such that they give rise to a unitary algebra $U(\Sigma(2j+1))$. Since we are dealing with fermions, the wavefunctions of this system must be totally antisymmetric. A possible chain for such a system is:

$$U(\Sigma(2j+1)) \supset Sp(\Sigma(2j+1)) \supset \cdots \supset O(3) \qquad (3)$$

where Sp is a symplectic algebra. Since the basis consist of fermion operators it can only "mix" (or transform) fermions together[9]. Likewise, for a bosonic system (e.g. IBM), the basis consist of boson bilinear operators and thus can only mix bosons together. The wavefunctions of the system must be totally symmetric. Since unitary algebras may be applied to bosons as well as fermions, I shall from now on denote them with the superscript B(F).

Next, we can build a "product" symmetry for bosons and fermions. An example of the product algebra would be $U^B(6) \times U^F(\Sigma(2j+1))$. The basis of this algebra still cannot mix the bosons and fermions. For those of you who are familiar with the odd-A nuclei's Interacting Boson-Fermion Model of Iachello and Scholten[10], you will notice that it possesses precisely such a "highest" symmetry for the model. Let me stress here that such product symmetries are not supersymmetries.

The basis of a superalgebra not only contain operators of the type a^+a and b^+b (b^+ and b are the conventional symbol for boson creation and annihilation operators) but a^+b and b^+a as well. For simplicity, I am dropping all the subscripts for a's and b's and also not worry about details like angular momentum coupling. Clearly, we will not be able to close these four types of operators under commutations since the a's and a^+'s by themselves, anticommute with one another. It turns out, however, that these four types of operators can close under a mixture of commutation and anticommutation relations. The study of the various kinds of "generic" closures (i.e. classifications of superalgebra) is still a hot topic in current mathematical physics. For our purpose, there are only two types of superalgebras which is of concern to nuclear physics, at least for now. Loosely speaking, the first (more important one) is the "generalization" of the unitary algebra, called superunitary algebra and the second is the so-called orthosymplectic superalgebra.

Since the basis of a superalgebra consists not only of operators like a^+a and b^+b but also a^+b and b^+a as well, then in general, the system will mix bosons and fermions (i.e. bosons transform into fermions and vice versa). The superunitary algebra, denoted in the literature as U(n/m), has a boson part with dimension n (in our case, n=6) and a fermion part with dimension m (in our case, m=Σ (2j+1)). On the other hand, the orthosympletic algebra is denoted as OSp(n/m). When it was first postulated that supersymmentry may play a role in nuclear structure, it was only through dynamical supersymmetry with the highest symmetry characterized by a superunitary algebra. The chain for such a dynamical supersymmetry is[8,11]:

$$U(n/m) \supset U^B(n) \times U^F(m) \supset \text{ and so on} \tag{4}$$

The reasons why such a chain was used to construct a dynamical supersymmetry can be summarized as follows:

(1) With U(n/m) as the highest symmetry, one needs to introduce its Casimir operators into the Hamiltonian. The Casimir operators for U(n/m) has the special property that they commute with **N**, the sum of the boson and fermion number operators, $\mathbf{N_b}$ and $\mathbf{N_f}$ respectively. Therefore, quantum mechanics tells us that at the supersymmetric level, the interger spin states for the even-even and odd-half interger states for the odd even nuclei in this model must be degenerate.

(2) Unlike all other superalgebras, the U(n/m) Casimirs has a very special property: it commutes not only with **N**, but with $\mathbf{N_b}$ and $\mathbf{N_f}$ separately. In this way, although the even-even and odd-even nuclei are "merged" at the supersymmetric level, as shown schematically in **Fig**.2, we still have baryonic conservation for these states. I would like to add that this property of separate commutations is indeed special and lucky, so-to-speak, for nuclear structure physics. Other superalgebras, e.g. OSp, will in general not have this property[12].

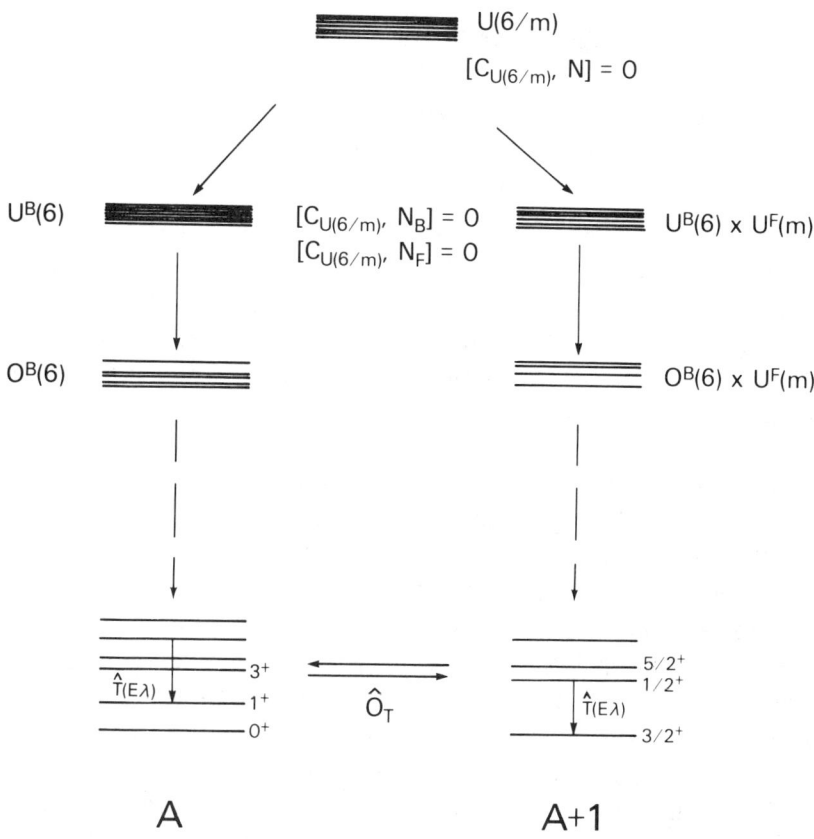

Fig. 2 Schematic diagram on the concept of dynamical supersymmetry

<u>In summary, supersymmetry is introduced in nuclear structure physics only because of the use of superunitary algebras as the highest symmetry in a dynamical supersymmetric chain.</u>

Before proceeding to the discussion of experimental verification of supersymmetry, let me, at this point, inject a few words about the orthosymplectic algebras $OSp(n/m)$. The chain given by eq.(4), which has the property that the supersymmetry is broken immediately by an ordinary symmetry, is by no means unique. There are of course other types of chains and some are only beginning to be understood mathematically. One such example is when the superalgebra $U(n/m)$ is broken by another superalgebra $OSp(n/m)$[12]:

$$U(n/m) \supset OSp(n/m) \supset \text{and so on} \tag{5}$$

In constructing a dynamical symmetry for chain (5), one encounters the difficulty that the Casimirs of the OSp superalgbra commute with **N** only and not, as in U(n/m), with $\mathbf{N}_b$ and $\mathbf{N}_f$ separately. Thus in the application to nuclear spectroscopy, states will admix the various nuclei together. Of course, this in itself may not be as fatal as it sounds especially in nuclear physics, one frequently encounters such models (e.g. BCS). Nevertheless, whether chains like (5) will prove to be useful in nuclear structure physics will depend entirely on whether there is experimental evidence. So far, it is too early to tell about this most intriguing problem.

There are two more points about supersymmetry which I like to make. First, I would like to say a few words about how, in my view, one may experimentally verify supersymmetry, or more generally, a higher symmetry in a chain (or dynamical symmetry). I have already discussed this point in a previous conference[13] but in view of its importance, I will briefly again reiterate it here. Let me begin by taking a closer look at the IBM. We know mathematically that given U(6) and O(3) as the highest and lowest symmetries respectively, there are only three chains, SU(3), U(5) and O(6). In constructing any one of the three dynamical symmetries, one needs, apart from the other Casimir invariants, the Casimir invariants for U(6). It is well known that such invariants are proportional to $\mathbf{N}_b$ only and has no manifestation of its effect in the spectra. Therefore, in hindsight, the detection of these three types of chains has the profound effect of pointing upstream, for the three chains, towards a common effective symmetry U(6). It is in this sense that U(6) may be regarded as the underpinning symmetry. From this discussion, we suggest that in order to demonstrate, in the sense of pointing upstream, the existence of some "global" symmetry, we must first study all the mathematical chains and find experimental verification for each, as in the IBM. This procedure is needless to say, out of reach at this stage. For this reason, the bulk of the work done thus far in supersymmetry is in answer of a slightly more modest question: given a complete chain with a superalgebra U(n/m) as its highest symmetry, can one find a set (or sets) of data which are consistent with its predictions.

The second point is that supersymmetry is now a general quantum mechanical scheme. This was developed by Witten in 1980. Although in supersymmetry quantum mechanics (SQM)[14], one needs no mention of bosons and fermions (although one can), it nevertheless relates two "physical" systems by the concept of a "supercharge", i.e. the Q, Q^+ operators of Witten. In our language, this operator would roughly correspond to our baryon non-conserving operators a^+b and b^+a. Supercharge is the operator which will bring a system to its supersymmetric partner. In the language of nuclear physics, this means an operator which will bring a system of A nucleons to a system of A+1 nucleons, or vice versa. To test such operators, one needs to carry out transfer reactions (one nucleon transfer, stripping or pickup). Therefore, <u>dynamical supersymmetry, or any supersymmetry concept, requires one to test the supercharge operators.</u> Thus, through supersymmetry, one relates not only the states(this is given by radioactivity) but also the process of transfer (this is given by accelerators). In nuclear physics, as in atomic physics[15], the transfer process is perhaps the most difficult to interpret.

4. Dynamical Supersymmetry U(6/20) - Proton Case

Although nuclear dynamical supersymmetry was introduced as a general concept in nuclear structure physics, the testing of this concept requires a chain which is closely related to and constructed from the physics of the problem. As I mentioned in Section 3, the first dynamical supersymmetry constructed has the highest symmetry U(n/m)(see eq. 4). To construct a chain, we need to discuss the inputs of the physics.

To begin our construction, we have to first determine the value of m, or rather the single particle levels to be included in the theory. Secondly, since the concept is so new, it was felt that we should rely on our previous knowledge of the IBM (or a chain scenario, as I called it previously). The most obvious chain scenario, with n=6 for eq.(4) are as follows:

$$U(6/m) \supset U^B(6) \times U^F(m) \supset \begin{cases} SU^B(3) \times U^F(m) \dots \\ O^B(6) \times U^F(m) \dots \\ U^B(5) \times U^F(m) \dots \end{cases} \qquad (6)$$

Eq. (6) clearly implies that such dynamical supersymmetries are best to be found in nuclei where the even-even component "preserves", as far as possible, its IBM dynamical symmetry. At the time when the concept of supersymmetry was introduced, the best known even-even nuclei which exhibited IBM dynamical symmetries were the platinum isotopes, therefore subsequent testing of dynamical supersymmetry revolved entirely around the platinum mass region. In this conference, Rick Casten has presented also some exciting results of nuclei in the Xe mass region as another example of O(6) dynamical symmetry[1a]. This could be potentially another very interesting mass region to test supersymmetry.

In the platinum mass region, the protons occupy the Z = 50 - 82 major shells. The positive parity states are $s_{1/2}$, $d_{3/2}$, $d_{5/2}$ and $g_{7/2}$ while the negative (intruder) level is $h_{11/2}$. For dynamical supersymmetry, we may link the even-even nucleus' positive parity states with the positive parity states of the even-odd partner, i.e. U(6/20) or with the negative parity states, i.e. U(6/12). For U(6/12) with $h_{11/2}$, there is no known way in group theoretic representation theory to carry out the chain decomposition (other than through the OSp chain). For the first supersymmetry test, it was assumed that only the $d_{3/2}$ level was dominent and therefore, instead of the U(6/20), the U(6/4) was used instead. The U(6/4) chain is[11]:

$$U(6/4) \supset U^B(6) \times U^F(4) \supset O^B(6) \times SU^F(4) \supset O^{B+F}(6)$$
$$\supset O^{B+F}(5) \supset O^{B+F}(3) \qquad (7)$$

Many pairs of supersymmetric partners are tested by this dynamical supersymmetry. They are ^{190}Os/^{191}Ir (n = 9), ^{192}Pt/^{193}Au (N=8) and so on. Experimentally, this includes spectra and E2 transition rates, this scheme seems to work remarkably well. Details can be found in ref.11. Although for the lighter masses, for example ^{193}Ir and ^{195}Ir, there is good agreement between theory and experiment for the spectroscopic factors for the one nucleon transfer reactions, this trend is not preserved as one moves into the heavier system, say, the Au masses.

Since the U(6/4) chain has only one fermion orbit in the theory and such isolated single orbit never quite seem to be realized in nature, therefore the development of a more realistic multi-j dynamical supersymmetric scheme is called for. This is where the U(6/20) enters. The development of the U(6/20) dynamical supersymmetry was prompted by the realization[16] that there is a 20 dimension representation of the algebra $SU^F(4)_{\underline{20}}$, namely $U^F(20) \supset SU^F(4)_{\underline{20}}$ which will satisfy this set of j values. With this, we can construct the following chain[17]:

$$U(6/20) \supset U^B(6) \times U^F(20) \supset O^B(6) \times SU^F(4)_{\underline{20}} \supset SO^{B+F}(6) \supset SO^{B+F}(5)$$

$$\supset Spin(3) \qquad (8)$$

The full chain decomposition is given in **Figure 3**. It should be noted that there is another chain which would involve the $O^B(6)$ symmetry. However, upon careful analysis of the spectrum, we find that the spectra generated by such a chain do not appear to describe the experimental situations.

The hamiltonian constructed from this chain is given as follows:

$$H = A_1 C^B_{206} + A_2 C^{B+F}_{206} + B C^{B+F}_{205} + C C^{B+F}_{203} \qquad (9)$$

where the symbol C^X_{Ng} in eq.(9) represents the Nth order Casimir invariants for the algebra g with character X(boson, fermion or coupled boson-fermion). The quantum numbers for the algebras $O^B(6)$, $O^{B+F}(6)$, $O^{B+F}(5)$ and Spin(3) are given in **Fig. 3**. are We have only included terms in the hamiltonian which will affect the spectra. In analogus to the O(3) O(2) chain of Section 2, the eigenvalues for the even-even and odd-even spectra for such a chain are given as follows:"

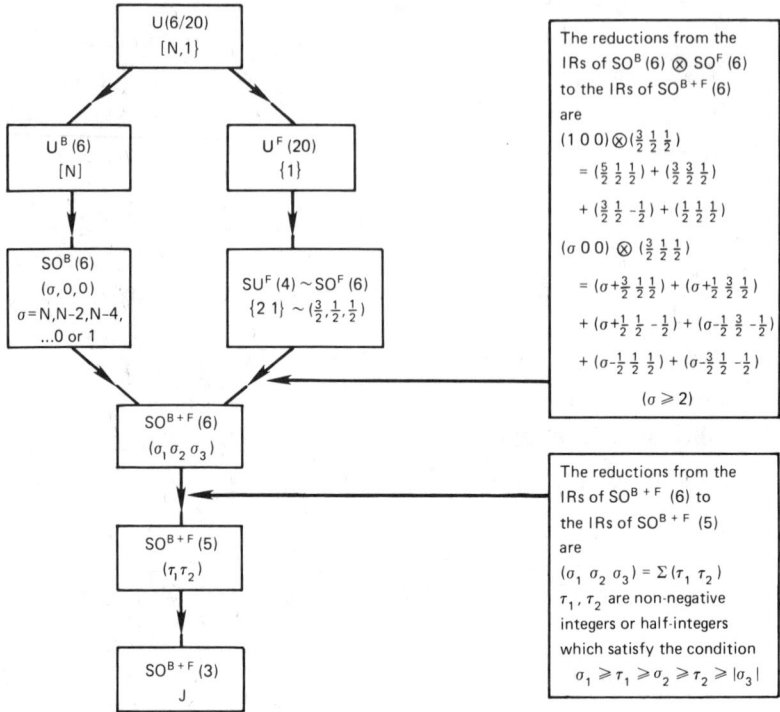

Fig.3 The U(6/20) Group Chain and its Reduction

$$E_{0-E} = (A_1/4)\sigma(\sigma+4) + (A_2/4)[\sigma_1(\sigma_1+4) + \sigma_2(\sigma_2+2) + \sigma_3^2]$$

$$+ (B/6)[\tau_1(\tau_1+3) + \tau_2(\tau_2+1)] + CJ(J+1) \tag{10}$$

$$E_{0-0} = (1/4)(A_1 + A_2)\sigma(\sigma+4) + (B/6)\tau(\tau+3) + CJ(J+1) \tag{11}$$

In a dynamical symmetry scheme, the coefficients A_1, A_2, B and C must be the same for the two expressions in eqs. (10) amd (11) since they are merely two of U(6/20)'s multiplets A noteworthy feature of the U(6/20) chain is the similarity with the U(6/4) chain. I will illustrate the use of this chain with the specific example of the nuclei ^{196}Pt(n_b=6,n_f=0) and ^{197}Au(n_b=5,n_f=1) by using all the known states below 1.5MeV in ^{196}Pt and 1.2 MeV in ^{197}Au. A similarly study with the U(6/4) chain was previously carried out Vervier et al[18]. The comparisons between the experimental and calculated spectra are presented in **Fig.** 4. In the same figure, we give also the results predicted by the U(6/4) chain. Clearly, for many of the low lying states, the two chains, U(6/20) and U(6/4) are virtually identical. There are some subtle differences. For example, the state $1/2^+$ (experimentally at 0.88MeV) for ^{197}Au is fairly well predicted by the U(6/20) scheme while it is not present in the U(6/4) chain. In fact, in the U(6/4) chain, this $1/2^+$ state cannot occur below the $3/2_3$ state, while in the U(6/20), it may.

To study the B(E2)s', we have chosen a simple form for the E2 operator which is just one of the generators of the $O^{B+F}(6)$ algebra. The results are given in ref. and because of length limitation, they will not be included here. For the ^{197}Au, there are very strong B(E2)s between the ground $3/2^+$ state and the set of states with $\tau_1 = 3/2$. The values are of the order of 30 W.U. This feature is well reproduced by the theory. On the other, there is one transition which is measured to be 10 W.U. but is strictly forbidden in this dynamical supersymmetry. All in all, however, the fit to the B(E2)s are very reasonable.

As I mentioned earlier, the concept of supercharge, or the operator to bring the system from one system to its supersymmetric partner is one of the crucial test of supersymmetry. To this end, Cizewski et al[19] have measured transfer reactions Hg(t,α)Au for both ^{195}Au, ^{197}Au and ^{199}Au. The simplest possible choice for the transfer operator for this reaction is

$$P_j = \text{(adjustable parameter)} \, a_j \qquad (12)$$

In **Fig.** 5, I present the spectrocopic factors for the three $3/2^+$ states transitions for all the Au isotopes as well as two Ir isotopes. It is clear that in general, the U(6/20) is a better representation of the experimental values although one glaring difficulty which haunts the U(6/4) chain persists here: the forbiddenness of the second 3/2 state transition. For the other states, the situation is not as good as the 3/2 states although, generally speaking the experimental situation is also less clean. Still, there exist some important contradictions between

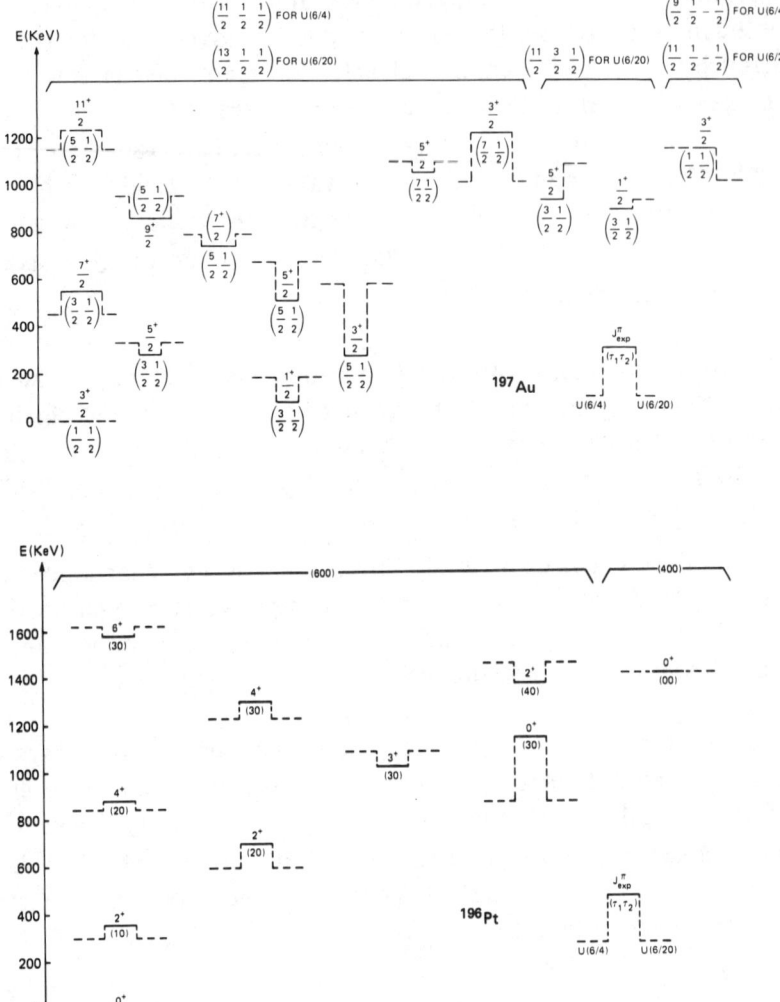

Fig.4 The SUSYs U(6/4) and U(6/20) fits for ^{197}Au/^{196}Pt

theory and experiment. An analysis of the transfer reactions within the context of the U(6/20) is the subject of a forthcoming paper[20].

5. Dynamical Supersymmetry U(6/12) - Neutron Case

For the platinum mass region, the neutron's negative parity states are $p_{1/2}$, $p_{3/2}$ and $f_{5/2}$ while the positive parity (intruder) state is $i_{13/2}$.

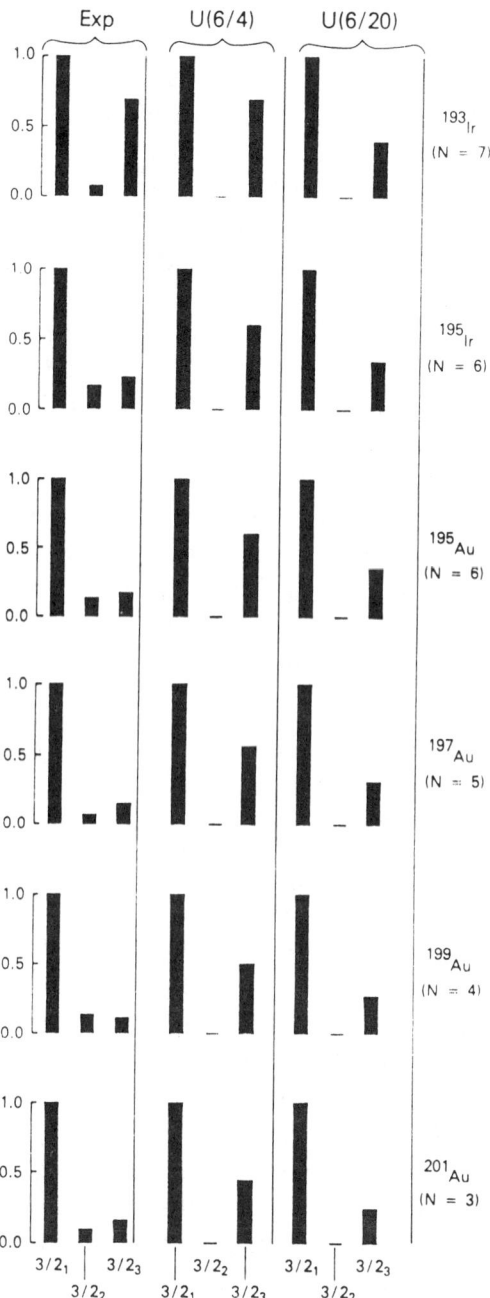

Fig.5 The Experimental, U(6/4) and U(6/20) Spectroscopic Factors for One Nucleon Transfer Reactions

For dynamical supersymmetry, we may link the even-even nucleus positive parity states with the negative parity states of the even-odd partner, i.e. U(6/12) or with the positive parity states, i.e. U(6/14). Once again, there is no known method to decompose the chain U(6/14) for such a single particle orbit other than via the OSp. Therefore, in this case, the U(6/12) dynamical supersymmetry was invoked. The subject of this supersymmetry test was the subject of several invited talks in international conferences, the lastest two were only published very recently[13,21]. I therefore will only be very brief in this discussions. I would like to remind you that there is the new work of Dave Warner, which is about the U(6/12) chain in the SU(3) limit. For the details of this exciting development, may I refer you to his paper in these proceedings[1b].

In the O(6) limit for the U(6/12), there are two chains only. One preserves the $O^B(6)$ symmetry for the even-even core and the other allows the odd neutron to polarize the core's $U^B(6)$ symmetry[22]. The energy expressions for these chains differ in a non-trivial way. The chain which does not preserve the $O^B(6)$ symmetry has the possibility of adjusting the side bands in the spectrum. This adjustment is demanded by the data. Nevertheless, fine-tuning the fit of the data led us to consider a more general hamiltonian which involves non-linear product of the standard Casimir operators. The effect of this more general hamiltonian allow separate intraband scaling for each band. Such "band stretching" is reminiscent of the Variable Moment of Inertia model in nuclear physics. In Fig. 6, I present a example of the supersymmetric fit for the nuclei ^{198}Pt and ^{199}Pt. There are two theoretical fits to the data. The first one, on the left hand side of the data (spin symmetry SPSY) is a separate least square fit for the even-even and even-odd nuclei while on the right hand side is a supersymmetric fit. The difference between these two fits are not significant. This signals the fact that the data do demand a dynamical supersymmetry interpretation. The SUSY (or SPSY) fit has a particularly notable feature in that the side band, which is generally higher in excitation energy than the ground band in lighter masses, is in fact lower in energy here. This feature of lowering side band as a function of masses can only be understood via the chain we have chosen. Furthermore, if we did not allow intraband "stretching", than bringing down the side band will cause a large number of states to "crash" into the picture and thus totally destroying the fit.

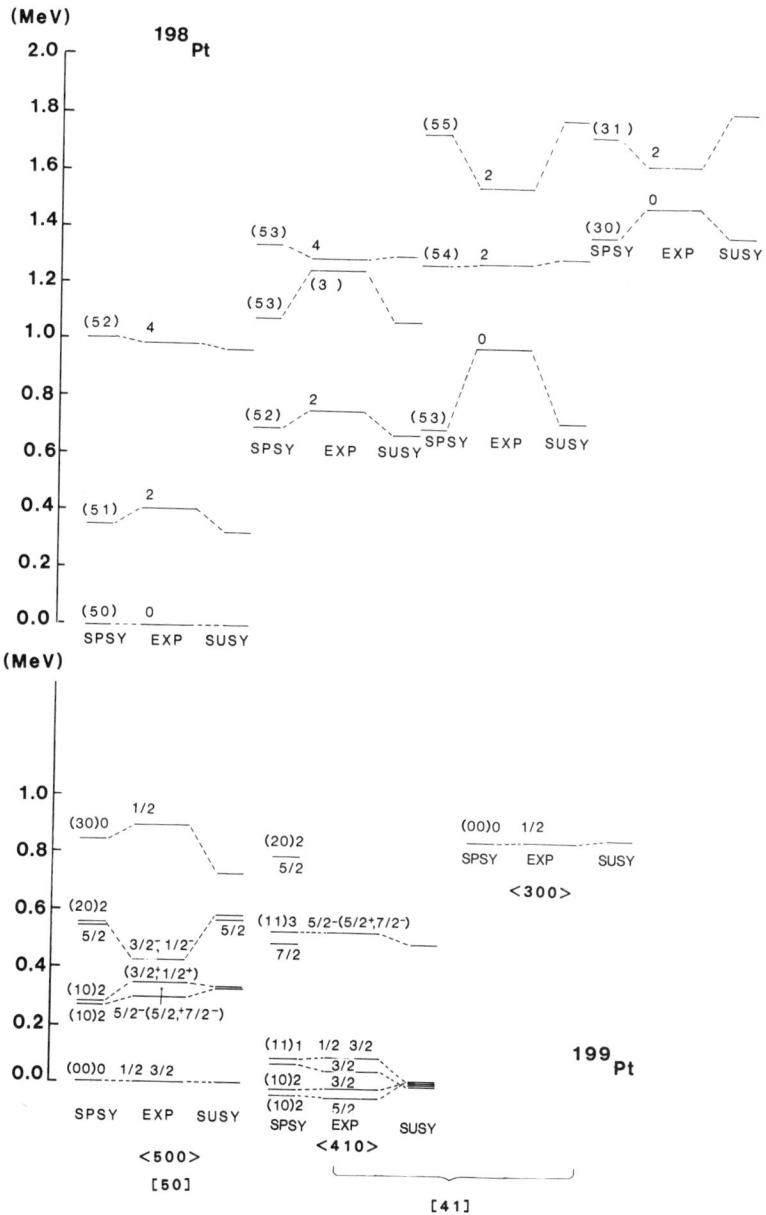

Fig.6 The SUSY and SPSY fits for ^{198}Pt/^{199}Pt

The number of B(E2)s are small. For this small sample, the theory seems to do fairly well. Nevertheless, far larger number of experimental data is needed here before more definitive statements can be made. As in the U(6/20) case, comparison of the supersymmetric predictions to neutron transfer data on the platinum isotopes shows good overall agreement; however, yet there exist a few transitions which are completely forbidden theoretically which have strong transitions. In some cases these forbidden transitions account for 20% to 30% of the observed strength. Details of the status of the transfer reactions can be found in a review talk by Vergnes[23] and the Ph.D. thesis of R. Bijker[24]. I urge the interested readers to consult those articles.

Summary and Conclusions

Clearly, the most serious disagreement with dynamical supersymmetry predictions is in the realm of transfer reactions. Yet it is precisely here that the concept of "supercharge" plays a vital role. The question is: do we understand how to construct such an operator? To try to understand this very point, it would be useful if one could analyse the problem in a model independent way. In transfer reactions, the best way to carry out this is via sum-rules. The work is in collaboration with Tom Pinkston of Vanderbilt University[25]. The operators which we use in computing spectroscopic factors are constructed from fermion and boson operators. For pickup and stripping, the transfer operators are

$$P^+_{jm} = \zeta_j a^+_{jm} + \sum_{j'} \zeta_{jj'} s^+ [da^+_{j'}]_{jm} \qquad (13)$$

$$Q^+_{jm} = \theta_j s a^+_{jm} + \sum_{j'} \theta_{jj'} [da^+_{j'}]_{jm} \qquad (14)$$

respectively. The quantities labeled by θ and ζ are constants. Frequently, only the first term in eqs. (13) and (14) are retained. The procedure followed in deriving the sum sules consists of writing first the spectroscopic factors S_j in terms of the matrix elements of the operators **P**'s and **Q**'s then using the unity operator is identified and removed from the sum. For example, the result for pickup is just

$$\Sigma S_j = \Sigma(2j+1)[\zeta_j^2 + \langle N_d(1+N_s)\rangle \Sigma \zeta_{jj}^2/5] \qquad (15)$$

A similar sum rule can be derived for stripping. The combined strength of strilpping and pickup is equal to the total occupancy $\Sigma(2j+1)$ of the active orbitals. This is a model independent test of internal consistency, independent of how well the theory agrees with experiment. Any fully microscopic theory must satisfy this tets. Although the total transfer strength is never observed experimentally, these sum rules provide rigorous consistency tests of the theory. In the application to the U(6/12) supersymmetry scheme, we find that there is a rather large discrependy in this test. Details of this work will appear soon.

At this point, it is still too early to pinpoint what is the source of this problem. It is important to bear in mind that for the spectra and electromagnetic transitions, we are dealing primarily with the bulk properties of nuclei and what our analyses seem to tell us are that the IBM and supersymmetry are excellent caricatures of such properties. Yet in single nucleon transfer reactions, we are probing the finer details of the nucleus (surface wavefunctions and so on). It is conceivable that our assumption that nucleon pairs behave as bosons may be too extreme when dynamical processes are important. Clearly, much more work is needed in this direction.

The dynamical supersymmetry, just as dynamical symmetry in the IBM for even-even nuclei, is the "bench-mark" symmetry of even-even and even-odd nuclei. As such, very few regions of the mass table will truly satisfy the dynamical supersymmetry. Nevertheless, they should be used carefully as the starting point to study how realistic nuclei may deviate from these bench-marks. This is in the same spirit as the IBM. For example, in the case of U(6/20), there is some concern as to whether the $g_{7/2}$ is close enough to the other levels so as to render the U(6/20) totally physical(although this in itself is still an open question). One could, nevertheless, break the U(6/20) chain symmetry by adding some additional single particle 7/2 energies to see its effect on the spectrum as well as other physical quantites. All in all, supersymmetry seems to give a fairly nice account of the structure of the nucleus. It is indeed exciting to think of the nucleus as a quantum system for us to locate such a intellectually profound new symmetry, albeit an effective one.

Acknowledgements

I am most grateful to Dr. Ramayya and Dave Warner of Brookhaven for their probing questions which led me to write the section on supersymmetry in this paper. I am also very grateful to A. Kostelecky for telling me recently about the interesting results in atomic physics and for teaching me about the difference between supersymmetry as I know it and the supersymmetry quantum mechanics of Witten. My collaborations with Hong-Zhou Sun(who taught me a great deal about graded Lie algebra) and Mei Zhang, both from Qing-Hua University in Beijing, have been most beneficial in my educational process. Finally, I thank Dr. S. Raman for organizing such a nice conference.

References

1. See, for example, (a) R. F. Casten, (b) D. D. Warner, (c) R. Hoff and (d) H. G. Borner in these proceedings.
2. D. D. Warner, R. F. Casten, M. L. Stelts, H. G. Borner and G. Barreau, Phys. Rev. **C26**, 1921(1982).
3. M. Mayer, Phys. Rev. 74, 235(1948); **75**, 1696(1949); O. Haxel, J. H. D. Jansen and H. E. Suess, Phys. Rev. **75**, 1766(1949).
4. A. Bohr and B. Mottelson, Mat. Fys. Dan. Vid. Selsk, **27**, 174(1953).
5. A. Arima and F. Iachello, see the most recent review article, Ann. Rev. of Nucl. Sci. **31**, 75(1981).
6. R. Casten and J. A. Cizewski, Nucl. Phys. **A309**, 477(1978).
7. J. N. Ginocchio and M. W. Kirson, Phys. Rev. Lett. **44**, 1744(1980); A. E. L. Dieperink, O. Scholten and F. Iachello, Phys. Rev. Lett. **44**, 1747(1980); D. H. Feng, R. Gilmore and S. R. Deans, Phys. Rev. **C23**, 1254(1981).
8. F. Iachello, Phys. Rev. Lett. **44**, 772(1980).
9. The details of this discussion is best found in the NORDITA publication of lecture notes by Ben F. Bayman. The title is <u>GROUP AND ITS APPLICATIONS TO SPECTROSCOPY</u>, 1960.
10. F. Iachello and O. Scholten, Phys. Rev. Lett. **43**, 679(1979).
11. A. B. Balantekin, I. Bars and F. Iachello, Phys. Rev. Lett. **47**, 19(1981);

12. I. Bars and A. B. Balantekin, see for example the lectures given by Bars in <u>SUPERSYMMETRY IN PHYSICS</u>, North Holland Publication ed. A. Kostelecky (1984).
    Q.-Z. Han, H.-Z. Sun, M. Zhang and D. H. Feng, Jour. of Math. Phys.(to be published).
13. D. H. Feng et al , Nucl. Phys. **A421**, 167c(1984), Proceedings of the Suzhou Conference Collective States in Nuclei, ed. L.-M. Yang et. al.
14. E. Witten, Nucl. Phys. **B188**, 513(1981).
15. A. Kostelecky and M. Nieto, LA-UR-84-2405, preprint, unpublished(1984).
16. M. Hamermesh, <u>Group Theory</u>, Addison-Wesley, Reading, Mass (1962).
17. Y.-S. Ling, M. Zhang, J.-M. Xu, M. Vallieres, R. Gilmore, D. H. Feng and H.-Z. Sun, Phys. Lett. (in press).
18. J. Vervier, Phys. Lett. **B100**, 383(1981).
19. J. A. Cizewski, see for example, Proceedings for the Oaxtepec Meeting, Mexico(1984).
20. M. Zhang, Y.-S. Ling, M. Vallieres, D. H. Feng and J. A. Cizewski (to be published).
21. R. Casten, Nucl. Phys. **A421**, 27c(1984);
    A. B. Balantekin, I. Bars, R. Bijker and F. Iachello, Phys. Rev. **C27**, 1761(1983)
22. D. H. Feng, H.-Z. Sun and M. Vallieres, Commun. of Theo. Phys. (Beijing, in English) submitted.
23. See, for example, M. Vergnes, in Proceedings for the Conference on Bosons in Nuclei, World Scientific Publication(Singapore), ed. O. Scholten (1984).
24. R. Bijker, Ph.D. thesis, University of Groningen (1984), unpublished.
25. W. T. Pinkston, in same volume in ref.23;
    W. T. Pinkston and D. H. Feng, Phys. Rev. C(in press).

NEW COUPLING LIMITS, DYNAMICAL SYMMETRIES AND MICROSCOPIC OPERATORS OF IBM/TQM

V. Paar[*]
Prirodoslovno-matematički fakultet, University of Zagreb,
Marulićev trg 19, 41000 Zagreb, Yugoslavia

ABSTRACT

A new particle-core basis having approximate supersymmetric (SUSY) features associated with SU(3) dynamical symmetry is introduced. The SUSY and CO-SUSY limits of IBFM/PTQM appear for the characteristic intermediate coupling strengths $\Gamma/\delta = \pm(\Gamma/\delta)_{SUSY}$. The CO-SUSY limit is a truncated analog of the Stephens rotation-aligned scheme. A paradox was found in the relation of the SUSY and truncated strong coupling (TSC) limits to the strong coupling limit of the Bohr-Mottelson model.

Microscopic Dyson and Holstein-Primakoff realizations of RPA collective quadrupole phonon operators are explicitly constructed. Employing this mapping procedure in conjunction with the leading RPA diagrams, various operators of IBM/TQM, IBFM/PTQM have been derived in the particle-hole channel: E2 operator, one-particle transfer operator, two-particle transfer operator etc. In addition to the standard terms, this derivation gives in the same diagrammatic order the additional terms also.

A new model was introduced for the odd-odd nuclei in the framework of IBM/TQM. For the SU(3) core the truncated analog of Gallagher-Moszkowski bands appears as the approximate SUSY pattern, of the same intrinsic structure as in the odd-even system.

The idea of boson-fermion dynamical symmetry and supersymmetry is extended to odd-odd nuclei and hypernuclei.

INTRODUCTION

The approach to nuclear structure based on the SU(6) boson symmetry in IBM/TQM[1-16], IBFM/PTQM[5,16-25] and IBOM/OTQM[51] for even-even, odd-even and odd-odd nuclei, respectively provides a very useful theoretical framework.

This paper contains three main parts which briefly summarize the results of our recent work on this problematics.

I. New coupling limits of IBFM/PTQM associated with the approximate supersymmetry SUSY for the SU(3) core.

II. The method of deriving new forms of IBM/TQM operators based on RPA in the particle-hole channel.

III. Dynamical boson-fermion symmetry and supersymmetry for odd-odd nuclei and for hypernuclei.

---

[*]This project was assisted by the U.S. National Science Foundation under Grant No. YOR 80/001.

IBFM/PTQM FRAMEWORK FOR PARTICLE-CORE COUPLING IN SU(3) LIMIT

The boson-fermion IBFM/PTQM Hamiltonian [16-18] for a particle $|j>$ coupled to the IBM/TQM boson core in the SU(3) limit is

$$H^{BF}(SU(3)) = H^{B}(SU(3)) + \Gamma(G_2^B G_2^F)_0 + V_{EXC}^{BF}, \qquad (1)$$

$$H^{B}(SU(3)) = -\frac{\alpha}{2} C_2^B + \delta I^B \cdot I^B, \qquad (2)$$

$$C_2^B = 2 G_2^B \cdot G_2^B + \frac{3}{4} I^B \cdot I^B. \qquad (3)$$

In the s,d- boson (IBM) representation there is [1]

$$H^{BF} = H_{IBFM} \qquad (4)$$

$$H^{B} = H_{IBM} \qquad (5)$$

$$G_{2\mu}^{B} = d_\mu^+ s + s^+ \tilde{d}_\mu + x(d^+\tilde{d})_{2\mu}, \quad x = \pm \sqrt{7}/2 \qquad (6)$$

$$I_\nu^{B} = \sqrt{10}\, (d^+ \tilde{d})_{1\nu}; \qquad (7)$$

in the quadrupole phonon (TQM) representation there is [2,16,18]

$$H^{BF} = H_{PTQM} \qquad (8)$$

$$H^{B} = H_{TQM} \qquad (9)$$

$$G_{2\mu}^{B} = b_\mu^+ (\hat{N-N})^{1/2} + (\hat{N-N})^{1/2} \tilde{b}_\mu + x(b^+\tilde{b})_{2\mu}, x=\pm\frac{\sqrt{7}}{2} \qquad (10)$$

$$I_\nu^{B} = \sqrt{10}\, (b^+ b)_{1\nu}. \qquad (11)$$

Here, b_μ^+ is the creation operator of the quadrupole phonon. Integer N denotes the total number of s and d bosons in IBM and the maximum number of quadrupole phonons in TQM.

The bilinear fermion operator is

$$G_{J\mu}^{F} = (c_j^+ \tilde{c}_j)_{J\mu}. \qquad (12)$$

In TQM the basis states are $|n \nu I>$, where n quadrupole phonons are coupled to the total angular momentum I and the additional quantum numbers to specify the state are denoted by ν. In IBM the corresponding basis state is $|n_s = N-n, n_d=n, \nu I>$.

Performing the angular momentum projection from the coherent state

$$|C> = \mathcal{N} \exp \{\beta b_{20}^+ (N - \hat{N})^{1/2}\}| 0 > \qquad (13)$$

with $\beta = \sqrt{2}$ $(-\sqrt{2})$ for a prolate (oblate) system, an analytical expression for the wave functions of the ground-state band of the SU(3) boson system is [16,26]

$$|I>_{GSB} = B_I \sum_{n\nu} A_{n\nu I} |n\nu I >. \qquad (14)$$

The expressions for the coefficients B_I and $A_{n\nu I}$ are given in ref. [16].

The coherent state decomposition is

$$|C> = \sum_I \frac{1}{B_I} | I >_{GSB}. \qquad (15)$$

For the odd system one projects from the direct product [16,22]

$$|j K > |C > \qquad (16)$$

obtaining

$$|KJM>_{SY} = \mathcal{N} \sum_I \frac{1}{B_I} < jKI0|JK > |(j\ I_{GSB})\ J\ M >_{TWC}. \qquad (17)$$

Here,

$$|(j\ I_{GSB})JM>_{TWC} = \sum_{m_1 m_2} < jm_1\ Im_2\ |J\ M>|jm_1> |Im_2>_{GSB} \qquad (18)$$

is referred to as the truncated weak coupling (TWC) basis, obviously appropriate for $\Gamma \gtrsim 0$, $v_{EXC}^{BF} \approx 0$ in (1).

In this section the exchange term in (1) is not considered. The results, therefore, correspond to the particle-rotor model. Let us focus our attention on the states in odd system built on the ground-state band of the core. For these states the Casimir term in (2) gives an overall energy shift only, and the relevant part of the Hamiltonian (1) is

$$\delta \cdot H' = \delta I^B \cdot I^B + \Gamma (G_2^B G_2^F)_0. \qquad (19)$$

Then the corresponding wave functions depend only on the ratio Γ/δ. It turns out that the states $|K = j\ J\ M>_{SY}$ (17) are the eigenvectors of (19) for a particular value of Γ/δ to be denoted by $(\Gamma/\delta)_{SUSY}$. [27]

Different K-states in (17) are not orthogonal and the Schmidt orthogonalization has to be used. Of physical significance are the two different orthogonalization schemes:

i) The upper orthogonalized (UO) scheme: For each J, starting with the state K=j, we orthogonalize the state K=j-1 to it, and so on, ending with K=1/2. The quantum numbers K obtained in this way are denoted by K_u and the corresponding UO basis states

$$|K_u \, J \, M> \, . \qquad (20)$$

By construction, the state $|K_u=j \, J \, M>$ coincides with the state $|K=j \, J \, M>_{SY}$.

ii) The lower orthogonalized scheme: For each J, starting with the state $K = \frac{1}{2}$ we proceed the orthogonalization up to the state K=j. The corresponding quantum numbers will be denoted by K_ℓ and the corresponding basis states

$$|K_\ell \, J \, M> \, . \qquad (21)$$

By construction, the state $|K_\ell = \frac{1}{2} \, J \, M>$ coincides with the state $|K = \frac{1}{2} \, J \, M>_{SY}$.

APPROXIMATE SUSY LIMIT

In this section we present a particular analytic solution for the eigenstates of IBFM/PTQM, obtained for a particular ratio of the particle-core coupling constant Γ and the core moment of inertia $\mathcal{J} = \hbar^2/2\delta$. This solution defines a special limit of the model, with the features of supersymmetry and therefore is referred to as SUSY limit.

A straightforward derivation [27] gives the matrix elements of the Hamiltonian H´ (19) in the SUSY basis (17):

$$_{SY}<K \, J \, M \, |H´|K´J \, M>_{SY} = \, _{SY}<K \, J \, M|I^B \cdot I^B|K´J \, M>_{SY} +$$

$$+ \, (\frac{\Gamma}{\delta}) \, _{SY}< KJM \, | \, (G_2^B G_2^F)_o \, |K´JM>_{SY} \, . \qquad (22)$$

The matrix elements of H´(19) in the UO SUSY basis (20)

$$< K_u \, J \, M \, | \, H´ \, | \, K_u´ \, J \, M > \qquad (23)$$

are obtained by transforming the matrix elements (22), using the coefficients obtained by the Schmidt orthogonalization.

In ref.[27] it was shown that the Hamiltonian matrix (23) is tridiagonal, i.e. the only nonvanishing matrix elements are diagonal ($K_u = K_u´$) and immediately off the diagonal ($|K_u - K_u´| = 1$). This fact is at the origin of special properties of the basis set (17). It should be pointed out that the tridiagonal pattern appears only in the UO SUSY scheme (20) and thus singles out that scheme from all the others. What is surprising in any case is the $\Delta K=1$ selection rule in the interaction matrix $(G_2^B G_2^F)_o$; one would expect the explicit presence of the fermion operator to spoil any selection rule valid in the collective subspace.

Because of tridiagonal form of (23), the ratio of the two terms in the Hamiltonian (19), given by Γ/δ, can be adjusted to make the only off-diagonal term connecting the $K_u=j$ column with the $K_u=j-1$ column vanish:

$$< K_u=j\ J\ M\ |I^B \cdot I^B + (\frac{\Gamma}{\delta})_{SUSY} (G_2^B G_2^F)_o\ |K_u=j-1\ J\ M > = 0 , \quad (24)$$

where the ratio Γ/δ for which (24) is fulfilled is called supersymmetric (SUSY).

Numerical values for $(\Gamma/\delta)_{SUSY}$ are independent both of N and J [27]. Exact values of $(\Gamma/\delta)_{SUSY}$, obtained from condition (24) are very well approximated by the approximate expression [27]

$$\Gamma^S_{SUSY} = \frac{\sqrt{5}}{2}\ \sqrt{j(2j+1)(2j+2)}\ \left[1+4\ \frac{(2j-2)(2j+4)}{(2j-1)(2j+3)}\right]^{-1/4} (1+\frac{2j-3}{36})^{-1} \cdot \frac{8}{3}\ \delta .$$

(25)

Particularly, for $j = 11/2$ there is

$$(\frac{\Gamma}{\delta})_{SUSY} = 46.3 . \quad (26)$$

Knowing that in the realistic cases $\delta \approx 0.015$MeV, the eq.(26) gives $\Gamma_{SUSY} \approx 1$ MeV, which corresponds to the intermediate coupling strength.

Here, a hole is coupled to a prolate core ($x = -\frac{\sqrt{7}}{2}$); the same solution is obtained for a particle coupled to an oblate core ($x = \frac{\sqrt{7}}{2}$).[27]

For the characteristic ratio of interaction strengths $(\Gamma/\delta)_{SUSY}$ the $K_u=j$ band is uncoupled from the others. The wave functions are exactly given by (17) for $K=j$

$$|K_u = j\ J\ M> = |K = j\ J\ M>_{SY} \quad (27)$$

and the energies are given by the energy formula

$$E_J (K_u = j) = \delta J(J + 1) \quad (28)$$

where $\delta = \hbar^2/2\mathcal{J}$ is the same parameter which appears in the core Hamiltonian (2). Thus, the same energy formula is valid for the $K_u = j$ band in the odd system and for the $K=0$ band of the core, with the same moment of inertia and the same weight function appears in the wave functions (27) and (15) for the boson-fermion and boson system, respectively. These features are characteristics of supersymmetry.

On the other hand, the approximate character of the present SUSY is revealed in the fact that the SUSY type of wave functions and energy formula is exact only for a subset of solutions ($K_u=j$ band).

Following this line of approach, in ref. [27] the supersymmetry was simulated by truncating the SU(2j+1) fermion algebra to SU(3); the problem was also investigated on the level of representations involving infinite-dimensional representations.

APPROXIMATE CO-SUSY LIMIT AND TRUNCATED ANALOG OF STEPHENS ROTATION-ALIGNED SCHEME

In this section we discuss the Hamiltonian (19) when a hole is coupled to an oblate SU(3) core ($x = \sqrt{\frac{7}{2}}$) or a particle to a prolate one ($x = -\sqrt{\frac{7}{2}}$). The effective way to change a hole into a particle is to reverse the sign of Γ. Particularly, we discuss the case

$$\Gamma = - \Gamma_{SUSY} . \qquad (29)$$

Since $(\Gamma/\delta)_{SUSY}$ was determined by demanding maximal destructive interference between the off-diagonal matrix elements, we now have a maximal constructive interference. Therefore, the corresponding limit is referred to as coherent-SUSY (CO-SUSY) limit.

The CO-SUSY limit was investigated in refs. [27,28].

In this case the appropriate band attribution is given by the LO SUSY basis (21). The $K_\ell = 1/2$ band is the lowest one and exhibits a strong signature effect, with the state $J = j$ at the lowest energy. The states of the $K_\ell = \frac{1}{2}$ band separate in two $\Delta J=2$ trajectories : the E2 $J+1 \to J$ transitions between these two trajectories are hindered and the $J+2 \to J$ transitions within the trajectories enhanced.

The lower branch of the $K_\ell' = 1/2$ band closely resembles the decoupled band pattern of the Stephens rotation-aligned scheme. [30]

The connection to the Stephens scheme was shown on the level of the wave functions [28]. Instead of the $d^j_{K\alpha}(\frac{\pi}{2})$ function which rotates the particle angular momentum projection in the Stephens transformation [30], the Clebsch-Gordan coefficient

$$< 2N\ 0\ j\ K\ |\ 2N + \alpha\quad K > \qquad (30)$$

is used to rotate the state $|K_\ell = \frac{1}{2}\ J\ M>$. The coefficient (30) coincides with $d^j_{K\alpha}(\pi/2)$ in the asymptotic limit $2N \to \infty$.

PARADOXICAL RELATION BETWEEN SUSY LIMIT AND BOHR-MOTTELSON STRONG-COUPLING LIMIT

While the CO-SUSY limit is an unambiguous analog of the Stephens rotation-aligned scheme, a paradox was found in the relation of the SUSY limit of IBFM/PTQM to the strong coupling limit of the Bohr-Mottelson model.

To elucidate the problem, let us introduce an alternative definition of the quantum number K (first considered in ref.[5]) in a straightforward analogy to the geometrical model.

The TWC basis (18) can be used as the basis for the IBFM/PTQM analogs of Nilsson bands, which are built from the ground-state band (GSB) of the core. Generally, one may write

$$|KJM> = \sum_I f_{NJ}(I) <jKIO|JK> |(j\ I_{GSB})\ J\ M>_{TWC} \qquad (31)$$

and the weight function $f_{NJ}(I)$ defines the quantum number K.

The SUSY form is

$$f_{NJ} = \mathcal{N}\ \frac{1}{B_I} \qquad (32)$$

corresponding to the SUSY states (17). On the other hand, in a straightforward analogy to the Bohr-Mottelson weak to strong coupling transformation there is

$$f_{NJ}(I) = \left(\frac{4I + 2}{2J + 1}\right)^{1/2} \qquad (33)$$

The basis (31) with the weight function (33) is referred to as the truncated strong-coupling basis (TSC).

In the TSC basis the Hamiltonian matrix does not have a simple tridiagonal form, and thus there is no finite value of the coupling constant for which the TSC basis states would be eigenstates of the Hamiltonian (19). On the other hand, by construction, TSC basis states are good approximation of the eigenstates of (19) in the limit $\Gamma \gg 1$ and/or $N \gg 1$.

We are now in position that we have two IBFM/PTQM limits, SUSY and TSC limits, which in the energy pattern bear resemblance to the strong coupling limit of Bohr-Mottelson model.

However, the SUSY limit corresponds to the particular intermediate coupling strength Γ_{SUSY}, which is, moreover, independent of N. For the same absolute value of Γ_{SUSY}, but with reversed sign, the resulting CO-SUSY limit is the truncated anlog of the Stephens rotation-aligned limit, a special intermediate coupling case of the rotational model. It is tempting to identify the SUSY limit with the strong coupling limit of Bohr-Mottelson model, but the SUSY coupling strength is not strong and it is independent of N.

Thus, there are two possibilities:

i) SUSY and TSC limits of IBFM/PTQM are two truncated analogs of the Bohr-Mottelson strong-coupling limit; the model parameters are, however, not expected to have the same meaning in the truncated and geometrical models.

ii) The TSC limit is the truncated analog of the strong-coupling limit and the SUSY limit corresponds to some new limiting case of the Bohr-Mottelson model, which was not pointed out to date.

MICROSCOPIC DERIVATION OF IBM/TQM OPERATORS IN PARTICLE-HOLE CHANNEL USING RPA FOR QUADRUPOLE PHONONS

Microscopic derivation of IBM and IBFM has been considered on the basis of generalized seniority approximation.[5,31,32]

Here we discuss the microscopic derivation of the operators (E2,M1,one-particle transfer, two-particle transfer) in the TQM and PTQM representation, employing RPA for quadrupole phonons.

The mapping of RPA quadrupole phonon creation and annihilation operators, defined in the fermion space, into the quadrupole boson space of TQM, can be done by using two methods.[33,34]

In the recent consideration [35] we employed the operator method, using as a starting point the generating state with the RPA quadrupole phonon operators Q_μ^+, Q_μ. This enables us:

i) to obtain closed microscopic expressions for Q_μ^+, Q_μ in terms of quadrupole boson operators b_μ^+, b_μ;

ii) to derive Holstein-Primakoff representation for the algebra underlying TQM;

iii) to recognize the RPA phonon creation operators Q_μ^+ as five raising operators of the Cartan-Weyl canonical form of $SU(6)$ algebra, corresponding to five positive simple roots. The recognition of the last point may be the key for the generalization of the operator method to deformed nuclei.

In the first step we obtain Dyson representation for Q_μ and Q_μ^+:

$$Q_\mu^{DR} = b_\mu \qquad (34)$$

$$Q_\mu^{+DR} = b_\mu^+ - \bar{c}_{RPA} \, b_\mu^+ \sum_\nu b_\nu^+ b_\nu . \qquad (35)$$

Here, b_μ^+ is the creation operator of the quadrupole boson and $\bar{c}_{RPA}$ denotes the norm of the two-RPA-phonon states.

Approximating the inverse two-RPA phonon norm by the closest integer N and using the orthogonalization procedure one obtains the microscopic Holstein-Primakoff (TQM) realization for RPA phonons

$$Q_\mu^{+HPR} = b_\mu^+ (1 - \frac{1}{N} \sum_\nu b_\nu^+ b_\nu)^{1/2} \qquad (36)$$

$$Q_\mu^{HPR} = (1 - \frac{1}{N} \sum_\nu b_\nu^+ b_\nu)^{1/2} b_\mu . \qquad (37)$$

The set of 35 operators { Q_μ^{HPR}, Q_μ^{+HPR}, $[Q_\mu, Q_\nu^+]^{HPR}$ } close the SU(6) algebra, automatically appearing in the Cartan Weyl canonical form.

Now, the physical operators of TQM and PTQM can be obtained microscopically employing Bohr-Mottelson microscopic treatment of vibrations in the framework of RPA (Nuclear field theory - NFT in the particle-hole channel)[29,36,37]. This treatment of the quadrupole vibrations is fully microscopic in the sense that the collective variables are expressed entirely in terms of the degrees of freedom of the individual particles.

In deriving TQM operators we proceed as follows:
i) The matrix elements are derived in NFT (with quasiparticle RPA phonons), including the leading quasiparticle-vibration diagrams;
ii) The RPA phonons (in the fermion space) are mapped into TQM phonons (in boson space) using the bosonization procedure (36),(37).

MICROSCOPIC FORM OF ONE-PARTICLE TRANSFER OPERATOR IN IBFM/PTQM (PARTICLE-HOLE CHANNEL)

The method i, ii) of the proceeding section has been applied to the one-particle transfer operator by including diagrams up to the first-order in the particle-vibration coupling. The resulting PTQM transfer operator, expressed in the IBFM representation is[38]

$$T_j^{IBFM} = u_j c_j^+ + \frac{v_j}{\sqrt{N}} \frac{5}{\sqrt{4\pi}} \{ \sum_{jj'} (u_j v_{j'} + u_{j'} v_j)^2 <j||Y_2||j'>^2 \}^{-1/2}$$

$$\cdot \sum_{j'} (-)^{j'-\frac{1}{2}} (2j'+1)^{1/2} \begin{pmatrix} j & 2 & j' \\ \frac{1}{2} & 0 & -\frac{1}{2} \end{pmatrix} (u_j v_{j'} + v_j u_{j'}) [s^+ (\tilde{d}c^+)_{j'j} +$$

$$+ (d^+ c^+)_{j'j} s]. \qquad (38)$$

The first two terms, which contain the operator c_j^+ and $s^+(\tilde{d} c^+)_{j'j}$, coincide with the standard microscopic form of IBFM transfer operator from ref.[5]. However, the third term in (38), which contains the operator $(d^+ c^+)_{j'j} s$ is missing in the standard transfer operator of IBFM[5,40,41].

Let us comment on the comparison between the microscopic derivations of transfer operator in IBFM and PTQM. In IBFM the transfer operator was obtained as the lowest-order approximation to the single-particle operator in the sense that its matrix elements in the boson-fermion space are equal to the matrix elements of the single-particle operator in the equivalent fermion space for states with lowest fermion seniority $v \leq 2$ [15]. (In the case of non-degenerate shells instead of usual seniority formalism the generalized seniority is used.) Obviously, in this approximation the new term $(d^+ c^+_{j'})_j s$ does not appear, because it would involve in the mapping procedure the states of fermion seniority $v = 3$ ($| S^{N-1}(Dj')_j >$). On the other hand, in our microscopic derivation of PTQM transfer operator the new term $(b^+ c^+_{j'})_j (N-\hat{N})^{1/2}$ appears in the lowest perturbation order on equal footing with the standard-type term $(N-\hat{N})^{1/2} (\tilde{b} c^+_{j'})_j$, because they arise from the first-order ground-state correlation and the first-order scattering term, respectively.

The microscopic derivations of IBFM and PTQM transfer operators are based on the lowest-order expansions with respect to the generalized seniority and to the quasiparticle-RPA-phonon coupling, respectively.

The term $s^+ (\tilde{d} c^+_{j'})_j$ which is included in IBFM calculations performed so far, and the term $(d^+ c^+_{j'})_j s$, which was not included in IBFM, appear in the same perturbation order in our microscopic derivation; therefore, they should both be taken into account.

The contributions from the new term in transfer operator can be comparable in some cases to those from the standard term. In particular cases the new term can even dominate. This can appear for the one-particle transfer $|0_1> \longrightarrow |J_1>$ if the target nucleus is close to SU(5) limit and the single-particle state $|\ell_{j=J}>$ does not appear in the valence shell. An illustration is given by a sizeable spectroscopic factor $S(3/2^-_1) = 0.15$ in ^{61}Co for the $\ell = 1$ transfer $|0^+_1> \rightarrow |3/2^-_1>$ in the reaction $^{62}Ni(d, ^3He)^{61}_{42}Co$. In this case the $|0_1>$ and $|3/2^-_1>$ wave functions are dominated by the zero-phonon component $|00>$ and the one-phonon multiplet component $|f^{-1}_{7/2}, 12; 3/2>$, respectively. Thus, the main contribution to the spectroscopic amplitude is due to the new term in the transfer operator (38), while the contributions due to the standard terms are very small.

Concluding, the present microscopic derivation of PTQM transfer operator gives in the lowest order an additional term and we propose to include this additional term into the transfer operator.

MICROSCOPIC FORM OF E2 OPERATOR IN IBM/TQM (PARTICLE-HOLE CHANNEL)

Using the same method as in the previous section, the microscopic expression for E2 operator has been derived, including diagrams up to the second order in the particle-vibration coupling.

Expressed in IBM representation, this microscopic E2 operator contains the standard IBM E2 operator and the additional term

$$M_\mu^{IBM}(E2)_{CORR} = e_v \times \left[(d^+ d^+)_{2\mu} ss + s^+ s^+ (\tilde{d}\tilde{d})_{2\mu} \right]. \quad (39)$$

It should be stressed that this $\Delta n_d = 2$ term arises in the microscopic RPA derivation of TQM in the same perturbation order as the standard $\Delta n_d = 0$ term. The new term relaxes the forbiddeness of $\Delta n_d = 2$ transitions, i.e. of those transitions for which the sizeable components in the wave functions of the initial and final state differ in two quadrupole bosons. Such situations appear, for example, for $2_2^+ \rightarrow 0_1^+$ transition in the SU(5) limit and for $2_\gamma^+ \rightarrow 0_g^+$ transition in (or near) the SU(3) limit if the maximum phonon number N is of a moderate value.

Employing standard TQM/IBM E2 operator in the SU(5) limit, the B(E2) values for the cross-over transitions $2_2 \rightarrow 0_1$, $4_2 \rightarrow 2_1$, $3_1 \rightarrow 2_1$ are equal to zero. Similarly, in the neighbouring odd system with anomalous low-lying I=j-2 state, lowered due to the dynamical and exchange force in PTQM/IBFM [5,44], the I=j-2 → I=j E2 transition is strongly hindered. However, due to the new $\Delta n=2$ term in the E2 operator, these hindrances are sizeably relaxed. Illustrative calculations are given in refs. [43,44].

The new form of E2 operator has also an interesting consequence on description of the inelastic electron scattering. Including the new term of the type (39), the quadrupole transition operator for the inelastic scattering reads [45]

$$\hat{\rho}_\mu^{(2)}(r)_{IBM} = \alpha(r)(d_\mu^+ s + s^+ \tilde{d}_\mu) + \beta(r)(d^+\tilde{d})_{2\mu} +$$
$$+ \gamma(r)\left[(d^+d^+)_{2\mu} ss + s^+ s^+ (\tilde{d}\tilde{d})_{2\mu}\right], \quad (40)$$

where $\alpha(r), \beta(r), \gamma(r)$ are the density functions.

In comparison to the standard transition operator[46-48] there appears the additional two-boson term with $\Delta n_d = 2$ d-boson changing density.

The transition operator (40) leads to a new form of transition densities for the $0_1^+ \rightarrow 2_i^+$ transitions

$$\rho_i(r) = A_i \alpha(r) + B_i \beta(r) + C_i \gamma(r) \quad (41)$$

with an additional coefficient given by $\Delta n_d = 2$ matrix element

$$C_i = \langle 2_i^+ || (d^+d^+)_2 \, ss + s^+s^+ \, (\tilde{dd})_2 ||0_1^+ \rangle . \tag{42}$$

Thus, the standard three-level density relation[46-48] does not hold any more and the three transition densities are not linearly dependent. Instead, there follows a four-level linear relation between the transition densities for inelastic excitations of four collective 2^+ states:

$$\rho_4(r) = C\rho_1(r) + D\rho_2(r) + E\rho_3(r). \tag{43}$$

Therefore, the violation of three-level density relation found in ref. 48 is not in contradiction with the new consistent form (40) of the transition operator.

We suggest that the new γ-term of transition operator be included in IBM codes for the analysis of inelastic scattering experiments (electron-, proton-, pion-, ... scattering).

The new forms of IBM/TQM operators have been investigated for the other operators also, for example for two-particle transfer [49] and for M1 transitions [50].

Generally, it seems that the form of the IBM/TQM Hamiltonian is on a higher level of sophistication than the forms of other physical operators. Therefore, more attention should be paid to the microscopic derivation of the operators and to the investigation of the effects of additional terms on the IBM/TQM calculations. Thus, some discrepancies between IBM and experiment might be attributed to additional terms in operators rather than to shortcomings of the wave functions.

TRUNCATED ANALOG OF GALLAGHER-MOSZKOWSKI BANDS IN ODD-ODD NUCLEI

A new Hamiltonian for description of odd-odd nuclei in the framework of IBM/TQM has been recently introduced and investigated[51]. The Hamiltonian reads

$$H^{BFF} = H_{IBFM}^{(\pi)} + H_{IBFM}^{(\nu)} - H_{IBM} + H_{RES}^{(\pi\nu)} \tag{44}$$

and

$$H^{BFF} = H_{PTQM}^{(\pi)} + H_{PTQM}^{(\nu)} - H_{TQM} + H_{RES}^{(\pi\nu)} \tag{45}$$

in the s,d-boson and quadrupole phonon representation, respectively. Here, $H_{IBFM}^{(\pi)}$ ($H_{PTQM}^{(\pi)}$) and $H_{IBFM}^{(\nu)}$ ($H_{PTQM}^{(\nu)}$) denote the IBFM (PTQM) Hamiltonian for odd-even nuclei, with an odd proton and odd neutron, respectively. H_{IBM} (H_{TQM}) denotes the IBM(TQM) Hamiltonian of the even-even core and $H_{RES}^{(\pi\nu)}$ is the residual proton-neutron interaction.

In ref. [51] the Hamiltonian H^{BFF} was diagonalized in the case of a proton particle j_p and a neutron particle j_n (the exchange force and residual interaction absent) coupled to the SU(3) core. (This is a proton-neutron-core coupling analog of the particle-core coupling discussed in the first part of the paper.) The resulting energy pattern exhibits two regular low-lying bands based on the states of angular momenta $J = j_p + j_n$ and $J = |j_p - j_n|$; the other bands are higher-lying.

It has been shown[51] that the wave functions of the state of angular momentum J belonging to the $(j_p + j_n)$-band can be presented in the same form as the SUSY wave functions (17) for odd-even system with the trivial replacements

$$| j > \rightarrow |(j_p j_n) j_{pn} > , \qquad (46)$$

halfinteger $K \rightarrow$ integer K,
halfinteger $J, M \rightarrow$ integer J, M.

The energies of the states of this band are given by

$$E(K = j_p + j_n, J) = \delta J(J + 1), \qquad (47)$$

with the same moment of inertia as for the core.

Let us now discuss the relations of these SU(3)-results to the rotational model. The IBOM/OTQM bands $|K = j_p + j_n, J>$ and $|K = |j_p - j_n|, J>$ are the truncated analogs of the Gallagher-Moszkowski bands based on $(\Omega_p = j_p, \Omega_n = j_n)$, expressed in the spherical particle-quadrupole phonon basis. Their precise relative energy in OTQM is sensitive to the residual spin-spin interaction. Analogously as in the case of the parabolic rule,[52] due to spin-spin interaction we have the lowering of the band heads $K = |j_p - j_n|$ or $K = j_p + j_n$ for $\nu = 0$ or $|\nu| = 1$, respectively, depending on the Nordheim number $\nu = j_p - \ell_p + j_n - \ell_n$.

Concluding, the results of this section can be considered as an extension of the approximate SU(3) SUSY for even-even and odd-even nuclei to encompass also odd-odd nuclei.

The truncated rotational-like bands

$$| K = 0 \; I \; M = 0 >_{IBM/TQM} = \mathcal{N} P^I_{00} \{| C > \} \qquad (48)$$

$$|\Omega = j_\rho \; J \; M >_{IBFM/PTQM} = \mathcal{N} P^J_{M\Omega = j_\rho} \{| j_\rho \Omega = j_\rho > | C > \}, \quad \rho = p, n \qquad (49)$$

$$|K = j_p + j_n \; J \; M >_{IBOM/OTQM} = \mathcal{N} P^J_{MK = j_p + j_n} \{| j_p \Omega_p = j_p > | j_n \Omega_n = j_n > | C > \} \qquad (50)$$

in even-even, odd-even and odd-odd nuclei, respectively, have the same structure of the wave functions and the same moment of inertia, revealing a more general SUSY pattern.

EXTENSION OF BOSON-FERMION DYNAMICAL SYMMETRY AND SUPERSYMMETRY TO ODD-ODD NUCLEI

The approach to odd-even nuclei based on exploiting the exact boson-fermion symmetries, or, more generally, supersymmetries, has received much attention in recent years [19-25].

We have recently extended the idea of boson-fermion symmetry to odd-odd nuclei [53].

The basic idea is to construct the maximal group of transformations for the group of off-core fermions (proton and neutron) and to explore the subgroup chains that lead to the well-defined spin-group embedding, i.e. that provide us with the correct spin-content of the system explored. When this structure is found, we analyze the possible scenarios of couplings of the off-core fermions and the core, and obtain the corresponding energy formulas. The group-theoretical analysis is done following refs. [54-56].

We have constructed the classifying schemes and the energy formulas for three choices of angular momenta of protons and neutrons:
(A) $j^\pi, j^\nu = 1/2$; (B) $j^\pi_i j^\nu = 3/2$; (C) $j^\pi_i j^\nu = 1/2, 3/2, 5/2$, and for the three standard dynamical symmetries of the even-even core.

For example, for $j^\pi_i j^\nu = 3/2$ and SO(6) boson core we construct two group-chain structures:

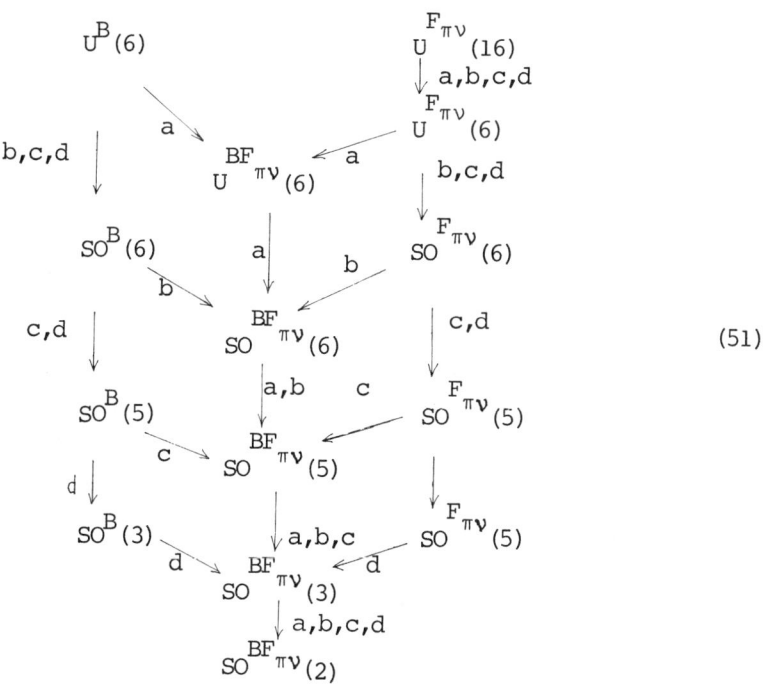

(51)

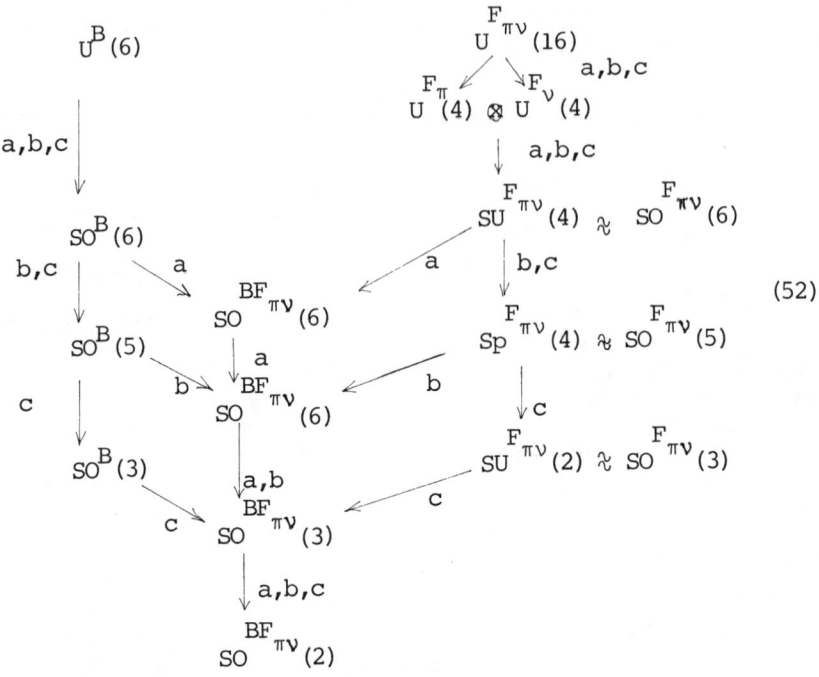

(52)

The energy formula which corresponds, for example, to the chain (51a) reads

$$E = aN + bN(N+5) + c(N_1^F + N_2^F) +$$
$$+ d\{N_1^F(N_1^F + 5) + N_2^F(N_2^F + 3)\} + e(N_1 + N_2 + N_3) +$$
$$+ f\{N_1(N_1+5) + N_2(N_2+3) + N_3(N_3+1)\} +$$
$$+ 2g\{\sigma_1(\sigma_1+4) + \sigma_2(\sigma_2+2) + \sigma_3^2\} + 2h\{\tau_1(\tau_1+3) + \tau_2(\tau_2+1)\} +$$
$$+ 2i\, J(J+1) \qquad (53)$$

where the quantum numbers of the state are $\{[N], [N_1^F, N_2^F], [N_1, N_2, N_3], (\sigma_1, \sigma_2, \sigma_3), (\tau_1, \tau_2), J, M\}$ labelling the irreducible representations (irreps) of the groups appearing in the corresponding group chain, $U^B(6)$, $U^{F\pi\nu}(6)$, $U^{BF\pi\nu}(6)$, $SO^{BF\pi\nu}(6)$, $SO^{BF\pi\nu}(5)$, $SO^{BF\pi\nu}(3)$ and $SO^{BF\pi\nu}(2)$, respectively. The parameters in the energy formula are $a, b, \ldots, i$. An algorithm for obtaining the quantum numbers associated with each group chain in (51),(52) is given in ref. 53.

Finally, let us comment on a possible extension of the present spinor symmetry to a supersymmetry, by imbedding the symmetry group $U^B(6) \otimes U^{F\pi\nu}(16)$ into a graded Lie algebra $U(6/16)$. This would present an extension of the supersymmetry of ref. 21 for even-even and odd-even nuclei, to encompass also odd-odd nuclei. In this case a graded Lie algebra would appear with the proton and neutron subsectors in its Fermi sector. A supermultiplet characterized according to the totally supersymmetric representation $|\mathcal{N}_0\}$, with $\mathcal{N}_0 = N + M_\pi + M_\nu$, would comprise the irreducible representations: ($N = \mathcal{N}_0 - 2i$, $M_\pi = 2i - 2j$, $M_\nu = 2j$), corresponding to even-even nuclei, ($N = \mathcal{N}_0 - 2i - 1$, $M_\pi = 2i - 2j + 1$, $M_\nu = 2j$) and ($N = \mathcal{N}_0 - 2i - 1$, $M_\pi = 2i - 2j$, $M_\nu = 2j + 1$), corresponding to odd-even nuclei and ($N = \mathcal{N}_0 - 2i - 2$, $M_\pi = 2i - 2j + 1$, $M_\nu = 2j + 1$), corresponding od odd-odd nuclei; here, $i = 1, 2, \ldots$ and $j = 1, 2, \ldots$ with $i \geq j$, $\mathcal{N}_0 \geq 2i$. It is desirable to explore this possible type of extended supersymmetry.

As an illustration, assuming that the nuclei ^{64}Zn, ^{63}Cu and ^{62}Cu are the ($N=4, M_\pi=0, M_\nu=0$), ($N=3, M_\pi=1, M_\nu=0$) and ($N=2, M_\pi=1, M_\nu=1$) irreps of an extended supersymmetry, respectively, we have recently obtained the spectrum of ^{62}Cu, without introducing any new parameter for odd-odd nucleus.[53]

EXTENSION OF BOSON-FERMION DYNAMICAL SYMMETRY AND SUPERSYMMETRY TO HYPERNUCLEI

We have recently considered the extension of the idea of boson-fermion dynamical symmetry to hypernuclei. The classifying schemes and the corresponding energy formulas have been constructed for the cases when the odd proton, odd neutron and a hyperon are coupled to the boson core in the cases (A) $j_1^\pi, j_1^\nu, j_1^\Lambda = 1/2$ and (B) $j_1^\pi, j_1^\nu, j_1^\Lambda = 3/2$, for three standard dynamical symmetries of the even-even core.

As an illustration, a boson-fermion group chain associated with $j^\pi = j^\nu = j^\Lambda = 3/2$ and $SO(6)$ core, is [57]

$$\begin{array}{c}
U^B(6) \otimes U^{F\pi\nu\Lambda}(64) \\
\downarrow \swarrow \searrow \\
U^B(6) \otimes U^{F\pi\nu}(16) \otimes U^{F\Lambda}(4) \\
\downarrow \qquad \downarrow \qquad \downarrow \\
U^B(6) \otimes U^{F\pi\nu}(6) \otimes U^{F\Lambda}(4) \\
\downarrow \qquad \qquad \downarrow \\
U^{BF\pi\nu}(6) \otimes U^{F\Lambda}(4) \\
\downarrow \qquad \qquad \downarrow \\
SO^{BF\pi\nu}(6) \otimes SU^{F\Lambda}(4) \\
\searrow \swarrow \\
\text{Spin}^{BF\pi\nu\Lambda}(6) \\
\downarrow \\
\text{Spin}^{BF\pi\nu\Lambda}(5) \\
\downarrow \\
\text{Spin}^{BF\pi\nu\Lambda}(3) \\
\downarrow \\
\text{Spin}^{BF\pi\nu\Lambda}(2) \ .
\end{array}$$

(54)

For a given total boson number N we have determined the allowed values of other quantum numbers in each consecutive step of the group chain. If one assumes that the Hamiltonian of the hypernucleus can be written in terms of the Casimir operators of the groups appearing in a boson-fermion group chain the corresponding hypernuclear energy formula arises.

Finally, we note a possible extension of this dynamical symmetry to a supersymmetry, by imbedding the irreps associated with the hypernuclear group chain into a graded Lie algebra U(6/64), involving hypernuclei in addition to even-even, odd-even and odd-odd nuclei. The parameters appearing in the hypernuclear energy formula could be taken from the corresponding nuclear levels belonging to the same U(6/64) supermultiplet; this opens a possibility to determine the corresponding energy levels in hypernuclei. The other two possibilities are to fit the parameters to hypernuclear levels (when available) or to attempt the microscopic derivation.

As an illustration, the energy spectrum of $^{61}_{\Lambda}$Cu has been derived [57], using the assumption that $^{61}_{\Lambda}$Cu belongs to a supermultiplet together with the nuclei ^{62}Cu, ^{63}Cu and ^{64}Zn, classified as (N=1, M_π=1, M_ν=1, M_Λ=1), (N=2, M_π=1, M_ν=1, M_Λ=0), (N=3, M_π=1, M_ν=0, M_Λ=0) and (N=4, M_π=0, M_ν=0, M_Λ=0), respectively.

REFERENCES

1. A. Arima and F. Iachello, Phys.Rev. Lett. 35,1069 (1975); Ann. Phys. 99, 253 (1976); 111, 201 (1978); 123,468 (1978).
2. D. Janssen, R. V. Jolos and F. Dönau, Nucl. Phys. A224,93 (1974).
3. J. P. Blaizot and E. R. Marshalek, Nucl. Phys. A309,422(1978).
4. V. Paar, in Interacting bosons in nuclear physics, ed. F. Iachello (Plenum, New York, 1979),p.163.
5. O. Scholten, Thesis (University of Groningen, 1980).
6. G. Kyrchev, Nucl. Phys. A349, 416 (1981).
7. A. Klein and M. Vallieres, Phys. Rev. Lett. 46, 586 (1981); A. Klein, Ch. T. Li and M. Vallieres, Phys. Scripta 25, 452 (1982); A. Klein, Ch. T. Li and M. Vallieres, Phys. Rev. C25, 2733 (1982).
8. M. Moshinsky, Nucl. Phys. A338,56 (1980).
9. J. N. Ginocchio and M. W. Kirson, Nucl. Phys. 350,31 (1980).
10. A. E. L. Dieperink, O.Scholten and F. Iachello, Phys. Rev. Lett. 44, 1747 (1980).
11. H. J. Assenbaum and A. Weigungy, Phys. Lett. 120B,257 (1983).
12. Y.Y. Tian and J. M. Irvine, J. Phys. G9,185 (1983).
13. R. L. Hatch and S. Levit, Phys. Rev. C25,614 (1982).
14. R. Gilmore and D. H. Feng, Phys. Rev. C26, 776 (1982).
15. N. Yoshida, A. Arima and T. Otsuka, Phys. Lett. 114B,86 (1982); J. Morrison, A. Faessler and C. Lima, Nucl. Phys. A372, 13 (1981).
16. V. Paar, S. Brant, L. F. Canto, G. Leander and M. Vouk, Nucl. Phys. A378, 41 (1982).
17. F. Iachello and O. Scholten, Phys. Rev. Lett. 43,679 (1979).

18. V. Paar et al, Verhandlungen der Deutschen Physikalischen Gesellschaft $\underline{3}$, 683 (1979);
    V. Paar, Inst. Phys. Conf. Ser. $\underline{49}$,53 (1980);
    V. Paar and S. Brant, Phys. Lett. $\underline{105B}$, 81(1981).
19. F. Iachello, Phys. Rev. Lett. $\underline{49}$,772 (1980);
    F. Iachello and S. Kuyucak, Ann. Phys. (N.Y.) $\underline{136}$,19 (1981).
20. A.B. Balantekin, I. Bars and F. Iachello, Nucl. Phys. $\underline{A370}$, 284 (1981).
21. A.B. Balantekin, I. Bars, R. Bijker and F. Iachello, Phys. Rev. $\underline{C27}$, 1761 (1983).
22. V. Paar, S. Brant and H. Kraljević, Phys. Lett. $\underline{110B}$,181 (1982).
23. H. Z. Sun, A. Frank and P. Van Isacker, Phys. Lett. $\underline{124B}$, 275 (1983).
24. M. Vallieres, H. Z. Sun, D. H. Feng and R. Gilmore, Phys. Lett. $\underline{135B}$, 339 (1984).
25. R. Bijker, Thesis (University of Groningen, 1984).
26. L. F. Canto and V. Paar, Phys. Lett. $\underline{102B}$,217 (1982).
27. D. K. Sunko, S. Brant, V. Paar, I. Dadić and H. B. Nielsen, submitted to Nucl. Phys. A.
28. D. K. Sunko and V. Paar, Phys. Lett. ___B, (1984).
29. A. Bohr and B. R. Mottelson, Nuclear Structure (W. A. Benjamin, Reading, Massachusetts, 1975), Vol. 2.
30. F. S. Stephens, Rev. Mod. Phys. $\underline{47}$, 43 (1975).
31. A. Arima, T. Otsuka, F. Iachello and I.Talmi, Phys. Lett. $\underline{66B}$, 205 (1977); T. Otsuka, A. Arima, F. Iachello and I. Talmi, Phys. Lett. $\underline{76B}$,141 (1978).
32. T. Otsuka, A. Arima and F. Iachello, Nucl. Phys. $\underline{A309}$,1 (1978).
33. G. Holzwarth, D. Janssen and R. V. Jolos, Nucl. Phys. $\underline{A261}$,1 (1976).
34. R. V. Jolos, F. Dönau and D. Janssen, Theor. Mat. Phys. $\underline{20}$, 112 (1974).
35. G. Kyrchev and V. Paar, to be published.
36. D. R. Bes and R. A. Broglia, Phys. Rev. $\underline{C3}$,2349 (1971).
37. I. Hamamoto, Phys. Rep. $\underline{10C}$, No 2 (1974).
38. V. Paar and S. Brant, Phys. Lett. ___B, (1984).
39. G. Kyrchev and V. Paar, Nucl. Phys. $\underline{A395}$, 61 (1983).
40. O. Scholten, Phys. Lett. $\underline{108B}$, 155 (1982).
41. O. Scholten and A. E. L. Dieperink, in Interacting Bose-Fermion systems in nuclei, ed. F. Iachello (Plenum Press, New York, 1981), p.343.
42. A. Marinov et al, submitted to Nucl. Phys. $\underline{A}$.
43. V. Paar and S. Brant, submitted to Nucl. Phys. A.
44. Y. Tokunaga et al, submitted to Nucl.Phys. $\underline{A}$.
45. V. Paar and G. Kyrchev, submitted to Phys. Lett. $\underline{B}$.
46. A. E. L. Dieperink, F. Iachello, A. Rinat and C. Creswell, Phys. Lett. $\underline{76B}$, 435 (1978).
47. M. A. Moinester et al., Nucl. Phys. $\underline{A383}$, 264 (1982).
48. F. W. Hersman et al., Phys. Lett. $\underline{132B}$, 47 (1983).
49. V. Paar, S. Brant, G. Kyrchev and R. A. Meyer, Z.Phys.A$\underline{\quad}$, (1984).
50. V. Lopac and V. Paar, to be published.

51. V. Paar, D. Sunko and D. Vretenar, submitted to Phys. Rev. Lett. and to be published.
52. V. Paar, Nucl. Phys. $\underline{A331}$,16 (1979).
53. T. Hübsch and V. Paar, submitted to Phys. Lett. $\underline{B}$ and to Ann. Phys. (N.Y.).
54. R. Slansky, Phys. Rep. $\underline{C79}$,1 (1981).
55. M. Fischler, J. Math. Phys. $\underline{22}$,637 (1981).
56. Ph. Combe, A. Sciarinnio and P. Sorba, Nucl. Phys. $\underline{B158}$,452 (1979).
57. T. Hübsch and V. Paar, submitted to Phys. Lett. $\underline{B}$ and to Z. Phys. $\underline{A}$.

SHELL-MODEL PREDICTIONS
FOR THE BROAD TRENDS AND FINE DETAILS OF
ELECTROMAGNETIC MATRIX ELEMENTS IN NUCLEI

B. H. Wildenthal
Drexel University, Philadelphia, PA 19104

ABSTRACT

Current techniques make it possible to carry out shell-model calculations which incorporate several single-particle orbits and hundreds to thousands of different combinations of particle distribution and spin coupling for a given nuclear state. We review briefly the current status of the "technology" of such calculations. We then consider in some detail one example of current work, in which wave functions for all sd-shell nuclear states have been obtained from a shell-model calculation which employs an empirical Hamiltonian and the complete set of $0d5/2-1s1/2-0d3/2$ basis vectors. We examine the electromagnetic matrix elements calculated from these wave functions in a variety of contexts to demonstrate how this theoretical approach works and to illuminate important issues in current nuclear structure research.

GENERAL COMMENTS ON SHELL-MODEL CALCULATIONS

The results to be discussed here have been obtained with a particular set of computer programs. The foundation of these programs is the Oak Ridge-Rochester Code developed by French, Halbert, McGrory and Wong.[1] Many small adjustments to the original codes have been made over the years at Michigan State by Chung and Kruse.[2,3] It is essential to realize and remember that in the kind of shell-model calculations we are going to be discussing (I like to refer to them as "neo-classical" calculations, in allusion to the fact that, while they differ by several orders of magnitude in dimensionality, they are similar to the original classic calculations conceptually) all codes which are operating correctly give precisely the same answers to a given problem.

That is to say, a given shell-model calculation is defined by the quantum numbers of the state in question, by the detailed specification of the basis space assumed for the model, and by the chosen Hamiltonian. Once these are prescribed, there is not an "Oak Ridge" answer, a "Glasgow"[4] answer and an "Oxford"[5] answer, there is only "the" answer. Different codes provide different routes to the same end. The choice of one over the others is largely a matter of personal or computer-installation history, although the differences between codes can make one code easier to use for one application and another almost essential for some other goal.

A COMPLETE CALCULATION FOR THE SD SHELL

The calculation[6] whose results[7] will be discussed here was designed to extract the best possible wave functions for sd-shell states with the minimum of theoretical preconceptions. To this end we assumed that the complete three-orbit sd-shell space is the appropriate model for "all" positive-parity states in the N,Z = 8 to 20 range and that isospin is a good quantum number. We assumed that the Hamiltonian could be characterized explicitly in terms of one-body and two-body interactions and that a single formulation of the Hamiltonian must suffice to generate the level structures of all the nuclei encompassed in the model space. We assumed, after hard-won experience experience, that the two-body matrix elements of the Hamiltonian must have magnitudes which decrease with increasing mass value. We chose to formulate this mass dependence by scaling the matrix elements by the factor of A to the -0.3 power.

The values of the matrix elements which define the Hamiltonian were determined by iteratively adjusting them so as to achieve a least-squares (non-linear) fit to 440 experimentally determined values[8] of level energies in A = 17-40 nuclei. The guiding philosophy was that, given the best "reasonable" approximation to the physically relevant "degrees of freedom" of the nuclear states, the "best" wave functions are correlated with the best match of model and experimental energies. In design, this calculation is more comprehensive and internally consistent than earlier work.[9] The theoretical energy levels which are obtained from the converged Hamiltonian parameters agree at least as well with experimental spectra as do the results of previous calculations, even though the latter have treated only fractions of the sd-shell region.

The rms deviation between calculated and experimental energies for more than 700 matched pairs of sd-shell levels is less than 140 keV. For nuclei in the middle of the shell, between say A=22 and A=34, the model levels match one to one with the experimentally known positive parity levels up to excitation energies of about 6 MeV. At higher excitations the densities of the model spectra necessarily fall below experimental values, but in A=26, one of the systems most thoroughly studied experimentally[10], obvious discrepancies between experiment and theory have still not emerged at 8 MeV. By this excitation, a dozen of more levels of each total angular momentum value J have typically been encountered. In the 0-5 Mev range of excitation energy in this region, the details of the observed spin sequences and their energy spacings are so closely reproduced by the model that experiment and theory are virtually interchangable as far as visual impressions of spectra are concerned.

The model spectra from this calculation begin to deviate from experiment progressively with proximity to the A=16 and A=40 shell closures. The deviations take the form of "extra" positive-parity states in the experimental spectra. We interpret these extra states as having their basic parentage in particle-hole excitations which cross over the shell closures. These "intruder" states (relative to the sd-shell basis) typically can be distinguished from the "native" states in the observed data, and appear to play the role of a background upon which the features of the sparse sd-shell spectra are displayed. Understanding the relationships between the intruder and native states, and the mechanisms by which the intruders move from lower to higher excitation energies with increasing distance from shell closures, is fundamental to improving our understanding of the shell model.

GENERAL COMMENTS ON ELECTROMAGNETIC MATRIX ELEMENTS

The results of the shell model calculation we have been discussing are, at least in so far as energies are concerned, so good that it is natural to consider throughly testing the wave functions associated with the eigenvalues. In this regard, primary attention in this review is focussed on M1 and E2 matrix elements. The preponderance of experimental data, in particular those from capture gamma-ray spectroscopy, concerns these multipolarities. In our examinations we attempt to distinguish between the effects of the model-space assumptions, the consequences of the definition of the electromagnetic operators, and the effects of the particular Hamiltonian used in the specific calculation. The high degree of internal consistency of the present wave functions makes it possible to examine their predictions for electromagnetic matrix elements with a minimum of concern about the effects of changing truncation schemes and different Hamiltonians. When such effects are present they can easily be misinterpreted as reflecting significant deviations between experimental observations and the fundamental assumptions of nuclear shell structure.

We consider first the average magnitudes of the model matrix elements relative to the averages determined from experiment. In more detail, we attempt to determine the degree to which the scale factors which match model values to experiment are dependent upon the mass number and "type" of the states. From comparisons of experiment with model we attempt to infer the values of "effective" (relative to the model space) one-body electromagetic operators and compare these values to the "free-nucleon" values, which are obtained by assuming that the neutrons and protons of the model calculation have the same properties they do in free space. Individual nuclei for which extensive experimental information is available are examined to determine the degree to which the shell-model formulation can provide a comprehensive accounting of low-lying electromagnetic properties.

It is useful to remind ourselves of several basic aspects of electromagnetic matrix elements which stem from the characteristics of the operators. In all contexts, the size scale of matrix elements is set by their single-particle values, those values obtained with states characterized by single terms corresponding to a particle in an orbit j coupled to a core of angular momentum zero. Large matrix elements for the electric quadrupole operator, a product of radial and spherical harmonic components, indicates deformation of shape together with a similarity in shape between initial and final states. Coherent effects from configuration mixing can enhance the magnitudes by factors of three to four.

Large matrix elements for the magnetic dipole operator, a sum of intrinsic and orbital angular momentum operators with the magnetic g factors as coefficients, indicate rather pure single particle structure in the initial and final states, with the two states being related either by a recoupling within the dominant orbit or by a spin flip. Selection rules dominate many aspects of M1 phenomena. The values of the proton and neutron magnetic moment combine to make the isovector M1 matrix elements roughly ten times larger than the isoscalar. Also, the properties of the angular momentum operators forbid transitions between states of different orbital angular momentum and many potential transitions are found to be approximately "l-forbidden". Coherent effects for M1 matrix elements tend to reduce their magnitudes, so "large" M1 values are those almost as big as the single particle values.

ELECTRIC QUADRUPOLE MOMENTS
AND GENERAL COMMENTS ABOUT EFFECTIVE CHARGE

The conceptually simplest evidence about electric quadrupole phenomena comes in the form of the quadrupole moments of ground (and a few excited) states, since only one rather than two systems are involved. Three categories of data are available. The "classical" atomic physics measurements of past decades provide values[11] for most of the stable ground states, but the systematic uncertainties in these numbers, stemming from the necessary corrections for the effects of the details of the electronic wave functions, are very difficult to evaluate and are probably large on our scale of single particle values. Quadrupole moment values have been obtained for 2+ first excited states in doubly-even nuclei, but these numbers suffer from both large statistical and large systematic uncertainties.[12] Only the third category of data is precise enough to thoroughly test the current wave functions. It is comprised by the values for stable ground states extracted from muonic atom experiments.[13] (The correction factors for muonic wavefunctions are negligible on our scale of concern.)

Comparison of theory and experiment for all the data of the sd shell[14] reveals that the relative aspects of the data, namely the "large-small" and "positive sign - negative sign" relationships, are all in good agreement with the predictions. But, the sizes of the theoretical numbers are, systematically, almost a factor of two too small if the theory incorporates the real charges of the free neutron and proton. This universal feature of nuclei has been recognized since the pioneering work of Arima and Horie.[15] The enhancement is found even for states which must be dominantly single particle in structure, such as 17F and 17O.

Explanation of this feature is completely outside the province of the one-major-shell model we are exploring here. Since in our model the A=17 and A=39 states are precisely single particle in structure, there is no mechanism by which alterations of the Hamiltonian could enhance (alter in any way) the predicted quadrupole moment values.

Hence, we are led quickly to the realization, to which a study of the underlying theory would have alerted us, that the shell model as we construe it here does not necessarily give quantitatively accurate absolute predictions for nuclear matrix elements and, indeed, should not do so in principle. This is because the truncation of the "true" degrees of freedom of a nuclear state to the dimensions of the vector spaces used in the model wave functions imply correspondingly massive renormalizations of the properties of the constituent particles. That is to say, in the shell model we are dealing with "quasi-particles", not raw neutrons and protons.

The major aspects of electric quadrupole phenomena which are omitted in a single-major-shell model are the particle-hole excitations between major shells of the same parity, such as sd-sdg and p-fp, which give rise to the giant E2 resonance. Mixing of these excitations into low-lying states gives rise to the pervasive enhancement of E2 matrix elements over bare proton, one-major-shell estimates. For the single-major-shell predictions to be able to have the same magnitudes as those measured, the effects of the important excluded excitations must be introduced into the model. The simplest scheme for this is associated with the term "effective charge". The properties of the operator or, equivalently, the particles are modified so as to scale up the values of the associated matrix elements in a uniform way. Most simply, the values of the proton and neutron charges in the expression for the matrix element are changed from 1.0e and 0e, respectively, to larger values.

Our study of a wide variety of sd-shell data[16-18] suggests that the empirically optimum values for these charges in this region are 1.35e and 0.35e. More precisely, these are the optimum values of E2 effective charge in the context of complete-space sd shell configurations combined with single-particle radial wave functions of harmonic oscillator form which yield fits to the measured ground state rms radii.

We see that the relevant tests of the shell-model wave functions focus on the accuracy with which the quantitative variations between a succession of experimental values are theoretically reproduced modulo some overall scale factor. Ideally, this scale factor can be theoretically calculated. In the practical event, it is more easily determined empirically, and this is the path we have followed in the case of E2 data. Having set the over-all scale empirically, and thus cut the "Gordian Knot" of "how" to renormalize the model for the effects of truncation, the active questions are ones of the agreement between the renormalized model values and those of corresponding experiments.

The degree to which the effective charge renormalization should be independent of state is not clear from the present theoretical standpoint. Empirically, we assume total state independence. Hence, discrepancies with experiment could imply the failure of this assumption rather than a local defect in the model wave function. However, this line of rationalization cannot be pursued too far. The entire utility of the shell model as we are using it rests implicitly on a basic state-independence of the key parameters.

While there would be nothing wrong theoretically with a model in which all the renormalizations for truncation effects were strongly state dependent (indeed, some calculations tend to suggest that they are), such a state of affairs would vitiate most of the practical utility of the model in understanding experimental data. Fortunately, from our standpoint, the evidence from the quadrupole moments suggests a surprizingly high degree of state-independence in the E2 effective charge. Given the uncertainties in most of the data, which can be of the order of an e-fm2 or greater, the agreement between the 1.35e-0.35e predictions and experiment is such that there is no justification for adding any additional complexity to the model. Only data of the precision of the muonic atom experiments can address this issue at the relevant level of detail, namely at the order of a tenth of an e-fm2. The few extant data in this category are very close to the global 1.35e-0.35e best-fit line.

E2 TRANSITION STRENGTHS

The matrix elements extracted from E2 transitions between low-lying nuclear states typically are characterized by much smaller uncertainties than are associated with most of the quadrupole moment measurements. The experimental values which involve ground state decays can be cross-checked with a variety of techniques, and the combination of statistical and systematic uncertainties for the best cases is of the ordr of a tenth of an e-fm2, or a few per cent. There are about three dozen experimental sd-shell values of E2 matrix elements which have uncertainties smaller than 5%. They range in magnitude from 1 to 25 e-fm2.

When these "best" data are compared with the matching model transition values we find no significant correlation of deviations with either nuclear mass or transition strength. The 1.35e-0.35e renormalization fits the model to the data with a chi-square of about 4, based strictly on the quoted experimental uncertainties. A comparison expanded to include somewhat less precisely measured E2 matrix elements, with uncertainties in the range 5%-10%, gives the same conclusions as reached with the best data, but, as with the quadrupole moment data, lack of precision vitiates much of the potential interest in the critique of the theory. Progress in establishing firmly the limits of simple, global renormalizations of the shell-model predictions for E2 phenomena awaits more, and particularly more precise, experimental measurements.

A different perspective on E2 predictions can be obtained by concentrating on the transitions within a particular nucleus rather than on a survey of the whole shell which samples only a few transitions per system. Several different issues come into play with this more "local" perspective. Each low-lying level is typically involved in several different transitions, so the characterization of its structure is more complete, and the model's reproduction of these features better established. The "band structure" of the nucleus can thus be elucidated, the dependence of the effective charge renormalization upon excitation energy potentially examined, and the issue of "intruder states" addressed anew. Again, the critical problem is the availability and precision of the data. As higher excited states are studied, the experimental techniques become fewer and more difficult. An unusually thorough study has been made of 26Al, with the result that about three dozen E2 matrix elements of transitions between levels in the first six MeV of excitation have been determined. The typical experimental uncertainties here are much greater than for the more tractable measurements we have considered so far, and the deviations between model and experiment are much greater as well. The overall correlation of the theoretical values with experiment is, however, strikingly strong. As we would expect, it is best for the larger matrix elements.

At this stage of complexity, comparison of experiment with theory involves a trivial but vital new uncertainty, namely the question of correctly matching the model states with the experimental levels. Matching model with experimental state is not a problem when we are considering the ground and first few excited states of a nucleus, but by the fourth, fifth, sixth or so level, the spacing between states of a given J value has typically become only a few hundred keV, namely of the same order as the "theoretical" uncertainty in the shell-model eigenvalues. Hence, as an example, the fourth experimental level of a given J in a particular nucleus may have a wave function corresponding to the fifth model state of that J in that system. Our approach to this would be that the wave functions (namely the transition properties) overrule the energy ordering, and we would criss-cross the lines connecting the levels in the comparison of spectra. However, until comparisons of matrix elements are done, we do not know the full extent of this problem.

Of course, the situation can deteriorate still further, to the point that the individual model wave functions no longer correlate with the individual experimental levels in any sense. This should be the dominant situation when the density of intruder states begins to overwhelm the "native", in the present example the sd-shell, states. We expect this to happen somewhere in the 5 to 20 MeV region, with a strong dependence upon the specific J-T value. But, even in the absence of intruder states, the spacing between levels will eventually be much smaller than the amount and accuracy of information which has been stored in the Hamiltonian can deal with, and the individual matchups of model states with experimental states will inevitably become meaningless.

What should remain meaningful in the higher excitation regions of high level density are the average properties of the model states. That is, while it is probably nonsense to try to find the experimental match for the thirteenth model state, for example, the amount of strength predicted from the model levels in the 9-10 MeV region of excitation could be expected to reproduce the corresponding measured quantity. Deviations in such average properties could not be excused on the grounds of expected imprecisions in the Hamiltonian, and would be significant indications of intruder effects or other fundamental breakdowns of the essential model hypotheses.

ELECTRON SCATTERING FORM FACTORS

An added dimension of electric quadrupole phenomena is available from the study of the form factors of inelastic electron scattering. The sampling of the E2 transition strength at a variety of finite momentum transfers can also be considered as a sampling of the radial dependence of the matrix element. As such, it is potentially possible to critically examine the detailed shapes of the single particle radial wave functions. The matrix element value at the photon point, obtained from a photon experiment, only allows a consistancy check with their rms radii.

The available data on sd shell nuclei are adequate for only qualitative examinations of wave functions. The comparisons of the predictions of the present shell model wave functions with the extant data on even-mass sd-shell nuclei[17] and on $27Al$[18,19], indicate that the basic features of observed form factors are reproduced out to momentum transfers of almost 3 fm-1. This agreement is in the context of an effective charge model renormalization, of course, in parallel to the case for the q=0 matrix elements. There seems a slight preference[17] for a radial dependence of the effective charge transition density which has a Tassie model shape, rather that a shape which follows the shape of the model space transition density. More and better data and continued analysis promises to yield important clues to understanding the microscopic foundations of the effective charge.

Electron scattering also offers an avenue to study of a corollary to E2 data, namely E4 phenomena. While E4 transitions are unobservable in gamma decay of light nuclei, the electron scattering process makes them essentially as amenable to study as E2 transitions. Questions about E4 phenomena arise in parallel to those we ask about the corollary E2 phenomena: "What are the effective charges?" "What are the radial dependence of the effective charges?" and "Do the wave functions correctly position and distribute the transition strength?" The answers at present seem to be that the E4 effective charges are 1.5e and 0.5e in the same context as the 1.35e and 0.35e for E2, that the Tassie model shape is again preferred for the effective charge transition density, and that the observed E4 strength is accurately predicted in all measured cases.[17,18,19]

MAGNETIC DIPOLE MOMENTS

The magnetic dipole moments of nuclear ground states offer the same advantages in principle as do the electric quadrupole moments, namely uniqueness to a single state and sign as well as magnitude information. Unlike the quadrupole moments, the dipole moments are measured with great precision, with uncertainties much smaller than the conventional nuclear theories can match with their own internal accuracies. The present calculations reproduce the known magnetic dipole moments of the sd shell very well when the values of the magnetic moments of the free neutron and proton are used as coefficients of the spin and orbital terms in the M1 operator.[20]

However, it is possible to improve the agreemeent between model predictions and experiment by adjusting the values of the single particle M1 matrix elements. While the adjustments are do not change the qualitative aspects of the agreement, the new values of the single-particle paramenters are meaningfully determined and the changes from the free-nucleon values are significant, even if small.[20] It is interesting to compare the changes in the single particle M1 matrix elements thus determined by fitting the shell model densities to data with the comparable numbers determined from a similar fit to Gamow-Teller beta decay.[20,21] The results of such a comparison clearly indicate a difference in the renormalizations of the M1 and GT spin operators which must be interpreted in the context of the non-nucleonic effects.[20]

SPIN-DOMINATED (M1 AND GT) TRANSITIONS

The correlations between M1 and GT phenomena are interesting in all contexts, in transition data as well as for moments. It is important to consider transitions as well as moments in these sorts of studies since the former tend to feature more involvement of the spin-flip single particle terms as opposed to the single-orbit dominance which typifies the moments. Transitions mediated by these two elated operators which occur between low-lying states are typically weak, i.e., small fractions of a single particle value. As such, the percentage accuracy of the shell model predictions will not be all that good. Many of the existing M1 data of this sort are not highly precise either, unlike the beta decay data, which generally have all the accuracy that the shell model can use.

The analysis via shell-model wave functions is useful even for these weak transitions, however, from the aspect of average properties. The totality of such M1 data is consistent with the magnetic moment values in that no appreciable overall renormalization of magnitudes from the free-nucleon estimates seems to be needed. The totality of sd-shell Gamow-Teller beta decay, on the other hand, unambiguously indicates that the free-nucleon shell-model estimates are too large.[22,21] The present wave functions[21] suggest that an overall quenching of the transition strengths of 0.60 +/- 0.02 (an 0.78 quenching of the matrix elements) serves as an accurate, mass and state independent renormalization for this process.

The process of determining the optimum empirical renormalizations of nuclear operators from shell model analyses of multiple-particle data is most reliable when the matrix elements are least sensitive to the details of the shell model calculations themselves. In practice, this means when the matrix elements are largest. The E2 renormalizations previously discussed are hence rather securely established, since the dominant data correspond to calculated matrix elements that are large and insensitive to small to moderate variations in the shell model wavefunctions. On the other hand, the results we have quoted for M1 and Gamow-Teller phenomena are based in large part on weak transitions, corresponding to highly cancelled shell model matrix elements. The latter are hence inevitably sensitive to the details of the model calculation and the conclusions about renomalizations which are based upon these analyses incorporate these uncertainties. It would be preferable to base conclusions on these issues on data which correspond to the dominant portion of the strength of a process rather than to a small residual.

For the spin-type processes such as M1 and Gamow-Teller, the dominant portion of transition strength is associated with spin flip and lies in the neighborhood of 10 MeV excitation energy. Experiments which survey a wide range of excitation energies are needed. The foundations of our knowledge in this area are the back-angle electron scattering experiments of Fagg and collaborators.[23] These studies yield a picture of the distribution and overall magnitude of M1 transition strength which is consitent with the present calculations and the free-nucleon value of the M1 operator. There is need to supplement these first existing data with newer values obtained from higher resolution experiments, such as those of reference [24].

An alternate experimental technique for obtaining accurate, high resolution M1 data in the "giant resonance" region is that of inelastic gamma ray scattering. This is a particularly powerful approach when coupled with polarization constraints. A recent survey of the even sd-shell nuclei with this technique[25] has provided an accurate and complete survey of M1 strength up to excitation energies at which these nuclear systems become unbound to particle emission, around 10 to 15 MeV. These energies are typically sufficient to cover most of the M1 giant resonance, although in a few cases the predictions indicate that significant strength lies just beyond the region of observability.

The comparison of shell model predictions to these results is instructive in many ways. First, the data establish that the predictions correctly place the preponderance of the model strength in the region of excitation energy at which experimental strength is observed. Second, the magnitudes of the observed strength are closely matched by the model magnitudes based on the free-nucleon parametrization of the M1 operator. That is to say, again, there is no significant evidence that observed M1 strength is quenched beyond the amount that can explained by intra-sd-shell configuration mixing. Third, it should be noted that the individual states sometimes have strengths which are in poor agreement with the model predictions, but that such disagreements are averaged out over small clusters of states, so that the total strength predicted for a small region of excitation energy is always in good agreement with the corresponding data. This aspect of the results demonstrates that it is unrealistic to expect the model predictions to be "better" than the rms uncertainty in the fit to energy levels. The 140 keV rms deviation in energies tells us that levels of the same spin which have eigenvalues separated by gaps of this order can be expected to have an uncertain mix of wave function components, even if the aggregate for the same levels is well determined.

Similar data on odd-mass nuclei can be even more informative. A recent study of 23Na with photon scattering[26] has identified a dozen M1 excitations of significant strength below 9 MeV. The corresponding shell model predictitions match the observations one for one over the complete range of the data. The aggregate magnitude predicted from the free-nucleon operator is in close agreement with the measured value.

CONCLUSIONS

We have illustrated here how a comprehensively systematic exploitation of the degrees of freedom of the sd-shell orbits suffices to account for a large body of information on level energies and the electromagnetic and weak decay properties of these levels. The fundamental point which should be drawn from the various comparisons of model predictions and experiment we have made is that the characteristics of the "multiparticle" (in the model context) nuclei are completely consistent with those of the "single particle" nuclei. The often profound enhancements and quenchings which are observed for multiparticle systems are seen to be the consequences of configuration mixing within the sd-shell orbits, rather than evidence for more exotic processes or strong state dependences in the renormalizations of the spectroscopic operators. All the E2 data in the region seem consistent with an additive effective charge of 0.35e. All of the M1 data seem consistent with the free nucleon magnetic moments. Finally, all the Gamow-Teller beta decay data seem consistent with a quenching of the strength by a factor of 0.6.

REFERENCES

1. J. B. French, E. C. Halbert, J. B. McGrory and S. S. M. Wong, Advances in Nuclear Physics, Vol. 3, Chapt. 3, Plenum Press (New York) (1969).
2. W. Chung, unpublished.
3. H. Kruse, unpublished.
4. R. R. Whitehead, A. Watt, B. J. Cole and I. Morrison, Advances in Nuclear Physics, Vol. 8, Chapt. 3, Plenum Press (New York) (1977).
5. W. D. M. Rae, A. Etchegoyen, N. S. Godwin and B. A. Brown, unpublished.
6. B. H. Wildenthal, B. A. P. S. 27, 725 (1982).
7. B. H. Wildenthal, Progress in Particle and Nuclear Physics, vol. 11, page 5 (1984).
8. P. M. Endt and C. van der Leun, Nucl. Phys. A310,1 (1978).
9. B. H. Wildenthal, E. C. Halbert, J. B. McGrory and T. T. S. Kuo, Phys. Rev. C4, 1266 (1971); W. Chung, PhD Thesis, Michigan State University (1976)
10. P. .M. Endt, private communication; H. Ropke, private communication.
11. C. M. Lederer and V. S. Shirley, Table of Isotopes, seventh edition, (Wiley, New York, 1978).
12. R. H. Spear, Physics Reports 73,369 (1981).
13. R. Weber et al., Physics Letters 98B, 343 (1981).
14. H. Horie and A. Arima, Phys. Rev. 99, 778 (1955).

15. B. A. Brown, W. Chung and B. H. Wildenthal, Phys. Rev. C22, 774 (1980).
16. B. A. Brown et al., Phys. Rev. C26, 2247 (1982).
17. B. A. Brown, R. Radhi and B. H. Wildenthal, Physics Reports 101, 314 (1983).
18. R. Radhi, PhD Thesis, Michigan State University (1983).
19. P. J. Hicks et al., Phys. Rev. C27, 2515 (1983).
20. B. A. Brown and B. H. Wildenthal, Phys. Rev. C28, 2397 (1983).
21. B. .A. Brown and B. H. Wildenthal, to be published in Atomic Data and Nuclear Data Tables.
22. B. A. Brown, W. Chung and B. H. Wildenthal, Phys. Rev. Lett. 40,1631 (1978).
23. L. W. Fagg, Rev. Mod. Phys. 47, 683 (1975)
24. R. Schneider et al., Nucl. Phys. ,A323,13 (1979).
25. U. E. P. Berg et al., Phys. Letters 140B, 196 (1984).
26. R. Vodhanel et al.,Phys. Rev. C29, 409 (1984).

LOCAL PROPERTIES OF INTERACTING BOSONS

V.R. Manfredi
Dipartimento di Fisica dell'Università, 35100 Padova, Italy
Istituto Nazionale di Fisica Nucleare, Sezione di Padova

ABSTRACT

By the use of Δ_3 and $\sigma^2(k)$ statistics a comparison is made between the fluctuation properties of a two-body random boson ensemble and the Gaussian orthogonal ensemble (GOE). With the two statistics no significant differences between the two ensembles have been found.

The aim of this work is to study the local property of an interacting boson system, in particular to see if and how the boson signature in the dense limit [1] affects the level density fluctuations. (For the global properties see references [2] and [3]).

In order to carry out this program, we chose a number-conserving Hamiltonian

$$H = \tfrac{1}{2} \sum_{i,j,k,l} \langle ij|V|kl \rangle a_i^+ a_j^+ a_k a_l, \qquad (1)$$

where the two-body matrix elements V_{ijkl} were taken as random numbers uniformly distributed over the $(-0.1, 0.1)$ interval. Forty different sets of V_{ijkl} were used in the following special case

Ensemble dimensionality (number of matrices)	$D=40$
Dimension of each matrix	$d=364$
Number of bosons	$M=11$
Number of single-boson levels	$N=4$
Number of independent V_{ijkl} (direct+exchange)	$=58$.

Starting from the "exact" eigenvalues $E_1^{(i)}, \ldots E_d^{(i)}$, obtained diagonalizing the Hamiltonian (1), we have calculated centroid $\varepsilon^{(i)}$, width $\sigma^{(i)}$, skewness $\gamma_1^{(i)}$ and excess $\gamma_2^{(i)}$, where i runs from 1 to 40. In order to avoid spurious fluctuations over the ensemble, each eigenvalue is renormalized by the transformation

$$E_j^{(i)} \to E_j^{,(i)} = (E_j^{(i)} - \varepsilon^{(i)})/\sigma^{(i)}. \qquad (2)$$

The level density for each eigenvalue distribution is quasi Gaussian: in fact $\langle \gamma_1 \rangle_e = 0.01 \pm 0.23$ and $\langle \gamma_2 \rangle_e = -0.16 \pm 0.24$, where $\langle \gamma_{j=1,2} \rangle_e$ is the mean value of γ_j over the ensemble. In order to remove the secular

Table I Comparison of calculated values of Δ_3 with prediction from Gaussian orthogonal ensemble (GOE)

$(-L,L)$	$\langle n \rangle_e$	$\langle \Delta_3 \rangle_e \pm (\text{Var}\Delta_3)^{\frac{1}{2}}$	$\Delta_3 \pm (\text{Var}\Delta_3)^{\frac{1}{2}}$ GOE
-10,10	19.97	0.2901±0.0839	0.2964±0.11
-20,20	40.22	0.3829±0.1546	0.3674±0.11
-30,30	60.20	0.4481±0.1547	0.4082±0.11
-40,40	80.37	0.5166±0.2026	0.4375±0.11

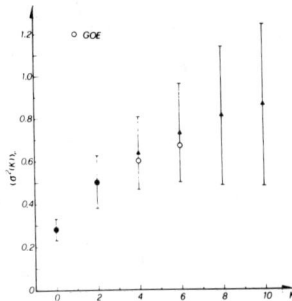

Fig. 1. Variance $\sigma^2(k)$ of the kth nearest-neighbour spacing versus k.

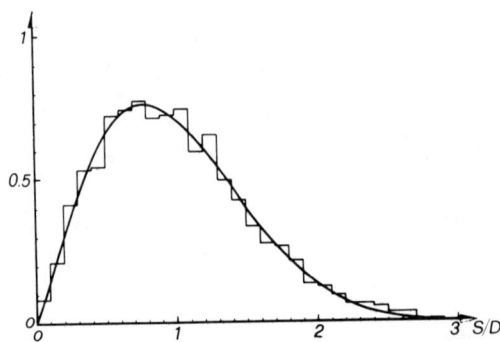

Fig. 2. Distribution of nearest-neighbour spacing S in unit of mean spacing D. The solid curve is the Wigner distribution.

variation of the density, we map each spectrum, by a numerical procedure [4] into one having a constant level density.

For each interval (-L,L) Table I shows the mean value of the number of levels n, of Δ_3 and, for comparison, of the GOE theoretical value [5]. As can be seen, for larger $\langle n \rangle_e$ there are significant discrepancies between the calculated values and the theoretical ones {see also ref.[6]}. These differences are connected with the sharp tails of the original level densities. In fact our smoothing procedure works well in the central region, giving uniform density spectra with the mean spacing very near to one and a constant number of levels, whereas at the end of the spectra the mean spacing is quite different from one and the mean number of levels is not at all constant [4].

Another statistic we have used is the variance of the kth nearest-neighbour spacing $\sigma^2(k)$. In fig. 1 $\sigma^2(k)$ is plotted versus k; Gaussian orthogonal ensemble (GOE) results are also indicated [7]. Fig. 2 shows the total spacing distribution in units of the mean spacing. The agreement with the Wigner distribution is astonishingly good.

We may conclude that the fluctuations of a two-body random boson ensemble, even in the dense limit, seem to be consistent with the GOE predictions.

REFERENCES

1. V. K. B. Kota and V. Potbare, Phys. Rev. C21, 2637 (1980).
2. V. R. Manfredi, Nuovo Cimento, A64, 101 (1981).
3. V. R. Manfredi, Global and Local Properties of Interacting Bosons to be published.
4. V. R. Manfredi, Lett. Nuovo Cimento, 40, 135 (1984).
5. F. J. Dyson and M. L. Mehta, J. Math. Phys. 4, 701 (1963)
   M. L. Mehta, Statistical Properties of Nuclei, edited by J. B. Garg (Plenum Press, N.Y., 1972), p. 179.
6. S. S. M. Wong, Statistical Properties of Nuclei, edited by J. B. Garg (Plenum Press, N.Y., 1972), p. 202.
7. M. L. Mehta and J. des Cloizeaux, Ind. J. Pure Appl. Math., 3, 329 (1972).

MIXED SYMMETRY STATES IN THE VIBRATIONAL LIMIT OF THE IBA MODEL

W.D. Hamilton, A. Irbäck and J.P. Elliott
Physics Division, University of Sussex, Brighton, BN1 9QH, England

ABSTRACT

It is shown that the 2^+ level at about 2 MeV in the N=84 isotones ^{140}Ba, ^{142}Ce and ^{144}Nd may be identified as the lowest state of mixed symmetry in the U(5) limit of the IBM-2.

A feature common to ^{140}Ba, ^{142}Ce and ^{144}Nd is the occurance and decay modes of the 2_3^+ level at ~2 MeV. The level, whose energy is largely independant of neutron number, decays by a predominantly M1 transition to the 2_1^+ state while there is also a significant branch to the groundstate. It is not possible to account for this level within the IBM1 description, which is adequate for the other low lying levels, and it is considered that the state may be of mixed symmetry in the IBM2 which distinguishes between neutron and proton bosons[1]).

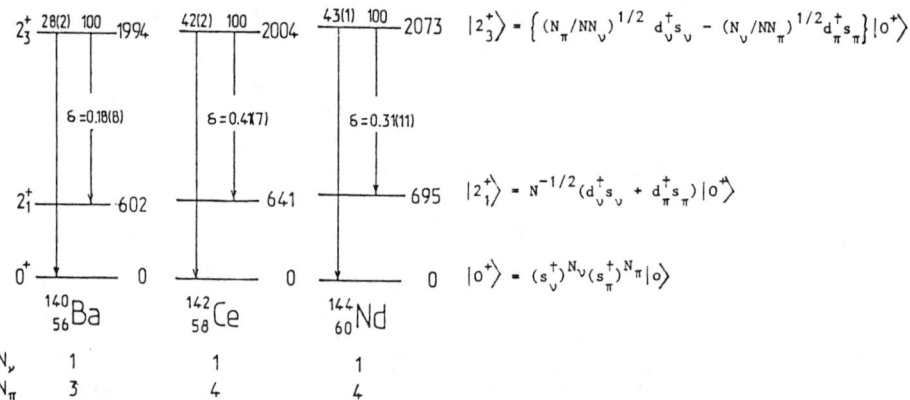

Fig. 1: The decays of the 2_3^+ level in ^{140}Ba, ^{142}Ce, and ^{144}Nd. The structure of the three states is given and also the neutron and proton boson numbers, N_ν and N_π. The shell closure at Z=64 has the effect of reducing N_π for ^{144}Nd [2]) from 5 to 4.

The decay schems of these isotopes were established by experiments carried out at the ILL Grenoble[3,4,5]). Gamma-ray directional correlation measurements were made on the decay of the fission products ^{140}Cs and ^{142}La which were selected by the on-line mass separator OSTIS while levels and transitions in ^{144}Nd were studied following n-capture by enriched ^{143}Nd.

Within the IBM the transtion operators are defined as

$$T^{E2} = e_\pi Q_\pi + e_\nu Q_\nu$$
$$Q = (s^\dagger d + d^\dagger s) + \chi(d^\dagger d) \quad (2)$$

$$T^{M1} = (3/4\pi)^{1/2} (g_p L_\pi + g_n L_\nu)$$

and in order to test the possible mixed symmetry description of these 2^+ levels we must derive quantities from the experimental data which depend on model predictions. We may write the branching ratio from the level as

$$\frac{T(2_3^+ \to 0^+)[E(2_3^+) - E(2_1^+)]^2}{T(2_3^+ \to 2_1^+)[E(2_3^+) - E(0^+)]^5} = \frac{1}{2.07(1+\delta^2)} \cdot \frac{N(e_\pi - e_\nu)^2}{g_p - g_n} \qquad (2)$$

where the E2 component has been eliminated from the $2_3^+ - 2_1^+$ transition.

Now the g-factors of 2^+ states have the values $g_p \simeq 1$ and $g_n \simeq 0$ and from the data in fig. 1 we obtain

$$|e_\pi - e_\nu| = 0.12 \text{ eb} \qquad (3)$$

Also we may write the transition probability as

$$B(E2; 2_1^+ \to 0^+) = (N_\pi e_\pi + N_\nu e_\nu)^2/N \qquad (4)$$

and re-arranging we obtain

$$[B(E2; 2_1^+ \to 0^+)N]^{1/2}/N_\pi = e_\pi + e_\nu N_\nu/N_\pi \qquad (5)$$

Thus a plot of the left-hand side of eq. (5) vs. N_ν/N_π should be linear and give e_π and e_ν. The results from B(E2) values of transitions in this region are shown in Fig. 2 and we obtain

$$e_\pi = 0.12 \quad \text{and} \quad e_\nu = 0.24$$

These values agree with the result given by eq. (3).

The mixing ratio may be written

$$\delta = \frac{E(2_3^+) - E(2_1^+)}{1.44} \cdot \frac{e_\pi \chi_\pi - e_\nu \chi_\nu}{g_p - g_n} \qquad (6)$$

If we take the mean value $\delta = 0.31$ and use the above values for g_p, g_n, e_π and e_ν we obtain

$$\chi_\pi - 2\chi_\nu = 2.6 \qquad (7)$$

This is consistent with the accepted values; $\chi_\nu = -1$ near the beginning of a shell and χ_π is small at about mid-shell. The effect of the partial shell closure at $Z = 64$ would imply $\chi_\pi \simeq 0$ for Ba, Ce and Nd with $\chi_\nu \simeq -1.3$.

The positive signs of the measured mixing ratios is crucial for the relation in eq. (7) to hold.

The quantities χ_π and χ_ν also occur in the expression for the quadrupole moment of the 2_1^+ state:

$$Q(2_1^+) = 1.7(N_\pi e_\pi \chi_\pi + N_\nu e_\nu \chi_\nu)/N \qquad (8)$$

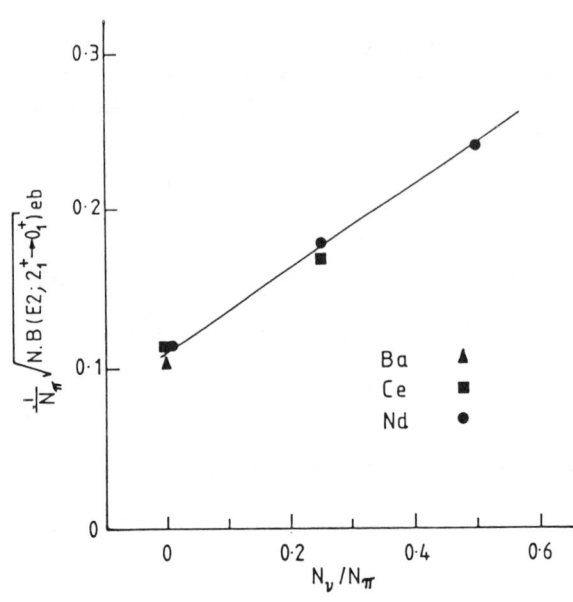

Fig. 2: The quantity $[B(E2; 2_1^+ \to 0^+)N]^{1/2}/N_\pi$ plotted versus N_ν/N_π for several nuclei. The straight line is a least-squares fit to the data taken from refs. 6,7 and 8. Experimental errors are insignificant compared to the size of the data symbols. The data are for ^{138}Ba, ^{140}Ce, ^{142}Ce, ^{142}Nd, ^{144}Nd and ^{146}Nd.

There are few precise values of $Q(2_1^+)$ but we have for ^{142}Ce the result $Q(2_1^+) = -0.12(9)$b [9] and with the above values for the quantities in eq.(8) we obtain $Q(2_1^+) = -0.11$b which confirms that χ_ν should be negative.

We conclude that the γ-decay properties of the 2_3^+ state in these nuclei are well described by the lowest mixed symmetry state in the vibrational limit of the IBM2.

REFERENCES

1. A. Arima and F. Iachello in Advances in Nuclear Physics 13(1948) 139, ed. M. Baranger and E. Vogt, pub. Plenum Press, New York
2. O. Scholten, Phys. Lett. 127B (1983) 144
3. S.J. Robinson, W.D. Hamilton, P. Hungerford, B. Pfeiffer, G. Jung and M. Snelling, to be published in J. Phys. G
4. E. Michelakakis, W.D. Hamilton, P. Hungerford, G. Jung, B. Pfeiffer and S.M. Scott, J. Phys. G 8 (1982) 111
5. D.M. Snelling and W.D. Hamilton, J. Phys. G 9 (1983) 763
6. G. Kindleben and Th. W. Elze, Z. Phys. A 286 (1978) 415
7. G. Engler, Phys. Rev. C1 (1970) 734
8. P.A. Crowley, J.R. Kerns and J.X. Saladin, Phys. Rev. C3 (1971) 2049
9. C.M. Lederer and V.S. Shirley, Tables of Isotopes, 7th Edition 1978, pub. Wiley, New York

Collective Magnetic Octopole Transitions

O. Scholten

Cyclotron Laboratory and Department of Physics-Astronomy
Michigan State University, East Lansing, MI 48824

ABSTRACT

The properties of the magnetic octopole operator are discussed in the framework of the neutron-proton Interacting Boson model. It is predicted that in deformed nuclei the $I^\pi=3^+$ member of a low-lying collective K=1$^+$ band can be excited with an appreciable strength.

In recent electron scattering experiments[1] strong M1 transitions have been observed in deformed medium heavy even-even nuclei at around 3 MeV excitation energy. In this energy regime there are several 1$^+$ states, but the exceptional features of this observation are that first of all the M1 strength of 1~2 μ_N^2 is much larger than one would expect for a 2 q.p. transition and secondly that the form factor shows an almost pure orbital excitation. These points indicate the excitation of a collective 1$^+$ state. Both in the geometrical[2] and in the Interacting Boson Model (IBA)[3] the existence of such states have been predicted and the experimental observation constitutes an important confirmation.

This finding has triggered the investigation of the existence of other collective magnetic transitions, specifically M3 transitions, in the IBA model. In the IBA model M3 transitions are calculated by introducing the operator

$$T^{M3}_{B\mu} = \sqrt{\frac{35}{8\pi}} \{ \Omega_\pi (d^\dagger_\pi \tilde{d}_\pi)^{(3)}_\mu + \Omega_\nu (d^\dagger_\nu \tilde{d}_\nu)^{(3)}_\mu \}$$

$$= \sqrt{\frac{35}{8\pi}} \{ \Omega_S [(d^\dagger_\pi \tilde{d}_\pi)^{(3)}_\mu + (d^\dagger_\nu \tilde{d}_\nu)^{(3)}_\mu]$$

$$+ \Omega_A \frac{2}{N} (N_\nu (d^\dagger_\pi \tilde{d}_\pi)^{(3)}_\mu - N_\pi (d^\dagger_\nu \tilde{d}_\nu)^{(3)}_\mu) , \quad (1)$$

where the parameters Ω_S and Ω_A are defined in terms of the magnetic octopole moments, $\Omega_\rho (\rho=\nu,\pi)$, of the neutron and proton d-boson as

$$\Omega_S = \frac{1}{N} (N_\pi \Omega_\pi + N_\nu \Omega_\nu) , \quad \Omega_A = \frac{1}{2} (\Omega_\pi - \Omega_\nu) , \quad (2)$$

and N_π and (N_ν) denote the number of proton (neutron) bosons (N=N_π +N_ν).[4] The first term on the right hand side of Eq. 1, isoscalar term, only excites states that are fully symmetric in the neutron and proton degrees of freedom, i.e. states that have maximal F-spin.

The second, the isovector term, connects the fully symmetric states with states that have $F=(F_{max}-1)$, hereafter referred to as antisymmetric (a.s.) states).

In the U(5) limit (spherical limit) of the IBA-2 model, the matrix elements of the M3 transition operator between the ground-state and lowest 1^- antisymmetric state vanish. In the SU(3) limit (the axially symmetric rotor limit), the picture is more complex. The operator of Eq. (1) leads from the ground state, $(\lambda,\mu)=(2N,0), I^{\pi}=0^+$ to $I^{\pi}=3^+$ states in both the gamma-band [i.e. the symmetric SU(3) representation $(\lambda,\mu)=(2N-4,2)$], in the $(\lambda,\mu)=(2N-2,1)$, $K^{\pi}=1^+$ band, and in the anti-symmetric $(\lambda,\mu)=(2N-4,2)$, $K^{\pi}=2^+$ band.

As an example of the M3 strength distribution in the transitional region between the SU(5) and the SU(3) limits of the IBA model, we will present the results of a calculation for the Sm isotopes. The numerical calculations were done the standard IBA-2 Hamiltonian as is given for example in ref 5, using the parameters given in ref 6. Only the strength of the Majorana force has been readjusted, $\xi_1=\xi_2=\xi_3=0.15$ MeV such that the energy of the 1^+ state is about 3MeV, the energy where it has been observed in ^{156}Gd. The B(M3 ↑) values, calculated separately for the isoscalar and isovector components of the operator, are given in Fig 1. For symmetric levels the isovector component is essentially zero, while anti-symmetric states are predominantly excited by the isovector part of the operator and the isoscalar part vanishes. As is expected, in the vibrational nuclei ^{148}Sm and ^{150}Sm the M3 excitation probability is very small. In the deformed nucleus ^{156}Sm three 3^+ levels are excited. It should be noted that although the energy difference between the $K=2_2^S$ and the $K=1_1^A$ levels in ^{154}Sm is small the selection rules are still rather well obeyed. This

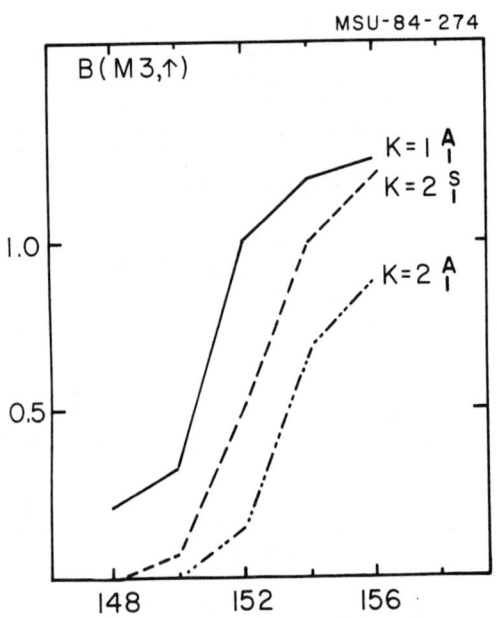

Fig. 1 - The calculated B(M3) values in the IBA model. For symmetric (antisymmetric) states only the isoscalar (isovector) part of the operator is considered.

implies that M3 transitions can be used to identify the position of the a.s. bands in deformed nuclei.

To obtain a microscopic estimate of the coupling constants Ω in Eq(1), the matrix elements between the lowest seniority states[7] in the boson space and the collective fermion space[8] (consisting of S_ρ and D_ρ pairs), are equated, i.e.

$$\Omega_\rho = \frac{1}{7} \sqrt{\frac{8\pi}{5}} \langle S_\rho^{N-1} D_\rho || T^{M3} || S_\rho^{N-1} D_\rho \rangle \quad , (\rho=\pi,\nu) \qquad (3)$$

The values of the M3 matrix element thus calculated[9] for ^{154}Sm (using a quenching factor of 0.7 for the neutron and proton spin magnetic moments) are $\Omega_\nu = -.13\mu_N b$ and $\Omega_\pi = .71\mu_N b$. In general the sign of Ω_ν is negative while that of Ω_π is positive. This implies that the M3 operator has a strong isovector component and thus predominantly excites a.s. states. The calculated B(M3) values for ^{154}Sm are given in Table 1, where the levels are labelled by their K values and the symmetry character (totally symmetric (S) or antisymmetric (A)). Note, however, that the IBA-2 Hamiltonian will in general lead to some mixing of both K and symmetry character and therefore the labels refer only to the dominant components. The $I_i^\pi = 3_2^+$ state which is a member of the $K^\pi = 1^+$ band is most strongly excited. The single particle value for a M3 transition is 0.13 $\mu_N^2 b^2$ while the typical strength of a transition to the first 2q.p. 3+ state, as calculated in the generalized seniority model[7], is only of the order of 0.03 $\mu_N^2 b^2$. The M3 transition probability to the collective a.s. states is thus large.

Table 1. Calculated excitation energies and $B(M3, 0^+ \to 3^+)$ values for the first four collective $I^\pi = 3+$ states in ^{154}Sm in units of $\mu_N^2 b^2$.

state	band	E_x[MeV]	B(M3↑)
3_1^+	2_1^S	1.51	0.23
3_2^+	2_2^S	2.46	0.001
3_3^+	1_1^a	2.99	0.56
3_4^+	2_1^a	3.26	0.31

M3 transitions offer an alternative way of exciting a.s. states. Microscopic calculations based on the generalized seniority[7] model indicate that the M3 operator is predominantly isovector in character. The predicted strength is of the order of 5 s.p. units.

REFERENCES

1. D. Bohle, A. Richter, W. Steffen, A.E.L. Dieperink, N. Lo Iudice, F. Palumbo, and O. Scholten, Phys. Lett. 137B (1984) 27.

2. N. Lo Iudice and F. Palumbo, Phys. Rev. Lett. 41 (1978) 1532; ibid, Nucl. Phys. A326 (1979) 193, T. Suzuki and D.J. Rowe, Nucl. Phys. A289 (1977) 461,

E. Lipparini and S. Stringari, Phys. Lett. 130B (1983) 139.

3. A.E.L. Dieperink, Prog. Part. Nucl. Phys. 9(1983) p121,
   M. Sambataro, O. Scholten, A.E.L. Dieperink, and G. Piccitto,
   Nucl. Phys. A423 (1984) 333.

4. A. Arima, T. Otsuka, F. Iachello and I. Talmi, Phys. Lett.
   66B (1977) 20.

5. Interacting Bose-Fermi Systems in Nuclei, ed. F. Iachello
   (Plenum Press, New York 1981).

6. O. Scholten, Ph. D. Thesis, Univ. of Groningen 1980.

7. O. Scholten, Phys. Rev. C28 (1983) 1783.

8. T. Otsuka, A. Arima, and F. Iachello, Nucl. Phys.
   A309 (1978) 1.

9. O. Scholten, A.E.L. Dieperink, K. Heyde, and P. van Isacker, to
   be published in phys. Lett.

GENERALIZED-SENIORITY MIXING IN SEMI-MAGIC NUCLEI

G. Bonsignori, M. Savoia
Ist. di Fisica dell'Università, Bologna
Via Irnerio n. 46; 40126 Bologna, Italy

K. Allaart
Natuurkundig Laboratorium, Vrije Universiteit
De Boelelaan 1081; 1081 HV-Amsterdam; The Netherlands

ABSTRACT

We have described the spectra and electromagnetic properties of even semi-magic nuclei using a generalized-seniority (v_g) scheme (broken-pair model; v_g = twice the number of broken pairs), with a model space including up to $v_g = 4$ states. We conclude that ground states are for about 90 percent $v_g = 0$ states, 2_1^+ and 3_1^- are for about 95 percent $v_g = 2$ states, while for other excited states one has at least 10-20 percent $v_g = 4$ admixtures in predominantly $v_g = 2$ states. Experimental evidence for these $v_g = 4$ admixtures comes mainly from E2 decay (lifetimes, E2/M1 mixing ratios). However also negative-parity spectra for $J \geqslant 5$, when compared with $v_g \leqslant 2$ and with $v_g \leqslant 4$ calculations, suggest rather strong (≈ 20 percent) $v_g = 4$ admixtures.

A v_g mixing mechanism which is of particle-phonon coupling type may explain qualitatively the $v_g = 4$ admixtures and their effect on electromagnetic properties.

The v_g mixing causes a strong fragmentation of "two-phonon" states as revealed by B(E2, $J^\pi \rightarrow 2_1^+$) values. For $J^\pi = 0^+$ this total E2 strenght is a factor two less than in (harmonic) vibrator or IBA models. The v_g mixing also causes a 30 percent reduction of $0_1^+ \rightarrow J^\pi$ excitation strenght (for the strongest excitations) for innatural parity states, but less reduction or even enhancement (2_1^+, 3_1^-) for natural parity states.

Identification of 1^+, 3^+ states as well as measurements of half-lives in the range of 10 to 100 ps would provide further tests of the model.

INTRODUCTION TO THE MODEL

A Shell Model state is said to have generalized seniority [1] (v_g) zero if it consists only of superfluid (coherent) Cooper pairs :

$$S^+ = \sum_{nlj} \frac{1}{2} \varphi_{nlj} (2j+1)^{1/2} (a^+_{nlj} a^+_{nlj})^{J=0} \tag{1}$$

A state has $v_g = 2$ if all but one pairs are of type (1), $v_g = 4$ if all but two pairs are of type (1), etc.. The basis states for a $v_g \leqslant 4$ model space are pictorially represented in Fig. 1. The coefficients φ_{nlj}, which determine the contribution of a certain orbit (n l j) to the S-pair, are obtained by minimization of the Shell Model energy for a state with only S-pairs ($v_g = 0$). Due to this procedure there is an approximate relation between these coefficients and the BCS parameters, as follows

$$\varphi_{nlj} \approx v_{nlj} (u_{nlj})^{-1} \quad ; \quad u^2_{nlj} + v^2_{nlj} = 1 \qquad (2)$$

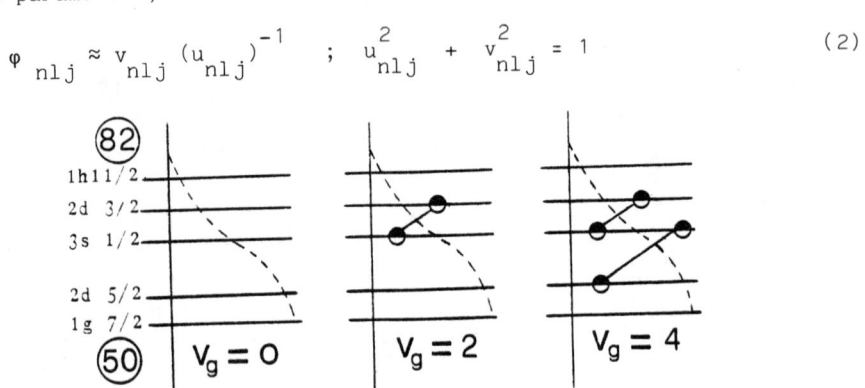

FIG.1. Types of basis states included.

Computational techniques for states up to $v_g = 4$ have been described in [2]. In a particle-number non-conserving approximation of the model one has the BCS quasiparticle model (with up to four quasiparticles). In qualitative discussions of the results one may use arguments derived from this approximate model. Thus one may consider the pairing factors

$$P_0(j_1, j_2) \approx (u_{j_1} u_{j_2} \mp v_{j_1} v_{j_2}) \quad ; \quad (\Delta v_g = 0) \qquad (3)$$

by which a single-particle transition matrix element is multiplied for transitions which do not change v_g, and the pairing factors

$$P_2(j_1, j_2) \approx (u_{j_1} v_{j_2} \mp v_{j_1} u_{j_2}) \quad ; \quad (\Delta v_g = 2) \qquad (4)$$

which apply when v_g is changed by two. The upper sign applies to electric operators, the lower to magnetic operators.

RESULTS; DEGREE OF v_g MIXING

The influence of $v_g = 4$ admixtures in states which are predominantly $v_g = 2$ is most clearly seen in the spectra and properties of negative parity states with $J \geqslant 5$ as shown in Fig. 2. One may notice that the relative spacing and order of the yrast levels is much improved by $v_g = 4$ admixtures. These also enhance, by factors three to five the B(E2) and B(E3) values and consequently they lead to shorter lifetimes (for electric decay) and larger E2/M1 mixing ratios. A Surface Delta Interaction (SDI) yields less v_g mixing than a finite-range force and too little improvement of the description of the data.

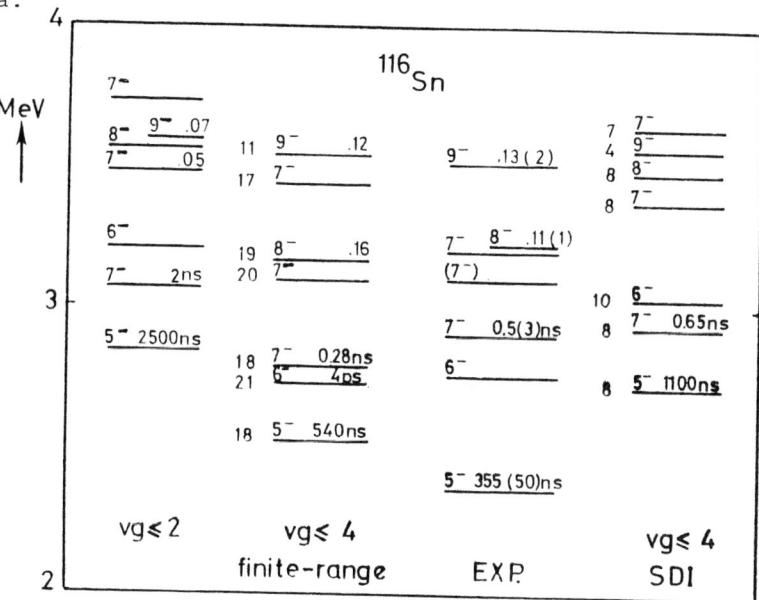

FIG.2. Negative-parity states, as calculated within a $v_g \leqslant 2$ and within a $v_g \leqslant 4$ model space are compared with experimental data. A finite-range force and a Surface Delta Interaction were used. Lifetimes and mixing ratios are indicated. The percentages of $v_g = 4$ admixtures are indicated besides the levels.

MECHANISM OF v_g MIXING

By inspection of the wave functions which resulted from our calculations within a $v_g \leqslant 4$ model space,[3] we observed that the main content of $v_g = 4$ admixtures in predominantly $v_g = 2$ states can be ascribed to a kind of particle-phonon coupling mechanism, as depicted in the left part of Fig. 3. This observation is confirmed by the fact that E2 and E3 transitions are often strongly enhanced in a $v_g \leqslant 4$ model as compared to a $v_g \leqslant 2$ model. Moreover this enhancement is governed by pairing factors $P_0(j_2, j'_2)$, eq. (3), as indicated in the Figure. It is interesting to note the effect on the $10^+ \rightarrow 8^+$ decay, the most important amplitudes of which are de-

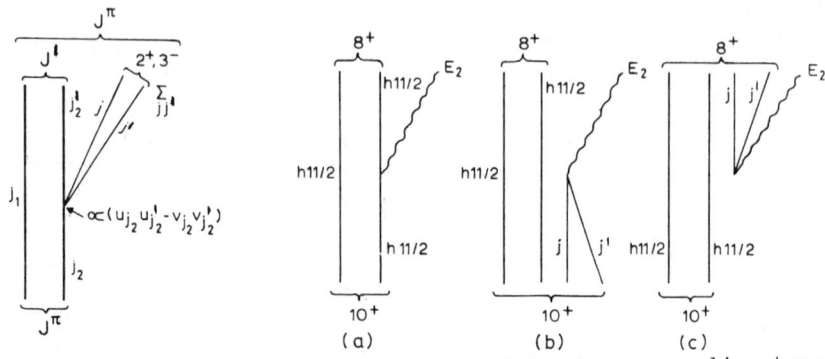

FIG.3. Main origin of v_g mixing is of particle-phonon coupling type (left figure). Diagrams (a), (b) and (c) depict three main contributions to the $10_1^+ \to 8_1^+$ decay.

picted on the right in Fig. 3. Diagram (a) is proportional to P_0 ($h^{11/2}, h^{11/2}$), diagrams (b) and (c) are proportional to P_2 (j, j') but there contribution is proportional to the size of the $v_g = 4$ admixtures, i.e. to the paring factor in the left part of the figure, which is also $P_0(h^{11/2} h^{11/2})$ in this case. Therefore the B(E2) becomes small for a half filled $h^{11}/2$ shell, also in a $v_g \leq 4$ model. These B(E2) values are listed in Table I.

Table I. B(E2, $10_1 \to 8_1$) in ^{A}Sn and paring factor P_0 ($h^{11}/2$, $h^{11}/2$), eq. (3).

A =	116	118	120	122	124	126
$v_g \leq 2$	15	11	4.4	.2	1.8	7
$v_g \leq 4$	52	40	21	2	8	44
exp	42(30)	37(2)				
$P_0 (h^{11}/2\ h^{11}/2)$	.80	.63	.37	.05	-.26	-.51

REFERENCES

1. I. Talmi, Nucl. Phys. A172, 1 (1971)
2. G. Bonsignori, M. Savoia, Nuovo Cimento 44A, 121 (1978)
3. G. Bonsignori, M. Savoia, K. Allaart, A. van Egmond, G. te Velde Nucl. Phys. (in press)

CONFIGURATION DEPENDENT PAIRING FROM THE EFFECTIVE DECOUPLING PICTURE

Ashok Kumar Jain and Kiran Jain
Department of Physics, University of Roorkee, Roorkee, India

ABSTRACT

The effective decoupling picture, proposed recently, enables us to estimate the amount of pairing correlations for Nilsson configurations corresponding to different odd-N and odd-Z bands. The configuration dependence of the pairing correlations is however found to be different from that pointed out by Garett et al. Further, the lower alignment frequency of certain odd-N bands, interpreted by Garett et al. as a sign of decreased pairing, may be explained in terms of a compression in transition energies which comes from the large Coriolis interaction due to the Fermi energy being located near low-K Nilsson orbital. Results from single quasi-particle plus band-mixing calculations seem to support this explanation.

INTRODUCTION

Our recent interpretation of the strongly coupled[1] and also the strongly perturbed bands[2] as effectively decoupled bands opens up the possibility of estimating the amount of pairing correlations for different Nilsson configuration upon which the given odd-N or, the odd-Z bands are based. In this paper we consider only the strongly coupled bands. We find that the pairing correlations are indeed configuration dependent but in a manner opposite to that found by Garett et al.[3,4]. It has been pointed by Garett et al. that the crossing rotational frequency, characterizing the crossing of the unaligned band with the aligned band, is lower for a number of negative parity bands of odd-N nuclei than in the neighbouring even-N isotopes. It has been argued[4] that the decreased pairing correlations lead to an early alignment and hence a decrease in the crossing frequency of these bands whose Nilsson configurations have also been found to possess a positive quadrupole moment, $q_2(\nu) > 0$. Other bands having $q_2(\nu) < 0$ do not show any shift in the crossing frequency. According to the effective decoupling picture, however, the bands having negative quadrupole moment are found to possess a decreased pairing. Those bands, which have $q_2(\nu) > 0$, may have full pairing but a compression in transition energies (due to large Coriolis interaction) which is reflected as a shift in the crossing frequency. Further, none of the odd-Z bands exhibit any significant lowering of the crossing frequency, and therefore should have according to Garett et al. the same amount of pairing. However, our model suggests a configuration dependent pairing in odd-Z bands too.

RESULTS AND DISCUSSION

In Fig.1, we compare the experimental transition energies of some typical odd-N bands having $q_2(\nu)<0$ with those of the respective even-even core ground band. The transition energies have been suitably displaced, in accordance with the effective decoupling picture to provide best matching. The bandhead angular momentum for the odd-N bands $I = R'+j'$, where j' is the effective aligned spin and R' is the core rotational angular momentum, directly given by the even-even core angular momentum. From the figure it is clear that the bandhead spin for different bands carries different amount of rotational angular momentum. We may emphasize here that these bands, according to Garett et al., are supposed to possess full pairing correlations.

In Fig.2, we show the results of a single quasi-particle plus rotor bandmixing calculations for the same bands. Proper Nilsson wavefunctions and a gap parameter $\Delta=1$ MeV were used. It is clear that, with the proper selection of the Fermi energy, trends identical to those exhibited by the experimental data appear. The bandheads for the $11/2^-[505]$, $5/2^-[512]$ and $7/2^-[514]$ bands may have a rotational angular momentum $R'=6,4, 6/4$, respectively. In Figs. 1 and 2 we have shown only one of the two possible sequences, because calculations as well as experimental data reveal that one of the sequences is closer to the core transition energies.[1]

As already pointed out by us[2], this may help us in determining the amount of pairing correlations in different odd-A bands. It is now well accepted that the

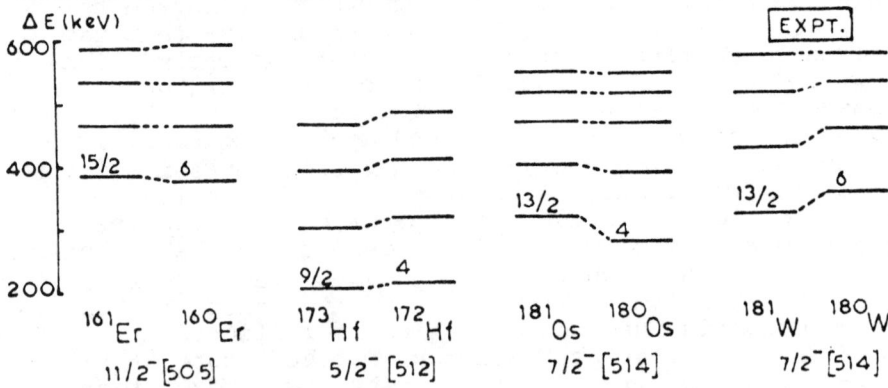

Fig.1 Comparison of transition energies $\Delta E(I \to I-2)$ of some odd-N bands with their respective core. The levels are labelled by I.

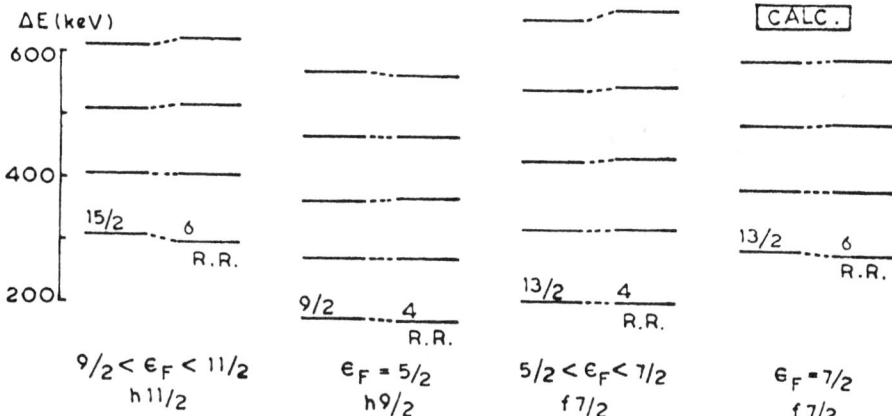

Fig.2 Calculated transition energies compared with the rigid-rotor core transition energies

slow rise of the moment of inertia at lower spins in even-even nuclei, is mainly due to a slow but gradual decrease in the pairing correlations. Since the strongly coupled bands discussed above closely follow the ground band transition energies of the respective core nuclei, it is safe to assume that a decrease of pairing correlations in a manner similar to the core is taking place in the odd-A bands too. Further, since different odd-A bands show a matching with the core band beginning from different core rotational angular momentum values, the Nilsson configurations corresponding to these bandheads have different pairing correlations.

Phenomenologically, the decrease in pairing for even-even nuclei may be given by the relation $\Delta = \Delta_0 (1-I/I_c)^{1/2}$, where I_c is the critical rotational angular momentum for zero pairing. Although, it is now known that the pairing does not go to zero even at relatively higher angular momenta, this relation approximates the low spin variation nicely.[5] If the odd-A bandhead, therefore, carries a rotational angular momentum $R' = 6$ or 4, the pairing would be lesser by 15% or 10% respectively, taking $I_c=20$. The $11/2^-[505]$ bands would thus have lesser pairing than the $5/2^-[512]$ bands. We may point out here that similar situation prevails in the odd-Z bands too. According to Garett et al. all the odd-Z bands should have same amount of pairing as no lowering of crossing frequency has been observed in any of these bands.

Difficulty arises when one considers the odd-N bands having $q_2(\nu) > 0$, which are found to exhibit a compression in the transition energies. This is reflected in an early 'backbending' which has been interpreted by Garett et al. as a lowering of the alignment frequency

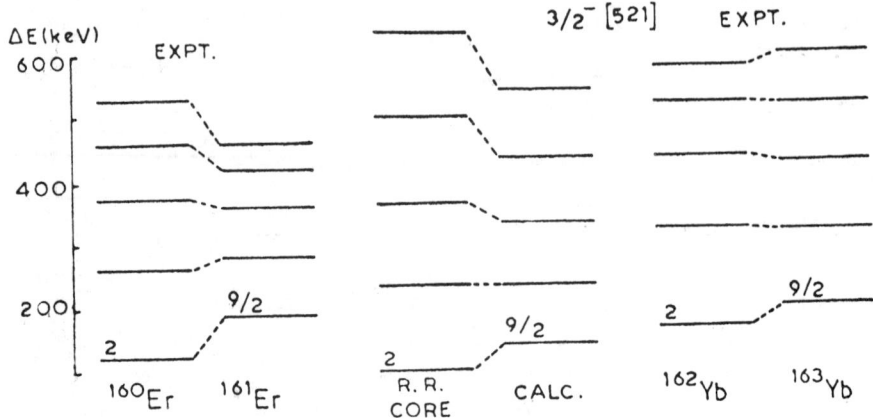

Fig.3 Experimental and calculated transition energies for $3/2^-[521]$ band of ^{161}Er. Also shows the data for ^{163}Yb

and hence a lesser pairing. We notice that this compression of the transition energies is similar to that observed in the unique parity decoupled bands[6]. The Fermi energy in both the cases lies near small-K Nilsson orbital, resulting in large Coriolis interaction and hence a compression in the transition energies. Results of a particle plus rotor bandmixing calculations for the $3/2^-[521]$ band, which is known to exhibit such a compression, are shown in Fig.3. The Fermi energy was chosen to lie slightly below the $3/2^-[521]$. Experimentally the $(9/2 \to 5/2)$ sequence has been found to exhibit greater compression than the $(7/2 \to 3/2)$ sequence. Accordingly we have shown only this sequence. It is clear that the transition energies of the $3/2^-[521]$ band are reduced compared to the core. Similar calculations for protons, however, reveal that there is a very little compression in the transition energies and therefore, no shift in the alignment frequency is expected in accordance with the experimental data. We may point out here that $3/2^-|521|$ band of ^{163}Yb (Fig.3) does not show any such compression in the transition energies. Such an exception may easily be explained by raising the Fermi energy.

One of us (KJ) acknowledges the award of a fellowship under the UGC special assistance programme.

REFERENCES

1. A.K.Jain, Z.Physik A (In press).
2. Kiran Jain and A.K. Jain, Phys.Rev.C (In press).
3. J.D.Garett et al., Phys.Rev.Lett.47, 75(1981).
4. J.D. Garett et al., Phys.Lett. B118, 297(1982).
5. P.C. Sood and A.K. Jain, Phys. Rev.C18, 1906(1978).
6. F.S. Stephens, Rev.Mod. Phys.47, 43(1975).

INTRINSIC EXCITATIONS IN DOUBLY ODD NUCLEI

P.C. Sood
Department of Physics, Banaras Hindu University,
Varanasi 221005, India

ABSTRACT

A procedure is outlined for predicting the bandhead energies of the two-particle (intrinsic) states of odd-odd deformed nuclei based on a quantitative evaluation of the zero range n-p residual interaction energy. We present our results for 250Bk, where many such levels are experimentally known, and for 236Np and 246Am, where the information is very scarce and that too uncertain, to illustrate the effectiveness of this approach.

On the one hand it may be true that 'highly deformed odd-odd nuclei are perhaps the most accessible both to experiment and to interpretation' [1], it is equally true that such nuclei are very sparingly investigated both by the experimentalists and the theorists. Only seven out of over fifty odd-odd actinides have confirmed spin-parity assignments to ground states. Capture gamma ray spectroscopy is one of the most effective methods of studying such nuclei as demonstrated by Hoff and collaborators at Livermore [2]. This method supplements traditional decay studies. However, in many such investigations one basic difficulty remains when low and high spin isomers are placed close to the ground state. In such cases levels are usually placed relative to one of the isomers which may or may not be the true ground state. Thus even to define the ground state properly and also to appropriately place all the expected intrinsic two-particle excitations, it is desirable to have some quantitative guidelines from the theory for relative placement of the bandheads of the various two-particle states expected to appear in the low energy spectra of odd-odd nuclei.

We have recently developed [3] a formulation for determining these bandhead energies based on quantitative evaluation of the zero range n-p residual interaction which explicitly includes a spin-spin interaction term in order to explain the Gallagher-Moszkowski (GM) rule. This rule places the spin-parallel state K_T lower in energy than the spin anti-parallel state K_S. Using the superposition principle we write the bandhead energy as

$$E_K(Z,A) = E_p(\Omega_p, Z, A-1) + E_n(\Omega_n, Z-1, A-1) + (\hbar^2/2\mathcal{J})(K-K^+)$$
$$+ A_0 W + A_\sigma (1-\mathcal{L})W + (-1)^I B\delta_{K,0}$$

where $K=K^\pm = |\Omega_p \pm \Omega_n|$ constitute a GM pair; W and $\mathcal{L}$ are interaction parameters derived for each configuration from atomic mass data and observed GM splitting energies, and A's are the matrix elements of the spin-independent and the spin-dependent terms in the interaction evaluated with appropriate two-particle Nilsson model wavefunctions. The last term denotes the odd-even shift observed for K=0 bands. This formulation has been successfully applied [3-5] for a wide range of nuclei both in the actinide and the rare earth regions. We illustrate the effectiveness of the method by citing results for three cases below and shown in the figure on next page.

Our first example is the nucleus 250Bk which presently is the heaviest nucleus for which somewhat detailed level information is available. Our calculations [3b] successfully describe the ten experimentally known intrinsic states and predict eight more two-particle states upto an excitation energy of 600 keV. A noteworthy feature of this spectrum is that the ground state of 250Bk corresponds to an excited state of the odd-particle (proton) rather than the ground state. This crossing of bands, which is outside the scope of routine predictions based on systematics only, is clearly expected in our formulation since the interaction matrix elements values for $4^+(p_0 n_0)$ are (-67, -29) whereas they are (-104, -51) for the $2^-(p_1 n_0)$ configuration, the latter thus appearing as the ground state. Considering the nature of agreement with known levels, we reasonably expect our predictions for as yet unobserved states to be borne by further experiments.

Our other two examples concern 236Np [5d] and 246Am [5c], both of which have isomer pairs and hence undefined ground states. The nucleus 236Np has a 1.15×10^5y and a 22.5h isomer known for over thirtyfive years but even today [6] their relative placing is undecided. The nucleus 246Am has a 39m high spin isomer and a 25m low spin isomer but with relative order not known. Using only the interaction parameters derived from our analysis of other Np and Am isotopes we predict not only the ground state character unambiguously but also predict the location of ten two-particle bandheads in each of these nuclei upto an excitation of 600 keV as shown in the figure.

These cases are typical of many other nuclei already studied by us [3-5] and the treatment is being extended

250Bk (97, 153)

CALC levels:
- $p_3 n_1$: 1^+, 6^+
- $p_3 n_0$: 2^+, 3^+
- $p_2 n_1$: 3^+, 4^+
- 5^+, 8^+
- $p_0 n_2$: 2^+
- $p_0 n_1$: 7^+
- $p_1 n_1$: 2^-, 5^-
- $p_2 n_0$: 1^+, 3^+
- $p_0 n_3$: 1^+
- $p_0 n_0$: 4^+
- $p_1 n_0$: 1^-, 2^-

EXPT levels: 1^+, 1^-, 3^+, 4^+, 2^-, 5^-, 2^+, 7^+, 5^+

$p_0 \frac{7}{2}^+ [633]$ $p_1 \frac{3}{2}^- [521]$ $p_2 \frac{1}{2}^+ [400]$ $p_3 \frac{5}{2}^+ [642]$

$n_0 \frac{1}{2}^+ [620]$ $n_1 \frac{7}{2}^+ [613]$ $n_2 \frac{3}{2}^+ [622]$ $n_3 \frac{9}{2}^+ [615]$

236Np (93, 143)   246Am (95, 151)

236Np CALC:
- $p_0 n_2$: 1^+, 5^+
- $p_1 n_2$: 5^-, 1^-
- $p_0 n_1$: 3^+, 2^+
- 6^+
- $p_1 n_1$: 2^-, 3^-
- $p_1 n_0$: 1^+
- 1^-
- $p_0 n_0$: 6^-

246Am:
- 3^+, 6^+, 7^+ ($p_2 n_0$)
- $p_0 n_1$: 1^+, 5^+
- $p_1 n_1$: 5^-, 1^-
- $p_0 n_0$: 2^-, 7^-
- $p_1 n_0$: 2^+ CALC

$n_0 \frac{7}{2}^- [743]$ $p_0 \frac{5}{2}^+ [642]$ $n_0 \frac{9}{2}^- [734]$
$n_1 \frac{1}{2}^+ [631]$ $p_1 \frac{5}{2}^- [523]$ $n_1 \frac{5}{2}^+ [622]$
$n_2 \frac{5}{2}^+ [622]$ $p_2 \frac{3}{2}^- [521]$

to cover various chains of isotopes systematically.

These investigations are supported by the University Grants Commission and the Department of Atomic Energy, Government of India.

REFERENCES

1. M.F.Slaughter et al.,Phys.Rev. C29, 114 (1984).
2. R.W.Hoff, invited talk at this conference.
3. P.C.Sood and R.N.Singh, Nucl.Phys. A373, 519 (1982).
4. P.C.Sood and R.N.Singh, Z.Phys. A314, 219 (1983); Nucl.Phys. A419, 547 (1984); in Nuclear Spectroscopy and Nuclear Interactions, eds. H.Ejiri and T.Fukuda (World Scientific, Singapore, 1984) in press.
5. P.C.Sood, Z.Phys. A310, 95 (1983); Physica Scripta 29, 540 (1984); Phys.Rev. C29, 1556 (1984); Z.Phys. A318 (1984) in press.
6. I.Ahmed, J.Hines and J.E.Gindler, Phys.Rev. C27, 2239 (1983).

SHELL EFFECTS ON THE E1 MOMENTS OF Ra-Th NUCLEI

G.A. Leander
UNISOR, Oak Ridge Associated Universities, Oak Ridge, Tn. 37830

ABSTRACT

Large systematic shell effects on intrinsic E1 moments are found, which should modulate any E1 moment induced by β_3 deformation. The calculated shell effects can explain an emerging trend for E1 data in Ra-Th nuclei, if and only if the gross β_3-induced polarization of finite nuclear matter goes in the same direction as the "lightning rod" effect.

INTRODUCTION

If the left-right symmetry of the intrinsic nuclear shape is broken, the nucleus can have an intrinsic dipole moment. The dipole moment, Q_{10}, induced by octupole distortions of a nuclear liquid drop was estimated many years ago as[1,2]

$$Q_{10}^{LD} = C_{LD} \, A \, Z \, \beta_2 \, \beta_3 \qquad (1)$$

The estimate for the numerical coefficient is $C_{LD} = +0.00069$ fm in Ref. 1 and -0.00052 fm in Ref. 2, about the same magnitude but with opposite signs (Fig. 1).

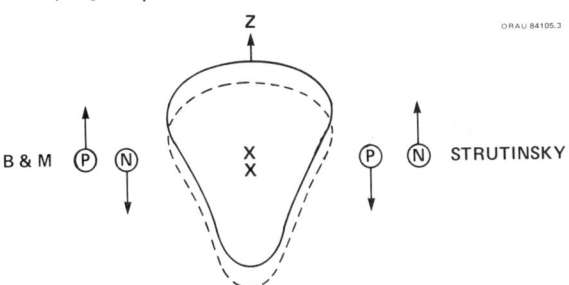

Fig. 1. Schematic drawing of an octupole shape with $\varepsilon_3 > 0$ ($\beta_3 < 0$). Strutinsky[1] and Bohr and Mottelson[2] obtained the neutron-proton polarization in opposite directions, because Strutinsky allowed the Coulomb potential to push the protons toward the surface where they are driven by the "lightning rod" effect towards regions of maximum curvature.

Recently, a number of experiments[3-10] on isotopes in the Ra-Th region have revealed unusually fast E1 transitions of apparently collective character. Theoretical potential-energy calculations had previously yielded octupole-deformed intrinsic equilibrium shapes in these nuclei.[11] However, the liquid-drop formula (1) provides at best an order-of-magnitude estimate for the E1 rates in this region. In particular, Eq. (1) does not explain why the E1 enhancement is absent in some heavier Ra and Ac isotopes[10,12-14], where β_2 is large and β_3 is non-zero according to both theory[11] and the spectroscopic evidence.[15] This paper presents a first investigation of shell effects on the E1 moments in an octupole-deformed single-particle potential. It will be seen that such shell effects do in fact lead to substantial systematic fluctuations around the liquid-drop value.

CALCULATIONS

Single-particle wave functions were calculated in the folded Yukawa single-particle potential with octupole deformation as described in Ref. 11. Let us first examine the center of mass for neutrons and protons separately, obtained by summing $\langle z \rangle_{sp}$ for the single-particle orbits at some fixed, representative deformation (Fig. 2). The z axis is defined as in Fig. 1, with the origin at the equivalent sharp-surface center of mass. Fig. 2 shows a systematic shell effect on the sum. Thus the contribution of successive orbitals does not give rise to random fluctuations around some average. For particle numbers in the Ra-Th region the single-particle contributions have coherent signs. The sum peaks at the closed shells and decreases smoothly towards mid-shell. Similar results are obtained at other relevant deformations.

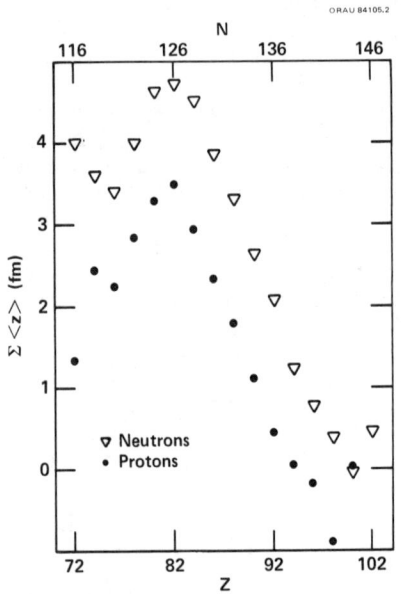

Fig 2. Sum of the single-particle expectation values of z as a function of the number of particles occupying the lowest orbits at $\varepsilon_2 = 0.10$, $\varepsilon_3 = 0.08$.

A nuclear E1 moment,

$$\delta Q_{10}^{sp} = \frac{N}{A} \sum_p v^2 \langle z \rangle - \frac{Z}{A} \sum_n v^2 \langle z \rangle \qquad (2)$$

was evaluated at the appropriate equilibrium deformations,[11] with BCS occupation factors v^2. This E1 moment from the independent-particle model also varies smoothly and systematically through the Ra-Th region. However, it is only a "raw" shell correction, to be renormalized, reduced by the restoring force of neutron-proton interaction and finally added to Q_{10}^{LD}. Work on these steps is in progress. In particular, the appropriate reduction might be obtained from that dipole-dipole interaction which also reproduces the giant dipole resonance.

COMPARISON WITH EXPERIMENT AND CONCLUSIONS

The δQ_{10} values from Eq. (2) increase as N decreases and Z increases from ^{226}Ra. Both these smooth trends can be represented by plotting vs. N-1.5Z: a single dashed curve goes through the theoretical points for Ra and Th in Fig. 3. The experimental data

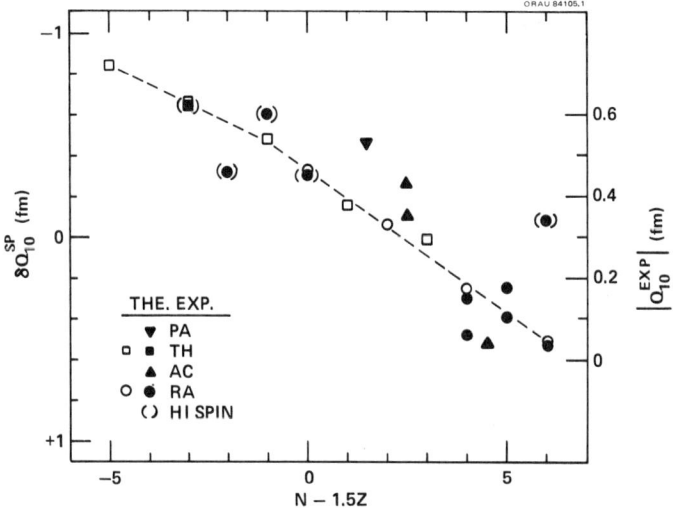

Fig. 3. El moments from independent-particle theory (δQ_{10}^{sp}) and absolute moments from experiment ($|Q_{10}^{exp}|$).

points in Fig. 3 have been extracted from a variety of experiments by the formula

$$|Q_{10}^{exp}| = \{ \frac{4\pi}{3} B(E1; I_i \to I_f) \}^{1/2} / \langle I_i K 10 | I_f K \rangle \qquad (3)$$

and using various assumptions about the rate of transitions by competing modes. The data exhibit an overall trend which could be accounted for by the El shell effects. Actually, theory and experiment in Fig. 3 seem to agree simply because they are plotted against different scales. Such judicious rescaling can be viewed as a phenomenological renormalization of δQ_{10}. The ad hoc reduction factor in Fig. 3 is 2, to be compared with a factor of about 3.6 that is estimated[16] to arise from a dipole-dipole interaction consistent with the giant dipole resonance.

The value of $|Q_{10}^{exp}|$ at $\delta Q_{10} = 0$ is 0.3 fm, which can be interpreted as the empirical value of $|Q_{10}^{LD}|$. Furthermore, since Q_{10}^{LD} appears to be cancelled by the positive shell corrections in nuclei around ^{226}Ra, and augmented by the negative shell corrections around ^{222}Th, if follows that the sign of Q_{10}^{LD} is negative when the nucleus is oriented as in Fig. 1. In conclusion, the still preliminary empirical trend of E1 rates in the Ra-Th region, and the trend of calculated shell corrections, suggests $C_{LD} \sim +0.001$ and would thereby confirm the presence of "lightning rod" effect[1] in nuclei.

Discussions with P. Vogel are gratefully acknowledged. UNISOR is a consortium of twelve institutions, supported by them and by the Office of Energy Research of the U.S. Department of Energy under Contract No. DE-AC05-76OR00033 with Oak Ridge Associated Universities.

REFERENCES

1. V. Strutinsky, Atomnaya Energiya 4, 150 (1956); J. Nucl. Energy 4, 523 (1957).
2. A. Bohr and B.R. Mottelson, Nucl. Phys. 4, 529 (1957); 9, 687 (1959).
3. I. Ahmad et al., Phys. Rev. Lett. 49, 1758 (1982).
4. J. Fernandez-Niello et al., Nucl. Phys. A391, 221 (1982); Munich Annual Report, 1983, p. 51.
5. D. Ward et al., Nucl. Phys. A406, 591 (1983).
6. W. Bonin et al., Z. Phys. A310, 249 (1983).
7. M. Gai et al., Phys. Rev. Lett. 51, 646 (1983).
8. C. Mittag et al., Munich Annual Report, 1983, p. 49.
9. A. Celler et al., Nucl. Phys. A, to be published.
10. I. Ahmad et al., Phys. Rev. Lett. 52, 503 (1984).
11. G.A. Leander et al., Nucl. Phys. A388, 452 (1982).
12. W. Kurcewicz et al., Nucl. Phys. A289, 1 (1977).
13. R. Zimmerman, Ph.D. thesis, Munich (1980).
14. C.W. Reich et al., to be published.
15. G.A. Leander and R.K. Sheline, Nucl. Phys. A413, 375 (1984).
16. G.F. Bertsch, private communication.

II. PHOTON-INDUCED REACTIONS

THE ABSORPTION AND SCATTERING OF PHOTONS BY THE Δ RESONANCE

Evans Hayward
National Bureau of Standards, Gaithersburg, MD 20899

ABSTRACT

Recently, the experiments on the total photonuclear absorption cross sections have been extended to encompass the Δ resonance in complex nuclei. These important experiments involve at least four different techniques and have been performed in European Laboratories. These results are compared with the total cross sections measured in the giant resonance region and extending up to the meson threshold. The Gell-Mann-Goldberger-Thirring sum rule provides a connection between the absorption cross sections in these two energy regions and the photopion cross sections of the nucleon.

The total photonuclear absorption cross sections are related to the forward coherent scattering cross sections through the optical theorem and dispersion relation. At backward angles, where measurements are possible, the scattering cross sections are strongly depressed by a form factor. The experimental cross sections do, however, exceed the prediction of a simple model.

THE PHOTONUCLEAR ABSORPTION MEASUREMENTS

In the last four or five years the measurements of the total photonuclear absorption cross sections for complex nuclei have been extended up to ~ 400 MeV to include the Δ-resonance region. These measurements encompass the range of nuclei from beryllium to uranium and were performed by four different methods in three European laboratories, Mainz, Bonn, and Saclay, and hence they deserve some attention.

The Mainz measurements are an extension of their attenuation experiments[1] described almost ten years ago. In these experiments a long absorber is interposed between two Compton spectrometers in a bremsstrahlung beam and the attenuation measured as a function of incidence photon energy. Photons are removed from the beam by the electronic processes, Compton scattering and pair production, and to a much lesser extent by nuclear interactions. To obtain the nuclear absorption cross section the electronic cross sections need to be well-known; the uncertainties need to be small compared to the nuclear cross section. For this reason the attenuation method has only been applied to light elements, A ≤ 40, where the pair production cross section is smallest. Up to the present time these measurements, extending up to ~ 380 MeV, have been performed for lithium,[2] beryllium,[2] oxygen,[3] and aluminum[3].

The Bonn experiment[4,5] used a tagged photon beam which impinged on the target viewed by an array of counter telescopes. These detectors registered charged pions and hadrons that emerged from the target; the electromagnetic debris was confined to the forward cone and was not counted. The measured cross sections were integrated over particle energy and angle. In order to convert these data into a total photonuclear absorption cross section, a correction needs to be made for the neutral particles that were not registered. This correction is large and unreliable for heavy targets so that this method succeeds only for the lightest targets.

The Saclay measurements are an extension of the photoneutron experiments that have been performed in that laboratory[6,7] for almost twenty years. In these experiments the nearly monochromatic photons produced by positron annihilation in flight impinge on a target located in a 4π neutron detector and the number and multiplicities of the photoneutrons determined. The total photonuclear cross section can be derived from these data on the assumption that the charged particle decay channels contribute neglible strength to the total absorption cross section; this is certainly true for heavy targets, with $A \gtrsim 100$. Leprêtre et al.[8] have shown that for photon energies above 30 MeV the total photonuclear cross section for heavy nuclei can most accurately be represented as the cross section for the production of at least two neutrons. Using this technique they have mapped out the absorption cross sections for cesium, tin, tantalum, and uranium in the quasideuteron region, 30-140 MeV, and up to 440 MeV for lead.[9]

Photofission measurements[10] made in the quasideuteron region showed that the fission cross section was the same as that obtained by neutron counting techniques. This result led a team from Bonn, Giessen, Mainz, and Saclay to measure the photofission cross sections[11] for ^{235}U and ^{238}U in the energy range 120-460 MeV using the tagged photon facility in Bonn. These cross sections have been equated to the total photonuclear cross section for these nuclides.

Figures 1 and 2 show the total photonuclear cross sections for beryllium and lead extending from the giant dipole resonance to ~ 400 MeV. The beryllium data in the giant resonance and quasideuteron regions were measured in Mainz[1]; those in the delta region come from Bonn[4]. The lead data come from three Saclay experiments,[8,9,12]; the high energy points[13] are included to show the trend at much higher energies. Note that the giant resonance is no longer the giant it once was; though the peak cross section near 350 MeV is less than that of the giant resonance, the area under the Δ resonance is much larger.

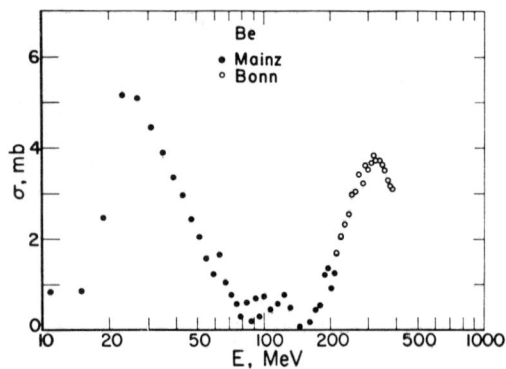

Fig. 1 The total photonuclear cross section for beryllium.

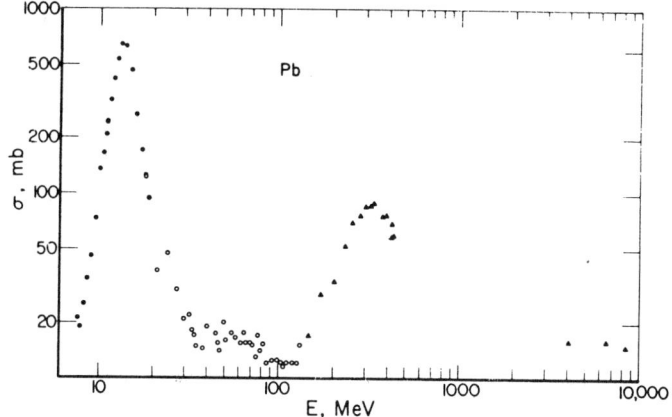

Fig. 2. The total photonuclear cross section for lead.

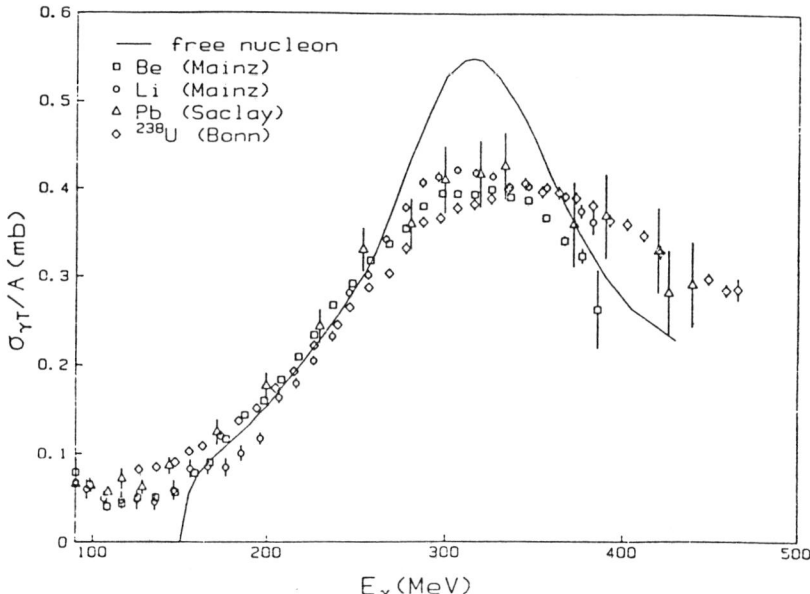

Fig. 3. The total cross sections per nucleon for Li, Be, Pb, and U along with that for the free nucleon.

It has already been pointed out[3,11] that when the absorption cross sections in the Δ region are divided by A, i.e., to obtain the cross section per nucleon, the curves for the different nuclides lie on top of one another. This is illustrated in Fig. 3 which shows the absorption cross sections for Li, Be, Pb, and U as well as that for a free nucleon. In a separate experiment the Saclay group measured[14] the total cross sections, i.e., the cross section for producing at least two neutrons, for eight different targets at 235 and 330 MeV. It is clear that for nuclei with A > 100, for which charged particle emission is not important, the cross section per nucleon at the peak, 330 MeV, is 425 μb/nucleon (see Fig. 4). The width of the resonance is more than 200 MeV. These data on complex nuclei may be compared with the total cross section[15] on the proton, i.e., $\sigma(\gamma,\pi^+) + \sigma(\gamma,\pi^0)$, shown in Fig. 5. The main peak in this cross section is 540 μb high and has a width of 150 MeV. The three bumps at higher energies have yet to be explored for complex nuclei.

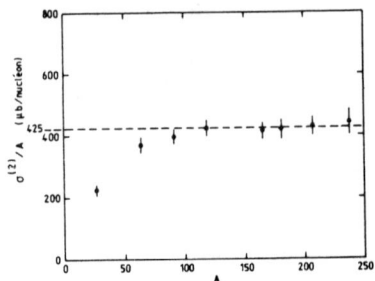

Fig. 4 The cross section per nucleon for producing at least two neutrons at 330 MeV as a function of A.

All of these results obtained with real photons strongly suggest that the nucleus is just a container for the A nucleons with which the photon interacts. This idea is strongly supported by the results of a related experiment[16] in which the inelastic scattering cross sections of 730 MeV electrons through the relatively small angle of 37.1° were measured for hydrogen, helium, beryllium, carbon, and oxygen. Here, too, the cross sections for the complex nuclei, helium, beryllium, carbon, and oxygen, are proportional to A (see Fig. 6). That the virtual photon interacts with a single nucleon is very evident here because in the analysis the appropriate excitation energy is obtained by treating the recoil as though it were quasi-elastic scattering. For both real and virtual photons, then, the cross section per nucleon for a complex nucleus can be represented by a universal curve.

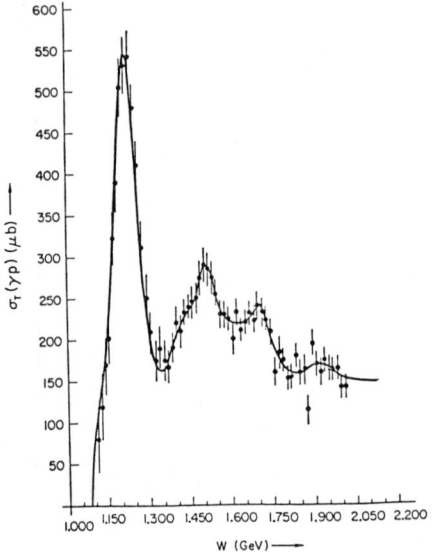

Fig. 5 The total cross section for the proton as a function of the total energy of the system.

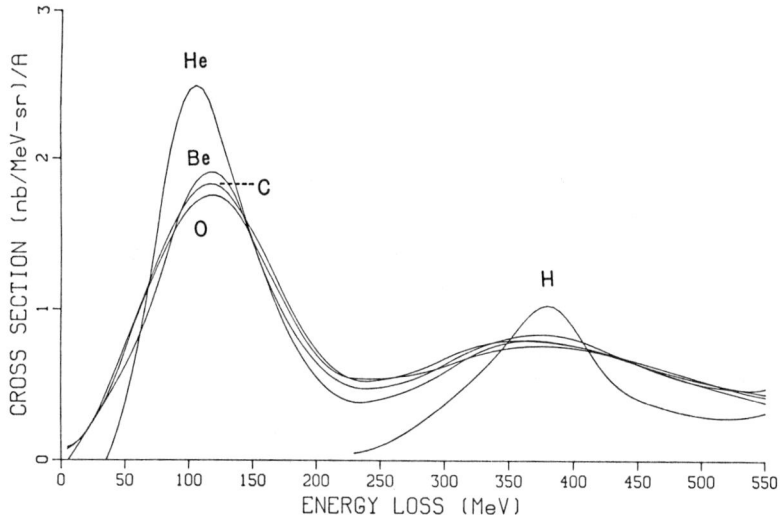

Fig. 6 The smooth curves that fit the data in the electron scattering experiment.

THE GGT SUM RULE

The Gell-Mann-Goldberger-Thirring (GGT) sum rule, derived using dispersion relations, includes all multipoles and retardation effects and relates the total integrated photonuclear cross sections above and below the pion threshold to the integrated photomeson production cross sections on the free nucleon:

$$\int_0^\pi \sigma(E) dE = \frac{2\pi^2 e^2 \hbar}{Mc} \frac{NZ}{A} + \int_\pi^\infty [\sigma_A(E) - \sigma(E)] dE . \qquad (1)$$

Here $\sigma(E)$ stands for the total photonuclear absorption cross section and $\sigma_A(E)$ is the photomeson production cross section on A free nucleons:

$$\sigma_A(E) = Z\sigma(\gamma,\pi^+) + N\sigma(\gamma,\pi^-) + A\sigma(\gamma,\pi^0)$$

$$\simeq A\sigma_N . \qquad (2)$$

The first term on the right-hand side of Eq. (1) may be recognized as the classical dipole sum. Equation (1) then states that the enhancement in the absorption cross section integrated to the pion threshold is exactly the amount by which the cross section on A free nucleons is reduced between the pion threshold and infinite energy. Meson exchange increases the photonuclear cross section below the pion threshold at the same time depleting the strength above.

As Weise[17,18] has pointed out this sum rule cannot be valid if the integral is to be carried out to truly infinite energies. At energies above about 2 GeV shadowing effects become important; then Eq. (2) is more accurately represented by

$$\sigma_A(E) = A^{0.9} \sigma_N . \tag{3}$$

At high energies, above 5 GeV, the photon decays into virtual hadronic states living long enough to interact strongly with the nucleons in the nucleus. The interaction is strong enough so that the photon sees only the surface nucleons.

At the present time we do not even have enough data on the photon absorption cross sections for complex nuclei to tell whether the GGT sum rule is obeyed if the integrals are made over the nucleon resonance region to ~ 2 GeV. In Fig. 3 the curve, representing the cross section for the free nucleon, crosses the data near 380 MeV. The integral of the free nucleon cross section from the pion threshold to 380 MeV exceeds the integral of the data by approximately half a dipole sum. Photonuclear absorption cross sections integrated up to the pion threshold amount to approximately 1.7 sums[7]. Therefore the GGT sum rule is approximately obeyed if 380 MeV is the upper limit of the integration.

On the other hand, the integral to 440 MeV, 90 MeV mb/nucleon, is the same as that for the proton. Total absorption data for complex nuclei extending over the entire resonance region are therefore badly needed to test the GGT sum rule.

THE PHOTON SCATTERING CROSS SECTION

Photon scattering is a very fundamental electromagnetic process in its own right. It is also of interest because of its relation to the total absorption cross section through the optical theorem and the dispersion relation.

$$\sigma_a(E) = 4\pi \lambdabar \, \text{Im} \, R(E,0°) \tag{4}$$

$$\text{Re} \, R(E,0°) = \frac{E^2}{2\pi^2 \hbar c} P \int \frac{\sigma_a(E')dE'}{E'^2 - E^2} . \tag{5}$$

These rules result because the coherent scattering cross section and the total absorption cross section stem from the same complex scattering amplitude. This amplitude refers only to those transitions that leave the nucleus in its initial state of energy and angular momentum, i.e., the same m state. Thus, nuclei having spins different from zero can have elastic scattering which is incoherent and not described by the above equations.

In the giant resonance region the scattered photons have a dipole angular distribution and good enough cross section measurements can, in principle, determine the absorption cross section. At higher energies where higher multipoles participate we simply do not know how to relate the forward scattering cross section to that at backward angles where measurements can be made.

The photon scattering cross sections for the proton have been measured extensively in the resonance region. A few measurements for the deuteron have also been made. In both cases the photon scattering angle is large and the important background, photons from π^0 annihilation, was circumvented by requiring a coincidence between the photon and the recoil proton or deuteron. Figure 7 compares[19] the forward scattering cross section for the deuteron with measurements[20] made at backward angles. It is clear from this comparison that a form factor is needed to describe this angular distribution. Using

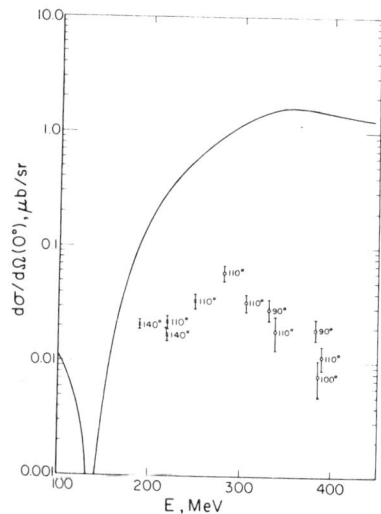

Fig. 7 The forward scattering cross section for the deuteron compared with measurements made at back angles.

$$\frac{d\sigma}{d\Omega}(\theta) = \frac{d\sigma}{d\Omega}(0°) \, F^2(\theta) \, \frac{(1+\cos^2\theta)}{2} \quad (6)$$

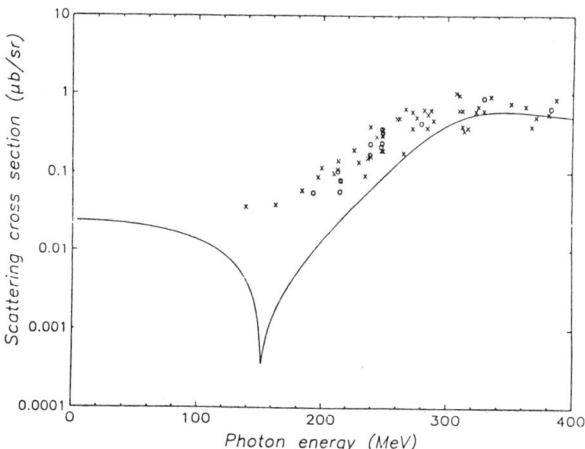

Fig. 8 The forward scattering cross section of the proton compared with the data adjusted according to Eq. (6).

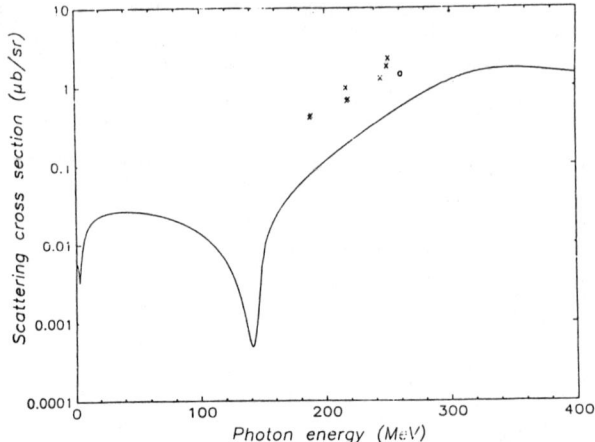

Fig. 9 The forward scattering cross section for the deuteron compared to the data adjusted according to Eq. (6).

Schelhaas et al.[21] have compared the forward cross sections for the proton and the deuteron with the existing data[22] (see Figs. 8 and 9). The form factor for elastic electron scattering was used for $F(\theta)$. The very sharp minimum near the meson threshold results from the cancellation of the real part of the forward scattering amplitude by the Thomson scattering amplitude. In both cases it can be seen that the measured cross sections are larger than those obtained from this formula. There are two possible explanations:
1) the elastic electron scattering form factor does not describe coherent photon scattering or
2) there is incoherent elastic scattering not described by the optical theorem and dispersion relation but which does contribute to the measurement. This scattering can result from the finite spins of the proton and deuteron.

For nuclei heavier than the deuteron it has not been possible to detect the recoiling nucleus in coincidence with the scattered photon. In a recent experiment[23] the "elastic" scattering cross sections for ^{12}C and ^{208}Pb were measured using the tip of the bremsstrahlung spectrum. The scattered photons were detected at an angle of 115° with an energy resolution of 10%. Figure 10 shows a linear plot of the data

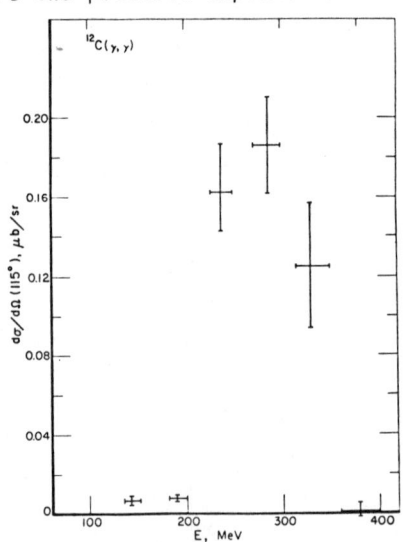

Fig. 10 The photon scattering cross section for ^{12}C in the delta region.

obtained for carbon using bremsstrahlung end-point energies varying from 150 to 400 MeV. The points may be seen to trace out the Δ resonance; in fact, the width appears to be only about 120 MeV, narrower than found in the absorption experiments. The lead data are much less reliable because of a very low counting rate. In Fig. 11 the carbon data are compared with the prediction of Eq. (6). The point shown here measured with 300 MeV bremsstrahlung includes some new data and therefore has a slightly smaller error than the one previously[23] reported. The forward scattering amplitudes used here are also a better estimate; they were obtained by multiplying the Be amplitudes shown in Fig. 12 by 12/9. We are using the A-dependence of the cross section and taking Be as a standard.

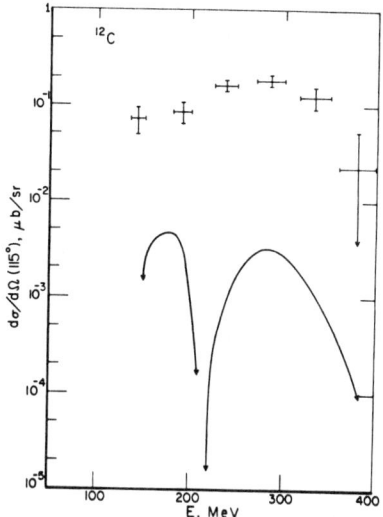

Fig. 11 The photon scattering cross section for ^{12}C compared to the prediction of Eq. (6).

The scattering cross sections for both ^{12}C and ^{208}Pb exceed the prediction of Eq. (6) by at least an order of magnitude. It may be that the form factor for elastic electron scattering is a bad choice to describe the charge distribution seen in photon scattering. On the other hand, since the experimental energy resolution includes inelasticities up to 10%, high energy photons emitted by the Δ resonance and populating low-lying excited states would certainly be registered in this experiment. All of these results point out the importance

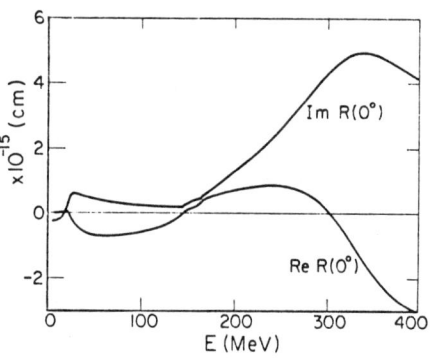

Fig. 12 The forward scattering amplitudes of Be.

of measuring the scattering cross section for helium in the Δ region as has been proposed by Booth and Miller. This nucleus has spin zero and no low-lying excited states.

Finally Fig. 13 shows the lead data (115°) measured[23] in the Δ region along with the previously measured[24] lower energy results obtained using positron annihilation radiation. The curves are meant to illustrate that a single form factor can be constructed to

represent the data. The form
factor applied here to the
quasi-deuteron and Δ
amplitudes is the Fourier
transform of the Fermi
distribution
$\rho(r) = \rho_0/[1 + e^{(r-c)/z}]$ with
c = 1 fm and z = 0.3 fm.

The availability of
photon energies extending to
2 GeV would permit the study
of the nucleon resonances in
complex nuclei, and a high
duty factor accelerator will
make the rate of data
acquisition more favorable.

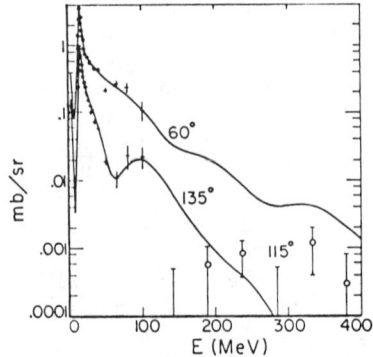

Fig. 13 The photon scattering cross section for ^{208}Pb.

ACKNOWLEDGMENTS

Thanks are due to both B. Ziegler and J. Ahrens for hospitality at the Max Planck Institute in Mainz and for providing much of the material discussed here.

REFERENCE

1. J. Ahrens, H. Borchert, K. H. Czock, H. B. Eppler, H. Gimm, H. Gundrum, M. Kröning, P. Riehn, G. Sita Ram, A. Zieger, and B. Ziegler, Nucl. Phys. A251, 479 (1975).
2. B. Ziegler, Lecture Notes in Physics 108, 148 (1979).
3. J. Ahrens, unpublished.
4. H. Rost, Thesis, Bonn University, 1980.
5. J. Arends, J. Eyink, A. Hegerath, K. G. Hilger, B. Mecking, G. Nöldeke, and H. Rost, Phys. Lett. 98B, 423 (1981).
6. H. Beil, R. Bergère, and A. Veyssière, Nucl. Instr. Methods 67, 293 (1969).
7. A. Veyssière, H. Beil, R. Bergère, P. Carlos, J. Fagot, A. Leprétre, and J. Ahrens, Nucl. Instr. Methods 165, 417 (1979).
8. A. Leprétre, H. Beil, R. Bergère, P. Carlos, J. Fagot, A. DeMiniac, and A. Veyssière, Nucl. Phys. A367, 237 (1981).
9. C. Choller, J. Arends, H. Beil, R. Bergère, P. Bourgeois, P. Carlos, J. L. Fallou, J. Fagot, P. Garganne, A. Leprétre, and A. Veyssière, Phys. Lett. 127B, 331 (1983).
10. H. Ries, U. Kneissel, G. Mank, H. Stroher, W. Wilke, R. Bergère, P. Bouirgeois, P. Carlos, J. J. Fallou, P. Garganne, and A. Veyssière, Phys. Lett. 139B, 254 (1984).
11. J. Ahrens, J. Arends, P. Bourgeois, P. Carlos, J. L. Fallou, N. Floss, P. Garganne, S. Huthmacher, U. Kneissel, G. Mank, B. Mecking, H. Ries, R. Stenz, and A. Veyssière, Proceedings of the International Conference on Nuclear Physics, Florence (Italy) August 29-September 3, 1983.

12. A. Veyssière, H. Beil, R. Bergère, P. Carlos, and A. Leprêtre, Nucl. Phys. A$\underline{159}$, 561 (1970).
13. D. O. Caldwell, V. B. Elings, W. P. Hesse, R. J. Morrison, F. V. Murphy, and D. E. Yount, Phys. Rev. D $\underline{7}$, 1362 (1973).
14. H. Beil, R. Bergère, P. Bourgeois, P. Carlos, C.Challet, J. Fagot, J. L. Fallou, P. Garganne, A. Leprêtre, A. De Miniac, and A. Veyssière, CEA Department de Physique Nucleaire, Compte Rendu d'Activité, 1982-1983.
15. M. Damashek and F. J. Gilman, Phys Rev. D $\underline{1}$, 1319 (1970).
16. J. S. O'Connell, W. R. Dodge, J. W. Lightbody, Jr., X. K. Maruyama, J-O. Adler, K. Hansen, B. Schroder, A. M. Bernstein, K. I. Blomqvist, B. H. Cottman, J. J. Comuzzi, R. A. Miskiman, and B. P. Quinn, to be published.
17. W. Weise, Phys. Rev. Lett. $\underline{31}$, 773 (1973).
18. W. Weise, Phys. Reports, $\underline{13C}$, 53 (1974).
19. E. Hayward in Highly Excited States in Nuclear Ractions, H. Ikegami and M. Muraoka, Eds. (Research Center for Nuclear Physics, Osaka University, Suita Osaka, Japan, 1980).
20. Heinz-Josef Weyer, Thesis, Bonn University (1977).
21. K. P. Schelhaas, B. Ziegler, and E. Hayward, Proceedings of the International Conference on Few Body Problems in Physics Karlsrube, W. Germany, August 1983.
22. H. Genzel and W. Pfeil, Landolt-Börnstein I/8, 1 (1973).
23. E. Hayward and B. Ziegler, Nucl. Phys. A$\underline{414}$, 333 (1984).
24. R. Leicht, M. Hammen, K. P. Schelhaas, and B. Ziegler, Nucl. Phys. A$\underline{362}$, 111 (1981).

PHOTON SCATTERING ABOVE THE GIANT DIPOLE RESONANCE

Alan M. Nathan
University of Illinois, Urbana, IL 61801

ABSTRACT

Several recent experiments which have investigated elastic photon scattering above the GDR are discussed. It is shown how these experiments can provide useful information on the size, energy dependence, and multipolarity of the total photoabsorption cross section. Several examples are given of how one uses the scattering data to study the distribution of E1 and E2 strength in the giant resonance region of light nuclei. For ^{12}C, ^{16}O, and ^{40}Ca, no compact E2 resonance is found between the GDR and 50 MeV, although the sum rule is exhausted below 50 MeV in ^{16}O by a broad distribution of E2 strength. The main concentration of E2 strength seems to lie above 50 MeV in ^{12}C. For ^{40}Ca, the data suggest that the integrated E1 absorption per nucleon is closer than previously thought to the nearly constant value found for medium and heavy nuclei. Finally, a brief account is given of the role of exchange effects in medium energy photon scattering, and our present understanding of the experiment on lead is discussed.

I. INTRODUCTION

In the past several years, the field of photon scattering has been an active and fruitful endeavor. At this conference, there are four invited talks devoted to various aspects of this subject, ranging in energy from a few MeV to a few hundred MeV. Across this vast energy range, there is wide variety in the kinds of nuclear physics information that can be learned from photon scattering. The present talk will focus on the energy range extending from giant resonance energies to the meson threshold.

The scope of this talk can best be appreciated by referring to Figure 1, where experimental values for the total photoabsorption cross section σ_γ on Pb is shown. Three distinct energy regions are indicated; the principal mechanism for photon-induced reactions is different in each region. At low energies ($E \lesssim 50$ MeV), σ_γ is dominated by the excitation of the highly collective giant dipole resonance (GDR) and possibly by the excitation of other giant resonances (GR) of higher multipolarity. Between 50 MeV and the threshold for meson production is the so-called quasideuteron region (QD), where the photon is believed to interact primarily with a pair of nucleons. Above meson threshold, the absorption proceeds predominantly by the interaction with a single nucleon. In particular, in the 200-400 MeV region the dominant process is the excitation of a single nucleon to the Δ resonance.

Just as the mechanisms for photoabsorption are different in the three regions, so too are the mechanisms for photon scattering. At photon energies corresponding to wave lengths larger than the

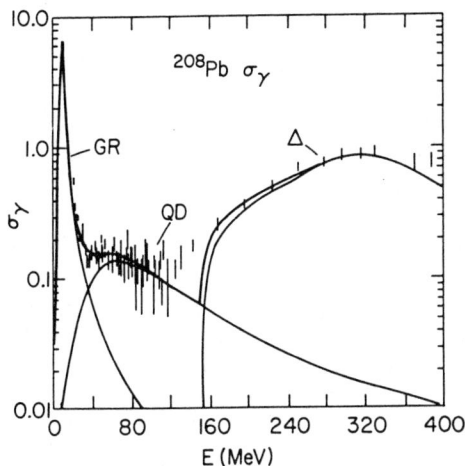

Fig. 1 The total photoabsorption cross section for lead.

nuclear size, one expects the scattering to be mediated primarily by the collective giant resonances. As I will show in this talk, if one has precise values for elastic scattering cross sections as a function of both energy and angle, one can extract reliable information about the size, energy dependence, and multipolarity of σ_γ. This in turn can be interpreted in terms of giant resonances, sum rules, widths, etc. The advantage of this technique over others is that photon scattering is a precise probe that requires little model dependence in the interpretation of the data. In Section III I will give several examples of how one has used elastic photon scattering to study the distribution of E1 and E2 strength in light nuclei.

At higher photon energies, when the photon wave length becomes small compared to the nuclear size, the collective giant resonances cease to become important and the scattering amplitude is dominated by the coherent scattering from individual scattering centers. These scattering centers are essentially the charged particles in the nucleus, whether they are protons or charged mesons exchanged between interacting neutron-proton pairs. The photon scattering cross section becomes characterized by strong diffractive effects which can be interpreted in terms of the spatial distribution of scattering centers in the nucleus. In Section IV I will discuss our present understanding of the only experimental study of this type that has been reported in the literature.

At still shorter wavelengths, the photon probes the spatial extent and structure of the scattering centers themselves. For example, in the 200-400 MeV region, the scattering amplitude is dominated by the excitation of a single nucleon into the Δ resonance. I will not discuss this energy regime, since it has already been presented at this conference by Dr. Hayward.

II. FORMALISM FOR PHOTON SCATTERING

The analysis of (γ,γ) data uses the close link between elastic scattering cross sections $d\sigma/d\Omega$ (E,θ) and total photoabsorption cross sections $\sigma_\gamma(E)$. In order to appreciate the kind of information one learns from (γ,γ), it is necessary to present a small amount of formalism in order to establish this link.[1,2] In the forward direction, one writes the elastic cross section as the square of the coherent sum of fundamental amplitudes:

$$\frac{d\sigma}{d\Omega}(E,\theta=0^0) = |f(E) + D_0|^2, \qquad (1)$$

where D_0 is the classical Thomson amplitude for the scattering of photons from a point object of charge Ze and mass AM:

$$D_0 = -\frac{(Ze)^2}{AMc^2} \equiv -\frac{Z^2}{A} r_0, \qquad (2)$$

and f(E) is the forward amplitude for scattering from the internal degrees of freedom of the nucleus. Other possible contributions to the coherent scattering amplitude (e.g., atomic Rayleigh and Delbrück) are small at the energies and scattering angles of interest here and will not be considered further. A low-energy theorem states that in the limit of zero energy, the total scattering amplitude is just D_0.[3] On the other hand, f(E) is linked to $\sigma_\gamma(E)$ through the optical theorem:

$$\text{Im}[f(E)] = \frac{E}{4\pi\hbar c} \sigma_\gamma(E) \qquad (3)$$

and a dispersion relation:

$$\text{Re}[f(E)] = \frac{E^2}{2\pi^2 \hbar c} P\int_0^\infty \frac{\sigma_\gamma(E')}{E'^2 - E^2} dE', \qquad (4)$$

where P denotes the principal value of the integral. Eqs. 1-4 imply that $\sigma_\gamma(E)$ uniquely specifies the forward elastic cross section $d\sigma/d\Omega$ (E,$\theta=0°$).

In order to predict the elastic cross section for arbitrary θ, it is first useful to decompose $\sigma_\gamma(E)$ and f(E) into multipoles:

$$\sigma_\gamma(E) = \sum_\lambda \sigma_\gamma^\lambda(E)$$

$$f(E) = \sum_\lambda f^\lambda(E) \qquad (5)$$

$$\lambda = E1, M1, E2, \dots.$$

Then one can show that in the long wavelength limit, the optical theorem and dispersion relation hold for <u>each multipole separately</u>:

$$\text{Im } [f^\lambda(E)] = \frac{E}{4\pi\hbar c} \sigma_\gamma^\lambda(E)$$

$$\text{Re } [f^\lambda(E)] = \frac{E^2}{2\pi^2\hbar c} P\int_0^\infty \frac{\sigma_\gamma^\lambda(E')}{E'^2 - E^2} dE'. \quad (6)$$

Straightforward angular momentum algebra is used to relate $d\sigma/d\Omega$ (E,θ) to the $f^\lambda(E)$:

$$\frac{d\sigma}{d\Omega}(E,\theta) = |\sum_\lambda f^\lambda(E) g_\lambda(\theta) + D g_{E1}(\theta)|^2, \quad (7)$$

where the $g_\lambda(\theta)$ are the angular factors appropriate to each multipole, with the normalization $g_\lambda(0) = 1$. In this expression I have anticipated the result that at finite energy and scattering angle, the amplitude D differs from the low energy Thomson limit D_0 given by eq. 2. The modification arises because part of the Thomson amplitude is the coherent sum of scattering amplitudes from Z protons. This term must be modified at finite momentum transfer $q = (2E/\hbar c) \sin(\theta/2)$ in order to properly account for the phase relation among waves scattered from different points in the nucleus. The correct expression is[4]

$$D_0 = -\frac{Z^2}{A} r_0 \to D = -Zr_0 F(q) + \frac{NZ}{A} r_0, \quad (8)$$

where $F(q)$ is the charge form factor of the nucleus. Since $F(0)=1$, D reduces to D_0 for $q=0$, i.e., at forward angles. There are additional modifications to D due to mesonic effects[2] and sub-nucleon structure.[5] However, for the giant resonance studies to be discussed in Sec. III, it will be assumed that these additional effects are small enough so that any uncertainty in our estimate of them will not appreciably affect our ability to extract information on the $\sigma_\gamma^\lambda(E)$ from the scattering data. In Sec. IV, these mesonic and subnucleonic effects will be further explored.

III. GIANT RESONANCE STUDIES

We want to use the formalism presented above in order to extract information on the photoabsorption cross section. Eq. 7, along with eq. 5, 6, and 8, imply that $d\sigma/d\Omega$ (E,θ) is completly specified if one knows the $\sigma_\gamma^\lambda(E)$ and $F(q)$. It is often the case, however, that one has partial knowledge of the $\sigma_\gamma^\lambda(E)$ which is used along with measurements of $d\sigma/d\Omega$ (E,θ) in order to deduce additional information about the $\sigma_\gamma^\lambda(E)$. I will give examples of the use of this technique to study the E1 and E2 strength distributions.

The first example is ^{16}O, for which elastic photon scattering data have been measured and analyzed by the NBS group.[1] Data were taken at several energies and angles; the points at 135° are shown

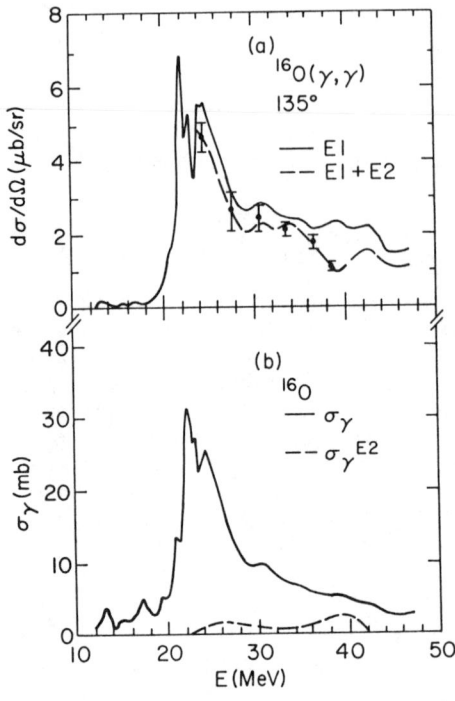

Fig. 2
(a) Elastic photon scattering and
(b) photoabsorption cross sections for ^{16}O.

in Figure 2(a). The curve labeled σ_γ in Figure 2(b) is a smooth line through experimental values of the total photoabsorption cross section $\sigma_\gamma^{expt}(E)$, which was measured at Mainz and which extends to about 400 MeV.[6] The analysis uses $\sigma_\gamma^{expt}(E)$ and the measured elastic scattering data in order to deduce a partitioning of $\sigma_\gamma^{expt}(E)$ into E1 and E2 parts. The curve labeled E1 in Figure 2(a) is obtained by applying the optical theorem and dispersion relation to $\sigma_\gamma^{expt}(E)$, appropriately modifying the Thomson amplitude using the measured form factor, and then applying the dipole angular distribution, assuming that $\sigma_\gamma^{expt}(E)$ is entirely E1. The deviation of this curve from the scattering data shows that this latter assumption is wrong. The curve labeled E1 + E2 is derived as follows: one assumes that σ_γ^{expt} is composed only of E1 and E2 parts, parameterizes the energy dependence of the E2 part of σ_γ in some convenient manner, and adjusts these parameters to best fit the scattering data, while still constraining the total photoabsorption to be consistent with the experimental value. In order to achieve an acceptable fit, it is necessary to slightly renormalize $\sigma_\gamma^{expt}(E)$ by an amount within the quoted systematic uncertainty. The curve labeled σ_γ^{E2} in Figure 2(b) shows the resulting E2 strength, which corresponds to $1.25^{+1.3}_{-0.9}$ total energy-weighted classical sums (TEWS). This strength does not appear to be in the form of a compact resonance but is spread over 20 MeV. Both the size and the energy distribution of this E2 strength are consistent with other experiments, such as $^{15}N(\vec{p},\gamma_0)^7$ and $^{16}O(\gamma,n_0)^8$.

One should remark that there is some model dependence to the analysis. In particular, it is not reasonable to expect that $\sigma_\gamma^{E2}(E)$ will be determined outside the energy range where $d\sigma/d\Omega(E,\theta)$ has been measured. Furthermore the dispersion relation implies that the scattering cross section at energy E is sensitive to the photoabsorption cross section <u>at all energies.</u> Therefore it is necessary to make reasonable assumptions about the energy dependence of $\sigma_\gamma^{E2}(E)$. In the preceeding analysis, it was assumed that $\sigma_\gamma^{E2}(E)$ is zero below 22 MeV and above 42 MeV and that the energy dependence is smooth between these energies. With elastic scattering cross sections at only 6 isolated energies, it is not possible to extract more information from the data.

Despite these limitations, the foregoing discussion demonstrates the power of this technique for determining E2 strength. To the extent that the formalism is valid and that the corrections to the Thomson amplitude are calculable, the data allow a precise and nearly model independent extraction of the total E2 strength function.

The next example is ^{12}C, for which elastic scattering data have been measured at Illinois[9] and NBS.[1] Since there exists experimental values of the total photoabsorption cross section on carbon extending up to 400 MeV, one could in principle analyze the scattering data in the same manner as was done for ^{16}O in order to obtain $\sigma_\gamma^{E2}(E)$. However, for reasons discussed below, scattering data alone have been used in order to extract both $\sigma_\gamma^{E1}(E)$ and $\sigma_\gamma^{E2}(E)$. Since the E2 contribution to the scattering occurs primarily through the interference with the dominant E1 contribution and since this interference is antisymmetric about 90°, it is possible to decouple the extraction of σ_γ^{E1} from that of σ_γ^{E2}. One first analyzes scattering data that have been averaged over angles symmetric about 90° in order to obtain $\sigma_\gamma^{E1}(E1)$. One then uses this $\sigma_\gamma^{E1}(E)$ and the un-averaged scattering data $d\sigma/d\Omega$ (E,θ) in order to obtain $\sigma_\gamma^{E2}(E)$. As with the analysis of ^{16}O, it is necessary to make some assumptions about the behavior of the $\sigma_\gamma^A(E)$ above 50 meV.

The Illinois scattering data at 135°, 45°, and the weighted mean of these two data sets are shown in Figure 3. The result of the fit to the angle-averaged data is the curve labeled E1, which was calculated using the E1 photoabsorption in Fig. 4. The fit to the angle-averaged data is excellent. Furthermore, below 35 MeV the curves calculated for the individual angles of 135° and 45° are in good agreement with the corresponding data. This strongly suggests that the scattering below 35 MeV is totally dominated by E1. However above 45 MeV these curves fall below the 135° data and above the 45° data in just the manner expected if there is E1-E2 interference. The sign of the interference pattern is what would be expected if the centroid of the E2 strength is above 52 MeV. For example, the dashed curve in Fig. 3 shows the result of adding roughly 1 TEWS of E2 strength centered at 53 MeV. For energies above the centroid, the sign of the interference changes. Unfortunately the existing data do not extend high enough in energy to observe this sign change. Thus the data do not provide very stringent constraints on $\sigma_\gamma^{E2}(E)$ except that much of the strength appears to be above 50 MeV. This result in itself is non-trivial, since it is at variance with microscopic calculations which suggest that the classical sum rule is already exhausted below 50 MeV.[10] Other than photon scattering, there are no other experiments which bear on the E2 strength in ^{12}C above 35 MeV.

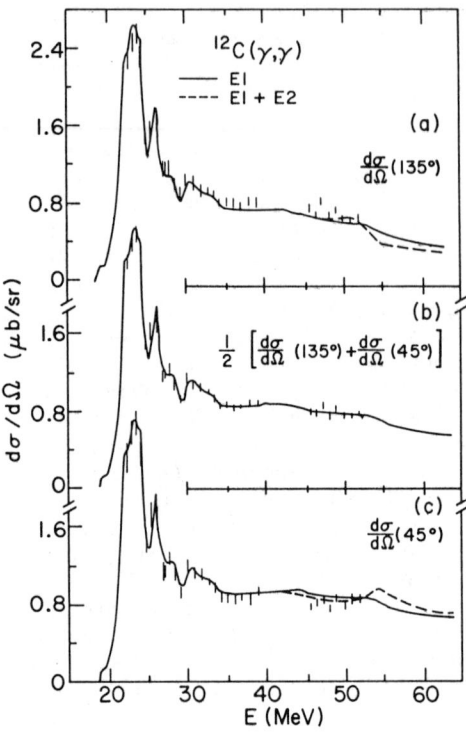

Fig. 3 Elastic photon scattering cross sections for ^{12}C.

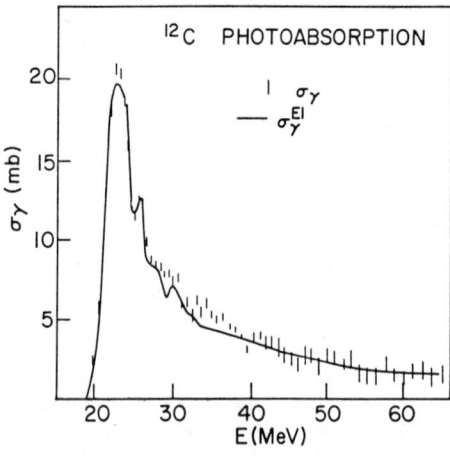

Fig. 4 Photoabsorption cross sections for ^{12}C.

In Figure 4, the extracted function $\sigma_\gamma^{E1}(E)$ is compared to the measured total photoabsorption cross section $\sigma_\gamma^{expt}(E)$, shown as points with error bars.[6] These two curves agree quite well with each other over the entire energy range shown. In fact if we compare the integrated cross sections up to 50 MeV, they agree to within 6 ± 2%. Nevertheless there is reason to believe that the small discrepancies between 27 and 40 MeV are real. As discussed above, the scattering data below 35 MeV are consistent with a purely E1 angular distribution. Therefore, if one would assume that the difference between $\sigma_\gamma^{expt}(E)$ and $\sigma_\gamma^{E1}(E)$ is due to E2 photoabsorption, the predicted scattering cross sections would be inconsistent with the measured data. Thus, there appears to be a small but nevertheless significant discrepancy between the elastic scattering cross sections and the total photoabsorption cross sections. This discrepancy is not particularly serious in terms of what one might like to know about the total photoabsorption cross section. However if one would interpret the small discrepancy between $\sigma_\gamma^{expt}(E)$ and $\sigma_\gamma^{E1}(E)$ as evidence for E2 strength, one would greatly overestimate the E2 strength in carbon between 30 and 40 MeV.[9] The essential point is that an interpretation of that type is extremely sensitive to the absolute cross sections for both $\sigma_\gamma^{expt}(E)$ and

$d\sigma/d\Omega$ (E,θ). The procedure outlined here, however, does not depend at all on previous measurements of σ_γ(E) and, in principle, results in the extraction of σ_γ^{E2}(E) whose detailed shape is quite insensitive to the absolute cross section scale for $d\sigma/d\Omega$ (E,θ). Unfortunately, as noted earlier, the goal of determining the detailed E2 strength distribution in ^{12}C was not accomplished. Nevertheless, the technique is feasible and reliable and should be extended to other light- and medium-weight nuclei. Finally, it is worth noting that if one can succeed in obtaining good absolute scattering cross sections over a wide range in energy and angle, the technique described here is an alternate means for determining the total photoabsorption cross section in the giant resonance region.

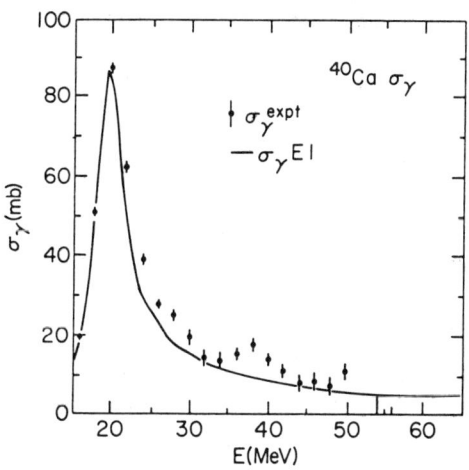

Fig. 5 Photoabsorption cross sections for ^{40}Ca.

As a final example, I will briefly discuss the analysis of the Illinois scattering data for calcium.[9] The essential results are shown in Figure 5. The points represent previous measurements of σ_γ from Mainz, whereas the curve represents σ_γ^{E1} derived from the scattering data in the same manner as was done for carbon. The discrepancies are considerably more significant than for carbon. Further, the scattering data suggest that there is not sufficient E2 strength between 25 and 50 MeV to resolve this discrepancy. In fact, there is no evidence for a compact E2 giant resonance between the GDR and 50 MeV. One should remark, however, that the Mainz total attenuation technique for measuring nuclear photoabsorption cross sections gets progressively more difficult as Z increases; calcium is the heaviest nucleus for which this technique has been used. On the other hand, the scattering cross section for calcium is somewhat more sensitive than that for carbon to model dependent modifications to the Thomson

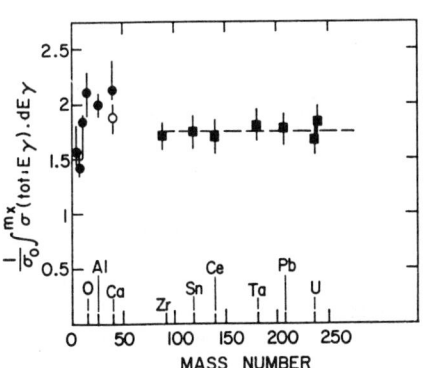

Fig. 6 Integrated photoabsorption cross sections.

amplitude due to mesonic effects. Nevertheless it is not expected that a more refined analysis of the photon scattering data will result in σ_γ^{E1} substantially different from that shown in Figure 5.

The consequences of the smaller photoabsorption derived from the scattering experiment are explored in Figure 6, where the world's data on the integrated photoabsorption cross sections up to 140 MeV are shown, in units of the classical dipole sum. For $A > 100$, the Saclay data[11] suggest a nearly constant value of 1.75, whereas the Mainz data[6] for $A \lesssim 40$ show considerable variation with A. The Mainz value for ^{40}Ca is 2.07, whereas if one uses the values of σ_γ^{E1} derived from the scattering data below 50 MeV and the Mainz values from 50 to 140 MeV, one obtains 1.82, shown by the open circle. This suggests that already for $A \approx 40$, the integrated photoabsorption per nucleon has stabilized at a roughly constant value.

One striking feature in Figure 6 is the vast region between A=40 and A=100 for which almost no total photoabsorption data exist, since neither the Saclay (γ,xn) technique nor the Mainz total attenuation technique are easily applicable. Perhaps it will be possible for the photon scattering technique, as outlined here, to fill in this gap in our knowledge.

IV. MESONIC EFFECTS IN PHOTON SCATTERING

As stated earlier, for energies large compared to giant resonance energies, the scattering amplitude is dominated by the scattering from individual scattering centers. For energies well below meson threshold, the scattering amplitude assumes the following approximate form:[2,12-14]

$$R(E,\theta) \propto -Zr_0 F_1(q) + \frac{NZ}{A} \kappa r_0 F_2(q), \qquad (9)$$

where κ is the enhancement to the classical sum rule for the unretarded dipole operator:

$$\int \sigma_{E1}^{unret}(E) \, dE = 2\pi^2 \frac{NZ}{A} r_0 (1+\kappa) \qquad (10)$$

The 1st term in eq. 9, the one-body contribution to the scattering, is the coherent scattering amplitude from Z point protons. The form factor $F_1(q)$ is related to the spatial distribution of protons in the nucleus, and is the same quantity one measures in elastic electron scattering. This same term also appears at lower energy as part of the Thomson amplitude (eq. 8) and is shown diagramatically in Fig. 7(a). The 2nd term in eq. 9, the two-body contribution to the scattering, can be interpreted as the coherent scattering amplitude from neutron-proton pairs. In a mesonic description of nuclear interactions, this amplitude results from photons which scatter from mesons exchanged between nucleons, as depicted in Figure 7(b).[2,15] The form factor $F_2(q)$ is related to the spatial distribution of neutron-proton pairs and is not easily

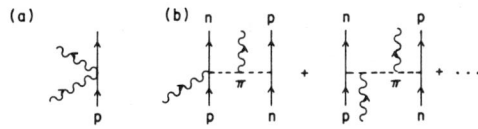

Fig. 7 (a) One-body and (b) two-body diagrams for photon scattering at medium energies.

measured by other means.[2] This quantity is the primary piece of information sought in medium energy photon scattering. In general both the one-[5] and two-body[2] amplitudes will also be energy dependent because of the finite size and structure of the scattering centers. However, for energies not too close to meson threshold ($\sim$ 100 MeV), the energy dependence is relatively weak, especially in comparison to the strong dependence on q through the form factors. This diffractive behavior is quite apparent in Fig. 8, where scattering data for ^{208}Pb are plotted in the form of backward angle-to-forward angle ratios.[13] The horizontal line at a ratio of 1.4 is the expected result for purely dipole scattering in the absence of form factors. The actual ratio falls below this by as much as a factor of 10, and the sharp diffraction pattern is indicative of a strong q dependence. For ^{208}Pb, all the terms in eq. 9 are known except for the pair form factor, so one might hope to extract $F_2(q)$ from these data. The three curves are calculated using quite different assumptions about the spatial distribu-

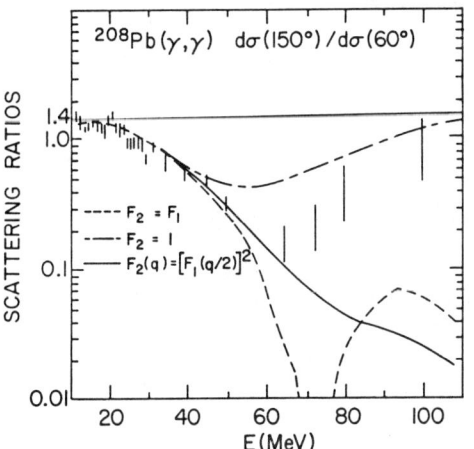

Fig. 8 Ratios of elastic scattering cross sections for ^{208}Pb.

tion of pairs. If this distribution is characterized by a uniform sphere of radius R_p, as compared to the charge radius R_c, then the three curves roughly correspond to $R_p = R_c$ (dashed), $R_p = R_c/\sqrt{2}$ (solid), and $R_p = 0$ (dash-dot). The solid curve is expected for fully correlated pairs, the dashed curve for uncorrelated pairs, and the dash-dot curve for no q dependence at all. For energies below 50 MeV, the data are not sensitive to extreme choices of $F_2(q)$. This gives us confidence that our previous analysis in the giant resonance region is on firm footing. For energies between 50 and 100 MeV, there is a marked sensitivity to $F_2(q)$, giving us hope that data such as these will teach us something about the pair distribution in nuclei. These data seem to suggest that the volume corresponding to $F_2(q)$ is significantly smaller than the nuclear volume. However, this conclusion depends critically on the q-dependence of other smaller contributions to the scattering amplitude not shown in eq. 9. For example, if there is a q-independent contribution to the amplitude which is negligible at low q, it will dominate the scattering at high q, where the form factors have damped out the dominant contributions. Therefore although there has been considerable theoretical progress in the last 2 years in the understanding of these data,^{2,12,14} an unambiguous interpretation is not yet possible.

ACKNOWLEDGEMENTS

It is a pleasure to acknowledge Dr. R. Bergère and Dr. P. Carlos of Saclay for providing a stimulating atmosphere for thinking about the analysis of photon scattering data, and my Illinois colleagues, particularly Dr. D. Wright, for their many contributions to this research. Many lively discussions with Drs. E. Hayward, B. Ziegler, and S. Fallieros are also gratefully acknowledged. This research was supported in part by the U.S. National Science Foundation under Grant No. NSF PHY83-11717.

REFERENCES

1. W. R. Dodge, Evans Hayward, R. G. Leicht, Miles McCord, and Richard Starr, Phys. Rev. C28, 8 (1983).
2. M. Weyrauch and H. Arenhövel, Nucl. Phys. A408, 425 (1983); Phys. Lett. 134B, 21 (1984).
3. J. L. Friar, Ann. Phys. (N.Y.) 95, 170 (1975).
4. T. E. O. Ericson and J. Hüfner, Nucl. Phys. B57, 603 (1973).
5. D. Drechsel and A. Russo, Phys. Lett. 137B, 294 (1984).
6. J. Ahrens, H. Borchert, K. H. Czock, H. B. Eppler, H. Grimm, H. Gundrum, M. Kroning, P. Riehn, G. Sita Ram, A. Zieger, and B. Ziegler, Nucl. Phys. A251, 479 (1975).
7. S. S. Hanna, H. F. Glavish, R. Avida, J. R. Calarco, E. Kuhlmann, and R. LaCarma, Phys. Rev. Lett. 32, 114 (1974).
8. T. W. Phillips and R. G. Johnson, Phys. Rev. C20, 1689 (1979).
9. D. H. Wright, A. M. Nathan, P. T. Debevec, and L. Morford, Phys. Rev. Lett 52, 244 (1984).
10. Giampaola Co and Siegfried Krewald, Phys. Lett. 137B, 145 (1984).
11. A. Leprêtre, H. Beil, R. Bergère, P. Carlos, J. Fagot, A. deMiniac, and A. Veyssière, Nucl. Phys. A367, 237 (1981).
12. W. M. Alberico and A. Molinari, Z. Phys. A309, 143 (1982).
13. R. Leicht, M. Hammen, K. P. Schelhaas, and B. Ziegler, Nucl. Phys. A362, 111 (1981).
14. M. Sanzone-Arenhövel, K. P. Schelhaas, and B. Ziegler, Proc. Intern. School of Intermediate Energy Nuclear Physics, eds. R. Bergere, S. Costa, and C. Schaerf (World Scientific, Singapore, 1982) p. 291.
15. J. Friar, Phys. Rev. Lett. 36, 129 (1980).
16. M. Weyrauch, diploma thesis (Mainz, 1980) unpublished.

NUCLEAR STRUCTURE STUDIES WITH PHOTOREACTIONS

F. Iachello

A. W. Wright Nuclear Structure Laboratory
Yale University
New Haven, Connecticut 06511

ABSTRACT

Recent work on nuclear properties which can be studied with photoreactions is briefly reviewed. It is pointed out that photon inelastic scattering cross sections are very sensitive tests of nuclear structure models. Experiments to test these models are suggested.

1. INTRODUCTION

Photoreactions have been used for many years to extract nuclear structure information. With the expected availability of high duty cycle electron accelerators, and, consequently, of intense photon beams, these studies may be further extended. In the mean time, nuclear structure models have evolved considerably, and detailed and extensive calculations of nuclear properties that can be tested with photoreactions are now available. In this article, I shall discuss what type of nuclear structure information can be obtained in experiments with photon beams. I will concentrate my attention mostly to the giant resonance region (E1 excitations), although towards the end of this article I will also briefly mention the excitation region around ≃3 MeV, where a new collective M1 mode has been recently discovered.

2. HIGH-LYING STATES

Photons provide a natural tool for analyzing properties of giant dipole resonances. I will discuss here only three experiments: (i) measurements of total photoabortion cross sections, $\sigma_a(E)$; measurements of (ii) elastic and (iii) inelastic scattering cross sections,

$$d\sigma(E,\theta;0^+_1 \to I^+_f)/d\Omega.$$

The way in which a giant dipole resonance appears in a given nucleus depends on the structure of that nucleus. For example, the giant dipole resonance is expected to appear as a single bump in photoabortion experiments on spherical nuclei but as a doubly-humped bump in experiments on axially deformed nuclei, Fig. 1. Classically, the energies of the two components, corresponding to oscillations parallel and perpendicular to the symmetry axis, are related to the deformation, δ, by[1]

$$E_\| = E(1 - \tfrac{2}{3}\delta),$$
$$E_\perp = E(1 + \tfrac{1}{3}\delta), \qquad (1)$$

while the ratio of the corresponding dipole strengths is

$$S_\perp/S_\| = 2. \qquad (2)$$

Since the shape of the photoaborption cross section depends on the underlying nuclear structure (for example spherical or deformed), one could hope that accurate measurements of the photoaborption cross section, $\sigma_a(E)$, may be used to extract nuclear structure information. The shapes in Fig. 1 are due to the coupling of the giant dipole resonance to the low-lying collective states. Already several years ago, Greiner et al approached the problem of coupling low-lying and high-lying collective states by using the dynamic collective model[2]. More recently, several authors[3-5] have suggested that the coupling of low-lying and high-lying collective states could be studied within the framework of the interacting boson model by introducing, in addition to (s,d) bosons describing the low-lying states, a p-boson that describes the giant dipole resonance. The corresponding Hamiltonian is written as

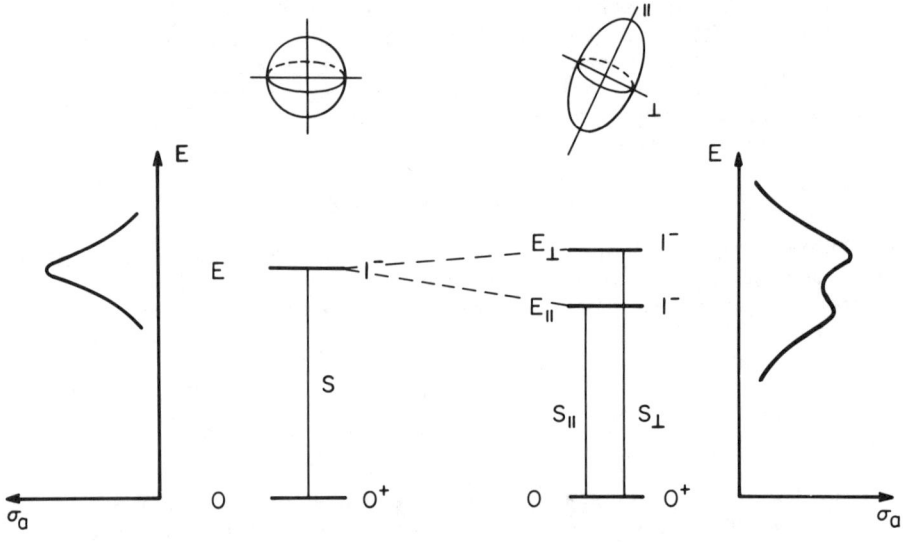

Fig. 1. Schematic representation of the splitting of the giant dipole resonance in nuclei with an axial deformation.

$$H = H_{sd} + H_p + H_{psd}, \qquad (3)$$

where H_{sd} is the usual interacting boson model Hamiltonian, H_p is the Hamiltonian describing the p-boson, $H_p = \varepsilon_p \hat{n}_p$, and H_{psd} is the coupling Hamiltonian. The approach based on the interacting boson model contains the classical result, Eqs. (1) and (2), as a limiting case. As shown in Ref.[5], for axially deformed nuclei, (SU(3) limit), one obtains the energies of the two components

$$\begin{aligned} E_\parallel &= E - \kappa(2\lambda+4) \\ E_\perp &= E + \kappa(\lambda-1) \end{aligned} \quad \xrightarrow[\lambda \to \text{large}]{} \quad \begin{aligned} &\simeq E - \kappa 2\lambda \\ &\simeq E + \kappa\lambda \end{aligned} \qquad (4)$$

where κ is related to the strength of the interacting and λ to the number of particles. The ratio of the dipole strengths is also given by

$$\frac{S_\perp}{S_\parallel} = \frac{2\lambda}{\lambda+3} \quad \xrightarrow[\lambda \to \text{large}]{} \quad \simeq 2, \qquad (5)$$

as in Eq. (2). However, the approach based on the interacting boson model is more general and allows one to calculate the splitting of the dipole strength for any situation, including that of transitional and triaxial nuclei.

Calculations of properties of the giant dipole resonances within the framework of the interacting boson model are done in the following way. One determines the parameters in H_{sd} by a fit to the low-lying states. Next one chooses a p-boson energy (usually taken to be given by the empirical law, $\varepsilon_p \approx 77A^{-1/3}$ MeV) and an interaction H_{psd} and calculates numerically the fragmentation of the E1 strength into states with energy E_n. Because of the coupling with the underlying more complex states (2p-2h,...), the strength of each dipole state at energy E_n is then spread with spreading width Γ_n. The Γ_n's are calculated using a power law dependence of Γ on E,

$$\Gamma(E) = \Gamma_0 \left(\frac{E}{E_o}\right)^\gamma. \qquad (6)$$

Results of calculations of this type for the transitional Nd and Sm isotopes[6] are shown in Figs. 2 and 3. The calculations appear to reproduce the change from a single- to a double-humped peak. However, the same agreement was obtained by Greiner et al, within the framework of the dynamic collective model[2], despite a rather different fragmentation pattern. One may thus conclude that the

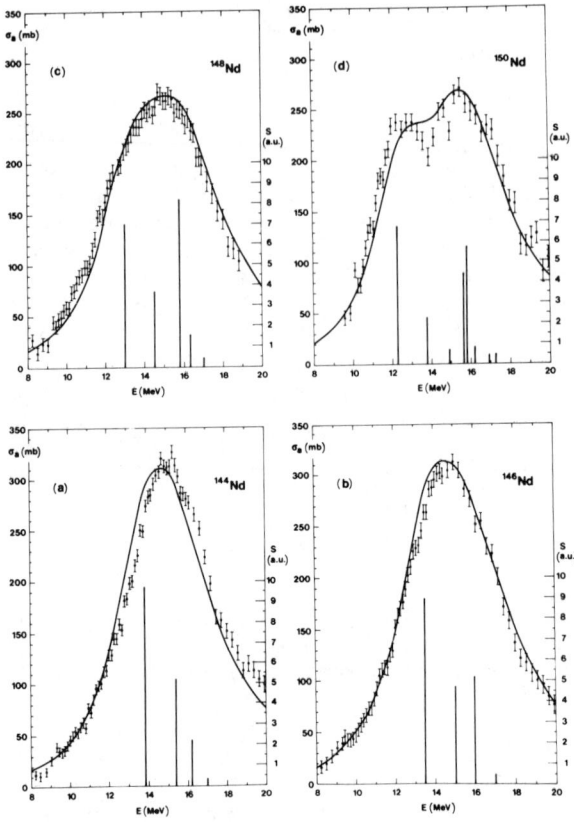

Fig. 2. Comparison between calculated and experimental[7] photoabsorption cross-section in the Nd isotopes. At the bottom of each graph the calculated fragmentation of the dipole strength is shown.

shape of the photoabsorption cross section is not a sensitive test of nuclear structure models. This is because the large spreading widths Γ tend to wash out all nuclear structure information. Only very accurate measurements of the photoabsorption cross section may reveal some details of nuclear structure. I mention two of these effects. In the classical oscillating drop model, the ratio $S_\perp/S_\parallel$ is expected to be $\simeq 2$. In the interacting boson model the ratio $S_\perp/S_\parallel$ appears to be always < 2, as one can see directly by summing the fragmented strengths in Figs. 2 and 3. The same result can also be seen from Eq. (5), if one replaces λ by its actual value $\lambda=2N$, twice the number of valence pairs. For example, for N=16, $S_\perp/S_\parallel$ = 1.83. If experiments could detect significant differences from the value 2, the results could be used to study the effect of the

finite number of particles in a nucleus. As a second point,
consider the fragmentation pattern of Figs. 2 and 3. It appears that
the perpendicular vibration ($\perp$) is systemically more fragmented than
the parallel vibration ($\parallel$). If the photoabsorption cross section is
analyzed in terms of two Gaussians with widths $\Gamma_\parallel$ and $\Gamma_\perp$, then
one expects always $\Gamma_\perp > \Gamma_\parallel$. The larger fragmentation of the
perpendicular vibration is related to an effective triaxiality of the
nucleus. Thus accurate measurements of $\sigma_a(E)$ may serve also to
elucidate this point.

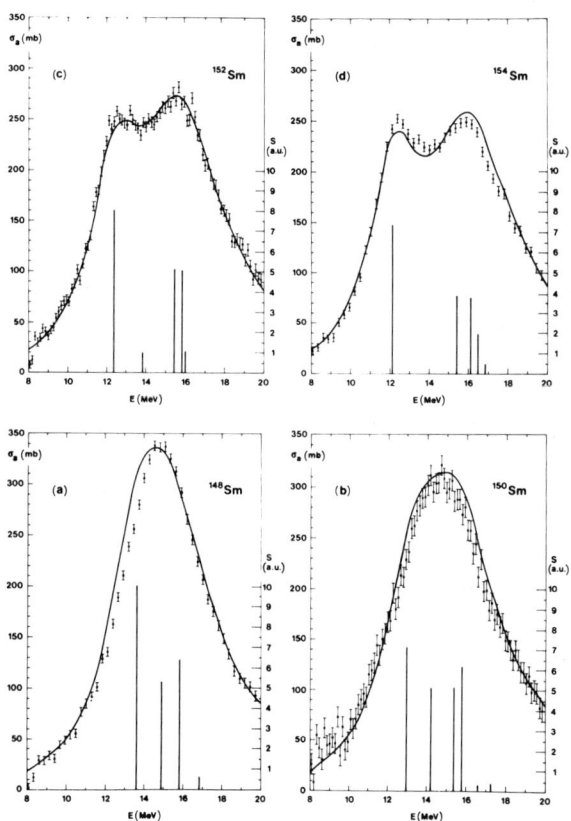

Fig. 3. Same as in Fig. 2 but for the Sm isotopes.

I have mentioned above that photoabsorption cross sections are not too sensitive to nuclear structure information. On the other side, it appears that elastic and inelastic photon scattering cross sections are very sensitive to the underlying nuclear structure. One can use the same interacting boson model to calculate these cross sections. As it is well known, several non-nuclear effects contribute to the elastic scattering amplitude, namely Rayleigh, Thomson and Delbruck scattering. In the calculations performed so far in terms of of the interacting boson model[4,8] only Thomson scattering has been included. These calculations are thus meaningful only for high photon energies ($\simeq$ 8 Mev) and large scattering angles ($\theta \simeq 90°$). The nuclear structure information is contained in the polarizabilities, P_J,

$$P_J = \delta_{JI_f} \frac{1}{\sqrt{3(2I_f+1)}} \left(\frac{e^2}{\hbar c}\right) EE' \sum_n \langle I_f^+ \| \hat{D}^{(1)} \| 1_n^- \rangle \langle 1_n^- \| \hat{D}^{(1)} \| 0_1^+ \rangle \times$$

$$\times \left(\frac{1}{E_n + E' + \frac{i\Gamma_n}{2}} + \frac{(-)^J}{E_n - E - \frac{i\Gamma_n}{2}} \right) - \delta_{J0} \delta_{JI_f} \sqrt{3} \frac{(Ze)^2}{Am_N c^2}, \quad (7)$$

where J denotes the nuclear angular momentum, I_f the angular momentum of the final state, E(E') the incident (scattered) photon energy, $|1_n^-\rangle$ the n-th dipole state at energy E_n and with width Γ_n and, in the Thomson amplitude, Z and A are the charge and mass number of the target nucleus and m_N the atomic mass unit. Absoption and scattering cross sections for unpolarized photons can then be written in the usual way, as

$$\sigma_a = \frac{4\pi}{\sqrt{3}} \left(\frac{\hbar c}{E}\right) \text{Im} P_0 = \frac{8\pi e^2}{3\hbar c} E^2 \sum_n \frac{E_n \Gamma_n |\langle 1_n^- \| \hat{D}^{(1)} \| 0_1^+ \rangle|^2}{(E_n^2 - E^2)^2 + \frac{\Gamma_n^2}{2} (E^2 + E_n^2 + \frac{\Gamma_n^2}{8})}, \quad (8) \quad (9)$$

$$\frac{d\sigma}{d\Omega}(E,\theta;0_1^+ \to I_f^+) = \frac{E'}{E} |P_J|^2 \delta_{JI_f} g_J(\theta),$$

where the functions $g_J(\theta)$ for J = 0 and J = 2 are given by

$$g_0(\theta) = \frac{1}{6}(1+\cos^2\theta); \quad g_2(\theta) = \frac{1}{12}(13 + \cos^2\theta). \quad (10)$$

Two calculations have been performed so far within the framework of the interacting boson model, one by Scholtz and Hahne [4] in ^{168}Er and one by Maino et al [8] in ^{232}Th and ^{238}U. Scholtz and Hahne have considered inelastic scattering to side (β and γ) bands, as shown schematically in Fig. 4. They found that their calculation was consistent with the data[9], while that based on the dynamic collective model was not. Maino et al[8] have considered elastic and inelastic scattering to the 2^+_1 state in ^{232}Th and ^{238}U. Their results are summarized in Fig. 5. It is interesting to note that, since the parameters in the collective Hamiltonian, H of Eq. (3), are fixed by fitting the energies of the low-lying states and the shape and magnitude of the photoabsorption cross section, the

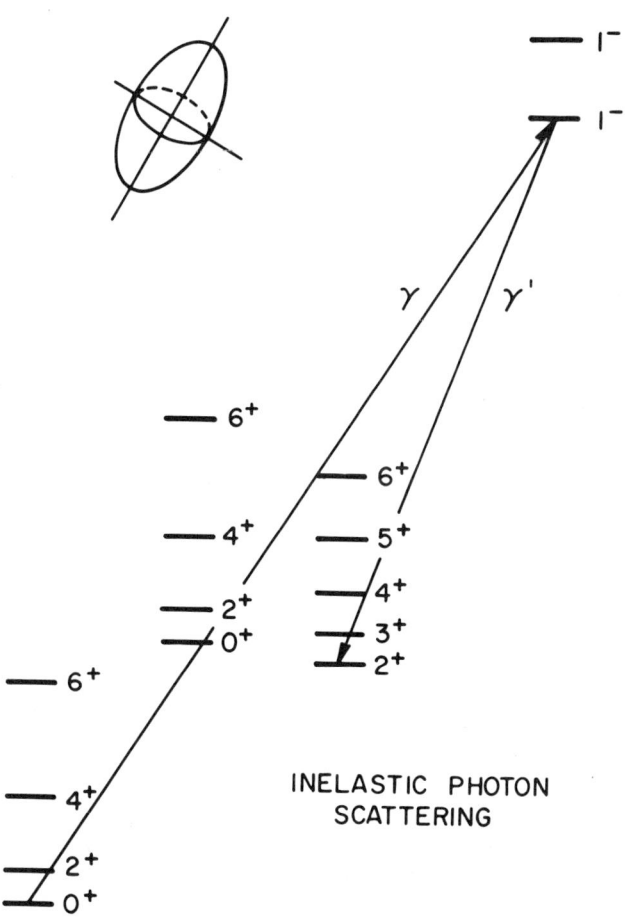

Fig. 4. Schematic representation of inelastic photon scattering to β and γ bands in deformed nuclei.

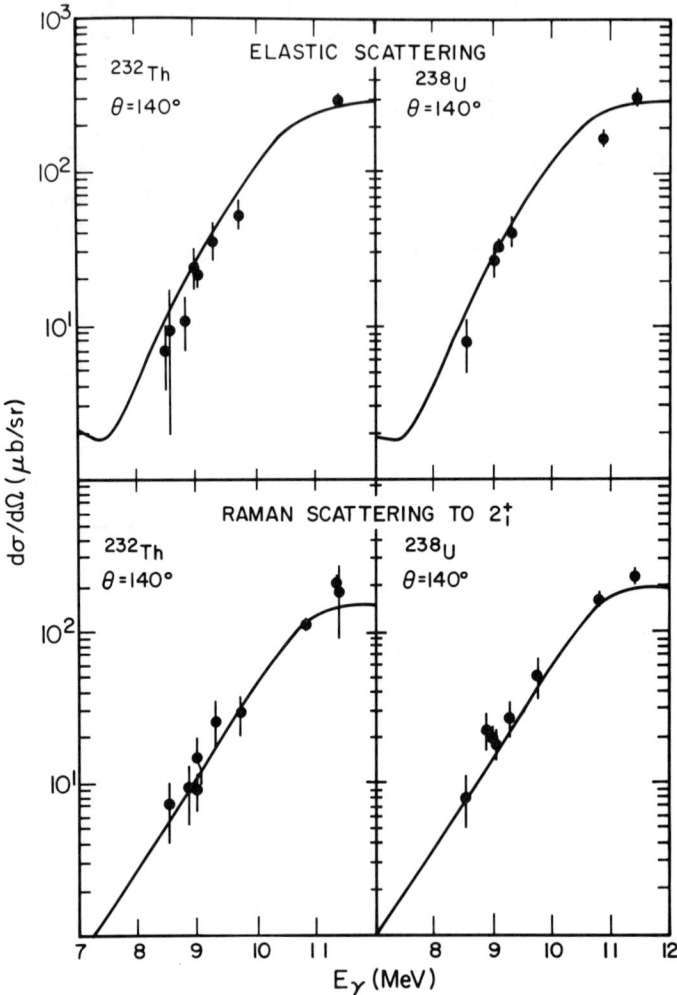

Fig. 5. Comparison between calculated and experimental[10] elastic and inelastic photon scattering cross sections at θ=140° in ^{232}Th and ^{238}U.

calculation of the scattering cross sections does not require additional parameters. A comparison with experiment provides thus a stringent test of the model calculation. The calculations shown in Fig. 5 appear to describe the data well. On the other side, direct non-resonant contributions were needed in order to obtain agreement between calculations based on the dynamic collective model and experiment[11]. This is in contrast to the case of the total photoabsorption cross section, for which both models gave similar results and indicates a sensitiveness of photon scattering cross sections to nuclear structure. Photoexperiments in this area could thus be particularly useful in elucidating nuclear structure aspects which are otherwise difficult to detect.

3. LOW-LYING STATES

Photons have traditionally been used for nuclear structure studies in the giant resonance region ($\simeq$ 10 MeV). This is because collective states with dipole (electric or magnetic) character were expected to occur only in the region. However, recently it has been suggested [12,13] that a new class of collective states should occur at much lower energy ($\simeq$ 3 MeV). Among these, particularly important is a magnetic dipole, 1^+, state expected to occur in deformed nuclei. This state corresponds classically to an angular oscillation of the proton and neutron deformations, Fig. 6.

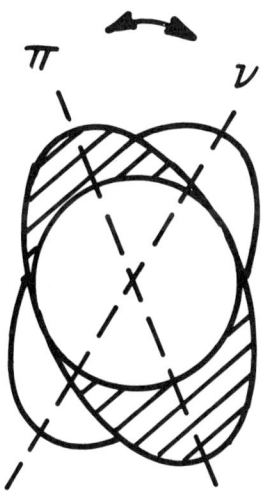

Fig. 6. Schematic representations of the new collective, M1, mode (twisting mode).

A state in the energy region (≈ 3 MeV) having the expected properties has been found recently by Bohle et al[14] using inelastic electron scattering. Because of its dipole, M1, character, this state (or states) should be reachable with photon beams. Indeed Berg et al[15] have observed these new collective states in (γ,γ') experiments. Furthermore, they have detected their decays to the 2_1^+ state, Fig. 7. This is seen in Fig. 8 where a portion of the (γ,γ') spectrum on a natural Gd target is shown. The 3069 keV peak, corresponding to the 1^+ state in ^{156}Gd appears to have a satellite peak 89.0 keV removed from it. This energy is precisely the energy of the 2_1^+ state.

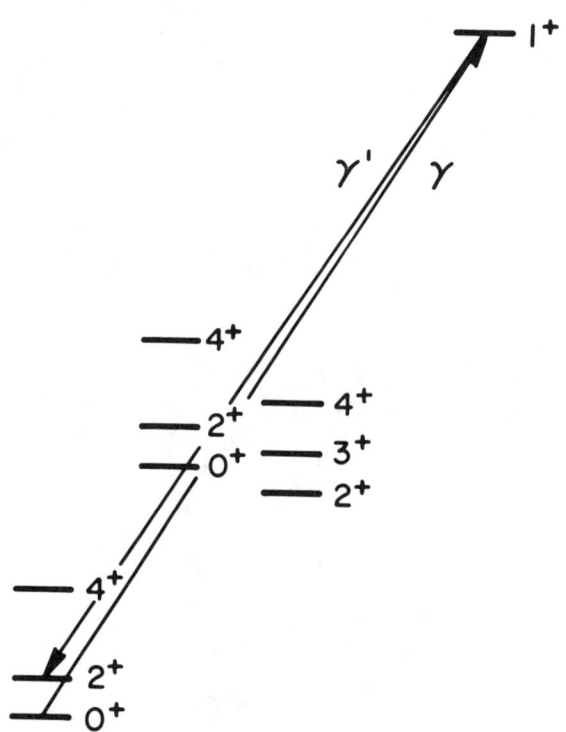

Fig. 7. Schematic representation of the inelastic excitation of the 2_1^+ state in (γ,γ') experiments.

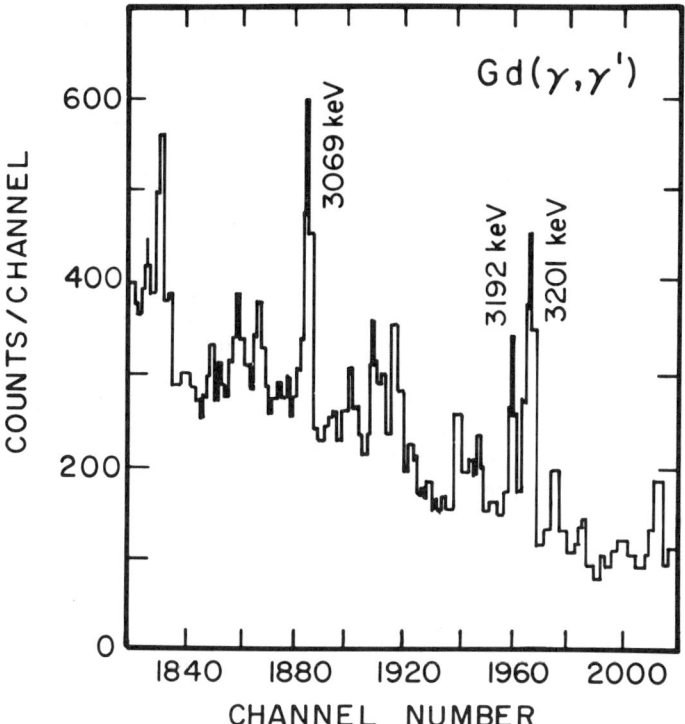

Fig. 8. Experimental observation of the 3069 keV line in ^{156}Gd and of the 3192 and 3201 keV lines in ^{158}Gd by Berg et al[15].

The experimental discovery of this new collective mode opens a new and interesting field for photoreaction studies. Experiments similar to those performed in the giant resonance region could be (and should be) repeated for this new region. They would provide a considerable amount of new nuclear structure information.

4. CONCLUSIONS

I have discussed in this article two aspects of nuclear structure that can be studied with photons. One, more traditional, is related to properties of the giant dipole (1⁻) resonance. Here, accurate photon experiments, especially inelastic scattering, may reveal important nuclear structure aspects which are otherwise difficult to detect. The photon in these experiments acts as a microscope that investigates the actual equibrium shape of a nucleus and its vibrations. The second, less traditional, is related to properties of a new collective, M1, mode, which has been predicted to occur in the energy region ≃ 3 MeV.

Comparison between experiments and theoretical expectations is facilitated by the development of elaborate nuclear structure models, in particular of the interacting boson model. This model can be used to describe the data available at present and to make predictions for properties hitherto unknown. For example, it predicted the occurence of the new class of collective states now observed in the energy region ≃ 3 MeV. These predictions can be qualitative, semi-quantitative or even quantitative. As an example, the expected elastic and inelastic photon scattering cross sections in ^{148,154}Sm are shown in Fig. 9. These are semi-quantitative predictions since the nuclear structure parameters on which they are based have not been optimized. They are shown here in order to stimulate further experiments.

In conclusion, the development of new and accurate nuclear structure models and of new experimental techniques, gives, in my opinion, unique possibilities to the field of photonuclear reactions.

This work was supported in part under the Department of Energy Contract No. DE-AC02-76 ER 03074.

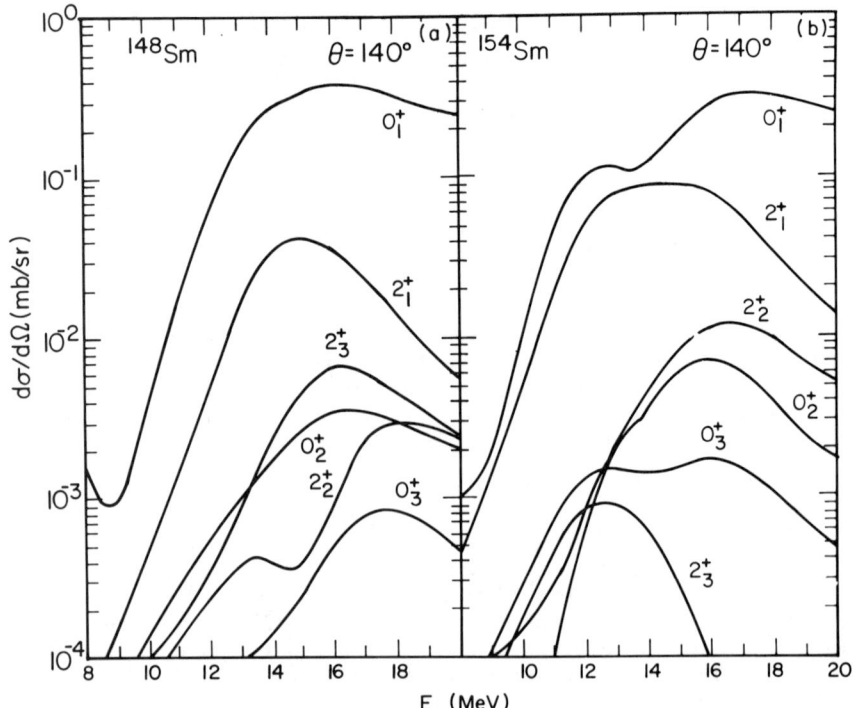

Fig. 9. Elastic and inelastic photon scattering cross sections expected in ^{148,154}Sm[16].

REFERENCES

1. M. Danos, Nucl. Phys. 5 (1958) 23; K. Okamoto, Phys. Rev. 110 (1958) 143.
2. See, for example, J. M. Eisenberg and W. Greiner, "Nuclear Theory", Vol. 1, North-Holland Publ. Co., Amsterdam, 1978, p.331.
3. I. Morrison and J. Weise, J. Phys. G8 (1982) 687.
4. F. G. Scholtz and F. J. W. Hahne, Phys. Lett. 123B (1983) 187.
5. D. J. Rowe and F. Iachello, Phys. Lett. 130B (1983) 231.
6. G. Maino, A. Ventura, L. Zuffi and F. Iachello, to be published; G. Maino, A. Ventura, L. Zuffi, These Proceedings.
7. P. Carlos, H. Beil, R. Bergère, A. Leprêtre, A. Veyssière, Nucl. Phys. A172 (1971) 437; P. Carlos, H. Beil, R. Bergère, A. Leprêtre, A. de Miniac, A. Veyssière, Nucl. Phys. A225 (1974) 171.
8. G. Maino, A. Ventura, L. Zuffi and F. Iachello, in preparation.
9. A. M. Nathan and R. Moreh, Phys. Lett. 91B (1980) 38.
10. M. Hass, R. Moreh, D. Salzmann, Phys. Lett. 36B (1971) 68; H. E. Jackson, G. E. Thomas, K. J. Metzel, Phys. Rev. C9 (1974) 1153; H. E. Jackson, G. E. Thomas, K. J. Metzel, Phys. Rev. C11 (1975) 1664; S. Kahane and R. Moreh, Nucl. Phys. A308 (1978) 88; P. Rullhusen, U. Zurmuhl, W. Muchenheim, F. Smend, M. Schumaker, H. G. Borner, Nucl. Phys. A382 (1982) 79; P. Rullhusen, U. Zurmühl, F. Smend, M Schumaker, H. G. Borner and S. A. Kerr, Phys. Rev. C27 (1983) 559.
11. H. Arenhövel, Phys. Rev. C6 (1972) 1449.
12. T. Suzuki and D. J. Rowe, Nucl. Phys. A289 (1977) 461; N. Lo Iudice and F. Palumbo, Phys. Rev. Lett. 41 (1978) 1532; N. Lo Iudice and F. Palumbo, Nucl. Phys. A326 (1979) 193.
13. F. Iachello, Nucl. Phys. A358 (1981) 89c; A. E. L. Dieperink, Progr. in Part. and Nucl. Phys. 9 (1983) 121.
14. D. Bohle, A. Richter, W. Steffen, A. E. L. Dieperink, N. Lo Iudice, F. Palumbo and O. Scholten, Phys. Lett. 137B (1984) 27.
15. U. E. P. Berg, C. Bläsing, J. Drexler, R. Heil, U. Kneissl, W. Naatz, R. Ratzek, S. Schennach, R. Stock, T. Weber, H. Wickert, B. Fischer, H. Hollick, D. Kollewe, to be published.
16. G. Maino, A. Ventura and L. Zuffi, private communication.

RECENT RESULTS OF THE GÖTTINGEN-GRENOBLE PHOTON SCATTERING COLLABORATION

M. Schumacher
II. Physikalisches Institut der Universität Göttingen
D - 3400 Göttingen, FRG

ABSTRACT

The present status of our photon-scattering studies using neutron capture γ-rays is reviewed. These studies are concerned with (i) the strengths of E1 transitions between bound nuclear levels, (ii) intermediate structure and its interpretation in terms of vibrations of the nuclear diffuseness, (iii) specific properties of Delbrück scattering like the Coulomb correction effect and scaling laws, and (iv) the exhaust of the TRK sum rule by giant resonances of heavy nuclei.

INTRODUCTION

Considerable progress has been made in elastic photon scattering using neutron-capture γ-rays since the last conference on this subject[1] in Grenoble 1981. As an introduction let me shortly remind the situation of 1981. At that time we had completed our high-precision studies on atomic Rayleigh and Delbrück scattering below 5 MeV using off-line γ-sources[2-4] and had started the work with neutron-capture γ-rays[5,6]. The main results of these previous investigations were: (i) The amplitudes for atomic Rayleigh scattering are predicted with a few-percent accuracy if the calculation is based upon the second order S-matrix of QED and appropriate self-consistent relativistic atomic wave-functions[3]. Form factor calculations are completely invalid at few MeV of energy and large scattering angles. (ii) The Feynman graph describing Delbrück scattering in lowest order is valid with an accuracy of at least 5%. This is the highest accuracy with which a single graph of the order $Z^2 e^6$ has ever been tested[4]. (iii) The Coulomb correction effect, as precisely examined by us at 2.7 MeV proved to be entirely due to the first graph of the order $Z^4 e^{10}$, and was shown[2] to modify the the amplitudes for high-Z elements by 100%. (iv) The amplitudes for nuclear Rayleigh scattering below particle threshold, which are known to be closely related to the dynamic electric polarizability of the nucleus, proved[6] to be in good agreement with the Lorentzian-shaped average photo-absorption cross-section of the giant-dipole resonance (GDR). This means that due to the large concentration of E1 strength in the GDR, the local E1 strength at the excitation energy of the incident photon has little influence on the dynamic polarizability at that energy. (v) First results of a systematic

---

* Work performed in cooperation with F.Smend, P.Rullhusen, U.Zurmühl (present address: MPI für Chemie, Mainz), H.G.Börner (ILL Grenoble). Supported by Deutsche Forschungsgemeinschaft.

study of the distribution of E1 strength below particle threshold using a dense series of neutron-capture γ-rays proved[1] to be in line with a Lorentzian tail of the GDR.

The last topic of our previous report will be the first of the present one. By now we have succeeded to understand in full the wealth of nuclear resonance-fluorescence (NRF) experiments using a dense series of γ-ray photons[7,8]. Furthermore, we were successful in disentangling[9] the different contributions to the elastic differential cross-section in the giant-dipole resonance region. Due to this progress, information is obtained on the Coulomb correction effect and on scaling laws in Delbrück scattering[10]. In addition it has become possible to scale the Lorentzian parameters of the GDRs, and thus to obtain an improved information on the exhaust of the TRK sum rule by the GDRs of heavy nuclei.

EXPERIMENTAL SET-UP

The GAMS through-channed of the HFR Grenoble was used to produce an intense γ-beam (fig.1).

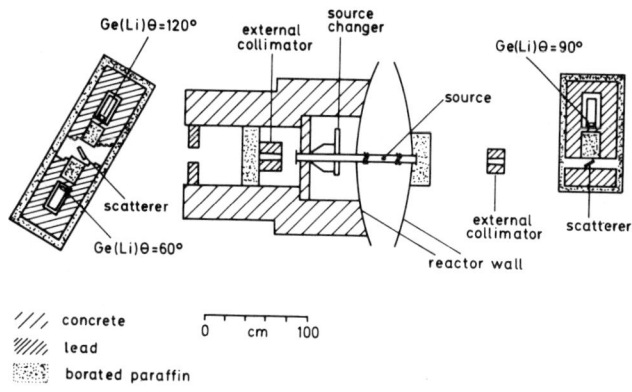

Fig.1 Experimental set-up at the HFR Grenoble

The widths of the γ-lines due to thermal Doppler broadening are typically between 5 and 10 eV FWHM. Due to the internal collimator system of the through-channel, only neutrons which are scattered by the internal neutron-capture target can escape from the reactor. This advantage and the hermetic shielding of the scattering chambers with borated paraffine lead to a comparatively small contamination of the useful γ-flux with neutrons (n/γ ~10^{-4}). Three large-volume, high-resolution Ge(Li) detectors have been used at the same time, thus providing three different scattering angles. In addition, a pair spectrometer has been applied for analysing and monitoring the beam. Absolute differential cross sections for photon scattering have been determined by measuring the direct beam with each of the Ge(Li) detectors placed behind a lead absorber of about 25 cm thickness. By interchangingly taking out part of the absorber, the transmission was calibrated with high precision. As an example we show the spectrum of photons from a ^{141}Pr neutron capture target scattered by ^{238}U (fig.2).

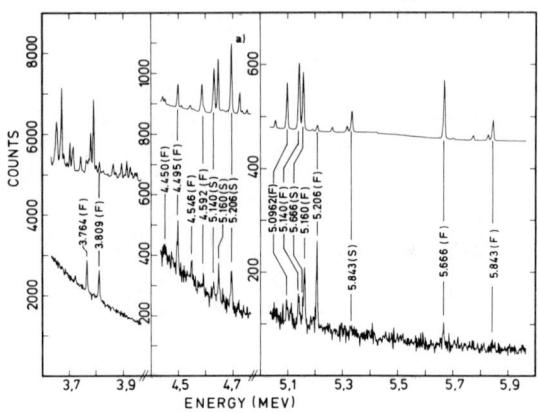

Fig.2 Upper part: Spectrum of photons from the ^{141}Pr(n,γ) reaction.
Lower part: Same photons scattered by ^{238}U.

NUCLEAR RESONANCE FLUORESCENCE AND THE E1 STRENGTH BELOW PARTICLE THRESHOLD

High-resolution studies of neutron cross-sections have lead to a deep insight into the statistical aspects of level spacings and widths. Unfortunately, this technique can only be applied in a very limited energy region just above the neutron binding energy of those nuclei which can be reached by the capture of slow neutrons. As stated in the previous section neutron-capture γ-rays are also of very high energy resolution. Therefore, in principle it should be possible to use neutron capture γ-rays for studies below particle threshold of stable nuclei, in a similar way as neutron resonances are used in the studies mentioned above. The basic difference, of course, in the two methods is, that it is very difficult to vary the γ-energy. Due to this drawback, the scattering of γ-rays does not lead to a complete information, but to a statistical sample of information about nuclear properties. Nevertheless, one may investigate the problem, to what extent the average and statistical properties of nuclear levels, as known from the study of neutron resonances, extend to lower energies. Following this idea NRF experiments have been carried out with different n-capture and γ-scattering targets in the A = 160-240 mass range. The most illustrative results are obtained for ^{238}U (fig. 3).

Fig.3 Predicted elastic scattering: C:coherent, B:coherent plus NRF for tagged photons, A: same as B for n-capture photons. Experiment:
o coherent elastic only,
■ off-line γ-sources,
⊕ n-capture γ-sources.

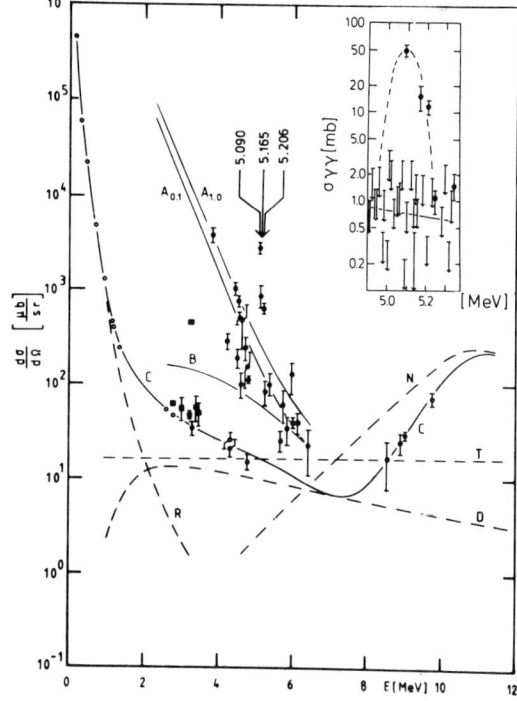

The predicted cross section (solid line) for coherent elastic scattering is well established by the experiment (open circles). This prediction includes atomic Rayleigh (R), nuclear Thomson (T), nuclear Rayleigh (N) and Delbrück (D) scattering. The D amplitudes have already been corrected for the Coulomb-correction effect (ref.[1] and next section). The closed circles and closed rectangles, lying above the coherent elastic cross section are shifted upward due to resonant excitation of one or several nuclear levels. No data are found below the coherent elastic cross section. This result reflects the smallness of interference effects due to phase mixing[11] so that special effort is required to observe interference at all[12].

As far as the angular distribution of the resonant radiation has been measured, an L = 1 photon angular momentum has been found, so that there is a good reason to assume an E1 multipolarity for all the resonances of fig.3. For an interpretation of the data we have made the following assumptions: (i) The E1 strength-function is given by the Lorentzian tail of the GDR. (ii) The density of levels is given by the back-shifted Fermi-gas model. (iii) The spacings between levels of the same spin-parity are distributed according to a Wigner distribution. (iv) The γ-width for transitions between a sequence of levels at about the same excitation energy to the same final state follows a Porter-Thomas distribution. (v) A GDR is built upon each nuclear level. Based upon these assumptions the three curves B, $A_{0.1}$ and $A_{1.0}$ have been calculated using the Monte Carlo technique[8]. Curve B is the predicted average resonant plus coherent-elastic cross-section obtained by averaging over an assembly of cross sections which reflects a full Porter-Thomas distribution of ground-state widths. This curve would have been the result of an experiment carried out with tagged photons. If we take the average only over cross sections lying above curve B or above 1/10 of curve B we arrive at $A_{1.0}$ and $A_{0.1}$, respectively. Apparently, the omission of small cross sections leads to predictions which are only weakly dependent on the out-off limit, which experimentally, for the closed circles, is given by curve B at the low-energy end and by 1/10 of curve B at the high-energy end of the curves $A_{1.0}$ and $A_{0.1}$. We see that there is a nice agreement between the closed circles and the predictions as given by the curves $A_{0.1}$ and $A_{1.0}$, except for the energy range between 5.0 and 5.3 MeV which will be discussed later. The importance of the foregoing investigation lies in the fact, that for the first time nuclear properties as known to be valid in the narrow energy interval above particle threshold where neutron resonances are observable, have been investigated and found valid for γ-resonances down to excitation energies of 3-4 MeV. Of course, the accumulation of a larger statistical sample of resonances and a direct measurement of the parity would be desirable, and appears feasible in a not too long beam time at the Grenoble reactor.

The success of the GDR and statistical models in explaining the properties of resonant cross sections in ^{238}U and the recent statement[13] that properties of the statistical model may be found valid at all mass numbers and all excitation energies has prompted us to apply this model to the histograms of E1 hindrance-factors as accumulated by P.M.Endt[14]. These histograms provide us with a very large sample of E1 transitions in the whole mass range between A = 6

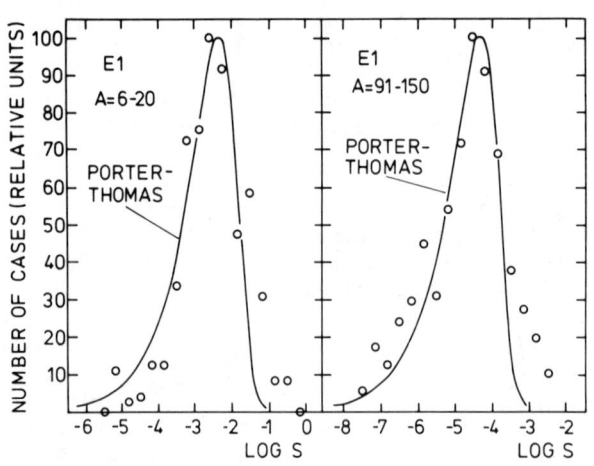

Fig.4 Histograms of E1 hindrance-factors compared with Porter-Thomas disbutions.

and 150. Our first observation is that these histograms are merely Porter-Thomas distributions on a logarithmic scale, as demonstrated by fig.4 for the mass ranges A = 6-20 and A = 91-150. The obvious explanation for this finding is that for each mass range there exists an average hindrance factor $\bar{S}$ corresponding to the maximum of the Porter-Thomas distribution on the log S-scale which is independent of the excitation energy U and independent of the transition energy E_γ. Fig.5 shows that this explanation is in line with the model outlined above. The curves calculated for different U/E_γ combinations are located in the same broad band as the crosses which represent the empirical average hindrance factor $\bar{S}$. For this calculation the GDR tail has been assumed to be Lorentzian down to 4 MeV and to decrease more steeply below this energy. The larger spreading of curves in the A = 91-150 mass range as compared to the vertical extension of the cross (fig.5) is in line with fig.4 where for the same mass range the empirical histogram is broader than the Porter-Thomas distribution. The foregoing consideration makes use of the supposition that the choice of E1 transitions incorporated in the histograms is random. Certainly, if by some procedure we select e.g. only the exceptionally large transition widths we would arrive at different conclusions.

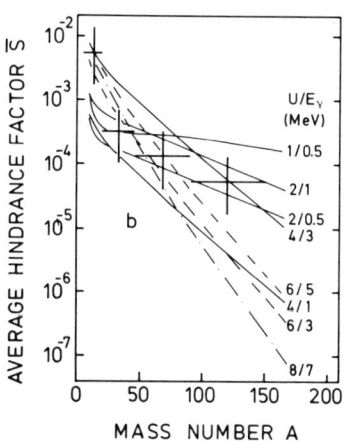

Fig.5 Predicted average hindrance factors for a Lorentzian tail of the GDR down to 4 MeV, and a steeper decrease below 4 MeV. U:excitation energy, E_γ: transition energy

NON-STATISTICAL STRUCTURE

We now return to fig. 3. The insert in this figure shows some exceptionally large cross sections in comparison with all the other data obtained in the same energy interval. Assuming an underlying Porter-Thomas distribution of ground-state widths, the statistical probability for finding three cross sections of this size is about 10^{-4}. This means that a non-statistical structure has been observed. An intermediate structure of E1 multipolarity is not unexpected at this energy. Satchler[15] has pointed out that a dipole-oscillation of the diffuseness of the Woods-Saxon potential (fig.6b) may lead to excited states at this energy. Furthermore, the inelastic scattering of α-particles has led to good evidence[16] for this type of excitation (fig.6a). In comparing inelastic α- and γ-scattering one has to remind that in heavy nuclei the nuclear states are of

mixed isospin. This facilitates an
isoscalar (α,α') and isovector (γ,γ')
excitation of this vibrational mode
(fig.6a).

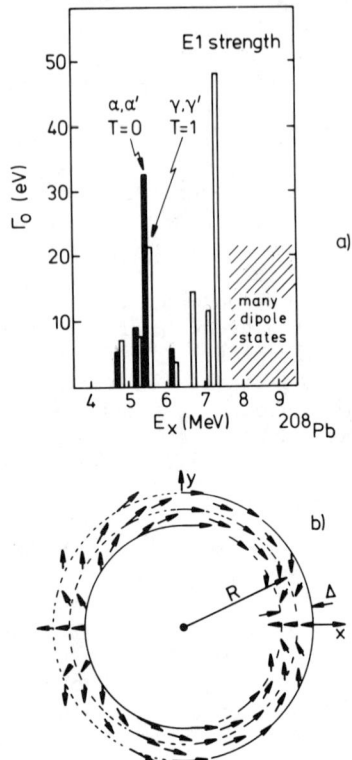

Fig.6 Oscillation of the nuclear
diffuseness. a) evidence from
α-scattering, b) flow-pattern

PROPERTIES OF DELBRÜCK SCATTERING IN THE GDR REGION AND ABOVE

Prior to our studies it had already been noticed[17] that the lowest-order D-amplitudes were not capable of fitting the angular dependence of the elastic differential cross-section measured for ^{238}U at 9.0 MeV. We have extended these investigations to ^{209}Bi and ^{232}Th and have made the same observation[9]. As an example fig. 7 shows the results obtained for ^{209}Bi: Though by proper scaling of the GDR peak cross-section the predicted elastic scattering cross-section has been brought into agreement with the experiment at $\Theta = 120°$ there still remains a discrepancy at $\Theta = 60°$. The reason for this behaviour has to be sought in a large contribution of Coulomb correction terms of the order of $Z^4 e^{10}$ which has not been calculated up to now and therefore is not included in the curves. We have developed[9] an empirical procedure to take Coulomb corrections into account by first reducing the four terms (fig.8) which contribute the Coulomb correction, viz. Re $B_{\parallel}$, Re $B_{\perp}$, Im $B_{\parallel}$ and Im $B_{\perp}$, to only one angle-dependent correction parameter. This was achieved by making use of the fact that for $\Theta \to 180°$, $B_{\parallel} \to -B_{\perp}$.
Furthermore, since for the lowest-order amplitudes

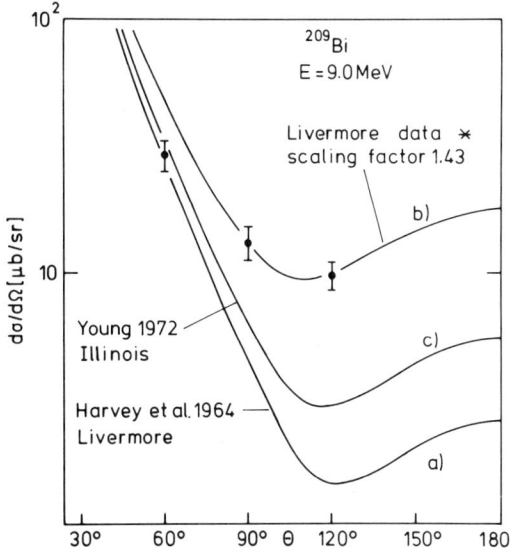

Fig.7 Coherent elastic scattering. Predictions for different GDR peak cross-sections not including the Coulomb correction effect.

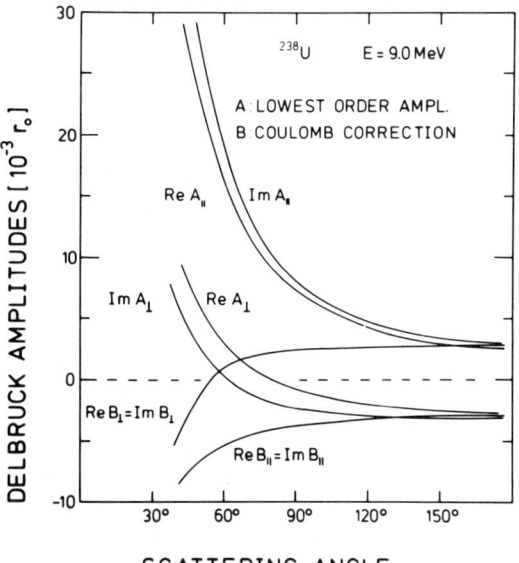

Fig.8 Amplitudes for Delbrück scattering: Predicted lowest-order amplitudes and empirical Coulomb correction terms.

we have Re $A_{\parallel} \approx$ Im $A_{\parallel}$ and Re $A_{\perp} \approx$ Im $A_{\perp}$, it is appropriate to assume that the same relations hold also for the Coulomb correction terms. By this procedure we have arrived at the Coulomb correction terms depicted in fig.8 and at the excellent fits to the experimental data shown in fig.9. Though the modifications of the differential cross sections due the Coulomb corrections are only of the order of at most 20 % (fig.9) the contributions of the Coulomb-correction terms to the total D-amplitudes are quite sizable (fig.8).

One major problem in these investigation was to arrive at appropriate parameters for the GDR. The data available in the literature differ little in resonance energies E_0 and widths Γ. But sizable discrepancies are found in the peak resonance

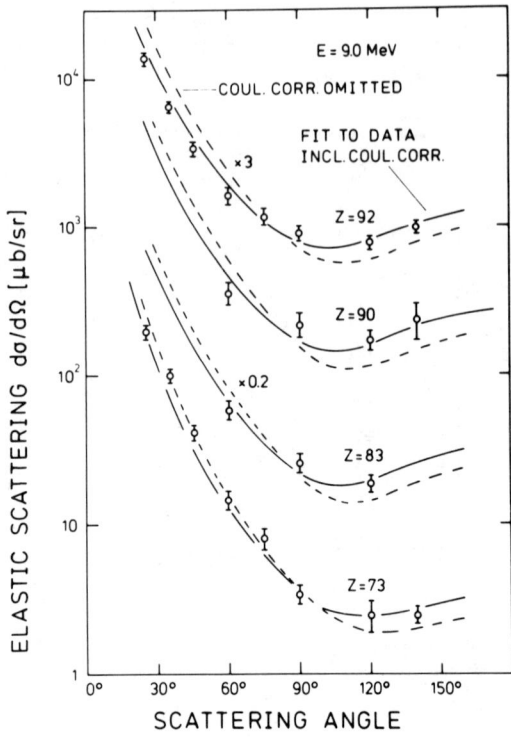

Fig.9 Coherent elastic scattering. Solid curves include the Coulomb correction effect. Part of data are from previous investigations (see ref.[9]).

cross-sections which therefore have been fitted to the experimental data. This was possible in an unambiguous way because elastic differential cross sections have also been measured at 11,4 MeV where Delbrück scattering is small. Furthermore, the amplitudes for nuclear Rayleigh scattering are predominantly real numbers at these energies and therefore only very slightly dependent on local structures of the photo-absorption cross-section.

The foregoing studies have provided us with a good knowledge of Delbrück scattering including the Coulomb correction effects. This puts us in position to test the scaling laws for D-amplitudes recently predicted by Cheng, Tsai and Zhu[18]. According to these authors[18] the lowest-order amplitudes and the Coulomb-correction terms should asymptotically become proportional to E^{-1} for fixed scattering angle, provided $\hbar\omega \gg mc^2$ and $|\hbar\vec{\Delta}| \geq mc$, where m is the electron mass. We have compared these scaling laws with predicted lowest-order amplitudes and empirical Coulomb-correction terms and have found a confirmation only for the lowest-order amplitudes (fig.

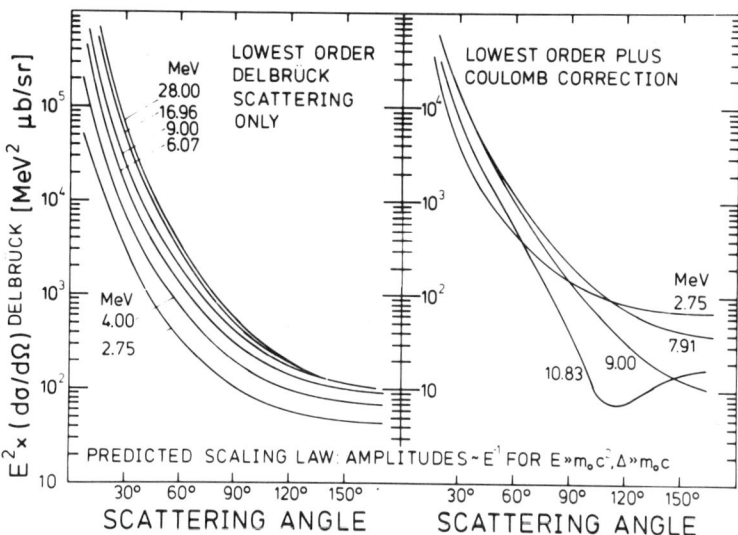

Fig.10 Test of scaling laws in D-scattering: Curves must be identical in case the scaling laws are fulfilled.

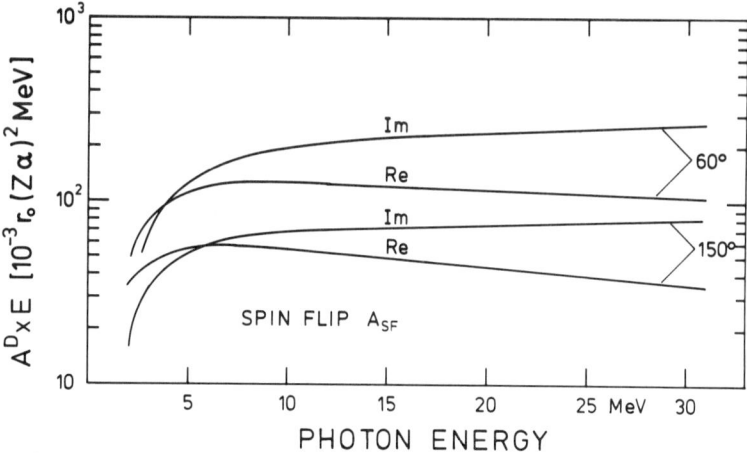

Fig.11 Lowest-order amplitudes: Curves must be horizontal in case the scaling laws are fulfilled.

10). However, by looking at the problem more carefully it appears uncertain that the scaling laws are precisely fulfilled even for the lowest-order amplitudes. This can be judged from fig.11 where the lines should become horizontal in the limit of high energies in case the scaling laws were fulfilled.

THE ENERGY WEIGHTED SUM RULE APPLIED TO THE GDR

The energy weighted sum rule relates the integrated photo-absorption cross-section to the number of absorbing charges in the many-body system. Studies in the whole range of mass numbers[19,20] have shown that the photo-absorption cross-sections integrated up to the pion threshold are almost the same for all nuclei namely

$$\int_0^{m_\pi c^2} \sigma(E)\, dE = 60 \text{ MeV mb } \frac{NZ}{A}(1+\kappa)$$

with $\kappa \approx 1$. The enhancement κ is usually attributed to exchange forces in the nuclear Hamiltonian though the detailed mechanism is not at all clear[21]. In attempts to understand the mechanism it appears advisable to look at the GDR and the quasi-deuteron region separately. This has recently been proposed by Brown and Rho[22]. According to these authors[22,23] the enhancement χ of the photoabsorption cross section integrated over the GDR separately, is related to δ_{gl} the correction from exchange currents to the orbital g-factor, whereas the large enhancement $\kappa \sim 1$ of the dipole sum when integrated up to the meson threshold may be understood theoretically to arise from the tensor part of the nucleon-nucleon interaction. Furthermore, for ^{208}Pb attempts have been made to understand the GDR in terms of 1 ℏω transitions between oscillator shells[24]. This latter aspect leads to the supposition that the integrated cross section measured for the GDR of ^{208}Pb or its neighbours may possibly be larger than the corresponding cross-sections measured for ^{181}Ta, ^{232}Th and ^{238}U where the nucleon configurations are far from closed shells and, hence, the the number of available 1 ℏω transitions may possibly be smaller. Apparently, there is more than one good reason for the experimentalist to separate the integrated cross-section of the GDR from that of the quasi-deuteron region. In heavy nuclei the integrated cross section attributed to the GDR is well defined by the areas underneath the Lorentzians[20]. This makes it possible to carry the separation of integrated cross sections through with little ambiguity.

Fig.12 shows the available data for the photo-absorption cross-sections integrated over the Lorentzians of the GDRs for the mass range between 181 and 238. Indeed, there is a suggestive support of the supposition that shell-effects may show up in the integrated cross-section of the GDR. The only data which clearly contradict the trend indicated by the dashed curve of fig.12 are the photo-absorption data obtained[25] by the Livermore group for ^{232}Th and ^{238}U. There are several[26] investigations which show that the peak cross-sections measured by the Livermore group [25] probably are too large for these nuclei, in agreement with our findings.

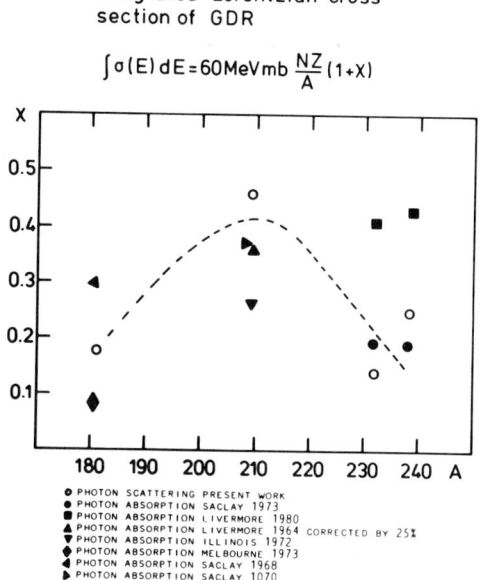

Fig.12 Enhancement χ for the Lorentzians of the GDR.

REFERENCES

1. M.Schumacher, F.Smend, P.Rullhusen, W.Mückenheim and H.G.Börner, Inst. Phys. Conf. Ser. No.62, 598 (1982)
2. P.Rullhusen, F.Smend, M.Schumacher, A.Hanser and H.Rebel, Z.Physik A 293, 287 (1979)
3. W.Mückenheim and M.Schumacher, J.Phys. G6, 1237 (1980)
4. P.Rullhusen, W.Mückenheim, F.Smend, M.Schumacher, G.P.A.Berg, K.Mork and L.Kissel, Phys.Rev. C23, 1375 (1981)
5. M.Schumacher, P.Rullhusen, F.Smend, W.Mückenheim and H.G.Börner, Nucl.Phys. A346, 418 (1980)
6. M.Schumacher, F.Smend, W.Mückenheim, P.Rullhusen and H.G.Börner, Z.Physik A300, 193 (1981)
7. M.Schumacher, P.Rullhusen, U.Zurmühl, F.Smend, H.G.Börner and S.A.Kerr, Internal Report Göttingen/Grenoble, 83 SC 06 T (1983)
8. U.Zurmühl, P.Rullhusen, F.Smend, M.Schumacher, H.G.Börner and S.A.Kerr, Z.Physik A314, 171 (1983)
9. P.Rullhusen, U.Zurmühl, F.Smend, M.Schumacher, H.G.Börner and S.A.Kerr, Phys. Rev. C27, 559 (1983)

10. P.Rullhusen, U.Zurmühl, F.Smend, M.Schumacher, H.G.Börner and S.A.Kerr, Phys. Rev. D$\underline{27}$, 1962 (1983)
11. P.Rullhusen, U.Zurmühl, W.Mückenheim, F.Smend and M.Schumacher, Nucl.Phys. A$\underline{382}$, 79 (1982)
12. S.Kahane, R.Moreh and O.Shahal, Phys. Rev. C$\underline{28}$, 1519 (1983)
13. T.A.Brody, J.Flores, J.B.French, P.A.Mello, A.Pandey and S.S.M.Wong, Rev. Mod. Phys. $\underline{53}$, 385 (1981)
14. P.M.Endt, Atomic Data and Nuclear Data Tables $\underline{26}$, 47 (1981)
15. G.R.Satchler, Nucl. Phys. A$\underline{100}$, 481 (1967)
16. P.Decowski, H.P.Morsch and W.Benenson, Phys. Lett. $\underline{101}$B, 147 (1981)
17. S.Kahane and R.Moreh, Nucl. Phys. A$\underline{308}$, 88 (1978)
18. H.Cheng, E.C.Tsai and X.Zhu, Phys. Rev. D$\underline{26}$, 908 (1982)
19. J.Ahrens, H.Borchert, K.H.Czock, H.B.Eppler, H.Gimm, H.Gundrum, M.Kröning, P.Riehn, G.Sita Ram, A.Zieger and B.Ziegler, Nucl. Phys. A$\underline{251}$, 479 (1975)
20. A.Leprêtre, H.Beil, R.Bergère, P.Carlos, J.Fagot, A.De Miniac and A.Veyssière, Nucl. Phys. A$\underline{367}$, 237 (1981)
21. W.Weise, Lecture Notes in Phys. $\underline{61}$, 484 (1977)
22. G.E.Brown and M.Rho, Nucl. Phys. A$\underline{338}$, 269 (1980)
23. J.I.Fujita and M.Ichimura, Mesons in nuclei, ed. M.Rho and D.H.Wilkinson (North-Holland, Amsterdam, 1979) p.627
24. J.Speth, Nucl. Sci. Res. Conf. Ser. $\underline{1}$, (Harwood 1979), p.33
25. J.T.Caldwell, E.J.Dowdy, B.L.Berman, R.A.Alvarez and P.Meyer, Phys. Rev. C$\underline{21}$, 1215 (1980)
26. E.W.Lees, Inst. Phys. Conf. Ser. $\underline{62}$, 613 (1982) and U.Kneissl, private communication

RECENT DEVELOPMENTS IN NUCLEAR SCATTERING OF CAPTURE GAMMA-RAYS

O. Shahal and R. Moreh
Nuclear Research Center-Negev, Beer-Sheva, Israel
Ben-Gurion University of the Negev, Beer-Sheva, Israel

ABSTRACT

Recent studies at the IRR-2 reactor using monoenergetic neutron capture γ-rays fall into three categories: 1) study the low energy GDR tail using the (γ,γ) and (γ,n) reactions; 2) study of interference effects between nuclear resonance fluorescence, Rayleigh and Delbruck scattering at small angles; 3) testing the 2D absorption structure of simple molecules on graphite using nuclear resonance scattering technique.

INTRODUCTION

The activities of the nuclear research group of the Negev reactor (IRR-2) is based mainly on the use of photon beams generated from neutron capture. These activities include (γ,n) angular distribution measurements on the deuteron and on heavy targets and are represented by a few contributions to this conference. Here we concentrate on new photon scattering studies which fall into three main categories: 1) Measurements which utilize the ultra-high monochromaticity of the photon beams ($\Delta E \sim 20$ eV) for probing the microstructure of the low-energy tail of the GDR in nuclei around A=208. In this mass region, nuclear structure effects are known to occur above the particle emission threshold e.g., the bumps at 9.0, 10.06, 11.3 MeV in ^{208}Pb. These bumps were studied using Ni capture γ-rays containing γ-lines with energies almost coinciding with those of the bumps. By combining the scattering information thus obtained with low resolution photon scattering data it was possible to set an upper limit to the widths of the 10.06 and 11.3 MeV bumps in ^{208}Pb. 2) Interference effects: such an effect was observed for the first time between Rayleigh-Delbruck scattering and nuclear resonance fluorescence from an isolated level. This was measured at forward angles ($\Theta = 1.0°$, $1.7°$) using the chance resonance scattering event in which a γ line of the Fe(n,γ) reaction overlaps the 7.28 MeV level in ^{208}Pb. The corresponsding theoretical expression was derived and an excellent agreement with measured values was obtained. 3) Molecular orientation: the nuclear resonance photon scattering from the 6324 keV level of ^{15}N was used for the determination of the spatial orientation of molecular groups containing nitrogen. This technique[1] senses the Doppler broadening of the ^{15}N level arising from the zero-point vibrational energy of the molecule. In diatomic molecules, the broadening is highly anisotropic being maximum along the molecular axis and minimum along the perpendicular direction. This fact was used for monitoring the orientation of adsorbed N_2 and NO molecules with respect to the adsorbing graphite planes. The results indicated that at low temperature ≤ 30°K the N_2 molecular axis align itself nearly parallel to the grafoil planes in fair agreement with measurements of neutron diffraction and low-energy electron diffraction. However, for the NO-grafoil system, a wide disagreement was observed. Our results indicate that at a coverage of ~ 0.24 monolayers (of NO on grafoil) and T < 50°K, the NO molecules (in dimer N_2O_2 form) <u>cannot all</u> lay parallel to the grafoil surface, in contrast to the findings of n-diffraction.[21]

The photon beam was generated from the (n,γ) reaction on metallic disks of (Fe, Cr, Ni) placed along a tangential beam tube near the core of IRR-2 reactor. The photon beam was collimated and neutron filtered. Details of the experimental system were given elsewhere.[1] In the present talk, we shall concentrate on the last three types of experiments.

1. Probing the Low Energy Tail of the GDR

The low energy tail of the GDR is known to have structure[2-5] above the particle emission threshold. This structure was observed in several independent studies. For example, the recent work of R. D. Starr, P. Axel and L. S. Cardman with (γ,γ') reaction and ~ 125 keV resolution in lead-208[3] small bumps were found at 10.04, 10.06, and 11.3 MeV. The Ni capture γ-rays spectra contain γ-lines with energies that coincide with those of the bumps (8.99, 10.06, 11.388 MeV).

The dipole character of the scattered radiation from those bumps was established by using (γ,γ')[3] reaction. The (e,e') measurements[4] have shown that those bumps are highly fragmented, with E1 and E2 strength. It is of interest to probe the structure using γ-lines with high energy resolution like those obtained from neutron capture γ-rays.

By combining the scattering information obtained with low energy resolution (varying between 140 keV for the (γ,n) reaction[2], 125 keV for the (γ,γ') reaction[3], ~ 30 keV for (e,e') reaction[4] and 5-50 keV using the threshold photoneutron technique[5] with the present data taken with the high resolution, neutron capture photon beams, it is possible to set an upper limit to the width of the 10.06, 11.3 MeV bumps in ^{208}Pb.

The main disadvantage of this technique however is the fact that the γ source is not variable in energy, so it is impossible to get a continuous scan. Experimentally, the photon beam hit a target of ~ 50 gr of ^{208}Pb (91% enrichment). Figure 1 shows the high energy spectra of the incident γ beam and that of the scattered photons at an angle of 145°. Note the strong scattered signal at 10.06 MeV. This line is almost absent in the scattered spectra from neighboring targets ^{206}Pb and ^{209}Bi. Its presence shows that there is a strong narrow bump in the GDR tail of ^{208}Pb. The dipole character of the scattered radiation was established by measuring the intensity ratio I(145°)/I(90°). For the 10.06 and 11.4 MeV line the obtained ratios are: 1.69 ± 0.17, 1.60 ± 0.08 compared to the theoretical value of 1.67 for a dipole transition. The result for the 8.999 MeV line was 1.43 which is also indicative of dipole scattering, but it is depressed because of the relatively large contribution of Delbruck scattering at 90°. At higher energies the effect of Delbruck scattering is much smaller and can be ignored. The absolute scattering cross sections obtained from the intensities of the incident beam and the scattered radiation are given in Fig. 2. The solid curve is the predicted cross section (calculated by S. Kahane) including Delbruck scattering, nuclear resonance scattering (the GDR parameters were obtained from ref. 2) and classical Thompson scattering.

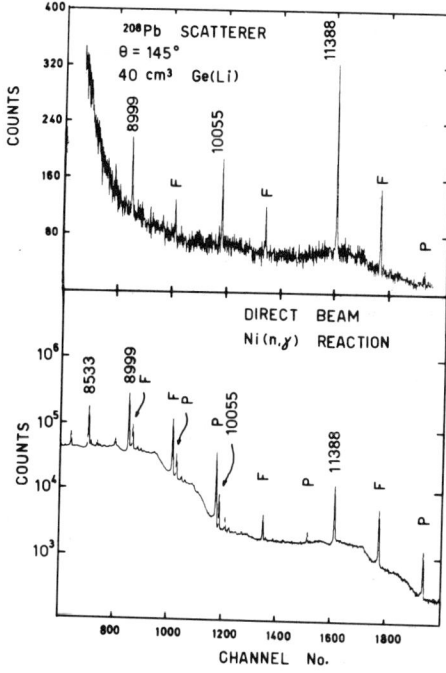

Fig. 1. Portion of the high-energy part of the scattered spectrum from a ^{208}Pb target at $\Theta=145°$ as measured with a 40 cm^3 Ge(li) detector, and the corresponding Ni(n,γ) spectrum of the incident beam as measured after passing through a 20 cm Pb absorber. P and F denote photo peaks and first escape peak.

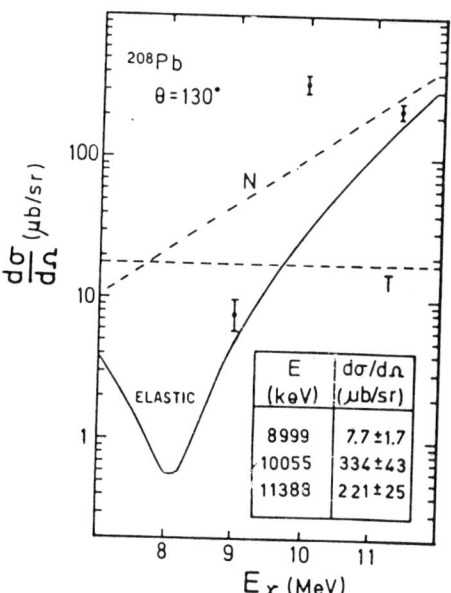

Fig. 2. Diffrential cross section (μb/sr) for elastic scattering of photons from ^{208}Pb of $\Theta=130°$: The dashed curves represent the calculated pure contributions of nuclear Thomson (T) and nuclear resonance (N) from the GDR. The solid curve represents the coherent contribution of the T,N and the Delbruck (D) scattering amplitudes.

E (keV)	dσ/dΩ (μb/sr)
8999	7.7 ±1.7
10055	334 ±43
11388	221 ±25

The Rayleigh scattering contribution was ignored. The present values are higher than predicted especially for the 10.06 MeV line. The present value at 10.06 MeV is a factor of 2 higher than that reported in the low resolution data. In order to explain our high measured scattering cross section compared to the experimental value of the low resolution data, it is necessary to assume the existence of a Lorentzian line at 10.06 MeV superimposed on the low energy tail of the GDR. If one uses a value of 21 mb-MeV for the integral $\int \sigma_a(E)dE$, our measured cross-section implies $\Gamma \leqslant 25$ keV for the Lorentzian width. A similar procedure was used to obtain an upper limit for the width of the 11.27, 11.45 MeV bumps. One finds that in order for the cross section to drop as low as 221 μb/sr at 11.388 MeV, the widths of the bumps at 11.27 and 11.45 MeV should be $\Gamma \leqslant 70$ keV if the predicted cross sections are to agree with present result and those of Illinois.[2]

The combination of our data with the earlier results imply a much narrower bump than was thought previously. It should be emphasized that the large measured value of the intensities ratio I(145°)/I(90°) at 8.999 MeV does not rule out the possibility of an E2 contribution because the (γ,γ') angular distribution is not sensitive to small E2 admixture in a region where dipole contribution predominate. A better search for some E2 admixture may be made by looking into photoneutron angular distribution, any asymmetry around θ=90° is indicative of interfering effects E1-E2 or E1-M1. Such measurements were already done.[6] The results of the 11.388 and 10.055 MeV lines have shown symmetry around 90°[6] while the 8.999 MeV state revealed strong neutron asymmetry around θ=90°.[6] Thus there is clear evidence for an E2 contribution at 8.999 MeV.

2. Studying the Interference Effects Between Nuclear Resonance Fluorescence and Rayleigh-Delbruck.

The use of a source with high energy resolution is essential for studying the interference effects between the elastic photon scattering processes. In the past few years, we already carried out a series of experiments, studying all the elastic processes separately by choosing the adequate scattering angles, Z, and energies in such a way as to enhance one process at a time while minimizing the others. With the advent of forward-angle elastic scattering measurements, where the cross sections are dominated by the R (Rayleigh) and D (Delbruck) contributions (in the energy range 4-10 MeV), it becomes feasible to observe possible interference effects between nuclear resonance fluorescences with R and D scattering. This is because the differential cross section of R and D scattering at θ ~ 1° and Z > 50 is about ~ 100 mb/sr and hence comparable to nuclear resonance fluorescence cross section. The contribution from the GDR together with that of Thomson scattering processes at E ~ 8 MeV is small and is thus ignored. In the present work we selected the Fe(n,γ) source together with lead-208 target because of four reasons:

1) ^{208}Pb is known to scatter resonantly the 7.28 MeV line obtained from the Fe(n,γ) reaction yielding the most intense nuclear resonance scattering signal.

2) Lead is a high Z element (Z=82) yielding intense R and D scattering cross sections at small angles.

3) It is possible to compare the elastic scattering of the 7.28 MeV line with other γ-lines, close in energy, where no nuclear resonance scattering process contributes.

4) The Pb scattering results can be compared with those from a Bi target (Z=83) which is nearly nonresonant at 7.28 MeV thus yielding more reliable information.

To achieve small forward scattering angles, we employed a shadow beam geometry shown in Fig. 3. In this geometry the incident γ beam is prevented from reaching the Ge(Li) detector directly by a long cylindrical absorber C and a collimator B. Ideally, the only radiation reaching the detector is that scattered by the ring target mounted on the absorber. The scattering angle could be varied by changing the diameter of the absorber, the scatterer-detector distance and the diameter and length of the colimator.

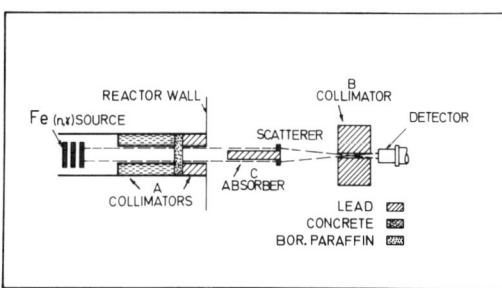

Fig. 3. Schematic diagram (not to scale) of the experimental system. The dimensions of the iron absorber C were 100 cm long, 4 cm in diameter (for θ ~ 1/0°). The dimensions of the lead collimator B were 2 cm diameter by 30 to 40 cm long (dependong on the value of θ). The scatterer-detector distance was ~ 135 cm.

In Fig. 4. the spectra of the high energy part of the scattered radiation from a Pb target at angles θ ~ 1° and 1.7° are shown. A broad Compton peak is present near each narrow elastic peak.

The appearance of the two peaks served three purposes:

1) To determine the absolute elastic differential cross section. This value is obtained by measuring the area ratios of the elastic to Compton peaks and employing the theoretically calculated Compton cross sections.

2) To find the scattering angle obtained from the energy separation between the elastic and Compton peaks.

3) To determine the angular spread $\Delta\theta$ obtained from the width and shape of the Compton peak.

The extraction of the elastic to Compton peak ratio was carried out using a computer code which fitted both the elastic peak and the

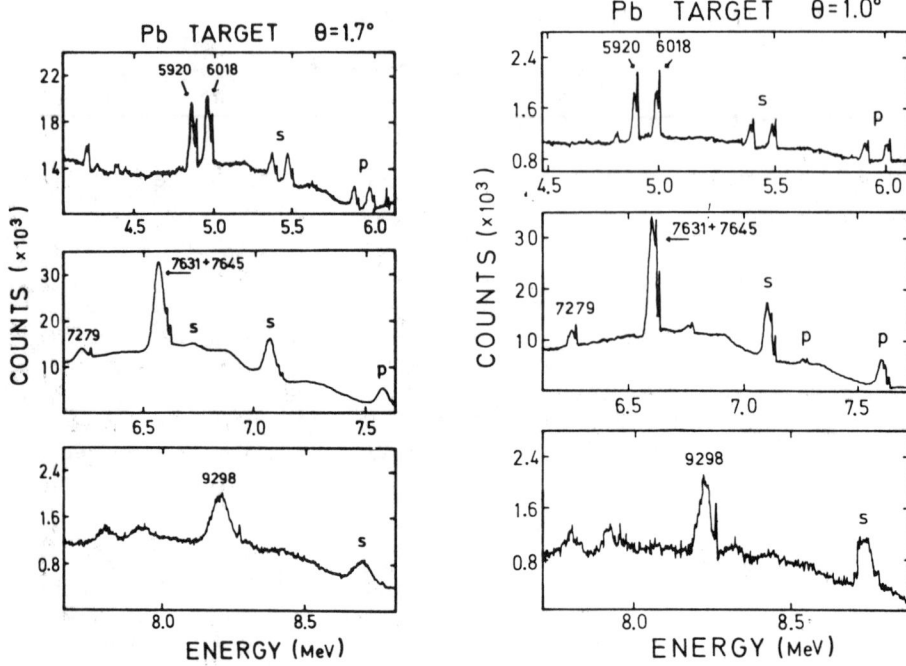

Fig. 4. Typical scattered spectrum from a Pb target at θ = 0.95 ± 0.07° and θ = 1.69 ± 0.02° showing a broad Compton peak near each narrow elastic peak. P and S denote photopeaks and first escape peaks. Other lines denote second escape peaks.

Compton peak. Figure 5 shows a typical result of such separation for the second escape peaks of the 5920 and 6018 keV lines scattered from a Bi target at θ=1.7°. The Compton cross section was calculated by multiplying the Klein-Nishina value by the incoherent scattering function S. The factor S takes into account the effect of bond electrons in cases where the recoil energy of the Compton scattered electrons is smaller than the binding energy.

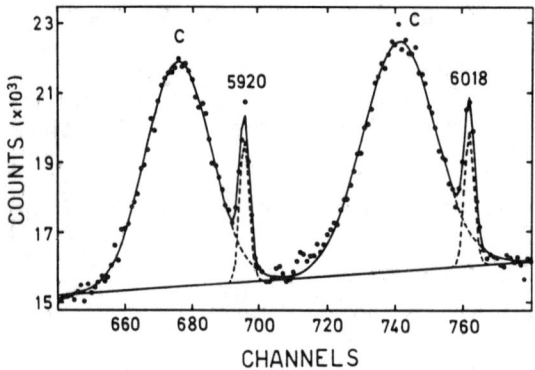

Fig. 5. Typical results of curve fittings of skew Gaussians to the elastic and Compton peaks. The procedure is shown for the double escape peaks of the 5920 and 6018 keV γ lines scattered at θ=1.7° from a Bi target.

The measuared elastic cross sections in the energy range 4.0-10.0 MeV are presented in Fig. 6.

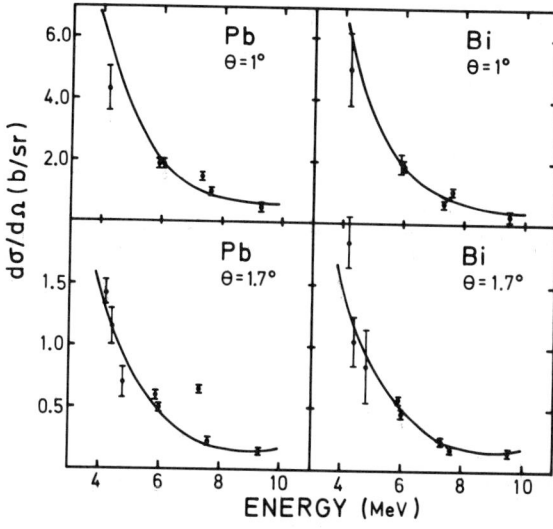

Fig. 6. Differential cross section for elastic scattering of photons from Pb and Bi at θ=1° and 1.7° the curves are obtained by reducing the R amplitudes by 5%, resulting in a better fit at the higher energies. The contribution of NRF scattering is not included.

The coherent differential elastic scattering cross section may be written in terms of linerly polarized amplitudes as

$$\frac{d\sigma}{d\Omega} = \frac{1}{2} r_0^2 \left(|A_\parallel|^2 + |A_\perp|^2 \right)$$

where $A_\parallel$ and $A_\perp$ are the scattering amplitudes polarized parallel to and perpendicular to the scattering plane. Each scattering amplitude is a superposition of D, R, and F (nuclear resonance fluorescence) scattering amplitudes

$$A_\parallel = A_D^\parallel + A_R^\parallel + A_F^\parallel = A_S^\parallel + A_F^\parallel$$

$$A_\perp = A_D^\perp + A_R^\perp + A_F^\perp = A_S^\perp + A_F^\perp \quad \text{where } A_S = A_D + A_R$$

In general each amplitude is complex. The Rayleigh scattering deals with the contribution of bound atomic electrons to the elastic scattering of photons. In the range of momentum transfers of the present work, namely, χ <10Å, the Rayleigh amplitudes are approximated by the <u>modified form factors</u> calculated using Hartree-Fock Slater relativistic wave functions of Liberman et al.[11]

$$g(q,Z) = 4\pi \int n(r) \frac{\sin(q,r)}{(qr)} \frac{mc^2}{mc^2 - E_B - V(r)} r^2 dr \quad (3)$$

where $q = (E/c)\sin(\theta/2)$ is the momentum transfer

E_B the electron binding energy

$\eta(r)$ density of the electron cloud

$V(r)$ H.F. mean potential of the electron

Delbruck scattering deals with the elastic scattering of photons from the Coulomb field of nuclei via real and virtual electron-positron production. Values of D amplitudes were calculated by Bar-Noy and Kahane[12] using the formulae given by Paptzacos and Mork[13] and by Constantini.[14]

The nuclear resonance fluorescence scattering amplitude may be obtained by considering a Breit-Wigner amplitude of a single photon of energy E incident on a nucleus of ground state spin J_0 with an excited level of peak energy E_r, total width Γ and spin J. In units of r_0 the classical electron radius, we have:

$$A_F^\parallel(E,\theta) = A_{F_1}^\parallel + iA_{F_2}^\parallel = C_0 \frac{\Gamma/2}{(E_r-E)-i\Gamma/2} f_\parallel(\theta) \tag{5}$$

where $C_0 = \sqrt{\sigma_0/r_0}$

$\sigma_0 = 2\pi\lambda^2 (2J+1)/(2J_0 + 1)$

for M1 radiation (7.28 MeV level in ^{208}Pb with spin 1^+)

$f_\perp(\theta) = \sqrt{3/8\pi}$ $f_\parallel(\theta) = \sqrt{3/8\pi} \cos\theta$ $A_F^\parallel = A_F^\perp \cos\theta$

for $\theta < 2°$ we may ignore the angular dependence

$f_\perp(\theta) = f_\parallel(\theta) = f(\theta)$ $A_F^\parallel = A_F^\perp$

The expression for the elastic scattering cross section is the coherent sum of the nuclear resonance fluorescence with the other scattering processes evaluated for each incident photon and then integrated incoherently over all photons, where A_F, A_S depend on the relative velocity v of the Fe(n,γ) emitting nucleus with respect to ^{208}Pb scatterer

$$|A_\parallel|^2 = |A_F^\parallel (E,\theta,v) + A_S^\parallel (E,\theta,v)|^2$$

An integration is done over all possible velocities v, with $\rho(v)$ the velocity distribution function, and $\int \rho(v)dv = 1$. The cross section may be written as:

$$\frac{d\sigma}{d\Omega} = \frac{r_0^2}{2} \int (|A_\parallel|^2 + |A_\perp|^2)\rho(\nu)d\nu$$

$$= \frac{r_0^2}{2} \Big\{ \int (|A_F^\parallel|^2 + |A_F^\perp|^2)\rho(\nu)d\nu + (|A_S^\parallel|^2 + |A_S^\perp|^2) +$$

$$2 \sum_{j=1}^{2} \Big[A_{S_j}^\parallel \int A_{F_j}^\parallel \rho(\nu)d\nu + A_{S_j}^\perp \int A_{F_j}^\perp \rho(\nu)d\nu \Big] \Big\} \quad (7)$$

The last expression contains three terms: the first is identifiable with the differential scattering cross section $d\sigma_F/d\Omega$ of the pure F process, as it contains an integration over all possible relative velocities ν. This term was also discussed in several earlier publications.[15] The second term is the combined coherent differential scattering cross section $d\sigma_s/d\Omega$ of the D and R process, while the third is the interference term between the F and the other two processes. The dependence of $A_s = A_R + A_D$ on ν is very weak and may be taken as a constant over the energy region of 20 eV.

Phase relations: The problem of the relative phases of the imaginary and the real parts of the amplitudes for the D, R, N, and T scattering processes was discussed in detail in earlier publications[13,16] and could be summarized as follows:

1) All the imaginary amplitudes at θ=0° are positive, as is immediately obvious from the optical theorem

$$\text{Im } A(E,\theta=0°) = \frac{E}{4\pi\hbar c} \sigma_a(E), \quad (8)$$

where $\sigma_a(E)$ is the absorption cross section of the corresponding absorption process, which is always positive.

2) The sign of the real part of the amplitude is positive or negative according to whether $\sigma_a(E)$ is an increasing or decreasing function of energy. For the F amplitude we use the same sign convention as that obtained for A_N, that being the nuclear resonance scattering amplitude[17] from the GDR. Using the Gaussin distribution for the thermal velocities of both the γ-emitting nuclei and the scattering nuclei and the phase relations the theoretical differential elastic cross section can be calculated.

The solid line in Fig. 6 was obtained for the nonresonant scattering events. It was necessary to reduce the Rayleigh amplitudes at θ=1° by 5% to achieve a better agreement with the experiment. The 5% reduction in the R amplitudes can be justified by noting that the contribution of the L- and M-shell electron is over-estimated in the modified form factor approximation and this deviation increases towards small momentum transfer, namely, smaller scattering angles. The contribution of Rayleigh scattering is large only at low energies (below 6 MeV). So this correction is relatively small at higher energies approaching the resonance at 7.28 MeV.

The magnitude of three terms constituting the elastic scattering cross section is given in Table 1.

Table 1. Measured and Theoretical Cross Section of the 7.278 MeV Line Scattered from Natural ^{208}Pb in (mb/sr)

	$\theta = 1°$	$\theta = 1.7°$
$d\sigma/d\Omega$ (measured)	1490 ± 80	663 ± 30 (mb/sr)
$d\sigma_S/d\Omega$	920	244
$Pd\sigma_F/d\Omega$	347	347
Interference	182	105
$d\sigma/d\Omega$ (calculated)	1449	696

P - relative abundance of the ^{208}Pb isotope

Only after including the interference term, a good agreement between the measured and predicted values is obtained. Hence, a constructive interference is obtained between nuclear resonance R and D scattering processes.

3. Molecular Orientation of Gases Adsorbed on Graphite

The last type of experiment is the studying of molecular orientation, by a technique that was developed in our Lab.[18] This type of experiment uses the very hgh monochromaticity of the incident beam and enables us to determine the orientation of small amounts of adsorbed gases such as NO or N_2, which have a concentration of ~ 1/150 that of the C-atoms constituting the adsorbing graphite surfaces. The n-diffraction method normally used for studying the structure of such systems[21] is not very sensitive to such small amounts of adsorbed gases. Thus we used our technique in which the 6.3 MeV line obtained from Cr capture γ-rays is scattered resonantly from the 6.324 MeV level of ^{15}N.[18] The scattering cross section σ_S is strongly dependent on the kinetic energy component of the N atom (including the zero point vibrational motion) along the incident photon beam direction. This dependence is due to the fact that σ_S is determined by the overlap between two Doppler-broadened shapes, the incident γ-line and the resonance level.[11] The dependence of the calculated σ_s is almost linear versus the effective temperature.[18] The kinetic energy along the molecular axis in both molecules NO and N_2 is higher than in the perpendicular direction because of the vibrational energy along the binding force of the two atoms. That means a broadened Doppler shape which implies a greater overlap and higher σ_s along the molecular axis. The effective temperature can be calculated in these cases if we consider that we have a system of only six degrees of freedom: three translational, two rotational and one vibrational. Using the vibrational frequency which is well known from infrared measurement (2281 cm^{-1} for $^{15}N_2$ and 1844 cm^{-1} for ^{15}NO) we find that T_{eff} is linear in T (and hence in σ_s):

$$T_{eff} = 5/6\, T + 274 \quad \text{for } ^{15}N_2 \text{ gas} \tag{9}$$

$$T_{eff} = 0.83T + 228 \quad \text{for } ^{15}NO \text{ gas}^{20} \tag{10}$$

If the molecules are forced to lay on the graphite planes we shall have two equations for T_{eff}. The effective temperature T_a parallel and T_c perpendicular to the graphite planes:

$$T_a = 3/4T + 411 \tag{11}$$

$$T_c = \sum_{j=1}^{2} \frac{h\nu_i}{2k} \frac{1}{e^{h\nu/kT}-1} + \frac{1}{2} \approx T \quad \text{for } ^{15}N_2 \tag{12}$$

where ν_1, ν_2 are explained elsewhere.[19]

Hence, a measurement of the scattering cross sections σ_a and σ_c (with the incident photon beam parallel or perpendicular to the graphite planes or respectively) can be used as a very sensitive monitor of the angle between the line joining the two atoms and the direction of the incident beam. Experimentally 350 mg of enriched $^{15}N_2$ gas was enclosed in a thin-walled stainless steel cylinder containing ~ 50g of parallel sheets of partially oriented graphite, grafoil, and corresponds to a coverage of ~ 1 monolayer at T = 78K. We measured the cross sections $\sigma_a(T)$, $\sigma_c(T)$ versus the temperature in the range of 16K to 300K using a Displex cryostat. The results are shown in Fig. 7. For T>160K σ_a and σ_c coincide and obey the linear relation (Eq. 9). The deviation between σ_a and σ_c begins at ~160K and increases towards lower T.

So even in the vapor phase, namely at 85K < T < 160K the adsorbed N_2 molecules have a prefered orientation with the molecular axes parallel to the graphite planes. The free rotation freezes and librational mode of vibration out of plane, is excited. At low temperatures, T ≤ 30°K, the results indicate that the N_2 molecular axis aligns itself nearly parallel to the grafoil plane in fair agreement with measurement of neutron diffraction and low-energy electron diffraction. The cross section ratio R = σ_a/σ_c at 16K was measured to be R=1.67 supporting the above conclusion.

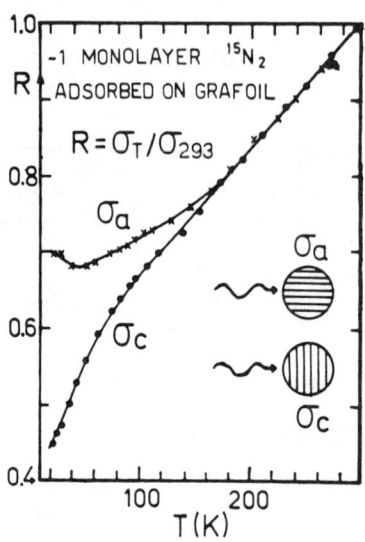

Fig. 7. Measured scattering cross sections σ_a and σ_c versus T for a one monolayer of $^{15}N_2$ adsorbed on Grafoil. The N_2 molecules from a commensurate $(\sqrt{3} \times \sqrt{3})30°$ structure on graphite.

Next we looked into the nitric oxide-adsorbed-graphite system. For this system a complicated phase diagram is suggested in the literature[21] based on neutron diffraction results. β, γ and δ solids were reported to be formed as function of coverage and temperature.[21] The orientation of NO axis in β and γ solid were reported to be parallel to the graphite planes (NO in the solid phase forms a dimer N^2O_2).

We used the nuclear scattering technique to test the suggested 2D crystallographic structure of nitric oxide adbsorbed on graphite. The orientation of NO in both the gaseous and liquid phases was also studied. A similar target container with grafoil was filled with enriched NO (in ^{15}N). The cross section ratio $R = \sigma_a/\sigma_c$ ws measured for the coverage 0.22-0.73 monolayers and in the temperature range 10-300K. Our results indicate that not all the NO molecules lay parallel to the surfaces since $R \leq 1$ for all the coverages tested meaning a wide disagreement with previous measurement.[21] The average vibrational energy per degree of freedom of nitrogen in N_2 and NO molecules can be deduced from the cross section indicating a much higher zero-point energy of the graphite binding potential in NO compared to N_2.

REFERENCES

1. R. Moreh, S. Shlomo and A. Wolf, Phys. Rev. C2, 1144 (1970).
2. A. Veyssiere, H. Beil, R. Bergere, P. Carlos, and A. Lepretre, Nucl. Phys. A159, 561 (1970).
3. R. D. Starr, P. Axel and L. S. Cardman, Phys. Rev. C25, 780 (1980).
4. G. Kuhner et al., Phys. Lett. B104, 189 (1981).
5. N. K. Sherman, H. M. Ferdinande, K. H. Lokan, and C. K. Ross, Phys. Rev. Lett. 35, 1215 (1975).

6. Y. Birenbaum, Z. Berant, S. Kahane, A. Wolf, and R. Moreh, Nucl. Phys. A369, 483 (1981).
7. S. Kahane and R. Moreh, Phys. Rev. C9, 2384 (1974).
8. Z. Berant, R. Moreh and S. Kahane, Phys. Lett. 69B, 281 (1977).
9. Z. Berant and R. Moreh, Phys. Lett. 73B, 142 (1978).
10. S. Kahane, R. Moreh and O. Shahal, Phys. Rev. C 18, 1217 (1978); Phys. Lett. 66B, 229 (1977).
11. D. A. Liberman, D. T. Cromer and J. T. Waber, Comput. Phys. Commun. 2, 107 (1971).
12. S. Kahane and T. Bar-Noy, Nucl. Phys. A288, 132 (1977).
13. P. Papatzacos and K. Mork, Phys. Rev. D 12, 266 (1975).
14. V. Constantini, B. de Tollis and G. Pistoni, Nuovo Cimento 2A, 733 (1971).
15. B. Arad and G. Ben-David, Rev. Mod. Phys. 45, 230 (1973).
16. R. Moreh, Nucl. Instrum. and Methods 166, 91 (1979).
17. E. G. Fuller and E. Hayward, Nucl. Phys. 30, 613 (1962).
18. O. Shahal and R. Moreh, Phys. Rev. Lett. 40, 1714 (1978).
19. R. Moreh and O. Shahal, Phys. Rev. Lett. 43, 1947 (1979).
20. R. Moreh and O. Shahal, to be published.
21. J. P. Coulomb, J. Suzanne, M. Bienfait, M. Matecki, A. Thomy, B. Corset, and C. Marti, J. Physique 41, 1155-1164 (1980).

PROTON-CAPTURE GAMMA RAYS AS A SPECTROSCOPIC TOOL

Ph.B. Smith
Laboratorium voor Algemene Natuurkunde, University of Groningen
9718 CM Groningen, The Netherlands

ABSTRACT

Proton-capture gamma rays can be used to induce nuclear reactions at a well-defined energy. The total intensity is considerably less (by a factor of at least a hundred) than the strong monoenergetic gamma rays from high-flux nuclear reactors exploited for this purpose. On the other hand the energy of the gamma rays, depending on the target mass and the energies, can be varied over a range from 20 to 80 keV by changing the angle of observation with respect to the proton beam. This means that specific levels can be excited at will. Experiments done to date using these gamma rays are reviewed. These are resonance self-absorption and fluorescence (capturing nucleus the same as resonating nucleus), resonance cross-absorption and fluorescence (where use can be made of the polarization of capture gamma rays) of bound levels, and resonance absorption on unbound levels. A discussion of the relative merits of these techniques, in comparison with other methods of obtaining the same information, is given. It is concluded that the greatest promise for future experiments lies in the last category mentioned.

I. INTRODUCTION

If we limit the discussion to discrete gamma-ray sources we may say that up until 1957 the resonance fluorescence of nuclear gamma rays had only been made observable by <u>line broadening</u> from either temperature motion or recoil motion from previous emissions or reactions and by <u>line shifting</u> by means of centrifuges. These methods and the cases studied are covered in older review articles[1,2].

Almost simultaneously in 1956/57 two other methods were found to provide exact resonance and complete overlap: the best-known is recoilless emission and absorption (Mössbauer), and the other is the unattenuated Doppler shift available after the two-body process of particle capture.

The energy resolution ($\Delta E/E$) of this second process is at best 10^{-5}. This is very good compared to other methods of gamma-ray spectroscopy but astronomically poor compared to Mössbauer spectroscopy. For this reason there have been few applications in condensed matter physics. The only solid-state parameter obtainable (from the resonance absorption integral) is $<v^2>$. Since this parameter is physically different from the parameters obtained in Mössbauer spectrometry, there is the possibility that certain solid-state experiments might be worthwhile. For lack of space they will not be discussed in this paper.

Resonance absorption and fluorescence where the emitting and

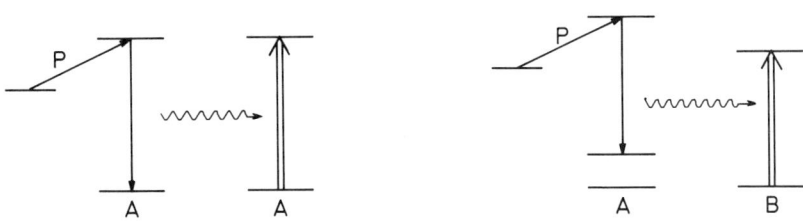

Fig. 1. Schematic representation of the processes discussed in this paper.

absorbing nucleus are the same (Fig. 1a) is limited to the small number of cases where the target and product nucleus are stable and the excited level has a reasonable partial decay width to the ground state (in practice $\Gamma_{\gamma_0} \gtrsim 0.1$ eV).

Only recently has the door been opened to more than incidental application of "cross-excitation", i.e. the excitation of a level in one nucleus by a gamma ray from another (Fig. 1b). This more general application has been stimulated by the greater precision in excitation energies and gamma-ray energies now available (unless both are known with keV precision finding an absorption dip is a "needle-in-the-haystack" search). A further stimulus is the systematic registration of intense gamma rays carried out in Utrecht by van 't Westende et al.[3]. A review of the use of resonance absorption to determine the transition widths of nuclear states is given by these authors. For cross excitation the gamma ray does not, of course, have to come from a ground-state transition, but for all practical purposes it must be a primary. "Coverage" is not complete, but van 't Westende[3] states that reasonably intense gamma rays are available for 90% of the energy region between 6 and 10 MeV.

In the present review the accent falls on the usefulness of particle-capture gamma rays in the excitation with high-resolution of specific interesting levels. Not only resonance absorption, but in particular, resonance fluorescence, will be treated. In the case of fluorescence the linear polarization of capture gamma rays makes the determination of the parity of the resonating level possible.

In the following section the question of precision and reliability of resonance absorption and fluorescence measurements will be treated in relation to measurement and analysis techniques, and in the last section a review of the applications of resonance absorption and fluorescence will be given.

II. PRECISION AND RELIABILITY

IIA. RESONANCE ABSORPTION

The charm of any resonance absorption method is that the results are in principle absolute, i.e. independent of background or of instrumental and non-resonant effects. This led to a spirit of euphoria among those who had first thought of the method (S.S. Hanna, L. Meyer-Schützmeister, and the present author). As time went on, however, it became apparent that it was quite possible to produce erroneous results. Now, a generation later, it is possible to take stock of the method. It may now be credibly claimed that absolute values of level parameters can be determined with a precision of approximately five percent for bound levels (e.g. Ref.4), if one keeps to the rules of the game. These rules are the subject of the following two paragraphs.

IIA.1. EXPERIMENTAL TRIBULATIONS

If trivial errors are excepted, the only things that can go wrong are in the background determination and the monitoring. Frequently intensity considerations dictate the use of a NaI detector. This means that the spectrum must be checked with a Ge-Li detector in order to know which gamma rays are being detected by the NaI detector. But even if a Ge-Li detector is used there is still danger of contamination <u>within</u> the peak by gamma rays from a nearby resonance (say 3 or so keV away). Even if at the beginning of a measurement this contamination is carefully ruled out, and even under the best conditions of target vacuum and proton energy stability, the bombardment of the target during a measurement (of the order of 1000 C/cm²) will lead to material damage and deterioration of the sharpness of the yield curve. Changing targets or target spots reduces this damage, but even 10 C/cm² causes serious damage. If a nearby resonance has a line of (practically) the same energy, the fuzziness of the yield curve, caused by the damage, will lead to a slow growth of contamination. A good example of such a case is to be found in the dissertation of J.W. Maas[5] in the resonance absorption measurement on the 11.44 MeV, $J^\pi = 1^+$, $T = 1$ level in ^{28}Si. Repetition of this measurement under better controlled conditions by Elsenaar*, eliminated this error (see Table II).

As far as monitoring is concerned, it has been the custom in Groningen to mount the monitor on a movable arm at 180° to the collimator (Fig. 2a), thus compensating reasonably well for the angular distribution. A serious error was introduced by the use of this technique in the thesis work of W. Biesiot because a thick piece of iron was placed before the monitor to reduce the counting rate. As it turned out there was a resonance in ^{56}Fe at the supplement of one of the angular regions covered, so that a broad <u>peak</u> appeared in the results. This led to an erroneous interpretation

* unpublished

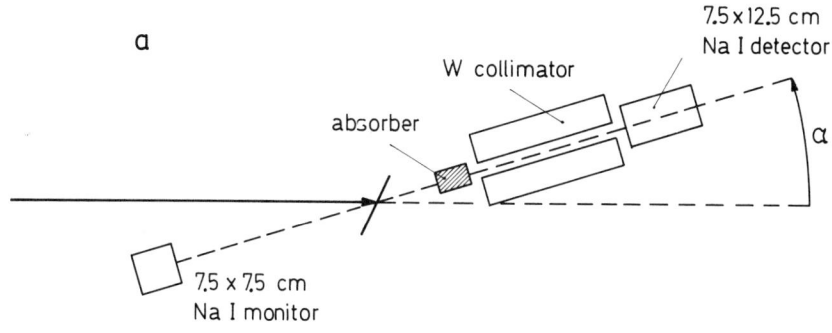

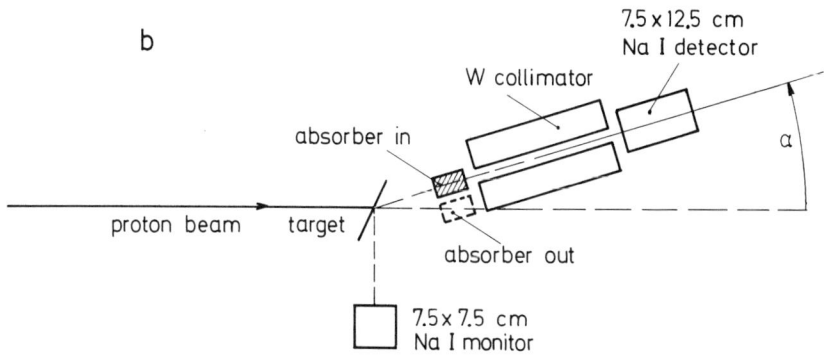

Fig. 2. Typical set-ups for resonant absorption experiments.

in the dissertation[6], since then corrected[7]. Incidentally, the work on ^{56}Fe mentioned earlier[4], was based on this "accident".

Both of the above mentioned errors were avoidable. But so are all errors. No one has yet encountered weird effects because of resonances in sodium, or iodine, or germanium, but one must always be on one's guard. The possibility of error in the monitoring has been practically eliminated in recent work at Groningen[7] by the adoption of the set-up shown in Fig. 2b. Each point is measured (in fact, many times) with the absorber in and with absorber out, as in neutron absorption work, to measure the true transmission of the absorber. This practice, however, increases the measurement time by 30 to 40% for the same statistical accuracy, which is not pleasant for measuring times already are of the order of 100 hours.

IIA.2. DIFFICULTIES IN ANALYSIS

IIA.2.a. THE RESONANCE ABSORPTION INTEGRAL

At first sight it appears that the determination of the resonance absorption integral is very easy. That this has frequently not been the case is due to inadequate curve fitting procedures, and too little attention to the obtainment of adequate statistics in the "background" region (far in the wings of the dip). This second point is one of experimental judgement, since the optimum experiment can only be designed after it has been done and the results have been analyzed.

The general equation for the resonance absorption at the angle α is:

$$A(\alpha) = \frac{\int dE\, f(E - E(\alpha))\{1 - \exp[-n\sigma_o \psi(x,\beta_T)]\}}{\int dE\, f(E - E(\alpha))} \quad (1)$$

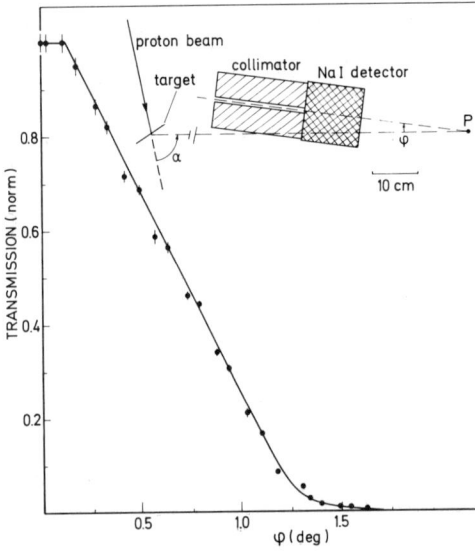

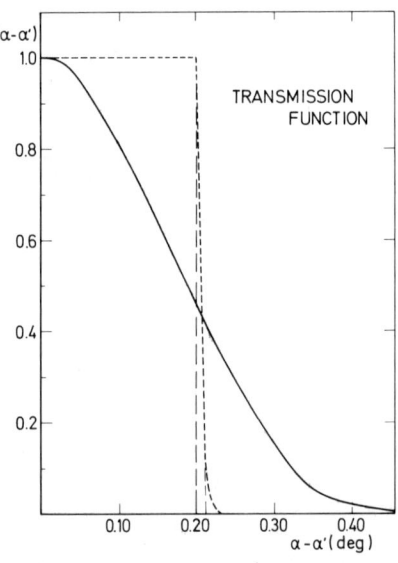

Fig. 3. Experimental and calculated results for the collimator transmission function, using the set-up shown in the inset.

Fig. 4. The dashed line gives the calculated transmission function of the collimator calculated as in Fig. 3. Two mathematically simpler variants are also shown which give undistinguishable fits to the data. The solid curve is a gaussian chosen such that the enclosed area equals the area under the trapezoid (see text).

Here $E(\alpha)$ is the nominal energy of the gamma rays passing through the collimator at angle α, $f(E - E(\alpha))$ the instrumental function, $x = (E - E_r)/(\Gamma_a/2)$, $\beta_T = 2\Delta/\Gamma_a$, and $\Delta = E_\gamma(2T_{eff}/Mc^2)^{\frac{1}{2}}$. The total width of the absorber level is Γ_a, the resonance energy is E_r, and $\sigma_0 = 2\pi \lambdabar^2 \{(2J + 1)/(2J_0+1)\}(\Gamma_{\gamma_0}/\Gamma_a)$. The ψ-function is the well-known convolution of a Breit-Wigner form and a temperature gaussian. It can be expressed as:

$$\psi(x,\beta_T) = \frac{\sqrt{\pi}}{\beta_T} \operatorname{Re} W(z) \tag{2}$$

with $z = x/\beta_T + i/\beta_T$ and $W(z)$ the complex error function, available in general, as a standard computer function.

There are many contributions to the instrumental function f. Biesiot[8] showed that, if we leave out the emitter function for the moment, a gaussian is better than the trapezoid which one would predict purely on the basis of the collimator transmission function. He was able to do this because we knew the other contribution (the Breit-Wigner width) very well from precision elastic proton scattering measurements. The gaussian form is not illogical, since other factors such as temperature motion of the target and (perhaps shifting) divergence of the beam will always tend to spread the trapezoidal function into a gaussian. In Fig. 3 is shown that the collimator transmission function agrees very well with what one calculates. In Fig. 4 are shown the trapezoidal function which would follow if the collimator were the only contributing factor to the instrumental function, and the gaussian which was finally used. The correct instrumental function is then the convolution of this gaussian and the Breit-Wigner form of the emitter line, again giving a complex error function.

In all of the work done up to the present on bound levels the instrumental function is at least ten times as broad as the (temperature spread) absorption function. The absorption can then be treated as a delta function and Eq. (1) takes the form:

$$A(\alpha) = \frac{1}{\sqrt{\pi}\,\gamma} A_\alpha(n\sigma_0, \Gamma_a, \beta_T) \operatorname{Re} W(z') \tag{3}$$

Here $z' = x/\beta_c + i/\beta_c$, $x = (E - E(\alpha))/\Gamma_e/2)$, $\beta_c = 2\gamma/\Gamma_e$, γ = FWHM of the instrumental gaussian (expressed in energy units) divided by $2\sqrt{\ln 2}$, and Γ_e the width of the emitter line. This last parameter must be known in order to get reliable results. It can in practically all cases be found from proton elastic scattering measurements. The resonance absorption integral A_α is the well-known integral:

$$A_\alpha = \Gamma_a \int_0^\infty dx\{1 - \exp(-n\sigma_0 \psi(x,\beta_T))\} \tag{4}*$$

In the fitting procedures now used in Groningen and Utrecht

---

* In Ref. 4 this equation is in error (the fore-factor is erroneously given as Γ_{γ_0}, instead of Γ_a).

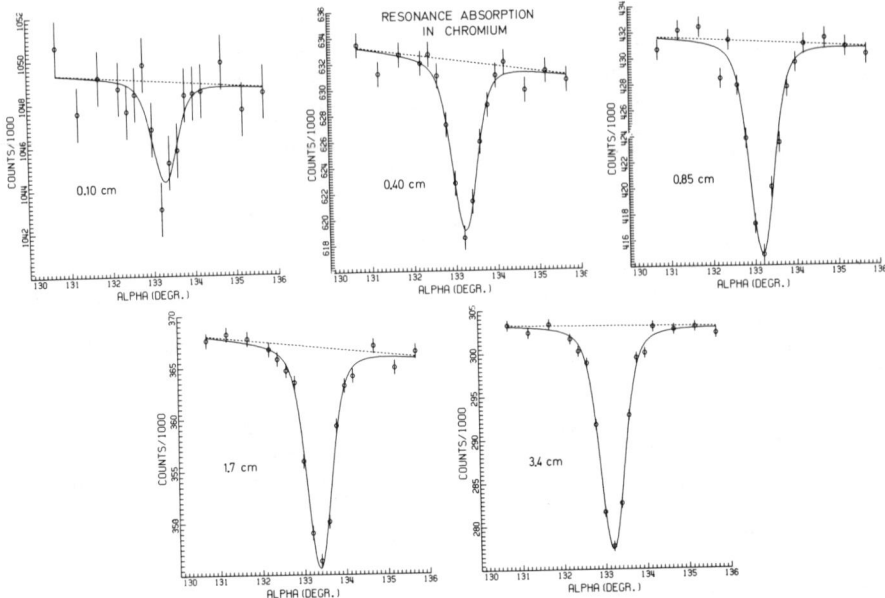

Fig. 5. Fitted results of 5 resonance absorption measurements on chromium. The curves are (single) convolutions of a Breit-Wigner emitter line and a gaussian instrumental function. The absorption for this bound level is treated as a delta function.

Eq. (3) is used for bound (i.e. narrow) levels. In much work in the past a gaussian form has been used. This can certainly not be justified with the argument that it is tedious to use the complex error function. For the programmer it makes no difference, and the computation time for simultaneous fitting of five transmission curves, such as shown in Fig. 5, with a total of 85 points is only 3 seconds. The fitted value of the gaussian width, γ, was constrained to be the same for all curves, but further the individual background parameters, A_α's and dip centres were left free.

Nor can the use of a gaussian form be justified with the argument that "you can't see the difference". The traitorous wings of a Lorentz curve contribute a great deal to the area, and if you know that they are there, there is no excuse to leave them out.

Even if the Doppler width is very small compared to the level width Γ_a, the analysis can be tedious, if the absorption is large. This explains why an almost factor-of-two discrepancy in the total width of the 9.17 MeV level in ^{14}N persisted more than 20 years in the literature. In 1959 Hanna and Meyer-Schützmeister[9] found a value of 77 ± 12 eV, and in 1981 Biesiot[8] a value of 135 ± 11 eV. This latter value agrees with the independent results of (p,p) work (135 ± 8 eV). In Fig. 6 are shown the transmission curves in the self-absorption measurement of Biesiot[8] on this level. The fits

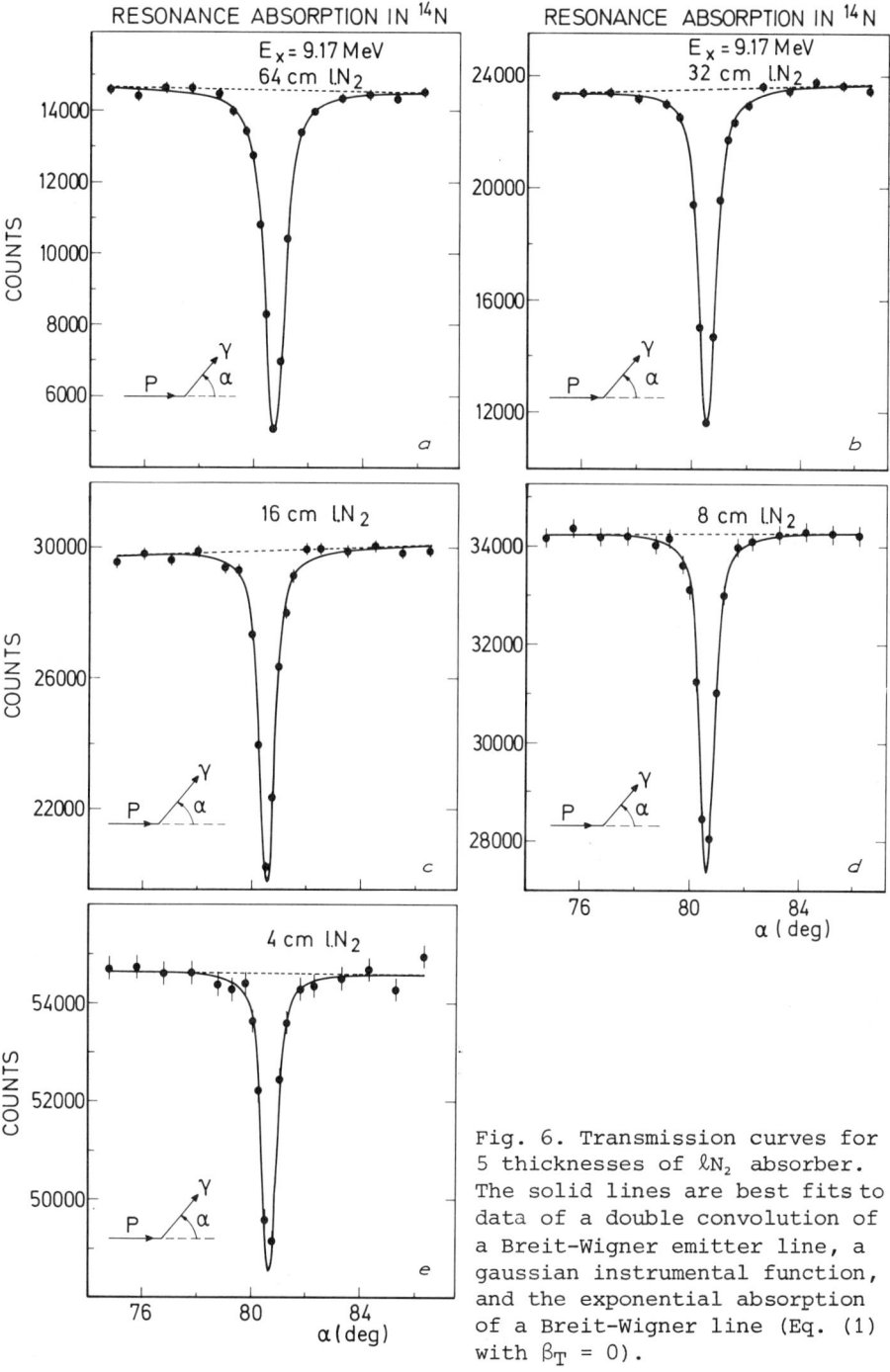

Fig. 6. Transmission curves for 5 thicknesses of ℓN_2 absorber. The solid lines are best fits to data of a double convolution of a Breit-Wigner emitter line, a gaussian instrumental function, and the exponential absorption of a Breit-Wigner line (Eq. (1) with $\beta_T = 0$).

were made using Eq. (1) in which the ψ function is replaced by its limiting value for $\beta = 0$, i.e. $(1 + x^2)^{-1}$. But even with this simplification the numerator is a double, numerical convolution, which hardly could have been carried out in 1959; even in 1980 the fitting process took hours of computing time.

Since the actual value found for the fitted width has a substantial effect on the value of σ_o and thus on the non-linearity of A_α vs n,[10] in this special case a check for internal consistency is possible. In ref. 8 it is shown that a width of around 75 eV cannot possibly lead to consistent results, whereas a width of 135 eV gives excellent consistency.

The point here is to underline that sophisticated analysis codes are a necessary part of the "equipment" needed to produce reliable and precise results in resonance absorption work.

IIA.2.b. ARRIVING AT THE FINAL RESULTS

Once a set of values of A_α for different absorber thicknesses has been found, as described in the previous section, Eq. (4) can be used to extract the value of Γ_{γ_0} and β_T. An example of this type of fit is shown in Fig. 7a for the 9.14 MeV $J^\pi = 1^+$ level in ^{52}Cr. The value of Γ_{γ_0}, 2.68 ± 0.16 eV ($B(M1)\downarrow = 0.302 \pm 0.018\ \mu_N^2$), was the principal result here. In Fig. 7b is shown that it is quite impossible to fit the absorption obtained with another isotope (^{53}Cr, with a natural abundance in the absorber of 9.5%). The ratio $\Gamma_{\gamma_0}/\Gamma_a$ must be known in order to get unique and reliable results. A separate fluorescence measurement is in general necessary for this

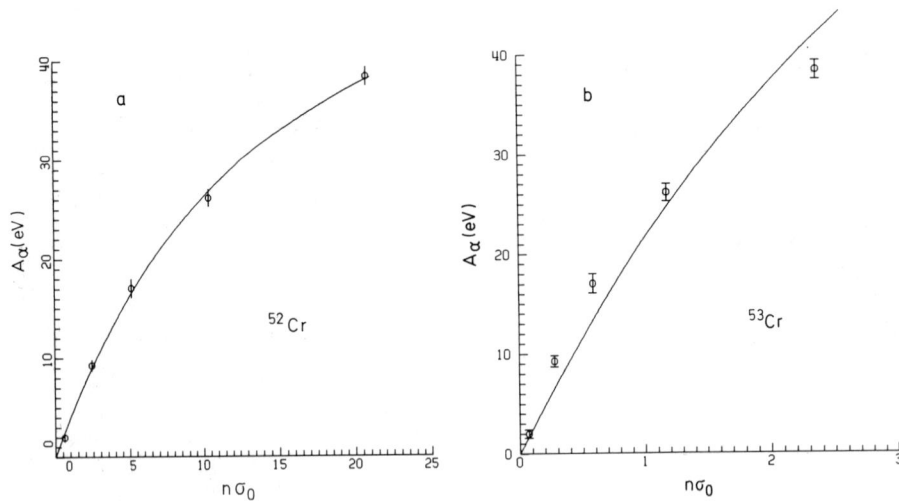

Fig. 7. Fits to Eq. (4) of the values of A_α found from the data of Fig. 5 to Γ_{γ_0} and β_T for ^{52}Cr (reduced χ^2 is 0.6) and ^{53}Cr (reduced χ^2 is 13) showing that ^{52}Cr is indeed the resonating nucleus.

purpose,[4] unless the level can be excited in a particle reaction.

The object of this section has been to outline the requirements which must be met in order to be able to rely on resonance absorption measurements at the level of 5%. It turns out that in order to reach this level much more work is necessary than has been spent on these measurements, in most cases, up to now. In particular, values of Γ_e and $\Gamma_{\gamma_0}/\Gamma_a$ should be known. Furthermore, for bound levels especially, a considerable range of absorber thicknesses should certainly be measured. Finally, the fitting codes must be quite sophisticated in order to extract precise results from precise data. These rigorous requirements can, of course, be relaxed in some cases, mostly unbound levels, where the absorption is so small that $n\sigma_0 \ll 1$. The absorption integral then becomes linear in Γ_{γ_0}, and usually the total width of the absorption curve, which then becomes $\Gamma_a + \Gamma_e$ is considerably greater than the gaussian part of the instrumental function so that the latter can be fixed in the analysis at a (not critical) value learned from experience with various levels. So both $(2J+1)\Gamma_{\gamma_0}$ and Γ_a are easy to determine. The precision depends entirely on counting statistics, and not much on the analysis (see section IIID).

IIB. RESONANCE FLUORESCENCE

It is assumed that the level parameters with the exception of the parity have been determined by resonance absorption and other spectroscopic methods. The sign of the azimuthal asymmetry of

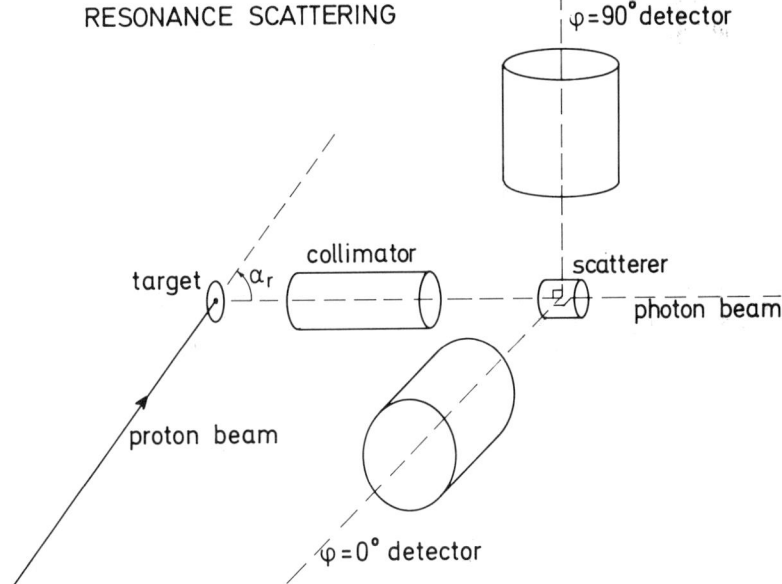

Fig. 8. Schematic drawing of the set-up for resonance fluorescence. The entire collimator and detection system rotates as a whole about the target.

resonantly scattered gammas is then the only quantity needed to determine the parity. A thorough discussion of the mathematical background of this type of measurement can be found in refs. 11 and 12. Our experimental set-up is shown in Fig. 8. Measurements are performed by measuring at α_r (the resonant angle), and two degrees to the left and to the right of α_r, where the gamma ray is no longer resonant. The chief experimental, and thus analytical, difficulty arises from the fact that there is no cylindrical symmetry about the axis of the scatterer (intensity considerations dictate the use of collimator <u>slit</u>, not a round hole as would appear from the figure. In measurements on the resonance scattering of the 9.17 MeV line in ^{14}N this was not important[8] because the electronic absorption in the BN scatterer was negligible. Later, when unique parity assignments were made[4] on levels in ^{52}Cr and ^{56}Fe a computer code was available which calculated the asymmetry. This was not available when resonance fluorescence was measured on the 4.84, 7.06 and 7.08 MeV lines of ^{208}Pb.[11] Fortunately we did have an unpolarized source of resonant gammas from a $J = \frac{1}{2}$ resonance with which to calibrate the asymmetry of the set-up. Resonance for both members of the doublet could be achieved with one gamma ray from the ^{34}S(p,γ) reaction at E_p = 2.54 MeV, at the widely distinct angles of 130° and 74°, where the polarization, P = 0.37. Both levels were clearly 1⁻, with more than three times the error bars separated from the 1⁺ solution. Although it had been clear from the work at Illinois in Axel's group[13] that even though they could not resolve the members of the doublet that both had to be 1⁻, it was reassuring to see that we could resolve the levels completely and prove the negative parity of each one independently.

The object of this paragraph is not, however, to tell how good the method is, but to define the requirements for achieving its potentialities of model-free, absolute, results. And that means that the failure to give a statistically significant answer to the parity of the 4.84 MeV level must be commented upon. Berg[12] has clearly shown that this level has negative parity, yet our results <u>indicated</u> (not at the level of an <u>assignment</u>, however) positive parity. The lesson to be learned from <u>this</u> is that although measured asymmetry factors may be reliable under identical conditions, they cannot be used too far away from the energy where they were measured. Furthermore the mistake was made of not taking sufficiently into account the uncertainty in P (calculated as 0.23). In general, with errors of the order of 50% it is probably unwise to present any results at all.

In conclusion, we believe that this technique can be made reliable. Its unique charm is the possibility of destroying the resonance condition by turning the whole system a couple of degrees away from the resonant angle so that even with a signal-to-noise ratio of 0.2 it is possible to be sure that you know what you are doing. But a good computer code to calculate the asymmetry of the set-up is absolutely necessary. As mentioned in Ref. 11 a modest expenditure on better passive shielding from the direct beam and active shielding from cosmic rays would greatly enhance the capabilities of the Groningen set-up.

Table I. Self-absorption measurements since 1964

nucleus	E_p (keV)	E_γ (MeV)	Γ_t (eV)	Γ_{γ_0} (eV)	Ref.
^{14}N	1750	9.17	135 ± 5	7.2 ± 0.4	8*
^{14}N			72 ± 10	8.9 ± 1.2	14
^{27}Al	1785	9.99	1.2 ± 0.3	0.6 ± 0.2	16
^{27}Al	1965	10.16	23 ± 3	0.6 ± 0.2	16
^{27}Al	1965	10.16	23 ± 5	0.8 ± 0.3	17
^{28}Si	733	12.33	11.2 ± 0.9	−	**†
^{31}P	1481	8.72	100 < Γ < 25	0.75 ± 0.05	18
^{31}P	1482	8.72	100 < Γ < 25	0.35 ± 0.06	18
^{35}Cl	1891	8.21	39 ± 5	0.70 ± 0.09	19
^{35}Cl	2791	9.08	65 ± 20	1.38 ± 0.16	20

* see also Refs. 9 and 15.
** R.J. Elsenaar, unpublished
† see also Ref. 10.

Table II. Cross-absorption measurements on bound levels

target	E_p (keV)	absorber	E_γ (MeV)	Γ_{γ_0} (eV)	Ref.
^{23}Na	1416	^{11}B	8.92	5 ± 0.5	*
^{30}Si	1830	^{23}Na	4.43	1.9 ± 0.4	21
^{30}Si	620	^{23}Na	7.89	3.0 ± 0.2	21
^{27}Al	1684	^{28}Si	11.44	9 ± 2	5
^{27}Al	1684	^{28}Si	11.44	23 ± 3	*
^{13}C	1750	^{52}Cr	9.14	2.68 ± 0.16	4
^{13}C	1750	^{56}Fe	9.14	1.28 ± 0.17	4
^{34}S	1680	^{208}Pb	4.84	$4.3^{+3.5}_{-1.4}$	11
^{34}S	1974	^{208}Pb	7.06	18 ± 3	22
^{34}S	1974	^{208}Pb	7.06	17 ± 3	11
^{34}S	2540	^{208}Pb	7.06	19.5 ± 1.7	11
^{34}S	2540	^{208}Pb	7.08	9.1 ± 1.3	11

* R.J. Elsenaar, unpublished.

III. APPLICATIONS UP TO THE PRESENT

The measurements performed up to the present using resonance reactions with capture gamma rays are summarized in this section.

IIIA. RESONANCE SELF-ABSORPTION

From simple non-relativistic kinematics the cosine of the angle of resonance of this process, shown in Fig. 1a, is:

$$\cos \alpha_r = E_o/(2E_p m_p c^2)^{\frac{1}{2}} \qquad (5)$$

Typical experimental set-ups are shown in Fig. 2. In an earlier review article[2] older work of this type is listed. In Table I is given a list of resonance absorption experiments published since 1964. A number of unpublished results are known to us, including the only example of this type of experiment ever done with the (d,γ) reaction. Hanna measured the resonance absorption of the 16.98 MeV level in ^{9}Be (second T = $^3/_2$) level using the ^{7}Li(d,γ) reaction at E_d = 365 keV. This has been repeated very recently in Utrecht in order to improve the accuracy, and the (preliminary) results will be reported at this conference.

IIIB. RESONANCE ABSORPTION ON BOUND LEVELS

As mentioned in the introduction this application should provide in most cases, much more interesting possibilities than the self-absorption experiments. Results of all work done are summarized in Table II. The problem of finding a gamma ray to excite a particular, interesting level has now become easier. In comparison with (e,e') and/or electron Bremsstrahlung this method has the advantage of being entirely free of model-dependent factors. The fact is, though, that one must conclude that this advantage is now almost academic, considering the extremely high reliability reached in contemporary electron work. The high-precision achieved in the measurement on the 9.14 MeV level in ^{52}Cr can serve, however, as an absolute calibration. Only in the case of very closely-spaced levels could our sub-keV resolution provide a unique approach.

IIIC. RESONANCE FLUORESCENCE ON BOUND LEVELS

In Table III are summarized the measurements done up to the present. As mentioned in section IIB, the separate assignment of the parity of the 7.08 MeV level is the only special achievement here. This was later confirmed by the Giessen group[23]; this shows that in the case of resonance fluorescence it is improbable that for bound levels this method can compete with the polarized Bremsstrahlung technique.

Table III. Resonance fluorescence measurements

target	E_p (keV)	absorber	E_γ (MeV)	unique parity assignment	Ref.
^{13}C	1750	^{14}N	9.17	*	8
^{13}C	1750	^{52}Cr	9.14	yes (+)	4
^{13}C	1750	^{56}Fe	9.14	yes (−)	4
^{34}S	1680	^{208}Pb	4.84	no	11
^{34}S	2540	^{208}Pb	7.06	yes (−)	11
^{34}S	2540	^{208}Pb	7.08	yes (−)	11

* Since this is resonance self-fluorescence the parity enters twice and thus cancels out.

IIID. RESONANCE ABSORPTION ON UNBOUND LEVELS

The only work done up to the present is shown in Fig. 9. It is the author's expectation that it is in this region that proton-capture gamma rays will deliver the most important contribution.

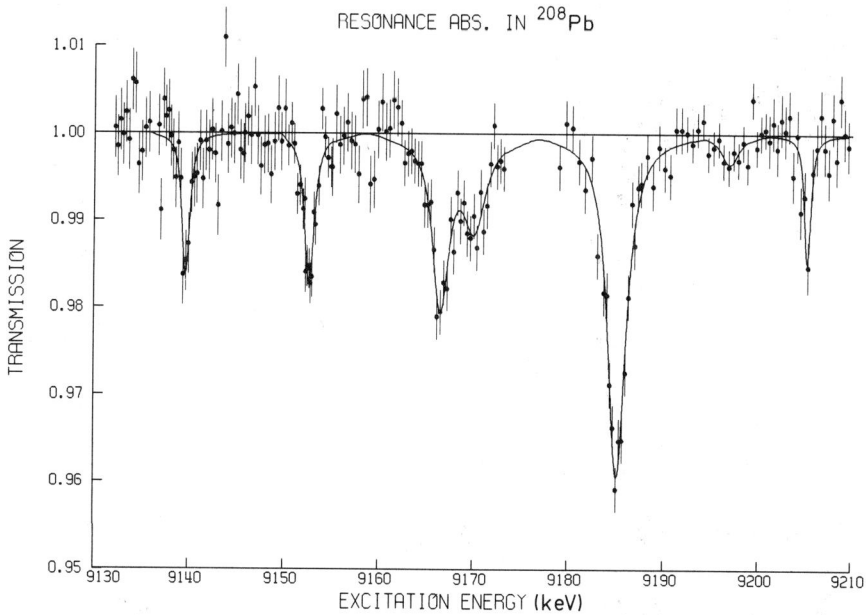

Fig. 9. Resonance absorption data on levels in ^{208}Pb in the excitation region from 1760 keV to 1830 keV equivalent ground-state neutron energy. The solid lines are fits to the data of a convolution of a Breit-Wigner line (with a width equal to the sum of the emitting and absorbing line width, since $n\sigma_0 \ll 1$) and a gaussian instrumental function.

Since resonance fluorescence is practically speaking impossible, one must look to (e,e') or hadron reactions for competing techniques. There are indications that the results from (e,e') and (p,p') do not agree[24]. This might mean that studies on specific levels with the high-resolution and model-free interpretation of proton capture gamma rays may provide answers to questions not obtainable otherwise.

A case in point is the data shown in Fig. 9. It is clear from the figure that there is little overlap in this region of excitation of ^{208}Pb. of levels with strong ground-state transitions (in fact, $D/<\Gamma> \sim 7$). There is, however, one clear case of overlapping levels and this can be identified with a doublet for which Laszewski[25,26] found interference in photoneutrons, which must come from an E1 resonance interfering with another multipolarity. This could, in principle, be M1 or E2. Independently of whether in this case it is M1 or E2, the fact remains that we can conclude that since in several cases there is evidence[26] of another multipolarity than E1, and yet there are some five to ten times as many non-overlapping levels for which no polarization would have been seen, we may expect with high-resolution work to uncover a great deal more non-E1 strength than has been seen up to the present.

How this can be done is a mute point. One suggestion is that if $(\vec{\gamma},n)$ could be measured with the polarized photons from capture reactions this could be as effective for unbound levels (and as model-independent) as $(\vec{\gamma},\gamma)$ is for bound levels. Some considerable development work would be required for this, however.

REFERENCES

1. F.R.Metzger, In: Progress in Nuclear Physics, vol. 7, pp. 53-87, Pergamon Press: London-New York-Los Angeles, 1959.
2. K.G. Malmfors, In: Alpha-, Beta- and Gamma-Ray Spectroscopy, (K. Siegbahn, ed.), North Holland Publ. Co. Amsterdam, 1965.
3. A.P.M. van 't Westende, H. Lancman, and C. van der Leun, Nucl. Instr. Meth. 151, 205 (1978).
4. Ph.B. Smith and W. Segeth, Nucl. Phys. A398, 397 (1983).
5. J.W. Maas, Thesis University of Utrecht, 1976.(unpublished)
6. W. Biesiot, Thesis University of Groningen, 1980. (unpublished)
7. Ph.B. Smith, submitted for publication in Phys. Rev. C.
8. W. Biesiot and Ph.B. Smith, Phys. Rev. C 24, 2443 (1981).
9. S.S. Hanna and L. Meyer-Schützmeister, Phys. Rev. 115, 986 (1959).
10. P.B. Smith and P.M. Endt, Phys. Rev. 110, 397 (1958).
11. W. Biesiot and Ph.B. Smith, Phys. Rev. C 24, 808 (1981).
12. U.E.P. Berg, Proc. Int. Symp. on Highly Excited States and Nuclear Structure, Orsay, 5-8 Sept. 1983.
13. A.M. Nathan, R. Starr, R.M. Laszewski, and P. Axel, Phys. Rev. Lett. 42, 221 (1979).
14. A. Luukko, Commentationes Physico-Mathematicae 31, nr. 6 (1965).
15. S.S. Hanna and L. Meyer-Schützmeister, Phys. Rev. 108, 1644 (1957).
16. P.R. de Kock, J.W. Koen and W.L. Mouton, Annals of Phys. 47, 481 (1968).

17. C. van der Leun and N.C. Burhoven Jaspers, Nucl. Phys. 88, 235 (1966).
18. H. van Rinsvelt and P.B. Smith, Physica 30, 59 (1964).
19. W. Biesiot, Ph.B. Smith, J.L. Stavast, P.B. Goldhoorn, and S. van der Hoek, Nucl. Phys. A359, 149 (1981).
20. R.J. Sparks, Nucl. Phys. A265, 429 (1976).
21. E.L. Bakkum, P.G. Bouwknegt, and C. van der Leun, submitted for publication in Nucl. Phys.
22. R.J. Sparks, H. Lancman, and C. van der Leun, Nucl. Phys. A259, 13 (1976).
23. K. Wienhard, K.A. Ackerman, K. Bangert, U.E.P. Berg, C. Bläsing, W. Naatz, A. Ruckelshausen, D. Rück, R.K.M. Schneider and R. Stock, Phys. Rev. Lett. 49, 18 (1982).
24. D. Bender, G. Eulenberg, A. Richter, E. Spamer, B.C. Metsch, and W. Knüpfer, Nucl. Phys. A398, 408 (1983).
25. R.M. Laszewski, R.J. Holt and H.E. Jackson, Phys. Rev. Lett. 38, 813 (1977).
26. R.J. Holt, H.E. Jackson, R.M. Laszewski and J.R. Specht, Phys. Rev. C 20, 93 (1979).

MEASUREMENT OF THE PHOTODISINTEGRATION OF DEUTERIUM

Y. Birenbaum, R. Moreh[*] and S. Kahane
Nuclear Research Centre - Negev, Beer-Sheva, ISRAEL

INTRODUCTION

Absolute cross sections for the photodisintegration of Deuterium in the 5-12 MeV energy range were measured in several occasions. Most of the measurements were carried out in the fifties[1-3] while the latest, published in 1972, was performed with limited accuracy merely in order to test an experimental arrangement.[4] The earlier measurements were done in rather complicated experimental techniques and suffered from large systematic errors. Most of the measurements had large uncertainty (generally greater than 15%) and with the exception of Ref. 3, largely deviate from theoretical predictions available today.

In this work we present the results of an experimental evaluation at the $D(\gamma,n)$ cross-section in the 6-11.4 MeV energy range, with an accuracy generally better than 3%. The present results are in very good agreement with the latest theoretical calculations which include also meson exchange-corrections (MEC).

THE MEASUREMENTS

The attenuation of a beam of monoenergetic photons in a highly enriched heavy water (D_2O) absorber was compared to that in an ordinary water (H_2O) absorber both contained in stainless steel tubes of comparable length. The lengths of the absorbers were chosen so as to satisfy (as far as possible) the relation $\ell_H/V_H = \ell_D/V_D$ with ℓ and V representing the length and molar volume of the D_2O and H_2O absorbers. Satisfying the above relation minimizes the dependence on "non $D(\gamma,n)$" attenuation being the same in both absorbers. The difference in atomic cross-sections for H and D was considered in a measurement of the $D(\gamma,n)$ cross-section in the 15-25 MeV energy range[5] and was found to be $2\mu b$ for 15 MeV photons, decreasing for lower energies. The characteristics of the two absorbers are listed in Table I.

Table I. Absorber characteristics

	D_2O	H_2O	uncertainty
Length (cm)	199.967	199.300	0.001
Molar Volume (cm³)	18.12863	18.06180	10^{-5}
Temperature (°C)	23.50	23.50	0.25
D_2 in D_2O	0.99778		10^{-5}
^{18}O in D_2O	0.0038		2.10^{-4}
σ_{tot}			2%

[*]Ben-Gurion University of the Negev, Beer-Sheva, ISRAEL

The ~ 2 meter length of the absorbers provided a 1.5-2.5% absorption ratio entirely due to $D(\gamma,n)$ attenuation. The absorption ratio is defined by

$$R \equiv \frac{I_D}{I_H} = \exp\{N_0[(\frac{\ell_H}{V_H} - \frac{\ell_D}{V_D})\sigma_{tot} - 2(1-p)\frac{\ell_D}{V_D}\sigma_D]\}$$

N_0 is the Avogadro number, I_D and I_H are the number of photons detected after passing the D_2O and H_2O absorbers, σ_{tot} and σ_D are the total absorption cross-sections in H_2O and the $D(\gamma,n)$ cross-section in D_2O, respectively. p is the H/D ratio in the heavy water. The term in brackets multiplying σ_{tot} appears due to incomplete fulfillment of the required ratio between absorber lengths.

The photon beams were obtained from thermal n-capture in Ni and Fe. The absorption spectra were taken with a 3"x3" NaI(Tl) detector. An interchanging facility was built which included a moving table carrying the two absorbers in a precise parallel position. By means of a linear stepping motor and an electronic control system, the two absorbers were alternately positioned in-and-out of the beam axis. In Figure 1 a schematic picture of the experimental setup is presented. Throughout the measurement, the two absorbers were interchanged every 5 minutes, thus ensuring total independence from variations of photon flux. The absorption spectra were separately accumulated in the M.C.A. memory. Typical photon spectra taken with the NaI detector are presented in Figure 2. Only the highest and the strongest γ-lines from each (n,γ) source were considered, thereby minimizing systematic uncertainties due to overlap of different γ-lines and due to background subtraction.

Preliminary tests were performed to insure: (a) negligible contribution from multiple scattering and pileups; (b) parallelism of the two absorbers, and (c) consideration of contributions from high energy γ-lines to lower ones. This last effect, existing in the case of the strong 7.64 and 9.00 MeV γ-lines (from Fe(n,γ) and Ni(n,γ) respectively) is smaller than the statistical uncertainty, but introduces a small systematic error which was properly taken care of. The reliability of the experimental setup was verified by measuring the total cross-section in H_2O (In this case both tubes were filled with H_2O.) for the 9.00 and 11.39 MeV γ-lines. The measured values (675±7 and 617±12 mb, respectively) overlap, within standard deviation, the cross-sections obtained from the tables of J.H. Hubble[6] (692±14 and 633±12 mb, respectively).

In addition to 4 high accuracy results at 7.64, 9.00, 9.30 and 11.39 MeV, the cross-sections at 8.53 and 5.97 MeV were also determined but with much larger uncertainties due to data analysis that involved response function measurements and subtraction of highly overlapping contributions.

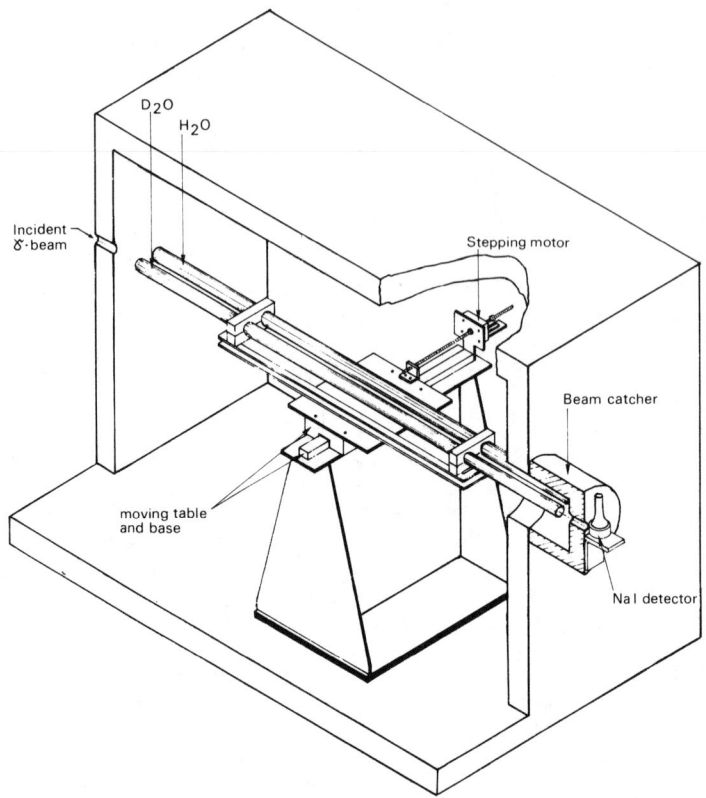

Fig.1. Experimental setup (not to scale) for the measurement of the D(γ,n) cross-section.

RESULTS

The results of the present work, together with different theoretical results are presented in fig.3. The lines marked SSC-B correspond to the calculations of M.L. Rustgi et al.[7] using the SSC-B potential without (I) and with (III) explicit consideration of MEC corrections. The line marked Paris corresponds to the calculations of H. Arenhovel[8] with the Paris potential and with the inclusion of MEC corrections. Obviously the agreement between the measured and calculated results is very good. It seems that the experimental results tend to prefer the calculated results obtained with MEC corrections. Yet it is not possible to determine which calculation follows the experimental results best since one can find differences in calculated results reported by different authors even with the same nuclear potential.

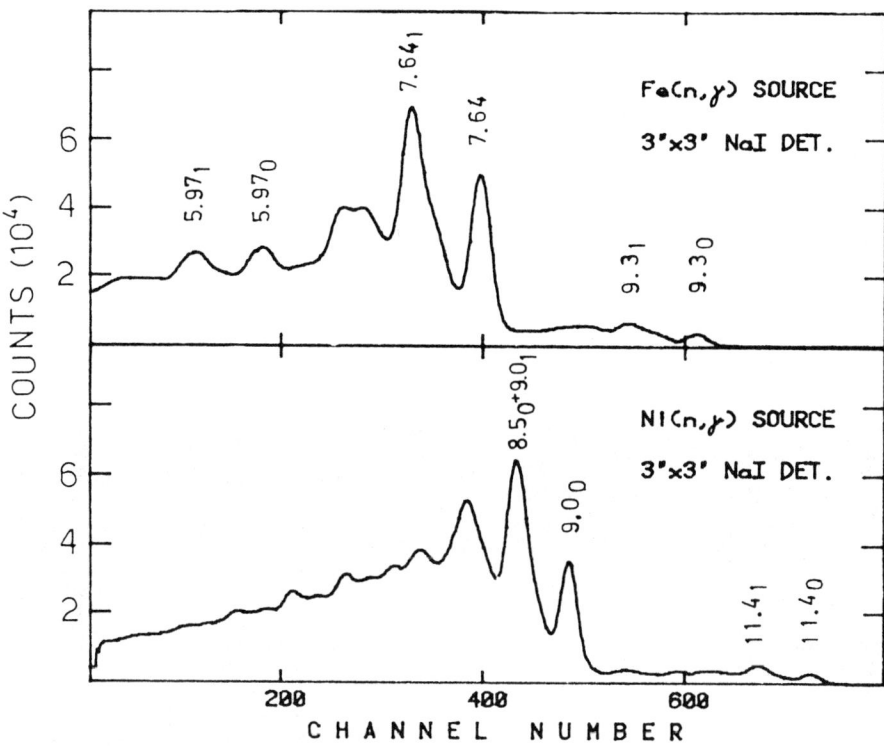

Fig. 2. Photon spectra obtained with the D_2O absorber.

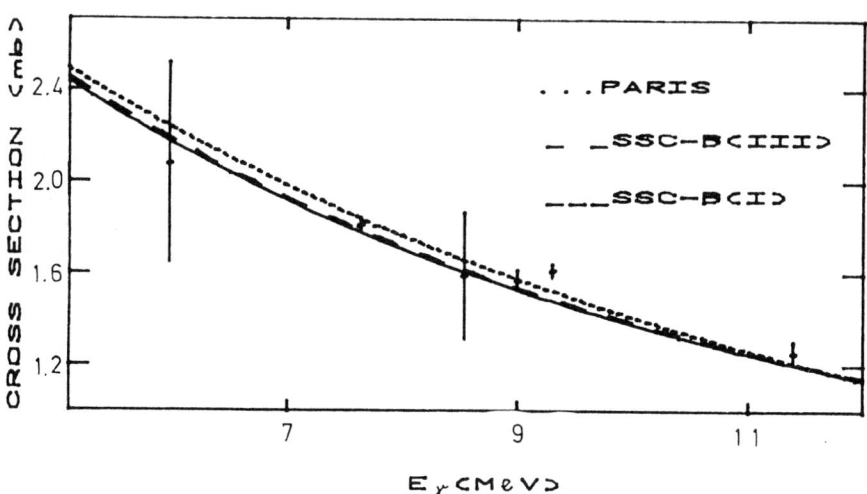

Fig. 3. $D(\gamma,n)$ cross sections

REFERENCES

1. A. Weststone and J. Halpern, Phys. Rev. 109, 2072 (1958).
2. J.A. Phillips, et al., Phys. Rev. 80, 326 (1950).
3. C.A. Barnes, et al., Phys. Rev. 86, 359 (1952).
4. O.Y. Mafra, et al., Nucl. Phys. A186, 110 (1972).
5. J. Ahrens, et al., Phys. Lett. 52B, 49 (1974).
6. J.H. Hubble, Int. J. App. Isot. 33, 1269 (1982).
7. H.L. Rustgi, et al., Phys. Rev. C29, 785 (1984).
8. H. Arenhovel, private communication (1984).

ANGULAR DISTRIBUTIONS FOR THE ^{2}H(γ,n) REACTION BELOW 10 MeV

A. Wolf[a], Z. Berant[a], Y. Birenbaum[a],
S. Kahane[a], and R. Moreh[a,b]

[a]Nuclear Research Centre Negev, P.O. Box 9001, Beer-Sheva, Israel
[b]Ben-Gurion University of the Negev, Beer-Sheva, Israel

ABSTRACT

Angular distributions of neutrons from the ^{2}H(γ,n) reaction were measured for six photon energies between 6-9 MeV. The results were fitted to a Legendre polynomials expansion up to and including $P_3(\cos\theta)$. The coefficients A_1/A_0 and A_3/A_0 were found to deviate considerably from theoretical calculations using different nucleon-nucleon potentials. These results indicate that the E1-E2 interference is considerably larger than expected.

INTRODUCTION

The photodisintegration of ^{2}H was extensively studied at γ energies above 10 MeV[1]. Below 10 MeV, the existing experimental data is sparse and with rather large uncertainties. Whetstone and Halpern[2] have measured angular distributions of photoprotons in the range E_γ = 9-23 MeV, using bremsstrahlung γ rays. Bosch et al.[3] have reported measurements of angular distributions of neutrons from the ^{2}H(γ,n) reaction using neutron-capture γ rays and liquid scintillation detectors. The results of these experiments were in general agreement with theoretical predictions available at that time. Since then, more refined calculations were performed, and more sophisticated experimental techniques became available. Thus, experimental results of higher accuracy can be obtained, enabling a more detailed comparison with theoretical predictions. Several more recent experiments[4,5] in the range E_γ = 17-43 MeV have indicated that angular distributions of protons from the ^{2}H(γ,p) reaction deviate to some extent from theoretical calculations.

In this work we studied the angular distributions of neutrons from deuteron photodisintegration for six photon energies in the range E_γ = 6-9 MeV. The results are compared with the calculations of Partovi[6], and also with the recent calculations of Rustgi, Vyas and Rustgi[7] in which meson exchange corrections were included.

EXPERIMENTAL TECHNIQUE

Incident γ beams were obtained from the Fe(n,γ) and Ni(n,γ) reactions, and contained monoenergetic γ lines with energies between 6-9 MeV and intensities of the order of 10^6 photons/cm^2/sec. The target consisted of D_2O in a 4 cm diameter, 6 cm high glass vial. The neutrons were detected at 12 angles between 40° and 150° using a high resolution ^{3}He spectrometer. The energy resolution was of the order of 30-60 keV, for neutrons between 2-4

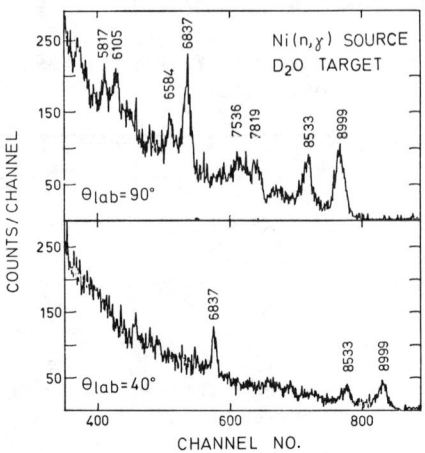

Fig. 1. Neutron spectra taken at 40° and 90°.

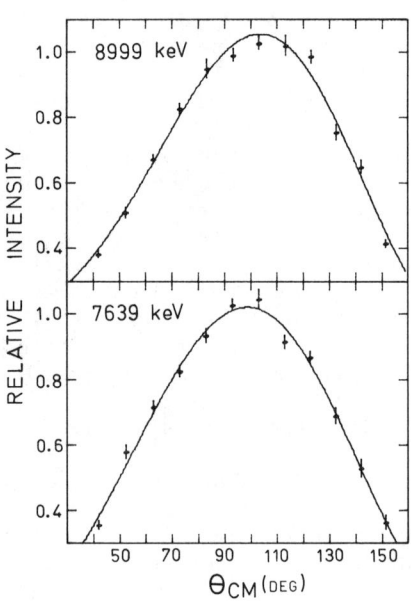

Fig. 2. Angular distributions at 8999 and 7639 keV.

MeV. The neutron spectra at 40° and 90°, obtained with the Ni(n,γ) source, are presented in Fig. 1. The neutron groups from the various lines in the incident beam are well resolved by the spectrometer. The numbers in Fig. 1 indicate the energy (in keV) of the respective γ line in the incident beam.

Angular distributions for six γ energies were obtained. The results were corrected for solid angle attenuation, for relative efficiency of the detector at the various angles, for neutron absorption and multiple scattering in the target, and were converted from laboratory to c.m. system.

RESULTS AND DISCUSSION

Two of the angular distributions are given in Fig. 2. A pronounced backward asymmetry is observed. The results of all the distributions were fitted to a Legendre polynomial expansion of the form:

$$\omega(\theta) = \sum_{i=1}^{3} A_i P_i (\cos\theta) \qquad (1)$$

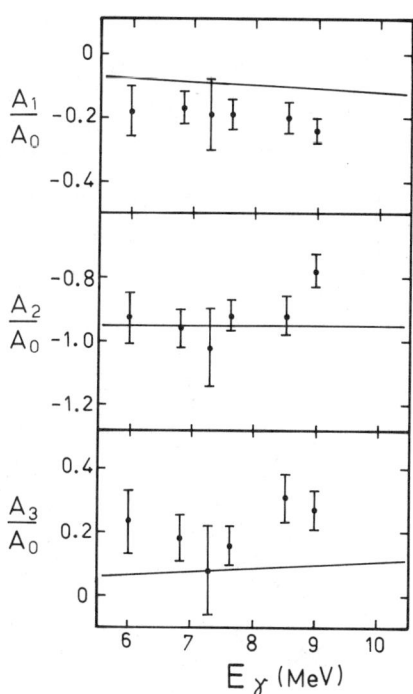

The contribution of $P_4(\cos\theta)$ was found to be negligible. The values of A_i/A_0 (i=1,3) are presented in Fig. 3 for all six γ energies measured in this work. The solid lines in Fig. 3 are from the calculations of Partovi[6]. We see that while the agreement between experiment and theoretical prediction is reasonably good for A_2/A_0, the values of A_1/A_0 and A_3/A_0 deviate considerably from the calculations. The data indicates a much larger asymmetry around 90° than expected. In the energy region considered in this work, this asymmetry is due to E1-E2 interference. Therefore, the

Fig. 3. The coefficients A_i/A_0 vs E_γ.
The solid lines are calculated from Partovi[6].

experimental data indicates a much larger contribution of E1-E2 interference than expected.

A comparison of our data with the recent calculations of Rustgi, Vyas and Rustgi[7] indicates that the above disagreement persists when different nucleon-nucleon potentials are employed and meson-exchange corrections are included.

REFERENCES

1. M. P. dePascale et al., Phys. Lett. 119B, 30 (1982).
2. A. Whetstone and J. Halpern, Phys. Rev. 109, 2072 (1958).
3. R. Bosch, J. Lang, R. Muller, and W. Wolfli, Helv. Phys. Acta. 36, 657 (1963).
4. B. Weissman and H. L. Schulz, Nucl. Phys. A174, 129 (1971).
5. J. E. E. Baglin et al., Nucl. Phys. A201, 593 (1973).
6. F. Partovi, Annals of Physics 27, 79 (1964).
7. M. L. Rustgi, R. Vyas and O. P. Rustgi, Phys. Rev. C29, 785 (1984).

RESONANT ABSORPTION OF DEUTERON-CAPTURE γ-RAYS

F. Zijderhand, S.W. Kikstra, S.S. Hanna* and C. van der Leun
Fysisch Laboratorium, Rijksuniversiteit Utrecht, The Netherlands

ABSTRACT

The experiment described in this paper shows that resonant absorption of deuteron-capture γ-rays is feasible. The E_x = 16.98 MeV, T = 3/2 level of ^{9}Be has been resonantly excited by γ-rays from the ^{7}Li(d,γ)^{9}Be resonance at E_d = 361 keV. Preliminary analysis of the data gives a total level width Γ(16.98 MeV) = 500 ± 50 eV, and a radiative width Γ_{γ_0} = 32 ± 5 eV. These data are relevant for the isospin mixing of this highly-excited and narrow state.

INTRODUCTION

The lowest T = 3/2 states of ^{9}Be, at excitation energies E_x = 14.39 and 16.98 MeV, have total widths of Γ = 0.38 and < 0.47 keV, respectively[1]. These remarkably narrow states fall in an excitation range where many broad (several hundred keV) T = 1/2 states are known. The small width of the T = 3/2 states is indicative of a high isospin purity. The ultimate aim of the present investigation is a determination of the isospin mixing of the E_x = 16.98 MeV, J^π = 1/2$^-$ state of ^{9}Be.

A simplified level and decay scheme of ^{9}Be, adapted from ref.[1], is presented in fig.1. Two isospin-forbidden decay channels of ^{9}Be(16.98) are energy-favoured (^{5}He + α, Q = 14.52 MeV and ^{8}Be + n, Q = 15.31 MeV), whereas the competing isospin-allowed decay channel, ^{8}Li + p, has a very low Q-value of Q = 0.09 MeV.

EXPERIMENTS

The first experiment is a ^{9}Be(γ,X) resonant absorption measurement, in which the γ-rays are provided by the E_d = 361 ± 2 keV resonance of the ^{7}Li(d,γ)^{9}Be reaction. Deuteron beam currents of up to 80 μA were available from the Utrecht 3 MV Van de Graaff accelerator. The resonant angle, at which the recoil energy loss upon emission and absorption of the γ-rays is compensated by the capture induced Doppler shift, can be calculated from the reaction kinimatics as α = 62°.

---

*Permanent address: Stanford University, Stanford, California, U.S.A.

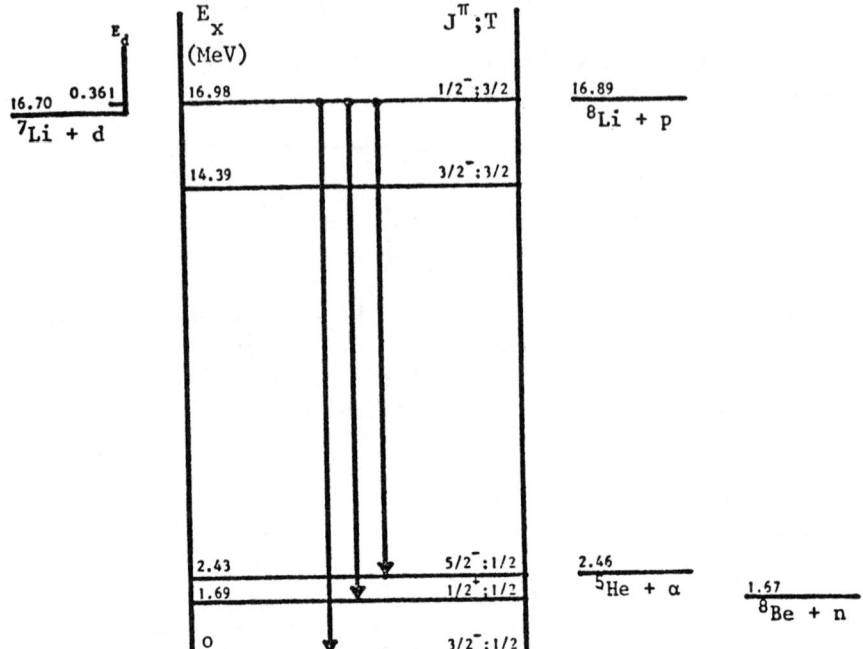

Fig.1. Simplified level and decay scheme of ^{9}Be; see ref.[1].

Berrylium absorbers of different lengths (L = 80, 200 and 300 mm) were placed in (and in front of) the 2.3 mm wide slit between two 15 cm long Pb-bricks. The high-energy part of a γ-ray spectrum measured with a Ge(Li) detector behind the slit is shown in fig.2. An example of an absorption dip, measured with a 200 mm Be absorber, is presented in fig.3.

The width of the dip, 1800 ± 400 eV, exceeds the instrumental width of 700 eV, indicating a total level width of Γ = 650 ± 200 eV. A preliminary analysis of the absorption integral, A_α = 0.39° ± 0.04°, yields a ground-state radiative width of Γ_{γ_0} = 32 ± 5 eV, appreciably larger than the adopted literature value of Γ_{γ_0} = 16 ± 2 eV [2]. For details about the method of analysis the reader is referred to ref.[3].

In a second experiment a ^{7}Li(d,γ)^{9}Be thick target yield curve was measured at the E_d = 361 keV resonance. The interquartile range (see fig.4) leads to a total resonance width $\Gamma_{c.m.}$ = 500 ± 50 eV, in good agreement with the value deduced from the resonant absorption experiment described above. An additional measurement is planned of the ^{7}Li(d,γ) resonance strength, that should yield in combination width the data presented here the ratio Γ_d/Γ, and thus gives a clue about the partial widths Γ_n and Γ_α of the isospin-forbidden decay channels, and thus of the isospin-purity of the E_x = 16.98 MeV level of ^{9}Be.

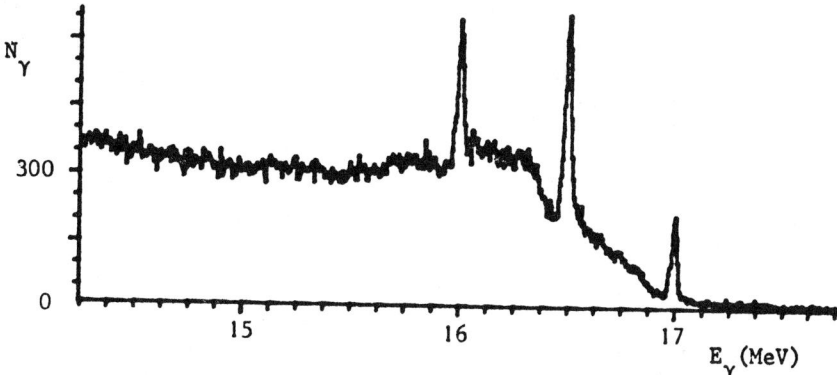

Fig.2. High-energy part of a ^{7}Li(d,γ)^{9}Be γ-ray spectrum at E_d = 361 keV, measured behind absorber and 2.3 mm collimator slit.

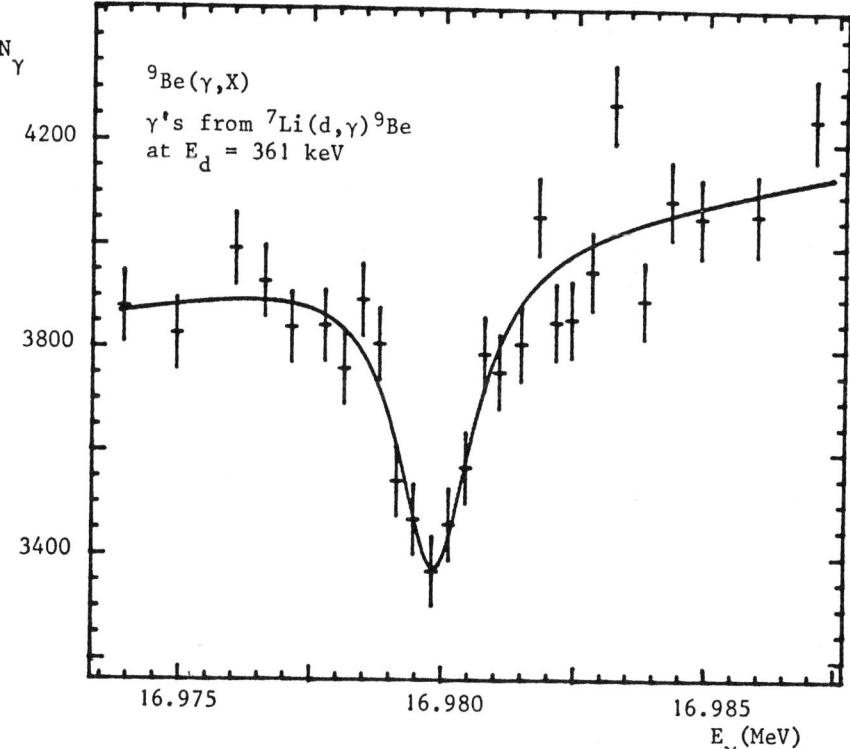

Fig.3. Absorption dip for 16.98 MeV γ-rays in ^{9}Be. The curve is a fit with a Breit-Wigner function and a linear background. The energy range shown corresponds to an angular range of 12°.

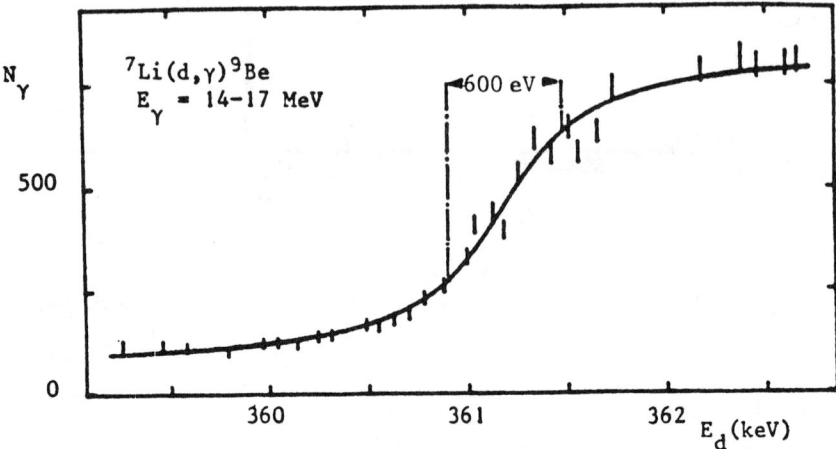

Fig.4. Thick-target yield curve of the ^{7}Li(d,γ)^{9}Be reaction around the E_d = 361 keV resonance. The interquartile range is indicated. Target material Li_2SO_4.

CONCLUSION

Resonant absorption by ^{9}Be of Doppler-shifted γ-rays from the ^{7}Li(d,γ)^{9}Be reaction provides information about the total and partial widths, and thus about the isospin mixing, of the narrow T = 3/2 state of ^{9}Be at E_x = 16.98 MeV. Preliminary results are presented.

REFERENCES

1. F. Ajzenberg-Selove, Nucl. Phys. A414, 1 (1984).
2. P.M. Endt, At. Data Nucl. Data Tables 23, 3 (1979).
3. P.B. Smith and P.M. Endt, Phys. Rev. 110, 397 (1958).

COMPARISON OF PHOTOABSORPTION BY ^{16}O AND ^{18}O

N.K. Sherman and W.F. Davidson
National Research Council, Ottawa, Canada K1A 0R6

S. Raman
Oak Ridge National Laboratory, Tennessee 37831

W. Delbianco and G. Kajrys
Université de Montréal, P.Q., Canada H3C 3J7

ABSTRACT

We are comparing the transmission of 3 to 38 MeV photons by ^{16}O-water and ^{18}O-water in order to measure the integral over the giant dipole resonance (GDR) of the photonuclear absorption cross section of ^{16}O.

INTRODUCTION

The sum of the integrated partial cross sections of ^{16}O for photoparticle emission from the GDR was reported[1] to be nearly 20% smaller than a careful measurement[2] of Σ, the integral over photon energy ω of the total photonuclear absorption cross section $\sigma_N(\omega)$. A preliminary[3] to the present experiment found that between 10 and 30 MeV Σ is 8% greater than the value given by the partial cross sections. The calculated atomic cross sections[4] of H and O were crucial to this result. The theoretical uncertainty of about ±0.2% in the molecular cross section σ_m of H_2O which dominates the total absorption introduces an uncertainty of about ±7% in Σ.

We are instead measuring $\sigma_m(\omega)$ by subtracting $\sigma_N(\omega)$ of ^{18}O from the total absorption cross section $\sigma_T(\omega)$ of heavy-oxygen water. We obtain σ_m in the presence of the same inscattering and higher order atomic processes as affect the light-oxygen water measurements. At a given energy,

$$\sigma_N(^{16}O) = \sigma_T(H_2{}^{16}O) - [\sigma_T(H_2{}^{18}O) - \sigma_N(^{18}O)], \quad (1)$$

and the background effects will cancel.

EXPERIMENTAL

The liquid deuterium (LD_2) neutron time of flight (TOF) spectrometer has been described before.[5] Absorbers of ^{16}O-water and ^{18}O-water and an identical empty container are exposed alternately and repeatedly under the same conditions to 42 MeV bremsstrahlung from the NRC linac. Transmitted photons are detected by photoneutrons they eject from a LD_2 target viewed by a TOF detector. Photon energies are calculated from the photoneutron energies. The absorption cross section for absorber i is given by

$$\sigma_T^i(\omega) = [\ln I_o(\omega)/I_i(\omega)]/n \quad (2)$$

where the neutron flux transmitted by the empty container is I_o and by the full container is I_i, and n is the number of molecules per unit area. An empty LD_2 target is also exposed to the unattenuated and attenuated photon beam during each cycle, from which we get the fluxes:

$$I(\omega) = F(\omega) - E(\omega) \tag{3}$$

where F, E refer to full and empty LD_2 target. Typical photon energy resolution is 0.6% at 6 MeV and 1.9% at 25 MeV.

RESULTS

Statistical uncertainty achieved to date in the $\sigma_T(\omega)$ of ^{16}O-water and ^{18}O-water is about ±0.4% in a 200-keV bin at 10 MeV and about ±1% at 20 MeV. The ^{16}O data are shown in Figure 1. Subtraction of values of $\sigma_N(^{18}O)$ interpolated from total photoparticle measurements[6] gives a smoothly varying $\sigma_m(\omega)$ which is 1.7% smaller than the theoretical value at 10 MeV and 0.9% at 30 MeV. Second order background processes account for part of these discrepancies. Up to about 16 MeV we expect that $\sigma_N(^{16}O)$ is nearly zero, in which case σ_T equals σ_m for light-oxygen water below this energy. Comparison of the two measured $\sigma_m(\omega)$ data sets between 3 and 16 MeV reveals a systematic ratio

$$R = \sigma_m(H_2{}^{16}O)/\sigma_m(H_2{}^{18}O) \tag{4}$$

of (1.0064 ± 0.0009) which within uncertainties is independent of ω. New data with the two water absorbers swapped in position on the absorber wheel will clarify the nature of this 0.6% shift from unity. Using normalized, fitted values of σ_m, and

$$\sigma_N(^{16}O) = \sigma_T(H_2{}^{16}O) - R \cdot \sigma_m(H_2{}^{18}O), \tag{5}$$

we find a preliminary value for

$$\Sigma(^{16}O) = \int_{10}^{30} \sigma_N(\omega)d\omega \tag{6}$$

of (182.5 ± 7.5) MeV-mb where the uncertainty is statistical. This gives, in units of the Thomas-Rieche-Kuhn sum rule applied to the nucleus,

$$\Sigma_{GDR}(^{16}O) = (0.76 \pm 0.15)\ \Sigma_{TRK}(^{16}O) \tag{7}$$

where the uncertainty is systematic. The value of Σ is the same as in the earlier experiment.[3] It does not however depend upon calculated cross sections.

Theoretical estimates of Σ including all multipoles integrated up to the meson threshold fall short of 2 sum rule units when based on models which explain other critical measurements such as the differential cross section at 0° for the $^2H(\gamma,p)$ reaction. The present measurement of the contribution by the GDR indicates that the integral for ^{16}O will not exceed this value.

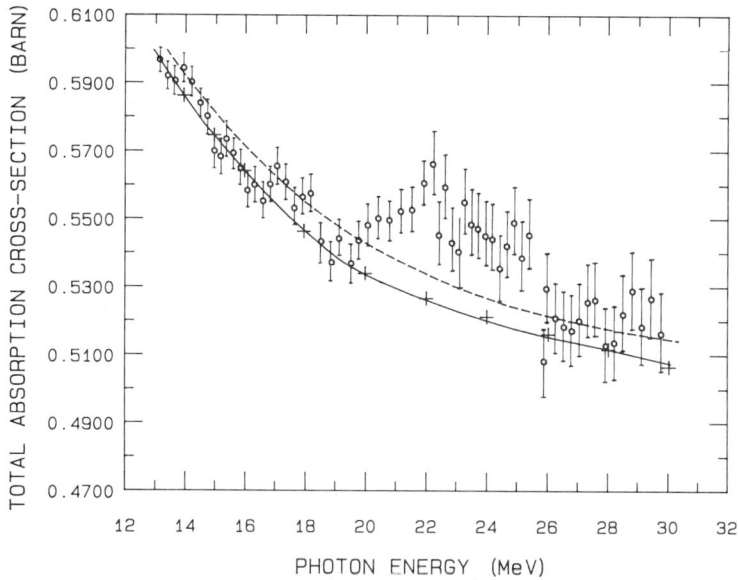

Fig. 1 Observed total cross section of ^{16}O-water for absorption of photons (circles) compared to the sum of the atomic cross sections (broken line) and the observed molecular cross section (crosses) for ^{18}O-water. The area between the solid line and the datum points is the photonuclear cross section.

ACKNOWLEDGEMENT

We are grateful to A. Nowak and M. Kosaki for support of linac and cryogenic target operations and to C.K. Ross for maintaining the computer system. We also thank D. Kleinbub, J. Bélanger and J. Stinson for technical help, J. Roy for computations and H. Matchett for word processing.
This work was supported by the National Research Council of Canada, by Oak Ridge National Laboratory (operated under DOE contract by Martin Marietta Energy Systems, Inc.) and by the Natural Sciences and Engineering Research Council of Canada.
One of us (S.R) is deeply grateful to T.R. Mills of the Los Alamos National Laboratory for facilitating the acquisition of the ^{18}O-water.

REFERENCES

1. B.L. Berman, R. Bergère and P. Carlos, Phys. Rev. C26, 304 (1982).
2. J. Ahrens, H. Borchert, K.H. Czock, H.B. Eppler, H. Gimm, H. Gundrum, M. Kroning, P. Riehn, G. Sita Ram, A. Zieger and B. Ziegler, Nucl. Phys. A251, 479 (1975).

3. N.K. Sherman, W.F. Davidson and A. Claude, J. Phys. G: Nucl. Phys. $\underline{9}$, 1519 (1983).
4. J.H. Hubbell, H.A. Gimm and I. Øverbo, J. Phys. Chem. Ref. Data $\underline{9}$, 1023 (1980).
5. N.K. Sherman, C.K. Ross and K.H. Lokan, Phys. Rev. $\underline{C21}$, 2328 (1980).
6. J.G. Woodworth, K.G. McNeill, J.W. Jury, R.A. Alvarez, B.L. Berman, D.D. Faul and P. Meyer, Phys. Rev. $\underline{C19}$, 1667 (1979).
7. B. Goulard and B. Lorazo, Can. J. Phys. $\underline{60}$, 162 (1982).

FORE-AFT ASYMMETRY IN POLARIZED PHOTOREACTIONS

F. Saporetti
ENEA, Centro "E.Clementel", Via Mazzini 2
I-40138 Bologna, Italy

R. Guidotti
Università di Bologna, Dottorato di Ricerca in Fisica
I-40126 Bologna, Italy

ABSTRACT

The forward-backward asymmetries of the proton angular distributions following the ^{141}Pr(γ,p_o) reaction induced by linearly polarized photons are discussed in the frame of the direct-semidirect model. The study points out the critical rôle played by the photon polarization state on the emission geometrical features.

As is well known[1], the polarization state of the photon initiating a photonuclear reaction seems to strongly affect the nucleon emission probability; e.g., at 90° the cross section obtained with γ-rays polarized parallelly ($\phi = 0°$) to the reaction plane is much larger than that with γ-rays polarized perpendicularly ($\phi = 90°$). The purpose of the present analysis is to show that the photon polarization state is equally crucial as regards the geometry of the angular distribution of the emitted nucleons.
In this connection we consider here the ^{141}Pr$(\gamma,p_o)^{140}$Ce reaction induced by linearly polarized photons. On the basis of the direct-semidirect (DSD) model for nuclear radiative capture and by the use of the microreversibility relation, we calculate the ratio between the difference and the sum of the differential cross sections at supplementary angles, viz.

$$I(\theta,\phi) = \frac{\frac{d\sigma}{d\Omega}(\theta,\phi) - \frac{d\sigma}{d\Omega}(\pi-\theta,\phi)}{\frac{d\sigma}{d\Omega}(\theta,\phi) + \frac{d\sigma}{d\Omega}(\pi-\theta,\phi)}, \qquad (1)$$

which expresses the forward-backward asymmetry of the proton angular distributions for the different polarization states of the incident photon. Calculations are carried out by considering the

DSD free-parameters suggested in ref.[2], assuming the real volume coupling form[3] for the nucleon-nucleus coupling interaction and using the multipole giant resonance parameters provided by the inelastic electron scattering experiments[4]. Calculations are performed at the photon energy E_γ = 26 MeV, just where the presence of the isovector quadrupole giant resonance shows itself more evident[1]. Assuming the beam entirely polarized and expanding the differential cross section in Legendre polynomials and associated Legendre functions of second order, we have

$$\frac{d\sigma}{d\Omega}(\theta,\phi) = A_o \left[1 + \sum_{L=1}^{4} a_L P_L(\cos\theta) + \cos 2\phi \sum_{L=2}^{4} d_L P_L^{(2)}(\cos\theta) \right]. \quad (2)$$

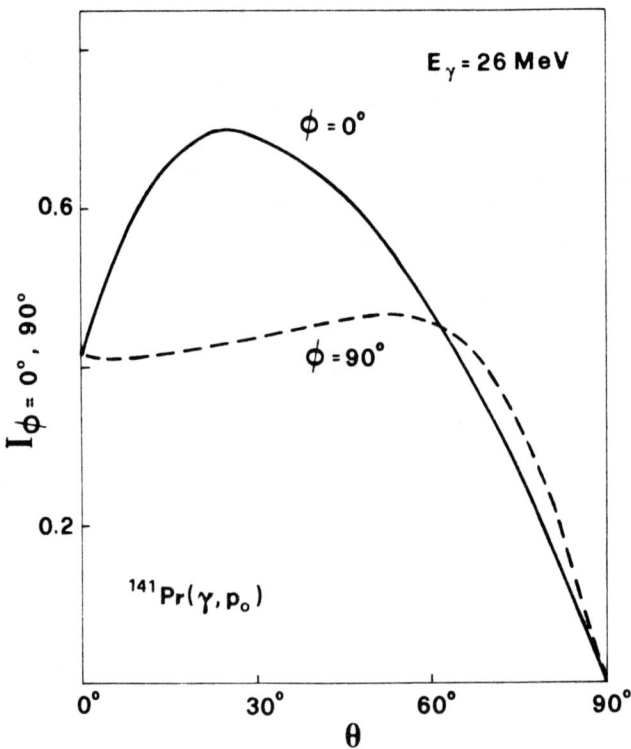

Fig. 1
Asymmetry ratios for different photon polarization states.

The a_L and d_L coefficients are related to the multipolarities involved in the photoreaction (in the present calculations only the E1 and E2 transitions are considered); the odd coefficients represent the E1-E2 interference between processes of opposite parity, while the even ones represent the strength of the E1 and E2 single processes. Since the dipole mechanism dominates, pure quadrupole effects are scarcely recognizable. From the expansion, it follows that

$$I(\theta,\phi) = \frac{a_1 P_1(\cos\theta) + a_3 P_3(\cos\theta) + \cos 2\phi \, d_3 P_3^{(2)}(\cos\theta)}{1 + a_2 P_2(\cos\theta) + a_4 P_4(\cos\theta) + \cos 2\phi [d_2 P_2^{(2)}(\cos\theta) + d_4 P_4^{(2)}(\cos\theta)]}. \quad (3)$$

It clearly appears that the forward-backward asymmetry ϕ-dependence expresses the weight of the E1-E2 interference for the different photon polarization states and model calculations can supply indications on the polarization state which gives the greatest prominence to the interference process.

A comparison of the results obtained with two different photon polarization states, i.e. $\phi = 0°$ (continuous line) and $\phi = 90°$ (dashed line), is given in Fig. 1. The latter shows how the geometrical features of the proton emission (which is steadily forward peaked) strongly depend on the polarization state of the incident γ-ray: in the $\phi = 0°$ case, the asymmetry exhibits a maximum close to $\theta = 25°$, while for $\phi = 90°$ the asymmetry remains practically constant between 0° and 70° before falling sharply to zero. Therefore, from the result we learn that in order to investigate the asymmetry around 25 MeV, it may be suitable to use photons polarized parallelly to the reaction plane and compare the proton yields at angles $\theta = 25°$ and $\theta = 155°$.

REFERENCES

1. F. Saporetti and R. Guidotti, Nucl. Phys. A404, 93 (1983).
2. F. Saporetti and R. Guidotti, Nucl. Phys. A390, 207 (1982).
3. G. Longo and F. Saporetti, Nucl. Phys. A199, 530 (1973)
4. R. Pitthan, H. Hass, D.H. Meyer, F.R. Buskirk and J.N. Dyer, Phys. Rev. C19, 1251 (1979).

INTERACTING BOSON DESCRIPTION OF THE GIANT DIPOLE RESONANCE IN THE LANTHANIDE REGION

L. Zuffi

Dipartimento di Fisica, Università di Milano and INFN, Sezione di Milano, via Celoria 16, I-20133 Milano, Italy

G. Maino and A. Ventura

ENEA, Divisione Fisica e Calcolo Scientifico, via Mazzini 2, I-40138 Bologna, Italy

ABSTRACT

Calculations of absorption, elastic and inelastic scattering of photons in the 8-20 MeV energy range by samarium isotopes have been carried out within the framework of the interacting boson model.

INTRODUCTION

In recent years, a number of Authors[1,2,3] have suggested that the interacting boson model (IBM)[4] could be a mathematical tool suited to the study of the coupling of low-lying collective modes, described by s, d bosons and high-lying collective excitations, or giant resonances. Realistic calculations are here presented for the giant dipole resonance, described by a p boson ($J^\pi = 1^-$), coupled to low energy 0^+, 2^+ states, or s-d boson excitations.

THE MODEL

The model of coupled s, d and p bosons has been described in refs. 1-3. The basis states are of the form $(s^+)^m (d^+)^n (p^+)^q |0\rangle$ with $m+n+q=N$, the boson number, and $q=0$, or 1. The Hamiltonian can be written in the form[1]:

$$\hat{H} = \hat{H}_{sd} + \hat{H}_p + \hat{H}_{psd} \qquad (1)$$

Here, $\hat{H}_{sd}$ is the standard IBM-1 Hamiltonian, written as a multipole expansion[4]:

$$\hat{H}_{sd} = \varepsilon \hat{n}_d + a_0 (\hat{P}^+ \cdot \hat{P}) + a_1 (\hat{L}\hat{L}) + a_2 (\hat{Q}\hat{Q}) + a_3 (\hat{T}_3 \hat{T}_3) + a_4 (\hat{T}_4 \hat{T}_4) \qquad (2)$$

$\hat{H}_p = \varepsilon_p \hat{n}_p$ is the free p-boson Hamiltonian, where the p-boson energy follows the empirical law $\varepsilon_p = 77.5\, A^{-1/3}$ MeV.

$\hat{H}_{psd}$ describes the coupling between the s, d and p bosons :

$$H_{psd} = b_0 (d^+ \tilde{xd})^{(0)} \cdot (p^+ x\tilde{p})^{(0)} + b_1 (d^+ \tilde{xd})^{(1)} \cdot (p^+ x\tilde{p})^{(1)} +$$
$$+ b_2 ((s^+ x\tilde{d} + d^+ x\tilde{s})^{(2)} + \chi (d^+ x\tilde{d})^{(2)}) \cdot (p^+ x\tilde{p})^{(2)} \qquad (3)$$

Here, use has been made of the standard notation for boson creators and annihilators and their tensor products. The Hamiltonian (1) is diagonalized by means of the PBOSON code[5].

For a given boson number N, two calculations, with q=0 and q=1, are performed. Once the wave functions have been obtained, the matrix elements of the electric dipole operator, $D^{(1)}$, are calculated by means of the PBOSON code[5]. $D^{(1)}$ is expressed in terms of s, d and p boson operators :

$$\hat{D}^{(1)} = D_0 (p^+ x\tilde{s} + s^+ x\tilde{p})^{(1)} + D_2 (p^+ x\tilde{d} + d^+ x\tilde{p})^{(1)} \qquad (4)$$

The strength of the transition from the nth dipole state, $|1_n^->$, to the ground state, $|0_1^+>$, is proportional to the quantity :

$$S_n = |<1_n^- \| \hat{D}^{(1)} \| 0_1^+>|^2 \qquad (5)$$

The energies, E_n, the widths, Γ_n, and the transition strengths S_n, of dipole states are the fundamental ingredients to be introduced in the photoabsorption cross section[6]. While E_n and S_n are calculated within the framework of the model, the widths Γ_n are assumed to be energy-dependent : $\Gamma(E) = \Gamma_0 (E/E_0)^\gamma$, with Γ_0 = 3.12 MeV, E_0 = 12 MeV and γ = 3/2 in the present work.

Once the parameters of H_{sd} have been adjusted on the experimental low-lying collective states and those of H_{psd} and of $D^{(1)}$ are fixed so as to reproduce the experimental photoabsorption cross section, it is possible to evaluate elastic and inelastic scattering cross sections[6] for unpolarized radiation without resorting to further adjustable parameters. In the case of elastic scattering a Thomson nuclear contribution is to be coherently added to the resonance contribution in the scattering amplitude, while limitation of the present analysis to large angle scattering makes it possible to neglect the Delbrück contribution, due to vacuum polarization effects, since it has a maximum at ϑ=0 and decreases rapidly with increasing scattering angle.

RESULTS

^{148}Sm and ^{154}Sm have been chosen as significant examples of spherical and deformed nuclei, respectively, in a transitional isotope chain.

Table I

IBM-1 parameters

	^{148}Sm	^{154}Sm
N	8	11
ε(keV)	770.30[a]	371.0[a]
a_0(keV)	52.30[a]	8.0[a]
a_1(keV)	4.00[a]	0.52[a]
a_2(keV)	-12.60[a]	-19.55[a]
a_3(keV)	29.60[a]	8.40[a]
a_4(keV)	19.70[a]	5.65[a]
χ	-0.65[a]	-1.20[a]
ε_p(MeV)	14.70	14.46
b_0(keV)	400.0	400.0
b_1(keV)	0.	0.
b_2(keV)	550.0	550.0
D_0(fm)	2.90	3.27

(a) IBM-1 projection of IBM-2 parameters in ref.11

The adopted IBM-1 parameters are given in Table I and the calculated photoabsorption cross sections are compared to the experimental ones[7] in fig. 1.

The agreement between calculations and experiment appears to be excellent and the transition from one-humped to two-humped shapes well reproduced. Extensive calculations in the Nd-Sm mass region[8] give results comparable to, and in some cases better than those obtained by means of the dynamic collective model[9].

Fig. 2 shows elastic and inelastic scattering cross sections at an angle $\vartheta = 140°$ as functions of the incident photon energy. These quantities, which cannot be compared to experimental data in the case of ^{148,154}Sm, are much more sensitive than the

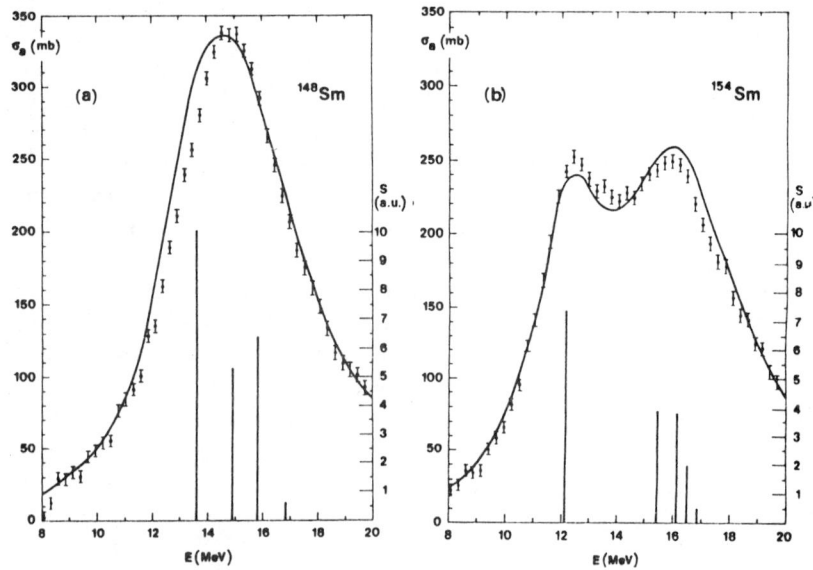

Fig. 1. Calculated and experimental photoabsorption cross sections: a) ^{148}Sm ; b) ^{154}Sm. Exp.data from ref.7.

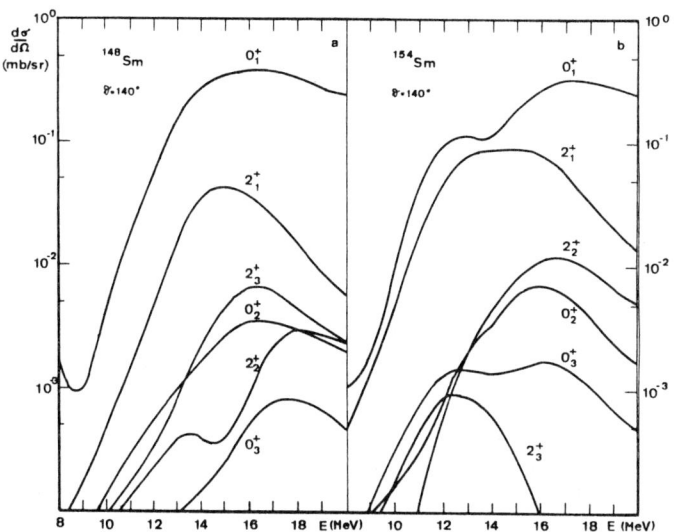

Fig. 2. Calculated elastic and inelastic scattering cross sections at $\vartheta = 140°$: a) ^{148}Sm ; b) ^{154}Sm. The curves are labelled with spin and parity of the final nuclear state.

absorption cross sections to the underlying nuclear model. IBM-1 calculations of photon scattering by Er^2 and U-Th^{10} isotopes compare quite well with experimental data and encourage us to carry out further analyses of giant resonances by means of the interacting boson model.

ACKNOWLEDGEMENTS

We are grateful to Prof. F. Iachello for many valuable comments and to Prof. K. Heyde and Dr. P. Van Isacker for permission to use PBOSON and PBOSONT.

REFERENCES

1. I.Morrison and J.Weise, J.Phys. G8, 687 (1982)
2. F.G.Scholtz and F.J.W.Hahne, Phys.Lett. 123B, 147 (1983)
3. D.J.Rowe and F.Iachello, Phys.Lett. 130B, 231 (1983)
4. A.Arima and F.Iachello, Ann.Rev.Nucl.Part.Sci. 31, 75 (1981)
5. P.Van Isacker, PBOSON and PBOSONT codes (1980)
6. V.Rezwani, G.Gneuss, H.Arenhövel, Nucl.Phys. A180, 254 (1972)
7. P.Carlos et al., Nucl.Phys. A225, 171 (1974)
8. G.Maino, A.Ventura, L.Zuffi, F.Iachello, subm.for publ. (1984)
9. J.M.Eisenberg and W.Greiner, Nuclear Theory (North Holland, 1978), vol. 1, p.331 fww.
10. F.Iachello, G.Maino, A.Ventura, L.Zuffi, subm.for publ. (1984)
11. O.Scholten, Thesis, University of Groningen (1980)

STUDY OF E1-E2 INTERFERENCE IN THE ^{159}Tb(γ,n)
AND ^{209}Bi(γ,n) REACTIONS

S. Kahane, Y. Birenbaum, Z. Berant, R. Moreh and A. Wolf

Nuclear Research Center-Negev, Beer-Sheva, ISRAEL

INTRODUCTION

The isoscalar giant E2 resonance, in medium and heavy nuclei, is localized at $E_o \sim 60\ A^{-1/3}$, close to the E1 giant resonance. Angular distributions of the (γ,n) reaction are sensitive to E1-E2 interference effects, between these two giant resonances, showing pronounced asymmetries arround 90°. The present work follows our previous investigations[1,2] on the spherical isotopes of lead by measuring ^{209}Bi and going further to the deformed nucleus of ^{159}Tb, being of interest to observe the E1-E2 interference in the case of a splitted giant dipole resonance. A modified DSD model is used to explain the qualitative features of the measured distributions.

EXPERIMENTAL METHOD

Incident photon beams were obtained from Fe(n,γ), Cr(n,γ) and Ni(n,γ) reactions and contained monoenergetic γ lines in the energy range 7-10 MeV, with intensities of the order of 10^6 photons/cm^2.sec. The neutrons were detected at 12 angles between 40° and 150° using a high resolution ^{3}He spectrometer having a resolution of $\sim$ 25 keV at 1 MeV.

The neutron spectra at 90° and 120° measured from ^{159}Tb, with a Cr(n,γ) incident beam, are presented in Fig. 1. Neutron groups leading to different excited states in the residual nucleus ^{158}Tb are clearly resolved in the spectra. Moreover, they are also differentiated according to the incident γ line producing them, at 8884 and 9720 keV respectively (the 90° spectrum is adnoted by the energies of the neutron groups).

RESULTS AND DISCUSSION

A number of angular distributions obtained in ^{209}Bi(γ,n_{o+1}) reaction are presented in Fig. 2. here n_{o+1} denotes neutron ground and first excited state transitions, not separated in the experiment. A pronounced asymmetry in respect with 90° is observed, this being an evidence for E1-E2 interference. The distributions were fitted with a Legeandre polynomial expansion

$$W(\theta) = \Sigma_{i=1}^{3} A_i P_i (\cos\theta)$$

The contribution of $P_4 (\cos\theta)$ was too small to be extracted.

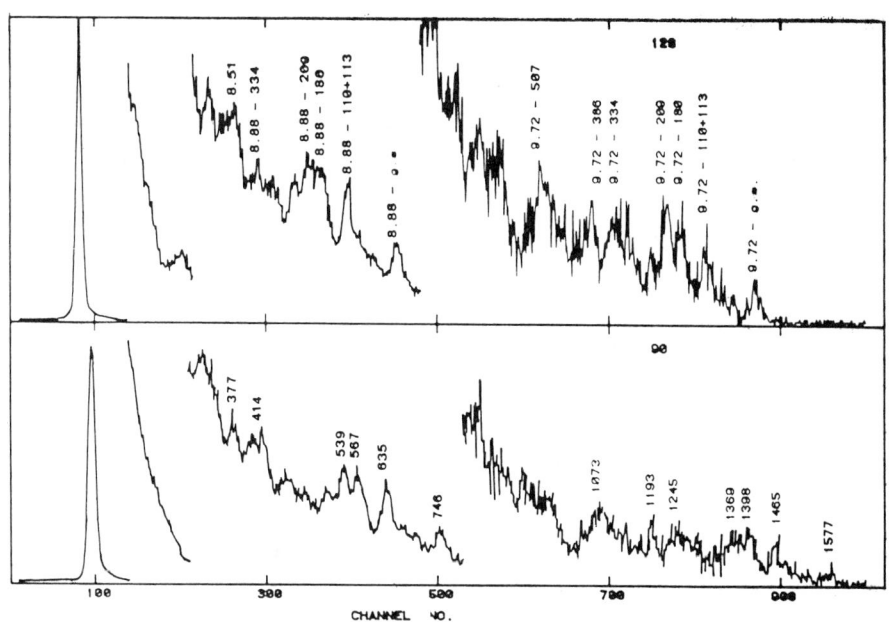

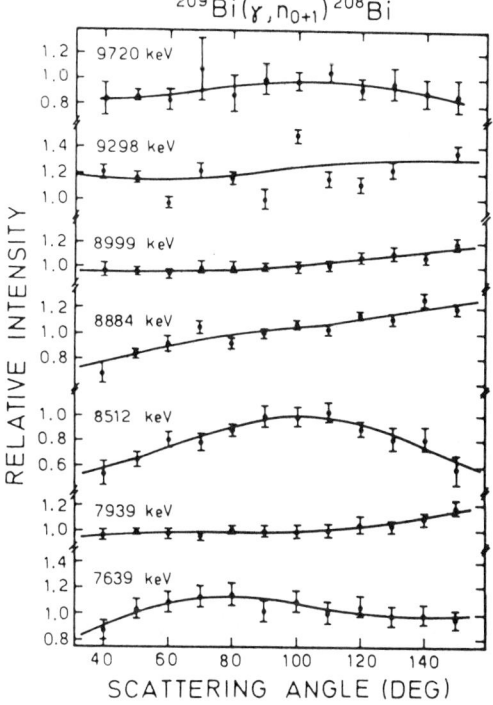

Fig. 1. Neutron spectra measured from ^{159}Tb(γ,n) reaction at 120^0 and 90^0. The upper spectrum is labeled according to the incided γ line in Cr(n,γ) source, and the final state in the residual nucleus ^{158}Tb. The lower spectrum is labeled by the neutron group energy.

Fig. 2. Angular distributions measured from ^{209}Bi(γ,n_{0+1}) reaction labeled according to the incident γ line. The fitted lines include Legeandre polynomials up to $P_3(\cos\theta)$.

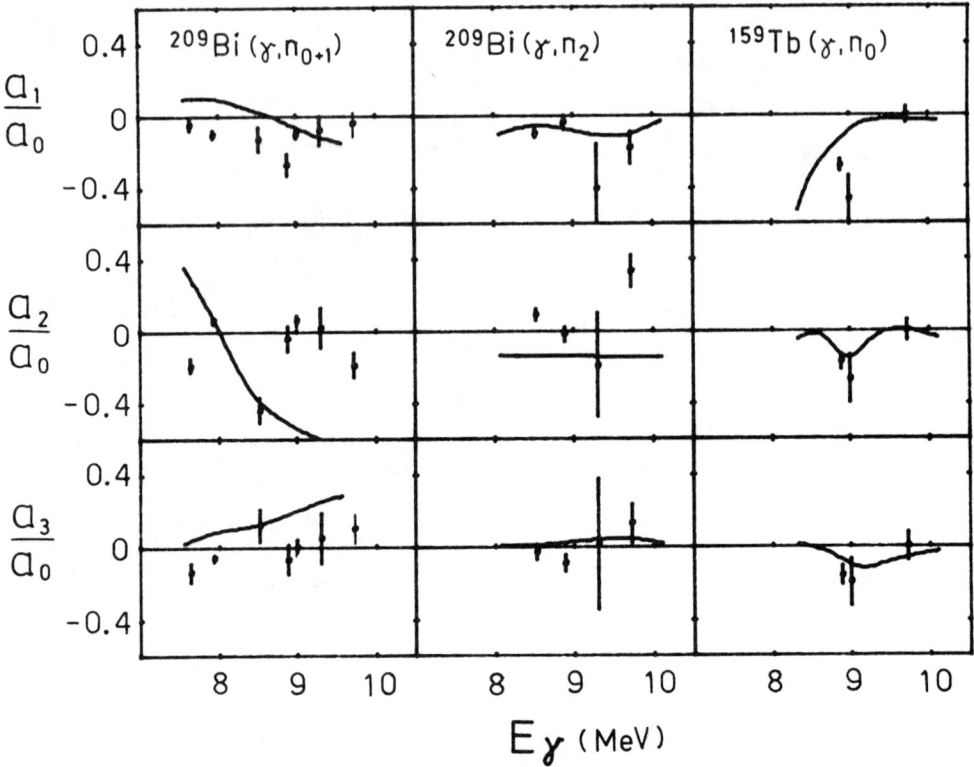

Fig. 3. The angular distribution coefficients A_i/A_0 for $^{209}Bi(\gamma,n_{0+1})$, $^{209}Bi(\gamma,n_2)$ and $^{209}Tb(\gamma,n_0)$ reactions. The solid curves are the modified DSD model calculations.

Values of A_i/A_o are presented in Fig. 3 for the transitions $^{209}Bi(\gamma,n_{0+1})$, $^{209}Bi(\gamma,n_2)$ and $^{159}Tb(\gamma,n_o)$, together with calculations based on the DSD model. This model was modified to accomodate targets with one proton outside an even-even core. It assumes in the initial state a valence neutron outside a core. To obtain such a state the proton was decoupled first, further the neutron was decoupled and the proton recoupled to the residual core:

$$|i> \sim \Sigma_x W(j_i, j_i, 0, j_p; 0\ X)\ W(j_i, j_p, 0, j_i, a\ X)\ |a>\ |j_i>$$

a, j_i j_p - are the spins of the core, valence neutron and the outside proton. These two Racah coefficients are multiplying the usual DSD operator, having a great influence on the calculated angular distributions.

For $^{209}Bi(\gamma,n_o)$ angular distribution expressions involving $s_{1/2}$, $d_{3/2}$, $p_{3/2}$ and $f_{5/2}$ partial waves, were obtained and found to be identical with those for ^{207}Pb, and ^{208}Pb [2,3] . The A_1/A_0 and A_3/A_0 are

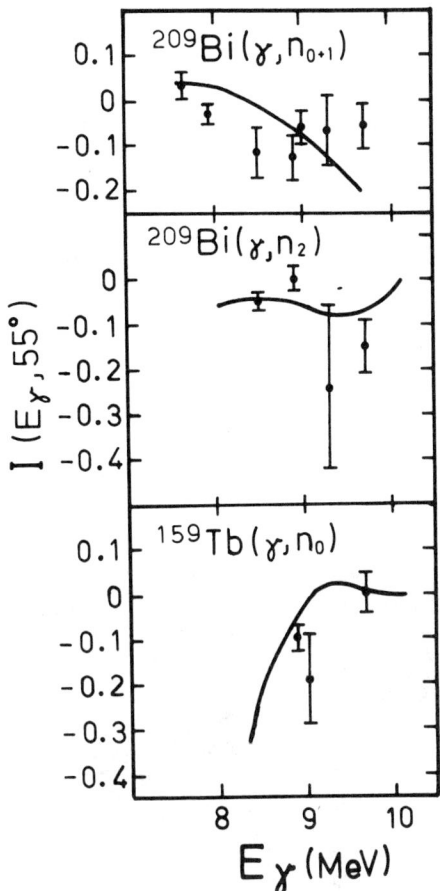

Fig. 4. Measured and calculated interference factor at $\theta=55°$. At this angle $P_2(\cos\theta)=0$; the dominant E1 contribution vanishes, allowing true E1-E2 interference effects via A_1 and A_3 coefficients. The measured points are obtained from $I=(N(\theta)-N(\pi-\theta))/(N(\theta)+N(\pi-\theta))$, where $N(\theta)$ is the intensity measured at the angle θ.

fairly well described by the calculations implying that the model is sensitive enough to the E1-E2 interference effects. This fact is further confirmed by the good agreement to the interference factor, defined by

$$I = (A_1 P_1 + A_3 P_3) / (A_0 + A_2 P_2 + A_4 P_4)$$

and presented in Fig. 4. As in ^{208}Pb[2,3] for A_2/A_0 only the general data trend is reproduced, but the vanishing region arround 9 MeV is missed.

In ^{159}Tb good agreement was obtained with all the measured coefficients, using larger than usual, particle vibration, real and imaginary coupling constants ($V_d=W_d=$ 300 MeV). They were especially needed in order to reproduce the isotropic distribution at 9.7 MeV. It is of interest to further investigate other deformed nuclei and look for a similar increase in the particle-vibration strength.

1. Y. Birenbaum, Z. Berant, A. Wolf and R. Moreh, Phys. Lett. <u>88B</u>, 239 (1979).
2. Y. Birenbaum, Z. Berant, S. Kahane, A. Wolf and R. Moreh, Nucl. Phys. <u>A369</u>, 483 (1981).
3. H.R. Weller, R.G. Seyler and R. Moreh, Phys. Rev. <u>C28</u>, 1959 (1983).

The ^{232}Th(γ,f) and (γ,n) reactions between 5 and 10 MeV

D. J. S. Findlay, G. Edwards, N. P. Hawkes, M. R. Sené

(AERE Harwell, England)

There is interest in the fission of thorium because the thorium isotopes seem to display characteristics which can be explained only in terms of a minimum in the outer hump of the double-humped fission barrier[1]). Because of this "thorium anomaly", several experiments on charged-particle, neutron and photon induced fission have been performed over the past few years. The present experiment was undertaken to measure simultaneously the (γ,f) and (γ,n) cross-sections of ^{232}Th and the mean number $\bar{\nu}$ of neutrons per fission at photon energies between ~5 and ~10 MeV; measurement of these three quantities provides more information on the fission process than the measurement of the (γ,f) cross-section alone.

The experiment was performed on the Low Energy beam line of the Harwell electron linear accelerator HELIOS. The electron beam energy was analysed to ±1.0% for one set of measurements and ±0.5% for a second set. Bremsstrahlung from a 0.1 g cm^{-2} Au radiator was collimated and intercepted by the Th photonuclear target; the bremsstrahlung dose was measured by an NBS P2 chamber. The Th

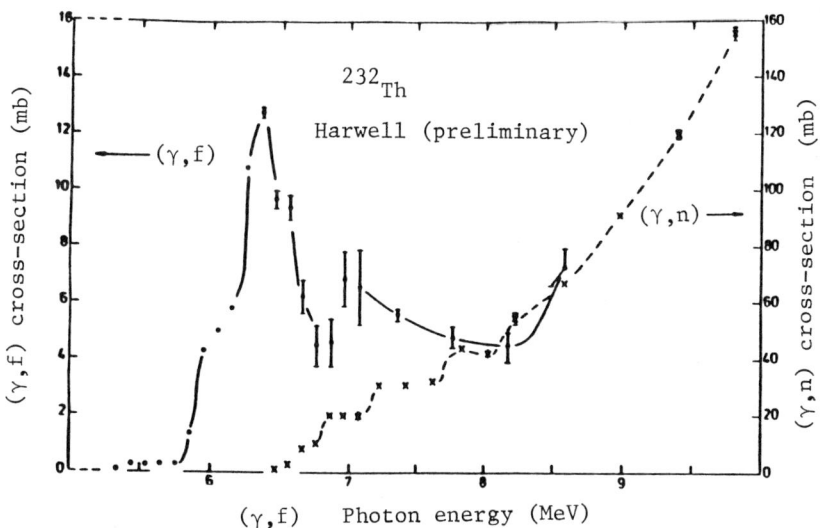

Fig. 1 The measured ^{232}Th(γ,f) and (γ,n) cross-sections (note separate vertical scales). The photon energy resolution is 130, 200 and 390 keV for the points spaced 100, 200 and 400 keV apart. The lines joining the points are only to guide the eye.

target thicknesses used ranged from 2 mg cm^{-2} to 4 g cm^{-2}. Neutrons from (γ,f) and (γ,n) reactions in Th were detected in a large oil-moderated assembly of fifty-six ^{10}BF$_3$ counters. Neutron multiplicity distributions were recorded, and photofission and photoneutron events were separated by using their different neutron multiplicities[3]). Cross-sections and some values of $\bar{\nu}$ were unfolded from the measured yields by specifically developed techniques[4]) using specifically calculated bremsstrahlung spectra[5]).

The ^{232}Th(γ,f) and (γ,n) cross-sections, measured with a photon energy resolution of 130 keV below 7 MeV, are shown in Fig. 1. These two cross-sections have not previously been measured simultaneously over such an energy range with such resolution, and so should prove useful in extracting the transmission through the fission barrier as a function of energy. The (γ,f) cross-section, measured with an energy resolution of 75 keV, is shown in Fig. 2. There is one plateau between about 5.45 and 5.75 MeV where two resonances at 5.48 and 5.63 MeV are evident, and there is a second plateau between about 5.95 and 6.10 MeV suggesting the presence of resonances at 5.94 and 6.05 MeV.

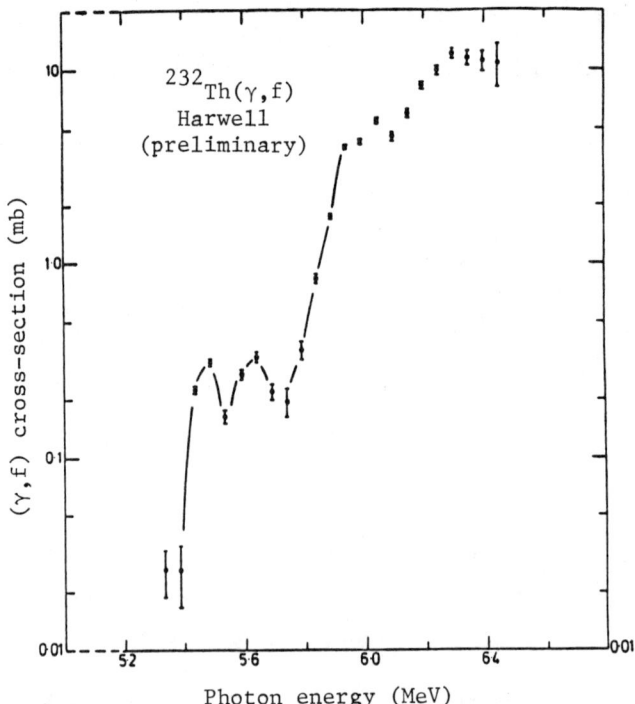

Fig. 2 The ^{232}Th(γ,f) cross-section measured with a photon energy resolution of 75 keV. The lines joining the points are only to guide the eye.

These data confirm resonances seen by Knowles et al[6]) using tagged photons, and demonstrate the value of using bremsstrahlung for making good resolution measurements at low photon energies where the cross-section is small.

It is tempting to interpret the resonances on the two plateaus in the (γ,f) cross-section as being resonances in two shallow "third" wells in the outer hump of the double-humped barrier. Recent calculations[7]) have suggested that there may be two separate paths to fission over the outer hump, about 300 keV apart. These separate paths may be the cause of the two plateaus separated by ∼450-500 keV seen in the present measurements; the resonances on these plateaus would then be resonances in two shallow wells, one in each separate path. Possible additional evidence for this interpretation may be provided by the energy dependence of the $\bar{\nu}$. values measured in the present experiment (Fig. 3). There is a

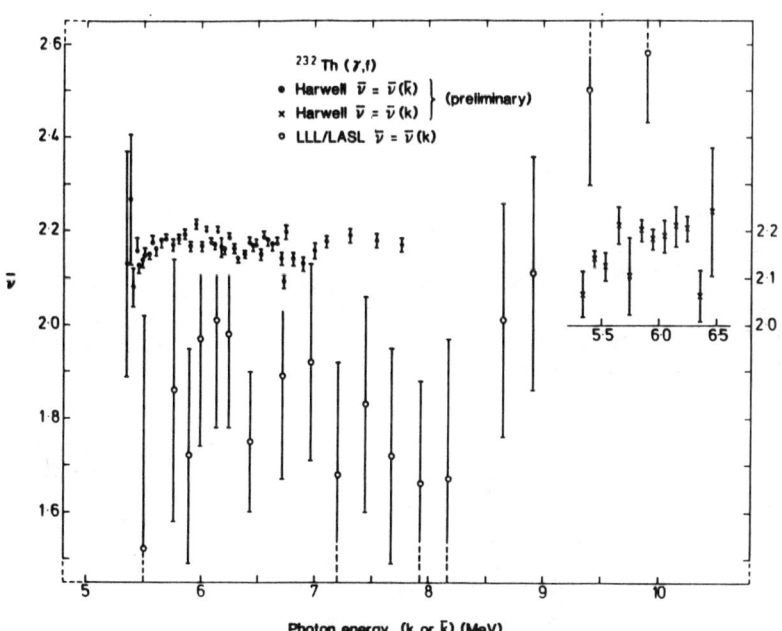

Fig. 3 Measured values of $\bar{\nu}$, the mean number of neutrons per fission. Most of the Harwell measurements (solid circles) are shown as a function of $\bar{k}$, a mean photon energy computed from the shapes of the (γ,f) cross-section and the bremsstrahlung spectrum, but some Harwell measurements (crosses) are also shown as a function of actual photon energy k below 6.5 MeV. The Livermore/Los Alamos values (open circles) are shown for comparison.

change in slope of $\bar{\nu}$ as a function of energy at ~5.9 MeV, the energy at which the second plateau in the (γ,f) cross-section begins. This change in slope could be due to the opening of an additional path to fission at ~5.9 MeV if this additional path corresponds to less internal excitation energy being given to the nascent fission fragments.

Support from Drs. J. E. Lynn and M. S. Coates is gratefully acknowledged. Assistance from and discussions with Dr. E. W. Lees are also acknowledged.

References

1) S. Bjornholm and J. E. Lynn Rev. Mod. Phys. 52 (1980) 725
2) E. W. Lees et al Nucl. Instr. Meth. 171 (1980) 29
3) G. Edwards et al Annals Nucl. Energy 8 (1981) 105
4) D. J. S. Findlay Nucl. Instr. Meth. 213 (1983) 353
5) D. J. S. Findlay Nucl. Instr. Meth. 206 (1983) 507
6) J. W. Knowles et al Phys. Lett. 116B (1982) 315
7) J. Dudek et al Nucl. Phys. A412 (1984) 61

HIGH RESOLUTION STUDIES OF PHOTOFISSION OF ^{232}Th and ^{238}U

H.X. Zhang and H. Lancman
Brooklyn College, CUNY, Brooklyn, N.Y. 11210

ABSTRACT

Photofission cross sections have been determined in several photon energy intervals with a photon energy resolution ~ 200 eV. The structure observed at 6.31 MeV in ^{232}Th is interpreted as a reflection of class II compound states in this nucleus.

Recently we have reported[1,2] the results of our measurements of intermediate structure in the photofission cross sections of ^{238}U and ^{232}Th. In both nuclei the structure was observed at slightly subbarrier energies. The aim of the experiments described in the present note was to extend those measurements to higher excitation energies in order to investigate the behavior of the structure above the fission barrier.

As before,[1] the gamma rays were produced in (p,γ) reactions at several proton resonances. Gamma rays emitted from a target bombarded by protons of a resonance energy E_p were allowed to fall on thin foils of uranium and thorium sandwiched between plastic films as shown in the inset in Fig. 1. The sandwiches were located on a

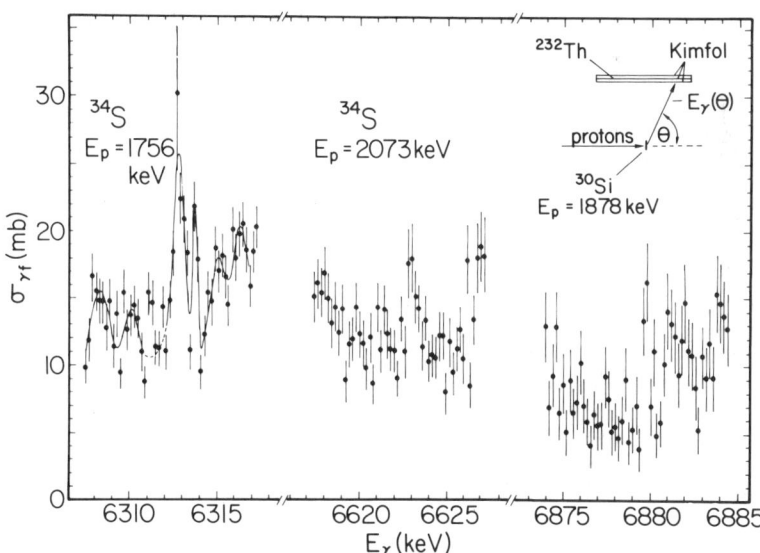

Fig. 1. Photofission cross section of ^{232}Th. The inset shows the experimental setup.

Table I. Average area, width and level spacing at 6.31 MeV

	Expected	Observed
A (b·eV)	11.4	8.2 ± 3.0
W (eV)	511	780 ± 300
D_{II} (keV)	1.2	1.6 ± 0.3

cylinder coaxial with the proton beam. The films were used as fission track detectors. The energy of the gamma rays, E_γ, varied continuously with the angle $\emptyset$ as a result of the Doppler shift. The average photon energy resolution was $\sim$ 200 eV. It was determined mainly by uncertainties in the proton beam position and the alignment of the sandwiches. The natural widths of the 6.62 MeV and 6.88 MeV resonances are 22 and 20 eV respectively.

The photofission cross sections of ^{232}Th measured in three different photon energy intervals are shown in Fig. 1. Structure can be seen in all three spectra, however only the spectrum at 6.31 MeV exhibits features similar to those seen at lower energy although the peaks in this case are less pronounced, reflecting the fact that the $K^\pi = 0^-$ channel is already almost entirely open. The average area, width and spacing of the observed peaks are listed in Table I. They are in good agreement with the computed values obtained with the barrier parameters determined previously.[1]

The structure observed in the spectra at the two higher energies could be caused by fluctuations of class II width and (or) level spacing. The expected value of the latter at these energies becomes comparable to the average width of the peaks.

The averaged photofission cross sections of ^{232}Th at each proton resonance are plotted in Fig. 2 versus the gamma ray energy. In addition to the values obtained from Fig. 1 we have used our results at lower photon energies reported previously. Our values are in good overall agreement with the results of Dickey and Axel[3] and those of Caldwell et.al.[4] It should be noted that our results are obtained by averaging over energy intervals an order of magnitude or more smaller than in these two experiments.

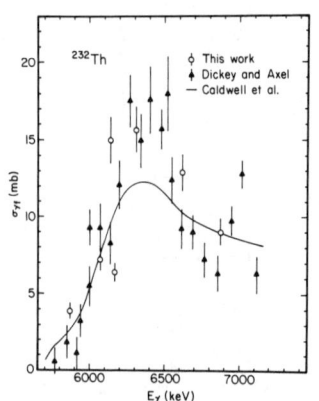

Fig. 2. Photofission cross sections of ^{232}Th determined in different experiments.

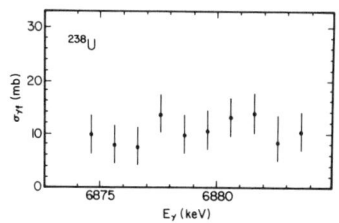

Fig. 3. Photofission cross section of 238U.

The background due to spontaneous fission of ^{238}U made it impossible to determine the photofission cross sections at the two lower photon energies. At 6.88 MeV the spectrum within the statistical accuracy of the experimental points shows no structure as can be seen from Fig. 3. The average value of the cross section at this energy is in good agreement with other determinations.

This work was supported in part by USDOE.

REFERENCES

1. H.X. Zhang, T.R. Yeh and H. Lancman, Phys. Rev. Lett. 53, 34 (1984).
2. H.X. Zhang, T.R. Yeh and H. Lancman, Proc. Int. Conf. Nucl. Phys., Florence, Italy 1983, edited by P. Blasie and R.A. Ricci (Tipografia Compositori, Bologna, Italy, 1984), Vol. 1, p. D121.
3. P.A. Dickey and A. Axel, Phys. Rev. Lett. 35, 501 (1975).
4. J.T. Caldwell, F.J. Dowdy, B.L. Berman, B.A. Alvarez and P. Mayer, Phys. Rev. C21, 1215 (1980).

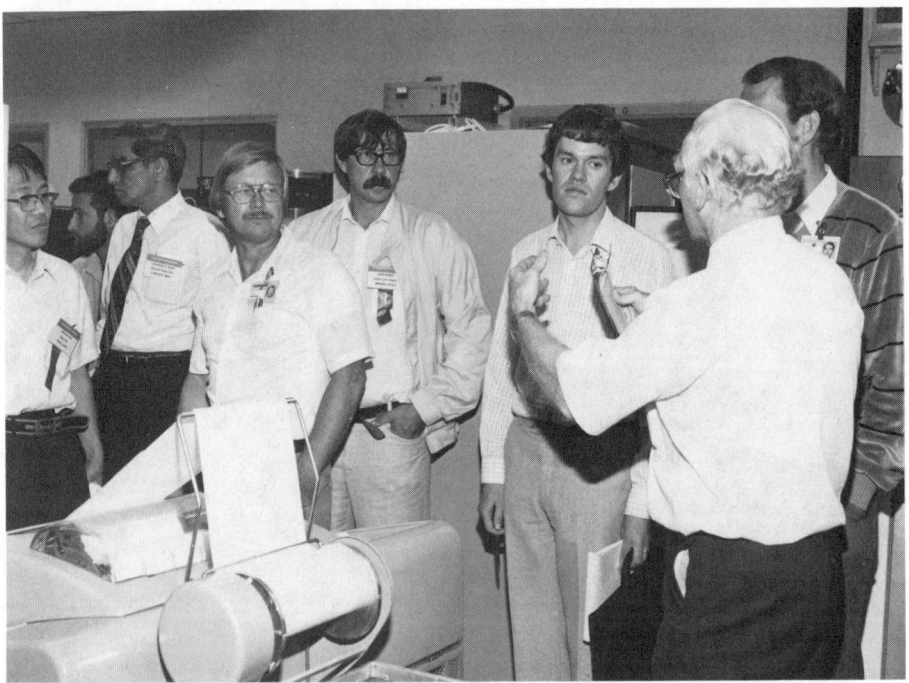

III. SLOW NEUTRONS

TESTS OF SUPERSYMMETRIES WITH THE (n,γ) REACTION

D. D. Warner
Brookhaven National Laboratory, Upton, New York, 11973, USA

ABSTRACT

The characteristics of the SU(3) and O(6) boson-fermion symmetries stemming from a U(6/12) group structure are discussed and compared with recent results of (n,γ) studies in the W-Pt region. The nuclei ^{185}W and ^{195}Pt are shown to represent the best empirical examples of the SU(3) and O(6) limits, respectively, and it is also shown that the Consistent Q Formalism can be extended to the odd A Hamiltonian to describe the transition between these two limits. Preliminary comparisons with the low lying structure of the odd Os nuclei are presented. The question of the empirical evidence for supersymmetry in this region is also discussed.

1. INTRODUCTION

The development of the Interacting Boson-Fermion Model[1], and the recognition of its associated symmetries[2], offers the chance to test our understanding of the collective structure of odd A nuclei over far broader ranges of mass and excitation energy than has hitherto been possible. This extension in scope stems principally from the inclusion of a core description which can run the full gamut of vibrational, rotational or asymmetric structure, and which incorporates essentially all collective excitations. Thus, in the region of well deformed nuclei, for instance, one can expect the model to generate an equally detailed description of both the low lying rotational structure which emerges from a Nilsson model treatment, and the subsequent vibrational modes which to date have, in general, been treated only qualitatively. Moreover, in regions outside those of axially symmetric deformation or near sphericity, the IBFM's capabilities should prove even more crucial, since here deficiencies in the core description can manifest themselves even at low excitation energies.

The validity of the IBFM can obviously first be examined in terms of its ability to reproduce odd-even structure in regimes which are already well understood. However its advantages, and the new insights which it may offer, must be searched for in regions which are as yet only poorly understood and, as a result, often neglected in experimental studies. The hallmark of the model, as explained above, is its implicit ability to generate the complete spectrum of collective modes in the odd A nucleus, and thus one of the most suitable probes to test it is the (n,γ) reaction. The statistical nature of the decay of the compound state formed after neutron capture ensures population of a broad range of final states, irrespective of their specific structural characteristics. The use of the Average Resonance Capture (ARC) technique is particularly important in this regard since it results in a reduction in the fluctuations inherent in the statistical origins of the final state population, to the point where typically all final states which can be attained by E1 primary radiation <u>must</u> be populated, up to an excitation energy dependent on the sensitivity of the measurement. Thus the completeness of the theoretical predictions is tested in absolute terms.

2. SYMMETRY IN ODD-EVEN NUCLEI

The development of the IBFM is following a similar course to that of its even-even counterpart and predecessor, the IBM. Thus the "basic" Hamiltonian appears relatively complex and holds little hope for physical intuition. For instance, the number of terms, and hence number of parameters, in the boson-fermion interaction alone is approximately n^3, where n is the number of single particle orbits considered. Nevertheless, as in the case of the IBM, it is possible to simplify the situation, and at the same time produce a Hamiltonian whose terms have a more physical interpretation, and which can be applied to broad ranges of nuclei in a comprehensible way. This simplification involves two steps. In the first, the various terms in the Hamiltonian are grouped according to their tensorial properties, so that, in the IBFM, only two significant composite contributions to the boson-fermion interaction remain, a quadrupole-quadrupole term and an "exchange" term. Note that essentially all descriptions of collective odd A structure incorporate a Q·Q interaction between core and particle, and hence it is the exchange term which is particular to the IBFM, in that it implicitly recognizes the underlying fermionic origins of the bosonic core.

The second step centers on the recognition of dynamical symmetries. These provide analytic solutions in certain specific limits of the general Hamiltonian and can be associated with a well defined geometrical structure appropriate to particular regions of nuclei. They thus define starting points, at least, in determining the parameterization of a specific nucleus, and they also provide insights into how to change that parameterization in a smooth and physically reasonable way to describe the transition regions between symmetries.

The group structure of a boson-fermion system is described by $U^B(6) \times U^F(m)$ where m specifies the number of states available to the odd fermion, and thus depends on the single particle space assumed. The ability to construct group chains corresponding to the symmetries SU(5), SU(3) or O(6) depends on the value of m, and this problem has already been discussed in detail in a separate contribution[3]. Of the structures studied in detail to date, the case of m=12 is the one with the broadest potential. The fermion is allowed to occupy orbits with $j = 1/2$, 3/2 and 5/2, so that the assumed single particle space corresponds to the negative parity states available to an odd neutron at the end of the N = 82-126 shell, namely, $p_{1/2}$, $p_{3/2}$ and $f_{5/2}$. The region of interest thus spans the W-Pt nuclei, and since one prerequisite for an odd-A symmetry is the existence of that same symmetry in the neighboring even-even core nucleus, the odd Pt nuclei around A = 196 offer the

obvious testing ground for the O(6) limit[4] of U(6/12). The heavier
even-even W nuclei, on the other hand, have the characteristics of
an axial rotor, and hence the negative parity structure of the
neighboring odd W isotopes offers the possibility to study the
validity of the SU(3) limit. Finally, given a definition and
understanding of these two limits, the construction of a simple
description of the transitional odd A Os nuclei can be considered.

The above discussion centers on symmetries in the boson-
fermion system. However, if a particular symmetry exists in neighboring even-even and odd-even nuclei, it is possible to ask whether
the two schemes stem from a common parent supersymmetric group
structure of the type U(6/m). The only way to examine such a question in the nuclear regime is to test whether the members of a
given supermultiplet, characterized by a constant total number of
bosons and fermions, can all be described by a single Hamiltonian.
In practice, this reduces to asking whether the odd-even and
appropriate even-even nucleus can be described with the same
parameters.

3. (n,γ) STUDIES IN THE W-Pt REGION

As pointed out earlier, the (n,γ) reaction in general, and the
ARC technique in particular represent ideal probes to test the
completeness of the predicted IBFM symmetry schemes. Such data
must, of course, be complemented by studies that probe the structure of the states via single particle transfer cross sections or
electromagnetic matrix elements. Nevertheless, the (n,γ) studies
represent the crucial first step in locating essentially all low-
lying, low spin states in the nuclei of interest.

A detailed discussion of the physical principles underlying
the ARC technique is presented in a separate contribution[5] to these
proceedings. However, it is worth summarizing the basic characteristics of the method here, and considering how they pertain to the
specific case of odd mass nuclei in the W-Pt region.

In single resonance, or thermal, neutron capture, the γ decay
of the compound state to low-lying final states is characterized by
intensities which, after correction for an energy dependence,
follow a Porter-Thomas distribution with one degree of freedom.
The form of such a distribution is illustrated by the $\nu=1$ curve in
Fig. 1a. Clearly, the probability for zero decay width is high; so
that not all final states need be populated and those that are will
be fed with widely ranging intensities. The ARC technique overcomes these problems by using neutron beams with a finite spread in
energy, such that a number of resonances are encompassed. This
results in a corresponding reduction in the intensity fluctuations,
as shown in Fig. 1a. As the number of resonances (or ν) becomes
large, the intensity distribution tends to a Gaussian with variance
$4/\nu$.

Given sufficient averaging, therefore, a complete set of
levels populated by E1 primaries can be established. In the case
of interest here, this corresponds to $J^\pi = 1/2^-$, $3/2^-$ states in
the odd A nuclei in the W-Pt region. However, the resonance level

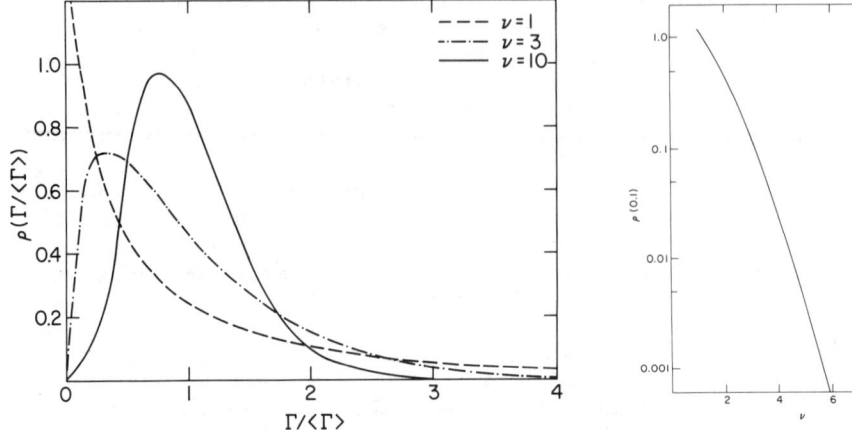

Fig. 1. a) Porter Thomas distributions for different numbers of degrees of freedom ν.
b) The probability function for $\Gamma/\langle\Gamma\rangle=0.1$.

density for these nuclei is low, so that on average only 5-10 resonances are involved in the averaging process. The question then arises as to whether the population of even $J^\pi = 1/2^-$, $3/2^-$ states can be guaranteed under such conditions. The answer can be seen in Fig. 1b, which shows the probability function for $\Gamma/\langle\Gamma\rangle = 0.1$, i.e., for primary intensities one tenth of the mean. It is clear that for $\nu=5$, the probability is already negligible. Since a reduced intensity of 10% of the mean is easily observable in these measurements, at least up to excitation energies $\approx$ 1200 keV, one can conclude that the ARC technique can still guarantee population of all $J^\pi = 1/2^-$, $3/2^-$ states in this region.

4. ODD Pt NUCLEI: THE O(6) LIMIT OF U(6/12)

As pointed out in Section 2, the well established O(6) symmetry in ^{196}Pt and its neighbors[4], coupled with the isolated $p_{1/2}$, $p_{3/2}$ and $f_{5/2}$ orbits available to an odd neutron in this region, implies that the odd Pt nuclei should offer the best opportunity to test the predictions of the O(6) group chain of U(6/12). The results of recent (n,γ) studies[6] of ^{195}Pt are summarized in the level scheme of Fig. 2. The scheme was constructed following ARC studies at Brookhaven National Laboratory, and also measurement of the secondary γ-ray spectrum with the GAMS curved crystal spectrometers at the Institut Laue-Langevin, Grenoble. It is worth noting that the completeness of the observed set of $J^\pi = 1/2^-$, $3/2^-$ states led to the establishment of a level at 222 keV which had hitherto escaped detection in other studies[7]. As will become apparent, this level plays a crucial role in the structural interpretation of this nucleus.

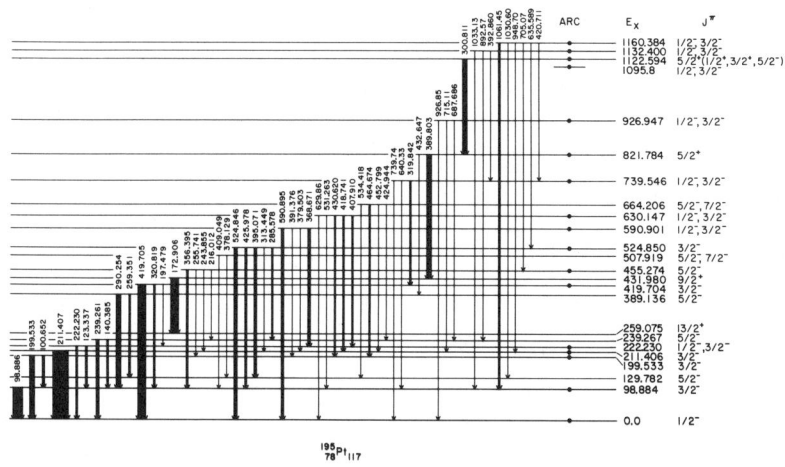

Fig. 2. The level scheme of ^{195}Pt (ref. 6).

It is evident that ^{195}Pt displays no obvious rotational structure. Indeed, simply the number of low-lying J^{π} = 1/2$^-$, 3/2$^-$ states precludes a simple Nilsson description, since a maximum of 8 such states can arise from the Nilsson orbits in this region. These features stem, of course, from the nature of the core nucleus which exhibits γ-unstable, rather than axially-symmetric structure. Moreover, earlier attempts[8] to allow for this difference by including the effects of coupling to the low-lying 2^+_γ core state also encountered severe problems, and appear unable to account for the newly discovered 222 keV state. Thus it appears that a more complete and realistic core description is necessary.

The comparison with the U(6/12) O(6) symmetry scheme is given in Fig. 3. The origins of this scheme, and its associated quantum numbers, have been discussed in detail elsewhere[9]. It is therefore sufficient to remark here that it offers an adequate description of the observed structure, at least below 600 keV. It is particularly encouraging that a one-to-one correspondence can be made between experimental and theoretical levels up to this energy. Moreover, subsequent (n,n',γ) studies[10] have removed a number of the ambiguities in the spin assignments of Fig. 2, and in all cases, the results confirm the association of states shown in Fig. 3. Data from Coulomb excitation studies[7,11] and single particle transfer studies[12] are also largely in agreement with the symmetry predictions, although some important discrepancies have been found[13] in the latter case for the reaction ^{195}Pt→^{196}Pt. However, it is possible, and indeed likely, that these stem from uncertainties in the form of the IBFM transfer operator itself.

A distinctive feature of the symmetry scheme is the existence of couplets of levels with J, J+1 separated by a constant J(J+1) spacing. This feature shows up clearly in the data and results

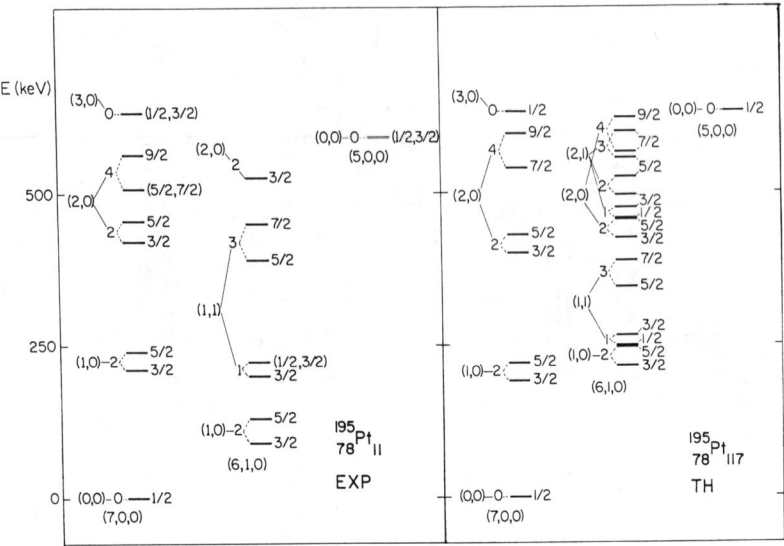

Fig. 3. A comparison of the O(6) symmetry scheme of U(6/12) with the levels of ^{195}Pt.

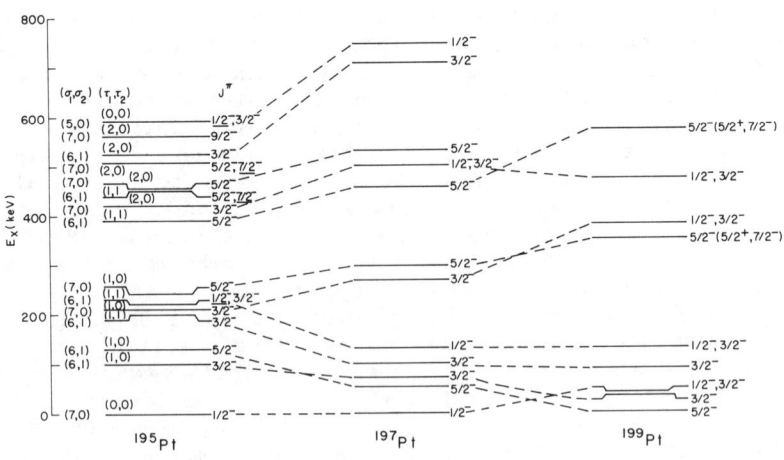

Fig. 4. Association of low lying levels in ^{195,197,199}Pt.

from the pseudo-spin symmetry inherent in all the group chains of U(6/12). There is, however, a clear discrepancy between theory and experiment at higher excitation energies in that the predicted states in the representation labelled [N,1] are two compressed, relative to the data. A modification to the original scheme which removes this problem, while maintaining the symmetry, is discussed in ref. 14.

It is clearly of interest to study if, and for how long, the symmetry structure persists in neighboring odd Pt nuclei, and to this end further ARC studies were made of ^{197,199}Pt. The results of these are discussed in detail in ref. 15, but can be summarized by reference to Fig. 4, which shows the evolution of the low lying states in this region. The number of low lying $J^\pi = 1/2^-, 3/2^-$ states found in ^{197}Pt was essentially identical to that in ^{195}Pt, and the couplet structure is also evident, although with less constancy in spacing. The situation for ^{199}Pt is, however, far less convincing. These basic conclusions are confirmed by a study[12] of single particle transfer cross sections, which indicate substantially increased symmetry breaking in ^{197}Pt.

5. ODD W NUCLEI: THE SU(3) LIMIT OF U(6/12)

The SU(3) limit of U(6/12) requires a rotational core structure, coupled to j = 1/2, 3/2 and 5/2 orbits. The odd W nuclei represent the best chance of observing characteristics of this symmetry since, in nuclei of lower mass in the well deformed rare earth region, the Fermi surface is progressively farther from the single particle orbits of interest. The predicted representations and their associated quantum number are illustrated in Fig. 5 for the case of coupling the boson and fermion degrees of freedom at the level of U(6), and again the reader is directed elsewhere[16] for a more detailed discussion of the origins of the scheme.

The SU(3) limit has the attractive advantage that its predictions can be compared with those of the Nilsson model for the same shell model states, so that a more physical interpretation of its structure can be formulated. To facilitate this type of comparison, the rotational band structure has been indicated in Fig. 5, in terms of the K quantum numbers of the bands contained within each SU(3) representation. Moreover, the results of a detailed study[16] have shown that in the lowest two representations, the only core states involved are those of the ground state rotational band, so that, in these cases, the bases of the two descriptions are identical. A more quantitatively-based link can then be established by means of the single particle structure of the wave functions. This is illustrated in Fig. 6, where the quantity $C_{j\ell}^{eff}$ is compared for the lowest three bands in the SU(3) scheme, and the lowest of the Nilsson orbits emanating from the $p_{1/2}$, $p_{3/2}$ and $f_{5/2}$ states, namely, the 1/2[521], 1/2[510] and 3/2[512] bands. The $C_{j\ell}^{eff}$ values in the Nilsson scheme correspond to the sum of $C_{j\ell}$ coefficients for a state I=j over the Coriolis mixed orbits. Coriolis coupling should be incorporated automatically in the equivalent IBFM scheme, and it is easy to show that the equivalent

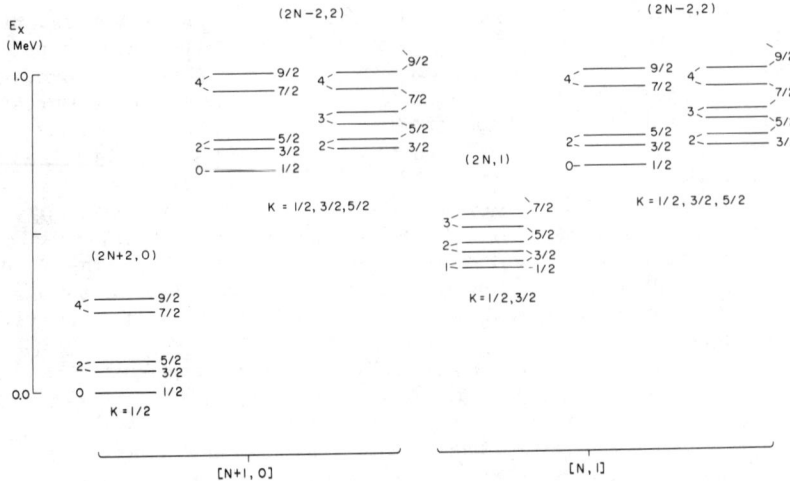

Fig. 5. The SU(3) scheme of U(6/12). Representations are labelled by the (λ,μ) quantum number of $SU^{BF}(3)$ and by the $[N_1 N_2]$ labels of $U^{BF}(6)$.

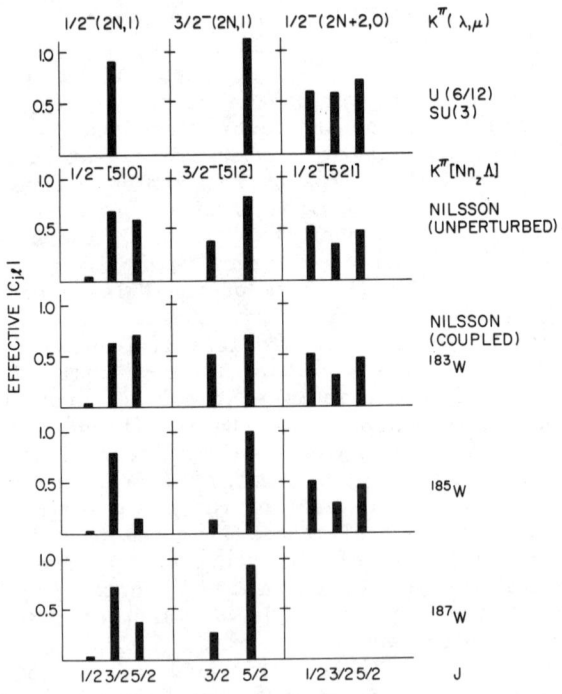

Fig. 6. Values of $C_{j\ell}^{eff}$ in the U(6/12) and Nilsson schemes. Those for the latter are calculated using the wave functions and Coriolis mixing amplitudes of ref. 17.

quantities correspond, with suitable normalization, to the amplitudes a_{j0}, for a single particle orbit of spin j coupled to the 0^+ ground state of the core.

Nilsson coefficients for both the unperturbed and Coriolis mixed orbits are shown in Fig. 6, the latter having been obtained from the results of ref. 17. The connection between the two frameworks is evident. The 1/2[521] and $(\lambda,\mu) = (2N+2,0)$ bands show an almost identical single particle structure throughout the W isotopes, although the absolute values in the latter case are larger because of the missing strength from the $f_{7/2}$ and $h_{9/2}$ orbits, which are not included in the U(6/12) basis. In the case of the K = 1/2 and 3/2 bands from the (2N,1) representation, the figure shows that the required structure corresponds to that of the 1/2[510] and 3/2[512] bands, after inclusion of a specific Coriolis interaction which is found empirically in ^{185}W.

Consideration of the predicted energy spectrum reinforces the above conclusions and the U(6/12) scheme is compared with the levels in ^{185}W in Fig. 7. Note that the $(\lambda,\mu) = (2N,1)$ representation of Fig. 5 can be made the ground state representation by a suitable choice of the strength of the Casimir operator of the group $U^{BF}(6)$. The Coriolis interaction between the 1/2[510] and 3/2[512] bands and the single particle structure of the 1/2[521] manifest themselves as a near degeneracy between states in the former pair, and a decoupling parameter near unity in the latter. Both features appear naturally in the symmetry scheme. However the figure also shows that recent ARC measurements[18] have identified five additional J^π = 1/2$^-$, 3/2$^-$ states in the region of 600-800 keV.

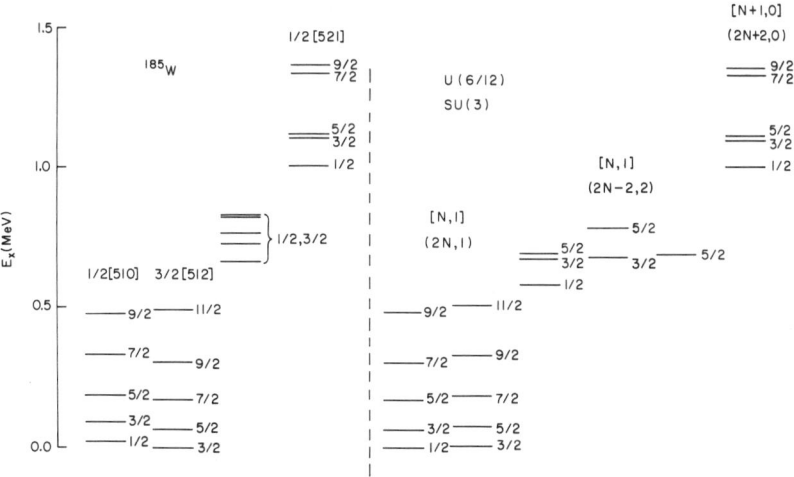

Fig. 7. Comparison of assigned Nilsson bands in ^{185}W with the SU(3) scheme of U(6/12). The location of five as yet unassigned J^π = 1/2$^-$, 3/2$^-$ states is also shown. Assigned bands stemming from the $f_{7/2}$ and $h_{9/2}$ orbits are not included.

Three of these were known from earlier work[17], but could not be associated with simple Nilsson configurations. The symmetry scheme, on the other hand, provides an interpretation for three of the five, but still leaves two unaccounted for. This draws attention to a fundamental inadequacy of the U(6/12) scheme in the deformed region, in that the single particle space is not sufficient. It has already been pointed out that significant components of the $f_{7/2}$ and $h_{9/2}$ orbits appear in the wave functions of the bands of interest, and in addition, the 9/2[505], 7/2[503] and 5/2[512] bands themselves have been identified below 1 MeV in ^{185}W. Thus a similar origin can be sought for the two "extra" $1/2^-$, $3/2^-$ states, in terms of the $1/2^-[770]$ and/or $3/2^-[761]$ bands from the $j_{15/2}$ state.

6. THE SU(3)→O(6) TRANSITION AND THE CQF IN ODD A NUCLEI

Despite its limited applicability, it is clear that the U(6/12) SU(3) scheme yields a basically valid description of the structure of ^{185}W. Thus there are now two benchmarks in the U(6/12) basis, ^{195}Pt and ^{185}W, which define the SU(3) and O(6) limits, respectively, and it is possible to consider a description of the transitional odd A nuclei in between. For the even-even nuclei in this region, a simple approach involves the Consistent Q Formalism[19], in which the variation of the single parameter χ in the boson quadrupole operator

$$Q_B = (s^+\tilde{d} + d^+s)^{(2)} + \chi/\sqrt{5}\ (d^+\tilde{d})^{(2)} \tag{1}$$

between its SU(3) and O(6) values ($-\sqrt{35}/2$ and 0) reproduces the gross structural changes across the region.

In the IBFM, the quadrupole operators, both boson and fermion, enter the symmetric Hamiltonian via the Casimir operator $C_{2SU^{BF}(3)}$, which generates a quadrupole-quadrupole interaction of the form

$$Q \cdot Q = (Q_B + Q_F) \cdot (Q_B + Q_F) \tag{2}$$

In fact, recent work[20] has shown that the fermion operator Q_F can also be parameterized by χ, such that when $\chi=0$

$$C_{2SU^{BF}(3)} = C_{2O^{BF}(6)} - C_{2O^{BF}(5)} \tag{3}$$

and the SU(3) Hamiltonian reduces to that of O(6), with the restriction that the O(6) and O(5) Casimirs are governed by the same constant. In fact, the first success of this approach can be taken as the fact that, in the best fit to ^{195}Pt of Fig. 3, the relevant two constants were indeed found to be almost equal (33.5 and 35.0 keV).

The transitional region in question spans the odd Os nuclei and unfortunately, our experimental knowledge of these nuclei is still, in most cases, inadequate to attempt a detailed fit. However, the low lying structure has been a subject of considerable interest for some time[21,22]. The situation can be summarized by noting that the four lowest low spin states have J^π = 3/2⁻, 1/2⁻, 5/2⁻ and 3/2⁻ in these nuclei, similar to ^{185}W. In fact, the accepted interpretation to date, based on the Nilsson model, is that these studies indeed stem largely from the Coriolis coupled 1/2[510] and 3/2[512] orbits, these being kept near the Fermi surface by the decreasing deformation in this region. However, the large single particle structure factors deduced for the second 3/2⁻ and first 5/2⁻ states in each case cannot be accounted for by these orbits alone, so that it has been assumed that fragments of the higher lying 3/2[501] and 5/2[503] also enter in the wave functions. The term fragment is used here to imply that mixing with

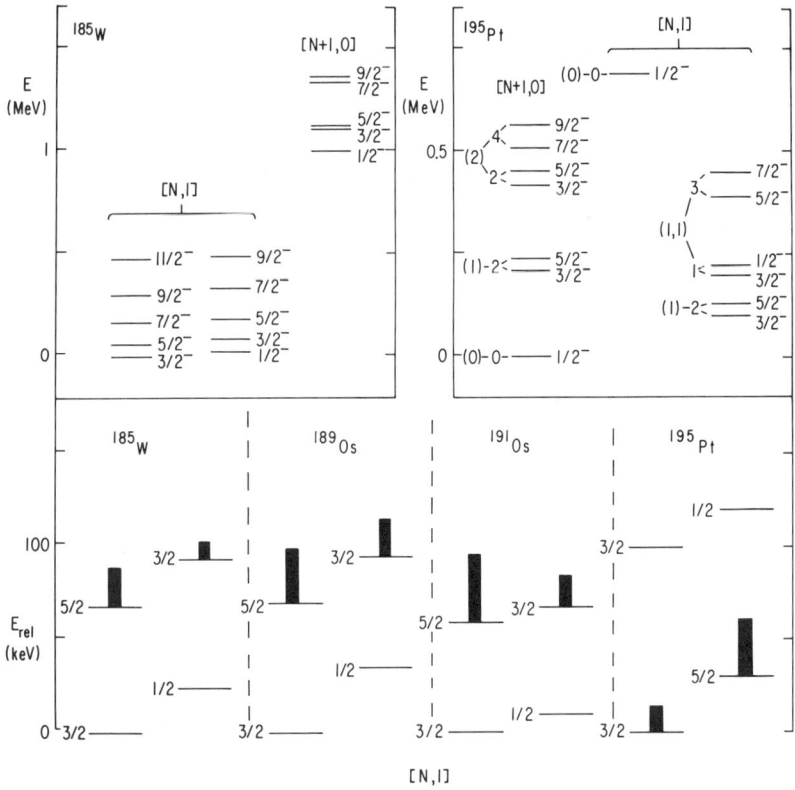

Fig. 8. (Top) The characteristic SU(3) and O(6) structure in ^{185}W, ^{195}Pt and (bottom) the single particle structure factors across the region in the low lying group of states shown (see text for details).

vibrational core excitations has to be assumed in order to bring the additional single particle strength sufficiently low in energy, and to account in general for the known spread of that strength across a large number of states in this region[23]. Such a conclusion is not, of course, surprising given the non-rotational structure of the even-even Os nuclei and, in particular, the low lying second 2^+ state in each.

The empirical situation is summarized in Fig. 8. At the top, the spectra of ^{185}W and ^{195}Pt are shown, displaying the characteristic features of the SU(3) and O(6) symmetries, respectively. Below, the four low lying states mentioned above are shown on a relative energy scale, and also their (d,t) structure factors. Note that in ^{195}Pt, the group of states no longer form the ground state, but appear at 99, 129, 199 and 222 keV instead. Figure 9 shows the simplest possible calculation within the CQF framework; in which all parameters, including the boson number, have been kept constant, except for χ. Figure 9a shows the SU(3) scheme and, for convenience, the various bands have been labelled with pseudo-K quantum numbers appropriate to the values of pseudo orbital angular momenta contained within each. The evolution of these bands as $\chi \to 0$ is then displayed in Fig. 9b. This part of the figure is necessarily schematic, since the exact details of the changing structure of states are extremely complex, and the rotational band structure disappears at some stage. Nevertheless, the most important feature is unambiguous. The $K_p = 2$ band descends rapidly in energy and eventually joins the $K_p = 1$ band to form the $(\sigma_1,\sigma_2,\sigma_3) = (N,1,0)$ structure of the O(6) limit. It is remarkable that it is precisely this band which has been shown[16] to contain the 3/2[501] and 5/2[503] Nilsson orbits, albeit mixed with β and γ vibrational core excitations. Thus the CQF predicts in this region, in a quantitative fashion, the same qualitative behavior deduced earlier.

The structure factors as a function of χ are shown in Fig. 9c. Two crucial points emerge. The predicted ratio S(5/2):S(3/2) is constant, and equal to 3:2. Empirically it ranges from 1.4 to 2.1. In addition, the effect of changing χ results in an increase in the <u>absolute</u> values and, again, this increase is seen in the data. Note that neither effect depends significantly on boson number.

Thus it must be concluded that the extension of the CQF to odd A nuclei can reproduce at least the low lying structure of the odd Os nuclei and therefore represents an attractively simple starting point for a general IBFM calculation in this region.

7. EVIDENCE FOR SUPERSYMMETRY

Up to this point, the validity of the U(6/12) schemes has been considered in the context of odd A nuclei only. However, as pointed out in Section 2, supersymmetry can be thought of as implying the simultaneous description of both even-even and odd-even

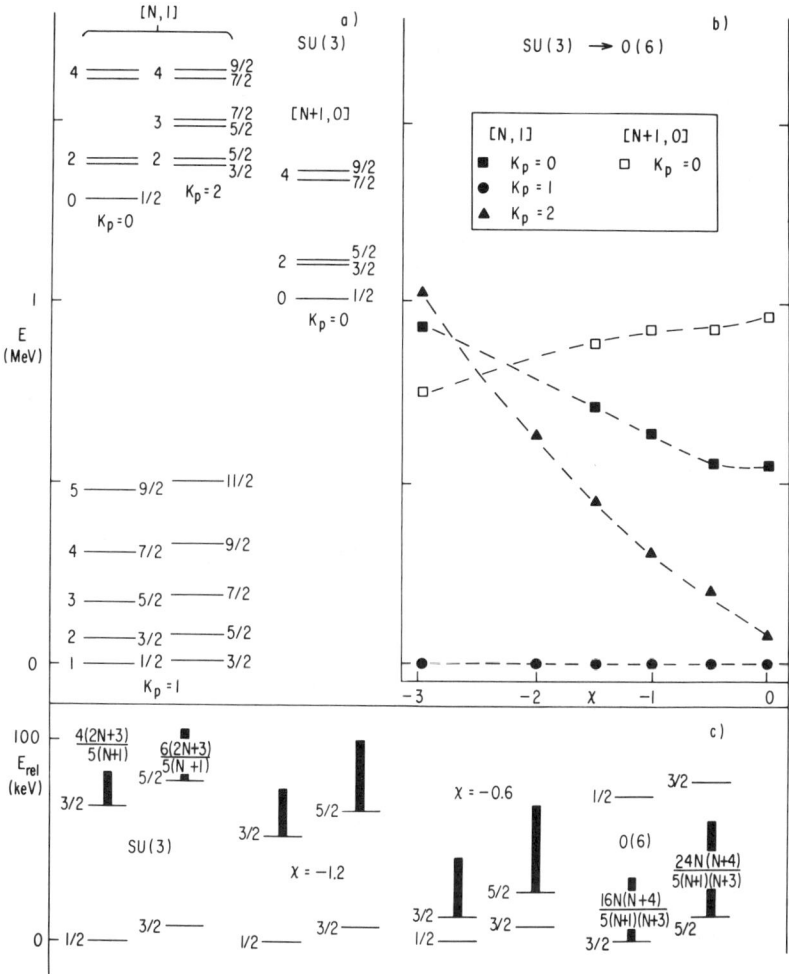

Fig. 9. a) The SU(3) limit; bands have been labelled by the pseudo projection quantum number K_p.
b) Schematic indication of the evolution of the bands as χ changes from its SU(3) value (-2.958) to O(6) (0).
c) The corresponding changes in the single particle structure factors of the low lying group of states shown in Fig. 8 (bottom).

partners in a multiplet characterized by a given total number of bosons plus fermions.

The eigenvalue expression for the O(6) limit can be written

$$AC_{2U BF(6)} + BC_{2O BF(6)} + CC_{2O BF(5)} + DC_{2O BF(3)} + EC_{2\ Spin(3)} \qquad (4)$$

where the various terms represent the relevant Casimir operators of the subgroups. The SU(3) expression is then obtained by replacing the second and third terms by $B'C_{2SU^{BF}(3)}$.

The supersymmetric partners of interest are ^{194,195}Pt and ^{184,185}W, and the parameters deduced for the even-even and odd-even nucleus in each case are compared in Table 1. The contribution from the $U^{BF}(6)$ Casimir is constant for all states in the even-even core, and hence does not affect the present discussion. Also, the last two groups of expression (1) combine in the even-even nucleus, so that it is the sum of the constants D and E which is relevant.

Table 1

Constant[a] in Exp (4)	O(6)		SU(3)	
	^{194}Pt[b]	^{195}Pt[c]	^{184}W	^{185}W
B	46.5	33.5		
C	42.0	35.0		
D+E	17.5	11.0	18.5	17.5
B'			>10	5.75

a) All parameter values are in keV.
b) From ref. 4.
c) From ref. 14.

The agreement for the O(6) case is reasonably good and, in fact, intermediate values of the two sets of parameters in Table 1 would produce a reasonable description of both ^{194}Pt and ^{195}Pt. In the SU(3) case, however, there is a clear problem. The strength of the Q·Q interaction, represented by the Casimir operator of SU(3), is required to be at least a factor of two stronger in ^{185}W than in its even-even partner and, in this case, there is no compromise which can be found to give acceptable fits in both cases. This result is perhaps not surprising in that it is known that the SU(3) symmetry in the even-even case does not extend to the E2 matrix elements, so that in the CQF, a χ different from the SU(3) value has to be adopted to obtain the best overall description. Since, of course, the same core quadrupole matrix elements enter into the IBFM problem, via the core-particle interaction of eq. (2), it is likely that this approach may represent a necessary symmetry breaking in ^{185}W also.

ACKNOWLEDGEMENT

The work reported here is the result of a number of collaborations, and the reader is directed to the original references cited in the text for a list of all participants. Particular thanks are due to A. M. Bruce, P. van Isacker and J. Jolie in connection with the as yet unpublished studies of ^{185}W and the extension of the CQF to odd A nuclei; also to R. Bijker for numerous discussions concerning the link between the general IBFM Hamiltonian and the symmetries; and to R. F. Casten in the case of the odd Pt studies.

Research has been performed under contract DE-AC02-76CH00016 with the United States Department of Energy.

REFERENCES

1. F. Iachello and O. Scholten, Phys. Rev. Lett. 43, 679 (1979).
2. F. Iachello, Phys. Rev. Lett. 44, 772 (1980).
3. D. H. Feng, contribution to this conference.
4. R. F. Casten and J. A. Cizewski, Nucl. Phys. A309, 477 (1978).
5. J. Kopecky, contribution to this conference.
6. D. D. Warner, R. F. Casten, M. L. Stelts, H. G. Borner, and G. Barreau, Phys. Rev. C26, 1921 (1982).
7. B. Harmatz, Nucl. Data Sheets 23, 607 (1978).
8. Y. Yamazaki and R. K. Sheline, Phys. Rev. C14, 531 (1976).
9. A. B. Balantekin, I. Bars, R. Bijker, and F. Iachello, Phys. Rev. C27, 1761 (1983).
10. A. P. Ghatak-Roy and S. W. Yates, Phys. Rev. C28, 2521 (1983).
11. A. M. Bruce, W. Gelletly, W. R. Phillips, J. Lukasiak, and D. D. Warner, Bull. Am. Phys. Soc. 29, 1049 (1984).
12. M. Vergnes, G. Berrier-Ronsin, and R. Bijker, Phys. Rev. C28, 360 (1983).
13. M. Vergnes, G. Berrier-Ronsin, G. Rotbard, J. Vernotte, J. M. Maison, and R. Bijker, Phys. Rev. C30, 517 (1984).
14. H. Z. Sun, A. Frank, and P. Van Isacker, Phys. Lett. 124B, 275 (1983), and Phys. Rev. C27, 2430 (1983), and R. Bijker, Ph.D. Thesis, University of Groningen (1984).
15. R. F. Casten, D. D. Warner, G. M. Gowdy, N. Rofail, and K. P. Lieb, Phys. Rev. C27, 1310 (1983).
16. D. D. Warner, Phys. Rev. Lett. 52, 259 (1984), and D. D. Warner and A. M. Bruce, Phys. Rev. C30, 1066 (1984).
17. R. F. Casten, P. Kleinheinz, P. T. Daly, and B. Elbek, Mat. Fys. Medd. Dan. Vid. Selsk. 38, No. 13 (1972).
18. A. M. Bruce and D. D. Warner, to be published.
19. D. D. Warner and R. F. Casten, Phys. Rev. C28, 1798 (1983).
20. D. D. Warner, P. van Isacker, J. Jolie, and A. M. Bruce, to be published.
21. D. Benson, P. Kleinheinz, and R. K. Sheline, Phys. Rev. C14, 2095 (1976); Z. Physik A281, 145 (1977) and A285, 405 (1978).
22. D. D. Warner, W. F. Davidson, H. G. Borner, R. F. Casten, and A. I. Namenson, Nucl. Phys. A316, 13 (1979).
23. R. F. Casten, D. D. Warner, and J. A. Cizewski, Nucl. Phys. A333, 237 (1980).

SPECTROSCOPY OF ODD NEUTRON ACTINIDE NUCLEI

H.G. Börner
Institute Laue-Langevin, F-38042 Grenoble, France

ABSTRACT

The combination of results obtained from complementary reactions such as particle transfer, (n,γ), and (n,e) has led to an extensive knowledge of the single particle- and vibrational structure of excited states in odd-neutron actinide nuclei. The current situation is illustrated in examples like ^{239}U and ^{231}Th where very complete level schemes have been obtained. Single particle-vibrational mixing has been observed in several cases including the $1/2^+[631]$, $3/2^+[631]$, $5/2[622]$, $5/2^-[752]$ and $7/2^-[743]$ configurations. For ^{231}Th the experimental results are compared to theoretical predictions.

INTRODUCTION

In recent years there has been considerable interest in studying the level schemes of the actinide nuclei. The heavy elements constitute one of the challenging regions of nuclear structure research. The study of the systematics of actinide nuclei has led to considerable advances in the understanding of alpha decay, fission, nuclear shapes and nuclear stability.

Detailed studies of odd-mass nuclei in this region can provide information on the locations of Nilsson orbitals and their energy systematics. Various techniques of experimental nuclear spectroscopy have been employed to determine excited levels. These data are interpreted in terms of single particle excitations, collective motion, such as rotations and vibrations and their interaction.

An extensive review of the single particle states in odd-neutron actinide nuclei was given by Chasman et al.[1] in 1977. Subsequent measurements have been carried out, using a variety of high resolution spectrometers. In the following we will concentrate mainly on the recent results obtained from the (n,γ) reaction.

The (n,γ) reaction is, in contrast to most other reactions, non selective and populates all levels with spin not too far from that of the capture state. Therefore it is well suited to identify vibrations in odd neutron nuclei whether or not they are based on the ground state band. Alternatively, particle transfer reactions proceed via single particle amplitudes and often provide a unique signature ('fingerprint') for the population of members of a rotational band which can be observed and used for the assignment of levels to rotational bands of specific single particle states. The combination of both types of reaction forms a very powerful tool for the construction of very complete level schemes and their interpretation.

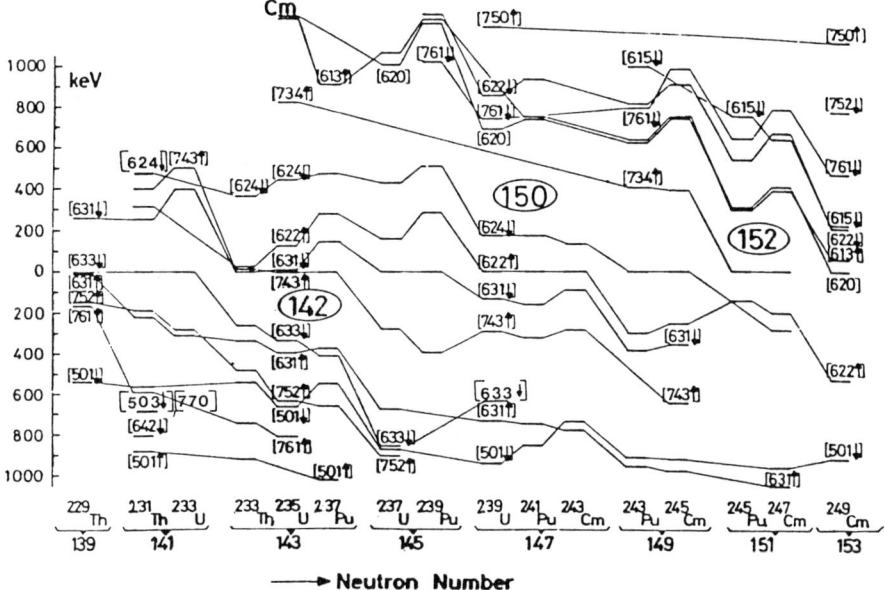

Fig. 1. Neutron single particle states in the actinide region. Most of these states have been identified via operator specific reactions.

EXPERIMENTAL SITUATION

Experimentally, there are certain difficulties inherent in this mass region. All target nuclei are unstable with respect to alpha decay or spontaneous fission and therefore in most cases special equipment is needed for the fabrication of such targets. In former years only a few (n,γ) experiments had been carried out because there often exists unfavourable ratio between the thermal neutron radiative capture cross section and the neutron-induced fission cross section. Moreover, γ radiation following the decay of the active targets provides an additional source of contaminant γ-ray lines superimposed by a γ-ray background stemming from fission (Fig. 2). It therefore follows, that meaningful γ-ray studies following neutron capture in actinides is only possible when using high thermal neutron fluxes. Since the excited states arise in a region of increasing level density, spectrometers with the best resolution are required. These criteria are fulfilled by the bent crystal spectrometers GAMS[2] and the electron spectrometer BILL[3] installed at the High-Flux Reactor of the Institute Laue-Langevin at Grenoble. For the past ten years studies on actinide ranging from ^{227}Ra to ^{249}Cm have been carried out using these instruments.

In many cases these studies were complemented by the application

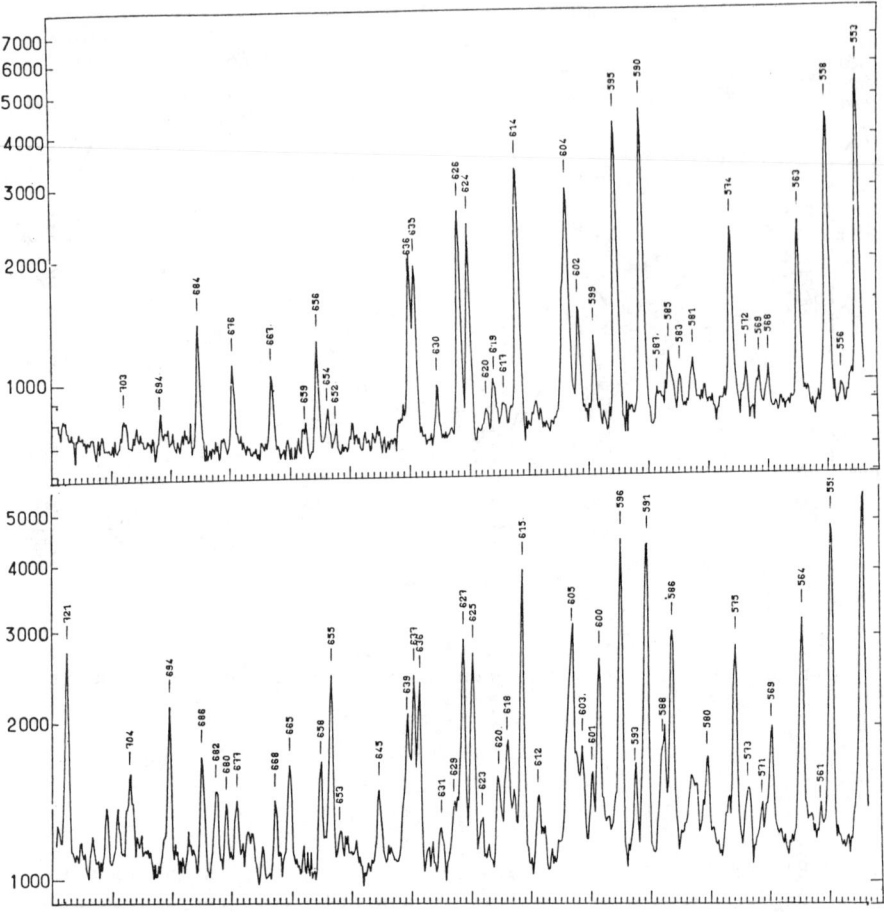

Fig. 2. Portion of a (n,γ) spectrum resulting from neutron capture in ^{240}Pu and measured with the Grenoble bent crystal spectrometers (the energies indicated are only approximate). Both figures show the same energy region. The upper part was measured after one <u>day</u> of irradiation, the lower part after one <u>week</u> of irradiation. Comparison shows the growth of fission product lines which, in fact, always constitute the highest number of γ-ray lines measured in these nuclei. One can also observe an increase in background which stems from thousands of low-intensity, unresolved fission-product lines.

of the average resonance neutron capture method[4,5] carried out at Brookhaven National Laboratory which allows the population of all capture states with $J \leqq 5/2$ (targets consist of even-even nuclei) in those cases where a sufficient amount of target material is available.

In the years prior to this new generation of spectrometers the most powerful tool in actinide structure studies proved to be the application of particle transfer reactions. The earliest attempt to apply this technique in this region was performed by Macefield and Middleton in 1964[6]. The assignment of more than 200 single particle states have been obtained or confirmed with this method[1] (Fig. 1). Its utility arises from the fact that the cross-sections depend on the single particle wavefunctions :

$$d\sigma/d\Omega = 2(C_j^k)^2 \Theta_j^{DW} P_K^2$$

which allows the direct inference of the structure of the states involved. For more details see ref. 1 and references therein.

VIBRATIONAL STATES

The energies of the vibrational states in the actinides show very interesting trends. In even-even nuclei, for instance, the first 1⁻ level is observed at 230 keV in ^{226}Th[6] and at 714 keV in ^{232}Th[7]. This indicates that vibrations depend very much on the single particle structure. Therefore the determination of the coupling of vibrations to single particle states in odd actinide nuclei may contribute to the clarification of the microscopic structure of these excitations.

Calculations show[8] that in deformed actinide nuclei there are virtually no pure single particle states and that phonon admixtures vary from a few percent to over ninety percent.

Most vibrationally-mixed states observed prior to the recent (n,γ) measurements were based on the ground state, since reactions such as Coulomb excitations and (d,d') selectively populate these vibrations. In the non-selective (n,γ) and (n,e) experiments, strong E0, E2 and E1 components in transitions indicate K=0⁺, 2⁺ and 0⁻ vibrations and can be used for the assignment.

In the contributions to the previous neutron capture symposium the most recent systematics of the quasi-particle-vibrational admixtures have been reviewed in the actinide region[9] and we will only briefly summarize the results.

The published results of actinide nuclei which have been studied using both high precision (n,γ) spectroscopy and particle transfer reactions involve the study of excited levels in ^{227}Ra, ^{231}Th, ^{233}Th, ^{235}U, ^{239}U and ^{249}Cm[10-15].

These investigations have shown evidence for E0 admixture in transitions deexciting levels in ^{231}Th, ^{233}Th, ^{235}U and ^{239}U. Most of these states are interpreted as $K^\pi = 0^+$ vibrational components coupled to low lying single particle states, namely 1/2[631] and 5/2[622]

Evidence for $K^\pi = 2^+$ phonons coupled to quasi-particle excitations was observed for ^{231}Th, ^{233}Th, ^{235}U and ^{239}U involving the gamma vibrational coupling on the [631↓], [622↑] and [743↑] states.

Octupole vibrational states with $K^\pi = 0^-$ phonons were assigned in ^{227}Ra, ^{233}Th, ^{235}U and ^{239}U. Here the octupole states are built on 1/2⁺[631], 3/2⁺[631], 3/2⁺[631̄] and 7/2⁻[743]. The systematics of these states is shown in Fig. 3.

New information, using the (n,γ) reaction, has been obtained for ^{239}U and ^{231}Th which is discussed in detail as follows.

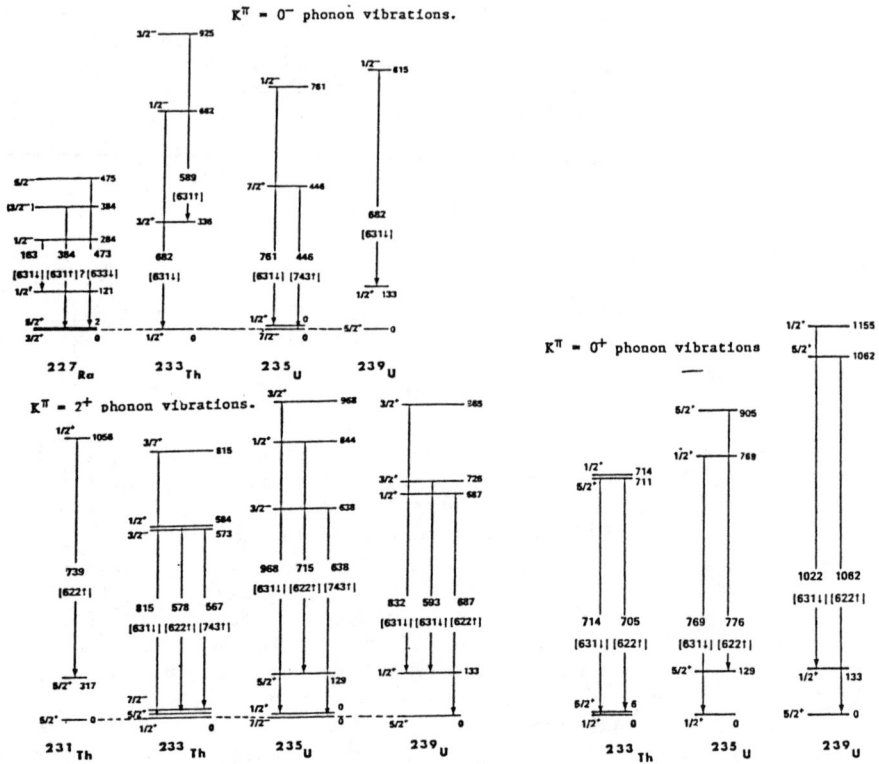

Fig. 3. Part of experimentally deduced levels in the actinide region involving vibrational excitations built on various quasi-particle excitations. This selection summarizes the situation reviewed in ref. 9.

^{239}U

A variety of experiments[6,16-19] have been undertaken to study ^{239}U and a considerable number of single particle states have been identified mainly from high resolution (d,p) measurements[19] and from high precision low energy γ-ray studies[14]. Based upon our understanding of the low-lying single-particle structure, we were able to identify vibrational states built on the [622↑] and [631↓] single particle states. More recently Chrien and Kopecky[20] have investigated γ-radiation resulting from average resonance capture and polarized thermal neutrons. As a result now all levels with J ≤ 5/2 below 1.4 MeV excitation energy are known in ^{239}U. These measurements have confirmed the proposed assignment[14] of a $K^\pi = 0^-$ octupole vibrational level at 539 keV built on the 5/2$^+$[622] ground state of ^{239}U (Fig. 4). This interpretation agrees well with the Alaga rule which predicts a ratio of 2.5 for the matrix elements of the transitions to the 5/2$^+$

ground state and 7/2⁺ of the 5/2[622] configuration, respectively. The experimental ratio is 2.3 ± 0.7.

This measurement also confirmed the existence of the levels belonging to the K=0⁺ states built upon the ground and the first excited state (Fig.4)

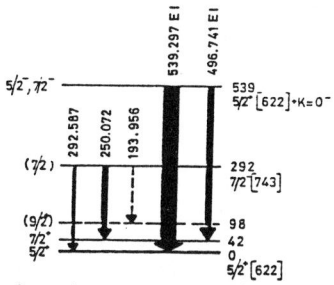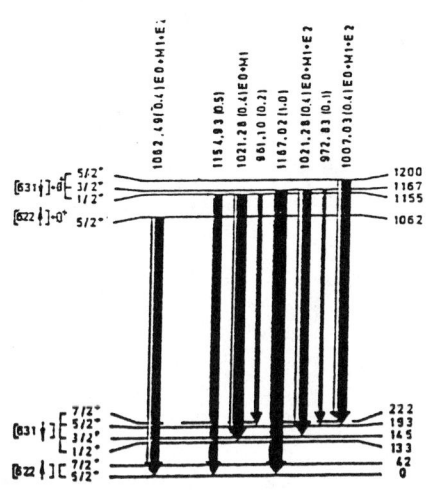

Fig. 4. The left figure shows the 5/2⁺|622|⁺ K=0⁻ state and its decay to the groundstate in ²³⁹U. The right figure shows the two K=0⁺ bands and their decay to the ground state band and first excited band.

Also the 5/2⁻ level, belonging to the K^π = 1/2⁻ band, assigned as 1/2⁻[631] + K = 0⁻, has been identified. This allows us, to calculate rotational constant A and decoupling parameter a which yield A = 4.88 keV and a = − 0.42, respectively, compared to A = 6.82 and a = − 0.415 for the 1/2⁺[631] base state.

Furthermore, it has been suggested[20] that a J^π = 5/2⁻ level at 1018 keV is the band head of a vibrational state with K^π = 2⁻ built on the 1/2⁺[631] Nilsson state and that levels at 1225, 1242 and 1295 keV are good candidates for the 5/2⁺[622] K = 2⁻ configuration in agreement with the predictions by Komov et al[23]. The study of ²³⁹U is a nice illustration how the combination of precisely measured primary and secondary γ-transitions together with high-resolution particle-transfer work can lead to a very complete level scheme, at least for levels with spin not too far from the capture spin (here J < 5/2). Currently ∿ 70 such levels are known in ²³⁹U.

²³¹Th

Very recently new measurements have been carried out to investigate the structure of ²³¹Th. These measurements comprise the determination of primary and secondary γ-rays and conversion electrons emitted after thermal neutron capture in ²³⁰Th, study of the (d,p) reaction on a ²³⁰Th target, and finally average resonance capture, carried out on a 2 keV neutron beam. These studies revealed the existence of ∿ 60 excited levels in ²³¹Th of which 80% where placed in rotational bands whose Nilsson configurations are assigned.

Newly identified Nilsson configurations include 5/2⁺[622]
7/2⁺[624], 3/2⁻[761], 1/2⁻[770] and 5/2⁻[503] (Fig. 5). Additionally
several vibrational states built on |631↓⟩ and |622↑⟩ are proposed.

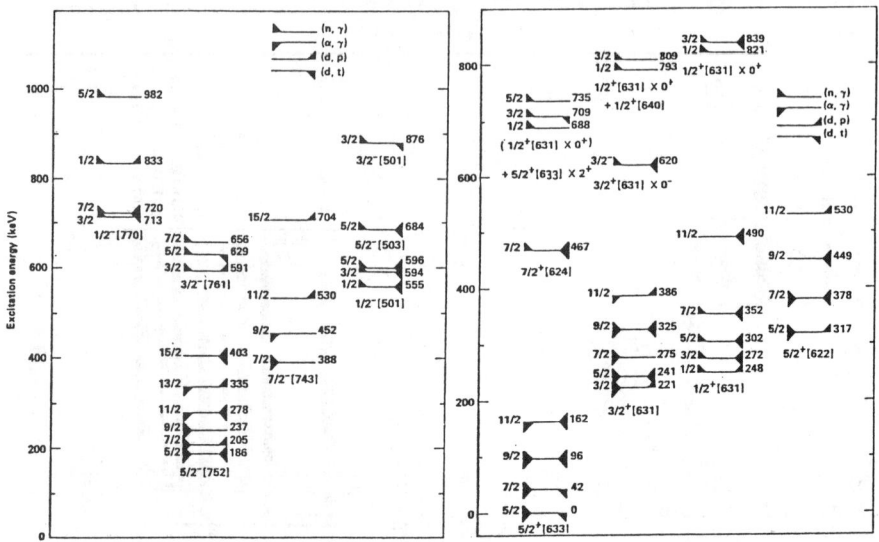

Fig. 5. Negative parity rotational (left) and positive parity
rotational (right) bands in ²³¹Th.

However the experimental situation regarding E0 admixtures in
²³¹Th transitions differs significantly from that in the other even
odd actinides. Generally three or four such transitions have been determined. In ²³¹Th there exist nine transitions with E0 admixtures (Fig. 6)
ranging from ∼ 20% to 90% and one transition with possibly 10% E0
admixture. Eight of these transitions have been placed in the decay
scheme of ²³¹Th (Fig. 7), and form part of the evidence for
the existence of new levels at 623, 634, 688, 709, 793, 809, 821
and 839 keV. With the exception of the first two levels, these transitions populate either the I = 1/2 or 3/2 level of the 1/2[631]
configurations. The levels of 623 and 634 keV decay to the 5/2 and
7/2 members of the [752↑] configuration. These new levels are interpreted as being the lowest-lying members of one new K = 5/2⁻ and
three new K = 1/2⁺ rotational bands.

All of the 1/2⁺ and 3/2⁺ levels just discussed also decay to
the ground state band. There is considerable E2 strength in the decay of the 688 and 709 levels. The E0 admixture in the 440 and
436 keV transitions populating the 1/2 and 3/2 members of the
1/2[631] band is 20% ± 10% and 10% ± 10%, respectively. This indicates that the main component in the wavefunction of this band is
[5/2⁺[633]-2⁺] with perhaps a small admixture of [1/2⁺[631]+0⁺].

Following the selection rules[21] one might expect to observe
mixing of the [3/2⁺[631]-2⁺] configuration into the 1/2⁺ bands.
Since most of the transitions detected that feed the levels of
the 3/2⁺[631] band at 221 keV are of pure M1 multipolarity, there

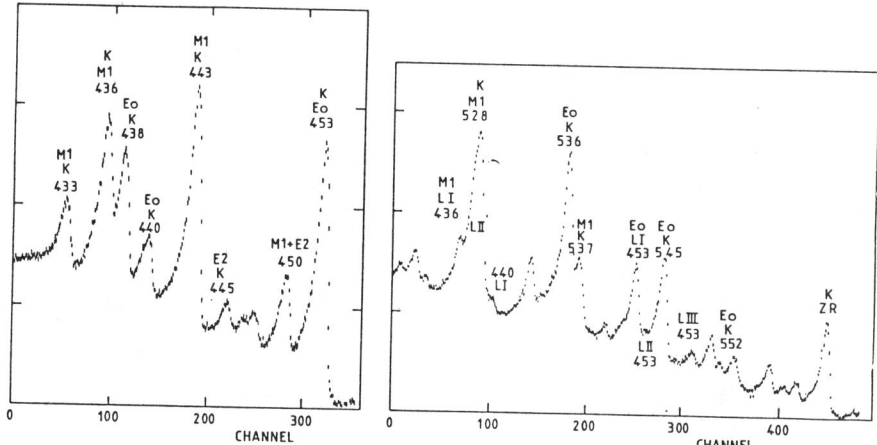

Fig. 6. Portion of the conversion electron spectra from the ^{230}Th(n,ē) reaction measured at Grenoble, show the K and L region of transitions containing E0-admixture.

is no evidence for coupling of this single particle state with a K = 2$^+$ vibration.

The theoretical calculations of Ivanova et al.[22] do not predict the presence of the quasiparticle-phonon mixed states just discussed (Fig. 8). In this energy range there is just one K=1/2$^+$ band at 610 keV with the components 50% 1/2$^+$[640] and 30% [1/2$^+$[640]+0$^+$]. If these new K = 1/2$^+$ band have some 1/2$^+$ [640] character they would be expected to decay strongly to the 3/2$^+$[631] band and some of the levels should be populated by the (d,t) reaction. These characteristics are found to some extent in each band, but there is no strong evidence. The best candidate would be the 793 keV band.

Thus, the predominant nature of the three K = 1/2$^+$ bands is assigned as follows: [1/2$^+$[631]+0$^+$], 793- and 821 keV bands; [5/2$^+$[633]-2$^+$], 688 keV band; 1/2$^+$ [640], 793 keV band, suggesting that the K = 1/2$^+$ states observed experimentally in ^{231}Th are more complex than expected from model calculations.

Two other 3/2$^+$ and 5/2$^+$ levels are observed at 960 and 980 keV, respectively which decay via E2 components to the 1/2$^+$ [631] band. This band is assigned [1/2$^+$[631]-2$^+$]. The energy separation of 713 keV between this band and the 1/2$^+$ [631] band can be compared with ΔE = 814 keV found for this case in ^{233}Th.

Levels at 554, 593 and 595 keV are assigned to be the 1/2$^-$, 3/2$^-$ and 5/2$^-$ members, respectively, of the 1/2$^-$[501] configuration. The strong decay with E2 transitions to the 5/2$^-$ and 7/2$^-$ members of the 5/2$^-$ [752] state suggest an admixture of 5/2$^-$[752] - K = 2$^+$ in this configuration. This is in agreement with the predictions of Ivanova et al[22] who propose a 19% admixture of 752↑ Q(22) in this state.

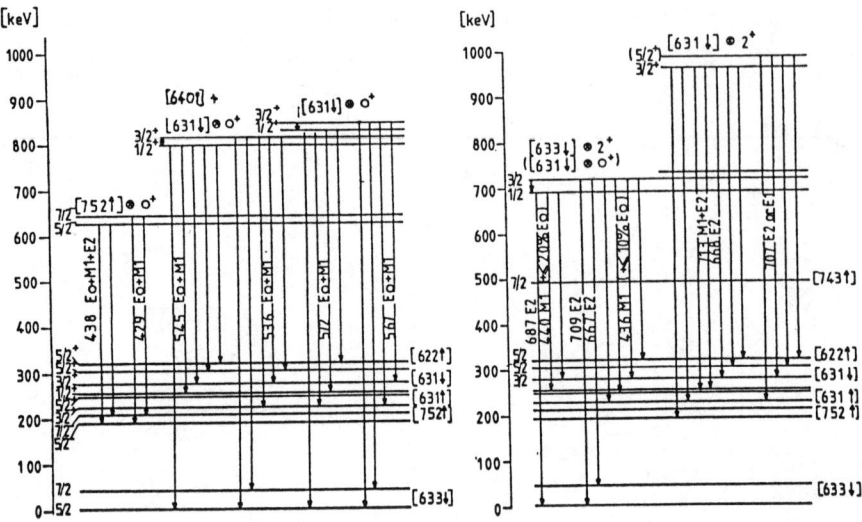

Fig. 7. Decay scheme of the levels in ^{231}Th involving K=0$^+$ and K=2$^+$ vibrational excitations built on 1/2[631], 5/2[633] and 5/2[752].

Table 1 : Transitions with E0-admixture in ^{231}Th				
E_γ [keV]	Shell	α^K_{exp}	α^K_{M1}	E0 Admixture
572	K,LI	0.448	0.124	73%
567	K,LI	0.614	0.127	80%
552	K,LI	>0.3	0.137	>60%
545	K,LI	>1.4	0.142	>90%
536	K,LI	0.948	0.148	85%
453	K,LI,LII	2.23	0.233	90%
(440	K,LI	0.32	0.255	20%)
438	K,LI,LII	>2.3	0.255	90%
(436	K,LI	0.28	0.255	10%
428	K,LI	0.799	0.271	75%

Concerning the observation of octupole vibrations with $K^\pi = 0^-$ in the ^{231}Th, the evidence is less definitive. A level at 620 keV whose spin, parity is uniquely determined to be 3/2$^-$ is a good candidate[11] for 3/2$^+$[631] + K = 0$^-$ and a 1/2$^-$ level at 833 keV assigned as probable band head of the 1/2[770] configuration[11] can also contain a significant 1/2$^+$[631] + K = 0$^-$ contribution. Yet, in both cases these levels seem to be populated also in the (d,p)-reaction and it seems not to be clear which single particle amplitudes are involved.

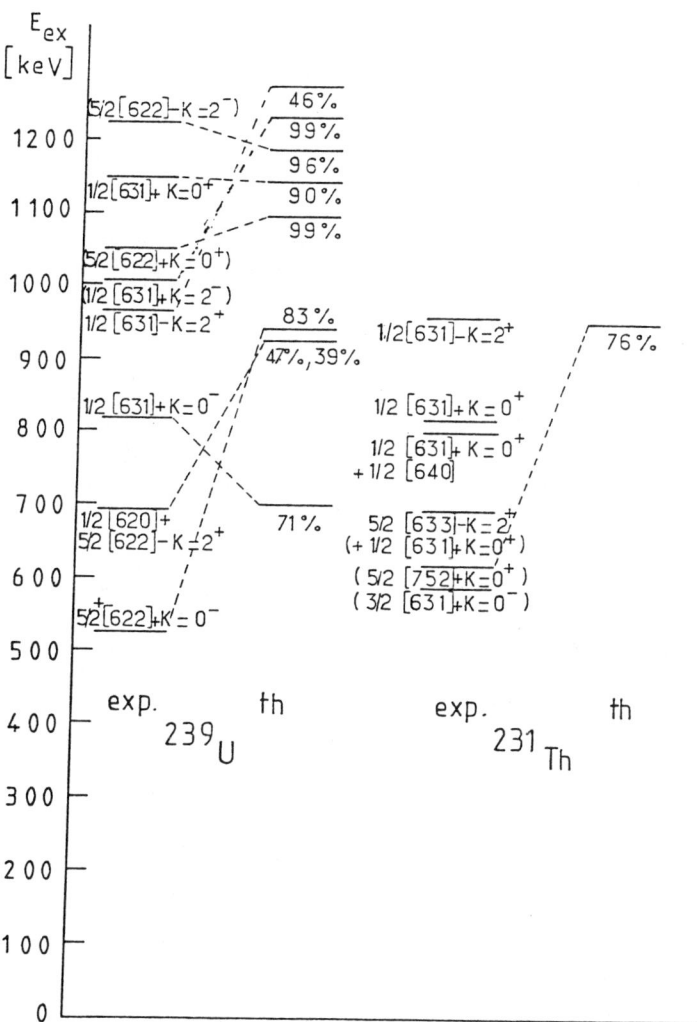

Fig. 8. Comparison of experimental and theoretical results in ^{239}U and ^{231}Th, respectively. Only such states are shown, where a correspondance between theory[22,23] and experiment[11,14,20] is observed.

Summarizing the new results obtained for ^{239}U and ^{231}Th one can conclude that the agreement between theory and experiment is remarkably good in one case (^{239}U) and remarkably bad in the other case (^{231}Th). As has been stated by Hoff et al[9] one can generally say that many of the experimental features in this region are not predicted by the calculations[22,23,8]. (Fig.8)

CONCLUSION

Whereas, for a certain number of years the theoretical predictions in the region of actinide nuclei seemed to be slightly ahead with respect to the experimental information the situation now seems to be reversed. The level structures which have been observed experimentally in the Thorium and Uranium region are certainly more complex than those calculated[22].

One should also bear in mind that, although much progress has been made on the experimental sector, there is still much to do :

In the region of Uranium and Plutonium isotopes recent measurements have been carried out on ^{237}U, ^{241}Pu and ^{245}Pu using the neutron capture techniques described in ref. 24. These data are under evaluation and should help to complete our knowledge of vibrational states in these nuclei. As already stated in the beginning, the (n,γ) reaction is an outstanding tool in the study of nuclear structure owing to its non-selectivity. But this is connected with an evident drawback concerning the interpretation of the observed states where often complementary techniques using operator specific reactions are extremely helpful.

A very nice example of how one can obtain a maximum of information by using many techniques is the study of the level scheme of ^{227}Ra[10] where the information from (n,γ), (d,p) and ($\vec{t}$,d) reactions and the β-decay of ^{227}Fr have been combined. It would be of outstanding interest to perform more single-particle transfer reaction studies with the best resolution possible, in order to probe on the new states observed in the (n,γ)-reaction.

1. R.R. Chasman, I. Ahmed, A.M. Friedman and J.R. Erskine, Rev. Mod. Phys. 49 (1977) 833
2. H.R. Koch, H.G. Börner, J.A. Pinston, W.F. Davidson, J. Faudou, R. Roussille and O.W.B. Schult, Nucl. Instr. and Methods 175 (1980) 401
3. W. Mampe, K. Schreckenbach, P. Jeuch, B.P.K. Maier, F. Braumandl, J. Larysz and T. von Egidy, Nucl. Instr. and Methods 154 (1978) 127
4. L.M. Bollinger and G.E. Thomas, Phys. Rev. C2 (1970) 1951
5. R.E. Chrien, Trans. N.Y. Acad. Sci. 44 (1980) 40
6. B.E.F. Macefield and R. Middleton, Nucl. Phys. 59 (1964) 561
7. Nuclear Data Sheets 20 (1977) 119, ibid 20 (1977) 165
8. F.A. Gareev, S.P. Ivanova, L.A. Malov and V.G. Soloviev, Nucl. Phys. A171 (1971) 134
9. R.W. Hoff, R.W. Lougheed, G. Barreau, H.G. Börner, W.F. Davidson, K. Schreckenbach, D.D. Warner, T. von Egidy and D.H. White, Inst. Phys. Conf. Ser. No 62 (1982) 250
10. T. von Egidy et al. Nucl. Phys. A365 (1981) 26
11. D. White et al. 3rd Int. Symp. Neutron Capture Gamma-Ray Spectr. (N.Y. Plenum Press) (1979) 802 and D. White et al., to be published

12. P. Jeuch, T. von Egidy, K. Schreckenbach, W. Mampe, H. Börner, W. Davidson, J.A. Pinston and R. Roussille, Nucl. Phys. A317 (1979) 363
13. J. Almeida, T. von Egidy, P. van Assche, H. Börner, W. Davidson, K. Schreckenbach and A. Namenson, Nucl. Phys. A315 (1979) 71
14. H. Börner, H.R. Koch, H. Seyfarth, T. von Egidy, W. Mampe, J.A. Pinston, K. Schreckenbach and D. Heck, Z. Physik A286 (1978) 31
15. R.W. Hoff, W.F. Davidson, D.D. Warner, H.G. Börner and T. von Egidy, Phys. Rev. C25 (1982) 2232
16. R.K. Sheline, W.N. Shelton, T. Ugadawa, E.T. Jurney, H.T. Motz, Phys. Rev. 151 (1966) 1011
17. L.M. Bollinger and G.E. Thomas, Phys. Rev. C6 (1972) 1322
18. B.P.K. Maier, Z. Physik 184 (1965) 143
19. J.R. Erskine, Phys. Rev. C17 (1978) 934
20. R.E. Chrien, J. Kopecky, Nucl. Phys. A414 (1984) 281-300
21. B.R. Mottelson and S.G. Nilsson, Mat. Fys. Skr. Dan. Vid. Selsk. 1 Nr.8 (1959)
22. S. Ivanova, A. Komov, L. Malov and V. Soloviev, Izv. Acad. Nauk SSSR Ser. Fiz. 39 (1975) 1612
23. A. Komov, L. Malov and V. Soloviev, Izv. Akad. Nauk SSSR Ser. Fiz. 35 (1971) 1550
24. H.G. Börner, G. Barreau, S.A. Kerr, K. Schreckenbach, Inst; Phys. Conf. Ser. No 62 (1982) 121

MODELING LEVEL STRUCTURES OF ODD-ODD DEFORMED NUCLEI*

R. W. Hoff
Lawrence Livermore National Laboratory, Livermore, California

J. Kern
Institut de Physique de l'Université, Fribourg, Suisse

R. Piepenbring and J. P. Boisson
Institut des Sciences Nucléaires
Université de Grenoble, Grenoble, France

ABSTRACT

A technique for modeling quasiparticle excitation energies and rotational parameters in odd-odd deformed nuclei has been applied to actinide species where new experimental data have been obtained by use of neutron-capture gamma-ray spectroscopy. The input parameters required for the calculation were derived from empirical data on single-particle excitations in neighboring odd-mass nuclei. Calculated configuration-specific values for the Gallagher-Moszkowski splittings were used. Calculated and experimental level structures for ^{238}Np, ^{244}Am, and ^{250}Bk are compared, as well as those for several nuclei in the rare-earth region. The agreement for the actinide species is excellent, with bandhead energies deviating 22 keV and rotational parameters 5%, on the average. Corresponding average deviations for five rare-earth nuclei are 47 keV and 7%. Several applications of this modeling technique are discussed.

* Work performed under the auspices of the U. S. Department of Energy by the Lawrence Livermore National Laboratory under Contract W-7405-ENG-48.

The method used in this paper to model the level structure of odd-odd deformed nuclei is based upon a technique first described in papers by Struble, Motz, et al.[1] and Scharff-Goldhaber et al.[2] They proposed that if the p-n residual interaction energy is small compared with the energy with which the odd nucleons are bound to the core, the excitation can be calculated by a simple extension of the odd-A model and the interaction energy can be treated later as a perturbation. Thus, the model can be described in terms of two simple concepts. The first is that in considering the coupling of two unpaired particles to a deformed core, the excitation energy of a given configuration can be described as the sum of each of the odd-nucleon excitations. The second concept is that the effective moment of inertia for a rotational band can be expressed as the sum of three components: the moment of inertia of the even-even core plus increments from the addition of each of the two odd nucleons.

The excitation of an odd nucleon in a deformed nucleus can be treated theoretically by various versions of single-particle potential theory. For the purposes of this paper, however, the excitations are obtained from experimental data for neighboring odd-mass nuclei. In the actinide region, these data have been systematically surveyed by Chasman et al.[3] The quasiparticle energy E_{qp} for a given orbital in an odd-mass nucleus can be found from the expression

$$E_I = E_{qp} + \hbar^2/2\theta[I(I+1)-K^2+\delta_{K,1/2}\, a(I+1/2)(-1)^{I+1/2}] , \quad (1)$$

where a is the decoupling parameter. The term $-K^2$ was assumed in this expression instead of the often-used form that contains a $-2K^2$ term. This modification has the effect of changing the magnitude (and even the sign) of the zero-point rotational energy E_{ZPR} for a given band. The quantity E_{ZPR} is the difference between E_K (defined by substituting K for I in Eq. 1), and E_{qp}. Thus, it can be expressed as

$$E_{ZPR} = \hbar^2/2\theta(K - a\delta_{K,1/2}) \cdot \quad (2)$$

A more precise definition of E_{ZPR} includes a term containing $\langle j^2 \rangle$, which is usually the major contributor and which consequently yields a larger zero-point energy than that given by Eq. 2. Nevertheless, several authors have chosen to incorporate this term in the expression E_{qp} and to evaluate E_{ZPR} using a simple expression, that given by Eq. 2. For example, Ogle et al.[4] adopted the use of a $-K^2$ term (as in Eq. 1) for their data analysis. They concluded that if one were to consider increasing the coefficient of this term, it could not be chosen much larger than 1.5 without destroying the qualitative agreement between Nilsson model predictions and quasiparticle level systematics.

The effective moment of inertia for a given rotational band, θ, is derived by the following method:

$$\theta_{odd-odd} = \theta_{even-even} + \delta\theta_p + \delta\theta_n$$

$$= \theta_{e-e} + (\theta_p - \theta_{e-e}) + (\theta_n - \theta_{e-e})$$

$$= \theta_p + \theta_n - \theta_{e-e}. \quad (3)$$

where θ_p and θ_n are the moments of inertia for the relevant rotational band in the neighboring odd mass nuclei and $\delta\theta_p$ and $\delta\theta_n$ are the increments to $\theta_{even-even}$ due to the unpaired proton and neutron. In the model being discussed here, effects due to Coriolis mixing that are explicit to odd-odd nuclei are not included. On the other hand, those experimentally-observed manifestations of Coriolis mixing in odd-A nuclei such as the compression or expansion of rotational spacing within a given band are included in the calculated effective moment of inertia for the odd-odd nucleus.

The excitation energies of levels in the odd-odd nucleus are calculated using the expression

$$E_I = E^p_{qp} + E^n_{qp} + \hbar^2/2\theta_{odd-odd}[I(I+1)-K^2] - (1/2-\delta_{\Sigma,0})E_{GM}$$
$$- \delta_{K,0}(-1)^I E_N \pi, \qquad (4)$$

where π denotes the parity of the states as introduced in Ref. 5 and is equal to ± 1 for positive or negative parity.

The E_{GM} and E_N terms in Eq. 4, which are designated as the Gallagher-Moszkowski splitting[6] and Newby shift,[7] respectively, are functions of the effective neutron-proton residual interaction. For the calculated Gallagher-Moszkowski splittings and Newby shifts reported in this paper, a zero-range central (δ) force between proton and neutron and a Nilsson-type potential were assumed. (For a detailed discussion of the method, see Ref. 5). The one adjustable parameter needed to describe the strength of the δ force is obtained from a global fit of G-M splittings in the actinide region, specifically the experimental values for the 12 configuration pairs shown in Figure 1. Calculated and experimental E_{GM} and E_N values are listed in Table 1.

In a few instances in Table 1, experimental data are reported for configurations occurring in more than one nuclide. Such an example is given in Fig. 2, which shows the excited levels of a K=0 band in ^{238}Np, ^{240}Am, ^{242}Am, and ^{244}Am. It can be seen that the level spacings within the band are very similar among these nuclei. The experimental and calculated bandhead energies and rotational parameters for each nucleus also show reasonably good agreement. On the other hand, the calculated matrix elements for the G-M splittings and Newby terms do not reproduce experiment in this particular case. Of the 15 separate configurations for which E_{GM} values are listed in Table 1, only in three cases do the calculated and experimental values show significant disagreement, i.e., the ratios fall outside the range $0.65 < E_{GM}(exp/calc) < 1.39$.

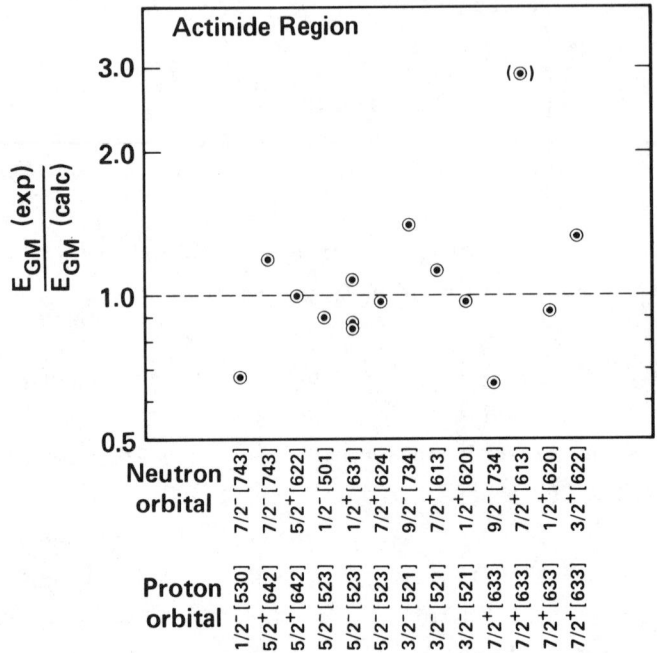

Figure 1. Comparison of experimental and calculated Gallagher-Moszkowski matrix elements for several configuration pairs in odd-odd actinide nuclei.

Several authors have published calculations of G-M splittings and Newby shifts made assuming simple central forces.[8] Boisson et al.[5] have worked with empirical data from rare-earth nuclei involving an initial sample of E_{GM} matrix elements for 43 two-quasiparticle configurations that was reduced to 23 empirical values of greatest reliability. They determined parameters for several forms of an assumed central-force effective interaction by fitting the calculated matrix elements to the empirical data. Sood and Singh[9] have calculated E_{GM} matrix elements for configuration pairs in the nuclei ^{238}Np, ^{244}Am, and ^{250}Bk with a zero-range proton-neutron residual interaction. For the adjustable parameter that determines the strength of the interaction, they have chosen to adopt a different

Table 1. Comparison of experimental and calculated E_{GM} and E_N values for configurations in some actinide nuclei.

Proton Config.	Neutron Config.	Nucleus	E(GM/N)[a] (keV) Exp.	Calc.	Exp/Calc
1/2⁻[530]	7/2⁻[743]	234Pa	77.5	117.	0.67
	1/2⁺[631]	234Pa	N−46.8	−44.2	1.06
		236Pa	N−50.2	−44.2	1.14
		238Np	N−37.9	−43.1	0.88
5/2⁺[642]	1/2⁺[631]	238Np	82.4	70.	1.18
5/2⁻[523]	1/2⁻[501]	242Am	39.	44.	0.89
	1/2⁺[631]	238Np	52.2	60.	0.87
		242Am	52.1	61.	0.85
		244Am	70.	61.	1.15
	1/2⁺[620]	242Am	21.	116.	0.18
		244Am	16.	116.	0.14
	5/2⁺[622]	238Np	1.	95.	0.01
		240Am	10.	96.	0.10
		242Am	5.	96.	0.05
		238Np	N+27.	−15.2	−1.78
		240Am	N+28.	−14.7	−1.90
		242Am	N+27.3	−14.6	−1.87
		244Am	N+25.7	−14.6	−1.76
	7/2⁺[624]	244Am	200.2	208.	0.96
3/2⁻[521]	9/2⁻[734]	248Bk	186.5	134.	1.39
	7/2⁺[613]	250Bk	66.4	60.	1.11
	1/2⁺[620]	250Bk	110.3	115.	0.96
7/2⁺[633]	7/2⁺[624]	244Am	N+33.1		
	9/2⁻[734]	248Bk	122.	189.	0.65
	7/2⁺[613]	250Bk	135.0	47.	2.87
		250Bk	N−25.0	−58.0	0.43
	1/2⁺[620]	250Bk	83.1	60.	1.39
	3/2⁺[622]	250Bk	91.2	69.	1.32
	1/2⁻[761]	250Bk	38.		

a - The values listed below are Gallagher-Moszkowski matrix elements except when indicated as a Newby term by an N in col. 4.

value for each nucleus; the value is usually adjusted to reproduce the G-M splitting that includes the ground state rotational band. For the 9 G-M splittings in these nuclei where comparison can be made with experiment, their calculations show approximately the same level of agreement as for those listed in Table 1. If one replots the data of Fig. 1 as a function of mass number, however, it appears there is no obvious trend, which weakens the case for adjusting the residual interaction strength for each nucleus.

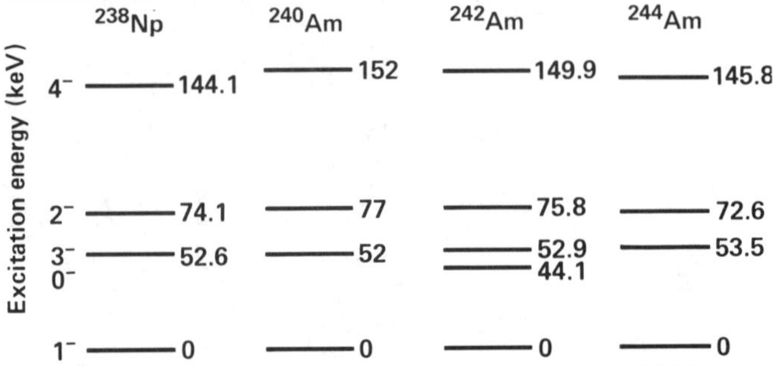

			^{238}Np	^{240}Am	^{242}Am	^{244}Am
		4⁻	144.1	152	149.9	145.8
		2⁻	74.1	77	75.8	72.6
		3⁻	52.6	52	52.9	53.5
		0⁻			44.1	
		1⁻	0	0	0	0
E_1(keV)	— exp		299	346	0	286.2
	— calc		279	264	0	315
A(keV)	— exp		5.1	5.4	5.29	5.29
	— calc		5.4	5.3	5.11	5.35
E_N(keV)	— exp		+26.9	+28	+27.3	+25.4
	— calc		−15.2	−14.7	−14.6	−14.6
E_{GM}(keV)	— exp		1	10	5	—
	— calc		95	96	96	96

Figure 2. Excited levels of the 5/2⁻[523]p - 5/2⁺[622]n rotational band in four actinide nuclei. Experimental and calculated values of the energy of the I=1 bandhead level(E_1), rotational parameter(A), Newby term(E_N), and Gallagher-Moszkowski splitting(E_{GM}) are compared.

Comparisons between the experimental and calculated bandhead energies and rotational parameters of ^{244}Am and ^{250}Bk are shown in Tables 2 and 3 and in Fig. 3. The calculated level energies, which have been obtained using Eq. 2, are given a zero-energy adjustment so that the calculated and experimental ground state energies match. The experimental data for these nuclei include information from recent measurements using the (n,γ) and (d,p) reactions[10,11], as well as data from earlier measurements.[12,13] The information on the level schemes of neighboring nuclei was taken largely from the Table of Isotopes.[13] The uncertainties on bandhead energy listed in the fourth columns of Tables 2 and 3 do not represent experimental error, but rather are derived from the spread in the two experimental values taken from the odd-mass nuclei. For example, the excitation of a given proton orbital in ^{250}Bk, relative to the ground-state orbital, is derived from observations in ^{249}Bk and ^{251}Bk. The two experimental observations are averaged and their spread (plus that from the odd-neutron observations) determines the uncertainty.

Table 2. Comparison of experimental and predicted band-head energies and rotational parameters for ^{244}Am.

K^π	Configuration	Bandhead Energy Exp. (keV)	Bandhead Energy Calc. (keV)	Rot. Par. A Exp. (keV)	Rot. Par. A Calc. (keV)
1^+	$5/2^+[642]$p - $7/2^+[624]$n	85	60+42	3.4	3.0
1^-	$5/2^-[523]$p - $7/2^+[624]$n	173	161+14	5.3	5.4
2^-	$3/2^-[521]$p - $7/2^+[624]$n	259	235+56	5.8	5.8
0^-	$5/2^-[523]$p - $5/2^+[622]$n	286(1^-)	319+31(1^-)	5.3	5.4
3^+	$1/2^+[400]$p - $7/2^+[624]$n	345	(336)	5.2	(5.4)
0^+	$7/2^+[633]$p - $7/2^+[624]$n	374(0^+)	362+69(0^+)	6.0	6.9
2^+	$5/2^+[642]$p - $1/2^+[631]$n	(416)	451+56	(2.7)	3.1
2^+	$5/2^-[523]$p - $9/2^-[734]$n	417	426+21	4.1	4.2
3^-	$5/2^-[523]$p + $1/2^+[631]$n	418	442+28	(5.6)	5.7
0^+	$5/2^+[642]$p - $5/2^+[622]$n	(475)(2^+)	463+59(2^+)	--	3.0
2^-	$5/2^-[523]$p - $1/2^+[631]$n	482	488+28	(6.6)	5.7
2^-	$5/2^+[642]$p - $9/2^-[734]$n	514	538+49	3.2	2.6
2^+	$3/2^+[651]$p - $7/2^+[624]$n	(612)	(395)	(5.8)	(6.2)
1^+	$5/2^-[523]$p - $7/2^-[743]$n	(668)	665+14	--	5.8
1^-	$3/2^-[521]$p - $5/2^+[622]$n	678	586+74	5.1	5.7
1^-	$5/2^+[642]$p - $7/2^-[743]$n		826+42		3.1
Average deviations:		19 (for 9 bands)		0.28(7.4%) (for 9 bands)	

Table 3. Comparison of experimental and predicted bandhead energies and rotational parameters for ^{250}Bk.

K^π	Configuration	Bandhead Energy Exp. (keV)	Bandhead Energy Calc. (keV)	Rot. Par. A Exp. (keV)	Rot. Par. A Calc. (keV)
2^-	$3/2^-[521]p + 1/2^+[620]n$	0	0	5.77	5.70
1^-	$3/2^-[521]p - 1/2^+[620]n$	104	99$\pm$4	5.35	5.70
4^+	$7/2^+[633]p + 1/2^+[620]n$	36	37$\pm$18	4.31	4.37
3^+	$7/2^+[633]p - 1/2^+[620]n$	115	88$\pm$18	4.14	4.37
5^-	$3/2^-[521]p + 7/2^+[613]n$	97	100$\pm$33	5.80	5.83
2^-	$3/2^-[521]p - 7/2^+[613]n$	146	137$\pm$33	5.59	5.83
7^+	$7/2^+[633]p + 7/2^+[613]n$	86	114$\pm$47	4.56	4.44
0^+	$7/2^+[633]p - 7/2^+[613]n$	175(1^+)	110$\pm$47(1^+)	--	4.44
2^+	$7/2^+[633]p - 3/2^+[622]n$	212	211$\pm$33	4.18	4.50
5^+	$7/2^+[633]p + 3/2^+[622]n$	316	287$\pm$33	4.43	4.50
6^+	$5/2^+[642]p + 7/2^+[613]n$	406	410$\pm$89	--	5.39
3^-	$7/2^+[633]p - 1/2^-[761]n$	526	513$\pm$18	3.4	3.30
4^-	$7/2^+[633]p + 1/2^-[761]n$	566	596$\pm$18	3.9	3.30
6^+	$7/2^+[633]p + 5/2^+[622]n$	552	557$\pm$25	--	4.41
Average deviations:		17 (for 13 bands)		0.20(4.7%) (for 11 bands)	

The agreement between experimental and calculated bandhead energies for these nuclides is excellent, the average deviations are $\pm$19 keV and $\pm$17 keV for ^{244}Am and ^{250}Bk, respectively. Similarly, the experimental rotational parameters agree extremely well with the calculated values; the average deviations are $\pm$7% and $\pm$5% for ^{244}Am and ^{250}Bk, respectively. This ability to predict rotational parameters proves to be useful in identifying the configuration of a rotational band. One can see in ^{250}Bk, for example, that there is a clear differentiation between configurations that contain the $3/2^-[521]p$ orbital and those that contain the $7/2^+[633]p$ orbital. For the former, the calculated rotational parameters have values of 5.7-5.8 keV. For the latter, where the proton orbital is subject to strong Coriolis interaction, the calculated values are smaller, 4.4-4.5 keV in several cases and even 3.3 keV when this proton is coupled with a 1/2⁻[761] neutron.

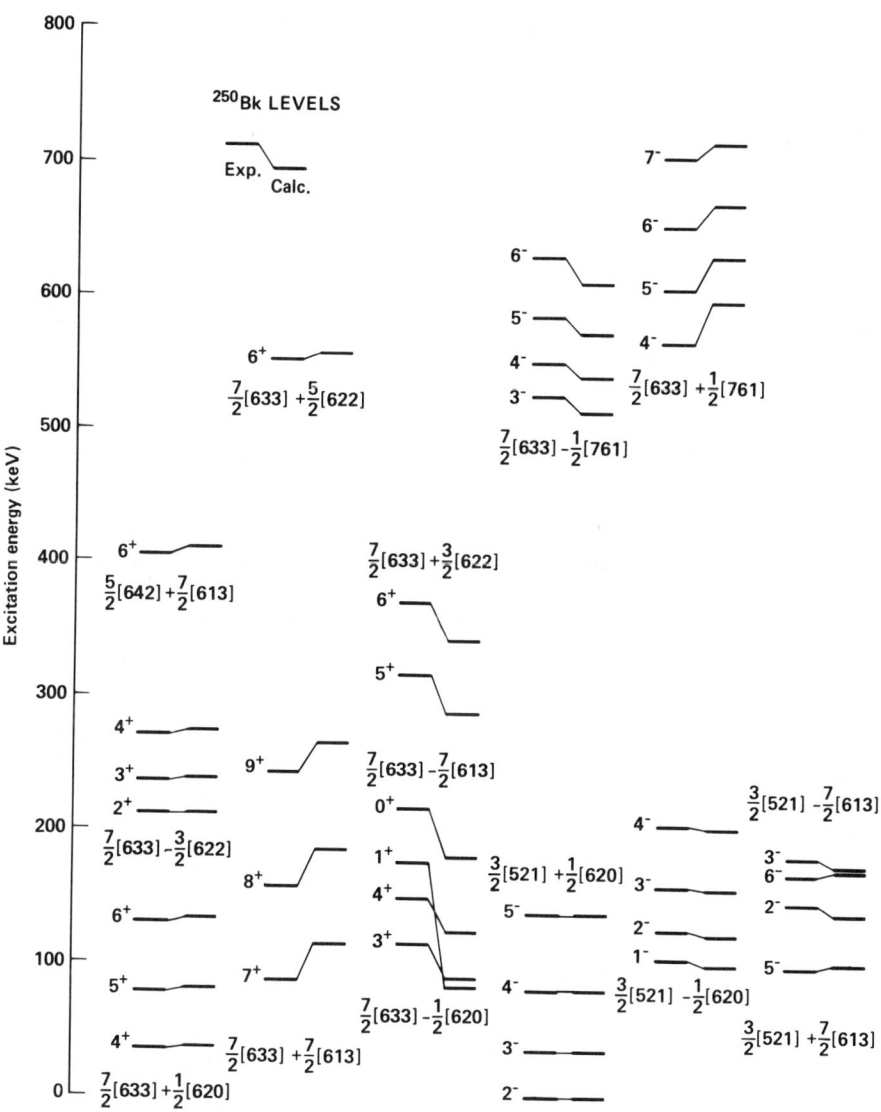

Figure 3. Experimental and calculated levels in ^{250}Bk. Thirty-seven levels, comprising 14 rotational bands, are shown.

In Table 4 the results of the comparisons are summarized for ^{250}Bk and ^{244}Am, along with data[14] for ^{238}Np, and for several rare-earth nuclides. For this latter set, the modeling techniques used was nearly identical to that for the actinides; one difference was that values for E_{GM} and E_N were obtained from Ref. 5 where these matrix elements were calculated employing a more complex force, namely, a central force with intrinsic-spin polarization and with long-range and tensor contributions which were determined from a fit to experimental data.

Table 4. Odd-odd nuclei in actinide and rare-earth regions: Comparison of experimental and calculated bandhead energies, rotational parameters, and G-M splittings.

Nucleus	Number of bands	Energy range (keV)	$\langle E_{exp}-E_{calc}\rangle$ (keV)	$\langle A_{exp}-A_{calc}\rangle$ (keV)	E_{GM} exp/calc
^{238}Np	9	0 - 345	29	0.14 (3.2%)	1.18, 0.87
^{244}Am	16	0 - 680	19	0.28 (7.4%)	1.15, 0.14, 0.96
^{250}Bk	14	0 - 570	17	0.20 (4.7%)	1.11, 0.96, 2.87, 1.39, 1.32
^{160}Tb	8	0 - 380	41	0.61 (8.1%)	1.03, 1.07, 1.13
^{166}Ho	10	0 - 560	47	0.74 (8.7%)	0.80, 1.08, 1.31
^{170}Tm	5	0 - 450	63	0.46 (5.2%)	2.04, 0.98
^{176}Lu	12	0 - 840	58	1.0 (9.2%)	1.14, 0.48, 1.01, 0.91, 0.39
^{182}Ta	7	0 - 270	24	0.47 (3.9%)	0.94, 0.97, 1.14

Thus, the evidence shows this modeling technique accurately reproduces experimental band-head energies and rotational parameters for these deformed nuclei. With this method, then, one can model all of the intrinsic single-particle excitations and rotational bands built on these excitations in any deformed odd-odd species where input data are available. This includes a capability to model structure in some odd-odd nuclei for which little or no experimental data exist, the modeling technique requires no specific knowledge of structure in the odd-odd nucleus of interest, other than calculated matrix elements arising from the proton-neutron interaction. Calculated level schemes can be extended to energies somewhat higher than the ranges given in Table 4, although it must be recognized that other kinds of excitations in these nuclei, e.g., vibrational motion and quasiparticle excitations involving more than two unpaired nuclei, are neglected in this limited approach.

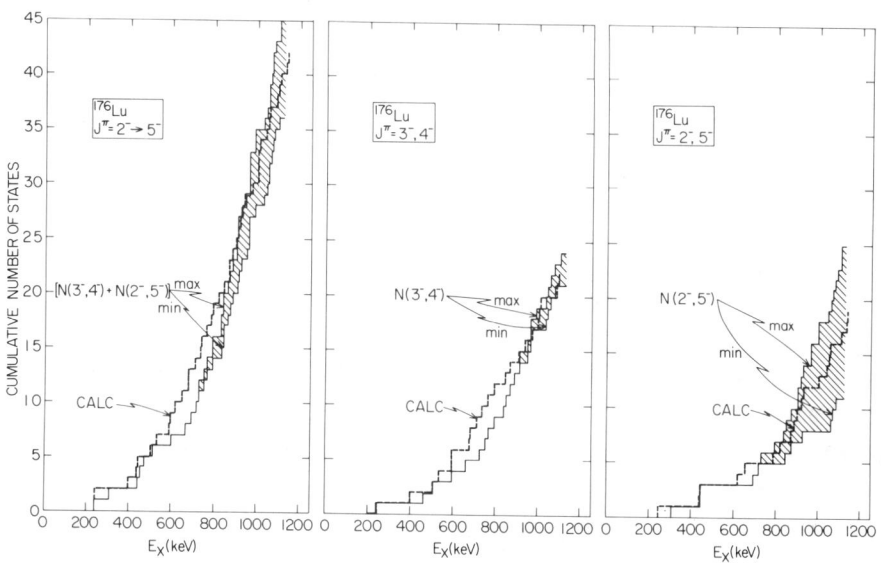

Figure 4. Comparison of experimental level structure data for ^{176}Lu from average resonance capture measurements with model calculation. Cumulative level number histograms for all states with spins $2^-,3^-,4^-,5^-$(left), $3^-,4^-$(center), and $2^-,5^-$(right). The cross-hatched regions show the range of the maximum and minimum numbers of states deduced from the ARC data. The thick dashed lines are the model calculations.

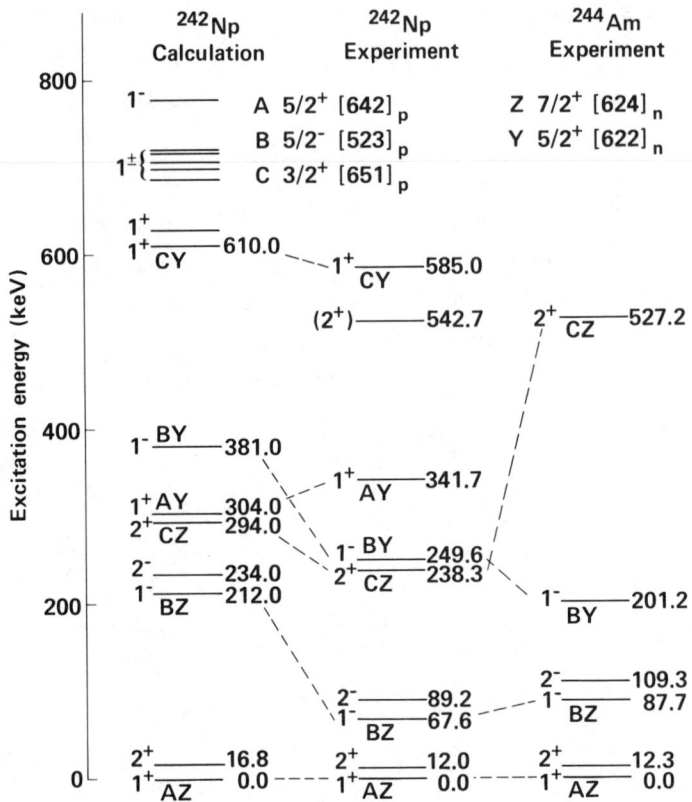

Figure 5. Comparison of the newly interpreted ^{242}Np level scheme with that predicted in a model calculation and with an experimental ^{244}Am level scheme. A key to notation for configuration assignments is shown in the figure. The present interpretation indicates that all levels shown for ^{242}Np are bandheads except for the I=2 levels at 12.0 and 89.2 keV.

Several applications of this modeling technique have proven to be useful. For example, an average resonance capture measurement[15] populating levels in ^{176}Lu has been performed using the filtered neutron beams available at the Brookhaven High Flux Beam Reactor. In this study, a complete set of $J^\pi = 2^- - 5^-$ levels up to an excitation of 1100 keV has been reported. A comparison of the cumulative number of levels in this spin and parity range obtained from the experiment with a corresponding calculated set shows that they are in generally good agreement (see Fig. 4). The comparison involves a total of 40 predicted rotational bands.

In another paper,[16] existing experimental data for ^{242}U beta decay which were measured by Haustein et al.[17] were given a new interpretation with the aid of the phenomenological model of this paper and by comparison with recent experimental data[10] for the levels of ^{244}Am. There has been developed an experimentally based level scheme for ^{242}Np (see Fig. 5) in which all predicted I=1 levels up to 585 keV have been identified and are being populated by ^{242}U beta decay. For the two lowest-lying K=1 bands, both I=1 and I=2 levels have been assigned. The agreement for the bandhead energies and rotational spacings of these bands in ^{242}Np and ^{244}Am is excellent. This behavior can be understood in terms of the near degeneracy of the $5/2^+[642]$ and $5/2^-[523]$ proton orbitals in these nuclei.

A final example of application of the model involves the calculation of cross-section ratios for the production of isomers in neutron-induced reactions, both capture and (n,2n), where the product nuclei are odd-odd deformed species. It has been found that the hundreds of discrete levels and their gamma-ray branching ratios provided by the modeling are necessary in order to achieve agreement between calculation and experiment. A more complete description of this is given in another paper[18] submitted to this conference.

The authors wish to thank G.L. Struble and R.G. Lanier for providing a computer code used to calculate some of the matrix elements listed in Table 1. It is a pleasure to thank H.G. Boerner, R.F. Casten, T. von Egidy, K. Schreckenbach, D.D. Warner, and D.H. White for many stimulating discussions, active assistance, and permission to quote unpublished experimental data. The support of C. Gatrousis in this work is gratefully acknowledged.

REFERENCES

1. G.L. Struble, J. Kern, and R.K. Sheline, Phys. Rev. 137B, 772(1965); H.T. Motz, E.T. Jurney, O.W.B. Schult, H.R. Koch, U. Gruber, B.P. Maier, H. Baader, G.L. Struble, J. Kern, R.K. Sheline, T. von Egidy, T. Elze, E. Bieber, and A. Backlin, Phys. Rev. 155, 1265(1967).
2. G. Scharff-Goldhaber and K. Takahashi, Bull. Acad. Sci. USSR 31, 42(1967).
3. R.R. Chasman, I. Ahmad, A.M. Friedman, and J.R. Erskine, Rev. Mod. Phys. 49, 833(1977).
4. W. Ogle, S. Wahlborn, R. Piepenbring, and S. Fredriksson, Rev. Mod. Phys. 43, 424 (1971).
5. J.P. Boisson, R. Piepenbring, and W. Ogle, Physics Reports 26C, 99(1976).
6. C.J. Gallagher, Jr. and S.A. Moszkowski, Phys. Rev. 111, 1282(1958).
7. N.D. Newby, Jr., Phys. Rev. 125, 2063(1962).
8. H.D. Jones, N. Onishi, T. Hess, and R.K. Sheline, Phys. Rev. C 3, 529(1971); N.I. Pyatov, Bull. Acad. Sci. USSR 27, 1409(1963); H. Massmann, J.O. Rasmussen, T.E. Ward, P.E. Haustein, and F.M. Bernthal, Phys. Rev. C 9, 2312(1974).
9. P.C. Sood and R.N. Singh, Nucl. Phys. A419, 547(1984); Nucl. Phys. A373, 519(1982).
10. T. von Egidy, R.W. Hoff, R.W. Lougheed, D.H. White, H.G. Boerner, K. Schreckenbach, D.D. Warner, G. Barreau, and P. Hungerford, Phys. Rev. C 29, 1243(1984).
11. R.W. Hoff, R.W. Lougheed, G. Barreau, H. Boerner, W.F. Davidson, K. Schreckenbach, D.D. Warner, T. von Egidy, and D.H. White, Inst. Phys. Conf. Ser. No. 62, p. 250(1981).
12. Z.M. Koenig, I. Ahmad, J. Milsted, and W.C. McHarris, Bull. Am. Phys. Soc. 27, 523(1982).
13. Table of Isotopes, Seventh Edition, C.M. Lederer and V. Shirley, Eds., (John Wiley, Inc., New York, NY, 1978), 7th ed.

14. V.A. Ionescu, J. Kern, R.F. Casten, W.R. Kane, I. Ahmad, J. Erskine, A.M. Friedman, and K. Katori, Nucl. Phys. A313, 283(1979).
15. R.W. Hoff, R.F. Casten, M. Bergoffen, and D.D. Warner, Brookhaven National Laboratory Report BNL-34811, May 1984, accepted for publication in Nucl. Phys. 1984.
16. R.W. Hoff, Lawrence Livermore National Laboratory Report UCRL-90794, May 1984; submitted for publication in Nuclear Physics.
17. P.E. Haustein, H-C. Hseuh, R.L. Klobuchar, E-M. Franz, S. Katcoff, and L. Peker, Phys. Rev. C 19, 2332(1979).
18. M.A. Gardner, D.G. Gardner, and R.W. Hoff, Lawrence Livermore National Laboratory Report UCRL-91091, August 1984; presented at the Fifth Intl. Symp. on Capture and Gamma-Ray Spectroscopy and Rel. Topics, Knoxville, TN, Sept. 10-14, 1984.

INVESTIGATION OF E0 TRANSITION RATES

G.G. Colvin and K. Schreckenbach
Institut Laue-Langevin, F-38042 Grenoble, France

ABSTRACT

Basic ideas on the nature of electric monopole transitions and their experimental realisation are presented. Some feeling for the sensitivity obtainable for $X = B(E0)/B(E2)$ ratios are discussed. Examples of measurements performed using the electron spectrometer BILL are described to demonstrate their relevance in testing nuclear models. These include even-even and odd-even nuclei such as nuclei close to $Z = 50$, rare-earth nuclei, Pt-Os isotopes and the actinides.

INTRODUCTION

From the many detailed measurements using the (n,γ) and (n,e^-) reactions, energy level spectra, branching ratios and multipole mixing ratios have been carefully compared with theoretical model predictions to show their respective strengths and weaknesses. One area which is not often emphasised (both theoretically and experimentally), but carries much nuclear structure information, is the investigation of electric monopole transition rates[1]. It is the aim of this article to high-light the investigations into E0 transitions performed at the ILL high flux reactor following thermal neutron capture. The properties of E0 transitions and how they can be measured by electron spectroscopy are outlines in sections 2 and 3 respectively. A brief review of some measurements performed using the electron spectrometer BILL is presented in section 4.

But first we will turn our attention to why E0 transitions may be of interest in nuclear spectroscopy. In the internal conversion process, to a good approximation, the rate of production of internal conversion electrons is directly proportional to the rate of gamma emission. However, after closer inspection, it is found that nuclear matrix elements do enter into the calculation of the rate of internal conversion. This is due to the finite size of the nucleus and that atomic wavefunctions may therefore penetrate into the nuclear volume. These nuclear matrix elements carry information about the structure of the nucleus, but in general are difficult to deduce from penetration calculations on the higher order multipoles. In the case of electric monopole transitions, which are themselves a direct consequence of the finite nuclear volume, ejection of the electrons proceeds solely by penetration of the atomic wavefunction into the nuclear volume, and thus carry with them directly, information on the nuclear matrix elements. It is this sensitivity to the nuclear structure which makes the study of E0 transitions useful for nuclear spectroscopy.

PROPERTIES OF E0 TRANSITIONS

Transitions between low lying nuclear levels occur by two main processes: Gamma-ray emission and internal conversion (IC). Internal pair production competes only at energies higher than $2m_ec^2$. These processes may comprise of a mixture of multipolarities as long as spin and angular momentum laws are conserved. Low energy electric monopole (E0) transitions occur by the internal conversion process, the emission of a single gamma-ray having this multipolarity is forbidden due to its transverse nature. Zero units of angular momentum are consequently transfered to the emitted electron. Therefore E0 transitions may occur between states where $I_i = I_f$ and $\pi_i = \pi_f$. In the case where $I \neq 0$, the E0 transitions will compete with higher order multipoles[1], whilst for $I = 0$, no higher multipoles of order λ are emitted due to the conservation law for angular momentum $I_i + I_f \leqslant \lambda \leqslant |I_i + I_f|$ for $I_i \rightarrow I_f$ transitions.

Electric monopole transitions occur via coulomb coupling between the nuclear protons and the atomic electrons. As the monopole moment is a constant outside of the nuclear volume, internal conversion leading to an E0 can only occur within the nucleus and is thus a pure penetration effect. It is this fact which gives the electric monopole its sensitivity to the details of the nuclear wavefunctions.

The E0 transition probability $\omega(E0)$ is defined as the product of an electronic factor Ω (which is spin and nuclear structure independent) and the so called 'strength parameter' ρ (E0) which contains the nuclear properties. Thus $\omega(E0) = \Omega|\rho(E0)|^2$. The strength parameter can be written

$$\rho(E0) = \Sigma \int \phi_f^* \left[\left(\frac{r_p}{R}\right)^2 - \sigma \left(\frac{r_p}{R}\right)^4 + \ldots \right] \phi_i dv \qquad (1)$$

where ϕ_i and ϕ_f are initial and final nuclear wavefunctions, r_p is the position of the pth proton and R is the nuclear radius. The constant σ is in most cases less than 0.1, thus to a good approximation only the first term of equation (1) is required. Inspection of equation (1) reveals the strong dependence of ρ on the nuclear radius and hence there is expected to be a strong Z dependence, as with increasing Z the nuclear radius also increases.

From the fact that E0 transitions are a consequence of a pure penetration effect, we can see that ρ will be dependent on the nuclear shape. It is the change in the radial extent of the charged field of the nucleus which gives rise to the E0 strength and not just a change in the charge distribution itself. Thus for instance, a proton pair excitation in the nucleus will favourably decay by an E0 transition, since the 'skin' of the nucleus (large proton radius) is involved. At this point it should be noted that the E0 transition matrix element depends on the individual wave functions of the initial and final state and not just on the change of the mean square radius $<r^2>$ of the nucleus. Thus a large $\Delta<r^2>$ in the nuclear deexcitation may enhance the E0 transition, but on the other hand the necessary microscopic structure change may hinder it. An outstanding illustration of this is the recently observed E0 transition from a shape isomer of ^{238}U back to the ground state, having the extremely small value of $\rho = 1.7 \times 10^{-9}$, despite the large $\Delta \langle r^2 \rangle$ from the change in deformation [5]. In comparison, the Mössbauer isomer shift is different, since it measures directly $\Delta \langle r^2 \rangle$ in that it is sensitive to the coulomb energy of the nuclear charge and the electron cloud.

A tabulation of Ω can be found in ref. 2. The way in which Ω changes with energy and atomic number can be seen from figure (1). If we assume that $\rho(E0) = 1$ (analogous to the Weisskopf estimate for higher multipoles), then we can make some interesting comparisons between the 'Weisskopf' estimates for E0, M1 and E2 transitions. Figure (2) shows the single-particle 'reduced monopole-conversion probability' ($\Omega = \omega/\rho^2 = \omega$) as a function of energy for the K-shell. The dotted line shows the single particle M1 estimate, whilst the dashed lines show the K-conversion probalities [3]. These figures clearly show how E0 transitions favourably compete with M1, E2 at lower energies and larger atomic number. In the regions of the nuclear chart where M1 transitions between 2^+ levels appear to be severely hindered relative to the single particle estimate, even this dominant multipole may effectively be absent so that the resulting E0 + E2 transition has a strong and readily measureable E0 component.

As will be described in section 3, shell and sub-shell ratios can be used to determine the E0 + M1 + E2 components of a mixed transition, so it would be instructive to compare such ratios for these multipolarities. In general E0 conversion is strongest in the $S_{1/2}$ shells (K, L_1, M_1, etc.) which is perhaps intuitively the expected result as the $S_{1/2}$ wavefunctions are strongest close to the nucleus, whilst for higher shells this becomes increasingly weaker. If we recall the classical figure-of-eight shape of the p-orbital, one would expect no E0 conversion at the nucleus, even for a finite nuclear volume. It is however the relativistic wave equations which gives rise to the finite conversion intensity. The L_2/L_3 ratio for E0 conversion ($p_{1/2}$, $p_{3/2}$ respectively) is of the order of 10^6 and thus no significant conversion in the L_3 shell is expected for E0 transitions. Figures (3) and (4) show the K/L_T, L_1/L_2 and L_1/L_3 internal conversion intensity ratios for E0, M1 and E2 multipolarities taken from the tabulations of references 3 and 4.

MEASUREMENTS OF E0 TRANSITIONS BY THE INTERNAL CONVERSION PROCESS

As was pointed out in section 2, electric monopole transitions can decay by internal conversion and internal pair production. More 'exotic' processes such as 2 gamma -decay are rather weak. Internal pair production can only occur at energies $> 2m_ec^2$. Important work has been done using this method[6,7], but we will limit our discussion here to the IC process which is able to measure low energy transitions. We hope in this section to briefly describe the measuring techniques available using an electron spectrometer and give some feeling for expected limits of sensitivity which can be obtained when measuring E0 transitions.

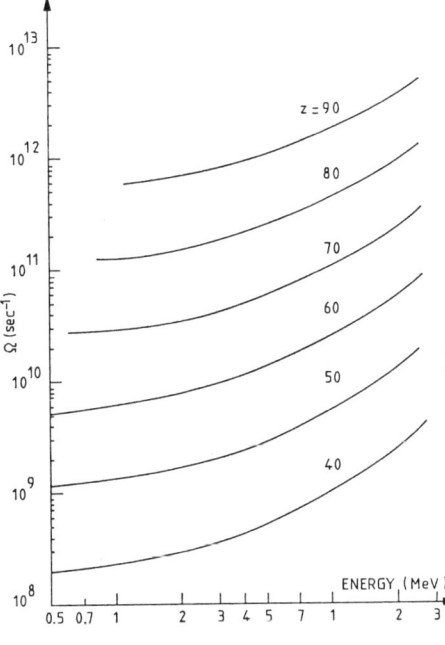

Figure(1)- The electronic factor Ω versus energy for various atomic number. An approx. parameterisation of Ω can be obtained from the equation $\log_{10} \Omega = \{5.8+0.0772Z-8.3\times10^{-5}Z^2\} \times \{0.96+0.1E-0.033E^2\}$ (E-KeV)

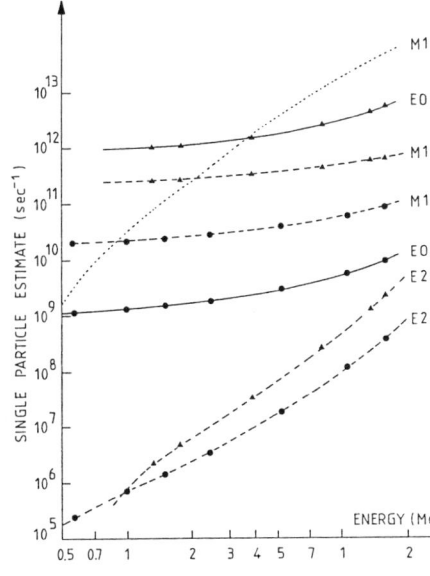

Figure(2)- The transition probability for transitions in the K-shell for Sn(●) and Pt(▲). For E0 transitions, $\rho=1$. Dashed lines give conversion probabilities in Weisskopf units whilst the dotted line is the single particle estimate for M1 transitions.

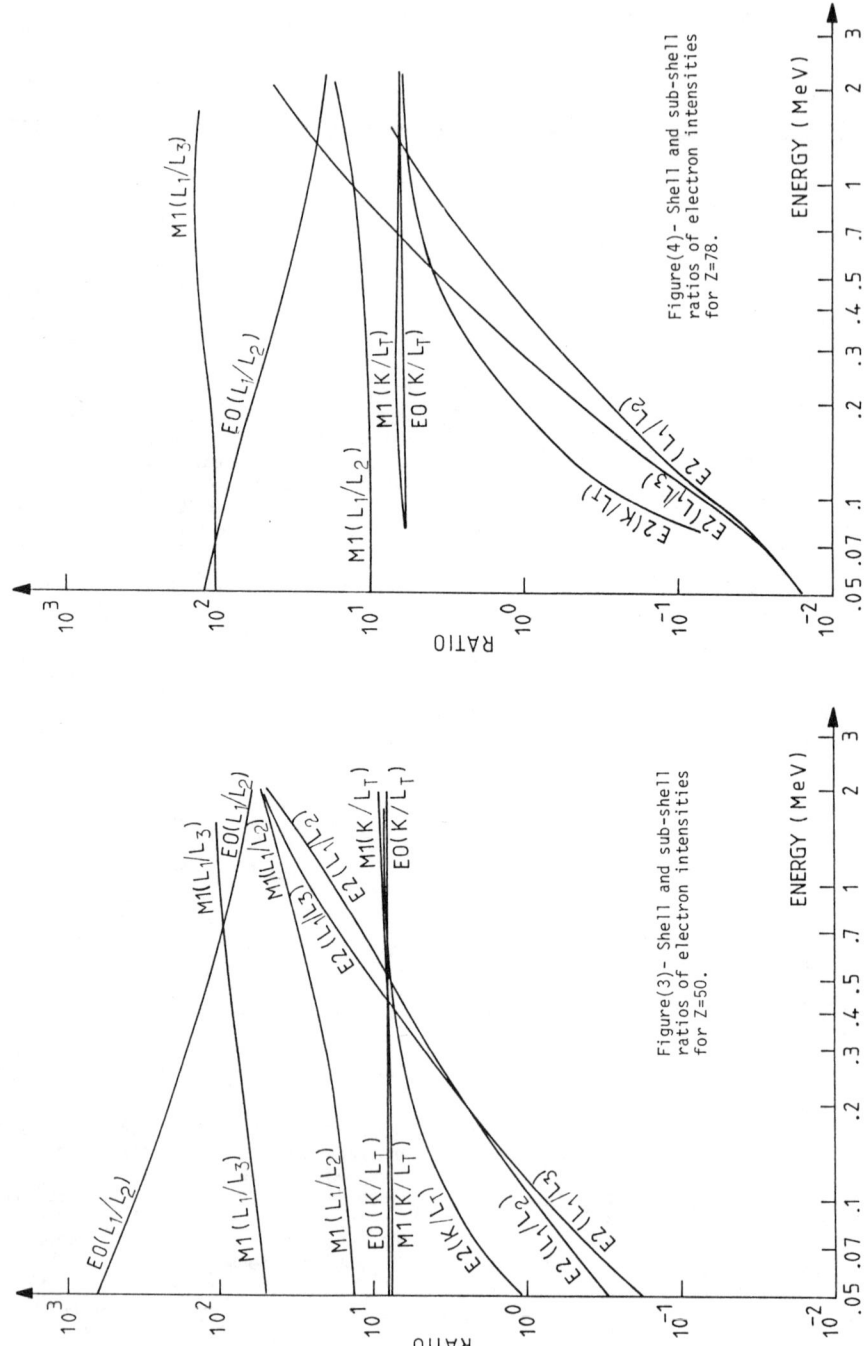

Figure(4) - Shell and sub-shell ratios of electron intensities for Z=78.

Figure(3) - Shell and sub-shell ratios of electron intensities for Z=50.

A measured series of electron lines which have no corresponding gamma-rays are of course a positive identification of pure E0 transition. In the case where there is mixed multipole radiation ($I_i=I_f$ and $I_i \neq 0$) then a direct consequence of an E0 component is an anomalous experimental internal conversion coefficient, not agreeing with any E2 + M1 admixture. Anomalous ICC's due to M1 penetration effects are not considered as this will only be visible when M1 lifetimes are retarded by $> 10^5$, and such cases are rare. For mixed multipole decays, a mixing ratio analogous to the E2/M1 mixing ratio δ^2 employed in the γ-ray case can be defined [8]:

$$q^2 = I_e(E0)/I_e(E2) \qquad (2)$$

where $I_e(E0)$, $I_e(E2)$ are the measured electron intensities for E0 and E2 respectively. As each component of the intensity of such a mixed transition is additive, then

$$q^2 = \frac{1}{\alpha(E2)} \left[(\alpha_e - \alpha(E2)) - \frac{1}{\delta^2} (\alpha(M1) - \alpha_e) \right] \qquad (3)$$

where α is the internal conversion coefficient. The subscript e refers to the experimental ICC.

The most fundamental measure of the E0 matrix elements is the nuclear strength parameter $\rho(E0)$. This can be determined when the partial life-times $T(E0)$ of the level in question is known and can be found from the relation

$$\rho^2(E0) = \ln 2 / T(E0)\Omega \qquad (4)$$

Tabulations of the electronic factor Ω can be found in reference 2. When such data is not available, a useful measure of the E0 strength that can be determined directly by IC spectroscopy is the value[9]

$$X(I_i, E_{ex}) = \frac{B(E0; I_i \to I_f)}{B(E2; I_i \to I_f)} \qquad (5)$$

where $B(E0; I_i \to I_f) = e^2 R^4 \rho^2(E0)$

This ratio can be written in terms of the E0 mixing ratio q, and is given by the relation[10],

$$X(I_i, E_{ex}) = 2.56 \times 10^9 A^{4/3} E_\gamma^5 (MeV) q^2 \alpha(E2)/\Omega \qquad (6)$$

The convention employed when defining X-values for $0^+ \to 0^2$ transitions is to take the ratio

$$X(0^+, E_{ex}) = B(E0; 0_2^+ \to 0_1^+)/B(E2; 0_2^+ \to 2_1^+) \qquad (7)$$

The experimental determination of q^2 can be done in two ways by IC spectroscopy.

i) In the case where δ^2 is known from other types of measurement (primarily γ-γ correlation, where precise values of δ^2 can be obtained[11]), the application of equation (3) is trivial once the experimental conversion coefficient has been determined.

ii) In the case where δ^2 is unknown and L-subshell ratios are resolved, then the following relation is true.

$$I_\gamma(E2) = I_e(L_3)/\alpha_{L_3}(E2) \qquad (8)$$

As indicated in section 2, the conversion of E0 in the L_3 subshell is of negligible intensity and reference to Fig. (4) also shows that the $\alpha_{L_3}(E2)/\alpha_{L_3}(M1)$ will always be large over a significant energy range. Therefore we can say that any conversion in the L_3 subshell is due solely to the E2 component. Thus from equation (8), δ^2 can be determined and hence q^2.

When both δ^2 is unknown, and L-subshells are not resolved, X-values cannot in general be determined from ICE alone. There exists however a small region of Z and energy where sensible limits can be calculated for the E0 strength of a mixed transition. In the region of Z around 50 and energies in the region > 500 keV, both the E2 and M1 conversion coefficients are similar in the $S_{1/2}$ shells. Hence a range of possible E0 components can be calculated by assuming both $\delta^2 = 0$ and $\delta^2 = \infty$. Such results are extremely useful if the lifetime of the decaying level is known as the partial lifetime for the E0 transition can then be calculated and hence $\rho(E0)$, which finally is a more sensitive measure of the E0 transition.

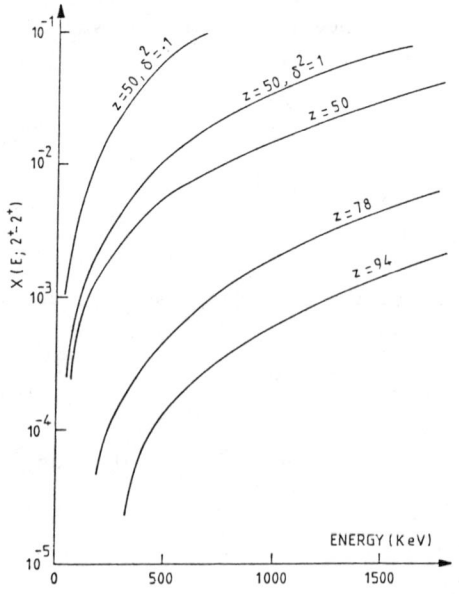

Figure(5)- Sensitivity limit for $X(E; 2^+-2^+)$ versus energy for various atomic number (Z) and E2/M1 mixing ratio (δ^2) taken from equation(6). For the calculation, a detection limit of 5% for the E0 component was assumed. Thus $q^2=0.05$ when $\delta^2 = \infty$.

When considering X-values, it is interesting to estimate the lower limits measurable by the IC process. If we assume that by careful experimentation, an error of only 5% could be achieved (such results have been obtained with the BILL spectrometer), then $\alpha_e/\alpha(E2+M1) > 1.05$. Figure (5) shows the relation between $X(I_i \neq 0, E_{ex})$ versus transition energy for ^{118}Sn, ^{196}Pt and ^{243}Pu for $q^2 = 0.05$. Pure E0 + E2 transitions ($\delta^2 = \infty$) are shown as solid lines whilst increasing M1 components ($\delta^2 = 1, .1$) are shown as dotted lines. It is clearly seen from this figure that the minimum measureable X-values rise steeply with increasing energy and decreasing Z. The effect of δ^2 is seen to strongly influence these values.

E0 MEASUREMENTS AT THE ELECTRON SPECTROMETER BILL

In the following section we will single out some typical experiments performed with the electron spectrometer BILL[12] concerning E0 transitions. The body of data already existing on E0 transitions have been reviewed in various compilations[10,11,13], but it is evident that further data taken at the ILL can play a significant role in increasing or clarifying certain areas. Appendix (1) tabulates the results of E0 measurements performed at the ILL for even-even nuclei. The even-odd actinide nuclei will be covered elsewhere in the proceedings[14]. Table (1) schematically show the expected X-values derived by various theoretical calculations and models. The reader should bare the results of this table in mind for the discussion of some of the experimental results outlined below.

A demonstration of the effect of an E0 component on an L-subshell measurement is shown in Figure (7). This figure[15,34] shows the 154.6 KeV, $2^+ \rightarrow 2^+$ transition in ^{114}Cd. The measured intensity and expected intensity (for an M1 + E2 transition of $\delta^2 = 3.8$) are clearly shown. The excess of electron intensity in the L subshells (and K-shell which is not shown) was attributed to an E0 admixture. In this case δ^2 was calculated directly from the L_3-subshell intensity and the γ-ray intensity[16] measured by the curved crystal spectrometer GAMS[17] as described in section 3. For this measurement the lifetime of the 2^+ state was known and hence a strength parameter of $\rho^2(E0) = 0.034$ was calculated.

The results of the E0 measurements of ^{114}Cd were used qualitatively to separate states arising from different nuclear deformations by the assumption that strong E0 transitions connect states of differing nuclear deformations. This latter assumption has the one important proviso that the strength of these transitions is only maintained if there is no mixing of the two sets of states involved. It was shown that some evidence for such mixing in the case of ^{114}Cd exists and does point out that such arguments must be carefully considered if E0 transitions are not to be misleading. The results of ^{114}Cd also emphasize that $\rho(E0)$ is a more reliable probe than

Table (1)   X-values derived from nuclear models.

Model	Ref.	X-value	comments	note
shell	27	0	$\rho(E0) = 0$	a
s.p. limit	28	4π	$\rho(E0) = 3/5$	-
vibrational (phonon)	29	β_o^2	$0_\beta^+ \to 0_g^+$	-
		∞	$2_\beta^+ \to 2_g^+$	b
rotational	11,29	$4\beta_o^2$	$0_\beta^+ \to 0_g^+$	
		$4\beta_o^2 \frac{2(2I-1)(2I+3)}{I(I+1)}$	$I_\beta = I = 2,4,..$	
		ν	$\gamma \to g$	c
	30	$4\beta_o^2(1+E_{2g}/E_{2\gamma})$	$0_\beta^+ \to 0_g^+$	
		$\frac{7}{2}X(2_\beta^+ \to 2_g^+)$	$2_\beta^+ \to 2_g^+$	
IBA	22,31	-	O(6)	d
		finite	SU(3)-rot	
		0	SU(5)-vib	
quasi-boson approx.	32	$.29 < X < .75$	rare earth	e
	33	$.29 < X < .33$	$0_\beta^+ \to 0_g^+$	e

Notes

a) Involves single nucleon between states of same j & l. Thus requires a jump of two shell closures and is unlikely for low lying states.

b) $B(E2; 2_\beta^+ \to 2_g^+) = 0$. However $\rho(E0)$ is of the same order for 0^+ & 2^+.

c) K-selection rule forbids $\gamma \to g$ E0 transitions. If γ,β,g mixing then[10]

$$\rho(2\gamma \to 2g) = -\frac{Ze}{\pi}\beta_o^2\left[1+\frac{5}{14}\sqrt{\frac{5}{\pi}}\beta_o\right]\frac{1}{b-a}\cdot\frac{1}{c}\cdot\frac{1}{\sqrt{2a-1}}$$

where $a = E_{2\gamma}/E_{2g}$; $b = E_{2\beta}/E_{2g}$; $c = E_{0\beta}/E_{2g}$

d) Selection rules are: E0 - $\Delta\sigma = \pm 2$, $\Delta\tau = 0$; E2 - $\Delta\sigma = 0$, $\Delta\tau = \pm 1$

e) See Figure (6)

X-values. Appendix (1) shows an X-value for the 0_3^+ state in ^{114}Cd which is 50 times greater than for the 0_2^+ state. Comparing $\rho(E0)$ however, this ratio is 0.05. This was found to be due to the lifetime of the 0_3^+ level possessing the large X-value, which is three orders of magnitude larger then the 0_2^+ level [34].

A recent measurement on ^{136}Ba performed with the BILL spectrometer demonstrates how some results can be trivially interpreted [18]. In this case the 1579.8 KeV level had been a candidate for a 0^+ level. The observation of an electron line but no corresponding gamma ray was thus the confirmation required. An X-value of 0.11 ± .02 was determined for this $0^+ \to 0^+$ transition. If this level is to be interpreted as a β-vibration for example, table (1) shows that $X = 4\beta_o^2$ is expected. Using a value of the deformation parameters $\beta_o = 0.139$ (calculated using the relation of Grodzins[19]) $X = 0.078$ can be calculated. This value is consistent with the experimental result and hence this state can be interpreted as

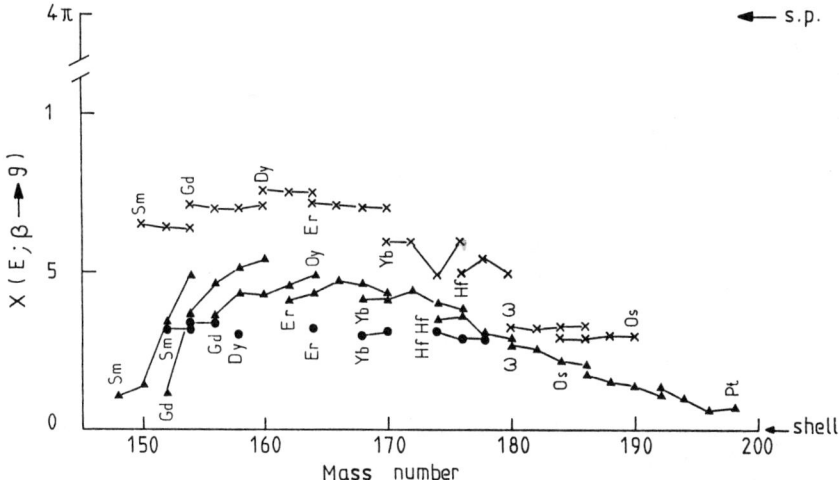

Figure(6)- Model predictions for $X(E;\beta \to g)$ versus mass number. Single-particle and shell model limits are indicated on the right. ▲ shows $4\beta_0^2$ using β_0 values of ref.39. × and ● show the calculations of Bess[32] and Silvestre-Brac[33] respectively.

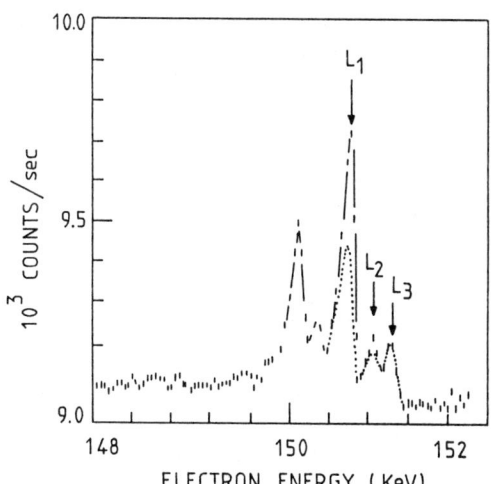

Figure(7)- Part of the ICE spectrum of ^{114}Cd. L_1,L_2,L_3 mark the E0+M1+E2 transition of 154.6KeV. The dotted line shows the expected spectrum if no E0 component were present.

a β-vibration. However, in the absence of lifetime measurements, this result cannot be taken as strong evidence.

Experimental X-values for β-vibrations sometimes show startling irregularities even within a limited region of the nuclear chart. For example, X-values within the rare-earth region have been measured with fluctuations of 3 orders of magnitude. A case where X-values have been measured of the order of the single particle estimate (see appendix) is in the ytterbium and hafnium nuclei[20,38]. For the case of ytterbium, both ^{172}Yb and ^{174}Yb are known to have several excited 0$^+$ states which show large differences in properties even though the two nuclei are only separated by two neutrons. It has been experimentally observed that ^{172}Yb has in general large X-values, with one transition at 1405.0 keV giving X = 2.81± 0.31, whilst in ^{174}Yb, no E0 decays could be observed[20] (see appendix). The interpretation of these results is not evident at first glance and it is perhaps instructive to list here some ways in which 0$^+$ states could be created : i) β-vibrations, ii) pair excitations, iii) pairing vibrations. Clearly the 1794 and 1895 levels in ^{172}Yb can be explained by β-vibrations ($\beta_0 \sim .3$ thus $X \sim .5$) but not the 1405 level. An attractive explanation of this large X-value is proton-pair excitations. A study of the systematics of the energies of the single particle levels in the neighbouring Lu isotopes[21] shows two levels which change sharply with deformation (these being the $\frac{1}{2}$ [411] and $\frac{1}{2}$ [541] levels). A rough calculation of X-values to be expected from such an explanation is that proton-pairs of large amplitude will give X > .9 and thus could explain the large value for the 1405 KeV transition. However, although these single particle levels can change rapidly, they lie close to the Fermi surface over a significantly wide region and would not be expected to give the abrupt variations in X-values observed. Again lifetime measurements may clarify the situation.

The interacting boson approximation model (IBA) can also be tested by a careful study of E0 transitions. The three limiting cases of the IBA are shown in table (1). For the 0(6) limit, tests have been performed at the ILL in the Os-Pt region[22]. In the IBA, the E0 transition operator can be expressed as
T(E0) = $\alpha(s^+s)° + \beta(d^+d)°$. Rewriting this in terms of the total boson number N = $n_s + n_d$ it can be easily shown that

$$T(E0) \propto \sum_i \alpha_i \beta_i n_{d_i} \qquad (9)$$

Thus relative E0 transition strengths can be readily estimated from an IBA calculation. The concept of X-values cannot be defined in the 0(6) limit as competing $0^+_1 \to 2^+_1$ transitions are forbidden by the selection rules and only occur through symmetry breaking terms which unfortunately are not easily calculated. Therefore comparison of E0 transitions between model and experiment are necessarily qualitative.

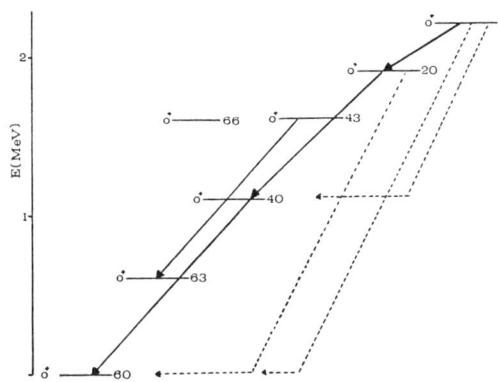

Figure(8)- Low lying 0^+ states from the O(6) limit of the IBA. Quantum numbers (σ,τ) are marked. Arrows connect levels where E0 transitions are expected, whilst dotted lines connect levels where the σ selection rule is broken.

Fig. (8) shows a typical spectrum of 0^+ states in the O(6) limit. The indicated E0 transitions are those allowed by the selection rules: $\Delta\tau=0$, $\Delta\sigma=\pm2$. The transitions indicated by the dotted lines are those which violate the σ selection rule. This selection rule arises by cancellation and is the most easily broken selection rule as only a small perturbation could cause a non-zero E0 matrix element. The τ selection rule should not however be so easily broken, as it arises through the fact that states with differing τ quantum numbers do not have any basis states in common and thus no E0 transitions can occur. To break this selection rule is therefore a more serious violation of the O(6) limiting case. Results for ^{188}Os and ^{196}Pt are in good agreement with the O(6) limit predictions in that the transitions allowed by the selection rules are observed with a significant intensity, although a broken O(6) calculation[23] was required in the ^{188}Os case in order to well reproduce the experimental results.

Finally a brief comment on E0 measurements performed at the ILL on actinide nuclei. The odd-neutron actinides will be covered elsewhere in these proceedings[14], but an example here is perhaps useful. In ^{233}Th, strong E0 admixtures were used phenomenologically to separate two β-vibrational bands coupled to Nilsson states[24]. That such a simple procedure was possible was due to the high resolution of the BILL spectrometer, which allowed the various E0 transitions to be resolved. Similar methods were used with ^{239}U to clearly identify rotational bands built on β-vibrational states[25].

An even-even actinide nucleus measured at BILL which showed some interesting E0 properties was ^{244}Cm following the β-decay of ^{244}Am isomers.[26] In this case the 984.9 KeV transition was identified as pure E0, with a large X-value equal to 1.5 ± 0.2. To interpret this state as a β-vibration cannot be successful, as the predicted X-value is of the order of 0.2. Again in this case, proton-pair excitation was invoked to explain this result.

SUMMARY

Properties and measuring techniques of E0 transitions have been discussed in order to show their use in the interpretation of nuclear spectroscopy data. Although not often emphasised, the E0 transition is a sensitive probe for nuclear spectroscopists. Attention has been payed to measurements involving the beta spectrometer BILL.

Further improvements in the use and interpretation of such data fall mainly into two areas. Firstly, more work is required in the theory of E0 transitions within particular nuclear models. Secondly, a different approach is required of the experimentalists. To explain this, it perhaps suffices to say that most of the E0 results described here have come from an experimental program of much larger proportions. A more systematic and careful approach to measurements of E0 transitions is required (along with lifetime measurements) in order to improve the precision and put on a firmer base the use of E0 transitions in nuclear spectroscopy.

REFERENCES

1. E.L. Church & J. Weneser, Phys. Rev., 103, 1035 (1956).
2. D.A. Bell, et al., Canadian J. Phys., 48, 2542 (1970).
3. R.S. Hager & E.C. Seltzer, Nucl. Data, A6, 1 (1969).
4. R.S. Hager & E.C. Seltzer, Nucl. Data, A4, 1 (1968).
5. J. Kantele et al., Phys. Rev. Lett., 51, 91 (1983).
6. A. Passoja et al., Nucl. Phys. A363, 399 (1981).
7. A. Passoya et al., Phys. Lett., 124B, 157 (1983).
8. E.L. Church & J. Weneser, Phys. Rev., 109, 1299 (1958).
9. J.O. Rasmussen, Nucl. Phys., 19, 85 (1960).
10. A.V. Aldushchenkov & N.A. Voinova, Nucl. Data, 11, 299 (1972).
11. J. Lange et al., Rev. Mod. Phys., 54, 119 (1982).
12. W. Mampe et al., Nucl. Inst. Meth., 154, 127 (1978).
13. D. Hageman, Internal Report KVI-54, (1982).
14. H. Börner, these proceedings
15. A. Mheemeed, Thesis, University of Grenoble, (1981).
16. A. Mheemeed et al., Nucl. Phys., A412, 113 (1984).
17. H.R. Koch et al., Nucl. Inst. Meth., 175, 401 (1980).
18. W. Gelletly et al., Annual Rep., Univ. Manchester, 31 (1980).
19. L. Grodzins, Phys. Lett., 2, 88 (1962).
20. J. Larysz et al., Nucl. Phys. A309, 128 (1978).
21. W. Ogle et al., Rev. Mod. Phys., 43, 424 (1971).
22. W.R. Kane et al., Phys. Lett., 117B, 15 (1982).
23. J. Cizewski et al., Nucl. Phys. A323, 349 (1979).
24. P. Jeuch et al., Nucl. Phys., A317, 363 (1979).
25. H.G. Börner et al., Z. Phys., A286, 31 (1978).
26. H.W. Hoff et al., Phys. Rev., C29, 618 (1984).
27. H.C. Pauli et al., in 'Nuclear Spectroscopy',ed.W.D.Hamilton (1975).
28. J.H. Hamilton et al., Phys. Rev., C10, 2540 (1974).
29. K.Kumar , in 'Nuclear Spectroscopy', ed. W.D. Hamilton, (1975).
30. A.S. Davidov & V.C. Rostovski, Nucl. Phys., 60, 529 (1964).
31. O. Scholten et al., Ann. Phys., 115, 325 (1978).
32. O.R. Bes, Nucl. Phys., 49, 544 (1963).
33. B. Silvestre-Brac & R. Piepenbring, Phys. Lett., 44B, 357 (1973).
34. K. Schreckenbach et al., Phys. Lett., 110B, 364 (1982).
35. W.D. Hamilton et al., to be published.
36. G.G. Colvin & K. Schreckenbach, Nucl. Inst. Meth., in press.

37. J.R. Larysz, Thesis, Univ. Manchester (1977).
38. A.M.I. Haque, Thesis, Univ. Koln, (1984).
39. P.H. Stelson & L. Grodzins, Nuc. Data Tables, 1, 21(1965).
40. F. Hoyler et al., to be published.

APPENDIX

Table(2) - E0 transitions in even-even nuclei measured by the electron spectrometer BILL

Nucleus	$I_i - I_f$	Level (KeV)	Trans-Energy (KeV)	X-Value	$\rho^2 \times 10^{-3}$	Ref.
^{114}Cd	0_2-0_1	1135	1135	.023(2)	27	16
	2_2-2_1	1210	651	<.01	<8	
	0_3-0_1	1306	1306	20(2)	1.3	
	0_3-0_2		171	-	.44	
	2_3-2_2	1364	806	>30	61	
	2_3-2_2		155	.032(6)	34	
	4_2-4_1	1732	449	.12(2)^a		
	2_4-2_1	1842	1283	<.2		
	2_4-2_2		632	.71(8)^a		
	2_4-2_3		478	<.014^a		
	0_4-0_1	1860	1860	.012(4)		
	4_3-4_1	1932	648	.057(22)^a		
	3_2-3_1	2204	340	.052(10)^a		
	0_5-0_1	2437	2437	.56(5)		
	0_6-0_1	2554	2554	.58(9)		
^{124}Te	0_2-0_1	1657	1657	0.012		35
	0_3-0_1	1883	1883	>3.5		
	0_3-0_2		226	~5		
	0_4-0_1	2390	2390	≥.6		
	2_2-2_1	1325	723	.059		
	4_2-4_1	1957	709	1.04		
	2_3-2_1	2039	1437	.21		
	2_3-2_2		714	.094		
	2_4-2_1	2092	1488	5.16		
	$2\ -2_1$	2183	1580	2.11^b		
	$2\ -2_1$	2323	1720	7.10		
^{136}Ba	0_2-0_1	1580	1580	.11(2)		18
^{152}Gd	0_2-0_1	615	615	.047(5)		36
^{156}Gd	$0\ -0_1$	1168	1168	.032(2)		40
^{172}Yb	0_2-0_1	1043	1043	.028(4)	2.3(3)	20,37
	2_2-2_1	1118	1039	<.106		
	4_2-4_1	1287	747	.016(4)^a		
	0_3-0_1	1405	1405	2.81(31)	<26	
	2_3-2_1	1477	1216	.015(3)^a		
	0_4-0_1	1794	1794	.34(4)		

Nucleus	$I_i - I_f$	Level	Trans-Energy	X-Value	$\rho^2 \times 10^{-3}$	Ref.
	2_4-2_1	1849	1771	.12(3)^a		
	0_5-0_1	1895	1895	.10(3)		
	2_5-2_1	1956	1878	.09(3)		
^{174}Yb	0_2-0_1	1478	1478	<.051		20,37
	0_3-0_1	1884	1884	<.175		
^{178}Hf	0_2-0_1	1199	1199	.28(3)		38
	2_2-2_1	1277	1184	.34(3)^a		
	4_2-4_1	1450	1144	.17(1)^a		
	6_2-6_1	1731	1099	.16(2)^a		
	0_3-0_1	1434	1434	>.15		
	2_3-2_1	1496	1403	.44(3)^a		
	0_4-0_1	1444	1444	.59(4)		
	2_4-2_1	1562	1468	.23(2)^a		
	0_5-0_1	1772	1772	>.2		
	2_5-2_1	1818	1725	.53(4)^a		
	4_3-4_1	1956	1650	.37(4)^a		
^{188}Os	0_2-0_1	1087	1087	.0006		22
	0_3-0_1	1478	1478	.0072		
	0_4-0_1	1704	1704	1.70		
	0_5-0_1	1765	1765	0.05		
	0_6-0_1	1825	1825	.15		
	0_7-0_1	1966	1966	>16		
^{196}Pt	0_2-0_1	1135	1135	<.005		22
	0_3-0_1	1403	1403	.092		
	0_4-0_1	1823	1823	<.03		
	0_5-0_1	1919	1919	.06		
^{244}Cm	0_2-0_1	985	985	1.5(2)		26

a - Assumes pure E0 + E2
b - Assumes 2^+ level

STRUCTURAL AND STATISTICAL ASPECTS OF EXTENSIVE LEVEL SCHEMES FROM (n,γ) AND TRANSFER REACTIONS

T. von Egidy, P. Hungerford, H.H. Schmidt, H.J. Scheerer,
A.N. Behkami*, G. Hlawatsch
Physik-Department, Technische Universität München, D8046 Garching,
Germany

B. Krusche, K.P. Lieb
II. Physikalisches Institut, Universität Göttingen,
D3400 Göttingen, Germany

H.G. Börner, S.A. Kerr, K. Schreckenbach
Institut Laue-Langevin, F38043 Grenoble, France

ABSTRACT

The very precise spectrometers for (n,γ) and (n,e) reactions at the ILL, Grenoble, and the high resolution Q3D spectrograph at the Munich Tandem Accelerator yield very detailed information on nuclear transitions and excitations. These results allow the establishment of level schemes which are rather complete in a given spin and energy region and which contain frequently more than 70 levels with spin and parity information. Recently the nuclei ^{20}F, ^{24}Na, ^{28}Al, ^{36}Cl, ^{40}K, ^{41}K, ^{42}K, ^{114}Cd, ^{134}Cs, ^{155}Sm, ^{154}Eu, ^{155}Gd, ^{161}Dy, ^{163}Dy and several actinide nuclei have been investigated. The limits of this method and the possibilities to identify nuclear structures will be discussed. In particular the mixing of single particle and vibrational excitations in deformed nuclei will be treated. A statistical analysis of these extensive level schemes gives information on the level density, on the strength of gamma transitions and on gamma multiplicities of levels. A systematic survey of gamma multiplicities of low spin levels shows that these multiplicities increase only slowly after the first 20 levels while the spread seems to decrease. Level density parameters for the constant temperature and Bethe formulae are determined. These formulae reproduce nicely the experimental level densities.
  The distribution of the E1 and M1 strength of primary transitions is compared with theoretical predictions. The energy dependence of dipole transitions in the Cl and K isotopes agrees better with an E^3 proportionality than with E^5 observed for heavier nuclei. Non-statistical distribution of M1 strength in the sd shell nuclei was seen and might be explained by nuclear structure effects.

*permanent address: Shiraz University, Shiraz, Iran

I. INTRODUCTION

The most detailed and most precise information on the atomic nucleus is obtained by nuclear spectroscopy. During the last decades (n,γ) and transfer reactions provided a large fraction of this spectroscopic information on nuclei close to stability at lower excitation energies. These results played a crucial role in the discussion of new theoretical approaches to the understanding of the nucleus. An important prerequisite for the test of nuclear models are complete level schemes in which all single particle excitations, predicted by the shell model, and all collective excitations are identified in a given energy and spin range. Only the combination of various non-selective and selective reactions, as for instance (n,γ) and transfer reactions, can guarantee the completeness and a certain classification of the level scheme. This completeness is also the basis for a statistical interpretation of the nucleus. We would like to report on a cooperative effort of physicists from many laboratories in Europe and America who established recently "complete" level schemes of ^{20}F (ref. 1), ^{24}Na (ref. 2), ^{28}Al (ref. 3), ^{36}Cl (ref. 4), ^{40}K (ref. 5), ^{41}K (ref. 6), ^{42}K (ref. 7), ^{114}Cd (ref. 8), ^{134}Cs (ref. 9), ^{154}Eu (ref. 10), ^{155}Gd (ref. 11), ^{161,163,165}Dy and several actinide nuclei.[12] Up to 100 levels with γ-decay properties and frequently with spins and parities were identified in each nucleus. The purpose of this contribution is to demonstrate what we can learn from these extensive level schemes.

II. EXPERIMENTAL METHODS

In order to construct schemes with a large number of well established levels, spectrometers are necessary which combine high sensitivity with high precision over a wide range of energies. Therefore, the best instruments presently available have been applied.

The basis of all level schemes discussed in this paper are secondary and primary γ-rays after the capture of thermal neutrons measured at the High Flux Reactor of the Institute Laue-Langevin at Grenoble, France. The bent crystal spectrometers GAMS1 and GAMS2/3 [13] are ideal tools in nuclear spectroscopy due to their outstanding resolution and sensitivity. Equally, the pair spetrometer [14] for primary γ-rays and the large Ge(Li) spectrometer for the intermediate energy range, both looking at an internal target of the High Flux Reactor, have excellent sensitivity. Conversion electrons from the (n,e) reaction observed with the very precise and sensitive beta spectrometer BILL [15] at Grenoble yield for nuclei with Z > 30 conversion coefficients and thus spins and in particular parities of levels, γ-γ coincidence or correlation measurements are also very useful.

Nuclei with sufficiently high neutron resonance density (>5/keV) can be studied with average resonance neutron capture spectroscopy (ARC)[16]. The best facilities for such experiments, filtered beams with 2 keV and 24 keV neutrons[17], are available at the Brookhaven High Flux Beam Reactor. These results are very valuable for the completeness of level schemes because all levels of given spin-parity groups are equally populated and identified.

Single particle transfer reactions such as (d,p) or (d,t) play an important role in the identification of neutron or proton particle or hole excitations, respectively. The Q3D spectrograph[18] at the Emperor Tandem Accelerator of the University and Technical University of Munich offers the best resolution (3-6 keV) and good sensitivity.

There are, of course, various other experimental methods which yield extensive level schemes. However, in most cases the completeness is not guaranteed.

III. LIMITS OF THE COMPLETENESS OF LEVEL SCHEMES

Frequently the test of nuclear models requires the complete set of all levels with special properties, for instance all 0^+ levels. Therefore, the question arrises how well and in which region this completeness can be established.

There are two kinds of completeness which can be discussed: 1. complete (n,γ) level schemes and 2. level schemes which contain all levels in a given energy and spin range.

As an example for a complete (n,γ) level scheme fig. 1 shows the scheme of ^{24}Na (ref. 2). About 99 % of the intensity depopulating the capture state is known. The population and depopulation of almost all levels is well balanced and about 100 % of the intensity arrives at the ground state. 216 out of 277 measured γ-rays are placed. The following conditions have to be fulfilled for such a level scheme: a) The level spacing also at high excitation energies has to be larger than several keV so that all primary transitions can be separated; b) High sensitivity and large dynamic range (smallest intensities <0.1 per 100 n captures); c) Good energy precision (ΔE<0.1 keV also for MeV transitions) for the reliable application of the Ritz combination principle; and d) Precise transfer reaction data which indicate

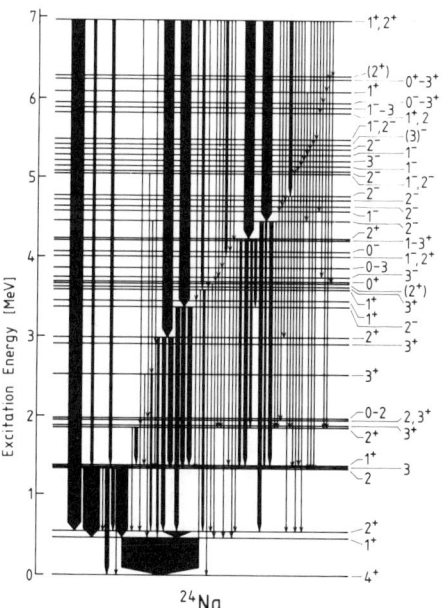

Fig. 1. (n,γ) level scheme of ^{24}Na with the strongest transitions. The widths of the transitions are proportional to the intensities[2].

level positions. Due to these restrictions complete (n,γ) schemes have been established only for nuclei with A<70 or close to closed shells.

Complete level schemes in a given energy and spin range require: a) A non-selective reaction such as (n,γ) or (n,n'γ) in order to populate all levels in question. ARC plays a special role because it guarantees equal population of certain spin-parity groups. b) Good energy precision (few eV) and high sensitivity of secondary γ-rays so that the Ritz principle gives reliable combinations also in the presence of very many levels and transitions. Consequently crystal spectrometer data are desirable. c) Additional information is necessary to confirm levels such as measured multipolarities, transfer reaction data and results from other experiments. d) One should have some idea of the expected level structure. The selectivity of transfer reactions and branching ratios gives the essential arguments for the classification of the levels which allows the comparison with model predictions. The increasing level density limits usually the energy range of the completeness. The reliable identification of levels becomes very difficult if the average level spacing is less than 5 keV. Complete level schemes of heavier nuclei (A>100) can be constructed up to about 2.5 MeV, 1.3 MeV and 0.8 MeV for even-even, odd an odd-odd nuclei, respectively.

A comparison of the sensitivities and spectroscopic information obtained in ARC, (d,p) and (d,t) experiments for ^{161}Dy is shown in fig. 2. These spectra demonstrate that most levels are observed in all four reactions. The selectivity of the (d,p) reaction is illustrated in fig. 3 which compares the "complete"

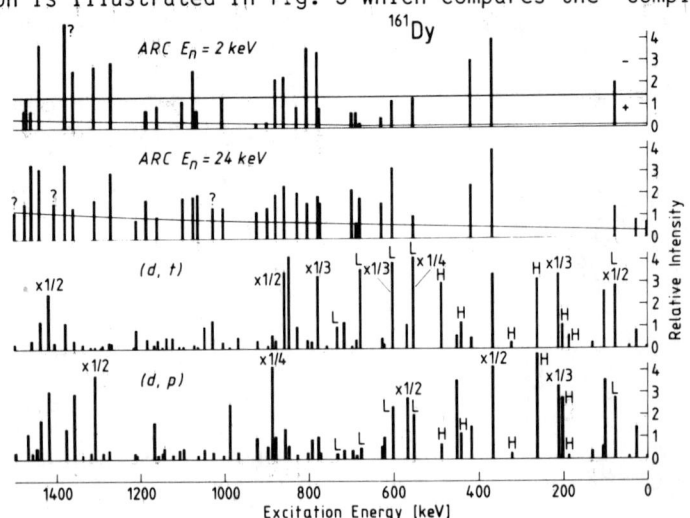

Fig. 2. Comparison of ARC spectra (2 keV and 24 keV neutrons) and (d,p) and (d,t) spectra of ^{161}Dy. The height of the lines corresponds to the populating intensity. (d,p) and (d,t) was measured at 35° with 14MeV d and at 45° with 22MeV d (shown). H and L indicate high and low spins deduced from the I(22MeV)/I(14MeV) ratios.

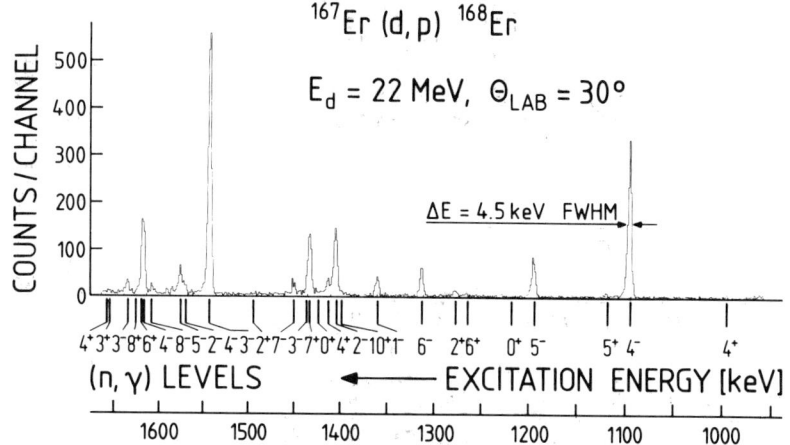

Fig. 3. Part of a (d,p) spectrum measured with the Munich Q3D spectrograph and the complete (n,γ) level scheme of ^{168}Er.

(n,γ) level scheme of ^{168}Er (ref. 19) with a (d,p) spectrum measured at Munich.

IV. LEVEL DENSITIES AND LEVEL SPACINGS

It has been pointed out that a statistical treatment of nuclear properties is meaningful not only because detailed spectroscopic information is not available at higher excitation energies, but because it reveals essentially new features of the nuclear system[20]. Therefore, the statistical interpretation of extensive and complete level schemes may shed new light on our understanding of the nucleus. Nuclear level energies contain two kinds of statistical properties: the level density and the level spacing distribution.

Two formulae are used to describe the <u>nuclear level densities</u> as a function of the excitation energy E:
constant temperature Fermi gas model[21]

$$\rho(E,J) = f(J)(1/T)\exp((E-E_0)/T) \quad \text{and}$$

Bethe formula

$$\rho(E,J) = f(J)\exp(2\sqrt{a(E-E_1)})/(12\sqrt{2}\, \sigma a^{1/4}(E-E_1)^{5/4}).$$

Both formulae use for the spin distribution

$$f(J) = \exp(-J^2/2\sigma^2) - \exp(-(J+1)^2/2\sigma^2)$$

with the spin cut-off parameter σ which is related to the parameters a and E_1 of the Bethe formula

$$\sigma^2 = 0.0888\, A^{3/2}\sqrt{a(E-E_1)} \quad (A = \text{atomic weight}).$$

Gilbert and Cameron[21] stated that the constant temperature formula gives a good fit to the experimental data over the first 10 MeV, while Dilg et al.[23] and Tielens et al.[24] find good agreement also with the Bethe formula if a and E_1 are fitted. Using our extensive level schemes we tried to answer the question which formula gives better agreement. The parameters T and E_0 or a and E_1 were adjusted to obtain a good fit to the number of levels at lower energies $N(E) = \int \rho(E)dE$ and to the neutron resonance densities[25].

Table I Level density parameters

Isotope	Bethe formula			constant temperature formula		
	a[MeV^{-1}]	E_1[MeV]	$\bar{\chi}^2$	T[MeV]	E_0[MeV]	$\bar{\chi}^2$
^{36}Cl	3.23±0.09	-1.96±0.25	0.6	2.08 ±0.08	-1.59±0.19	0.5
^{41}K	5.31±0.06	-0.59±0.06	0.9	1.46 ±0.02	-1.62±0.07	0.8
^{114}Cd	14.28±0.11	1.15±0.02	2.5	0.704±0.004	-0.06±0.02	0.6
^{134}Cs	12.82±0.11	-1.38±0.03	2.1	0.776±0.007	-2.87±0.03	1.4
^{154}Eu	18.55±0.08	-1.36±0.01	1.8	0.615±0.005	-2.96±0.02	2.9
^{155}Gd	19.56±0.21	-0.36±0.04	10.4	0.554±0.006	-1.69±0.04	5.1

The energy dependent σ of the Bethe formula was also used in the constant temperature Fermi gas formula with the Bethe parameters. Fig. 4 and Table I show that in most cases both formulae give good results; the constant temperature formula seems to be better for all isotopes examined except for ^{154}Eu.

The spacing distribution of nuclear levels with the same spins and parities reflects the number of good quantum numbers[20]. The spacing follows an exponential distribution if levels with different spins or parities are considered. The spacing of levels with the same spin and parity and no additional good quantum number is expected to be described by a Wigner distribution due to the level repulsion:

$P(S/D) = (\pi S/2D)\exp(-\pi S^2/4 D^2)$

with the spacing S and the average spacing $D = 1/\rho$.

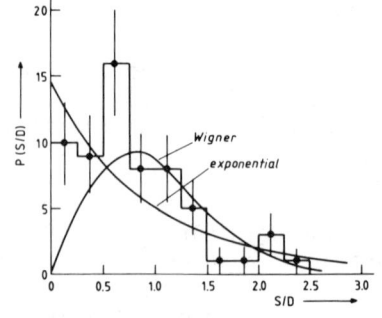

Fig. 5 Spacing distribution P(S/D) of levels with the same spin-parity below 500 keV in ^{154}Eu compared with Wigner and exponential distributions.

Our extensive level schemes which contain very many spin-parity assignments offer the unique possibility to determine the spacing distribution experimentally. It is demonstrated in

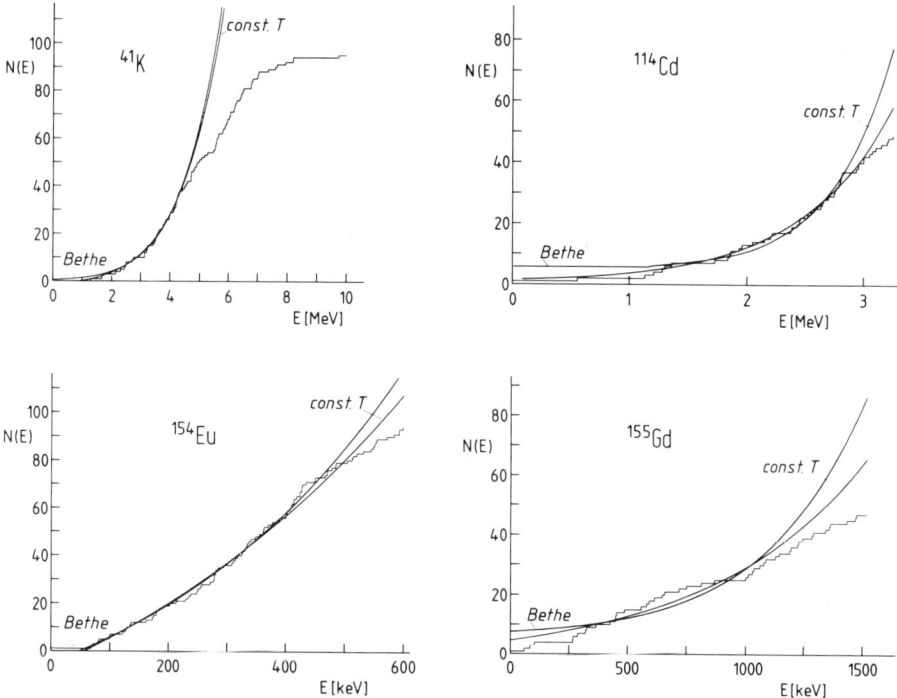

Fig. 4 The number of levels N(E) up to the energy E as a function of E, compared with calculated functions from the level density formulae.

Fig. 5 that the experimental spacing distribution of ^{154}Eu lies between a Wigner and an exponential distribution with a slightly better χ^2 for the exponential curve. This could be an influence of the K quantum number and the different proton-neutron pairs.

V. GAMMA RAY STRENGTH FUNCTIONS OF sd SHELL NUCLEI

The primary transitions depopulating the neutron capture level contain information on the <u>energy dependence of the average γ-intensities</u>. It is assumed that these transitions have preferentially dipole character. Fig. 6 shows the summed primary γ-ray yield Y(E) as a function of the energy of the populated levels. This experimental result can be compared with the theoretical prediction derived from the level density $\rho(E)$ and the energy dependence of the average intensity which is E_γ^3 according to the single particle model and E_γ^5 according to the giant dipole resonance model:

$$Y(E) = \Sigma_{E' \leqslant E}\ I(E\gamma) \propto \int_0^E \rho(E')\ E_\gamma^n\ dE'\ ,\quad E_\gamma = E_B - E'$$

with the neutron binding energy E_B and the energy exponent n = 3 or 5. In contrast to heavier nuclei such as ^{46}Sc, ^{64,66}Cu (refs. 26 and 27) where n = 5 gives better agreement, n = 3 was found for ^{36}Cl and ^{40,41,42}K and n ≃ 1 for ^{20}F, ^{24}Na and ^{28}Al (refs. 1 - 3). This remarkable result indicates that strong M1 and E1 single-particle components govern the primary spectrum for A = 30 to 40. For A < 30 the statistical approach seems to fail.

The distribution of reduced γ-intensities $r = I_\gamma/E_\gamma^3$ can be compared with Porter Thomas distributions $P_{\nu/2}(r)$ where ν denotes the degrees of freedom. Only transitions were included to levels where the scheme is still complete. The two examples in fig. 7 show that $\nu(^{36}Cl) = 1.8\pm0.4$ and $\nu(^{41}K) = 1.9\pm0.4$.

In the energy region where the level scheme is not any more complete, more and more weak primary transitions escape observation. Consequently the level schemes become less complete and the observed level densities seem to decrease. The number of missing primary transitions and of missing levels can be estimated with the known Porter Thomas distribution and the detector response function. Fig. 8 confirms

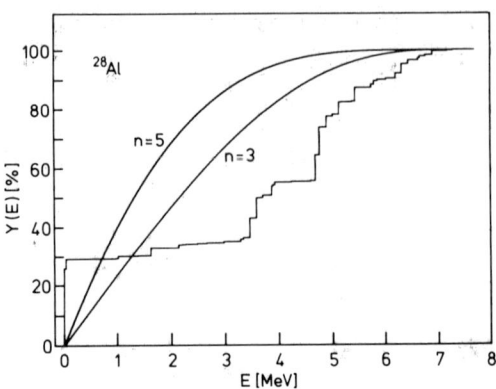

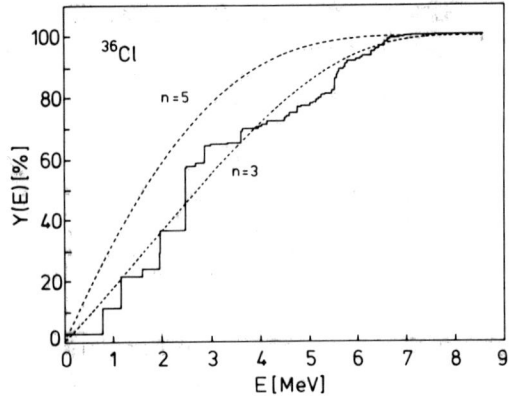

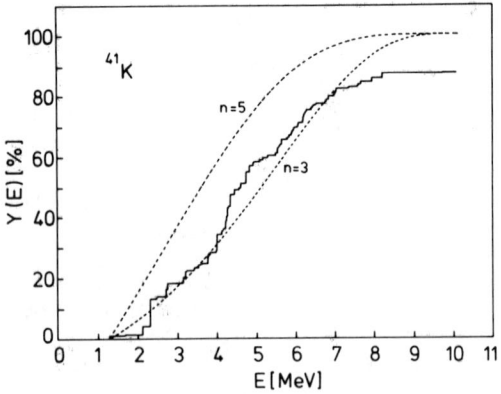

Fig. 6 Integrated γ-ray yield of primary dipole transitions leading to levels with $E \leq E_B - E_\gamma$. The lines correspond to predictions with the intensities αE_γ^3 and αE_γ^5.

that this estimate is correct.

The measured average M1 strengths $\bar{S}$ in most odd-odd nuclei studied is displayed in fig. 9 as a function of the excitation energy. The individual strengths S_i are calculated from the γ-intensities with the expression[28]:
$S_i = 4.8 \cdot 10^7 \, \Gamma_t D^{-1} \cdot E_{\gamma i}^{-3} I_{\gamma i}$ [Wu/MeV].
Γ_t denotes the total width of the capture state and D the average resonance spacing. While the measured values of $\bar{S}$ are in many cases close to the average value given by Mc Cullagh et al.[28] $\bar{S}$ = 1.4 Wu/MeV, it is evident from fig. 9 that in ^{24}Na, ^{28}Al, and ^{36}Cl systematic deviations occur between 2 and 5 MeV excitation energy.

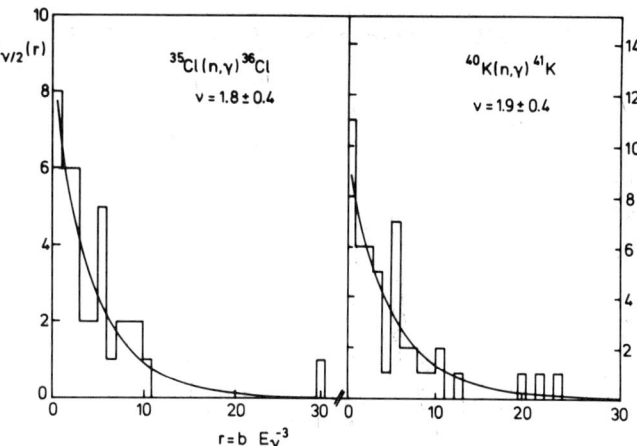

Fig. 7 Measured distributions of primary dipole transitions and Porter Thomas fits $P_{\nu/2}(r)$.

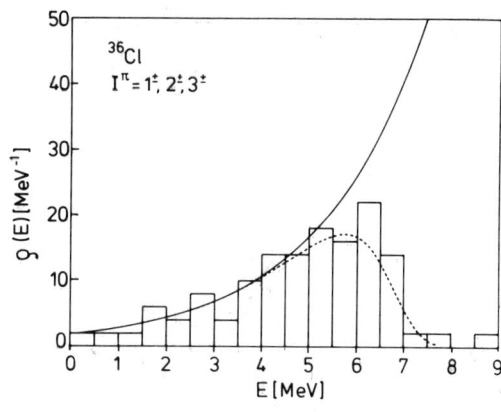

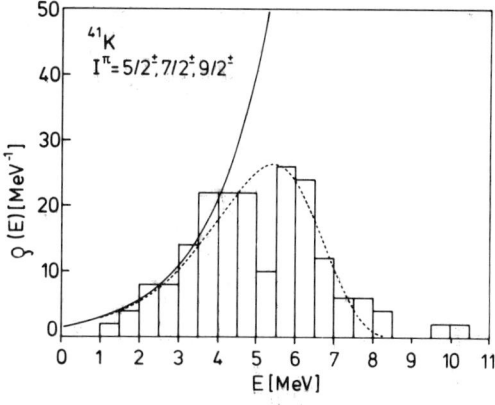

Fig. 8 Measured level density ρ(E) together with the constant temperature density fit (full line) and with the calculated density corrected for missing levels (dashed line).

Fig. 9 Strength functions of primary M1 transitions as a function of excitation energies, averaged over 1 MeV intervals. The hatched areas indicate the corrections for missing lines due to Porter Thomas fluctuations.

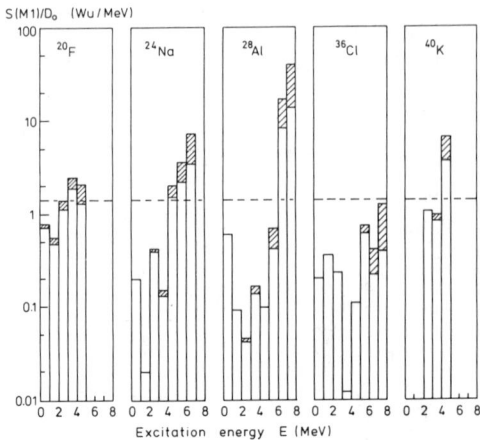

VI. GAMMA MULTIPLICITIES OF LOW SPIN LEVELS

Gamma ray multiplicities are frequently used to obtain information on nuclear properties. The measurement is relatively simple, because they are proportional to the ratio of coincident to single rates of two γ detectors looking at a γ-source. Large multiplicities are usually indications of high spin states. Recently, a Scandinavian group determined γ-multiplicities of levels with spins $I \leqslant 10$ via the (^{3}He,αγγ) reaction[29,31]. They tried to deduce from this very selective reaction with the help of crude γ-multiplicity distributions information on level densities, γ-strength functions and nuclear structure changes. The complete level schemes established by our group contain the γ-multiplicities of very many individual levels. In order to provide a detailed understanding we studied the increase of γ-multiplicities of low spin levels with increasing excitation energies and the multiplicity distributions, and compared them with theoretical predictions based on level densities and average transition rates.

Gamma multiplicities of individual levels in four isotopes are displayed in fig. 10. Average multiplicities of groups with 20 levels and the standard deviations of the multiplicities in these groups are given in Table II. The new information from this investigation is: 1. The average multiplicities of low spin levels increase only slowly with increasing energy after the twentieth level; and 2. The standard deviations of the multiplicities within a group of 20 levels seem to have the tendency to decrease with increasing energy. This second point is an indication of an increasing configuration mixing.

γ-multiplicities have also been calculated with a cascade program[32] which uses the level densities (Bethe Formula) and average E1 and M1 transition rates[28]. The calculated multiplicities are lower for $I_\gamma \propto E_\gamma^5$ than for $I_\gamma \propto E_\gamma^3$, as expected. These curves in fig. 10 show reasonable agreement with the experimental

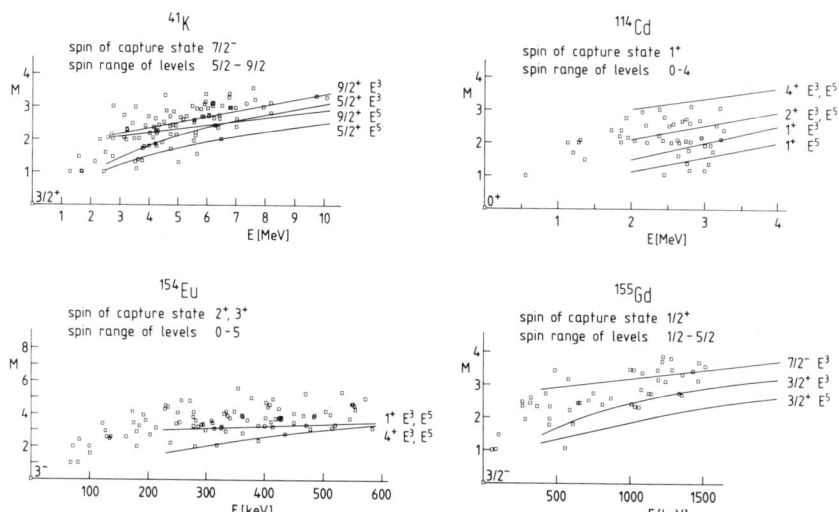

Fig. 10 Gamma ray multiplicities of individual levels. The lines correspond to calculated multiplicities for given I^π and with γ-intensity dependence E_γ^3 or E_γ^5.

TABLE II Average γ-multiplicities for groups of 20 levels together with the standard deviations in these groups (in brackets).

Isotope	Spin range	\multicolumn{4}{c}{level group}
		1 - 20	21 - 40	41 - 60	61 - 80
^{36}Cl	1 - 3	1.78(0.64)	2.03(0.50)	2.34(0.46)	2.53(0.52)
^{41}K	5/2-9/2	1.56(0.53)	2.02(0.49)	2.35(0.39)	2.58(0.51)
^{114}Cd	0 - 4	2.10(0.51)	2.16(0.54)	2.13(0.49)	
^{134}Cs	2 - 5	2.14(0.65)	2.86(0.61)	3.33(0.50)	
^{154}Eu	0 - 5	2.62(0.84)	3.62(0.74)	3.57(0.79)	3.90(0.62)
^{155}Gd	1/2-5/2	2.14(0.64)	3.02(0.51)	3.22(0.36)	

values and support the conclusion that multiplicities of low spin levels increase only slowly at higher energies.

VII. MIXING OF SINGLE PARTICLE AND VIBRATIONAL EXCITATIONS IN ODD DEFORMED NUCLEI

The lowest excitations in odd deformed nuclei are rotational bands built on Nilsson configurations. These single particle states can couple with vibrational excitations which are observed in the adjacent even-even nuclei. The information on vibrations in odd deformed nuclei is scarce, only in the actinide region systematic results are available[33]. The identification of these vibrations is very difficult because they appear in the region of increasing level density and increasing fragmentation of specific nuclear structures. Vibrations on the ground state can be observed by Coulomb excitation. The combination of (n,γ) and transfer reactions is a good method to identify the vibrations because they are supposed to be weakly populated by transfer reactions and decay preferably to the basis bands by E0, E1 and E2 transitions which indicate β, octupole and γ-vibrations, respectively.

Fig. 11 gives an example of such vibrations observed in ^{155}Gd (ref. 11) together with the vibrations in ^{154}Gd and ^{156}Gd. The distance of the vibrations to the basis bands is different for different bands and depends on the mixing of this band with other configurations and on the specific properties of the vibrations. Therefore special information on the microscopic structure of vibrations can be obtained by a careful analysis, because the con-

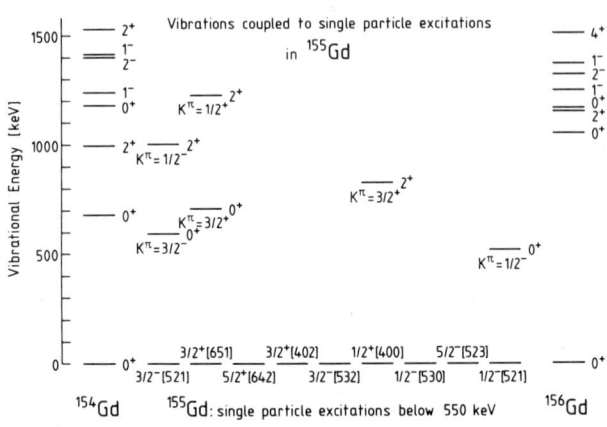

Fig. 11 Band heads in ^{154}Gd and ^{156}Gd compared with vibrational excitations in ^{155}Gd. The spacings of the Nilsson band heads and the vibrations built on them are given for ^{155}Gd.

figuration of the basis band is blocked in the coupled vibration. Due to the completeness of the ^{155}Gd level scheme until 1100 keV, the non-observation of vibrations indicates that they lie at higher excitation energies. Another contribution[34] to these proceedings will compare vibrations with theoretical predictions, in particular with the particle phonon coupling model of Soloviev et al.[35].

VIII. CONCLUSION

A wealth of information can be deduced from extensive and complete nuclear level schemes. The detailed knowledge of very many individual levels leads to a deeper understanding of the statistical properties of nuclei and to the identification of special structures in nuclear excitations. The construction of extensive level schemes is hard and tedious work and in many cases the result of a larger international cooperation. The progress in nuclear science requires precise and comprehensive experimental information and justifies the effort.

We wish to thank F. Fridgen and P. Stoeckel for computational and graphical help, respectively.

REFERENCES

1. P. Hungerford et al., Z. Physik A313, 339 (1983)
2. P. Hungerford et al., Z. Physik A313, 325 (1983)
3. H.H. Schmidt et al., Phys. Rev. C25, 2888 (1982)
4. B. Krusche et al., Nucl. Phys. A386, 245 (1982)
5. T. von Egidy et al., J. Phys. G10, 221 (1984)
6. B. Krusche et al., Nucl. Phys. A417, 231 (1984)
7. B. Krusche et al., to be published
8. A. Mheemeed et al., Nucl. Phys. A412, 113 (1984)
9. M. Bogdanovic et al., to be published
10. M. Balodis et al., to be published
11. H.H. Schmidt et al., to be published
12. H.G. Börner, contribution to these proceedings
13. H.R. Koch et al., Nucl. Instr. Meth. 175, 251 (1979)
14. D.D. Warner et al., J. Phys. G5, 1723 (1979)
15. W. Mampe et al., Nucl. Instr. Meth. 154, 127 (1978)
16. R.E. Chrien, Inst. Phys. Conf. Ser. 62, 342 (1982)
17. R.C. Greenwood and R.E. Chrien, Nucl. Instr. Meth. 138, 125 (1976)
18. M. Löffler et al., Nucl. Instr. Meth. 111, 1 (1973)
19. W.F. Davidson et al., J. Phys. G7, 455 and 843 (1981)
20. T.A. Brody et al., Rev. Mod. Phys. 53, 385 (1981)
21. A. Gilbert and A.G.W. Cameron, Can. J. Phys. 43, 1446 (1965)
22. H.A. Bethe, Phys. Rev. 50, 332 (1936)
23. W. Dilg et al., Nucl. Phys. A217, 269 (1973)
24. T.A.A. Tielens et al., Nucl. Phys. A403, 13 (1983)
25. G. Rohr, private communication (1984)
26. T.A.A. Tielens et al., Nucl. Phys. A376, 421 (1982)
27. M.G. Delfini et al., Nucl. Phys. A404, 225 and 250 (1983)
28. C.M. Mc Cullagh et al., Phys. Rev. C23, 1394 (1981)
29. J. Rekstad et al., Physica Scripta T5, 45 (1983)
30. R.M. Aasen et al., Phys. Rev. C28, 1857 (1983)
31. T. Ramsøy et al., Nucl. Phys. A414, 269 (1984)
32. T. von Egidy in "Neutron Capture Gamma-Ray Spectroscopy", IAEA, Vienna 1969, p. 541
33. T. von Egidy et al., Physics Lett. 81B, 281 (1979)
34. H.H. Schmidt et al., contribution to these proceedings
35. V.G. Soloviev et al., Bull. Akad. Sci. USSR, Phys. Ser. 31, 515 (1967)

AVERAGED NEUTRON CAPTURE DATA AND THEIR APPLICATIONS
TO NUCLEAR STRUCTURE AND REACTION MECHANISM

J. Kopecky
Netherlands Energy Research Foundation
P.O. Box 1, 1755 ZG Petten, The Netherlands

ABSTRACT

Techniques employing the averaging over compound nuclear capturing states have been used for investigations of nuclear structure since 1970. Recently, however, the statistical basis of the averaging process was formulated, allowing a quantitative formulation of the fluctuation properties. Further, this quantitative basis allows the investigation of some other nuclear parameters and properties related to the capture mechanism. Several examples based on the recent data are presented.

INTRODUCTION

The neutron capture reaction forms a highly excited state, which can decay by electromagnetic radiation. In general this is a narrow compound state, highly complex in structure, although many examples of simple configurations have been reported[1]. Discrete resonance capture may be studied by neutron time-of-flight methods, but the statistical accuracy of this method is limited by the low intensity of pulsed neutron beams. Another limitation stems from the distribution of the partial radiative widths for the statistical decay. It is normally assumed that these follow the chi-square distribution with $\nu=1$. The features of this distribution are responsible for the finding that many final states will be weakly or not at all populated.

There is, however, an advantage intrinsic in the statistical nature of the capture process. The decay of a compound resonance is non-selective, i.e., not dependent on the nature of the final state. Both collective or single particle states can be equally populated, because the matrix elements of transitions are statistically distributed and carry no information on the structure of the final states. In order to employ this feature fully, the accumulation of data over many resonances is required.

In the limit as the number of resonances approaches infinity one expects the distribution of intensities to approach a Gaussian form with a variance of $2/n$. The averaged radiation width is proportional to the sum of partial widths of contributing resonances and the dynamic range of the distribution is drastically reduced.

The method of instrumental resonance-averaging was first proposed by Bollinger and Thomas[2,3]. In their setup an internal target at the reactor was surrounded by a ^{10}B absorber to suppress the large thermal contribution. The resulting neutron spectrum was centered around 100-150 eV and had an 1/E high energy tail out to several keV. Because of the target geometry and the use of a thick

2. Their number can be derived from the expression

$$D_J = \frac{D_{J=0}}{(2I+1)} e^{1/2\left(\frac{J+1/2}{\sigma}\right)^2}, \qquad (2)$$

where σ is the spin cut-off factor.

3. The neutron strength functions, S_0 and S_1, are spin and channel spin independent.
4. The total radiative width, Γ_γ, is spin, parity and energy independent.
5. Only dipole transitions are considered.

Under these simplifications and expressing Γ_n with the help of S_0 and S_1 eq. (1) can be rewritten[9,10] as:

$$\langle \sigma_{\gamma f} \rangle = \frac{2\pi \lambda^2 E_n^{1/2} S_0 D_{J=0}}{2(2I+1)} \left\{ \sum_i^{s\text{-waves}} \sum_J \frac{e^{1/2\left(\frac{J+1/2}{\sigma}\right)^2}}{\Gamma(E_n, J, \pi_i)} \frac{\Gamma_{\gamma f}(E1, M1)}{D_J} F_0 \right.$$

$$\left. + \frac{S_1}{S_0} \frac{(ka)^2}{(1+ka)^2} \sum_i^{p\text{-waves}} \sum_J \frac{e^{1/2\left(\frac{J+1/2}{\sigma}\right)^2}}{\Gamma(E_n, J, \pi_i)} \frac{\Gamma_{\gamma f}(E1, M1)}{D_J} F_1 \right\}, \qquad (3)$$

where I is the target spin, E_n is the neutron energy and the penetrability of p-waves relative to s-waves is described by $(ka)^2/(1+ka)^2$. The factors F represent the replacement of the $\langle \Gamma_n \Gamma_\gamma / \Gamma \rangle$ term in eq. (1) either by introducing[9] the averages of the individual widths in terms of Lynn's fluctuation factors[16] or generating them as uncorrelated variables from the Porter-Thomas distributions[10].

The next step is the separation of E1 and M1 radiation and the introduction of the photon strength functions, S(E1) and S(M1). They are defined as:

$$S(E1) = \Gamma_{\gamma f}(E1) \cdot f(E_\gamma, A, E1) D_J^{-1} \text{ and}$$
$$S(M1) = \Gamma_{\gamma f}(M1) \cdot f(E_\gamma, A, M1) D_J^{-1}. \qquad (4)$$

The function f reflects the energy and mass dependence, which generally differs from E1 and M1 radiation.

At this point the previous derivation[9,10] assumed the same E_γ proportionality for the E1's and M1's. The present discussion shall, however, concentrate on a general treatment. It will be shown that this simplifying assumption may be after all applied to the analytical derivation but it should be taken into account in the comparison to experimental data. This has been very recently pointed out by Chrien et al. in refs. 17 and 18.

In order to proceed further, some model predictions for the function $f(E_\gamma, A, L)$ have to be introduced.

^{10}B absorber, the energy distribution of the neutrons was not well defined and the fluctuation analysis was difficult to carry out quantitatively.

To overcome this problem quasi-monoenergetic filtered beams applied on external targets were introduced. The idea of beam filters, making use of the minimum in the cross section due to destructive interference between resonance and potential scattering, originates by Simpson and Miller[4] from 1968. The present most commonly used beams have the average energy at 2 keV and 24 keV based on the transmission through scandium and iron filters. The historical development and parameters of filtered beams are described in several papers and the reader is referred to the most significant of them[5-7]. The application of filtered beams to neutron capture spectroscopy was discussed during the (n,γ) symposia in Petten[5], Brookhaven[8] and Grenoble[9].

A new development in this field appeared in 1981, when R. E. Chrien and M. Stelts together formulated the analytical basis for the theoretical treatment of the averaging process[9,10]. This formulation enables more accurate data analysis and broadens the scope of applications of averaged data substantially. In several recent papers[11-15] the new analysis method has been partly applied, but certainly not all possibilities have been exploited to date.

The instrumental part and the basis of the average resonance capture (ARC) method have been well documented in previous surveys. This paper primarily aims to describe the principle of the statistical analysis and to illustrate its connection with several nuclear or reaction parameters.

AVERAGED RESONANCE CAPTURE

The resonance averaged cross section for a γ transition to a final state f is given by[16]

$$\langle \sigma_{\gamma f} \rangle = 2\pi \lambdabar^2 \sum_{\ell,J} g_J D_J^{-1} \langle (\frac{\Gamma_n \Gamma_{\gamma f}}{\Gamma}) \rangle, \qquad (1)$$

where ℓ is the angular momentum of neutrons, J is the resonance spin and D_J is the average level spacing for resonances with the spin J, $\lambdabar$ is the neutron wave length, g_J is the statistical factor, Γ_n, Γ and $\Gamma_{\gamma f}$ are the neutron, total and partial radiative widths, respectively. The summation is over all capture states (resonances) in the averaging interval.

M. Stelts[9] and R. E. Chrien[10] applied several assumptions in order to make the analytical solution of eq. (1) possible. We shall briefly repeat the basics of their approach and concentrate in detail on the properties of E1 and M1 radiation and their impact on the expression for $\langle \sigma_{\gamma f} \rangle$ and its applicability.

The assumptions, which deal primarily with the properties of initial states, were listed in ref. 10 as follows:

1. Only s- and p-resonances are considered.

the comparison to experimental data. This has been very recently pointed out by Chrien et al. in refs. 17 and 18.

In order to proceed further, some model predictions for the function $f(E_\gamma, A, L)$ have to be introduced.

In the case of the statistical decay the E1 strength distribution is well described by the Brink-Axel model, which gives $f(E_\gamma, A, E1) = E_\gamma^5 A^{8/3}$. Various neutron and proton capture experimental values show an excellent fit with the E_γ^5 dependence, which can be attributed to the tail of the giant dipole resonance. Recent reviews[18,19] of photon strength functions have shown, however, that the mass dependence in the original parameterization of Axel is slightly overestimated, and that it follows rather $A^{6/3}$.

In contrast to the E1's, the behavior of the M1's is still uncertain. There is at this moment no solid evidence for a significant departure from the phase shift prediction, E_γ^3. A weak mass dependence has been proposed in ref. 19 based on the fit to the M1 strength functions, being approximately linear.

It is therefore somewhat surprising that almost all previous papers have assumed that E1 and M1 radiation have the same energy dependence, namely E_γ^5. This conclusion was based on a belief that the ratio between E1 and M1 strengths is independent of the γ-ray energy and for heavier nuclei (A>100) also of the mass[20]. This conclusion can now, however, be seen as derived from incomplete or inaccurately interpreted data. If the presently accepted models for the E1 and M1 strength functions are applied, the E1/M1 ratio becomes energy dependent and can be expressed, in case of averaging over the initial states, as

$$\frac{\langle \Gamma_{\gamma f}(E1)\rangle_i}{\langle \Gamma_{\gamma f}(M1)\rangle_i} = \frac{S(E1)\, E_\gamma^5 A^{6/3} D_J}{S(M1)\, E_\gamma^3 A\, D_J} = \frac{S(E1)}{S(M1)} E_\gamma^2 A = R(E_\gamma, A). \quad (5)$$

In the recent survey of the E1/M1 ratios[21], the radiative widths were averaged not only over the initial but also over the final states. From traditional and practical reasons the same reduction factors were applied for both multipolarities. The ratio R(A) derived in this way represents a value averaged over the γ-ray energies to the particular final states and can be written for a given nucleus as:

$$\frac{\langle\langle \Gamma_{\gamma f}(E1)\rangle_i\rangle_f}{\langle\langle \Gamma_{\gamma f}(M1)\rangle_i\rangle_f} = \frac{S(E1)}{S(M1)} = R(A). \quad (6)$$

R(A) is to a large extent energy independent, if the E_γ intervals are the same for the E1's and M1's and are broad enough. Such values are suitable to study the global mass dependence of the E1/M1 ratio, if small variations of the E_γ intervals among nuclei are considered.

For the final simplification of eq. (3) the partial radiative widths can be replaced by the γ-ray strength functions, but we have to make a choice between the two solutions. The "exact" one, taking into account the E_γ dependence of the E1/M1 ratio in the form of eq. (5), will make, however, the evaluation as well as its comparison with the data rather tedious and complicated. The "approximate" solution, which uses eq. (6), enables the analytical derivation in a rather simple way and can rely on the well documented R(A) values[19,21]. If this approach (identical to refs. 9 and 10) is chosen, which is the only practical way, eq. (3) can be written as:

$$\langle \sigma_{\gamma f} \rangle \approx \frac{2\pi \lambdabar^2 E_n^{1/2} S_0 D_{J=0} S(E1)}{2(2I+1) R(A)} \left\{ \sum_i^{\text{s-waves}} \sum_J C(E_n, J^\pi) \cdot \Delta(J_f, J) \right.$$

$$\left. \cdot F_0[\delta_0(\pi,\pi_f) \cdot R(A) + \delta_1(\pi,\pi_f)] + \frac{S_1(ka)^2}{S_0(1+ka)^2} \sum_i^{\text{p-waves}} \sum_J C(E_n, J^\pi) \right.$$

$$\left. \cdot \Delta(J_f, J) \cdot F_1[\delta_0(\pi,\pi_f) \cdot R(A) + \delta_1(\pi,\pi_f)] \right\}, \tag{7}$$

where $C(E_n, J^\pi) = \dfrac{e^{1/2(\frac{J+1/2}{\sigma})^2}}{\Gamma(E_n, J, \pi_i)}$;

$\Delta(J_f J)$, $\delta_0(\pi, \pi_f)$ and $\delta_1(\pi, \pi_f)$ are functions which give 1 or 0 following the dipole selection rules.

In this derivation we have applied the very crude simplification omitting the factor E_γ^2 from eq. (5). Therefore in the comparison between the evaluated $\langle \sigma_{\gamma f} \rangle$ and the data, this missing factor has to be considered.

The substitution of F_0 and F_1 in eq. (7) by the Lynn's fluctuation factors[16] is a reasonable approximation when the number of initial states is high and only the mean values of $\langle \sigma_{\gamma f} \rangle$ for different J_f should be evaluated. However, the correct way to study the dispersion of the distribution for different J_f values is to carry out a Monte Carlo analysis of eq. (7). This can be performed by replacing F_0 and F_1 with uncorrelated Porter-Thomas variables y_n, y_γ and y'_n, y'_γ, respectively. Both these solutions of eq. (7) have been written as a computer code (SPARC[9] and RACA[10]) at the Brookhaven National Laboratory and adopted for the use at the CYBER computer in Petten.

COMPARISON WITH EXPERIMENT (HEAVY NUCLEI)

One of the most extensively covered set of ARC data belongs to the ^{168}Er nucleus studied by Davidson et al.[22] We shall use this data to demonstrate some of the basic and some of the new features of the analysis, in particular in combination with the supporting

SPARC and RACA evaluations. Further we shall try to point out some new possibilities to determine selected radiative capture parameters.

1. Discussion of the fluctuation properties

This subject was discussed in detail by R. E. Chrien[10] during the Grenoble symposium. Therefore we restrict ourselves only to a few complementary remarks.

Both evaluation methods, SPARC and RACA, as outlined in the previous section, give identical results for the relative mean values of the distributions. Of special interest is, however, the evaluated standard deviation of the distribution and how this value is related to the crude estimation based on the properties of Porter-Thomas distribution and the number of capturing states on the averaging interval. It has been pointed out already in ref. 10 that the dispersion obtained from the Monte Carlo calculation may differ from the estimation for several reasons. Three main factors can be mentioned, namely the contributions of the opposite multipolarity for a given γ ray, the presence of p-wave capture and the effects of correlations between neutron and total widths. The importance of these effects increases with the neutron energy, as can be seen from Table 1, in which a comparison of the relative standard deviations is shown for accessible final states of ^{168}Er at 2 and 24 keV. The neutron resonance parameters used for this calculation were the same as quoted in ref. 10, ν is the estimated number of resonances in the averaging intervals.

Table 1. Relative dispersions of the ARC data for the ^{167}Er$(n,\gamma)^{168}$Er reaction.

	2 keV				24 keV			
	s waves		s+p waves		s waves		s+p waves	
J_f	calc.	$2/\sqrt{\nu}$	calc.	$2/\sqrt{\nu}$	calc.	$2/\sqrt{\nu}$	calc.	$2/\sqrt{\nu}$
2^-	.194	.200	.197	.141	.119	.133	.109	.080
3^-	.135	.137	.135	.098	.081	.092	.077	.054
4^-	.133	.137	.133	.098	.078	.092	.073	.054
5^-	.188	.189	.187	.134	.109	.127	.101	.073
2^+	.194	.200	.171	.141	.119	.133	.077	.080
3^+	.135	.137	.116	.098	.082	.092	.056	.054
4^+	.133	.137	.120	.098	.080	.092	.053	.054
5^+	.188	.189	.161	.134	.109	.127	.074	.073

The relative standard deviation of an averaged $\langle\sigma_{\gamma f}\rangle$ distribution can be categorized into three limiting cases:
(i) For the low-energy limit (p/s capture $\ll 1$ and $\Gamma_\gamma \gg \Gamma_n$) the relative standard deviation is given[23] as $\sigma/m = \sqrt{2/\nu}$.

(ii) In the intermediate limit (p/s capture $\ll 1$ but $\Gamma_\gamma \approx \Gamma_n$) the standard deviation[10] becomes $\sigma/m = 2/\sqrt{\nu}$, because the cross section is a product of two uncorrelated χ^2-square distributions.

(iii) For the high-energy limit (p/s capture ≥ 1, $\Gamma_\gamma \tilde{<} \Gamma_n$) the approximation of $2/\sqrt{\nu}$ underestimates the dispersion and the Monte Carlo simulation has to be used.

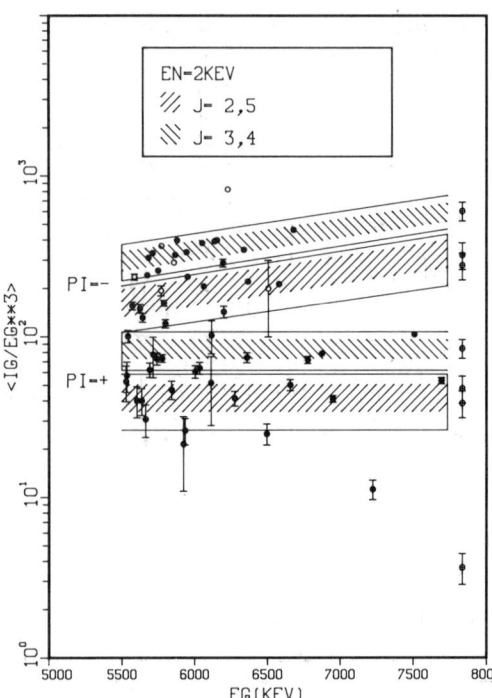

The results of Table 2 show that the ^{168}Er nucleus belong to the last limit for both neutron energies.

The application of evaluated dispersion to experiment is demonstrated in Fig. 1, in which the 2 keV data of ^{168}Er are shown. Calculated mean values for different spin groups together with their standard deviations are given at the right hand side of the plot. The shadowed (straight line) regions indicate the intensity fluctuations, based on the calculation, corresponding to one (two) standard deviations. Many spin assignments can now be made on the 96% or better confidence level.

Fig. 1. A plot of $I_\gamma E_\gamma^{-3}$ in ^{168}Er at E_n = 2 keV.

2. Energy dependence of γ-radiation

The E_γ dependence should be directly observable if the reduced intensities are plotted vs E_γ, which has been the customary way to display and analyze the ARC data. However, if the J_f^π groups are well discriminated, the dependence can be straightforwardly derived by a conventional least-squares fit.

The limiting factor of such procedure is, however, not only the statistical uncertainty but in the inability to distinguish experimentally between opposite multipolarities, especially for the M1's. One should also choose either even targets or $J_f = I\pm 1/2$ states in odd targets, because the mean values for these spin groups are identical, as proved by the RACA calculation.

Most previous papers have applied only a qualitative fit, not taking the above mentioned effects into account, which is probably the reason that for several nuclei the E_γ^5 proportionality has been suggested for M1 radiation.

In the current work four nuclei have been selected to try to fix the energy exponent by means of the least-squares fit. The SPARC evaluation permits a correction for opposite multipolarities from the p-wave capture quite reliably. The 24 keV data, however, were disregarded for the M1's, because the E1 component was almost of the same order. Only those intensities have been used, for which the corresponding J_f assignments were based on 96% confidence level.

The fitted values are given in Table 2.

Table 2. The E_γ exponents from ARC data

Nucleus	Ref.	E_n	E1	M1	E2
^{155}Sm	24	Sc,Fe	4.85(35)		
		Sc		3.67(78)	5.79(84)
^{162,164}Dy	11	Sc	5.02(35)	3.43(37)	
^{168}Er	2, 22	B,Sc,Fe	5.36(45)		
		B,Sc		3.32(25)	

The results support the assumption that the $E_\gamma^5(E_\gamma^3)$ factor should be generally applied for E1(M1) radiation, respectively. It further shows that the ARC data can be used to study this problem systematically.

The final remark concerns the data analysis. In our opinion the E_γ^3 reduction factor applied to I_γ is preferable, especially if the energy interval of the M1's is broader than the E1's. A good example is given in Fig. 1. The E_γ^5 dependence for E1 radiation is nicely demonstrated.

3. Neutron energy as a parameter

The quasi-monoenergetic beams at different neutron energies bring an important feature into the capture process, namely the competition of the p-wave against the s-wave neutrons. In order to demonstrate this, the calculated mean intensities for different J_f^π groups of ^{168}Er are displayed in Fig. 2 as a function of neutron energy. The error bars at the scandium and iron energies indicate the distribution dispersions as obtained from the RACA calculation.

At low energies (p/s wave capture <<1) the discrimination between negative and positive parity states is very clear and reflects accurately the E1/M1 ratio. With the increasing neutron energy, due to the p-wave contributions, the parity discrimination

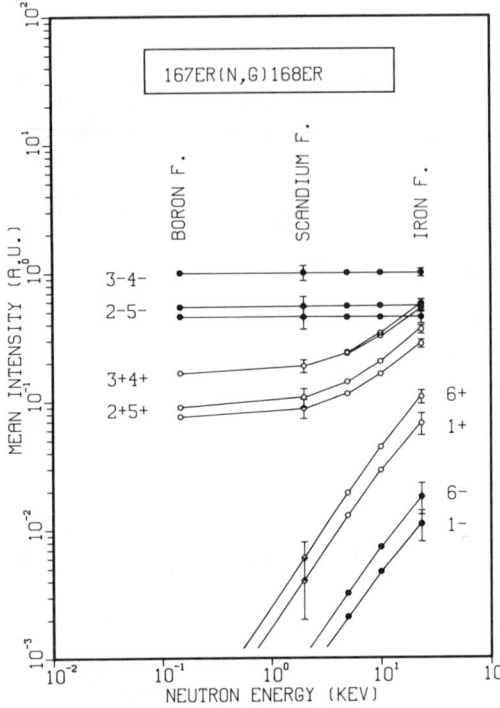

Fig. 2. Mean I_γ's for different I_f^π as a function of E_n.

is completely washed out. However, states with $J_f = I\pm 5/2$, not populated at lower energies, are now observable through the dipole components from the p-wave resonances.

The presence of the p-wave contributions can be used to determine the parity of final states. Since the positive parity states are relatively more strongly populated at 24 keV than at 2 keV, relative to the negative parity states, the ratio between these two measurements is a sensitive measure of the parity. A plot of ratios of γ-ray intensities of ^{168}Er at 2 and 24 keV is displayed in Fig. 3. The data are nicely separated into two groups. The corresponding mean values have been determined using the data points for which the parity was known from previous studies. The shadowed (solid lines) regions indicate the dispersion of one (two) standard deviations, respectively, as obtained from the RACA calculations. The discrimination between both groups is for most states sufficient to assign the parity on 96% confidence level. The present assignments are indicated by full or empty points.

For the predicted region for the $5/2^+$ states, based on the calculation, the data (two points) show slightly stronger intensities.

4. The E1/M1 ratio

We have so far only applied the standard analytical procedure as described in several earlier publications. There is, however, more information hidden in this data. The 2/24 keV ratio for the $J_f = I\pm 3/2$ states is primarily determined by the E1/M1 and S_0/S_1 values. The analytical expression of eq. (7) allows us to use them as variables in the calculation, and if one of them is known, to extract the other one by comparing the results with the experiment. Let us define the quantity P as eq. (8):

$$P = \frac{\langle I_\gamma(J_f^-)_2 / I_\gamma(J_f^-)_{24}\rangle}{\langle I_\gamma(J_f^+)_2 / I_\gamma(J_f^+)_{24}\rangle} , \qquad (8)$$

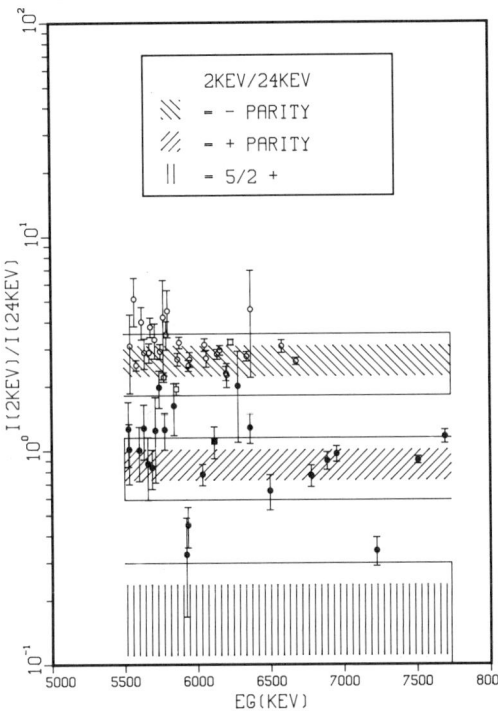

Fig. 3. The 2/24 keV data at ^{168}Er.

where the averaging takes place over the available energy intervals. The nominator in this expression is E_γ independent, because E1 radiation is dominant. The denominator may, however, carry a slight E_γ dependence, due to the presence of E1 components in the $I_\gamma(J^+_f)_{24}$ term. Since the intervals for a given nucleus are approximately equal, the ratio has the same averaging property as the R(A) values from eq. (6). In the case of an intercomparison between more nuclei, the averaging intervals may differ in energy and the proper correction should be applied to the P values.

The parameter P can be calculated by means of eq. (7) as a function of R(A) if all other parameters, in particular S_0 and S_1, are known. For a comparison with the experiment one or more J_f values can be selected, the most suitable being the $J_f = I\pm 1/2$ values, because their mean I_γ (theor.) are equal.

An example is given in Fig. 4, where the theoretical values of P vs $R(A)^{-1}$ are plotted for ^{168}Er (see the curve). The shadowed band corresponds to the dispersion of one standard deviation in P_{exp}, as obtained from the RACA calculation ($S_0 = 1.8 \times 10^{-4}$ and $S_1 = 1.38 \times 10^{-4}$ from Ref. 9). The deduced value of R(A) is 6.25±1.02.

If the S_1 strength function is not known, the situation becomes more complicated. If some information about the S_1 value may be obtained from the general trend of the S_1's, an upper and lower limit of the R(A) can be estimated with the accuracy almost comparable to that from the previous surveys.

To illustrate this possibility the P ratio for ^{162}Dy is compared to the theoretical P vs R(A) curves calculated for different

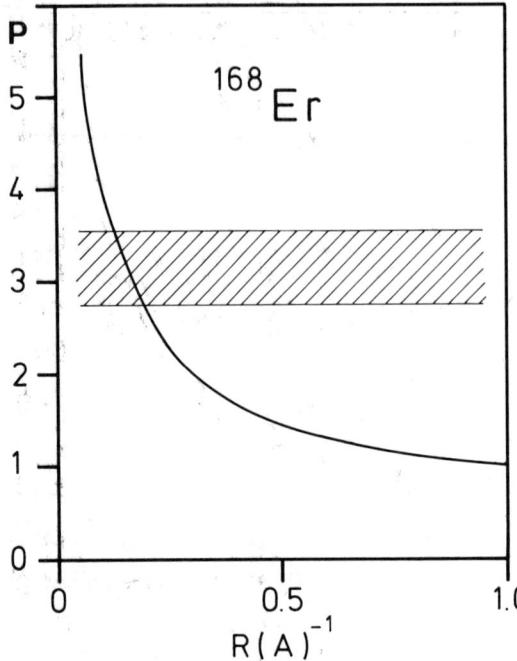

Fig. 4. The P vs R(A) plot.

values of S_1 (S_1 = 0.1, 1, 2 and 5×10^{-4}) and the estimate of 5.7±3.0 was made for R(A), assuming[26] that the value of S1 can have a value between $\approx 1-2 \times 10^{-4}$.

The R(A) values have been derived for several other ARC nuclei, in order to test the applicability of this method to a wide range of nuclei and to make a comparison with the comprehensive survey in ref. 21.

The results of the present analysis are given in Table 3.

The errors associated with the P values have been derived from the RACA calculations carried out separately for each individual $I_\gamma(J_f^\pi)$.

To account for differences among the E_γ intervals, the correction factor of $(\langle E_\gamma \rangle_i / 7 \text{ MeV})^2$ was applied and thus the values in the last column are related to energy of 7 MeV. The mean energy of the interval for a given nucleus i is denoted by $\langle E_\gamma \rangle_i$.

Table 3. R(A) values from the 2/24 keV data.

	Ref.	P	J_f	$R(A)_{\langle E_\gamma \rangle}$	$R(A)_{7 \text{ MeV}}$
^{114}Cd	13	2.57(13)	2,3	2.41(16)	2.25(15)
^{155}Sm	24	2.26(24)	1/2, 3/2	2.56(42)	5.42(89)
^{162}Dy	11	2.90(15)	2,3	5.70(306)	5.57(300)
^{164}Dy	11	3.52(18)	2,3	5.88(175)	6.72(200)
^{168}Er	22	3.06(16)	3,4	6.25(100)	7.25(115)
^{195}Pt	25	3.86(39)	1/2, 3/2	5.12(76)	8.22(122)
^{239}U	14	3.63(17)	1/2, 3/2	4.35(28)	12.53(80)

Finally let us make a comparison with the E1/M1 file of ref. 21. The easiest way is to incorporate the present data (full circles) into this file and the result is shown in Fig. 5. The plotted curve corresponds to a least-squares fit to the original values of ref. 21, denoted by open data points.

An inspection of this figure allows us to make the following conclusions:

(a) The accuracy of the ARC data as compared to the previous set, based on discrete resonance data[18], is much better owing to the improved averaging.

(b) Relating the data to the same $\langle E_\gamma \rangle_i$ certainly removes the energy fluctuations. It is therefore worthwhile to apply such correction generally, which may smooth the previous data too.

(c) The present values fall reasonably close to the fitted curve and support therefore the conclusion of refs. 19 and 21 that the E1/M1 ratio is dependent on A.

5. The M1/E2 ratio

The presence of pure E2 transitions in the primary decay has been established in several previous thermal and ARC measurements. Their intensities were found to be much weaker than that of M1 radiation and were therefore usually neglected compared to the M1's.

The information on the E2 strength functions S(E2) is rather scarce, due to the small number of observed pure E2 transitions. Recent compilations[27,28] suggested that the general behavior of the S(E2) can be approximately explained in terms of the giant resonance model, applying the Brink-Axel hypothesis and using the systematics of the E2 giant resonances.

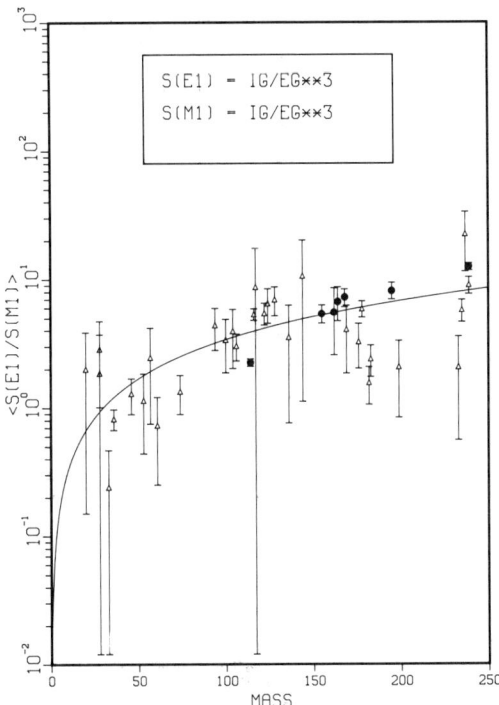

Fig. 5. The E1/M1 ratio.

Since the ARC data give only the relative I_γ intensities, the additional absolute normalization for a derivation of the γ-ray strength function is needed. This can be done either by relating $\langle \Gamma_{\gamma f} \rangle$ for several transitions to the value extracted from the isolated resonance data (e.g., as shown in ref. 15) or assuming ΣI_γ for a certain energy region equal to thermal data (as done in ref. 28).

Thus it seems to be practical to try to determine the M1/E2 ratio from the ARC data, and to relate the absolute S(E2) values to more accurately known S(M1)'s.

The M1/E2 ratio may be defined in the similar way to eq. (6). Although the available E2 γ-ray intervals are often rather narrow, the advantage of the resonance-averaged E2 intensities compensates for this and favors this approach against the thermal data of ref. 27.

The other complication may arise from the E1 components from the p-wave capture, which could be mixed with the E2's. Therefore the low energy neutron beams are preferable, since the p/s-capture is ≪1, and the E1 components are small.

These components can be evaluated from the SPARC code, but the calculation requires accurate E1/M1 input values. The performed tests showed, that the corrections at 2 keV still can play an important role and thus the boron data seem to be the most suitable.

We have surveyed the existing ARC data for well assigned E2 transitions and listed the results, together with the deduced M1/E2 values, in Table 4.

The number of observed E2's is denoted by N and the values corrected for the E1 p-wave components are given in the last column. As expected, for the boron data, the corrections have been small, less than 5%.

For the ^{168}Er the accuracy of the M1/E2 ratio was estimated from the RACA calculations, which reflects the averaging dispersion

Table 4. The M1/E2 ratios.

Nucleus	Ref.	$\bar{E}_n$	$\bar{E}_\gamma$ (E2) (MeV)	N	$\frac{<I_\gamma(M1)>}{<I_\gamma(E2)>}_{exp.}$	$\frac{<I_\gamma(M1)>}{<I_\gamma(E2)>}_{corr.}$
^{114}Cd	13	Sc	6.7	3	4.76	6.00
^{151}Sm	31	B	5	10	3.10 a)	3.24
^{153}Sm	31	B	5	18	5.75 a)	5.90
^{155}Sm	24	Sc	4.8	8	4.00	5.88
^{168}Er	22	Sc	7.2	1	3.79(90)	4.83 (113)
^{180}Hf	29	B	7.3	1	3.56 a)	3.67
^{233}Th	30	Sc	4.4	5	3.06	5.64
^{239}U	14	Sc	4.5	3	1.89	2.00
		B		3	2.46	2.50

a) Estimated only from the figures.

and a relative value of ≈25% was obtained. A safe upper limit of this relative error for the remaining nuclei of Table 4, taking into account their different spacings, may have a value of ≈50%. The achieved accuracy does not allow the determination of the detailed trend with the mass number.

Let us consider whether the present M1/E2 ratios are reasonable. If we apply the following mean values from Table 4, $\langle E_\gamma \rangle$ = 5.1 MeV, $\langle A \rangle$ = 176 and $\langle M1/E2 \rangle$ = 4.7, the simple estimate can be made and compared to existing models.

We know that the single particle M1 strength approximately (within 30%) agrees[18] with the experiment. For the SP model, estimates of $\langle \Gamma_{\gamma f} \rangle / D$ give

$$\langle \Gamma_{\gamma f}(M1) \rangle_{SP}/D \approx 2.1 \times 10^{-8} E_\gamma^3 = 2.8 \times 10^{-6} \text{ and}$$

$$\langle \Gamma_{\gamma f}(E2) \rangle_{SP}/D \approx 4.8 \times 10^{-14} E_\gamma^5 A^{4/3} = 1.7 \times 10^{-7}, \quad (9)$$

while the giant resonance model[32] gives:

$$\langle \Gamma_{\gamma f}(E2) \rangle_{GR}/D \approx 2.6 \times 10^{-14} E_\gamma^5 A^{8/3} = 8.7 \times 10^{-7}.$$

These values can be compared with the experimentally derived ones, which gives $\langle \Gamma_{\gamma f}(E2) \rangle/D \approx 6.0 \times 10^{-7}$ which accounts for 0.70 of the GR prediction and is approximately 3.5 times larger than the SP value. The first conclusion agrees rather well with ref. 28, in which the value 0.62 of the giant resonance prediction was quoted.

COMPARISON WITH EXPERIMENT (LIGHT NUCLEI)

For low A targets, only a few resonances may be present in the filtered neutron intervals and thus little or no averaging is possible. For this reason the ARC method has not been applied to light nuclei until now.

Very recent studies[33,34], however, indicate that useful spectroscopic information may be obtained from the ARC data even for nuclei with A << 100. The iron filter is more useful than the scandium one because it has a greater probability ($\Delta E \approx 3.8$ keV) of encompassing more resonances.

Since only a small number of resonances is involved in the averaging process, the J_f^π grouping of primary I_γ's is obscured by statistical fluctuations. However even a small number of contributing resonances produces a significant reduction in the fluctuations of the intensities. This can be understood from the properties of the chi-square distribution with a few degrees of freedom. For $\nu = 3$ or greater, the most probable intensity values are non-zero. This is demonstrated in Fig. 6, in which the reduced intensities $I_\gamma E_\gamma^{-3}$ are plotted for the

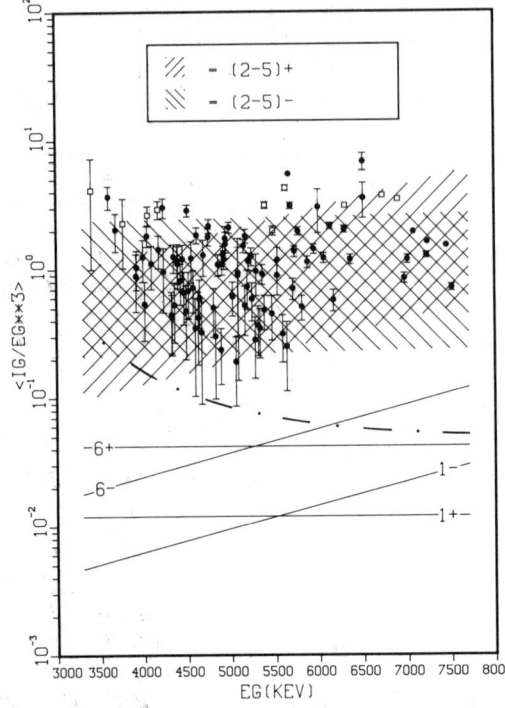

Fig. 6. The $I_\gamma E_\gamma^{-3}$ plot in ^{60}Co at E_n = 24 keV.

^{59}Co(n,γ)^{60}Co reaction[34] at 24 keV. An estimated number of 8 s-wave and p-wave resonances[26] contribute to the averaging.

The dashed bands reflect the dispersion of two standard deviations, obtained from the RACA calculations, with the E_γ^5 (E_γ^3) assumed dependence for E1(M1) radiation, respectively. For the J_f = 1,6 population, only the mean values have been plotted, to indicate that they lay below the detection limit (the dashed-dotted curve), and thus are not observable.

From the displayed data the following spectroscopic information, complementary to thermal capture, may be gained.

1. The kinematic γ-ray energy shift between thermal and 24 keV intensities identifies uniquely the primary transitions. The average energy at which capture occurs can be calculated from the relation[11]

$$\langle E \rangle = \int y(E) E dE / \int y(E) dE, \qquad (10)$$

where y is the total capture yield of a resonance given as y ≈ $\Gamma_n \Gamma_\gamma / \Gamma$.

This calculation resulted for ^{60}Co in a value of 22.30 keV, which nicely agrees with the average experimental shift $E_\gamma(24) - E_\gamma(th)$ = 22.42(17) keV. The achieved accuracy allows the determination of the E_x of levels, not populated in thermal capture, with a good precision.

2. The γ-rays with intensities significantly above the calculated regions are secondary transitions or doublets. The doublets are denoted by open squares in Fig. 6.

3. Even a low number of resonances ensures a reasonable population of <u>all</u> states with J_f = 2-5 and $E_x \leq$ 2 MeV. This is due to the fact that the peak in the chi-square distribution with ν=8 is well above the instrumental sensitivity.

The levels not populated in the 24 keV capture, but solidly established from other experiments, are thus the candidates for the J_f = 1,6 assignments.

ACKNOWLEDGEMENT

The author wishes to express his appreciation for the long and fruitful collaboration with Dr. R. E. Chrien on this subject.

REFERENCES

1. S. F. Mughabghab and R. Chrien, in Neutron Capture Gamma-Ray Spectroscopy (Plenum Press, N.Y. and London, 1979), p. 265.
2. L. M. Bollinger and G. E. Thomas, Phys. Rev. Lett. $\underline{18}$, 1143 (1967).
3. L. M. Bollinger and G. E. Thomas, Phys. Rev. $\underline{C2}$, 1951 (1970).
4. O. D. Simpson and L. G. Miller, Nucl. Instr. Meth. $\underline{61}$, 245 (1968).
5. R. G. Greenwood, in Neutron Capture Gamma-Ray Spectroscopy (RCN, Petten, 1974), p. 323.
6. R. G. Greenwood and R. E. Chrien, Nucl. Instr. Meth. $\underline{138}$, 125 (1976).
7. R. M. Brugger and R. C. Block, in Neutron Sources for Basic Physics and Applications (Pergamon Press, N.Y., 1982), p. 177.
8. C. W. Reich, in Neutron Capture Gamma-Ray Spectroscopy (Plenum Press, N.Y. and London, 1979), p. 105.
9. M. L. Stelts, in Nuclear Cross Sections for Technology (Nat. Bur. of Stds. Sp. Publ. 594, 1980), p. 936.
10. R. E. Chrien, in Neutron Capture Gamma-Ray Spectroscopy and Related Topics (Inst. of Physics, Bristol, 1981), p. 342.
11. R. E. Chrien, in Neutron Induced Reactions, Physics and Applications, Vol. 10 (Institute of Physics, Bristol and London, 1982), p. 189.
12. D. D. Warner et al., Phys. Rev. $\underline{C27}$, 2292 (1983).
13. A. Mheemeed et al., Nucl. Phys. $\underline{A412}$, 113 (1984).
14. R. E. Chrien and J. Kopecky, Nucl. Phys. $\underline{A414}$, 281 (1984).
15. R. E. Chrien et al., to be published.
16. J. E. Lynn, Theory of Neutron Resonance Reactions (Oxford Univ. Press, London, 1968).
17. P. Axel, Phys. Rev. $\underline{126}$, 671 (1962).
18. C. M. McCullagh, R. E. Chrien and M. L. Stelts, Phys. Rev. $\underline{C23}$, 1394 (1981).
19. J. Kopecky, Neutron Capture Gamma-Ray Spectroscopy and Related Topics (Institute of Physics, Bristol and London, 1981), p. 423.
20. L. M. Bollinger, in Proc. of Int. Conf. on Photonuclear Reactions (Springfield, 1972), p. 783.
21. J. Kopecky, ECN Report ECN-81-040 (1981).
22. W. F. Davidson et al., J. Phys. $\underline{G7}$, 455 (1981).
23. C. E. Porter and R. G. Thomas, Phys. Rev. $\underline{104}$, 483 (1956).
24. K. Schreckenbach et al., Nucl. Phys. $\underline{A376}$, 149 (1982).
25. D. D. Warner et al., Phys. Rev. $\underline{C26}$, 1921 (1982).
26. S. F. Mughabghab, M. Divadeenam and N. E. Holden, Neutron Cross Sections, Vols. I and II (Academic Press, New york, 1981).
27. J. Kopecky, ECN Report ECN-99 (1981).

28. W. V. Prestwich, M. A. Islam and T. J. Kennett, Z. Phys. A315, 103 (1984).
29. D. L. Bushnell, D. J. Buss and R. K. Smither, Phys. Rev. C10, 2483 (1974).
30. P. Jeuch et al., Nucl. Phys. A317, 363 (1979).
31. R. K. Smither, in Neutron Capture Gamma-Ray Spectroscopy (RCN, Petten, 1974), p. 358.
32. A. Fubini, P. R. Oliva and D. Prosperi, Lett. Nuovo Cim. 3, 401 (1972).
33. M. G. Delfini et al., Nucl. Phys. A404, 225 (1983).
34. J. Kopecky, M. G. Delfini and R. E. Chrien, Nucl. Phys. A427, 413 (1984).

SOME OPEN PROBLEMS IN NEUTRON CAPTURE γ-RAYS IN ROTATIONAL NUCLEI

M. Stefanon

ENEA - C.R.E. "E. Clementel", Bologna, Italy

The study of neutron resonances and, in particular, of their γ-decay, is of fundamental importance for an understanding of the statistical features of compound nucleus reactions in complex nuclei. The possibility to resolve several resonances and measure their individual γ-decay spectra seems to represent a quasi ideal situation for the application and testing of statistical models, both concerning fluctuations and average properties. Indeed it was the measurement of high-resolution neutron cross sections which stimulated the famous works of Wigner, Porter, Thomas and Dyson[1] on the fluctuation properties of complex quantum systems. At any rate, though medium-heavy nuclei, at an excitation of 6-8 MeV, are certainly far too complicated to be described in a deterministic way, they often seem to be still too simple to completely average out all the residual effects of simple structures in (n,γ) reactions.

One should consider that in the observed γ-decay of compound nucleus states, simple configurations describing low-energy final levels are coupled essentially via electric dipole operator to the extremely complex wave functions of neutron resonances. On the other hand resonances are coupled by neutron-nucleon interaction to the initial, simple, scattering state. It follows that resonance neutron-capture measurements project out of the complex compound-nucleus wave function, simple components related to simple states. It is then possible in this way to point out even very weak simple structures and search for deviations from a complete statistical description. All resonance capture works are indeed divided into those confirming the overall validity of a statistical picture and those giving evidence for non statistical effects.

A statistical approach to neutron capture reactions may concern three different items: fluctuations, correlations and average values. We very briefly recall the first two points before discussing in more detail a non-statistical effect concerning average partial γ-strengths observed in some deformed even-even nuclei.

The fluctuation theory based on random hamiltonian matrices from the Gaussian Orthogonal Ensemble (Goe), has very general quantum-mechanical and statistical foundations. Sensible deviations appear to be possible only if very large correlations exist between the hamiltonian matrix elements as they arise, e.g., from almost good hidden symmetries. Also, the presence of shall-model correlations or a doorway state mechanism does not appreciably change the level energy statistics or the effective number of degrees of freedom of partial width distribution. In fact, theoretical calculations and numerical simulations showed that both level energy distribution and the Porter-Thomas (P.T.) law are approximate general properties of large hamiltonians under much more general conditions than assumed by Goe^2. In spite of this, while no reliable evidence of deviations from Goe predictions was found in the case of level energies, several experiments showed a narrower distribution for E1 radiation than that predicted by the P.T. law^3. The observed number of degrees of freedom, $\nu \approx 1.4$-1.6, if real, would imply a complete failure of the statistical model. Deviations could perhaps be justified by the presence of unknown biases in measurements or in the complicated data analysis procedures, but this is in contrast with the fact that M1 radiation, which should be affected by equal or stronger biases, was always found to be consistent with P.T. law even in the same experiments which gave $\nu > 1$ for E1 transitions^3. The problem is then completely unsolved.

Another long-debated point is that of correlations$_1$. While correlations are allowed by the statistical model between different partial radiation widths, no correlation should exist between neutron channel and partial radiation widths if compound nucleus formation and γ-decay were completely independent. This is, however, more a surmise than a fundamental requirement of the model; in fact a simple doorway state capture mechanism would justify positive correlations without essentially changing fluctuation properties.

Positive correlations were indeed found in different mass regions and were interpreted with different models, giving a valuable contribution to better understanding of (n,γ) reactions. In particular, strong correlations were observed in the mass regions A = 35-65 and A = 90-112 where valence capture is expected. In most cases, also correlations with (d,p) spectroscopic factors were observed so that one can conclude that for such nuclei, direct-like capture processes and single particle configurations play an important role in neutron capture. Indeed the valence-capture mechanism provides both correlations and enhancement of particular transition probabilities; it gives a direct-like contribution which adds to the statistical part.

It is then particularly interesting to see whether deviations from a statistical picture of resonance capture also affect excitations in nuclei with no strong single-particle components, e.g. rotational levels. In this connection we quote the results discussed by Becvar in this Conference giving evidence of correlations in deformed nuclei not explainable with valence capture.

An important contribution to the understanding of compound nucleus processes can arise from the study of a possible systematic dependence of high energy transition probabilities on the collective structure of the final state.

Only a few experiments were aimed at this. In general the partial E1 γ-ray strength function (averaged over neutron resonances) was assumed to depend smoothly on energy and mass number according to the well-known Axel formula

$$\langle \Gamma_{\gamma ij}^{E1}/DE^5 \rangle A^{-8/3} = \text{cost} .$$

This, in fact, gave an overall satisfactory set of the average data and allowed reduction of the whole set of measured intensities to give a common average reduced strength. As is well known, this formula is based on the extrapolation of the photoabsorption Giant Dipole Resonance (GDR) with the further hypothesis, suggested in the Brink thesis, that the E1 photoabsorption cross section in excited states is the same as in the ground state [4].

This kind of decoupling of the E1 giant excitation and low energy excited states can hold only approximately, and observation of systematic departures was not general, probably only because the set of measurable resonances and transitions is too limited to put weak effects into evidence, due to the wide dispersion introduced by P.T. fluctuations. However, we will recall in the following some experiments which suggest that in deformed even-even compound nuclei, in the mass region $A \approx 170$, transitions to $K=0^+$ states are hindered with respect to transitions to levels belonging to other rotational bands, based on more complex excitations like $K = 2^+$ γ-vibrations.

The most convincing and direct evidence of this effect was obtained [5] in ^{177}Hf(n,γ). In this nucleus it was possible to analyze 38 γ-spectra and measure single primary transition intensities down to 29 final states with reliable spin and parity assignment. Extreme care was employed both in the measurement and in the data analysis. The results are probably the most significant set of neutron resonance capture γ-rays in this mass region. Transition intensities were reduced with Axel formula and expectation values were estimated separately for groups of final states having the same projection quantum number K. The results are reported in fig. 1. In this connection we mention a preceding work[7] in which a K-dependence of E1 transitions was also suggested for the odd Z ^{177}Lu.

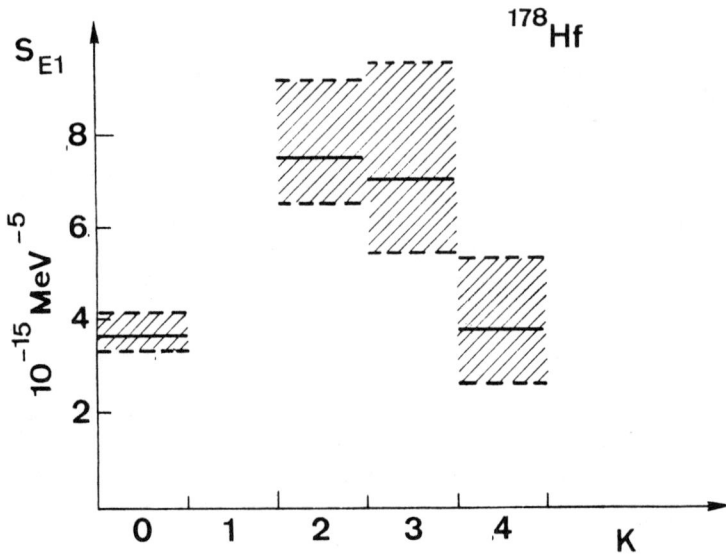

Fig. 1. E1 average reduced strengths of ^{178}Hf, versus K-value of the final state.

The data clearly show a "bell-shaped" K-dependence, the most convincing effect being the difference between K=0 and K=2 strengths for which a larger number of transitions were available. An attempt to explain this behaviour was given in ref.[8] but it seems to have a weak theoretical basis. It is to be noted that in the same experiment a value $\nu = 1.38^{+0.18}_{-0.13}$ was obtained for the number of degrees of freedom of the partial width distribution. There is no apparent relationship between the two effects; in particular the ν value becomes slightly larger if the data are reduced taking into account the systematic K-dependence. A more recent measurement[9] on ^{173}Yb(n,γ) showed the same ratio between γ-strengths to K=2 and K=0 but with larger uncertainties due to the lower number of well-resolved transitions. Five transition intensities, to K=0 and K=2 final states, averaged over 32 resonances up to 200 eV neutron energy, are reported in fig. 2.

Intensities were reduced with a conservative $1/E^3$ factor instead of $1/E^5$, in order to avoid overestimating K=2 transitions which have much lower energy.

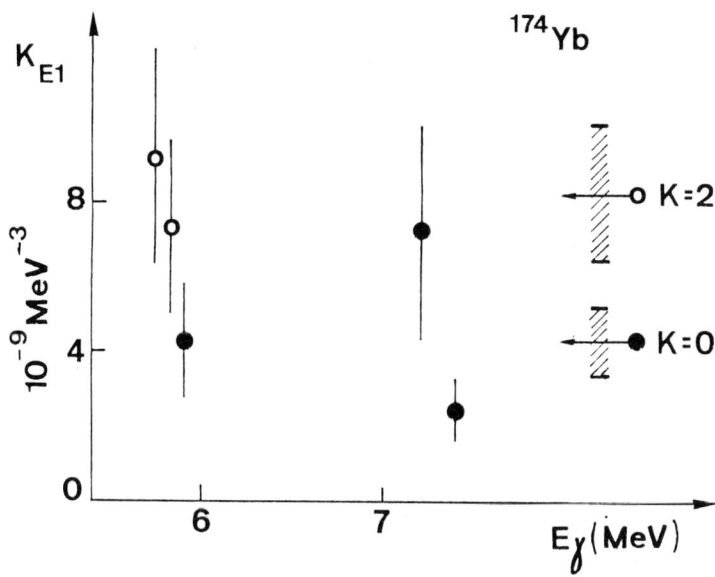

Fig. 2. E-1 single particle strengths of ^{174}Yb direct transitions. Full circles correspond to K=0 while open circles correspond to K=2 final states.

The ratio between K=2 and K=0 strengths is R = 1.9+0.5 ; if E^5 energy dependence is assumed to reduce the data, the ratio becomes R = 3.5±0.7 . Gamma spectra were measured also from other 17 resonances up to 500 eV. However the statistical quality of the spectra was too poor to allow individual fit of transition intensities; the intensities were then estimated from the analysis of a cumulative spectrum obtained by a weighted average^{10} over all resonances. In this case the data reduced with E^3 dependence show no K-effect, while a weak effect remains assuming E^5 dependence.

It is not completely clear whether the partial washing out of the K-dependence is real or due to the worsening quality of the data at higher neutron energy and to the less detailed analysis procedure. Anyway it must be clear that it is possible to detect this kind of effect in resonant neutron capture only by pushing measurements and data analysis up to the highest possible quality.

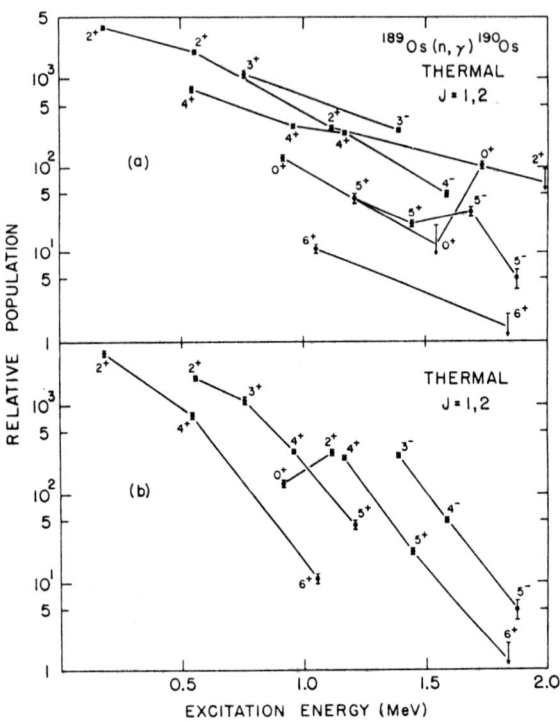

Fig. 3. Relative population of states in the ^{189}Os(n,γ) reaction from ref. [11].

Dependence of average properties of the whole γ-cascade was also observed [11]. The low-energy level populations measured in ^{189}Os(n,γ) are reported in fig. 3. The upper part represents the data connecting points corresponding to the same spin, while in the lower part the same data are given connecting points belonging to the same rotational band. A regular behaviour clearly appears in the second case, the curves being nearly parallel, while in the upper figure the curves are intersecting, showing that populations do not depend only on spin and energy[12], but that a band effect is present. In a recent experiment on ^{167}Er(n,γ), primary transitions and low-energy γ-rays were measured. The low-energy part of the sum spectrum of 56 resonances with spin J=4 was used to obtain population probabilities of low-lying states. The results display the same general features of fig. 3.

In order to bring into evidence structure effects on the γ-population probability, the data were corrected for the systematic spin dependence as predicted by the statistical simulation of the γ-cascade process. The N-step model[13] extensively employed to assign the spin of low-lying states[14], was tested by means of the experimental population ratios between different spin resonances. The results reported in tab. I show fair agreement for a number of N=5 steps. The same model, with N=5, was used to calculate the relative spin dependence P_c of population probability from J=4 resonances. This is given by:

$$P_c = \frac{P_{ac}}{P_{ac'}} = \frac{\sum \prod_{n=1}^{N-1} k(I_n)}{\sum \prod_{n-1}^{N-1} k(I'_n)}$$

where a indicates the initial state (J=4 resonance), c the final state having spin I, c' the final reference state spin I', and $k(I_n)$, is the normalized spin dependent part of the level density of the intermediate state with spin I_n. The summation extends over all spin paths compatible with dipole selection rules.

Table I. Observed and calculated population ratios in ^{168}Er from J=4 and J=3 resonances

R	Final state spins				
	J = 2	J = 3	J = 4	J = 5	J = 6
Experiment	0.65±0.06	0.89±0.10	1.12±0.10	1.48±0.21	2.08±0.14
N = 2	0.49	0.66	1.47	1.96	....
N = 3	0.49	0.84	1.14	1.95	2.92
N = 4	0.61	0.82	1.15	1.55	2.42
N = 5	0.66	0.86	1.10	1.45	1.93
N = 6	0.71	0.87	1.08	1.34	1.72

The experimental population probabilities were then divided by the systematic spin dependence P_c. The results are reported in fig. 4. If the γ-decay cascade proceeded in a completely statistical way, the corrected populations would display only a smooth energy variation due to the energy dependent part of the level density, not included in the N-step cascade calculation. Indeed the data show systematic energy behaviour but they spread considerably according to different rotational bands. The most remarkable feature is that 0^+ bands are strongly depressed, especially the ground state one. Unfortunately it was not possible to measure the direct effect in ^{167}Er, i.e. the corresponding ratio of primary transition strengths, because the positive parity of neutron resonances allows primary E1 transitions only to negative parity final states. However the evidence of fig. 4 definitely

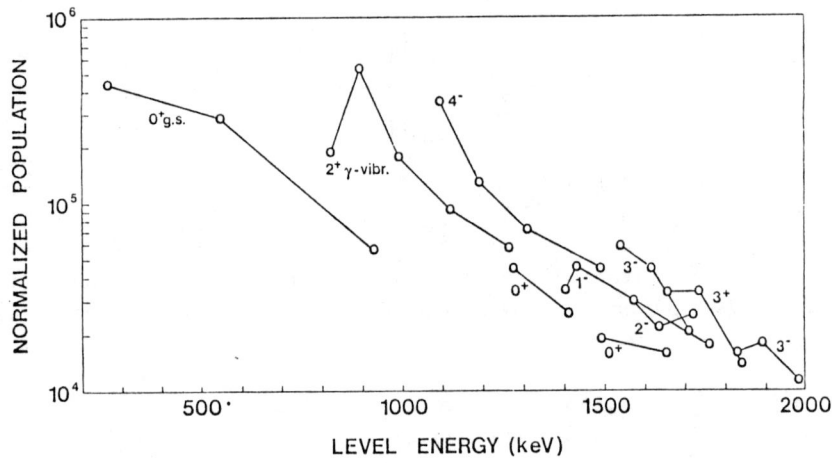

Fig. 4. Relative populations corrected for systematic spin dependence in the ^{167}Er(n,γ) reaction.

substantiates the idea that neutron resonances are more weakly coupled to states of $K=0^+$ bands than to others. The effect is of particular interest because it is not explainable, as in valence capture, by direct-like transition mechanisms from a part of quasi-bound compound state wave function and involves also high-energy bound states. A final state dependence of E1 reduced partial γ-strengths acquires an interesting physical interpretation if transferred to dipole photoabsorption giant resonance. In the Brink-Axel model, E1 photoabsorption giant resonances based on excited states are the same as the GDR of the ground state but for an energy shift equal to the excitation energy. The final state dependence of partial γ-strengths can be qualitatively understood if it is assumed that the photoabsorption resonance depends on the coupling of dipole giant excitation with nuclear shape vibrations. In particular, the resonance should be broader, or have a larger low-energy tail, if it is based on k=2 γ-vibrations. Of course, photoabsorption cross sections cannot be directly measured for excited states and neutron capture data are confined to a restricted energy region and are too scarce and uncertain to give a clear indication.

Some justification of Brink hypothesis, i.e. of the shift of giant resonances based on excited states, was given by Rosenzweig[15]. He showed, assuming the constancy of dipole sum rules, that the mean energy of the GDR is shifted upward by approximately the low-lying state excitation energy. However, in order that the shape of the resonance remain unchanged, higher order energy weighted sum rules should be independent of the low-energy state, and this does not seem to have any foundation. Especially in the low-energy tail does it seem reasonable that the GDR may be affected by the coupling with low-energy excitations and that wider damping occurs if strongly-correlated collective vibrations are involved.

In any case, theoretical calculations accounting for effects of this kind are difficult as it would not be sufficient to calculate the fragmentation of E1 excitation of different low-lying states in the bound eigenstates of a collective Hamiltonian, but the coupling with continuum states should be included to account for the total resonance width. In particular, damping in the neutron resonance region should be correctly described. It is also not clear whether there is any connection with other non-statistical effects observed in the γ-decay of neutron resonances such as, e.g., deviation from the P.T. law.

One can conclude by noting that the intrinsic difficulties one encounters in resonance neutron capture experiments such as the worsening of resolution and intensity with increasing neutron energy, the large P.T. fluctuations and, not least, the need of cumbersome data analysis if one tries to exploit the data correctly, have probably impaired a more systematic detection of deviations from the overall average behaviour predicted by a complete statistical picture. Furthermore, it is always difficult to have theoretical schemes to deal with mixed statistical and systematic behaviour. The consequence is that up to now, in spite of the large amount of work performed in this field, the quality of resonant capture data, especially in the region of complex rotational nuclei, is sufficient to suggest the presence of deviations from usually assumed behaviour, but it does not allow us to obtain definite conclusions. The possibility to detect and study effects like the K-dependence of the γ-decay of resonances more quantitatively, is strongly related to an improvement in the quality of the experiments, which should be possible, thanks to the continuous technical progress in measurement equipment. We think this programme is worth pursuing, as neutron capture is probably the most convenient experiment to get information on coupling the giant dipole mode with low-energy excitations.

REFERENCES

1. C.E. Porter, Statistical Theories of Spectra: Fluctuations (Academic Press, New York and London).
2. T.A. Brody et al., Rev. Mod. Phys. 53, 385 (1981).
3. C. Coceva, Proc. Bologna 1980, Lecture Notes in Physics 137, 339 (1981) (ed. H. Arenhövel and A.M. Saruis).
4. P. Axel, Phys. Rev. 126, 671 (1962).
5. M. Stefanon and F. Corvi, Nucl. Phys. A281, 240 (1977).
6. V.G. Soloviev, Phys. Lett. 36B, 199 (1971).
7. L. Aldea et al., Proc. Int. Conf. on Nucl. Phys., ed. De Boer and H. Hang (North Holland - Amsterdam, 1973)
8. P. Giacobbe and M. Stefanon, Int. Conf. on Selected Topics in Nuclear Structure, Dubna 1976.
9. O. Shahal et al., Phys. Rev. C 25, 1283 (1982).
10. S. Raman, O. Shahal and G.G. Slaughter, Phys. Rev. C 23, 1794 (1981).
11. M.R. Macphail, R.F. Casten and W.R. Kane, Phys. Lett. 58B, 39 (1975).
12. S. Kahane et al., to be published in Phys. Rev.
13. K.J. Wetzel and G.E. Thomas, Phys. Rev. C1 (1970) 1501.
14. C. Coceva et al., Nucl. Phys. A218, 61 (1974).
15. N. Rosenzweig, Nucl. Phys. A118, 650 (1968).

NONSTATISTICAL EFFECTS OF NEUTRON RADIATIVE CAPTURE IN DEFORMED NUCLEI

F. Bečvář
Charles University, Faculty of Mathematics and Physics
180 00 Prague 8, Czechoslovakia

ABSTRACT

The present review summarizes experimental results on neutron radiative capture in deformed nuclei at isolated resonances. For ^{154}Gd, ^{167}Er, ^{173}Yb, ^{176}Lu and ^{185}Re target nuclei a statistically significant correlation between the partial radiation widths and the reduced neutron widths of neutron resonances has been observed. These results are compared with predictions of the theory of Lane and Lynn, as well as with qualitative predictions following from Soloviev's model. Some consequences of a large effect of correlation are discussed in detail. In case of neutron radiative capture in ^{173}Yb a strong correlation between various partial radiation widths has been observed. It is concluded that a doorway state common to radiation channels is responsible for this effect.

INTRODUCTION

A prevailing part of this talk will concern one of the most relevant nonstatistical phenomena of highly-excited states of nuclei, the correlation between the partial radiation widths and the reduced neutron widths of neutron resonances.

The effects of width correlation represent a challenging problem whose study is extremely useful in investigating the structure of neutron resonances and the mechanism of their decay.

The effects of width correlation have been found especially distinct and statistically significant in the p-wave neutron capture in spherical nuclei with the mass number $A = 90-100$, see Refs. 1,2. These effects were to a large extent accounted for by the theory of Lane and Lynn[3,4] in terms of transitions between neutron single-particle components of the wave functions of the initial and final states.

In case of s-wave neutron capture in the nuclei with $150 < A < 190$ the most significant neutron single-particle transition is $4s \to 3p$. However, owing to nuclear deformation, occurring in the given mass region, the role of this transition is strongly reduced. Following a rough estimate, the contribution of the single-particle transition $4s \to 3p$ to partial radiation widths should be as small as 1%. Despite this, the width correlation in the nuclei in question was subject to an extensive experimental study. As a result, statistically significant effects were recorded. This effort, however, underwent a rather dramatic development.

At the beginning, startling cases of correlation were reported for the neutron capture in ^{159}Tb (Ref. 5), ^{163}Dy (Ref. 6), ^{169}Tm (Ref. 7) and ^{173}Yb (Ref. 8) nuclei. However, subsequent studies[8,9] threw strong doubts on the existence of correlation in cases of ^{159}Tb and ^{163}Dy, while in case of ^{169}Tm new data[10,11] even denied the original claim for correlation. Naturally, it was then suspected that a similar situation might develop also in the case of the ^{173}Yb nuclei and threaten the correlation effect.

A disturbing problem in the previous approach is the way in which the partial radiation widths were selected when analyzing a correlation. No firm *a priori* rule for this selection was kept. Instead, in an attempt to reveal a correlation effect, the individual partial radiation widths were selected *ad hoc*. It is evident that under such conditions the analysis of correlation might be biased.

An important step in clarifying the problem of width correlation for the deformed nuclei was undertaken at Dubna. For several nuclei positive results were obtained, including the crucial case of ^{173}Yb. A significant novelty was that the analysis of correlation has been performed not for arbitrarily selected partial radiation widths, but for the widths which are anticipated by Soloviev's quasiparticle-phonon model[12] as favourable objects for observing the correlation.

In this talk emphasis will be chiefly placed upon the results of Dubna measurements.

PREDICTIONS FROM THEORIES

In accordance with theory of Krieger and Porter,[13] a partial radiation width $\Gamma_{\lambda\gamma f}$ corresponding to the transition from a resonance λ to a final level f of the residual nucleus can be expressed as a square of the following sum of width amplitudes:

$$\Gamma_{\lambda\gamma f}^{1/2} = \Gamma_{\lambda\gamma f}^{(r)1/2} + \delta\Gamma_{\lambda\gamma f}^{1/2}. \qquad (1)$$

Here, $\Gamma_{\lambda\gamma f}^{(r)1/2}$ is a random amplitude fluctuating for fixed f and fixed spin J of resonances *independently* and randomly according to the normal distribution with a zero mean. For s-wave resonances the square of the amplitude $\delta\Gamma_{\lambda\gamma f}^{1/2}$ can be expressed via the reduced neutron width $\Gamma_{\lambda n}^o$

$$\delta\Gamma_{\lambda\gamma f} = q_{Jf}\,\Gamma_{\lambda n}^o/(1\text{ eV})^{1/2}, \qquad (2)$$

where a factor q_{Jf} is independent of λ. This factor is responsible for the correlation between the partial radiation widths and the reduced neutron widths.

If the contribution of the internal nuclear region ($r<R$) of the configuration space to the radiative process is neglected, then in case of $E1$-transitions, according to Lane and Lynn,[3] the factor q_{Jf} can be expressed as

$$q_{Jf} = \frac{e^2\hbar^2}{9\sqrt{2}\,c^3 M^{5/2} R^2}\left(\frac{Z}{A}\right)^2 S_{dp,f}^{(1)}\,\frac{2J_f+1}{2J+1}\,y^4\left(\frac{y+2}{y+1}\right)^2. \qquad (3)$$

Here, M is the reduced mass of neutron and $S_{dp,f}$ is the neutron spectroscopic factor for the neutron orbital momentum $\ell = 1$ and for a given level f with spin J_f. The quantity y is equal to $k_f R$, where k_f is the wave number of the neutron, residing at the bound level f.

As in the external region $(r>R)$ of configuration space the neutron wave functions of the initial and final states are known analytically, the expression in Eq. (3) should be regarded as a good approximation.

On the other hand, if one tries to include the internal region, model concepts have to be adopted. Within the valence neutron model of Lynn[4] the neutron in the initial and final states is assumed to be orbiting the inert core not only out of core, but also inside it $(r<R)$. The wave functions for these states can be calculated, using the optical model potential. The calculations available for Mo nuclei[2] show that the internal region contributes to $\delta\Gamma_{\lambda\gamma f}^{1/2}$ only by $\approx 20\%$. This weak contribution is due to a small overlap of strongly oscillating radial neutron wave functions for the initial and final states in the internal region.

Utilization of the valence neutron model for nuclei with $150<A<190$ *is seriously obstructed by the deformation*. In fact, no calculations are available, so far. However, it is expected that for the same reasons valence model predictions of $\delta\Gamma_{\lambda\gamma f}$ will not essentially differ from those according to Eqs. (2) and (3).

The wave function of a neutron resonance contains a large set of various components, but only a limited number of them is responsible for the emission or absorption of a neutron. For the case of s-wave neutrons and deformed even-even residual nuclei these components are of a specific two-quasiparticle structure. In terminology of Soloviev's model[12] this structure is of the type $(s_o\sigma_o, s\sigma)$. Here, $s_o\sigma_o$ denote a set of all quantum numbers which characterize the neutron single-quasiparticle configuration of the target ground state, while $s\sigma$ are quantum numbers of some other neutron quasiparticle state. The symbols σ and σ_o stand for the signs of the quasiparticle total angular momentum projection on the symmetry axis. Parity π and total angular momentum projection Ω of the second quasiparticle should satisfy the additional condition $\Omega^\pi = 1/2^+$. The number of various two-quasiparticle components $(s_o\sigma_o, s\sigma)$ that are responsible for the neutron emission or absorption is strongly restricted, as the energy of the corresponding two-quasiparticle states should be situated near the neutron threshold.

According to Soloviev's model,[12] a necessary condition for correlation between partial radiation widths and reduced neutron widths is the occurrence of such a two-quasiparticle component in the wave functions of resonances that is responsible for the emission of a neutron and a primary γ-ray simultaneously. It implies that the most suitable candidates for this correlation are transitions to final levels with a two-quasiparticle structure of the type $(s_o\sigma_o, s_f\sigma_f)$ or to levels with a large admixture of such structure. It can be seen that these transitions satisfy the general quasiparticle selection rule, as one of the quasiparticles remains passive upon transition.

If one takes into account the Ω-forbiddenness, then for

E1-transitions the following condition should be satisfied

$$|\sigma\Omega - \sigma_f\Omega_f| \leq 1, \tag{4}$$

where σ_f and Ω_f refer to the "active" quasiparticle, belonging to the final level f. It is evident that such a condition restricts the number of candidates for correlation. It is noteworthy that the restricted E1-transitions populate exclusively those final levels which may carry the $\ell = 1$ neutron single-particle strength. The size of this strength depends on a particleness of the "active" quasiparticle.

In analogy with the even-even residual nuclei, Soloviev's model predicts the width correlation also for the remaining types of nuclei. For the odd-odd residual nuclei the most suitable candidates for the correlation are primary transitions to the levels with the two-quasiparticle structure $(r_o\rho_o, s_f\sigma_f)$, where $r_o\rho_o$ is a set of all quantum numbers, characterizing the proton quasiparticle configuration of the target ground state. In case of the odd-even residual nuclei the most favourable structure is $(r_o\rho_o, s_o\sigma_o, s_f\sigma_f)$, where $(r_o\rho_o, s_o\sigma_o)$ is a proton-neutron quasiparticle configuration of the odd-odd target.

Finally, in case of the even-odd residual nuclei the best candidates for the correlation are transitions to the neutron single-quasiparticle levels.

EXPERIMENTAL RESULTS

The data on width correlation which are discussed in this talk are based on the (n,γ) experiments undertaken at the Pulsed Fast Reactor at Dubna by the time-of-flight technique. In these experiments the following target nuclei were subjected to the study: ^{152}Gd (Ref. 14), ^{154}Gd (Ref. 15), ^{167}Er (Refs. 16, 17, 18), ^{171}Yb (Refs. 17, 19), ^{173}Yb (Refs. 17, 20, 21, 22), ^{175}Lu (Ref. 23), ^{176}Lu (Ref. 24) and ^{185}Re (Ref. 25).

For most of these nuclei capture γ-ray spectra, serving as a primary source of information, could be obtained from a large number of well-isolated resonances with a firm spin assignment. Totally 141 γ-ray spectra have been employed for analysis. Examples are given in Figs. 1 and 2.

Among the resolved neutron resonances were those with a large "local" value of the s-wave neutron strength function $S^o(\text{local}) = g_J \Gamma_{\lambda n}^o/D_J$. Here, g_J is the spin factor and D_J — the average spacing between resonances of a given spin J. In many instances the quantity $S^o(\text{local})$ exceeded the value 10^{-3}. Occurrence of resonances with large values of $S^o(\text{local})$ is crucial for efficient detection of a correlation effect, masked by "noise" originating from fluctuations of random amplitudes $\Gamma_{\lambda\gamma f}(r)^{1/2}$.

Quantum numbers J, π and K of a majority of low-lying levels of all residual nuclei studied were known, as well as a structure of these levels.

The average coefficient of correlation, defined as

$$R = \sum_J \sum_f \omega_J \, r_{Jf}, \qquad (5)$$

served as a quantitative measure of the correlation between $\Gamma_{\lambda\gamma f}$ and $\Gamma_{\lambda n}^o$ for a given set of resonances and a set of final levels. Here, ω_J is a proper statistical weighting factor and r_{Jf} is the coefficient of linear correlation for a sample formed by pairs $\{\Gamma_{\lambda\gamma f}, \Gamma_{\lambda n}^o\}$ with fixed J and f.

Similarly, in some cases another average coefficient of correlation, defined as

$$T = \sum_J \sum_{f<f'} \tilde{\omega}_J \, t_{Jff'}, \qquad (6)$$

has been used as a measure of the correlation between *various* partial radiation widths. Here, $\tilde{\omega}_J$ is a corresponding weighting factor and $t_{Jff'}$ is a sample coefficient of linear correlation for pairs

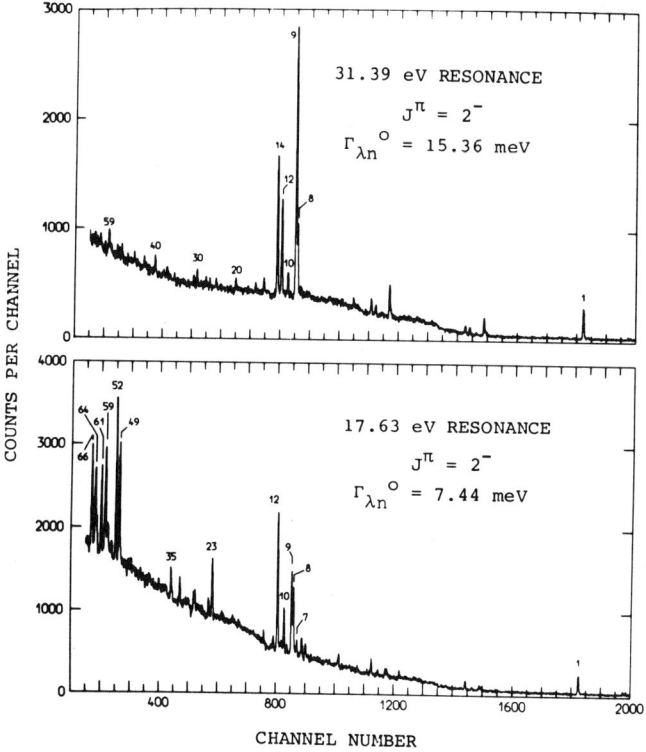

Fig. 1. Examples of γ-ray spectra from isolated $J^\pi = 2^-$ resonances in the ^{173}Yb$(n,\gamma)^{174}$Yb reaction. Peak numbers 7-10, 12, 14, and 20 correspond to primary $E1$-transitions to the bands in ^{174}Yb listed in Table I.

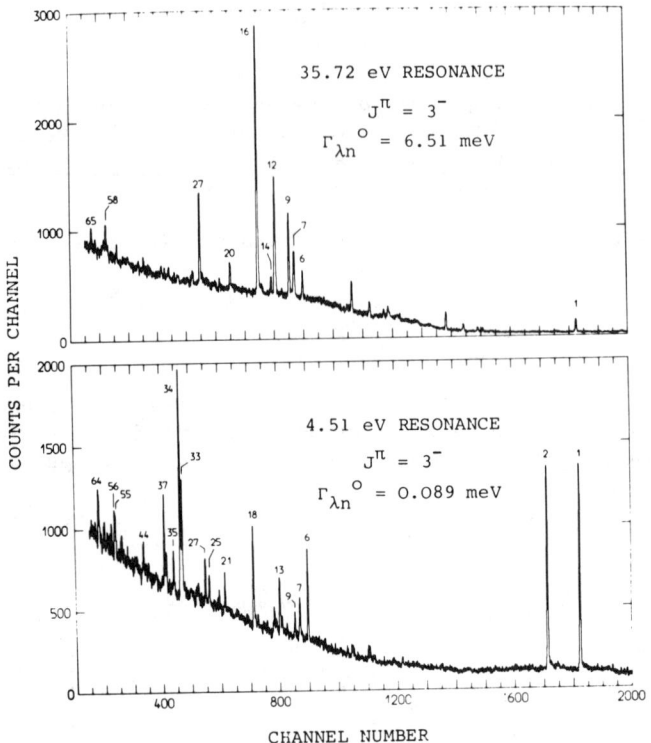

Fig. 2. Examples of γ-ray spectra from isolated $J^\pi = 3^-$ resonances in the ^{173}Yb(n,γ)^{174}Yb reaction. Peak numbers 7, 9, 12, 14, 16, 18, 20 and 27 correspond to primary $E1$-transitions to the bands in ^{174}Yb listed in Table I.

$\{\Gamma_{\lambda\gamma f}, \Gamma_{\lambda\gamma f'}\}$ with fixed J, f and f'.

To distinguish between two types of correlation, from now on the terms "R-correlation" and "T-correlation" will be frequently used.

Utilization of the Monte Carlo method enabled calculating the statistical significance with which a given experimental value of R, denoted further by R_{exp}, rejects the hypothesis that the widths $\Gamma_{\lambda n}^o$ and $\Gamma_{\lambda\gamma f}$ are not correlated. The value of statistical significance is understood as the probability $P(R<R_{exp})$ of finding an average coefficient of correlation R which is lower than R_{exp} under the assumption that widths $\Gamma_{\lambda n}^o$ and $\Gamma_{\lambda\gamma f}$ are not correlated. Similarly, the Monte Carlo method has been used to calculate statistical significance $P(T<T_{exp})$ related to T-correlation.

The main results of analysis of R-correlation are given in Table I. It should be stressed that these results are based only on an analysis of partial radiation widths that correspond exclusively to primary $E1$-transitions from s-wave resonances. In almost all cases the individual values R_{exp} in Table I belong to primary transitions to levels of a single rotational band. All the bands involved

Table I. R-correlation for deformed nuclei.

Target	Band in residual nucleus[a]	Number of widths	R_{exp}	$P(R<R_{exp})$ (%)
^{152}Gd	$n521\downarrow$, $n521\uparrow$, $n530\uparrow$	20	−0.107	36.1
^{154}Gd	$n521\uparrow$	11	0.086	68
	$n532\downarrow$	11	−0.290	19
	$n521\downarrow$[b]	22	0.664	99.80
^{171}Yb	$\underline{n521\downarrow}-n512\uparrow$[c]	22	−0.315	5.6
^{173}Yb	$n512\uparrow+n521\downarrow$	37	0.178	85.4
	$\underline{n512\uparrow}-n514\downarrow$[c]	69	0.395	99.90
	$\underline{n512\uparrow}-n510\uparrow$[b]	60	0.302	98.80
	$\underline{n512\uparrow}-n512\uparrow$[b,c]	37	0.243	91.1
^{167}Er	$n633\uparrow-n521\downarrow$	25	0.407	94
	$n633\uparrow+n521\downarrow$	25	0.246	90
	$n633\uparrow-n512\uparrow$[c]	51	0.168	82
^{175}Lu	$\underline{p404\downarrow}-n514\downarrow$[c]	45	−0.100	41
	$\underline{p404\downarrow}-n512\uparrow$[c]	45	0.366	97
^{176}Lu	$\underline{p404\downarrow+n514\downarrow}\pm n510\uparrow$[d,e]	36	0.599	99.94
	$\underline{p404\downarrow+n514\downarrow}-n521\downarrow$	36	−0.034	43
	$\underline{p404\downarrow+n514\downarrow}+n521\downarrow$	26	0.021	53
	$\underline{p404\downarrow+n514\downarrow}+n512\downarrow$[b]	18	0.042	56
^{185}Re	$\underline{p402\uparrow}-n510\uparrow$	48	0.054	68.9
	$\underline{p402\uparrow}+n510\uparrow$	38	0.502	99.7
	$\underline{p402\uparrow}-n512\downarrow$	72	0.074	72.2
	$\underline{p402\uparrow}+n512\downarrow$	14	−0.194	28.5
	$\underline{p402\uparrow}-n503\uparrow$[c]	48	−0.039	42.0
	$\underline{p402\uparrow}-n505\downarrow$[c]	62	−0.099	25.0

[a] The underlined part represents the structure of the target.
[b] A strong admixture of this structure.
[c] The R-correlation should be Ω-forbidden.
[d] Only the transitions to the band heads are included.
[e] Two bands.

are of a quasiparticle structure favourable for observing the R-correlation, as predicted by Soloviev's model.[12] In many cases the structure satisfies the condition for Ω-allowed R-correlation.

In case of the ^{173}Yb$(n,\gamma)^{174}$Yb reaction, information concerning the factors q_{Jf} has been deduced. Table II lists estimates of q_{Jf} summed over f and compares them with corresponding theoretical values calculated according to Eq. (3). As spectroscopic factors entering Eq. (3) are not known experimentally, their values were calculated instead. A modified expression of Satchler[26] has been used —

$$S_{dp,f}^{(\ell)} = \frac{2J+1}{2J_f+1} \left[\sum_j (jI \pm \Omega_f K | J_f K_f) \, C_{j\ell}^{\Omega_f} \, u_f \right]^2. \quad (7)$$

This expression is valid for cases when $K_f = K \pm \Omega_f$. Here, I and J_f are the target spin and the spin of a final level, respectively; corresponding projections on the symmetry axis are K and K_f, respectively; j stands for the total angular momentum of the captured neutron. Factor $C_{j\ell}^{\Omega_f}$ is the amplitude in an expansion of the deformed neutron wave function in terms of sherical limits; factor u_f is the amplitude of the particleness of the neutron quasiparticle. A correction for the overlap of the initial vibrational state with the final state has been neglected. Values of $C_{j\ell}^{\Omega_f}$ and u_f were taken from calculations of Gareev et al.[27]

A detailed analysis of T-correlation has been done only for the ^{173}Yb$(n,\gamma)^{174}$Yb reaction. A search for a statistically significant effect has been undertaken for primary $E1$-transitions populating three separate groups of ^{174}Yb levels: (i) the levels with a known structure, (ii) the remaining known levels in the interval of excitation energy 2000-2550 keV, and (iii) all the known levels in the energy interval 2550-2740 keV.

Although statistically significant values T_{exp} were found for the first group of levels, the results obtained turned to be physically uninteresting, as $T_{exp} \simeq R_{exp}^2$ in these cases. The observed T-correlation could be thus easily accounted for by a common dependence of various pairs of partial radiation widths on the reduced neutron width, in accordance with Eqs. (1) and (2).

The results for the remaining groups of final levels are presented in Table III. These results display two highly significant cases of T-correlation. On the other hand, in both these cases values R_{exp} were fully compatible with the absence of correlation between $\Gamma_{\lambda\gamma f}$ and $\Gamma_{\lambda n}^o$.

Table II. Estimates of q_{JF} for the ^{173}Yb$(n,\gamma)^{174}$Yb reaction.

J^π	$\sum_f q_{Jf}$, (eV$^{1/2}$)	
	Experiment[a]	Eq. (3)[b]
2^-	0.36±0.11	0.010
3^-	0.59±0.25	0.008

[a] Levels of all bands specified in Table I are included.
[b] Only the $n512\uparrow$-$n510\uparrow$ and $n512\uparrow$+$n521\downarrow$ bands contribute.

Table III. T-correlation in the ^{173}Yb$(n,\gamma)^{174}$Yb reaction.

J^π	Number of resonances	Number of final levels^a	T_{exp}	$P(T<T_{exp})$ (%)
2^-	9	15^b	+0.047	92
3^-	14	14^b	+0.037	90
2^-	9	16^c	+0.266	99.9988
3^-	14	15^c	+0.024	86
2^-	9	6^d	+0.748	>99.9997
2^-	8^e	6^d	−0.122	6

^{a}The levels with a specified excitation energy which are accessible by $E1$-transitions from s-wave resonances.
^{b}Excitation energies 2000-2550 keV.
^{c}Excitation energies 2550-2740 keV.
^{d}Levels at 2581.1, 2599.4, 2656.3, 2679.9, 2712.2, and 2732.0 keV with the spin assignment $J=1$, $J=1(2,3)$, $J=2,3$, $J=2,3$, $J=2,3$, and $J=1,2,3$, respectively.
^{e}The resonance at 17.63 eV is excluded from analysis.

Figure 3 shows a comparison of the Monte Carlo generated values of T with the experimental value T_{exp} for primary transitions from nine $J^\pi = 2^-$ resonances to six ^{174}Yb levels. The levels are specified in footnote "d" of Table III.

R-CORRELATION

Evidence for R-correlation

As is apparent from Table I, in case of the ^{173}Yb$(n,\gamma)^{174}$Yb reaction the high values of statistical significance belong to radiation widths, associated with transitions to the levels of the $n512\uparrow$-$n514\downarrow$ band and the $K^\pi = 2^+$ band at 1634 keV, whose ≈50% admixture has the structure $n512\uparrow$-$n510\uparrow$. When the transitions to the levels of all four bands listed in Table I are analyzed as a whole, the value $R_{exp} = 0.300$ is obtained with a significance of 99.99%. This constitutes strong evidence for R-correlation in the ^{173}Yb$(n,\gamma)^{174}$Yb reaction.

The existence of R-correlation seems corroborated by a result of Mughabghab,[28] who observed in the same reaction a large value $r_{jf} = 0.85$ for transitions from nine $J^\pi = 2^-$ resonances to the head of the $K^\pi = 2^+$ band at 1634 keV. On the other hand, Shahal et al.[29] have found that this distinct effect displays a marked instability, as it virtually disappears if a transition from just one resonance at 31.39 eV to the 1634 keV level is excluded from analysis. Despite this, it

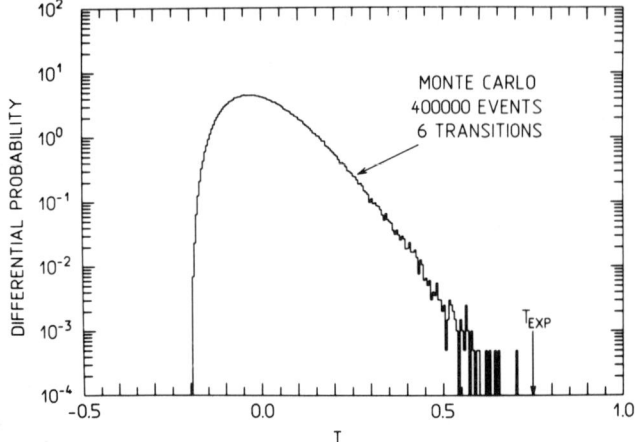

Fig. 3. A comparison of the generated values of T with the experimental value T_{exp} for the ^{173}Yb(n,γ)^{174}Yb reaction.

is evident that the existence of R-correlation in the ^{173}Yb(n,γ)^{174}Yb reaction is well established. Indeed, if all transitions from $J^\pi = 2^-$ resonances to the incriminated 1634 keV level are excluded, the original value $R_{exp} = 0.300$ decreases only to 0.274 and corresponding value of statistical significance remains still high, $P = 99.96\%$. Moreover, it should be stressed that there are no evident *a priori* reasons to exclude the transitions to the 1634 keV level from analysis.

From the data on the ^{167}Er(n,γ)^{168}Er reaction a statistically significant result can be drawn when transitions to the n633↑±n521↓ bands in ^{168}Er are analyzed simultaneously.[18] In this case, the analysis yields a value $R_{exp} = 0.349$ with a significance of 99.7%. It is to be noted that the neutron two-quasiparticle structures of both rotational bands involved satisfy the conditions for the Ω-allowed R-correlation.

As Table I shows, a highly significant result has been obtained for the ^{176}Lu(n,γ)^{177}Lu reaction for partial radiation widths that correspond to primary transitions to the heads of the p404↓+n514↓±n510↑ bands in ^{177}Lu. On the other hand, a detailed inspection of the data[24] shows that no significant effect of R-correlation is observed for the transitions to the remaining members of these bands, neither to the levels of the p404↓+n514↓±n521↓ bands, nor to the levels of the $K^\pi = 11/2^+$ band, whose admixture has a structure p404↓+n514↓-n512↓. These important results will be discussed later.

In spite of the occurence of several statistically significant values R_{exp} in Table I, a remaining large number of values is fully compatible with the assumption that the partial radiation widths involved are not correlated with reduced neutron widths. It may be argued that partial radiation widths are not correlated *at all* and that the occurrence of a few exceptional cases of statistically sig-

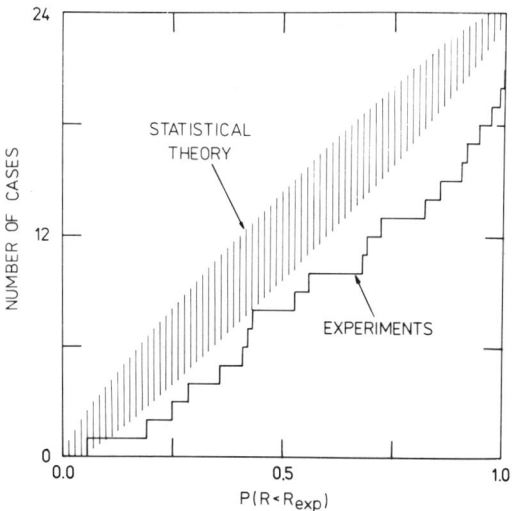

Fig. 4. A cumulative plot of twenty-four values of statistical significance $P(R<R_{exp})$ from Table I. If partial radiation widths involved were statistically independent of reduced neutron widths, then the step-like curve of this plot would be situated approximately within the hatched area.

nificant values R_{exp} is a mere manifestation of uninteresting random fluctuations of the average coefficient of correlation. This objection calls for a closer inspection of the results.

If the R-correlation does exist, the values of statistical significance P for all twenty-four sets of partial radiation widths in Table I should behave as a sample drawn from a population of uniformly distributed numbers in the interval (0, 1). However, Fig. 4 shows that the cumulative plot of values of P deviates appreciably from what is expected for the uniform distribution. The actual distribution clearly exhibits a systematic deficit of low values. Besides that, the highest values of statistical significance are densely clustered at the very end of the interval (0, 1). In particular, five values are higher than, or equal to 0.988, while for four of these the lower limit is even as high as 0.997. In case of completely absent R-correlation for the deformed nuclei, a probability for such clustering is equal to 8.4×10^{-6} and 8.1×10^{-7}, respectively. If any critical value of significance is removed from analysis, the probability for observation of clustering of remaining values will increase at most to 4.5×10^{-5}, remaining still negligible. The results of this analysis constitute a firm proof that the R-correlation really exists for the deformed nuclei with $150<A<190$.

Role of quantum number K

Shahal et al.[29] in their study of the ^{173}Yb$(n,\gamma)^{174}$Yb reaction noticed that an average single-particle strength for transitions to the levels of the $K^\pi = 2^+$ band with the structure $n512\uparrow-n514\downarrow$ is substantially higher than that for the transitions to the levels of the $K^\pi = 0^+$ bands. A ratio of the average transition strengths $<k_{E1}>_{K^\pi=2^+}/<k_{E1}>_{K^\pi=0^+} = 1.92$ can be deduced from the data in Ref. 29.

A closer inspection of the data from the Dubna experiment[17,20] confirms such a conclusion when strong resonances are taken into account. As Table IV shows, in this case the reduced intensities for the transitions to the levels of the $K^\pi = 2^+$ band are enhanced.

Table IV. Ratios of average reduced intensities for the transitions to the $K^\pi = 2^+$ and $K^\pi = 0^+$ levels in the ^{173}Yb$(n,\gamma)^{174}$Yb reaction. Values of statistical significance for accepting a hypothesis that transitions to the $K^\pi = 2^+$ levels are enhanced are also given.

J^π	$\langle I^0_{\lambda f}\rangle_{K^\pi=2^+} / \langle I^0_{\lambda f}\rangle_{K^\pi=0^+}{}^a$	
	Resonances with $\Gamma^0_{\lambda n} > 2$ meV^b	Resonances with $\Gamma^0_{\lambda n} < 2$ meV^c
2^-	5.48±0.22 (99.99%)	2.15±0.19 (87.8%)
3^-	2.14±0.07 (97.2%)	1.19±0.04 (70.5%)

^{a}Included transitions to the $K^\pi = 0^+$ levels at 77.1, 253.7, 1561.2, 1715.2, 1958.7, 2123.0, 2172.5, and 2335.9 keV together with transitions to the levels of the 1633.8 keV band with $K^\pi = 2^+$.
^{b}Six resonances with $J^\pi = 2^-$ and five resonances with $J^\pi = 3^-$.
^{c}Three resonances with $J^\pi = 2^-$ and nine resonances with $J^\pi = 3^-$.

However, in weak resonances the sensitivity of reduced intensities to K^π becomes vague and cannot be regarded as statistically significant.

In view of the evidence for R-correlation this finding suggests that the effect reported in Ref. 29 is a manifestation of the following properties of the partial radiation widths involved: (i) the widths for the transitions to the n512↑-n514↓ band are strongly correlated with $\Gamma^0_{\lambda n}$ and (ii) the correlated parts $\delta\Gamma_{\lambda\gamma f}$ are at least in some degree decoupled from the giant dipole electric resonance (GDR) and lead to enhancement of partial radiation widths. It suggests that the earlier observed enhancement of single-particle transition strength is not connected with a particular value of K, but with a favourable structure leading to the R-correlation. In view of these arguments it seems to be premature to interpret the reported effect[29] as a dependence of k_{E1} on K.

It is to be noted that sensitivity of partial radiation widths to K has been tested in the ^{176}Lu$(n,\gamma)^{177}$Lu reaction at 17 isolated resonances for a much broader interval of K, ranging from 1/2 to 15/2,[30] but no significant effect has been observed.

Discussion of results

The reported results for the deformed nuclei pose a question whether the strong R-correlation persists in s-wave resonances of other nuclei with a substantially different mass number. It can be speculated that the size of the overall sum $\sum_f q_{Jf}$ is roughly the same as in the case of the ^{173}Yb(n,γ)^{174} reaction, irrespectively of how the sum is distributed among individual resonances.

If the correlation persists, tails of distant s-wave resonances produce a non-resonant "background" capture. A partial cross section for this process can be expressed approximately as follows

$$\sigma_{\gamma f}^{(b)} = \frac{4\pi}{E_n^{1/2}} \sum_J g_J q_{Jf} (R-R')^2, \qquad (8)$$

where R' is the potential scattering amplitude.

The difference $(R-R')$ reaches a local maximum of $\simeq 2.2$ fm for nuclei with $A \simeq 130$. These nuclei are thus favourable for observing the "background" capture.

Neglecting dependence of q_{Jf} on J and on γ-ray energy, the experimental values of the sum $\sum_f q_{Jf}$ in Table II can be used and a total effect of the "background" capture at thermal energies can be estimated. For $(R-R') = 2.2$ fm Eq. (8) gives $\sum_f \sigma_{\gamma f}^{(b)} \simeq 1.9$ b. This value is by an order of magnitude higher than the values of the thermal total capture cross section[31] in the considered mass region. It is thus evident that for the nuclei with $A \simeq 130$ the R-correlation is substantially suppressed.

In accordance with the mechanism of the GDR some fraction of $E1$-strength of any simple nuclear state should be relocated into the GDR and subsequently spread over individual resonances. Gyarmati et al.,[32] who analyzed the problem of relocation, concluded that for the case of a neutron single-particle s-state the relocation of $E1$-strength to the GDR is effective if this state is situated in an unbound region, while for a bound s-state the relocated fraction should be small. In addition, it is believed that the process of spreading is of random nature, so that the size of a contribution of the GDR to the $E1$-strength of each resonance is not correlated with the reduced neutron width. Thus, a strong relocation of $E1$-strength to the GDR should lead to diminution of R-correlation.

For the nuclei with $A \simeq 130$ the neutron single-particle 4s-state is fully unbound. Therefore, the threshold-dependent effect of relocation accounts for the suppressed R-correlation for these nuclei. The discrepancy concerning the "background" capture thus seems to be easily clarified.

The experimental values of q_{Jf} in Table II are too high to be consistent with the values expected from the simplified version of the theory of Lane and Lynn,[3] in which only the *external* (channel) region of the configuration space is taken into account.

As for the *internal* region, according to the valence neutron model calculations for spherical nuclei,[2] it plays only a minor role in the neutron single-particle transition mechanism. Unfortunately, no quantitative conclusions from the valence model are available for the deformed nuclei. There are, however, no apparent reasons to expect that a rigorous inclusion of the deformation into the model would radically change the situation. It is therefore likely that also the valence model would not be able to account for the large experimental values of q_{Jf}.

Setting aside the question of quantitative interpretation of the experimental data, it is of interest to treat a more general problem, namely whether the observed cases of R-correlation originate from a neutron single-particle transition. If that is the case, the correlated parts of partial radiation widths are proportional to neutron spectroscopic factors, $\delta\Gamma_{\lambda\gamma f} \propto S_{dp,f}^{(1)}$. The given problem is connected with the question of validity of the Ω-forbiddenness. As the configuration $512\uparrow$-$n514\downarrow$ does not carry the $\ell = 1$ neutron single-particle strength $(S_{dp,f}^{(1)} \equiv 0)$, it may seem that the large effect of the Ω-forbidden R-correlation for the transitions to the levels of $n512$-$n514\downarrow$ band in ^{174}Yb violates the above-mentioned proportionality. Unfortunately, as follows from a recent work on ^{174}Yb (Ref. 33), the assignment of the structure $n512\uparrow$-$n514\downarrow$ to the rotational band in question cannot be considered to be absolutely firm. A reasonable alternative for the given band seems to be the structure $n512\uparrow$-$n512\downarrow$. This structure is favourable for observing the Ω-allowed R-correlation and leads to a non-zero value of the neutron spectroscopic factor. So, at this point the question of proportionality between $\delta\Gamma_{\lambda\gamma f}$ and $S_{dp,f}^{(1)}$ remains open.

Further information on this problem can be drawn from the ^{176}Lu(n,γ) ^{177}Lu reaction. In this connection it is important to note that the residual ^{177}Lu nuclei possess the following specific features:

(1) Due to an extremely large spin of the ^{176}Lu target ($I = 7$), Clebsh-Gordan coefficients in Eq. (7) lead to a strong concentration of the $\ell = 1$ neutron single-particle strength on the heads of ^{177}Lu bands. In particular, the heads of the $p404\downarrow+n514\downarrow\pm n510\uparrow$ bands exhaust approximately 87% of the overall sum of spectroscopic factors. The heads of the $p404\downarrow+n514\downarrow\pm n521\downarrow$ bands behave similarly.

(2) The neutron quasiparticle state $n510\uparrow$ is characterized by a strong particleness, $u_f^2 = 0.98$, in the region $A \simeq 177$, while the state $n521\downarrow$ is of a hole character with $u_f^2 = 0.05$.

(3) In case of the remaining ^{177}Lu band which satisfy the condition for the Ω-allowed R-correlation the pertinent structure, $p404\downarrow+n514\downarrow-n512\downarrow$, is present only as an admixture.

If the proportionality $\delta\Gamma_{\lambda\gamma f} \propto S_{dp,f}^{(1)}$ holds, then, in view of the features outlined, the only reasonable candidates for the R-correlation are the transitions to the heads of the $p404\downarrow+n514\downarrow\pm n510\uparrow$ bands and, possibly, to the head of the $p404\downarrow+n514\downarrow-n512\downarrow$ band.

The fact that a strong effect of R-correlation has been observed only for the transitions to the heads of the $p404\downarrow+n514\uparrow\pm n510$ bands, while no significant trace of the effect was recorded for any of the remaining transitions, speaks strongly in favour of a neutron single-particle origin of the R-correlation.

T-CORRELATION

The effects of T-correlation were considered by Lane[34] and Beer[35] on the basis of the common doorway concept. If the role of the common doorway is played by a simple nuclear state which is *different* from the neutron single-particle state, the T-correlation becomes interesting, as it reflects a qualitatively new situation. In such an instance, the R-correlation should be missing.

So far, the only case of T-correlation of this type was reported many years ago by Beer.[35] This author analyzed the experimental data of Saclay group for $J^\pi = 0^+$ resonances in the $^{187}W(n,\gamma)^{188}W$ reaction and deduced a value $T_{exp} = 0.37$.

As is apparent from the data in Table III, the most distinct effect of the considered type of T-correlation is observed for the case of $E1$-transitions from nine $J^\pi = 2^-$ resonances to six $J^\pi = 1^+$, 2^+, 3^+ levels in ^{174}Yb; these transitions correspond to peak numbers 49, 52, 59, 61, 64, and 66 in Fig. 1.

Excitation energies as well as spin values of all six levels are compatible with the assumption that the levels form 1^+, 2^+, and 3^+ members of two individual $K^\pi = 1^+$ rotational bands. This points to the physical nature of the observed effect.

The fact that T-correlation is missing in the $J^\pi = 3^-$ resonances is not suprising, as a common doorway involved may not be necessarily spread over the s-wave resonances of both spin values 2^- and 3^-.

Besides an unusually strong effect of correlation $T_{exp} = 0.748$, Table III shows another startling feature — a marked instability of this effect with respect to exclusion of the 17.63 eV resonance. This behaviour is in a sharp disagreement with the theory[13] which describes general properties of fluctuations and correlations of partial widths. As shown in Ref. 22, validity of this theory for the case of the $^{173}Yb(n,\gamma)^{174}$ reaction can be rejected with a statistical significance of 99.95%.

As a very tentative explanation of the instability of T-correlation it can be assumed that a process of spreading of the corresponding common doorway over resonances is not quite random in nature. It may be that fluctuations of the amplitude of common doorway contribution to a resonance are not governed by the normal distribution with a zero mean, as it is implicitly assumed by the general theory of Krieger and Porter.[13]

CONCLUSION

The vast amount of data on neutron capture at isolated resonances now clearly demonstrates that distinct and statistically significant effects of width correlation occur also in nuclei with 150<A<190.

The available data lead to a highly probable conclusion that the observed effects of correlation between the partial radiation widths and the reduced neutron widths are of neutron single-particle origin.

The extraordinary large size of correlated parts of partial radiation widths seems to contradict the theory of Lane and Lynn.[3,4]

In this respect, there is also a similar unsatisfactory situation in case of spherical nuclei with $A = 90-100$,[2] where the theory of Lane and Lynn is able to account for only ≈50% of the size of the observed correlated parts of partial radiation widths. Nevertheless, in case of nuclei with 150<A<190, a full verification of the theory of Lane and Lynn[3,4] requires a rigorous inclusion of nuclear deformation into the formalism.

Unprecedentedly strong effect of correlation between various partial radiation widths in the ^{173}Yb$(n,\gamma)^{174}$Yb reaction points to the existence of a doorway, common to various radiation channels and different from the neutron single-particle state.

REFERENCES

1. O. A. Wasson and G. G. Slaughter, Phys. Rev. C8, 297 (1973).
2. R. E. Chrien, G. W. Cole, G. G. Slaughter, and J. A. Harvey, Phys. Rev. C13, 578 (1976).
3. A. M. Lane and J. E. Lynn, Nucl. Phys. 17, 586 (1960).
4. J. E. Lynn, The Theory of Resonance Neutron Reactions (Clarendon Press, Oxford, 1968), p. 333.
5. A. P. Jain, B. Couvin, and H. Lotin, Nucl. Phys. A223, 509 (1974).
6. S. F. Mughabghab, R. E. Chrien, and O. A. Wasson, Phys. Rev. Lett. 25, 1670 (1970).
7. M. Beer, M. A. Lone, R. E. Chrien, O. A. Wasson, M. R. Bhat, and H. R. Muether, Phys. Rev. Lett. 20, 340 (1968).
8. S. F. Mughabghab in Nuclear Structure Study with Neutrons, edited by J. Ero and J. Szucs (Plenum Press, London and New York, 1974), p. 167.
9. P. Ribon, R. E. Chrien, and G. W. Cole, Bull. Am. Phys. Soc. 18, 1402 (1973).
10. R. E. Chrien, in Statistical Properties of Nuclei, edited by J. B. Garg (Plenum, New York, 1972), p.223.
11. B. W. Thomas, ibid., p. 251.
12. V. G. Soloviev, Part. and Nucl. 3, 770 (1973).
13. T.J. Krieger and C. E. Porter, J. Math. Phys. 4, 1272 (1963).
14. F. Bečvář, J. Honzátko, M. Králík, N. D. Nhuan, T. Stadnicov, and S. A. Telezhnikov, JINR Report No. R3 12516 (1979).
15. F. Bečvář, J. Honzátko, M. Králík, N. D. Nhuan, and S. A. Telezhnikov, Yad. Fiz. 34, 1158 (1981).
16. L. Aldea, F. Bečvář, J. Honzátko, V. Mateiciuc, and S. A. Telezhnikov, in Proceedings of International Conference on Selected Topics in Nuclear Structure, vol. 1, JINR Report No. D-9682 (1976), p. 116.
17. F. Bečvář, J. Honzátko, M. Králík, N. D. Nhuan, T. Stadnicov, and S. A. Telezhnikov, Yad. Fiz. 33, 3 (1981).
18. M. Králík, Thesis (Technical University of Prague, Prague, 1982).
19. F. Bečvář, J. Honzátko, M. Králík, N. D. Nhuan, T. Stadnicov, and S. A. Telezhnikov, in Neutron Capture Gamma-Ray Spectroscopy and Related Topics, edited by T. von Egidy, F. Gonenwein, and B. Maier (Institute of Physics, Bristol and London, 1982), p. 230.
20. L. Aldea, F. Bečvář, J. Honzátko, S. Pospíšil, and S. A. Telezhnikov, Czech J. Phys. B 27, 1002 (1977).
21. F. Bečvář, in Neutron Induced Reactions, vol.10 (Institute of Physics, Bratislava, 1982), p. 171.
22. F. Bečvář, J. Honzátko, M. Králík, M.-E. Montero-Cabrera, N. D. Nhuan, contributed paper to the VI Soviet National Conference on Neutron Physics, Kiev, Oct. 2-6, 1983 (in press).

23. F. Bečvář, H. T. Hiep, M.-E. Montero-Cabrera, S. Pospíšil, and S. A. Telezhnikov, *ibid*.
24. F. Bečvář, H. T. Hiep, M.-E. Montero-Cabrera, S. Pospíšil, and S. A. Telezhnikov, *ibid*.
25. F. Bečvář, J. Honzátko, M. Králík, N. D. Nhuan, T. Stadnicov, Yad. Fiz. $\underline{37}$, 1357 (1983).
26. G. R. Satchler, Ann. Phys. $\underline{3}$, 275 (1958).
27. F. A. Gareev, S. P. Ivanova, V. G. Soloviev, and S. I. Fedotov, Particles and Nuclei $\underline{4}$, 357 (1973).
28. S. F. Mughabghab, *in* Lectures at the Third International School on Neutron Physics, Alushta, USSR, JINR Report No. D-11787 (1978), p. 328.
29. O. Shahal, S. Raman, C. Coceva, and M. Stefanon, Phys. Rev. $\underline{C25}$, 1283 (1982).
30. F. Bečvář, *in* Lectures at the Second International School on Neutron Physics, Alushta, USSR, JINR Report No. D3-7991 (1974), p. 294.
31. S. F. Mughabghab and D. I. Garber, Neutron Cross Sections, vol. 1, Resonance Parameters (National Neutron Cross-Section Center, Upton, 1973).
32. B. Gyarmati, A. M. Lane, and Zymányi, Phys. Lett. $\underline{50B}$, 316 (1974).
33. R. C. Greenwood and C. W. Reich, Phys. Rev. $\underline{C23}$, 153 (1981).
34. A. M. Lane, Ann. Phys. $\underline{63}$, 171 (1971).
35. M. Beer, Ann. Phys. $\underline{65}$, 181 (1971).

CHANNEL WAVE FUNCTION AND CHANNEL RADIATIVE CAPTURE CROSS SECTIONS

Yu-Kun Ho
Physics Department, Zhengzhou University, Honan Province, China

M.A. Lone
Chalk River Nuclear Laboratories, Chalk River, Ont., Canada K0J 1J0

ABSTRACT

Based on the intermediate interaction model, an approximate wave function for the entrance channel in the nucleon-nucleus reaction is derived. It is valid in the exterior region as well as the region near the nuclear surface, and is expressed in terms of the wave function and reactance matrix of the optical model and of the near-resonance parameters. Using this formalism the following processes are calculated: 1) resonance averaged channel radiative neutron capture cross section; 2) radiative capture in the inelastic channel of compound nucleus; 3) interference effect in the channel radiative neutron capture process; 4) laser induced neutron radiative capture reaction.

1. INTRODUCTION

In the twenty years since the Lane and Lynn[1,2] theory of channel neutron capture was presented, much effort has been devoted to the experimental and theoretical investigation of this reaction mechanism[3-9]. Quantitative and qualitative verification of the theory has been achieved in several cases, although there also exist some examples that need further clarification.

A better test of the theory would be to calculate the absolute values of the cross sections with more realistic nuclear interactions and compare these with the measured cross sections. Calculations of the channel capture cross sections at sharp energies, in addition to the energy-averaged values, would permit a more detailed test of the theory, including features such as the interference between potential and valence capture.

These calculations require knowledge of the entrance-channel wave function, not only in the exterior asymptotic region, where the wave functions are usually expressible in terms of the scattering matrix and spherical Bessel functions, but also in the interior region, at least in the vicinity of the nuclear surface where the nuclear force is not negligible.

In section 2, we derive such a wave function for the neutron entrance channel using the intermediate interaction model. With this formalism a number of processes were calculated and the results are presented in sections 3 to 6.

2. CHANNEL WAVE FUNCTION

The channel capture cross section for radiative neutron transition from a scattering partial wave (ℓ, j) to a final state with spectroscopic factor S_f for a bound single particle orbit (ℓ_f, j_f) is[10])

$$\sigma^{ch}_{\ell j, f} = \frac{4}{3} \frac{k_\gamma^3}{\hbar v} \frac{\pi}{k^2} \sum_J \frac{(2J+1)}{2(2I+1)} <\ell j J \| D_I \| \ell_f j_f J_f>$$

$$(2J_f+1) S_f \, \overline{e}^{\,2} \frac{4}{\left|1-i K^J_{\ell j}\right|^2} \left| \int_R^\infty r \, U^{+J}_{\ell j}(r) \, U_{\ell_f j_f}(r) dr \right|^2, \qquad (1)$$

where

$$<\ell j J \| D_I \| \ell_f j_f J_f> = (2\ell+1)(2j+1)(2j_f+1)$$

$$\left[C^{\ell_f 0}_{\ell 0 1 0} \, W(\ell \, j \, \ell_f j_f, \frac{1}{2} \, 1) \, W(j \, J \, j_f J_f, I \, 1) \right]^2 ; \qquad (2)$$

$U_{\ell_f j_f}(r)$ is the radial wave function of single particle in state f. The reader is referred to the original article for nomenclature[10].

For the region outside the zone of nuclear interaction, two slightly different forms of the radial wave function for the entrance channel are in current use. These are

$$U^J_{\ell j}(r) = P_{\ell j}(r) - S^J_{\ell j} \, Q_{\ell j}(r), \qquad (3)$$

and

$$U^{+J}_{\ell j}(r) = J_{\ell j}(r) + K^J_{\ell j} \, N_{\ell j}(r). \qquad (4)$$

They are related by

$$U^J_{\ell j}(r) = \frac{-2i}{1 - i K^J_{\ell j}} U^{+J}_{\ell j}(r), \qquad (5)$$

where $S^J_{\ell j}$ is the scattering matrix and $K^J_{\ell j}$ the reactance matrix; $J_{\ell j}(r)$, $N_{\ell j}(r)$, $P_{\ell j}(r)$, $Q_{\ell j}(r)$ are related to spherical Bessel, Neumann and Hankel functions. In the resonance region we have

$$K^J_{\ell j} = K^{hs}_{\ell j} + \frac{1}{2} \sum_{i(J)} \frac{\Gamma^{+\ell j}_{ni}}{E^+_i - E} + \frac{1}{2} \sum_{\lambda(J)} \frac{\Gamma^{+\ell j}_{n\lambda}}{E^+_\lambda - E - \frac{i}{2} \Gamma^+_\lambda} \qquad (6)$$

there hs, i and λ designate respectively hard sphere, distant resonances and near resonances. The superscript + designates resonance parameters in the K representation[4,10]. For calculations of cross section we will use experimentally measured values of these parameters and consequently in the rest of the paper we drop this superscript. We note from the relationship

$$\left\langle \sum_{\lambda(J)} \frac{\Gamma^{\ell j}_{n\lambda}}{E_\lambda - E - \frac{i}{2}\Gamma_\lambda} \right\rangle = i\pi \frac{\langle \Gamma^{\ell j}_{n\lambda} \rangle_J}{D^J}, \qquad (7)$$

that

$$\text{Re}\langle K^J_{\ell j} \rangle = K^{hs}_{\ell j} + \frac{1}{2} \sum_{i(J)} \frac{\Gamma^{\ell j}_{ni}}{E_i - E}. \qquad (8)$$

Then[10]

$$U^{+J}_{\ell j}(r) = \text{Re}\langle U^{+J}_{\ell j}\rangle + \frac{1}{2} \sum_{\lambda(J)} \frac{\Gamma^{\ell j}_{n\lambda}}{E_\lambda - E - \frac{i}{2}\Gamma_\lambda} \frac{\text{Im}\langle U^{+J}_{\ell j}(r)\rangle}{\text{Im}\langle K^J_{\ell j}\rangle} \qquad (9)$$

Now we turn to calculate the wave function in the interior region. Inside the nucleus, the wave function $\phi^J_c(E)$ for a state of total energy E, spin J for incident channel c ($c \equiv \ell, j$) is written as

$$\phi^J_c(E) = \hbar^{1/2} \sum_q \frac{\Gamma^{1/2}_{qc} \chi^J_q}{E_q - E - \frac{i}{2}\Gamma_q}, \qquad (10)$$

where χ^J_q is the wave function of the compound state q. We can expand χ^J_q in a set of configuration wave functions

$$\chi^J_q = \sum_{\nu'c'} a^q_{\nu'c'} v_{\nu'}(r) \phi^J_{c'}, \qquad (11)$$

where $\phi^J_{c'}$ is the intrinsic channel wave function containing the internal excitation and $v_{\nu'}(r)$ the radial wave functions defined by the average potential between the incident particle and the target nucleus. The index ν' denotes single particle states of a specific ℓ. Substituting eq. (11) in eq. (10), and for the elastic channel c, the radial wave function for the interior region is

$$U^{+J}_c(r) = A_c \sum_{q(J)} \sum_{\nu'} \frac{\Gamma^{1/2}_{qc} a^q_{\nu'c} v_{\nu'}(r)}{E_q - E - \frac{i}{2}\Gamma_q}, \qquad (12)$$

where A_c is a normalization factor; the superscript + is used just for similarity with the notation used in eq. (9).

According to the intermediate interaction model[12], the contribution of the distant resonances to the nuclear wave function at a sharp energy corresponds to the real part of the optical model wave function, and the distribution for different levels ν' are essentially non-overlapping. Thus summation of eq. (10) can be separated into two parts. One over distant resonances with net sum equal to $\text{Re}\langle U_c^{+J}(r)\rangle$ and the other over local resonances of spin J. Then using the relationship[1]

$$\Gamma_{\lambda c}^{1/2} = \left(\frac{\hbar^2 R P_c}{M}\right)^{1/2} a_{\nu c}^{\lambda} v_{\nu}(R) , \qquad (13)$$

and eq. (7), we obtain

$$U_c^{+J}(r) = \text{Re}\langle U_c^{+J}(r)\rangle + \frac{1}{2} \sum_{\lambda(J)} \frac{\Gamma_{\lambda c}}{E_\lambda - E - \frac{i}{2}\Gamma_\lambda} \frac{\text{Im}\langle U_c^{+J}(r)\rangle}{\text{Im}\langle K_c^J\rangle} \qquad (14)$$

For a channel $c = (\ell,j)$ the partial neutron width $\Gamma_{\lambda c}$ in eq. (14) and $\Gamma_{n\lambda}^{\ell j}$ in eq. (9) are identical. Thus eq. (9) is formally the unified expression for the channel wave function applicable to all regions.

Figure 1 shows the variation of the function

$$N_{\ell j}(r) = \frac{\text{Im}\langle U_{\ell j}^{+J}(r)\rangle}{\text{Im}\langle K_{\ell j}^J\rangle} \qquad (15)$$

with r for ^{40}Ca target at neutron energy of E = 1 MeV. The optical model parameters of ref. 11 were used in these calculations. For comparison, the asymptotic functions $-kr\,\eta_\ell(kr)$ are also shown in the same figure.

3. AVERAGED ELASTIC CHANNEL RADIATIVE NEUTRON CAPTURE CROSS SECTION

To calculate the average of eq. (1) over incident energy, we define the fluctuation of the wave function as[10]

$$\Delta U_{\ell j}^J(r) = U_{\ell j}^J(r) - \langle U_{\ell j}^J(r)\rangle = \Delta S_{\ell j}^J Q_{\ell j}(r) , \qquad (16)$$

where $Q_{\ell j}(r)$ is defined by[10]

$$Q_{\ell j}(r) = \frac{\text{Im}\langle U_{\ell j}^{+J}(r)\rangle}{\text{Im}\langle K_{\ell j}^J\rangle} + i\left[\text{Re }\langle U_{\ell j}^{+J}(r)\rangle - \frac{\text{Re}\langle K_{\ell j}^J\rangle}{\text{Im}\langle K_{\ell j}^J\rangle}\text{Im}\langle U_{\ell j}^{+J}(r)\rangle\right] \qquad (17)$$

In terms of the statistical theory[13] of nuclear reactions

$$\langle |\Delta S^J_{\ell j}|^2 \rangle = \frac{T^J_{\ell j} T^J_{\ell j}}{T^J} W_{\ell j \ell j}, \quad (18)$$

where T^J and $T_{\ell j}$ are the total and neutron partial transmission coefficients, respectively, and the $W_{\ell j \ell j}$ are the width fluctuation correction factors. Then we obtain the resonance-averaged elastic channel radiative neutron capture cross section

$$\langle \sigma^{ch}_f \rangle = \frac{4}{3} \frac{k^3_\gamma}{\hbar v} \frac{\pi}{k^2} \sum_{\ell j J} \frac{(2J+1)}{2(2I+1)} \langle \ell j J \, \|D_I\| \, \ell_f j_f J_f \rangle$$

$$(2J_f+1) S_f \, \bar{e}^2 \, (R^{J(B)}_{\ell j \ell_f j_f} + R^{J(V)}_{\ell j \ell_f j_f} + R^{J(C)}_{\ell j \ell_f j_f}). \quad (19)$$

In eq. (19)

$$R^{J(B)}_{\ell j \ell_f j_f} = \frac{4}{|1-i\langle K^J_{\ell j}\rangle|^2} \left| \int_R^\infty r \, \mathrm{Re}\langle U^{+J}_{\ell j}(r)\rangle \, U_{\ell_f j_f}(r) dr \right|^2 \quad (20)$$

contains the hard sphere and distant resonance contributions. The second term

$$R^{J(V)}_{\ell j \ell_f j_f} = \frac{4}{|1-i\langle K^J_{\ell j}\rangle|^2} \left| \int_R^\infty r \, \mathrm{Im}\langle U^{+J}_{\ell j}(r)\rangle \, U_{\ell_f j_f}(r) dr \right|^2 \quad (21)$$

involves the contribution from the averaged near resonance wave functions, and the fluctuation term is

$$R^{J(C)}_{\ell j \ell_f j_f} = \langle |\Delta S^J_{\ell j}|^2 \rangle \left| \int_R^\infty r \, Q_{\ell j}(r) \, U_{\ell_f j_f}(r) dr \right|^2. \quad (22)$$

By using the terminology of the optical model, the sum of the first two terms in eq. (19) can be regarded as due to the capture of the shape elastic scattering wave, and the third (a fluctuation term) as due to the radiative capture of the compound elastic scattering wave.

The energy dependence and the relative magnitudes of the three terms are as following:

at $kr < 1$,

$$\langle \sigma^{ch(B)}_{\ell j,f} \rangle \approx \langle \sigma^{ch(V)}_{\ell j,f} \rangle \propto k^{2\ell-1}, \quad (23)$$

$$\langle \sigma_{\ell j,f}^{ch} \rangle^{(C)} \propto k^{2\ell-1}, \quad \text{for } T_\gamma \gg T_{\ell j}, \tag{24}$$

$$k^{-2}, \quad \text{for } T_\gamma \ll T_{\ell j},$$

$$\frac{\langle \sigma_{\ell j,f}^{ch} \rangle^{(C)}}{\langle \sigma_{\ell j,f}^{ch} \rangle^{(B)}} \approx \frac{D^J}{\langle \Gamma_\lambda \rangle_J} \tag{25}$$

where D^J is the average level space.

Figures 2 and 3 give calculated results for the averaged elastic channel radiative capture cross sections for E1 transitions in ^{56}Fe, ^{74}Ge, ^{90}Zr, ^{138}Ba, and ^{200}Hg. Those results show clearly the dominant role of the fluctuation term in the capture cross section, and the energy dependence of the cross sections as discussed above.

4. RADIATIVE NEUTRON CAPTURE IN THE INELASTIC CHANNEL OF COMPOUND NUCLEUS

It is of interest to extend the idea and approach developed above to calculate the resonance averaged inelastic channel radiative neutron capture cross section. The result is[14]

$$\langle \sigma_f^{ich} \rangle = \frac{4}{3} \frac{k_r^3}{\hbar v'} \frac{\pi}{k^2} \sum_{\ell' j' J} \frac{(2J+1)}{2(2I+1)} \langle \ell' j' J \| D_I \| \ell'_f j'_f J_f \rangle$$

$$(2J_f+1) \, S_f' \, \overline{e}^2 \, \left| Q_{\ell' j', \ell'_f j'_f} \right|^2 \, \frac{T_{\alpha \ell j} T_{\beta \ell' j'}}{T^J} \, W_{\alpha \ell j \beta \ell' j'}^J \tag{26}$$

We use the superscript prime to designate a quantity in the inelastic channel. The inelastic channel wave function can be written as

$$\phi_{\ell' j' I'}^{JM} = \left[Y_{\ell' j'} \times \phi_{iI'} \right]^{JM} \tag{27}$$

It consists of the intrinsic excited state wave function, $\phi_{iI'}$, of the residual nucleus coupled to the neutron angular momentum wave function $Y_{\ell' j'}$ to give total spin J and projection M. The final state wave function can be written as

$$\psi_f = \frac{U_{\ell'_f j'_f}(r)}{r} \phi_{\ell'_f j'_f I'}^{J_f M_f} \tag{28}$$

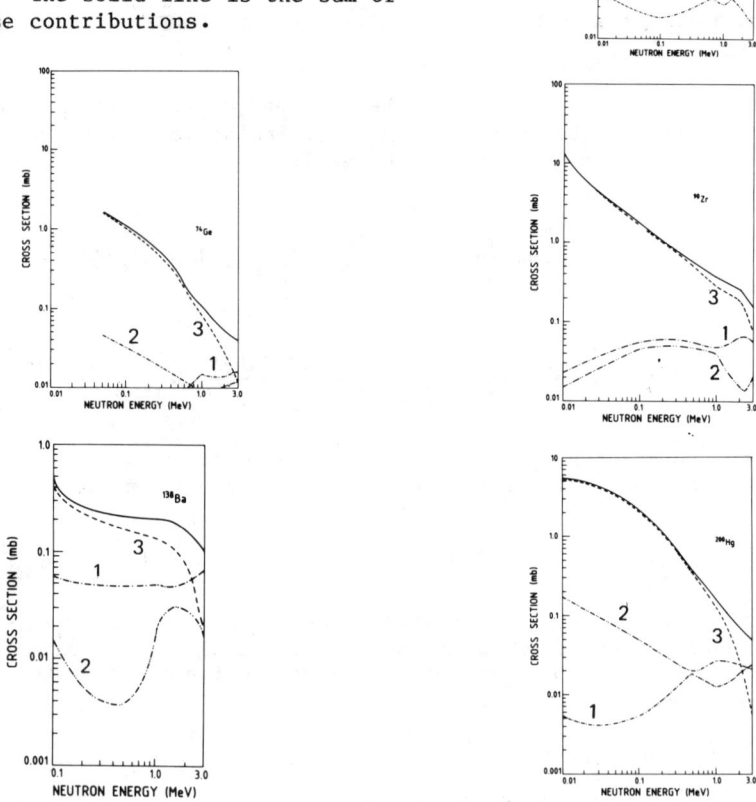

Fig. 1. The solid lines show the function $N_{\ell j}(r)$ (eq.15), for $(\ell,j) = (0,1/2),(1,1/2),(2,3/2)$ for the target nucleus ^{40}Ca at E=1 MeV. The dash lines give its asymptotic form $-kr\eta_\ell(kr)$.

Fig. 2. Averaged elastic channel capture cross sections for E1 transition in ^{56}Fe as a function of neutron energy. The line marked 1 is the background (eq. (20)). Line 2 is the contribution from eq. (21). Line 3 is the fluctuation term (eq. 22). The solid line is the sum of these contributions.

Fig. 3. The same conventions as in Fig. 2, but for the target nuclei ^{74}Ge, ^{90}Zr, ^{138}Ba and ^{200}Hg.

$U_{\ell'_f j'_f}(r)$ is the radial wave function of single particle bound state in a potential relevant to the excited residual nucleus. The radial integral in eq. (26) is

$$Q_{\ell'j'\ell'_f j'_f} = \int_R^\infty Q_{\ell'j'}(k'r) \, U_{\ell'_f j'_f}(r) \, rdr \tag{29}$$

where $Q_{\ell'j'}(k'r)$ has the form of eq. (17). Formally, $\langle U_{\ell'j'}^{+J}(r) \rangle$ and $\langle K_{\ell'j'}^{J} \rangle$ should be calculated in an optical model potential corresponding to the excited target. However, there is no conclusive experimental evidence that this potential is much different from the potential for the ground state. Thus following the general practice we use the optical model potential for the ground state with an effective neutron kinetic energy of $E-E_i$, where E_i is the excited-state energy.

In deriving eq. (26) we have ignored the pre-equilibrium process. So the cross section given by eq. (26) can be regarded as due to the radiative capture of the compound inelastic scattering wave.

Calculated results of the inelastic channel capture cross sections (eq. (26)), together with the elastic channel capture cross sections (eq. (19)) and the compound nucleus radiative capture cross sections for ^{56}Fe, ^{74}Ge, ^{90}Zr, ^{138}Ba and ^{200}Hg are given in Figs. 4 and 5. We note that in heavier nuclei the inelastic channel capture cross sections at higher neutron energies may be larger than the elastic channel capture cross sections.

5. INTERFERENCE EFFECT IN THE CHANNEL CAPTURE PROCESS

The channel wave function in eq. (9) was derived for a sharp energy. The first term in eq. (9) contains contributions from hard-sphere and distant-resonances. This is usually called the potential elastic scattering wave. The second term in eq. (9) corresponds to the nearby resonances, and can be regarded as the resonance scattering wave. It is of interest to explore the interference effect between the two components.

First we rewrite eq. (1) with different parameters[15]

$$\sigma_{\ell j,f}^{ch} = \sigma_{\ell j,f}^{po} \left[1 + \frac{\sum_{\lambda(J)} \frac{\Gamma_{n\lambda}^{\ell j}}{E_\lambda - E - \frac{i}{2}\Gamma_\lambda} \int_R^\infty rN_{\ell j}(r) U_{\ell_f j_f}(r) dr}{2 \int_R^\infty r \, \mathrm{Re}\langle U_{\ell j}^{+J}(r) \rangle \, U_{\ell_f j_f}(r) dr} \right]^2, \tag{30}$$

where

Fig. 4. Predicted radiative capture cross sections as a function of neutron energy in ^{56}Fe. The solid line is the inelastic channel capture cross section (eq. 26) calculated for energies above the single particle inelastic channel threshold. The dash-dot line is the elastic channel capture cross section (eq. 19). The dash line is the prediction of statistical theory of compound nucleus. The dash-dot-dot line is the sum of these three cross sections.

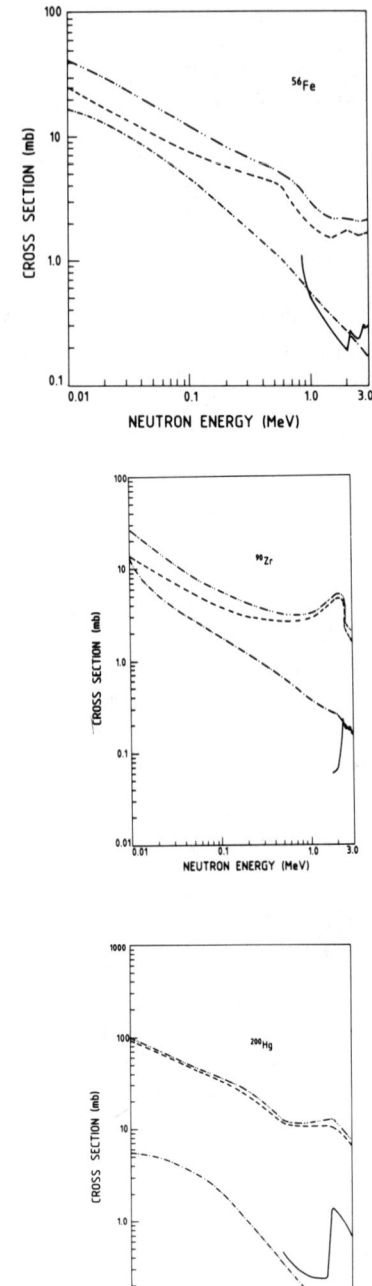

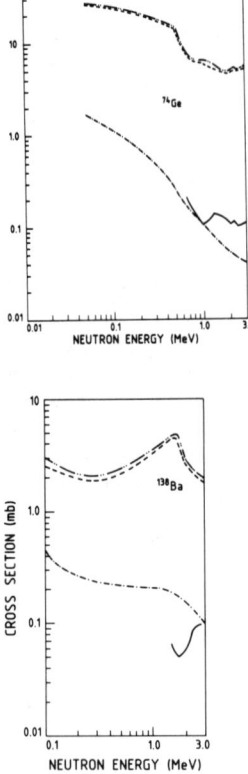

Fig. 5. The same conventions as in Fig. 4, but for the target nuclei ^{74}Ge, ^{90}Zr, ^{138}Ba, ^{200}Hg.

$$\sigma^{po}_{\ell j, f} = \frac{16}{3} \frac{k_\gamma^3}{\hbar v} \frac{\pi}{k^2} \frac{(2J+1)}{2(2I+1)} \langle \ell j J \| D_I \| \ell_f j_f J_f \rangle \cdot$$

$$(2J_f+1) S_{dp,f} \overline{e}^2 \left[\int_R^\infty r \ \text{Re} \langle U^{+J}_{\ell j}(r) \rangle \ U_{\ell_f j_f}(r) dr \right]^2 \tag{31}$$

is the potential capture cross section.

At thermal energy we are interested in the capture of an s-wave incident neutron into a bound p-wave orbit. From the relation

$$a^J = R' - \frac{1}{2k} \sum_{\lambda(J)} \frac{\Gamma_{n\lambda}^{\rho \ 1/2}}{E_\lambda - E - \frac{i}{2} \Gamma_\lambda} \quad,$$

where a^J is the coherent scattering length and R' the potential scattering length, we obtain

$$\sigma^{ch}_{0\frac{1}{2},f} = \sigma^{po}_{0\frac{1}{2},f} \left[1 - (a^J - R')k \frac{\int_R^\infty r \ N_{0\frac{1}{2}}(r) \ U_{\ell_f j_f}(r) dr}{\int_R^\infty r \ \text{Re} \langle U^{+J}_{0\frac{1}{2}}(r) \rangle U_{\ell_f j_f}(r) dr} \right]^2 . \tag{32}$$

Equation (32) has the same structure as the well-known Lane-Lynn formula[16] with the difference that in eq. (32) the contribution of distant resonances is contained in $\sigma^{po}_{0\frac{1}{2},f}$, which can be calculated using the optical model wave function; the interference factor $a^J - R'$ may be determined from the known parameters of nearby resonances.

We present two examples. One, thermal neutron capture in ^{12}C, shows destructive interference; the other, thermal neutron capture in ^{40}Ca, shows constructive interference. Both predicted and measured results are given in Table 1.

Let us examine the cause of these interference phenomena further. In Fig. 6 we show the radial integrands of the potential capture, $\left(r \ \text{Re} \langle U^{+J}_{\ell j}(r) \rangle \ U_{\ell_f j_f}(r) \right)$ and the channel capture, $\left(r \ U^{+J}_{\ell j}(r) \ U_{\ell_f j_f}(r) \right)$ in ^{12}C and ^{40}Ca. In ^{12}C, which lies above the 2s size resonance, the nearby strong resonance at negative energy (-2.02 MeV with reduced neutron width 990 ev) pushes the crossing point of the wave function along the abscissa from R' to a^J. This leads to the areas above and below the abscissa enclosed by the integrand being approximately equal, resulting in a drastic cancellation in the radial integral. In contrast, in the case of ^{40}Ca, which lies below the 3s size resonance, the nearby neutron s-wave resonances are at positive energy. Therefore $a^J < R'$, and the areas above and below the abscissa become unequal, resulting in an enhancement of the integral.

Since the channel capture cross section comes mainly from the contribution of the radial integral near nuclear surface, it is generally believed that calculations with a realistic nuclear interaction provide better results than the square-well potential model. However, there is difficulty in using eq. (1) to calculate a partial channel capture cross section to a final state whose observed binding energy is much different from the binding energy of the associated single particle orbit calculated in the potential model. This stems from the problem of determining the single particle wavefunction for the physical final state.

Table 1. Comparison of theoretical and measured radiative capture cross sections and scattering cross sections for ^{12}C and ^{40}Ca at thermal energy.

	^{12}C	^{40}Ca
Experimental (n,γ) cross sections (mb)	2.38 ± 0.05	410
Calculated (n,γ) cross sections (mb)	2.1	414
Experimental (n,n) cross section (b)	4.7	3.01
Calculated (n,n) cross section (b)	4.75	2.73

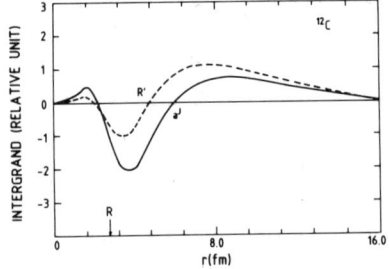
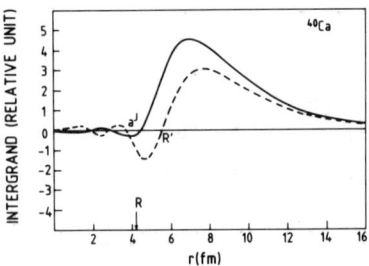

Fig. 6. Radial integrands in the radiative capture of an s-wave neutron into the $P_{1/2}$ orbit in ^{12}C and ^{40}Ca at thermal energy. The dashed lines are the integrands of potential capture $(r \, \text{Re}\langle U^{+J}_{\ell j}(r)\rangle \, U_{\ell_f j_f}(r))$ in eq. (31). The solid lines are the integrands of channel capture $(r \, U^{+J}_{\ell j}(r) \, U_{\ell_f j_f}(r))$ in eq. (1).

6. LASER INDUCED RADIATIVE NEUTRON CAPTURE PROCESS

There is current interest in investigating laser-stimulated nuclear reactions and transition rates. Many suggestions and proposals on this topic have been put forward. Here we examine the probability of p-wave radiative capture of a low energy neutron in the presence of a strong laser field[17,18].

In the giant resonance region of the p-wave neutron strength function (e.g. 3p region near A ≈ 90) there are most likely nuclei with p-wave resonances near the neutron binding energy with large reduced widths. However these are not observable because of the angular momentum barrier. Furthermore, in some of these nuclei there are levels near the ground state with large s-wave neutron single particle components. Then in the presence of a strong laser radiation field, an incident low-energy s-wave neutron may populate a nearby p-wave resonance resulting in an enhanced E1 transition to the low lying s-wave neutron bound state. This process can be identified by comparing the intensity of the primary γ-ray transition in the presence of the laser field (E1) with the intensity in the absence of the laser field (M1).

We have investigated this process theoretically using second-order perturbation theory to calculate the cross sections. A set of discrete resonances with large p-wave reduced widths was used as the intermediate states. The partial capture cross section to a final state f can be written in the form

$$\sigma_{if}^{(LVR)} = \frac{\pi}{k^2} \sum_{\lambda' J} \frac{(2J+1)}{2(2I+1)} \frac{\Gamma_{\lambda' n}^{(L)} \Gamma_{\lambda' \gamma f}^{(V)}}{(E_{\lambda'} - E \pm \hbar\omega)^2 + \frac{\Gamma_{\lambda'}^2}{4}}, \qquad (33)$$

where $\Gamma_{\lambda' vf}^{(V)}$ is the valence radiative capture width of an intermediate state, λ', to the final state $f^{10)}$.

$$\Gamma_{\lambda' n}^{(L)} = \frac{4}{9} \frac{1}{\hbar v} e_f^2 \varepsilon^2 \left| A_{\ell' j'}^{\lambda'} \right|^2 \frac{1}{\left|1 - i K_{\ell j}^J\right|^2} (2j'+1)(2J'+1) \cdot$$

$$W^2(j'J'jJ, I1) \left| Q_{\ell' j' \ell j}^{(B)} + \frac{1}{2} \sum_{\lambda(J)} \frac{\Gamma_{n\lambda}}{E_\lambda - E - \frac{i}{2}\Gamma_\lambda} Q_{\ell' j' \ell j}^{(R)} \right|^2 \qquad (34)$$

can be regarded as the laser induced effective p-wave neutron width of λ'. ε is the electric field strength of laser radiation field, and ω the laser photon circular frequency. We use superscript primes to designate quantities belonging to the intermediate state λ', and λ for the s-wave resonance in the entrance channel. In eq. (34) $Q_{\ell' j' \ell j}^{(B)}$ and $Q_{\ell' j' \ell j}^{(R)}$ are radial integrals corresponding to the two terms in eq. (9) and

$$A_{\ell' j'}^{\lambda'} = \left[\frac{M}{\hbar^2} \frac{\sqrt{E_{\lambda'}(eV)}}{k_{\lambda'}} \Gamma_{\lambda' n}^{\ell'}\right]^{1/2} ; \qquad (35)$$

where M is the neutron mass.

We define a laser induced enhancement factor as the ratio of the probability of excitation of an intermediate p-wave state via two step process (capture of s-wave neutron plus E1 laser photon) to the probability of p-wave capture of the incident neutron itself. This factor can be estimated by

$$F = \sim \frac{\Gamma^{(L)}_{\lambda'n}}{\Gamma_{\lambda'n}} = \frac{4}{9} \frac{1}{\hbar v} e_f^2 \varepsilon^2 \frac{M}{\hbar^2} \frac{\sqrt{E_{\lambda'}(eV)}}{k_{\lambda'}} \frac{1}{|1-i K^J_{\ell j}|^2} \cdot$$

$$(2j'+1)(2J'+1)W^2(j'J'jJ,I1) \frac{1+(kR)^2}{(kR)^2} \frac{1}{\sqrt{E(eV)}} \cdot$$

$$\left| Q^{(B)}_{\ell'j'\ell j} + \frac{1}{2} \sum_{\lambda(J)} \frac{\Gamma_{n\lambda}}{E_\lambda - E - \frac{i}{2}\Gamma_\lambda} Q^{(R)}_{\ell'j'\ell j} \right|^2 . \quad (36)$$

where

$$\Gamma_{\lambda'n} \underset{(\text{for } \ell'=1)}{=} \frac{(kR)^2}{1+(kR)^2} \sqrt{E(eV)} \; \Gamma^{\ell'}_{\lambda'n} \quad (37)$$

is the p-wave neutron width.

The following data were used to estimate the order of magnitude of eq. (36): E_f = 5 MeV, E = 0.024 eV, $\hbar\omega$ = 1 eV, A = 100, Z = 42, S_f = 1. Then by ignoring $Q^{(R)}_{\ell'j'\ell j}$ in eq. (36), we get

$$F^{(B)} = \frac{2}{9}(2J'+1) \frac{e_f^2 \varepsilon^2}{\hbar^2 v k^4} \left[\frac{M}{2 E_{\lambda'}} \frac{E_{\lambda'}(eV)}{E(eV)}\right]^{1/2}$$

$$\sim 5.2 \times 10^{-16} [\varepsilon(V/cm)]^2 . \quad (38)$$

If the incident energy just matches an s-wave resonance, in other words, if there exists a p-wave and an s-wave resonance separated by an amount approximately equal to the laser photon energy, then the enhancement factor becomes

$$F^{(R)} \simeq F^{(B)} \left(\frac{\Gamma_{n\lambda}}{\Gamma_\lambda}\right)^2 \left[\frac{Q^{(R)}_{\ell'j'\ell j}}{Q^{(B)}_{\ell'j'\ell j}}\right]^2 \simeq 1 \times 10^{-10} [\varepsilon(V/cm)]^2 \quad (39)$$

The above results show that no appreciable enhancement is obtained until the laser radiation field exceeds a value in the range 10^5-10^8 V/cm. The lower value is for the rare case where the energies of an s-wave and a p-wave resonance differ by the laser photon energy. For cases with no resonance in the entrance channel one needs a radiation field of greater than 10^8 V/cm. In view of the fact that the breakdown threshold for ionization of an ordinary gas target is usually of the order of 10^5 V/cm, experimental investigation of this phenomena may be very difficult.

REFERENCES

1. A.M. Lane & J.E. Lynn, Nucl. Phys. 17(1960)563;586.
2. J.E. Lynn, The theory of neutron resonance reaction (Clarendon, Oxford, 1968).
3. B.J. Allen & A.R. de L. Musgrave, Advances in nuclear physics, Vol. 10, (Plenum, New York, 1978) p.129.
4. J. Cugnon & C. Mahaux, Ann. of Phys. 94(1975)128.
5. A.M. Lane & S.F. Mughabghab, Phys. Rev. C10(1974)412.
6. S.F. Mughabghab & R.E. Chrien, Neutron capture gamma-ray spectroscopy (Plenum, New York, 1978) p.265.
7. G.A. Bartholomew, E.D. Earle, A.J. Ferguson, J.W. Knowles & M.A. Lone, Advances in nuclear physics, Vol. 7, (Plenum, New York, 1973) p.24.
8. M.A. Lone, Neutron capture gamma-ray spectroscopy (Plenum, New York, 1978) p.161.
9. S. Raman, R.F. Carlton, J.C. Wells, E.T. Jurney & J.E. Lynn, to be published in Phys. Rev. C.
10. Y.K. Ho and M.A. Lone, Nucl. Phys. A406(1983)1.
11. F.D. Becchetti & G.W. Greenless, Phys. Rev. 182(1969)1190.
12. A.M. Lane, G.R. Thomas & E.D. Wigner, Phys. Rev. 98(1955)693.
13. E. Vogt, Advances in Nuclear Physics, Vol. 1, (Plenum, New YOrk, 1968) p.261.
14. J.F. Liu, Y.K. Ho et al., this conference.
15. Y.K. Ho and M.A. Lone, Nucl. Phys. A406(1983)18.
16. S.F. Mughabghab, Phys. Lett. 81(1979)93.
17. D.F. Zaretskii & V.V. Lomonosov, Sov. Phys. JETP 54(1981)229; JETP Letter 30(1979)508.
18. Y.K. Ho, F.C. Khanna & M.A. Lone, this conference.

STUDY OF THERMAL NEUTRON CAPTURE BY ^{32}S

Guo Taichang, Shi Zongren, Zeng Xiantang,
Li Guohua and Ding Dazhao

Institute of Atomic Energy, Academia Sinica
P.O.Box 275, Beijing, China

The thermal neutron capture by ^{32}S has been studied using a 140 cm^3 Ge(Li) detector at the thermal column of the heavy water moderated reactor at Institute of Atomic Energy. The thermal neutron flux at sample is 2×10^6 n/cm^2-sec and the cadmium ratio (for gold) is 200.

Natural S is used as the sample and in order to determine the absolute value of transition cross sections, the well known strong primary transition intensities of Au(n,γ) is used as the standard, so the sandwich sample Au-S-Au-S-Au is used for this purpose. The thickness of the Au foil is 14 micron. The ^{14}N(n,γ) and ^{35}Cl(n,γ) is used for calibration of the detector efficiency in energy range 1.6-10.8Mev and a ^{152}Eu source is used for lower energy region down to 200Kev. The common line observed on S, sandwich, N, Cl samples and the empty teflon container were recognized as background.

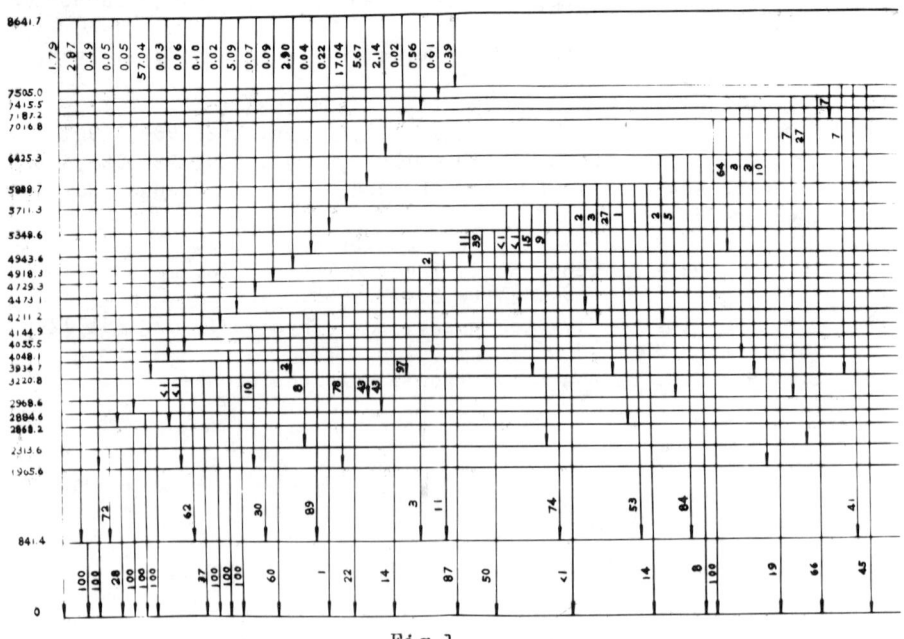

Fig.1

86 transitions from $^{32}S(n,\gamma)$ reaction were identified, among them 23 wre recognized as primary transitions. By normalization to the $Au(n,\gamma)$ cross section 98.8±0.3 barn, the sum of the partial cross sections of 23 primary transitions is 515mb and the sum of the partial cross sections of transitions to the ground state is 529mb,which is in good agreement with each other. The minimum intensity observed in this experiment is 0.1mb.

The level scheme of ^{33}S consisted of 25 levels shown as in Fig.1 is deduced from the transition energy corrected for the recoil correction using the Ritz principle. Among these 25 levels, three at 7505.0, 4473.1 and 2884.6Kev were not reported priviousely. (1) The decay branching ratio is in agreement with Endt's compilation in general except the 5888.7Kev level, which mainly decays to ground state($3/2^-$), first excited state(841.4, $1/2^-$) and the 3220.8Kev($3/2^-$) level with the branching ratio 14%, 53% and 27% respectively.

The neutron separetion energy of ^{33}S is determined as 8641.73Kev by averaging six two step cascade transitions.

Seven of the 23 primary transitions decaying to 3220.8, 4211.2, 4918.3, 5711.3, 5888.7, 6425.3 and 7187.2Kev levels are the E1 transitions and consiste of 90% of the cross section of $^{32}S(n,\gamma)$ reaction. The Lane-Lynn formula for direct capture is used to analyse these seven transitions. The result is shown in Tab.1. In calculation the spectroscopy factor of (d,p) reaction is taken from ref.(2), coherent scattering length a_{coh} is taken as 2.80fm and the nuclear radius is taken as $1.35A^{1/3}$fm. It is seen from the table, that the agreement is quite notable. It is conclude that the $^{32}S(n,\gamma)$ reaction is a good representative of the direct capture machenism at thermal neutron energy.

Table 1

transition energy (Kev)	level (Kev)	cross section (mb)	
		measured	calculated
5420.4	3220.8	302	238
4430.2	4211.2	26.9	29.9
3723.2	4918.3	15.3	7.2
2930.3	5711.3	90.1	67.2
2752.9	5888.7	30.0	26.1
2216.3	6425.3	11.3	15.1
1454.5	7187.2	3.0	5.7

References
(1) P.M. Endt and C.Von Der Leun   Nucl Phy.A310,(1978)358
(2) M.C. Mermaz et al   Phys Rev.C4,(1971) 1778

^{74}Ge: TRANSITIONS AND LEVELS EXCITED IN THERMAL-NEUTRON CAPTURE

C Hofmeyr, C Franklyn
NUCOR, Pelindaba, Transvaal, South Africa

G Barreau, H Börner, R Brissot, H Faust, K Schreckenbach
Institut Laue-Langevin, Grenoble, France

ABSTRACT

Gamma-ray transitions due to thermal-neutron capture in ^{73}Ge were measured at ILL, Grenoble, using the curved-crystal spectrometers GAMS 1, 2 and 3, a pair spectrometer and a Ge(Li) spectrometer. Some 750 transitions were identified, of which 450 were placed in a level and decay scheme with the aid of an interactive program. Selected energy regions were scanned with the internal-conversion electron spectrometer BILL, yielding 18 transitions corresponding to γ-rays and sixteen unmatched candidates. The levels up to 4 MeV are presented together with the degree of corroboration obtained from published (p, t) and (t, p) results and β-decay data.

INTRODUCTION

In the previous NDS compilation, A=74 of 1976[1], the ^{74}Ge levels were based heavily on ^{74}Ga β^--decay, ^{74}As β^+-decay, some earlier (n, gamma) studies, and to some degree on various charged-particle reactions. Since then the main thrust in the spectroscopic study of even Ge nuclei has been from (p, t)[2,3] and (t, p)[4,5,6] reaction studies and DWBA fits to the data. ^{74}Ge is the only even-even Ge isotope that can be studied by (n, gamma). A comprehensive high resolution study was motivated in an attempt to consolidate and expand the decay scheme.

EXPERIMENTAL

The experiment, performed at the Institut Laue-Langevin, Grenoble, comprised studies with GAMS1 (20-400 keV), GAM2/3 (200-1500 keV), a 20 % Ge(Li) detector (100 keV - 3 MeV) and a pair spectrometer (1.8-10 MeV). The internal sample of some 20 mg ^{73}GeO$_2$ was enclosed in very pure aluminium, for which precise calibration energies have become available[7]. Some 750 transitions were observed in the energy region up to 10 MeV in ^{74}Ge. Al and C were used as internal calibration lines.

Selected scans were undertaken with the internal-conversion

electron spectrometer BILL, which produced 18 transitions with corresponding gamma rays and 16 unmatched E0 candidates. Due to the low Z (32) only very strongly converted lines could be discerned, with a corresponding influence on the precision.

ANALYSIS

All spectra were analysed using ILL in-house programs. An interactive mini-computer program GRAP, which has been developed at NUCOR, is capable of handling up to 1000 transitions and 256 levels and their respective experimental and statistical errors. This program gives one the ability to edit a level scheme and to monitor the effect of any changes on the level and decay scheme as a whole, including intensity balances. Automatic search routines can scan any energy region for levels in defined energy steps. Optimum level values are calculated during trial fits. An important feature is the high speed of execution.

RESULTS

In Table 1 are presented the levels up to 4 MeV excitation energy in ^{74}Ge observed in the present experiment. The table contains the level energies in keV and relative feed and decay intensities for each level. Levels also observed in β-decay are indicated by a '1' in the 'Beta' column. The column marked 'This' indicates levels fed directly from the capture state (C) or decaying directly to the ground state (0). In levels fed directly from the capture state at 10196.0 keV the spin is limited to $7 \geqslant J \geqslant 2$. Levels through which simple cascades from the capture state, $J^{\pi} = (4^+, 5^+)$ go to the ground state (0^+), are 2^+ with a high probability, as is mostly borne out by spin and parity allocations from charged-particle reactions which are indicated in Columns a[2], b[3], c[4] and d[5]. The normal uncertainty in the association of these levels pertains. Levels in other reactions which have not been seen in our work are not recorded in this contribution. For the 2165 keV level our work supports the 4^+ allocation against 1^- and 0^+. A level scan produced no other candidate levels nearby.

The balance between feed and decay intensities is satisfactory for most levels since the accuracy of the intensity calibration is only 10 - 20 % at best. The observed decay intensity of the capture state is only about 1/3 of the intensity reaching the ground state. Out of a total of 750 observed transitions about 450 representing $\approx$98 % of the observed intensity have been fitted to the decay scheme. Due regard was given to the proper placing of transitions observed in β-decay and in coincidence measurements[1].

74Ge LEVELS UP TO 4 MeV.

Level (keV)		Int Feed / Decay		Beta	This		a	b	c	d
0.000	(.000)	55.97 /	0.00	1			0+	0+	0+	0+
595.850	(.006)	56.39 /	49.90	1	C	0	2+	2+	2+	2+
1204.206	(.008)	15.93 /	16.24	1	C	0	2+	2+	2+	2+
1463.755	(.009)	23.55 /	29.52	1	C		4+	4+		
1483.048	(.059)	.26 /	.16	1			0+	0+	0+	0+
1697.140	(.010)	9.68 /	12.40	1	C					
1724.952	(.014)	.24 /	.27							
2165.256	(.012)	6.19 /	10.78	1	C		4+	1-	0+	
2197.921	(.037)	1.15 /	1.00	1	C	0	2+	2+	2+	2+
2227.758	(.117)	.36 /	.98						0+	0+
2536.199	(.094)	.97 /	2.20	1	C		3-	(3-)	3-	3-
2569.324	(.015)	1.05 /	4.11		C		4+			
2600.324	(.092)	.69 /	.31				(1-)		0+	
2669.690	(.049)	1.12 /	2.22		C		4+	3-	4+	4+
2693.730	(.100)	1.13 /	1.76	1	C					
2696.920	(.015)	.97 /	3.82							
2828.504	(.014)	1.13 /	2.84		C					
2835.900	(.065)	.49 /	.64		C		2+	(2)	2+	2+
2935.500	(.017)	3.60 /	5.78		C		2+	2+		(2+)
2949.121	(.105)	.11 /	.29	1					4+	
2973.440	(.050)	1.32 /	2.76		C	0				
3033.948	(.100)	.55 /	.56	1	C					
3048.573	(.050)	.70 /	1.04		C	0	4+		4+	4+
3081.340	(.045)	1.47 /	2.09		C					
3104.492	(.016)	1.69 /	2.30		C		5-	4+	5-	
3139.534	(.015)	.46 /	.50	1			3-	(2)	3-	
3175.460	(.099)	.77 /	.43	1	C					
3271.260	(.090)	.59 /	.55		C					
3315.739	(.039)	.11 /	.79		C					
3358.515	(.016)	.22 /	.23				5-	3-	0+	
3381.767	(.051)	.64 /	.69		C			5-		
3392.610	(.020)	.12 /	.75		C		3-		2	
3409.940	(.074)	.79 /	.95		C		2+			
3478.450	(.094)	.94 /	1.28	1	C					
3515.448	(.019)	.65 /	.69		C			4+		
3578.948	(.033)	.11 /	.20			0	2+	2+	2+	
3651.883	(.039)	.82 /	.62		C		2+			
3685.430	(.060)	.13 /	.12		C				5-	
3691.770	(.090)	.19 /	.17							
3707.103	(.100)	.21 /	.15		C		(2+)			
3720.790	(.110)	.20 /	.31	1	C					
3743.344	(.020)	.28 /	.29				2+			
3771.750	(.060)	.84 /	1.01				0+		0+	
3783.419	(.073)	.23 /	.29							
3790.720	(.095)	.31 /	.39		C					
3806.782	(.028)	.53 /	.58	1	C					
3831.978	(.060)	.38 /	.58		C				2+	
3835.277	(.065)	.82 /	.49		C					
3889.689	(.030)	.09 /	.31							
3898.060	(.096)	.28 /	.88							
3932.970	(.030)	.16 /	.12		C			3-		
3941.078	(.030)	.32 /	1.07		C					
3957.892	(.117)	.10 /	.30		C			3-	2+	
3975.832	(.065)	.18 /	.48	1						
3995.820	(.060)	.75 /	.76	1	C					
4022.968	(.060)	.47 /	.72		C		5-	5-	2+	

The sensitivity obtained improves appreciably on previous (n, gamma) studies. Below 4 MeV all but 20 transitions observed in β-decay of ^{74}Ga were seen in this experiment, although sometimes with very different intensities (e.g. the intensity of the 2353 keV transition is 50 % of the 595 keV transition intensity in β-decay, but only 0,5 % in (n, gamma).

The excited 0^+ levels posed a particular problem, being weakly excited in (n, gamma). The position regarding the fitting of the E0 candidate transitions has not been resolved satisfactorily, whereas all but one of the conversion electron lines matched by gamma-transitions (non-E0) have been placed satisfactorily.

REFERENCES

1. R Kocher, Nucl. Data Sheets 17, 519 (1976).

2. F Guilbault, D Ardouin, J Uzureau, P Avignon, R Tamisier, G Rotbard, M Vergnes, Y Deschamps, G Berrier, Phys Rev C16, 1840 (1977).

3. A Rester, J Ball, R Auble, Nucl Physics A346, 731 (1980).

4. S Lafrance, S Mordechai, H Fortune, R Middleton, Nucl Physics A307, 52 (1978).

5. C Lebrun, F Guilbault, D Ardouin, E Flynn, D Hanson, S Orbesen Phys Rev C19, 1224 (1979).

6. S Mordechai, H Fortune, M Carchidi, R Gilman, Phys Rev C29, 1699 (1984).

7. H Schmidt, P Hungerford, H Daniel, T von Egidy, S Kerr, R Brissot, G Barreau, H Börner, C Hofmeyr, Phys. Rev C25, 2888 (1982).

LOW-LYING STATES OF ^{94}Nb

M. Bogdanović
Boris Kidrič Institute of Nucl. Sciences, 11001 Belgrade,
Yugoslavia
H. Seyfarth,
Institut für Kernphysik, KFA Jülich, D-5170 Jülich, F.R. Germany
H.R. Börner, S. Kerr, F. Hoyler, K. Schreckenbach, G. Colvin
Institut Laue-Langevin, 38042 Grenoble, France

ABSTRACT

A preliminary level scheme of ^{94}Nb$_{53}$ has been established using the results from conversion electron and γ-ray measurements following thermal neutron capture. An attempt has been made to interpret part of the states in terms of proton-neutron multiplets using the lowest order one phonon exchange. The staggering is attributed to higher order multipoles.

The measurements of secondary γ-rays from the thermal neutron ^{93}Nb(n,γ) reaction have been performed with the bent crystal spectrometers GAMS 1 in the energy range $40 < E_\gamma < 500$ keV and GAMS 2/3 in the region $150 < E_\gamma < 1500$ keV. The conversion electron lines of the strongest transitions in ^{94}Nb have been measured with the electron spectrometer BILL. At the DIDO reactor in the KFA Jülich two singles gamma spectra from ^{94}Nb have been measured with 1.4 cm^3 Ge and 60 cm^3 Ge(Li) detectors for gamma energies up to 550 keV and 1350 keV, respectively.
Combining all results from the new measurements we have been able to revise the level scheme previously summarized in the Nucl. Data Sheets[1] and also obtained from a ^{94}Zr(p,nγ)^{94}Nb reaction study[2]. The level scheme of ^{94}Nb is shown in Figs. 1 and 2.
From the values of the level energies in ^{94}Nb and the primary gamma transition energies taken from ref. 1, a more precise value was obtained for the neutron binding energy: $B_n = 7229.13 \pm 0.12$ keV.
It was expected that the low-lying states of ^{94}Nb could be interpreted in terms of a quasiproton-quasineutron (cluster) multiplet structure. The proton-neutron multiplet structure is based on the quasi-proton $\tilde{g}_{9/2}^{+1}$, $\tilde{p}_{1/2}^{-1}$, $\tilde{p}_{3/2}^{-1}$ and $\tilde{f}_{5/2}^{-1}$ and on the neutron-cluster $(d_{5/2})^3{}_{5/2^+}$, $s_{1/2}(d_{5/2})^2{}_{1/2^+}$, $g_{7/2}(d_{5/2})^2{}_{7/2^+}$ configurations.
The 6^+, 3^+, 4^+, 7^+, 5^+ and 2^+ states at energies 0, 40.9, 58.7, 78.7, 113.4 and 334.1 keV are confirmed as the states of the $[\pi \tilde{g}_{9/2}^{+1}, \nu(d_{5/2})^3{}_{5/2^+}] \; 2^+, 3^+, 4^+, 5^+, 6^+, 7^+$ multiplet.

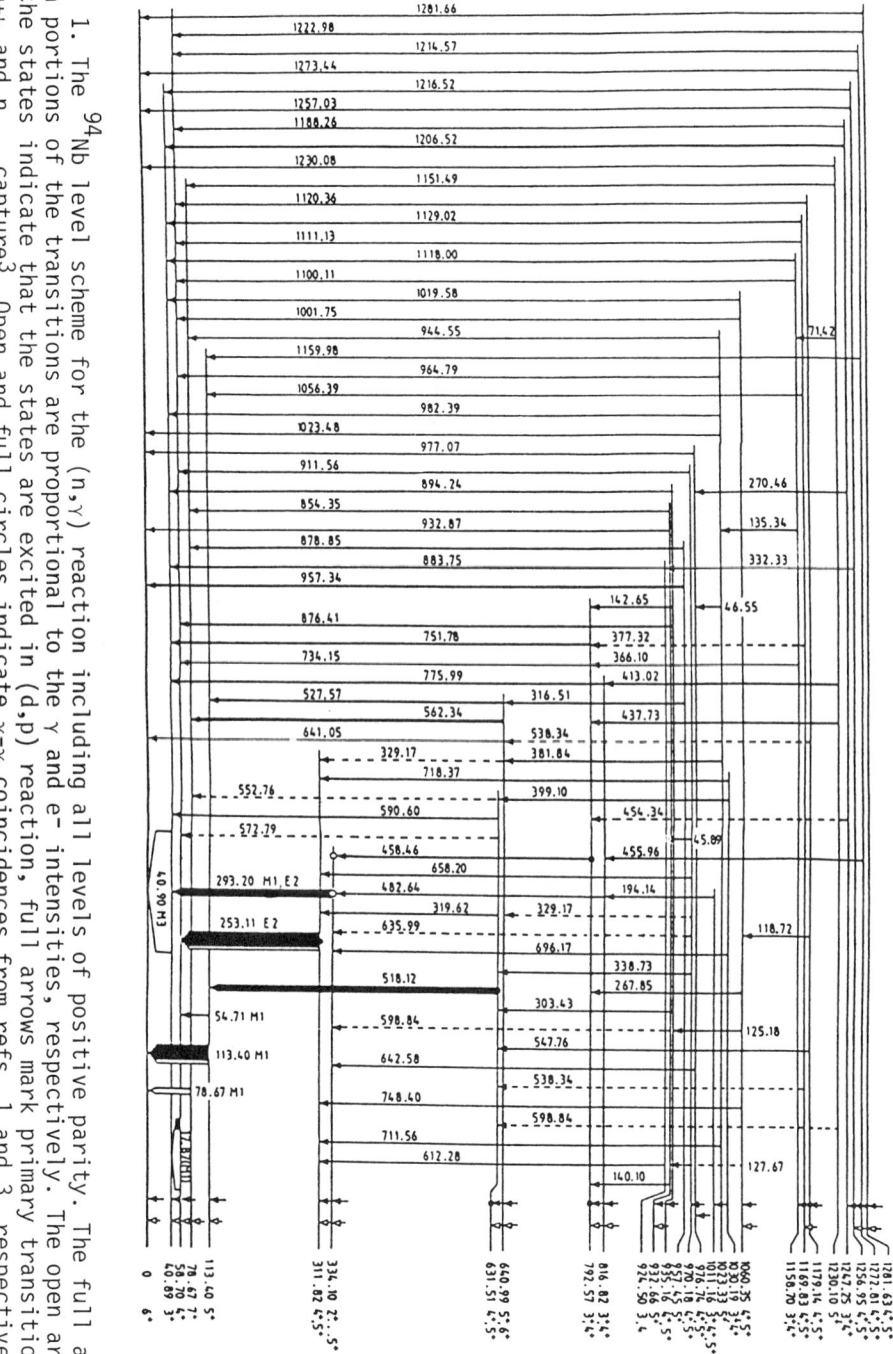

Fig. 1. The ^{94}Nb level scheme for the (n,γ) reaction including all levels of positive parity. The full and open portions of the transitions are proportional to the γ and e- intensities, respectively. The open arrows to the states indicate that the states are excited in (d,p) reaction, full arrows mark primary transitions in nth and nres capture3. Open and full circles indicate γ-γ coincidences from refs. 1 and 3, respectively.

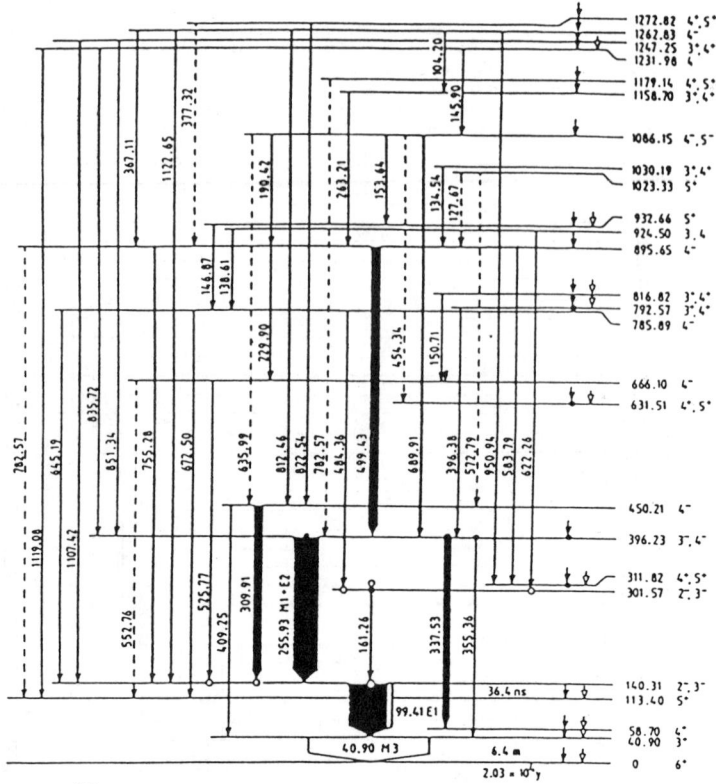

Fig.2. The ^{94}Nb level scheme from the (n,γ) reaction including all levels of negative parity (for explanations see caption of fig.1)

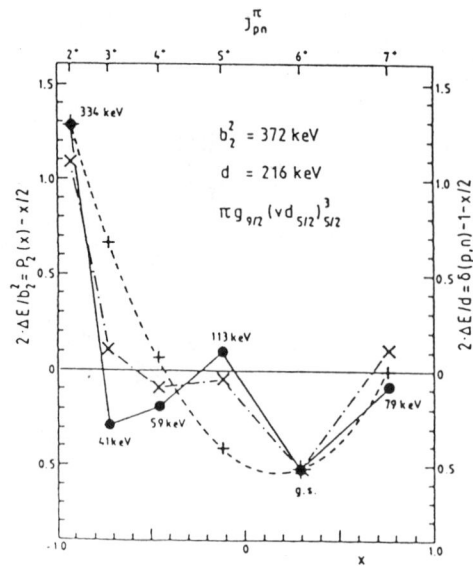

Fig.3. The g.s. multiplet in ^{94}Nb (● experimental energies, + lowest order quasiparticle-vibrational model, × short range residual interaction).

$$x = \frac{J_{pn}(J_{pn}+1) - j_p(j_p+1) - j_n(j_n+1)}{2(j_p(j_p+1)j_n(j_n+1))^{1/2}}$$

The 2^- and 3^- states at 140.3 and 396.2 keV seem to form the $[\pi\tilde{p}_{1/2}^{-1},\nu(d_{5/2})^3{}_{5/2}+]$ 2^-, 3^- doublet. There are some ambiguities in resolving and confirming the other multiplets. The resolved and confirmed ground state multiplet on average shows a parabolic structure of excitation energies with the parabola open upwards. The deviations from the parabola are pronounced exhibiting a staggering pattern.

In Fig. 3 we present the experimental and the calculated states of the $[\pi\tilde{g}_{9/2}^{+1},\nu(\tilde{d}_{5/2})^3{}_{5/2}+]$ multiplet. In the first calculation we use the lowest order quasiparticle vibrational coupling model and in the second the delta function reduced by the contribution of the linear term $\vec{\sigma}_1 \cdot \vec{\sigma}_2$ as a residual two body interaction.

The parabola resulting from the first calculation is open upwards and smoothly varying as a function of angular momentum. The curve from the second calculation shows a staggering pattern. The staggering pattern thus appears as a contribution of the higher multipoles.

The most promising procedure for the future is to take a suitable linear combination of the contributions of the particle vibrational and residual two-body interaction. The particle vibrational coupling even in the lowest order produces moderate polarization charges and accounts for the enhanced E2 transitions and quadrupole moments, while the residual interaction (higher multipoles) produces the staggering and the mixing of the multiplets allowing for M1 transitions and one particle transfer reactions.

1. Nucl. Data Sheets 14, 191 (1975)
2. I.D. Fedorits, Yu. P. Antufjev, I.I. Zabjubovskii, A.I. Popov, V.E. Storizhko, Izv. Akademii Nauk SSSR, Ser. Fiz. 40, 1260 (1976)
3. E.J. Jurney, H.T. Motz, R.K. Sheline, E.B. Shera and Jean Vervier, Nucl. Phys. A111, 105 (1968)

PROPERTIES OF LOW-LYING STATES IN ^{134}Cs

M. Bogdanović[1], R. Brissot[2], G. Barreau[2], K. Schreckenbach[2], S. Kerr[2], H. Börner[2], I.A. Kondurov[3], Yu.E. Loginov[3], V.V. Martynov[3], P.A. Sushkov[3], H. Seyfarth[4], T. von Egidy[5], P. Hungerford[5], H.H. Schmidt[5], H.J. Scheerer[5], A. Chalupka[6], W. Kane[7]

[1] Boris Kidrič Institute of Nucl. Sciences, 11001 Belgrade, Yugoslavia
[2] Institut Laue-Langevin, 38042 Grenoble, France
[3] Konstantinov Leningrad Nucl. Physics Institute, Gatchina 188350, USSR
[4] Institut für Kernphysik, KFA Jülich, D-5170 Jülich, F.R. Germany
[5] Physik Department, Techn. Univ. München, D-8046 Garching, F.R. Germany
[6] Institut für Radiumforschung und Kernphysik, Österreichische Akademie der Wissenschaften, Wien, Austria
[7] Brookhaven National Laboratory, Upton, New York 11973, USA

ABSTRACT

The low - lying states in ^{134}Cs have been investigated using the ^{133}Cs$(n,\gamma)^{134}$Cs and ^{133}Cs$(d,p)^{134}$Cs reactions. The level scheme is interpreted by model calculations based on two quasiparticles coupled to a vibrator.

The level scheme of ^{134}Cs$_{79}$ in the range 0 - 1250 keV has been established on the basis of the previously known data[1,2,3] and the present experimental results. The measurements of the ^{133}Cs(n,γ) reaction have been made at the High-Flux Reactor of the ILL in Grenoble with the bent crystal spectrometers GAMS1 and GAMS2/3, the magnetic beta spectrometer BILL and the pair-spectrometer. The additional experiment with the γ-γ coincidence spectrometer has been performed at the WWR-M reactor of LNPI in Gatchina. At the Brookhaven National Laboratory 5.9 eV and 22.6 eV resonance neutron-capture has been investigated. The ^{133}Cs$(d,p)^{134}$Cs reaction has been studied with the Q3D spectrograph at the tandem accelerator of the University and Technical University at München. The resulting scheme of $^{134}_{55}$Cs$_{79}$ is presented in Figs. 1 and 2. It includes 63 states 30 of which are new and about 250 secondary transitions. The intensity of these transitions sums up to 80 % of the total intensity of all measured transitions. The total intensity of transitions populating the ground state is 100 % within the error of 15 %. From the primary transition energies and

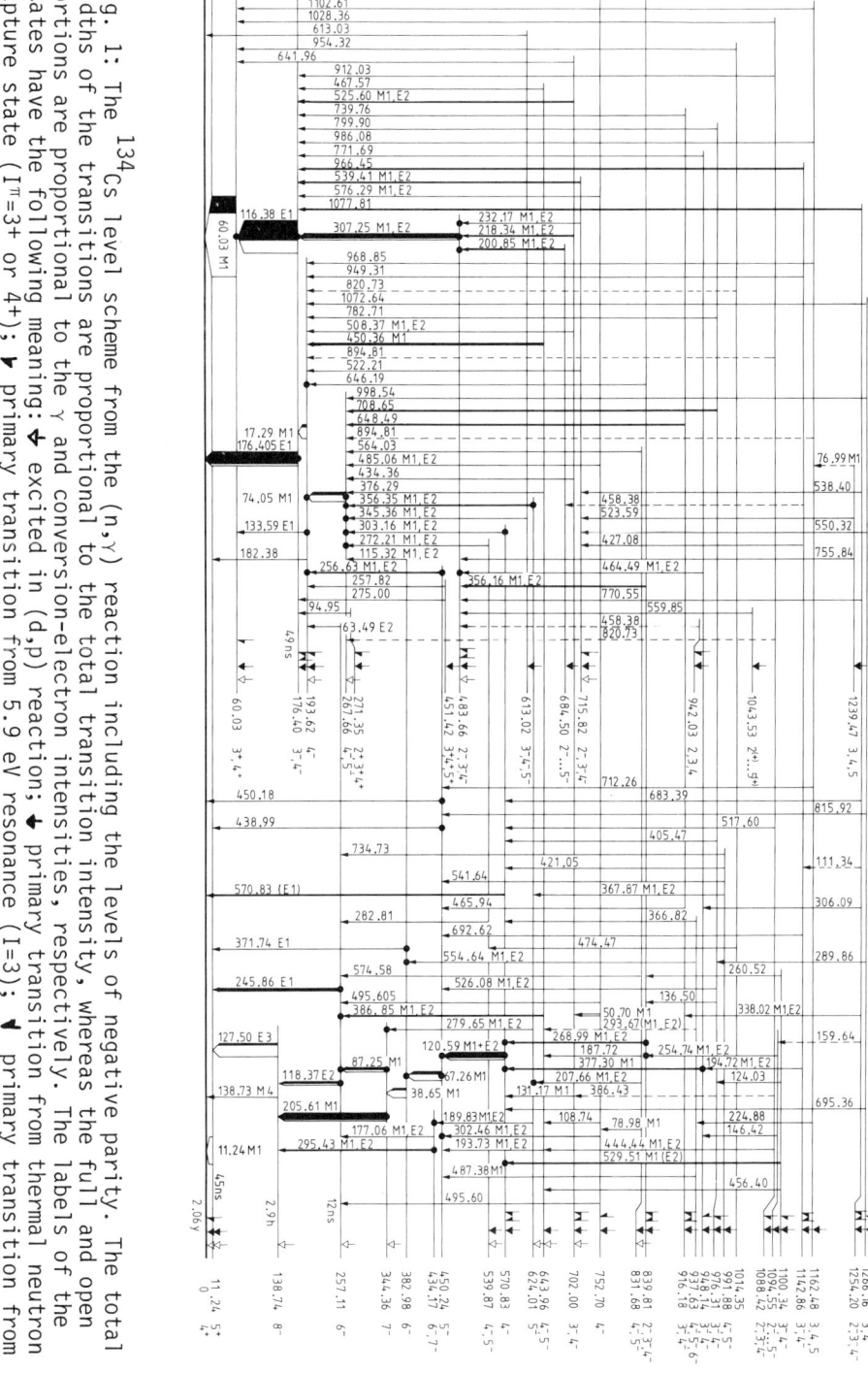

Fig. 1: The ^{134}Cs level scheme from the (n,γ) reaction including the levels of negative parity. The total widths of the transitions are proportional to the total transition intensity, whereas the full and open portions are proportional to the γ and conversion-electron intensities, respectively. The labels of the states have the following meaning: ↮ excited in (d,p) reaction; ▼ primary transition from thermal neutron capture state (I^π=3+ or 4+); ▼ primary transition from 5.9 eV resonance (I=3); ▼ primary transition from 22.6 eV resonance (I=3).

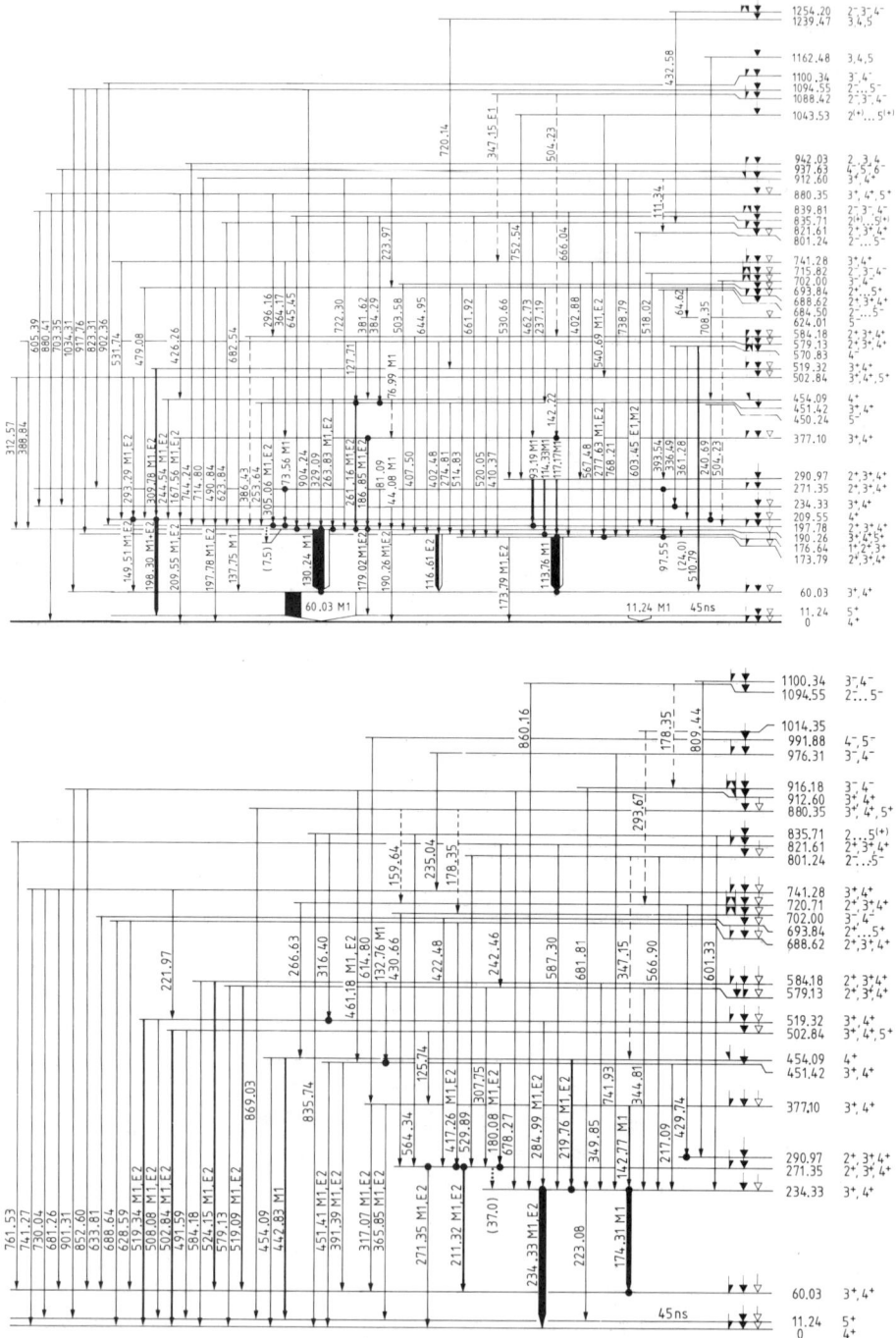

Fig. 2: The ^{134}Cs level scheme from the (n,γ) reaction including all levels of positive parity (further explanations see caption of fig.1).

the derived level energies the neutron binding energy was deduced
to be 6891.70 ± 0.13 keV.

An attempt has been made to interprete part of the levels in terms
of quasiproton-quasineutron multiplets[4]. The negative parity states
at 484(2^-), 176.4(3^-), 194(4^-), 268(5^-), 257(6^-), 344(7^-), 139 keV
(8^-) and an until now missing 9^- state, predicted at about 600 keV,
are assigned to the ($\pi\tilde{g}_{7/2}^{+1}$, $\nu\tilde{h}_{11/2}^{-1}$) multiplet. The states at 840
(3^-), 571(4^-), 450(5^-), 383(6^-), 434(7^-) and an until now missing
8^- state, predicted at 400 - 500 keV, are assigned to
the ($\pi\tilde{d}_{5/2}^{+1}$, $\nu\tilde{h}_{11/2}^{-1}$) multiplet. The average spectroscopic factor
$S[7/2 \to (g_{7/2}, h_{11/2})]$, being $\sim u^2(h_{11/2})$, yields $u^2(h_{11/2}) = 0.169$
with use of the experimental value $S = 0.22 \pm 0.04$ [3] for the well
separated 139 keV (8^-) state. From the multiplet splitting and the
transfer reaction data we deduce $v^2(h_{11/2}) = 0.831$, $v^2(g_{7/2}) = 0.435$, $v^2(d_{5/2}) = 0.427$ and the quasiparticle energies $E(g_{7/2}) = 0$,
$E(d_{5/2}) = 199$, $E(h_{11/2}) = 337$ keV.

The positive parity states at 174(2^+), 60(3^+), 0(4^+) and 11 keV(5^+)
are assigned to the ($\pi\tilde{g}_{7/2}^{+1}$, $\nu\tilde{d}_{3/2}^{-1}$) multiplet, those at 177(1^+),
198(2^+), 191(3^+) and 210 keV (4^+) to the ($\pi\tilde{d}_{5/2}^{+1}$, $\nu\tilde{d}_{3/2}^{-1}$) multiplet,
and those at 234(3^+) and 377 keV (4^+) to the ($\pi\tilde{g}_{7/2}^{+1}$, $\nu\tilde{s}_{1/2}^{-1}$) 3^+, 4^+
doublet. Here $v^2(d_{3/2}) = 0.638$, $v^2(s_{1/2}) = 0.799$ and $E(d_{3/2}) = 0$,
$E(s_{1/2}) = 242$ keV are deduced.

This classification is in agreement with magnetic moments and
magnetic transitions within multiplets but in disagreement with the
observed strong M1 transitions between the multiplets and
strong $\Delta I = 2$ E2 transitions within the lowest negative parity
multiplet. The spectroscopic factor for exciting the two highest
states of the ($\pi\tilde{g}_{7/2}^{+1}$, $\nu\tilde{d}_{3/2}^{-1}$) multiplet is almost twice as large as
expected in disagreement with the classification considered for
this multiplet, indicating much stronger mixing of the multiplets
than anticipated by the lowest order particle vibrational model.

1. Alexeev et al., Nucl. Phys. A248, 249 (1975) ; A297, 373 (1978)
2. Nuclear Data Sheets 34, 475 (1981)
3. A.G. Lee and R.G. Summers-Gill, McMaster Accelerator
   Laboratory, Annual Report 1980, p. 37; Annual Report 1982, p. 10
4. V. Paar, Nucl. Phys. A331, 16 (1979)

MULTIPOLARITY OF GAMMA TRANSITIONS IN ^{140}La PRODUCED IN THE ^{139}La (n,γ) ^{140}La REACTION

M.P.Stojanović, J.Simić
Boris Kidrič Institute - Vinča, Belgrade, Yugoslavia

K.Schreckenbach, G.Colvin
Institut Laue-Langevin, Grenoble, France

ABSTRACT

The conversion electron lines of ^{140}La following thermal neutron capture in ^{139}La were studied by the magnetic spectrometer and the (n,γ) lines by the Ge(Li) spectrometer. Forty conversion lines in ^{140}La have been measured. The conversion coefficients of all transitions and the multipolarities for some of them were determined. A new transition of 538.8 keV was identified by coincidence measurement.

INTRODUCTION

The doubly odd ^{140}La has one neutron out of the 82-neutron closed shell-core and 7-protons out of the 50-proton closed-shell core. Previous investigations of ^{140}La have been studied by ^{140}Ba β-decay[1], as well as (d,p)[2] and (n,γ)[3] reaction spectroscopy. These studies have shown that the low lying states are well explained by mixtures of two low-lying pure configurations: $(\pi 1g^0_{7/2} \nu 2f_{7/2}; J=0^-,...7^-)$ and $(\pi 2d^0_{5/2} \nu 2f_{7/2}; J=1^-,...6^-)$. All 14 states ranging from 0 to about 600 keV excitation energies have been confirmed [2,3]. A quantitative comparison between experiment and theory was made using an odd-odd quasiparticle model[4].

The higher levels in ^{140}La may contain components of the configurations: $(\pi g^0_{7/2} \nu p_{3/2}; J=2^-,...5^-)$ and $(\pi d^0_{5/2} \nu p_{3/2}; J=1^-,...4^-)$, as was suggested from the (d,p)[2] experiments. Due to the lack of conversion electron data, unique spin and parity assignments to these higher states were not possible until now.

391

Table

Run No.	$E_j(I_1) - E_j(I_2)$ (keV)	E_γ (keV)	I_γ	α_i exp.	shell	α_i theor. M1	Hager-Seltzer E2	Multipolarities exp. / theor. M1 E2
1	28.537	28.537	0.17(6)	3.2(12)	L1	4.51(0) M1	4.64(-1) E1	
					L2	3.87(-1) M1	2.08(-1) E1	
					L1	2.75(+2) M2	9.44(-1) E2	
					L2	2.31(+1) M2	9.94(+1) E2	
				1.52(58)	M1	9.26(-1) M1	9.26(-2) E1	
					M2	8.38(-2) M1	3.91(-2) E1	
					M1	6.16(+1) M2	2.50(-1) E2	
					M2	5.45(0) M2	3.31(+1) E2	
2	29.9653(2^-)-0.0(3^-)	29.9653	1.9(7)	2.9(11)	L1	3.91(0) M1	9.52(-1) E1	M1 7.9 11.7 0.01
				3.7(14)x10^{-1}	L2	3.33(-1) M1	7.82(+1) E1	L1/L2 2.8 4.9 0.4
				1.04(38)	M1	8.01(-1) M2	2.37(-1) E2	L2/M1 0.4 0.4 330
3	34.659(5^-)-0.0(3^-)	34.659	0.29(9)	2.79(90)	L1	2.54(0) M1	9.52(-1) E2	E2
				4.4(14)x10^1	L2	2.14(-1) M1	3.84(+1) E1	L1/L2 0.06 11.9 0.02
				6.0(20)x10^1	L3	4.68(-2) M1	5.25(+1) E1	L1/L3 0.05 54.3 0.02
				1.02(34)x10^1	M1	4.66(-2) M2	3.18(+1) E1	L2/L3 0.73 4.6 0.73
				1.57(51)x10^1	M3	1.01(-2) M1	1.15(+1) E1	
4	318.214(3^-)-272.314(4^-)	45.9	0.18(6)	7.8(28)x10^{-1}	L1	1.12(0) M1	6.84(-1) E1	M1
				v.w.	L2	9.13(-2) M1	9.79(0) E1	
				2.4(11)x10^{-1}	M1	2.28(-1) M1	1.45(-1) E1	
5	103.803(6^-)-48.865(6^-)	54.9382	2.0(5)	5.8(16)x10^{-1}	L1	6.57(-1) M1	4.90(-1) E1	M1
				4.1(13)x10^{-2}	L2	5.31(-2) M1	4.11(0) E1	L1/L2 14.2 10.4 0.1
				1.26(34)x10^{-1}	M1	1.35(-1) M1	9.98(-2) E1	L1/M1 4.6 4.9 4.9
								L2/M1 0.3 0.4 41.2
6	658.15(3,4)$^-$-601.89(4,5)$^-$	56.238	0.17(4)	5.4(15)x10^{-1}	L1	6.13(-1) M1	4.66(-1) E1	M1
				v.w.	L2	4.95(-2)	3.68(0)	

Table (Continued)

1	2	3	4	5	6	7	8	9	
7	63.171(4⁻)-0.0(3⁻)	63.171	3.8(10)	1.92(52)	K	3.52(0)	4.28(0)	M1	
				4.0(11)x10⁻¹	L1	4.37(-1)	3.61(-1)	L1/M1	12.6 12.6 0.2
				3.18(88)x10⁻²	L2	3.48(-2)	2.10(0)	L1/M1	2.8 4.9 5.0
				1.43(39)x10⁻¹	M1	8.98(-2)	7.26(-2)	L2/M1	0.2 0.4 28.9
8		69.198	0.28(8)	9.1(29)x10⁻¹	K	2.70(0)	5.51(-1)	E1	
					L1	3.36(-1)	5.27(-2)	E1	
					K	3.35(+1)	3.43(+1)	E2	
					L1	6.38(0)	2.91(-1)	E2	
9	162.656(2⁻)-29.9653(2⁻)	132.68	0.21(4)	3.31(79)x10⁻¹	K	4.20(-1)	5.22(-1)	M1	
				v.w.	L1	5.21(-1)	4.84(-2)		
10	318.214(3⁻)-162.656(2⁻)	155.557	2.3(2)	2.41(34)x10⁻¹	K	2.70(-1)	2.65(-1)	M1	
				3.06(52)x10⁻²	L1	3.33(-2)	3.01(-2)	K/L1	7.9 8.1 8.8
				v.w.	L2	2.33(-3)	3.08(-2)		
11	162.656(2⁻)-0.0(3⁻)	162.656	5.8(6)	2.38(24)x10⁻¹	K	2.38(-1)	2.24(-1)	M1	
				3.03(27)x10⁻²	L1	2.95(-2)	2.63(-2)	K/L1	7.8 8.1 8.5
				v.w.	L2	2.04(-3)	2.51(-2)	K/M1	39.5 39.3 42.8
12	771.310(5,4)⁻-601.89(4,5)⁻	169.394	0.45(6)	6.03(54)x10⁻³	M1	6.05(-3)	5.24(-3)	L1/M1	5.0 4.9 50.2
				2.44(43)x10⁻¹	K	2.13(-1)	2.01(-1)		
13	272.311(4⁻)-63.171(4⁻)	209.14	0.48(5)	1.14(17)x10⁻¹	K	1.20(-1)	1.26(-1)		
				v.w.	L1	1.48(-2)	1.22(-2)		
					L1	2.63(-2)	2.32(-2)		
14	322.003(5⁻)-103.803(6⁻)	218.200	9.1(9)	1.07(11)x10⁻¹	K	1.07(-1)	1.10(-1)	M1	
				1.22(14)x10⁻²	L1	1.32(-2)	1.08(-2)	K/L1	8.8 8.1 10.2
				v.w.	L2	8.57(-4)	6.73(-2)	K/M1	24.5 38.5 51.2
				4.36(52)x10⁻³	M1	2.71(-3)	2.15(-3)	L1/M1	2.8 4.9 5.0
15	284.634(7⁻)-48.856(6⁻)	235.765	1.3(2)	8.5(13)x10⁻²	K	8.69(-2)	8.36(-2)	M1	
				2.00(33)x10⁻²	L1	1.07(-2)	8.51(-3)	K/L1	4.2 8.4 11.0
				v.w.	L2	6.82(-4)	4.78(-3)		

Table (Continued)

1	2	3	4	5	6	7	8	9
16	272.311(4⁻)-34.659(5⁻)	237.66	3.8(4)	8.1(12)x10⁻²	K	8.50(-2)	8.12(-2)	M1
				1.36(21)x10⁻²	L1	1.05(-2)	8.31(-3)	K/L1 6.0 8.1 9.8
				v.w.	L2	6.66(-4)	4.62(-3)	
17	322.003(5⁻)-63.171(4⁻)	258.83	0.34(5)	4.6(10)x10⁻²	K	6.76(-2)	6.01(-2)	
18ᵃ	272.311(4⁻)-0.0(3⁻)	272.303	6.5(7)	5.25(75)x10⁻²	K	5.91(-2)	5.09(-2)	
				8.3(11)x10⁻³	L1	7.29(-3)	5.51(-3)	K 304.85keV addm.
				2.6(4)x10⁻³	M1	1.50(-3)	1.11(-3)	
19	601.89(4,5)⁻-322.003(5⁻)	279.93	0.79(9)	4.97(76)x10⁻²	K	5.50(-2)	4.66(-2)	K 310.13keV addm.
				9.4(19)x10⁻³	L1	6.77(-3)	5.08(-3)	
					L2	4.12(-4)	2.27(-3)	
20	1054.93(4,5)⁻-771.6(5,4)⁻	283.615	0.50(6)	5.42(89)x10⁻²	K	5.31(-2)	4.47(-2)	M1 0.0≤M1≤1.0
21	318.214(3⁻)-29.9653(2⁻)	288.249	9.0(9)	4.94(70)x10⁻²	K	5.09(-2)	4.25(-2)	M1+≤69%E2
22ᵃ	467.505(1⁻)-162.656(2⁻)	304.85	0.25(4)	5.82(83)x10⁻³	L1	6.27(-3)	4.66(-3)	K/L1 8.5 8.1 9.1
23ᵃ	912.07(2,3,4)⁻-601.89(4,5)	310.13	0.27(4)	2.15(39)x10⁻¹	K	4.39(-2)	3.59(-2)	L1 272.303keV addm.
24		389.18	0.37(5)	6.26(19)x10⁻²	K	4.20(-2)	3.41(-2)	M1 272.303keV addm.
				2.68(54)x10⁻²	K	2.34(-2)	5.34(-3)	E1
25		396.94	0.52(7)	2.28(43)x10⁻²	K	2.23(-2)	1.75(-2)	E2
						8.46(-2)	1.65(-2)	E1
26	744.61(4,5)⁻-322.003(5⁻)	422.63	4.8(5)	1.58(24)x10⁻²	K	7.96(-2)	1.85(-3)	E2
				3.00(38)x10⁻³	L1	1.90(-2)	1.38(-2)	M2
				v.w.	L2	2.32(-3)	1.55(-3)	
27	769.09(2⁻)-318.214(3⁻)	477.92	0.31(9)	2.28(84)x10⁻²	K	1.24(-4)	4.02(-4)	K/L1 5.3 8.2 8.9
28	658.15(3,4)⁻-162.656(2⁻)	495.80	0.87(17)	1.27(33)x10⁻²	K	1.40(-2)	9.76(-3)	M1
						1.27(-2)	8.84(-3)	M1
								0.15≤M1≤0.99

Table (Continued)

1	2	3	4	5	6	7	8	9
29[b]	601.89(4,5)⁻-63.171(4⁻)	538.8	0.37(8)	9.1(33)x10⁻³	K	1.04(-2) M1 7.09(-3) M2	2.49(-3) E1	0.62M1+0.38E2 0.0≤M1≤1.0 E2
30	711.6(2,3,4)⁻-162.656(2⁻)	548.92	1.1(2)	1.18(25)x10⁻²	K	9.89(-3)	6.75(-3)	M1 0.8≤M1≤1.0
31	601.89(4,5)⁻-48.865(6⁻)	553.0	0.67(14)	7.5(21)x10⁻³	K	9.71(-3)	6.63(-3)	0.38M1+0.72E2 0.0≤M1≤0.96
32	601.89(4,5)⁻-34.659(5⁻)	567.3	3.7(7)	9.9(21)x10⁻³	K	9.12(-3)	6.20(-3)	M1 0.52≤M1≤1.0
33	658.15(3,4)⁻-63.171(4⁻)	594.9	1.1(2)	7.0(15)x10⁻³	K	8.12(-3)	5.49(-3)	0.58M1+0.42E2 0.04≤M1≤0.82
34	658.15(3,4)⁻-34.659(5⁻)	623.3	0.54(11)	7.2(22)x10⁻³	K	7.24(-3)	4.88(-3)	M1 0.05≤M1≤0.96
35	658.15(3,4)⁻-0.0(3⁻)	658.3	1.2(2)	7.8(16)x10⁻³	K	6.34(-3)	4.26(-3)	M1 0.93≤M1≤1.0
36	771.71(5,4)⁻-103.803(6⁻)	667.6	0.63(13)	8.0(21)x10⁻³	K	6.13(-3)	4.12(-3)	M1 0.93≤M1≤1.0
37	771.71(5,4)⁻-63.171(4⁻)	708.3	1.4(3)	5.2(13)x10⁻³	K	5.31(-3)	3.57(-3)	M1 0.19≤M1≤0.93
38	744.61(4,5)⁻-34.659(5⁻)	710.4	0.76(15)	6.1(19)x10⁻³	K	5.28(-3)	3.54(-3)	M1 0.37≤M1≤1.0
39	771.71(5,4)⁻-48.865(6⁻)	722.8	2.4(5)	6.1(14)x10⁻³	K	5.06(-3)	3.40(-3)	M1 0.65≤M1≤1.0
40	1054.93(4,5)⁻-63.171(4⁻)	992.1	0.46(9)	5.6(20)x10⁻³	K	2.40(-3)	1.67(-3)	

a. When possible addmixtures are cleared off new experimental values are obtanied: $\alpha_{L_1}(272.303) = 6.7(9) \times 10^{-3}$, $\alpha_{M_1}(272.303) = 1.21(18) \times 10^{-3}$, $\alpha_K(304.85) = 4.05(74) \times 10^{-2}$, $\alpha_K(310.13) = 3.03(57) \times 10^{-2}$.

b. A new transition of 538.8 ± 0.1 keV.

RESULTS

In this work the conversion electron lines of ^{140}La following thermal neutron capture in ^{139}La were studied by the magnetic spectrometer BILL at the Laue-Langevin Institute-Grenoble.

Two targets of 100 μ/cm^2 and 800 μ/cm^2 have been used in the electron energy ranges 20 to 165keV and 150 to 1000keV respectively. The targets are produced by evaporating La$_2$O$_3$ on the Al foil. The (n,γ) lines were studied by the Ge(Li) spectrometer at the Boris Kidrič Institute - Belgrade.

The results of conversion electron spectra analyses are presented in the Table. For conversion coefficients determination in the energy region 20 to 500keV the γ-rays intensities measured by curved-crystal spectrometer[3] have been used. For the energy range 500-1000keV we used our measured γ-rays intensities. By comparinge the experimental conversion coefficients or their ratios with the calculated from tables the multipolarities for some of the transitions have been assigned. The conversion coefficients were calculated in such way that the ratio of theoretical and measured K conversion coefficients for the 162.656keV (M1) and 218.200keV (M1) transitions was unity.

The γ-γ coincidence were measured by Ge(Li) detectors at the Boris Kidrič Institute. A new transition of 538.8keV was identified by coincidence measurement with the 4559.2keV.

REFERENCES

1. J.Kern and G.Mauron, Helv.Phys. Acta. 43, 272 (1970).
2. J.Kern, G.L.Struble and R.K.Sheline, Phys.Rev. 153,1331 (1967).
3. E.T.Jurney, R.K.Sheline, E.B.Shera, H.R.Koch, B.P.K.Maier, U.Gruber, H.Baader, D.Breitig, O.W.B.Schult, J.Kern and G.L. Struble, Phys. Rev. C2, 2323 (1970).
4. Gordon L.Struble, Phys.Rev. 153, 1347 (1967).

^{144}Nd, ^{165}Dy AND ^{175}Yb LEVEL SCHEMES FROM THE (n, 2γ) REACTION

V.A.Khitrov, Yu.P.Popov, A.M.Sukhovoj, Yu.S.Yazvitsky
Laboratory Of Neutron Physics, Joint Institute for
Nuclear Research, Dubna, USSR, Head Post Office,
P.O. Box 79

ABSTRACT

The (n, 2γ) reaction on thermal neutrons has been studied by the method of summation of amplitudes of coinciding pulses from two Ge(Li)-detectors on ^{143}Nd, ^{164}Dy and ^{175}Yb nuclei. Intensities of two-quanta cascades, populating low-lying levels of ^{144}Nd, ^{165}Dy, ^{175}Yb and γ-transition energies in these cascades have been determined.

The ^{143}Nd(n, 2γ)^{144}Nd, ^{164}Dy(n, 2γ)^{165}Dy, ^{174}Yb (n, 2γ)^{175}Yb reactions on thermal neutrons were studied by the method of summation of amplitudes of coinciding pulses from two Ge(Li)-detectors. More detailed description of the spectrometer and measuring and calculating methods are given in refs /1-3/. The investigation has been performed on the IBR-30 reactor. The example of the resulting spectrum of the sum of amplitudes of coinciding pulses for the ^{174}Yb(n, 2γ) reaction is given in Fig.1. The peaks corresponding to the registration of the total energy of two-quanta cascades in two detectors can be seen there. The figures on the peaks are the cascade energy in kev.

If a spectrum is accumulated from only these amplitude codes, which being summed up with the codes of the other detector give a certain separate peak of a summary spectrum, it is called a differential spectrum. The two-quanta cascades in such spectrum are located symmetrically in relation to the spectrum centre. They correspond to the case of total absorption of γ-quanta energy in the detectors. The method of removing the background unfluence in summary spectrum /1/ and method of improvement in resolution without the loss of efficiency/3/ were used by processing differential spectra. The corrected differential spectrum for a total absorption peak 5266,5 kev is given in Fig.2. Analysis of positions and areas of separate peaks in differential spectra permitted to determine the γ-transitions energy in cascades and the intensity of these cascades. The order of γ-quanta in cascade was defind by comparison of differential spectra for different final states. If there are γ-transitions with equal energy in different spectra, one may suppose these transitions to be primary and the population of final states must undergo the same intermediate state. Correspondingly, the secondary transitions to various final states differ in energy by the difference of excitation energies of final states.

Five differential spectra for total absorption peaks with energies of 5607, 5556, 5534, 5176, 5142 kev corresponding to two-quanta transitions to levels with excitation energy less than 780 kev for ^{165}Dy have been analysed. During the experiment 180 cascades have been measured and 113 of them have been placed in the level scheme. The value of 43 level energies with an error less than 3 kev within 1.1 - 4 MeV have been determined. The 13 new levels have been found at excitation energy $E_f > 3025$ kev; it could not be accomplished by other methods earlier.

A total of 155 cascades populating nine low-lying levels of ^{175}Yb with excitation energy $E_f < 1$ MeV have been determined. Total energies of investigated cascades are: 5822, 5307, 5266, 5219, 5183, 5010, 4951, 4902 and 4831 kev. The obtained level scheme includes 105 cascades exciting 33 levels up to excitation energy 3578 kev. The excitation level scheme for this nucleus has been defined up to $E_f = 2737$ kev, and we have determined 14 new levels above this energy.

In ^{144}Nd 39 cascades have been investigated populating 5 low-lying states with $E_f < 1570$ kev. Total energies of these cascades are: 7819, 7120, 6504, 6305 and 6256 kev. The above mentioned method of identification of γ-quanta order in cascade allows one to place 2o cascades in a level scheme exciting 12 levels with $E_f < 5723,5$ kev. We give 18 more states as tentative (we attribute a larger energy to primary quanta). Seven new levels have been determined above the excitation energy of 3,5 MeV, four of them are given as tentative.

Level energies lying above the known level schemes for all three nuclei are listed in Table 1.

Table 1. Energy of new levels ^{144}Nd, ^{165}Dy and ^{175}Yb determined above known level schemes.

^{144}Nd	^{165}Dy		^{175}Yb	
3563,7	3051,2	3454,8	2770,3	3223,5
(3750,1)	3123,7	3474,0	2861,3	3266,7
(3915,8)	3193,4	3539,7	2877,5	3327,2
(3931,7)	3256,3	3586,7	2903,0	3414,9
(4006,4)	3379,0	3650,7	3022,6	3440,3
5389,9	3421,5	3848,4	3159,9	3545,1
5723,5	3443,1		3198,8	3578,1

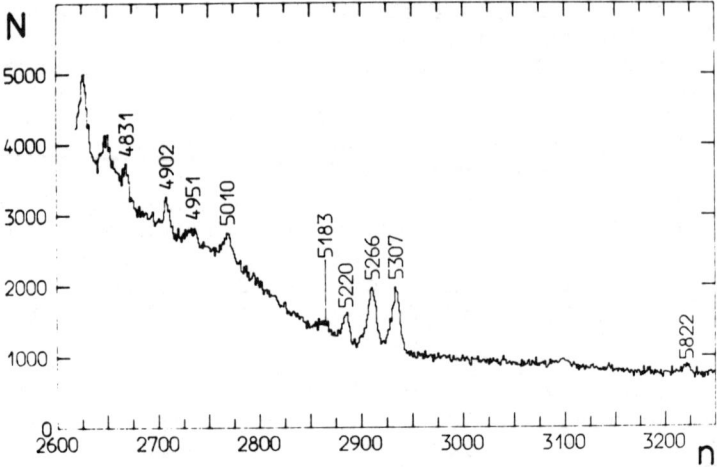

Fig. 1.

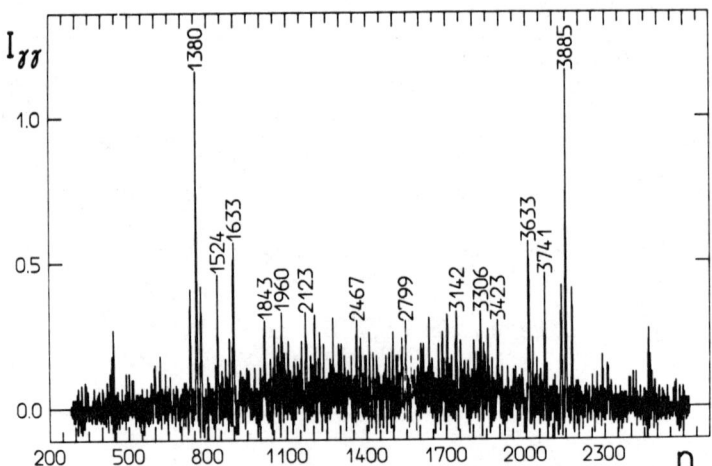

Fig.2.

REFERENCES

1. A.A.Bogdzel et al. , JINR, P15-82-706, Dubna, 1982.
2. Yu.P. Popov et al. , JINR, P6-83-316, Dubna, 1983.
3. A.M. Sukhovoj, V.A. Khitrov, JINR, P13-82-905, Dubna, 1982.

AVERAGE INTENSITIES OF TWO-QUANTA CASCADES IN ^{144}Nd, ^{165}Dy AND ^{175}Yb AFTER THE CAPTURE OF THERMAL NEUTRONS

V.A.Khitrov, Yu.P.Popov, A.M.Sukhovoj, Yu.S.Yazvitsky
Laboratory Of Neutron Physics, Joint Institute for
Nuclear Research, Dubna, USSR, Head Post Office,
P.O. Box 79

ABSTRACT

Two-quanta cascades in three nuclei are investigated. The sum of their intensities makes 0,5-0,7 Γ_γ .

The method of summation of amplitudes of coinciding pulses from two Ge(Li) detectors was used for the investigation of γ-decay cascades of compound states. Some characteristics of the method used as well as the peculiarities of the method for the improvement of coinciding pulses spectra are described in /1,2/. For the ^{144}Nd, ^{165}Dy and ^{175}Yb isotopes 19 spectra were obtained from one detector. The sum of amplitude codes of both detectors were equal to total absorption energy of two γ-quanta of the cascade between the compound state and the fixed low-lying state. The probability of populating the low-lying levels by 2 γ-quanta cascades through any intermediate state was determined. The data are listed in Table 1.

Table 1. Experimental $I_{\gamma\gamma}^E$ and calculated $I_{\gamma\gamma}^W$, $I_{\gamma\gamma}^A$ intensities of 2 γ-quanta cascades for total energies $E_f + E_s$ (E_f and E_s being the energies of quanta forming the cascade).

Compound nucleus	$E_f + E_s$ keV	J^π	$I_{\gamma\gamma}^E$	$I_{\gamma\gamma}^W$	$I_{\gamma\gamma}^A$
^{144}Nd	7819	0^+	3,7±1,2	1,2	3,8
	7120	2^+	32 ±3	5,2	13,6
	6504	4^+	7,3±1,4	2,2	5,6
	6308	3^-	3,9±1,0	8,0	5,7
	6256	2^+	(2,8)	1,2	2,8
	all 5		49,7	17,8	31,5
^{165}Dy	5715	$7/2^+$	(1,6)	0,002	0,02
	5607	$1/2^-$	11,0±1,9	2,8	7,7
	5556	$3/2^-$	11,5±2,5	2,6	6,9
	5534	$5/2^-,5/2^-$	9,8±2,3	2,3	6,4
	5176	$5/2^-,3/2^-$	7,4±2,5	4,8	3,4
	5142	$1/2^-,3/2^-$	12,1±4,8	2,1	5,3
	all 6		54,2	14,6	29,7

Compound nucleus	$E_f + E_s$ kev	J^π	$I^E_{\gamma\gamma}$	$I^W_{\gamma\gamma}$	$I^A_{\gamma\gamma}$
^{175}Yb	5822	7/2$^-$	2,9±0,4	0,6	1,1
	5307	1/2$^-$	17,2±4,0	3,0	7,6
	5266	1/2$^-$	18,1±4,7	2,8	6,9
	5220	1/2$^-$	9,0±1,6	1,3	3,2
	5183	5/2$^-$	2,3±1,6	1,2	2,9
	5010	3/2$^-$	11,5±5,9	1,5	3,7
	4951	3/2$^-$	2,3±1,3	0,7	1,6
	4902	1/2$^-$	6,3±3,7	1,2	2,9
	4831	1/2$^-$	3,4±1,3	1,0	2,4
	all 9		72,8	12,7	32,3

The comparison of experimental and theoretical yields of cascades populating the low-lying levels shows that the experimental data regularly exceed the values predicted by statistical theory, especially in case of ^{165}Dy and ^{175}Yb, where the total population through these cascades of the low-lying levels shown in the table makes 54% and 73% of a total radiative width, respectively. Spectra from one detector can be calculated /3/ theoretically, since the γ-quanta yield for ΔE_γ interval is proportional to:

$$\varphi(E_\gamma) = \sum^n \frac{\Gamma_{\lambda g}(E_f)\Gamma_{gf}(E_s)}{\Gamma_\lambda \Gamma_g} + \sum^m \frac{\Gamma_{\lambda h}(E'_f)\Gamma_{hf}(E'_s)}{\Gamma_\lambda \Gamma_h} \quad (1)$$

Here λ, f represent the initial and final states, respectively; g and h are the intermediate states, Γ is the total (one letter index) and partial (two letter index) γ-width of corresponding levels or transitions. The number of levels in ΔE_γ interval excited by primary trasitions with energies of $E_f \div (E_f + \Delta E_\gamma)$ equals n and in case of $E'_f \div (E'_f + \Delta E_\gamma)$ the number of levels equals m, where $E_f + E_s = E'_f + E'_s$, $E_f = E'_s$ and $E_s = E'_f$. The sum (1) over the energy interval from (B_n-511) kev to the excitation energy level E_f+511 kev determines the total cascade yield of two γ-transitions. In cases where energy dependence of partial γ-widths is $\Gamma_{\lambda g} \sim E^{2l+1}$ (l denotes multipolarity of transition) the resulting sum is denoted "W" in the Table. Index "A" means that the Lorentz curve has been used for calculating γ-transitions. The dependence of level densities on an excitation energy has been determined by /4/.

The experimental and predicted spectra values can be compared if their total areas are assumed to be equal. As the spectra fluctuate and are symmetrical in relation to their centres it is recommended to compare the increasing sums up to the half of a spectrum. Fig.1 gives diagrams of ^{144}Nd and ^{165}Dy spectra, respectively. Fig.2 gives diagrams of ^{175}Yb spectra.

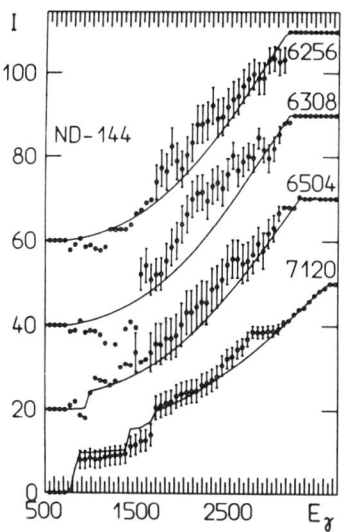
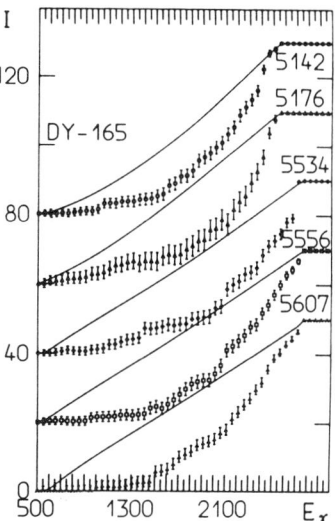

Fig. 1.

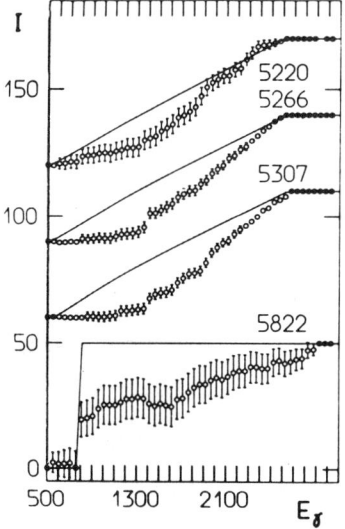
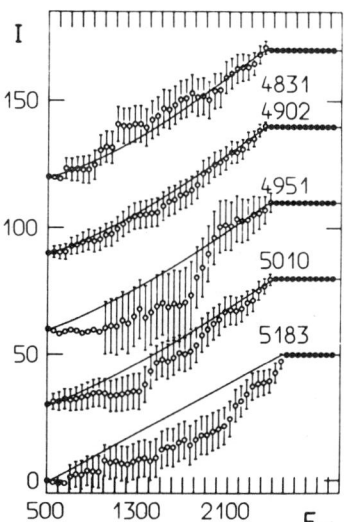

Fig.2.

Dots in figures represent experimental values and lines represent the values calculated by (1). It can easily be seen that spectra dependence for ^{144}Nd may be described by (1) with sufficient precision. The agreement of experimental and calculated values for ^{165}Dy and ^{175}Yb is a good deal worse. A more detailed study of ^{165}Dy spectra has shown /3/ that experimental intensity values of cascades passing through levels with excitation energy of 2-4 MeV essentially exceed the values calculated according to the statistical model. The result of ^{175}Yb experiments are similar. In this region of excitation energies there exist single particle neutron states $3/2^-[512]$ $1/2^-[510]$ and proton states $3/2^-[532]$, $1/2^-[530]$ for the given nuclei. The contribution of these components to intermediate levels structure can result in increasing both primary E1 - transitions from compound states and secondary M1-transitions to the low-lying single particle levels.

REFERENCES

1. A.A. Bogdzel et al. JINR, P15-82-706, Dubna, 1982.
2. A.M. Sukhovoj, V.A. Khitrov, JINR, P13-82-905, Dubna, 1982.
3. Yu.P. Popov et al. JINR, P3-83-651, Dubna, 1983.
   Yu.P. Popov et al. JINR, P3-84-94, Dubna, 1984.
4. W. Dilg et al. Nucl.Phys., A217, p.269, 1974.

A COMPLETE IDENTIFICATION BY THE PARTICLE-ROTOR MODEL OF ^{153}Gd STATES UP TO 1 MeV

A.M.J. Spits, P.H.M. Van Assche
SCK/CEN, 2400 Mol, Belgium

H.G. Börner, W.F. Davidson, D.D. Warner, K. Schreckenbach, G. Colvin
ILL, 38042 Grenoble Cedex, France

R.C. Greenwood, C.W. Reich
EG&G Idaho, Inc., Idaho Falls, Idaho 83415, U.S.A.

ABSTRACT

Measurements performed at the Institut Laue-Langevin in Grenoble regarding gamma rays and conversion electrons following thermal-neutron capture in ^{152}Gd together with measurements of 2 keV neutron capture in the same nucleus at the High Flux Reactor in Brookhaven have resulted in a 100-level ^{153}Gd scheme. For some 200 transitions in ^{153}Gd conversion coefficients have been calculated. This enabled the determination of transition multipolarities and spin and/or parity restrictions for many levels.

A Coriolis analysis has been performed which, due to the increase of information regarding level spins, could take into account a larger number of interacting bands than was hitherto possible, both regarding positive as well as negative parity levels. In this way some 95 levels including all levels observed up to an excitation energy of 1 MeV could be identified as to their energies within 30 keV. Good agreement as to other experimental parameters such as life times and branching ratios, especially regarding the lower-lying levels, is also achieved by this analysis.

- - - - -

This paper encompasses an effort to interpret the results of a threefold experiment on neutron capture in ^{152}Gd by an extensive analysis in the framework of the Nilsson particle-rotor model. These experiments, all performed with high-resolution instruments, involve a study of 1) the ^{152}Gd(n_{th},γ) reaction by the curved-crystal spectrometers GAMS 1,2,3 and pair spectrometer of the ILL at Grenoble, 2) the ^{152}Gd(n_{th},e^-) reaction by the conversion electron spectrometer BILL also at the ILL and 3) average-resonance capture in ^{152}Gd at the Sc-filter facility in the HFBR at BNL, Brookhaven.

The results of these experiments can be summarized as follows: some 450 capture γ-rays, 220 of which having known type of multipolarity, or in some cases, multipole mixing ratio, were incorporated in a 100-level ^{153}Gd scheme up to 1.7 MeV. Spins and parities could be determined or restricted for many levels and it is believed that all (1/2,3/2)$^-$ states in the relevant region were observed.

This experimental set of levels was subjected to a theoretical interpretation on the basis of the Nilsson particle-rotor model. In principle, the calculation was carried out the same way as performed by e.g. Tuurnala[1] in his description of low-lying states in ^{153}Gd,

0094-243X/85/1250403-03 $3.00 Copyright 1985 American Institute of Physics

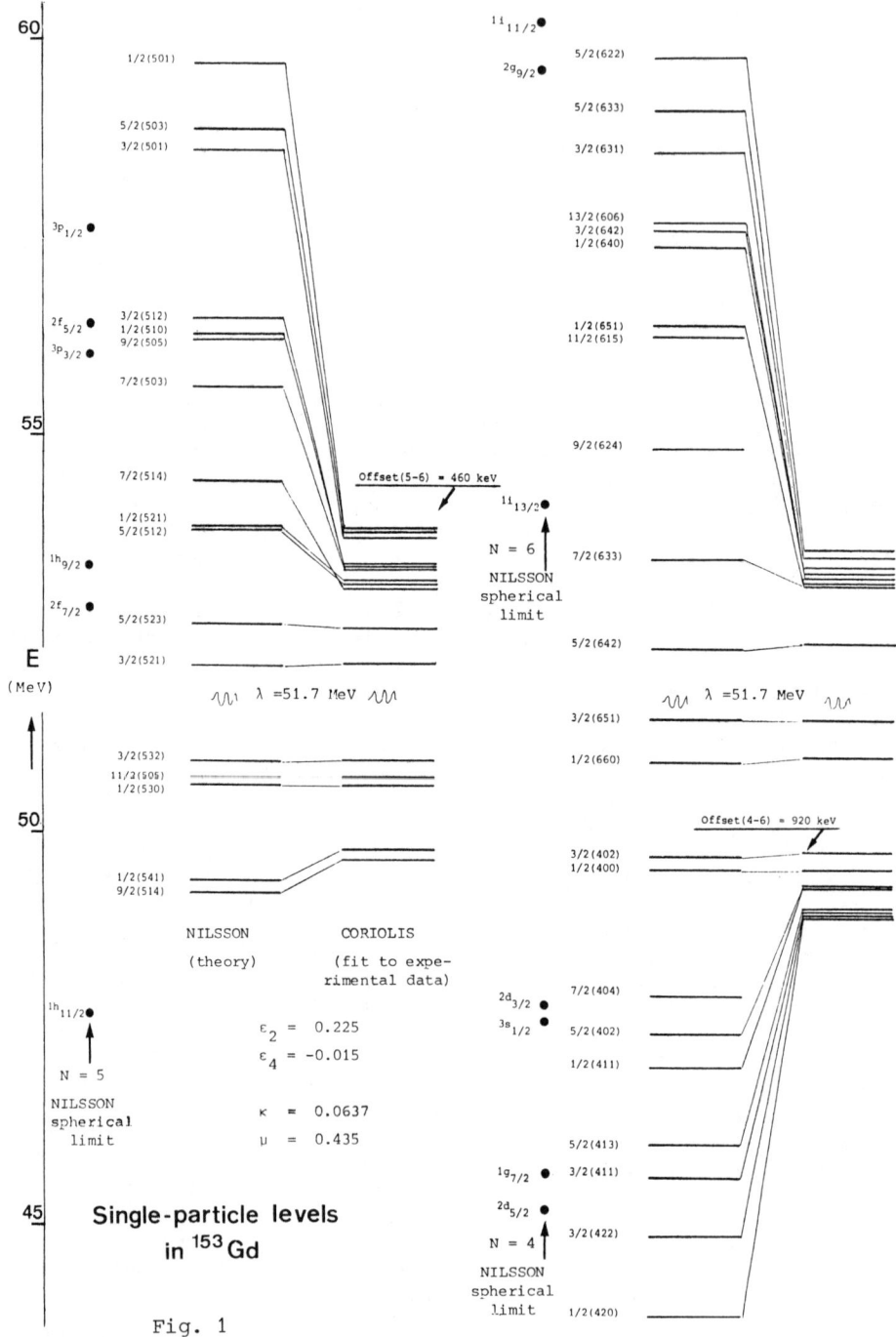

Fig. 1. Single-particle levels in ^{153}Gd

except that a larger number (33) of interacting bands was included in the analysis as well as that some levels were interpreted differently as compared to previous work. As a starting point, the set of single-particle levels belonging to the N=4,5 and 6 oscillator shells were calculated. The potential parameters κ and μ were assigned the shell-independent values of κ=0.0637 and μ=0.435, the deformation parameters ε_2 and ε_4 were initialized at values of ε_2=0.20 and ε_4=0. A BCS calculation was performed on this set of states, in which each state succesively was blocked. A value of λ=52.0 MeV for the Fermi level resulted. Now for each orbital the single-particle energies were allowed to vary such as to minimize the residues. Common inertial parameters A=h^2/2I were adopted for the set of positive- and negative-parity states (A=13 and 19 keV, resp.) with an exception for the 11/2⁻(505) band for which a value of A=16 keV gave the best fit to the data. Also other parameters such as the decoupling parameters were varied. A coupling of the 5/2⁻(523) and the 3/2⁻(532) bands expressed in a matrix element of 80 keV was taken into account in order to reproduce the correct order of the second 3/2⁻ and 5/2⁻ levels (129 keV and 110 keV, resp.). The set of bands was extended step by step to bands further away from the Fermi level to finally include 18 and 17 for positive- and negative-parity states, resp., with the constraint that the order of the bands should be preserved as much as possible. In order to account for an observed shift in energies between single-particle levels originating from the N=6 and N=4 shells, and to further improve the overall agreement for N=5 shell levels, an "offset" of 460 keV each was introduced between N=5 and N=6 as well as between N=4 and N=5 shells.

After that, in the first calculation the deformation parameters ε_2 and ε_4 and the Fermi level λ were allowed to vary to reproduce the latter set of single-particle energies centered around the Fermi level as closely as possible. This cycle was repeated several times, finally resulting in values of ε_2=0.225, ε_4=-0.016 and λ=51.7 MeV.

In such a way, all states observed up to ∼1 MeV (including a few high-spin states from other reactions) could be identified as to their energies within 30 keV. With the given configuration- a few 5/2⁺ levels were predicted, but not observed, however. On the average other physical quantities such as life times and branching ratios are surprisingly well reproduced, especially for the lower-lying levels. These results will not be reproduced here.

The final results are shown in Fig. 1. It is seen that there exists nice agreement for levels not far from the Fermi surface, but for levels further away a compression effect, which has been observed earlier, e.g. in ref.²), becomes apparent. It will be the subject of future investigations in how far the new features of the model will be reproduceable for other nuclei.

REFERENCES

1. T. Tuurnala, Z. Phys. **268**, 371 (1974)
2. B. Elbek and P.O. Tjøm, Adv. Nucl. Phys. **3**, 301 (1969)

SINGLE PARTICLE AND VIBRATIONAL BANDS in ^{155}Gd, ^{161}Dy, and ^{163}Dy

H.H. Schmidt, P. Hungerford, T.v. Egidy, H.J. Scheerer
Physik-Department, Technische Universität München, D8046 Garching,
Germany

H.G. Börner, S.A. Kerr, K. Schreckenbach, F. Hoyler, G. Colvin
Institut Laue-Langevin, F38043 Grenoble, France

R.F. Casten, D.D. Warner, W. Kane
Brookhaven National Laboratory, Upton, New York, 11973, USA

Abstract

The nuclei ^{155}Gd, ^{161}Dy and ^{163}Dy have been investigated by means of the (n,γ), (n,e) and transfer reactions. Level schemes have been constructed up to about 1.5 MeV, which include spin and parity assignments for many levels. New rotational bands based on single particle states (Nilsson) as well as on vibrational states have been identified. The results are discussed.

I. Introduction

Though the low lying levels of odd-A nuclei are well understood, due to the increasing level density and to increasing configuration mixing the available information decreases rapidly with increasing excitation energy. In particular, not many vibrational excitations have been identified. A combination of the non selective (n,γ) reaction and the selective single particle transfer reactions using high resolution instruments allows the extension of the level schemes up to 1.5 MeV and the identification of the main cofigurations for many levels.

II. Experiments

Primary and secondary γ-rays following thermal neutron capture have been studied at the ILL, Grenoble, using the pair spectrometer[1] and the crystal spectrometers GAMS[2]. Primary γ-rays following average resonance neutron capture (ARC)[3] with 2 keV and 24 keV neutrons[4] have been measured at the BNL, Brookhaven. The (n,e) reaction has been studied at the spectrometer BILL[5] at the ILL. (d,p) and (d,t) spectra have been recorded at the Q3D Spectrograph[6] at the Munich Tandem accelerator with deuteron energies of 14 MeV and 20 MeV.

III. Results and Discussion

From these extensive experimental data, level schemes have been constructed up to about 1.5 MeV, which are rather complete in the spin region 1/2 to 5/2 up to about 1 MeV and include spin and parity assignments for many levels. About 15 rotational

bands in each nucleus could be identified, which are based on
single particle excitations (Nilsson) as well as on vibrational
excitations coupled to single particle excitations. The
rotational bands are shown in Fig.1 and Fig.2 for ^{155}Gd, in Fig.3
for ^{161}Dy and in Fig.4 for ^{163}Dy. As can be seen from Fig.1 and
Fig.2 most of the levels are populated in the (d,p) and the (d,t)
reaction. Similar patterns of the population of the members of a
band based on a vibrational excitation and the members of a band
based on a single particle excitation indicate mixing between the
two configurations and allow the approximate determination of the
degree of the mixing. For example, the {3/2$^-$[521]-2$^+$} band in
^{155}Gd is estimated to have an admixture of approximately 20% of
the 1/2$^-$[521] band. Fig.5 and Fig.6 show a comparison of the
band head energies with the predictions of Soloviev et al.[7] for
^{155}Gd and ^{163}Dy, respectively. Fig.7 and Fig.8 show the
systematics of the band head energies of the bands based on
Nilsson configurations for the odd Gd isotopes and the odd Dy
isotopes, respectively. Remarkable is the similarity even in
details (for example, 5/2$^-$[523] band and 1/2$^-$[521] band).

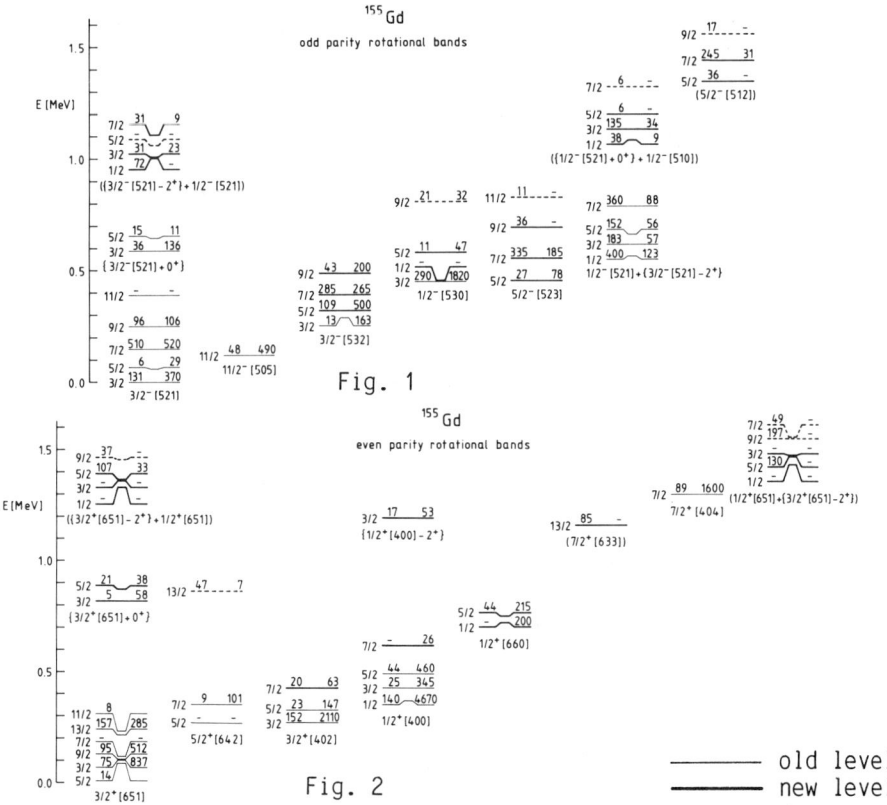

Fig. 1 and Fig. 2 Rotational Bands in ^{155}Gd. The numbers above
the levels are the (d,p) (left number) and
(d,t) (right number) populations in μb/sr.

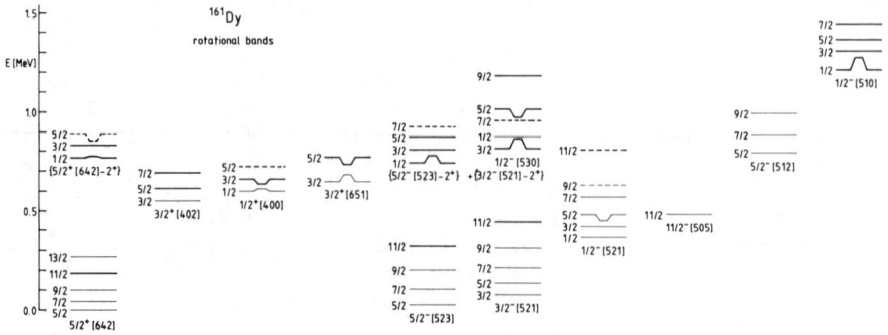

Fig. 3 Rotational Bands in ^{161}Dy

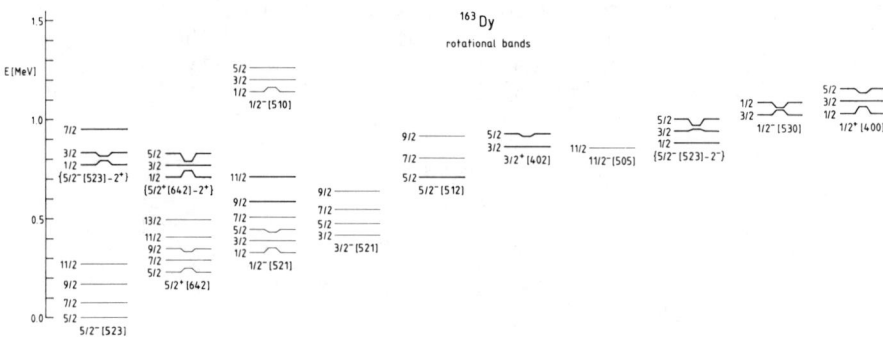

Fig. 4 Rotational Bands in ^{163}Dy

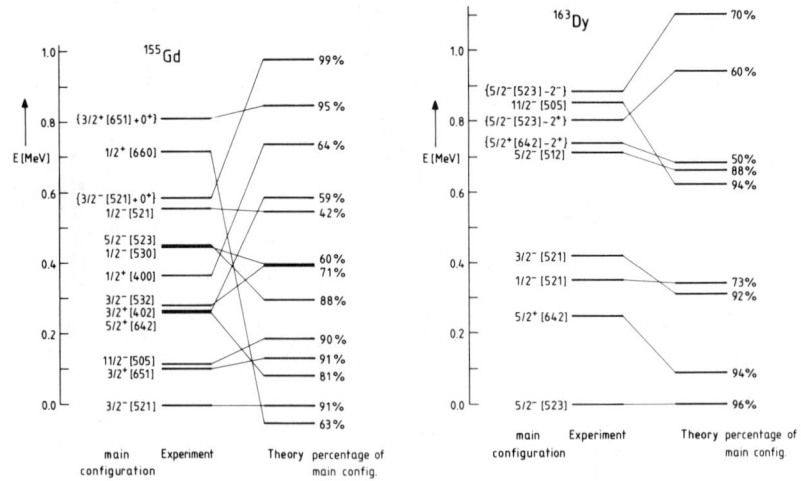

Fig. 5 and Fig. 6   Comparison of the band head energies in ^{155}Gd (Fig. 5) and ^{163}Dy (Fig. 6) with the Theory of Soloviev et al.

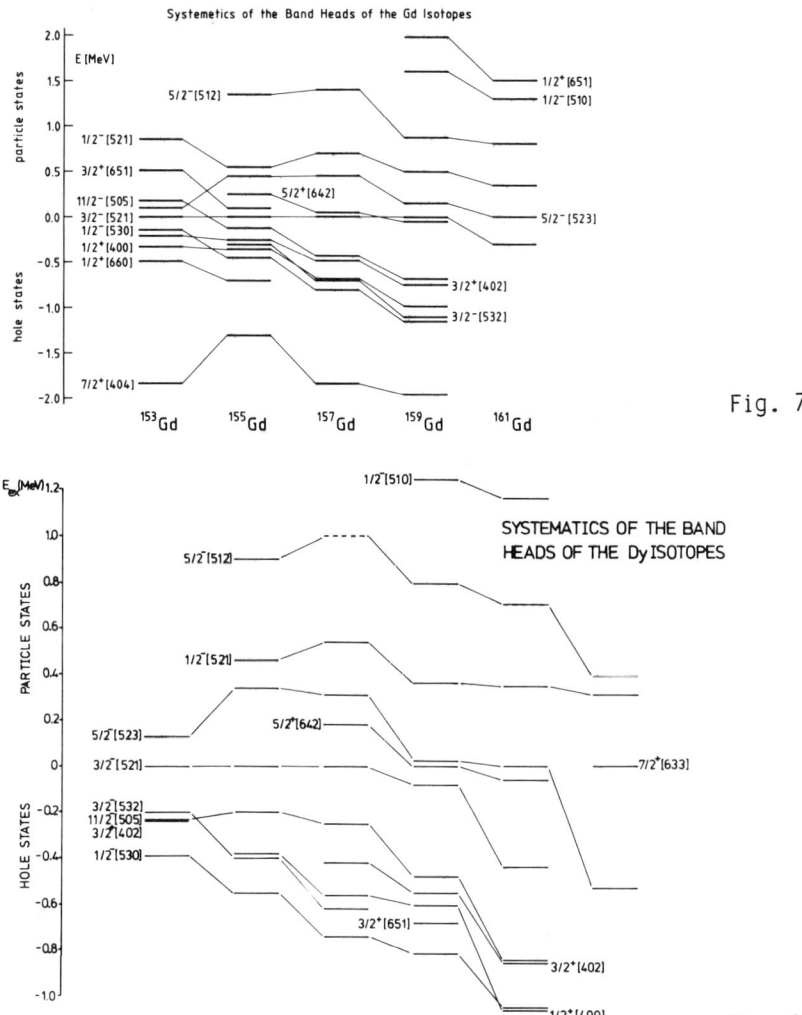

Fig. 7 and Fig. 8 Systematics of the band head energies in the odd Gd- (Fig. 7) and the odd Dy- (Fig. 8) Isotopes.

References
1) D.D. Warner et al., J.Phys. G5, 1723 (1979)
2) H.R. Koch et al., Nucl.Instr. Meth. 175, 251 (1979)
3) R.E. Chrien, Inst. Phys. Conf. Ser. 62, 342 (1982)
4) R.C. Greenwood and R.E. Chrien,
   Nucl. Instr. Meth. 138, 125 (1976)
5) W. Mampe et al., Nucl Instr. Meth. 154, 127 (1978)
6) M. Löffler et al., Nucl. Instr. Meth. 111, 1 (1973)
7) V.G. Soloviev et al.,Bull. Akad. Sci. USSR,
   Phys. Ser. 31, 515 (1967)

INVESTIGATION OF INTER- AND INTRABAND TRANSITIONS OF THE 0_2^+, 2_1^+, 0_3^+ BANDS IN ^{156}Gd

F. Hoyler, K. Schreckenbach, H.G. Börner, G. Colvin
I.L.L., F-38042 Grenoble, France

ABSTRACT

A search for inter- and intraband transitions of the 0_2^+, 2_1^+, 0_3^+ bands in ^{156}Gd was performed by means of the ^{156}Gd(n,γ) reaction. The curved crystal spectrometers GAMS 1,2/3 and the conversion electron spectrometer BILL at the I.L.L. high flux reactor were used for this investigation. A sensitivity of $3 \cdot 10^{-6}$ per neutron capture at 100 keV transition energy was obtained. With this sensitivity it was possible to measure the ratio of the B(E2)-values of the 2_γ-0_β and the 2_γ-0_g transition as $B(E2, 2_\gamma$-$0_\beta)/B(E2, 2_\gamma$-$0_g)=0.97(50)$. This result is in good agreement with the prediction of the IBA for deformed nuclei. An E0 branch from the 0_3^+ to the 0_2^+ state was observed.

INTRODUCTION

The ^{155}Gd(n,γ) reaction has been extensively studied by Bäcklin et al.[1] using a curved crystal spectrometer, Ge(Li) detectors and an electron spectrometer. This study resulted in a level scheme comprising more than 50 levels up to 2.3 MeV excitation energy. Particularly the low spin members of the 0_2^+ band (β band) at 1049 keV, the 0_3^+ band (i-band) at 1168 keV and the 2_1^+ band (γ band) at 1154 keV are well established. From the theoretical point of view ^{156}Gd is the widely used example of the almost complete realisation of the SU(3) symmetry in nuclear physics[2]. On the other hand there are also investigations on configuration mixing in an extended IBA frame to explain the occurence of the low lying 0_3^+ intruder band, which cannot be explained in the conventional IBA model[3].

0094-243X/85/1250410-04 $3.00 Copyright 1985 American Institute of Physics

EXPERIMENTAL PROCEDURE

The (n,γ) reaction on ^{155}Gd was measured using the 5.6m bent crystal spectrometer GAMS1 for the low transition energies. Due to the prior knowledge of the level scheme it was possible to measure specific lines with very high sensitivity (3 10^{-6}/per capture at 1oo keV). The targets consisted of 5 mg highly enriched ^{155}Gd (99.8%). In the high flux (ϕ_n=5.5 10^{14}ncm^{-2}s^{-1}) the half life of ^{155}Gd was about 24h, including self shielding. The impurity of ^{157}Gd was burned-up much faster. The conversion electrons were measured using the β - spectrometer BILL. The half life of the 90 μg/cm^2 Gd targets was in the flux of 3.3 10^{14}ncm^{-2}s^{-1} about 17h. Fig. 1 highlights the enormous complexity of the electron spectrum and the high sensitivity of the measurement. The steep increase of the background to the low energies is due to the strong 89 keV transition (2_g-0_g), which masks several of the interesting transitions.

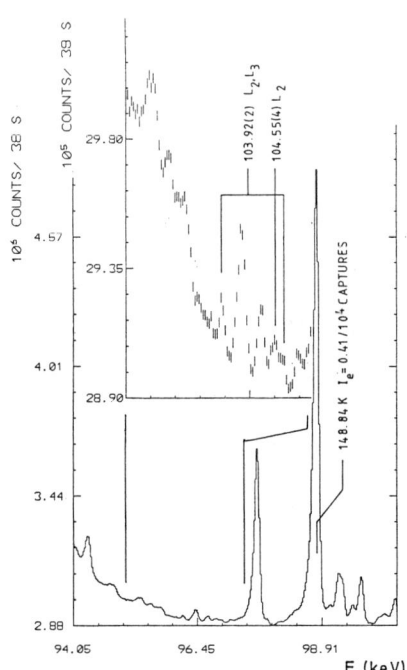

Fig. 1 Part of the electron spectrum measured with BILL

RESULTS AND DISCUSSION

The measured inter- and intraband transitions are summarised in Fig. 2. The assignment is based on energy considerations and the comparison of the γ - ray intensities with the electron intensities. Due to the high density of of the spectrum, interference with transitions between other high lying levels cannot be excluded.

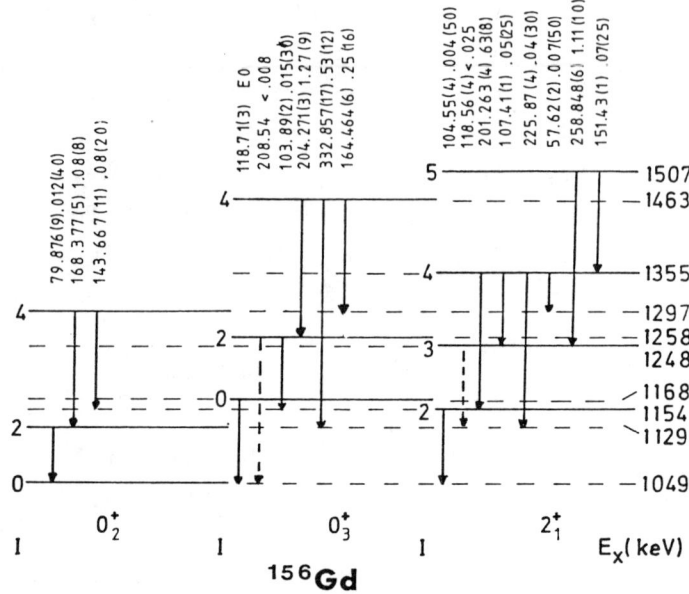

Fig. 2 Summary of the measured transitions in ^{156}Gd (Intensities in quanta/10^4 captures with errors in percent)

Intraband transitions

Except the 2_i-0_i and the 3_γ- 2_γ transition, which are in the vicinity of the very strong 2_g-0_g line, up to spin 5 all possible transitions have been observed. Several of these transitions have already been placed in the previous measurement[1]. From the ICC the E2 character of most of these transitions could be confirmed.

Interband transitions

Great care was taken to measure the transitions between the γ- and β- band. The 2_γ-0_β transition at 104.6 keV (Fig.1) could be seen in the GAMS and the BILL measurement with consistent intensities for a pure E2 multipolarity. The branching ratio

$$\frac{B(E2,2_\gamma-0_g)}{B(E2,2_\gamma-0_\beta)} = \frac{I_\gamma(2_\gamma-0_g)E_\gamma^5(2_\gamma-0_\beta)}{I_\gamma(2_\gamma-0_\beta)E_\gamma^5(2_\gamma-0_g)}$$

was found to be 0.97 (50). This result is in agreement with the prediction of the IBA in a consistent Q framework.[4] In this model values between 0.7 and 1.1 are calculated for deformed nuclei. The $4_i - 4_\beta$ and the $4_i - 2_\beta$ transitions were already tentatively placed[1]. Their E2 multipolarity found in this measurement confirms this placement. The limit for the $2_i - 0_\beta$ transition was taken from the GAMS measurement. A possible candidate for the $2_i - 2_\gamma$ transition could only be seen in the electron spectrum (Fig.1). The $4_\beta - 2_\gamma$ transition was already measured in the study of Bäcklin et al.[1]. The intensity of the electron line seen in our measurement is consistent with E2 multipolarity. The E0 transition $0_i - 0_\beta$ at 118.7 keV is relatively strong ($I_e = 0.18 / 10^4$ captures) and no transition is found at this energy in the γ measurement. The measurement of the $0_i - 0_g$ E0 transition enabled us to determine the branching ratio $B(E0, 0_i - 0_g)/B(E0, 0_i - 0_\beta) = 0.021$ (6). Thus the E0 branch to the 0_β^+ state is strongly dominating over the E0 transition to the ground state.

REFERENCES

1. A.Bäcklin et al. Nucl.Phys.A380 (1982) 189
2. A.Arima, F.Iachello, Ann.ofPhys.111 (1978)201
3. P.Van Isacker et al. Nucl.Phys.A380 (1982) 383
4. D.D.Warner, R.F.Casten Phys.Rev.C28 (1983) 1798

ENERGY LEVELS OF ^{164}Dy FROM THE REACTION ^{163}Dy(n,γ)

T.J.Al-Janabi, A.K.Mheemeed, S.S.Kamoon and S.T.Ahmed
Department of Nuclear Physics, Nuclear Research Centre,
Tuwaitha - Baghdad - Iraq

ABSTRACT

The nuclide ^{164}Dy has been studied by several reactions and results up to September 1973 were summarised by Buyrn[1]. There has also been a recent investigation using (n,n'γ) reaction[2], gamma-gamma directional correlation measurements following thermal neutron capture[3] and average resonance capture[4] measurements.

The present measurements were carried out using the thermal neutron beam of the IRT-5000 reactor in Baghdad[5]. The ^{163}Dy target used in these measurements consisted of 2g of Dy_2O_3 enriched to 92.8% in ^{163}Dy. The 4.5% impurity of ^{164}Dy represents a capture probability of 49.9% due to its large cross-section. Therefore, a separate ^{164}Dy target enriched to 96.99% ^{164}Dy has been studied as well.

The γ-rays following thermal neutron capture in ^{163}Dy target have been investigated with a Ge(Li) detector operated in anti-Compton and pair modes. Gamma radiations in the energy range from 966.0 to 5195.9 keV were investigated for the first time. The present measurements have also provided accurate values for the high energy γ-transitions with precision better than 0.6 keV using the intense lines of ^{36}Cl for the energy calibration[6]. Approximately 320 γ-rays were observed out of which, 113 were attributed to the reaction ^{163}Dy(n,γ)^{164}Dy and 70 have been incorporated into the level scheme of ^{164}Dy with 29 excited states. The remaining 207 γ-transitions were ascribed to the reaction ^{164}Dy(n,γ)^{165}Dy.

Table I shows the level energies obtained from the present investigation. Energies close to 1809.9, 1949.4, 2078.3, 2124.5, 2152.4, 2207.6, 2348.2 and 2531.2 keV have not been observed in thermal neutron capture studies prior to the present work. The present results were compared with the previously published values of Warner et al.[4] and the agreement was found to be satisfactory.

The neutron binding energy was determined to be 7657.68(11) keV where the error is statistical. This value is consistent with the value 7655.3(13) keV of ref.(1).

Table I Levels in ^{164}Dy from the ^{163}Dy(n,γ)^{164}Dy reaction

Level energy keV	Level energy keV	Level energy keV
0	1123.36(18)	1979.32(15)
73.64(10)	1226.02(21)	2078.30(21)
242.50(15)	1588.59(41)	2124.49(34)
501.54(20)	1716.15(16)	2152.36(18)
762.22(11)	1796.64(14)	2207.62(31)
828.57(13)	1809.98(30)	2248.55(28)
916.56(13)	1840.60(17)	2348.25(44)
976.99(13)	1921.49(31)	2459.60(17)
1024.78(42)	1933.40(15)	2531.18(16)
1039.83(16)	1949.38(32)	

REFERENCES

1. A.J. Buyrn; Nucl. Data Sheets 11 327(1974)
2. H.R. Hooper, J.M. Davidson, P.W. Green, D.M. Sheppard and G.C. Neilsen; Phys. Rev. C15 1665(1977)
3. P. Hungerford, W.D. Hamilton, S.M. Scott and D.D. Warner; J. Phys. G. Nucl. Phys. 6 741(1980)
4. D.D. Warner, R.F. Casten, W.R. Kane and W. Gelletly; Phys. Rev. C27 2292(1983)
5. T.J. Al-Janabi, J.D. Jafar, S.J. Hassan, K.M. Mahmood and M.A. Al-Amili; to be published
6. B. Krushe, K.P. Lieb, H. Daniel, T. Von Egidy, G. Barreau, H.G. Borner, R. Brissot, C. Hofmeyr and R. Rasher; Nucl. Phys. A386 245(1982)

THE LEVEL SCHEME OF ^{166}Dy OBTAINED BY DOUBLE NEUTRON CAPTURE

S.A. Kerr*, F. Hoyler, K. Schreckenbach, H.G. Borner, G. Colvin
ILL, F38042 Grenoble, France

P.H.M. Van Assche, E. Kaerts
SCK-CEN, B2400 Mol, Belgium

ABSTRACT

The level scheme of $^{166}_{66}$Dy has been obtained from measurements of the gamma rays and conversion electrons from the ^{164}Dy$(2n,\gamma)^{166}$Dy reaction. The γ-vibration bandhead occurs at 857.151(12)keV and the neutron separation energy of ^{166}Dy is 7043.5(4)keV. The results of this work show that, within the systematics of the even-even Dy isotopes, maximum deformation occurs at N = 98 (A = 164).

INTRODUCTION

The level scheme of $^{166}_{66}$Dy has been investigated by the two-neutron capture reaction on a target of isotopically enriched ^{164}Dy. As far as the authors are aware, no information exists on the excited states of ^{166}Dy apart from the preliminary results of this study which have been reported previously[1]. This nucleus is of particular interest since it is the most neutron rich Dy isotope studied to date and therefore extends systematic investigations of such features as nuclear deformation further from stability and nearer to the middle of the N = 82 to 126 neutron shell.

EXPERIMENTS

The double neutron capture reaction provides one of the few possibilities for the study of ^{166}Dy :

$$^{164}\text{Dy} \xrightarrow{2700b} {}^{165}\text{Dy} \xrightarrow{3900b} {}^{166}\text{Dy}$$
$$\downarrow 2.35h \qquad\qquad \downarrow 81.5h$$
$$^{165}\text{Ho} \xrightarrow{63b} {}^{166}\text{Ho}$$

The β-decay of ^{165}Dy competes with the double capture process. In a constant neutron flux, ϕ_n, and in saturation of the 2.35h activity, the double to single capture rate is proportional to ϕ_n^2. This strong ϕ_n dependence implies that it is only feasible to observe ^{166}Dy by irradiation in a high neutron flux facility. However, the spectrum of gamma-rays from such an irradiation, even with an isotopically enriched (95.7%) ^{164}Dy target, is extremely complex, with contributions from several different isotopes present either as target impurities (e.g. ^{161,2,3}Dy).

---

*Present address : Physics Department, University of Surrey, Guildford, Surrey GU2 5XH, U.K.

or as reaction products (e.g. ^{165,6}Ho, ^{166,7}Er). A series of experimental techniques have therefore been employed in order to unambiguously assign observed lines to ^{166}Dy :

1.) Signals from multiple capture or decay reactions can often be identified by their particular time-dependent intensity behaviour during irradiation. Fig. 1 shows this behaviour for a number of the more dominant processes in this study. However, difficulty arises in differentiating ^{165}Dy from ^{166}Dy since, due to the short half-life of ^{165}Dy, the single and double capture rates reach secular equilibrium within ~ 1/2 day, after which time they both burn-up in an identical manner. Nevertheless, this time dependence has been observed with the conversion electron spectrometer, BILL (ϕ_n ~ 3 x 10^{14}n cm^{-2}s^{-1}) at the ILL by repeatedly cycling, with a 24h period, a ^{164}Dy target between the in-pile measurement position and a low-flux 'cooling' position.

2.) Such time-dependent measurements are not possible with curved-crystal spectrometers (CCS) due to the longer time required for target adjustments before measurements can commence. Therefore gamma spectra were measured at CCS facilities with substantially different neutron fluxes, namely GAMS 1,2,3 at the ILL (ϕ_n ~ 5.5 x 10^{14}n cm^{-2}s^{-1}) and at CEN, Mol (ϕ_n ~ 2.7 x 10^{13}n cm^{-2}s^{-1}). The double to single capture rates in ^{164}Dy at these two facilities were approximately 0.025 and 0.0013, respectively, as shown in Fig. 1, with the result that all gamma-rays from ^{166}Dy were below the sensitivity limit of the Mol instrument. Thus comparison of spectra from these two measurements enabled a number of gamma lines to be assigned to ^{166}Dy.

3.) Finally, the primary gamma-rays from a ^{164}Dy target were recorded with the ILL pair spectrometer. The difference in neutron binding energies (~ 1.3 MeV) between ^{165}Dy and ^{166}Dy enabled several primary transitions to be assigned to ^{166}Dy.

RESULTS

Results from these experiments have permitted the construction of a level scheme for ^{166}Dy, which is summarised in Table 1. The level scheme comprises the ground state rotational band up to spin 6^+ and the rotational band built on the γ-vibration up to spin 4^+, together with the capture-state. The reported intensities have been normalised to a total transition intensity of 100% for the 76.6 keV $2_1^+ \to 0_1^+$ transition. All measured conversion coefficients or L-subshell ratios are consistent with pure E2 multipolarities. The neutron binding energy of ^{166}Dy has been determined as 7043.5(4) keV which is in agreement with, and improves, the earlier Wapstra value[2].

DISCUSSION

Fig. 2 shows the ground state band energy systematics of the even-even Dy isotopes. The ratio $E(4_1^+)/E(2_1^+)$ shows ^{166}Dy to be a well-deformed rotor, as expected, whilst the behaviour of the 2_1^+ energy demonstrates that maximum deformation occurs at N = 98 (A = 164)[1] which is before the neutron shell N = 82 to 126 is half-filled.

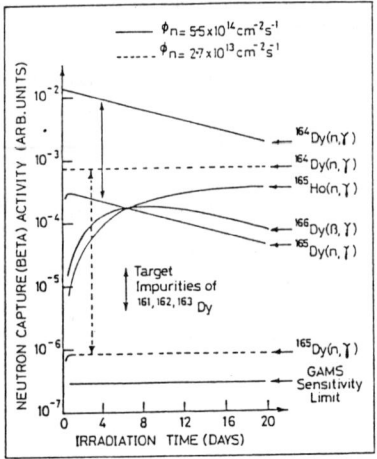

Fig. 1. Multiple neutron capture in a ^{164}Dy target calculated for different neutron fluxes

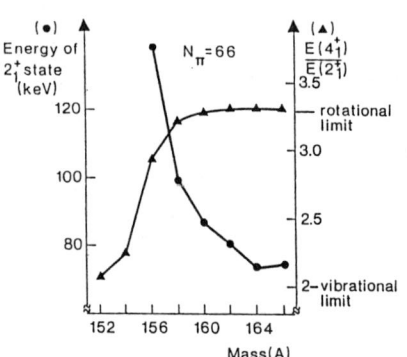

Fig. 2. Systematics of even-even Dy-isotopes showing that maximum deformation occurs at $N = 98$ ($A = 164$)

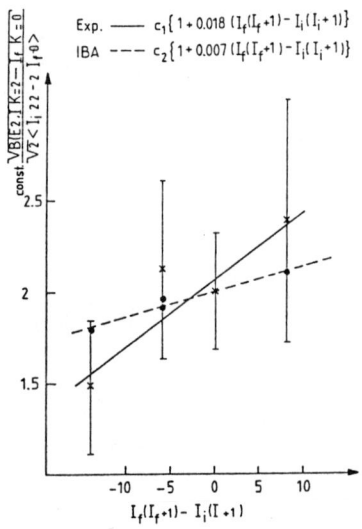

Fig. 3. "Mikhailov plot" of the relative reduced transition probabilities from the γ-band to the g-band in ^{166}Dy

Finally, consideration has been given to the observed branching ratios of the γ → g transitions and their deviations from the simple rotational values as predicted by the Alaga rules. A Mikhailov plot[3] (Fig. 3) has therefore been made of those branching ratios which could be obtained from the level scheme. These data fall on a straight line, within the errors, the slope and intercept of which can be used to extract the single parameter, Zγ, commonly employed as a measure of the ΔK = 2 band mixing effects. For the case of ^{166}Dy, Zγ was found to be 0.026 (12), which, in the context of the systematics of the rare earth nuclei[3], is consistent with a maximum Zγ at the mid neutron shell and is indicative of weak γ-g mixing. The slope of this line was compared with the prediction of the IBA in the exact SU(3)-limit using the analytic expressions for E2 transitions given by Van Isacker[4]. As for ^{168}Er the model underestimates the experimental value. The discrepancy is due to the over simplified exact SU(3)-Hamiltonian as has been shown in a numerical study of this problem[3].

Table 1 : Summary of the level scheme of ^{166}Dy

E_x(keV)	J^π	E_γ(keV)	I_γ	L_1/L_2 exp.	theor.(E2)	Transition g → g
g.s.-band						
76.583(1)	2^+	76.583(1)	12	.095(12)	.091	2 → 0
253.514(7)	4^+	176.939(1)	30	.56(2)	.56	4 → 2
526.942(12)	6^+	273.438(2)	7.4	1.04(8)	1.13	6 → 4
γ-band				α_K(x 10^{-4})		γ → g
857.151(12)	2^+	857.148(35)	8.5	35.35(360)	norm.	
		780.566(11)	6.7	43(5)	43	2 → 2
928.711(19)	3^+	852.118(9)	17	37(4)	36	3 → 2
		675.198(21)	2.7	64(11)	60	3 → 4
1023.391(19)	4^+	946.781(61)	3.8	24(4)	29	4 → 2
		769.902(12)	7.2	50(5)	45	4 → 4
capture state			I_γ rel.			c → g
7043.5(4)	3,4$^+$	6968.0(10)	0.4			c → 2
		6789.6(4)	1.0			c → 4

REFERENCES

1. S.A. Kerr, F. Hoyler, K. Schreckenbach, H.G. Borner, G. Colvin, Proc. Int. Conf. Nucl. Phys. Florence, Sept. 1983, Vol. 1, pp.166
2. A.H. Wapstra, K. Bos, At. Dat. and Nucl. Dat. Tab., 20, 3 (1977)
3. R.F. Casten, D.D. Warner, A. Aprahamian, Phys. Rev. C28, 894 (1983)
4. P. Van Isacker, Phys. Rev. C27, 2447 (1983)

M1-E2 MIXING RATIOS OF GAMMA-GROUND TRANSITIONS IN ^{168}ER

W Gelletly
University of Manchester, Manchester M13 9PL, U.K.

D D Warner
Brookhaven National Laboratory, Upton, Long Island, N.Y., U.S.A.

G C Colvin and K Schreckenbach
Institut Laue-Langevin, 38042 Grenoble, France

ABSTRACT

The L subshell ratios of four γ-g transitions in ^{168}Er have been measured with the high resolution, iron, β-spectrometer at the Institut Laue-Langevin. The reduced M1-E2 mixing ratios are not consistent with the values predicted by Warner on the basis of IBA-1. It may be possible to explain the results in IBA-2 in terms of the ΔK=1 mixing of the gamma-band and the pure antisymmetric states such as the 1+ isovector mode.

INTRODUCTION

M1 transitions in even-even nuclei are forbidden in the simplest form of collective models because the leading term of the collective M1 operator is proportional to the total angular momentum, which is a good quantum number. Such transitions are observed however in the decay of the gamma band in deformed nuclei. To date their origins are not fully understood.

Schreckenbach and Gelletly[1] studied the M1 components in the intraband transitions within the gamma band of ^{168}Er by measuring the L subshell conversion line intensity ratios for these transitions. A comparison with the less precise measurements on neighbouring nuclei revealed that the experimental E2/M1 mixing ratios for the ΔI = -1 transitions within the γ-band are remarkably constant in the even-even, rare-earth nuclei. On the assumption that IBA-1 gives the E2 decay rates for levels in the γ-band Warner et al[2] showed that B(M1) ≈ 8 x 10^{-4} s.p.u. for all these M1 transitions. The constancy and small size of these M1 components in a range of nuclei argue that these transitions are collective.

On the basis of these measurements of mixing ratios in intraband transitions Warner[3] derived values of

$$\Delta(^{E2}/_{M1}) = \frac{\delta(^{E2}/_{M1})}{0.835\, E_\gamma\, (\text{MeV})}$$

for the γ-g transitions in a series of rare-earth nuclei on the assumption that the M1 matrix element is effectively proportional to the E2 matrix element. Overall he found passable agreement with experiment although the errors on the experimental values, which are mainly derived from angular correlations, are large. γ-γ correlations are a powerful tool for measuring spins, parities and mixing ratios but they are subject to systematic errors and ambiguities where the spectra are complex.

MEASUREMENTS

Appropriate sections of the conversion electron spectrum from the ^{167}Er (n,e$^-$) ^{168}Er reaction were scanned with the BILL spectrometer. The target (3 × 10 cm) consisted of ≈ 100 μg cm^{-2} enriched ^{167}Er evaporated on to an Al backing. A ten wire proportional

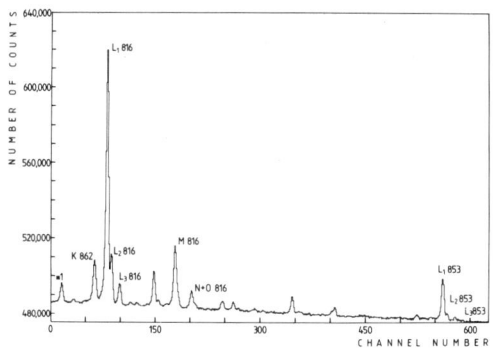

Fig. 1. A section of the ^{167}Er (n,e$^-$) ^{168}Er spectrum.

counter was used to detect the electrons. Each set of lines was scanned at least three times. An example of a section of the spectrum is shown in fig. 1.

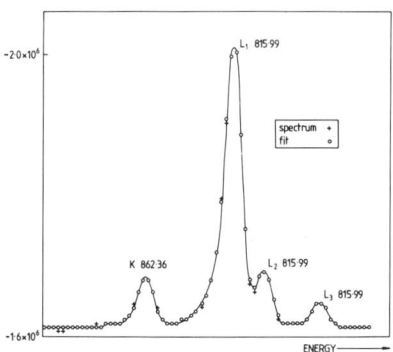

Fig. 2. A fit to the L lines of the 815.99 keV, $3^+_\gamma - 2^+_g$ transition

Fig. 2 shows a fit to the L Lines of the 815.99 keV, $3^+_\gamma - 2^+_g$

transition. In all it was possible to extract L subshell ratios for four of the γ-g transitions studied. Overlapping lines, identified from our previous, extensive study[4] of this reaction, interfered in the other three cases.

RESULTS

The results are summarised and compared with previous results and Warner's predictions in Table 1. In the only case where Δ had been measured previously with high precision the agreement is excellent.

The measured values are less than half of Warner's predicted values, which suggests that we require a more sophisticated description of the M1 matrix element in IBA-1.

A better explanation may be found in IBA-2, where neutron and proton degrees of freedom are included explicitly. Here M1 transitions result from admixtures of states which are not fully symmetric with respect to the neutron and proton degrees of freedom. These admixtures depend on the positions of the pure antisymmetric states. A principal component of any admixture in the rare-earth nuclei is expected to be the 1^+ isovector mode which was recently identified [5] in ^{156}Gd. If this state is fairly constant in energy in the rare-earth region it would explain the observed constancy of the M1 components in the γ-band, intraband transitions. At the same time the observed M1 components provide a clue to the portion of the 1^+ isovector mode.

TABLE 1

Preliminary results of measurements of M1-E2 mixing in gamma-ground transitions in ^{168}Er.

I_i	I_f	IBA^2	$\Delta(E2/M1)$ Previous expts.[3]	Present Work
2^+	2^+	-26.5 ± 1.0	$|\Delta| > 140$	--
3^+	2^+	-25.0 ± 1.0	$+24.7^{+3.7}_{-4.7}$	$7.5^{+1.8}_{-1.0}$
3^+	4^+	-18.1 ± 0.7	-9.3 ± 0.6	$9.6^{+10.6}_{-2.4}$
4^+	4^+	-13.8 ± 0.5	$-164^{+98}_{-\infty}$	--
5^+	4^+	-14.2 ± 0.5	--	$5.1^{+3.3}_{-1.1}$
5^+	6^+	-11.9 ± 0.5	--	--
6^+	6^+	-9.6 ± 0.4	--	$5.0^{+2.6}_{-1.0}$

REFERENCES

1. K Schreckenbach and W Gelletly, Phys. Rev. Letters 94B (1980) 298.
2. D D Warner et al, Phys. Rev. Letters 45 (1980) 1761; Phys. Rev. C24 (1981) 1713.
3. D D Warner, Phys. Rev. Letters 47 (1981) 1819.
4. W F Davidson et al, J. Phys. G: Nuclear Physics 7 (1981) 455.
5. D Bohle et al, Phys. Letters 137B (1984) 27.

A STUDY OF THE LOW-LYING STATES IN ^{178}Hf THROUGH THE NEUTRON CAPTURE REACTION *

A.M.I. Haque, R. Richter, A. Gelberg, I. Förster, R. Rascher and
P. von Brentano, Institut für Kernphysik, Universität Köln, FRG
H.G. Börner, K. Schreckenbach, S.A. Kerr, G. Barreau and R.
Brissot, Institute Laue-Langevin, Grenoble Cedex, France and
R.F. Casten and D.D. Warner, Brookhaven Nat. Lab.,Upton,N.Y., USA

The decay of the low-lying states of ^{178}Hf was investigated using: (1) High-resolution curved crystal spectrometry of the secondary γ-rays using the GAMS-1 and GAMS 2/3 facilities at the ILL, (2) Measurements of the secondary (n,e$^-$) transitions using the Electron Spectrometer BILL at the ILL, (3) Measurements of the primary γ-transitions following thermal neutron capture with the pair-spectrometer at the ILL and (4) Average Resonance Capture (ARC) measurements at the neutron energies of 2 keV and 24 keV, using the tailored beam facilities at BNL.

A level scheme including 69 levels and 270 transitions up to an excitation energy of 2.1 MeV was constructed. Most of the levels were ordered in 18 different rotational bands. The levels assigned to rotational bands along with the deexcitation modes of the γ-band (inset), are displayed in Fig. 1. The level scheme of ^{178}Hf seems to be complete below 2 MeV for spins between 2 and 5.

In order to determine which, if any, of the numerous 0^+ bands are collective, approximate absolute B(E2) values to the ground state band were estimated from the measured inter to intra band transition intensities using the plausible assumption that the intraband intrinsic matrix elements were similar to the ground

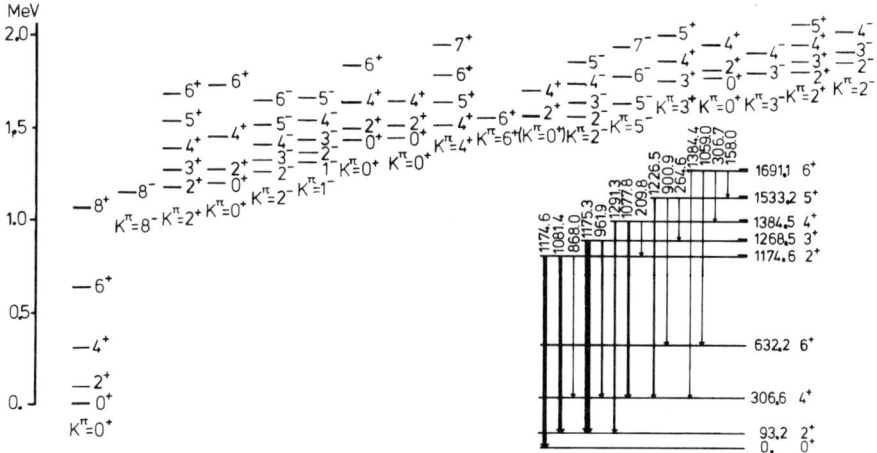

Fig. 1   Experimental rotational bands in ^{178}Hf (inset;γ-band decay)

Table - 1 ESTIMATED ABSOLUTE B(E2) VALUES (IN W.U.) FOR TRANSITIONS FROM EXCITED K=0⁺ BANDS IN ^{178}Hf TO THE GROUND STATE BAND[a]

$I_i \rightarrow I_f$	1199 keV Band	1434 keV Band	1443 keV Band	1500 keV Band
$2^+ \rightarrow 0^+_g$		0.0032(4)	0.00084(7)	
$4^+ \rightarrow 2^+_g$	0.124(28)		0.0218(22)	0.0033(4)
$4^+ \rightarrow 6^+_g$	1.243(75)			
$6^+ \rightarrow 4^+_g$		0.019(2)		
$6^+ \rightarrow 8^+_g$	0.195(16)			

a) $\dfrac{\langle I_i K=0 || E2 || I_g \rangle}{\langle I_i K=0 || E2 || I_f K=0 \rangle_{Intra}} \cdot \langle I_g || E2 || I_g-2 \rangle_{Intra}$ in W.U.

band. The results are shown in Table-1. It is apparent that the only candidate for a ß-band is the band at 1199 keV.

The extensive nature of the data provides an opportunity for a detailed test of the IBA model[1]. The Consistent Q Formalism (CQF) (Ref. 2) was used. The Hamiltonian has the form

$$H = -\varkappa Q \cdot Q - \varkappa' L \cdot L$$

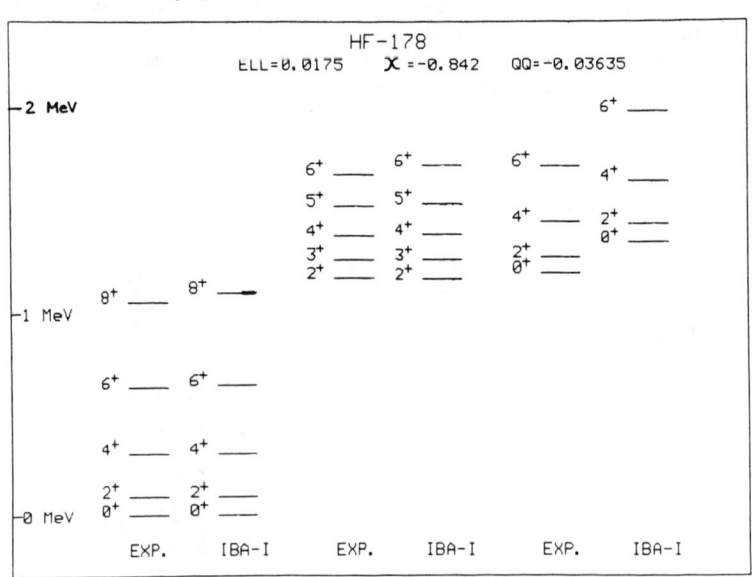

Fig. 2 Energy fit for the ground state, γ and ß bands (for X =-0.62, the IBA-energies of the beta-band lie ∿200 keV higher)

Table - 2 DECAY PROPERTIES OF THE GAMMA AND BETA BANDS IN ^{178}Hf

E_i	$I_i, K_i=2$	E_f	I_f, K_f	I_γ (rel.)[a]	B(E2) Expt.	Alaga	B(E2) IBA[b] $X=-0.62$	$X=-0.842$
				γ - band				
1174.6	2_2^+	0.0	$0_g^+,0$	25.02 ±2.46	100	100	100	100
		93.2	$2_g^+,0$	18.87 ±0.41	145±20	143	178	168.2
		306.6	$4_g^+,0$	0.369±0.020	9.34±1.3	8	13.2	11.6
1268.5	3_1^+	93.2	$2_g^+,0$	29.53 ±3.28	100	100	100	100
		306.6	$4_g^+,0$	5.95 ±0.21	56±8	40	63.4	57.2
1384.5	$4_2^!$	93.2	$2_g^+,0$	6.36 ±0.21	0.821±0.162	1.55	0.888	1.033
		306.6	$4_g^+,0$	13.95 ±0.41	4.56±0.84	4.56	4.56	4.56
		1174.6	$2_2^+,2$	0.082±0.021	100	——	69.4	174.5
1533.2	$5_1^!$	306.6	$4_g^+,0$	9.02 ±0.41	4.33±0.78	6.78	3.29	3.86
		632.2	$6_g^+,0$	1.723±0.123	3.87±0.75	3.87	3.87	3.87
		1268.5	$3_1^!,2$	0.185±0.041	100	——	130.1	347.1
1691.1	6_2^+	306.6	$4_g^+,0$	0.882±0.082	0.531±0.10	0.93	0.362	0.473
		632.2	$6_g^+,0$	1.497±0.123	3.45±0.63	3.45	3.45	3.45
		1384.5	$4_2^+,2$	0.082±0.021	100	——	102.9	256.6
				β- band				
1276.7	2_3^+	0.0	0_g^+	1.477±0.082	0.17±0.01	0.38	0.29	0.30
		93.2	2_g^+	7.998±0.615	——[c]	0.56	0.342	0.350
		306.6	4_g^+	2.256±0.205	1	1	1	1
1450.4	4_3^+	93.2	2_g^+	0.677±0.082	0.063±0.004	0.63	0.352	0.387
		306.6	4_g^+	10.66 ±0.82	——[c]	0.57	0.127	0.125
		632.2	6_g^+	0.861±0.082	1	1	1	1
		1276.7	2_3^+	0.103±0.010	(0.27±0.03)*10^3	——	5.58*10^3	21.6*10^3
1731.1	6_3^+	632.2	6_g^+	0.287±0.041	——[c]	0.59	0.009	0.005
		1058.6	8_g^+	0.041±0.004	1	1	1	1
		1450.4	$4_3^!$	1.025±0.062	(2.12±0.29)*10^3	——	5.41*10^3	18.8*10^3

a) Normalised to 100 for I_γ ($2^+ \to 0^+$)
b) For $X=-0.62(-0.842)$; ELL=0.01378(0.0175), QQ=-0.0463(-0.03635)
c) For $\Delta I=0$ transitions the M1 components could not be extracted

where $Q = (s^\dagger \tilde{d} + d^\dagger s)^{(2)} + \chi(d^\dagger \tilde{d})^{(2)}$. The corresponding parameters for the IBA code PHINT1 (ref. 3) are QQ=-4$\varkappa$, ELL=-2$\varkappa'$. The E2 transition operator is $\underline{T(E2)} = \alpha Q$, where α determines the absolute scale. X varies from $-\sqrt{7/2}$ to 0 associated with the symmetry limits SU(3) and O(6).

Two calculations were performed. 1) $X = -0.62$: this gives the best results for the relative B(E2) values for the inter and

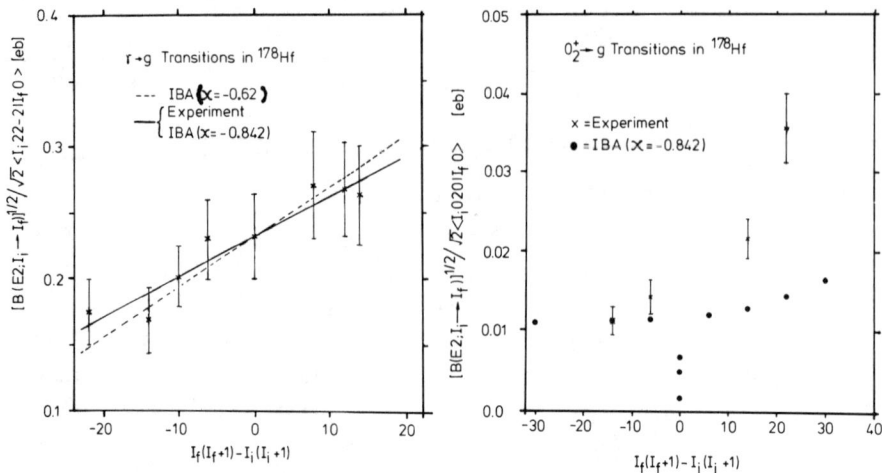

Fig. 3 Mikhailov plots for the $\gamma \rightarrow g$ and $\beta \dashrightarrow g$ transitions. The straight lines are least square fits to the data and the IBA

intraband transitions for the γ-band. 2) $\chi = -0.842$: this gives a straight line on a Mikhailov plot of $\gamma \rightarrow g$ transitions which is identical to the experimental one. The wave functions depend only on χ. $\varkappa$ and $\varkappa'$ were required to fit the excitation energies only. The energy fit is shown in fig. 2 for calculation 2. Table 2 gives the results for B(E2) values and fig. 3 gives Mikhailov plots for the $\gamma \rightarrow g$ and $\beta \dashrightarrow g$ transitions. The results show excellent agreement for the ground and γ-bands, qualitative accord for the energy of the lowest 0^+ band, and little or no agreement for the B(E2) values of this band. Both the data and the IBA calculations for this band suggest that its decay is more complicated than can be accounted for by simple 2-band mixing. From the $\gamma \rightarrow g$ Mikhailov plots one can extract the following values for the bandmixing parameter Z_γ (ref. 4) :

Z_γ : Exp. 0.0249(17) ; IBA: 0.0249 ($\chi = -0.842$) & 0.0314 ($\chi = -0.62$)

The experimental value is consistent with regional systematics[2].

* Work supported in part under contract DE-AC02-76CH00016 with the United States Dept. of Energy. Work supported by the Ministerium für Wissenschaft und Forschung NRW, FRG. RFC acknowledges support from the von Humboldt foundation, FRG.

References :

(1) A. Arima and F. Iachello ; Ann. Phys. (NY) 99, 253 (1976)
(2) D.D. Warner and R.F. Casten, Phys. Rev. C28, 1798 (1983)
(3) O. Scholten ; Program PHINT1, private communication
(4) L.L. Riedinger et al, Phys. Rev 179, 1214(1969)

HIGH PRECISION NEUTRON CAPTURE GAMMA-RAY AND CONVERSION ELECTRON MEASUREMENTS OF ^{181}Ta

I.Förster°[*], H.G.Börner[*], P.v. Brentano°, G. Colvin[x], A.M.I. Haque°
S.A.Kerr[*^], R.Rascher°, R.Richter°, K. Schreckenbach[*]

° Institut of Nuclear Physics, University of Köln, West Germany
[*] ILL, Grenoble, France
[^] Present address: Physics Department, University of Surrey, U.K.

As a part of neutron capture γ-ray studies in the rare earth region, the deexcitation spectra of the reactions ^{180}Ta(n,γ)^{181}Ta and ^{180}Ta(n,e$^-$)^{181}Ta have been measured at the ILL high flux reactor, using bent crystal, pair and electron spectrometers. Due to the exceptional high target spin ($I^\pi=9^-$) levels with spin as high as 21/2 were populated, a region normaly reserved for heavy ion reactions. About 73% of the measured intensity was placed in ^{181}Ta level scheme, defining 45 excited states, most of which were previously unknown. Spin-parity assigments and γ-ray branching ratios were also

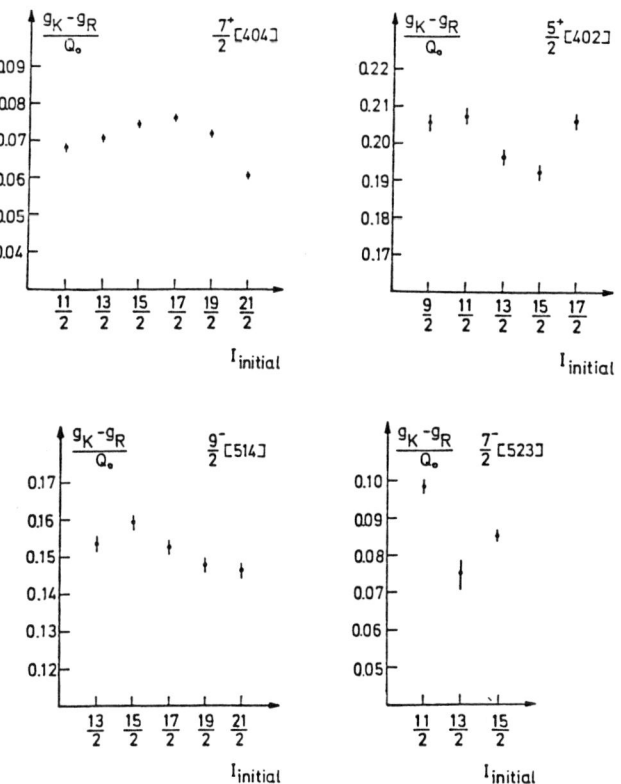

Fig.1: The $(g_K-g_R)/Q_o$ ratio for the intra-band transitions, calculated with the aid of the Alaga rules.

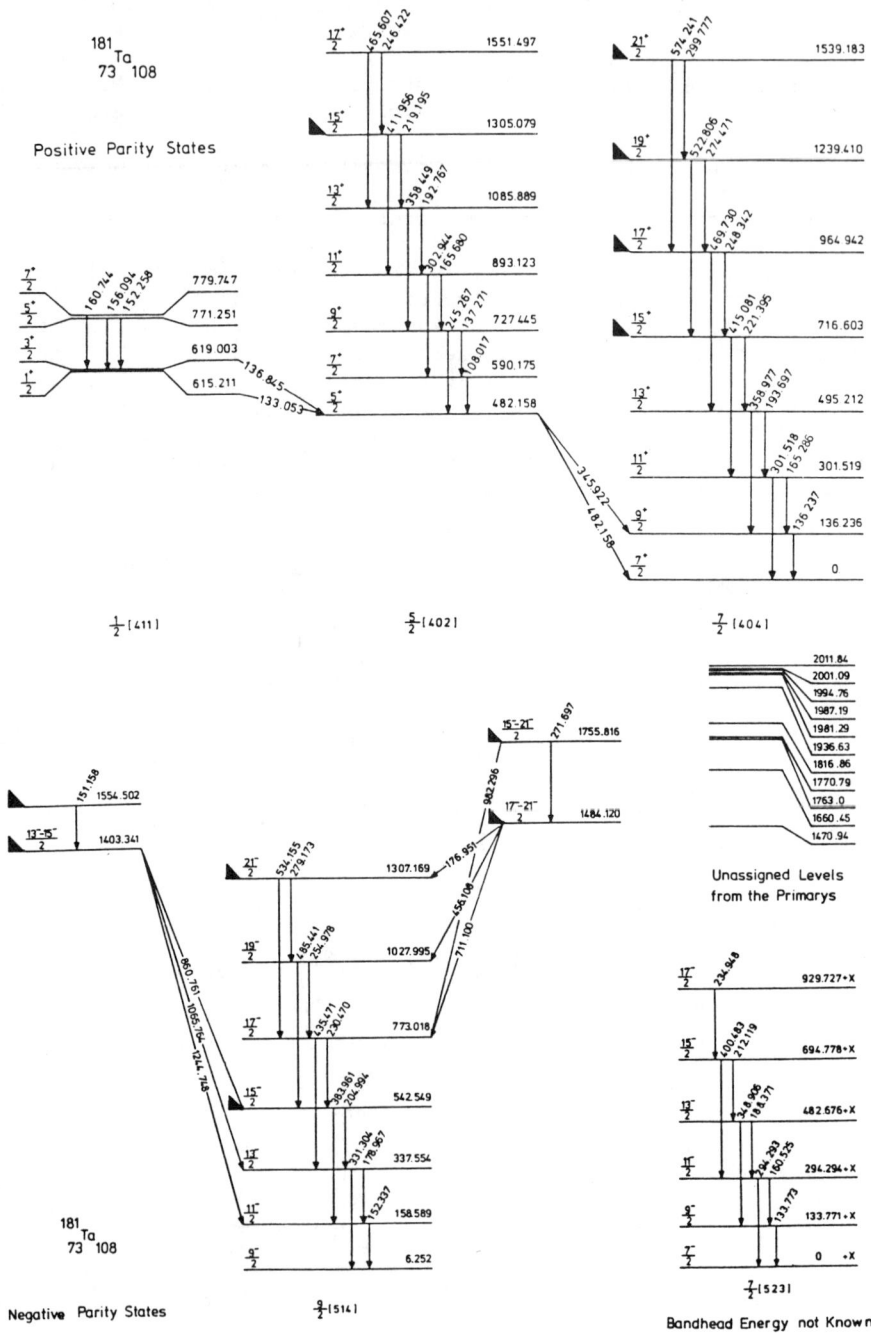

Fig. 2: The level scheme of ^{181}Ta

determined. The neutron binding energy was found to be
Q_n = 7652.18(20) keV and a thermal neutron capture cross section
σ = 420 ± 200 b was calculated. The $7/2^+$ |404| ground state band and
the $9/2^-$ |514| first excited band are in agreement with previous
data[1] up to the 21/2 levels. A new rotational band has been constructed on the $5/2^+$ |402| Nilsson state (see fig.2). Another 7/2
band, thought to be built on the $7/2^-$ |523| state, has also been observed, but the band head energy is not yet known. Only the first
two level of the $1/2^+$ |411| band had been established so far.

For the positive parity $7/2^+$ |404| and $5/2^+$ |402| bands and the
negative parity $9/2^-$ |514| and $7/2^-$ |523| bands a two band Coriolis
mixing calculation was performed.

Assuming a rigid axially symmetric deformed core the Hamiltonian can be written as

$$H = H'_p + \frac{\hbar^2}{2\Theta} I(I+1) + H_C \qquad (1)$$

$$H_C = -\frac{\hbar^2}{2\Theta}(I_+j_- + I_-j_+) \qquad (2)$$

where H'_p describes the particle-core coupling and H_C the particle-rotation coupling, so called Coriolis coupling.

The Coriolis matrix element <11/2 5/2|H_C|11/2 7/2> between the
11/2 state of the 5/2 |402| and 7/2 |404| bands was calculated to
be -24.6 keV. The calculated mixing between the bands, which are
separated by ~500 keV, was therefore very small and, thus, only
weak interband E2 transitions are expected. For example, the
branching ratio is

$$\frac{B(E2, 11/2\ 5/2 \to 7/2\ 7/2)}{B(E2, 11/2\ 5/2 \to 7/2\ 5/2)} = 0.0006$$

This is in good agreement with our experiment, which gave an upper
limit of 0.008.

The value for $(g_K-g_R)/Q_0$ were found to be constant within
±8% in each of this bands comparing up to 6 branchings in the
|404|, |402| and |514| bands respectively (see. fig.1). The
constancy of $(g_K-g_R)/Q_0$ of the two bands calculated from the
intra-band branchings with the aid of the Alaga rules supports
the mixing predictions.

The transition rate for the cross-over transition in a
rotational band is

$$I(E2) = a\ E_\gamma^5 Q_0^2\ <I_i K20|I_i\,-2K>^2 \qquad (3)$$

and for the stop-over transition

$$I(E2'+M1') = b\ E_\gamma'^5 Q_0^2\ <I_i K20|I_i\,-1K>^2 + c\ E_\gamma'^3 (g_K-g_R)^2 <I_i K10|I_i\,-1K>^2 \qquad (4)$$

where a, b and c are constants. $(g_K-g_R)/Q_0$ can be calculated from 3 and 4. The values should be constant in an unturbed band.

Although both the 9/2|514| and 7/2 |523| bands are known, the excitation energy of the latter was not determined. A Coriolis calculation of the mixing of these two bands compresses the energies of one and expands the other. From the empirical difference in inertial parameters an upper limit of 1400 keV for the bandhead energy difference was obtained. The constancy of $(g_K-g_R)/Q_0$ also implies small mixing and therefore a large energy separation. The weak population[2] of the 7/2⁻ |523| band in the (n,γ) reactions puts another limit on its energy. Combining these results, one estimates that it lies 600 - 1400 keV above the 9/2⁻ |514| band. This is consistent with an estimate from the Nilsson model with pairing. For δ = 0.3, with the Fermi energy assumed to be at the position of the 9/2⁻|514| band, and with a pairing gap calculated from the odd-even mass difference, a quasi-particle energy for the 7/2⁻ |523| band was calculated from

$$E_{qp} = \sqrt{(\varepsilon_{7/2}-\lambda)^2 - \Delta^2} - \Delta$$

A value of 900 keV for E_{qp} (7/2 |523|) was obtained in agreement with the above empirical limits.

Acknowledgement

We thank Dr. R.F. Casten for fruitful discussions.

Work supported by BMFT.

References:

1. L.G. Mann, R.G. Lanier, J.T. Larsen, W.J. Richards, 2nd International Symposium of Neutron capture γ - Ray Spectroscopy, Petten 1974, p.578
2. I. Förster, Diplomarbeit an der Universität Köln, 1983

E2/M1 MIXING RATIOS IN ^{196}Pt FROM THE ^{195}Pt$(n,\gamma)^{196}$Pt REACTION

A. M. Bruce
Schuster Lab., Univ. of Manchester, Manchester, M13 9PL, England

D. D. Warner
Brookhaven National Laboratory, Upton, New York, 11973, USA

ABSTRACT

Angular correlations in ^{196}Pt have been measured using a new two detector correlation system at Brookhaven National Laboratory (BNL). The properties of ^{196}Pt can be well described in terms of the O(6) limit of the IBA-1 formalism[1] and the empirical determination of mixing ratios offers a further refinement to that comparison. The measured M1 components may provide information concerning the location of the isovector mode in the O(6) region. In addition, known correlations can provide a test of the new correlation system.

EXPERIMENT

A target of 6.3 gms enriched to 97.3% in ^{195}Pt was placed in the external thermal neutron beam at the monochromater beam facility of the High Flux Beam Reactor at BNL. Angular correlations were measured using a new two detector system which comprises one fixed detector and one moving. The latter is attached to a lead screw by a sliding rod and the screw is driven by a stepping motor. Microswitches at each end serve to reverse the direction of the motor and, in one case, to define the first counting position at 180°. For this experiment the two detectors were situated at 11.6 and 14.6 cm from the target position. Angles ranging between 90° and 180° can be chosen by setting a fixed number of motor steps and the resultant precision in defining the angle is better than 0.1°. The cycle can be repeated with the same degree of accuracy. At each position data is collected for a pre-determined interval before the detector moves the prescribed number of steps to the next counting position. Each time the motor stops, it increments a binary counter which is read by the data acquisition computer and 'tagged' onto the coincidence event to indicate the angle between the detectors. A singles spectrum is also taken at each of the angles and, in this case, the binary signal is used to route the counts into different parts of memory to create a 1K spectrum for each of the measuring stations. These spectra can then be used to normalize for different counting times, neutron beam fluxes, etc.

RESULTS

J^π assignments for the low lying levels in ^{196}Pt have been well established in an earlier study[1]. The present experiment can therefore be used not only to unambiguously determine mixing ratios but also to test the new correlation system using known decay

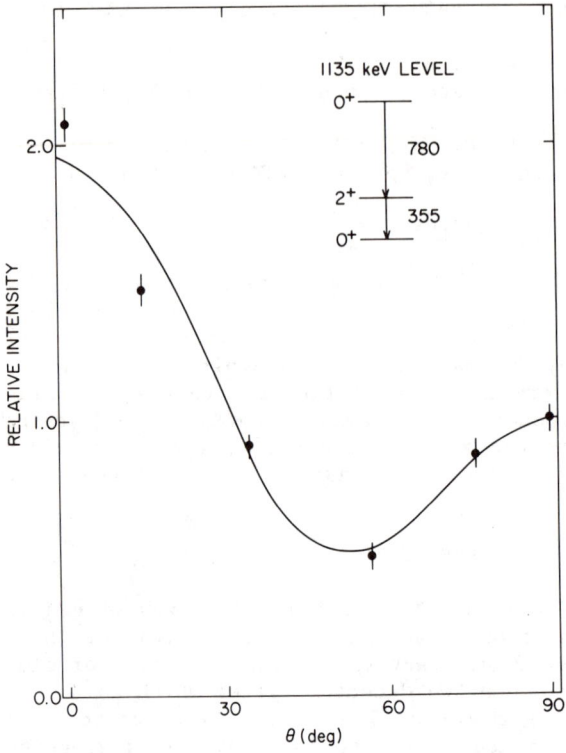

Fig. 1. Gamma-ray intensity as a function of angle for the 780→355 cascade. The solid line represents a theoretical $0^+ \to 2^+ \to 0^+$ correlation.

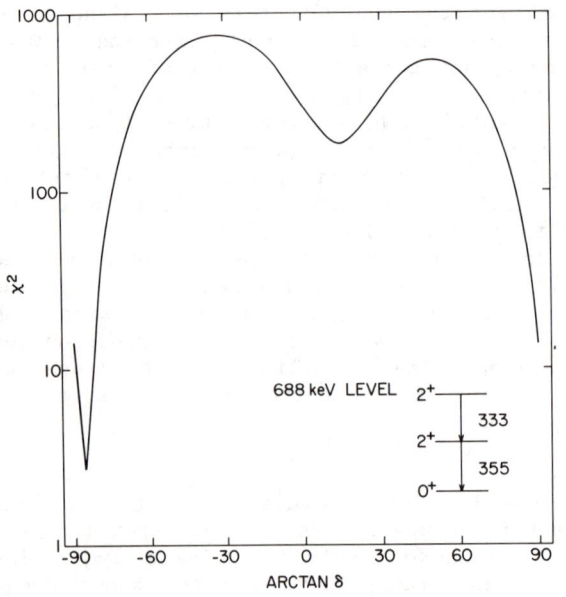

Fig. 2. Plot of χ^2 versus arctan δ for the 333→355 cascade.

Table of measured mixing ratios.

E_γ(keV)	Spin Sequence	O(6) Quantum Numbers $(\sigma, \tau, \nu_\Delta)$	Mixing Ratio δ
326	$3^+ \to 2^+$	$(6,3,0) \to (6,2,0)$	<-19.08 >7.12 or $-0.09^{-0.33}_{+0.14}$
333	$2^+ \to 2^+$	$(6,2,0) \to (6,1,0)$	$-14.30^{-4.78}_{+2.87}$
673	$2^+ \to 2^+$	$(6,4,1) \to (6,2,0)$	$-0.62^{-0.61}_{+0.41}$
1006	$2^+ \to 2^+$	$(6,4,1) \to (6,1,0)$	<-19.08 >8.14
1091	$3^- \to 2^+$		$0.25^{+0.15}_{-0.13}$

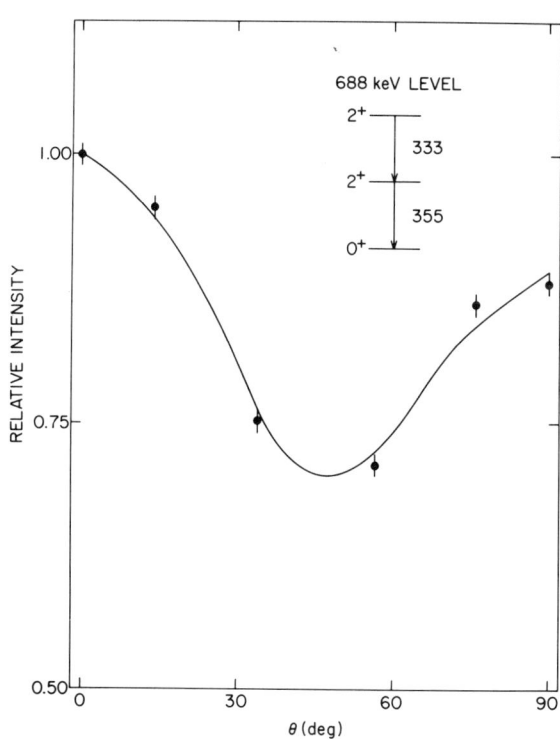

Fig. 3. Gamma-ray intensity as a function of angle for the 333→355 cascade. The solid line represents the fit for the assigned value of δ.

patterns. One such test is shown in Fig. 1 where the relative intensity of the 780-355 keV cascade is plotted against angle. The 1135 level has been identified as a 0^+ level and indeed the data shows the distinctive pattern of a $0^+ \to 2^+ \to 0^+$ correlation. In addition, values of E2/M1 mixing ratios (δ) for four transitions have been determined and are given in the table. The criteria for setting limits on δ was taken from ref. 2. Figure 2 shows a plot of χ^2 versus arctan δ for the 333-355 keV cascade and Fig. 3 shows the relevant fit to the data points as a function of angle.

DISCUSSION

A successful description of the properties of ^{196}Pt has been made in terms of the O(6) limit of the IBA-1 formalism[1]. In this limit, states are characterized by quantum numbers $(\sigma, \tau, \nu_\Delta)$ and the selection rules for E2 are $\Delta\sigma=0$, $\Delta\tau=\pm 1$. The $\Delta\tau$ rule arises from the form of the E2 operator and is therefore strictly applied. However, the $\Delta\sigma=0$ rule arises from a numerical cancellation and is therefore more susceptible to a slight change in the wavefunction when the strict symmetry is broken. Both the 326 and 333 keV transitions obey these selection rules and the table shows that these are predominantly E2 transitions, although in the case of the 326 keV transition, the possibility of a large M1 component cannot be eliminated. The decay of the 1361 keV (6,4,1) level to the (6,1,0) level via the 1006 keV transition has been determined to be more than 98% E2 even though it breaks both E2 selection rules. This may arise because the strict O(6) symmetry is slightly broken, allowing mixing between states which have non-zero E2 matrix elements with the (6,1,0) state.

In the IBA-1, the M1 matrix element consists of two parts[3], one of which yields the same selection rules as for E2 transitions whilst the other gives $\Delta\sigma=2$, $\Delta\tau=0$. While the best test of these rules would involve transitions from the $\sigma=4$ states, the existence of M1 components in transitions between states with $\sigma=6$ cannot be explained by this approach. In IBA-2, where neutrons and protons are treated separately, M1 transitions result from admixtures of states which are not fully symmetric with respect to the n and p degrees of freedom. Therefore, the large M1 admixture in the 673 keV is a measure of the antisymmetric component in the 1361 keV level and could be used, in conjunction with a numerical IBA-2 calculation of ^{196}Pt, to provide an indication of the likely position of the lowest fully antisymmetric state, which is characterized by a large B(M1) strength to the ground state, and in this region should have $J^\pi=1^+$. Evidence for such an isovector mode has recently been obtained in the deformed region[4], via electron scattering.

ACKNOWLEDGEMENT

Research performed under contract DE-AC02-76CH00016 with the USDOE.

REFERENCES

1. J. A. Cizewski et al., Nucl. Phys. A323, 349 (1979).
2. A. N. James et al., N.I.M. 115, 105 (1974).
3. D. D. Warner, Phys. Rev. Lett. 47, 1819 (1981).
4. D. Bohle, Phys. Lett. 137B, 27 (1984).

AVERAGE RESONANCE CAPTURE STUDIES OF ^{102}Ru

Z.-R. Shi[†] and R. F. Casten
Brookhaven National Laboratory
Upton, New York, 11973, USA

J. Stachel
State University of New York
Stony Brook, New York, 11974, USA

A. M. Bruce
University of Manchester
Manchester M13 9PL, England

ABSTRACT

The ^{102}Ru nucleus has been investigated via the ARC technique which ensures a complete set of $J^\pi = 0^+$, $1^\pm$, $2^\pm$, $3^\pm$, $4^\pm$, and 5^+ levels up to 2 MeV. The results are discussed in the framework of the IBA-1 with Consistent Q. The calculations show good agreement with the empirical data especially for the 0^+_2 state, suggesting that it can be described in terms of collective degrees of freedom.

INTRODUCTION

The structure of Ru isotopes has received much attention because they exhibit properties of different types of deformation. The first excited 0^+ state has been interpreted as an intruder state[1-2]. The present experiment has been performed by means of the Average Resonance Capture technique (ARC) at the Brookhaven High Flux Beam Reactor (HFBR) using external neutron beams of mean energy 2 and 24 keV. IBA-1 calculations have been performed in the framework of the Consistent Q Formalism[3] (CQF) where the parameter χ in the quadrupole operator has the same value in both the E2 transition operator and the Hamiltonian. There are only four parameters in the Hamiltonian. Good agreement has been obtained for both energies and absolute B(E2) values of some low-lying states, in particular for the 0^+_2 state.

EXPERIMENTAL TECHNIQUE

The ARC technique has been used and a detailed description of the filtered beam facility has been given elsewhere[4]. A target of 7.2g enriched to 97.8% in ^{101}Ru was used and the γ rays were detected in a three-crystal pair spectrometer.

[†]Permanent address: Institute of Atomic Energy, Beijing
P.O. Box 275, People's Republic of China

Table 1: ^{101}Ru(n,γ)^{102}Ru ARC Results

E_x(keV) a)	I_R(2) 2 keV	I_R(24) 24 keV	$\frac{I_R(2 \text{ keV})}{I_R(24 \text{ keV})}$	J^π ARC	J^π d)
475.1(2)	100(3)	100(3)	1.00(4)	2^+ c)	2^+
943.8(3)	7.5(12)	15(1)	.53(9)	0^+	0^+
1102.3(2)	81(12)	87(9)	.93(16)	$2^+,4^+(3^+)$	2^+
1106.3(2)	100(12)	74(9)	1.35(22)	$2^+,4^+$	4^+
1521.4(2)	107(8)	94(17)	1.14(22)	3^+ c)	3^+
1580.8(4)	98(11) b)	79(3)	1.24(16)	2^+ c)	2^+
1798.3(3)	69(5)	49(3)	1.40(14)	$1^+,4^+,5^+$	4^+
1836.7(10)	9(3)	12(2)	0.75(26)	0^+	0^+
1966.9(15)	4(4)	8(3)	0.5(5)	$0^+,5^-$	0^+

a) The energies are from the average of lines in 2 keV and 24 keV.
b) The intensities are corrected for contaminants.
c) J^π values from literature used for normalization.
d) J^π values are from ref. 8.

RESULTS AND DISCUSSION

J^π values of levels can be determined by comparing experimental primary γ ray reduced intensities $I_R = I_\gamma/E_\gamma^5$ and the ratio $I_R(2 \text{ keV})/I_R(24 \text{ keV})$ with those predicted by a Monte Carlo calculation[5]. The resulting excitation energy and J^π assignments for levels up to 2 MeV are given in Table 1.

The Hamiltonian and E2 transition operator are as follows:

$$H = \varepsilon n_d - \kappa Q \cdot Q - \kappa' L \cdot L$$

$$Q = (s^+\tilde{d}+d^+s)^{(2)} + (\chi/\sqrt{5})(d^+\tilde{d})^{(2)}$$

$$T(E2) = \alpha Q$$

^{102}Ru is considered[6] as a triaxially deformed nucleus with $\langle\gamma\rangle$ = 26°. The relationship[7] between χ and $\langle\gamma\rangle$ gives the value of χ as −0.3 and κ' is empirically taken as 0.0137 MeV. The remaining two parameters ε and κ are then determined by the least square method such that theoretical excitation energies agree with the four well known levels at E_x = 475.1(2^+_1), 1102.3(2^+_2), 1106.3(4^+_1) and 1521.4(3^+_1). This yields ε = 0.56 MeV and κ = −0.035 MeV. The deduced band structure of positive parity states is shown in Fig. 1. A comparison of predicted B(E2) values and their ratios and experimental ones[8] is shown in Tables 2 and 3. The parameter α in the E2 operator is fixed by adjusting the theoretical B(E2:$2^+_1 \rightarrow 0^+_1$) value to the experimental one.

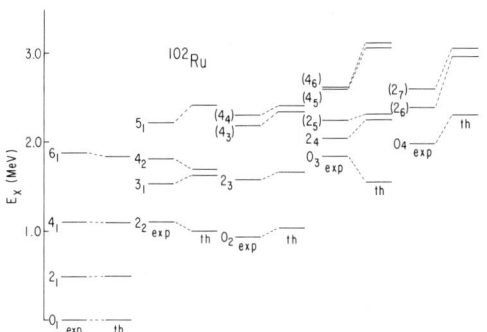

Fig. 1. Comparison of the experimental band structure for even-parity states with the predictions of IBA-1. J^π values in parentheses are tentative. The $J^\pi=6^+$ level at 1.9 MeV is from ref. 8.

As seen from Fig. 1, Tables 2 and 3, theoretical B(E2) values, branching ratios and excitation energies are in good agreement with experimental ones. In particular it is possible to describe the properties of the 0^+_2 level. From the properties of the g-band a β-deformation of 0.2 was deduced[6] and B(E2) ratios for the γ band yielded $\langle\gamma\rangle$ = 26° with a large γ softness. The present experiment proves that the 0^+_2 state has to be a collective excitation since it is the only excited 0^+ state at low excitation energy (<1800 keV) and the collective strength of its transition to 2^+_1 cannot be due to strong mixing with a higher lying 0^+ state. The good agreement of the experimental 0^+_2 properties with the present IBA-1 calculation can be understood as an indication that the nucleus ^{102}Ru is also soft in the β direction. Previous IBA[6] calculations discussing properties of the ground and γ bands correspond to a β-deformed, β-rigid and γ-soft potential, whereas the present calculation represents a spherical β- and γ-soft potential. These different approaches should not be seen as contradictory, but rather, each of them bears out a special aspect of the complex properties of a β- and γ-deformed and β- and γ-soft nucleus. However, the empirical energy difference between the 3^+_1 and 4^+_2 states is larger than the theoretical value (i.e., odd-even staggering effect of γ band), suggesting that a cubic term describing a rigid triaxial rotor could be necessary in the IBA-1 Hamiltonian.

Table 2. B(E2) Values.

$J^\pi_i - J^\pi_f$	$2^+_1 - 0^+_1$	$0^+_2 - 2^+_1$	$2^+_2 - 2^+_1$	$2^+_2 - 0^+_1$	$4^+_1 - 2^+_1$
B(E2)$_{exp}$ w.u.[8]	43.9(40)	35(6)	27(3)	1.09(20)	66(10)
B(E2)$_{th}$ w.u.	43.5	38.9	66.0	0.021	66.9

Table 3. B(E2) Ratio.

B(E2) Ratio	Exp.[8]	Th.
$B(0^+_2-2^+_1)/B(2^+_1-0^+_1)$	0.81	0.98
$B(2^+_2-2^+_1)/B(2^+_1-0^+_1)$	0.62	1.52
$B(4^+_1-2^+_1)/B(2^+_1-0^+_1)$	1.52	1.54
$B(2^+_2-0^+_1)/B(2^+_2-2^+_1)$	0.04	0.00033
$B(4^+_1-2^+_1)/B(2^+_2-2^+_1)$	2.44	1.01
$B(3^+_1-2^+_1)/B(3^+_1-4^+_1)$	0.16	0.0016
$B(4^+_2-4^+_1)/B(4^+_2-2^+_2)$	0.58	0.90

ACKNOWLEDGEMENTS

It is a pleasure to acknowledge discussions with D. D. Warner. Research has been performed under contract DE-AC02-76CH00016 with the United States Department of Energy.

REFERENCES

1. P. Van Isacker and G. Puddu, Nucl. Phys. A348, 125 (1980).
2. A. M. Vandenberg, R. Bijker, N. Blasi, M. Sambataro, R. H. Siemssen, and W. A. Sterrenburg, Nucl. Phys. A422, 61 (1984).
3. D. D. Warner and R. F. Casten, Phys. Rev. C28, 1798 (1983).
4. R. G. Greenwood and R. E. Chrien, Nucl. Instr. Meth. 138, 125 (1976).
5. R. E. Chrien, Proc. 4th Int. Symp. on Neutron Capture Gamma-Ray Spectroscopy and Related Topics, Grenoble (1981), p. 342.
6. J. Stachel, Invited talk presented at the "Workshop on Electromagnetic Properties of High Spin Nuclear Levels", Ein Bokek, Israel.
7. R. F. Casten, A. Aprahamian and D. D. Warner, Phys. Rev. C29, 356 (1984).
8. P. de Gelder, D. de Frenne and E. Jacobs, Nucl. Data Sheets 39, Vol. 4, 443 (1982).

ON SIMULATION OF NONSTATISTICAL EFFECTS IN NEUTRON RESONANCE GAMMA-DECAY

V.A. Knat'ko
Institute of Physics, BSSR Academy of Sciences,
Minsk 220602, USSR

ABSTRACT

The properties of the γ-widths of s-neutron resonances of ^{64,66,68}Zn are analysed in terms of a quasiparticle-cluster-vibration model.

The main contribution to the nonstatistical effects observed in resonance neutron capture is made by primary high-energy γ-transitions. In [1], it was proposed to analyse such transitions in terms of a model used to describe low-lying levels in a product-nucleus. Knowing the wave functions of the final states and generating the resonance wave function components, one may attempt to simulate widths for high-energy γ-transitions and estimate their role in nonstatistical effects. In the present work (see also [2]), this approach is used to analyse the properties of γ-widths of s-neutron resonances of zinc isotopes with A=64, 66 and 68.

The partial γ-width was written as

$$\Gamma_{\gamma if} = \left[(\Gamma_{\gamma if}^v)^{1/2} + \sum_{m,n} C_m^i a_n^f \Gamma_{m \to n}^{1/2} \right]^2 = \left[(\Gamma_{\gamma if}^v)^{1/2} + \sum_m C_m^i (\Gamma_m^f)^{1/2} \right]^2$$

where C_m^i (a_n^f) are the expansion coefficients for resonance (final level) wave functions, $\Gamma_{\gamma if}^v$ is the valence capture γ-width [3], Γ_m^f is the width for the γ-transition from the configuration m in the resonance wave function. The coefficients C_m^i were expressed as $C_m^i = b_m^i [f_m(E_i)]$. Here, b_m^i is the random number drawn from a normal distribution with parameters $\mu = 0$ and $\sigma = 1$. The factor $f_m(E_i)$ characterizing the fragmentation of a simple component over resonances was taken in the Lorentz form $f_m(E_i) = (D_0 \cdot W_m / 2\pi) / [(E_i - E_m)^2 + W_m^2 / 4]$ where D_0 is the average resonance spacing, E_i and E_m are the resonance and the configuration m energy, respectively, is the spreading width for the configuration m.

The $\Gamma_m^\downarrow$ values were calculated in terms of a quasiparticle-cluster-vibration model which satisfactorily describes low-lying levels in odd-mass Zn isotopes [4]. Depending on the structure of the populated p-levels, the three-quasiparticle (3QP), 3QP-plus-phonon, and 1QP-plus-phonon configurations were taken into account in the resonance wave functions. The E1 γ-transitions to five low-lying p-levels of ^{65,67}Zn were considered. Attempts to describe the level scheme of ^{69}Zn were hindered by the lack of experimental data. Therefore, transitions only to three lowest p-levels of ^{69}Zn (at 0, 0.835 and 1.007 MeV) were included in calculations.

It should be noted that the fragmentation of simple components may differ from the Lorentz one [5]. To take into account the uncertainty of the factor $f_m(E_i)$, the values of E_m and W_m were arbitrarily varied within the limits $E_m^o \pm 0.5$ MeV and $W_m^o \pm 0.5$ MeV, respectively. Here, $W_m^o = 0.3$ MeV and 1.5 MeV are for 3QP- and phonon-containing configurations, respectively. The energy E_m^o was approximated as $\sum_j E_j + E_2$ where E_j is the quasiparticle excitation energy and E_2 is the phonon energy.

Simulated $\Gamma_{\gamma if}$ values were used to calculate the correlation coefficients $T_{ff'} = [2/K(K-1)] \sum_{f \neq f'}^{K} r(\Gamma_{\gamma if}, \Gamma_{\gamma if'})$ and $r(\Gamma_{\gamma i}, \Gamma_{ni}^o)$, the resonance-averaged γ-widths $\langle \Gamma_{\gamma i} \rangle = \sum_f^K \langle \Gamma_{\gamma if} \rangle_i$, (K is the number of populated p-levels) and to determine the degree of freedom parameter ν for χ^2-distribution describing theoretical γ-widths. The process of generation of $\Gamma_{\gamma if}$ simulating the physical sample was iterated 100 times. The calculated values of the above quantities averaged over 100 sets are compared with the experimental values [6] in Fig. 1.

The results show that the contribution of the transitions in question to the total experimental γ-width increases as the mass number increases. It can be seen, taking into account this contribution, that there is a qualitative agreement between calculated and experimental values of the parameter ν and the coefficient $r(\Gamma_{\gamma i}, \Gamma_{ni}^o)$. For more detailed comparison measurements of partial γ-widths are needed. In particular, a comparison between predicted and experimental values would be of interest.

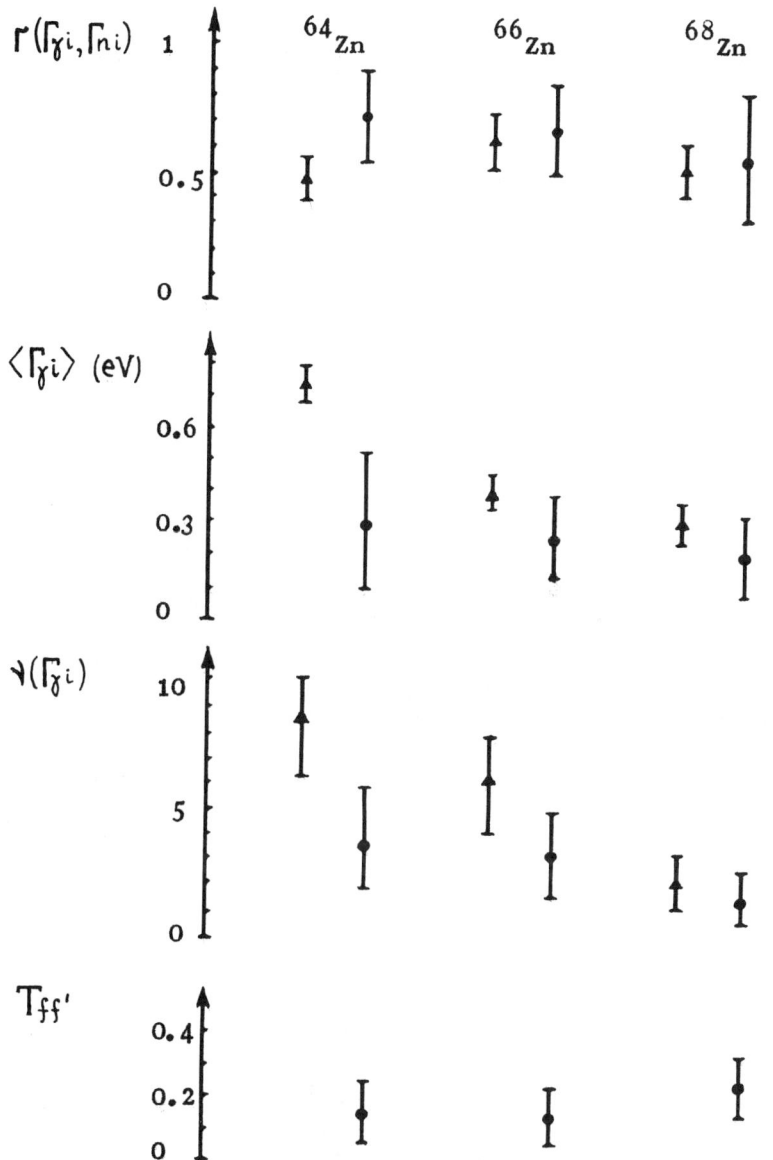

Fig.1. Results of experimental (triangles) and theoretical (points) γ-widths analysis. For theoretical values of $r(\Gamma_{\gamma i}, \Gamma_{ni})$, $v(\Gamma_{\gamma i})$, $T_{ff'}$, $\langle \Gamma_{\gamma i} \rangle$ and experimental values of v errors indicative 10 to 90% confidence limits. For experimental values of $r(\Gamma_{\gamma i}, \Gamma_{ni})$ and $\langle \Gamma_{\gamma i} \rangle$ the errors were taken from [6].

REFERENCES

1. V.A. Knat'ko, E.A. Rudak, Nucl. Phys. A194, 458 (1972).
2. V.A. Knat'ko, E.A. Shimanovich, Proc. of the 33-th All-Union-Conf. on Nuclear Spectroscopy (Nauka, Leningrad, 1983), p. 425.
3. B.J. Allen, A.R. del Musgrove, in Adv. Nucl. Phys. 10, 129 (1978).
4. K. Allaart, P. Hofstra, V. Paar, Nucl. Phys. A366, 384 (1981).
5. V.G. Soloviev, Proc. Intern. School on Nuclear Structure (JINR, Dubna, 1980), p. 57.
6. J.B. Garg et al., Phys. Rev. C23, 683 (1981); C24, 1922 (1981); C25, 1808 (1982).

IV. FAST NEUTRONS

RADIATIVE CAPTURE OF NUCLEONS IN THE GIANT-DIPOLE RESONANCE REGION

F. S. Dietrich
Lawrence Livermore National Laboratory, Livermore, CA 94550

ABSTRACT

Several features of radiative capture in the GDR region are discussed in terms of the direct-semidirect (DSD) and pure-resonance (PRM) models. The relationship of these models is clarified by results of a derivation that explicitly allows for E1 strength decoupled from the GDR. A microscopic folding-model estimate of the coupling form factor is presented, and its consistency with the (p,n) isobaric-analog reaction discussed. A Coulomb-excitation contribution is found to be significant for proton capture. The correlation between the nucleon escape widths and the particle-hole spectroscopy of the GDR is discussed in terms of the single-particle and rearrangement escape amplitudes in the PRM formulation. Applications are made to radiative capture near A=208.

I. INTRODUCTION

The general properties of the giant-dipole resonance (GDR), such as its location, width, and electromagnetic strength, have been well determined from systematic studies with inclusive photonuclear reactions[1]. On the other hand, understanding the detailed spectroscopic properties of the GDR, such as the amplitudes of the various particle-hole components, requires the use of other reactions, such as (p,γ), (n,γ), (α,γ), and their inverses. Extracting information from these reactions requires knowledge of the reaction mechanisms. The GDR, although the best studied of the giant multipole resonances, is still interesting in this regard, since the reaction-mechanism problem has not yet been completely solved. The radiative-capture mechanism for the GDR is also of importance because other multipolarities, such as E2, manifest themselves primarily through interference with the dominant E1 radiation, and therefore the magnitude and phase of the E1 amplitudes must be well known if quantitative information on the higher multipolarity giant resonances is to be extracted from radiative capture measurements.

The most complete treatment of the radiative-capture mechanism, at least at low energies when only nucleonic degrees of freedom are considered, is the coupled-channels formalism[2,3], in which the structure and reaction-mechanism problems are treated on the same footing. Coupled-channels calculations have been useful in reproducing the cross sections, angular distributions, and fine structure of the GDR in light nuclei, but become impractical for heavy nuclei because of the large number of channels that must be treated simultaneously. Other approaches introduce various further approximations and phenomenological ingredients. In one type of calculation that focusses on the structure of the GDR, (e.g. Wang and Shakin[4]; Castel and Micklinghoff[5]), the nuclear Hilbert space is divided with projection operators into an "internal" part in which a diagonalization is performed to yield the GDR structure, and the internal space is connected to the continuum by using a projection-operator reaction formalism. These calculations have been successful in explaining the fine structure of the GDR in ^{16}O and ^{29}Si. This paper is concerned with two closely-related models, the direct-semidirect model (DSD) and the pure-resonance model (PRM), which are particularly suited to systematic studies because the nuclear structure is described rather simply by a form factor, exactly as in the DWBA treatment of inelastic scattering.

In the DSD model[6,7] there are interfering direct-capture and resonant core-polarization amplitudes; the magnitude and phase of the latter depends on the properties of the form factor. An intuitive, semi-classical description of the physics underlying this model has been given by Halpern[8]. The DSD model has been extensively developed[9,10] and has been qualitatively successful in reproducing the general features of (n,γ) and (p,γ) excitation functions and angular distributions. It has also been extended to other multipolarities[11-14]. However, the microscopic origin of the form factor is not adequately understood, and phenomenological parameterizations of the form factor have not led to satisfactorily predictive results when the model is extrapolated from one nucleus to another.

The development of the PRM[15] was motivated by the observation that the GDR contains nearly all the E1 strength built upon the final state after capture, and consequently the nonresonant direct term in the DSD should be canceled almost exactly by some identifiable feature of the semidirect term. This feature was found to be the single-particle resonances that occur in both terms. Removing these resonances led to an expression with only a resonant term, containing the same form factor as in the DSD model, but with altered continuum wave functions. The hope that removing an unstable feature of the capture model by requiring the cancellation of the direct component would improve reliability has been borne out through decreased sensitivity to the imaginary part of the form factor for capture cross sections in the region of the GDR peak, but extensions to E2 capture[16] do not show such decreased sensitivity.

In the remainder of this paper, three further investigations of the DSD and PRM models are made. First, an additional resonance is introduced that shows explicitly the near cancellation of the direct capture strength that is assumed to be exact in the original formulation of the PRM; this term significantly clarifies the relation between the two models. Next, a calculation of the real part of the form factor for ^{208}Pb is made by folding a transition density with a density-dependent effective interaction, and found to be consistent with a similar treatment of the (p,n) reaction to the isobaric analog state; a Coulomb-excitation contribution to the form factor is found to be significant for proton capture. Finally, the extended PRM is used to study the connection between the nucleon escape widths for the GDR and the schematic-model description of the particle-hole amplitudes. In all of these investigations, applications are shown for nucleon capture near A=208.

II. EXTENSIONS OF THE DSD AND PRM FORMULATIONS

In this section, the results of a rederivation of the DSD and PRM using the Feshbach reaction theory are presented, in which the original <u>a priori</u> assumption that all the E1 strength lies in the GDR is relaxed.

We define a spatially localized configuration built on the final state $|G\rangle$ after capture as $|D\rangle = R|G\rangle$, where R is proportional to the E1 operator, and the configuration is normalized to unity. For capture of a nucleon from a channel $|lj\rangle$ to a bound state $|u\rangle$, it is essential to identify the part of $|D\rangle$ that corresponds to a hole in the state $|u\rangle$ and a particle in the configuration $|w\rangle$ having the same quantum numbers (l,j) as the channel. Physically, $|w\rangle$ corresponds to the configuration in the channel (l,j) reached by raising a particle in $|u\rangle$ to a higher state by absorption of a photon; it is defined by $|w\rangle = \eta P_{lj}(1-P_{occ})R|u\rangle$, where $(1-P_{occ})$ removes occupied states, P_{lj} projects onto the correct channel quantum numbers, and η symbolizes the normalization.

The configuration $|D\rangle$ may then be written as

$$|D\rangle = \frac{\mu}{\sqrt{1+\mu^2}} |wu^{-1}\rangle + \frac{1}{\sqrt{1+\mu^2}} |B\rangle,$$

in which the normalized configuration $|B\rangle$ is the part of the coherent state $|D\rangle$ <u>not</u> associated with the capture channel. The quantity μ is a small, real, parameter whose value is close to that of the schematic-model[17] amplitude of $|wu^{-1}\rangle$ in the GDR. The orthogonal configuration,

$$|\emptyset\rangle = -\frac{1}{\sqrt{1+\mu^2}} |wu^{-1}\rangle + \frac{\mu}{\sqrt{1+\mu^2}} |B\rangle,$$

carries no electromagnetic strength. Because the first term in $|\emptyset\rangle$ dominates, the energy of $|\emptyset\rangle$ is close to the unperturbed particle-hole energy ($1\hbar\omega$ in a harmonic-oscillator model). The relationship between $|D\rangle$ and $|\emptyset\rangle$ resembles that between analog and anti-analog states.

In the original Feshbach-theory[18,19] derivations of the DSD and PRM, it was assumed that $|D\rangle$ corresponds to the physical GDR, and $|\emptyset\rangle$ plays no role. Relaxing this assumption leads to a mixing of the two states via coupling to that part of the space <u>not</u> spanned by $|D\rangle$ and $|\emptyset\rangle$ (principally, the continuum). The physical states then become

$$|d\rangle = \frac{\lambda}{\sqrt{1+\lambda^2}} |wu^{-1}\rangle + \frac{1}{\sqrt{1+\lambda^2}} |B\rangle$$

and

$$|\phi\rangle = -\frac{1}{\sqrt{1+\lambda^2}} |wu^{-1}\rangle + \frac{\lambda}{\sqrt{1+\lambda^2}} |B\rangle.$$

The coefficient λ is complex, and is calculable without introducing new parameters; λ should be fairly close to μ, since the coupling is weak and the schematic model is a useful first approximation to the GDR. The state $|\phi\rangle$ may radiate, and this leads to an additional resonant term in the capture amplitudes, which for the two types of models are

$$T_{DSD}' = W_D - \lambda\mu \frac{1 - (\lambda/\mu)}{1 + \lambda^2} \frac{W_{SD}}{E - \epsilon_\phi} + \frac{1 + \lambda\mu}{1 + \lambda^2} \frac{W_{SD}}{E - \epsilon_d} \text{, and}$$

$$T_{PRM}' = \frac{1 - (\lambda/\mu)}{1 + \lambda^2} \frac{W_S - \lambda\mu W_R}{E - \epsilon_\phi} + \frac{1 + \lambda\mu}{1 + \lambda^2} \frac{(\lambda/\mu) W_S + W_R}{E - \epsilon_d}.$$

The complex energies ϵ_d, ϵ_ϕ contain the position and width of the GDR and "anti-GDR," respectively; ϵ_d is given the phenomenological value $E_{GDR} - i\Gamma_{GDR}/2$, whereas ϵ_ϕ is calculated (with no new free parameters), and is very nearly equal to the expectation value of the optical-model Hamiltonian in the configuration $|w\rangle$. Typically the position and width in ϵ_ϕ are $1\hbar\omega$ and 10 MeV, respectively; the width is dominated by the damping due to the imaginary part of the optical Hamiltonian. The quantities W_D, W_{SD}, W_S, W_R are:

$W_D = \langle u|r|\chi^{(+)}\rangle$, direct capture;
$W_{SD} = M_{GDR} \langle u|h'(r)|\chi^{(+)}\rangle$, semidirect capture;
$W_S = \langle u|r|w\rangle\langle w| H^{OPT}|\phi^{(+)}\rangle$, single-particle escape;
$W_R = M_{GDR} \langle u|h'(r)|\phi^{(+)}\rangle$, rearrangement escape.

In these expressions, M_{GDR} and $\langle u|r|w\rangle$ are the E1-decay matrix elements of the GDR and the particle-hole configuration $|wu^{-1}\rangle$, respectively; $|\chi^{(+)}\rangle$ is the usual optical-model wave function, whereas $|\phi^{(+)}\rangle$ is the optical-model wave function obtained after projecting the configuration $|w\rangle$ out of the same optical-model Hamiltonian H^{OPT} that is used to generate $|\chi^{(+)}\rangle$. The form factor $h'(r)$ describes the coupling of the incident channel (l,j) to the part of the GDR that is not in that channel, and is therefore in principle channel dependent. The usual approximation of a channel-independent form factor is likely to be satisfactory for heavy nuclei for which the GDR of the projectile-plus-target system has so many particle-hole components that the amplitude in any particular channel (l,j) is small. In the limit $\lambda = \mu$, the original DSD and PRM amplitudes are recovered:

$$T_{DSD} = W_D + \frac{W_{SD}}{E - \epsilon_D} \text{ ; } T_{PRM} = \frac{W_S + W_R}{E - \epsilon_D}.$$

The generalized DSD and PRM expressions have the property that they are precisely identical, both formally and after making the approximations necessary for numerical calculations. The purely resonant form is useful for identifying the particle widths of the GDR, as will be discussed is Section IV. The "anti-GDR" term in the PRM expression exhibits a near cancellation via the factor $1-(\lambda/\mu)$. The analogous term in the DSD expression also has the same factor, but this term is very small because of the additional factor $\lambda\mu$. Because both $|\lambda\mu|$ and $|\lambda^2|$ are small ($\lesssim 0.1$), the practical result is that the two terms of the extended PRM description are nearly equivalent to the original DSD expression. The "anti-GDR" term in the PRM exhibits explicitly the near-cancellation that was assumed to be exact in the original PRM. This cancellation is analogous to the cancelling of the $1\hbar\omega$ strength in the schematic model.

The extended model (PRM or DSD) goes beyond the schematic-model description in predicting a specific amount of strength decoupled from the GDR. The same issue of decoupled strength has been addressed in a somewhat different way by Potokar[20]. However, the correct amount of decoupled strength may be difficult to calculate accurately because the cancellation depends on the form factor and the optical potential, which may not be sufficiently well known to yield an accurate value for the factor $1-(\lambda/\mu)$. The original PRM is a useful approximation in circumstances in which the decoupled strength is expected to be small, and one wishes to avoid calculating the cancellation, either explicitly in the "anti-GDR" or implicitly in the original DSD model.

A comparison of the models with a body of neutron and proton capture data[21-23] in the region near A=208 is shown in Figs. 1-3. The solid curves represent the extended model, and the dashed curve the original PRM model. The real part of the form factor was calculated by a folding model, as described in the next section

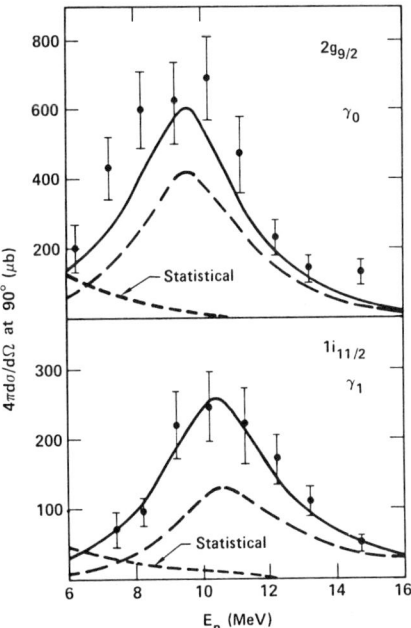

Fig. 1   PRM (dashed line) and extended model (solid line) calculations for ^{208}Pb(n,γ). An estimate of the compound (statistical) contribution is also shown. Data from Ref. 21.

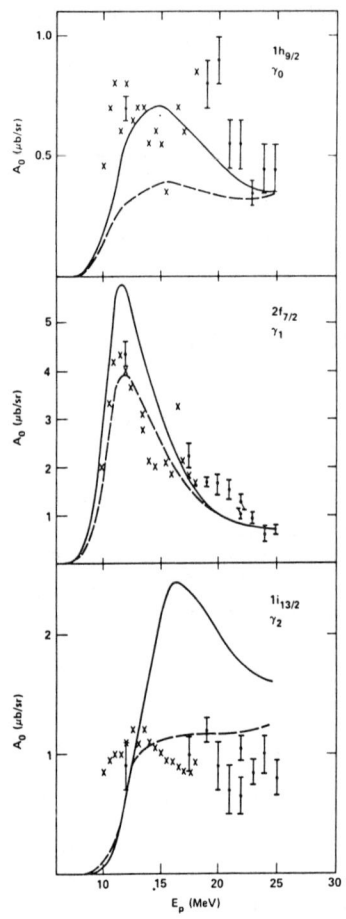

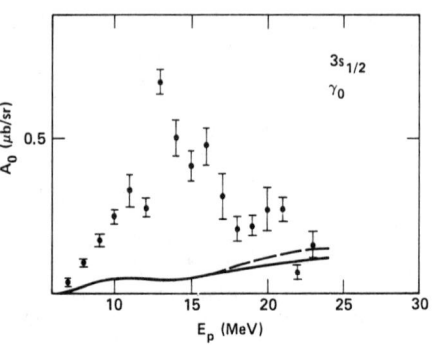

Fig. 2 (at left). Same as Fig. 1, for ^{208}Pb(p,γ). Data from Ref. 22,23.

Fig. 3 (above). Same as Fig. 1, for ^{205}Tl(p,γ). Data from Ref. 22.

(curves labeled JLM, Fig. 4). A phenomenological form was chosen for the imaginary part[9,24], with parameters (using the normalization of Ref. 12) W =33 MeV, r =1.32 fm, a =0.56 fm. The calculations were performed with a nonlocal optical potential that had been fit to elastic neutron scattering in the range 7-26 MeV and proton scattering between 21 and 45 MeV; the use of a nonlocal potential reduces ambiguities associated with the energy dependence of the real potential, particularly for proton scattering near the Coulomb barrier[25]. An additional feature of the calculation of the continuum wave functions is the fact that occupied-state orbitals, which can give spurious contributions to radiative capture, have been projected out of the optical Hamiltonian. In the illustrations shown, this has proved to be important for the ^{208}Pb(p,γ_2) reaction. The calculations yield qualitatively reasonable results, except for the ^{205}Tl(p,γ_0) reaction; this case seems to be particularly susceptible to small changes in the bound-state potential parameters, as a DSD calculation with slightly different parameters has given better agreement with the data[22].

III. MICROSCOPIC CALCULATION OF THE REAL PART OF THE FORM FACTOR

It is well known that the GDR form factor should be related to the isovector terms in the nucleon optical potential, and to the $\vec{\tau}_1 \cdot \vec{\tau}_2$ part of an effective two-body interaction. Although these relationships are invoked in deriving the form factor from hydrodynamic models for the GDR transition density[24], a satisfactory quantitative correspondence between the isovector strengths required for capture, elastic scattering, and charge exchange reactions has yet to be established. This problem is illustrated in Fig. 4, in which the curve labeled B-G is the real part of the form factor for ^{208}Pb, calculated with the Becchetti-Greenlees potential[26] and the Steinwedel-Jensen GDR model[27], assuming unit energy-weighted sum rule (EWSR). The magnitude is roughly a factor of two too low to reproduce the GDR peaks in the capture measurements, even though the strength of V_1 (24 MeV) and the ratio V_1/V_0 (about 0.5) in the Becchetti-Greenlees potential are in reasonable accord with elastic scattering and (p,n) measurements.

Progress during the last few years in developing energy- and density-dependent effective interactions[28] and optical potentials[29] starting from free nucleon-nucleon potentials suggests that it should be useful to calculate at least the real part of the form factor by folding a transition density with the isovector part of such an effective interaction, and to compare the results with a similar treatment for elastic scattering and charge exchange. The curves labeled JLM in Fig. 4 represent the form factor obtained by folding the schematic-model transition density (normalized to unit EWSR, and shown in the lower portion of the figure) with an effective interaction obtained by dividing the real isovector part of the Jeukenne-Lejeune-Mahaux optical potential[29] by the ground-state density, and ascribing a range to the interaction (≈ 1 fm) that reproduces elastic and inelastic scattering[30,31]. The effect of density dependence in the effective interaction is evident in the peaking of the form factor about 1 fm beyond the peak in the transition density, and in the modification of the form factor toward a shape that is more surface-peaked than in the "volume-like" phenomenological treatment based on the Steinwedel-Jensen model. The strength of the interaction increases by about a factor of three from the center of the nucleus to the region of one-tenth the central density. The overall strength of the form factor is too weak, and the nuclear interaction has been multiplied by 2.5 in calculating the form factors shown in Fig. 4, which are those used in the capture calculations of Figs. 1-3. However, approximately the same normalization of the interaction is needed to agree with the magnitude of the charge exchange reaction, as shown in Fig. 5. Such an upward shift of the interaction strength for both reactions is reasonable, since the ratio of isovector to isoscalar strengths in the JLM potential is $V_1/V_0=0.23$ in the surface region, which is smaller by about a factor of two than phenomenological values.

We conclude tentatively that the strength of the real part of the form factor is not anomalous from the point of view of the microscopic folding-model approach. It will be interesting to see whether this conclusion is supported as further progress is made in understanding effective interactions in the nuclear medium.

Two additional effects that have been included in the folding-model treatment of the form factor are the Coulomb and two-body spin-orbit interactions. The Coulomb interaction is significant, and in fact nearly all of the difference between the proton and neutron form factors shown in Fig. 4 is due to Coulomb excitation of the GDR. On the other hand, the additional form factors resulting from folding the spin-orbit interaction with the isovector matter density have very small effects, even on the angular-distribution and analyzing-power observables.

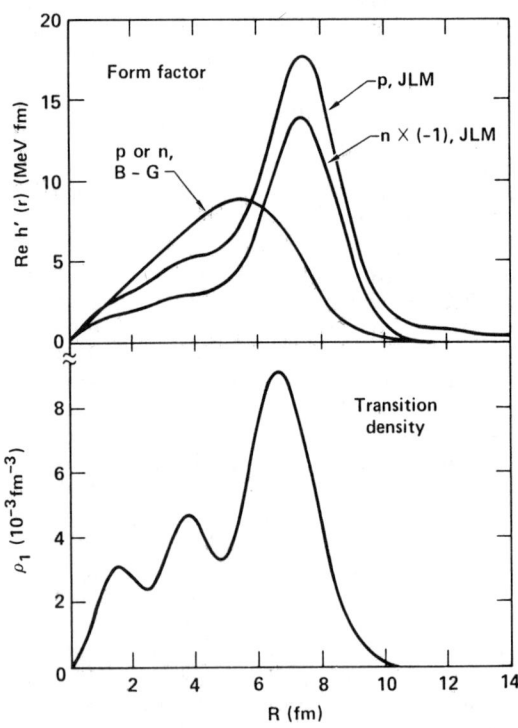

Fig. 4. Folding-model calculation of the real part of the GDR form factor for ^{208}Pb, using a schematic-model transition density and the JLM[29] isovector interaction normalized by a factor of 2.5. A typical phenomenological form factor (B-G) is shown for comparison.

The microscopic origin of the imaginary part of the form factor[10], which appears to be required in most cases for optimal agreement with the data, is not yet well enough known to allow a meaningful calculation. The one-step folding approach, using the imaginary part of an effective interaction, is unlikely to describe the physics adequately at low energies, and has not been attempted. This remains one of the outstanding problems in understanding radiative capture in the GDR region.

IV. ESTIMATES OF GDR SINGLE-PARTICLE WIDTHS FROM THE PRM

Since the GDR is commonly described as a coherent particle-hole excitation, it is natural to seek a correlation between the cross sections for nucleon capture involving a particular combination of incident channel (l,j) and final bound orbital, and the amplitude of the corresponding particle-hole component in the GDR state vector. Such a correlation has been reported by Dowell et al.[32] and Snover[33], who have presented experimental evidence for a proportionality between photonucleon cross sections integrated over the giant-resonance peaks (i.e., integrated capture cross sections, converted by detailed balance), and the corresponding spectroscopic factors for single-nucleon transfer. The remarkable feature of this correlation, which has been established for a number of light nuclei ($A \leq 40$), is that the ratio of integrated cross section to spectroscopic factor appears to be nearly independent of the nuclear state. These results can be explained[32,33] either by a direct-capture model, or by

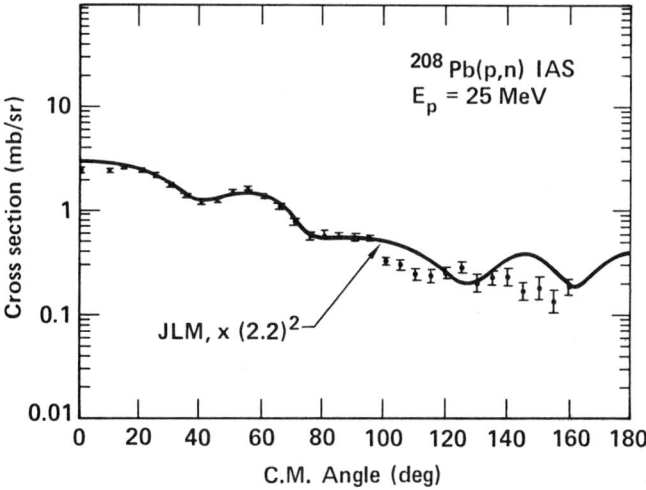

Fig. 5 DWBA calculation of the (p,n) reaction to the isobaric analog state, using a folding model. The normalizing factor (2.2) required for the JLM interaction is similar to that for the GDR form factor. Data from Ref. 34.

capture through the GDR, if the GDR is described by the schematic model and rearrangement processes are neglected. The empirical correlation was established without factoring out barrier penetration.

The PRM and its extended version are a useful framework in which to discuss the connection between the capture reaction and the spectroscopy of the GDR, since the resonance numerators contain quantities whose absolute square is the particle escape width. The escape-width amplitude has components that correspond to both single-particle escape (which is the quantity discussed by Dowell et al.) and rearrangement escape, as noted in Section III. The barrier penetration is included naturally through the use of optical-model wave functions.

We discuss the spectroscopic problem using as examples the calculations in the A=208 region shown in the previous section. For each final-state orbital, there are three continuum channels obeying the E1 selection rules (except for the $s_{1/2}$ final state, for which there are only two). These channels may be separately identified in the calculations, even though they are not separated in the experiments.

Viewed as a photo-ejection reaction, the single-particle escape part of the reaction is a two-step process in which the photon first raises the bound nucleon to the configuration $|w\rangle$ with an amplitude determined by μ in the original PRM, or λ in the extended version. Next, the excited, spatially localized configuration decays into the appropriate channel ($|\phi\rangle$, having the same l,j as $|w\rangle$), with an amplitude $\langle w| H^{OPT} |\phi^{(+)}\rangle$ that depends only on the optical-model parameters, and includes the barrier-penetration effects.

The absolute squares of the spectroscopic amplitudes μ and λ extracted from the original and extended PRM calculations are shown in Fig. 6 as solid and open bars, respectively. These may be compared with the squared particle-hole amplitudes for the schematic model based on harmonic-oscillator wave functions, represented by the hatched bars. The correlation between μ and the schematic-model amplitudes is very strong, showing that the result of Dowell et al. is embodied in the original PRM, at least in the single-particle escape contribution to the particle widths. For the extended model, the correlation is less impressive in detail than for the original PRM, but it is still clearly evident. For each particle-hole component shown in Fig. 6, the energy-dependent quantities were determined by choosing the energy at the GDR peak position (E =13.47 MeV). Because of the proximity to the Coulomb barrier in the case of the proton components, and high angular momenta for many of the particle states, there is an enormous variation (three orders of magnitude) in the penetration factors necessary to obtain the actual particle widths from the squared amplitudes.

In the rearrangement-escape process, the photon first excites the part of the GDR that does not have the particle-hole quantum numbers of the channel. The subsequent decay requires the intervention of a residual effective interaction, which in the present case is included in the form factor in the rearrangement-escape matrix element $<u|h'|\phi^{(+)}>$. There is no obvious reason why the rearrangement-escape amplitude should be proportional to the particle-hole spectroscopic amplitudes of the GDR in the same way as the single-particle escape amplitude. In the present example, the rearrangement-escape amplitudes are comparable in magnitude to the single-particle escape; the two amplitudes must be added coherently. The ratios of the total particle width to the single-particle escape width are shown at the bottom of Fig. 6 for the extended-model calculations. It is apparent that the ratios vary strongly from configuration to configuration. On average, the ratio is smaller for protons than for neutrons, since the Coulomb barrier inhibits the rearrangement-escape term more than the single-particle escape; this is because the form factor h' cuts off sharply with radius, whereas H^{OPT} does not. Evidently, inclusion of the rearrangement-escape term impairs the correlation, but does not destroy it entirely, as the ordering of the various particle-hole components according to intensity is reasonably well preserved when the rearrangement-escape terms are added.

To further investigate the correlations between the particle widths and the GDR spectroscopic amplitudes, it would be desirable to extend the present calculations to the light nuclei in which the

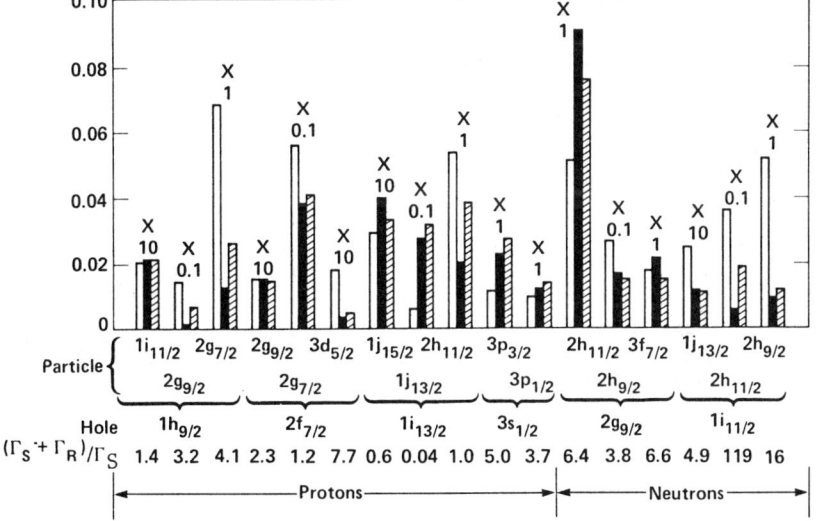

Fig. 6   Comparisons of schematic-model intensities (hatched bars) for various particle-hole components of the ^{208}Pb GDR, with quantities μ^2 in the PRM (solid bars) and $|\lambda|^2$ in the extended model (open bars). Also shown are the ratios of the total particle width to the single-particle escape width (see text).

correlations have been observed, and in which the rearrangement-escape amplitude is expected to be less important relative to the single-particle escape because there are fewer particle-hole components in the GDR state vector. It would also be interesting to use the extended PRM to estimate the total nucleon escape width of the GDR in ^{208}Pb, since the calculations presented in this paper show reasonable agreement with the data, at least for the strong transitions.

V. SUMMARY AND CONCLUSIONS

A rederivation of the direct-semidirect and pure-resonance capture models using the Feshbach reaction formalism has been made, explicitly allowing for the possibility of E1 strength decoupled from the GDR. This results in an additional resonance-like term that explicitly displays the approximate cancellation of low-lying strength that had previously been assumed to be exact in the PRM. The extended versions of the DSD and PRM are identical, and in fact the extended model is numerically very close to the original DSD. Folding-model calculations of the real part of the capture form factor for ^{208}Pb and of the (p,n) reaction show that both can be reconciled with the same effective interaction; this interaction must be much stronger than that implied by the JLM microscopic optical potential. Coulomb excitation is significant for the proton-capture form factor, but form factors related to the spin-orbit interaction were found to be unimportant. The PRM formulation was used to study the particle widths of the GDR, and to investigate the correlation between the widths and the particle-hole spectroscopy of the GDR. A strong correlation is exhibited by the single-particle escape widths, but this correlation is reduced by the rearrangement-escape terms.

VI. ACKNOWLEDGMENTS

I would like to acknowledge the collaboration of A. K. Kerman and F. Petrovich in various aspects of this work, as well as interesting conversations with K. A. Snover, particularly concerning the use of the PRM to discuss escape widths. This work was performed under the auspices of the U.S. Department of Energy by the Lawrence Livermore National Laboratory under contract number W-7405-ENG-48.

REFERENCES

1. B. L. Berman and S. C. Fultz, Rev. Mod. Phys. 47, 713 (1975).
2. B. Buck and A. D. Hill, Nucl. Phys. A95, 271 (1967).
3. M. Marangoni, P. L. Ottaviani, and A. M. Saruis, Nucl. Phys. A277, 239 (1977), and references therein.
4. W. L. Wang and C. M. Shakin, Phys. Rev. C5, 1898 (1972).
5. B. Castel and M. Micklinghoff, in Neutron Capture Gamma-Ray Spectroscopy, R. Chrien and W. Kane, eds. (Plenum, N.Y., 1979), p. 285.
6. G. E. Brown, Nucl. Phys. 57, 339 (1974).

7. C. F. Clement, A. M. Lane, and J. A. Rook, Nucl. Phys. $\underline{66}$, 273, 293 (1965).
8. I. Halpern, in Photonuclear Reactions and Applications, B. Berman, ed., (USAEC Office of Information Services, Oak Ridge, 1973), p. 909.
9. G. Longo and F. Saporetti, Phys. Lett. $\underline{42B}$, 17 (1972).
10. M. Potokar, Phys. Lett. $\underline{46B}$, 346 (1973).
11. K. Snover et al., Phys. Rev. Lett. $\underline{37}$, 273 (1976).
12. F. Dietrich, D. Heikkinen, K. Snover, and K. Ebisawa, Phys. Rev. Lett. $\underline{38}$, 156 (1977).
13. A. Likar, M. Potokar, and F. Cvelbar, Nucl. Phys. $\underline{A280}$, 49 (1977).
14. G. Longo, F. Saporetti, and R. Guidotti, Nuovo Cimento $\underline{A46}$, 509 (1978), and references therein.
15. F. S. Dietrich and A. K. Kerman, Phys. Rev. Lett. $\underline{43}$, 114 (1979).
16. A. Likar and R. Martincic, Nucl. Phys. $\underline{A350}$, 74 (1980).
17. G. E. Brown, Unified Theory of Nuclear Models (North-Holland, Amsterdam, 1967).
18. H. Feshbach, Ann. Phys. (N.Y.)$\underline{19}$, 287 (1962).
19. N. Auerbach, J. Hufner, A. Kerman, and C. Shakin, Rev. Mod. Phys. $\underline{44}$, 48 (1972).
20. M. Potokar, Phys. Lett. $\underline{92B}$, 1 (1980).
21. I. Bergqvist, D. Drake, and D. McDaniels, Nucl. Phys. $\underline{A191}$, 641 (1972). These data are consistent with the more recent results of S. King et al., Nucl. Phys. $\underline{A384}$, 129 (1982).
22. K. A. Snover, in Neutron Capture Gamma-Ray Spectroscopy, R. Chrien and W. Kane, eds. (Plenum, N.Y., 1979), p. 319.
23. K. A. Snover, J. F. Amann, W. Hering, and P. Paul, Phys. Lett. $\underline{37B}$, 29 (1971).
24. G. R. Satchler, Nucl. Phys. $\underline{A195}$, 1 (1972).
25. J. S. Eck and W. J. Thompson, Nucl. Phys. $\underline{A237}$, 83 (1975).
26. F. D. Becchetti and G. W. Greenlees, Phys. Rev. $\underline{182}$, 1190 (1969).
27. H. Steinwedel and J. H. D. Jensen, Zeit. fur Naturforsch. $\underline{5}$, 413 (1950).
28. F. A. Brieva and J. R. Rook, Nucl. Phys. $\underline{A291}$, 299, 317 (1977).
29. J. P. Jeukenne, A. Lejeune, and C. Mahaux, Phys. Rev. $\underline{C16}$, 80 (1977), and references therein.
30. Ch. Lagrange and J. C. Brient, J. Phys. (Paris) $\underline{44}$, 27 (1983).
31. S. Mellema, R. Finlay, F. Dietrich, and F. Petrovich, Phys. Rev. $\underline{C28}$, 2267 (1983).
32. D. H. Dowell et al., Phys. Rev. Lett. $\underline{50}$, 1191 (1983).
33. K. A. Snover, in proceedings of International Symposium on Highly Excited States and Nuclear Structure, Orsay, France, 1983; p. C4-337.
34. R. R. Doering, D. M. Patterson, and A. Galonsky, Phys. Rev. $\underline{C12}$, 378 (1975).

GIANT ISOVECTOR E2 RESONANCES OBSERVED IN (n,γ) AND (γ,n) REACTIONS

L. Nilsson
Tandem Accelerator Laboratory,
S-751 21 Uppsala, Sweden

ABSTRACT

A summary of the gross properties of isovector E2 resonances in nuclei and a survey of recent experimental efforts to study these resonances are presented. The main emphasis is on applications of radiative capture and photonuclear reactions, which have proved to be valuable tools for these studies. Planned and ongoing activities in this field are discussed.

INTRODUCTION

Since the first experimental observation of the isoscalar giant quadrupole resonance (GQR) more than a decade ago [1-3], the study of giant resonance phenomena in nuclei has been one of the central themes in nuclear structure research. During the initial period in this new area of research, studies of inelastic scattering of hadrons (and to some extent electrons) were the main sources of new information. This implied that the great majority of new data concerned isoscalar resonances, in particular the isoscalar GQR. As the experimental techniques for studying inelastic scattering at extreme forward angles (including 0°) developed, it became possible to extract information on the isoscalar giant monopole resonance (the nuclear breathing mode) [4,5]. Simultaneously, radiative nucleon capture [6] and photonuclear reactions [7] turned out to be useful complementary tools in these studies. Other techniques that have recently been applied to obtain information on giant multipole resonances involve charge-exchange reactions, i.e. (p,n), (n,p), (^{3}He,t), (π^+,π^0), (π^-,π^0), etc., through which the isobaric analogues of the isovector giant resonances in the target nucleus can be studied [8-11].

Parallel to the experimental development in the study of giant resonance characteristics, the theoretical understanding of these phenomena has improved substantially. In particular, extensive calculations based on the random-phase approximation have been able to account for many new observations and to predict new phenomena [12-14]. Long before the experimental discovery of the "new" giant resonances at the beginning of the 1970's, excitation energies and other properties of various giant multipole resonances had been predicted by the hydrodynamic and other macroscopic nuclear models [15-17].

Several excellent review papers and conference talks have been devoted to various aspects of this exciting development in the study of giant resonance phenomena during the past decade [13,18-21].

This paper is aimed at a discussion of the properties of the isovector E2 resonance, the study of which has been a challenge to experimentalists, in particular for light nuclei. In the next section, we will make an attempt to summarize the present knowledge concerning the isovector GQR and to present various efforts which are being made to reveal its properties. The following section will deal in some detail with one of the methods utilized for these studies, namely the radiative neutron capture reaction and its inverse the photoneutron reaction. In the final section, the results from various techniques are discussed and some suggestions for further studies in this field are put forward.

THE ISOVECTOR E2 RESONANCE

In heavy nuclei, evidence for the isovector GQR has been obtained from studies of inelastic scattering of electrons [22]. The summary in ref [22] shows that the excitation energies are about $130A^{-1/3}$ MeV, but that the values in the low mass region (A = 50 - 100) tend to approach $120A^{-1/3}$ MeV. The widths range from about 10 MeV at A = 60 to about 5 MeV for the heaviest nuclei (Pb and U). The resonance exhausts, roughly independent of mass number, 60 to 100 % of the isovector energy-weighted sum rule (EWSR) based on the Goldhaber-Teller model [15].

In light nuclei, important information on the E2 strength has been obtained from polarized and unpolarized proton capture experiments, in which - as in the electron scattering experiments - it is possible to distinguish between isoscalar and isovector excitations only by means of the excitation energy. This work has been excellently summarized by Snover [6]. The proton capture technique has also been applied to study isovector E2 strength in heavy nuclei [23], where the results have been interpreted in terms of the direct-semidirect [24] (DSD) and pure-resonance [25] (PR) models.

Other interesting techniques used to study the E2 strength in light nuclei involve photonuclear and photon scattering experiments. From measurements of angular distributions of neutrons from the $^{16}O(\gamma,n_0)^{15}O$ reaction in the γ-energy region 25 - 45 MeV, and with the assumption that multipoles higher than E2 can be neglected, Phillips and Johnson [26] report a lower limit for the E2 strength in this energy region of 23 % of the EWSR for the isovector GQR. By combining their data with data at the low-energy end on the polarization of the photoneutrons, the authors estimate that 68 % of the EWSR is observed in their experiment.

Elastic photon scattering experiments on ^{12}C and ^{16}O performed at 90° and 135° at NBS [27] have been analyzed together with total absorption cross section data [28]. The scattering cross section is related to the total absorption cross section through the optical theorem and the dispersion relation after correction for Thomson scattering. To obtain consistency between these data, the authors introduce, in the 20 - 45 MeV region, (in addition to the E1 cross section) an E2 cross section corresponding to 1.9 and 1.3 times the total (isoscalar plus isovector) E2 EWSR (TEWS) in ^{12}C and ^{16}O, respectively. The interpretation of these results is, however, in contradiction to some recent photon scattering data of Wright et. al.[29], who have performed measurements for carbon and calcium. The carbon data agree with those from NBS, i.e. at 90° and 135°. Introducing the E2 strength reported by Dogde et.al. [27] in the data analysis does not, however, lead to a satisfactory description of the 45° data of Wright et.al. [29]. The authors conclude that the scattering data and the total absorption cross sections are mutually inconsistent. Analysing the scattering data for carbon independently of the absorption data indicates that no E2 strength larger than 50 % of the TEWS or concentrated in a resonance less than 5 MeV wide is required in the 20 - 35 MeV region to explain the data. The data are, however, compatible with a 20 MeV wide E2 resonance at 58±4 MeV exhausting 150 % of the TEWS.

The recently reported E2 strength of 35 % of the isovector EWSR (i.e. 100 % of the strength of the lower isospin ($T_<$) component) at 32 MeV in calcium from an (n,γ_o) experiment is below the sensitivity limit of the photon scattering experiments. We shall return to this and other (n,γ) and (γ,n) experiments later.

During the last few years single-charge-exchange reactions have been applied to the study of isovector excitations in nuclei. The excitations which become available for study are isobaric analogues of $\Delta T = 1$ states in the target nucleus. The most spectacular example is the Gamow-Teller resonance which has been studied extensively using the (p,n) reaction [31]. Particularly simple cases to study are $T_z = 0$ (self-conjugate) targets, because only one isospin component in the final nucleus is available; $T = 1$, $T_z = -1$ in (p,n), (^{3}He,t) and (π^+,π^o) and $T = 1$, $T_z = 1$ in (n,p) and (π^-,π^o) reactions. An interesting feature but also a complication in the studies of charge-exchange reactions is that spin-flip ($\Delta S = 1$) transitions occur in addition to $\Delta S = 0$ transitions. Several experimentalists have observed the analogue of the giant dipole resonance [8-11] and some have attempted to localize the corresponding isovector GQR [10, 11]. As in hadron scattering the continuum background is considerable, which means that the detection limit for the isovector GQR is a substantial fraction of the EWSR. From a recent (^{3}He,t) experiment [10] at a beam energy of 197 MeV in self-conjugate nuclei (^{12}C, ^{24}Mg, ^{28}Si and ^{40}Ca), the authors state that the isovector GQR would have been seen if it were no more than 10 MeV wide with at least 60 % of the EWSR strength. A similar negative result was obtained from a recent (π^-,π^o) experiment [11], where again the detection limit is well below the sum-rule strength.

NEUTRON CAPTURE IN THE ISOVECTOR GQR REGION

The use of radiative capture (and its inverse reaction) in studies of the isovector E2 strength is based on the interference between radiation amplitudes of opposite parity (E1 and E2). This interference gives rise to asymmetries around 90° in the angular distribution. The asymmetry, defined by

$$A_1 = \frac{I(55°) - I(125°)}{I(55°) + I(125°)} \tag{1}$$

is expected to change dramatically in the vicinity of the isovector E2 resonance in neutron capture from around zero below the resonance to close to unity above, depending on the strength of the resonance. This result has been obtained qualitatively by simply adding a resonant dipole amplitude and a resonant quadrupole amplitude coherently [32]. Similar results were obtained many years ago by applying the DSD capture model [33]. It was also pointed out several years ago [34] that there is a distinct difference between the (p,γ) and (n,γ) reactions in this respect, namely that the direct E2 component in neutron capture is very weak compared with that in proton capture. This comes about because the effective charge of a neutron in an E2 transition is about 1/A times that of a corresponding proton. The E2 effective charge of the proton implies that the fore-aft asymmetry in proton capture increases continuously above the GDR, which makes the extraction of the E2 strength more complicated. Nevertheless, the proton capture reaction has been successfully applied to studies of the isovector E2 resonance in lead [23].

The neutron capture reaction has been used to study the isovector E2 strength in ^{209}Pb (ref [32]) and ^{41}Ca (ref [30]). The (γ,n) reaction has recently been used [35] with the same objective for Cd and Pb. In all these cases the asymmetry shows the expected behaviour in the region of the isovector GQR. From the (n,γ) experiment in Pb a value for the excitation energy of the isovector E2 resonance of 22.5 MeV is obtained in agreement with other observations. The data are compatible with a full exhaustion of the EWSR, but the data do not allow a precise extraction of the strength as it was not experimentally possible to extend the measurements to more than 1 - 2 MeV above the peak of the resonance. The asymmetry data for Pb have been compared with calculations based on the direct-semidirect model. The comparison shows that the asymmetry increases even faster than predicted. The most extensive use of the DSD model in connection with studies of the isovector E2 strength was recently reported for the ^{40}Ca(n,γ$_0$)^{41}Ca reaction. The application of the DSD model in this particular case will be described here together with the basic ingredients of the model. For a more detailed account of this work the reader is referred to ref [30] and references therein.

The original formulation of the DSD model [24] has been modified to include spin-orbit coupling and interference between direct and semidirect capture amplitudes. The transition amplitude can be written

$$T_{\lambda\tau} = \langle \Psi_{nlj} | d_{\lambda\tau}(r) | \chi^{(+)} \rangle + \langle \Psi_{nlj} | \frac{h_{\lambda\tau}(r)}{E_\gamma - E_{\lambda\tau} + \frac{1}{2}i\Gamma_{\lambda\tau}} | \chi^{(+)} \rangle \qquad (2)$$

where $\chi^{(+)}$ is the scattering wavefunction (calculated with an optical model potential) and Ψ_{nlj} the final bound-state wave-function (calculated with a real Woods-Saxon potential). The first term in eq. (2) is the direct capture amplitude, where $d_{\lambda\tau}(r)$ is the single-particle multipole operator for multipolarity λ and isoscalar (τ=0) or isovector (τ=1) excitations. The second term is the semidirect capture amplitude, which represents capture via the giant resonance state (with energy $E_{\lambda\tau}$ and width $\Gamma_{\lambda\tau}$).

The interaction between the incoming neutron and the target nucleus is contained in the form factor $h_{\lambda\tau}(r)$. The explicit forms of the form factors are [36]

$$h_{11}(r) = -10\frac{NZ}{A^2} Y_{1v}(\bar{r})\frac{\hbar^2}{2m}(1+0.8x)\frac{p_{11}}{R^2 E_{11}} r[\frac{1}{4}V_1 f_r(r) - \frac{i}{4}W_1 4b\frac{df_i(r)}{dr}] \quad (3)$$

for the giant dipole resonance,

$$h_{21}(r) = -28\frac{NZ}{A^2} Y_{2v}(\bar{r})\frac{\hbar^2}{2m}\frac{p_{21}}{R^2 E_{21}} r^2[\frac{1}{4}V_1 f_r(r) - \frac{i}{4}W_1 4b\frac{df_i(r)}{dr}] \quad (4)$$

for the isovector quadrupole resonance and

$$h_{20}(r) = -\frac{Z}{A} Y_{2v}(\bar{r})\frac{\hbar^2}{2m}\frac{p_{20}}{E_{20}} r[V_0\frac{df_r(r)}{dr} + iW_0 4b\frac{d^2 f_i(r)}{dr^2}] \quad (5)$$

for the isoscalar quadrupole resonance. In eqs. (3) to (5), V_0 and W_0 are the depths of the real and imaginary parts of the central potential, respectively, and V_1 and W_1 are the corresponding depths of the symmetry potential. The functions $f_r(r)$ and $f_i(r)$ are Woods-Saxon form factors for the real and imaginary potentials, respectively. The factors $p_{\lambda\tau}$ in front of the radial form factors are the respective fractions of the EWSR exhausted by the capture reaction. These fractions are particularly important in neutron capture, because isospin conservation requires that only the $T_<$ part of the isovector resonances is excited. The fractional strength distribution between the isospin components is given by

$$\frac{S_>}{S_<} = \frac{1}{T_0}\frac{1 - 1.5T_0/A^{2/3}}{1 + 1.5/A^{2/3}} \quad (6)$$

for the giant dipole resonance [37] and by

$$\frac{S_>}{S_<} = \frac{1}{T_0}\frac{1 - 2T_0/A}{1 + 2/A} \quad (7)$$

for the isovector quadrupole resonance [38]. In eqs. (6) and (7) T_0 is the isospin of the ground state in the final nucleus. For ^{41}Ca ($T_0 = 1/2$) the $T_>$ strength is about twice the $T_<$ strength in both resonances.

In the calculations the sensitivity of the results to optical model parameters was investigated. Universal optical model parameter sets as well as parameter sets from elastic neutron scattering on calcium were used and it was found that the calculated 90° cross sections and fore-aft asymmetries were roughly the same for the four optical model parameter sets used in this study. The calculations also nicely describe angular distribution data from TUNL [39] in the neutron energy range 8 - 13 MeV. The giant resonance and potential parameters used in the calculations are summarized in Table I.

Table I Giant resonance and interaction strength parameters.

Resonance	$E_{\lambda\tau}$ (MeV)	$\Gamma_{\lambda\tau}$ (MeV)	EWSR %	V_τ (MeV)	W_τ (MeV)
Isovector dipole $\lambda=1, \tau=1$	18.1	4.0	38	120	40
Isovector quadrupole $\lambda=2, \tau=1$	32.0	5.0	35	120	40
Isoscalar quadrupole $\lambda=2, \tau=0$	18.4	4.2	60	50	10

The measured cross sections and fore-aft asymmetries are shown in Figs. 1 and 2, respectively, compared with calculations where

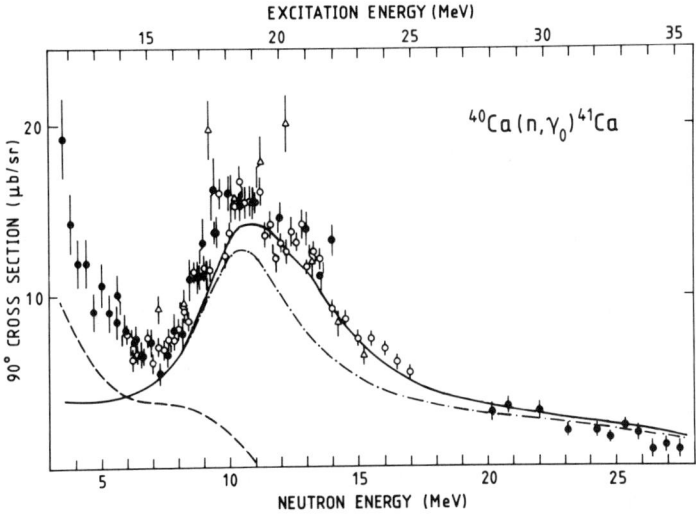

Fig. 1. The 90° cross section for the $^{40}Ca(n,\gamma_0)^{41}Ca$ reaction [30,39,44]. The full and dot-dashed curves represent cross sections calculated using the DSD model and the dashed curve the compound-nucleus contribution to the cross section.

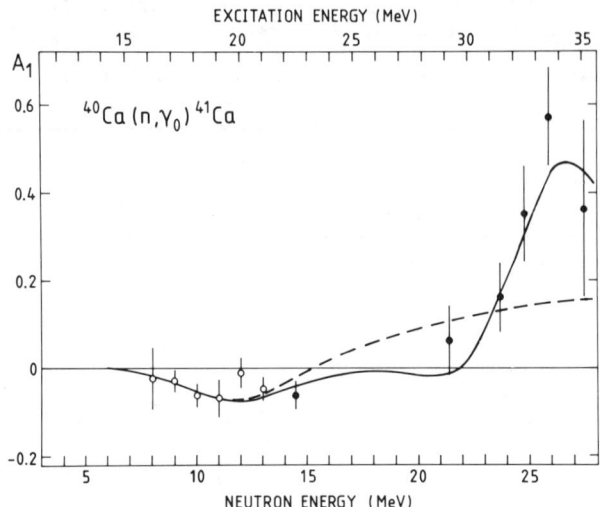

Fig. 2. The fore-aft asymmetry of the $^{40}Ca(n,\gamma_0)^{41}Ca$ reaction [30,39]. The full curve is calculated using the DSD model including an isovector GQR with parameters given in Table I. The dashed curve represents the case where no isovector GQR was included in the calculation.

the optical model parameters of Tornow et.al. [40] are used. The depths, V_1 and W_1, of the symmetry potentials, are treated as free parameters and adjusted to describe the 90° cross section in the region of the giant dipole resonance. The obtained values of V_1 = 120 MeV and W_1 = 40 MeV are slightly larger than those obtained from quasi-elastic (p,n) reactions. The use of the parameters given in Table I for the $T_<$ component of the isovector GQR implies that the strength for this component of the resonance is fully exhausted. The uncertainty in the excitation energy has been estimated to ±1 MeV. The uncertainties in the width and strength are coupled and are difficult to estimate.

To give an impression of these uncertainties, calculated asymmetries for different sets of values for the width and strength are presented in Fig. 3. The asymmetry curves calculated with widths of 2.5 and 5 MeV give equally good fits to the measured data points, whereas the 1 MeV curve is too steep. It is not likely that very narrow structures occur at excitation energies above 30 MeV. The figure also shows that a constant ratio between the width and the strength gives rise to the same total change in the asymmetry. This would imply that, if the width is larger than 5 MeV more than 100 % of the $T_<$ strength is observed. The conclusion is that the width is between 2.5 and 5 MeV and that 20 - 35 % of the isovector EWSR is exhausted.

The comparison in Fig. 3 clearly illustrates the desirability to extend the measurements of fore-aft asymmetries to energies well

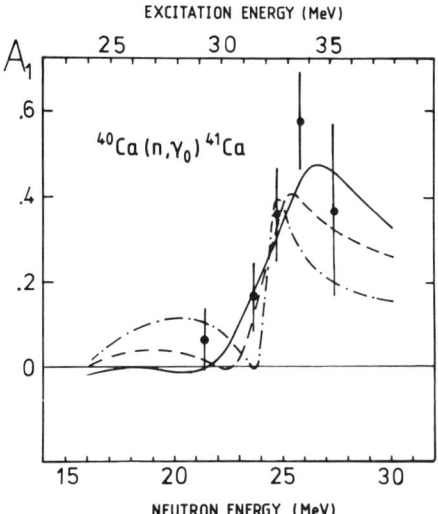

Fig. 3. The fore-aft asymmetry of the $^{40}Ca(n,\gamma_0)^{41}Ca$ reaction [30]. The curves are calculated on the basis of the DSD model with various values for the width (Γ) and fractions of the $T_<$ EWSR (full curve – Γ = 5 MeV and 100 %, dashed curve – Γ = 2.5 MeV and 50 %, dot-dashed curve – Γ = 1 MeV and 20 %).

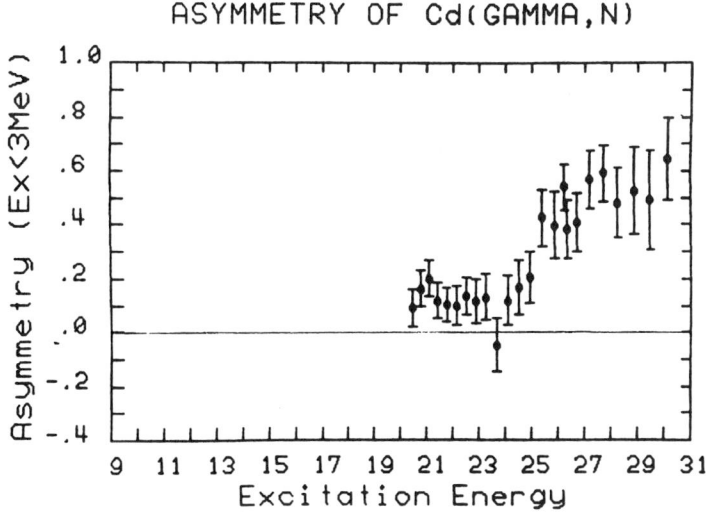

Fig. 4. Fore-aft asymmetry data for the $Cd(\gamma,n)$ reaction in the region of the isovector GQR [35]. Photoneutrons to levels below 3 MeV in the final nucleus are included.

above the peak of the resonance. Photoneutron experiments of this kind have already been performed [35] and preliminary data for lead and cadmium have been reported. Results for cadmium from this work are shown in Fig. 4. These asymmetry data include photoneutrons to final levels below 3 MeV. It is not presently understood why the asymmetry below the resonance is slightly positive, but isotope effects might play a role.

DISCUSSION AND CONCLUSIONS

As mentioned above studies of the isovector E2 resonance have been very difficult, in particular in light nuclei. As a matter of fact, several attempts to reveal its properties have failed. One of the reasons for this is that in the capture and photonuclear reactions the E2 amplitude is weak compared with the E1 amplitude even well above the giant dipole resonance. In hadron scattering the cross section for the excitation of isovector states is small in comparison with that for isoscalar excitation. This is illustrated by the weak excitation of the giant dipole resonance. A further difficulty is the high excitation energy (and large width) of the isovector GQR which is well up in the continuum where the "background" is considerable. This is also a serious problem in charge-exchange reaction studies. The electron scattering experiments have been fairly successful in heavy nuclei in spite of the background from the radiation tail. We are presently not aware of any electron scattering experiment in nuclei lighter than Ni, where there is clear evidence for isovector E2 strength.

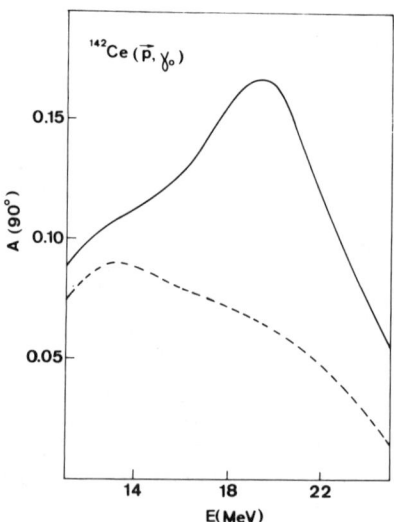

Fig. 5. The analysing power at 90° for the ^{142}Ce($\vec{p},\gamma_0$) reaction [43]. The dashed curve is calculated with E1 and direct E2 capture, whereas the full curve also includes resonant E2 capture.

The question is what can be done to increase our knowledge of the isovector GQR, in particular in light nuclei, where the situation now is far from settled. Below follows a list of suggestions, which are aimed at improving the understanding of the isovector GQR.

* Electron scattering experiments should be extended to nuclei in the mass region below A = 60. It is desirable to have electron scattering data as a complement to other information in this mass region, and attempts should be made even if the resonances are expected to be rather wide.

* Neutron capture studies seem to give a clear signature for E2 strength in the region of the isovector GQR. The experiments with monoenergetic neutrons (from the $^3H(d,n)^4He$ reaction) are, however, very difficult and time-consuming and cannot easily be extended to higher energies well above the peak of the isovector GQR. On the other hand, neutron capture experiments at the WNR facility at LAMPF [41] look promising and should be continued. The recent attempts to improve the energy resolution of bismuth germanate detectors [42] form an essential part in this program.

* The preliminary results from the photoneutron work at Illinois are encouraging and further experiments can be expected to give important information. One of the advantages in this technique is that the measurements can be extended to energies well above the resonance. The time resolution in these experiments should be improved to allow separate observation of the ground-state transition.

* Photon scattering is a natural means of studying E2 strength. From the contradictory interpretations of recent data, there are, however, reasons to question whether the sensitivity limit can be pushed down sufficiently.

* It would be interesting to study charge-exchange reactions in nuclei where the isovector GQR has been observed by other techniques, e.g. in Ni isotopes.

* In a recent paper Saporetti and Guidotti [43] pointed out that calculations based on the DSD model indicate that the analysing power at 90° in polarized proton capture gives a clear signature for resonant E2 capture. (See Fig. 5.)

In conclusion, the study of isovector E2 resonances in nuclei has been and will certainly continue to be a very exciting and challenging field of research. There are indeed several methods to choose between, which means that laboratories with various types of equipment can contribute to future advances in this field.

ACKNOWLEDGEMENTS

It is a great pleasure for me to thank my collaborators for many years Drs. I. Bergqvist, A. Likar and A. Lindholm for many enlightening discussions and valuable comments to the manuscript. I would also like to express my sincere gratitude to Prof. I. Halpern who kindly supplied me with information on the recent photoneutron experiments at the University of Illinois.

REFERENCES

1. R. Pitthan and Th. Walcher, Phys. Letters 36B, 563 (1971)
2. S. Fukuda and Y. Torizuka, Phys. Rev. Letters 29, 1109 (1972)
3. M.B. Lewis and F.E. Bertrand, Nucl. Phys. A196, 337 (1972)
4. M.N. Harakeh et.al., Phys. Rev. Letters 38, 676, (1977)
5. D.H. Youngblood et.al., Phys. Rev. Letters 39, 1188 (1977)
6. K.A. Snover, in ref. 19, p. 229
7. E. Hayward, in ref. 19, p. 275
8. F. Brady et.al., Phys. Rev. Letters 51, 1320 (1983)
9. S.Y. van der Werf et.al., Journal de Physique 45, C4-471 (1984)
10. S.L. Tabor et.al., Nucl. Phys., to be published
11. A. Erell et.al., Phys. Rev. Letters 52, 2134 (1984)
12. J. Speth, in ref. 19, p. 33
13. J. Speth, Nucl. Phys. A396, 153c (1983)
14. N. Auerbach and A. Klein, Nucl. Phys. A395, 77 (1983)
    N. Auerbach and A. Klein, Phys. Rev. C27, 1818 (1983)
    N. Auerbach and A. Klein, Phys. Rev. C28, 2075 (1983)
    N. Auerbach and A. Klein, Nucl. Phys. A422, 501 (1984)
15. M. Goldhaber and E. Teller, Phys. Rev. 74, 1046 (1948)
16. H. Steinwedel and J.H.D. Jensen, Zeits für Naturforschung 5A, 413 (1950)
17. J.D. Walecka, Phys.Rev. 126, 653 (1962)
18. F.E. Bertrand, Ann. Rev. Nucl. Sci. 26, 457 (1976)
19. Giant Multipole Resonances, Proc. of the Giant Multipole Resonance Topical Conference, Oak Ridge, Tennessee, October 1979, ed. F.E. Betrand, (Harwood Academic Publishers, New York, 1980)
20. F.E. Betrand, Nucl. Phys. A354, 129c (1981)
21. J. Speth and A. van der Woude, Rep. Prog. Phys. 44, 719 (1981)
22. R. Pitthan, in ref. 19, p. 161
23. K.A. Snover, in Neutron Capture Gamma-ray Spectroscopy, eds. R.E. Chrien and W.R. Kane (Plenum Press, New York, 1979), p. 319
24. G.E. Brown, Nucl. Phys. 57, 339 (1964)
    C.F. Clement, A.M. Lane and J.R. Rook, Nucl. Phys. 66, 273, 293 (1965)
25. F.S. Dietrich and A.K. Kerman, Phys. Rev. Letters 43, 114 (1979)
26. T.W. Phillips and R.G. Johnson, Phys. Rev. C20, 1689 (1979)
27. W.R. Dodge et.al., Phys. Rev. Letters 44, 1040 (1980)
    W.R. Dodge et.al., Phys. Rev. C28, 8 (1983)
28. J. Ahrens et.al., Nucl. Phys. A251, 479 (1975)

29. D.H. Wright et.al., Phys. Rev. Letters $\underline{52}$, 244 (1984)
30. I. Bergqvist et.al., Nucl. Phys. $\underline{A419}$, 509. (1984)
31. C.D. Goodman, in The (p,n) reaction and the nucleon-nucleon force, eds. C.D. Goodman, S.M. Austin, S.D. Bloom, J. Rapaport and G.R. Satchler, (Plenum Press, New York, 1980), p. 149
    C. Gaarde, Nucl. Phys. $\underline{A396}$, 127c (1983)
    C. Gaarde, Journal de Physique $\underline{45}$, C4-405 (1984)
32. D.M. Drake et.al., Phys. Rev. Letters $\underline{47}$, 1581 (1981)
33. F. Saporetti, G. Longo and R. Guidotti, Phys. Letters $\underline{76B}$, 15 (1978)
34. I. Halpern, Proc. of the Int. Conf. on Photonuclear Reactions and Applications, vol. 2, Asilomar, California, ed. B.L. Berman, (1973), p. 909
35. T. Murakami et.al., Bull. Amer. Phys. Soc. $\underline{29}$, 707 (1984)
    I. Halpern, private communication
36. A. Likar and R. Martincic, Nucl. Phys. $\underline{A350}$, 74 (1980)
37. S. Fallieros and B. Goulard, Nucl. Phys. $\underline{A147}$, 593 (1970)
38. J. Krumlinde, private communication
39. S.A. Wender et.al., Phys. Rev. Letters $\underline{41}$, 1217 (1978)
40. W. Tornow et.al., Nucl. Phys. $\underline{A385}$, 373 (1982)
41. S.A. Wender et.al., Nucl. Instr. Meth. $\underline{220}$, 371 (1984)
42. S.A. Wender, paper presented at this conference and private communication
43. F. Saporetti and R. Guidotti, Nucl. Phys. $\underline{A390}$, 207 (1982)
44. I. Bergqvist, D.M. Drake and D.K. McDaniels, Nucl. Phys. $\underline{A231}$, 29 (1974)

RADIATIVE CAPTURE IN FEW NUCLEON SYSTEMS

H. R. Weller
Triangle Universities Nuclear Laboratory and Duke University
Durham, North Carolina 27706

ABSTRACT

This talk will constitute an attempt to report recent low energy capture experiments which are providing new information on small components of the nuclear wavefunctions in the 2, 3 and 4 body nuclear systems. Recent polarized n-capture data on ^{1}H which address the question of whether these data require meson exchange current effects will be presented. Several new results will be described in the case of the three body problem; new polarized proton capture data, analyzed with D-state effects included, yield new and lower E2 strengths which are in reasonable agreement with inverse reaction measurements and calculations. Polarized $\vec{d}$ on p and $\vec{p}$ on d data will be presented which exhibit the sensitivity of these measurements to the presence of s=3/2 M1 strength just above threshold in ^{3}He. And tensor analyzing power measurements of the ^{1}H($\vec{d},\gamma$)^{3}He reaction which are sensitive to D-state effects in ^{3}He will be described and compared to calculations which employ recent Faddeev generated wavefunctions. Finally, new data on the tensor analyzing powers for the $^2_4$H($\vec{d},\gamma$)^{4}He reaction which provide new information on the D-state of ^{4}He will be described. A simple model calculation will be used to demonstrate that the observed isotropic $T_{20}(\theta)$ of -0.22 ± 0.014 is consistent with a 4.8% D-state admixture in the two deuteron wavefunction used to describe ^{4}He.

There have been a number of experimental and theoretical developments in the last few years which have provided new information on the two-three and four body systems which have involved capture and inverse photonuclear reactions. In this talk I will attempt to describe a few of these studies to indicate the power of the capture reaction when applied to the study of these fundamental nuclear systems.

It has been slightly more than 10 years since it was shown that the effects of meson exchange currents (MEC) are essential in understanding the cross section for radiative n-p capture at thermal neutron energies.[1] Calculations were also performed at that time for the cross section and polarizations of the outgoing nucleons in the d(γ,n)p reaction up to E_γ=22.2 MeV.[2] While the effects on the cross section were small, more pronounced effects were observed in the polarization of the outgoing nucleons. It is the sensitivity of the polarizations to the presence of M1 radiation, especially at 90°, which is responsible for the size of these effects.

Unfortunately, discrepancies have persisted in the results of the measurements of the photonucleon polarization in the energy region below 30 MeV. The situation as of about a year and a half

0094-243X/85/1250470-13 $3.00 Copyright 1985 American Institute of Physics

ago is shown in Fig. 1. The two curves shown here are the results of an impulse approximation calculation[3], and the calculation which includes MEC effects[2]. Unfortunately, one could not make a definitive statement as to whether or not MEC effects are properly included from this situation.

Recently, a new experiment (Holt et al.) was performed to remeasure the neutron polarization for the $^2H(\gamma,\vec{n})H$ reaction at 90° between $E_\gamma=6$ and 13 MeV.[4] These results are also shown in Fig. 1 and appear to be in

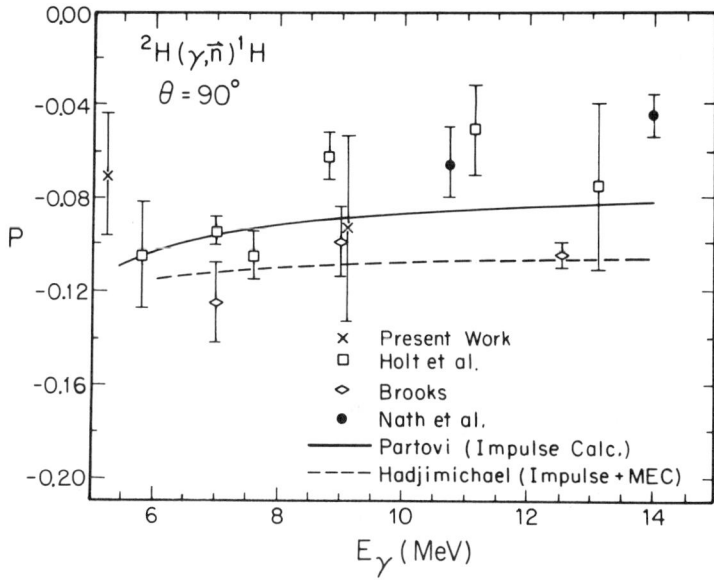

Fig. 1 Comparison of theory and experiment. The "present work" refers to the recent capture data of Ref. 6. See Ref. 4 (Holt et al.) for further details and additional references.

better agreement with the impulse calculation of Partovi[3] than with the calculation which includes MEC effects[2], suggesting the possibility that the MEC effects are either absent or improperly included. A new series of calculations[5] which include two body charge and current density effects in the traditional theory leads to a similar conclusion when compared with the data of Fig. 1. This work also shows that the result does not depend on the choice of potential. The Yale, Hamada-Johnston, Paris and supersoft-core potentials give essentially the same results.

The first <u>polarized neutron capture</u> data on hydrogen in this energy region are now becoming available. Due to the rather low γ-ray energy and the high backgrounds, this is a very difficult experiment.

It was performed at the University of Wisconsin[6] by using a liquid scintillator as the target and requiring a coincidence between the γ-rays and the recoiling deuterons detected in the scintillator. Although the preliminary data shown in Fig. 1 do not allow us to distinguish between the different calculations, this technique, which is being pursued at Wisconsin and at TUNL, promises to provide precision data in the near future. At the present time the best data set for polarization measurements at low energies is probably that of Jewel et al.[7] obtained at Eγ=2.75 MeV. These data indicate a basic disagreement between theory and experiment which is made even worse when two-body charge and current effects are included. This observation has led Rustgi et al.[5] to conclude that the present theory of the $^2H(\gamma,n)^1H$ process is inadequate.

Data on the $^1H(\vec{n},\gamma)^2H$ reaction have also been obtained at neutron energies of 180 and 270 MeV.[8] The results of the 270 MeV experiment are shown in Fig. 2. The solid curve shown here is from Partovi[3] and does not include MEC; the dashed curve is due to Arenhövel[9] and included MEC and isobar corrections. Here again, a discrepancy persists suggesting an inadequacy in the theory, although at this energy the corrections push the theory in the proper direction. Hopefully new theoretical calculations, inspired by these new experimental results, will lead to a better understanding of the two body problem in the near future.

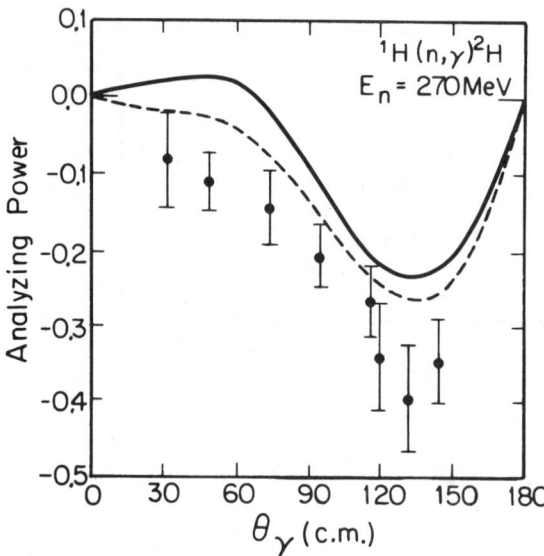

Fig. 2 Analyzing powers in $\vec{np}\rightarrow d\gamma$ at E_n=270 MeV from Ref. 8. Solid curve (Ref. 3) is without MEC, dashed curve includes MEC (Ref. 9).

There are three new results in the mass 3 systems which I would like to mention. The first of these concerns a comparison between the angular distributions of the $^3H(\gamma,n)^2H$ and $^3He(\gamma,p)^2H$ reactions. The work of Skopik et al.[10] on the ^3H case indicated that the behavior of the fore-aft asymmetry in the differential cross sections of these two reactions:

$$a_s = \frac{Y(55°)-Y(125°)}{Y(55°)+Y(125°)},$$

did not behave as expected from the simple theory. This follows if we consider the problem as an effective 2-body problem. In that case the only relevant coordinate is the relative coordinate between the nucleon and the deuteron. Since the fore-aft asymmetry in the excitation energy region of 7 to ~20 MeV is expected to arise from E1-E2 interference, it follows from simple effective charge arguments that the values of a_s in the case of ^{3}H should be -1/5 times those found in the ^{3}He case. The relevant photonuclear data have been combined to form the ratio $a_s(^3\text{H})$-to-$a_s(^3\text{He})$. The results are shown in Fig. 3, where we see that the ratio is about -0.5 rather than the predicted value of -0.2 for this reaction.

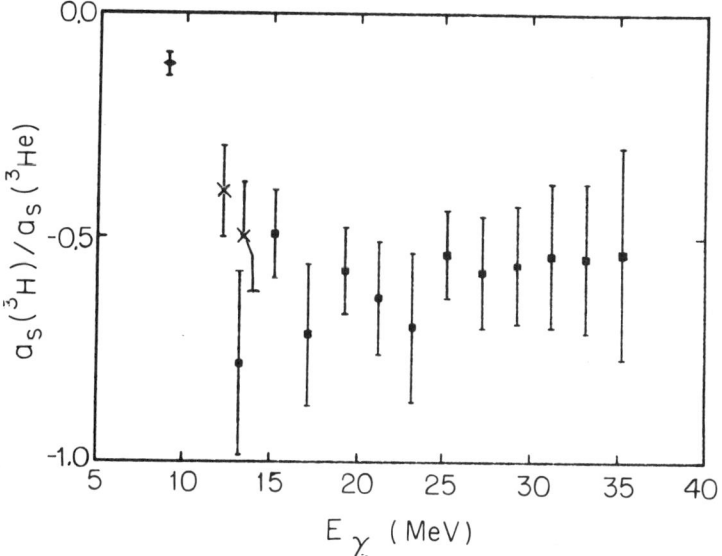

Fig. 3 Asymmetry ratios: (x) TUNL data(Refs. 11 and 12); (■) Skopik (Ref. 10)/Kundu (Ref. 13); (♦) Büsch (Ref. 14)/Wölfli (Ref. 15)

In order to verify this somewhat surprising result we have performed the capture reactions ^{2}H(p,γ)^{3}He and ^{2}H(n,γ)^{3}H in this energy region. Data for the ^{2}H(p,γ)^{3}He reaction was taken using a gas target cell whose entrance and exit windows were shielded from the NaI detector assemblies. A pulsed proton beam provided a time-of-flight condition which lead to extremely clean spectra. The experimental arrangement is shown in Fig. 4.[12,16]

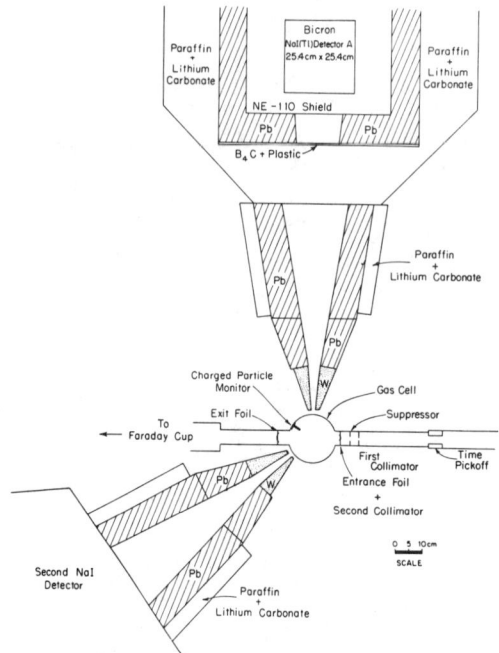

The $^2\text{H}(n,\gamma)^3\text{H}$ reaction was much more difficult to measure because of serious background problems, not to mention low count rates. However, high quality spectra were obtained by employing a deuterated liquid scintillator as the target and using the response of this scintillator to the recoiling tritons as a coincidence requirement on the γ-ray detectors. This technique worked quite well and, incidentally, allowed us to determine the response of our deuterated NE232 liquid scintillator to tritons.[11] A typical γ-ray spectrum is shown in Fig. 5.[11]

Fig. 4. Experimental arrangement for $^2\text{H}(p,\gamma)^3\text{He}$ measurements at TUNL.

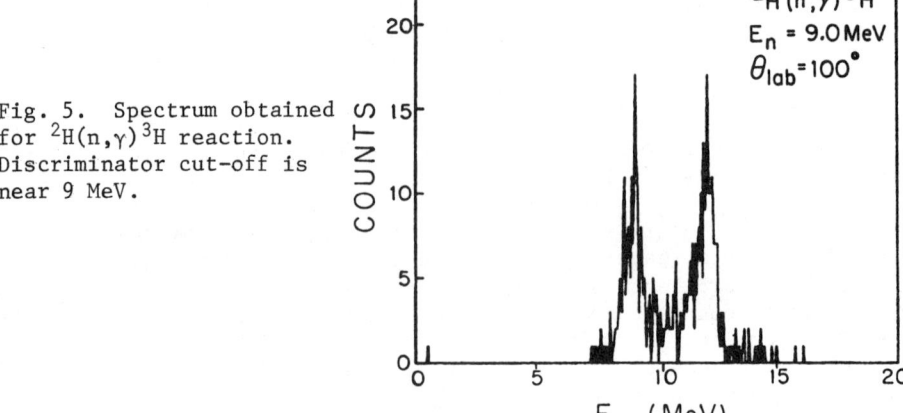

Fig. 5. Spectrum obtained for $^2\text{H}(n,\gamma)^3\text{H}$ reaction. Discriminator cut-off is near 9 MeV.

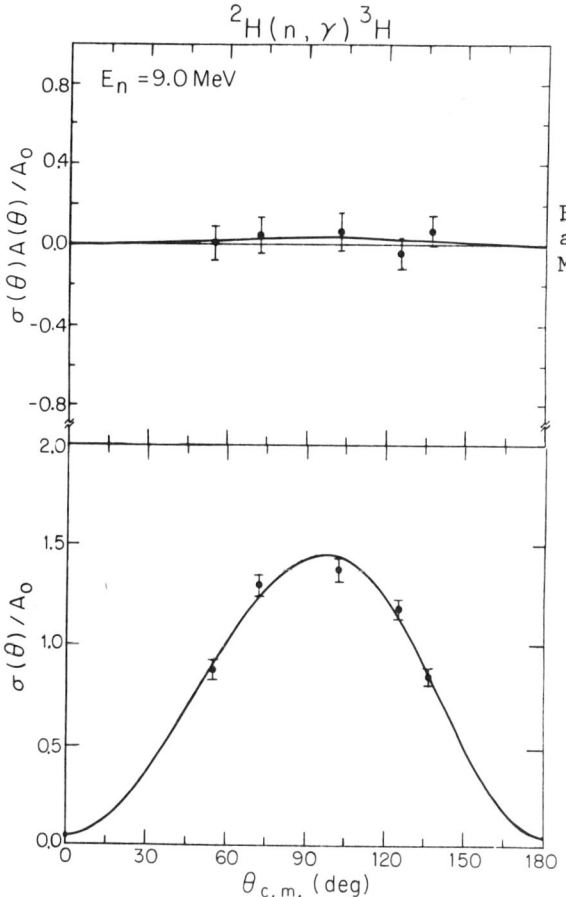

Fig. 6 Cross section and analyzing powers at E_n=9.0 MeV (E_x=12.25 MeV).

The data at E_x=12.25 MeV are shown in Fig. 6. The resulting a_s values obtained at E_x=12.25 and 13.45 MeV are shown in Fig. 7. Note that these values are based on "full" angular distribution measurements rather than just two angles. Our ratios confirm the photonuclear results! At E_x=12.25 we obtain $a_s(^3H)/a_s(^3He)$=0.4±0.1 while at E_x=13.4 MeV we obtain -0.5±0.12, as shown in Fig. 3.

In order to account for this result we began by investigating the two body assumption which leads to the -1/5 result. If the full operator is considered, it can be shown[19] that, if ρ is the coordinate between the center-of-mass of the deuteron and the third nucleon, and $\vec{r}$ is the coordinate from the n-to-the-p (or p-to-n) of the deuteron for ^{3}He (or ^{3}H), the E1+E2 operator has several terms:

For ^{3}He: $\varepsilon \cdot (\frac{\vec{\rho}}{3} + \frac{\vec{r}}{2}) + \frac{i}{2} [\frac{5}{9} (\varepsilon \cdot \vec{\rho})(\vec{k} \cdot \vec{\rho}) - \frac{1}{6} (\varepsilon \cdot \vec{\rho})(\vec{k} \cdot \vec{r})$

$- \frac{1}{6} (\varepsilon \cdot \vec{r})(\vec{k} \cdot \vec{\rho}) + \frac{1}{4} (\varepsilon \cdot \vec{r})(\vec{k} \cdot \vec{r})]$

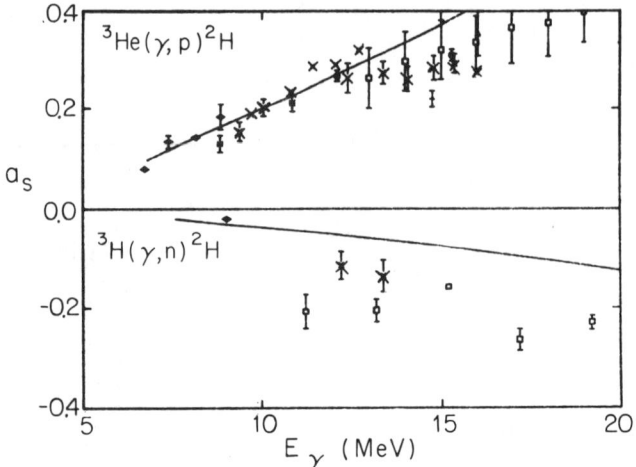

Fig. 7 Asymmetry data and results of two-body direct capture calculation (solid curves - see Ref. 16) which predict -1/5 for the ratio. (x): TUNL results; □: Refs. 10, 13; ◆: Refs. 14, 15; ● Ref. 17; ▲: Ref. 18

while for ^{3}H:

$$\varepsilon \cdot (-\frac{\vec{\rho}}{3} - \frac{\vec{r}}{2}) + \frac{i}{2} [\frac{1}{9} (\varepsilon \cdot \vec{\rho})(k \cdot \vec{\rho}) + \frac{1}{6} (\varepsilon \cdot \vec{\rho})(\vec{k} \cdot \vec{r})$$
$$+ \frac{1}{6} (\varepsilon \cdot \vec{r})(\vec{k} \cdot \vec{\rho}) + \frac{1}{4} (\varepsilon \cdot \vec{r})(\vec{k} \cdot \vec{r})]$$

The result of -1/5 follows for the assumption of point deuterons when the observed asymmetry is attributed to E1-E2 interference. What happens if realistic deuteron wavefunctions are used? Although we see that the additional terms above could, in principle, change this ratio away from the -1/5 value, it turns out to be not so easy. This is mainly because of the fact that the E2 matrix elements which correspond to transitions from the D-state to the S-state component of either the A=2 or the A=3 system do <u>not</u> interfere with the S-state to S-state transition matrix elements which dominate the (E1) cross section. Although the S-D E1 matrix elements <u>will</u> interfere with the S-D and D-S E2 matrix elements, the contribution to the asymmetry due to these terms is expected to be quite small. Detailed calculations using realistic wavefunctions for ^{3}He (Ref. 20) and ^{2}H (Ref. 21) are in progress, but we do not expect the results to change the -1/5 result by any significant amount. Is there an obvious alternative explanation? What about other multipoles? The behavior of a_1 and a_3 suggest that this is an E1-E2 effect! We conclude that this result may indicate a basic flaw in the way in which this problem has been traditionally formulated.

The second A=3 result concerns the study of D-state effects in ^{3}He. The same ^{2}H(p,γ)^{3}He data which generated the a_S coefficients also provided a very accurate and systematic set of a_2 coefficients (from expanding $\sigma(\theta)$ as $A_0[1+\sum_k a_k P_k(\theta)]$) as shown in Fig. 8.[16]

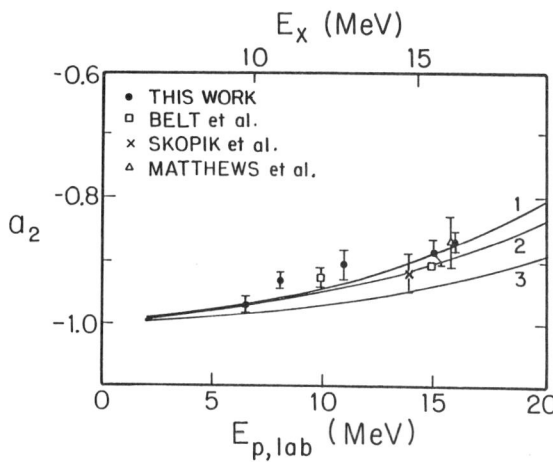

Fig. 8 Measured a_2 coefficients. See Ref. 16 for further details. Curves 1, 2 and 3 are 9.12%, 5.08% and 0% D-state in ^{3}He.

Although previous claims suggested no D-state sensitivity in this reaction at these energies[22], our careful measurements seem to indicate otherwise. The solid curves represent 2-body calculations using three different Faddeev generated wave functions for ^{3}He. One of these had no D-state, one had 5.08% for the total D-state component and one has 9.12% for the D-state probability in ^{3}He. The data are clearly fit better when a reasonable amount of D-state is included.[16] It has also been shown[16] that the inclusion of D-state effects in the analysis of the polarized and unpolarized capture data reduces the amount of E2 strength necessary to describe the measurements. Whereas previous analyses required ∼10% E2 strength, the inclusion of D-state reduces this to about 2%, which is close to theoretical predictions.[22]

A new experiment has just been completed at McMaster University.[23] In this work a tensor polarized deuteron beam was used to measure the tensor analyzing power $T_{20}(\theta)$ in the $^1H(d,\gamma)^3He$ reaction at E_d=19.8 MeV. This quantity is of interest because of the fact that it is zero if there is only s=1/2 capture. Since the radiation is mainly E1 and E2, spin flip transitions should be negligibly small. Therefore, the presence of s=3/2 capture strength should imply s=3/2 ground state components which would arise from D-state admixtures in the ground state of ^{3}He.

Precision results were obtained by using an Enge spectrometer to detect the recoiling ^{3}He nuclei in the $^1H(d,\gamma)^3He$ reaction at E_d=19.8 MeV (E_x=12.1 MeV). Since a one-to-one relationship exists between the energy of the ^{3}He particles and the angle of the γ-ray, and since the ^{3}He nuclei are confined to a cone about the beam axis with a maximum angle of 2.6 degrees, the entire γ-ray angular distribution can be observed at one time in the focal plane of the spectrometer. Two position sensitive solid-state counters were used to detect the ^{3}He particles. Consecutive runs were made for an m=1 and an m=0 beam, where the spin direction was set along the beam axis. The results for T_{20} as a function of θ are shown in Fig. 9.

The same model and the identical parameters used to calculate the a_2 coefficients of Fig. 8 were used to calculate $T_{20}(\theta)$ at $E_x=12.1$ MeV. The results for the ^{3}He wavefunctions which were calculated for an assumed deuteron D-state probability of 4 and 7% are essentially identical and are represented by the solid curve in Fig. 9. The result for no D-state is $T_{20}(\theta) \equiv 0$ in this model.

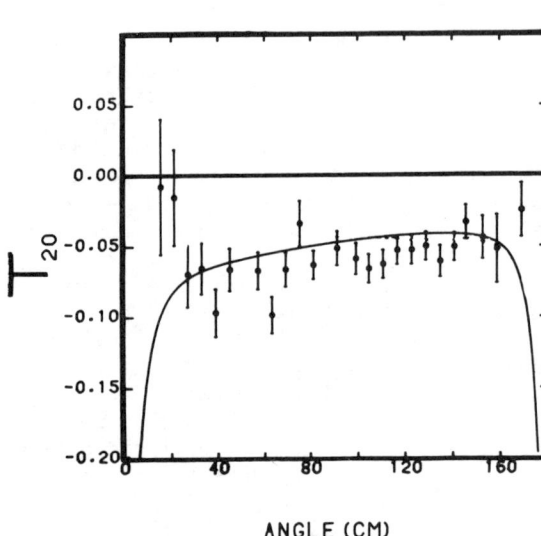

Fig. 9 T_{20} for the ^{1}H($\vec{d},\gamma$)^{3}He reaction. Solid line is the theoretical calculation which has 5.08% or 9.12% D-state in ^{3}He.

The remarkably good agreement shown in Fig. 9 provides confirming evidence for the validity of the wavefunctions generated in Ref. 24. A somewhat surprising result, however, is the apparent insensitivity to the D-state strength present in the ^{3}He wavefunction. The two calculations performed ($P_D(d)=4$ and 7%) correspond to a total-D-state probability in ^{3}He of 5.08 and 9.12%, respectively. Furthermore, the ratios of the asymptotic normalization constants (C_S'/C_D') for the two calculations are 0.05 and 0.038, respectively, so one might expect a significant difference for the two calculations. However, 70% of the transition amplitude strength arises from the ^{3}He wavefunction in the region between 2.5 and 6.5 fm. It has been found that the two-body projected wavefunctions in this region of space are almost identical. A study similar to this has been recently reported by Arriaga and Santos[25], who compared their calculation with the unpublished data of Baumgartner et al.[26] Their work showed that the Sasakawa wavefunction gave a good description of the data. However, the sensitivity which they suggest to the asymptotic D/S state ratio does not appear to be realistic since the asymptotic region of the wavefunction does not appear to make an important contribution to the capture strength.

The final new result in A=3 which I want to mention is based on an experiment in which the vector analyzing powers were measured for both the ^{1}H($\vec{\mathrm{d}},\gamma$)^{3}He and the ^{2}H($\vec{\mathrm{p}},\gamma$)^{3}He reactions.[27] The proton and deuteron energies were chosen to yield the same excitation energy (6.0 MeV) in ^{3}He. At this energy (0.5 above threshold), the E2 strength is predicted to be essentially zero by the model of Ref. 16, and M1 effects may become observeable. If the product of the cross section ($\sigma(\theta)$) and the analyzing powers ($A(\theta)$) are expanded in b_K coefficients according to

$$A(\theta)\sigma(\theta) = A_o[\sum_{k=1} b_K P_K^1(\theta)],$$

then it can be shown that if the reaction is proceeding via a particular set of s and s' (s and s' are the channel spins of 1/2 and/or 3/2, then

$$\frac{b_k(d,\gamma)}{b_k(p,\gamma)} = \frac{-3}{2} \frac{W(1s1s';1/2\ 1)}{W(1/2\ s\ 1/2\ s';1\ 1)} .$$

This implies that if s=s'=1/2 (pure doublet terms) this ratio is -3.0; if s=1/2, s'=3/2, the ratio is 3/4, while for s=s'=3/2 (pure quartet terms), the ratio is 3/2.

The experimental results are shown in Fig. 10. The b_k coefficients are presented in Table 1. It is seen that the (p,γ) data

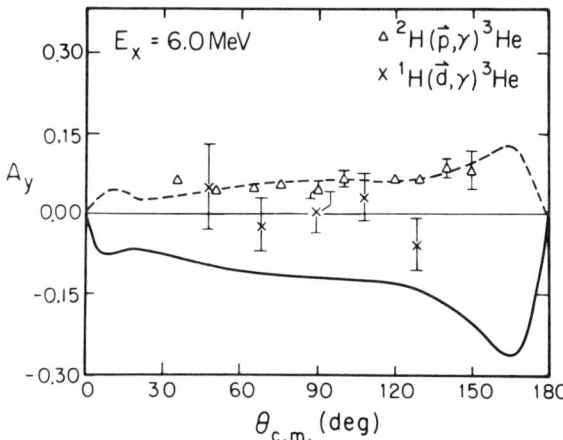

Fig. 10 The ($\vec{\mathrm{p}}$,γ) and ($\vec{\mathrm{d}}$,γ) analyzing powers. The solid curve is the predicted (d,γ) result based on the ($\vec{\mathrm{p}}$,γ) data and the assumption that only s=1/2 terms contribute.

Table 1

Expansion coefficients for the $^1H(\vec{d},\gamma)$ and $^2H(\vec{p},\gamma)$ reaction at $E_x = 6.0$ MeV

$^2H(p,\gamma)^3He$	$^1H(\vec{d},\gamma)^3He$ (using p-on-d angles)
$b_1 = 0.08 \pm 0.007$	$b_1 = -0.006 \pm 0.045$
$b_2 = -0.005 \pm 0.006$	$b_2 = 0.018 \pm 0.036$
$b_3 = -0.006 \pm 0.006$	$b_3 = -0.0.8 \pm 0.036$
$b_4 = -0.001 \pm 0.006$	

contain a finite b_1 coefficient suggesting M1-E1 interference. And b_1 for the $(\vec{d},\gamma)$ data is not -3.0 times this, implying a significant quartet contribution. If the model of Ref. 16 is used, b_1 is predicted to be zero (this is an E1+E2+E3 calculation). If as little as 1% (and less than 8%) M1 strength is added, however, the present data are accounted for. Furthermore, the minimum M1 strength was found to be predominantly s=3/2 type strength. This example points to the power of polarized capture; not only does it identify possible M1 strength but it provides, in this case, a sensitivity as to whether this is s=1/2 or s=3/2 type capture strength. A full three body calculation is, of course, needed to properly evaluate these results.

Finally, I will discuss our recent new result in ^4He. What we have done is along the lines of the d+p measurement of $T_{20}(\theta)$ which showed D-state effects in ^3He, but for the $^2H(d,\gamma)^4He$ reaction. We have measured $T_{20}(\theta)$ at six angles at $E_d=9.7$ MeV($E_x=28.7$ MeV) and found the result to be isotropic with a value of $T_{20}=-0.22\pm0.014$. Since the colliding deuterons are identical bosons, only states with L+S even are allowed. In the case of E1 radiation, this requirement is met only by the transition matrix element which corresponds to the incident channel having (L=1,S=1)1$^-$. Besides being inhibited by the isospin selection rule in self conjugate nuclei ($\Delta T=\pm 1$), E1 transitions to the ground state (S=0 or S=2) will be further diminished because they have $\Delta S=1$.

In fact, as pointed out by Mandl and Flowers[28], to the extent that the magnetic multipole operators depend only upon the spin coordinates and the electric operators only upon the spatial coordinates of the nucleon, the $^2H(d,\gamma)^4He$ reaction should be dominated by E2 radiation. Previous experimental work has shown that the angular distribution of the cross section is consistent with this assumption.[29] Our results agree with this, as shown in Fig. 11.

If the ground state of ^4He is pure (L=0,S=0) and the radiation is pure E2, $T_{20}(\theta)$ is identically zero. If we allow for a small ground state admixture having (L=2,S=2), then it can be shown that, keeping only first order terms, $T_{20}(\theta)$ is given by $T_{20}(\theta)=-0.594A+0.79B+0.497C$ where A, B and C are given by

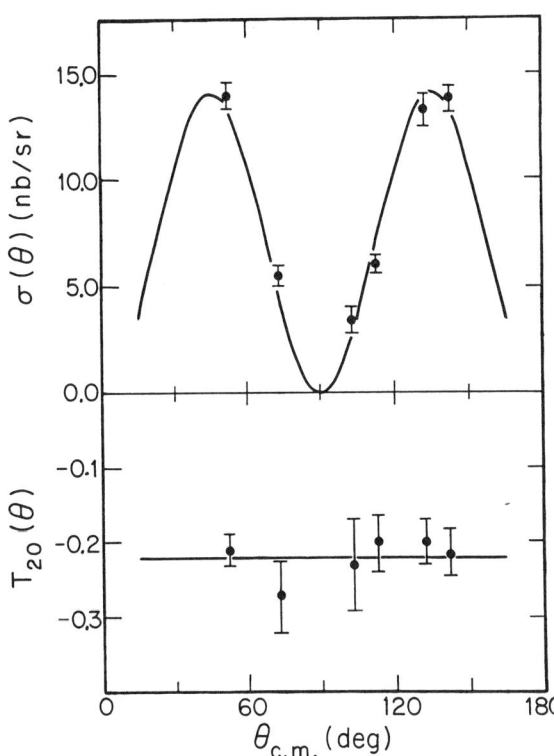

Fig. 11 The measured $\sigma(\theta)$ and $T_{20}(\theta)$ for the $^2H(\vec{d},\gamma)^4He$ reaction at E_d=9.7 MeV. The curve for $\sigma(\theta)$ is of the form $\sin^2\theta\cos^2\theta$ while for $T_{20}(\theta)$ it results from fitting the data to a constant.

$$A \equiv {}^0D_2{}^2D_2 \cos\Phi({}^0D_2 - {}^2D_2),$$

$$B \equiv {}^0D_2{}^2G_2 \cos\Phi({}^0D_2 - {}^2G_2),$$

$$C \equiv {}^0D_2{}^2S_2 \cos\Phi({}^0D_2 - {}^2S_2),$$

and ${}^S L_J$ represents the E2 transition matrix element from the continuum partial wave having these quantum numbers to the ground state. The four terms contributing to the cross section were normalized according to

$$\frac{5}{9}[{}^0D_2{}^2 + {}^2D_2{}^2 + {}^2G_2{}^2 + {}^2S_2{}^2] = 1.0$$

The results of our measurements of $T_{20}(\theta)$ are shown in Fig. 11. The fit shown was obtained by fitting the data to a constant and gives T_{20}=-0.220±0.014. The observed isotropy strongly supports the pure E2 view of this reaction.

This rather remarkably large value of T_{20} can be put in perspective if we assume pure D-wave capture and a ${}^0D_2 - {}^2D_2$ phase difference of $0°$. This allows us to solve for the ${}^2D_2$ contribution to the cross section. We find that the observed T_{20} value implies that 3% of the cross section arises from S=2 type strength.

A simple heuristic model has been used to estimate the D-state in ^4He required to account for the observed T_{20}. The ground state wavefunction was constructed from two Wood-Saxon potentials which were adjusted to bind the two point deuterons at 23.84 MeV with L=0 and L=2, respectively. The scattering wavefunctions were generated from these same potentials. Siegert's form of the E2 operator was used (no longwavelength approximation) to calculate the four previously mentioned E2 transition matrix elements, from which T_{20} was found. A 4.8% admixture of L=2 strength in the ground state of ^4He was needed to reproduce the observed T_{20} value. It seems clear that the present measurements provide an important new observable which should be of great value in studying tensor force effects in ^4He.

Measurements involving polarized capture into the two, three and four body systems have only begun to be exploited. I am confident that we will see more complete and precise data of this type in the near future. I hope that my examples have convinced you that such data are worthwhile.

Work supported by the U.S. Department of Energy Director of Energy Research, Office of High Energy and Nuclear Physics, under Contract No. DE-AC05-76ER01067.

REFERENCES

1. D.O. Riska and G.E. Brown, Phys. Lett. 38B, 193 (1972).
2. E. Hadjimichael, Phys. Lett. 46B, 147 (1973).
3. F. Partovi, Ann. Phys. (NY) 27, 79 (1964).
4. R.J. Holt, K. Stephenson, and J.R. Specht, Phys. Rev. Letts. 50, 577 (1983).
5. M.L. Rustgi, Reeta Vyas, and Manoj Chopra, Phys. Rev. Lett. 50, 236 (1983).
6. J.P. Soderstrum and L.D. Knutson, Bull. Am. Phys. Soc. and private communication.
7. R.W. Jewell, W. John, J.E. Sherwood, and D.H. White, Phys. Rev. 139, B71 (1965).
8. J.M. Cameron et al., Phys. Letts. 137B, 315 (1984).
9. H. Arenhövel, private communication.
10. D.M. Skopik, D.H. Beck, J. Asai, and J.J. Murphy II, Phys. Rev. C24, 1791 (1981).
11. G. Mitev, Ph.D. thesis, Duke University (1984); see also G. Mitev, H.R. Weller, D.R. Tilley, NIM 224, 324 (1984).
12. S. King, Ph.D. thesis, Duke University (1983).
13. S.E. Kundu, Y.M. Shin and G.D. Wait, Nucl. Phys. A171, 384 (1971) and Ph.D. thesis, University of Saskatchewan (1972).
14. R. Bösch, J. Lang, R. Müller and W. Wölfli, Phys. Lett 8, 120 (1964).
15. W. Wölfli et al., Phys. Lett. 22, 75 (1966).
16. S. King, N.R. Roberson, H.R. Weller, and D.R. Tilley, Phys. Rev. Lett. 51, 877 (1983); and Phys. Rev. C30, 21 (1984).
17. B.D. Belt et al., Phys. Rev. Lett. 24 (1970) 1120.
18. J.L. Matthews et al., Nucl. Phys. A223 (1974) 221.
19. D. Lehman, private communication.
20. D.O. Riska, Nucl. Phys. A350, 227 (1980).
21. Ronald J. Adler, Phys. Rev. C16, 1231 (1977).
22. S. Aufleger and D. Drechsel, Nucl. Phys. A364, 81 (1981).
23. M. Vetterli et al., private communication.
24. B.F. Gibson and D.R. Lehman, Phys. Rev. C29, 1017 (1984).
25. A. Arriaga and F.D. Santos, Phys. Rev. C29, 1945 (1984).
26. M. Baumgartner, J. Jourand, G.R. Plattner, W.D. Ramsay, H.W. Roser, and I. Sick (unpublished).
27. S. King, N.R. Roberson, H.R. Weller, D.R. Tilley, H.P. Englebert, H. Berg, E. Huttel and G. Clausnitzer, to be published in Phys. Rev. C.
28. B.H. Flowers and F. Mandl, Proc. R. Soc. (London) A206, 131 (1950).
29. Walter E. Meyerhof et al., Nucl. Phys. A131, 489 (1969).

"Fast Neutron Capture with a White Neutron Source"
S.A. Wender and G. F. Auchampaugh
Los Alamos National Laboratory
Los Alamos, New Mexico 87545

Abstract

A system has been developed at the Los Alamos National Laboratory to measure gamma-rays following fast neutron reactions. The neutron beam is produced by bombarding a thick tantalum target with the 800 MeV proton beam from the LAMPF accelerator. Incident neutron energies, from 1 to over 200 MeV, are determined by their times of flight over a 7.6-m flight path. The gamma-rays are detected in five 7.6 x 7.6-cm cylindrical bismuth germanate (BGO) detectors which span an angular range from $45°$ to $145°$ in the reaction plane. With this system it is possible to simultaneously measure the cross section and angular distribution of gamma-rays as a function of neutron energy. The results for the cross section of the $^{12}C(n,n'\gamma=4.44$ MeV) reaction at $90°$ and $125°$ show good agreement with previous measurements, while the complete angular distributions show the need for a large a_4 coefficient which was not previously observed. Preliminary results for the $^{12}C(n,n'\gamma=15.1$ MeV) reaction have also been obtained. The data obtained for the $^{40}Ca(n,\gamma_o)$ reaction in the region of the giant dipole resonance demonstrate the unique capabilities of this system. Future developments to the neutron source which will enhance the capabilities of the system will be presented.

1. Introduction

In this paper we will describe a system recently constructed at the Los Alamos National Laboratory to study γ-rays following fast neutron induced reactions. The system consists of a pulsed spallation neutron source and an array of five bismuth germanate (BGO) scintillators. The basic system has been described previously[1] so we will concentrate mainly on the modifications implemented during the past year. We will present some preliminary results on the $^{12}C(n,n'\gamma=4.44$ MeV), $^{12}C(n,n'\gamma=15.1$ MeV) and the $^{40}Ca(n,\gamma_o)$ reactions that show the capabilities of this system. We will conclude the paper with a description of a proposed upgrade to the neutron production source that will significantly enhance the performance of the system.

2. Neutron Source

The neutron beam is produced by a spallation reaction of the 800 MeV proton beam from the Los Alamos Meson Physics Facility (LAMPF) on a thick tantalum cylinder. The time structure of the beam consists of macro-pulses with a repetition rate of 12 Hz and micro-pulses within each macro-pulse. Each macro-pulse is approximately 700 µsec wide and is made up of sharp (less than 300 psec wide) micro-pulses whose separation can be varied from 1 μsec to over 500 μsec. This produces a maximum pulse rate of approximately 8,400 proton pulses/second. The number of neutrons/sr/sec for one μamp of proton beam is plotted as a function of neutron energy in Fig. 1. Typical proton beam currents are 70 namps.

The collimation of the neutron beam from the tantalum production target to the sample is shown in Fig. 2. The major changes to the system have been: 1. the removal of the concrete shielding wall behind the detector stand, 2. the removal of a significant amount of lead shielding from around the magnet trough and, 3. the installation of a fission chamber in the neutron beam. The first two improvements have reduced the backgrounds in the detectors. The fission chamber monitors the neutron flux by measuring the fission yield from a thin ^{235}U foil. The fission chamber, located at 5.9 m from the neutron production target, is shielded from the detectors so that it does not significantly contribute to the background. The sample is located 7.6 m from the neutron production target. With this flight path the neutron energy resolution, which is dominated by the time resolution of the detector system, is approximately 200 kev at 10 MeV. There are no filters or absorbers in the neutron beam, and the beam is stopped 80 m downstream from the sample.

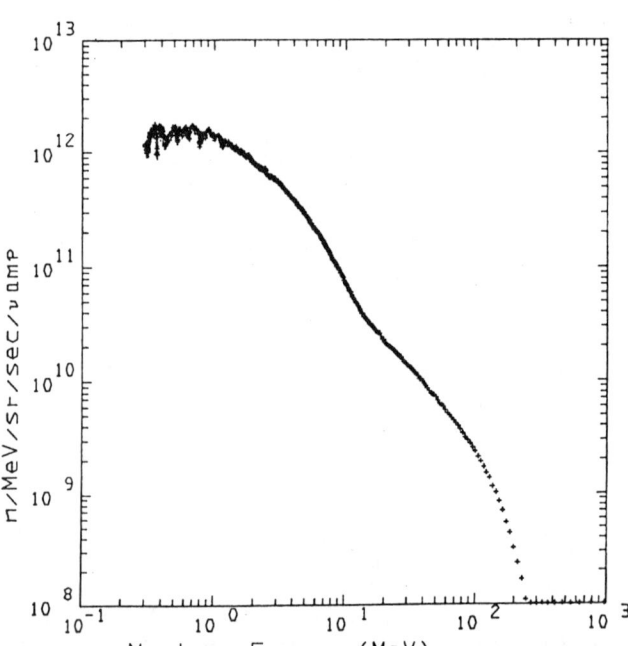

Figure 1. The neutron flux (neutrons/sr/MeV/μamp) as a function of neutron energy. Typical proton beam currents are 70 namps.

3. Detector System

The detector system consists of five 7.6 x 7.6-cm cylindrical BGO scintillators. A more complete description of the characteristics of BGO detectors is given in ref. 1. The detectors were place 35.5 cm from the center of the sample at angles of 145, 125, 90, 55, and 45 degrees. Approximately 12 cm of ^{6}LiD was placed in front of each detector to scatter neutrons away from the BGO crystal, and each detector was surrounded by approximately 1 cm of lead. The time resolution of the system was measured to be 1.6 nsec for gamma-ray energies greater than 2.0 MeV.

Data from each detector were stored in a 512-by 512-channel two dimensional array of neutron time-of-flight versus gamma-ray pulse height. The 1.25 megawords of data were recorded on magnetic tape at four hour intervals during each run to monitor the progress and stability of the run.

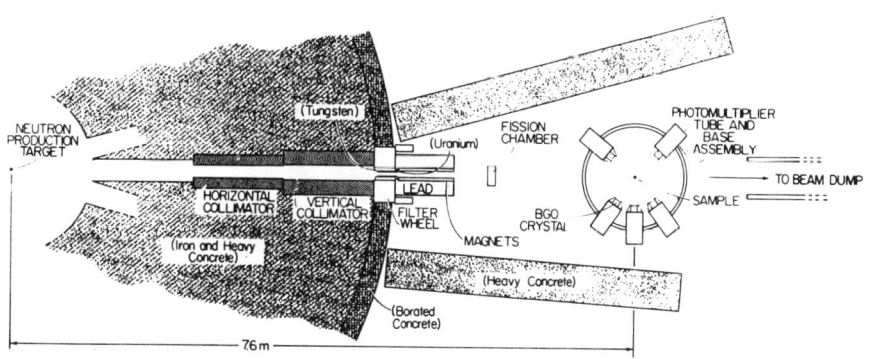

Figure 2. Experimental setup showing collimation of neutron beam and location of detector system.

4. Data analysis

The processing of the data is performed in several stages. The time independent background (i.e., thermal neutron capture in the detector, gamma-rays following thermal capture, or cosmic rays) was removed by subtracting out the pulse height spectrum associated with neutron flight times preceeding the gamma flash. The data were corrected for time correlated backgrounds (i.e., neutrons scattered from the sample directly into the detector) by measuring the pulse height spectrum of a Be sample and subtracting it from the data. Be was chosen because of its small gamma-ray production cross section. The data were corrected for the energy dependence of the incident neutron flux and the variable neutron energy bin widths. The data were binned into appropriate neutron and gamma-ray bins. Monte-Carlo techniques were used to determine the effective sample thickness including neutron absorption. No corrections were made for gamma-ray attenuation in the sample.

The detector response was removed from the pulse height spectrum using the unfolding code "FERD" [2]. The response matrix required by the unfolding code for each detector was calculated using the Monte-Carlo radiation transport code "CYLTRAN" [3] and was measured using the mono-energetic tagged photon facility at the University of Illinois. The measured line shape for a 10 MeV gamma-ray in our detector is shown in Fig. 3. The solid line represents the calculated response function [4]. The full width at half maximum is approximately 10%. The response matrix also contains the efficiency and solid angle for each detector. The unfolded spectrum was fit to a sum of gaussian functions. The area under each gaussian is the differential cross section at that gamma-ray and neutron energy.

To calibrate the absolute effieiency of the fission chamber neutron monitor, we normalized our $^{12}C(n,n'\gamma=4.44$ MeV) at $90°$ and $125°$ to data taken at the Oak Ridge Electron Linear Accelerator (ORELA) by Dickens et. al.[5] . The relative flux is based on the $^{235}U(n,f)$ cross section given in the ENDF/B-IV data set.

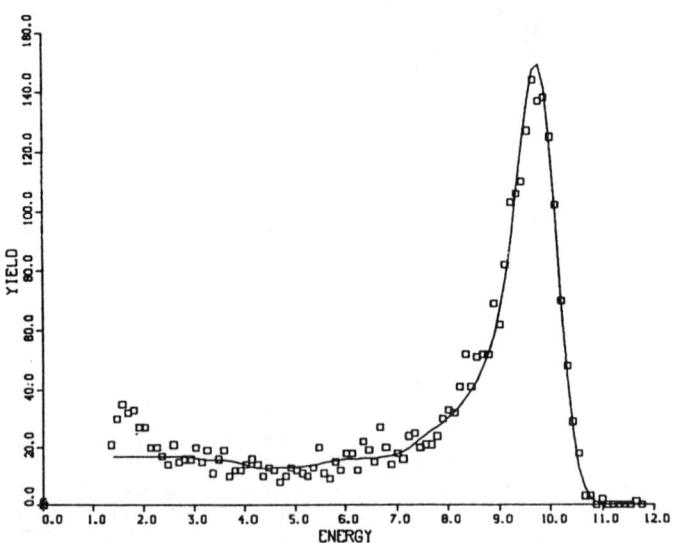

Figure 3. Response function for a 10 MeV gamma-ray in our $125°$ 7.6 x 7.6-cm BGO detector. The squares are the measured data, the line represents the calculated line shape.

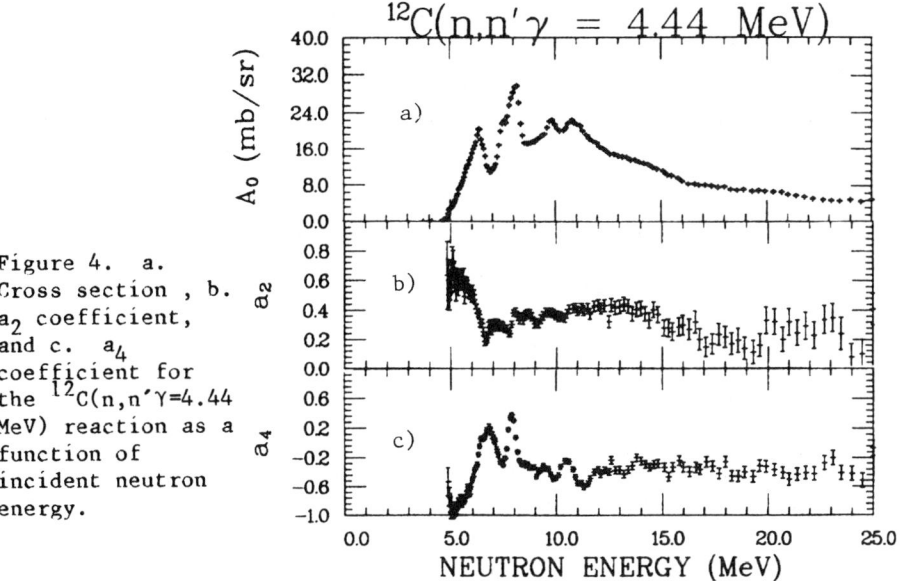

Figure 4. a. Cross section, b. a_2 coefficient, and c. a_4 coefficient for the $^{12}C(n,n'\gamma=4.44$ MeV) reaction as a function of incident neutron energy.

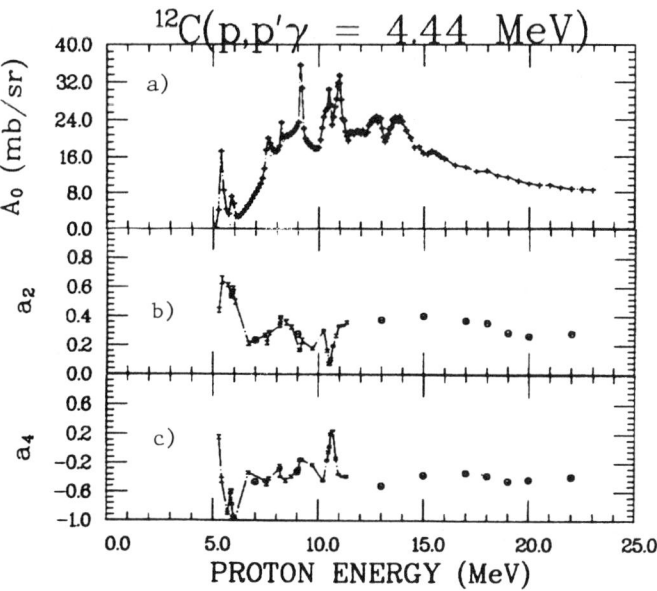

Figure 5. a. Cross section (ref 7), b. a_2 coefficient, and c. a_4 coefficient for the the $^{12}C(p,p'\gamma=4.44$ MeV) reaction as a function of incident proton energy. The connected points are from ref. 8, and the circles are from ref. 9.

5. Results and Discussion

Data were obtained over a 12 hour period for the $^{12}C(n,n'\gamma)$ reaction. The sample consisted of a 2.54-cm-diam, 5.08-cm-long cylinder of natural carbon weighing 48 grams. Fig. 4a shows the integrated cross section for the 4.44 MeV gamma-ray for neutron energies from threshold up to 25 MeV. The shapes of the 90° and 125° cross sections agree very well with the ORELA results. The data were fit to an expansion in Legendre polynomials including terms to a_4. The a_2 coeffieients are shown in Fig. 4b and the a_4 coefficients are shown in Fig 4c.

Our results for the a_2 coefficients are different from the evaluated [6] ORELA results because we measured enough angles to fit the angular distributions to terms including a_4. The previous data on this reaction were obtained at only two angles so the determination of a_4 was not possible. The present results do not give the unphysically large values for the a_2 coefficients that were present in the ORELA data. As seen from Fig 4c the values of the a_4 coefficients are large, ranging from -1.0 to +0.5, confirming the problem with the ORELA results.

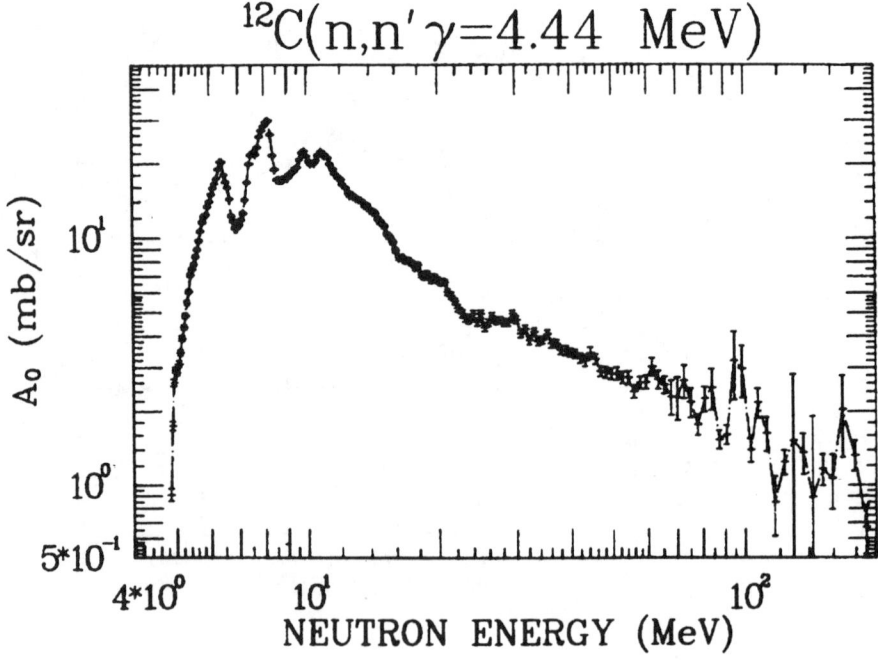

Figure 6. The cross section for the $^{12}C(n,n'\gamma=4.44$ MeV) reaction for incident neutron energies from threshold to 200 MeV.

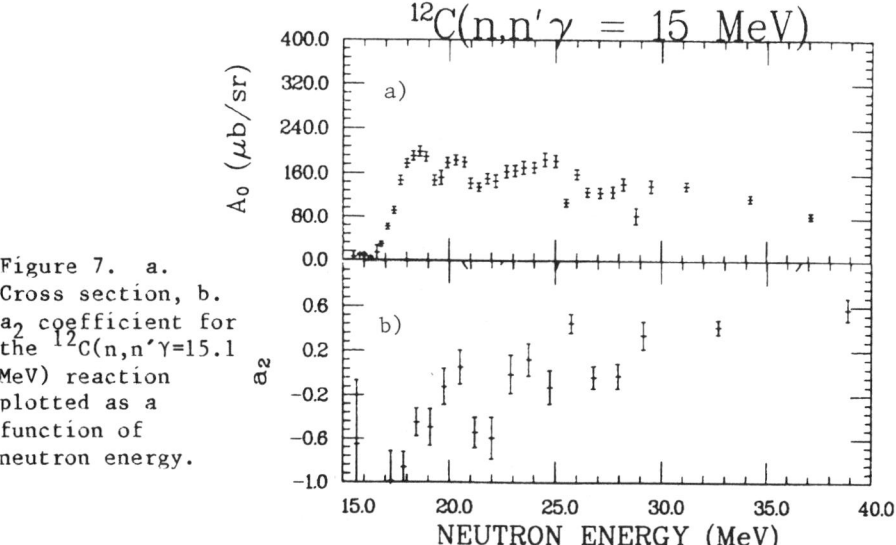

Figure 7. a. Cross section, b. a_2 coefficient for the $^{12}C(n,n'\gamma=15.1$ MeV) reaction plotted as a function of neutron energy.

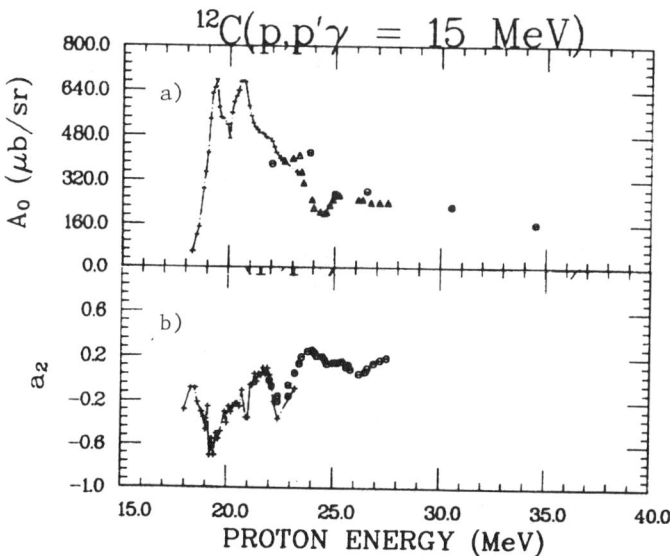

Figure 8. a. Cross section, b. a_2 coefficient for the $^{12}C(p,p'\gamma=15.1$ MeV) reaction. the connected point are from ref. 11, the triangles are from ref. 12 and the circles are from are the 90° differential cross section from ref. 13.

It is interesting to compare our results with those for inelastic proton excitation of the 4.44 MeV level. The $^{12}C(p,p'\gamma=4.44$ MeV) results are shown in Fig. 5 for: a. the cross section,[7] b. the a_2,[8,9] and c. the a_4[8,9] coefficients. If we shift the proton energy scale up 3 MeV to account for the difference in binding energies between ^{13}C and ^{13}N, we see good agreement for the compound resonances in these nuclei. In particular, the cross sections are remarkably similar and much of the structure observed in the cross section and angular distribution coefficients is present in both reactions.

The excitation function for the $^{12}C(n,n'\gamma=4.44$ MeV) reaction from threshold up to 200 MeV is plotted in Fig. 6. Suggestions of broad structures are evident around 20 MeV. Similar structure has been observed in proton inelastic scattering to this level.[10]

The $^{12}C(n,n'\gamma=15.1$ MeV) results are shown in Fig. 7, and the $^{12}C(p,p'\gamma=15.1$ MeV)[11,12,13] results are shown in Fig 8. In contrast to the 4.44 MeV data, where the cross section over the low energy resonances has similar magnitudes, the 15 MeV cross section is approximately a factor of four less in the neutron reaction than in the proton reaction. Only above 25 MeV are the cross sections similar. Also shown in Fig. 7b are the a_2 coefficients obtained from fitting our data. The a_2 coefficients for the proton reaction are shown in Fig. 8b. In this case, as in the case of the 4.44 MeV transition, the angular distribution coefficients are in good agreement when the proton energy scale is shifted by approximately 1.5 MeV.

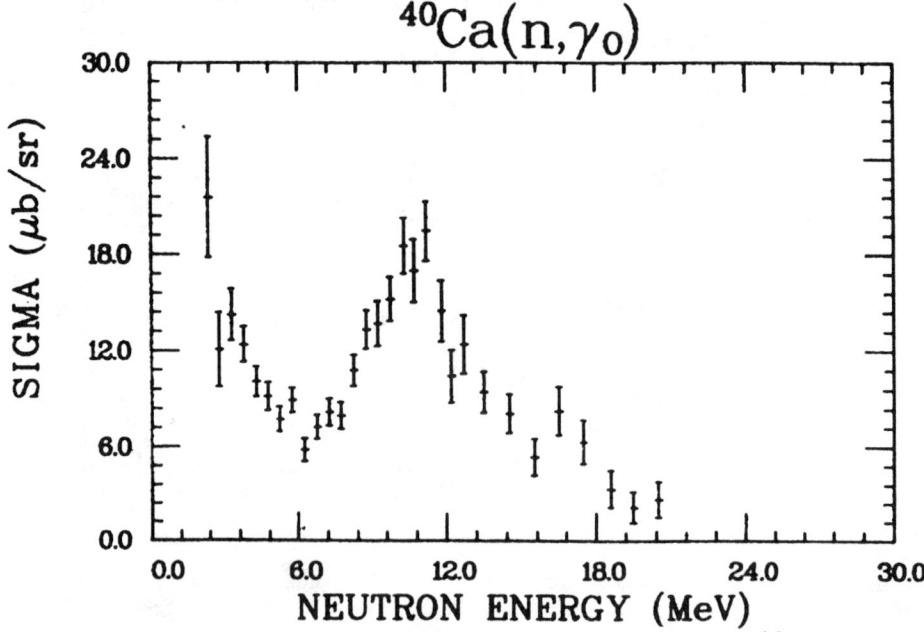

Figure 9. The $90°$ differential cross section for the $^{40}Ca(n,\gamma_0)$ reaction.

The $^{40}Ca(n,\gamma_o)$ reaction was chosen for study because it has been previously investigated [14] and would allow us to evaluate our system for fast neutron capture. Because the cross section for this reaction at the peak of the giant dipole resonance is approximately four orders of magnitude smaller than the cross section for the inelastic excitation of the 4.44 MeV state in ^{12}C, a larger (159 gram) sample was used and the opening of the neutron beam collimator was increased to raise the neutron flux at the sample. The 90° differential cross section is plotted in Fig. 9 where we have normalized our data to previous measurements.[14] The data shown were acquired in approximately 48 hours. The shape agrees very well with previous measurements.

6. Future plans

At present the major limitation of this system is the low pulse rate. To acquire data in a reasonable amount of time, a large instantaneous count rate is required which results in pulse pile-up problems. A new beam transport system and a new neutron production target are being planned that will allow a factor of five increase in the number of macro-pulses/sec and raise the maximum micro-pulse rate to approximately 42,000 proton pulses/sec. This will increase our data rate a factor of five without increasing pulse-pile up. In addition, this new target position will allow viewing the production target at angles more forward to the incident proton beam. This will result in a greater flux (approximately a factor of two at 50 MeV) of high energy neutrons with a correspondingly smaller increase in low energy neutrons. The intensity in each micro-pulse will be increased by a factor of five which will allow us to increase our flight path with a corresponding improvement in neutron energy resolution. At present the beam time at the facility is shared with condensed matter experiments, with nuclear physics experiments receiving somewhat less than 20% of the available time. In the new configuration it should be possible to multiplex the proton beam and run simultaneously with the condensed matter experiments. This will increase the available running time for fast neutron physics by a factor of five. Therefore, the over all improvement in the system including pulse rate, beam intensity, and available running time should be at least a factor of 50.

7. Conclusions

The use of a white neutron source in conjunction with an array of detectors has shown itself to be a powerful tool for studying neutron induced gamma-ray reactions. We have shown that high quality data can be obtained in a short time over a wide energy range on inelastic gamma-ray cross sections and angular distributions. We are presently able to perform fast neutron capture measurements but, with the planned upgrade to the neutron source, the ability of doing these types of measurements will be greatly enhanced.

References

1. S.A. Wender, G.F. Auchampaugh, J.F. Wilkerson, N.W. Hill, L.R. Nilsson and N.R. Roberson, Nucl. Instr. and Meth. 220(1984) 371.

2. Bert W. Rust, Daniel T. Ingersoll and Walter R. Burrus, Oak Ridge National Laboratory, ORNL/TM-8720 (1983).

3. J.A. Halbleib, Sr., W.H. Vandevender, Sandia Laboratories, SAND 74-0030(1974).

4. S.A. Wender, G.F. Auchampaugh, P.T. Debevec, S. Hoblit and S. LeBrun, Nucl. Instr. and Meth. to be published.

5. J.K. Dickens, G.L. Morgan, G.T. Chapman, T.A. Love, E. Newman, and F.G. Perey, Nucl. Sci. and Eng. 124(1977)515, and G. L. Morgan, T.A. Love, J.K. Dickens and F.G. Prery, Oak Ridge National Laboratory, ORNL-TM-3702 (1972).

6. J. Lachkar, F. Cocu, G. Haouat, P. LeFloch, Y. Patin, J. Sigaud, "An Evaluation of the Neutron-Induced Scattering, Reaction and Photon-production Cross Sections of Carbon", NEANDEC(E) 168 "L" (1975).

7. P. Dyer, D. Bodansky, A.G. Seamster, E.B. Norman, and D.R. Maxton, Phys. Rev. C23(1981)1865.

8. H.S. Adams, J.D. Fox, N.P. Heydenberg, and G.M. Temmer, Phys. Rev. 124 (1961) 1899.

9. Private communication P. Dyer

10. S. Ferroni, G. Ricco, G.A. Rottigni, M. Sanzone and G. LO Bianco, Nuovo Cem. 35(1976)15.

11. D. Berghofer, M.D. Hasinoff, R. Helmer, S.T. Lim, D.F. Measday, and K. Ebisawa, Nucl. Phys. A263(1976)109.

12. R.H. Howell, F.S. Dietrich, and F. Petrovich, Phys. Rev. C21(1980)1158.

13. D.F. Measday, P.S. Fisher, A. Kalmykov, F.A. Nikolaev and A.B. Clegg, Nucl. Phys. A44(1963)98.

14. I. Bergqvist, R. Zorro, A. Hakansson, A. Lindholm, L. Nilsson, N. Olsson, and A. Likar, Nucl. Phys. A419(1984)509.

OBSERVATION OF EXTREMELY LOW s-WAVE STRENGTH IN THE REACTION ^{136}Xe + n

B. Fogelberg*, J. Harvey, M. Mizumoto[†] and S. Raman
Oak Ridge National Laboratory, Oak Ridge, TN 37831

The neutron cross section of ^{136}Xe has been investigated using transmission measurements at the Oak Ridge electron linear accelerator. A sample of xenon gas, enriched to 93.6% in ^{136}Xe, was used as target. Measurements were made at a flight path of 80 m with an energy resolution ≈0.1%. Thirty-five resonances were found in the 0-500 keV region. Considerations of the experimental sensitivity suggest that another 3-6 resonances may have escaped detection. All strong resonances could be assigned as p-wave from the absence of interference between potential and resonance scattering. Only four very weak resonances can possibly originate from s-wave neutron interactions, but other ℓ-values are not excluded for these resonances.

The very weakness of the four unassigned resonances is actually an indication that they are probably not s-wave. The total probability of finding resonances with reduced widths up to and including the experimental values can be simply estimated from the Porter-Thomas law provided that a mean reduced width for a distribution of s-wave resonances has been determined. This quantity is of course unknown for ^{137}Xe, but can be assumed to have a magnitude similar to the mean reduced width for the p-wave resonances which was found to be about 3.5 eV in the current work. The above assumption is guided by data[1] for other valence nuclei in this mass region. In Table I, the mean of the fictitious distribution of s-wave resonances was varied within the range 1.5-6 eV. It is evident from this table that none of the four weak resonances is particularly likely to be

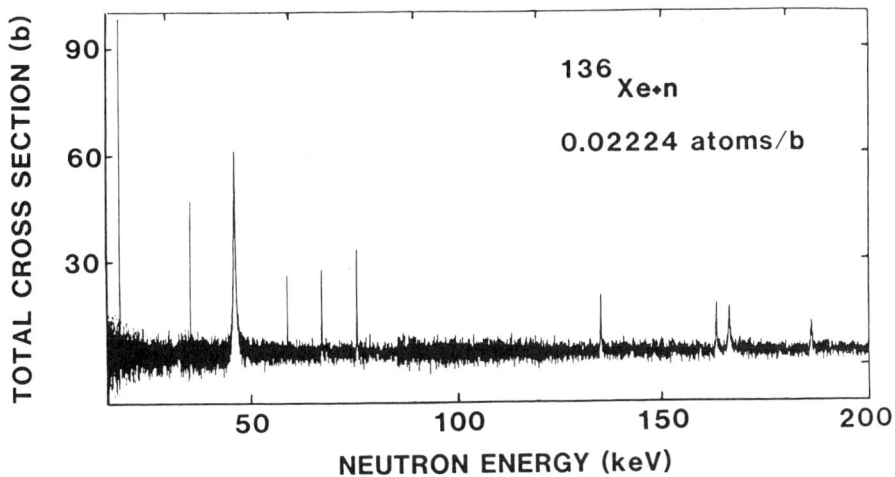

Fig. 1. A portion of the cross section curve derived from the present transmission measurements at ORELA.

s-wave. It is thus quite possible that ^{137}Xe is completely lacking s-wave strength in the first 500 keV of unbound levels. In any event, a conservative upper limit of the s-wave strength function in this region can be derived as $S_0 < 1.0 \times 10^{-6}$. Such a low value of s-wave strength is highly unusual in a heavy nucleus, and is matched only by the reaction ^{208}Pb+n.

In contrast, the p-wave strength function was found to be similar to what is observed for other nuclei in the mass region near ^{137}Xe. We derived a value of $S_1=(8.7\pm2.4)\times10^{-5}$ for a radius of 6.96 fm from the reduced widths of resonances with definite $\ell=1$ assignment. This value would not be significantly altered by inclusion of strength from the few weak unassigned resonances.

The reason for the differences in s- and p-wave strength is that few, if any, levels with $J^\pi=1/2^+$ are present in the region 4.03-4.53 MeV in ^{137}Xe, while a substantial number of $p_{1/2}$ and $p_{3/2}$ resonances were identified there, corresponding to levels with $J^\pi=1/2^-$, $3/2^-$. In fact, the $1/2^-$ levels were found to be more abundant than the $3/2^-$ levels. This shows that the spin and parity distribution of levels in the neutron resonance region (below 500 keV) of ^{137}Xe is not in accord with statistical theories. A detailed discussion of the properties of high lying levels in ^{137}Xe will be given in a forthcoming publication.

Table I

Estimated probabilities for s-wave assignments of weak resonances in ^{137}Xe. The average reduced widths $g\Gamma_n^0$ has been assumed to be in the range 1.5-6 eV.

E_n (keV)	$g\Gamma_n^0$ (eV)	$P(\ell = 0)$
79.30	0.030	< 0.05-0.12
232.24	0.056	< 0.08-0.16
252.62	0.098	< 0.10-0.21
253.47	0.083	< 0.10-0.19

Research was sponsored by the Division of Basic Energy Sciences, U.S. Department of Energy, under Contract No. DE-AC05-84OR21400 with the Martin Marietta Energy Systems, Inc. and the Swedish Natural Science Research Council.

*Permanent address: The Studsvik Science Research Laboratory, 61182 Nyköping, Sweden.
†Permanent address: JAERI, Tokai-mura, Naka-gun, Ibaraki-ken, Japan.
1. S. F. Mughabghab, M. Divadeenam and N. E. Holden, Neutron Cross Sections, Vol. I, Part A (Z=1-60) (Academic Press, New York, 1981).

LOCATION OF A DOORWAY STATE USING THE CHANNEL n + ^{207}Pb

L. C. Dennis*
Florida State University, Tallahassee, Fla. 32306

S. Raman†
Oak Ridge National Laboratory, Oak Ridge, Tennessee 37831

ABSTRACT

The location of a doorway state in the n + ^{207}Pb channel is established through a statistical analysis of the observed partial widths for gamma-rays and neutrons. Several statistical tests developed to help locate doorway states are presented. The statistical analysis focuses on the strong correlation between large partial widths in the two exit channels. Widths in both exit channels exhibit extremely large values in the energy region near E_n = 120 keV. This clustering of large widths, even when considered seperately for each exit channel, is relatively unlikely to occur in a statistical sample. The strong correlation between channels decreases the likelyhood for this clustering of large widths to occur in a statistical sample to less than 0.0003.

INTRODUCTION

The concept of doorway states, though theoretically well founded, has been difficult to verify experimentally except in special cases, such as with analog states and fission doorway states[1]. One possible candidate for a doorway state is an apparent 1$^+$ doorway state, common to both neutrons and ground state photons in n+^{207}Pb p-wave data at 120 keV[2,3]. This apparent doorway state is characterized by a clustering of large widths for levels near 120 keV in both exit channels. This paper compares the gamma-ray and neutron partial widths with statistical model predictions. To make such a comparison it is neccesary to use statistical test that precisely address the question of how likely are the large observed widths and correlations. The statistical tests and their results are discussed in detail.

DISCUSSION

Figure 1 shows the accumulated widths for the neutron and gamma exit channels. A qualitative indication of the degree of correlation between widths in the two exit channels can be obtained from this figure. The shaded areas in the figure show where the sum of the widths jumps markedly in both exit channels. Such

---

* Work supported in part by the National Science Foundation and The Florida State University.
† Work supported by the U.S. Department of Energy under Contract No. DE-AC05-84OR21400 with the Martin Marietta Energy Systems, Inc.

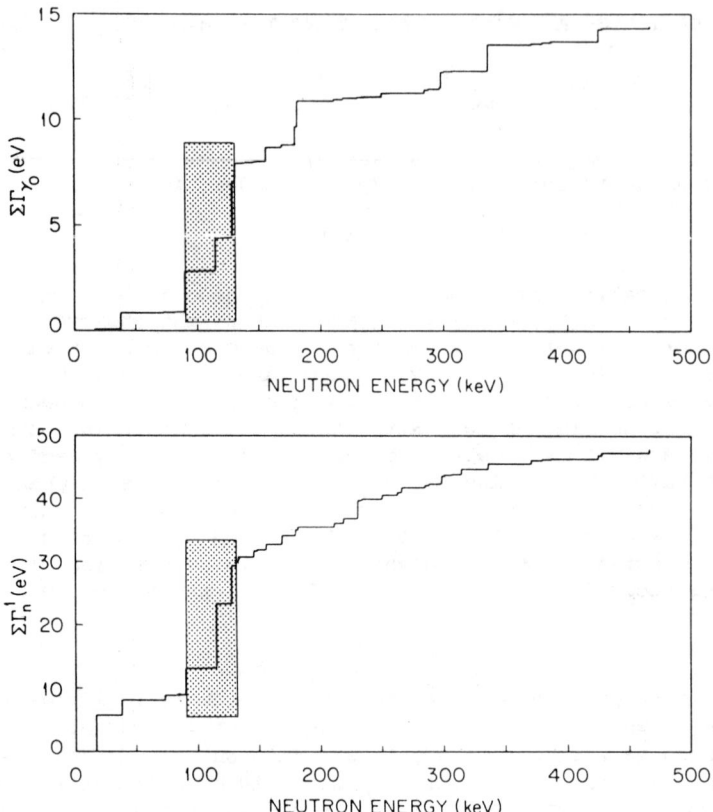

Figure 1. Cummulative sums of neutron and gamma-ray partial widths.

qualitative indications of correlations, though a good starting point are not necessarily convincing demonstrations of a doorway state's existence.

The statistical model as used to explain fluctuations[4] requires that resonance partial widths for different compound nucleus states should be uncorrelated. The actual distribution of widths will vary according to the type of exit channel. The statistical model also predicts that partial widths for different exit channels should be uncorrelated. However, when a common doorway state is present then the resonance widths for different levels in the doorway and different exit channels should be correlated and the widths may be larger than those expected from a statistical distribution of widths. There are many statistical tests that can be used to search for narrow energy regions where strong correlations exist within or between exit channels.

The correlation coefficient defined below is commonly used in

$$\rho = \frac{1}{N} \sum_{i=1}^{N} \left(\frac{\Gamma_{ni} - \bar{\Gamma}_n}{\sigma_n} \right) \left(\frac{\Gamma_{\gamma i} - \bar{\Gamma}_\gamma}{\sigma_\gamma} \right) \qquad (1)$$

statistical tests. In eq. 1 Γ_{ni} and $\Gamma_{\gamma i}$ are the neutron and gamma widths for the ith level in the compound nucleus, $\overline{\Gamma}$ and σ are the average widths and the variances and N is the number of widths. The value of ρ obtained using the data of ref. 2 and 3 is 0.64 (between 0.35 and 0.83 at the 95% confidence level). Thus there is a correlation between the gamma-ray and neutron widths. The value of ρ obtained after removing the four largest widths in the region of the suspected doorway state is ρ = 0.13 (-0.27 to 0.5 at the 95% confidence level). Thus the strong correlation is due to levels near E_n = 120 keV.

A statistical test that is useful for locating large widths within a single exit channel is based on the expected value for the sum of N randomly selected widths. If the distribution of widths is known it can be used in a Monte-Carlo calculation to predict the sum of N randomly selected widths. In the present case the probability distribution of the Γ_γ values is well reproduced by the predicted distribution[4] given by eq. 2 with $\overline{\Gamma}$ = 0.109 eV.

$$P(\Gamma)d\Gamma = \frac{1}{\sqrt{\pi}} \left(\frac{\Gamma}{2\overline{\Gamma}}\right)^{-1/2} e^{-(\Gamma/2\overline{\Gamma})} \frac{d\Gamma}{2\overline{\Gamma}} \quad (2)$$

The incomplete gamma function above $\Gamma(1/2,\overline{\Gamma})$ can be used to predict the distribution of the sum of N widths, which should have a distribution of the form $\Gamma(N/2,\overline{\Gamma})$[5]. Using these distributions and the tables of ref. 6 we find that the probability of four Γ_γ values adding to 7.1, as do those near E_n = 120 keV, is 0.0005. The Γ_n distribution is not well fit by the predicted incomplete gamma distribution. Consequently a Monte-Carlo approach using the observed width distributions was adopted to try to predict the distribution of the sum of the four Γ_n values. The Monte-Carlo calculation showed that the probability of getting a sum of 21.4, which was obtained for the four widths near E_n = 120 keV, is 0.0009. This result is clearly at odds with statistical preditions.

The distribtuion of runs tests developed by Mood[7] are well suited for locating doorway states. In this test one counts the number or runs, i.e. groups of consecutive widths above or below the average. The probability of observing N1 runs whose members are all above the average and N2 whose members are all below the average when there are A (B) widths above (below) the average width is given by eq. 3.

$$P(N1, N2) = \frac{\binom{A-1}{N1-1}\binom{B-1}{N2-1}}{\binom{A+B}{A}} F(N1, N2) \quad (3)$$

where $\binom{A}{B}$ denotes the standard binomial coefficient and $F(N1,N2)$ is 2, 1 and 0 for N1=N2, |N1-N2|=1 and |N1-N2|>1, respectively.

In the measured partial widths the number of runs in both exit channels is lower than expected because the large widths are clustered. The probabilities for the observed run distributions are 0.019, 0.25 and 0.032 in the neutron, gamma-ray and combined width distributions. The later distribution was obtained by counting as above only those resonances for which both the neutron and gamma-ray widths were above average. What is even less likely is the length of the runs in these three distributions. The probability distribution for the length of runs is given by eq. 4.

$$P(N_{ij}) = \frac{\begin{bmatrix} N1 \\ N1i \end{bmatrix} \begin{bmatrix} N2 \\ N2j \end{bmatrix} F(N1, N2)}{\begin{pmatrix} A+B \\ A \end{pmatrix}} \quad (4)$$

where the square brakets denote the standard multinomial coefficient. In the above equation $N1i$ is the number of runs whose i members are all above the average width and $N2j$ is the number of runs whose j members are all below the average. Summing $N1i$ ($N2j$) over all i (j) values gives $N1$ ($N2$). The probabilities of observing the runs distributions seen in the data of ref. 2 and 3 is 0.00019, 0.00041 and 0.00029 for the neutron, gamma-ray and combined widths, respectively.

CONCLUSION

The clustering of large widths correlated in the neutron and gamma exit channels which is observed in the n + ^{207}Pb reaction is clearly non-statistical. Of the three tests used the distributions of runs test seems to be the most sensitive to the presence of large, correlated widths. The other two tests worked well in this case but do not test for all of the characteristics of doorway states and thus may be less sensitive to their presence than the runs tests.

REFERENCES

1. A. M. Lane, Conference on the neutron and its applications (Cambridge, 1982), p. 125.
2. D. J. Horen, etal., Phys. Rev. C18, 722 (1978).
3. S. Raman, etal., Phys. Rev. Lett. 39, 598 (1977).
4. R. O. Stephen, Clarendon Laboratory Report, Ph.D. thesis (1963).
5. M. Dwass, Probability Theory and Applications, (Benjamin, 1970), p. 285.
6. M. Abramowitz and I. A. Stegun, Handbook of Mathematical Functions (National Bureau of Standards, 1972), p. 978.
7. A. M. Mood, Ann. Math. Stat. 11, 367 (1971).

NEUTRON STRENGTH FUNCTIONS OF ^{205,209}Pb

V.G.Soloviev, V.V.Voronov
Joint Institute for Nuclear Research, Dubna, USSR

Ch.Stoyanov
Institute of Nuclear Research and Nuclear Energetics of the Bulgarian Academy of Science, Sofia

ABSTRACT

The s-, p-, and d-wave neutron strength function are calculated for ^{204}Pb + n and ^{208}Pb + n resonances.

Value of neutron strength functions is determined by the strength distribution of one-quasiparticle components of the wave functions near the neutron binding energy. In ref. [1] the s-, p- and d-wave neutron strength functions for some Pb isotopes have been calculated with the quasiparticle-phonon model of a nucleus. Our calculations provide rather a good description of energy dependence of the strength functions observed in experiment [2]. In the present paper we have calculated the neutron strength functions for ^{205,209}Pb. The scheme of calculation and the Hamiltonian parameters are given in ref. [1]. Our calculations predict the existence of a strong substructure in ^{209}Pb at $E_n = 0.5$ MeV for s-wave neutrons. The strength distribution of the $4s'_{1/2}$ subshell has a pronounced local peak due to the coupling with the configuration $2g_{9/2} \otimes 4^+_1$, which appears as a substructure in the strength function (see fig. 1). In ^{205}Pb this substructure is smeared due to a large density of levels and it manifests itself very weakly. Our calculations show the existence of substructures for d-wave strength functions in ^{209}Pb as well. In ^{205}Pb these substructures are smoothed. The calculated strength functions for the averaging interval of 1 MeV are shown in the Table. While averaging we neglected the presence of substructures.

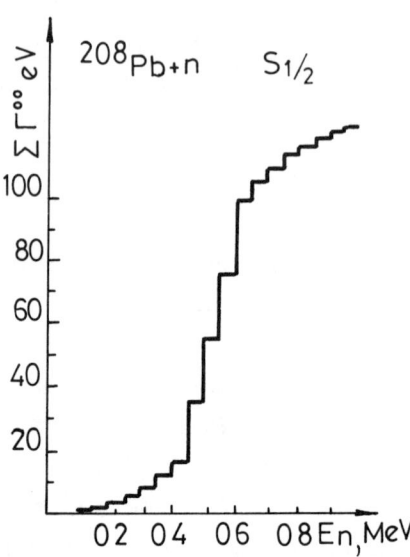

Fig. 1

Plots of the sum of the reduced neutron widths for the ^{208}Pb + n resonances versus neutron energy

Table 1. Calculated values of the strength functions S^{ℓ} in ^{204,208}Pb + n

Target	Partial wave	$S^{\ell} \times 10^4$
^{204}Pb	s	1,9
	p	0,2
	d	2,2
^{208}Pb	s	1,3
	d	0,7

References

1. V.G.Soloviev, Ch.Stoyanov, V.V.Voronov, Nucl.Phys. A399, 141 (1983).
2. D.J.Horen, Y.A.Harvey, W.W.Hill, Phys.Rev. C18, 722 (1978):, C20, 478 (1979); C24, 1961 (1981).

VALENCE NEUTRON CAPTURE IN s- AND p-WAVE RESONANCES IN ^{32}S

B.J. Allen
Australian Atomic Energy Commission Research Establishment
Lucas Heights Research Laboratories
Private Mail Bag, Sutherland, NSW 2232, Australia

F.Z. Company
University of Wollongong, Wollongong, NSW 2500, Australia

ABSTRACT

Neutron capture γ-ray spectra from resonances with large reduced neutron widths have been measured in ^{32}S. Spectra were obtained with a large NaI detector with time-of-flight discrimination of resonance capture events. Partial radiation widths are deduced and found to be consistent with a major valence contribution in both the 102.7 keV s-wave and 202.6 keV p-wave resonances. The observed and calculated valence partial radiation widths are found to be highly correlated, supporting an important role for the valence model in both s- and p-wave resonances in ^{32}S.

INTRODUCTION

The valence model of resonance neutron capture describes the transition of a valence neutron outside a closed shell nucleus. The strength of the transition is dependent, inter alia, on the single particle character of the resonance and final state. For the resonance state, the reduced neutron width is a measure of the single particle strength, with maximum values occurring for s-wave resonances in the A ∿50 region and p-wave resonances at A ∿90. However, in light A nuclides, both s- and p-wave resonances with large reduced widths can sometimes occur in the same nuclide. This is the case for ^{32}S for which Halperin et al.[2] have measured s- and p-wave radiation widths for comparison with valence calculations using the optical model approach.

We have further investigated the role of the valence model by the measurement of capture γ-ray spectra from neutron resonances at <100> and <200> keV, so as to compare the widths with the calculated partial valence widths.

EXPERIMENT AND ANALYSIS

Neutrons were produced by the ^{7}Li(p,n) Be reaction using the 3 MV Van de Graaff accelerator. The pulsed and bunched proton beam delivered an average beam current of about 5 μA on the 20 keV thick ^{7}Li target. The beam pulse repetition rate was 2 MHz and the bunched pulse width typically about 3.5 to 4.5 ns (FWHM). The proton energy was calibrated by resonance yields in the ^{27}Al(p,γ) reaction. A well-shielded NaI detector (20 x 15 cm) set at 125° to the beam, measured γ-ray spectra from the 1151 g sulphur sample (0.123 atom b^{-1} of 95%

^{32}S) at a flight path of 20 cm.

After background subtraction from the γ-ray spectra, the γ-ray response function of the detector[3] was removed by line shape analysis to give the γ-ray intensity histogram (Fig. 1). The γ-ray absorption in the shielding materials and in the sample was calculated. After Monte Carlo neutron self-shielding and multiple scattering corrections, summed partial radiation widths $\Sigma\Gamma_{\lambda\mu}$ in a specified neutron energy range were calculated using $\Sigma\Gamma_{\lambda\mu} = I_{\lambda\mu}\Sigma\Gamma_\gamma$ where $I_{\lambda\mu}$ is the normalised γ-ray intensity to final state μ, $\Sigma\Gamma_\gamma$ is the summed radiation strength (taken from ref. 2) over the neutron energy dispersion.

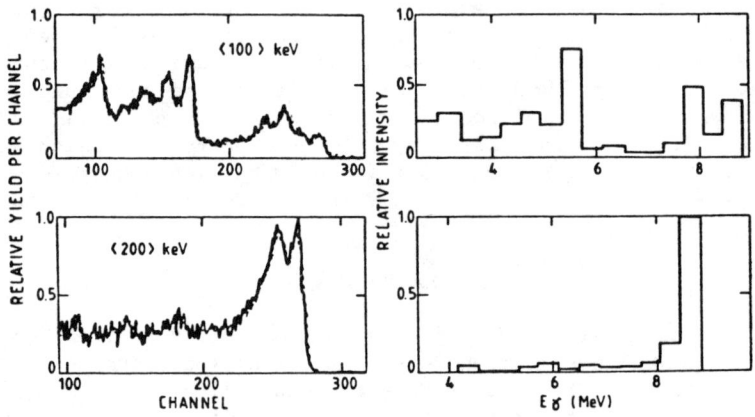

Fig. 1. Fitted γ-ray spectra and intensity histograms

RESULTS

Table 1 shows the relative γ-ray intensities of the primary transitions I_γ (I_γ is normalised to 100 primary transitions) and the partial radiative widths of ^{32}S at <100> and <200> keV.

The partial E1 widths of valence transitions were calculated following Allen and Musgrove[1] using the spectrosopic factors[4] for ^{32}S(d,p)^{33}S and the well established level scheme. Partial E1 widths for the valence transitions $\Gamma^V_{\lambda\mu}$ were calculated at <100> and <200> keV and compared with the observed partial widths in Table 1. Relevant resonance parameters are also shown.

<100> keV. There is a large error (±2 eV) in the total radiation width measurement of the 102.7 keV resonance (because of the uncertainty relating to the contribution of scattered neutrons to the capture yield[2]) which is three times higher than the actual value of the radiation widths of the p$^3/_2$ resonances. After subtraction of the calculated p$^3/_2$ valence component to the ℓ_n = 0,2 states and assuming M1 transitions to the ℓ_n = 1 group can be neglected (cf. 202.6 keV γ-ray spectrum), the spectrum can be ascribed to the single s$^1/_2$ wave resonance without a significant change in the uncertainties.

s$^1/_2$ wave resonance. The calculated and experimental widths are

Table 1 Partial Valence and Observed Radiation Widths at E_n = <100> and <200> keV

					$s_{1/2}$ 102.7 keV Γ_γ = 7 eV Γ_n^0 = 48.3 eV Γ_n = 15 keV		$p_{3/2}$ 97.5, 112.2 keV $\Sigma\Gamma_\gamma$ = 0.66 eV $\Sigma\Gamma_n^1$ = 12.1 eV		$p_{1/2}$ 202.6 keV Γ_γ = 0.9 eV Γ_n^1 = 24.77 eV Γ_n = 3 keV	
E_μ MeV	ℓ	J_μ^π	θ_μ^2	$(E_\gamma-E_n)$ MeV	I_γ	$\Sigma\Gamma_{\lambda\mu}$ meV	$\Sigma\Gamma_{\lambda\mu}^V$ meV	I_γ	$\Gamma_{\lambda\mu}$ meV	$\Gamma_{\lambda\mu}^V$ meV
0.000	2	$3/2^+$	0.93	8.640	13±2	995±230	35	75±10	675±115	740
0.842	0	$1/2^+$	0.32	7.799	19±3	1460±335	126	18±4	160±50	260
1.968	(2)	$5/2^+$	0.00	6.671	2±1	150±90	0			0
2.313	2	$3/2^+$	0.07	6.325	2±1	150±90	0			22
2.869	2	$\begin{pmatrix}3/2\\5/2\end{pmatrix}^+$	0.10	5.771	3±1	230±150	<10	7±4	63±35	24
3.221	1	$3/2^-$	0.48	5.419	27±5	2070±560	2160			0
4.43	1	$3/2^-$	0.07	4.427	8±2	610±190	210			0
4.920	1	$1/2^-$	0.04	3.720	6±2	460±170	90			0
5.715	1	$1/2^-$	0.53	2.925	12±3	920±290	570			0
5.894	1	$3/2^-$	0.11	2.746	8±3	610±370	75			0

Fig. 2. Comparison of measured and valence partial radiation widths for E1 transitions from the 102.7 keV s-wave and 202.6 keV p-wave resonances.

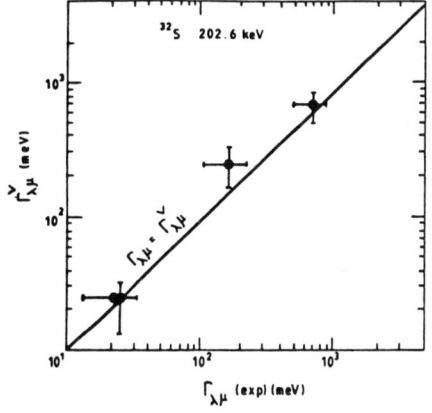
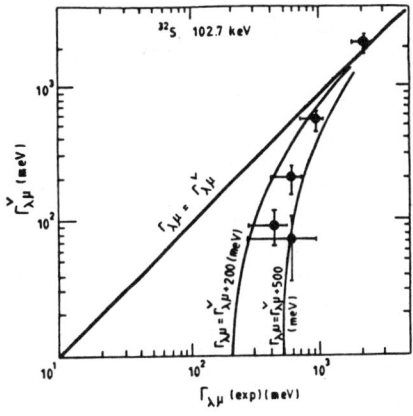

compared in Fig. 2 where a 200 to 500 meV non-valence width contribution is deduced. The correlation between the observed and valence widths, $\rho(\Gamma_{\lambda\mu},\Gamma^V_{\lambda\mu}) = 0.89$ (SD = +0.07, -0.21) is consistent with a major role for s-wave valence capture in the 102.7 keV resonance. The error is the standard deviation of a correlated distribution calculated for the appropriate sample size using Fisher's transformation. The uncertainty associated with the error in $\Gamma_{\lambda\mu}$ is calculated to be ±0.33.

For the E_γ = 5.5 to 7.0 MeV region, $\Gamma_{\lambda\mu} \simeq \Gamma^V_{\lambda\mu} \simeq 0$. The corresponding final states have very low spectroscopic factors and the results are as predicted in the theory of valence capture.

$p^3/2$ wave resonances. The sum of the partial widths of the ground and first excited state at <100> keV is about four times larger than the total radiation widths of the $p^3/2$ waves. These transitions, after subtracting the calculated p-wave valence component, give an upper limit for M1 transitions from the $s^1/2$ resonance. It is found that the $<I_\gamma(E1)E_\gamma^{-3}>$ value (averaged over the ℓ_n = 1 group) is about ten times higher than the $<I_\gamma(M1)E_\gamma^{-3}>$ value (average of the ground and first excited state). This result is very much greater than the interpolated value of 1.5 from the systematics of Kopecky[5].

<200> keV. For the ground and first excited state, $\Gamma_{\lambda\mu}$ equals $\Gamma^V_{\lambda\mu}$ within the experimental error. A transition to the 1.968 MeV level would have M2 multipolarity and is not observed. Transitions to levels at 2.313 and 2.869 MeV are E1 and have $\Gamma_{\lambda\mu} \simeq \Gamma^V_{\lambda\mu} \simeq$ 25 meV. The valence and observed partial radiation widths are shown in Fig. 2, and their correlation is $\rho(\Gamma_{\lambda\mu},\Gamma^V_{\lambda\mu}) = 0.96$ (SD = +0.11, -0.34).

A large sample size can be achieved by combining the five E1 transitions at <100> keV with the four E1 transitions at <200> keV. For this group the valence correlation is $\rho(\Gamma_{\lambda\mu},\Gamma^V_{\lambda\mu}) = 0.88$ (SD = +0.12, -0.34), including the error (±0.24) arising from uncertainties in $\Gamma_{\lambda\mu}$.

These results provide confirmation of the valence theory for a limited set of partial radiation widths from the strongest s- and p-wave reduced neutron width resonances in ^{32}S.

REFERENCES

1. B.J. Allen and A.R. Musgrove, Advances in Nuclear Physics, 10, 129 (1978) (and references therein).
2. J. Halperin, C.H. Johnson, R.R. Winters and R.L. Macklin, Phys. Rev. C21, 545 (1980)
3. R.F. Barrett, K.H. Bray, B.J. Allen and M.J. Kenny, Nucl. Phys. A278, 204 (1977).
4. M.C. Mermaf, C.A. Whitten Jr., J.W. Chaplin, A.J. Howard and D.A. Bromley, Phys. Rev. C4, 1778 (1971).
5. J. Kopecky, Inst. Phys. Conf. Ser. No. 62, 434 (1982).

EVIDENCE FOR VALENCE TRANSITIONS IN
NEUTRON CAPTURE GAMMA-RAY SPECTRA IN ^{88}Sr

B.J. Allen
Australian Atomic Energy Commission Research Establishment
Lucas Heights Research Laboratories
Private Mail Bag, Sutherland, NSW 2232, Australia

F.Z. Company
University of Wollongong, Wollongong, NSW 2500, Australia

ABSTRACT

Neutron capture γ-ray spectra have been measured at 11 average neutron energies from 10 to 530 keV in ^{88}Sr using a 20 x 15 cm NaI detector with time-of-flight discrimination of background events. The partial radiation widths and the calculated partial valence widths are compared for the strong p-wave resonances at 287 and 321 keV and found to be highly correlated. At these energies, the spectra are dominated by strong transitions to low-lying single particle states, in confirmation of the role of valence capture in the 3p region. However, the data do not support this mechanism at <508> keV.

INTRODUCTION

The ground state of ^{88}Sr has a closed shell of 50 neutrons and the ground and first excited states of ^{89}Sr have strong $d^5/_2$ and $s^1/_2$ single particle configurations[1]. Further, the p-wave strength function is large and intermediate structure in the $p^3/_2$ strength function is observed at ∿300 keV. These conditions are ideal for the manifestation of valence resonance capture. Indeed, large initial state width correlations $\rho(\Gamma_{\lambda n}, \Gamma_{\lambda \gamma})$ have been reported[1,2] which support the valence theory. However, the contribution of prompt resonance scattered neutrons to the observed yields may not have been adequately determined, and measurements of capture γ-ray spectra are required. To this end, we have measured resonance γ-ray spectra with a NaI detector[3], and report on partial width correlations, particularly in the 300 keV region.

MEASUREMENTS AND ANALYSIS

A pulsed 5 μA beam of protons produced neutrons via the ^{7}Li(p,n) reaction. A well shielded 20 x 15 cm NaI detector measured capture γ-ray spectra in a 241.55 g Sr(NO₃)₂ sample, enriched to 99.84% ^{88}Sr (9.666 x 10⁻³ atom b⁻¹). The sample was provided on loan by the US Department of Energy. Time of flight discrimination allowed the simultaneous measurement of resonance capture and background γ-rays. The low capture cross section of ^{88}Sr required a short flight path of 10 to 20 cm, with the effect of reducing the timing resolution and increasing the neutron energy range over the sample area. Average incident neutron energies and the spread in energy (Δ) were calculated from kinematics, and are given in Table 1. The incident neutron flux per keV was obtained from time of flight measurements with a ^{6}Li glass. After background subtraction, γ-ray intensities were deduced by line shape analysis. Monte Carlo neutron self-shielding and multiple scattering corrections were made, as well as corrections for γ-ray attenuation in the sample and detector shielding. Relative γ-ray intensities per incident neutron per keV are listed in Table 1. Primary transitions are identified by their energy shift with neutron energy, whereas the energies of secondary transitions are constant.

DISCUSSION

Although the reported initial state correlations[1,2] are sensitive to scattered neutrons, our measurements do not suffer from this effect. We derive correlations between the summed reduced neutron widths in each energy range Δ and γ-ray intensities per incident neutron per keV to five sets of final states. Results are given in Table 1 with standard deviations derived from Fisher's transformation, and are consistent with zero correlation. However, correlations are observed for transitions to the ground, first and fourth excited states if resonances above 400 keV are excluded. These final states are strongly single particle in nature, and the correlations are consistent with a valence effect.

The deduced γ-ray intensities I_γ over the excitation region (Δ) are proportional to the integrated partial capture cross section over resonances λ in that region, which reduces to $\Sigma g_\lambda \Gamma_{\lambda\mu} \Gamma_{\lambda n}/\Gamma_\lambda$. For many resonances, g is unknown and, at higher energies, $\Gamma_{\lambda\gamma}$ has yet to be measured. Further, the level density decreases at high energies indicating that resonances have been missed in the capture cross section measurements. The observed γ-ray spectra are sums over s-, p- and d-wave resonances, and undoubtedly weak transitions and primary transitions below $E_\gamma = 3$ MeV are missing. To make the problem more tractable, we assume that
- only E1 transitions from $p^1/_2$, $p^3/_2$ resonances contribute,
- missed resonances would not significantly affect the observed γ-ray intensities because they would have small capture areas,
- missed primary γ-ray transitions comprise 20% of the summed radiation widths.

The deduced γ-ray intensities $I_{\gamma\mu}$ are related to the summed partial widths by $\Sigma g_\lambda \Gamma_{\lambda\mu} = I_{\gamma\mu} \Sigma g_\lambda \Gamma_{\lambda\gamma}$. This equation cannot be accurately evaluated, so we use the approximation $<g>\Sigma\Gamma_{\lambda\mu} = I_{\gamma\mu} <g>\Sigma\Gamma_{\lambda\gamma}$.

The radiation widths given in refs. 1,2 are generally in good agreement, except for those resonances with large reduced neutron widths at ∼300 keV. Both experiments inadequately analyse these resonances, and large uncertainties are associated with their deduced radiation widths. In this paper, resonance parameters are taken from ref. 1, and it can be assumed that $\Gamma_{\lambda\mu}$ values are upper limits because of the neutron scattering effect.

Partial widths are deduced for spectra taken at <287>, <321> and <336> keV, and are compared with valence widths calculated after Allen and Musgrove[4]. Valence widths to the higher excited states are an order of magnitude too low, explaining the lack of significant correlations for these states. However, the ground and first excited state

Table 1. Relative γ-ray yield (I_γ) per incident neutron per keV at the sample angle per $\Delta E_\gamma = 0.3$ MeV interval

E_μ (MeV)	ℓ	J^π	θ_μ^2	E_γ (MeV)	E_n keV: 13; Δ: 2; $\Sigma\Gamma_n^1$ eV: 10	29; 15; 25	54; 20; 3.5	122; 43; 36	150; 59; 48	287; 48; 58	321; 47; 62	336; 40; 58	399; 47; 52	423; 48; 57	508; 50; 167	$\rho(\Sigma\Gamma_n^1, I_\gamma)$ (SD) <400 keV	$\rho(\Sigma\Gamma_n^1, I_\gamma)$ (SD) <530 keV
0.000	2	$5/2^+$	0.79	6.360	1.4	1.8	9.0	2.2	14.0	16.5	6.0	3.4	0.3	1.7		0.61(+0.20−0.33)	0.00(+0.35−0.35)
1.032	0	$1/2^+$	0.90	5.328	3.8	1.2	0.5	6.0	2.9	11.0	10.5	4.5	3.0	0.3	1.0	0.65(+0.17−0.29)	0.00(+0.33−0.33)
1.473		$(7/2)^+$		4.887													
1.940	2	$5/2^+$	0.09	4.420													
2.008	2	$3/2^+$	0.46	4.352	1.7	1.5	1.4	1.5	1.0	3.8	4.5	1.8	3.0	1.2	0.6	0.55(+0.19−0.30)	−0.09(+0.34−0.32)
2.061		$(9/2)^+$		4.299													
2.079	5	$(11/2)^-$		4.281													
2.880	(1)			4.080													
2.452	2	$3/2^+$	0.34	3.908	0.7	1.3	1.5	2.0	0.8	1.5	0.6	2.2	0.7	1.4		0.09(+0.36−0.39)	0.08(+0.32−0.34)
2.570	1	$3/2^-$		3.790													
2.671	4	$7/2^+$	0.73	3.689													
2.805	(2)			3.442													
2.918	2	$(3/2, 5/2)^+$		3.386													
2.974	2	$9/2^+$		2.974													
3.070	2																
3.128	2	$(3/2)^+$	0.05	3.232	0.6	0.5	0.9	2.0	0.9	1.3	1.5	0.6	4.0	2.0	3.1	0.30(+0.29−0.38)	0.47(+0.23−0.31)
3.228		( )−		3.132													
3.245	2	$(3/2, 5/2)^+$	0.06	3.115													
3.386		$(7/2^+, 9/2^+)$		2.974													
3.388		$(11/2^-, 12/2^-)$	0.08	2.972													

widths are in good agreement with valence expectations.

There are similar numbers of resonances in the <287>, <321> and <336> keV groups and $\Sigma\Gamma_{\lambda\mu}$ is comparable for the higher excited states. We can therefore assume that these widths represent the summed statistical contribution $\Sigma\Gamma^S_{\lambda\mu}$ and determine $k = \Sigma\Gamma^S_{\lambda\mu}(nE^3_\gamma)^{-1}$. The curve $\Sigma\Gamma_{\lambda\mu} = \Sigma\Gamma^V_{\lambda\mu}+(knE^3_\gamma)$ can then be obtained which fits the data in Fig. 1 rather well.

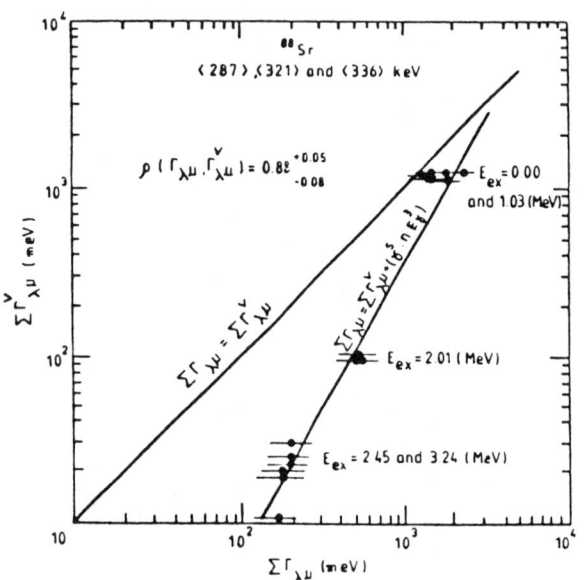

Fig. 1 Correlation of measured and valence summed partial radiation widths

If the adopted $p^3/_2$ resonance radiation widths are too large because of neutron scattering effects, the calculated valence widths would also become too large. However, the partial width correlation is independent of the magnitude of the total radiation widths. We find this correlation to be large and highly significant, $\rho(\Sigma\Gamma_{\lambda\mu},\Sigma\Gamma^V_{\lambda\mu}) = 0.88$ (SD = +0.05,-0.08), and in confirmation of the role of the valence capture mechanism for these strong $p^3/_2$ resonances in ^{88}Sr. Similar results pertain to other resonance groups, particularly at <13> and <122> keV.

Above 400 keV, valence capture should be dominated by $p^1/_2$ resonances, so a reduction in the $d^5/_2$ transition strength would be expected. The $s^1/_2$ strength should increase because of the very strong resonances at 513 and 520 keV ($p^1/_2$) and 521 keV ($p^3/_2$), but this does not occur. Instead, the strength to the weak $d^3/_2$, $^5/_2$ states at ~3.2 MeV increases. These observations are consistent with the radiative decay of a p-wave doorway state[5] such as $3^- \times d^3/_2$ to final states of vibrational character.

REFERENCES

1. J.W. Boldeman et al., Nucl. Phys. A269, 397 (1976).
2. B.J. Allen et al., Inst. Phys. Conf. Series 62, p. 404 (1982).
3. B.J. Allen and F.Z. Company, ibid., p. 398.
4. B.J. Allen and A.R. de L. Musgrove, Adv. Nucl. Phys. 10, 129 (1978).
5. J.G. Malan et al., Ann. Phys. 89, 284 (1975).

GAMMA-RAY STRENGTH FUNCTIONS IN ^{139}La AND ^{141}Pr

B.J. Allen
Australian Atomic Energy Commission Research Establishment
Lucas Heights Research Laboratories
Private Mail Bag, Sutherland, NSW 2232, Australia

F.Z. Company
University of Wollongong, Wollongong, NSW 2500, Australia

ABSTRACT

Neutron capture γ-ray spectra at average neutron energies <180> and <270> keV in ^{139}La and at <35>, <270>, <725> and <1075> keV in ^{141}Pr have been measured with a NaI detector. The corrected energy level distributions are obtained and γ-ray strength functions investigated. The data fit well within the framework of the Lorentzian giant dipole resonance. The non-statistical capture mechanism hypothesis is ruled out as a dominant capture mechanism in these nuclides in the threshold region.

INTRODUCTION

Lanthanum-139 and ^{141}Pr are both magic isotones with N = 82 neutrons. Low lying states in the compound nuclides ^{140}La, ^{142}Pr correspond to proton spin fragmentation of the ℓ_n = 3 and, at higher energies, the ℓ_n = 1 neutron configurations.

High resolution thermal capture γ-ray spectra in ^{139}La and ^{141}Pr, 10-70 keV and 210 keV neutron capture in ^{139}La all showed apparently anomalous γ-ray strength to the low-lying ℓ_n = 3 states. Both initial and final state resonance correlations were small and no evidence of valence capture was found. It appeared that neither statistical nor valence models could account for these data[1]. Because the energies of the enhanced transitions are comparable to unperturbed p-h energies for E1 transitions, a 2p-1h mechanism was proposed.

To investigate this hypothesis, further measurements of capture γ-ray spectra were made at average neutron energies of <180> and <270> keV in ^{139}La and at <35>, <270>, <725> and <1075> keV in ^{141}Pr, preliminary results being reported at the 1981 Grenoble Capture Gamma-Ray Symposium[2]. In that paper, the need for more accurate level density data was recognised.

EXPERIMENT AND ANALYSIS

Neutron capture γ-ray spectra were measured with 0.00796 atom b^{-1} ^{139}La in the form La$_2$O$_3$ and 0.0289 atom b^{-1} ^{141}Pr at the Australian Atomic Energy Commission's 3 MV Van de Graaff accelerator. Pulse width was 3 ns and pulse rate 2 MHz. Neutrons were produced by the ^{7}Li(p,n) reaction, with samples at a flight path of 15 to 20 cm. A well shielded 20 x 15 cm NaI detector was placed at 125° to the beam, 60 cm from the sample. After background subtraction, the pulse height spectra were analysed by a non-linear least squares program, using a

Table 1. Gamma-ray intensities in ^{139}La and ^{141}Pr

E (MeV)	L	n_c	n_o	Thermal I_γ	<180> I_γ	<270> I_γ
5.0	4.45	10	10	23.7	15.0	17.0
4.7	4.4	5	5	19.4	4.5	4.0
4.4	4.35	9	6	12.7	6.8	8.2
4.1	4.3	17	7	2.5	4.8	5.3
3.8	4.25	28	9	6.0	9.8	8.5
3.5	4.2	60	13	6.6	15.8	14.3
3.2	4.17	80	12	10.7	12.8	13.0
2.9	4.15	150	12	10.3	16.0	15.0
2.6	4.12	250	14	7.7	14.3	14.0

E (MeV)	L	n_c	n_o	Thermal I_γ	<35> I_γ	<270> I_γ	<725> I_γ	<1075> I_γ
5.7	4.60	11	11	11.4	9.6	16.0	12.4	10.0
5.4	4.55	1	1	0.7	2.0	2.1	2.0	1.3
5.1	4.45	6	6	10.0	4.8	5.7	4.9	5.4
4.8	4.40	11	11	9.0	3.1	4.7	3.8	4.8
4.5	4.35	12	12	4.0	5.5	4.7	6.2	7.6
4.2	4.30	28	28	9.6	11.5	9.5	11.5	8.7
3.9	4.25	42	42	9.8	7.8	8.7	8.6	10.5
3.6	4.20	75	57	11.4	14.2	9.4	12.5	11.2
3.3	4.17	140	67	11.0	10.4	11.3	10.5	13.0
3.0	4.12	200	74	10.8	13.4	12.4	13.3	12.0
2.7	4.10	380	95	12.0	17.5	16.0	14.3	15.0

L = Lorentzian
n_c = corrected number of energy levels/0.3 MeV
n_o = observed number of energy levels/0.3 MeV
I_γ = normalised γ-ray intensity/0.3 MeV

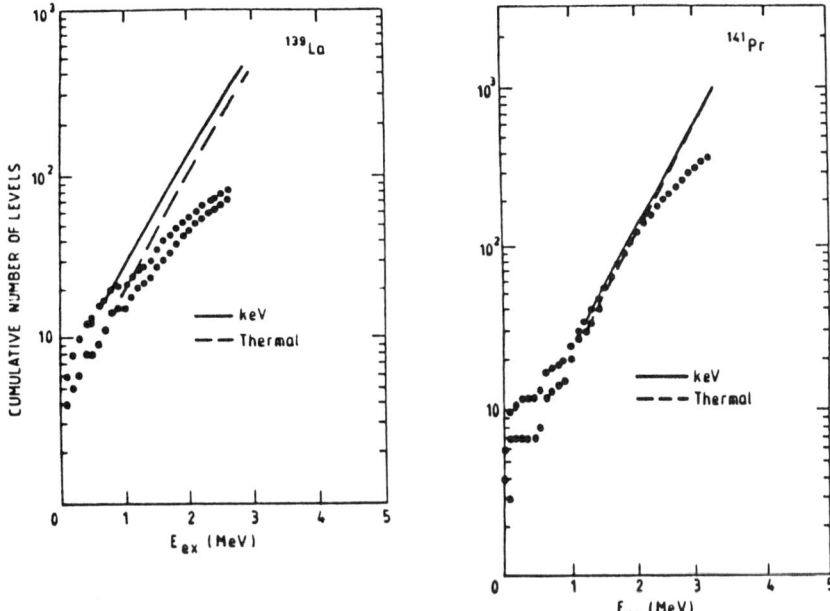

Fig. 1. Cumulative level distributions for final states in thermal and keV capture in ^{139}La and ^{141}Pr

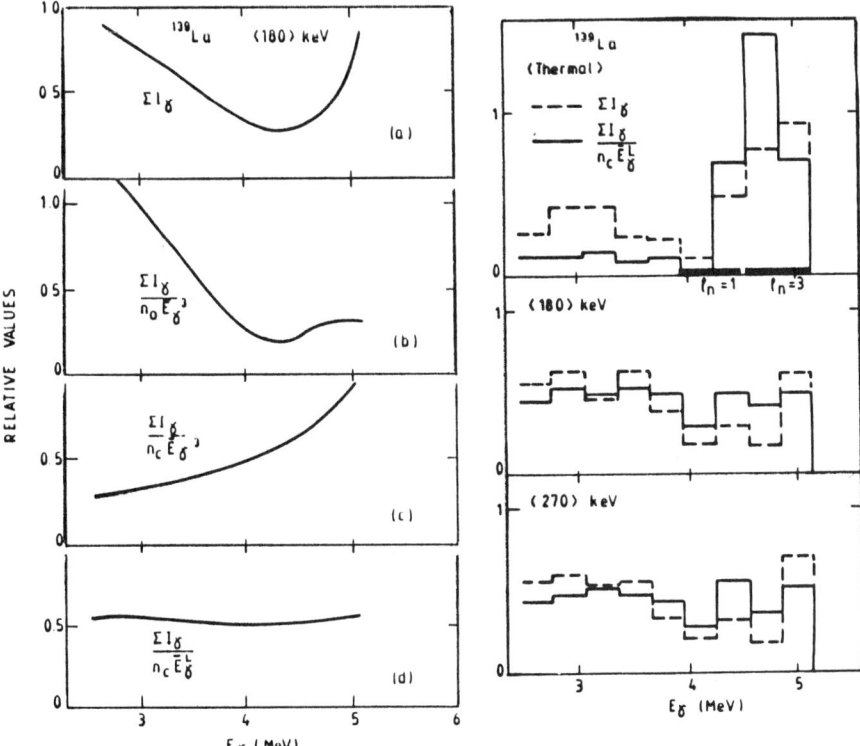

Fig. 2. Energy dependence of various strength functions in ^{139}La

library of NaI response functions[3] to yield intensity histograms in
300 keV intervals. Results are summarised in Table 1.

LEVEL DENSITY AND GAMMA-RAY STRENGTH FUNCTION

The cumulative level count rate for ^{139}La and ^{141}Pr (see Table 1)
reveals increasing numbers of missed levels (dotted curves in Fig. 1).
The corrected level distribution can be determined using the constant
temperature Fermi gas model[4] evaluated at 1 MeV and at the neutron
binding energy to determine the temperature T and pairing energy E_o.
The cumulative number of levels up to excitation energy E and the s-
wave resonance density at binding energy B_n are given by

$$N(E) = \sum_J f(J)\exp[(E-E_o)/T]+c; \qquad \rho_o = \frac{1}{2} \sum_{J=I-\frac{1}{2}}^{I+\frac{1}{2}} \frac{1}{T} f(J)\exp[B_n-E_o)/T]$$

For s-wave capture in ^{139}La at thermal energy and the level data of
Kern[5], T = 0.6 and E_o = -1.3 and, for s-, p- and d-wave capture at keV
energies, T = 0.7, E_o = -1.8. Using these values, the cumulative
number of levels versus excitation energy is calculated for thermal
and keV capture and plotted in Fig. 1. Note that for s-wave thermal
capture, E1 transitions to the final states with J_u^π = 2$^-$ to 5$^-$ are
possible. Transitions to all levels are considered in keV capture.
For ^{141}Pr, T and E_o values were obtained from the recent data of
Kennett et al.[6]

We define the γ-ray strength function, $S_\gamma = \Sigma I_\gamma (n_c \bar{E}^n)^{-1}$, where
ΣI_γ is the E1 transition strength in a 300 keV γ-ray energy interval,
n_c is the corrected number of levels in 300 keV, $\bar{E}_\gamma$ is the average
γ-ray energy and n the exponent characterising the strength function.
The influence of the GDR on the E1 strength function can be described
by the classical Lorentzian line shape, for which the Axel approximation is made. However, a more accurate estimate of the GDR contribution in the threshold region is obtained with an energy dependent
exponent, $L = n(E_\gamma)$, calculated from the GDR parameters for La and Pr
and given in Table 1. The different strength functions for La are
shown in Fig. 2, each curve being normalised to the same area. Overall, the Lorentzian strength functions fit the keV capture γ-ray distributions in La and Pr quite well, but the thermal result for La
remains anomalous (Fig. 2).

The data are now adequately accounted for by the influence of the
GDR in the threshold region and there is no need to invoke a nonstatistical capture mechanism.

REFERENCES

1. B.J. Allen and A.R. de L. Musgrove, Neutron Capture Gamma-Ray Spectroscopy, Plenum Press, 538 (1979).
2. B.J. Allen and F.Z. Company, Inst. Phys. Conf. Series No. 62, 401 (1982).
3. R.F. Barrett et al., Nucl. Phys. A278, 204 (1977).
4. A. Gilbert and A.G.W. Cameron, Can. J. Phys. 43, 1446 (1965).
5. J. Kern, Phys. Rev. 152, 1331 (1967).
6. T.J. Kennett, Can. J. Phys. 59, 1212 (1981).

ABSOLUTE DIPOLE GAMMA-RAY STRENGTH FUNCTIONS FOR ^{176}Lu*

D. G. Gardner, M. A. Gardner and R. W. Hoff
Lawrence Livermore National Laboratory, Livermore, CA 94550

ABSTRACT

We have derived absolute dipole strength-function information for ^{176}Lu from an average resonance capture study of ^{175}Lu with 2-keV neutrons, and from neutron capture cross-section measurements with neutrons from 30 keV to about 1 MeV. We found that we needed to increase our previous estimate of the relative M1/E1 strengths near 5 MeV by a factor of 3, and to revise downward the absolute magnitude of our E1 strength function. We accomplished the latter, while still maintaining continuity with the photonuclear data, by adjusting the one free parameter in our line shape. The present E1 and M1 strengths now seem correct both near the neutron separation energy and also around 1 MeV.

METHOD

The results from two types of experiments were used here. The first was an average resonance capture (ARC) study[1] of ^{175}Lu using 2-keV neutrons, while the second type consisted of the neutron capture cross-section measurements of Macklin, et al.,[2] and Beer, et al.,[3] over the neutron energy range of 30 keV to about 1 MeV.

All cross-section and primary capture γ-ray intensity calculations were made with our version of the STAPRE statistical model code.[4] We employed the spherical equivalent neutron optical-model potential that had been derived from a deformed potential for ^{169}Tm.[5] Gilbert-Cameron level density parameters were used, as modified by Cook, et al.,[6] and further adjusted to agree with the discrete levels of ^{175}Lu and to reproduce a value of D_{ob} for ^{176}Lu of 3.35 eV. For ^{175}Lu, we used 21 discrete levels up to 0.69 MeV; for ^{176}Lu we used a modeled set of 291 levels, comprising 62 rotational bands up to an energy of 1.5 MeV.[7]

Dipole γ-ray transitions were modeled according to our E1 and M1 systematics,[8] where the E1 strength function is taken to have an energy-dependent Breit-Wigner (EDBW) line shape and the M1 is a constant.

$$f_{E1}(E_\gamma) = 3.32 \times 10^{-6} \, (NZ/A)(F_{SR}/E_\gamma \Gamma_R) \, G_R(E_\gamma, E_R, \Gamma_R) \, \text{MeV}^{-3}. \quad (1)$$

Here $G_R(E_\gamma, E_R, \Gamma_R)$ is the functional line shape taken to describe the energy dependence of the giant dipole resonance (GDR), with peak energy, E_R, and peak width, Γ_R. For a deformed nucleus, the GDR can better be represented by a sum of two peaks. Each peak in the EDBW formulation has the functional form

---

*Work performed under the auspices of the U.S. Department of Energy by the Lawrence Livermore National Laboratory under Contract W-7405-ENG-48.

$$G_R(E_\gamma, E_R, \Gamma_R) = [1 + 4(E_\gamma - E_R)^2/\Gamma_R \Gamma(E_\gamma)]^{-1} \,, \quad (2)$$

with $\quad \Gamma(E_\gamma) = \Gamma_R[2E_\gamma/(E_x + E_R)]^2 \,; \quad E_\gamma \leq (E_x + E_R)/2 \,,$

$\qquad\qquad = \Gamma_R \qquad\qquad\qquad\qquad ; \quad E_\gamma > (E_x + E_R)/2 \,.$

The remaining terms in Eq. 1, as well as the systematics for determining the values of E_R and Γ_R for each peak, have been described elsewhere.[8] The only free parameter in Eq. 2 is E_x, the energy at which the EDBW function has the same value as a Lorentzian line shape with the same E_R and Γ_R parameters. Previously we found that a value of $E_x = 5$ MeV worked well in all cases studied, under the assumption that the M1 radiation width was about 15-20% of the total s-wave radiation width at the neutron separation energy.

Using the total radiation width calculated from our original dipole systematics, along with the other input quantities described above, we were able to calculate a ^{175}Lu$(n,\gamma)^{176m,g}$Lu excitation function that agreed very well with the experimental measurements.[2,3] We then calculated the relative primary γ-ray intensities to all of the 291 discrete levels in ^{176}Lu following capture of 2-keV neutrons. Comparison with the ARC measurements[1] showed that our calculated M1/E1 strengths were too small by a factor of 3.

PRESENT RESULTS

Since we wished to preserve the total radiation width value which had given agreement with experiment while at the same time utilizing the M1/E1 ratio from the ARC measurements, both the absolute E1 and M1 strength functions had to be modified. The f_{M1} value was increased by a factor of 3 from 5.8×10^{-9} MeV^{-3} to 1.65×10^{-8} MeV^{-3}. To decrease the E1 strength function in the region around 5-6 MeV while still maintaining continuity with the photonuclear data above 10 MeV, the value of the parameter E_x in Eq. 2 was increased from 5 MeV to 11 MeV.

In Fig. 1 we show the newly derived E1 and M1 strength functions for ^{176}Lu, as a function of the γ-ray energy (the full curves). For comparison, we also show the results predicted by our original systematics (the dashed curves) and from a Lorentz E1 strength function (the dotted curve) using the same double-peak GDR parameters. The arrows indicate the E_x value of 11 MeV determined in the present study and the 5-MeV value used previously.

In the E1 case, it is convenient to compare with experiment the function $S_{E1} = f_{E1}(E_\gamma) A^{-8/3} E_\gamma^{-2}$. The McCullagh et al., compilation[9] of f_{E1} values expressed in this form is plotted in Fig. 2. For ^{176}Lu at $E_\gamma = 5$ MeV, our new E1 strength function has a value of 5.1×10^{-8} MeV^{-3}, which converts to $S_{E1} = 2.1 \times 10^{15}$. The shaded rectangle in Fig. 2 shows this value, along with the estimated error limits of $\pm$ 40%. The set of curves in the figure were predicted by our original systematics. Figure 3 shows our M1 result, again with $\pm$ 40% error limits, compared with data from the compilation of McCullaugh et al. The M1/E1 ratios themselves are probably much less uncertain than the $\pm$ 40% assigned to the absolute values, since the ARC

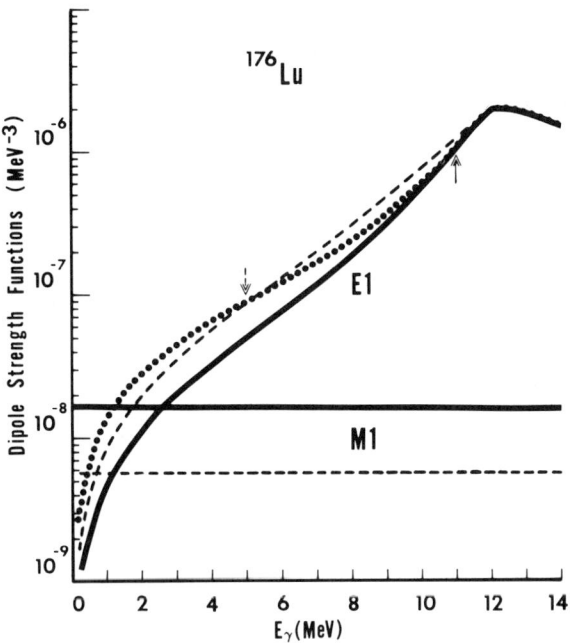

Fig. 1. Absolute E1 and M1 strength functions for the ^{176}Lu nucleus. See text for details.

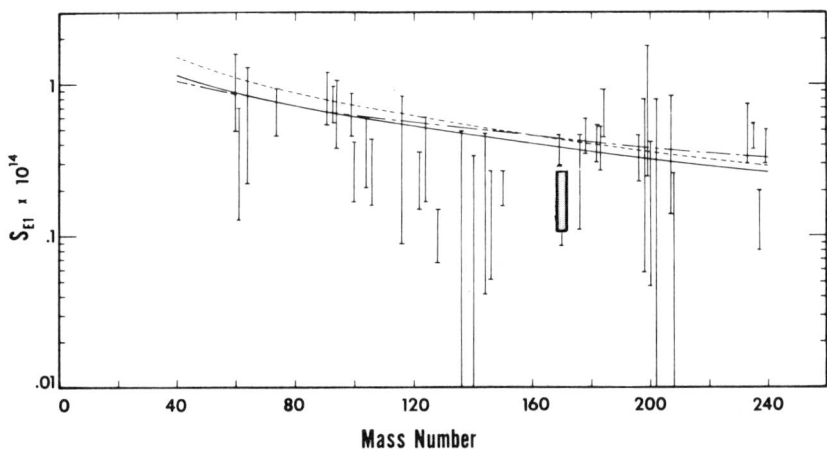

Fig. 2. Comparison of E1 strength function value derived in this work for ^{176}Lu (shaded rectangle includes error estimate) with the data from the compilation of McCullagh et al.[9] The curves are S_{E1} values predicted by our systematics[8] for γ-ray energies of 3, 5 and 7 MeV.

measurements determine this ratio in the 5-6 MeV energy region. For γ-ray transitions in the 0-1 MeV range, our average M1/E1 ratio is about 6.4, which is in general agreement with literature values for this mass region.[7] It does not seem possible to represent this entire energy range, up to the photopeak region, with a Lorentz E1 line shape.

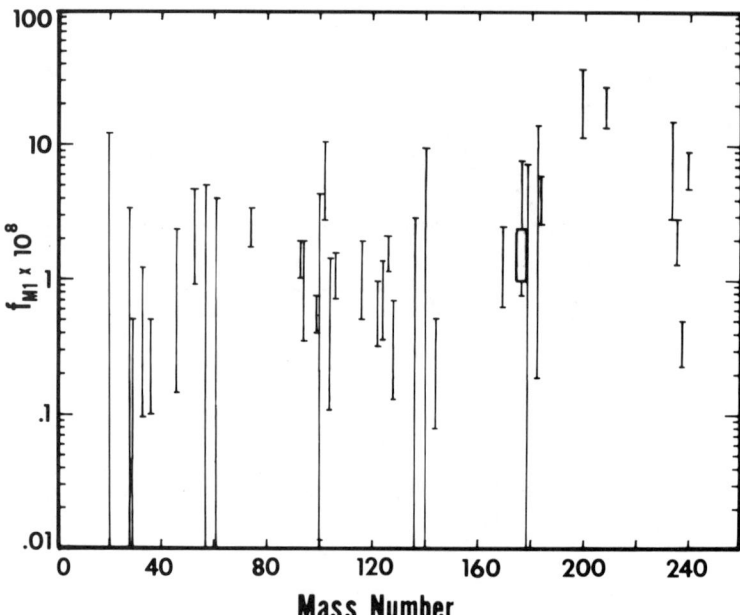

Fig. 3 The present M1 strength function value in units of MeV^{-3} (shaded rectangle includes error estimate) compared with the data from the compilation of McCullagh et al.[9]

REFERENCES

1. R. W. Hoff, R. F. Casten, M. Bergoffen and D. D. Warner, BNL-34811 (1984).
2. R. L. Macklin, D. M. Drake and J. J. Malanify, LA-7479-MS (1978).
3. H. Beer, F. Kappeler and K. Wisshak, Proc. Int. Conf. Nuclear Cross Sections for Technology, Knoxville, NBS Special Publication 594 (1980), p. 340.
4. M. Uhl, Acta Phys. Austriaca 31, 245 (1970).
5. M. Collin and E. D. Arthur, LA-9647-PR (1982), p. 17.
6. J. L. Cook, H. Ferguson and A. R. Musgrove, AAEC/TM-392 (1967).
7. T. Von Egidy et al., Phys. Rev. C29, 1243 (1984).
8. D. G. Gardner, Proc. NEANDC/NEACRP Specialist's Meeting on Fast-Neutron Capture Cross Sections, Argonne, IL, ANL-83-4 (1983), p. 67.
9. C. M. McCullagh, M. L. Stelts and R. E. Chrien, Phys. Rev. C23, 1394 (1981).

GAMMA RAYS FROM 565-keV p-WAVE NEUTRON RESONANCE
AND OFF-RESONANCE CAPTURE BY ^{28}Si

M. Shimizu, M. Igashira, H. Komano, and H. Kitazawa
Research Laboratory for Nuclear Reactors,
Tokyo Institute of Technology,
2-12-1 O-okayama, Meguro-ku, Tokyo 152, Japan

ABSTRACT

We have measured gamma-ray spectra from the 565-keV $p_{3/2}$-wave neutron resonance capture by ^{28}Si and from the off-resonance capture at 485 keV with an anti-Compton NaI(Tl) detector. The resonance capture gamma-ray spectrum cannot be reproduced by the valence model. On the other hand, the off-resonance neutron capture gamma-ray spectrum is successfully explained by the direct capture model.

INTRODUCTION

In the nucleus ^{28}Si, both proton and neutron $1d_{5/2}$-subshells are closed in the extreme single-particle shell model. Therefore the strong single-particle interaction between the incident neutron and target nucleus is expected to play an important role in the neutron capture process. In addition, since the p-wave neutron strength function reaches a peak in the vicinity of the nuclei A=30, enhanced E1 valence transitions from the resonance to the ground and first excited states in ^{29}Si which have respectively large components of $2s_{1/2}$- and $1d_{3/2}$-orbits may occur following p-wave neutron resonance capture.

EXPERIMENTAL METHOD

Experiments have been performed for the 565-keV $p_{3/2}$-wave neutron resonance capture and for the off-resonance capture at the neutron energy of 485 keV where no resonance of silicon isotopes have been observed, using a time-of-flight (TOF) technique. Pulsed neutrons were produced from the ^{7}Li(p,n)^{7}Be reaction with a 1-ns bunched proton beam from the 3.2-MV Pelletron accelerator. Neutron target thicknesses for resonance and off-resonance capture experiments were 28 and 30 keV in FWHM, respectively. The sample was a 6-cm diam by 1.5-cm disk of natural silicon, which was placed at the distance of 15.5 cm from the neutron source. Gamma rays were observed with a 76-mm diam by 152-mm NaI(Tl) detector which was centered in a 254-mm diam by 286-mm NaI(Tl) hollow anti-coincidence detector surrounded by an optimized heavy shield. The gamma-ray spectrometer located at the distance of 80 cm from the sample made an angle of 125° with respect to the proton beam direction.

RESULTS AND DISCUSSION

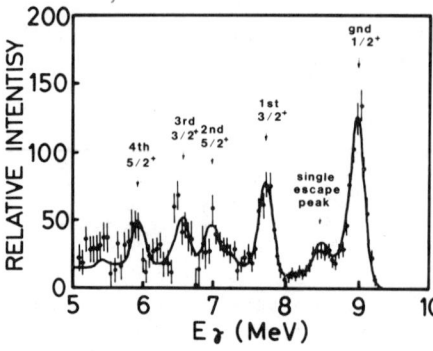 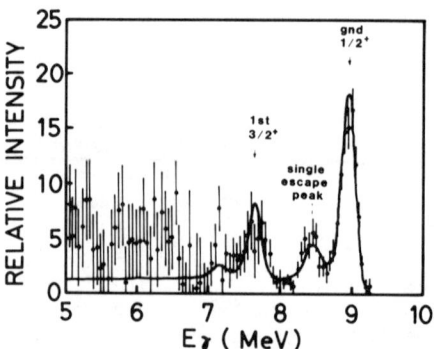

Fig. 1. Gamma-ray pulse height spectrum from the 565-keV $p_{3/2}$-wave neutron resonance capture by ^{28}Si. The solid line shows a spectrum fitted with response functions of the anti-Compton NaI(Tl) spectrometer.

Fig. 2. Gamma-ray pulse height spectrum from the off-resonance capture by ^{28}Si at the neutron energy of 485 keV. See caption of Fig. 1.

The observed gamma-ray pulse-height spectra from resonance and off-resonance capture are shown in Figs. 1 and 2. The partial radiative widths derived from the resonance capture spectrum are given in Table 1 in comparison with the data of Joly et al.[1] As a whole, our values are lager than those of Joly et al. The theoretical values in Table 1 have been predicted by the VALENCE computer code[2] based on the valnce capture formalism[3]. The spectroscopic factors of final states in ^{29}Si were cited from the work of Peterson et al.[4] The optical potential parameters were taken from the work of Moldauer[5] and the neutron escape width used in present calculations to be 10.4 keV from the compilation of Mughabghab et al[6].

The calculated partial radiative width for the ground state transition agrees well with the observed one. For other states, however, the theoretical values, particularly for both $3/2^+$ states, are much smaller than the experimental ones. These discrepancies could not be removed in the framework of the valence model, even if any available parameters are used.

The experimental values of off-resonance partial radiative cross sections presented in Table 2 were obtained from the whole partial radiative cross sections observed at 485 keV by subtracting the contribution of tails of the 188-, 565- and 813-keV resonances which have a large total width. The partial cross sections originated from these resonance tails were within 20% of the whole partial cross sections at 485 keV.

Theoretical off-resonance partial radiative cross sections were calculated by the HIKARI computer code[7] based on the direct-semidirect capture model. It is well known that the direct capture cross section is quite sensitive to the optical potential parameters used in the calculation. Therefore, we chose the Moldauer

Table 1. Experimental and theoretical partial radiative widths for 565-keV $p_{3/2}$-wave resonance on ^{28}Si.

J^π	Ex(MeV)	Eγ(MeV)	Partial radiative widths (eV)		
			present	Joly et al.[1]	valence model
$1/2^+$	0.0	9.02	0.85±0.10	0.60±0.15	0.73
$3/2^+$	1.27	7.75	0.40±0.06	0.18±0.05	0.05
$5/2^+$	2.03	6.99	0.18±0.04	0.13±0.08	0.08
$3/2^+$	2.43	6.59	0.18±0.04	0.12±0.04	0.00
$5/2^+$	3.07	5.94	0.14±0.04	0.07±0.03	0.02

Table 2. Experimental and thoretical off-resonance partial radiative cross sections of ^{28}Si at the neutron energy of 485 keV.

J^π	Ex(MeV)	Eγ(MeV)	Partial radiative cross sections (μb)	
			present	direct capture model
$1/2^+$	0.0	8.95	45±9	39.0
$3/2^+$	1.27	7.67	20±5	16.8
$5/2^+$	2.07	6.92		2.0
$3/2^+$	2.43	6.52		0.0
$5/2^+$	3.07	5.88		0.53

optical potential which well reproduced s- and p-wave neutron strength functions of ^{28}Si. The results show that neutron capture reactions in off-resonance are successfully explained by the direct capture model.

REFERENCES

1. S. Joly, G. Grenier and J. Voignier, Nucl. Phys. A334, 269 (1980).
2. K. Hida, Master Thesis of Tokyo Institute of Technology (1980).
3. A. M. Lane and S. F. Mughabghab, Phys. Rev. C10, 412 (1974).
4. R. J. Peterson, C. A. Fields, R. S. Raymond, J. R. Thieke and J. L. Ullman, Nucl. Phys. A408, 221 (1983).
5. P. A. Moldauer, Nucl. Phys. 47, 65 (1963).
6. S. F. Mughabghab, M. Divadeenam and N. E. Holden, Neutron Cross Sections, Vol. 1 (Academic Press, N. Y., 1981).
7. H. Kitazawa, not published.

EXCITATION OF ISOVECTOR M1 STATES IN p-WAVE NEUTRON
RESONANCE CAPTURE REACTIONS ON ^{56}Fe

H. Kitazawa, H. Komano, M. Igashira, and M. Shimizu
Research Laboratory for Nuclear Reactors,
Tokyo Institute of Technology,
2-12-1 O-okayama, Meguro-ku, Tokyo 152, Japan

ABSTRACT

Strong, anomalous M1 transitions leading from the p-wave neutron resonance states at 34.20, 38.40 and 59.20 keV to either ground($1/2^-$) or first excited state($3/2^-$) in ^{57}Fe have been observed. These transitions are comprehensible in the framework of the semidirect model, assuming excitation of isovector M1 states in the target nucleus.

INTRODUCTION

In the closed shell nuclei with N=28, the l·s particle-hole pairs, $1f_{5/2}-1f_{7/2}^{-1}$, are expected to produce M1 states a few MeV above the l·s splitting energy (∿6.6 MeV), because of the repulsive nature of the spin-dependent part of the particle-hole interaction. However, the M1 strength in ^{57}Fe in which neither proton nor neutron shells are closed must be greatly fragmented, and therefore some fragments of the strength located near the neutron separation energy might be observed in neutron resonance capture reactions on ^{56}Fe.

EXPERIMENTS

Experiments[1] were performed at the neutron energies of 1-110 keV, using a time-of-flight (TOF) technique. The 3.2-MV Pelletron accelerator available at our laboratory produced pulsed neutrons by bombarding a Li-evapolated copper disk with the pulsed proton beam of width 1.0 ns and repetition rate 2 MHz. The average beam current was 3-5μA. The sample was a natural iron disk of 6 cm in diameter and 0.5 cm in thickness, which was placed at the distance of 40 cm from the neutron source in the direction of proton beam. Gamma rays for the transitions from resonance states to the excited states below 1.5 MeV in ^{57}Fe were measured with a 60-cm^3 pure germanium detector placed at the distance of 11 cm from the sample in the direction of 90° with respect to the incident neutron beam.

RESULTS AND DISCUSSION

The capture gamma-ray spectra for 34.20-, 38.40- and 59.20-keV p-wave resonances on ^{56}Fe are shown in Fig. 1. As seen from the figure, the strong, anomalous transitions have been observed which led from these resonance states to either ground ($1/2^-$) or first excited state ($3/2^-$) in ^{57}Fe.

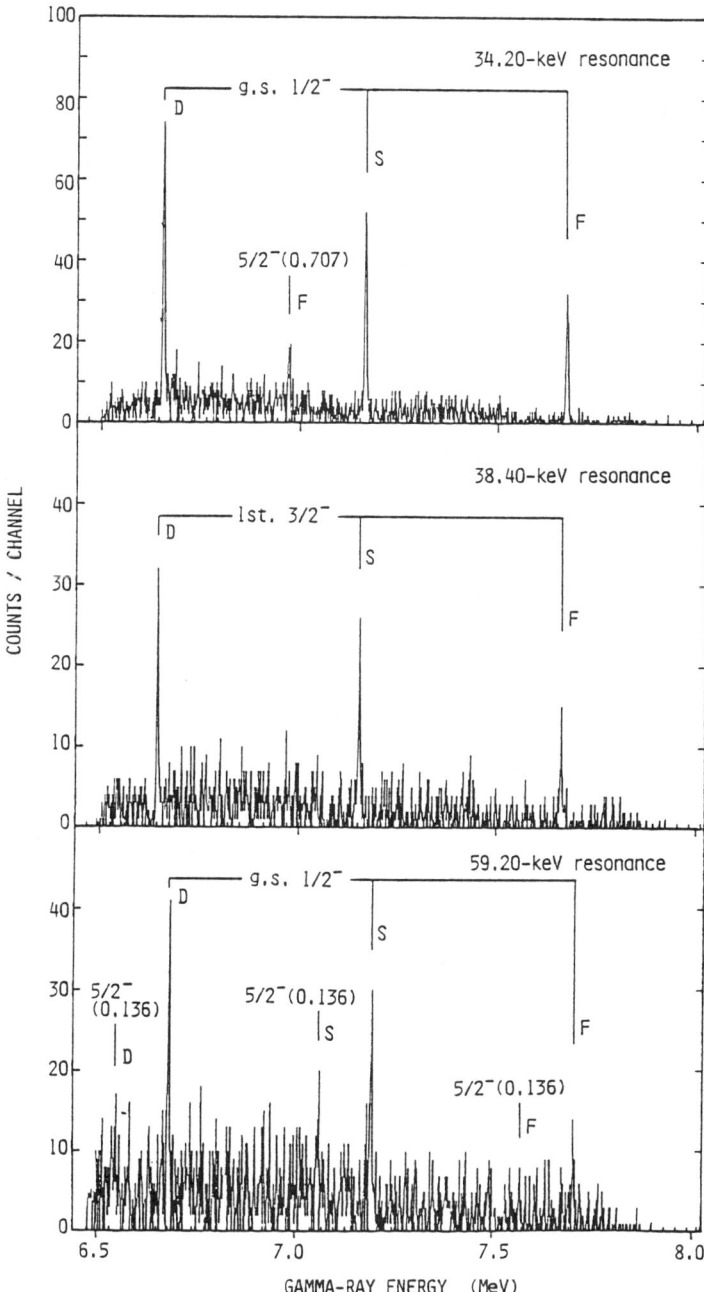

Fig. 1. Gamma rays for the transitions to the excited states below 1.5 MeV in ^{57}Fe following p-wave neutron resonance capture.

Table 1. Experimental and theoretical reduced M1 strengths.

Eres (keV)	Ex (keV)	J_i^π	J_f^π	B(M1) (n.m.^{2}) experimental	theoretical
34.20	0	$3/2^-$	$1/2^-$	0.047	0.031
38.40	14	$3/2^-$	$3/2^-$	0.036	0.008
59.20	0	$3/2^-$	$1/2^-$	0.022	0.033
sum				0.105	0.072

Perey et al.[2] have assigned to be $J^\pi=3/2^-$ for these resonance states with improved measurements of neutron resonance scattering on ^{56}Fe. Using their resonance parameters, the partial radiation widths of the resonance states were obtained for the ground and first excited state transitions. Assuming the most probable M1 transition from the resonance states to the ground or first excited state, the experimental reduced M1 strengths of these resonance states in Table 1 were obtained from the observed partial radiation widths. The theoretical values were calculated by the semidirect model so as to reproduce the observed resonance peak cross section. In this model, the incident neutron interacts with the target nucleus ^{56}Fe to be captured in the ground or first excited state in ^{57}Fe, while target nucleons in the $1f_{7/2}$-orbit make coherent spin-flip transitions to the $1f_{5/2}$-orbit to excite isovector M1 states in the target nucleus. Then, the annihiration of particle-hole pairs is followed by gamma-ray emission. Based on the Bohr and Mottelson estimate[3] for observed magnetic moments in the neighbourhood of ^{208}Pb, we have 40 MeV for the coupling strength between the incident neutron and the isovector M1 state. Using this value, each M1 strength of three resonances and sum of these strengths are in good agreement with experiments.

In conclusion, we showed that the strong, anomalous transitions observed in ^{57}Fe are explained by assuming excitation of the isovector M1 states in the target nucleus. However, the observed M1 strength exhausts about 1% of the sum rule strength for isovector M1 transitions which arises from the core particle transition $f_{7/2}^n \rightarrow (f_{7/2})^{n-1} f_{5/2}^1$. Much stronger M1 strength might exist at higher excitation energy.

REFERENCES

1. H. Komano, M. Igashira, M. Shimizu, and H. Kitazawa, Phys. Rev. C29, 345 (1984).
2. F. G. Perey, G. T. Chapman, W. E. Kinney, and C. M. Perey, Neutron Data of Structural Materials for fast Reactors, (Pergamon, N. Y., 1977), p. 530.
3. A. Bohr and B. R. Mottelson, Nuclear Structure (Benjamin, Mass., 1975), p. 638.

NEUTRON CAPTURE GAMMA-RAY SPECTRA OF NUCLEI WITH N = 82 - 118
AT THE NEUTRON ENERGIES OF 10 - 800 keV

M. Igashira, M. Shimizu, H. Komano,
H. Kitazawa, and N. Yamamuro
Research Laboratory for Nuclear Reactors,
Tokyo Institute of Technology,
2-12-1 O-okayama, Meguro-ku, Tokyo 152, Japan

ABSTRACT

We have measured capture gamma-ray spectra of Pr, Tb, Ho, Ta, and Au with keV neutrons and observed a pygmy resonance in all these spectra. Remarkable features of the pygmy resonance were found to be that the resonance energy and the electric dipole strength exhausted in the resonance increase with neutron number.

INTRODUCTION

Many workers have observed an anomalous bump around 5.5 MeV in gamma-ray spectra from (n,γ) and $(d,p\gamma)$ reactions on several nuclei 110 < A < 140 and 180 < A < 210 at the neutron energies of -1.5 - 4.0 MeV[1-3].

Joly et al.[4] found a bump around 3.5 MeV in the neutron capture gamma-ray spectra of Tm at the neutron energies of 0.5 - 2.5 MeV. In addition, they have shown that the gamma-ray spectra can be expressed by the Brink-Axel gamma-ray strength function[5] with an additional small resonance of the width 1.0 MeV at 3.5 MeV.

These bumps are generally referred to as pygmy resonance. Great interest is to know whether or not these resonances have a common physical origin. From this point of view, we have measured keV-neutron capture gamma-ray spectra in the mass region 140 < A < 200.

EXPERIMENTAL PROCEDURES AND DATA PROCESSING

Neutrons were generated by the $^7Li(p,n)^7Be$ reaction using the pulsed proton beam from the 3.2-MV Pelletron accelerator in Tokyo Institute of Technology. Measurements of capture gamma-ray spectra were performed by a time-of-flight technique at several neutron energies between 10 and 800 keV for Tb, Ho, and Au, and for Pr and Ta respectively at 540 and 420 keV. Gamma rays from the sample were detected by a 76 mmϕ x 156 mm NaI(Tl) detector centered in an annular NaI(Tl) crystal. Both detectors operated as an anti-Compton gamma-ray spectrometer, which was placed in a heavy shield of borated paraffin, lead, and cadmium. Gamma rays were observed at an angle of 125° with respect to the proton beam direction.

Net capture gamma-ray pulse height spectra were unfolded by the computer code FERDOR[6], using the response matrix of the gamma-ray detector. Correction for the gamma-ray attenuation in the sample was made by a Monte-Carlo calculation.

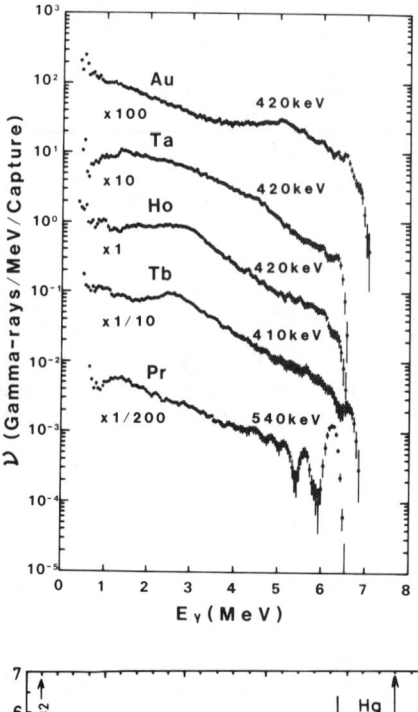

Fig. 1. Unfolded capture gamma-ray spectra of Pr, Tb, Ho, Ta, and Au. Incident neutron energies are shown in the figure.

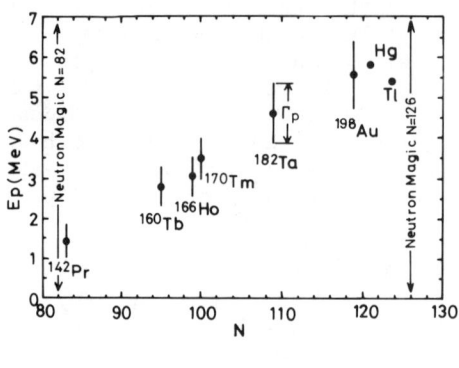

Fig. 2. Energies and widths of the pygmy resonance derived from capture gamma-ray spectra by spectrum fitting calculations.

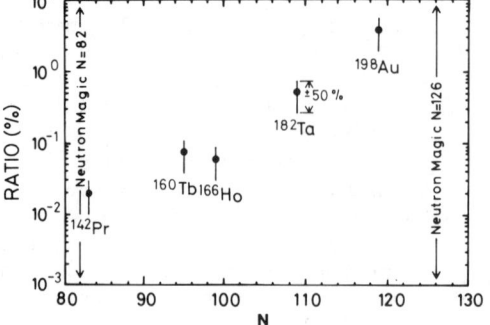

Fig. 3. Ratios of the electric dipole strength of the pygmy resonance to the classical sum-rule value.

RESULTS AND DISCUSSION

Typical unfolded capture gamma-ray spectra of Pr, Tb, Ho, Ta, and Au are shown in Fig. 1. In the capture gamma-ray spectra except for Ta, a distinct bump is found around 1.5, 2.5, 3.0, and 5.5 MeV. A slight change of the gradient of the gamma-ray spectrum of Ta, which could be attributed to a bump by the present analysis, can be recognized around 4.5 MeV. The bumps for Tb, Ho, and Au were observed for all neutron energies between 10 and 800 keV. The energy of the bump is independent of the incident neutron energy, namely, its behavior is resonance-like. Therefore, the characteristics of the bump can be reflected phenomenologically in the gamma-ray strength function as a pygmy resonance.

The gamma-ray strength functions for residual nuclei were derived from capture gamma-ray spectra, using a spectrum fitting method[7]. The composite formula proposed by Gilbert and Cameron[8] was used for a nuclear level density distribution in the spectrum fitting calculation. A Brink-Axel type of function with an additional small resonance was used as a trial function for the gamma-ray strength function.

The energies E_p and widths Γ_p of the pygmy resonance derived from spectrum fitting calculations are shown with neutron number N in Fig. 2. The length of bar shows the resonance width. In the figure, the value for ^{170}Tm, Hg, and Tl are taken from other works[3]. The figure shows that the resonance energy increases with neutron number $82 < N < 126$.

In addition, we investigated the electric dipole strength exhausted in the pygmy resonance. The ratio of the electric dipole strength of the pygmy resonance to the classical sum-rule value is expressed by $(\pi/2)\sigma_p \Gamma_p/(60NZ/A)$, where σ_p is the peak cross section of the pygmy resonance, Z is the proton number, and A is the mass number. These ratios for ^{142}Pr, ^{160}Tb, ^{166}Ho, ^{182}Ta, and ^{198}Au are shown in Fig. 3. The figure shows that the ratio, a few % at most, increases with neutron number.

REFERENCES

1. I. Bergqvist and N. Starfelt, Nucl. Phys., **39**, 353 (1962).
2. G. A. Bartholomew, I. Bergqvist, E. D. Earle and A. J. Ferguson, Can. J. Phys., **48**, 687 (1970).
3. R. F. Barrett, K. H. Bray, B. J. Allen and M. J. Kenny, Nucl. Phys., A**273**, 204 (1977).
4. S. Joly, D. M. Drake and L. Nilsson, Phys. Rev., C, **20**, 2072 (1979).
5. P. Axel, Phys. Rev., **126**, 671 (1962).
6. H. Kendrick and S. M. Sperling, GA-9882 (1970).
7. G. A. Bartholomew, E. D. Earle, A. J. Ferguson, J. W. Knowles and M. A. Lone, Adv. Nucl. Phys., **7**, 229 (1973).
8. A. Gilbert and A. G. W. Cameron, Can. J. Phys., **43**, 1446 (1965).

INVESTIGATION OF COMPOUND AND DIRECT-SEMIDIRECT INTERFERENCE EFFECTS IN THE $^{89}Y(n,\gamma_0 + \gamma_1)^{90}Y$ REACTION

S. Joly
Service de Physique Neutronique et Nucléaire
B.P. n° 12
91680 Bruyères-le-Châtel, France.

ABSTRACT

The partial radiative capture cross section of the $^{89}Y(n,\gamma_0 + \gamma_1)^{90}Y$ reaction was analyzed to look for possible interference effects between compound-nucleus (CN) and direct-semidirect (DSD) processes in the capture reaction. The contribution of these mechanisms has first to be estimated as well as possible. The SPRT method has been used to get the optical potential representing the interaction of protons and neutrons with ^{89}Y. A good overall agreement is found between the measured and the calculated compound-nucleus (inelastic and capture) cross sections. A good agreement is also found for the capture cross section in the giant dipole resonance (GDR) region using the complex isovector potential we have deduced and a volume type coupling of the incident neutron to the GDR. As the sum of the CN and DSD contributions accounts for the observed cross section in the 3-7 MeV energy region, it is concluded that the interference effects should be small as confirmed by estimates.

INTRODUCTION

Capture cross sections below 3 MeV are known to be mainly due to the compound-nucleus (CN) process. On the other hand, the direct and semidirect (DSD) mechanism was introduced to account for the excitation functions observed at higher energies, in the region of the giant dipole resonance (GDR). In the intermediate region, both processes are contributing by about the same amount but effects due to interference between them are ignored.

In this contribution, we investigate this interference effect for the $^{89}Y(n,\gamma_0 + \gamma_1)^{90}Y$ reaction as data are available[1,2] between 0.5 and 16 MeV for comparison with theoretical predictions. For this purpose, the CN capture cross sections have to be calculated between 0.5 and about 6 MeV as well as DSD cross sections from 16 MeV down to 3 MeV.

We determine first the optical potential representing the interaction of a nucleon with the nucleus ^{89}Y as this potential is used to determine the neutron transmission coefficients in CN calculations, the incident neutron wave function χ_E^+ and the particle-vibration coupling function $h(r)$ which depends on the isovector component $U_1(r)$ in the DSD model. Then the CN and DSD capture cross sections are calculated and a simple expression for the interference effect between these processes is given.

THE NUCLEON-^{89}Y NUCLEUS OPTICAL POTENTIAL

The determination of the neutron optical potential parameters was made by means of the SPRT method developed at Bruyères-le-Châtel[3]. The parameters are first defined by fitting the scattering length R' and the s and p-wave neutron strength functions S_o and S_1 to the data defined at 10 keV. The energy dependence of the potentials is then determined by the neutron total cross section in the energy range 10 keV-15 MeV. Final adjustements of these parameters are obtained by fitting the neutron elastic scattering data.

As far as the proton optical potential parameters are concerned, the geometrical parameters for neutrons are used and the different potential depths were deduced by fitting experimental elastic scattering data obtained between 18.9 and 24.5 MeV and analyzing powers measured at 21.1 MeV.

The quasielastic (p,n) reaction is known to give the best information on the isovector part of the optical potential but this reaction has not been studied for the ^{89}Y target and we had to rely on the data obtained from the analysis of the ^{93}Nb(p,n)^{93}Mo reaction, which is a good approximation. The agreement between the data and the corresponding quantities calculated with this unique potential is very good.

COMPOUND-NUCLEUS CROSS SECTION CALCULATIONS

Inelastic cross sections corresponding to excitation of low-lying levels in ^{89}Y were calculated to test the neutron transmission coefficients deduced from the above optical potential. These were obtained for discrete levels up to 2.22 MeV and for different treatments of the level width fluctuation correction :
a) no correction, i.e. the usual Hauser-Fesbach term $<\sigma^{HF}>$,
b) correction according to the method of Hofmann et al.[4] (the so-called HRTW method) introducing the elastic enhancement factor W_n,
c) the technique developed by Moldauer[5] using the concept of effective number of degrees of freedom,
d) this number of degrees of freedom is expressed in terms of W_n.
The best agreement with the inelastic cross section data is obtaining with method c). The effect is larger for the first excited level at 0.91 MeV when few channels are open as expected.

As the capture cross sections are calculated from 10 keV to 6 MeV, level densities in the compound nucleus ^{90}Y as well as in ^{89}Y were estimated using discrete low-lying excited states, s-wave neutron spacing or inelastic neutron spectrum shapes.

The measured total radiative width[6] at B_n and the γ-ray strength function deduced from our capture spectra were used to define the γ-ray transmission coefficient T(Eγ). From the deduced γ-ray branching ratio, the partial cross section for ($\gamma_o + \gamma_1$) was obtained from the total capture cross section. These are compared to the data in fig.1 up to 6 MeV.

DIRECT-SEMIDIRECT CROSS SECTION CALCULATIONS

In the DSD theory, the transition amplitude for capture of the incident neutron from the continuum optical model state χ^+_E to the single-particle state $\psi_{n\ell j}$ is given by[7] :

$$T_{DSD} = c \langle \psi_{n\ell j} | r^L | \chi^+_E \rangle + \langle \psi_{n\ell j} | \frac{h_L(r)}{E_\gamma - E_R + \frac{1}{2} i \Gamma_R} | \chi^+_E \rangle$$

where the first term represents the direct capture and the second, the semidirect capture through the excitation of a collective state (energy E_R, width Γ_R and multipolarity L). In the Steinwedel-Jensen model, the interaction coupling function is given by[8] :

$$h_L(r) = \frac{\hbar^2}{m} \frac{f_L}{E_R} L(2L+1) \frac{NZ}{A^2} \frac{\langle r^{2L-2} \rangle}{\langle r^{2L} \rangle} r^L U_1(r)$$

where f_L represents the fraction of the energy-weighted sum rule exhausted in the reaction, the averages $\langle r^M \rangle$ are taken over a sphere of uniform density. When using the isovector potential determined previously, the DSD cross sections are too low. Increasing $U_1(r)$ by 43 % along with f_L = 100 %, E_R = 16,9 MeV and Γ_R = 3.69 MeV give the results shown in fig.1. As suggested by Dietrich and Kerman[9], an RPA correction term c = 0.75 was included in the direct term ; using c = 1.0 slightly changes the cross section. The continuous curve represents the sum of CN and DSD cross sections as calculated above. The agreement with the data between 0.5 and 16 MeV is fairly good as predicted cross sections are always within quoted error bars for the observed values.

INTERFERENCE BETWEEN CN AND DSD CONTRIBUTIONS

Kawai et al.[10] have shown that the Hauser-Feshbach formulation was not correct in the presence of direct reactions. The fluctuation cross section has to be expressed in terms of generalized transmission coefficients and can be approximated by :

$$\sigma_{f\ell} = (\sum_c T_{cc})^{-1} (T_{nn} T_{\gamma\gamma} + T_{n\gamma} T_{\gamma n})$$

We make further approximations and the cross section reduces to the simple expression :

$$\sigma_{f\ell} = \Pi \lambda^2 \frac{T_n T_\gamma}{\sum_c T_c} [1 + \frac{|\langle S_{nn} \rangle|^2}{T_n} \frac{|\langle S_{n\gamma} \rangle|^2}{T_\gamma}] = \langle \sigma^{HF} \rangle W_{n\gamma}$$

We estimated this factor at 4 and 6 MeV where the interference effect is expected to be larger. The γ-ray transmission coefficient was obtained by extrapolating the $f(E_\gamma)$ deduced function at higher energies but this procedure is somewhat uncertain. Angular momentum and parity selection rules restrict the possible values involved in the evaluating the $W_{n\gamma}$ factor. We obtain $W_{n\gamma}$(4 MeV) ≈ 1.015 and W(6 MeV) ≈ 1.008, i.e. very small corrections. These values correspond to the good agreement obtained between the observed partial capture cross sections and the sum of the predicted CN and DSD cross sections eventhough the uncertainties are relatively large.

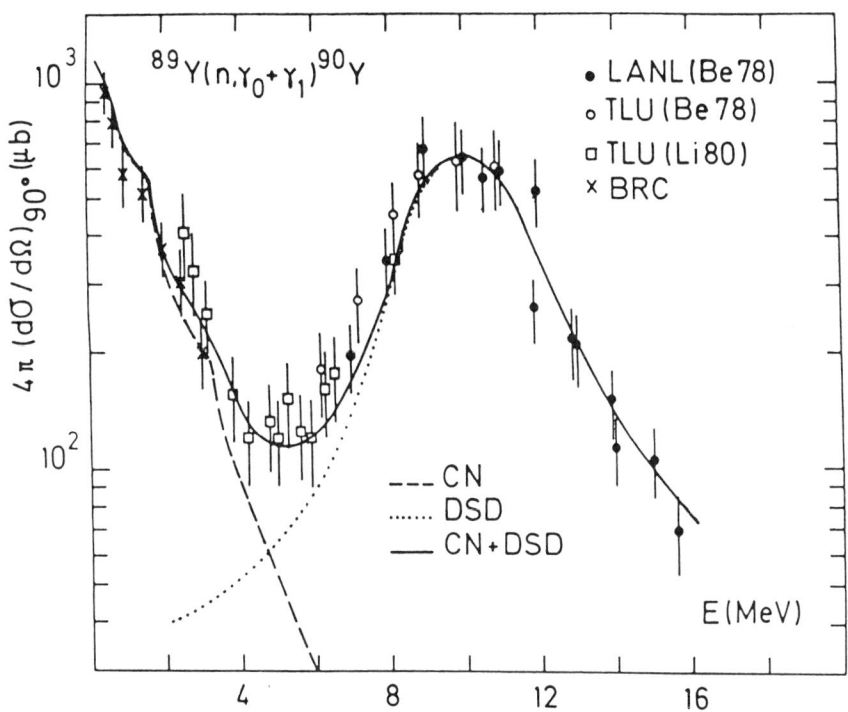

Fig.1. Comparison of data with CN and DSD calculations.

REFERENCES

1. I. Bergqvist et al., Nucl. Phys. A295, 256 (1978).
2. A. Lindholm et al., Nucl. Phys. A339, 205 (1980).
3. J.P. Delaroche et al., Nuclear Theory in Neutron Data Evaluation IAEA-190 Report (1976).
4. H.M. Hofmann et al., Ann. Phys. 90, 403 (1975).
5. P.A. Moldauer, Phys. Rev. C11, 426 (1975); Phys. Rev. C12, 744 (1975).
6. J.W. Boldeman et al., Nucl. Sci Eng. 64, 744 (1977).
7. M. Potokar et al., Nucl. Phys. A277, 29 (1977).
8. F.S. Dietrich et al., Phys. Rev. Lett. 38 156 (1977).
9. F.S. Dietrich and A.K. Kerman, Phys. Rev. Lett. 43, 114 (1979).
10. M. Kawai, A.K. Kerman and K.W. McVoy, Ann. Phys. 75, 156 (1973)

ANALYSIS OF n + ^{197}Au CROSS SECTIONS FOR E_n = 0.01-20 MeV

P. G. Young and E. D. Arthur
Los Alamos National Laboratory, Los Alamos, NM 87545

ABSTRACT

An analysis of n + ^{197}Au reactions has been completed for incident neutron energies between 0.01 and 20 MeV. The analysis involves use of a deformed optical model to calculate neutron transmission coefficients, a giant-dipole-resonance model and experimental data to determine gamma-ray transmission coefficients, and Hauser-Feshbach statistical theory to calculate partial reaction cross sections. Particular emphasis was given to obtaining gamma-ray strength functions that are consistent with spectral measurements of gamma-ray emission between E_n = 0.2 and 20 MeV by Morgan and Newman,[1] while at the same time requiring agreement with (n,γ) and (n,xn) cross-section data.

THEORETICAL CALCULATIONS

The deformed optical model parameterization by Delaroche[2] was utilized in our analysis. This parameterization gives good agreement with neutron total cross-section measurements on ^{197}Au between 10 keV and 27 MeV, elastic (n,n) angular distributions near 5 MeV, and, making use of isospin relationships, ^{197}Au(p,p) elastic scattering angular distributions at 13.8 and 55 MeV. The coupled-channel code ECIS[3] was used for our deformed optical model calculations. The lowest three states of the ^{197}Au ground state rotational band were coupled in the calculation (J^π = 3/2$^+$, 5/2$^+$, 7/2$^+$ at E_x = 0, 279, 548 keV, respectively). Neutron transmission coefficients were calculated to 20 MeV with ECIS and were collapsed to a form depending only on incident neutron energy and orbital angular momentum for use in Hauser-Feshbach calculations.

The Hauser-Feshbach statistical theory calculations were performed with the COMNUC[4] and GNASH[5] reaction theory codes. The COMNUC cross section calculations include width-fluctuation corrections, important at lower energies, and the GNASH calculations incorporate preequilibrium effects, which become significant at higher energies. COMNUC was used to calculate all cross sections below 3 MeV, whereas GNASH was used for cross sections above 3 MeV and for all spectra calculations. Both codes use the Gilbert and Cameron[6] level density formulation and the Cook[7] tabulation of level density parameters, modified slightly as described below. A maximum amount of experimental information on discrete energy levels was incorporated into the calculations, and the constant temperature part of the Gilbert and Cameron level density was matched to the discrete level data for each residual nucleus in the analysis.

Gamma-ray transmission coefficients were calculated from E1 and M1 strength functions. A giant-dipole-resonance shape[8] with the parameters E_R^{M1} = 8 MeV and Γ_R^{M1} = 5 MeV[9] was used for the M1 strength function. For radiative capture the shape of the E1 strength function was determined for $E_\gamma < 8$ MeV by trial-and-error calculation of the ^{197}Au(n,γ) spectrum measurements of Morgan and Newman[1] at $E_n = 0.2$-0.6 MeV. A second E1 strength function was similarly determined from the measured (n,n'γ) spectrum at $E_n = 6$-7 MeV. Above $E_\gamma = 8$ MeV, the empirically determined E1 strength functions were joined to a giant-dipole-resonance shape.

The normalizations of the strength functions were determined with the relationship

$$\frac{\langle \Gamma_\gamma \rangle}{\langle D_0 \rangle} = \int_0^{B_n} [f_{E1}(E_\gamma) + f_{M1}(E_\gamma)] E_\gamma^3 \rho(B_n - E_\gamma) dE_\gamma$$

where $f_{x\ell}(E_\gamma)$ are the gamma-ray strength functions, B_n is the neutron binding energy of ^{198}Au, $\langle \Gamma_\gamma \rangle$ is the average gamma-ray width (= 0.122 eV),[10] and $\langle D_0 \rangle$ is the mean s-wave resonance spacing (= 16.2 eV).[10] Based on the review by Lone,[9] the ratio $\Gamma_{M1}/(\Gamma_{M1} + \Gamma_{E1}) = 0.12$ was assumed in the calculations.

The level density parameter 'a' for ^{198}Au was taken from the Cook tabulation.[6] Cook's values were slightly modified (within ± 15%) for ^{197}Au and ^{196}Au to concurrently optimize agreement with higher energy measurements of neutron emission spectra, (n,2n) and (n,3n) cross sections, and the gamma-ray emission spectra. At 14 MeV a preequilibrium fraction of 33% was used in the analysis.

RESULTS

The E1 gamma-ray strength functions that resulted from this analysis are compared in Fig. 1 with values inferred from experiments by Joly et al.,[11] Loper et al.,[12] and Veyssiere et al.[13] The present curve is quite similar to one also obtained from Morgan's data[1] by Kitazawa[14] using different neutron transmission coefficients and level density parameters. The two strength functions obtained in our analysis of the (n,γ) and (n,n'γ) measurements are quite similar, differing mainly at the shoulder near $E_\gamma = 5$-6 MeV. While this difference is not large, significantly improved agreement with the higher energy data was obtained with the (n,n'γ) E1 strength function. This strength function was also used in the (n,2nγ) and (n,3nγ) calculations.

The resulting ^{197}Au(n,γ)^{198}Au cross section from $E_n = 0.01$-20 MeV is compared with a selection of experimental data and with the ENDF/B-V evaluation[15] (with resonance structure averaged out) in Fig. 2. The pronounced peak in the theoretical cross section near $E_n = 12$ MeV results from inclusion of a semi-direct component,[16] which is necessary to reproduce the data near 14 MeV.

The calculated gamma-ray emission spectra are compared in Fig. 3 with Morgan and Newman's data at two neutron energies not involved in extracting the gamma-ray strength functions. The left side of Fig. 3 illustrates the calculated and measured data at $E_n = 1.0$-1.5 MeV. Above $E_\gamma = 1.5$ MeV, of course, the spectrum results entirely from radiative capture, and the agreement is seen to be reasonably good. The results for $E_n = 14$-17 MeV are shown in the right side of Fig. 3. In this case the spectrum is due mainly to to (n,2nγ) reactions, with a smaller but appreciable mixture of (n,n'γ) present. Again, within the accuracy of the experiment, the agreement is reasonable.

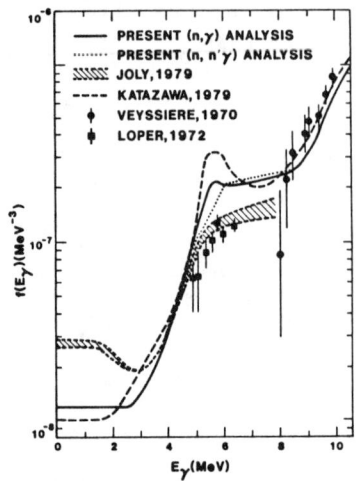

Fig. 1. E1 γ-ray strength functions inferred from ^{197}Au(n,γ) ^{198}Au and ^{197}Au(n,n'γ) ^{197}Au measurements.

Fig. 2. Measured and calculated ^{197}Au(n,γ) ^{198}Au cross sections from 10 keV to 20 MeV. The dotted curve is ENDF/B-V.

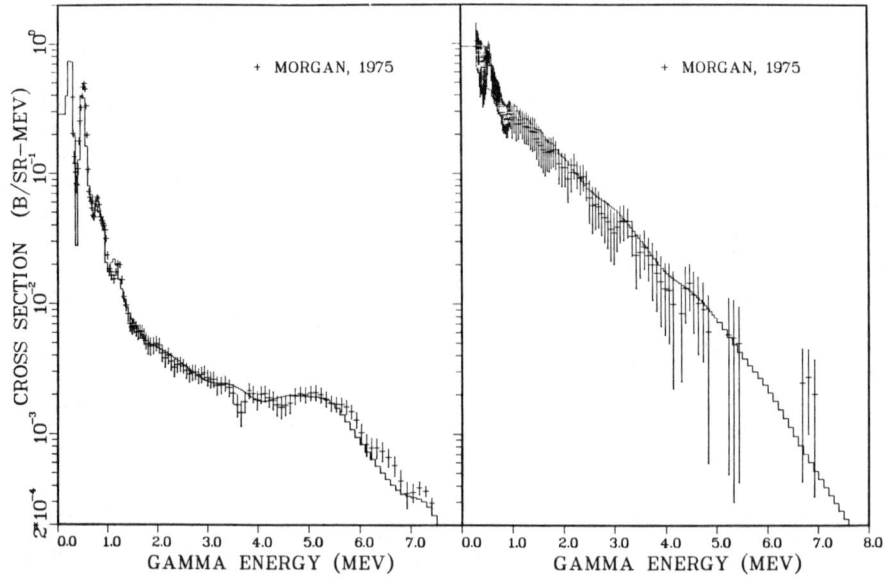

Fig. 3. γ-ray emission spectra for n + ^{197}Au reactions with E_n = 1-1.5 MeV and 14-17 MeV. The histograms are the present calculations.

CONCLUSIONS

The deformed optical model and Hauser-Feshbach statistical theory calculations and gamma-ray strength function formulations described here adequately represent available neutron-induced gamma-ray measurements on ^{197}Au between E_n = 0.01 and 20 MeV. Although not presented here, the calculations also reproduce measurements of the neutron total, (n,2n), and (n,3n) cross sections as well as neutron emission spectra from 14-MeV neutron reactions.

REFERENCES

1. G. L. Morgan and E. Newman, "The Au(n,xγ) Reaction Cross Section for Incident Neutron Energies between 0.2 and 20.0 MeV," Oak Ridge Nat. Lab. report ORNL-TM-4973 (1975).
2. J. P. Delaroche, "Potential Optique Nucleon ^{197}Au Entre 10 keV et 57 MeV," Int. Conf. Neutron Physics and Nucl. Data for Reactors and Other Applied Purposes, Harwell, England (1978), p. 366.
3. J. Raynal, "Optical Model and Coupled-Channel Calculations in Nuclear Physics, Int. At. Energy Ag. report IAEA SMR-9/8 (1970).
4. C. L. Dunford, "A Unified Model for Analysis of Compound Nuclear Reactions," Atomics Int. report AI-AEC-12931(1970).
5. P. G. Young and E. D. Arthur, "GNASH: A Preequilibrium-Statistical Nuclear Model Code for Calculations of Cross Sections and Emission Spectra," Los Alamos Sci. Lab. report LA-6947 (1977).
6. A. Gilbert and A. G. W. Cameron, Can. J. Phys. $\underline{43}$, 1446 (1965).
7. J. L. Cook, H. Ferguson, and A. R. Musgrove, Aust. J. Phys. $\underline{20}$, 477 (1967).
8. P. Axel, Phys. Rev. $\underline{125}$, 671 (1962); see also D. M. Brink, Nucl. Phys. $\underline{4}$, 215 (1957).
9. M. A. Lone, "Photon Strength Functions," Proc. Third Int. Symp. Neutron Capture Gamma-Ray Spectroscopy and Related Topics, Brookhaven, N.Y., Sept. 18-22, 1978 (Plenum Press, N.Y, 1979), p. 161.
10. S. F. Mughabghab and D. I. Garber, "Neutron Cross Sections, Vol. 1, Resonance Parameters," Brookhaven Nat. Lab. report BNL-325, 3rd Ed., Vol. 1 (1973).
11. S. Joly, D. M. Drake, and L. Nilsson, Phys. Rev. C $\underline{20}$, 2072 (1979).
12. G. D. Loper, L. M. Bollinger, and G. E. Thomas, "Search for the Pygmy Resonance in ^{197}Au(n,γ)^{198}Au," in Argonne Nat. Lab. report ANL-7971, Phys. Div. Ann. Review, 1 April 1971-1 March 1972 (1972), p. 7.
13. A. Veyssiere, H. Beil, R. Bergere, P. Carlos, and A. Leprêtre, Nucl. Phys. A $\underline{159}$, 561 (1970).
14. H. Kitazawa, Y. Harima, H. Yamakoshi, Y. Sano, T. Kobayashi, and M. Kawai, "Gamma-Ray Production Cross Sections for MeV Neutrons," Proc. Int. Conf. on Nucl. Cross Sections for Technol., Oct. 22-26, 1979, Knoxville, Tenn. (NBS Special Pub. 594, 1980), p. 775.
15. S. F. Mughabghab, ENDF/B-V Data File for ^{197}Au (MAT 1379), described in BNL-NCS-17541 (ENDF-200), R. Kinsey, Ed., Nat. Nucl. Data Center, Brookhaven Nat. Lab., Upton, N.Y. (July 1979).
16. E. D. Arthur, "Calculation of Neutron Cross Sections on Isotopes of Yttrium and Zirconium," Los Alamos Sci. Lab. report LA-7789-MS (April 1979) p. 24.

LEVEL STRUCTURE OF QUASI-MAGIC ^{96}Zr

G. Molnár, B. Fazekas, T. Belgya and Á. Veres
Institute of Isotopes of the Hungarian Academy of
Sciences, Budapest, H-1525, Hungary

ABSTRACT

The ^{96}Zr nucleus is located in the A=100 region where an extremely sharp transition from spherical to deformed shape takes place. Due to the double subshell closure at Z=40 and N=56, however, this nucleus is reminiscent of magic ^{90}Zr rather than of transitional ^{98}Mo, its isotone. Nevertheless, the presence of many additional states with regard to the spectrum of the former and the strong excitation of the first excited 0$^+$ state in various particle transfer reactions imply more complex structure than that of a simple magic nucleus.

In order to reveal this structure the ^{96}Zr nucleus was investigated through the (n,n'γ) reaction, induced by reactor fast neutrons. All known levels up to 3500 keV energy were observed via their γ-decay and for most of them spin assignments could be made on the basis of angular distribution measurements. In addition, several strong γ-rays were discovered which are believed to connect still unknown low-lying levels. On the basis of the present results a possibility of shape coexistence in ^{96}Zr is discussed.

INTRODUCTION

The importance of doubly closed shell (or magic) nuclei in nuclear structure physics is quite obvious. The recent discovery of the magic character of ^{146}Gd [1] with closed major neutron shell but with only a closed subshell (Z=64) for protons focussed attention also on closed subshell nuclei. The only example of this class of nuclei known previously is ^{90}Zr which has a closed proton subshell at Z=40.

The ultimate generalization of magicity would be the doubly closed subshell nuclei. A possible candidate is ^{96}Zr with closed proton and neutron subshells Z=40 and N=56, respectively. This nucleus resembles ^{90}Zr in the sense that the first 2$^+$ state has very high energy and, as in some other magic nuclei, the first excited state is a 0$^+$ state. The presence of many additional states with respect to the spectrum of ^{90}Zr imply, however, that the N=56 subshell closure is not complete. Unfortunately, the scarcity of the experimental data (spin-parities of only the first three excited states are established firmly) has so far prevented detailed analysis [2].

THE ^{96}Zr (n,n'γ) REACTION

The (n,n'γ) reaction induced by reactor fast neutrons is highly suitable for studying low-lying, low spin states because it combines the nonselectivity of statistical excitation with high precision of γ-spectroscopic methods. The ^{96}Zr nucleus was studied via this reaction for the first time. Since isotopic enrichment of the target amounted only to 60 per cent, all measurements were repeated using a natural target with very low ^{96}Zr abundance. The use of spectrum subtraction made possible the unambiguos identification of the ^{96}Zr γ-rays, as demonstrated in Fig.1.

Prior to the present experiment only 20 γ-ray transitions between the ^{96}Zr levels were observed in the beta decay of 9.8 s ^{96}Y [3]. Of these, 13 were also detected in the present work in which a total of 31 γ-rays was observed. Levels up to 3500 keV energy were excited. The 3150 keV level is a new observation. Although many of the levels known only from particle spectroscopy could unambiguously be identified via their γ-decay it was not possible to connect any of the observed γ-rays with the

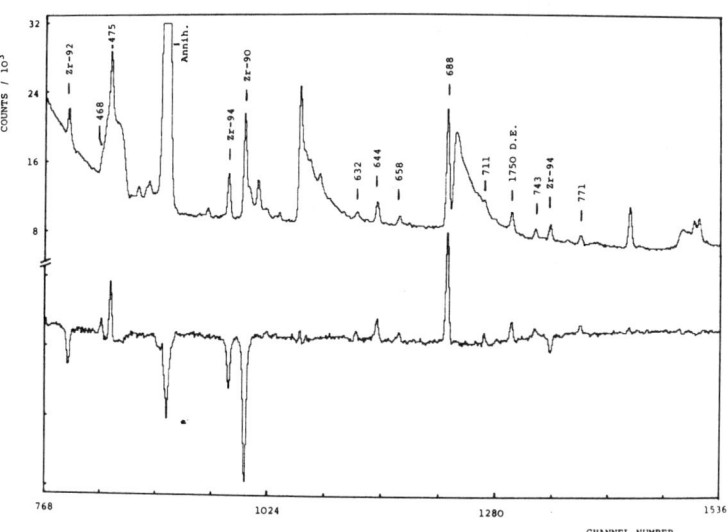

Fig.1. Part of the ^{96}Zr (n,n'γ) spectrum. The lower curve is obtained by subtracting a similar spectrum of natural Zr.

2330 keV 2+ state. Since 2+ states are strongly excited in inelastic neutron scattering we conclude that this level cannot belong to ^{96}Zr, as has also been noted by other authors [4]. On the other hand, the existence of the 2438 keV level has been confirmed by observing its γ-decay for the first time. There are several strong γ-rays, such as the 644 keV, 658 keV 743 keV and 1279 keV lines that most probably de-excite still unknown low-lying levels.

For one third of the observed transitions angular distributions could be measured that made possible new spin assignments. The assignment of 3⁻ to the 2225 keV state is in accordance with a tentative proposal made in the beta decay[3]. Another interesting result is that the spin of the 3081 keV state cannot be 5⁻ as has been suggested [4]. Hence another candidate should be sought for the first 5⁻ state. The experimental results are summarized in Fig. 2 where the level schemes obtained in various shell model calculations [5,6] are also included.

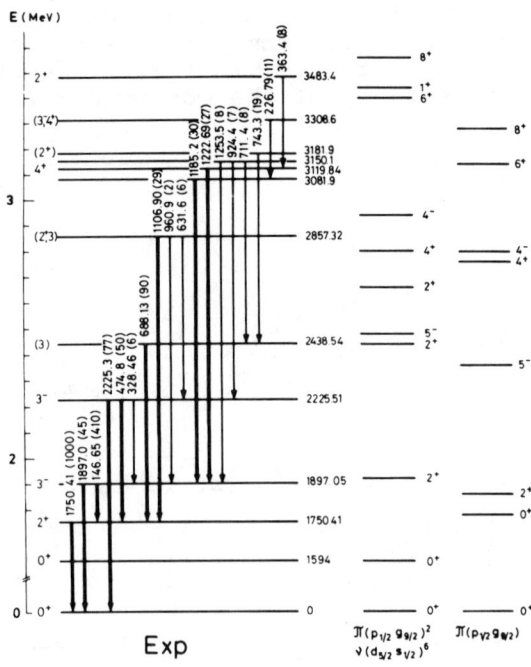

Fig. 2. Partial level scheme of ^{96}Zr, compared with various shell model calculations.

DISCUSSION

Shell model studies of ^{90}Zr usually assume an ^{88}Sr inert core with active protons distributed among the $2p_{1/2}$ and $1g_{9/2}$ orbitals. If the N=56 subshell is supposed to be closed, the resulting spectrum for ^{96}Zr [5] is of the ^{90}Zr type. It is clear from Fig. 2 that neither this nor a more sophisticated calculation with inclusion of the $2d_{5/2}$ and $3s_{1/2}$ neutron orbitals [6] agree with the experimental level pattern. The only exceptions are perhaps the first 2^+ and the first excited 0^+ states. On the other hand, the identification of several levels with spin 3 above the first (octupole)3^- state suggest similarity between ^{96}Zr and ^{94}Sr, which is an isotone of the former and has 38 protons. Hence, the properties of both ^{90}Zr and ^{94}Sr can be traced in the level spectrum of ^{96}Zr.

In order to envisage submagicity in the case of ^{96}Zr, a unified view of subshell closures is proposed. It was pointed out earlier [7] that the N=56 subshell closure effect is enhanced by the presence of neutron orbitals of low degeneracy, such as the $3s_{1/2}$ and $2d_{3/2}$ orbitals above the subshell gap. Since Z=40 is in the upper part of the 28-50 major proton shell it can be viewed as a subshell closed for proton holes. The $2p_{1/2}$ and $2p_{3/2}$ orbitals lying below the subshell will now enhance the subshell closure effect for proton holes. It is obvious that this symmetric picture readily incorporates the breaking of submagicity when high spin orbitals, such as $\nu g_{7/2}$ and $\pi f_{5/2}$, come into play. If this picture is correct, deviations from submagicity in the case of ^{96}Zr can be explained only by taking into account these high spin orbitals for protons and neutrons simultaneously.

Pair excitations across closed shells and their role in producing intruder configurations leading to coexistence or mixing, are well understood [8]. However, the regions around closed subshells are less investigated [9]. In submagic (or quasi-magic) ^{96}Zr the 0_2^+ state is strongly excited in particle transfer reactions [10-12] which is an unambiguous feature of intruder states. Since no γ-rays decaying to this state have been observed so far [3] it is difficult to decide whether a coexistent band associated with this 0^+ state exists, as it was proposed long time ago [13]. The strong γ-ray transitions discovered in the present work may resolve the puzzle. To find their place in the level scheme further experiments are being carried out.

We wish to thank K. Heyde, K. Sistemich and J.L. Wood for most useful discussions.

REFERENCES

[1] P. Kleinheinz, R. Broda, P.J. Daly, S. Lunardi, M. Ogawa and J. Blomqvist, Z. Phys. A290 (1979) 279
[2] H. W. Moeller, Nucl. Data Sheets 35 (1982) 281
[3] G. Sadler, T. A. Khan, K. Sistemich, J. W. Güter, H. Lawin, W. D. Lauppe, H. A. Selic, M. Shaanan, F. Schussler, J. Blachot, E. Monnand, G. Bailleul, J. P. Bocquet, B. Pfeiffer, H. Schrader and B. Fogelberg, Nucl. Phys. A252 (1975) 365
[4] E. R. Flynn, D. D. Armstrong and J. G. Beery, Phys. Rev. C1 (1970) 703
[5] J. Vervier, Nucl. Phys. 75 (1966) 17
[6] D. H. Gloeckner, Nucl. Phys. A253 (1975) 301
[7] F. Tondeur, Nucl. Phys. A359 (1981) 278
[8] K. Heyde, P. Van Isacker, M. Waroquier and R. A. Meyer, Phys. Reports 102 (1983) 291; K. Heyde, P. Van Isacker, J. Moreau and M. Waroquier, Proc.Int. Symp. on In-Beam Nuclear Spectroscopy, Debrecen, May 14-18, 1984 (to be published)
[9] J. L. Wood, ibid
[10] E. R. Flynn, J. G. Beery and A. G. Blair, Nucl. Phys. A218 (1974) 285
[11] A. Saha, G. D. Jones, L. W. Put and R. H. Siemssen, Phys. Lett. 82B (1979) 208
[12] R. S. Tickle, W. S. Gray and R. D. Bent, Phys. Lett. 92B (1980) 283 and Nucl. Phys. A376 (1982) 309
[13] R. K. Sheline, I. Ragnarsson and S. G. Nilsson, Phys. Lett. 41B (1972) 115

DECAY SCHEME OF ^{116}Sn FROM (n,n'γ) AND (n,γ) RESULTS

Z. Gácsi, J. Sa and J.L. Weil[†]
University of Kentucky, Lexington, KY 40506

E.T. Jurney[+]
Los Alamos National Laboratory, Los Alamos, NM 87545

S. Raman[*]
Oak Ridge National Laboratory, Oak Ridge TN 37831

ABSTRACT

A decay scheme for ^{116}Sn containing approximately 190 γ-rays and 112 levels below 6.0 MeV has been constructed by combining the results of ^{116}Sn(n,n'γ) and ^{115}Sn(n,γ) experiments.

INTRODUCTION

The most detailed information about the level structure and decay scheme of ^{116}Sn available to date has come from radioactive decay studies[1] and the ^{115}Sn(n,γ) study of McClure and Lewis.[2] More limited results on the level structure have come from particle transfer and scattering experiments.[1]

EXPERIMENTAL DETAILS

For the ^{116}Sn(n,n'γ) study, monoenergetic neutrons in the energy range of 1.9 to 4.5 MeV were scattered from a highly enriched (95.7%) 45 gram sample. The gamma rays following inelastic scattering were detected in a Gamma-X Ge detector with 21% relative efficiency and 1.9-keV energy resolution at $E_γ$=1.33 MeV. The neutrons were produced in the T(p,n)^{3}He reaction by a pulsed beam so that time-of-flight methods could be used to reduce the background. A massive Cu collimator and borated polyethylene shield for the detector were also very effective in reducing the background. Gamma-ray yields were determined from spectra measured at 50-keV intervals in neutron bombarding energy and excitation functions for approximately 140 γ-rays were thus obtained. From the threshold for production of a particular γ-ray, it can often be uniquely placed in the decay scheme, and if not unique the energy assignment of the emitting level is usually correct to within 50-100 keV. The (n,n'γ) measurements were performed at the University of Kentucky.

The ^{115}Sn(n,γ) study was made on a 13 mg target enriched to 97% in ^{115}Sn using the thermal (n,γ) facility at the Los Alamos Omega West reactor. Other details were reported previously.[3] Approximately 500 γ-rays were identified with energies in the range 0.1-9.6 MeV, but

---

[†]Work supported in part by National Science Foundation.
[+]Work supported by the U.S. Department of Energy under Contract No. W-7405-eng-36 with the University of California.
[*]Work supported by the U.S. Department of Energy under Contract No. DE-AC05-84OR21400 with the Martin Marietta Energy Systems, Inc.

TABLE I. Energy Levels in ^{116}Sn

E(level) (keV)	J^π [a]	E(level) (keV)	J^π [a]	E(level) (keV)	J^π [a]
0.0	0^+	3303.5		3748.0	
1293.6	2^+	3309.1		3777.1	
1756.8	0^+	3315.0		3778.5	
2027.5	0^+	3326.4		3806.0	
2112.3	2^+	3327.6		3812.1	
2225.5	2^+	3332.5		3840.4	
2266.2	3^-	3333.8		3843.7	
2366.1	5^-	3344.4		3884.2	
2390.9	4^+	3349.6		3904.9	
2496.6		3371.4		3917.0	
2529.3	4^+	3416.6		3946.5	
2545.7	$(0)^+$	3417.4		3952.4	
2585.6		3426.8		4001.1	
2650.5	(2^+)	3427.3		4013.2	
2773.6	6^-	3436.1		4015.4	
2790.6	$0,1^+$	3469.7		4026.9	
2801.3	4^+	3507.3		4037.2	
2843.9	(2^+)	3508.4		4113.9	
2908.8	7^-	3514.1		4131.1	
2960.0		3515.2		4142.8	
2960.1	$1,2^+$	3520.9		4162.4	
2996.3		3552.1		4170.7	
3032.6		3573.0		4190.5	
3046.4	4^+	3576.2		4200.2	
3088.6	(2^+)	3586.6		4201.4	
3096.9	(4^+)	3593.9		4211.5	
3105.3		3616.0		4217.5	
3142.7		3621.5		4235.3	
3147.0		3622.9		4236.9	
3157.9		3640.9		4238.2	
3179.7		3647.8		4251.7	
3194.4		3651.7		4278.0	
3209.9	7^-	3658.6		4297.1	
3227.5		3673.4		5401.1	
3228.1		3701.8		5427.8	
3236.0		3711.9		5449.5	
3248.2		3730.8		5461.1	
3257.8		3742.1		5931.4	
3288.7		3743.0		9563.5	

[a]Taken from Ref. 1

many of them have not yet been placed in the decay scheme. The unique contributions of the (n,γ) work to the present results are the accuracy and precision of the γ-ray energies because of the careful calibration of the detector[4] and good statistics of the spectrum.

541

RESULTS

In the range of excitation energy up to 4.3 MeV, we have now identified approximately 110 levels, compared to 50 levels for the same region given in Ref. 3. The combination of very precise γ-ray energies from the (n,γ) study and accurate placement of the decays from excitation function thresholds has eliminated the ambiguity of most of the γ-ray placements. From the excitation of levels of known spin, it appears that both the (n,γ) and (n,n'γ) reactions are able to non-selectively excite levels through J=6, and for neutrons several hundred keV above threshold, even 7^- states are excited in the (n,n'γ) reaction. We believe that essentially all levels with J≤6 below an excitation energy of 3.5 MeV have been located, and in addition we have found some 65 levels in the region $3.5 \leq E_x \leq 5.931$ MeV which have J≤4. The levels identified in this work are given in Table I.

REFERENCES

1. Table of Isotopes, 7th Ed., edited by C.M. Lederer and V.M. Shirley, (John Wiley, 1978), p. 548; J. Blachot, J.P. Husson, J. Oms, G. Marguier, and F. Haas, Nuclear Data Sheets 32, 287 (1981).
2. D.A. McClure and J.W. Lewis, III, Phys. Rev. C5, 922 (1972).
3. R.F. Carlton, S. Raman, and E.T. Jurney, in Neutron Capture Gamma-Ray Spectroscopy, Ed. by R.E. Chrien and W.R. Kane, (Plenum Press, 1979).
4. S. Raman, E.T. Jurney, D.A. Outlaw and I.S. Towner, Phys. Rev. C 27 1188 (1983).

LOW-LYING LOW-SPIN STATES IN ^{136}Ba STUDIED VIA (n,n'γ) REACTION

I. Diószegi, Cs. Maráczy and Á. Veres

Institute of Isotopes of the Hungarian Academy of Sciences, H-1525 Budapest, P.O.B. 77, Hungary

ABSTRACT

Low-spin levels in ^{136}Ba have been studied by means of the (n,n'γ) reaction using reactor fast neutrons. Excited states up to an excitation energy 2.8 MeV and with spin values up to 7 were populated. Energies, intensities and angular distribution of γ-rays attributed to ^{136}Ba have been measured. Preliminary results are reported here.

INTRODUCTION

In ^{136}Ba nucleus there are six protons outside the Z=50 magic proton core, and N=80 neutrons, i.e. two neutrons are missing to complete the N=82 shell. In consequence the spectrum of the low-lying excited states of this nucleus is an interesting coexistence of the pure proton excitations and several two neutron-hole states[1].

Experimental information on level scheme of ^{136}Ba is available from studies[1,2,3] of the decay of ^{136}Cs and ^{136}La, furthermore from thermal neutron capture investigations[4,5,6].

Continuing the study of the barium isotopes[7] we investigated the ^{136}Ba(n,n'γ) reaction. The aim of the present work was to resolve the several existing ambiguities in the low-energy level scheme of this nucleus[8].

RESULTS

The γ-ray spectra from a 15 g sample enriched to 93.0% in ^{136}Ba have been measured with a shielded Ge(Li) detector at the neutron beam facility at the 5 MW research reactor of the Central Research Institute for Physics, Budapest.

The level scheme of the ^{136}Ba nucleus, as obtained from the present experiment is given in fig.1. All known states in ^{136}Ba up to an excitation energy 2.8 MeV and with spin values up to 7 were observed by their γ-decay.

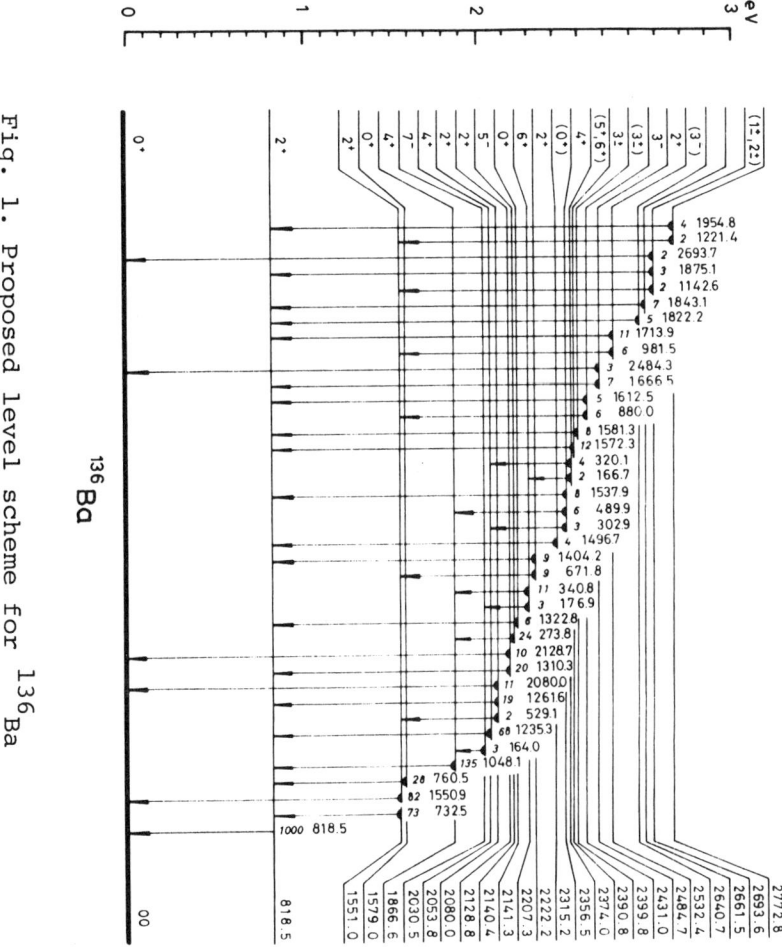

Fig. 1. Proposed level scheme for ^{136}Ba

In addition to confirming known level characteristics[8] the angular distribution data resulted in new unique spin assignments for 2080.0 keV 2^+, 2128.8 keV 2^+, 2141.3 keV 0^+, 2390.8 keV $3^\pm$, 2399.8 keV $3^\pm$, 2431.0 keV 3^- and 2484.7 keV 2^+ levels. Furthermore E2/M1 multipole mixing ratios were determined for the most intensive mixed γ-transitions (Table I.).

Table I.

E2/M1 multipole mixing ratios of ^{136}Ba γ-transitions

E_γ (keV)	732.5	1261.6	1310.3	671.8	1666.5
$J_i \to J_f$	$2_2^+ \to 2_1^+$	$2_3^+ \to 2_1^+$	$2_4^+ \to 2_1^+$	$2_5^+ \to 2_2^+$	$2_6^+ \to 2_1^+$
δ(E2/M1)	−1.5(1.0)	−1.5(1.5)	0.03(18) or 2.2(1.6)	−0.01(12) or 2.5(1.2)	0.07(36) or 1.9(2.0)

REFERENCES

1. R.D. Griffioen, R. Gunnink and R.A. Meyer, Z. Physik A274, 391 /1974/.
2. G.M. Julian and T.E. Fessler, Phys. Rev. 172, 1208 /1968/.
3. R.A. Meyer and R.D. Griffioen, Phys. Rev. 186, 1220 /1969/.
4. W. Gelletly, J.A. Moragues, M.A. Mariscotti and W.R. Kane, Phys. Rev. 181, 1682 /1969/.
5. R.E. Crien, G.W. Cole, J.L. Holm and O.A. Wasson, Phys. Rev. C9, 1622 /1974/.
6. K. Schreckenbach, H. Faust, S. Blakeway, W. Gelletly, W.F. Davidson, R.F. Casten, D. Warner and M.F. Stelts, Neutron-Capture Gamma-Ray Spectroscopy and Related Topics 1981, Inst. Phys. Conf. Ser. No. 62, p.200.
7. I. Diószegi, Á. Veres, W. Enghardt and H. Prade, J. Phys. G: Nucl. Phys. 10, 969 /1984/.
8. L.K. Peker, Nuclear Data Sheets 26, 473 /1979/.

ANGULAR DISTRIBUTION STUDIES OF GAMMA RAYS FROM THE ^{172}Yb(n,n'γ) REACTION

H.M. Youhana, M.A. Al-Amili, S.R. Al-Obeidi and H.E. Abid

Department of Nuclear Physics, Nuclear Research Centre,
Tuwaitha, Baghdad, Iraq

ABSTRACT

Energy levels of ^{172}Yb have been studied using the (n,n'γ) reaction with fast reactor-neutrons. The de-excitation γ-rays from a target (5.68g of Yb_2O_3 enriched to 95.0% in ^{172}Yb) were observed with a 30 cm^3 coaxial Ge(Li) detector located at a distance of 40 cm from the target and coupled to a 4096 channel analyser. The system resolution was 2.5 keV FWHM for the 1.33 MeV γ-rays.

Gamma ray spectra, recorded at five angles (θ = 90, 114, 125, 135 and 145°) relative to the beam direction, were analysed by using the peak-finding, peak-fitting computer program BABIL[1]. An additional measurement was also performed at each angle with a 19g plate of Cu placed close to the Yb target. The angular distribution data, corrected for the attenuation in the target and normalised using the isotropic 669.7 keV γ-transition in ^{63}Cu, were computer fitted to an even order Legendre polynomial series.

Spin assignment to levels in ^{172}Yb were carried out by comparing the measured a_2 and a_4 coefficients for each γ-transition with those computed, for a range of spin sequences as a function of δ-mixing ratio, using the computer code CINDY[2]. The a_2 coefficient was used to deduce the allowed values of δ. The phase convention used for the definition of δ is that of Krane and Steffen[3]. In cases where one of the γ-transitions from a particular level is expected to be pure, a model-independent procedure[4], hereafter called "complete alignment method CAM", was employed to calculate the δ-mixing ratios.

The measured angular distribution coefficients, spin values and the relevent δ-values, compared where possible with the values obtained by CAM and by other authors, are presented in Table I. The agreement is in general satisfactory apart from the δ-values obtained for the 90.5 and 1002.5 keV γ-transitions.

Table I Angular distribution results for γ-transitions in ^{172}Yb

E_i keV	E_γ keV	J_i^π	J_f^π	a_2	a_4	δ CINDY	δ CAM	δ Previous
260	182	4^+	2^+	+0.32(5)	−0.04(6)	+0.02(7)		
540	280	6^+	4^+	+0.31(5)	−0.14(5)	−0.03(5)		
1118	857	2^+	4^+	+0.06(2)	−0.03(2)	+0.03(4)	E2	
	1039	2^+	2^+	+0.18(4)	−0.06(4)	$+2.3^{+0.5}_{-0.3}$	$+2.0^{+0.9}_{-0.6}$	$+5.0^{+2.5a}$

Table I (cont.)

E_i keV	E_γ keV	J_i^π	J_f^π	a_2	a_4	δ CINDY	δ CAM	δ Previous
1172	912	3^+	4^+	+0.25(4)	+0.06(5)	-1.5(4)		-1.7(2)b -1.9(1)c
1198	1120	2^-	2^+	+0.19(4)	-0.04(5)	+0.03(9)		
1222	1143	3^-	2^+	-0.19(4)	+0.01(5)	+0.02(3)		
1263	91	4^+	3^+	+0.39(10)	+0.18(10)	+0.50(12)		-2.4(2)c
						$+3.6^{+1.6}_{-1.0}$		-1.7(2)d
	1003	4^+	4^+	+0.24(7)	-0.09(8)	-0.17(11)		$-17^{+9}_{-\infty}$ b
						+1.4(3)		$+50^{+\infty}_{-29}$ c
1287	1027	4^+	4^+	+0.38(8)	-0.10(9)	$+0.77^{+0.33}_{-0.42}$		+0.87(13)a +0.75(17)c
1331	1070	4^-	4^+	+0.37(10)	-0.01(11)	$+0.06^{+?}_{-0.17}$		$+0.0^{+0.3}_{-0.9}$a
1466	1466	2^+	0^+	+0.32(7)	-0.04(8)	E2	E2	
	1388	2^+	2^+	-0.20(4)	-0.06(4)	$-0.9^{+0.2}_{-0.5}$	$-0.8^{+0.2}_{-0.4}$	$-9.3^{+2.5}_{-5.0}$b
						$-3.0^{+1.2}_{-1.5}$	$-4.0^{+1.8}_{-3.9}$	$-2.4^{+1.5}_{-0.8}$c
1477	1398	2^+	2^+	+0.31(7)	-0.07(8)	+1.3(6)		$+1.0^{+\infty}_{-0.6}$b
1641	378	4^-	4^+	+0.37(5)	-0.03(5)	+0.04(10)		
1663	400	3^+	4^+	-0.13(2)	+0.03(2)	+0.04(2)		+0.00(4)c
1701	528	3^+	3^+	+0.34(3)	-0.05(4)	$+0.11^{+0.8}_{-0.6}$		+0.05(4)c

a) Ref 5   b) Ref 6   c) Ref 7   d) Ref 8

REFERENCES

1. A. Hussain; M.Sc. Thesis, University of Baghdad (1977)
2. E. Sheldon and V.C. Rogers; Comp. Phys. Commun. 6 699(1973)
3. K.S. Krane and R.M. Steffen; Phys. Rev. C2 724(1970)
4. E. Sheldon and D.M. Van Patter; Rev. Mod. Phys. 38 143(1966)
5. J.R. Cresswell, P.D. Forsyth, D.G.E. Martin and R.C. Morgan; J. Phys. G. Nucl. Phys. 7 235(1981)
6. K.S. Krane; At. Data Nucl. Data Tables 16 383(1975)
7. K.S. Krane; Phys. Rev. C13 1295(1976)
8. R.M. Walker, W.R. Faber, W.H. Bentley, R.M. Ronningen and R.B. Firestone; Nucl. Phys. A343 45(1980)

ISOMER RATIO CALCULATIONS USING MODELED DISCRETE LEVELS*

M. A. Gardner, D. G. Gardner, and R. W. Hoff
Lawrence Livermore National Laboratory, Livermore, CA 94550

ABSTRACT

We have calculated isomer ratios for the ^{175}Lu(n,γ), ^{175}Lu(n,2n), ^{237}Np(n,2n), ^{241}Am(n,γ), and ^{243}Am(n,γ) reactions using modeled level structures in the deformed, odd-odd product nuclei. We find: that the hundreds of discrete levels and their gamma-ray branching ratios provided by the modeling are necessary to achieve agreement with experiment, that many rotational bands must be included in order to obtain a sufficiently representative selection of K quantum numbers, and that the levels of each band must be extended to appropriately high values of angular momentum.

METHOD

We have made isomer ratio calculations for the reactions: ^{175}Lu(n,γ)^{176m,g}Lu, ^{175}Lu(n,2n)^{174m,g}Lu, ^{237}Np(n,2n)^{236m,g}Np, ^{241}Am(n,γ)^{242m,g}Am, and ^{243}Am(n,γ)^{244m,g}Am. All of the cross-section calculations were made with our version of the STAPRE statistical model code,[1] with no fission competition. Appropriate neutron potentials were used for the different targets and the Cook-modified, Gilbert-Cameron level density parameters were employed, further adjusted to match discrete level information or to reproduce D_{ob} values where known. Below about 1.5 MeV in each deformed, odd-odd product nucleus, we replaced the level density expression with a discrete level set, modeled as described elsewhere in this conference.[2] Table I summarizes the modeled sets used and compares each with the experimental level data available for that nucleus.

The gamma-ray cascades leading to the ground-state and isomeric products are modeled as follows. The continuum bins are depopulated according to our E1 and M1 strength function systematics, where the E1 strength function has an energy-dependent Breit-Wigner line shape and the M1 strength function is a constant.[3] The depopulation of the discrete levels proceeds as described in Ref. 2. The use, among the discrete levels, of the same E1 and M1 strength functions as described for the continuum led to calculated isomer ratios for the ^{175}Lu(n,γ) and ^{237}Np(n,2n) reactions that were equal to those obtained using a constant E1/M1 ratio of 0.167.

RESULTS

In Fig. 1a we show the m/g ratios vs E_n that we calculated for the ^{175}Lu(n,γ) reaction, compared with values derived from experimental data at about 30 keV.[4,5] The use of 291 modeled discrete levels for ^{176}Lu(Set A in Table I) yields the lowest calculated band. If

*Work performed under the auspices of the U.S. Dept. of Energy by the Lawrence Livermore National Lab. under Contract No. W-7405-ENG-48.

Table I Comparison of modeled and experimental level sets

Level-set designation		Energy range (MeV)	Number of levels	Number of rotational bands	Range of quantum nos. K	I
^{174}Lu	A	0-1.50	433	82	0-9	0-12
	Expt.	0-0.97	57	10	0-7	0- 7
^{176}Lu	A	0-1.50	291	62	0-9	0-12
	B	0-1.04	109	33	0-9	0-11
	C	0-0.77	49	15	0-9	0-10
	D	0-0.62	27	8	0-8	0-10
	Expt.	0-0.99	38	7	0-7	0-10
^{236}Np	A	0-1.48	998	94	0-6	0-20
	B	0-0.91	453	71	0-6	0-15
	C	0-0.36	50	15	0-6	0-11
	D	0-0.20	14	5	1-6	0- 9
	Expt.	0-0.22	5	3	1-6	1- 6
^{242}Am	A	0-1.50	788	107	0-7	0-13
	Expt.	0-1.10	30	9	0-5	0- 7
^{244}Am	A	0-1.40	769	101	0-8	0-15
	Expt.	0-0.78	44	15	0-6	0- 6

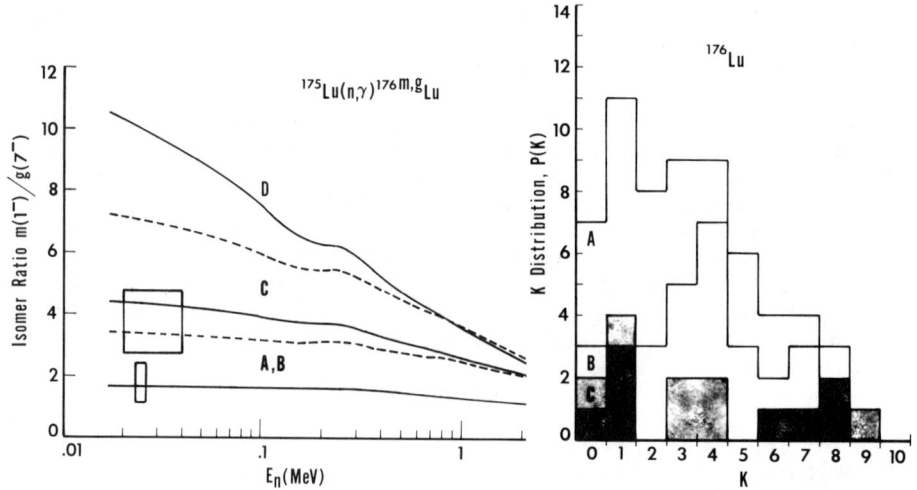

Fig. 1 a) Calculated ^{175}Lu(n,γ) m/g ratios vs E_n using ^{176}Lu level Sets A-D, compared with data ($\square$).[4,5] b) The distribution of K quantum numbers, P(K), for Sets A-D.

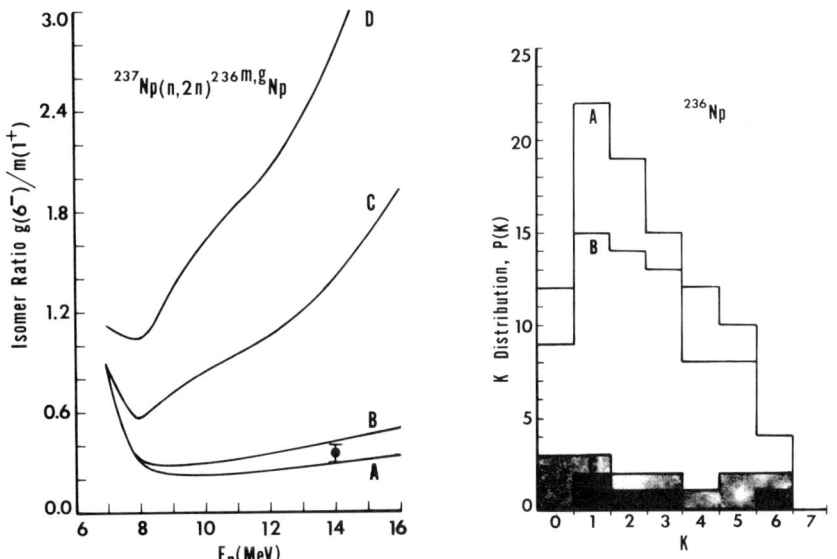

Fig. 2 a) Calculated ^{237}Np(n,2n) g/m ratios vs E_n using ^{236}Np level Sets A-D, compared with data (●).[6] b) The P(K) for Sets A-D.

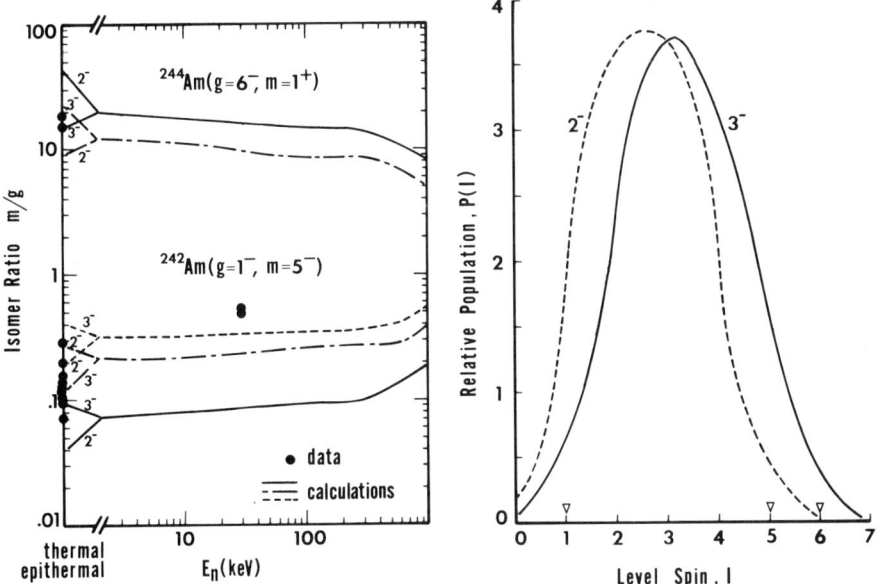

Fig. 3 a) Calculated ^{241,243}Am(n,γ) m/g ratios vs E_n compared with data[7,8] (——— present work, ----- Ref. 7, -——— Ref. 8). b) Spin distributions among levels calculated after γ-ray cascade from continuum for 2⁻ and 3⁻ resonance cases.

only dipole transitions are allowed during the depopulation of the discrete levels, the lower bound of the band is obtained; if allowed E2 transitions are considered to take place from some of the band heads, then the upper bound is calculated. Sensitivity studies of the computed isomer ratio to the number of discrete levels describing ^{176}Lu also were made. Set A was truncated several times just below a new band head to yield Sets B-D (see Table I). Set B led to about the same calculated isomer ratios as Set A. No E2 transitions were considered to begin at or below 0.62 MeV, therefore Set D results do not involve a band. The distribution of the K quantum numbers, P(K), associated with Sets A-D are shown in Fig. 1b. Although the P(K) of Sets A and B are quite different, each has a sufficiently representative sampling of K values to yield equal isomer ratios. Sets C and D do not sample all of the K values, leading to isomer ratios that are too high.

The calculated g/m ratios for the ^{237}Np(n,2n) reaction are shown in Fig. 2a, compared with a 14-MeV measurement.[6] Similar sensitivity studies were carried out with truncated ^{236}Np level sets. Again the corresponding P(K) for the sets are shown in Fig. 2b.

In Fig. 3a we show our calculated m/g ratios for the ^{241}Am(n,γ) and ^{243}Am(n,γ) reactions compared with measurements.[7,8] We also computed isomer ratios for s-wave neutrons leading to the 2⁻ and 3⁻ resonances for comparison with the thermal and epithermal data. Other calculational results from Refs. 7 and 8 are shown too. For each reaction, our s-wave 3⁻ resonance result agrees well with data. Fission competition should not have a significant effect at thermal energies for target ^{241}Am and throughout the neutron energy range shown for target ^{243}Am. However, the lack of fission competition in our calculations may be the reason why our ^{241}Am isomer ratio is not in good agreement with the 30-keV data. For the ^{241}Am(n,γ) reaction, our isomer ratio for the 3⁻ resonance is larger than that for the 2⁻ resonance, in agreement with the Ref. 7 calculation. In Fig. 3b we show the spin distributions among the discrete levels that we calculate after the gamma-ray cascade from the continuum for both the 2⁻ or 3⁻ resonance cases. These distributions show why the high spin state is favored by the 3⁻ resonance and why we obtain the relative order of the isomer ratios for each case.

REFERENCES

1. M. Uhl, Acta Phys. Austriaca 31, 241 (1970).
2. R. W. Hoff, et al., "Modeling Level Structures of Odd-Odd Deformed Nuclei," this Conference; UCAR 10062-83/1 (1983), p. 218.
3. D. G. Gardner, ANL-83-4 (1983), p. 67.
4. B. Allen, et al., Proc. 4th Int. Symp. on Neutron-Capture Gamma-Ray Spectroscopy and Related Topics, Grenoble (1981), p. 573.
5. H. Beer and F. Kappeler, Phys. Rev. C21, 534 (1980).
6. W. A. Myers, et al., J. Inorg. Nucl. Chem. 37, 637 (1975).
7. K. Wisshak, et al., Nucl. Sci. Eng. 81, 396 (1982).
8. F. M. Mann and R. E. Schenter, Nucl. Sci. Eng. 63, 242 (1977).

AN ESTIMATE OF THE INELASTIC CHANNEL NEUTRON RADIATIVE CAPTURE CROSS SECTIONS

J.F. Liu and Yu-Kun Ho
Physics Department, Zhengzhou University, Honan Province, China

M.A. Lone
Chalk River Nuclear Laboratories, Chalk River, Ont., Canada K0J 1J0

ABSTRACT

A formalism to calculate the resonance-averaged inelastic channel radiative neutron capture cross sections is presented. The relative contribution of these cross sections in the predicted total (n,γ) cross sections is estimated for a number of nuclei.

Since Lane & Lynn's channel neutron capture model was presented, most efforts have been devoted to the investigation of the valence and potential capture mechanism with the assumption of unexcited core. Some calculations have been made on the channel capture with excited target states, which can be called inelastic channel capture or, in some cases, doorway states contributions.[1] Since inelastic channel capture is related to the contributions from three or more quasiparticles configurations in the initial state wave function, it is difficult to investigate it theoretically and identity its contributions experimentally. Here we generalize the idea and approach developed in our previous paper[2] to calculate resonance-averaged inelastic channel capture cross sections.

The inelastic channel radiative neutron capture cross section for an E1 transition is

$$\sigma_f^{ich} = \frac{16\pi}{9} k_\gamma^3 \frac{1}{\hbar v} \frac{1}{2(2I+1)} \sum_{m\mu} \left| \langle \psi_f | \bar{e} r Y_{1\mu} | \psi_i \rangle \right|^2 \qquad (1)$$

where k_γ is the photon wave number, $\bar{e}$ the neutron effective charge.

$$\psi_i = \sum_{\ell j J} -\frac{i\sqrt{(2\ell+1)\pi}}{k} C_{\ell\, 0\, i\, m_i}^{j\, m_j} C_{j\, m_j\, I\, m_I}^{JM}$$

$$\sum_{\ell' j'} \phi_{\ell' j' I'}^{JM} \frac{v}{v'} S_{\beta\ell' j'\, \alpha\ell j}^{J} \frac{1}{r} Q_{\ell' j'}(r) \qquad (2)$$

is the initial state inelastic channel wave function. We use superscript prime to designate a quantity belonging to the inelastic channel. In eq. (2)

$$\phi_{\ell' j' I'}^{JM} = [\phi_{iI'} \times Y_{\ell' j'}]^{JM} \qquad (3)$$

is the channel wave function consisting of the intrinsic excited state wave function of the target nucleus $\phi_{iI'}$ coupled to the neutron angular momentum wave function $Y_{\ell'j'}$ to give total spin J and projection M; $S^J_{\beta\ell'j'\alpha\ell j}$ is the S matrix elements corresponding to the inelastic scattering. The function

$$Q_{\ell'j'}(r) = \frac{\operatorname{Im}\langle U^{+J}_{\ell'j'}(r)\rangle}{\operatorname{Im}\langle K^J_{\ell'j'}\rangle} + i[\operatorname{Re}\langle U^{+J}_{\ell'j'}(r)\rangle - \frac{\operatorname{Re}\langle K^J_{\ell'j'}\rangle}{\operatorname{Im}\langle K^J_{\ell'j'}\rangle} \operatorname{Im}\langle U^{+J}_{\ell'j'}(r)\rangle]$$

(4)

is the radial wave function of the inelastic channel[2], and has the asymptotic form $ik'rh^{(1)}_{\ell'}(k'r)$ in the external region. In eq. (4), $\langle U^{+J}_{\ell'j'}(r)\rangle$ and $\langle K^J_{\ell'j'}\rangle$ are respectively the optical model wave function and reactance matrix corresponding to the excited target state, and Re and Im denote real and imaginary parts respectively.

In eq. (1) the final state wave function has the form

$$\psi_f = \sqrt{S'_f} \frac{U_{\ell'_f j'_f}(r)}{r} \phi^{J_f M_f}_{\ell'_f j'_f I'} ,$$

(5)

where $U_{\ell'_f j'_f}(r)$ is the radial wave function of single particle bound state in a potential corresponding to excited residual nucleus, and s'_f its coefficient.

Substituting eqs. (2) and (5) into eq. (1), averaging over the incident energy, we get the average inelastic channel radiative capture cross section

$$\langle \sigma^{ich}_f \rangle = \frac{4}{3} \frac{k_\gamma^3}{\hbar v'} \frac{\pi}{k^2} \sum_{\ell'j'J} \frac{(2J+1)}{2(2I+1)} \langle \ell'j'J \| D_I \| \ell'_f j'_f J_f \rangle (2J_f+1)$$

$$S'_f \bar{e}^2 |Q_{\ell'j'\ell'_f j'_f}|^2 \langle |S^J_{\beta\ell'j'\alpha\ell j}|^2 \rangle$$

(6)

where $\langle \ell'j'I \| D_I \| \ell'_f j'_f J_f \rangle$ is the reduced matrix element[2], and

$$Q_{\ell'j'\ell'_f j'_f} = \int_R^\infty r \, Q_{\ell'j'}(r) \, U_{\ell'_f j'_f}(r) \, dr$$

(7)

is the radial integral. The value of $\langle |S^J_{\beta\ell'j'\alpha\ell j}|^2 \rangle$ can be calculated by means of the statistical theory of nuclear reactions. Thus

$$\langle |S^J_{\beta\ell'j'\alpha\ell j}|^2 \rangle = \frac{T_{\beta\ell'j'} T_{\alpha\ell j}}{T^{J\pi}} W^{J\pi}_{\alpha\ell j \beta\ell'j'} \tag{8}$$

$T_{\beta\ell'j'}$ is the inelastic scattering neutron transmission coefficient, and $W^{J\pi}_{\alpha\ell j \beta\ell'j'}$ is the width fluctuation correction factor.

Eqs. (6) and (8) can be regarded as contributions from radiative capture of the compound inelastic channel scttering wave.

Furthermore, we rewrite eq. (6) as

$$\langle \sigma^{ich}_f \rangle = \frac{\pi}{k^2} \sum_{\ell j J \ell'j'} \frac{(2J+1)}{2(2I+1)} \frac{T_{\alpha\ell j} T^J_{f,\beta\ell'j'}}{T^{J\pi}} W^{J\pi}_{\alpha\ell j \beta\ell'j'} \tag{9}$$

where

$$T^J_{f,\beta\ell'j'} = \frac{4k^3_\gamma}{3\hbar v'} \langle \ell'j'J \| D_I \| \ell'_f j'_f J_f \rangle (2J_f+1) S'_f \bar{e}^2 \left| Q_{\ell'j',\ell'_f j'_f} \right|^2 T_{\beta\ell'j'} \tag{10}$$

can be considered as the radiative transmission coefficient in the compound inelastic channel (ℓ', j').

At MeV region of neutron incident energy, there are four processes that are expected to contribute to the average radiative neutron capture cross sections, namely compound nucleus radiative capture, elastic channel radiative capture, inelastic channel radiative capture and semidirect capture.[3] The last one is important only at E > 3-5 Mev. Table 1 shows the fractions of each process in the calculated total cross sections in ^{56}Fe, ^{74}Ge, ^{90}Zr, ^{138}Ba and ^{200}Hg at neutron energies of 0.1 MeV, 1.0 MeV and 2.0 MeV respectively.

These results show that the non-statistical processes (including elastic and inelastic channel radiative capture) may account for about one third of the total (n,γ) cross sections[1] for target nuclei in the 3s giant resonance regions of the neutron strength function. In addition, for heavier nuclei the inelastic channel capture cross section at high energies may be larger than that of the elastic channel.

Table 1

Fractions of each process in the calculated (n,γ) cross sections

	Elastic Channel Capture			Inelastic Channel Capture			Compound Nucleus Capture		
Target	0.1 MeV	1 MeV	2 MeV	0.1 MeV	1 MeV	2 MeV	0.1 MeV	1 MeV	2 MeV
^{56}Fe	0.348	0.176	0.160	0.09	0.174	0.079	0.560	0.650	0.760
^{74}Ge	0.048	0.017	0.011	0.0	0.016	0.022	0.952	0.967	0.967
^{90}Zr	0.312	0.114	0.051	0.0	0.0	0.012	0.688	0.886	0.937
^{138}Ba	0.148	0.060	0.051	0.0	0.0	0.019	0.852	0.94	0.93
^{200}Hg	0.058	0.015	0.006	0.0	0.023	0.11	0.942	0.962	0.884

REFERENCE

1. B.J. Allen & A.R. de L. Musgrave, Advances in Nuclear Physics, Vol. 10, Plenum, New York, 1978.
2. Y.K. Ho & M.A. Lone, Nucl. Phys. A406(1983)1.
3. G.E. Brown, Nucl. Phys. 57(1964)339.
   C.F. Clement et al., Nucl. Phys. 66(1966)273.

V. CHARGED PARTICLES

USE OF CAPTURE REACTIONS TO MEASURE SHORT LIFETIMES BY THE DSA METHOD[*]

J. Keinonen
University of Helsinki, Accelerator Laboratory
Hämeentie 100, SF-00550 Helsinki 55, Finland

ABSTRACT

The experimental techniques of measuring the short mean lifetimes (1-100 fs) of excited nuclear states by the Doppler-shift-attenuation method through the capture reactions are discussed. Emphasis is put on the most recent development in the preparation and use of implanted targets and in the utilization of the Monte Carlo method and the experimental stopping power in the analysis. Lifetime results are compared with values obtained from other methods. Latest isoscalar E2 transition strengths in the nuclear mass triplets are used to illustrate the relevance of short lifetimes in understanding the structure of excited states.

INTRODUCTION

The aim of this talk is to discuss the status of the measurements of short lifetimes (1-100 fs) by the use of charged particle capture reactions coupled with the Doppler-shift attenuation (DSA) method and to demonstrate the relevance of short lifetimes in understanding the structure of excited states of the 1p and sd shell nuclei. General reviews of lifetime measurements have recently been given in two extensive articles[1,2].

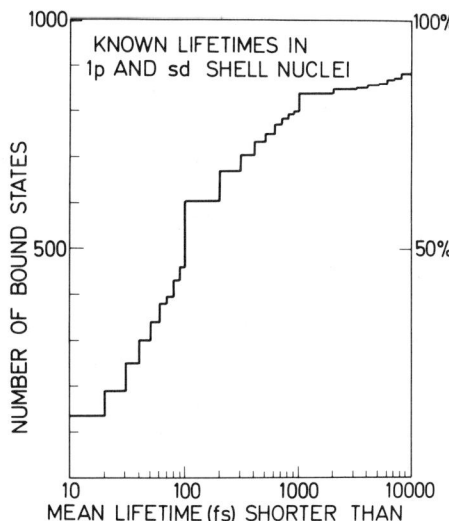

Fig. 1. Distribution of the known lifetimes of bound states in the 1p and sd shell nuclei. The data are taken from Refs. 3-10.

As shown in Fig. 1, the main part of the lifetimes of bound states in the 1p and sd shell nuclei lies in the region 1 to 1000 fs. The only way to measure lifetimes in this region is often by the DSA method. In connection with capture reactions it offers a unique method for systematic study of short lifetimes. The use of the capture reactions has the following advantages:
(i) The proton and α-particle capture reactions populate most of the bound levels in the product nuclei.
(ii) Disturbing effects due to the reaction kinematics are negligible.
(iii) The selective decay of

[*] The studies on which this talk is based have been carried out in collaboration with A. Anttila, M. Bister, M. Hautala and A. Luukkainen.

each resonance to a relatively small number of bound states gives the possibility to avoid disturbing feeding transitions and ensures γ-ray spectra with well separated lines. Lifetime measurements can be performed in singles experiments.

(iv) The primary γ-ray transitions (presenting short lifetimes $0.01 \lesssim \tau \lesssim 1$ fs, $1 < \Gamma < 100$ eV) can be used to experimentally correct for the solid angle attenuation of the observed Doppler shifts and to check on the stability of the detection system.

In the DSA method the lifetime of an excited state is compared with the slowing down time of the recoiling excited nucleus in a backing material, the main drawback being in insufficient knowledge of this time. This problem has especially been connected to the use of the capture reactions[11-13], where small initial recoil velocities ($\beta \sim 0.1$-1%) are produced. The techniques which remove this ambiguity are discussed in the following section.

EXPERIMENTAL PROCEDURE

Target preparation. In typical DSA measurements, relatively high currents (10-100 μA) of proton or α-particle beams at energies 0.5 to 3 MeV, are focused to a beam spot of about 3 mm in diameter. Thus, the targets are required to withstand high proton or α-particle fluxes and doses.

The implantation of the target atoms into a heavy backing material is the only practical way to prepare targets from rare and gaseous isotopes and targets with the high effective stopping power necessary in the measurements of short lifetimes. Tantalum is the most suitable backing material for the following reasons: (i) It has low sputtering yield and high saturation value for implanted atoms[89]. (ii) Contaminations which are disturbing in the capture reactions, can be completely removed by outgassing at temperatures above 1000°C (melting point of Ta is 2996°C). (iii) Tantalum has relatively high Z value (73) and high density (16.6 g/cm^3). (iv) Because of efficient hydrogen diffusion in Ta[14], the implanted targets withstand high proton beam intensities and doses. The blistering of Ta backings in the α-particle bombardments can be avoided by roughening the surface[15]. (v) As the diffusion of the implanted target atoms is slow in Ta (Ref. 90 and our own tests), high beam intensities do not deteriorate the targets.

Depending on the cross section of the resonance to be used, the implanted doses vary between 10^{19} to 10^{21} ions/m^2. For a typical implantation energy of 60 keV, the corresponding maximum concentrations are 1 to 30 at.%. The concentration profile of the implants can be studied accurately by the use of the Rutherford backscattering (RBS) and nuclear resonance broadening (NRB) method[16]. Typical depth profiles of ^{22}Ne implanted at 60 keV into Ta and Ta$_2$O$_5$ are shown in Figs. 2 and 3.

The maximum concentrations obtained for different implanted targets are illustrated in Figs. 4 and 5. It can be seen that the saturation atomic concentration of ^{22}Ne in Ta$_2$O$_5$ (12 at.%) is only about half of that in Ta (28 at.%).

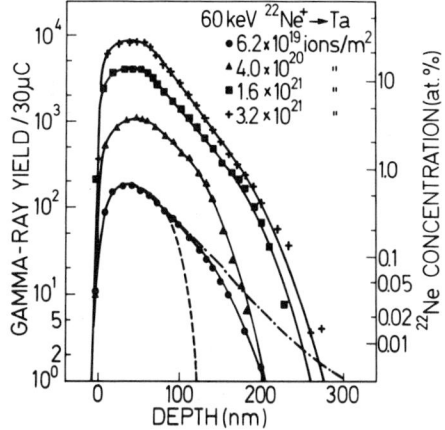
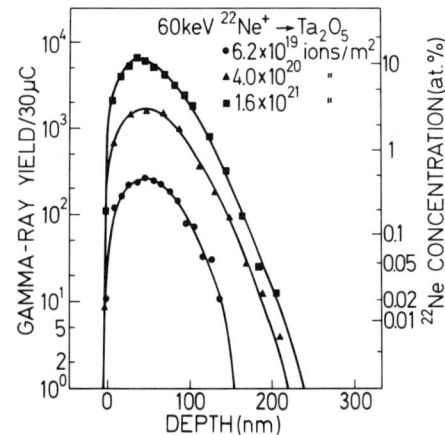

Fig. 2. The γ-ray yield curves from the ^{22}Ne$^+$ implantations into Ta backings: In converting the energy scale into the depth scale (nm), the proton stopping power[85] has to be used and the correction due to the Ne concentration should be taken into account. The scale is shown for the lowest dose. The solid lines are drawn to guide the eye. In the case of the lowest dose, the dashed line is the Monte Carlo calculated range profile assuming an amorphous structure for the backing material and the dot-dashed line is the range simulation for a polycrystalline backing material.

Fig. 3. As for Fig. 2, but for Ta_2O_5. In the case of the lowest dose, the solid line shows also the Monte Carlo simulated range profile.

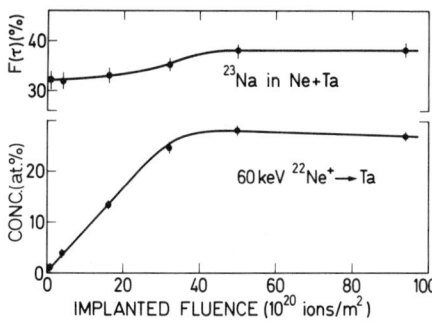
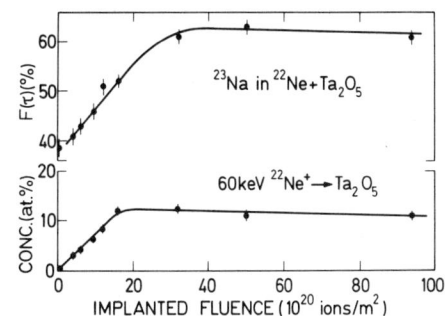

Fig. 4. The collection curve obtained in the preparation of ^{22}Ne targets in Ta backings (lower part) and the $F(\tau)$ values measured for different targets (upper part). The solid lines are drawn to guide the eye.

Fig. 5. As for Fig. 4, but for Ta_2O_5.

Stopping power. The use of implanted targets in DSA measurements demands (1) experimental knowledge on the correction for the cross section of the atom-atom scattering (potential and scattering angle) and (2) experimental knowledge on the slowing down conditions as affected by the structural effects of the backing material and as modified during the preparation of the target.

In the slowing down at low reduced-recoil energies ($\varepsilon \sim 0.1-1$), the total stopping power

$$\left(\frac{d\varepsilon}{d\rho}\right)_{corr} = f_n \left(\frac{d\varepsilon}{d\rho}\right)_n + f_e \left(\frac{d\varepsilon}{d\rho}\right)_e^{LSS} \tag{1}$$

where energy ε and range ρ are given in the reduced units of the LSS theory[18], is dominated by the nuclear (n) part. In the DSA analysis performed in our laboratory, the analytical form given in the LSS theory for the uncorrected nuclear stopping cross section is replaced by Monte Carlo calculations, where the scattering angles of the recoiling ions are directly derived from the classical scattering integral and the inter-atomic interaction is derived from the Thomas-Fermi potential. The test of different potentials by comparing with experiment the calculated ranges and range distributions of several ions in various amorphous materials, has indicated that the ranges agree to within 10% over the reduced energy interval $\varepsilon = 0.1-1$[19-20]. Thus, there is no large difference in the stopping calculated from various potentials[19,21]. The uncorrected electronic stopping power (e) is calculated according to the LSS theory.

The experimental correction factors f_n and f_e are determined in the following way[25]: (i) The range values of typically 20 to 100 keV atoms, corresponding to the recoil energies of the compound nucleus, measured with the NRB method for a semi-infinite medium are compared with theoretical ranges. Monte Carlo calculations are used to take into account the number of reflected ions and the influence of the surface

Table I. Summary of stopping parameters.

Backing	Recoiling atom	f_n a)	f_e	Ref.
Mo	Na	0.78 ± 0.07	1.00 ± 0.17	16
Ta	N	0.93 ± 0.05	1.0 ± $^{0.4}_{0.3}$	22
	F	0.85 ± 0.05	1.0 ± 0.16	23
	Na	0.83 ± 0.05	1.15 ± $^{0.6}_{0.3}$	24
	Al	0.67 ± 0.08	1.0 ± $^{0.8}_{0.2}$	25
	Cl	0.68 ± 0.06	1.00 ± 0.19	26

a) The values given here do not any more include the corrections which are needed if the analytical, approximative form of the nuclear stopping cross section[18] is used.

to the range distribution. The f_n, f_e-combinations producing the experimental range values are obtained. (ii) The experimental line shape of a Doppler broadened γ-ray peak obtained at angles 90° and 0° relative to the beam are then compared with the corresponding line shapes calculated with the Monte Carlo method. The f_n, f_e, τ-combinations producing the experimental F-value and line shape at 0° are obtained. The unique f_n, f_e, τ-values are obtained from the comparison of the line shapes at 90°. At this angle the broadening of the γ-ray peak is most sensitive to the velocity components perpendicular to the beam direction. Correction factors obtained for different ions recoiling in Mo and Ta are summarized in Table I. In the Monte Carlo calculations the target material has been assumed to be amorphous. The fact that the correction factors for the nuclear stopping power are smaller than one is due to the following reasons: (i) The Thomas-Fermi potential is not quite accurate in describing the collisions of low Z_1-high Z_2 atoms. This is supported by the experimental observation[27] that in the case of low Z_1-low Z_2 atoms, where the reduced energy at the same energy of Z_1-atoms is relatively high and therefore collisions with smaller impact parameters are dominant, the Thomas-Fermi potential is more accurate. A further indication is that the scattering angle determined from the line shape of the γ-rays obtained at 90° are smaller than calculated. (ii) The polycrystalline structure of the backing material causes microchanneling of the implants yielding longer ranges. This can be seen in comparing ranges in Ta and Ta_2O_5 (Figs. 2 and 3) where the broadening of the range profile in Ta is due to this effect. The microchanneling can be included in the Monte Carlo calculations[17], but cannot be considered to be the whole explanation for the reduction of the nuclear stopping power. The fact that the portion of channeled atoms decreases with the implanted dose indicates that the first channeled atoms block the channels.

It should be pointed out that by the above method the stopping parameters are determined under the circumstances where the lifetime measurements are performed and they thus fully describe the slowing down of the recoiling nuclei needed in the DSA analysis. The collisions of the recoiling atoms with the target atoms are taken into account in the analysis. In the calculations, the atomic density of the slowing down material is corrected for the experimentally known concentration distribution of the implanted target nuclei[16]. At low concentrations (∼1 at.%) the effect of the target atoms on the stopping power is negligible. However, in DSA analysis where the $F(\tau)$ values and line shapes are measured with the targets having high implant concentrations, the stopping parameters should also be known for the compound-nucleus atom - target atom collisions. The concentrations less than 20 at.% are, however, so low that the uncertainty of the stopping parameters in this description has a negligible effect on the calculated values.

An important factor in the case of the implanted targets with relatively high doses, is possible changes in the local subsurface density due to the implants and this effect should be included in the DSA analysis. Recent studies with the Mo[16] and Ta[16,28] backings show that the atomic density first increases with the increasing dose, but at higher values of implanted doses the amorphization of the host lattice takes place and the density decreases. How different amounts of implanted target material ^{22}Ne effect the F(τ) value measured with the Ta and Ta$_2$O$_5$ backings is illustrated in Figs. 4 and 5. Typical DSA measurements are shown in Figs. 6 and 7.

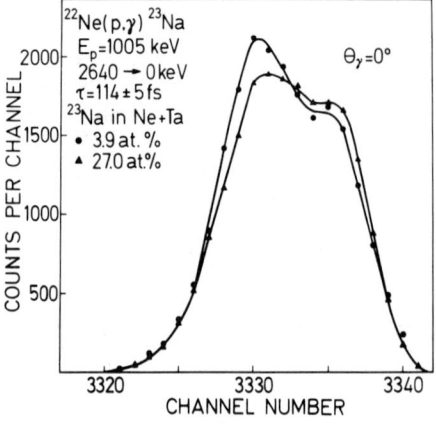
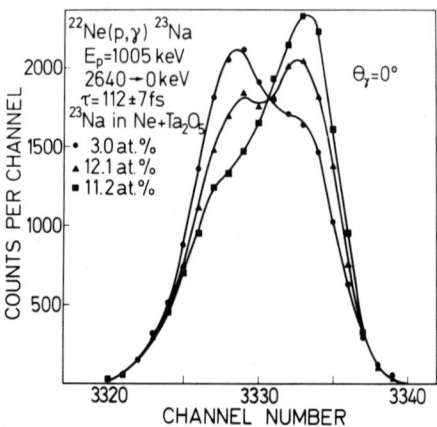

Fig. 6. The peaks of the 2640 keV γ-rays from ^{23}Na recoiling in a Ta backing. The solid curves are the best fits for the 0° line shapes with $f_n=0.83$ and $f_e=1.15$. The dispersion is 0.791 keV/ch.

Fig. 7. As for Fig. 6, but for for Ta$_2$O$_5$. The stopping parameters were $f_n=1.00$ and $f_e=1.00$. The dispersion is 0.794 keV/ch.

The F(τ) values indicate that the lattice structure can withstand higher doses than the amorphous structure without decrease of the effective stopping power and further the efficient stopping power of Ta seems not to be strongly affected even by relatively high doses of implants.

LIFETIME RESULTS

Examples shown in Table II have been chosen to illustrate how lifetime values measured at low recoil velocities relate to those obtained under widely varying conditions. Recent development of the DSA measurements by the Chalk River, Brookhaven and Utrecht groups, where light-ion implanted targets are bombarded by a beam of heavy ions, where experimental data on the characteristics of the targets are used and where an experimentally known stopping power has been utilized in the DSA analysis[66,72,73,86-88], gives a possibility to compare short lifetimes from the precision DSA measurements at low and high recoil velocities. Unfortunately, the results of the high recoil velocities thus far reported are mainly for relatively longer lifetimes ($\tau \gtrsim 50$ fs). The consistency of the lifetime values obtained with the techniques described in the present talk with the weighted average values indicates that no serious sources of systematic errors exist and that the error limits quoted are realistic.

Table II. Comparison of lifetime values determined by the techniques described in this talk (*) with results obtained also by other methods.

Method	τ(fs)	Ref.	Reaction	v/c(%)	Slowing-down medium
The 1.04 MeV level in ^{18}F:					
DSA	*2.6$^{+0.7}_{-0.4}$	41	^{18}O(p,n)	0.50	Ta+implanted ^{18}O
	*2.7±0.9	41			Ta$_2$O$_5$+implanted ^{18}O
	4 ±2	42			Ta$_2$O$_5$
DSA HI[d]	2.7±0.7	43	^{3}He(^{16}O,p)	4.1	Al,Zr,Nb and Au+ implanted ^{3}He
weighted average	2.7±0.4				
The 1.35 MeV level in ^{19}F:					
DSA	*3700±700	23	^{15}N(α,γ)	0.66	Ta+implanted ^{15}N
	5200±900	29	^{19}F(α,α'γ)	0.95	CaF$_2$, SrF$_2$
	5300$^{+1500}_{-900}$	31	^{19}F(p,p'γ)	0.55	SrF$_2$
	3500±1300	32		0.70	SrF$_2$+Au
RDM[e]	4700±600	30	^{19}F(α,α'γ)	0.95	
weighted average	4500±400				
The 2.64 MeV level in ^{23}Na:					
DSA	*114±5	28	^{22}Ne(p,γ)	0.20	Ta+implanted ^{22}Ne
	*112±7	28		0.20	Ta$_2$O$_5$+implanted ^{22}Ne
	*112±6	24		0.20	Ta+implanted ^{22}Ne
	*114±11	24		0.20	Ta$_2$O$_5$+implanted ^{22}Ne
	*119±5	16		0.20	Mo+implanted ^{22}Ne
	94±14[a]	33		0.16	Mo,Ta+implanted ^{22}Ne
	146±12[a]	34		0.16	Mo,Ta+implanted ^{22}Ne
	100±10[a]	35		0.21	Ta+implanted ^{22}Ne
	363±60	36	^{23}Na(p,p'γ)	0.37-0.51	evapor. Na
	100$^{+80}_{-40}$	37		0.39-0.44	NaOH
	200±80	38		0.44	evapor. Na and NaI
DSA HI[d]	95±35	39	^{12}C(^{12}C,pγ)	4	C+Au
res. fluor.	136±40	40	^{23}Na(γ,γ)		
weighted average	115±3				
The 2.07 MeV level in ^{26}Al:					
DSA	*730±100	8	^{25}Mg(p,γ)	0.16	Ta+implanted ^{25}Mg
	530±200	44		0.13-0.23	^{25}Mg

Table II. cont.

	560±280	45	^{24}Mg(^{3}He,p)	0.17-0.23	MgO
	390±50	46		0.67	evaporated ^{24}Mg+Ni
	460±130	47	^{26}Mg(p,n)	0.51	^{26}Mg
	460±120	48	^{23}Na(α,n)	0.76-1.0	evaporated Na
RDM[e)]	1000±100	49	^{23}Na(d,n)	0.9	
weighted average	530±40				

The 3.51 MeV level in ^{26}Al:

DSA	*26±8	8	^{25}Mg(p,γ)	0.18	Ta+implanted ^{25}Mg
DSA HI[d)]	23±8	50	^{12}C(^{16}O,d)	3	C with and without Ni backing
weighted average	25±7				

The 2.21 MeV level in ^{27}Al:

DSA	*38.5±2.0	6	^{26}Mg(p,γ)	0.20	Ta+implanted ^{26}Mg
	55±9	51	^{27}Al(p,p'γ)	0.36,0.42	Al
DSA HI[d)]	36±19	52	^{27}Al(^{37}Cl,^{37}Cl)	3.3	Al
res.fluor.	38.3±1.3	53	^{26}Mg(γ,γ)		
	49±11	54			
	41±6	55			
	44±6	56			
	40±2	57			
	37±7	58			
	37±8	59			
el.scatt.	38±7	60[b)]	^{26}Mg(e,e')		
Coulex[f)]	19±7	61[b)]	^{14}N(Al,Al)		
	29±7	62[b)]	^{14}N,^{12}C(Al,Al)		
weighted average	38.6±0.9				

The 2.94 MeV level in ^{30}P:

DSA	*90±11	5	^{29}Si(p,γ)	0.13,0.20	Ta+implanted ^{29}Si
	35±5	63		0.13,0.20	SiO$_2$
	70±10	64		0.13,0.20	SiO$_2$
	115±20	3[c)]	^{29}Si(^{3}He,p)	-	-
	95±30	65	^{27}Al(α,n)	0.83	Al
DSA HI[d)]	90±12	66	^{3}He(^{28}Si,p)	3-4	Al,Zr,Nb and Au+ implanted ^{3}He
adopted value	90±8				

[a)] For the inclusion of the experimental stopping parameters in the reanalysis of the original F(τ) value, see Ref. 24. [b)] Computed from B(E2) values with the aid of branching and/or mixing ratios from Ref. 3. [c)] Private communication given in Ref. 3. [d)] DSA measurement with a heavy-ion-induced reaction. [e)] Recoil distance method.
[f)] Coulomb excitation measurement.

ISOSCALAR E2 TRANSITION STRENGTHS

The E2 transition strengths in the nuclear mass triplets with A=4n+2 have previously been used to test the isospin dependence of electromagnetic transitions[67]. The decomposition of the E2 matrix element into isoscalar (M_0) and isovector (M_1) or, equivalently, proton and neutron matrix elements, i.e.[67,68]

$$M_{n/p}(T_z) = \frac{1}{2}[M_0(T_z) + T_z M_1(T_z)] \qquad (2)$$

has recently [68,91] been shown to give a possibility for the determination of the neutron and proton polarization charges in the sd shell. As the E2 transition strength is given by the quadrature of eq. (2), a unique solution for the matrix elements is only found if the transition strength in the self-conjugate nucleus is known. Summary of the isoscalar E2 transition strengths obtained in our laboratory for sd shell nuclei is given in Table III. The strengths

Table III. Isoscalar E2 transition strengths of the A=22,26,30 and 34 nuclei.

Nucleus	E_i (MeV)	J_i^π	E_f (MeV)	J_f^π	τ(fs)	Branching (%)	$\|M(E2)\|^2_{IS}$ (W.u.)
^{22}Na	1.95	2_1^+	0.66	0^+	11±4	0.30±0.07 [a)]	16±7 [b)]
^{26}Al	2.070	2_1^+	0.23	0^+	22±3	3.6 ±0.7	13.8±3.3 [c)]
	3.16	2_2^+	0.23	0^+	9±2	0.44±0.17	0.40±0.18 [c)]
^{30}P	2.94	2_1^+	0.68	0^+	90±11	33.2±1.0	9.2±1.2 [d)]
	4.18	2_2^+	0.68	0^+	3.1±0.9	1.3±0.3	1.2±0.4 [d)]
^{34}Cl	2.16	2_1^+	0	0^+	47±4	16.3±0.5	9.2±0.8 [e)]
	3.38	2_2^+	0	0^+	7.5±1.7	2.1±0.7	0.8±0.3 [e)]

a) Ref. 71. b) Ref. 69. c) Ref. 7. d) Ref. 5. e) Ref. 9.

obtained from inelastic hadron scattering data (9.2 W.u.[82] and 14.3 ± 1.9 W.u.[83] for the $2_1^+ \to 0^+$ transition in ^{26}Al, 1.3 W.u.[82] and 2.57 ± 0.35 W.u.[83] for the $2_2^+ \to 0^+$ transition in ^{26}Al, 10.5 ± 0.8 W.u.[74] and 0.44 ± 0.10 W.u.[74] for the $2_1^+ \to 0^+$ and $2_2^+ \to 0^+$ transitions, respectively, in ^{34}Cl) are in agreement with the electromagnetic data. In all cases, the comparison with the E2 strengths of the analogue transitions in the mirror nuclei (Table IV and Fig. 8) show that $|M_1| \ll |M_0|$ and that the relative sign of the neutron and proton matrix elements is positive. This is in agreement with very recent inelastic hadron scattering data obtained for both the first 2^+ states in ^{26}Mg, ^{30}Si and ^{34}S [74] and for the 2_2^+ states in ^{26}Mg and ^{30}Si [75]. The negative sign has, however, also been obtained[75] for the 2_2^+ state in ^{34}S.

Table IV. $\Delta T=0$, E2 transition strengths for A=22, 26, 30 and 34 nuclei and their comparison with shell-model calculations.

Nucleus	T_z	Transition	$|M(E2)|^2$ (W.u.)					
			Experiment				Theory	
^{22}Ne	+1	$2_1^+ \to 0_1^+$	12.2±0.5 a)	13.1 i)			12.1 o)	
^{22}Na	0		16 ±7 b)	15.3 i)				
^{22}Mg	-1		24 ±10 a)				17.1 o)	
^{26}Mg	+1	$2_1^+ \to 0_1^+$	13.9±0.6 c)	12.3 j)	5.4 l)		12.5 o)	
^{26}Al	0		13.8±3.3 d)	9.9 j)	4.2 l)		10.1 o)	
^{26}Si	-1		15.4±1.5 e)	7.9 j)	3.9 l)		8.0 o)	
^{26}Mg	+1	$2_2^+ \to 0_1^+$	0.40±0.05 c)	0.70 j)	0.10 l)		0.72 o)	
^{26}Al	0		0.40±0.18 d)	2.2 j)	0.53 l)			
^{26}Si	-1		1.58±0.37 e)	4.2 j)	1.4 l)		4.3 o)	
^{30}Si	+1	$2_1^+ \to 0_1^+$	7.3 ±0.4 f)	4.5 k)			9.7 o)	
^{30}P	0		9.2 ±1.2 g)	5.4 k)	3.1 m)		10.1 o)	
^{30}S	-1		10.9±0.9 f)	6.4 k)			10.4 o)	
^{30}Si	+1	$2_2^+ \to 0_1^+$	1.42±0.13 f)	1.8 k)	2.5 m)		0.75 o)	
^{30}P	0		1.2 ±0.4 g)	0.8 k)			0.11 o)	
^{30}S	-1		0.38±0.09 f)	0.2 k)			-0.04 o)p)	
^{34}S	+1	$2_1^+ \to 0_1^+$	6.0 ±0.2 a)	8.4 k)	5.3 m)	5.0 n)	7.2 o)	
^{34}Cl	0		9.2 ±0.8 h)	8.3 k)	6.1 m)	5.7 n)	7.3 o)	
^{34}Ar	-1		14 ±4 a)	8.3 k)		6.4 n)	7.5 o)	
^{34}S	+1	$2_2^+ \to 0_1^+$	0.73±0.05 a)	0.1 k)		0.69 n)	0.31 o)	
^{34}Cl	0		0.8 ±0.3 h)			0.27 n)		
^{34}Ar	-1		0.23±0.09 a)	-0.3 k)p)		0.25 n)	-0.42 o)p)	

a) Ref. 3. b) Ref. 69. c) Refs. 71,72. d) Ref. 7. e) Ref. 73.
f) Ref. 66. g) Ref. 5. h) Ref. 9. i) Ref. 80. j) Ref. 78. k) Ref. 76.
l) Ref. 79. m) Ref. 77, FPSDI calculations. n) Ref. 81. o) Ref. 91.
p) The negative sign indicates the negative relative sign of the proton and neutron matrix elements.

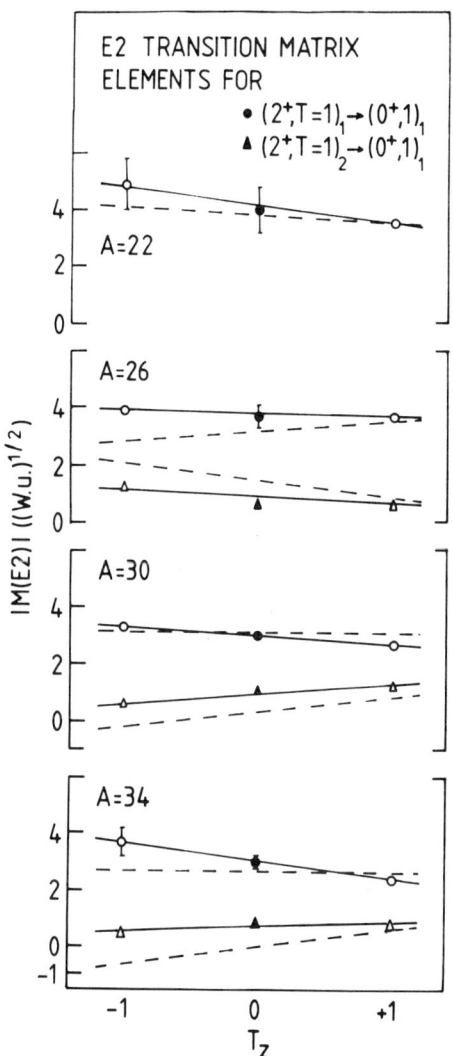

The comparison of the experimental and theoretical E2 transition strengths (Table IV) shows that the current shell-model calculations have tendency to underestimate the isoscalar effective charge of the $2_1^+ \rightarrow 0^+$ transition but no systematic deviation exists for the $2_2^+ \rightarrow 0^+$ transition. Matrix elements from recent calculations[91] where empirical isoscalar effective charge of $e_p + e_n = 1.70 e$ and isovector charge of $e_p - e_n = 1.0 e$ were used are in Fig. 8 compared with the E2 matrix elements obtained from the electromagnetic transition strengths. It can be seen that the isoscalar transition strengths are well reproduced but in the cases where the isovector strength is included deviations between the experimental and theoretical values seem to be more state dependent. A recent observation[84] indicates that the isoscalar effective charge of $1.65 e$ and the isovector effective charge of $0.63 e$ would better reproduce the data and indicates thus enhanced neutron polarization charge ($\varepsilon_n/\varepsilon_p \sim 3.5$ instead of 1.0 used in Ref. 91).

Fig. 8. The square roots of the E2 transition strengths in the A=22, 26, 30 and 34 nuclei plotted as a function of T_z. The solid lines illustrate the fits to the experimental data and the dashed lines the values obtained from a recent shell-model calculation[91].

SUMMARY

The experimental techniques presented allow measurements of precise short lifetimes by the DSA method through capture reactions. Such data should provide a testing ground for any new shell-model wave functions and also for effective charges and g-factors which may differ markedly from state to state and cannot be sufficiently taken into account by using average values.

REFERENCES

1. P.J. Nolan and J.S. Sharpey-Schafer, Rep. Prog. Phys. $\underline{42}$, 1 (1979).
2. T.K. Alexander and J.S. Forster, in Advances in Nuclear Physics, Vol 10, edited by M. Baranger, E. Vogt (Plenum Press, N.Y.-London, 1980), p. 197.
3. P.M. Endt and C. van der Leun, Nucl. Phys. $\underline{A310}$, 1 (1978).
4. F. Ajzenberg-Selove, Nucl. Phys. $\underline{A413}$, 1 (1984).
   F. Ajzenberg-Selove and C.L. Busch, Nucl. Phys. $\underline{A336}$, 1 (1980).
   F. Ajzenberg-Selove, Nucl. Phys. $\underline{A360}$, 1 (1981).
   F. Ajzenberg-Selove, Nucl. Phys. $\underline{A375}$, 1 (1982).
   F. Ajzenberg-Selove, Nucl. Phys. $\underline{A392}$, 1 (1983).
5. A. Anttila and J. Keinonen, Phys. Rev. $\underline{C21}$, 1196 (1980).
6. A. Anttila et al., Nucl. Phys. $\underline{A385}$, 194 (1982).
7. J. Keinonen at al., Nucl. Phys. $\underline{A385}$, 461 (1982).
8. J. Keinonen et al., Nucl. Phys. $\underline{A403}$, 45 (1983).
9. J. Keinonen et al., Nucl. Phys. $\underline{A412}$, 101 (1984).
10. R. Lappalainen et al., Nucl. Phys. $\underline{A}$, in print.
11. W.M. Currie et al., Nucl. Phys. $\underline{A135}$, 325 (1969).
12. C. Broude et al., Phys. Lett. $\underline{39B}$, 185 (1972).
13. M. Toulemonde and F. Haas, Phys. Rev. $\underline{C15}$, 49 (1977).
14. J. Völkl and G. Alefeld, in Diffusion in Solids, edited by A.S. Norwick and J.J. Burton (Academic, N.Y., 1975), p. 231.
15. A.Z. Kiss et al., Nucl. Instrum. $\underline{203}$, 107 (1982).
16. J. Keinonen et al., Nucl. Instrum. $\underline{216}$, 249 (1983).
17. M. Hautala, Phys. Rev. $\underline{C}$, in print.
18. J. Lindhard, M. Scharff and N.E. Schiøtt, Kgl. Dan. Vid. Selsk. Mat. Fys. Medd. $\underline{33}$, no. 14 (1963).
19. M. Bister et al., Radiat. Eff. $\underline{42}$, 201 (1979).
20. M. Hautala and M. Bister, Radiat. Eff. $\underline{54}$, 191 (1981).
21. B.M. Latta, Can. J. Phys. $\underline{58}$, 1783 (1980).
22. M. Bister et al., Phys. Rev. $\underline{C16}$, 1303 (1977).
23. A. Anttila et al., Nucl. Phys. $\underline{A334}$, 205 (1980).
24. A. Anttila et al., Z. Physik $\underline{A274}$, 227 (1975).
25. M. Bister et al., Phys. Lett. $\underline{53A}$, 471 (1975).
26. J. Keinonen et al., Nucl. Phys. $\underline{A412}$, 101 (1984).
27. A. Anttila et al., Radiat. Eff. $\underline{33}$, 13 (1977).
28. J. Keinonen et al., to be published.
29. P. Paul et al., Phys. Rev. $\underline{173}$, 1063 (1968).
30. K.W. Jones et al., Phys. Rev. $\underline{178}$, 1773 (1969).
31. A.R. Poletti et al., Phys. Rev. $\underline{182}$, 1054 (1969).
32. K. Bharuth-Ram et al., Nucl. Phys. $\underline{A269}$, 327 (1976).
33. M.A. Meyer and J.J.A. Smit, Nucl. Phys. $\underline{A205}$, 177 (1973).
34. Z.B. DuToit et al., Z. Physik $\underline{246}$, 170 (1971).
35. M. Bister et al., Ann. Acad. Scient. Fenn. Serie A, VI Physica $\underline{349}$, 1 (1970).
36. J.L. Durell et al., Phys. Lett. $\underline{29B}$, 100 (1969).
37. N.J. Meier et al., Nucl. Phys. $\underline{A141}$, 99 (1970).
38. A.R. Poletti et al., Phys. Rev. $\underline{184}$, 1130 (1969).
39. G.G. Frank et al., Can. J. Phys. $\underline{51}$, 1155 (1973).
40. V.K. Rasmussen, Nucl. Phys. $\underline{A169}$, 166 (1971).
41. J. Keinonen et al., Phys. Rev. $\underline{C22}$, 351 (1980).

42. A.E. Blaugrund et al., Phys. Rev. 158, 893 (1967).
43. J. Keinonen et al., Phys. Rev. C23, 2073 (1981).
44. E.O. De Neijs et al., Nucl. Phys. A230, 490 (1974).
45. O. Häusser and N. Anyas-Weiss, Can. J. Phys. 46, 2809 (1968).
46. C.R. Gould et al., Phys. Rev. C7, 1068 (1973).
47. O. Häusser et al., Can. J. Phys. 46, 1035 (1968).
48. J.L. Durell et al., J. of Phys. A5, 302 (1972).
49. Z. Berant et al., Nucl. Phys. A218, 324 (1974).
50. D. Branford and I.F. Wright, Nucl. Instrum. 106, 437 (1973).
51. P.J.M. Smulders et al., Can. J. Phys. 46, 261 (1968).
52. D. Schwalm et al., Nucl. Phys. A293, 425 (1977).
53. F.R. Metzger, Nucl. Phys. A182, 213 (1972).
54. V.J. Vanhuyse and G.J. Vanpraet, Nucl. Phys. 45, 602 (1963).
55. E.C. Booth et al., Nucl. Phys. 57, 403 (1964).
56. L.A. Schaller and W.C. Miller, Bull. Amer. Phys. Soc. 9, 666 (1964).
57. N.A. Khan and V.K. Rasmussen, Phys. Rev. 138, B1385 (1965).
58. S.W. Robinson et al., Phys. Rev. 174, 1320 (1968).
59. J.H. Haugh et al., Nucl. Phys. A132, 110 (1969).
60. R.M. Lombard and G.R. Bishop, Nucl. Phys. A101, 601 (1967).
61. D.G. Alkhazov et al., Izv. Akad. Nauk. (ser. fiz.) 27, 211 (1963).
62. O.F. Afanin et al., Sov. J. Nucl. Phys. 6, 160 (1968).
63. G.I. Harris et al., Phys. Rev. 187, 1413 (1969).
64. M. Bini, et al., Nuovo Cimento 4A, 45 (1971).
65. J.F. Scharpey-Schafer, Nucl. Phys. A167, 602 (1971).
66. T.K. Alexander et al., Phys. Rev. Lett. 49, 438 (1982).
67. E.K. Warburton and J. Weneser, in Isospin in Nuclear Physics, edited by D.H. Wilkinson (North-Holland,Amsterdam, 1969), p. 173.
68. A.M. Bernstein et al., Phys. Rev. Lett. 42, 425 (1979).
69. M. Bister et al., Nucl. Phys. A306, 189 (1978).
70. J. Görres et al., Nucl. Phys. A385, 57 (1982).
71. K. Dybdal et al., Nucl. Phys. A359, 431 (1981).
72. G.C. Ball et al., Nucl. Phys. A377, 268 (1982).
73. T.K. Alexander et al. Phys. Lett. 113B, 132 (1982).
74. A. Saha et al., Phys. Rev. Lett. 52, 1876 (1984).
75. A.M. Bernstein et al., private communication.
76. P.W.M. Glaudemans et al., Ann. Phys. (N.Y.) 63, 134 (1971).
77. B.H. Wildenthal et al., Phys. Rev. C4, 1708 (1971).
78. B.A. Brown and B.H. Wildenthal, Phys. Rev. C21, 2107 (1980).
79. G.A. Timmer et al., Z. Physik A288, 83 (1978).
80. B.M. Preedom and B.H. Wildenthal, Phys. Rev. C6, 1633 (1978).
81. B.H. Wildenthal, Bull. Amer. Phys. Soc. 27, 725 (1982).
82. K. van der Borg et al., Nucl. Phys. A365, 243 (1981).
83. A.M. Berstein, in Advances in Nuclear Physics, edited by M. Baranger and E. Vogt, vol. 3 (Plenum, N.Y., 1969).
84. T.K. Alexander et al., private communication.
85. H.H. Andersen and J.F. Ziegler, The Stopping Powers and Ranges of Ions in Matter, vol. 3 (Pergamon, N.Y., 1977).
86. J.S. Forster et al., Nucl. Phys. A313, 397 (1979).
87. E.K. Warburton et al., Phys. Rev. C20, 628 (1979).
88. L.P. Ekström et al., Nucl. Phys. A295, 525 (1978).
89. O. Almén and G. Bruce, Nucl. Instrum. 11, 257 (1961).
    O. Almén and G. Bruce, Nucl. Instrum. 11, 279 (1961).
90. A.L. Bragoo, At. En. Rev., Special Issue 3, 131 (1972).
91. B.A. Brown et al., Phys. Rev. C26, 2247 (1984).

MEDIUM ENERGY PROTON AND HELIUM-3 CAPTURE
IN LIGHT NUCLEI

S.L. Blatt
Department of Physics, The Ohio State University
Columbus, Ohio 43210

ABSTRACT

Since our initial report of unexpectedly strong capture of 40 to 80 MeV protons to highly excited states in ^{12}C, ^{13}N, and ^{28}Si, the Ohio State - Indiana University Cyclotron collaboration has continued to investigate the characteristics of medium energy radiative capture reactions. Among these are the existence of $2\hbar\omega$ giant resonances built on $1\hbar\omega$ excited states, as well as $3\hbar\omega$ resonances on $2\hbar\omega$ states; and the striking similarities of angular distributions and analyzing powers for similar single-particle transitions in closed-subshell and neighboring closed-subshell-plus-one-proton nuclei. We have also been impressed with the generally good description of the γ_0 and γ_1 observations provided by a direct-semi-direct reaction picture toward the higher energies investigated. Recently, however, we have observed an anomalously strong capture in ^{7}Be to a state which appears to require a multi-step process to populate. Radiative capture of ^{3}He particles has also been studied by our group over the past few months. Early results for captures into ^{12}C and ^{16}O show significantly different populations of final states than in the same nuclei excited by proton capture. Another reaction, ^{12}C(^{3}He,γ)^{15}O, appears to populate the same set of states observed in ^{12}C(^{6}Li,t)^{15}O. Selectivity for 3-particle configurations is suggested in these cases.

INTRODUCTION

In our earliest report of radiative capture observations above the giant dipole energy region,[1] we pointed out what, at that time, were unexpectedly large transitions to high-lying excited states in the nuclei studied. In particular, the 19 MeV region in ^{12}C, including 4$^-$ "stretched" 1p-1h states, was an outstanding feature of ^{11}B(p,γ)^{12}C spectra recorded above E_p=40 MeV; similarly, the 1p-1h cluster of states in ^{28}Si, including the 6$^-$ stretched configurations, was found to be a major feature of the ^{27}Al(p,γ)^{28}Si spectra. Furthermore, as pointed out by Arnold,[2] the main features of the ^{11}B(p,γ)^{12}C spectra were also present in those we measured on $^{12}C(p,\gamma)^{13}N$; in effect, there were similarly strong captures at the same E_γ in these neighboring nuclei. An understanding of the dominant states as consisting primarily of good 1p-1h (in the closed-subshell cases) or single-particle (in the ^{13}N case) states came from consideration of a direct-capture reaction mechanism.[2,3] Subsequent higher-resolution spectra[4,5] confirmed the presence of the many states postulated, and studies in additional nuclei pairs[4,6] ($^{16}O-^{17}N$, $^{28}Si-^{29}P$) showed that the similarity of (p,γ) spectra for these neighboring nuclei seems to be a general characteristic, due to the similar energies of their single particle configurations.

An early suggestion[1] that the captures to states with one proton at an excitation of $1\hbar\omega$ might go through a giant resonance at $2\hbar\omega$ was followed up with measurements on $^{11}B(p,\gamma)$ between 24 and 40 MeV,[4,7,8] and such a resonance, built on the 19 MeV cluster in ^{12}C, was found. Subsequently, the work of Anghinolfi et al.[9] in ^{12}C and of Dowell et al.[10] in ^{28}Si showed that there are in fact giant resonances built on <u>all</u> resolvable states from the ground state to the vicinity of the stretched configurations. More recently, we have established evidence for a resonance at $3\hbar\omega$, built upon $2\hbar\omega$ states in ^{12}C; a preliminary report was given a few years ago;[11] a more detailed analysis, based on more complete data, is being readied for publication.

The relationships between transitions in neighboring nuclear pairs, mentioned above, have been explored in greater detail over the past several years. Results for $^{12}C-^{13}N$ have recently been presented[12], along with a generalized description of the radiative capture mechanism which includes the usual direct and semi-direct mechanisms as primary contributors, but also allows more complex mechanisms while still preserving interesting angular and energy-dependent characteristics. The transitions which were compared were: a) $^{11}B(p,\gamma)^{12}C^*$(19 MeV) and $^{12}C(p,\gamma)$ $^{13}N^*$(3.6 MeV), both of which have the final proton configuration primarily $d_{5/2}$ (with a $p_{3/2}$ hole also coupled, in the ^{12}C final state); and b) $^{11}B(p,\gamma)^{12}C^*$ (4.4 MeV) and $^{12}C(p,\gamma)^{13}N$(g.s.), where the proton is captured into a $p_{1/2}$ orbital. In both cases, the angular distributions of cross section and analyzing power for the <u>corresponding</u> transitions were found to be very similar. By way of contrast, the analyzing powers for the transitions of case a) had both a smaller magnitude and opposite sign from those of case b), indicating the sensitivity of these measurements to <u>different</u> configurations.

In the following sections, more recent measurements on other neighboring pairs will be described. We also report on the good success we have found in describing many of the higher-energy (p,γ) data with a simple direct-semidirect approach, as well as one case in which this approach fails completely. Finally, we review briefly some of our newest measurements, involving radiative capture of ^{3}He-particles in some of the same nuclei we have been investigating with the (p,γ) reaction.

RECENT PROTON-CAPTURE RESULTS

A. Closed-Subshell/Closed-Subshell-Plus-One-Proton Comparisons

In our study[12] of the ^{12}C-^{13}N pair noted above, we were struck by the experimental observation of similar angular distributions and analyzing powers for captures which leave the proton in the same orbital, regardless of whether it was coupled to a residual hole or simply outside a closed core. Such an effect comes naturally out of the standard direct capture picture, where the target nucleus acts as a "spectator" in the reaction; however, in the case we were studying, some of the captures went through resonances, definitely not a simple direct process. We therefore worked, together with our theory colleagues L.G. Arnold and R.G. Seyler, to see if a less restrictive picture of the capture mechanism could be constructed, which, in some limit, would explain our observations.

The more generalized capture description assumes: a) that the initial and final states can be constructed from orthogonal one-nucleon configurations; and b) that the electromagnetic operator can be represented as a sum of one-nucleon operators. Using projection operator notation, where P projects from the nuclear wave function the part which representes one nucleon coupled to the target ground state and Q is its complement, we can write the transition amplitude for radiative capture as

$$M = \langle \phi_f | H^\lambda | \phi_i \rangle = \langle P\phi_f | H^\lambda | P\phi_i \rangle + \langle P\phi_f | H^\lambda | Q\phi_i \rangle + \langle Q\phi_f | H^\lambda | P\phi_i \rangle + \langle Q\phi_f | H^\lambda | Q\phi_i \rangle. \quad (1)$$

The first term, where both the initial and final states look like the target ground state plus a nucleon, is the usual direct capture amplitude. The next two terms are, respectively, initial-state and final-state semidirect amplitudes. If $Q\phi_j$ were approximated by a single nucleon coupled to a coherent 1p-1h excitation of the target ground state, the second term would be the conventional semidirect capture amplitude. The terms with final states $Q\phi_f$ have generally not been taken into account. However, if those components of $Q\phi_f$ which are 1p-1h excitations of the target ground state are considered, the third term does not have to be zero; indeed, this represents the very real (although generally small) probability that the radiative transition is by a nucleon in the core (i.e., target nucleus), while the incoming proton plays the role of a spectator.

For captures dominated by the first three terms of eq. 1, we have a generalized direct-semidirect picture which predicts the following three relationships for closed-subshell and adjacent 1-particle nuclei: a) The cross section angular distributions have the same shape; b) their analyzing powers (at any given energy and angle) are equal; and c) their cross sections have magnitudes related by

$$\frac{\sigma(J_i,j_f,J_f)}{\sigma(0,j_f,j_f)} = \frac{(2J_f+1)}{(2J_i+1)(2j_f+1)} \frac{C^2S(J_i,j_f,J_f,T_f)}{C^2S(0,j_f,j_f,t_f)}. \quad (2)$$

In the latter expression J_i, J_f refer to the initial and final spins of the states in the closed-subshell case, and j_f is the total angular momentum of the nucleon in its final configuration in either case. (It should be noted that the above results are only valid if a single j_f dominates.) The examples studied earlier[12] showed all three of these characteristics.

Recently, Rackers[13] has investigated corresponding transitions in ^{27}Al(p,γ)^{28}Si and ^{28}Si(p,γ)^{29}P, and, to a more limited extent, in ^{15}N(p,γ)^{16}O and ^{16}O(p,γ)^{17}F. In the former pair, the following transitions were compared: I) ^{28}Si first excited state (2^+), 88% [$2s_{1/2}$, $1d_{5/2}^{-1}$] configuration,[14] and ^{29}P ground state (1/2$^+$), [$2s_{1/2}$]; II) ^{28}Si second excited state (4^+), [$1d_{3/2}$, $1d_{5/2}^{-1}$] and ^{29}P first excited state (3/2$^+$), [$1d_{3/2}$]; and III) the cluster of negative-parity states from 11.6 to 14.4 MeV in ^{28}Si, including the 6^- states, primarily [$1f_{7/2}$, $1d_{5/2}^{-1}$], and the fifth excited states of ^{29}P, a 7/2$^-$ state with configuration [$1f_{7/2}$].

For the ^{16}O - ^{17}F pair, only one comparison could readily be made, namely the transitions to the 3^- second excited state of ^{16}O, [$1d_{5/2}$, $1p_{1/2}^{-1}$] and to the 5/2$^-$ ground state of ^{17}F, which has the configuration [$1d_{5/2}$].

Fig. 1 shows the first-mentioned transitions in the ^{28}Si-^{29}P pair. The angular distributions have been normalized for direct comparison of their shapes; the

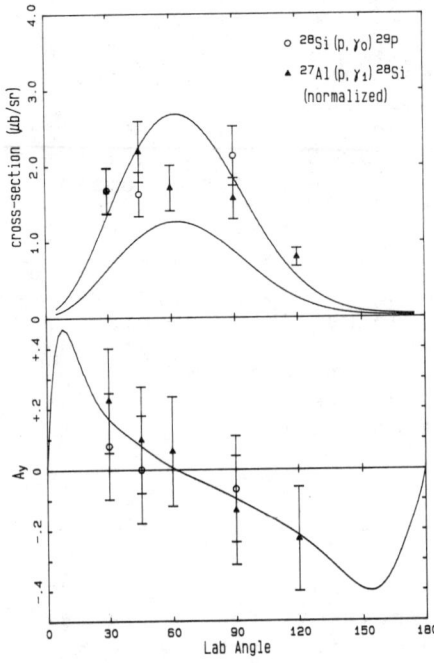

Fig. 1. Comparison of ^{28}Si(p,γ_0) and ^{27}Al(p,γ_1). Both capture to $s_{1/2}$ orbits.

analyzing powers are shown directly as derived from the data. Also shown is a theoretical prediction generated utilizing the program HIKARI,[14] which includes the conventional direct-semidirect terms (I and II of eq. 1). (Since the third term produces identical predictions on the angular dependences, no additions are needed for the comparison being made.) Similarly good comparisons, between data on corresponding transitions and between experiment and theory, are found for the other cases, with the exception of the negative-parity cluster in ^{28}Si. In this latter case, too many components are present to be experimentally resolvable, and the total contribution from the entire cluster averages the analyzing power to essentially zero.

B. <u>Energy Dependence of (p,γ_0) Reactions above the GDR</u>

During the course of our experimentation over the past several years, we have acquired a great deal of data on ground-state captures in the 26-100 MeV range. Proton captures leading to the following nuclei are included: ^{7}Be, ^{12}C, ^{13}N, ^{16}O, ^{17}F, ^{28}Si, and ^{29}P. Recently, S. Jensen[16] and H.J. Hausman have collated and analyzed these data, and made comparisons with the standard DSD capture model.[15] The approach taken was to normalize the calculated direct-capture component to the highest-energy data points for a given reaction, thereby providing a rough determination of the ground state spectroscopic factor; next, reasonable GDR parameters were used to add the semi-direct contribution, with the amplitude relative to the direct component a free parameter, varied to obtain a best fit to the lower-energy data. Wherever possible, comparisons were then made to known GDR cross sections in the region below our own data; these comparisons, in general, have been found to be quite good. One

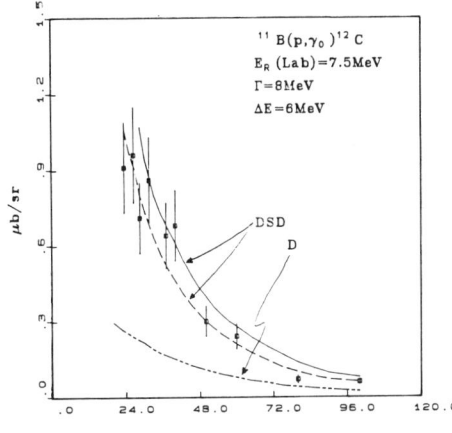

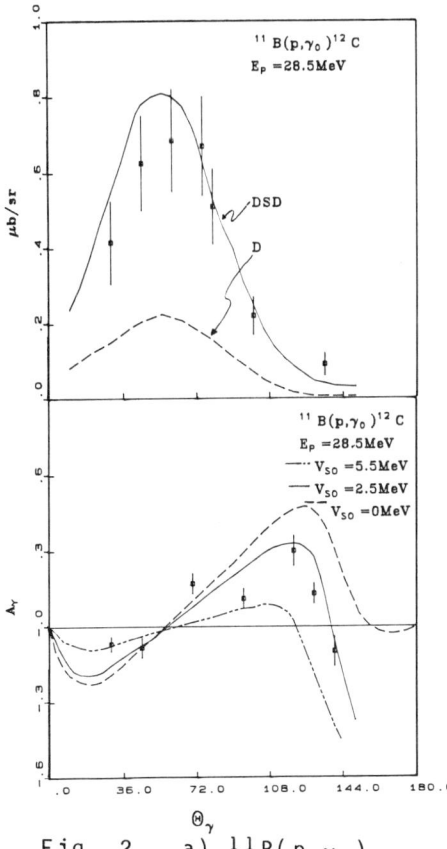

Fig. 2. a) $^{11}B(p,\gamma_0)$ cross section vs. E_p. b) Angular distribution at 28.5 MeV. c) Analyzing power measurements.

example is shown in Fig. 2. The energy dependence for $^{11}B(p,\gamma_0)^{12}C$ is indicated, along with the direct calculation and two possible DSD fits, in Fig. 2a. The data and predicted angular distributions at 28.5 MeV are shown on Fig. 2b. Fig. 2c shows analyzing power measurements at this same energy; while the other fits were all made with a global set of optical-model parameters, variations in the spin-orbit potential make large differences in the predictions for this quantity, and the various curves in the figure point out this sensitivity.

The spectroscopic factors obtained as described above are displayed in Table 1. Rough as our numbers are, they are not very different from values either calculated[17] or derived from (^{3}He,d) or (d,n) measurements.[18]

We note, however, that our numbers from the (p,γ) measurements tend to lie on the low side of the previously measured or theoretical values.

C. "Non-Direct" Captures

In the previous two sections, the fourth term in eq. 1 has been ignored. This term, which may be designated as a "non-direct" amplitude, represents a more complex set of transitions than the other three terms. As shown in ref. 12, captures to final states with <u>any</u> given target-state parentage can occur, so long as components of the initial state with identical parentage are involved; these can be excited through inelastic scattering, for example.

Table 1. Spectroscopic factors from direct-semidirect calculations on proton capture data.

Reactions	$S(p,\gamma)$	$S(Theo.)^a$	$\dfrac{S(expt.)}{[(^3He,d),(d,n)]}^b$
$^{11}B(p,\gamma_0)^{12}C$	4 - 5.7	5.69	3.3 - 8.0
$^{12}C(p,\gamma_0)^{13}N$	0.31 - 0.49	0.61	0.48 - 0.68 (0.7 - 1.25)
$^{15}N(p,\gamma_0)^{16}O$	1.0 - 1.32		1.3 - 3.7
$^{16}O(p,\gamma_0)^{17}F$	0.5 - 0.8		0.85 - 1.25
$^{27}Al(p,\gamma_0)^{28}Si$	1.0-2.0		3.0 - 5.3
$^{28}Si(p,\gamma_0)^{29}P$	0.52 - 0.81		1.0
$^6Li(p,\gamma_0)^7Be$	0.59	0.43	

[a] ref.17.
[b] ref.18.

    Inclusion of the non-direct amplitude may be crucial in some radiative capture reactions. We consider here captures to the $7/2^-$ second excited state of ^7Be, in the reaction $^6Li(p,\gamma)$. This state, at an excitation of 4.57 MeV, is not a single particle configuration outside of a ^6Li (g.s.) core. The 1p orbitals have too-small angular momenta, the s-d shell has incorrect parity, and the $1f_{7/2}$ shell is much too far away in energy; the $7/2^-$ analog in ^7Li has been shown to have zero spectroscopic strength for the $f_{7/2}$ single-particle configuration.[19]

    The more complex configuration of the $7/2^-$ state could not, therefore, be populated in a simple direct or semi-direct capture, since only a single nucleon transition occurs in these mechanisms. However, a multistep process, allowed by the non-direct amplitude, is possible. For example, the ^6Li could be excited to its 3^+ first excited state in the incident channel; the proton could then be captured into a 1p orbital to produce the $7/2^-$ state of ^7Be.

    A gamma ray spectrum of $^6Li(p,\gamma)^7Be$, taken[20] at 44.4 MeV, is shown in Fig. 3. Transitions to the ground- and

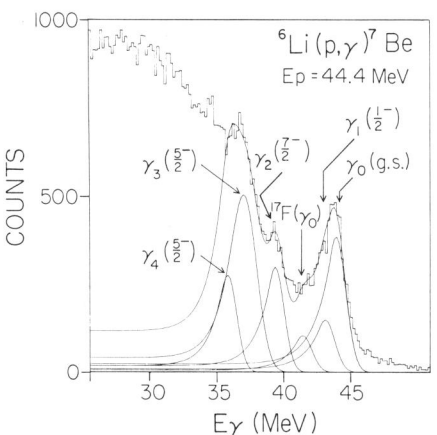

Fig. 3. ^{6}Li(p,γ)^{8}Be spectrum, showing strong transition to 7/2$^-$ state.

first four excited states are present. A small component from oxygen contamination in the target also appears; spectra taken with a natural oxygen gas cell verify that no additional background from this source can be found in the vicinity of the 7/2$^-$ peak. It is striking to note, then, that the 7/2$^-$ state is produced with an intensity comparable to those of the other low-lying ^{7}Be states, even though these others can be produced via the supposedly much more probable direct and semidirect mechanisms.

The strength of this multi-step capture provides a warning for all of the cases which have been studied over the years and analyzed from the simple direct-semidirect point of view: more complex mechanisms many well be operating, at a non-negligible level, in those cases, as well. Conclusions drawn from the comparison of data with the DSD picture, including the extraction of spectroscopic factors reported in the previous section of the present paper, must be re-examined to determine if the presence of non-direct amplitudes produce significant corrections.

CAPTURE OF ^{3}He-PARTICLES

If there is a sizable probability for direct and/or semidirect capture of projectiles more complex than single nucleons, it should be possible to obtain information about the importance of, say, 2-particle configurations via (d,γ) reactions, 3-particle configurations via (^{3}He,γ) or (t,γ) reactions, etc. For heavy projectiles, statistical effects have been shown to dominate the observed capture spectra.[21] Snover[22] has shown that, for medium-weight nuclei and low energies, such effects are also important in ^{3}He capture. We have, thus far, limited our own (^{3}He,γ) studies to very light nuclei, and at somewhat higher energies; we have also concentrated on transitions to the ground and low-lying excited states, which are at values of E_γ far above the region of importance in the statistical process.

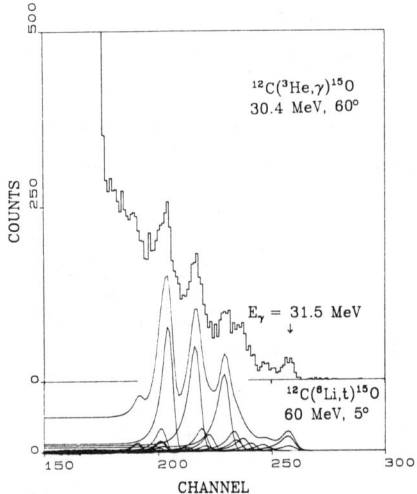

Fig. 4. $^{12}C(^3He,\gamma)$ spectrum compared with $^{12}C(^6Li,t)$ data folded with our γ-ray line shape.

A spectrum taken on $^{12}C(^3He,\gamma)^{15}O$, at 30.7 MeV, is shown in Fig. 4a. The components shown in Fig. 4b, however, are not fit to this spectrum; they are instead taken from the spectrum for $^{12}C(^6Li,t)^{15}O$ obtained at $E(^6Li) = 60$ MeV and $\theta_t = 5°$ by Bingham et al.[23] with our experimental gamma-ray line shape folded in. The similarity of the spectra is striking. Analysis of the 3-particle transfer reaction[23] indicates direct transfer is occurring, and states with significant 3-particle components are being selected. In our $(^3He,\gamma)$ spectra, the same states also dominate, and it is reasonable to assume direct (or semidirect) capture to these 3-particle states is the predominant mechanism.

Data on $^9Be(^3He,\gamma)^{12}C$ and $^{13}C(^3He,\gamma)^{16}O$ have recently been obtained. Marchlenski[24] has analyzed these measurements and compared the population of the final states seen in these reactions with those in $^{11}B(p,\gamma)^{12}C$ and $^{15}N(p,\gamma)^{16}O$, respectively. As expected, the populations are quite different. Ratios for producing states of ^{12}C via ^3He and proton capture are shown in Table 2. The states with small σ $(^3He,\gamma)$ are clearly not 3p-3h in nature.

Can we say that the stronger $(^3He,\gamma)$ transitions do pick out such configurations? The evidence in the $^{12}C(^3He,\gamma)$ case of Fig. 4 suggests that this is so. A purely statistical mechanism, for example, might be expected to populate both the 0^+ ground state and the 0^+ second excited states equally; instead, γ_2 is quite a bit stronger. (It is still weak in an absolute cross section sense; thus a large 3p-3h component in the 0^+ excited state is still not indicated.) The importance of 3p-3h configurations in the 19 MeV region, however, where the (p,γ) cross sections are also quite strong, is very interesting. Such configurations, in this region and higher, have been predicted[25] in attempts at theoretical understanding of the fine structure of the ground-state GDR.

Table 2. Capture cross section ratios for $^9\text{Be}(^3\text{He},\gamma)^{12}\text{C}$ and $^{11}\text{B}(p,\gamma)^{12}\text{C}$; $E_x \sim 45$ Mev, $\theta_\gamma = 60°$ (preliminary values).

E_f	$J\pi$	$\sigma(^3\text{He},\gamma)/\sigma(p,\gamma)$
0.0 MeV	0^+	0.02
4.44	2^+	0.04
7.66	0^+	0.84
9.64	3^-	0.24
~19 MeV	$4^-, 3^-$	0.32

Work is continuing in this area. Angular distributions have been measured, and a preliminary analysis indicates that they can be understood in the context of DSD calculations. Data have also been obtained on the energy dependence of the $(^3\text{He},\gamma)$ cross sections, and analysis of these measurements is in progress.

Based on results such as those reported here and at other laboratories throughout the world, radiative capture reactions appear to be as fascinating in the variety of phenomena they display, and as potentially valuable in determining nuclear structure information at intermediate energies as they have been at lower energies.

The author acknowledges the major contributions of many collaborators, either referred to explicitly in the above text or in the references to our published work. Support for this work from the U.S. National Science Foundation is also gratefully acknowledged.

REFERENCES

1. M.A. Kovash, S.L. Blatt, R.N. Boyd, T.R. Donoghue, H.J. Hausman, and A.D. Bacher, Phys. Rev. Lett. <u>42</u>, 700 (1979).
2. L.G. Arnold, Phys. Rev. Lett. <u>42</u>, 1253 (1979).
3. S.-F. Tsai and J.T. Londergan, Phys. Rev. Lett. <u>43</u>, 576 (1979).
4. S.L. Blatt, Los Alamos National Laboratory Report LA-8303-C (1980), p. 90.

5. S.L. Blatt, T.R. Donoghue, M.A. Kovash, R.N. Boyd, H.J. Hausman, A.D. Bacher, and C.C. Foster, Bull. Am. Phys. Soc. $\underline{25}$, 603 (1980).
6. S.L. Blatt, M.A. Kovash, T.R. Donoghue, R.N. Boyd, H.J. Hausman, and A.D. Bacher, Bull. Am. Phys,. Soc. $\underline{24}$, 843 (1979).
7. S.L. Blatt, M.A. Kovash, H.J. Hausman, T.R. Donoghue, R.N. Boyd, A.D. Bacher, and C.C. Foster, in Giant Multipole Resonances (Harwood, Chur, Switzerland, 1980) p.435.
8. H.R. Weller, H. Hasan, S. Manglos, G. Mitev, N.R. Roberson, S.L. Blatt, H.J. Hausman, R.G. Seyler, R.N. Boyd, T.R. Donoghue, M.A. Kovash, A.D. Bacher, and C.C. Foster, Phys. Rev. C$\underline{25}$, 2921 (1982).
9. M. Anghinolfi, P. Corvisiero, G. Ricco, M. Taiuti, and A. Zucchiatti, Nucl. Phys. A$\underline{399}$, 66 (1983).
10. D.H. Dowell, G. Feldman, K.A. Snover, A.M. Sandorfi, and M.T. Collins, Phys. Rev. Lett. $\underline{50}$, 1191 (1983).
11. S.L. Blatt, H.J. Hausman, T.R. Donoghue, R.N. Boyd, P. Koncz, M.A. Kovash, A.D. Bacher, and C.C. Foster, Bull. Am. Phys. Soc. $\underline{26}$, 1128 (1981).
12. S.L. Blatt, H.J. Hausman, L.G. Arnold, R.G. Seyler, R.N. Boyd, T.R. Donoghue, P. Koncz, M.A. Kovash, A.D. Bacher, and C.C. Foster, Phys. Rev C $\underline{30}$, 423 (1984).
13. T. Rackers, S.L. Blatt, H.J. Hausman, T.R. Donoghue, D. Marchlenski, M. Wiescher, M.A. Kovash, A.D. Bacher, and C.C. Foster, Bull. Am. Phys. Soc. $\underline{28}$, 650 (1983); T.W. Rackers, Ph.D. dissertation, The Ohio State University (1984), unpublished.
14. R.W. Barnard and G.D. Jones, Nucl. Phys. A$\underline{108}$, 641 (1968); ibid., p. 655.
15. Hideo Kitazawa, Duke University, (1980).
16. S.M. Jensen, M.S. Thesis, The Ohio State University (1984), unpublished.
17. S. Cohen and D. Kurath, Nucl. Phys. A$\underline{101}$, 1, (1967).
18. F. Ajzenberg-Selove, Energy Levels of Light Nuclei. Nucl. Phys. A248, 1 (1975) and references therein.
19. E.W. Hamburger and J.R. Cameron, Phys. Rev. $\underline{117}$, 781 (1960).
20. H.J. Hausman, S.L. Blatt, R.N. Boyd, T.R. Donoghue, L.G. Arnold, R.G. Seyler, M.A. Kovash, D.G. Marchlenski, T.W. Rackers, M. Wiescher, A.D. Bacher, and C.C. Foster, Int'l Conf. on Nuclear Physics, Florence, Italy (1983), and to be published.
21. J.O. Newton, B. Herskind, R.M. Diamond, E.L. Dines, J.E. Draper, K.H. Linderberger, C. Schuck, S. Shih, and F.S. Stephens, PHys. Rev. Lett. $\underline{46}$, 1383 (1981); E.F. Garman, K.A. Snover, S.H. Chew, S.K.B. Hesmondhalgh, W.N. Catford, and P.M. Walker, Phys. Rev. C $\underline{28}$, 2554 (1983).

22. K.A. Snover, private communications (1983, 1984); see also paper by this author in the present conference proceedings.
23. H.G. Bingham, M.L. Halbert, D.C. Hensley, E. Newman, K.W. Kemper, and L.A. Charlton, Phys. Rev. C $\underline{11}$, 1913 (1975).
24. D.G. Marchlenski, S.L. Blatt, H.J. Hausman, T.R. Donoghue, T.W. Rackers, M. Wiescher, M.A. Kovash, A.D. Bacher and C.C. Foster, Bull. Am. Phys. Soc. $\underline{28}$, 650 (1983), and to be published.
25. C.M. Shakin and W.L. Wang, Phys. Rev. C $\underline{5}$, 1898 (1972); see also H.D. Shay, R.E. Peschel, J.M. Long, and D.A. Bromley, Phys. Rev. C $\underline{9}$, 76 (1974).

MEDIUM ENERGY GAMMA RAYS FOLLOWING
RADIATIVE CAPTURE OF POLARIZED PROTONS ON LIGHT NUCLEI

H. Ejiri, T. Shibata, Y. Nagai, T. Kishimoto*,
H. Ohsumi, N. Kamikubota, and T. Satoh

Dept. Physics, Osaka Univ., Toyonaka, Osaka, 560, JAPAN
*Dept. Physics, Tokyo Inst. Tech., Tokyo, JAPAN

ABSTRACT

Medium energy gamma rays with E_γ=40-100 MeV following radiative capture of polarized protons on ^{11}B and ^{12}C were well studied by means of the HERMES. The angular distribution shows a uniform bell-shape pattern, being independent of the final state. On the other hand the analyzing power depends on the spin direction with respect to the orbital angular momentum of the captured orbit. These results are well reproduced in terms of the stretched E1-E2 radiations with the large E2 contributions for $E_\gamma \gtrsim 60$ MeV. A simple geometrical consideration is given to account for the spin dependence of the analyzing power. Experimental results are compared with the Siegert theorem calculation and the model calculation including both the impulse term and the pion exchange term.

1. INTRODUCTION

The present work is concerned with medium energy gamma rays in the energy range of 40-100 MeV. This energy range is beyond giant resonance and below the pion threshold energies. The wave length is an order of $\lambda \approx$ 2~4 fm. Such energetic gamma rays produced by radiative capture process of medium energy protons give rise to large momentum transfers of $q \approx 0.5$ GeV/c.

Radiative capture gamma rays in the giant resonance regions of 10~30 MeV have been extensively studied[1-5], and the radiative process has been analyzed in terms of direct and semi-direct processes[6]. Radiative capture processes in the medium energy region far above giant resonances, however, have not been well explored so far. Pioneer works of the radiative process of medium energy protons have recently been made by the IUCF, OHIO and TUNL groups[7].

This work reports first extensive measurements of medium energy gamma rays following radiative capture of polarized protons on light nuclei. Since the excitation energy is well above the discrete single particle levels and giant resonances the electromagnetic radiation may be rather independent of individual proton orbits associated with the radiative process. Then it may elucidate dynamic properties of the radiative process and the electromagnetic multipole strengths in the medium energy region. In this region with the large momentum transfer the giant resonance process and the direct (one body) process become less important and the exchange current (two-body process) may contribute significantly in addition to the one body process. Diagrams for the radiative processes are shown in Fig. 1. The exchange current contribution has been dis-

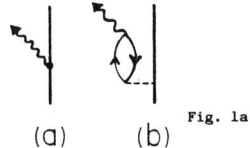
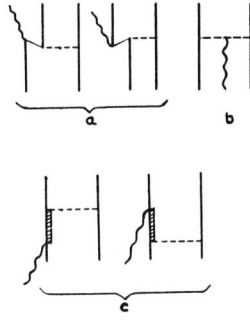

Fig. 1a
Schematic diagram of radiative capture processes. a) Simple direct process (one body process). b) Semi-direct process.

Fig. 1b
Pair current (a), pion current (b) and isobar contribution (c). See ref. 11.

cussed to understand the deviation κ of the photo absorption process from the TRK sum rule. The radiative capture process is an inverse process of the particular channel (γ,p_o) of the photo nuclear reaction, and it deals with the matrix element $<\psi_f|T_\gamma|\psi_i>$. Thus one can study the wave functions asociated with the radiative process, in particular the components responsible for such medium energy radiation, as well as the electromagnetic interactions far beyond giant resonances. It is also noted that radiative capture gamma rays to highlying states may reveal the properties of these states.

2. EXPERIMENTAL PROCEDURES

Medium energy gamma rays following radiative capture reactions of 40-80 MeV polarized protons on ^{11}B and ^{12}C were studied. The polarized protons were provided from the Osaka Univ. RCNP cyclotron. The gamma rays in the range of 35-100 MeV were measured by means of the HERMES (High Energy Radiation Measuring System)[8,9]. It consists of a 6" × 11" central NaI crystal and four-segment 11" × 11" annular NaI crystal. The energy and time signals from these crystals were recorded in a list mode on a magnetic tape. The energy spectrum is constructed by summing up all energy signals from the crystals. Unique point of the HERMES is excellent energy resolution (2.6 % for 60 MeV γ rays) with the low background counts. The radiative capture process in the medium energy region is a very rare process. The cross-section at backward angles goes down even below 10 nb/sr for 80 MeV protons. Thus one has to select such rare events among huge neutrons and charged particles with the cross-sections of an order of 1b. Selection of gamma rays were nicely made by measuring the time of flight (TOF) with the flight path of 400-1200 mm. The experimental arrangement is schematically illustrated in Fig. 2. An example of the TOF spectra is given in Fig. 3.

Angular distributions and analyzing powers for gamma rays were measured for the ^{11}B target at E_p=40, 50, 65 and 80 MeV and for the ^{12}C target at E_p=40 MeV. The targets used were the self supporting enriched ^{11}B with 31 mg/cm^2 and 62 mg/cm^2.

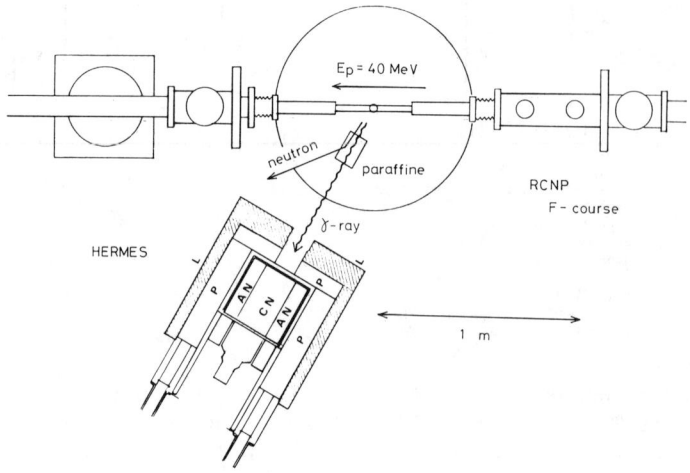

Fig. 2. Top view of the experimental arrangement.
CN: central NaI detector, AN: annular NaI detector,
P: plastic scintillator, L: lead shield.

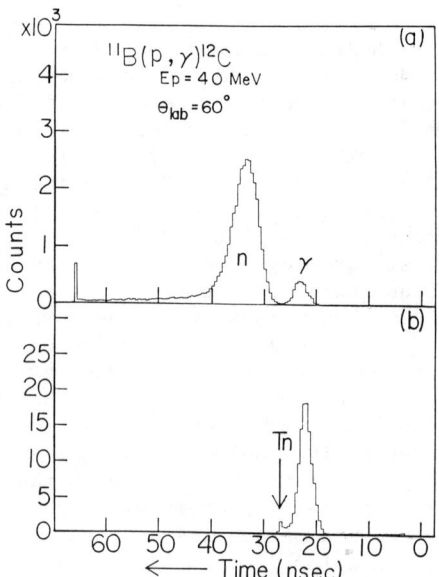

Fig. 3. Time spectra of neutrons and γ-rays following the $^{11}B(p,\gamma)^{12}C$ reaction.

3. RESULTS

Gamma ray spectra for the ^{11}B($\vec{p},\gamma$) and the ^{12}C($\vec{p},\gamma$) reactions at E_p=40 MeV are shown in Fig. 4. Medium energy gamma rays feeding

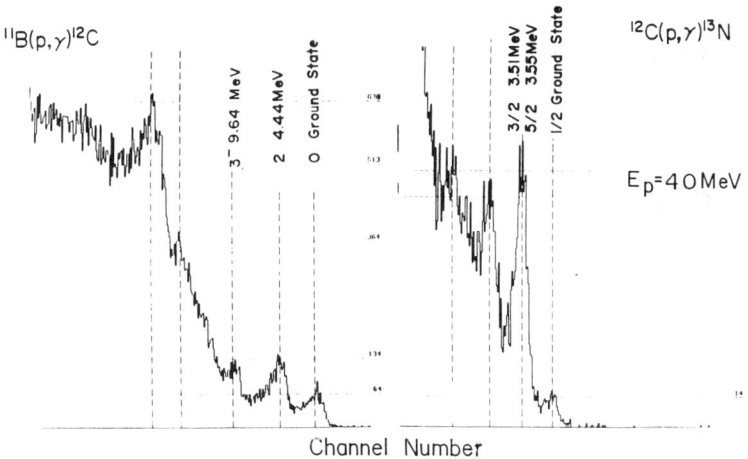

Fig. 4. Gamma ray spectra for the ^{11}B(p,γ)^{12}C reaction at θ_γ=95° and for the ^{12}C(p,γ) reaction at θ_γ=150°.

low-lying levels are clearly observed. The energy resolution including the target thickness is about 3 % for 40-90 MeV gamma rays. Interesting is to note in the spectrum of the ^{11}B(p,γ) the strong gamma feeds to higher excitation region of 19-23 MeV.

Angular distributions both for the ^{12}C($\vec{p},\gamma$) reaction at E_p=40 MeV and the ^{11}B($\vec{p},\gamma$) reactions at E_p=40, 50, 65 and 80 MeV are shown in Figs. 5, 6 and 7. The analyzing powers for these reactions are shown in Figs. 8, 9 and 10.

Interesting is to find the following uniform features in the angular distributions and the analyzing powers.
  i) The observed angular distributions show bell-shape distributions with peaks around θ_m=60~30 deg. The angular distributions of the (p,γ_i) reactions feeding various low-lying states ψ_i are nearly the same as shown in Figs. 5 and 6. The shape for the same proton energy does not depend on the final state ψ_i.
 ii) The angular distributions are all asymmetric with respect to θ_γ=90°, showing a strong forward peak. The peak of the distribution moves forward as the incident proton energy increases as shown in Fig. 7. These features clearly indicate strong admixture of positive parity transitions into the E1 transition.
iii) The analyzing powers are classified into two types. Those for the final states with $j_>=\ell+1/2$, namely the orbital angular momentum parallel to the spin, decrease as the angle increases, while those for the $j_<=\ell-1/2$ increase as the angle increases.

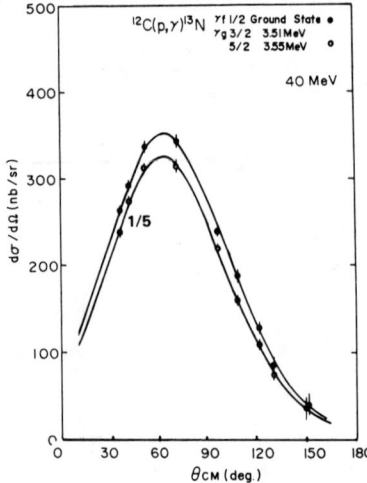

Fig. 5. Angular distributions of gamma rays following the $^{12}C(p,\gamma)^{13}N$ reaction at $E_p=40$ MeV.

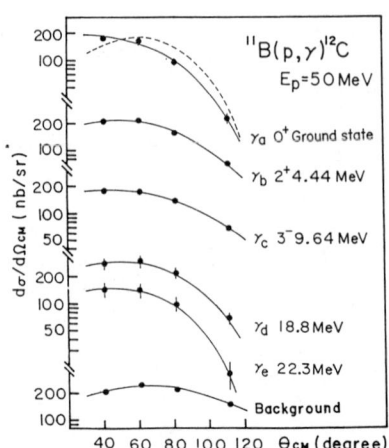

Fig. 6. Angular distributions of gamma rays following the $^{11}B(p,\gamma)^{13}N$ reaction at $E_p=50$ MeV.

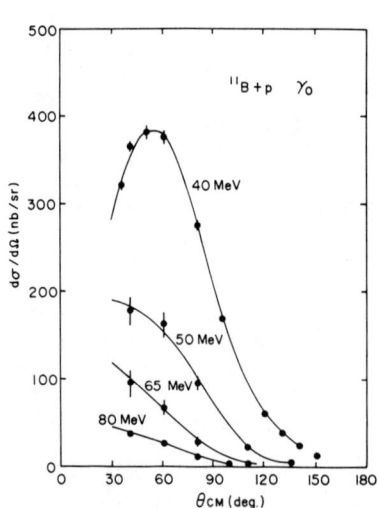

Fig. 7a. Angular distributions of gamma rays following the $^{11}B(p,\gamma_0)^{12}C$ reactions at $E_p=40-80$ MeV. γ_0 is the γ rays to the ground state.

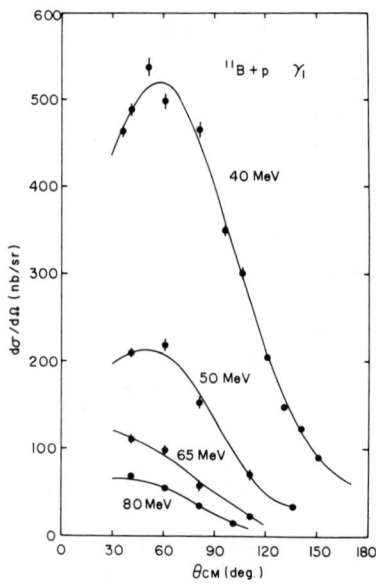

Fig. 7b. Angular distributions of gamma rays following the $^{11}B(p,\gamma)^{12}C$ reactions at $E_p=40-80$ MeV. γ_1 is the γ-transition to the first excited state.

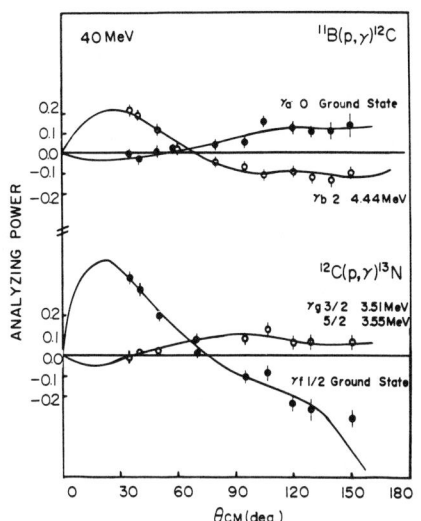 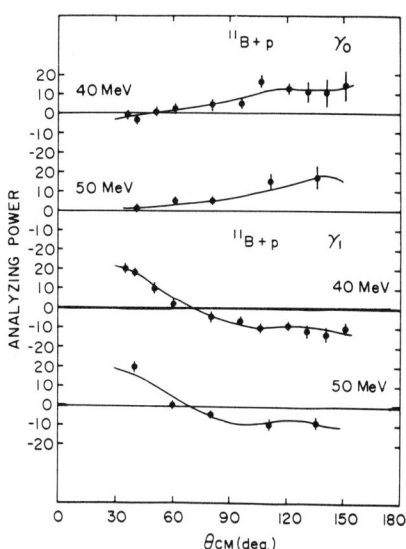

Fig. 8. Analyzing powers for the ^{11}B(p,γ)^{12}C and ^{12}C(p,γ)^{13}N reactions at E_p=40 MeV.

Fig. 9. Analyzing powers for the ^{11}B(p,γ)^{12}C reactions at E_p=40 MeV and 50 MeV. γ$_0$: γ-rays to the ground state, γ$_1$: γ-rays to the first excited state.

These features are clearly seen in Figs. 8-10.
The differential cross-section for the radiative capture gamma ray is expressed in terms of the Legendre polynomials P_k and P_k^1 as follows.

$$\frac{d\sigma(\theta)}{d\Omega} = \frac{\sigma_0}{4\pi} [1 + \Sigma a_k P_k(\cos\theta) + \vec{P}\vec{n}\Sigma b_k P_k^1(\cos\theta)], \quad (1)$$

where $\vec{p}$ is the polarization vector of the incident proton and $\vec{n}=(\vec{k}_p \times \vec{k}_\gamma)/|\vec{k}_p \times \vec{k}_\gamma|$. The coefficients a_k and b_k are derived from the observed angular distributions and the analyzing powers. They are plotted against the gamma ray energies E_γ in Figs. 11 and 12. The data at lower bombarding energies have been obtained at TUNL, and are shown also for comparison. The a_1 coefficients coming mainly from the E1-E2 interference increase with the increasing E_γ. The a_2 coefficients increase from negative values to positive values as the E_γ increases from 20 MeV to 90 MeV. The b_2 coefficients for the $j_>=\ell+ 1/2$ final states are negative, while those for the $j_<=\ell- 1/2$ are positive. The values become small as the E_p or E_γ increases. It should be noted that the ^{12}C(p,γ) data fit better to the general trend of the ^{11}B(p,γ) data when they are plotted as a function of the E_γ rather than those plotted as a function of the E_p.

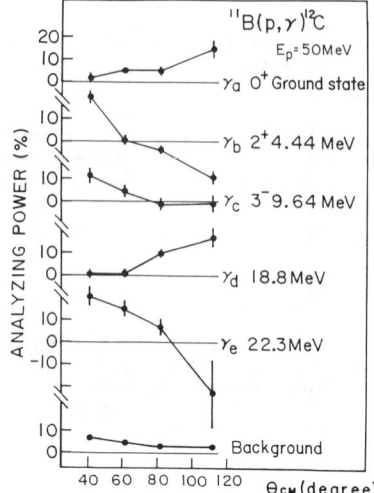

Fig. 10.
Analyzing powers for the $^{11}B(p,\gamma)^{12}C$ reaction at $E_{\bar{p}}$=50 MeV.

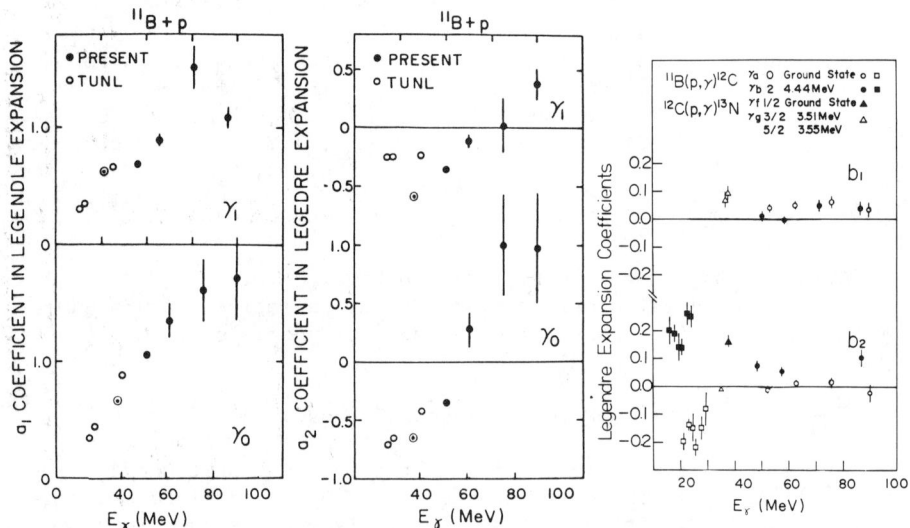

Fig. 11.
Legendre polynomial coefficients for the $^{11}B(p,\gamma_0)$ and the $^{11}B(p,\gamma_1)$ reactions. The values for the $^{12}C(p,\gamma_0)$ and the $^{12}C(p,\gamma_1$ 3.5 MeV excited state) reactions are shown by double circles; see ref. 7, Weller et al.

Fig. 12.
Legendre expansion coefficients of the analyzing powers for the $^{11}B(p,\gamma)$ and $^{12}C(p,\gamma)$ reactions. Squares are taken from ref. 2.

4. ANALYSIS

4.1 SCHEMATIC E1-E2 MODEL

The uniform features found in the angular distributions and the analyzing powers indicate some general mechanism of the radiative capture process in the medium energy region. Analysis of the observed data in terms of a simple schematic model is interesting in order to get some transparent views for the radiative process.

For simplilcity we consider the streched E1 and E2 gamma rays following direct radiative capture reactions. The major components of the radiative capture into the final orbit of the $\psi_f(j,\ell)$ are the stretched E1 radiations of $A_1(\ell+1\rightarrow\ell)$ and $A_1^1(\ell-1\rightarrow\ell)$ and the stretched E2 radiation of $A_2(\ell+2\rightarrow\ell)$. The transition matrix elements for these transitions are large because of the large angular overlap integrals. The main parts of the $A_1(\ell+1\rightarrow\ell)$ and $A_2(\ell+2\rightarrow\ell)$ components are $(n,\ell+1) \rightarrow (n,\ell)$ and $(n,\ell+2) \rightarrow (n,\ell)$ transitions, where n is the number of radial nodes. Thus the A_1 and A_2 have large radial integrals. The component A_1^1 is necessarily taken into account to get a finite b_2 coefficient. The E2 transitions of the $A_2(\ell\rightarrow\ell)$ and $A_2(\ell-2\rightarrow\ell)$ are not taken into account because of the small overlap integrals and the small spin factor of $2\ell_i+1$ for the $\ell-2$ wave. Since the electric transition has no spin operator, we take into account only the spin non-flip processes of $j_i=\ell_i \pm 1/2 \rightarrow j_f=\ell_f \pm 1/2$ in the analysis. Actually the spin flip processes are smaller by a factor $\approx 1/(2\ell)$ compared with the non spin flip transition.

In the framework of this simple schematic model the asymmetry coefficient a_1 arises from the interference between the $A_1(E1)$ and the $A_2(E2)$, and the analyzing power coefficient b_2 arises from the interference of the two E1 radiations with different orbital angular momenta, A_1 and A_1^1. The a_2 coefficient is the sum of those for the A_1, A_1^1 and A_2. Then the a_1, a_2 and b_2 coefficients are written as

$$a_1(j\ell) = NC_1(j\ell)R_1R_2\cos(\delta_2 - \delta_1), \qquad (2)$$

$$a_2(j\ell) = N[a_2(A_1) + a_2(A_1^1) + a_2(A_2) + C_2(j\ell) R_1^1 R_1 \cos(\delta_1 - \delta_1^1)], \qquad (3)$$

$$b_2(j\ell) = Nd_2(j\ell) R_1^1 R_1 (-\sin(\delta_1 - \delta_1^1)), \qquad (4)$$

where $a_2(A_i)$ is the coefficient of the Legendre polynomial $P_2(\cos\theta)$ for the A_i radiation and N is the normalization factor given by $N=1/[a_0(A_1)+a_0(A_1^1)+a_0(A_2)]$. The R_1, R_1^1 and R_2 are the radial matrix elements for the A_1, A_1^1 and A_2 radiations, respectively, and δ_1, δ_1^1 and δ_2 are corresponding phases. The angular integral is included in the coefficients $c_1(j\ell)$ and $d_2(j\ell)$. The phase of the transition matrix element depends on the phase shift of the incoming proton, which is obtained from the optical potential.

The phase difference $\delta_2 - \delta_1$ is written as $\delta_2 - \delta_1 \approx \delta_{\ell+2} - \delta_{\ell+1} + \pi/2$, where $\delta_{\ell+2}$ and $\delta_{\ell+1}$ are the phase shifts of the incoming waves with $\ell+2$ and $\ell+1$, and the term $\pi/2$ comes from the difference $i^{\ell+2} - i^{\ell+1}$. The difference $\delta_{\ell+2} - \delta_{\ell+1}$ is considered to be due to the different centrifugal potentials $V^{cf}\ell(\ell+1)$ and the spin orbit potentials $V_{so}\vec{\ell}\cdot\vec{s}$ acting on the different partial waves of $\ell+2$ and $\ell+1$. Thus one gets

$$a_1(j\ell) = NC_1(j\ell)R_1R_2\cos(\frac{\partial\delta_\ell}{\partial V^{cf}}\cdot 2(\ell+2)V^{cf} \pm \frac{\partial\delta_\ell}{\partial V_{so}} + \frac{\pi}{2}). \quad (5)$$

The ± spin for the V_{so} term is positive for the $j_<=\ell-1/2$ and negative for the $j_>=\ell+1/2$. The derivative $\partial\delta_\ell/\partial V$ is negative since a repulsive interaction gives a negative phase-shift, and V^{cf} and V_{so} are positive. Thus the term $\frac{\partial\delta_\ell}{\partial V}2(\ell+1)V^{cf}$ is negative and does not exceed much $\pi/2$ in the present energy region where no single resonance pole exists. The spin-orbit potential itself is much smaller than the centrifugal potential, and the effect may be neglected for the a_1 coefficient. Then one gets a positive value for the $\cos(\delta_2 - \delta_1)$. The R_1R_2 is positive for the major component of the n, $\ell+1 \to n,\ell$ and n, $\ell+2 \to n,\ell$ transitions. The coefficient $C(j\ell)$ is also positive. Then we get finally the positive coefficient a_1,

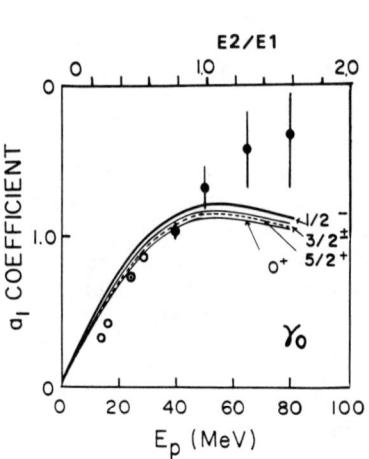 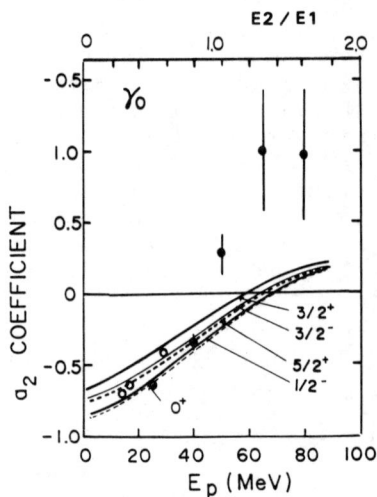

Fig. 13.
The a_1 coefficient for the $P_1(\cos\theta)$ term of the angular distribution. The curves are calculated for radiative capture processes on the 0^+ target into the final orbits of $1P1/2^-$, $1P3/2^-$, $1d3/2^+$ and $1d5/2^+$. The curve 0^+ is for the radiative capture on the $1P3/2^-$ target into the ($1P3/2$, $1P3/2$) 0^+ final state. See the caption of Fig. 11 for the date points.

Fig. 14.
The a_2 coefficient for the $P_2(\cos\theta)$ term of the angular distribution. See the caption of the Fig. 13.

which increases from 0 as the E2 component A_2 increases. The calculation is made for the final orbits of $(n\ell j)$ = 1d5/2, 1d3/2, 1P3/2 and 1P1/2 assuming $\cos(\delta_2 - \delta_1) = 1$ for simplicity. The value for the radiative capture on the 1P3/2 target feeding 0^+ state with the (1P3/2 1P3/2) 0^+ configuration is also obtained. The calculated values reproduce the general trends of the observed ones as shown in Fig. 13. Here the E2 contribution is assumed to increase as $a_0(A_2)/[a_0(A1) + a_0(A_1^1) + a_0(A_1)] \sim (E_\gamma - 12 \text{ MeV})/50 \text{ MeV}$. The a_2 coefficient is calculated in a similar manner as the a_1 coefficient. The phase factor for the A_1^1 and A_1 interference is $\cos(\delta_1 - \delta_1^1)$ = $\cos(\delta_{\ell+1} - \delta_{\ell-1} + \pi)$. Assuming $\delta_{\ell+1} - \delta_{\ell-1} < \pi/2$, one gets $\cos(\delta_1 - \delta_1^1) < 0$. For simplicity we assumed $\cos(\delta_1 - \delta_1^1) = -1$. The calculated value is generally in accord with the observed values as shown in Fig. 14. Both the observed a_1 and a_2 coefficients exceed the calculated values for the higher E_γ region. This suggests the contribution of the E3 component.

The b_2 coefficient for the analyzing power is written as

$$b_2 = Nd_2(j\ell)R_1^1R_1\sin(\delta_{\ell+1} - \delta_{\ell-1})$$
$$= Nd_2(j\ell)R_1^1R_1\sin[\frac{\partial\delta_\ell}{\partial V^{cf}}V^{cf} \cdot (4\ell+2) \pm \frac{\partial\delta_\ell}{V_{so}} \cdot 2V_{so}]. \quad (6)$$

The coefficient $d(j\ell)$ is positive for $j_> = \ell + 1/2$ and negative for $j_< = \ell + 1/2$. The $\pm$ signs of the V_{so} term is same as in Eq. (3), namely + for $j_<$ and - for $j_>$. Noting that the V^{cf} term plays a major role in the $\delta_{\ell+1} - \delta_{\ell-1}$ phase difference and $R_1^1R_1>0$, one gets a positive b_2 value for $j_<$ and a negative b_2 value for $j_>$. These signs are just in good agreement with the observed values. The calculated b_2 values for the $\sin(\delta_{\ell+1} - \delta_{\ell-1}) = 1$ are shown in Fig. 15. They reproduce the general feature of the observed values. The difference between the b_2 values for the $j_<$ and $j_>$ decreases as the proton (γ-ray) energy increases. This is considered to be partly due to the increase of the E2 admixture and partly due to the decrease of the absolute value of the $\partial\delta_\ell/\partial V$.

The analyzing power for the electric radiative capture reaction can be schematically understood in terms of a geometrical consideration. Since the electric gamma transition does not flip spin, initial protons with $j_>^i = \ell_i + 1/2$ contributes to the radiative capture into the final orbit of $j_> = \ell + 1/2$, and protons with $j_<^i = \ell_i - 1/2$ to the capture into the $j_< = \ell - 1/2$ orbit. Thus the incident spin-up proton captured into the $j_> = \ell + 1/2$ final orbit must interact with the right-hand side of the nucleus because the orbital angular momentum is oriented upwards. On the other hand the incident spin-down proton must interact with the left-hand side to have downward orbital angular momentum. In case of the $j_< = \ell - 1/2$, the right-left relation is just reversed. These are schematically illustrated in Fig. 16. The analyzing power, which is the difference between the spin-up cross-section and the spin-down one, is thus given by $A_R - A_L$ for $j_>$, while it is given by $A_L' - A_R'$ for $j_<$. The b_2 coefficient, which arises from the interference between the proton waves of $\ell+1$ and $\ell-1$, changes the sign for the change from the $j_>$ state to $j_<$

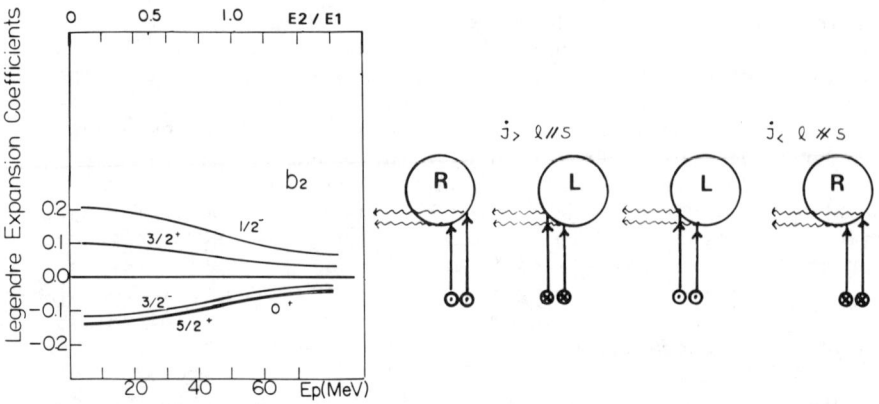

Fig. 15.
The b_2 coefficient for the $P_2^1(\cos\theta)$ term of the analyzing power. See the caption of the Fig. 13.

Fig. 16.
Schematic diagrams of the analyzing power for the E1 radiative capture process. R (L) indicates the case of interaction of the incident protns with the right (left) hand side of the nucleus. Asymmetry for the $j_>$ is R-L, while it is L-R for the $j_<$.

state provided that $A_R \sim A_R'$ and $A_L \sim A_L'$. This is so in the geometrical consideration of the proton-nuclear interaction provided that the effect of the spin orbit interaction is much smaller than the effect of the centrifugal potential as illustrated by the different impact parameters in Fig. 16. With the increasing proton energy, both the impact parameter and the phase shift difference between the $\ell+1$ and $\ell-1$ waves become small, resulting in reduction of the difference between the b_2 values for the $j_<$ and $j_>$ final states.

4.2 THEORETICAL CALCULATIONS

Theoretical analysis of the radiative capture reactions in the medium energy region have been made by various ways[10-13]. Ohtsubo and Hommura[13] have calculated the $^{11}B(p,\gamma_0)$ reactions by two methods, one by using the Siegert theorem and another by calculating directly the impulse term and the pion exchange term. The wave functions for the target state of the ^{11}B and the final state of the ^{12}C are those given by Cohen-Kurath[14]. The incoming proton waves in the 40 - 80 MeV region were obtained from the optical potential. The parameters of the optical potential were obtained from the elastic scattering data so as to change smoothly with the increasing proton energies.

Calculated values are shown in Fig. 17. The Siegert theorem calculation predicts larger cross-sections by a factor two than the impulse + exchange model calculation does. Here the non-Siegert term which is about 3 ~ 10 % of the the Siegert term, is included in the calculation. The model calculation with the impulse term + pion exchange current (pionic + pair) term is in accord with the gross shape of the observed angular distribution at E_p = 40 MeV, as shown in Fig. 17b. Fig. 17c shows the calculated angular distributions in terms of the impulse term, the pionic current term and the pion pair current term[13]. Here the optical potential used is somewhat different from that used in Figs. 17a and 17b. The final state (^{12}C 0^+) and the target state (^{11}B) are the simple closed shell state and the 1P3/2 hole state. Interesting is to find that the pair current contributes constructively with the impulse term, while the pion current contributes distructively. Consequently the two exchange currents cancel each other, and the final result with both the pair and exchange currents is close to the impulse term calculation. The large excess of the Siegert theorem calculation over the experimental data indicates improper wave functions used in the calculation. Since the overlap integral for the medium energy proton wave ψ_i and the bound proton waves ψ_f is very small, the matrix element in the Siegert theorem is sensitive to the mixing of higher shell orbits into the ψ_f induced by the tensor interaction and the behavior of the ψ_i inside the nucleus. Since the exchange current contributions are properly included in the Siegert theorem calculation the ψ_i and ψ_f have to be adjusted so as to reproduce the experimental data. Then it is interesting to check how the model impulse term + pion exchange term) calculation by using such ψ_i and ψ_f may deviate from the data or not. The deviation might indicate contributions of heavier meson exchange current. Both the Siegert and model calculations give small dependence of the cross-section on the incident proton energy (or gamma ray energy), while the observed data decrease exponentially with the increasing proton energy as $\sigma \propto \exp(-E_p/15 \text{ MeV})$.

Gari and Hebach[11,12] calculated differential cross-sections of (γ,P) reactions in terms of the shell model, the correlation (giant resonance) and the Rosenfeld interaction, as shown in Fig. 17d. The calculated values are much larger than the observed ones. Similar calculation with the Rosenfeld interaction was made by Tsaj and Londergan[10]. The calculated value is generally close to the obsered ones. It is not clear in these calculations what the deviation or the agreement means.

5. HIGHER EXCITATION CONFIGURATION

It is interesting to note that the radiative capture of medium energy protons feeds higher excited states as well as low-lying states. Strong gamma transitions feeding the 19 MeV excitation region and the 22 MeV excitation region in ^{12}C have been found in the radiative capture of 40 MeV protons on ^{11}B as shown in Fig. 18. The former may correspond to the 19.2 ± 0.6 MeV state with the $[(d5/2) (P3/2)^{-1}]_{4-}$ as observed before[7]. The observed analyzing powers for these two groups of the gamma rays (see Fig. 10) suggest the $j_>$ and $j_<$ configurations for the 19 MeV and 22 MeV excitation regions, respectively. Actually the spectroscopic strengths for the d5/2 and 1d3/2 configurations[15-17] are concentrated in the two excitation regions as shown in the insert of Fig. 18.

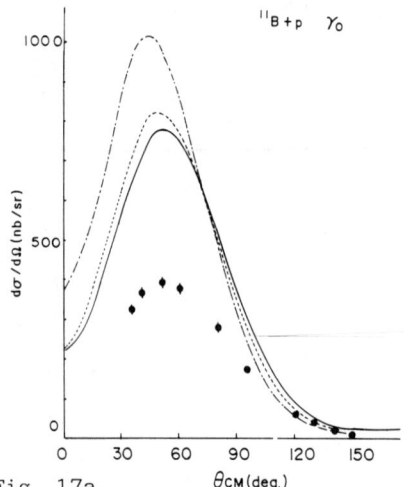

Fig. 17a.
Angular distributions for the $^{11}B(p,\gamma_0)^{12}C$. Solid, dashed and dot-dashed lines are calculated for the E_γ=55, 65 and 100 MeV, respectively, in terms of the Siegert theorem including the non Siegert term (ref. 13). The data points at E_p=40 MeV correspond to the E_γ=52.3 MeV.

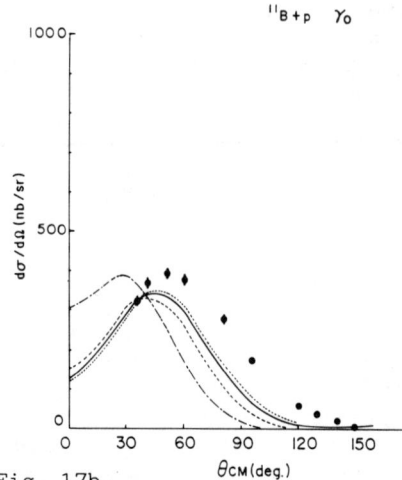

Fig. 17b.
Angular distributions for the $^{11}B(p,\gamma_0)^{12}C$. Solid, dashed, dot-dashed and dotted lines are calculated for the E_γ=55, 65, 100 and 52.3 MeV in terms of the impulse + pion exchange terms (ref. 13). The data points are for the E_γ=52.3 MeV.

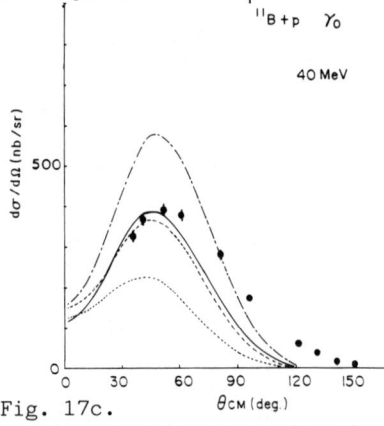

Fig. 17c.
Angular distributions for the $^{11}B(p,\gamma_0)^{12}C$ at E_p=40 MeV. Dashed, dotted, dot-dashed and solid lines are calculated in terms of the impulse term, the impulse + the pion-current terms, the impulse + pion pair current terms, the impulse + the pion current and pair current terms.

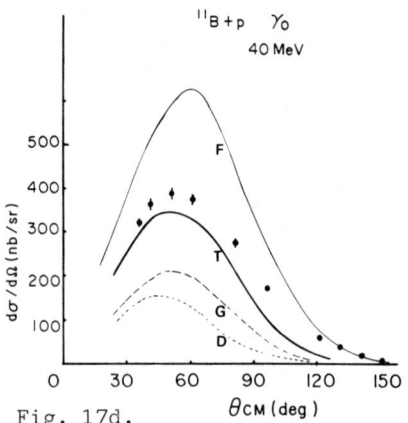

Fig. 17d.
Angular distributions for the $^{11}B(p,\gamma_0)$ reaction at E_p= 40 MeV. The lines D, G and F stand for the calculations of the direct shell model term, the D + the giant resonance term, and the G + the exchange term in refs. 11,12. The line T stands for the calculation[10] with the exchange term.

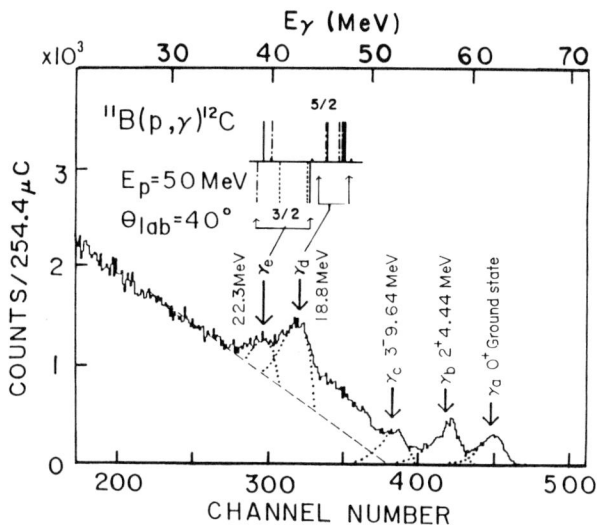

Fig. 18. Gamma ray spectrum for the $^{11}B(p,\gamma)^{12}C$ at E_p=50 MeV. The insert shows the states with the 1d5/2 proton configuration and those with the 1d3/2 one as found in refs. 15-17.

6. CONCLUDING REMARKS

1. The HERMES, which consists of a central and four-segment annular NaI crystals surrounded by the plastic anti-counters, is shown to give good energy resolution spectra with low-background counts for 40 - 80 MeV polarized protons.
2. Angular distributions of radiative capture gamma rays are rather independent of the final state. They show a strong forward-peaking pattern, indicating the large admixture of positive parity transitions.
3. Analyzing powers depend on the spin direction with respect to the orbital angular momentum of the final proton orbit for the radiative capture process.
4. Schematic model calculations with stretched E1-E2 radiations reproduce well the angular distributions and the analyzing powers. The analyzing power can be simply understood by the geometrical consideration in the framework of the schematic model of the stretched E1-E2 radiations.
5. Analyses of the Legendre polynomial coefficients in the schematic model lead to the $<E2>^2/<E1>^2 \approx (E_\gamma - 12 \text{ MeV}) \cdot 2 \cdot 10^{-2} \text{ MeV}^{-1}$. This is an order of magnitude larger than the single particle estimates. The large a_1 and a_2 coefficients may suggest the existence of E3 and higher multipole radiations in the 80 - 100 MeV gamma transitions. Magnetic transitions are also interesting. The linear polarization measurement of these gamma-rays is necessary.

6. Theoretical analysis in terms of the Siegert theorem gives much larger cross-sections than the observed value. This may indicate improper wave functions used for the initial proton and/or the final state. Comparison with the model calculation including the impulse + pion-exchange currents, together with the Siegert theorem calculation, suggests heavier meson contributions provided that the pion current contribution and the pion pair current one cancel each other.

ACKNOWLEDGEMENT

The authors thank Prof. H. Ohtsubo for valuable discussions, and T. Honmura for the calculation.

REFERENCES

1. C. Brassard, et al., Phys. Rev. C6, 53 (1972).
2. H.F. Glavish, et al., Phys. Rev. Lett. 28, 766 (1972).
3. K.A. Snover, P. Paul, H.M. Kuan, Nucl. Phys. A285, 189 (1977).
4. S. Manglos, Phys. Rev. C24, 2378 (1981).
5. F. Saporetti, R. Guidotti, Nucl. Phys. A330, 53 (1979).
6. G.E. Brown, Nucl. Phys. 57, 339 (1964).
7. M.A. Kovash, S.L. Blatt, R.N. Boyd, T.R. Donoghue and H.J. Hausman, Phys. Rev. Lett. 42, 700 (1979).
   H.R. Weller et al., Phys. Rev. C25, 2921 (1982).
8. T. Kishimoto, T. Shibata, M. Sasao, M. Noumachi and H. Ejiri, Nucl. Inst. Meth. 198, 269 (1982).
9. M. Noumachi, Ph.D. thesis, Osaka Univ. 1983 (unpublished).
   M. Noumachi, T. Shibata, K. Okada, T. Motobayashi, F. Ohtani, H. Ejiri and T. Kishimoto, to be published in Phys. Rev. C, (1984).
10. S.F. Tsai and J.T. Londergan, Phys. Rev. Lett. 43, 576 (1979).
11. M. Gari and H. Hebach, Phys. Rev. C18, 1071 (1978).
12. H. Hebach, A. Wortberg and M. Gari, Nucl. Phys. A267, 425 (1976).
    M. Gari and H. Hebach, Phys. Lett. 49B 29 (1974).
13. T. Honmura and H. Ohtsubo, private communication (1984).
14. S. Cohen and D. Kurath, Nucl. Phys. 73, 1 (1965).
15. A. Yamaguchi, T. Terasawa, K. Nakahara and Y. Torizuka, Phys. Rev. C3, 1750 (1971).
16. T.W. Donnelly, et al., Phys. Rev. Lett. 21, 1196 (1968); Ann. Phys. (N.Y.) 60, 209 (1970).
17. M. Buenerd, P. Martin, P.de Saintignon and J.M. Loiseaux, Nucl. Phys. A286, 377 (1977).
18. G.M. Reynolds, D.E. Rundquist and R.M. Poichar, Phys. Rev. C3, 442 (1971).

RECENT RESULTS FROM PROTON CAPTURE TO GIANT RESONANCES
BUILT ON HIGHLY-EXCITED STATES

D. H. Dowell
Brookhaven National Laboratory
Upton, New York, 11973, USA

ABSTRACT

New measurements of the (p,γ) reaction on natural targets of K and Ca at intermediate proton energies (E_p = 11 to 36 MeV) reveal excitation functions that are dominated by excited-state giant dipole resonances (ESGDR's) built upon single-particle states in ^{40}Ca and ^{41}Sc. These ESGDR's all peak near E_γ = 20 MeV and have widths which increase monotonically with the final state's excitation energy. The integrated, inverse (γ,p_0) cross sections are shown to be proportional to the known single-particle spectroscopic strengths indicating a semi-direct mechanism for the capture process.

1. INTRODUCTION

Since its discovery the giant dipole resonance (GDR) has been studied extensively for almost all the nuclei throughout the periodic table. This large body of work has shown that the GDR mode of excitation is common to all nuclei and its properties vary smoothly from one nucleus to the next. Generalizing this concept, it was suggested by Brink and Axel[1] that these same qualities should hold for GDR's built upon excited states; that is, the excited state-giant dipole resonance (ESGDR) should be common to all nuclear levels and its properties should vary smoothly from one state to the next. Earlier experimental evidence showed that this may be the case. In a measurement of γ-ray spectra from the (p,γ) reaction on medium-mass nuclei[2], it was found that the simple direct capture process underestimated the data. The excitation of an intermediate GDR state, sometimes called a 'semi-direct' process, was needed to explain the data.

Our (p,γ) experiments have been motivated by the work of Blatt and co-workers[3] in which they observed strong γ-ray transitions to the group of final states near 19 MeV in the $^{11}B(p,\gamma)^{12}C^*$ reaction. However this and other work[4] have concentrated on the behavior of the 19 MeV states which are not resolvable with a NaI(Tl) spectrometer and which did not allow the systematic study of ESGDR properties since only one final state energy was being investigated. Therefore it has been our intent to choose cases where the γ rays are reasonably resolved for final state energies up to at least 10 MeV.

To this end, we have studied the $^{27}Al(p,\gamma)^{28}Si^*$ reaction, and the results have been published in ref. 5. We report here new measurements for the $^{Nat}K(p,\gamma)Ca^*$ and $^{Nat}Ca(p,\gamma)Sc^*$ reactions. The properties of the ESGDR energy, width and integrated strength observed in these three reactions are presented. Also by using the direct semi-direct model, a single constant is shown to relate the proton capture strength to single particle transfer strength. The validity of this relation is demonstrated and later used to estimate previously unmeasured high-lying spectroscopic strength.

2. THEORETICAL BACKGROUND

In Brown's formulation of the direct, semi-direct model[6], the emission amplitude for (γ,p) includes not only the usual direct term but also a resonance-like term. This resonance term was introduced to describe an intermediate giant dipole resonance state which decays by emitting a proton. These two processes are shown in Fig. 1 for photons being absorbed onto a 1-particle, 1-hole (1p-1h) excited state, $|n\ell\rangle$, at an excitation energy of E_x^f. The resonance reaction is considered to be a two-step process in which the photon raises the proton into the collective GDR state in the first step and then decays by emitting this proton.

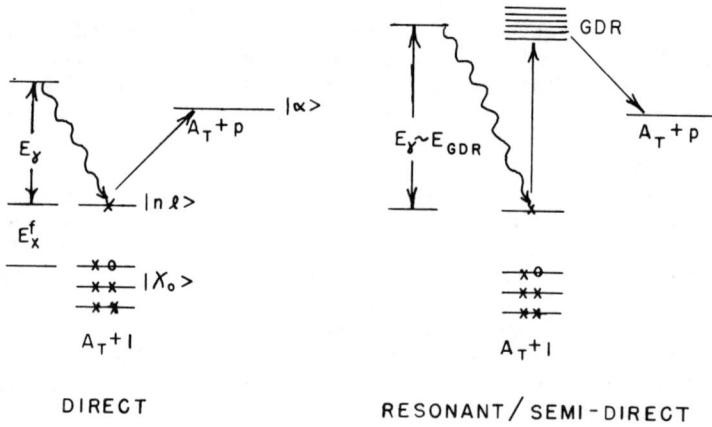

Fig. 1. The direct and resonant or semi-direct processes for the (γ,p) reaction on a 1p1h excited state with an excitation energy of E_x^f.

In this model, the emission amplitude is given by

$$S_{\alpha,n\ell} = \left[1 + \frac{\Delta E}{E_\gamma - E_{GDR} + \frac{i\Gamma}{2}}\right] \langle \chi_o;\alpha|D|\chi_o;n\ell\rangle \qquad (1)$$

This equation expresses the essential features of the direct, semi-direct model. Starting with the matrix elements, $|\chi_0\rangle$ is the core's wavefunction which is explicitly shown to be unaffected by the one-body dipole operator, D, which only operates on the valence proton's wavefunction |nℓ>, exciting it to the continuum state |α>. The core-excitation (i.e., the intermediate GDR state) has been factorized into the second term in the brackets. This second bracket term is the semi-direct part of the interaction with a resonance energy of E_{GDR} and a width of Γ. The strength of this semi-direct term, noted by ΔE, is proportional to the residual interaction. That is, it is the energy difference between the actual dipole state and the unperturbed, particle-hole energy obtained in the simple one-body shell model. The first term in the brackets is the direct part; and since the direct matrix element multiplies the resonant part, it is easily seen that semi-direct nucleon emission or capture is proportional to the spectroscopic strengths measured in direct reactions.

This can be made more quantitative by keeping only the second, semi-direct term in the brackets of eq. 1, which is valid for energies near E_{GDR} and squaring. Then the cross section can be written as the product of the direct cross section and an energy-modulating GDR resonance,

$$\sigma_{exp}(\gamma,p) = \frac{\Delta E^2}{(E_\gamma - E_{GDR})^2 + \Gamma^2/4} C^2 S(n,\ell) \sigma^{n,\ell}(\gamma,p) \qquad (2)$$

Here $\sigma^{n,\ell}(\gamma,p)$ is the theoretical cross section for direct emission of the proton, n and ℓ are the principle and angular momentum quantum numbers of the initial 1p1h state, and $C^2 S(n,\ell)$ is the spectroscopic factor. C^2 is the Clebsch-Gordon coefficient coupling the proton and residual nuclear isospins. The γ-ray energy dependence can be removed by integrating eq. 2 to obtain

$$\int \sigma_{exp}(\gamma,p) dE_\gamma = C^2 S(n,\ell) \int \frac{\Delta E^2}{(E_\gamma - E_{GDR})^2 + \Gamma^2/4} \sigma^{n,\ell}(\gamma,p) dE_\gamma \qquad (3)$$

$$\int \sigma_{exp}(\gamma,p) dE_\gamma = K_{n,\ell} C^2 S(n,\ell) \qquad (4)$$

where the integral is denoted by $K_{n,\ell}$.

Following a derivation by K. Snover[7], $K_{n,\ell}$ can be calculated using the schematic model by "building" a GDR upon the excited state of interest. This is done by summing all of the allowed E1 transitions between the 1p1h excited state, n,ℓ>, and the states, j₂>, which make up the ESGDR,

$$K_{n,\ell} = \sum_{j_2} |<j_2|D|n,\ell>|^2 \qquad (5)$$

Here j_2 is the total angular momentum of the final state. In terms of the radial matrix elements and Clebsch-Gordon coefficients this is[8]

$$K_{n,\ell} = \frac{4\pi}{\hbar c} \sum_{j_2} \frac{q^2}{3} E_{12} (2j_1+1) (j_1\tfrac{1}{2}10|j_2\tfrac{1}{2})^2 \langle r\rangle^2 \qquad (6)$$

where q is the effective charge of the nucleon making the transition, E_{12} is the energy of the transition, and j_1 is the total angular momentum of the initial 1p1h state, $j_1 = \ell \pm 1/2$.

By considering the three allowed E1 transitions, $j_2 = j_1+1$, j_1, and j_1-1, and using harmonic oscillator wavefunctions to calculate $\langle r\rangle^2$, the following simple relation is obtained,

$$K_{n,\ell} = \frac{4}{3}\pi^2 \frac{e^2\hbar}{Mc} (N/A)^2 (n + \ell/2 + 1/2). \qquad (7)$$

Here M is the mass of the nucleon and N/A is the effective charge of the proton. In terms of convenient units, eq. 7 becomes,

$$K = 39.48\ (N/A)^2\ (n + \ell/2 + 1/2)\ \text{MeV·mb} \qquad (8a)$$

or

$$K = 9.87\ (n + \ell/2 + 1/2)\ \text{MeV·mb} \qquad (8b)$$

for N=Z nuclei.

The value of $K_{n,\ell}$ is constant within each shell since $n+\ell/2$ is constant, and increases with increasing shell number. These values of $K_{n,\ell}$ are later shown to agree quite well with our experimental results.

3. EXPERIMENTAL DETAILS

Our $^{Nat}K(p,\gamma)Ca^*$ and $^{Nat}Ca(p,\gamma)Sc^*$ measurements were made using the two-stage and three-stage tandem accelerators at Brookhaven National Laboratory (BNL) for protons from 11 to 36 MeV. γ rays were detected at $\theta_\gamma=90°$ in the BNL-MK III NaI(Tl) γ-ray spectrometer. This detector system has been discussed in detail elsewhere[9]. In addition to the usual lead shielding a paraffin and Li_2CO_3 hardener approximately 50 cm in length was placed between the target and the detector to moderate and remove fast neutrons from the data. Also a meter thick wall of paraffin and borated-paraffin was placed on either side of the spectrometer's usual lead housing to shield against neutrons coming from both the beam dump and the upstream beam line components. This shielding made the neutron backgrounds in the high-energy portion ($E_\gamma > 15$ MeV) of the spectra negligible, as verified by pulsed-beam time-of-flight measurements.

During the experiment we also collected pulse-pileup routed spectra which were later used in the data analysis to correct for pileup events. The pileup efficiency of our electronics was determined by two methods. One was by a difference of high- and low-current runs and the other was obtained from the high-energy tail of an electronic pulser placed near $E_\gamma = 50$ MeV in the spectra. These two methods agreed within errors. All the analyzed spectra were corrected for pulse pileup.

The targets were self-supporting foils of natural isotopic abundance. The ^{Nat}K target was hand-rolled in a dry box with an atmosphere of argon containing less than 1% oxygen. The metallic potassium was first cut into a thin wafer, approximately 1 mm thick and then placed between two sheets of teflon and rolled. The rolling was done in several small steps with the potassium foil removed from between the teflon sheets between rolls and lubricated with kerosene. After rolling, the target was rinsed in hexane, mounted on a target frame and transferred to the scattering chamber with a vacuum interlock chamber. Once placed in the vacuum, the target had the blue-grey appearance of slightly oxidized lead. A comparison using the proton beam with an Al_3O_2 of known thickness showed that the target contained approximately 500 μg/cm² of ^{16}O. The target's thickness was determined by comparison of our γ_0 yields with those of an earlier $^{39}K(p,\gamma)$ experiment[10] at E_p = 11, 12, 13, 14, 15, 16 and 17 MeV. The target thickness was found to be 15.4 ± 1.mg/cm².

The ^{Nat}Ca target was a sandwich of two evaporated foils. Each foil was 2.4 mg/cm² thick giving a total target thickness of 4.8 mg/cm². After evaporation the calcium foils were also mounted on a target frame in the dry box and the vacuum interlock chamber used to move it to the scattering chamber. The calcium target contained very little oxygen and had the appearance of shiny aluminum foil.

4. DATA ANALYSIS

Knowledge of the NaI(Tl) spectrometer's line shape was critical to the data analysis and was obtained from the $^{11}B(p,\gamma)^{12}C$ reaction at E_p = 7.25 MeV, which is near the peak of the GDR in ^{12}C. Figure 2 shows the line shape fit to the γ_0 and γ_1 lines at

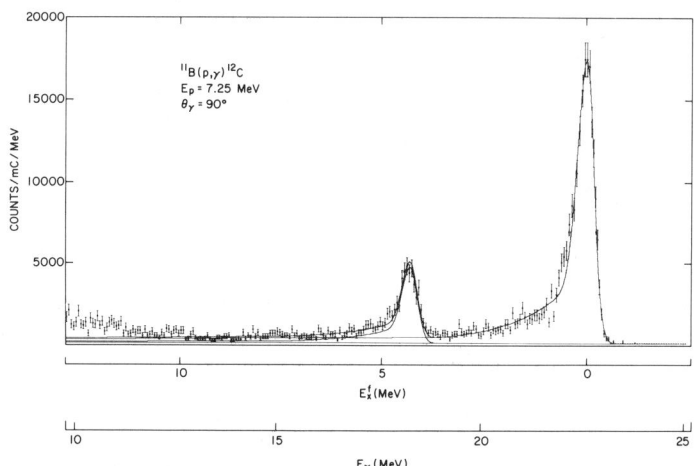

Fig. 2. $^{11}B(p,\gamma)^{12}C$ spectrum used to determine BNL-MK III line shape.

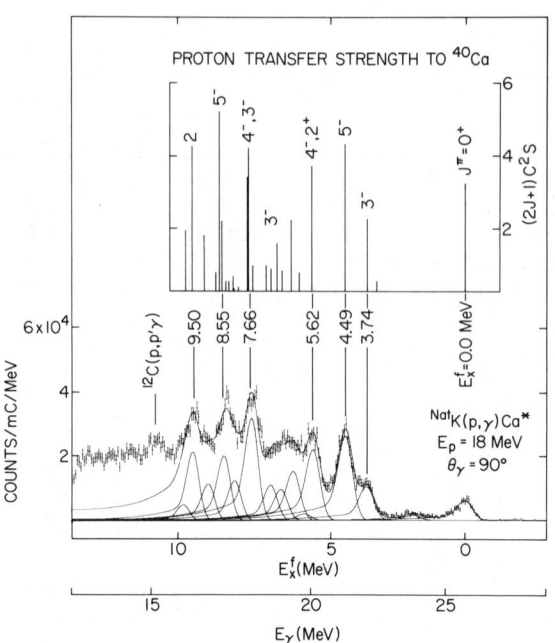

Fig. 3. Bottom: ^{Nat}K+p γ-ray spectrum at E_p = 18 MeV with line shape decomposition. Top: Known proton transfer spectroscopic strength to states in ^{40}Ca.

E_γ = 22.6 and 18.16 MeV. The parameterized line shape of ref. 9 was used in the peak region with a straight-line tail matched on at about 3 MeV below the full energy peak position and extrapolated to zero counts at zero energy. This tail is larger than that given in ref. 9 because of the 50 cm of paraffin between the target and detector. The line shape was assumed to have a constant percentage energy resolution at all γ-ray energies.

A typical γ-ray spectrum for the ^{Nat}K(p,γ)Ca* reaction is shown in Fig. 3. In a manner similar to our previous study of the ^{27}Al(p,γ)^{28}Si* reaction[5] we have fit this and the other spectra with a sum of line shapes of variable amplitudes whose positions are fixed at energies corresponding to states of known proton transfer strength. All of the structures in the spectrum are observed to shift in γ-ray energy according to E_γ = 39/40E_p + 8.33 - E_x^f, as expected for radiative capture to final states in ^{40}Ca.

The upper portion of Fig. 3 shows the location and values of known proton transfer strength from the ^{39}K(d,n)^{40}Ca and ^{39}K(^{3}He,d)^{40}Ca reactions[11]. The known direct transfer strength accounts for all of the structure in the spectrum for final state energies from E_x^f = 0 to 9.7 MeV.

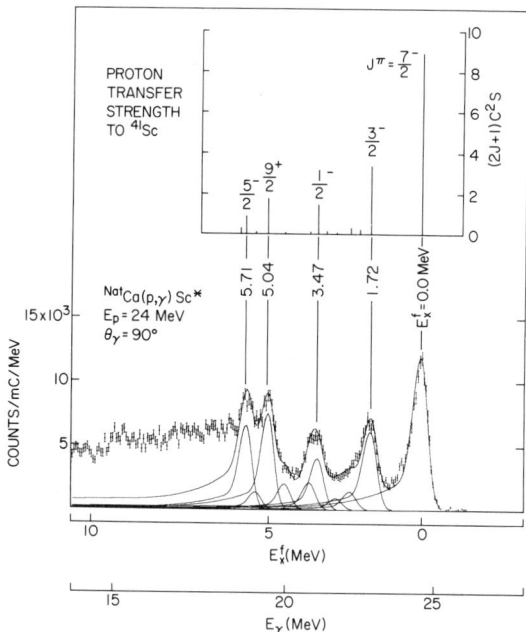

Fig. 4. Bottom: $^{Nat}Ca+p$ γ-ray spectrum at E_p = 24 MeV with line shape fit. Top: Known proton transfer spectroscopic strength in ^{41}Sc.

The γ-ray spectrum in Fig. 4 for the $^{Nat}Ca(p,\gamma)Sc^*$ reaction shows a similar line shape fit and comparison with the known proton spectroscopic strength[12] in ^{41}Sc. The Q-value for this reaction is low (Q = 1.09 MeV) and this is the first measurement made on this nucleus in the region of the GDR. Again line shapes have been placed in the spectrum at the positions of known spectroscopic strength.

Figures 5, 6 and 7 show our cross sections at θ_γ = 90° for the $^{Nat}K(p,\gamma)Ca^*$, $^{Nat}Ca(p,\gamma)Sc^*$ and $^{27}Al(p,\gamma)^{28}Si^*$ reactions. The $^{27}Al(p,\gamma)^{28}Si^*$ is our older work reported in ref. 5. The cross sections have been plotted vs E_γ where

$$E_\gamma = \frac{A_T}{A_T+1} E_p + Q - E_x^f \qquad (9)$$

Here A_T is the atomic number of the target nucleus, E_p is the lab proton energy, Q is the (p,γ_0) reaction-Q value and E_x^f is the final state energy in the $A_T + 1$ nucleus. The Q-values for these three reactions are 8.33 MeV for $^{39}K + p$, 1.09 MeV for $^{40}Ca + p$ and 11.58 MeV for $^{27}Al + p$.

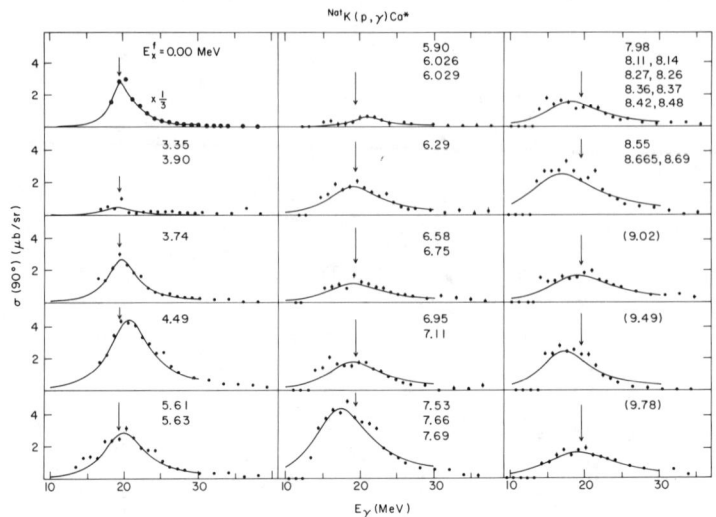

Fig. 5. $^{Nat}K(p,\gamma)Ca^*$ cross sections at $\theta_\gamma = 90°$ vs E_γ for various final state energies in ^{40}Ca. The curves are Lorentzian line shapes fit to the data. Arrows indicate energy of ground state GDR.

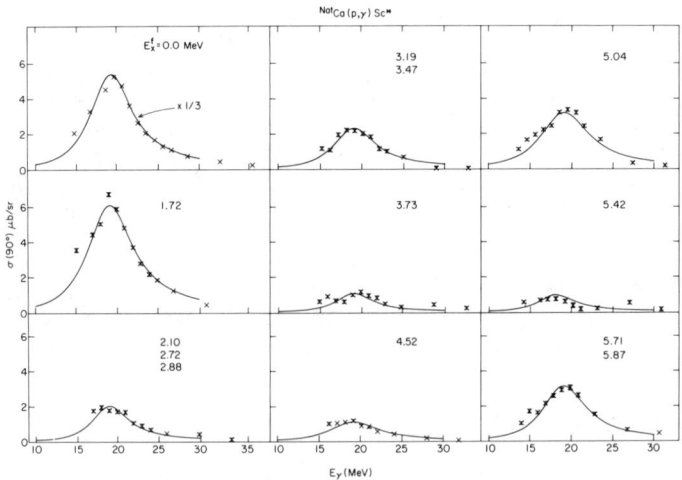

Fig. 6. $^{Nat}Ca(p,\gamma)Sc^*$ cross sections at $\theta_\gamma = 90°$ vs E_γ for various final state energies in ^{41}Sc. The curves are Lorentzian line shapes fit to the data.

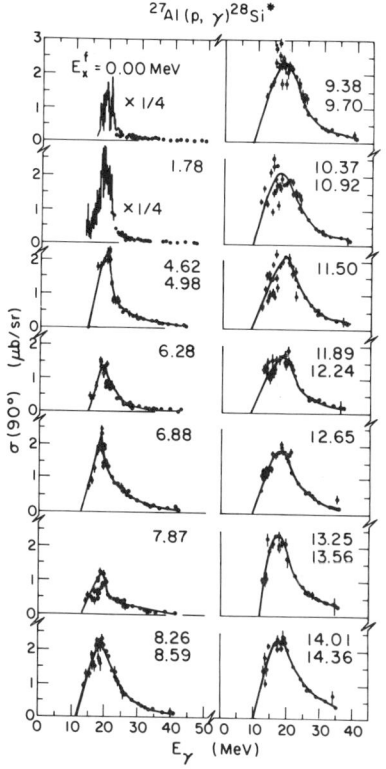

Fig. 7. $^{27}Al(p,\gamma)^{28}Si^*$ cross sections vs E_γ at $\theta_\gamma = 90°$ for various final state energies, E_x^f, in ^{28}Si. The curves are drawn to guide the eye. See ref. 5.

5. DISCUSSION

The cross sections for all three reactions exhibit resonance-like behavior whose energies peak at or very near the ground state GDR energy. For final state energies greater than E_x^f = 6 MeV the width increases from about 4 MeV (the width of the ground state GDR) to nearly 12 MeV for ESGDR's at 10 MeV. This is shown in Figs. 8, 9 and 10 where the width vs. final state energy is plotted for the three reactions. The similar increase in width for ESGDR's in ^{40}Ca and ^{28}Si suggests a simple mechanism. Perhaps this is just due to increasing escape width. The data for ESGDR's in ^{41}Sc does not extend high enough in E_x^f to make any similar comparison.

We have used detailed balance to convert our (p,γ) cross sections to (γ,p_0) cross sections. Specifically

$$(2J+1) \frac{d\sigma}{d\Omega}(\gamma,p_0) = (2I_{A_T}+1) \left[\frac{A_T}{A_T+1}\right]^2 2Mc^2 \frac{E_p}{E_\gamma^2} \frac{d\sigma}{d\Omega}(p,\gamma) \qquad (10)$$

Where J is the spin of the final state in the $A_T + 1$ nucleus, I_{A_T} is the ground state spin of the (p,γ) target nucleus and M is the mass of the nucleon. The smooth curves shown in Figs. 5, 6 and 7 were used in eq. 10 to obtain the (γ,p_0) cross sections for the various excited states. These (γ,p_0) cross sections were then integrated in E_γ and related to the spectroscopic strengths by an expression similar to eq. 4,

$$4\pi/f \ (2J+1) \int d\sigma/d\Omega \ (\gamma,p_0) \ dE_\gamma = \Sigma(2J+1) \ KC^2S(n,\ell) \qquad (11)$$

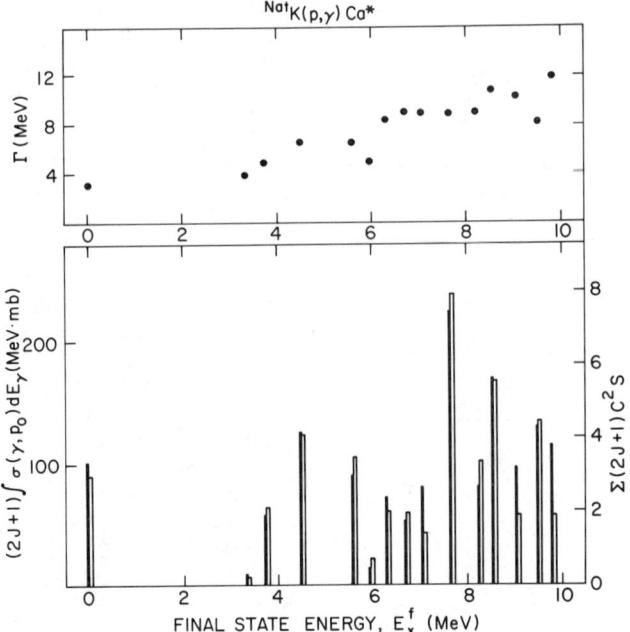

Fig. 8. Top: ESGDR width vs final state energy for the $^{Nat}K(p,\gamma)Ca^*$ reaction. Bottom: Comparison of integrated (γ,p_0) cross sections with proton transfer strengths for various final state energies. Solid bars are the $\int\sigma(\gamma,p_0)dE_\gamma$ cross sections and the open bars are the $^{39}K(^3He,d)^{40}Ca$ spectroscopic strengths.

Here f is a factor to account for any anisotropy in the angular distribution and K is the experimental constant of proportionality.

Since we have not measured any angular distributions, f has been taken from the literature for proton capture to the ground state GDR and this value used for all of the excited states. The values of C^2, f and K used in the above relationships to generate the bar graphs in Figs. 8, 9 and 10 are given in Table I along with the references used to obtain f.

Figure 11 shows the comparison between $K_{n,\ell}$ calculated using Eq. 8 and the experimental values listed in Table I. The agreement is very good, especially considering the simple model used to calculate $K_{n,\ell}$. This agreement gives us confidence in the resonance reaction model described above. That is, the (p,γ) reaction selectively picks out single-particle strengths and the mechanism proceeds through an ESGDR at intermediate proton energies.

So far the discussion has concentrated on understanding the reaction mechanism of proton capture at intermediate energies. To do this we have used the available spectroscopic information for discrete, resolvable levels to investigate the use of an

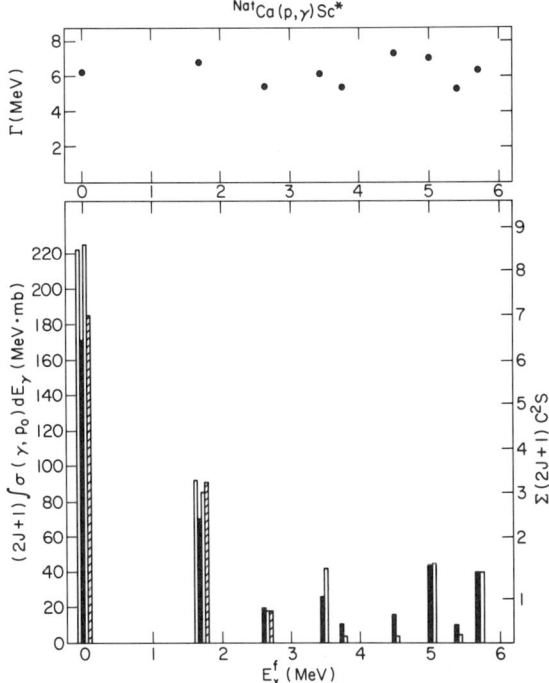

Fig. 9. Top: ESGDR width vs final state energy, E_x^f, for the $^{Nat}Ca(p,\gamma)Sc^*$ reaction. Bottom: Comparison of integrated (γ,p_0) cross sections with known proton transfer strengths. Solid bars are the $\int\sigma(\gamma,p_0)dE_\gamma$ strengths, open bars to the left of solid bars are results from (d,n) reaction, open bars to the right of solid bars correspond to the $(^3He,d)$ results, and the cross-hatched bars are (α,t) spectroscopic factors.

intermediate resonance model, i.e., the semi-direct model. Having justified its validity, this model should allow the determination of high-lying proton spectroscopic strength using the (p,γ) reaction which may be difficult to study using transfer reactions. However, at these higher excitation energies there is no certainty that the capture strength continues to be only single-particle in character. In fact, one would expect transitions to more complex final states ranging from two-particle states to the statistical states of the compound nucleus. To what extent the (p,γ) capture process remains single-particle requires more study.

Figures 12 and 13 show γ-ray spectra for $^{Nat}K(p,\gamma)Ca^*$ at E_p = 26 MeV and $^{Nat}Ca(p,\gamma)Sc^*$ at E_p = 28 MeV. The line shape fits shown have already been discussed. What is observed in both cases is the persistence of capture strength to even higher excitation energies. Also, what is especially evident in Fig. 12 is a GDR resonant shape centered near E_γ = 20 MeV. For a smooth, continuous level density the γ-ray spectrum will have the shape of

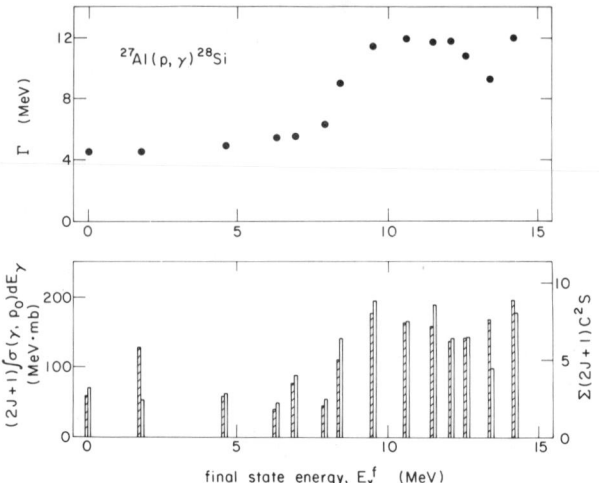

Fig. 10. Top: ESGDR width vs final state energy, E_x^f, for $^{27}Al(p,\gamma)^{28}Si$ reaction. Bottom: Comparison of integrated (γ,p_0) cross sections with known proton transfer strengths vs E_x^f. Cross hatched bars are $\int \sigma(\gamma,p_0)dE_\gamma$ strength and open bars are $^{27}Al(^3He,d)^{28}Si$ spectroscopic strengths.

Table I. Values of the constants used in evaluating eq. 11 to make the comparisons with spectroscopic factors shown in Figs. 8, 9 and 10. C^2 is the isospin Clebsch-Gordon coefficient, f is the angular anisotropy at $\theta_\gamma = 90°$ and K is the experimental value of $K_{n,\ell}$.

Reaction	C^2	a) f	K (MeV·mb)	b) Ref
$^{39}K(p,\gamma)^{40}Ca$	1/2	1.165	30	A
$^{40}Ca(p,\gamma)^{41}Ca$	1	1	25	None
$^{27}Al(p,\gamma)^{28}Si$	1/2	1.14	22	B

A: E. M. Diener et al., Phys. Rev. C7, 695 (1973).
B: P. P. Singh et al., Nucl. Phys. 65, 577 (1965).
a. $f = 1 - a_2/2$ where $W(\theta_\gamma) = 1 + a_2 P_2(\cos \theta_\gamma)$.
b. These references are the sources of the values of a_2 used to compute f.

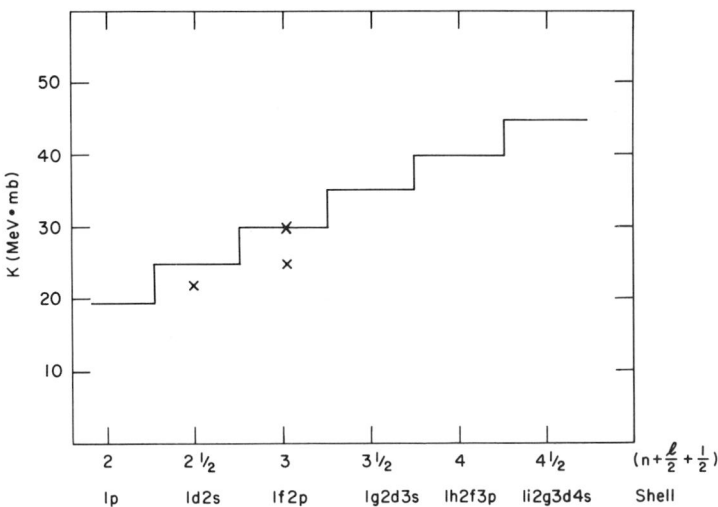

Fig. 11. Dependence of $K_{n,\ell}$ upon the shell occupied by the 1p1h excited state. $K_{n,\ell}$ is calculated in the schematic model using harmonic oscillator wavefunctions. The crosses are the experimental values, K. See Table I and text.

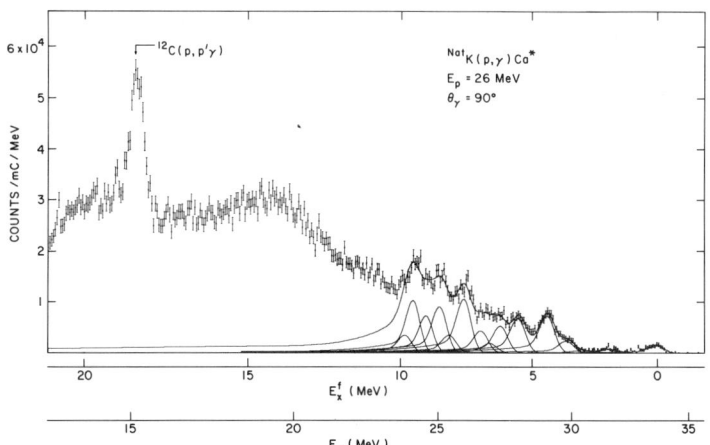

Fig. 12. $^{Nat}K(p,\gamma)Ca^*$ γ-ray spectrum at E_p = 26 MeV showing capture strength near E_x^f = 15 MeV in ^{40}Ca.

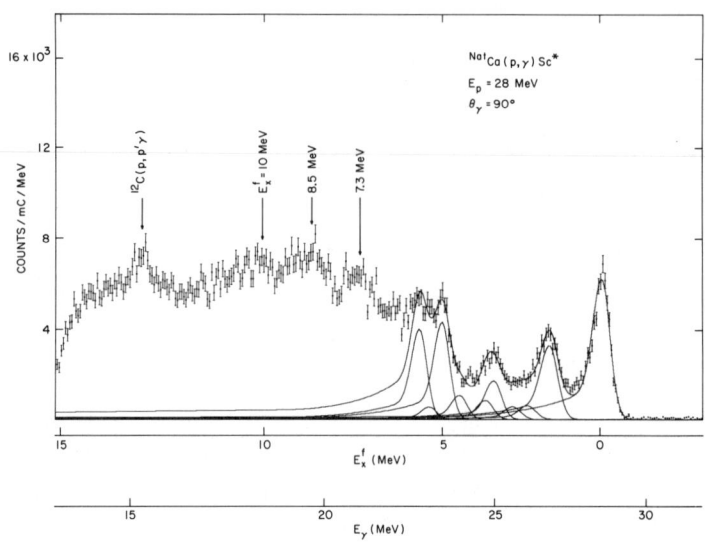

Fig. 13. $^{Nat}Ca(p,\gamma)Sc^*$ γ-ray spectrum for E_p = 28 MeV. Arrows indicate positions of clustered single-particle strength at E_x^f = 7.3, 8.5 and 10 MeV in ^{41}Sc.

the GDR convoluted with the detector's response function. The spectroscopic strength shown in Fig. 12 near E_x^f = 15 MeV is estimated to be about 5.5 to 7.5 units of $(2J+1)C^2S$ per MeV of excitation energy.

Figure 13 indicates that there is some clustering of strength for states in ^{41}Sc at E_x^f = 7.3, 8.5 and 10 MeV. Similar clustering is not seen in the $^{Nat}K(p,\gamma)Ca^*$ spectra since the coupling of the valence proton with the $d_{3/2}^{-1}$ hole in the ^{39}K core would result in a multiplet of states, spreading the strength out in energy. In addition to the three clusters there is a sizable, continuum of unresolved strength. There is approximately 2.5 to 3.1 units of $(2J+1)C^2S$ per MeV near E_x^f = 8.5 MeV.

6. SUMMARY

We have used the (p,γ) reaction at intermediate proton energies to study excited-state giant dipole resonances (ESGDR's) for discrete states up to E_x^f = 9.78 MeV in ^{40}Ca, E_x^f = 5.87 MeV in ^{41}Sc and E_x^f = 14.36 MeV in ^{28}Si. A comparison of the γ-ray spectra with proton transfer strength clearly shows a preference for the (p,γ) reaction to populate single-particle states. The cross sections for γ-ray transitions to each final state all show resonance-like shapes when plotted as functions of E_γ. The energies of these resonances peak at or very near E_γ = E_{GDR}, the energy of the ground state GDR, and show increasing widths for final state energies greater than about 5 MeV. Using detailed balance to convert from (p,γ) to (γ,p_0) cross sections, the energy-integrated (γ,p_0) strength, $\int\sigma(\gamma,p_0)dE_\gamma$, was compared with the proton transfer spectroscopic factor for each final state. A simple constant, $K_{n,\ell}$, was found to relate the integrated (γ,p_0) cross sections to the spectroscopic factors. $K_{n,\ell}$ was also shown to be accurately calculated using harmonic oscillator wave functions.

These results firmly establish that the (p,γ) reaction mechanism at these energies is proceeding through an intermediate GDR, and that it should be a viable probe of single-particle strength in nuclei.

ACKNOWLEDGEMENTS

I am pleased to acknowledge the efforts of my colleages who have contributed to this work. In particular, I am indebted to K. A. Snover with whom I began this work while I was at the University of Washington, Seattle, and A. M. Sandorfi who collaborated with us on the BNL experiments. I also wish to acknowledge the help of C. Gossett, L. Ricken, and G. Feldman.

Research has been performed under contract DE-AC02-76CH00016 with the United States Department of Energy.

REFERENCES

1. D. M. Brink, D. Phil. Thesis, Univ. of Oxford, 1955 (unpublished), reported in B. B. Kinsey, Handbuch der Physik, ed. S. Flugge (Springer-Verlag, Berlin, 1957), Vol. XL, p. 314ff; P. Axel, Phys. Rev. 126, 671 (1962).
2. D. M. Drake et al., Nucl. Phys. A203, 257 (1973).
3. M. A. Kovash et al., Phys. Rev. Lett. 42, 700 (1979); H. R. Weller et al., Phys. Rev. C25, 2921 (1982).
4. M. Anghinolfi et al., Nucl. Phys. A399, 66 (1983); J. T. Londergan and L. D. Ludeking, Phys. Rev. C25, 1722 (1982); R. J. Philpott and Dean Halderson, Nucl. Phys. A375, 169 (1982).
5. D. H. Dowell et al., Phys. Rev. Lett. 50, 1191 (1983).
6. G. E. Brown, Nucl. Phys. 57, 339 (1964).
7. K. A. Snover, private communication and Proc. of the Intern. Symp. on Highly Excited States and Nuclear Structure, Orsay, France, Sept. 5-8, 1983, Editions de Physique, to be published.
8. E. Hayward, Photonuclear Reactions, National Bureau of Standards Monograph 118, August 1970.
9. A. M. Sandorfi and M. T. Collins, NIM 222, 479 (1984).
10. E. M. Diener et al., Phys. Rev. C7, 695 (1973).
11. H. Fuchs et al., Nucl. Phys. A129, 545 (1969); P. M. Endt and C. Van der Leun, Nucl. Phys. A310, 1 (1978), p. 556ff and references therein.
12. D. H. Youngblood et al., Phys. Rev. C2, 477 (1970); P. M. Endt and C. Van der Leun, Nucl. Phys. A310, 1 (1978), p. 5933ff and references therein.

NUCLEAR SPECTROSCOPY USING TESSA FOLLOWING HEAVY ION FUSION REACTIONS

P. J. Nolan
Oliver Lodge Laboratory, University of Liverpool,
Liverpool, L69 3BX, U.K.

ABSTRACT

The total energy suppression shield array (TESSA) has been used to make a wide range of nuclear spectroscopy measurements following heavy ion fusion reactions. The experiments have been carried out at the Nuclear Structure Facility at Daresbury Laboratory using beams from ^{4}He to ^{82}Se at energies just above the Coulomb barrier. TESSA currently consists of six escape suppressed germanium detectors and a bismuth germanate ball designed to measure the total energy-multiplicity of the γ-radiation.

Discrete line studies have yielded levels with very high spins (e.g. $89/2^+$ in ^{159}Er), evidence for the collapse of neutron pairing (^{168}Hf) and highly deformed nuclear shapes (^{84}Zr and ^{132}Ce). TESSA can also be used to study gross effects in nuclei by studying the statistical and rotational continuum radiation. These studies have yielded extremely deformed nuclear shapes (e.g. ^{152}Dy), nuclear shapes and properties up to spin 60ℏ and information on the statistical γ-decay processes following heavy ion fusion reactions. Recent studies using a symmetric target-beam (^{80}Se + ^{80}Se) have shown evidence for higher than expected yields in the 2n channel.

INTRODUCTION

The use of arrays of escape suppressed spectrometers has greatly extended the work that can be done using gamma-ray spectroscopy especially in the study of high spin states in nuclei. In this talk I will present some of the work that has recently been carried out at the Nuclear Structure Facility (NSF) at Daresbury using the gamma-ray spectrometer TESSA2. The work has included a wide range of studies using heavy ion reactions to investigate the high spin structure of nuclei using both discrete line and continuum γ-ray spectroscopy. The nuclei investigated range from A = 80 to A = 236, but in this talk I will concentrate on just a few to illustrate the kind of work that can be done.

HEAVY ION FUSION EVAPORATION REACTIONS

The work I will talk about involves heavy ion fusion evaporation reactions. An example of such a reaction is shown in figure 1. The fusion of ^{98}Mo (target) and ^{36}S (beam) leads to the compound nucleus ^{134}Ce. At a beam energy of 150 MeV this compound

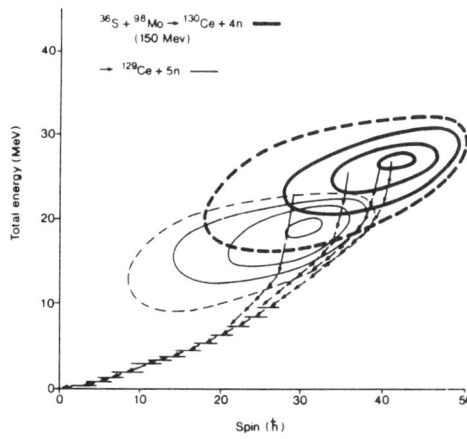

Fig. 1. Entry contours for ^{130}Ce and ^{129}Ce following the ^{98}Mo + ^{36}S reaction. Representative decay paths and the yrast states in ^{130}Ce are shown.

system decays by evaporating neutrons, protons and alphas, the main final product nuclei are ^{131}Ce (3n), ^{130}Ce (4n), ^{129}Ce (5n), ^{129}La (p4n), ^{127}Ba (α3n). Entry contours in excitation energy versus spin space are shown in figure 1 for the 4n and 5n products with some representative decay paths. Two distinct areas of study are apparent. Discrete line γ-ray spectroscopy allows the structure near the yrast line to be determined. This is carried out using high resolution germanium detectors, the final product and entry point being selected by measuring the total energy and number of gamma rays (related to spin). At high spin or away from the yrast line discrete transitions cannot be seen so the continuum gamma-ray spectrum has to be studied to gain further information. The continuum gamma-rays may be studied in high resolution using germanium detectors or in low resolution and higher efficiency using scintillation detectors (e.g. sodium iodide (NaI), bismuth germanate (BGO)). The total energy suppressed spectrometer array (TESSA2) has been designed to carry out both types of investigation.

THE TOTAL ENERGY SUPPRESSION SHIELD ARRAY (TESSA)

The use of escape suppression shields (ESS) to improve the quality of the γ-ray spectra from germanium detectors is well established. Until recently ESS were used in small numbers to carry measurements of γ-ray singles and γ-γ coincidences. The use of multiple ESS arrays is now a few years old, TESSA being the first example of such a system. TESSA was first used in 1980-1982 at the Niels Bohr Institute, Risø, Denmark by a Liverpool/ Manchester/NBI collaboration. Since then the apparatus has evolved to its present form (TESSA2) which has been operational at the NSF at Daresbury for almost two years.

The layout of TESSA2 is shown in figure 2. It consists of 6 ESS, employing sodium iodide suppression shields and n-type Hp germanium detectors. Also included is a 50 element bismuth germanate (BGO) total energy/multiplicity detector. The performance of TESSA2 has been described in detail by Twin et al [1].

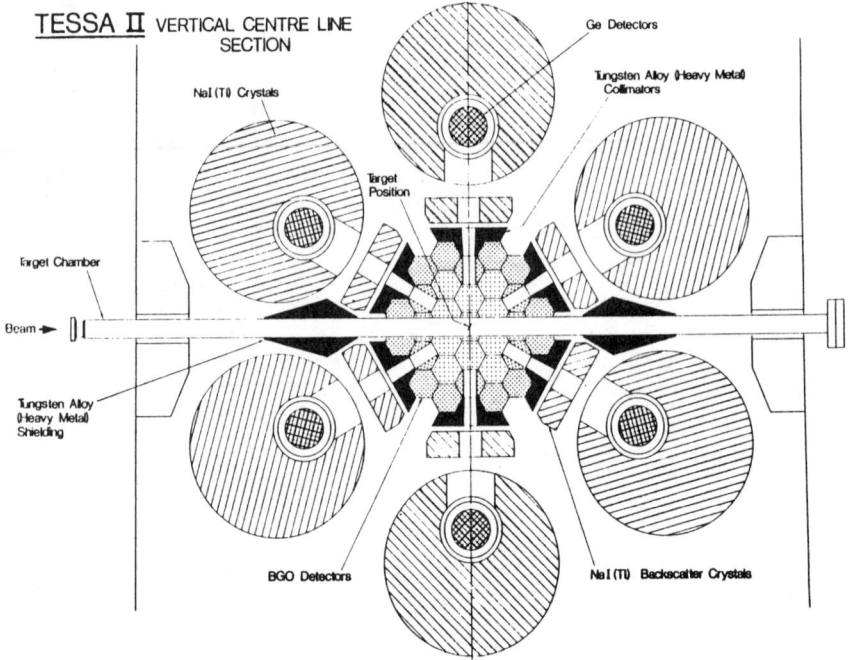

Fig. 2. The TESSA2 spectrometer consisting of 6 Compton suppressed germanium detectors and a 50 element bismuth germanate sum energy/multiplicity detector.

The important features will be summarised briefly here. The ESS give excellent peak to total fractions of 70% and 60% for γ-rays from ^{137}Cs and ^{60}Co sources. This is important when considering γ-γ coincidence measurements where it is the (peak/total)^2 that is important. So, compared to an unsuppressed Ge detector, the photopeak-photopeak coincidence fraction increases from 0.04 to 0.36 for 1.2 MeV γ-rays. The BGO detector used in TESSA2 is compact fitting into a cylinder < 25cm in diameter. This allows the ESS to be placed close to the target. The detection efficiency and response of the BGO ball has been measured using the technique described by Jääskaläeinen et al [2]. For 1.2 MeV gamma-rays, the detection efficiency is 93%. The sum energy and multiplicity resolutions are about 35% for 20 1.2 MeV gamma-rays. These results compare well with the much bigger NaI crystal balls at Oak Ridge [2] and Heidelberg [3] for E_γ < 1.2 MeV. The large photo-electric contribution in BGO reduces the scattering between elements. For 1.2 MeV the ratio of single to multiple hits is 0.17 (c.f. ~ 0.5 for the NaI balls). The neutron response of BGO has been investigated by Lone et al [4] and found to be small. On average in our BGO ball one detector fires in response to the neutrons from a (HI, 4n) reaction (c.f. 6-7 for a NaI ball).

DISCRETE LINE STUDIES

TESSA2 has been used to study high spin states in many nuclei. Some published examples are ^{84}Zr [5], ^{130}Ce [6], ^{157,158,159}Er [7] and ^{168}Hf [8]. I will give two examples of how TESSA2 data has allowed discrete line studies to be extended to high spin.

The deformed nucleus ^{158}Er has previously been established up to $J^{\pi} = 38^+$ in the yrast band by Burde et al [9]. This nucleus has been studied at Daresbury using the ^{114}Cd(^{48}Ca, 4n) reaction at 200 MeV [10]. The spectrum for the high spin part of the yrast band is shown in figure 3. The yrast band changes structure

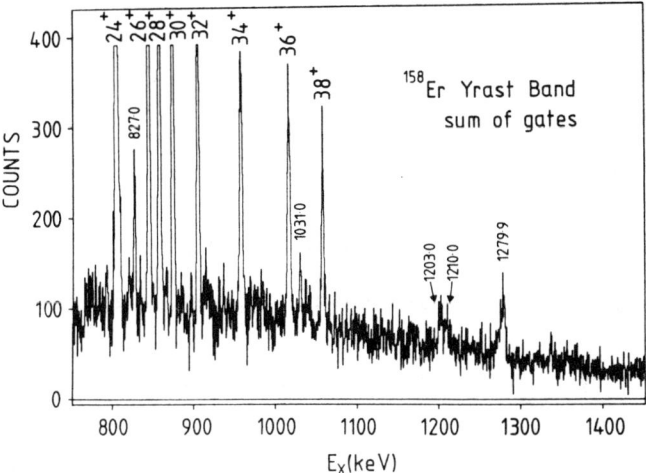

Fig. 3. Part of the gamma ray spectrum in coincidence with the yrast transitions in ^{158}Er. The data are from the ^{114}Cd + ^{48}Ca reaction at a beam energy of 200 MeV. The transitions marked with their energies lies above the 38^+ state.

abruptly above spin 38ℏ, the transitions from higher levels not following the expected rotational pattern. The decay scheme is shown in figure 4. This is evidence for the change in shape predicted by several calculations. One example is that described by Leander et al [11] who predict a shape change from prolate through triaxial towards oblate above spin 40ℏ. The highest spins seen in discrete line spectroscopy in TESSA2 data in deformed rare earth nuclei all come in nuclei near N = 90. In ^{159}Er the highest state seen has $J^{\pi} = 89/2^+$ [12], while in ^{160}Er [12] the yrast band is seen to $J^{\pi} = 40^+$. In heavier rare earth nuclei, the yrast band is only seen in discrete line spectroscopy to a lower spin e.g. ^{162}Hf 28^+, ^{168}Hf 34^+ [8].

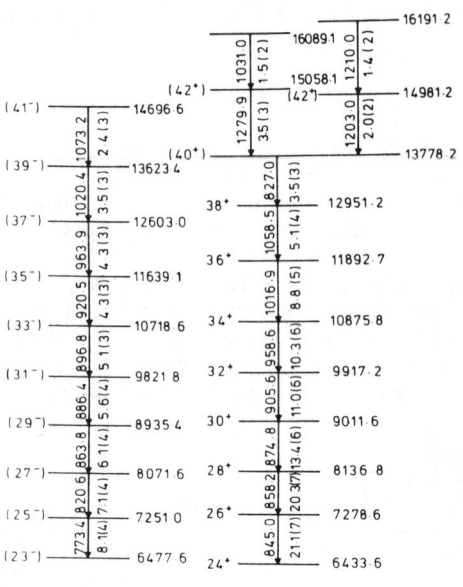

^{158}Er

Fig. 4. Partial decay schemes for ^{158}Er states in the yrast and one of the negative parity bands are shown. Transition energies (in keV) and relative intensities ($4^+ \rightarrow 2^+ = 100$) are given.

In ^{130}Ce the yrast band is only seen to $J^\pi = 28^+$ in discrete line spectroscopy [6], corresponding to a gamma-ray energy of 1320 keV. These data come from the ^{98}Mo(^{36}S,4n)^{130}Ce reaction at 150 MeV. In ^{132}Ce the situation in the yrast band is similar. Again the states are seen to $J^\pi = 28^+$, a gamma-ray energy of 1240 keV in the ^{100}Mo(^{36}S,4n)^{132}Ce reaction at 150 MeV. In ^{132}Ce a new high spin sequence is seen. Figure 5 shows the gamma-ray spectrum. A series of regularly spaced gamma-rays is seen from 807 keV to 1484 keV. This band feeds into the yrast band at spin 16 and 18 and hence, assuming the transitions are quadrupole, lies in the spin 20-40$\hbar$ region. This band has a high moment of inertia, $\mathcal{J}_{band}^2$ 58-62 MeV$^{-1}\hbar^2$. The spherical rigid body value for A = 132 is 48 MeV$^{-1}\hbar^2$, so this data may indicate a deformed shape. A rigid prolate ellipsoid (with a sharp surface) with $\varepsilon_2 \sim 0.35$ has a moment of inertia near 60 MeV$^{-1}\hbar^2$. A change in structure of the yrast band is expected in these nuclei in this spin region [13] and this could be the first indication of this in discrete line spectroscopy.

There are many other examples of new data at high spin which I do not have time to include. TESSA also allows non-yrast bands

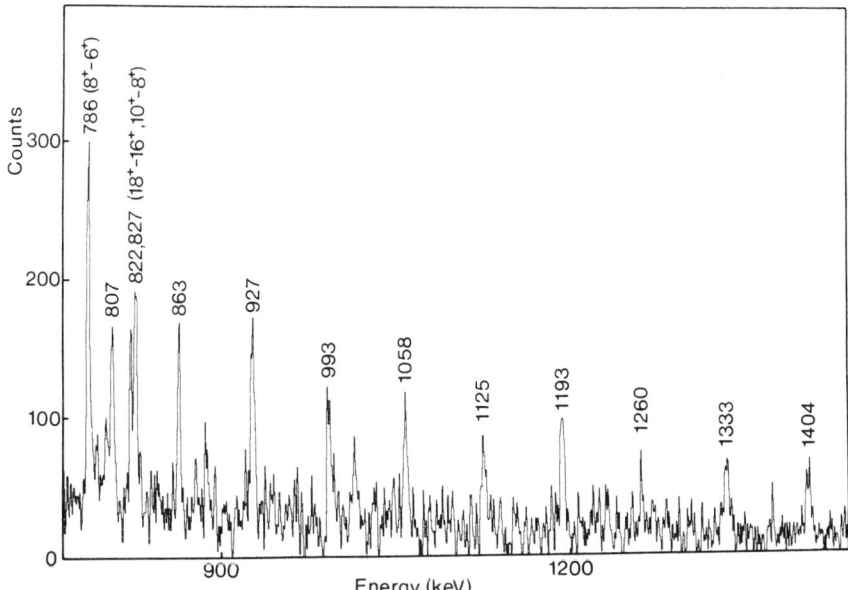

Fig. 5. Transitions in the new high spin, high moment of inertia band in ^{132}Ce. The spectrum shown is in coincidence with the sum of the 807, 863, 993, 1057, 1125 and 1260 keV transitions. The data are from the ^{100}Mo + ^{36}S reaction at 150 MeV.

to be studied and many are found in most nuclei (five or six being typical e.g. ^{159}Er [12], ^{130}Ce [14], but up to 10 or 12 in some cases e.g. ^{156}Dy [15], ^{180}Os [16]). Arrays of ESS are a rich source of data in discrete line spectroscopy allowing complex decay schemes to be assembled and much new nuclear structure information to be obtained.

CONTINUUM GAMMA-RAY STUDIES

The study of the continuum using E_γ-E_γ correlation spectroscopy is a well established technique [17]. Most experiments in the past have used NaI detectors ref. [17], a few have used germanium detectors e.g. ref. [18]. All these experiments have required numerical techniques to remove the uncorrelated background so that the ridge structure characteristic of rotational behaviour can be seen. Coincidence data from ESS with a good peak/total ratio allow these ridges to be seen in the raw data. This was first reported in ^{130}Ce by Nolan et al [6]. Data from TESSA2 have also been used to establish superdeformation in ^{152}Dy [19].

The nucleus ^{152}Dy was populated at high spin using the ^{108}Pd(^{48}Ca, 4n)^{152}Dy reaction at 205 MeV. Discrete lines are seen up to about spin 40ℏ, the decay scheme being typical of a spherical/oblate shape. Rotational structure is characterised at high spin by regularly spaced transitions where ideally the difference in gamma-ray energies is the same. Thus a plot of ΔE_γ ($= |E_{\gamma 1} - E_{\gamma 2}|$) v ($E_{\gamma 1} + E_{\gamma 2}$)/2 should show a ridge if rotational behaviour is present. This plot is shown in figure 6 and such a

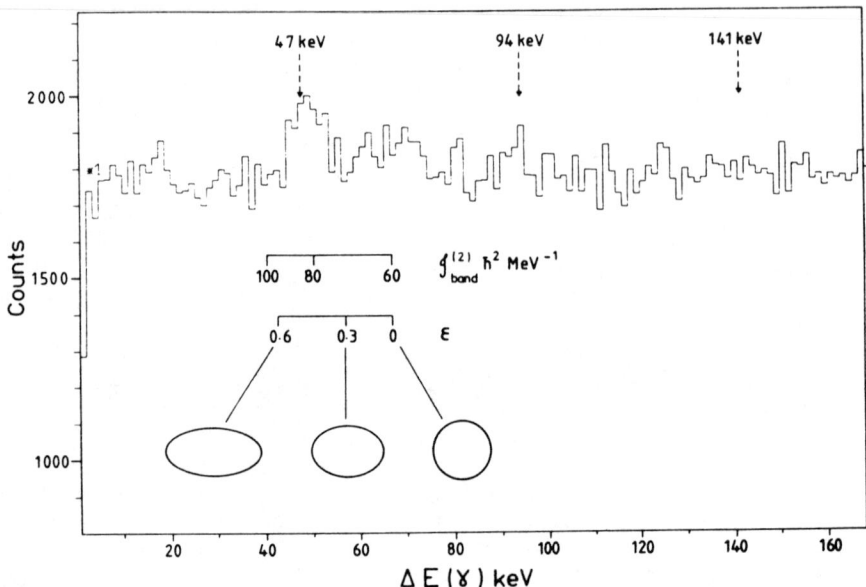

Fig. 6. A spectrum of $\Delta E_\gamma (= |E_{\gamma 1} - E_{\gamma 2}|)$ for ^{152}Dy. The data are from the ^{108}Pd + ^{48}Ca reaction at 205 MeV and correspond to ($E_{\gamma 1}$ + $E_{\gamma 2}$)/2 = 785-1330 keV. The ridge at ΔE_γ = 47 keV indicates rotational behaviour. Values of $\mathcal{J}^{(2)}_{band}$ ($= 4\hbar^2/\Delta E_\gamma$) and ε (for a rigid axially symmetric ellipsoid with sharp surfaces) are also shown.

ridge is seen at ΔE_γ = 47 $\pm$ 1 keV. The moment of inertia $\mathcal{J}^{(2)}_{band}$ ($= 4\hbar^2/\Delta E_\gamma$) obtained from the data is 85 $\pm$ 2 MeV$^{-1}\hbar^2$. This is 1.4 times the rigid spherical body value for A = 152. Such a large moment of inertia implies a high deformation. For a rigid axially symmetric prolate shape (with sharp surfaces) this corresponds to ε_2 = 0.51. This is in excellent agreement with recent predictions [20-22], which indicate a shape of $\varepsilon_2 \sim 0.5$, $\gamma = 0°$. The ridges are also narrow (figure 6) which indicate a constant moment of inertia and a stable shape. The FWHM of the ridge is 6.8 keV compared to the instrumental resolution of about 6 keV. This structure is seen in the data between 800 keV and 1350 keV, which corresponds to a spin 34-55$\hbar$ (assuming spin = $\mathcal{J}^{(2)}_{band} \times E_\gamma/2$).

The continuum γ-ray spectra also allow the gross structure of the continuum to be probed. The response function of the ESS allows it to be stripped from the data easily. The spectrum is assumed to consist of a narrow peak and a constant background, the ratio of which has been measured experimentally. The results of this procedure are shown in figure 7. Having removed the instrumental response, the ridges characteristic of rotational bands are still weak. The data shown are for ^{130}Ce from the ^{98}Mo(^{36}S, 4n)^{130}Ce reaction at 150 MeV. The BGO sum energy/multiplicity detector has been used to select the ^{130}Ce channel, but there are still contributions from the other reaction channels (mainly 3n, 5n, p4n and α3n). The results indicate that in the spin 30-50 region 10-20% of the intensity decays down rotational bands. The remaining intensity is uncorrelated thus indicating the high probability of out of band decay. This is perhaps not surprising as the many bands expected above the yrast line down which the nuclei decay will mix, hence the probability for many in bands decays is low.

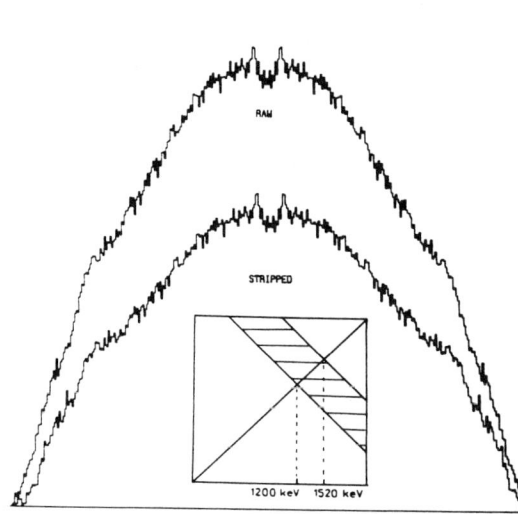

Fig. 7. Cuts perpendicular to to the $E_{\gamma 1} = E_{\gamma 2}$ for the ^{130}Ce $E_{\gamma 1}$ versus $E_{\gamma 2}$ matrix. The two spectra shown correspond to the raw data and that with the detector response stripped away.

Similar results are found for nuclei in the A = 160-170 deformed rare earths. This data would be improved by the ability to select the final nucleus more cleanly (e.g. triple coincidences).

FUSION OF SYMMETRIC SYSTEMS

It has been established for some time that the energy dependence of fusion evaporation cross-sections cannot be reproduced by statistical model decay calculations [23, 24]. For example, when light erbium or ytterbium systems are formed with an excitation energy of 55 MeV about 1% of the reaction cross section is observed in the 1n evaporation residue, this being about two orders of magnitude greater than expected. Work has been carried out to measure near barrier fusion cross sections [25, 26] where enhancements of up to three orders of magnitude are observed relative to single dimensional barrier penetration calculations.

This enhancement has recently been attributed to coupling to inelastic and transfer reaction channels [27]. This coupling leads to enhanced cross sections for evaporation residues for a few emitted neutrons and to greatly enhance partial wave cross sections for large angular momenta [28]. Recently Khoo et al [29] have studied the decay of the ^{156}Er compound system and found an enhanced yield in the two neutron evaporation residue. They suggest that the fusion of a symmetric (or near symmetric) target and beam system populates superdeformed states. Multiple detector arrays, including TESSA2, are ideal for studying such systems.

Preliminary results will be presented from the ^{80}Se + ^{80}Se system at beam energies between 260 and 310 MeV. The evaporation residues are identified by their characteristic gamma-rays detected in the 6 ESS. The resulting coincident multiplicity distribution can be measured in the BGO detector. Figure 8 shows

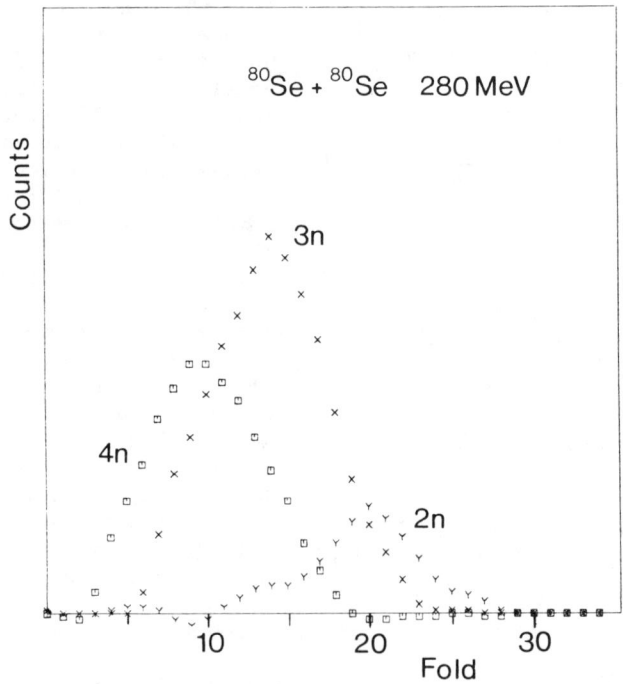

Fig. 8. Fold spectra for the 4n(^{156}Er), 3n(^{157}Er) and 2n(^{158}Er) products from the ^{80}Se + ^{80}Se reaction at 280 MeV. The channel were selected using known γ-rays. (Multiplicity ~ 1.3 x Fold).

the multiplicity distribution at a beam energy of 280 MeV for the 2n, 3n and 4n products. High multiplicity (and presumably high angular momentum) components are found especially in the 2n channel. This data analysis is continuing along with data taken on the ^{82}Se + ^{82}Se, ^{76}Se + ^{82}Se, ^{82}Se + ^{78}Se systems.

SUMMARY

The examples given illustrate the power of the TESSA2 system. The majority of the work carried out to date has concerned high spin states investigated using both discrete and continuum gamma-ray spectroscopy. The 6 ESS in the system have been the dominant feature in most of this work, the BGO detector providing mainly channel selection. The spectra from the 4π BGO detector have been used to investigate multiplicity distributions in the near barrier fusion of ^{80}Se + ^{80}Se, this time the ESS providing the channel selection. TESSA2 is a versatile gamma-ray spectrometer which has provided much new information during its two years of operation.

The main limitation of TESSA2 is the counting efficiency of 6 ESS with the germanium detectors being 25-27cm from the target. The size of the NaI shields makes it impractical to increase the number in the array beyond six. A new compact ESS (figure 9) using a BGO shield has been designed and tested [30].

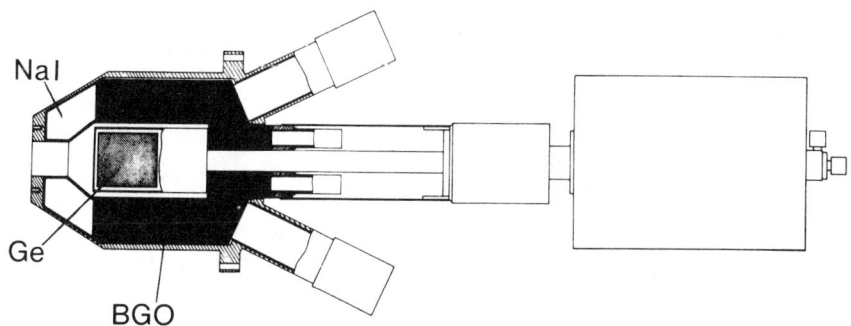

Fig. 9. The prototype BGO escape suppression shield and germanium detector system for future use in TESSA. The shield diameter is 170 mm.

This will allow the detector to be placed much closer to the target (~ 15cm) and up to 30 to be placed around the target, thus increasing the counting efficiency for γ-γ and γ-γ-γ coincidences.

ACKNOWLEDGEMENTS

The work I have described is the result of much labour by very many people. Space does not permit all to be named individually. I would like to thank the gamma-ray groups at Liverpool and Manchester Universities, Daresbury Laboratory and the Niels Bohr Institute whose data I have used in this talk. The work was made possible by the excellent range of beams available from the NSF and the invaluable technical support from the Physics

Department at Liverpool University in the design, construction and maintenance of TESSA2. This work was financially supported by the U.K. Science and Engineering Research Council.

REFERENCES

1. P. J. Twin et al., Nucl. Phys. A409, 343c (1984).
2. M. Jääskeläeinen et al., Nucl. Instr. Meth. 204, 385 (1983).
3. V. Metag et al., Nucl. Phys. A409, 331c (1984).
4. M. A. Lone et al., Proc. Int. Workshop BGO, Princeton University 237 (1982).
5. H. G. Price et al., Phys. Rev. Lett. 51, 1842 (1983).
6. P. J. Nolan et al., Phys. Lett. 128B, 285 (1983).
7. M. A. Riley et al., Phys. Lett. 135B, 275 (1984).
8. R. Chapman et al., Phys. Rev. Lett. 51, 2265 (1983).
9. J. Burde et al., Phys. Rev. Lett. 48, 530 (1982).
10. J. Simpson et al., Phys. Rev. Lett. 53, 648 (1984).
11. G. A. Leander et al., Physica Scripta 24, 164 (1981).
12. J. Simpson et al., Proc. 5th Nordic Meeting on Nuclear Physics, Jyväskylä, March 1984.
13. T. Bengtsson and I. Ragnarsson, Nucl. Phys. to be published.
14. D. M. Todd et al., J. Phys. G. 10, (1984).
15. M. A. Riley et al., Proc. 5th Nordic Meeting on Nuclear Physics, Jyväskylä, March 1984.
16. R. M. Lieder, Private communication.
17. O. Andersen et al., Phys. Rev. Lett. 43, 687 (1979).
18. M. A. Deleplanque et al., Phys. Rev. Lett. 45, 172 (1980).
19. B. Nyakó et al., Phys. Rev. Lett. 52, 507 (1984).
20. I. Ragnarsson et al., Nucl. Phys. A347, 287 (1980).
21. S. Åberg, Physica Scripta 25, 23 (1982).
22. Y. Schutz et al., Phys. Rev. Lett. 48, 1534 (1982).
23. J. Gilat et al., Phys. Rev. C7, 1973 (1973).
24. S. Della Negra et al., Z. für Physik A282, 75 (1977).
25. M. Beckerman et al., Phys. Rev. Lett. 45, 1472 (1980) Phys. Rev. C25, 837 (1982).
26. R. Stockstad and E. Gross, Phys. Rev. C23, 281 (1981).
27. R. Broglia et al., Phys. Lett. 133B, 34 (1983).
28. S. Landowne and C. H. Dasso, Phys. Lett. 138B, 32 (1984).
29. T. L. Khoo et al., 'High angular momentum properties of nuclei' ed. N. R. Johnson, (Harwood Academic Press, 1982), p.179.
30. P. J. Nolan et al., Nucl. Instr. Meth. (to be published).

CRYSTAL BALL STUDIES OF GIANT RESONANCE GAMMA DECAY

J. R. Beene, F. E. Bertrand, and M. L. Halbert*
Oak Ridge National Laboratory,† Oak Ridge, Tennessee 37831

ABSTRACT

We have carried out coincidence experiments to investigate the photon and neutron emission from the giant resonance region in ^{208}Pb and ^{90}Zr using the ORNL Spin Spectrometer, a 72-segment NaI detector system. States in ^{208}Pb and ^{90}Zr were excited by inelastic scattering of 380-MeV ^{17}O. We have determined the total gamma-decay probability, the ground-state gamma branching ratio, and the branching ratios to a number of low-lying states as a function of excitation energy in ^{208}Pb to ~15 MeV. Especially interesting observations include the absence of a significant branch from the giant quadrupole resonance to the 3⁻ state at 2.6 MeV, a strong branch from this resonance to a 3⁻ state at 4.9 MeV, and the dominance of decays to various 1⁻ states at 5-7 MeV from the region around 14 MeV of excitation (E0 resonance). Comparable but less complete data were also obtained on ^{90}Zr.

INTRODUCTION

A large amount of systematic data on the average properties of isoscalar giant electric multipole resonances has been obtained over the last decade, primarily from inelastic scattering experiments using a variety of probes.[1-3] Many of the questions now being asked about the giant resonances (GR) will require more detailed experimental data, which can only be obtained from coincidence (i.e., decay) experiments.[1,4,5] Such experiments offer the possibility of probing details of the structure of the GR not addressed by existing systematics.

The GR are described microscopically as a coherent superposition of one-particle one-hole excitations relative to the ground state.[1,6,7] This coherent state is connected — by definition — to the ground state by a strong electromagnetic matrix element. Observation of the corresponding electromagnetic decay deexciting the GR is of great importance, because of its direct relationship to the concept of a GR, since it offers the possibility of a determination of the resonance strength independent of that provided by analysis of

---

*Collaborators on the experimental work include D. C. Hensley, R. L. Auble, D. J. Horen, R. L. Robinson, T. P. Sjoreen, and R. O. Sayer.

†Operated by Martin Marietta Energy Systems, Inc., under contract DE-AC05-84OR21400 with the U.S. Department of Energy.

inelastic scattering data with reaction models. Unfortunately, the electric GR lie above particle emission thresholds, with the consequence that the γ decay, in heavy nuclei, occurs for only about one in 10^4 decays.

The particle-hole states that make up the resonance can decay directly into the continuum, producing a free particle and the A-1 nucleus in the corresponding hole state. This is considered a direct decay process, and the corresponding width $\Gamma^{\uparrow}$ is called the escape width. Observation of the distribution of hole states left behind after such decays would provide detailed information about the microscopic structure of the resonances. Unfortunately, in a heavy nucleus such direct particle decays are also rare and difficult to isolate from more common processes.[8,9] The resonances in a heavy nucleus typically lie in a region of very high level density. The simple 1p-1h states of the resonance are consequently mixed or damped into the more complex np-nh states which exist at the same excitation energy.

This mixing or damping can be thought of as an alternative decay process for the coherent state.[5,8,9] From this point of view, the GR is excited as a primary doorway state in the inelastic scattering process. This state decays directly via $\Gamma^{\uparrow}$ or by gamma emission, or it "decays" into the continuum of more complex (compound) states. These states then decay statistically, usually by particle emission. A width $\Gamma^{\downarrow}$, the spreading width, is associated with this decay into the continuum. The observed width of the GR state is thus $\Gamma_T = \Gamma^{\uparrow} + \Gamma^{\downarrow}$ (we can safely neglect Γ_γ). For heavy nuclei, $\Gamma_T \sim \Gamma^{\downarrow}$ (Refs. 8,9). A microscopic understanding of this damping process is the focus of current theoretical work on giant resonances.[5] Decay studies can provide insight into this process too, if, as has been suggested, the most important states involved in the mixing process are the 2p-2h states formed by coupling the 1p-1h states of the resonance to low-lying surface vibrations.[4] Evidence for the importance of such couplings should appear in the particle or gamma decay to the low-lying collective states.

EXPERIMENT

We have recently carried out an experiment at the HHIRF at ORNL, designed to study both the γ decay and particle decay of the giant resonance region (~9 to 20 MeV of excitation) in ^{208}Pb.

The resonance region of ^{208}Pb was excited by inelastic scattering of 381-MeV ^{17}O from ^{208}Pb. Oxygen-17 was chosen as a projectile because the low neutron binding energy (4.1 MeV) minimizes interference from gamma rays from projectile excitation. The inelastically scattered ^{17}O was detected in six cooled Si surface-barrier telescopes arranged symmetrically around the beam at an angle θ = 13° subtending Δθ = 3° and Δφ = 9° each and a total solid angle of 22.6 msr. The telescopes consisted of two elements of thickness ~500 μm and ~1000 μm, respectively. The energy resolution was ~800 keV, and the mass resolution was sufficient to separate ^{17}O from adjacent oxygen isotopes. A singles spectrum of inelastically scattered ^{17}O is shown

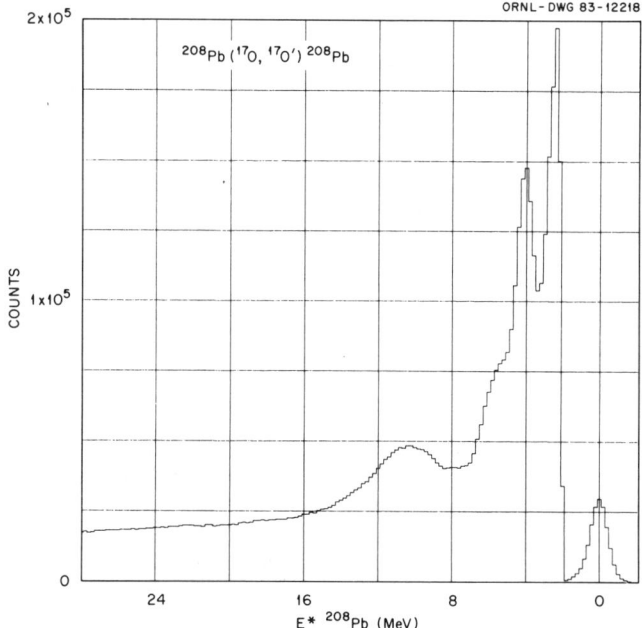

Fig. 1. Spectrum of 380-MeV ^{17}O scattered by ^{208}Pb. The elastic peak has been prescaled by a factor 512. The large bump between 9 and 15 MeV results from excitation of giant resonances.

in Fig. 1. Strong excitation of the giant resonance region, centered at ~11 MeV, is evident. Decay products were detected in 70 elements of the ORNL Spin Spectrometer. The spectrometer, which has been described in detail elsewhere,[10] consists of a spherical shell of NaI 17.8-cm thick, divided into 72 independent modules surrounding the target chamber. For the present experiment, two modules at 0° and 180° were removed to allow the beam to enter and leave the chamber. A photograph showing the experimental setup, including a portion of the Spin Spectrometer, is shown in Fig. 2. The response of the spectrometer to high-energy photons was determined by using the ^{12}C(p,p')^{12}C reaction with 24-MeV protons, which produces 4.43-, 12.71-, and 15.11-MeV gamma radiation. The response at lower energies was obtained from a variety of radioactive sources.

Figure 3 shows some of the levels of ^{207}Pb and ^{208}Pb relevant to the present experiment. The goal of the experiment is to study, as completely as possible, the decay of states in the 9- to 16-MeV region of ^{208}Pb. Only two decay modes are important. Neutron emission accounts for >99% of decays, while gamma rays are emitted with a probability of ~10^{-3} to 10^{-4}. This report focuses on the gamma decay. Copious information on n decay was also obtained and will be reported elsewhere.

Fig. 2. Six-telescope array in the Spin Spectrometer. The exit hemisphere and one part of the spherical reaction chamber were removed for the photograph. The beam enters from the right, and the target is at the center.

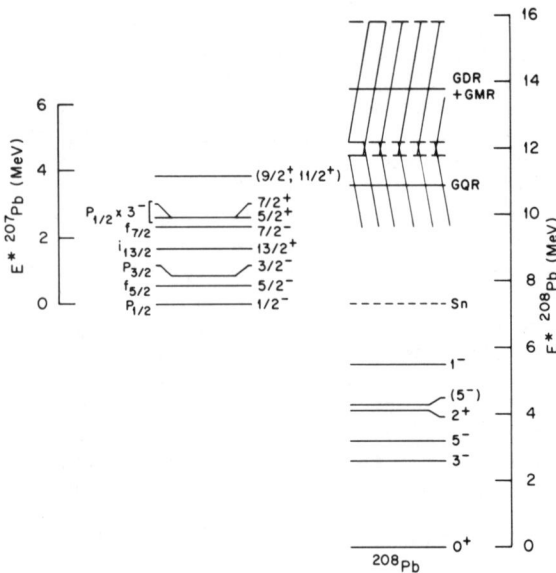

Fig. 3. Selected levels in ^{208}Pb and ^{207}Pb. The configuration labels on the ^{207}Pb states refer to neutron hole states.

ANALYSIS AND DISCUSSION

The experiment posed a number of difficulties for which the Spin Spectrometer, with its very large efficiency and multiple segments, proved almost ideal. The chief experimental problems were, first, isolating gamma decays from the $>10^3$ times more frequent n decays in the GR region; second, distinguishing direct gamma transitions to the ground state from multiple or cascade decays; and, third, isolating decays which directly populated low-lying states of interest (e.g., the 3^-, 2.61-MeV state) by a single gamma ray from the GR region.

The raw data obtained from the spectrometer consisted of pulse heights from the individual NaI elements and times of these pulses relative to the inelastically scattered ^{17}O with which they were in coincidence. A number of derived parameters were obtained which were used to address the questions raised earlier. The total gamma-ray pulse height, $H = \Sigma h_i$, was constructed by summing all those pulses which occurred within a prompt time window. This window (which was a function of pulse height) was narrow enough to eliminate pulses resulting from detection of neutrons with energies less than ~5 MeV, due to their longer flight time to the NaI. Single high-energy gamma rays are extremely unlikely to trigger a single NaI detector. Consequently, the number of detectors triggered is not very useful for isolating single gammas. A more useful quantity can be constructed by considering each pulse height observed in an element of the spectrometer as a vector quantity, $\vec{h}_i$, with direction determined by the location of the element, from which the quantity $V = |\Sigma \vec{h}_i|/H$ is formed. For a single high-energy gamma ray, $V \sim 1$, while for multiple gamma rays a smaller value of V is much more likely. Other useful quantities are the cluster sum pulse height and the cluster multiplicity. They are constructed for each event as follows. First, the largest pulse height is found, and a cluster sum is created by adding to it all the pulse heights in the five or six nearest neighboring detectors. Then the next largest pulse height not yet included in a sum is found, and a cluster sum is calculated from its nearest neighbors (not including those already used). This process continues until all the NaI pulses which satisfy the time gate are used. The number of clusters found is called the cluster multiplicity. For events such as those encountered in ^{208}Pb decay, in which a small number of gamma rays (usually fewer than four) are emitted, the cluster sums are a much better reflection of individual gamma-ray energies than the separate NaI pulse heights.

The separation of neutron decays from purely gamma decays is illustrated in Fig. 4. The horizontal axis measures excitation energy in ^{208}Pb obtained from the kinetic energy of the inelastically scattered ^{17}O ions. The vertical axis is the sum gamma-ray energy. The upper solid line in the figure is the line which would be occupied by events for which these two quantities are equal. Purely gamma decays were isolated by placing a gate around this line, as in Fig. 5a (the width of the gate in each direction reflecting instrumental resolution). Another line is drawn 7.4 MeV (the n binding energy in ^{208}Pb) below this line in Fig. 4. Events in which a neutron was emitted should lie below this line in the figure.

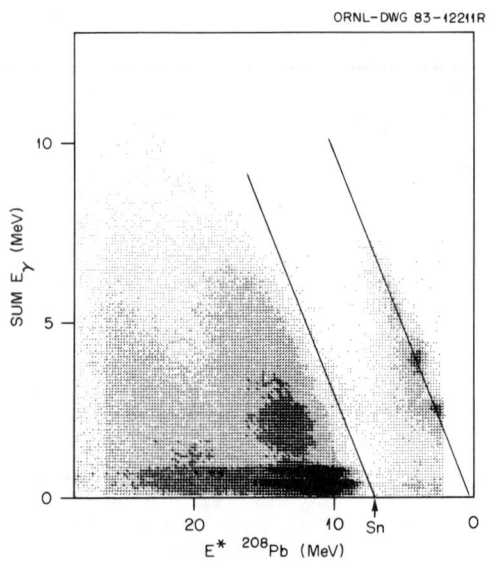

Fig. 4. Two-parameter density plot of events from ^{208}Pb(^{17}O,^{17}O') in which one or more NaI detectors registered a delayed pulse. These should be due to neutron decays. The abscissa is derived from the energy lost by the inelastic ^{17}O. The ordinate is the sum of the γ-ray energies seen in the NaI detectors and is equal to the excitation energy in the residual nucleus.

Direct single-step transitions to the ground state were isolated by requiring that the cluster multiplicity be one and the parameter V ≥ 0.98. (This value was arrived at experimentally using the 15.1-MeV ^{12}C calibration data.) Figure 5b shows the result of imposing this requirement. Figure 6 shows spectra obtained by projecting the gates in Figs. 5a and 5b onto the sum gamma-energy axis. Taking the ratio of spectra such as these produces the ground-state branching spectrum shown in Fig. 7. The peaks in this spectrum below the neutron binding energy are at the position of states in ^{208}Pb known to have large ground-state branches. Above the neutron binding energy, the ground-state branching falls off rapidly until the vicinity of the giant quadrupole resonance (GQR) is reached. The large peak in the branching spectrum between ~9 and 15 MeV nicely illustrates the strong localization of electromagnetic strength to the ground state in this region. The bump contains contributions from both the GQR at 10.6 MeV and the giant dipole resonance (GDR) at 13.4 MeV, which is very weakly excited in the reaction. We have not

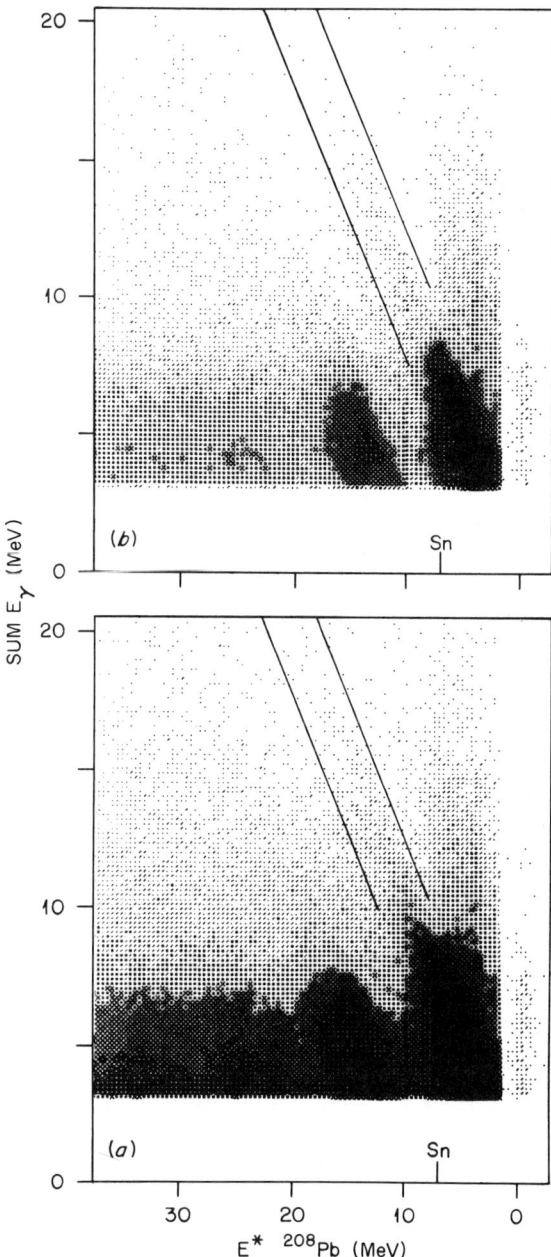

Fig. 5. Two-parameter density plot of events from ^{208}Pb(^{17}O,^{17}O') in which no NaI detector registered a delayed pulse. The axes are the same as for Fig. 4. Events falling between the pairs of lines are due to γ-decay events: (a) all events; (b) events satisfying the additional requirement $V > 0.95$ to select ground-state transitions.

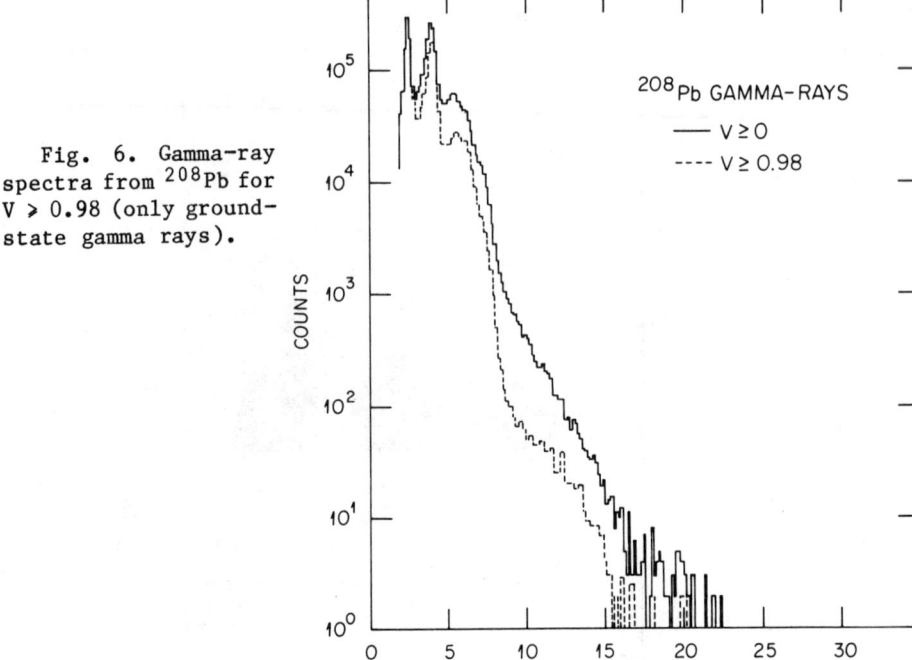

Fig. 6. Gamma-ray spectra from ^{208}Pb for V > 0.98 (only ground-state gamma rays).

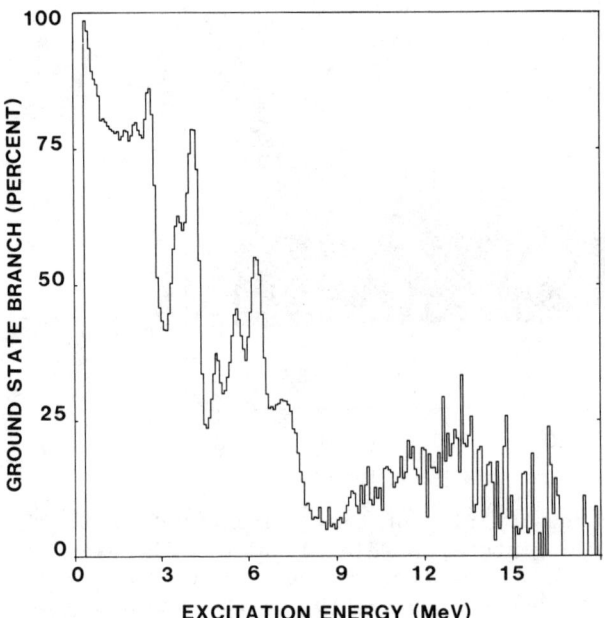

Fig. 7. Ratio of ground-state γ-decay events to total gamma yield as a function of ^{208}Pb excitation energy.

been able to decompose the ground-state gamma spectrum into quadrupole and dipole components as a function of energy; however, we can establish that the region of 9.5 to 11.5 MeV consists of (70 ± 10)% quadrupole radiation. Contribution of multipolarities, other than L = 1 and 2, to the ground-state decay is extremely unlikely. The spectrum of ground-state gamma rays was fit using the resonance parameters in Table I. By dividing the ground-state gamma-ray yield by the singles yield of scattered particles populating the GQR, obtained from a fit to the ^{17}O singles spectrum, we obtain

$$\frac{\Gamma_{\gamma 0}}{\Gamma_T} = (3.27 \pm 0.45) \times 10^{-4} .$$

This value has been corrected for instrumental efficiency and for the fraction of quadrupole radiation in the region obtained from fits to the photon angular distribution. This result can be used to obtain an absolute value for $\Gamma_{\gamma 0}$ if we identify the Γ_T with the spreading width of the GQR, which, in turn, is identified with the experimental width of the resonance from Table I.

TABLE I. Properties of states above 8 MeV in ^{208}Pb observed in ^{208}Pb(p,p'), from Ref. 14. The last two columns refer to the present ^{208}(^{17}O,^{17}O') experiment. Expected σ is the cross section expected for ^{17}O scattering based on the proton results. Observed σ is the cross section which we observe. Uncertainties of about 15% apply to both the observed and expected cross sections.

Excitation energy (MeV)	L	Γ (MeV)	EWSR fraction (%)	For (^{17}O,^{17}O') Expected σ (mb/sr)	Observed σ (mb/sr)
8.11	4	0.4	3	9 }	13
8.35	3	0.4	4	5 }	
8.86	2	0.4	7	8	6
9.34	2	0.4	5	5	13
10.6	2	2	70	50	60
12.0	4	2.4	10	17	18
13.6	1	4.0	100	~4 }	20
13.9	0	2.9	100	10 }	

$$\Gamma_{\gamma 0} = \frac{\Gamma_{\gamma 0}}{\Gamma_T} \times \Gamma_{EXP} = 654 \pm 91 \text{ eV} \quad (GQR).$$

This implies, taking $E_{\gamma 0} = 10.6$ MeV,[11]

$$B(E2\downarrow) = (5.81 \pm 0.81) \times 10^3 \text{ } e^2 fm^4.$$

This should correspond with the energy-weighted sum-rule value,[12,1]

$$\sum EB(E2\uparrow) = 49.9 \text{ } A^{5/3} \text{ } e^2 fm^4$$

or

$$B(E2\downarrow)_{EWSR} = 49.9 \text{ } A^{5/3}/5E_{\gamma 0} = 6.86 \times 10^3 \text{ } e^2 fm^4.$$

Therefore,

$$\frac{B(E2\downarrow)}{B(E2\downarrow)_{EWSR}} = 0.85 \pm 0.12$$

for the GQR.

A significant yield of dipole ground-state gamma transitions is observed above 12 MeV. It is not possible to establish the excitation cross section for the GDR from the spectrum of inelastically scattered ^{17}O since the GDR is so weakly excited and so broad. Calculations assuming the GDR is exclusively Coulomb excited predict a cross section of ~4 mb. Taking this value and assuming a ground-state decay width exhausting the E1 sum rule, we find the observed yield of dipole transitions to be ~60% of the predicted value, with an uncertainty of ~30%.

It is also of great interest to see if gamma-decay branches other than the ground-state decay can be identified. In particular, direct decays to the low-lying collective states, the 3⁻ state at 2.61 MeV and the 2⁺ state at 4.085 are of interest,. Figure 8 shows the relative strength of gamma-ray branches to a number of low-lying states. Figure 8a is for ground-state transitions, and Figs. 8b and 8c are for direct decays to the 3⁻, 2.61 and 2⁺, 4.08 states, respectively. Multistep cascades are ruled out in these cases by requiring that the cluster multiplicity discussed earlier be precisely two. Figure 8d is the relative strengths for decays populating the 4.97-MeV, 3⁻ state. The yield distributions in Fig. 8, other than the ground-state yield, must be considered semiquantitative, especially where they indicate very small strengths, since adequate background subtraction has not been done. Nevertheless, they are valuable to indicate general features. A few of the more striking aspects include the marked absence of strength to the 2.61 and 4.08 MeV states across the resonance region. Another interesting feature is the strong yield of decays to the 2.61-MeV state at ~5.2 MeV of excitation energy. This might be an indication of the long sought two-phonon octupole vibrational states. A strong yield of decays to the 3⁻ state at 4.97 MeV (thought to be a noncollective state dominated by a single 1p-1h configuration) is seen to appear at ~9 MeV and remains significant across the GQR region. This is in marked contrast to the absence of decays to the lower lying collective 3⁻ state. A very similar, though weaker, strength distribution to that shown in

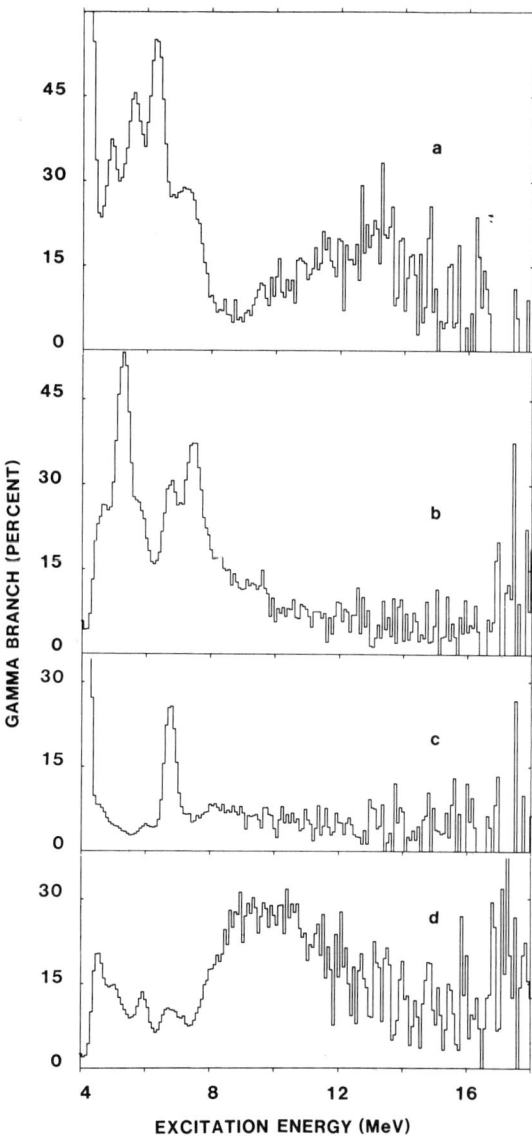

Fig. 8. Relative gamma-decay strengths for transitions to a number of low-lying levels in ^{207}Pb: (a) for ground-state decays; (b) for transitions to the 2.61-MeV, 3^- state; (c) the 4.08-MeV, 2^+ state; (d) the 4.97-MeV, 3^- state.

Fig. 8d is seen for decays to a 5⁻ state at 3.9 MeV. This indicates the existence of high-spin strength underlying the GQR. A more quantitative treatment of decay branches from the GQR region (i.e., a bin from 9.5 to 11.5 MeV) is shown in Table II. It should be noted that the absence of decay to the 2.6-MeV, 3⁻ state, which appears remarkable at first sight, agrees with a recent calculation by Bortignon, Broglia, and Bertsch.[13]

TABLE II. Relative gamma branching to low-lying states in ^{208}Pb from an excitation energy region 9.5-11.5 MeV [E(GQR) ± Γ(GQR)/2]. The 5-7 MeV, 1⁻ states refers to a group of 1⁻ states in that region known from (γ,γ') experiments.

Energy	J^π	Relative gamma branch (%)
0	0⁺	20 ± 2
2.61	3⁻	0.8 ± 0.8
3.97	5⁻	~5-10
4.08	2⁺	$0.3^{+1.0}_{-0.3}$
4.97	3⁻	36 ± 5
5-7	1⁻	23 ± 9

We hope we have been able to give an indication of the richness of detailed data which these experiments have opened up. We have not even mentioned the neutron decay information. We believe that as the results are refined and more and improved data are obtained, we will make a significant contribution to the understanding of the microscopic properties of the giant resonances.

REFERENCES

1. F. E. Bertrand, Annu. Rev. Nucl. Sci. 26, 457 (1976).
2. Giant Multipole Resonances, ed. F. E. Bertrand (Harwood, Academic, New York, 1980).
3. F. E. Bertrand, Nucl. Phys. A354, 129c (1981).
4. P. F. Bortignon and R. A. Broglia, Nucl. Phys. A317, 405 (1981).
5. G. F. Bertsch, P. F. Bortignon, and R. A. Broglia, Rev. Mod. Phys. 55, 287 (1983).
6. G. R. Satchler, Phys. Rep. 14, 99 (1974).
7. K. Goeke and J. Speth, Annu. Rev. Nucl. Sci. 32, 65 (1982).
8. G. J. Wagner in Giant Multipole Resonances, ed. F. E. Bertrand (Harwood Academic, New York, 1979).
9. L. S. Cardman, Nucl. Phys. A354, 173c (1981).
10. M. Jääskeläinen et al., Nucl. Instrum. Methods Phys. Res. 204, 385 (1983).
11. A. Bohr and B. R. Mottelson, Nuclear Structure, Vol. I (Benjamin, Reading, Mass., 1969).
12. A. Bohr and B. R. Mottelson, Nuclear Structure, Vol. II (Benjamin, Reading, Mass., 1975).
13. P. F. Bortignon, R. A. Broglia, and G. F. Bertsch, to be published.
14. F. E. Bertrand, to be published.

OBSERVATION OF RADIATIVE CAPTURE IN ^{90}Zr-INDUCED FUSION REACTIONS

H.-G. Clerc, C.-C. Sahm, E. Tschöp, W. Schwab
Institut für Kernphysik, Technische Hochschule,
6100 Darmstadt

K.-H. Schmidt, R.S. Simon, J.-G. Keller, W. Reisdorf,
F. Heßberger, G. Münzenberg, B. Quint
Gesellschaft für Schwerionenforschung, GSI,
6100 Darmstadt
Federal Republic of Germany

ABSTRACT

In the fusion reactions ^{90}Zr + ^{90}Zr, ^{92}Zr, and ^{94}Mo the radiative fusion process, that is the deexcitation of the compound nucleus only by γ-radiation, was detected. The radiative fusion was identified in all three cases by the observation of the ground state α-decay of the compound nucleus. For the reaction ^{90}Zr (358 MeV) + ^{90}Zr, the γ-rays were detected in coincidence with evaporation residues. Gamma rays from radiative fusion and from particle evaporation channels were separated by establishing a time correlation between the evaporation residues and their subsequent radioactive α-decay. The γ-rays were observed in 20 NaI detectors covering a total solid angle of 90 % of 4 π. This set-up allowed a determination of the γ multiplicity distribution as well as of the spectral distribution of single γ-transitions, of the γ sum energy, and of the angular distribution of γ-rays. The results do not support the assumption that specific direct processes associated with high energy γ rays are responsible for the large cross sections of up to 50 μb observed for radiative fusion. The low excitation energy of the compound nuclei, which is achievable with these heavy, symmetric systems seems to favour radiative fusion. The excitation functions for radiative fusion and for particle evaporation were reproduced by statistical model calculations.

INTRODUCTION

The fusion process of two heavy nuclei involves a large rearrangement of nuclear matter, and it usually leads to a highly excited compound nucleus. This compound nucleus may either cool down by the emission of nucleons and α-particles and form evaporation residues, or it may fission. Recently, however, the compound nucleus itself in its ground state was observed[1] as a final reaction product in the reaction $^{90}Zr + ^{90}Zr$, see fig. 1. Obviously, in this process the compound nucleus must have emitted γ-rays only. The cross section for this radiative fusion process is rather large and may reach up to 50 μb, see fig. 2. Following this observation, the radiative fusion process was also found[2] in the reactions $^{90}Zr + ^{92}Zr \to ^{182}Hg$ and $^{90}Zr + ^{94}Mo \to ^{184}Pb$. In all three cases, the γ-channel was identified by the observation of the α-decay of the compound nucleus. Due to their Q-values, very low excitation energies of the compound nuclei can be achieved with these systems. This may play an important role for enabling the radiative fusion by reducing the probability for particle evaporation. A comparison of the excitation functions for radiative fusion and the neutron evaporation channels in the system $^{90}Zr + ^{90}Zr$ as shown in fig. 2 demonstrates that radiative fusion becomes the dominating channel for excitation energies below about 16 MeV.

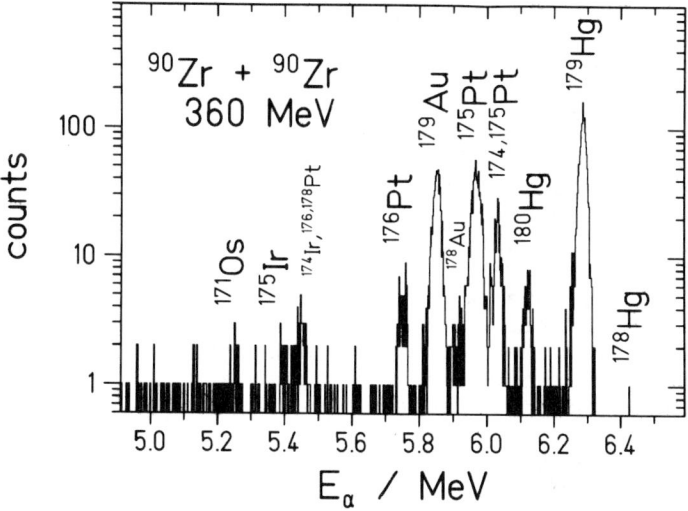

Fig. 1. α-spectrum of fusion products and their daughter nuclei in the reaction $^{90}Zr + ^{90}Zr$ at E_{LAB} = 360 MeV.

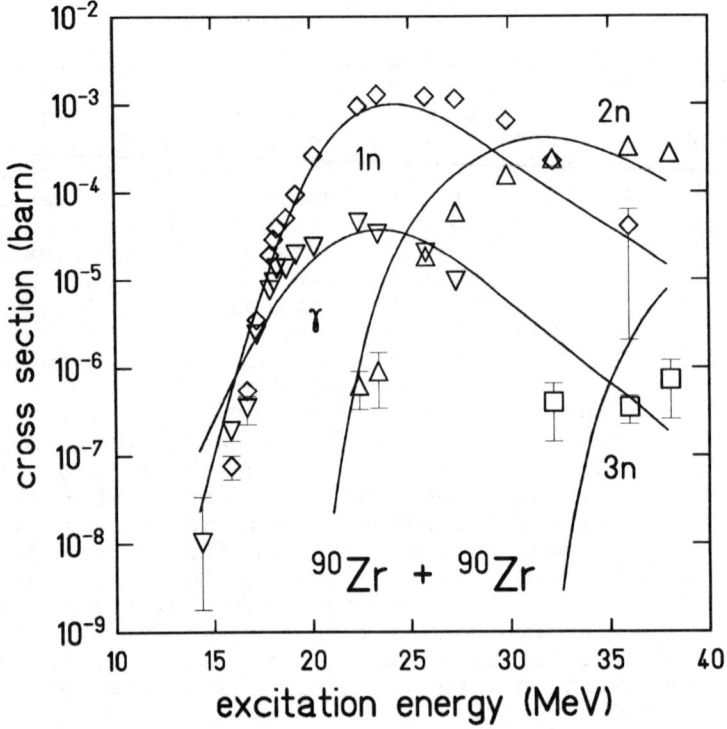

Fig. 2. Data points: Measured excitation functions of the reaction ^{90}Zr (^{90}Zr, xn) $^{180-xn}Hg$. Curves: Result of an evaporation calculation with an E1 γ-strength function equal to the giant dipole resonance strength function[10] multiplied by a factor of 4.6.

In order to understand the physical nature of this process, a direct observation and spectroscopy of the γ-rays emitted during the radiative fusion process seems to be necessary. The γ ray energy distributions and also the angular distributions may give information about the character of the electromagnetic transitions involved. The γ ray multiplicity distributions are connected to the spin distributions of the emitting nuclei. Results of a test experiment have been reported recently[11].

In the present paper we report for the first time the clear separation of the γ rays associated with the radiative fusion process from those associated with the competing one neutron evaporation channel. Gamma rays from ^{90}Zr + ^{90}Zr were detected in coincidence with the evaporation residues, thus suppressing in a very efficient way the intense γ-ray background from compound nucleus fission and other, non-compound reaction channels. Gamma rays from radiative fusion and from particle evaporation channels were separated by establishing time correlations between the impinging evaporation residues and their radioactive α-decay. The analysis of the data is still in a preliminary stage. Nevertheless, however, it is possible to recognize important general features of the radiative fusion process.

EXPERIMENTAL SET-UP

A target of about 60 μg/cm^2 monoisotopical ^{90}Zr on a 130 μg/cm^2 Au-backing was irradiated with a ^{90}Zr beam from the UNILAC heavy ion accelerator at GSI Darmstadt. The evaporation residues were separated from the primary beam and transported to a detector telescope by the velocity filter SHIP[3]. The detector telescope consisted of two thin time-of-flight detectors, a thin secondary electron detector, and an array of three position sensitive surface barrier detectors. The evaporation residues could be distinguished from scattered projectiles passing through SHIP by registrating their time-of-flight, their ΔE-signal from the secondary electron detector, and their energy. The evaporation residues implanted into the detector array were individually identified by their subsequent radioactive α-decay. With a resolution of about 0.3 mm FWHM, the position sensitive detector allowed to establish that the implantation of an evaporation residue and an α-decay occurred at the same position. In the analysis of the time differences between evaporation residues and α-decays, the area of the detector array was divided into narrow strips, thus reducing the counting rate accordingly. This allowed to establish time correlations even for the α-decay of ^{180}Hg with its rather long halflife of 2.9 s.

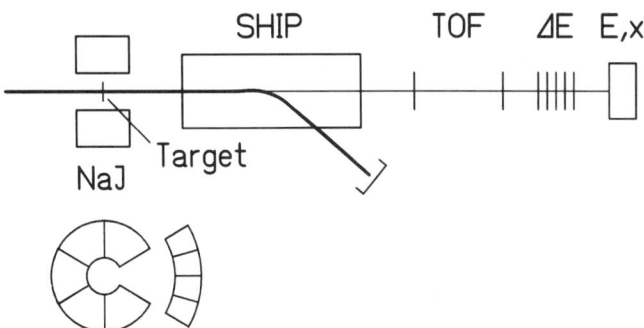

Fig. 3. Schematical drawing of the experimental set-up.

The γ-rays from the target were observed in a combined arrangement (see fig. 3) of a cylindrical NaI sum crystal[4] with one of the six segments removed, and a section of the Darmstadt-Heidelberg crystal ball[5,6] consisting of 15 NaI detectors each covering about 0.6 % of 4π. The coincidence requirement between γ-events from the target and evaporation residues identified by the detector telescope behind SHIP reduced the immense background rate in the NaI-detectors (up to one hundred thousand per second instantaneous rate per NaI cylinder segment) to a negligible portion. The 15 detectors of the crystal ball section served to measure the spectral distribution of the γ-rays. All NaI detectors together, which covered about 90 % of 4π, were used to determine the γ sum energy and the γ ray multiplicity.

The α-spectra and the excitation functions shown in figs. 1 and 2, respectively, were measured in a slightly modified set-up which is described in ref.[7].

OBSERVATION OF γ-RADIATION

The γ-radiation was detected at a projectile energy of 358 MeV corresponding to an excitation energy of about 21 MeV. At this energy the cross section of the radiative fusion channel is near its maximum value and reaches about 40 μb, see fig. 2. By reducing the angular acceptance of SHIP from its normal value of ± 1.5° to about ± 0.4° (both horizontally and vertically), the number of detected evaporation residues from charged particle evaporation could be reduced strongly. This is due to the angular distribution of the evaporation residues, which for charged particle emission is distinctly wider than for the γ-channel. On the basis of the time correlation of the evaporation residue with the subsequent α-decay, about 700 events were attributed to the γ-channel, and about 5400 events to the 1n-channel. The total contamination of the events assigned to the γ-channel with events from other channels (in particular the 1n-channel) was estimated to be about 10 %.

Fig. 4 shows spectra of the total γ-ray energy deposited in all NaI detectors in a single fusion event. The detectors were calibrated by the photopeaks produced by γ lines at 0.344 MeV (^{152}Eu), 0.570 MeV (^{207}Bi), 1.17 MeV (^{60}Co) and 2.754 MeV (^{24}Na). The average sum energy for the 1n-channel is about one half of the value for the γ-channel. This is to be expected, since the neutron carries away about half of the total excitation energy. The energy for the γ-channel reaches up to about 20 MeV, in agreement with the excitation energy of the compound nucleus. The spectrum for the 1n-channel has a tail towards higher energies. This may be due to a contribution of the evaporated neutron which was not corrected for.

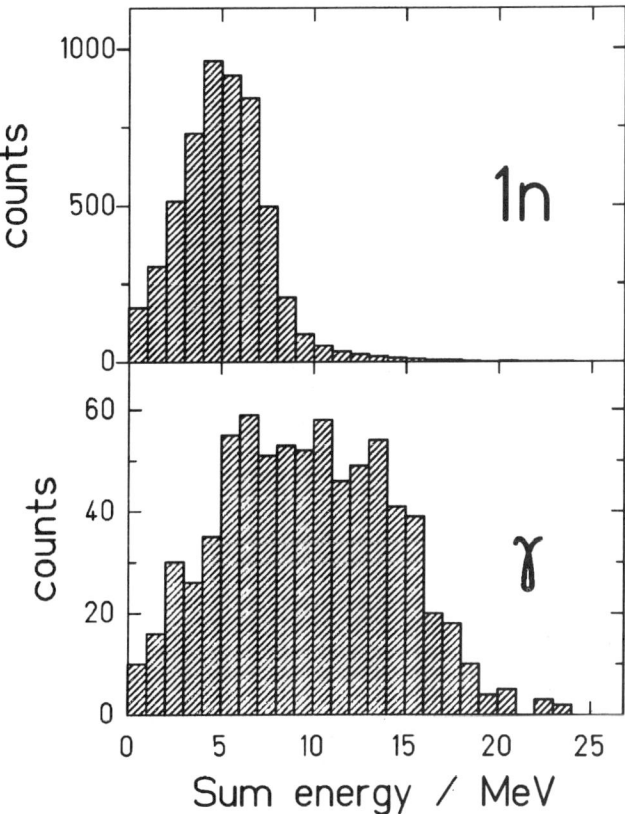

Fig. 4. Measured spectrum of the γ sum energies, not corrected for the detector response functions, for the radiative fusion process and for the 1n evaporation channel in the reaction ^{90}Zr (358 MeV) + ^{90}Zr.

Fig. 5 shows the raw spectra of the individual γ rays observed in the 15 NaI detectors of the crystal ball segment. As in fig. 4, no unfolding was performed to recover the original γ ray spectrum. Above about 3 MeV the shapes of the spectra for the 1n-channel and for the γ-channel are quite different. In this energy range the spectrum associated with the γ-channel has a much larger relative intensity. This may be explained by the larger initial excitation energy of the γ-ray emitting nucleus. However, above 10 MeV, not a single count was detected. If every radiative fusion event would be accompanied by the emission of a γ-ray in the energy region above 10 MeV, it can be estimated that about 15 counts should appear in the spectral range above 10 MeV. It may be concluded that a significant influence of a specific direct capture process associated with a high energy γ ray does not seem to be present.

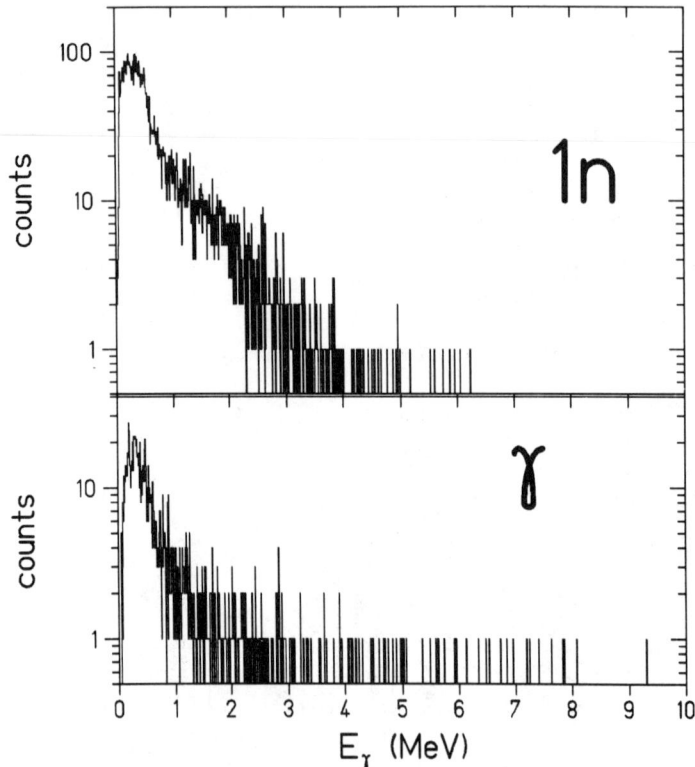

Fig. 5. γ transition energy spectrum of individual γ-rays observed in the 15 NaI detectors from the crystal ball segment, not corrected for the detector response function, for the γ-channel and for the 1n-channel in the reaction ^{90}Zr (358 MeV) + ^{90}Zr.

Fig. 6 shows the measured angular distributions of the γ-rays. The anisotropy observed for the 1n-channel corresponds to a ratio

$$W(0°)/W(90°) = 1.5 \pm 0.3,$$

if the function

$$W(\theta) = 1 + A_2 P_2 (\cos \theta)$$

is used for the extrapolation to 0°. This value is similar to the values found for different systems emitting several neutrons[8]. For the γ-channel, a more isotropic angular distribution is observed. This is expected, if the γ-ray emission is dominated by statistical transitions.

Another important experimental information is the number of γ-detectors firing simultaneously in one event. The response of the detector arrangement has been determined with the 1.17 MeV transition of a ^{60}Co source. By

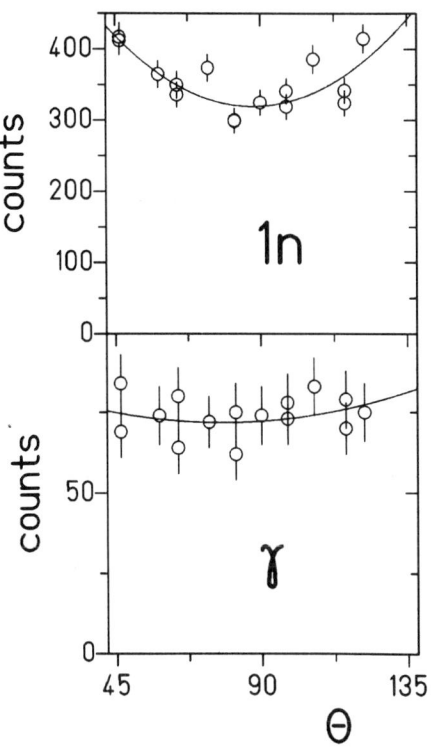

Fig. 6. Angular distribution of γ-rays associated with radiative fusion and 1n evaporation, respectively. The distribution for the 1n-channel was not corrected for the influence from neutron emission. The lines are drawn to interpolate between the measured points (^{90}Zr(358 MeV)+ ^{90}Zr)

adding a given number of events, the multiplicity distributions for long γ-cascades have been simulated. For the analysis of the experiment, a Gaussian function was assumed to represent the γ multiplicity distribution. The γ multiplicity distributions fitted to the measured detector multiplicity distributions are shown in fig. 7, and the corresponding mean values and standard deviations are listed in table I. The high multiplicity observed for radiative fusion is consistent with the emission of many low and medium energy γ-rays during this process.

Table I: Moments of the γ multiplicity distributions in the reaction ^{90}Zr (358 MeV) + ^{90}Zr

Channel	mean value	standard deviation
1 n	6.9 ± 0.1	3.3 ± 0.2
γ	10.2 ± 0.2	5.3 ± 0.4

STATISTICAL MODEL CALCULATION

Fig. 2 shows that it is possible to reproduce the measured excitation functions for the radiative fusion process and for the xn-channels by an evaporation calculation. The γ-strength used for E1-transitions was that predicted by the giant dipole resonance as given in ref. [10], multiplied by a factor of 4.6. The calculated excitation functions were found to be rather insensitive to modifications of the strength of E2-transitions which in the calculation seem to compete only in the latest stage of the deexcitation process. The relatively large E1-strength was not fitted to this specific case alone, but was also found necessary to reproduce the shape of excitation functions for particle evaporation in other systems[9]. It must be kept in mind, however, that γ-transitions have no threshold energy and consequently may lead to levels of higher excitation energy than particle evaporation. Therefore the competition between γ-deexcitation and particle evaporation is sensitive to the excitation energy dependence of the nuclear level density.

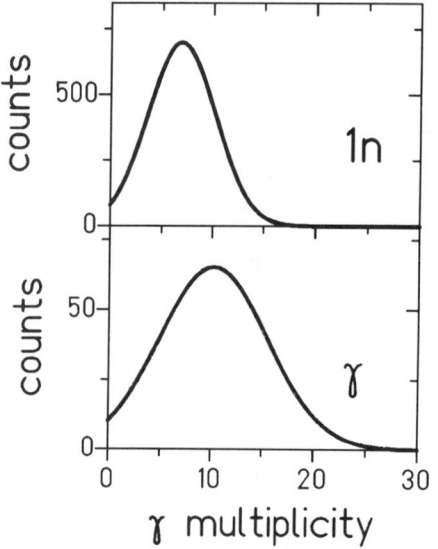 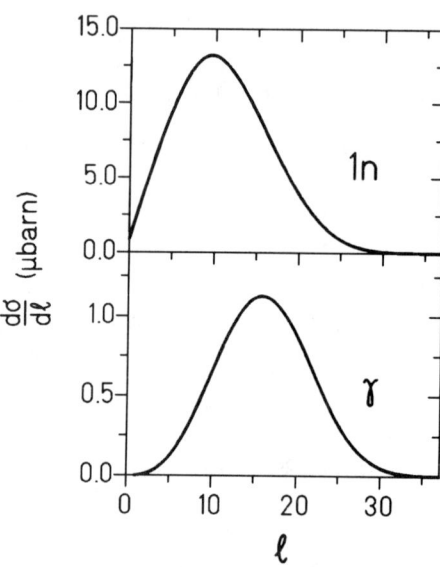

Fig. 7. γ multiplicity distributions determined from the measured detector multiplicities under the assumption of Gaussian functions for the shape (^{90}Zr(358 MeV) + ^{90}Zr).

Fig. 8. Angular momentum distribution underlying the evaporation calculation shown in fig. 2 for the reaction ^{90}Zr + ^{90}Zr at 358 MeV.

The compound nucleus angular momentum distributions underlying the fusion-evaporation-calculation fitting the measured excitation functions of fig. 2 are shown in fig. 8. As can be seen, the calculated average angular momenta contributing to the 1n-channel and to the γ-channel are not very different. Although the detailed relationship between the angular momentum and the γ-multiplicity is not known for the present case, this seems to be compatible with the measured average multiplicities for the 1n-channel and for the γ-channel, which are not very different either.

A more refined data analysis, including the recovery of the original γ ray energy spectra from the measured NaI pulse height spectra will be performed in the near future. Also it is planned to calculate γ-ray spectra and γ-multiplicities in the frame work of the statistical model. This will allow a more detailed comparison of the present γ-ray data with the predictions of the statistical model.

CONCLUSION

The present γ ray measurements do not support the assumption that specific direct processes associated with high energy γ rays above 10 MeV are responsible for the large cross sections observed for radiative fusion. The low excitation energy of the compound nuclei, which is achievable in heavy ion fusion reactions with a symmetric entrance channel, seems to favour this process. There are indications that the present data are compatible with the statistical model for the deexcitation of the compound nucleus.

ACKNOWLEDGEMENT

One of the authors, H.-G. C., wishes to thank Dr. K.A. Snover for fruitful discussions.

This work was supported by the German Bundesministerium für Forschung und Technologie.

REFERENCES

1. J.G. Keller, H.-G. Clerc, K.-H. Schmidt, Y.K. Agarwal, F.P. Heßberger, R. Hingmann, G. Münzenberg, W. Reisdorf, C.-C. Sahm, Z. Phys. A 311, 243 (1983)
2. J.G. Keller, H.-G. Clerc, F.P. Heßberger, G. Münzenberg, W. Reisdorf, C.-C. Sahm, K.-H. Schmidt in: GSI Annual Report 1983, p. 3
3. G. Münzenberg, W. Faust, S. Hofmann, P. Armbruster, K. Güttner, H. Ewald, Nucl. Instrum. Methods 161, 65 (1979)
4. P. Oblozinsky, R.S. Simon, Nucl. Instrum. Methods 223, 52 (1984)
5. R.S. Simon, Journal de Physique Coll. 41 C 10-281 (1980)
6. V. Metag, R.D. Fischer, W. Kühn, R. Mühlhans, R. Novotny, D. Habs, U.v. Helmolt, H.W. Heyng, R. Kroth, D. Pelte, D. Schwalm, W. Hennerici, H.J. Hennrich, G. Himmele, E. Jaeschke, R. Repnow, W. Wahl, E. Adelberger, A. Lazzarini, R.S. Simon, R. Albrecht, B. Kolb,

Nucl. Phys. A 409, 331 C (1983)
7. J.G. Keller, K.-H. Schmidt, H. Stelzer, W. Reisdorf, Y.K. Agarwal, F.P. Heßberger, G. Münzenberg, H.-G. Clerc, C.-C. Sahm, Phys. Rev. C 29, 1569 (1984)
8. R.M. Diamond, F.S. Stephens, Ann. Rev. Nucl. Part. Sci. 30, 85 (1980)
9. W. Kühn, P. Chowdhury, R.V.F. Janssens, T.L. Khoo, F. Haas, J. Kasagi, R.M. Ronningen, Phys. Rev. Lett. 51, 1858 (1983)
10. M.A. Lone in: Neutron Capture Gamma-Ray Spectroscopy, Editors R.E. Chrien, W.R. Krane, Plenum Press N.Y. 1979, p. 161
11. K.-H. Schmidt, R.S. Simon, J.-G. Keller, W. Reisdorf, H.-G. Clerc, C.-C. Sahm, E. Tschöp, W. Schwab, Int. Conf. on Fusion below the Coulomb Barrier, June 13-15, 1984, Cambridge, MA, USA

THE ROLE OF GIANT RESONANCES IN HEAVY-ION RADIATIVE CAPTURE

A. M. Sandorfi
Physics Department, Brookhaven National Laboratory
Upton, N.Y. 11973

INTRODUCTION

The capture reactions discussed at this conference have dealt almost exclusively with the radiative capture of light particles (n,p,He). These reactions have been used in a variety of ways for many years as spectroscopic tools. In contrast, much less is known about the radiative capture of heavier ions. This, and the following review by Snover, attempt to summarize the main features and physics of heavy-ion capture.

Very generally, the deexcitation process in such a reaction can follow one of the routes shown schematically in Figure 1. In case (a) the compound nucleus loses energy by the emission of a high-energy gamma ray to states below particle-emission threshold (t_p). Subsequent gamma decay then leads to the fused nucleus in its ground state. It is the experimental information on this process that is discussed here. Type (b) radiative transitions, involving high-energy gamma decay to specific states above t_p, become increasingly difficult to separate from the background as the gamma-ray energy decreases. In fact this process (b) has only been observed when decays to many different final states coincide in transition energy. Such is the case for the decay of large numbers of giant resonances, one built on each excited level that is statistically populated in a heavy-ion reaction. This latter process is reviewed by Snover.

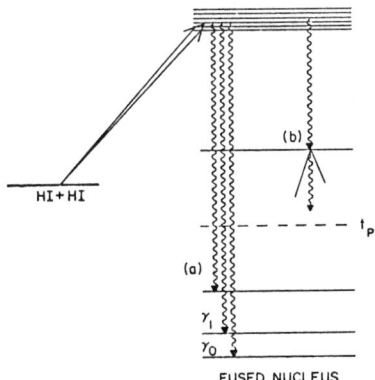

Fig. 1. Possible radiative decay routes following heavy-ion capture.

GENERAL CONSIDERATIONS FOR DECAYS TO BOUND LEVELS

The most interesting characteristic of radiative decays to levels below t_p (Fig. 1) is the presence of resonances that exhibit nonstatistical heavy-ion partial widths. Most of these structures are not correlated with previously identified features in other heavy-ion reaction channels. However, there are three factors that strongly affect the likelihood of observing such resonances. First of all, the multipolarity of high-energy gamma-ray transitions is very unlikely to be greater than 2. [1] Since the spins of states below t_p are generally low ($J \lesssim 5$), these transitions can be observed only if the spins of the capturing states are relatively low. With heavy-ions, these states can be excited only in the vicinity of the Coulomb barrier, which is much more well-defined than it is in light-ion reactions. As the heavy-ion bombarding energy is increased above the barrier, the cross section for forming the compound nucleus peaks at increasingly higher spin, with the result that at energies above about twice the barrier high-energy gamma decay to bound levels is greatly diminished. A second consideration is that, the average capture yield is always indirectly dependent upon the location of giant electromagnetic resonances. The dependence arises simply because the capture cross section is proportional to Γ_γ, the radiative width, and sum rules restrict the total radiative width of a given multipolarity. If nearly all of the strength is contained in a giant resonance (GR) at an energy E_{GR}, then the magnitude of Γ_γ, and hence the capture cross section, cannot help but be greatly reduced at energies significantly different from E_{GR}. However, this is a dependence imposed by sum rules and need not necessarily reflect any structural similarity between the capturing states and giant resonances. Finally, the behavior of the Coulomb barrier, and the sum-rule restrictions on the total Γ_γ available, effectively limit heavy-ion capture measurements to s-d shell nuclei. For heavier projectiles and targets, the Coulomb barrier is significantly higher, while the giant resonances, which change in position with $A^{-1/3}$, are shifted to lower energies. The farther the barrier is from the GR, the more difficult radiative capture will be to observe. (This, of course, ignores the possibility of high-lying pockets of multipole strength, about which little experimental information is available.)

These general considerations are indeed born out by the available experimental data. But apart from these, it is difficult to isolate trends that persist through the various systems that have been studied. The decays to the ground state of the compound nucleus, or to the members of its rotational band, should be the easiest to understand, since the wavefunctions of these final states are fairly well understood and since the giant multipole strength built upon them has been measured. This information has been used to deduce interesting and unexpected characteristics of resonances in the decays to low-lying levels. However, these characteristics vary considerably among different heavy-ion systems. Most of the capture data is in this general category. However, in at least one case, decays have been observed to higher-lying levels that are shape-isomers of the compound nucleus, and these results are particularly simple to interpret. Data from three reactions –

$^{12}C(^{12}C, \gamma)^{24}Mg$, $^{14}C(^{12}C, \gamma)^{26}Mg$, $^{12}C(^{16}O, \gamma)^{28}Si$ –

are discussed here as examples of heavy-ion capture. A more extensive review is given in reference 2. Large anticoincidence-shielded NaI spectrometers were used for these measurements, and these are described in reference 3. In addition to the obvious requirements of good energy resolution and efficient cosmic-ray rejection, the small cross sections encountered in these reactions necessitate using high beam currents and maintaining a stable gain at high counting rates for periods of typically 6 hours in duration. The experimental problems peculiar to heavy-ion captive are discussed in some detail in reference 2.

DECAYS TO THE GROUND STATE AND ITS ROTATIONAL BAND

One of the more extensively studied heavy-ion capture reaction is the fusion of two ^{12}C nuclei leading to low-lying levels in ^{24}Mg. A typical spectrum is shown in Fig. 2. The decays to the ground state of ^{24}Mg, 1.37MeV(2^+) first excited state, and 4.1–4.2MeV(4^+–2^+) second and third excited states are clearly visible. The peaks are superimposed upon a smooth background due primarily to the pileup of many lower-energy events. Spectra taken at energies below the Coulomb barrier often show additional lines corresponding to decays to higher lying levels.

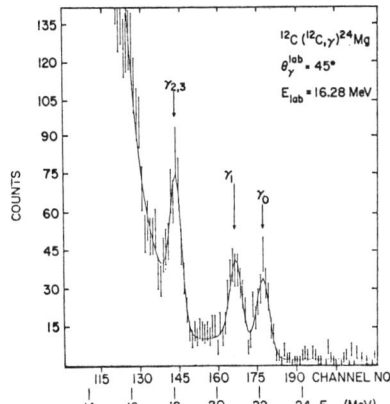

Fig. 2. A high-energy gamma-ray spectrum from the $^{12}C(^{12}C, \gamma)^{24}Mg$ reaction.

The $\theta\gamma = 45°$ excitation functions for these transitions are shown in Fig. 3. There are two striking features of these data. First, as we have already anticipated, the yield is localized in the vicinity of the Coulomb barrier (6.6 MeV c.m.), especially in the γ_0 channel where it extends from 19 MeV excitation up to about 24 MeV. Second, within this gross structure, there is considerable finer structure in the form of narrow ($\Gamma < 0.3$ MeV) resonances, several of which are correlated in all three γ-decay channels.

For identical bosons in the entrance channel, the J^π of the compound state is limited to 0^+, 2^+, 4^+,.... The observation of a photon decaying to the 0^+ ground state rules out 0^+ assignments for structures appearing in this channel, and makes multipolarities greater than 2 extremely unlikely.[1] Thus, all of the yield in the

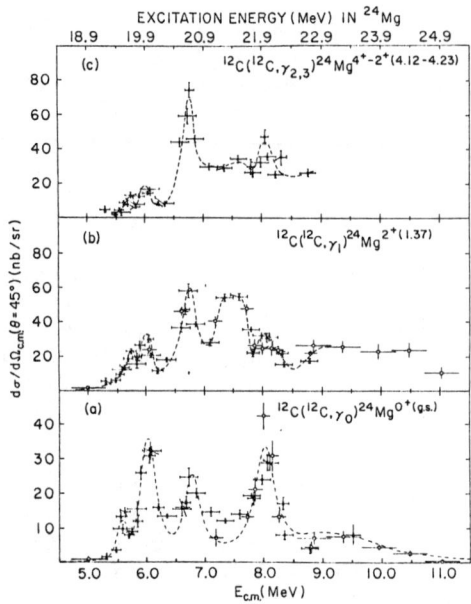

$^{12}C(^{12}C, \gamma_0)$ reaction arises from 2^+ states in ^{24}Mg and exhibits a $\sin^2(2\theta\gamma)$ angular dependence. The dashed line through the γ_0 excitation function is the result of a fit to an incoherent sum of Breit-Wigner resonances.[4] The fit falls below the data near 20.3 and 21.3 MeV excitation, which may indicate contributions from additional less prominent structures. The dashed lines through the γ_1 and $\gamma_{2,3}$ excitation functions are merely to guide the eye.

Fig. 3. Excitation functions for $^{12}C + ^{12}C$ radiative capture to the low lying levels of ^{24}Mg.

A great deal of structure has been observed in $^{12}C + ^{12}C$ reactions ranging from below the Coulomb barrier up to more than six times the barrier. A large number of resonances have been reported in elastic, inelastic, and a variety of reaction channels, and most of these are viewed as resulting from the formation of some kind of nuclear molecule. In Fig. 4 the $^{12}C(^{12}C, \gamma_0)$ excitation function (Fig. 4b solid curve) is compared with the previously identified 2^+ $^{12}C + ^{12}C$ "quasimolecular" resonances above 5.0 MeV c.m. (Fig. 4c). The γ_0 peak at 5.6 MeV corresponds to a 2^+ quasimolecular resonance. However, the more prominent peaks at 6.0, 6.8, and 8.0 MeV c.m. do not correlate with any known $^{12}C + ^{12}C$ 2^+ structures. Furthermore, the average width of a 2^+ level between 19 and 23 MeV excitation in ^{24}Mg is about 20 KeV, while the total widths of the capture resonances of Fig. 4b are an order of magnitude larger. An Ericson fluctuation phenomenon [5] can thus be ruled out.

The best available data on the distribution of E2 strength built on the ^{24}Mg ground state have come from several high-energy ^{24}Mg (α, α') experiments. If this E2 strength would entirely fission into two ^{12}C nuclei, then the corresponding capture cross section would be given by the curve in Fig. 4a.[4] Upon comparing Figure 4a with Figure 4b, solid curve, we see that the small 5.6-MeV c.m. resonance seems to be correlated with a peak in the E2 strength function. However, the 6.0- and 6.8-MeV c.m. resonances line up with valleys in the E2 distribution, while the 8.0 MeV c.m. resonance appears near the middle of a much broader structure. On the whole, the pronounced features of the capture yields do not directly reflect

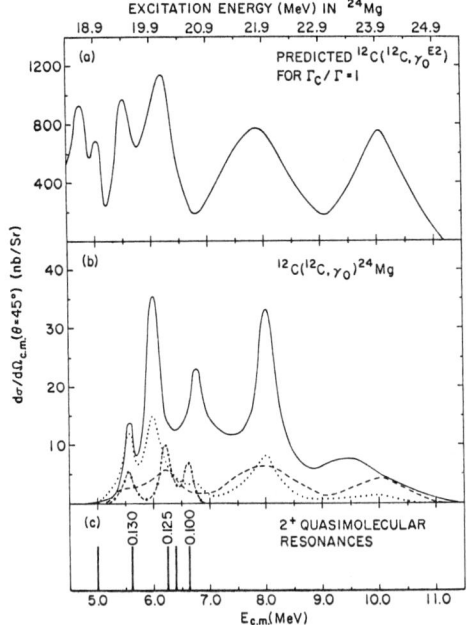

Fig. 4 (a) Predicted cross sections for capture through the giant quadrupole resonance in ^{24}Mg.
 (b) The ^{12}C(^{12}C, γ_0) excitation function (solid line) and various calculations - see text.
 (c) The locations of 2^+ quasimolecular states. Total widths of resonances are indicated in MeV.

the structure in Γ_γ. Before trying to interpret the unusual features of this capture reaction, it is useful to know the extent of the contributions to the cross section from the process of compound nucleus formation followed by competitive statistical decay.[2,4] If the carbon widths were purely statistical, then a Hauser-Feshbach calculation would predict a capture cross section given by the dashed curve in Fig. 4b. Viewing the reaction backwards, this would correspond to the process in which a photon excites the ^{24}Mg nucleus into its giant quadrupole resonance, which then mixes into the compound levels and decays statistically into two ^{12}C nuclei. This calculation reflects the structure of Fig. 4a. However, over most of the excitation function, it falls far below the data. Nonetheless it does succeed in explaining the drop in cross section below 5.5 MeV c.m. as resulting from the Coulomb barrier.

The statistical model can also be used to estimate statistical photon decay, that is the decay of a resonance excited in the ^{12}C + ^{12}C channel (i.e., a resonance in Γ_c) which then mixes into the underlying compound levels, and emits a photon by virtue of the E2 strength of these underlying levels. If each of the resonances observed in ^{12}C(^{12}C, γ_0) were due to resonances in Γ_c then, assuming reasonable values for the elastic width, statistical photon decay would produce the dotted curve in Fig. 4b. This curve peaks wherever the data does, but of course that is by construction. What is important here is the relative magnitude. The γ_0 decay of the 5.6-MeV c.m. resonance (the only one that has been observed in other channels) is completely consistent with the process of statistical mixing into the compound nucleus followed by E2 γ decay. The magnitude of the 6.0-MeV resonance predicted by this calculation is about a factor of 2 below the data. This process could account for the 6.0-MeV γ_0 peak if Γ_c/Γ were as large as 0.30. However, this would be grossly inconsistent with elastic-scattering measurements.

The other peaks in the γ_0 yield are far above the dotted curve.

Since statistical photon decay causes the 5.6-MeV c.m. resonance to appear in the $^{12}C(^{12}C, \gamma_0)$ excitation function, the other 2^+ quasimolecular resonances of Fig. 4c would also be expected, at some level, and this is given by the circle-dashed curve in Fig. 4b. These contributions are always small compared with the dominant features of the γ_0 yield. The conclusion from this analysis is that the photon widths of the capture resonances are, for the most part, significantly enhanced over statistical widths.

We can draw two conclusions from these statistical calculations. First, at least one and possibly all of the previously identified quasimolecular resonances (Fig. 4c) are present in the radiative capture yields, but only at a very low level consistent with a statistical γ decay to ^{24}Mg. Second, the $^{12}C + ^{12}C$ and the γ_0 decay probabilities of the dominant resonances observed in radiative capture are significantly greater than statistical probabilities for decay from the compound nucleus. These capture resonances must then reflect states with large $^{12}C + ^{12}C$ parentage that also have a close link with the structure of the ground state of ^{24}Mg. The wave functions of the levels in the ground-state rotational band (GSB) of ^{24}Mg are very similar, and thus 2^+ levels that decay strongly to the ground state would also be expected to have significant decay branches to the 2^+ and 4^+ members of the GSB. This is exactly what is observed in Fig. 3. The dominant 2^+ peaks at 6.0, 6.8 and 8.0 MeV are present in the γ_1 and $\gamma_{2,3}$ excitation functions as well as in γ_0. (In contrast, the 5.6-MeV quasimolecular resonance appears only in the γ_0 excitation function. However, the γ decay of this resonance arises from statistical photon emission from an overlapping peak in the distribution of ground-state E2 strength. The E2 strength built on excited states of ^{24}Mg may be very different, and thus γ decay to the ground state via statistical photon emission does not guarantee a comparable decay rate to other members of the GSB.) These capture resonances are undoubtedly present in elastic scattering at some level due to their nonstatistical ^{12}C widths, but they have evidently been hidden by the much more dominant quasimolecular structures.

From these considerations it would seem that the heavy-ion capture reaction is a sensitive way of picking out unusual states of a nucleus that like to fission after absorbing a photon. This is an exciting prospect, which seems to fall completely apart when one looks at other heavy-ion systems. Recent results from the $^{14}C(^{12}C, \gamma)^{26}Mg$ reaction are reported in a contribution to this conference.[6] The excitation functions for decay to the low-lying states of ^{26}Mg are shown in Fig. 5. Some of these yields do indeed exhibit peaks with cross sections and widths comparable to the resonances of $^{12}C(^{12}C, \gamma)$, but these do not appear at the same c.m. energy. In particular, the yields to the 0^+ ground state, and to the 2^+ and 4^+ members of its rotational band at 1.81 MeV and 4.32 MeV, respectively, show essentially no structual similarities.

Although forbidden by the symmetry of the $^{12}C + ^{12}C$ system, E1 dipole radiation is both symmetry and isospin allowed in the

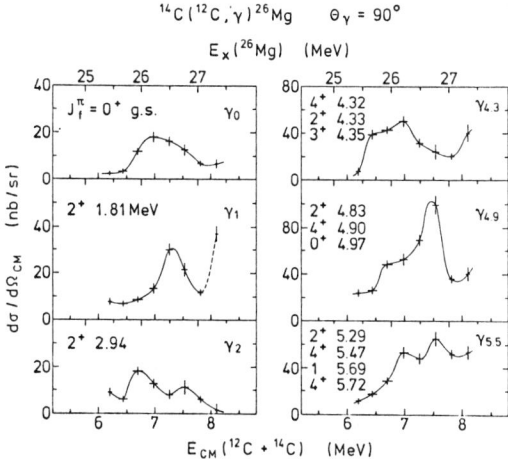

Fig. 5. Excitation functions for $^{12}C + ^{14}C$ radiative capture to the low lying levels of ^{26}Mg.

$^{14}C(^{12}C, \gamma)$ reaction, and indeed the radiation patterns indicate strong dipole components. Since E1 transitions can have much larger radiative widths than E2 decays, the excitation functions of Fig. 5 might have significant contributions from statistical carbon decay of the giant dipole resonance (GDR.) This question has been investigated in Ref. 6. Elastic scattering is almost devoid of any structure in this energy range,[7] suggesting that the carbon widths are largely statistical. With this assumption the photon absoption cross sections infered from the capture data are factors of 2 to 10 less than those observed in the ground state GDR. This suggests either that the GDR's built on excited states of ^{26}Mg are very different, which seems unlikely,[12] or that the photon widths are nonstatistical. The nature of the intermediate structure that gives rise to the yields of Fig. 5 is evidently quite complicated.

As a final example of heavy-ion capture, leading to the ground state band of the fused system, the γ_0 and γ_1 excitation functions of the $^{12}C(^{16}O, \gamma)^{28}Si$ reaction are shown in Fig. 6.[8] Again, as anticipated, relatively narrow resonances are present, which appear with cross sections comparable to those of $^{12}C(^{12}C, \gamma)$, and these are localized in the vicinity of the Coulomb barrier (8.7 MeV c.m.). However, the resonances do not appear simultaneously in γ_0 and in γ_1 and the cross sections for γ_0 decay are, for the most part, surprisingly small.

Since the ground state of ^{28}Si is oblate while the incoming heavy-ion channel is inherently prolate, the small γ_0 yield may be viewed as resulting from the inability to connect these two intrinsic shapes with a simple electromagnetic operator. Unfortunately, the same argument can be made for the γ_1 yield since the wave functions of the ground- and first-excited states of ^{28}Si look very much alike.[9] Alternatively, the resonances could be of spin higher than 2^+, 3^- or 4^+, for example. Since the multipolarity of the γ transitions is effectively restricted to 1 or 2, decays to the 1.78 MeV 2^+ first excited state could be observed, while decays to the ground state would be precluded. Angular distributions were measured at the positions indicated by the arrows in Fig. 6.[2] For the γ_1 peaks, these are not strictly definitive because of the $J^\pi = 2^+$ final state. Nonetheless, several of the peaks in Fig. 6 are quite well isolated and a unique spin and γ-ray multipolarity might well dominate each.

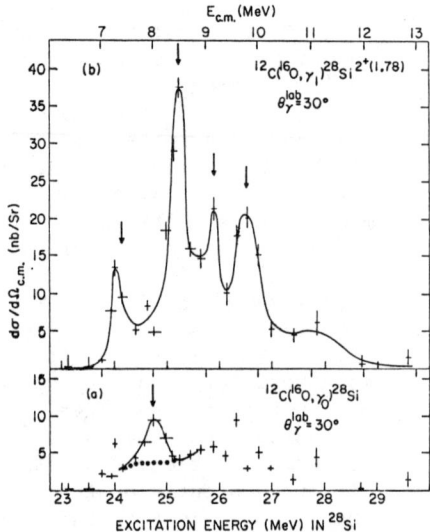

Fig. 6. Excitation functions for $^{12}C + ^{16}O$ radiative capture to the low-lying levels of ^{28}Si.

The 9.8-MeV resonance is likely a $4^+ \rightarrow 2^+$ decay, and so its absence in the decay to the 0^+ ground state is not surprising. However, the distributions for the 7.3-MeV c.m. and 8.5-MeV c.m. resonances suggest $J^\pi = 2^+$ assignments. These could decay to the ^{28}Si ground state, but do not, for as yet, unknown reasons.

An analysis of statistical decay probabilities is a bit more difficult here, since the more pronounced resonances occur in the γ_1 excitation function and the multipole strength built on the first excited state of ^{28}Si is unknown. Nevertheless, under the assumption that, on the average, the distribution of E2 strength built on the 2_1^+ state is just shifted up by 1.78 MeV from that built on the ground state, the calculated capture cross sections still fall far below the resonances of Fig. 6. This largely due to the increased Q value and Coulomb barrier which pushes the structures to higher excitation energies, far above the quadrupole resonance in ^{28}Si. However, the GDR still has appreciable strength at these energies, and for the ground state transitions, dipole radiation is allowed by symmetry, although still inhibited by isospin. The γ_0 cross sections of Fig. 6a may well be accounted for by assuming statistical heavy-ion widths and γ decay through a small isospin-split component of the GDR.[2] This still leaves the dominant peaks of the γ_1 yield which may have very nonstatistical carbon widths, as we have inferred. However, since they do not appear in the γ_0 excitation function, they would not be expected in $^{12}C + ^{16}O$ elastic scattering, which indeed shows almost no structure in this energy range.[10]

In summary, the $^{12}C(^{12}C, \gamma)$ data and analysis provided the tantalizing suggestion that heavy-ion capture was a very sensitive tool for picking out unusual highly-deformed states of a nucleus that are simply connected to the ground state, and to its rotational band. The sensitivity to unusual highly-deformed structures is undoubtedly there, but it would seem that the connection, the intermediate structure leading to large photon and heavy-ion partial widths, is far from simple. Except for $^{12}C(^{12}C, \gamma)$ which seems to be somewhat anomalous, these structures do not resonate in the yields to all of the members of the GSB. It is more generally the case that the excitation functions to different members of the GSB are very different.[2]

CAPTURE THROUGH GDR'S BUILT ON SHAPE-ISOMERIC STATES

One might have thought that if capture to the ground state was difficult to understand, capture to a highly-excited state would be hopeless. Delightfully, there are at least some transitions for which this is not the case. In the course of the $^{12}C(^{16}O,\gamma)$ measurements described above, unusually strong transitions were observed to the 0_3^+ at 6.69-MeV excitation in ^{28}Si, with cross sections at least five times larger than the decays to the low-lying levels.[8] Although the ground and low-lying states of ^{28}Si are oblate, this 0_3^+ has been identified as the bandhead of a K = 0 prolate shape-isomer.[9,11] Enhanced high-energy gamma-ray transitions are usually characteristic of giant electromagnetic resonances. Some examples of such structures built upon excited states have been discussed by Dowell,[12] although none of these states involve a shape change. Capture through the heavy-ion channel would normally be a rather improbable way of looking for such giant resonance strength. However, since the incoming heavy-ion channel has inherently a large prolate distortion, heavy-ion capture will be much more sensitive to prolate shape-isomers than light-ion reactions, most of which involve relatively minor deformations.

The possibility of comparing giant resonances built upon different intrinsic shapes within a single nucleus is quite exciting. However, measurements of transition rates to excited final states, where the gamma-ray energy is significantly less than the maximum (the γ_0 transition to the ground state), are extremely difficult because of large backgrounds. These come mainly from the pileup of the very large numbers of low-energy signals, and increase very rapidly with bombarding energy. Essentially all of these backgrounds can be removed by detecting the fused compound nucleus in coincidence with the high-energy photon. A crossed $\vec{E}$ and $\vec{B}$ field velocity selector has been used for just such measurements and, as described in ref. 13, this technique reduces pileup backgrounds by up to four orders of magnitude.

These coincidence measurements revealed strong transitions, not only to the 0_3^+, but also to the 2_p^+ and 4_p^+ members of this prolate band at 7.41-MeV and 9.16-MeV in ^{28}Si, respectively. Their excitation functions are shown in Fig. 7. The decay to the 6.7-MeV 0_p^+ is substantially stronger than to the 7.4-MeV 2_p^+. The 4_p^+ is very near the 6_1^+ at 8.54-MeV which was strongly populated at a few bombarding energies, and it was not always possible to clearly separate the two. For this reason, only the sum of their yields is plotted in Fig. 7c. All three excitation functions display prominent peaks which are uncorrelated with center-of-mass energy. (This, again, is in contrast to $^{16}O + ^{12}C$ elastic scattering which shows very little structure below 12-MeV c.m.[16]) However, the three lowest energy peaks in the 0_p^+ and 2_p^+ yield curves occur at the same excitation energy above the final state (i.e. at the same gamma-ray energy). This is evident in Fig. 8 where the $\sigma(\gamma)$ deduced for these decays is plotted against E_γ (see below). The second narrow structure in the 6^+-4_p^+ yield of Fig. 7c also lines up with one of these

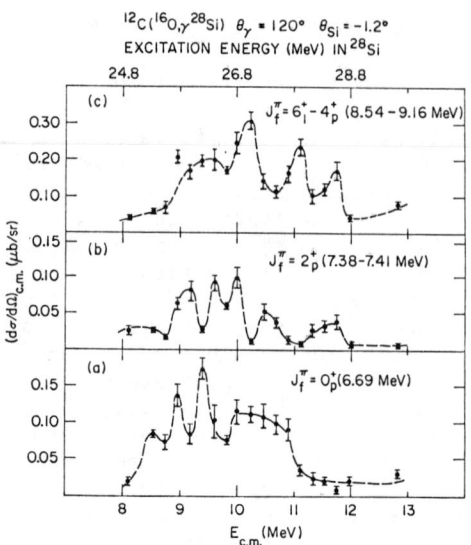

Fig. 7. Excitation functions for ^{12}C + ^{16}O radiative capture to the members of the excited prolate band in ^{28}Si.

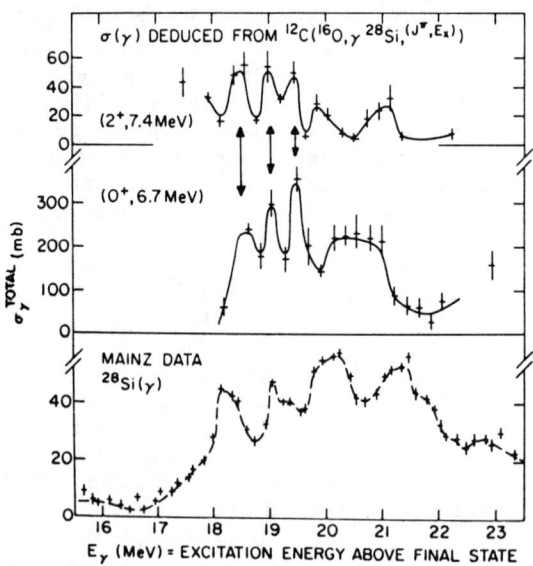

Fig. 8. Total absorption cross sections deduced from Fig. 7 under the assumption of statistical heavy-ion widths, and (bottom) the GDR of the ^{28}Si ground state.

peaks at $E\gamma$ = 19-MeV, although the ambiguity of the final state makes the comparison with the 0^+_p and 2^+_p yields difficult.

The capture yields as displayed in Fig. 8 are remarkably similar to the GDR built on the oblate ground state of ^{28}Si. The photoabsorption data, showing the latter, are reproduced at the bottom of Fig. 8.[14] The envelopes of these three excitation curves are very nearly the same. Much of the intermediate structure in the gamma-ray strengths built on the 0^+_p and on the 2^+_p is the same, and the peak at $E\gamma$ = 19-MeV even persists in the ground-state photoabsorption data. The angular dependence of the transitions to the 6.7MeV 0^+_p state, measured at several beam energies, is always strongly peaked at 90° indicating a dominant dipole component. The capture data of Fig. 8 are strongly suggestive of GDR's built upon the members of the prolate band in ^{28}Si.

The transition strength to the 0^+_p level shown in Fig. 7a exhausts 0.14/< Γ_{16_0}/Γ > percent of the energy-weighted classical dipole sum rule. A Hauser-Feshbach calculation, taking into account isospin mixing in ^{28}Si, has been used to estimate the branching ratio < Γ_{16}/Γ >, [13] and it is these results that are plotted in Fig. 8. In the calculation of $\sigma(\gamma)$ for the 2^+_p state, radiative capture through 3^- levels was assumed. The calculated branching fraction is a smoothly varying function of $E\gamma$ and the structure of Fig. 8 reflects that of the capture data. The resulting absorption cross sections of Fig. 8 account for 1.9 ± 0.3 and 0.3 ± 0.1 classical sum rules built on the 6.7-MeV 0^+_p and 7.4-MeV 2^+_p levels, respectively.

These data and calculations are consistent with giant dipole excitations of the prolate intrinsic shape of ^{28}Si. However, it is surprising that the prolate-GDR appears narrower than the oblate ground-state GDR, since the widths of GDR's built on excited states increases with energy.[12] A lack of significant broadening of the giant resonance might be reasonable if the 6.7-MeV 0^+ is viewed as the ground state of the prolate shape, and indeed the mixing between prolate and oblate states in ^{28}Si is very small.[9] Nonetheless, the quadrupole moment of the prolate states (+876mb) is twice as big as that of the oblate states (-480mb),[11] and all models coupling deformation degrees of freedom to the GDR would predict a significantly broader prolate giant resonance.

Prolate shape isomers have been predicted in other s-d shell nuclei, noteably ^{24}Mg, ^{32}S and ^{40}Ca, [15,16] and it would be extremely interesting to study the giant resonances built upon such states. Isomers in these nuclei have yet to be identified. On the other hand, the inherently prolate entrance channel makes heavy-ion radiative capture a rather selective tool in the search for these unusual states. There are in fact, specific predictions for the ^{16}O(^{16}O, γ)^{32}S reactions leading to an isomer at 8-MeV in ^{32}S.[16] We have investigated this reaction. However, no unusually strong transitions were observed in the predicted energy range (11 to 13-MeV c.m.).[17] The symmetry of the entrance channel does preclude dipole radiation, but a giant quadrupole resonance might still be observable. Nonetheless, the capture cross sections where no larger than the ^{12}C(^{12}C, γ) data shown in Fig. 3. Since the available dipole radiative width will always be much larger than that for quadrupole radiation, capture through a $T \neq 0$ heavy-ion channel would be a much more sensitive probe for such an isomer search. As a possible example of this, a spectrum from the $^{16}O(^7Li, \gamma)^{23}Na$ reaction is shown in Fig. 9.[2] Here the γ_0 and γ_1 transitions are completely dwarfed by the much stronger decays to a doublet of levels at 7.1- and 7.7-MeV in ^{23}Na. These data are suggestive of a GDR built upon a prolate shape-isomer, although as yet there are no calculations for such levels in this nucleus. If such calculations can be extended to ^{23}Na and other s-d shell nuclei, heavy-ion radiative capture may prove an exceptionally well-suited probe.

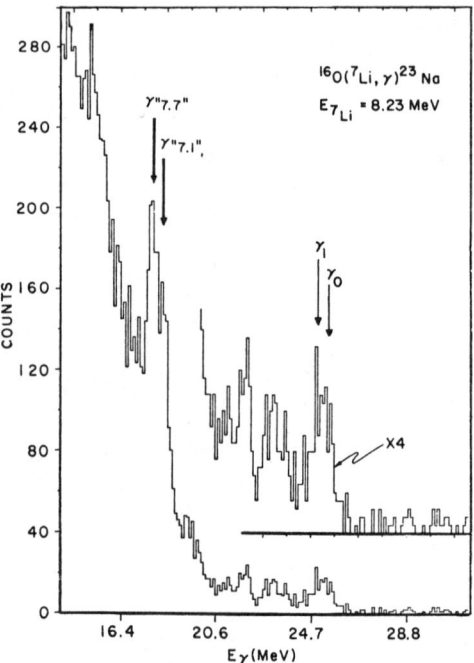

Fig. 9. A high-energy gamma-ray spectrum from the $^{16}O(^7Li, \gamma)^{23}Na$ reaction.

This work was supported by the U. S. Department of Energy under contract No. DE-AC02-76CH00016.

REFERENCES

1. P.M. Endt and C. van der Leun, At. Data Nucl. Data Tables 13, 67 (1974).
2. A.M. Sandorfi, Heavy-Ion Science, Vol. II (Ed. D. Allan Bromley), Plenum Press (1984).
3. A.M. Sandorfi and M.T. Collins, Nucl. Inst. and Meth. 222, 479 (1984).
4. A.M. Nathan, A.M. Sandorfi, and T.J. Bowles, Phys. Rev. C24, 932 (1981).
5. T. Ericson, Adv. Phys. 9, 425 (1960).
6. L. Ricken, A.M. Sandorfi, D.H. Dowell, and P. Paul, contributed paper to this conf.; and to be published.
7. H.G. Bohlen, M. Feil, A. Gamp, B. Kohlmeyer, N. Marquardt, and W. Von Oertzen, Phys. Lett. 41B, 425 (1972).
8. A.M. Sandorfi and M.T. Collins, Lecture Notes in Physics 156, 264 (1981).
9. S. Das Gupta and M. Harvey, Nucl. Phys A94, 602 (1967); S.S.M. Wong and G.D. Longheeed, ibid. A295, 289 (1978).
10. H. Spinka, and H. Winkler, Nucl. Phys. A233, 456 (1974), and references therein.
11. F. Galtz, P. Betz, J. Siefert, F. Heidinger and H. Röpke, Phys. Rev. Lett. 46, 1559 (1981).
12. D.H. Dowell, invited paper to this conf.
13. M.T. Collins, A.M. Sandorfi, D.H. Hoffmann, and M.K. Salomaa, Phys. Rev. Lett. 49, 1553 (1982); A.M. Sandorfi and M.T. Collins to be published.
14. J. Ahrens, H. Borchert, K.H. Czock, H.B. Eppler, H. Gimm, H. Gundrum, M. Kroning, P. Reihm, G. Sita Ram, A. Zieger and B. Ziegler, Nucl. Phys. A251, 479 (1975).
15. H. Schultheis and R. Schultheis, Phys. Rev. C27, 1367 (1983).
16. K. Langanke, Phys. Rev. C28, 1574 (1983).
17. L. Ricken, A.M. Sandorfi and P. Paul, unpublished.

HIGH ENERGY GAMMA RAYS FROM COMPLEX PARTICLE COLLISIONS

K.A. Snover
Nuclear Physics Laboratory, University of Washington
Seattle, WA 98195

Abstract

High energy γ-rays ($E_\gamma \sim 10$-30 MeV), from the decay of Giant Dipole Resonances built on excited nuclear states (e.s.-GDRs) are a common feature of all energetic nuclear collisions. Current results indicate that for bombarding energies below 5-6 MeV per nucleon the statistical emission of high energy γ-rays predominates in complex particle collisions. Statistical GDR properties have now been studied over a wide range of mass, energy and angular momentum. Recent results include 1) in light compound nuclei considerably broadened e.s.-GDRs are observed, with strengths that indicate relatively pure compound nuclear isospin, 2) in medium mass nuclei (A~60-80) e.s.-GDRs are found with substantially reduced strength (<0.5 of the classical dipole sum rule) and 3) in rare-earth deformed nuclei a splitting of e.s.-GDRs is apparent, indicating the persistence of deformation at elevated temperature. At higher bombarding energies (6-9 MeV per nucleon) for ^{3}He and ^{4}He strong enhancements are seen for $E_\gamma \gtrsim 15$ MeV, due to nonstatistical effects. These enhancements are associated with strong forward-backward asymmetries, arising from (phase coherent) E1-E2 interference, which peaks at energies near the isovector e.s.-GQR.

INTRODUCTION

Until three years ago the emission of high energy ($E_\gamma \gtrsim 10$ MeV) γ-rays in heavy ion collisions was considered to be an unusual process occuring with relatively small cross section in special cases, in the form of resonances near the Coulomb barrier. Then it was discovered at Berkeley that high energy γ-rays are emitted in the decay of compound nuclei copiously produced in heavy ion collisions.[1] In light ion reactions (mostly ^{3}He, α, and some d-induced reactions)[2] discrete radiative capture transitions to low-lying states had been studied earlier in a number of light nuclei and, although it had been realized in a few cases that continuum high-energy γ-ray emission originated from the decay of the compound nucleus, the continuum spectra that result from such a process had never been analyzed quantitatively.[3]

Most of you involved in neutron capture studies have known for years that slow neutron capture proceeds mainly via compound nucleus formation, and the γ-ray strength function derived from decay studies reflects the energy dependence of the low-energy tail of the Giant Dipole Resonance. Just as the GDR in medium and heavy nuclei decays (or is believed to decay) mainly via statistical particle emission, so also the statistical emission of high energy γ-rays should (and does) reflect the average properties of the GDR built on the states populated in the γ emission process. Energetic nuclear collisions produce highly excited compound nuclei for which γ-decay strength function studies are possible for energies spanning the GDR and up as high as the isovector Giant Quadrupole Resonance (IVGQR). In contrast to γ-emission following slow neutron capture, the high energy γ-rays in energetic reactions must be emitted in direct competition with energetic particle emission - this occurs for $E_\gamma \sim E_G$ (E_G ≡ the resonance energy of the GDR) with an absolute probability $\Gamma_\gamma / \Gamma \sim 10^{-4}$ to 10^{-5} per MeV, a small number, but it leads to respectably large γ-ray production cross sections $\sigma_\gamma \sim \sigma_c \Gamma_\gamma / \Gamma \sim$ 10-100 μb/MeV for typical above-barrier compound nucleus formation cross sections $\sigma_c \sim$ 1 barn.[4]

The statistical emission of high energy γ-rays appears to dominate in all complex [A (projectile) > 2, probably also for the deuteron] particle collisions at sufficiently low bombarding energies. For these reactions, the statistical emission mechanism is believed to be understood, and the interesting physics lies in the properties of the GDR in highly excited nuclei - how are they modified and what can this teach us about new aspects of nuclear behavior. Two new aspects are discussed below: compound nuclear isospin, and the deformation of highly excited "hot" nuclei.

At higher bombarding energies I present here new results which demonstrate intriguing nonstatistical effects in ^{3}He and α-induced reactions. These include spectrum shape deviations from statistical expectations, along with strong apparent E1-E2 interference effects. At this early stage, the emphasis lies in a qualitative understanding of the reaction mechanism (direct and/or semidirect emission?). The results presented here, plus what little is known for heavier ion (A>4) induced reactions at high energies, suggest that very roughly the onset of appreciable nonstatistical effects occurs at bombarding energies E/A ~ 6-7 MeV per nucleon.

GENERAL CONSIDERATIONS

In general, we may write the cross section for photoexcitation of a nuclear level followed by particle emission as

$$\sigma(\gamma,x) = \sigma_{abs} (\Gamma_x / \Gamma) = \sigma_{abs} f \Gamma_x\!\downarrow / \Gamma + \sigma_{abs} (1-f) \Gamma_x\!\uparrow / \Gamma$$

where σ_{abs} is the photoabsorption cross section and Γ_x/Γ is the decay branching ratio for emission of particle x. In general, both statistical and nonstatistical processes may contribute to the decay widths, and we may write $\Gamma_x = \Gamma_x\!\downarrow + \Gamma_x\!\uparrow$ and $\Gamma = \Gamma\!\downarrow + \Gamma\!\uparrow$ where $\Gamma\!\downarrow$ and $\Gamma\!\uparrow$ are the total spreading (or statistical) width and the total escape (or nonstatistical) width, and similarly for the partial widths. Then $\sigma(\gamma,x)$ separates into two terms as shown above, the first term being the statistical part and the second the nonstatistical part, where $f = \Gamma\!\downarrow/(\Gamma\!\uparrow+\Gamma\!\downarrow)$ is the fraction of the total decay strength which is statistical and 1-f the fraction which is nonstatistical.

The first term along with the Brink hypothesis[5] leads to the dipole strength function for the time-reversed process of statistical γ-emission from a (compound) nuclear level

$$f(E_\gamma) = \Gamma_\gamma^J \rho_J(E_i)/E_\gamma^3 = 3.32 \times 10^{-6}\, S_{E1}^S(NZ/A)\, \Gamma\, E_\gamma\, [(E_\gamma^2-E_G^2)^2+E_\gamma^2\Gamma^2]^{-1}$$

where Γ_γ^J is the E1 radiative width between individual levels for a decay $J \to J_0$ where $J_0 = J$, $J \pm 1$, $\rho_J(E_i)$ is the density of initial states, E_G and Γ are the resonance energy and width of the GDR and $S_{E1}^S = S_{E1} \cdot f$ where S_{E1} is the total GDR strength (in units of the classical sum rule) and f is given above. S_{E1}^S is thus the fraction of the GDR strength which decays statistically. The γ-emission spectrum from the decay of the compound nucleus is then calculated in the usual manner[6] using transmission coefficients and level densities to obtain the contribution from the parent (initial) compound nucleus and all its daughters populated by particle emission. By the hypothesis of independence of formation and decay, the identity of the entrance channel does not matter except insofar as it determines the compound nucleus formation cross section at a given energy, spin, and perhaps isospin. One should also keep in mind the approximation involved in this type of application, that the strength function given above holds with the same parameter values for all excited states which contribute in a given reaction. A primary goal then is to look for GDR parameter dependencies on excitation energy, spin, isospin and nuclear species.

The second term in $\sigma(\gamma,x)$ above contains all other (nonstatistical) processes. For the inverse (x,γ) reaction, direct and/or semidirect (one-step doorway) amplitudes may contribute at an early stage in the time evolution of the excited nuclear configuration, as well as pre-equilibrium γ-emission at some later stage before statistical equilibration is reached. In contrast to nucleon capture, very little is known about such processes in complex particle collisions.

EXPERIMENTAL TECHNIQUES

The measurement of reliable continuum γ-ray spectra requires careful consideration of backgrounds. Several different techniques have been employed. For most of the results shown here, (inclusive) singles spectra are measured with a large shielded NaI spectrometer. For light targets (A<40), passive neutron shielding is adequate to allow DC beam measurements; for heavier targets pulsed beam measurements are required to eliminate neutron backgrounds by time-of-flight. Random pulse pileup must be suppressed. In heavy targets, light contaminants must be kept small, and their contribution measured and corrected for if necessary. Good gain stability is required and, since there are no spectral lines present at high energies, an accurate gain calibration must be measured using other reactions.

Another common technique employs the use of a trigger

detector to detect low energy γ-rays emitted near the end of the decay cascade, providing time-of-flight for the high energy γ ray which is measured in coincidence in a separate detector. Crystal ball studies of excited-state GDR decays are also underway at Heidelberg.[7] These measurements may reveal detailed information on GDR decay properties as a function of nuclear spin; on the other hand, neutron discrimination is more difficult in such a close array of detectors.

STATISTICAL DECAYS IN LIGHT NUCLEI

The example[8] of ^{3}He + ^{25}Mg illustrates many of the general features of high energy γ-ray production in complex particle collisions, including both statistical and nonstatistical behavior. Fig. 1 shows γ-ray spectra from this reaction measured at three different, relatively low bombarding energies. These spectra are all very similar in shape. The cross section is very large below particle threshold (E_γ < 10 MeV) while at higher energies a broad structureless yield is apparent. The solid curves, statistical model calculations using CASCADE,[9] are in remarkably good agreement with the experimental data over a factor of 10^6 variation in intensity. The shape of the calculated spectrum at energies below particle threshold (E_γ < 10 MeV) is insensitive to the GDR strength function parameters (energy, width and strength), while its magnitude depends on the compound nuclear formation cross section, σ_c. At higher energies (E_γ > 10 MeV), the yield scales with the product $\sigma_c \cdot S^s_{E1}$ and the spectrum shape

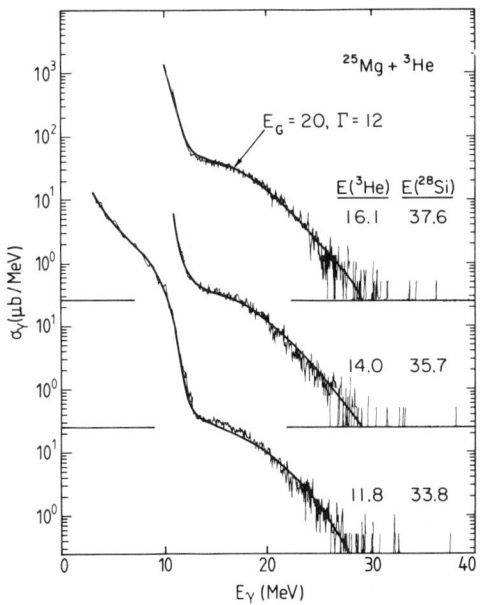

Fig. 1 Gamma-ray production cross section for ^{3}He + ^{25}Mg. The smooth curves are statistical model calculations with the indicated GDR parameters (Ref. 8).

depends on E_G and Γ; a comparison of the calculated and the experimental cross sections for different parameter values determines $E_G \sim$ 20 MeV and $\Gamma \sim$ 12 MeV.

Fig. 2 shows a decomposition of the calculated cross section into components arising from γ-decay of the initial parent nucleus and from all of the daughters populated by particle emission. At high E_γ, the γ-decay of the initial compound nucleus dominates. This is required for $E_\gamma >$ E (initial, ^{28}Si) - 10 MeV, where energy conservation rules out contributions from any of the daughters populated by particle decay. At higher bombarding energies, calculations show that the daughters become relatively more important at high E_γ, but that the γ-decay of the initial compound nucleus is always expected to be the strongest single channel. This follows[1] from level density considerations for the circumstance where the γ-decay energy exceeds the mean particle decay energy (binding plus kinetic energy), which is generally the case for $E_\gamma \sim E_G$ (for continuum γ-decay of energy less than the mean particle decay energy the situation reverses and daughter decays are more important). Several different experiments in heavier systems have provided experimental information supporting this feature.[10,11]

The value $E_G \sim$ 20 MeV extracted from the ^{3}He + ^{25}Mg analysis is comparable to the value for the GDR built on the ground-state, while the width $\Gamma \sim$ 12 MeV is considerably broader. Similar resonance energies and widths have been found[12] in an ^{27}Al(p,γ)^{28}Si study of GDRs built on 1 particle-1 hole states in this same energy region of ^{28}Si. The large width (relative to the width $\Gamma \sim$ 4.5 MeV of the ground-state GDR) is not understood theoretically.

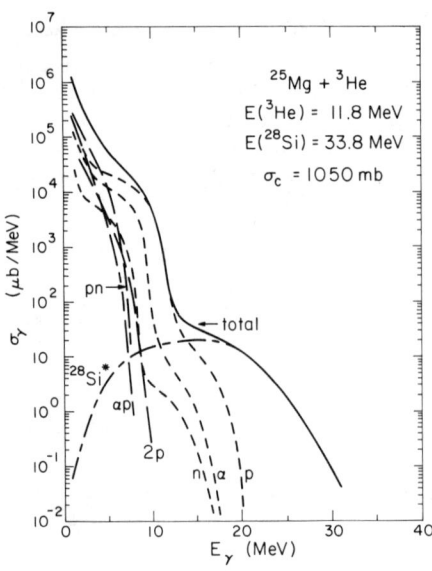

Fig. 2 CASCADE calculation of the partial and total cross section for ^{3}He + ^{25}Mg. Each curve is labelled by the particle(s) emitted preceding the γ-ray (Ref. 8).

For N=Z compound nuclei the isospin of the entrance channel can have a strong effect on the high energy γ-ray yield, as is illustrated in the case[13] of $^{12}C + ^{16}O$ (Fig. 3), which shows a strong inhibition of high energy $E_\gamma \sim E_G \sim$ 20 MeV emission relative to that observed in $^{25}Mg + ^{3}He$. This effect follows from the well-known selection rule that $\Delta T=\pm 1$ (no $\Delta T=0$) for E1 decays in a self-conjugate nucleus. Thus E1 decays from T=0 initial states must go to T=1 final states and vice-versa. Since the density of T=1 final states at moderate excitation energies in an N=Z nucleus is much less that the density of T=0 final states, the high energy statistical γ-yield from $^{12}C + ^{16}O$, which has a T=0 entrance channel, should be much less than the yield from $^{25}Mg + ^{3}He$ which has an equal mixture of T=0 and T=1 in the entrance channel, as is observed. At low E_γ the spectra from the two reactions are similar, since the emission of a particle preceding the γ-ray tends to wash out the effects of isospin. Shown in Fig. 3 are calculations[14] for both pure and completely mixed isospins, along with a calculation which fits the data well, in which isospin increases linearly with energy, with a value of 3% for the initial ^{28}Si compound nucleus at 34 MeV. Actually, the high energy γ-yield is sensitive to the sum of the isospin mixing intensity in initial and final states; this quantity is 5% for the best-fit calculation, and is insensitive to the assumed energy dependence of the isospin mixing, as long as it varies slowly. Thus compound nuclear isospin is still relatively pure! Here E2 emission is predicted to be about 15% in the high energy, which is not negligible (Fig. 3).

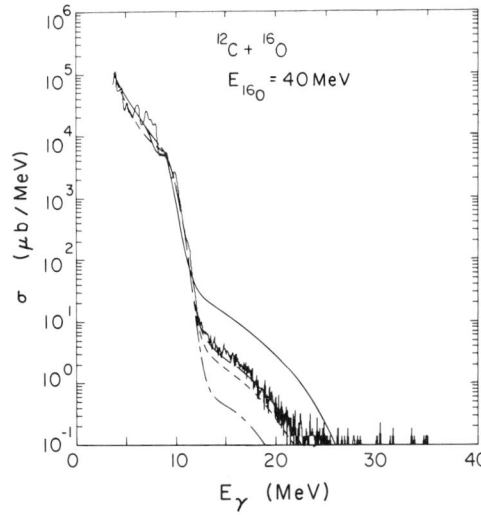

Fig. 3 Gamma-ray production cross section for $^{12}C + ^{16}O$, $E(^{16}O) = 40$ MeV, $E(^{28}Si) = 34$ MeV. CASCADE calculations: solid line-completely mixed isospin, all curves E1 + E2; long-short dash, E2 only (Ref. 13).

STATISTICAL DECAYS IN MEDIUM MASS NUCLEI

As an example of statistical decays in medium mass nuclei, I turn to a recent study[15] of γ-ray spectra from $\alpha + ^{59}Co$ and $^{12}C + ^{51}V$ reactions, both of which form the ^{63}Cu compound nucleus. In an earlier study,[16] limits were placed on a possible spin dependence of the GDR strength function in ^{63}Cu, ^{71}Kr, and ^{127}Cs. The present measurements span a range of excitation energies E_x(initial) of 17 to 52 MeV (see Fig. 4). In all but the lowest energy α-spectrum,

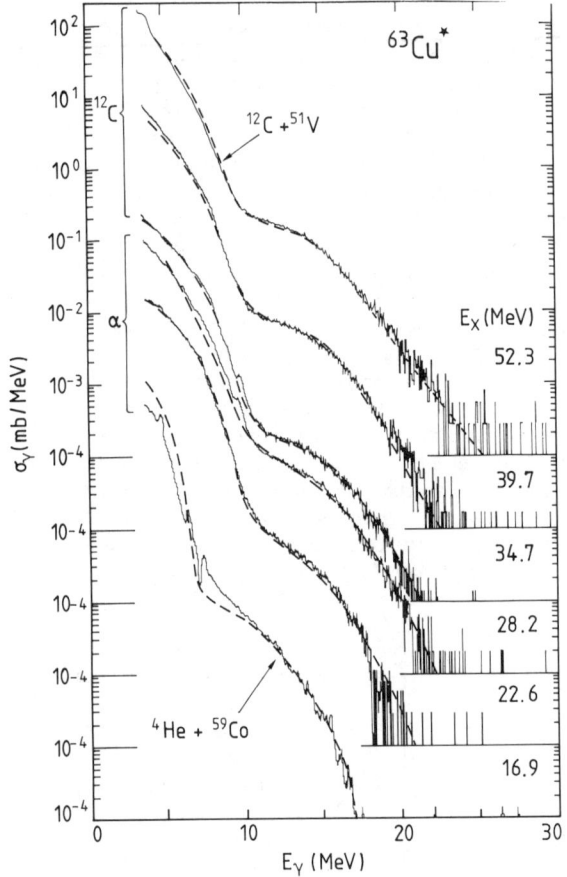

Fig. 4 Gamma-ray cross sections from ^{12}C + ^{51}V (top 3 graphs) and α + ^{59}Co (bottom 3 graphs) reactions forming the ^{63}Cu compound nucleus at six different excitation energies. The dashed curves are fitted CASCADE calculations described in the text (Ref. 15).

contributions from light contaminants are negligible. These spectra clearly demonstrate that the probability of high energy γ-emission increases with increasing excitation energy of the initial compound nucleus. This is just a level density effect as discussed above. The similarity in the shapes of the 24 MeV α + ^{59}Co spectrum (E_x = 28.2 MeV) and the 26 MeV ^{12}C + ^{51}V spectrum (E_x = 34.7 MeV) provides support for the assumption that the compound nucleus reaction mechanism dominates. These spectra have been fitted with a chi-squared minimization routine based on the CASCADE code. Preliminary fitting results indicate E_G values between 16 and 17.5 MeV, widths Γ ~ 5-8 MeV and strengths S_{E1}^s ~ 0.5. For comparison, the ground-state GDR has E_G = 17.5 MeV, Γ = 4.7 MeV and

$S_{E1} \sim 1.0$.[17] The small value of S^S_{E1} needed to fit the data, in agreement with recent $^{16}O + {}^{58,61,64}Ni$ results (J. Gundlach et al., unpublished), is puzzling. We are currently looking into whether isospin considerations are important here. Final analysis of these data will help to answer the question raised by earlier experiments of whether there exists a systematic deviation of E_G for excited-state GDR decays compared to ground-state GDR decays.

STATISTICAL DECAYS IN HEAVY NUCLEI

One of the most interesting applications of excited-state GDR decay studies to the field of nuclear structure is the use of the GDR strength function to reveal the deformation of highly excited nuclei, in an analogy with the well-known splitting of the ground-state GDR into two components in prolate-deformed rare-earth nuclei. This application is particularly exciting since, in contrast to rapidly rotating "cold" nuclei near the yrast line, relatively little is known about "hot" deformed nuclei, in which most of the excitation energy resides in internal degrees of freedom. It has been proposed[18] that a hot nucleus should retain memory of its low energy deformation up to a temperature $T \sim \delta \; 40$ MeV $A^{-1/3}$ where $E_x \cong aT^2$, $a \cong A/8$ (MeV^{-1}) and δ is the nuclear deformation. Furthermore, at sufficiently high angular momentum (I $\sim 70 \; \hbar$ for $A \sim 160$) centrifugal forces should drive a nucleus into an oblate shape regardless of its initial (low-energy) deformation.[19]

Recent high-energy γ-decay studies[20] of the ^{166}Er compound nucleus formed at $E_x = 61.5$ MeV in the 84 MeV $^{16}O + {}^{150}Nd$ reaction indicate a spectrum shape which cannot be described by a GDR strength function with a single Lorentzian. A description of the data in terms of a two-component GDR strength function yields a strength ratio of 2:3 for the upper component relative to the lower component, suggesting an average oblate deformation (in the hydrodynamic model, a strength ratio of 0.5 is expected for oblate deformation and 2.0 for prolate deformation) with $\delta \sim 0.27$ estimated from the energy splitting. For comparison, a study of ^{108}Sn decays at comparable temperature is consistent with a one-component GDR strength function. The finding of Ref. 20 of an oblate average deformation for the excited Er nucleus is very surprising, since ^{166}Er at low energies has a large prolate deformation ($\delta \cong 0.32$)[21] and the mean angular momentum of the initial compound nucleus, $\sim 22 \; \hbar$, is much too small to cause the shape change by rapid rotation.

We[22] have studied decays of the ^{166}Er compound nucleus at $E_x = 49.6$ MeV with a mean initial angular momentum of $\sim 15 \; \hbar$ formed in 61.5 MeV $^{12}C + {}^{154}Sm$. Shown in Fig. 5 is the measured γ-ray spectrum at $\theta_\gamma = 90°$, and also the spectrum multiplied by $\exp(E_\gamma / T_E)$ with $T_E = 1.5$ MeV. This multiplicative factor is arbitrarily adjusted to roughly cancel the rapid energy dependence of the cross section due to level density variation, allowing, for display purposes, data and calculations to be compared on a linear scale. The solid and dashed curves are (preliminary) best fit χ^2 minimization results for a two-component and a one-component GDR strength function, respectively, in which the energy, width and strength was allowed to vary independently for each component. The single component strength function is clearly inadequate, while the two-component strength function provides a good description of the data below E_γ = 19 MeV. Above E_γ = 19 MeV there is an excess of

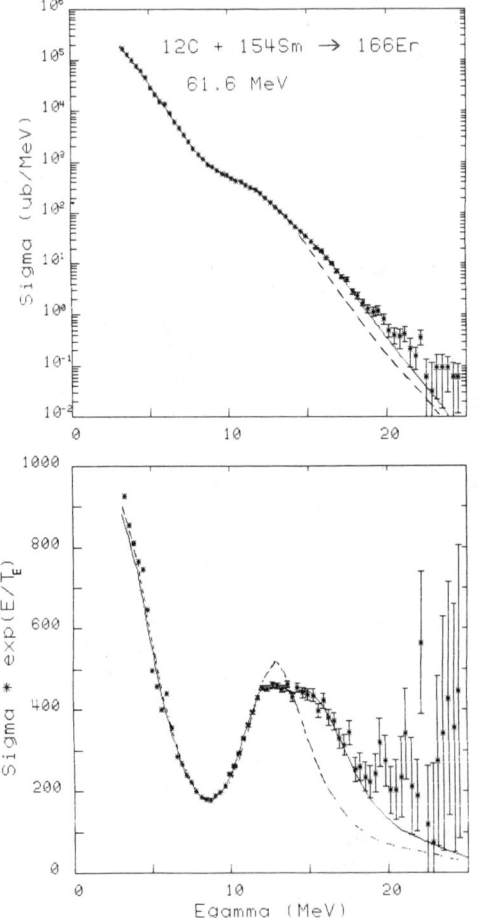

Fig. 5 Top: γ-ray production cross sections from ^{12}C + ^{154}Sm at $\overline{E}_c$ = 61.5 MeV. The curves are CASCADE fits described in the text. Bottom: display of both the data and the fits multiplied by $\exp[E_\gamma / T_E]$ with T_E = 1.5 MeV for ^{12}C + ^{154}Sm (Ref. 22).

yield above the calculation, whose origin is not understood; however, it does not significantly affect the fitted parameters. Results for the two-component fit are E_{G1} = 12.07, Γ_1 = 3.66, E_{G2} = 15.75, Γ_2 = 5.77 MeV, $S^s_{E1}(2)/S^s_{E1}(1)$ = 1.64, $S^s_{E1}(2) + S^s_{E1}(1)$ = 1.5

and $\chi^2/N = 1.22$. The strength ratio of 1.64 indicates <u>prolate</u> deformation and the energy ratio $E_{G2}/E_{G1} = 1.31$ implies an average deformation[23] $\delta = 0.32$ for the excited ^{166}Er nucleus (plus neighboring daughters) at $E_x \sim 30$ MeV (this is the mean residual energy <u>following</u> ~14 MeV γ-ray emission), corresponding to a temperature $T \sim 1.2$ MeV. This deformation is essentially identical to that of ^{166}Er in the ground state,[21] indicating no appreciable change with excitation energy. In fact, the excited-state GDR strength ratio and individual component energies and widths are remarkably similar to ground-state GDR values measured for nuclei of similar mass and deformation; e.g., ^{154}Sm(g.s.) which also has $\delta = 0.32$, has[24] $E_{G1} = 12.35$, $\Gamma_1 = 3.35$, $E_{G2} = 16.1$, $\Gamma_2 = 5.25$, $S_{E1}(2)/S_{E1}(1) = 1.67$ and $S_{E1}(2) + S_{E1}(1) = 1.21$. The width <u>ratio</u> Γ_2/Γ_1 is 1.67 for both the excited ^{166}Er and the ^{154}Sm(g.s.) GDR, while the individual widths are roughly 10% broader in the former case, pehaps due to increased damping. The difference between the present results and those of Ref. 20 for ^{166}Er* are not fully understood, but appear to be partly differences in data and partly differences in analysis (Ref. 20 did not do a χ^2-fit to their data). These results clearly indicate that statistical GDR studies are a powerful tool for studying the deformation of highly excited nuclei.

Finally I show in Fig. 6 the γ-ray spectrum from 27 MeV α + ^{154}Sm leading to ^{158}Gd, which has a large deformation comparable to ^{154}Sm and ^{166}Er, and the spectrum from 27 MeV α + ^{148}Sm leading to

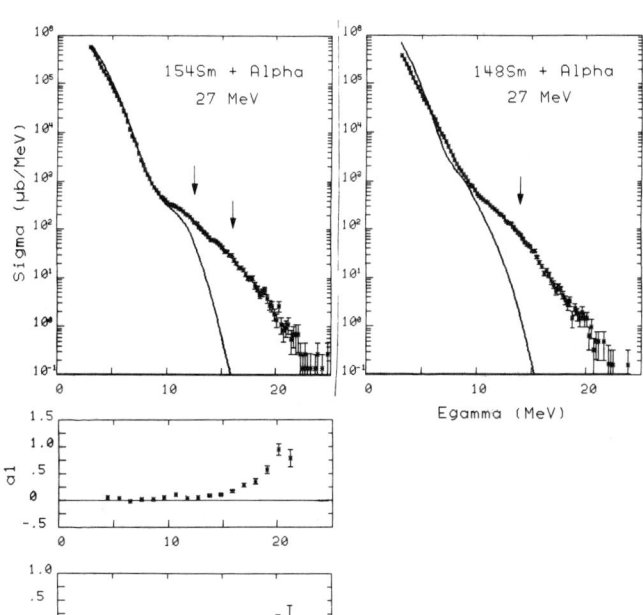

Fig. 6 Gamma ray production cross sections for α + ^{148}Sm and α + ^{154}Sm at $E_\alpha = 27$ MeV. For α + ^{148}Sm, a_1 and a_2 coefficients were extracted from $\theta_\gamma = 40°$, $90°$, and $140°$ (Ref. 22).

the substantially less-deformed ^{152}Gd nucleus.[21] The effect of deformation on the α + ^{154}Sm spectrum shape is clearly evident. Here we are unable to calculate the observed spectrum shape – I discuss this; along with the measured angular distributions in more detail below.

NONSTATISTICAL EFFECTS

To begin the discussion of nonstatistical high energy γ-ray production in complex particle collisions, I go back to the case of ^{28}Si. At high bombarding energies $E > 20$ MeV in the ^{25}Mg + ^{3}He reaction, there is clear evidence[4,8] for nonstatistical contributions to the γ-ray spectrum at high E_γ. Fig. 7 shows

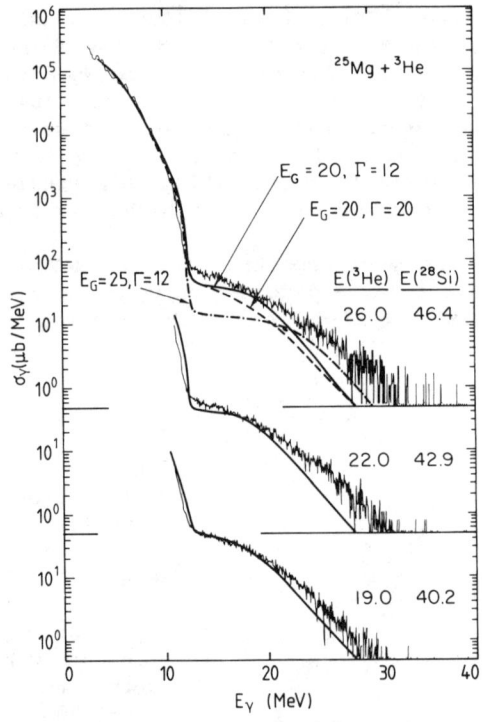

Fig. 7 γ-ray cross sections for ^{3}He + ^{25}Mg at E(^{3}He) = 19, 22, and 26 MeV. The curves are CASCADE calculations with the indicated parameters (Ref. 8).

spectra measured at E(^{3}He) = 19, 22 and 26 MeV. There is a clear deviation of the experimental data from the statistical model calculation for $E_\gamma > E_G$, which grows with increasing bombarding energy. Angular distributions show a forward-backward asymmetry (even perhaps at low E(^{3}He) – see Fig. 8) which grows with E_γ for a given E(^{3}He) and with E(^{3}He) at a given E_γ. For these data the NaI gain was adjusted at each angle to keep the low-energy spectrum (E_γ < 10 MeV) fixed in position. Since these γ-rays come primarily from fusion-evaporation, this procedure eliminates the angle-dependent Doppler-shift from the spectra. The a_1 coefficients (the coefficients of $P_1(\cos\theta)$ in the Legendre Polynomial expansion of the angular distribution of the γ-decay) deduced from the resulting spectra, as shown in Fig. 8, are thus to a good approximation representative of those values which would be observed in the center-of-mass of the radiating system, and thus must arise from the interference of multipoles of opposite parity, most likely E1 and E2. This type of effect is well-known in (p,γ) and (n,γ) reactions to discrete states in and above the GDR. In ^{25}Mg + ^{3}He, however, the smoothness of the spectra indicate that many different final states participate in the radiation process, too closely spaced to be resolved. Radiative emission populating any particular final state must, of course, arise from <u>different</u> (opposite parity) initial states for E1 and E2 emission. Hence the large a_1 coefficients, which are obtained by integrating over 1 MeV bins, indicate that a remarkable degree of coherence persists when one sums over many states of differing angular momenta (both final and initial).

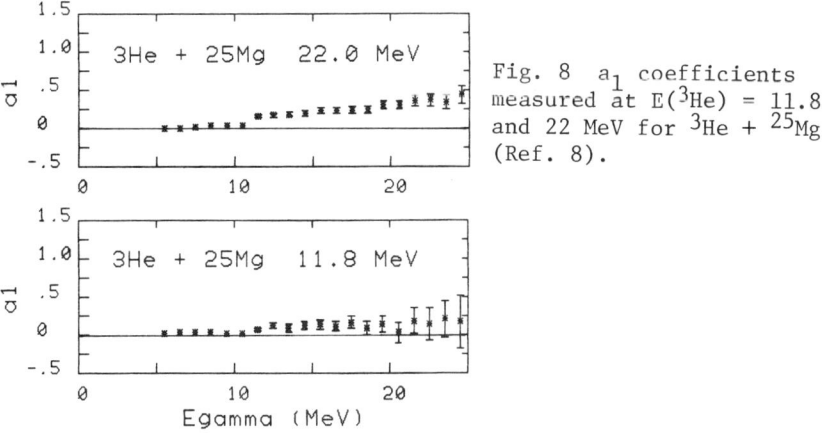

Fig. 8 a_1 coefficients measured at $E(^3He) = 11.8$ and 22 MeV for ^{3}He + ^{25}Mg (Ref. 8).

Fig. 9 shows the spectrum shape observed for 27 MeV ^{3}He colliding with various medium to heavy targets.[25] The solid curves are CASCADE statistical model calculations using default parameters for level densities. In all cases the magnitude of the observed γ-ray spectrum exceeds the calculation by several orders of magnitude or more at high E_γ. For 27 MeV ^{3}He + ^{146}Sm a similar result is seen[25] (Fig. 10); in addition the a_1 coefficient (estimated from data at $\theta_\gamma = 40°$ and $140°$) rises rapidly at high E_γ to a very large value.

Fig. 6 shows results analogous to Fig. 10, but for the α + Sm entrance channel at 27 MeV.[22] Again, one observes spectrum shape deviations at high E_γ relative to statistical model calculations, and very large a_1 coefficients. For these α and ^{3}He data the NaI gain was stabilized at a constant value independent at angle; here, however, effects of the lab-cm transformaton are small.

In addition, we have searched for angular distribution asymmetries in ^{12}C + ^{148,154}Sm and ^{12}C + ^{103}Rh at $E_c = 63$ MeV. Our results at present indicate values consistent with zero once the effects of light contaminants and the lab-cm transformation have been accounted for.

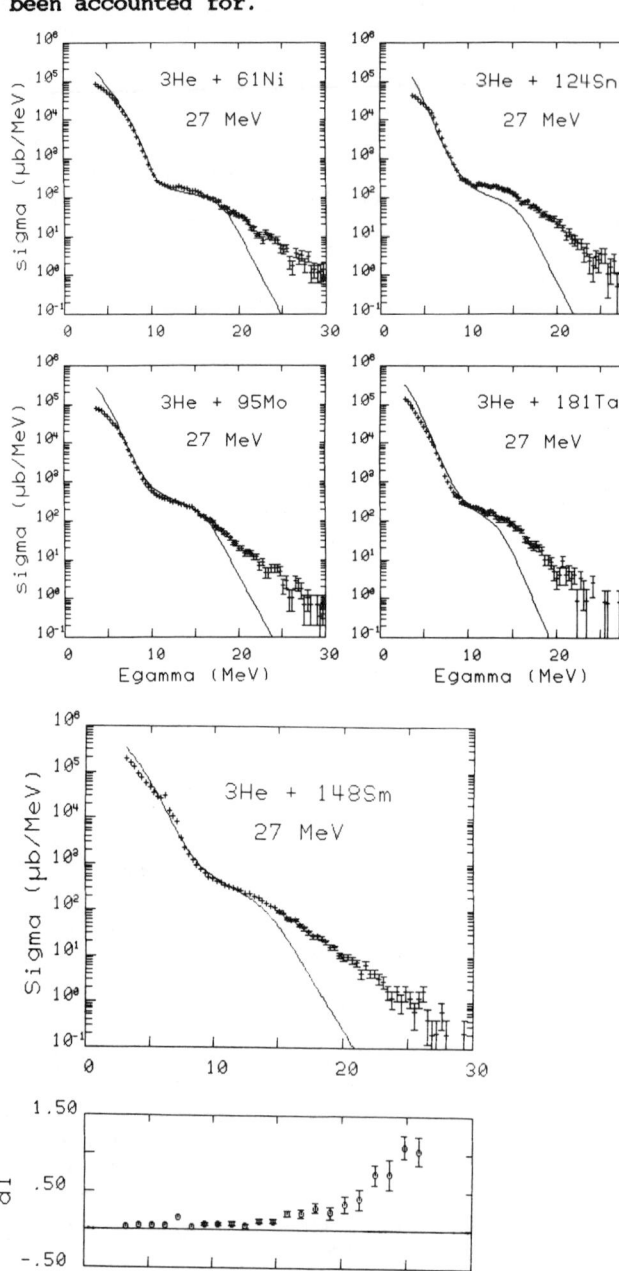

Fig. 9 90° spectra for 27 MeV ^{3}He + various targets. Solid curves are CASCADE calculations (Ref. 25).

Fig. 10 Averaged 40° + 140° spectra for 27 MeV ^{3}He + ^{148}Sm. The solid curve is a CASCADE calculation. The bottom half shows the a_1 coefficient deduced from $(Y_{40}-Y_{140})/(Y_{40}+Y_{140})$ (Ref. 25).

NONSTATISTICAL REACTION MECHANISM

The very strong degree of phase coherence required to produce the large a_1 coefficients in α and ^{3}He-induced reactions suggests that the nonstatistical γ-ray production mechanism involves radiative emission at an early stage of the reaction process, such as direct or semidirect (one-step doorway resonance) emission. In analogy with proton and neutron capture, we expect E1 emission to dominate at all energies except perhaps where the forward-backward cross section asymmetry is very large. Also from such reactions, we know that interference between either direct or resonant E1 and E2 amplitudes leads to large positive forward-backward asymmetries (positive a_1 coefficients) in the region above the GDR. Thus the asymmetries seen in composite particle reactions offer the possibility of studying the giant isovector E2 resonance built on excited states.

Both the direct and the semidirect E1 and E2 amplitudes should scale with the spectroscopic factor for target + projectile $\rightarrow$ final state, so that the observed nonstatistical cross sections should in principle be related to this spectroscopic strength distribution in the final (excited) nucleus, as is now well-established for proton capture.[8,12] These reactions may prove to be a powerful tool for studying composite-particle strength distributions at high energies. The lack of structure in these cross sections suggests that the ^{3}He and α spectoscopic strength distributions vary smoothly with energy in the nuclei studied.

The direct amplitude also contains the well-known effective charge factor $e_{eff}(L) = \mu^L[Z_1/M_1^L + (-)^L Z_2/M_2^L]$ for electric radiation of multipolarity L, while the semidirect amplitude, corresponding to one-step excitation of the collective giant resonance, involves the strength of the effective interaction for coupling the (composite) projectile to the collective vibration. For isovector resonances this coupling interaction strength is related to the strength of the $\vec{\tau} \cdot \vec{T}$ term in the projectile-target interaction, where $\vec{\tau}$ is the projectile isospin and $\vec{T}$ the target isospin. In addition both amplitudes involve (different) radial overlap integrals.

Since ^{3}He has isospin - 1/2, both direct and semidirect amplitudes may be present by analogy with proton capture. However, the semidirect amplitude should be weaker in ^{3}He vs. p capture since the ^{3}He contains only one nucleon which is unpaired (in isospin) and hence available to contribute to the $\vec{\tau} \cdot \vec{T}$ interaction. This is consistent with experiments which show that nonstatistical (^{3}He,γ) is much weaker than (p,γ).

The α-particle has isospin zero and hence has, in the usual formulation, no semidirect E1 or isovector E2 amplitude. The α-nucleus effective charge is zero for N=Z targets, but is appreciable ($\sim$0.4) for heavy targets. This may explain why at comparable bombarding energies, nonstatistical spectrum shape deviations are much larger in α + Sm (Fig. 6), than in α + ^{59}Co

(Fig. 4) for which $e_{eff}(E1) = 0.16$. If direct E1 and E2 dominates for nonstatistical (α,γ) reactions, the quantitative analysis of (heavy nucleus)(α,γ) experiments would be attractively simple, including extraction of α-spectroscopic factors for highly excited states. Further theoretical development of these issues is badly needed.

Nonstatistical effects, besides being interesting in their own right, represent an upper limit to the energy range over which the γ-ray production cross section can be approximated by the statistical model. Data on nucleon, ^{3}He and α capture suggests very crudely that the onset of significant nonstatistical effects occurs at bombarding energies (per nucleon) of E/A ~ 6-7 MeV/nucleon. Our negative searches for nonstatistical effects in 63 MeV ^{12}C + Sm and ^{12}C + ^{103}Rh (E/A = 5.6 MeV/nucleon) are consistent with this estimate. It is intriguing to look at higher bombarding energies for nonstatistical effects in reactions initiated by projectiles with A>4. Recent observations (Ref. 26) of spectrum shape deviations from statistical model calculations in 140 MeV ^{20}Ne + ^{91}Zr may be such an example.

SUMMARY AND CONCLUSIONS

The statistical emission of high energy γ-rays from the GDR built on excited states is prevalent in composite particle collisions. These reactions provide quantitative information about the properties of the GDR in highly excited nuclei, which leads to new information on isospin purity in light nuclei and deformation in highly excited nuclei. Nonstatistical effects are seen in both spectral shapes and angular distributions for α and ^{3}He - induced reactions. These effects present an opportunity to study spectroscopic strength distributions for composite particle transfer to highly excited states, and possibly the properties of giant isovector E2 resonances built on excited states. They also represent an upper limit to the energy range over which the γ-ray production cross section can be approximated by the statistical model.

ACKNOWLEDGMENT

I am indebted to my colleagues who have shared in this work, including J. Behr, D.H. Dowell, G. Feldman, E.F. Garman, J. Gundlach, M.N. Harakeh, R. Loveman, T. Murakami, and J.L. Osborne. I am particularly indebted to C.A. Gossett for a very rewarding and enjoyable collaboration.

REFERENCES

1. J.O. Newton, B. Herskind, R.M. Diamond, E.L. Dines, J.E. Draper, K.H. Lindenberger, C. Schück, S. Shin, and F.S. Stephens, Phys. Rev. Lett. 46, 1383 (1981).
2. Fast nucleon capture, the most extensively studied radiative capture reaction in earlier years, proceeds mainly via a nonstatistical process. This is the subject of several other papers at this conference and hence will not be discussed here (see also Ref. 4).
3. D. Drake, S.L. Whetstone, and I. Halpern, Phys. Lett. 32B, 349 (1970); Nucl. Phys. A 203, 257 (1973).
4. K.A. Snover, Proc. of the Internat. Symp. on Highly Excited States and Nuclear Structure, Orsay, France, Sept. 1983, Journal de Physique 45, C4-337 (1984).
5. D. Brink, D. Phil. Thesis, University of Oxford, 1955 (unpublished).
6. F. Pühlhofer, Nucl. Phys. Phys. A 280, 267 (1977) and private communication.
7. H. Hennerici, V. Metag, J.J. Hennrich, R. Repnow, W. Wahl, D. Habs, K. Helmer, U.V. Helmolt, H.W. Heyng, B. Kolb, D. Pelte, D. Schwalm, R.S. Simon, and R. Albrecht, Nucl. Phys. A 396, 329c (1983); D. Habs, private communication.
8. K.A. Snover, D.H. Dowell, E.F. Garman, G. Feldman, and M.N. Harakeh, to be published.
9. A computer code (Ref. 6) modified to include the GDR strength function. Although some earlier experiments were interpreted using approximate expressions based on effective temperature, it is clear that complete statistical model decay calculations are necessary for quantitative data analysis.
10. J.J. Gaardhoje, O. Andersen, R.M. Diamond, C. Ellegaard, L. Grodzins, B. Herskind, Z. Sujkowski, and P.M. Walker, Phys. Lett. 139B, 273 (1984).
11. A. Lazzarini, et al., Phys. Rev. Lett., to be published.
12. D.H. Dowell, G. Feldman, K.A. Snover, A.M. Sandorfi, and M.T. Collins, Phys. Rev. Lett. 50, 1191 (1983).
13. K.A. Snover, M.N. Harakeh, D.H. Dowell, G. Feldman, and J.L. Osborne, to be published.
14. Figs. 3-6, 9, 10 all show the measured data compared with CASCADE calculations folded with the measured detector response. The absolute cross section scale is exactly correct at E_γ = 15.1 MeV, and is approximate for other E_γ values.
15. E.F. Garman, K.A. Snover, M.N. Harakeh, D.H. Dowell, J. Gundlach, G. Feldman, R. Loveman, and J.L. Osborne, Proc. of the IOP Conference, Bradford, England, April 1984.
16. E.F. Garman, K.A. Snover, S.H. Chew, S.K.B. Hesmondhalgh, W.N. Catford, and P.M. Walker, Phys. Rev. C 28, 2554 (1983).
17. B.L. Berman and S.C. Fultz, Rev. Mod. Phys. 47, 713 (1975).
18. S. Bjornholm, A. Bohr, and B.R. Mottelson, Proc. Conf.

Phys. and Chem. of Fission, Rochester 1973, IAEA Vienna 1974, Vol. I, p. 367.
19. A. Bohr and B.R. Mottelson, Physica Scripta 10A, 13 (1974); G. Andersson et al., Nucl. Phys. A 268, 205 (1976).
20. J.J Gaardhoje, C. Ellegaard, B. Herkind, and S. Steadman, Phys. Rev. Lett. 53, 148 (1984).
21. V. Gotz, H.C. Pauli, K. Adler, and K. Junker, Nucl. Phys. A 192, 1 (1972).
22. C.A. Gossett, K.A. Snover, J.A. Behr, G. Feldman, T. Murakami, and J.L. Osborne, to be published.
23. See e.g. Ref. 17, and A. de Shalif and H. Feshbach, Theoretical Nuclear Physics, Vol. 1, John Wiley and Sons, 1974.
24. P. Carlos, H. Beil, R. Bergere, A. Lepretre, A. de Miniac, and A. Veyssiere, Nucl. Phys. A 225, 171 (1974).
25. J. Behr, G. Feldman, C.A. Gossett, and K.A. Snover, to be published.
26. J.J. Gaardhoje, Proc. of the XXII Int. Conf. on Nucl. Structure, Bormio, Italy, 1984.

THE (p,γ) REACTION ON THICK TARGETS AS A SPECTROSCOPIC TOOL

T. PARADELLIS

Tandem Accelerator Laboratory, N.R.C. Demokritos
15341 Agia Paraskevi, Greece

ABSTRACT

When thick even-even targets around mass A-60 are bombarded with low energy protons the (p,γ) reaction excite a large number of resonances in the compound nucleus with a strong initial alignment. The statistical gamma decay of these states to the low lying discrete levels of the compound nucleus introduces some attenuation of the initial alignment. It is shown experimentally that the resulting alignment of the low discrete states is strong enough to permit usefull spectroscopic study of these states, since the resulting attenuation do not depend on the target or the bombarding energy being function of the spin of the level. This type of spectroscopy can be extended also to measure life-time of levels through Doppler-shift.

INTRODUCTION

During the proton bombardment with E_p=2-4 MeV of even-even targets, 3-5 mg/cm^2 thick, around the mass A=60, thousands of resonances are excited predominantly of low spin in the compound nucleus. The γ- decay of these resonances feed the low lying discrete states of the even-odd compound nucleus. Thus γ-ray spectra from this reaction will contain discrete γ-rays which originate from the deexcitation of the low lying states, superimposed on a continuum of γ-rays which extend up to the reaction threshold.

The angular distribution of the discrete γ-rays provide useful information about the spin, the lifetime of the levels as well as their multipolarity, since as it is shown, the alignment of these states is known experimentally.

ANGULAR DISTRIBUTIONS

As it is well known the angular distribution of a γ-ray of multipolarity δ proceeding between states $J_i \to J_f$ is given by

$$W(\theta) = \Sigma \ Q_\nu \ A_\nu \ P_\nu(\cos\theta) \quad \nu = 0,2,4 \quad (1)$$

where Q_ν are the solid angle correction factors and P_ν the Legendre polynomials. The factor A_ν is written

$$A_\nu = \rho_\nu(J_i) \ L_\nu(J_i, J_f, \delta)$$

where $\rho_\nu(J_i)$ are the statistical tensors for alignment[1] and L_ν are the γ transition linking parameters[1].

More appropriatelly the $\rho_v(J)$ parameter may be written in terms of the statistical tensor for complete alignment[1] $B_k(J_i)$ as

$$\alpha_v(J_i) = \frac{\rho_v(J_i)}{B_k(J_i)}$$

where $\alpha_v(J_i)$ is the attenuation coefficient.

As it has been shown previously[2] the α_v have been determined for a number of levels and transitions in several nuclei by measuring the angular distribution of γ-rays deexciting levels with well known spin, parity and mixing ratio. The compound nuclei studied in ref.2 were ^{67,69}Ga, ^{63}Cu and ^{57}Co. Thus by measuring W(θ) through eqs 1,2,3 the α_v's have been determined for several levels. Additional measurements have been obtained from ^{65}Ga and ^{65}Cu. The following table summarizes the results.

TABLE I: Average values of $\alpha_v(J_i)$ at E_p=3MeV (ni refers to the total number of levels examined. Errors are statistical at the 95% confidence limit

J_i	n_2	$\alpha_2(J_i)$	n_4	$\alpha_4(J_i)$
3/2⁻	9	0.24(2)	-	-
5/2⁻	17	0.42(2)	9	0.08(3)
7/2⁻	13	0.62(3)	10	0.17(4)
9/2⁻	7	0.64(3)	4	0.29(10)

The constancy of the α_v's have also been checked against changes in bombarding energy. Thus by changing the energy of the protons from 2.5 to 3.5 MeV no change has been observed within 5% in the angular distribution of the γ-rays.

The derived attenuation parameters have been used succesfully in the spectroscopy of ^{65}Ga via the ^{64}Zn(p,γ) reaction[3] and ^{65}Cu via the ^{64}Ni (p,γ) reaction[4], where levels up to 3 MeV have been identified, assigned spin and parity and δ of the corresponding transitions have been determined

LIFE - TIME OF LEVELS

Although the recoil of the compound nucleus is small, excited states with τ=10-500 fs will exhibit measurable Doppler-shift. The main problem in analyzing such data is the unknown effect of the side feeding from the continuum. In Fig. 1. the F(τ) curves are plotted assuming different average mean lives for the continuum side feeding. Also marked are experimental F(τ) for a number of states of known τ in ^{57}Co and ^{63}Cu. As it can be seen data are reproduced by an average τ of 20 to 30 fs. Similar results have been obtained in ^{67}Ga and ^{69}Ga. Thus observed F(τ) in experiments are analyzed with τ_s=25 ±5 fs to obtain mean lives in other nuclei as for example in the case[2,4] of ^{69}Ga and ^{65}Cu

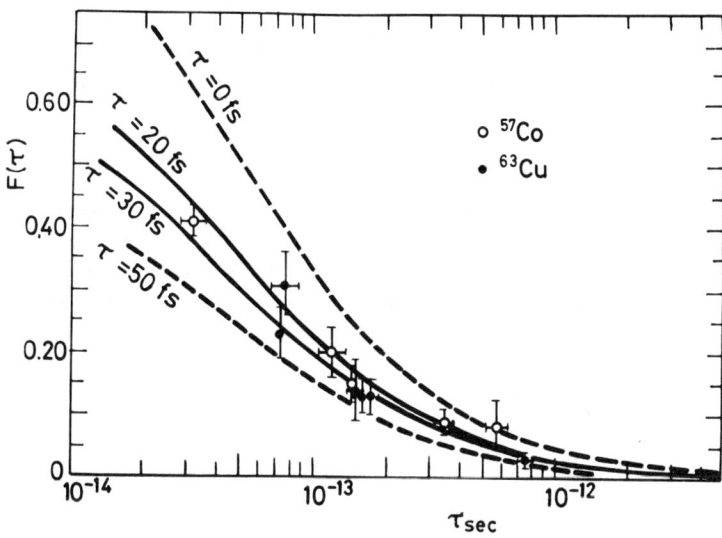

Fig. 1 F(τ) curve for different effective side feeding times.

REFERENCES

1. T. Yamazaki, Nuclear Data, A3, 1 (1967).
2. T. Paradellis, G. Vourvopoulos, G. Costa and E. Sheldon
                                    Phys. Rev C24,398 (1981).
3. T. Paradellis, under preparation
4. E. Paleodimopoulos, T. Paradellis: work under progress.

ANALYSIS OF THE $^{11}B(p,\gamma_0)^{12}C$ REACTION
IN TERMS OF THE DIRECT-SEMIDIRECT MODEL

M. Kicińska-Habior, T. Matulewicz

Institute of Experimental Physics, University of Warsaw, Poland

ABSTRACT

The experimental data for the $^{11}B(p,\gamma_0)^{12}C$ reaction in the proton energy range of 0.5-14 MeV were analysed within the framework of the direct-semidirect capture model. This analysis reveals an enhanced coupling of the GDR with the incident proton $d_{5/2}$ wave consistent with the microscopic structure of the GDR in the ^{12}C nucleus.

INTRODUCTION

In the interaction of a low energy proton with an odd-Z target nucleus the total energy of the system is often high enough to allow significant coupling of the reaction entrance channel to the nuclear collective giant modes and it is natural to expect that such a coupling will in some way manifest itself in different exit channels. Therefore, a particularly important role is expected to be played in the radiative capture reaction by virtual excitation of the giant dipole resonance (GDR) mode.

Our recent experiments have shown that, indeed, the γ-ray yields observed in the nonresonant radiative capture of low energy protons by odd-Z s-d shell targets cannot be satisfactorily understood in terms of pure direct capture mechanism and that one has to take into account virtual excitation of the GDR.[1,2] In these experiments the extracted coupling strengths of the GDR to various proton partial waves were found to be related to the microscopic structure of the GDR involved, built on the ground and excited states of the final nucleus.

It seems that the sensitivity of the direct-semidirect (DSD) model to the microscopic structure of the GDR should be higher for the light nucleus for which the GDR is far off collectivity and consists of a small number of components separated in energy. The ^{12}C nucleus was chosen to test this hipothesis. It has been suggested in theoretical calculations [3,4] and also found experimentally [6,8] that the dipole strength in this nucleus is distributed in four peaks, one of which at excitation energy 23 MeV exhausts about 70% of the dipole strength and is identified as the $(1p_{3/2})^{-1}(1d_{5/2})$ state. The other peaks at 17.6 MeV, 19.1 MeV and 25.6 MeV are identified as

$(1p_{3/2})^{-1}(2s_{1/2})$ and $(1p_{3/2})^{-1}(1d_{3/2})$ states, respectively.

The existing experimental data on the excitation functions and the angular distributions of the $^{11}B(p,\gamma_0)^{12}C$ reaction in the proton energy range of 0.5-4 MeV [5] and 4-14 MeV [6,7] are reanalysed in the present paper in terms of two versions of the DSD model describing the capture cross section.

ANALYSIS AND RESULTS

The experimental data for $^{11}B(p,\gamma_0)^{12}C$ reaction were analysed in terms of the traditional approach based on combined direct plus R-matrix calculations.[9] The compound nucleus resonances at 675 keV, 1388 keV, 2620 keV, 3500 keV and 3700 keV were accounted for by using appropriate Breit-Wigner amplitudes. Resonance parameters were taken from the literature.[5] Calculations of the nonresonant E1 direct capture amplitudes based on the direct-semidirect model followed closely the scheme described earlier.[1] The coupling with the GDR was taken into account via an effective charge factor. The proton-target interaction in the entrance channel was approximated by Coulomb plus hard sphere potentials with the radius parameter r_0=1.3 fm for proton energy range of 0.5-4 MeV, and by the Woods-Saxon potential with parameters proposed by Perey [10] for proton energy range of 4-14 MeV. In the final state the interaction was taken as Coulomb plus real Woods-Saxon potential with the same radius parameter and the depth adjusted to the binding energy of the captured proton. The proton spectroscopic factor value C^2S=3.0 was chosen. The energy of the GDR and its width were assumed as E_{GDR}=23.5 MeV and Γ_{GDR}=3 MeV for the $(1p_{3/2})^{-1}(1d_{5/2})$ and $(1p_{3/2})^{-1}(1d_{3/2})$ components. The complex coupling constants v for the dominant incident partial wave l=2 ($d_{5/2}$, $d_{3/2}$) were treated in the analysis as free parameters adjusted to fit to the experimental data. The $(1p_{3/2})^{-1}(2s_{1/2})$ component of GDR was accounted for by using the appropriate Breit-Wigner amplitudes.

The importance of coupling with the GDR in the low proton energy range is most pronounced for $E_p \gtrsim 3$ MeV, where the total cross section and the angular distribution coefficients, calculated assuming the compound resonance mechanism only, differ from the experimental ones. Four parameter fits to the experimental data for both energy ranges were performed independently because of the relatively larger values of the coupling constants in low energy proton range related to the damping of the hard sphere wave functions close to the nuclear surface where the coupling formfactor reaches maximum. Results of these calculations are shown by solid lines in Fig.1.

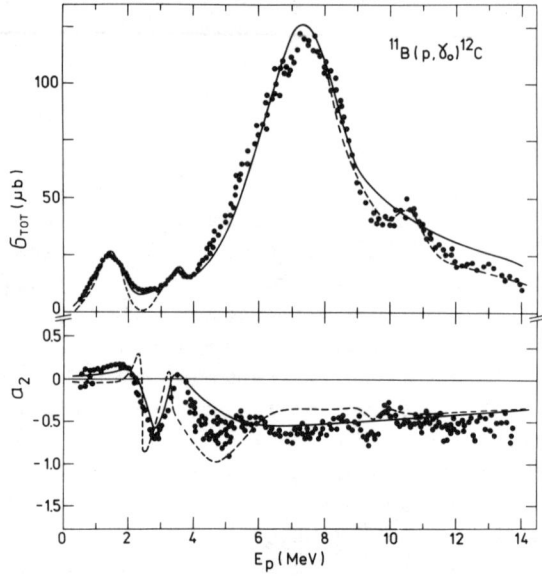

Fig. 1. Total cross section and a_2 coefficient for $^{11}B(p,\gamma_0)^{12}C$ reaction. Experimental data taken from Refs.5 and 6 (corrected as recommended in Ref.7). The solid lines show the fit obtained in traditional DSD model. The dashed lines correspond to the results of the modified DSD calculations.

It is worth noting that the angular distribution coefficient a_2 is in good agreement with the experimental value over a large proton energy range. The complex coupling constants v necessary to fit the experimental data have shown the same enhancement for the $d_{5/2}$ partial wave in both energy ranges:

$$v_{d_{5/2}} / v_{d_{3/2}} = 10 \pm 2 \qquad (1)$$

Enhanced coupling of the GDR with the incident proton $d_{5/2}$ wave is consistent with the microscopic structure of the GDR in the ^{12}C nucleus.[3,4]

Small collectivity of the GDR built on the ground state of the ^{12}C nucleus suggests that experimental data of the $^{11}B(p,\gamma_0)^{12}C$ reaction should be reanalysed in terms of a modified direct-semidirect capture model which accounts for the presence of broad shape (single-particle) resonances in the entrance channel. The reaction excitation function exhibits four broad resonances in the proton energy range of 0.5-14 MeV at E_p = 1.4 MeV, 3.5 MeV, 7.2 MeV and 10.3

MeV what correspond to excitation of the individual GDR components. The first two resonances can be regarded as two components of the $2s_{1/2}$ proton wave shape resonance split by coupling with the bound states embedded into the continuum. The remaining two resonances can be regarded as two components of the 1d proton wave shape resonance. These resonances were used to determine the single-particle potential depths for both s and d waves and the strength of the spin-orbit force in the reaction entrance channel. Comparison of the experimental widths of these resonances with the single-particle ones points to the almost single-particle nature of the former. The energy-dependent effective potential method ^{11} was applied to split the single-particle states in order to reproduce the positions and widths of the experimentally observed resonances. Analysis yielded the spectroscopic factor of the final state, $C^2S=3.2$. Agreement (see Fig.1) of the theoretical excitation curves and the experimental data supports the concept of the predominantly single-particle nature of the GDR in the ^{12}C nucleus.

CONCLUSION

The presented analysis has shown that the DSD model can be applied to proton capture even by nuclei as light as ^{11}B, if the microscopic structure of the GDR is taken into consideration. It seems that in the case of light nuclei the coupling constants, obtained from the fit of the DSD model calculations to the experimental values of the total cross section and angular distribution coefficient a_2, may give some information about the dominant components of the GDR.

REFERENCES

1. M. Kicińska-Habior et al., Z. Phys. A312, 89 (1983)
2. M. Kicińska-Habior et al., Z. Phys. submitted
3. M. Marangoni and A. M. Saruis, Nucl. Phys. A132, 649 (1969)
4. V. Gillet and N. Vinh-Mau, Nucl. Phys. 54, 321 (1964)
5. R. E. Segel et al., Phys. Rev. B139, 818 (1965)
6. R. G. Allas et al., Nucl. Phys. 58, 122 (1964)
7. M. T. Collins et al., Phys. Rev. C26, 332 (1982)
8. K. A. Snover et al., Nucl. Phys. A285, 189 (1977)
9. C. Rolfs, Nucl. Phys. A217, 29 (1973)
10. F. G. Perey, Phys. Rev. 131, 745 (1963)
11. J. Toke, T. Matulewicz, Acta Phys. Pol. B14, 609 (1983)

DESCRIPTION OF SUBBARRIER RESONANCES IN RADIATIVE PROTON CAPTURE REACTION BY THE EFFECTIVE POTENTIAL APPROACH

T. Matulewicz, P. Decowski, M. Kicińska-Habior, B. Sikora
Institute of Experimental Physics, University of Warsaw
Hoża 69, 00681 Warsaw, Poland

and J. Tõke
GSI, 6100 Darmstadt, W. Germany

ABSTRACT

An energy-dependent effective potential approach was applied to generate the relative motion continuum wave functions in the region of subbarier resonances in radiative proton capture reactions. The analysis of the (p,γ) reactions on ^{28}Si, ^{20}Ne and ^{12}C target nuclei provided information about the parameters of the interaction potential at low energies as well as spectroscopic factors of resonant and bound states in the final nuclei.

INTRODUCTION

Direct radiative proton capture reaction has proved in the past its usefulness as a tool of nuclear spectroscopy [1]. The non-resonant part of the interaction cross section is related to the DC model calculations by the value of the spectroscopic factor of the final state. However, in the presence of strong and broad resonances, the extraction of direct part of the cross section becomes potentially ambiguous.

We have proposed a modification of the theoretical analysis scheme, which treats the direct capture and the capture via broad resonances on equal basis [2]. The idea is based on employing an energy-dependent single-particle potential for generating the relative motion continuum (scattering) wave function. The potential accounts for effects of coupling of single-particle continuum states with bound states embedded into the continuum (BSEC). The energy dependence of the effective potential (in a simplified form) consists in variation of nuclear (Saxon-Woods) potential according to the formula:

$$U_{eff}(E) = U_o + \Sigma_{i=1}^{N} [aV_{ci}^2 / (E-\varepsilon_i)] \qquad (1).$$

Energy dependence (1) results in splitting of a single-particle resonance into N+1 resonances, while preserving the total single-particle strength. The action of the energy-dependent potential can be understood from Fig.1, which assumes N=1.

Variation of the well depth with the particle energy produces shape resonances at two different energies (Fig.1c) which result from equation:

$$U_{eff}(E) = U_{res}(E) \qquad (2).$$

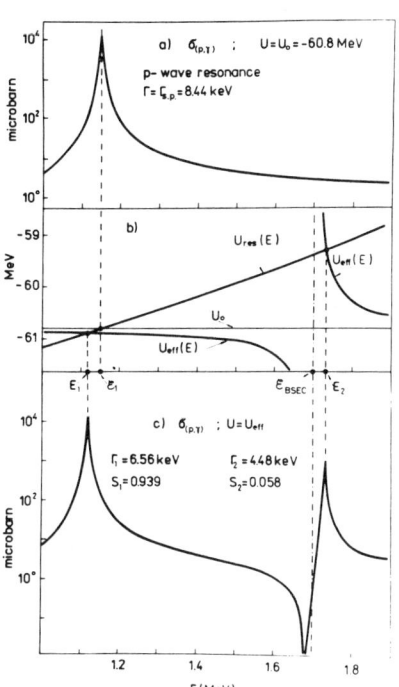

Fig.1. Excitation function for the (p,γ) reaction on a hypothetic A=28 Z=14 nucleus generated by using in the entrance channel a constant depth potential (section a) and the energy dependent effective potential with $U_{eff}(E)$ shown in the middle section (section c). The $U_{res}(E)$ line in the middle section represents the dependence of the single-particle resonance energy on the well depth for a particular well geometry. Figure taken from [2].

ANALYSIS AND RESULTS

The excitation function of the ^{28}Si(p,γ) reaction exhibits two broad resonances at 1.65 MeV and 2.88 MeV. These resonances can be regarded as two spin components of the 2p wave shape resonance. The effective potential and the spin-orbit potential parameters were determined by requesting the reproduction of positions and widths of both resonances[3]. The semi-direct mechanism[4] was also taken into account. In the case of the ^{28}Si(p,γ) reaction leading to the first excited state of the ^{29}P nucleus it was necessary to assume much stronger coupling to the GDR of the incident f-wave compared with the p-wave, which is consistent with the results of similar analysis carried out for odd-A target nuclei[5].

The ^{20}Ne(p,γ) reaction was studied in the same way as the ^{28}Si(p,γ) one. In this case two broad resonances at 1.83 MeV and 2.68 MeV were analyzed[6], although comparison with experimental data was possible only for $E_p < 2.2$ MeV.

The excitation function of the ^{12}C(p,γ) reaction is dominated by a strong resonance located at 457 keV. Our numerical studies have shown, that the position of this resonance is compatible with a combination of the depth V_o and the radius parameter r_o of the Saxon-Woods interaction potential which satisfy the equation:

$$V_o \times r_o^{1.7} \simeq 89.6 \text{ MeV} \times \text{fm}^{1.7} \qquad (3).$$

The width of a pure single-particle resonance generated at 457 keV depends with good accuracy linearly on r_o:

$$\Gamma_{s.p.} \text{ (keV)} = 35.5 + 15.7 \times (r_o - 1.2) \qquad (4)$$

when r_o is varied in a physically meaningful range[7]. Confronting eq. (4) with the experimentally measured value $\Gamma_{exp} = 36.5 \pm 1.0$ keV[8] allows to assume that this resonance is of pure single-particle nature. The final fit, which required also accounting for the semidirect mechanism, is compared with the experimental data on Fig.2.

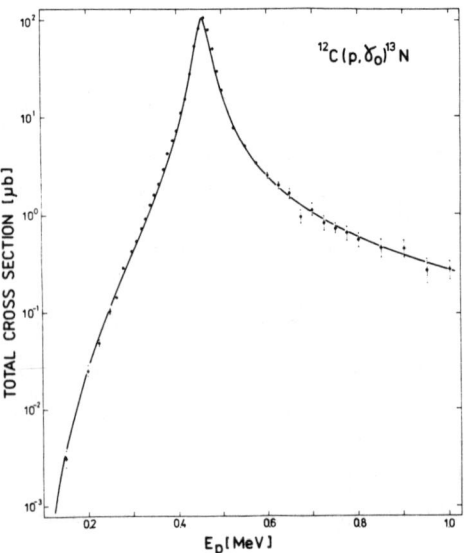

Fig.2. Comparison of the experimental total cross section values with the theoretical excitation curve (from ref.[7]).

The normalization of the theoretical excitation curves to the experimental data yielded the values of the spectroscopic factors of bound states of the final nuclei. The spectroscopic factors of broad shape resonances were extracted through comparison of their experimental widths with appropriate single-particle values. Table I presents the results of the analysis.

Table I. Spectroscopic factors from the modified DC analysis

Nucleus	E^* (MeV)	J^π	S
^{13}N	0.0	$1/2^-$	0.88
	2.37$^{(r)}$	$1/2^+$	1.00
^{21}Na	0.33	$5/2^+$	0.71
	4.17$^{(r)}$	$3/2^-$	0.95
^{29}P	0.0	$1/2^+$	0.36
	1.38	$3/2^+$	0.10
	4.34$^{(r)}$	$3/2^-$	0.85
	5.53$^{(r)}$	$1/2^-$	0.66

$^{(r)}$ resonances

REFERENCES

1. C. Rolfs, Nucl.Phys. A217,29 (1973)
   C. Rolfs, R.E. Azuma, Nucl.Phys. A227,291 (1974)
   C. Rolfs et al., Nucl.Phys. A241,460 (1975)
   H.P. Trautvetter, C. Rolfs, Nucl.Phys. A242,519 (1975)
   F. Terrasi et al., Nucl.Phys. A324,1 (1979)
2. J. Tõke and T. Matulewicz, Acta Phys.Pol. B14,609 (1983)
3. T. Matulewicz et al., Acta Phys.Pol. B14,617 (1983) and to be published
4. C.F. Clement, A.M. Lane, J.R. Rook, Nucl.Phys. 66,273 (1965)
5. M. Kicińska-Habior et al., Z.Phys. A312,89 (1983) and to be published in Z.Phys.A
6. K. Czerski et al. NPL Annual Report, U.Warsaw, p.19 (1981)
7. T. Matulewicz and J. Tõke, to be published in A.Phys.Pol.B15
8. F. Ajzenberg-Selove, Nucl.Phys. A360,1 (1981)

FRAGMENTATION OF THE $1g_{9/2}$ ISOBARIC ANALOG RESONANCE IN ^{53}Mn

J. Sziklai[x], and J.A. Cameron
Tandem Accelerator Laboratory, McMaster University,
Hamilton, Ontario, Canada L8S 4K1

I.M. Szöghy
Département de Physique, Université Laval, Québec City,
Québec, Canada G1K 7P4

ABSTRACT

Fragments of the $g_{9/2}$ isobaric analog resonance in ^{53}Mn corresponding to the 3.715 MeV state ($S_n=0.57$) in ^{53}Cr have been located using the ^{52}Cr(p,$p_1\gamma$), ^{52}Cr(p,$p_2\gamma$) and ^{52}Cr(p,γ)^{53}Mn reactions. Excitation curves were measured in the 4.04 - 4.35 MeV proton energy region. The excitation function of the (p,$p_2\gamma$) reaction is a sensitive tool for locating resonances with higher, $\frac{5}{2} \leq J_R \leq \frac{13}{2}$ spins and helped to locate the candidates for the $g_{9/2}$ resonance fragments. Since the (p,γ) channels are relatively strong for this case differential excitation functions for different primary transitions to low lying levels of ^{53}Mn were also measured. The spins of the resonances were found from (p,$p_1\gamma$), (p,$p_2\gamma$) and (p,γ) angular distribution measurements.

INTRODUCTION

The $g_{9/2}$ isobaric analog resonance (IAR) was first identified by Gunn et al.[1] and Galès et al.[2] in the ^{52}Cr(^{3}He,d)^{53}Mn reaction. Later Fodor et al.[3] localized nine fragments with the aid of differential excitation functions and (p,γ) and (p,$p_1\gamma$) angular distribution measurements. In a recent paper on the $g_{9/2}$ IAR's in ^{51}Mn[4] the method of normalized angular distributions and the (p,$p_2\gamma$) excitation function helped to identify the $g_{9/2}$ fragments. The use of these methods for studying the decay of the $g_{9/2}$ IAR in the ^{52}Cr+p system was very promising because the population of the 4^+ state in the ^{52}Cr(^{3}He,d$\bar{p}$) reaction[2] in the decay of the $g_{9/2}$ IAR had already been observed.

[x]Permanent address: Central Research Institute for Physics,
Budapest 114, P.O.Box 49, Hungary, H-1525

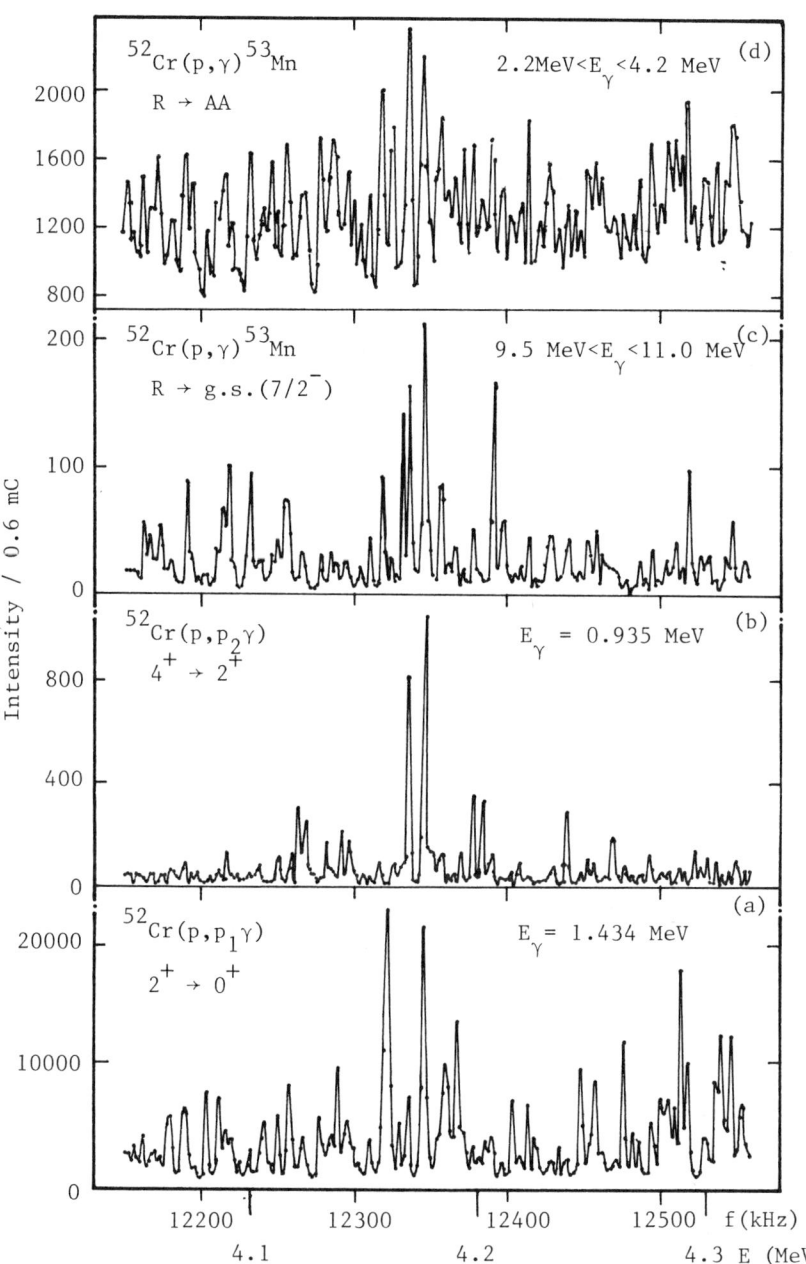

Fig.1 Excitation functions for the ^{52}Cr(p,p$_1\gamma$), ^{52}Cr(p,p$_2\gamma$) and ^{52}Cr(p,γ)^{53}Mn reactions in the 4.04 - 4.35 MeV proton bombarding energy region measured in 1.3 keV steps.

EXPERIMENTAL PROCEDURE

The measurements were carried out at the FN Tandem Accelerator of McMaster University where recently installed spiral-inclined-field accelerator tubes by DOWLISH and an NEC pelletron charging system with the earlier optical-fast-feedback stabilizer made it possible to achieve an overall beam resolution of ~ 700 eV even for extended runs.

The targets were prepared by evaporating isotopically enriched (98.8%) $^{52}Cr_2O_3$ onto thick, previously degassed tungsten backings. Degassed tungsten was also used for beam collimators and the small target chamber with diameter of 5 cm was lined with degassed tungsten too. The use of degassed tungsten enabled us to collect γ spectra with almost negligible fluorine contamination.

The γ rays were collected by two ORTEC GammaX detectors. The one with 29% efficiency and 1.68 keV resolution for the 1.33 MeV line was used for the excitation function measurements and as the moving detector for the angular distributions. The other one with 23% efficiency and 1.78 keV resolution, at $-90°$ to the beam direction was used as monitor. Both detectors had extremely good resolution at high energies i.e. 6.5 keV and 7.0 keV at 10.67 MeV.

The spectra were collected by a TRACOR NORTHERN TN-1710 multichannel analyzer. The data were transferred to a VAX 11/750 computer for later analyses.

For the angular distribution measurements the centering of the beam spot with respect to the detector circle was adjusted to be better then 1%.

RESULTS

The excitation functions measured in the 4.04 - 4.35 MeV proton bombarding energy region in 1.3 keV steps are shown in Fig.1. Figure 1(a) shows the excitation function of the $^{52}Cr(p,p_1\gamma)$ reaction. Several intense resonances arround 4.17 MeV show the presence of the $g_{9/2}$ IAR in this channel, however other resonances with spins of $\frac{1}{2} - \frac{9}{2}$ are expected to populate the first excited 2^+ state in ^{52}Cr. Figure 1(b) shows the excitation function of the $^{52}Cr(p,p_2\gamma)$ reaction. The structure of this curve is the simplest of the curves on Fig.1. The background is very low and only a few stronger resonances can be seen. Using the same consideration for the penetrabilities for resonances of different spins, as in Ref.4 it can be expected that all of the $g_{9/2}$ fragments are to show up in this excitation curve. We may find some $5/2^-$ levels but no $5/2^+$ resonances are expected to populate the 4^+ state. Figure 1(c) shows the excitation function of the ground state decay. Since the ground state has $7/2^-$ spin-parity and a large single particle strength resonances of spins 5/2 - 9/2 and retaining some single particle character are to appear in this curve. Figure 1(d) shows the integral excitation function for 2.2MeV<E_γ<4.2MeV. This is the region where the possible analog to antianalog transition was expected to appear. According to Ref.1

and Ref.2 important $g_{9/2}$ single particle strengths are located around the proton emission threshold. The main component at $E_x=6.54$ MeV ($S_p=0.42$) is considered as possible antianalog state. The primary γ rays from the $g_{9/2}$ IAR fragments were expected to appear in this excitation curve. The structure of this curve is rather complex due to the relative low energy where different secondary transitions, originating from other than $g_{9/2}$ resonances also appeared. There are ,however, some important correlations with other channels like the $(p,p_2\gamma)$ and (p,γ) where mainly the $g_{9/2}$ fragments are expected. Unfortunately no individual analog → antianalog transition has been identified yet.

The work to select the real $g_{9/2}$ fragments from the 24 candidates with the help of different angular distributions and decay properties is in progress.

REFERENCES

1. G.D.Gunn, J.D.Fox, and G.J.KeKelis, Phys. Rev. C13,595(1976).
2. S.Galès, S.Fortier, H.Laurent, J.M.Maison, and J.P.Schapira, Phys. Rev. C14,842(1976).
3. I.Fodor, J.Sziklai, P.Kleinwächter, H. Schobbert, and F.Herrmann, J.Phys.G 5,1267(1979).
4. J.Sziklai, J.A.Cameron, and I.M.Szöghy, Phys. Rev. C30,490(1984).

TOTAL (p,n), (p,γ), (p,p'γ) AND DIFFERENTIAL (p,p) CROSS-SECTION MEASUREMENTS FOR ^{61,64}Ni

R.L. Hershberger and F. Gabbard[+]
University of Kentucky
Lexington, Kentucky 40506

and

C.E. Laird
Eastern Kentucky University
Richmond, Kentucky

ABSTRACT

Absolute total (p,n) and differential elastic (p,p) cross sections have been measured for ^{61,64}Ni in the energy range of E_p = 2 to 7 MeV. The (p,γ) and (p,p'γ) cross sections were measured from as low an energy as feasible to approximately one MeV above the (p,n) threshold. Standard optical potentials have been used with a Hauser-Feshbach model to analyze the data. The adopted model values are used to deduce a total proton strength function which displays features of the 3s single particle resonance.

INTRODUCTION

The proton absorption cross section, measured for nuclides in the mass 60 region, should be strongly dominated by the 3s single particle resonance at proton energies near the Coulomb barrier. For neighboring nuclides, such as ^{61,64}Ni, the absorption cross sections are expected to be similar and hence described by similar sets of optical model parameters. Even so, widely different partial cross sections in the outgoing channels are to be expected because of the large differences in level densities in the target and residual nuclides. In fact, the total ^{64}Ni(p,n) cross section is approximately twice that of ^{61}Ni(p,n).

The objective of this paper was to make a statistical model fit of new data on the (p,γ) and (p,p'γ) for ^{61,64}Ni cross sections, together with existing (p,n) and (p,p) data. The analysis was done using similar optical model parameter sets for both ^{61}Ni and ^{64}Ni.

ANALYSIS

A major problem with an optical model analysis in this mass region is that a large fraction of the absorption cross section goes into the compound elastic channel, which cannot be directly measured. In addition, the (p,γ) channel, in general, cannot compete for decay of compound nuclear states with the compound

[+]Supported by the National Science Foundation.

elastic channel at energies near the (p,n) threshold. Hence there is no overlap region of the (p,γ) and (p,n) channels to show the competition of the two and thereby give a relative normalization.

In ^{61}Ni however there is a large (p,p') cross section which was measured by (p,p'γ). This cross section not only is calculated directly from the proton potential but also competes for a large fraction of the total absorption with both the (p,γ) and (p,n) channels. This provides a bridge for the relative normalization of the (p,γ) and (p,n) calculations and added confidence in the deduced proton absorption cross sections.

The form of the OM potentials used, and the experimental techniques for the (p,n) and (p,p) measurements are described in Ref. 1). The prescription for the γ-ray strengths was that of Johnson[2]. The (p,γ) measurements are described in Ref. 3).

The procedure used was to first vary parameters in order to fit the (p,γ), (p,p'γ) and (p,n) cross sections for ^{61}Ni. Then the same parameters were used to calculate the cross sections for ^{64}Ni. The only change necessary in order to fit the ^{64}Ni data was to increase W_D by 50%.

The obtained parameter sets were:

V_R = 64 MeV $W_D(61)$ = 20 MeV $W_D(64)$ = 30 MeV

a_R = 0.73 fm a_D = 0.15 fm

r_R = 1.20 fm r_D = 1.30 fm

the imaginary diffuseness, a_D, is quite small, however the product $W_D a_D$ = 3 – 4.5 MeV fm is consistent with measurements made in the mass 100 region [1,4-7]. However the γ-ray strength had to be reduced by a factor of fifty over what was used in the mass 100 region. The cause is not clear but was necessary to fix the relative competition of the (p,γ) and (p,p'γ) channels in the ^{61}Ni analysis.

The (p,p) differential elastic scattering was predicted quite well for ^{64}Ni, however the large compound elastic cross section and a contamination by the 67 keV first excited state in ^{61}Ni made comparison with the data difficult.

CONCLUSION

In conclusion the cross sections for protons on ^{61,64}Ni were well described with similar sets of parameters. The widely differing partial cross sections were predicted by the differing nuclear structure of the target and residual nuclides in the frame work of the statistical model.

REFERENCES

1. R.L. Hershberger, D.S. Flynn, F. Gabbard and C.H. Johnson, Phys. Rev. C 21, 896 (1980).
2. C.H. Johnson, Phys. Rev. C 16, 2238 (1977).

3. C.E. Laird, D.S. Flynn, R.L. Hershberger, F. Gabbard, "The proton-^{90}Zr interaction at sub-Coulomb energies," to be published.
4. D.S. Flynn, R.L. Hershberger, and F. Gabbard, Phys. Rev. C 20, 1700 (1979).
5. R. Schrils, D.S. Flynn, R.L. Hershberger, and F. Gabbard Phys. Rev. C 20, 1706 (1979).
6. D.S. Flynn, R.L. Hershberger, and F. Gabbard, Phys. Rev. C 26, 1744 (1982).
7. C.H. Johnson, A. Galonsky, R.L. Kernell, Phys. Rev. C 20, 2052 (1979).

THE PROTON-^{90}ZR INTERACTION AT SUB-COULOMB PROTON ENERGIES

C.E. Laird
Eastern Kentucky University
Richmond, Kentucky 40475

David Flynn, Robert L. Hershberger and Fletcher Gabbard[+]
University of Kentucky
Lexington, Kentucky 40506

ABSTRACT

Measurements have been made of proton elastic scattering differential cross sections for proton scattering at 135° and 165° from 2 to 7 MeV, of inelastic scattering cross sections for proton scattering from 3.9 to 5.7 MeV, and of the radiative capture cross sections from 1.9 to 5.7 MeV detecting primary and cascade gamma rays. Optical potentials with Hauser-Feshbach and coupled-channel models have been used to analyze the data. This analysis yields an energy dependent absorptive potential of $W = 2.63 + .73$ whose mean value of 5 MeV at $E_p = 4$ MeV is consistent with previously reported, but anomalously small values. The diffuseness of the real potential is .54 fm, which is consistent with values found for ^{92}Zr and ^{94}Zr. The adopted model values are used to deduce a total proton strength function which displays the features of both the 3s and the 3p single particle resonances.

Precision measurements of total (p,n) cross sections at sub-Coulomb proton energies have proven to be a valuable aid to understanding properties of proton-nucleus interactions at low bombarding energies. For example, the first clear observation of a single-particle or potential resonance in an excitation function for a medium or heavy-mass nucleus was made in very accurate measurements of total (p,n) cross sections for isotopes of tin[1]. Subsequent measurements of (p,n) cross sections in the mass-100 region have utilized the effects of the single-particle resonances to study the systematic behavior of the proton absorption from one nucleus to the next[2-7].

The reason for the special status of total (p,n) cross sections in this mass region is that the highly excited compound nucleus finds neutron decay--when it is energetically possible-- the dominate mode of decay for energies less than the Coulomb

[+]Work supported in part by the National Science Foundation.

barrier. Since direct reactions are negligible at these energies, measuring (p,n) cross sections essentially yields total proton-absorption cross sections. It is in measurements of the total proton-absorption cross sections that the effects of the single-particle resonances are most evident. Thus, fitting the detailed shapes produced by the single-particle resonances makes possible sensitive studies of the proton-absorptive potential.

A nucleus with a (p,n) threshold energy near the energy of the peak of the Coulomb barrier height does not lend itself to study by (p,n) cross section measurements. In cases, such the (p,n) excitation function cannot be measured over a sufficiently large energy range to clearly resolve the shapes due to the single-particle resonances. There are several nuclei around mass-90 whose absorptive properties are not amenable to study by (p,n) measurements because their (p,n) thresholds are too high. An objective of the present work was to extend the study of the sub-Coulomb proton absorption of nuclei in the mass-100 region by studying proton-induced reactions-(p,p), (p,p'), and (p,γ)- on ^{90}Zr. (Other channels may be open, but the cross sections are always negligible at these energies.)

To accomplish this experimental measurements have been made of elastic proton differential cross sections at 135° and 165° angles from 2 to 7 MeV, of inelastic cross sections for proton scattering to levels above the first excited state from an effective threshold at 3.9 MeV to 5.7 MeV, and of the radiative capture cross section by measuring the primary gamma rays to the ground and first excited states and the cascade gamma rays to these levels from levels up to 4.2 MeV excitation energy. These cross sections have been analyzed with a Hauser-Feshbach model using transmission coefficients from an Isospin Coupled-Channels code for elastic protons, from a standard optical model for inelastic protons, and from a modified Giant Dipole Resonance strength function for gamma rays.

This analysis has shown several improtant properties of model analysis parameters. First, the absorptive potential is energy dependent (i.e.,2.63 + 0.73 Ep) but has an energy-averaged value consistent with the anomalous behavior previously reported[2,3]. Second, the real diffuseness (.54 fm) is abnormally small, as it is for other Zr isotopes[4]. The diffuseness appears to be decreasing as one approaches the closed neutron shell at N=50. Finally, the deduced total proton strength function is similar to those found in the (p,n) studies of ^{92}Zr and ^{94}Zr but reduced from those strength functions by a factor of about 2. This seems due both to the reduced absorptive potential and to the reduced real diffuseness for protons in the entrance channel. This strength function shows a slow rise at low energy, indicative of the 3s single particle resonance and a high energy rise indicative of the 3p SPR and consistent with previous studies[1,3]. This gives further evidence of the existence of an s-wave single particle resonance in the mass-90 region.

Coupled-Channel and Optical Model Parameters Used in the Analysis of Experimental Data (Notation Defined in Refs. 3,4,6,7)

$V_R(E) = V_o + .45 \, Z/A^{1/3} - .32E$

$V_o = 55.6$ MeV, $1 \neq 2$ $V_o = 54.2$, $1 = 2$

$a_R = .54$ fm, $R_R = 1.20$ fm, $V_i = 24$ MeV

$V_{so} = 6.1$ MeV $R_{so} = 1.03$ fm $a_{so} = .63$ fm

$R_c = 1.231$ fm,

$W_D = 2.63 + 0.73 \, E$ $\Delta_{coul} = 11.868$ MeV

1. C.H. Johnson, J.K. Blair, C.M. Jones, S.K. Penny, and P.W. Smith, Phys. Rev. C15, 196 (1977).
2. C.H. Johnson, Phys. Rev. Letts. 39, 1604 (1977) and C.H. Johnson, Phys. Rev. C20, 2052 (1979).
3. D.S. Flynn, R.L. Hershberger, and F. Gabbard, Phys. Rev. C20, 1700 (1979).
4. R. Schrills, D.S. Flynn, R.L. Hershberger, and F. Gabbard, Phys. Rev. C20, 1706 (1979).
5. D.S. Flynn, R. L. Hershberger, and F. Gabbard, Bull. Am. Phys. Soc. 27, 461 (1982).
6. D.S. Flynn, R. L. Hershberger, and F. Gabbard, Phys. Rev. C26, 1744 (1982).
7. R.L. Hershberger, D.S. Flynn, and F. Gabbard, Phys. Rev. C21, 896 (1980).

HIGH RESOLUTION IN-BEAM γ-RAY SPECTROSCOPY

J. Kern, J.-Cl. Dousse, M. Gasser, B. Perny and Ch. Rhême
Physics Department, University, 1700 Fribourg, Switzerland

ABSTRACT

An in-beam curved crystal facility has been installed at the SIN variable energy cyclotron. Using the (110) planes of a 3.0 mm thick quartz lamina bent at 3.15 m, diffraction peaks typically 6 arcsec wide (FWHM) are obtained. The energy resolution is thus, for instance, 110 eV at 170 keV in 3rd order. Due to a sophisticated detector system and heavy shielding, the sensitivity of the instrument is quite good. The facility proves quite useful in (p,xnγ) reaction studies whenever the γ-ray spectrum is very complex, e.g. in the study of odd-odd deformed nuclei. Complicated multiplets appearing in the ^{176}Yb(p,3nγ)^{174}Lu spectrum could be successfully resolved. From the results we derive that the g-factors of the 142 d, $J^\pi = 6^-$ isomer, take anomalous values.

Spectroscopy using (n,γ) and (p,xnγ) reactions presents a number of differences but also of similarities. In particular (p,xn) reactions have also a limited selectivity, so that the decay of the entry states populates a large fraction of the low lying levels. Whenever the level density in the final nucleus is large, the spectrum is very complex. This problem has been solved in (n,γ) spectroscopy by using curved crystal spectrometers.

We have installed such a spectrometer at the SIN variable energy cyclotron. The crystal radius of curvature is 3.15 m and the angle of observation is $\Theta = 55°$. Using the (110) diffraction planes of a 3.0 mm thick quartz lamina, we obtain diffracted peaks about 6 arcsec wide corresponding to an energy resolution FWHM $\Delta E \cong 0.0115\, E^2/n$, where n is the diffraction order, ΔE is expressed in eV and E in keV. The target is gas cooled. Beam currents of a few μA can be used, the limit being given in general by the heat resistance of the material. Good background conditions are achieved by use of a sophisticated detector system and heavy shielding. In some energy regions, it can be as low as a few counts in 100 s (Fig. 1). The sensitivity permits one to observe transitions with an intensity of the order of 0.1 photon per 100 reactions in the energy range 40 to 300 keV within a reasonable measuring time.

A first application is a study of the ^{176}Yb(p,3n)^{174}Lu reaction. The $I^\pi = 1^-$ ground state configuration is { π 7/2$^+$ (404) - ν 5/2$^-$ (512) }. The configuration of the 6$^-$ 142 d isomer involves the same orbitals, reversely coupled. The magnetic moments of these states have been measured by Krane et al.[1]. It has been shown by Kern and Struble[2] that if, for a rotational band of a deformed doubly odd nucleus, the magnetic moment of one level and an intraband branching ratio are known, the value of the collective g_R factor and that of a linear combination of g_{Ω_p} and g_{Ω_n} can be obtained.

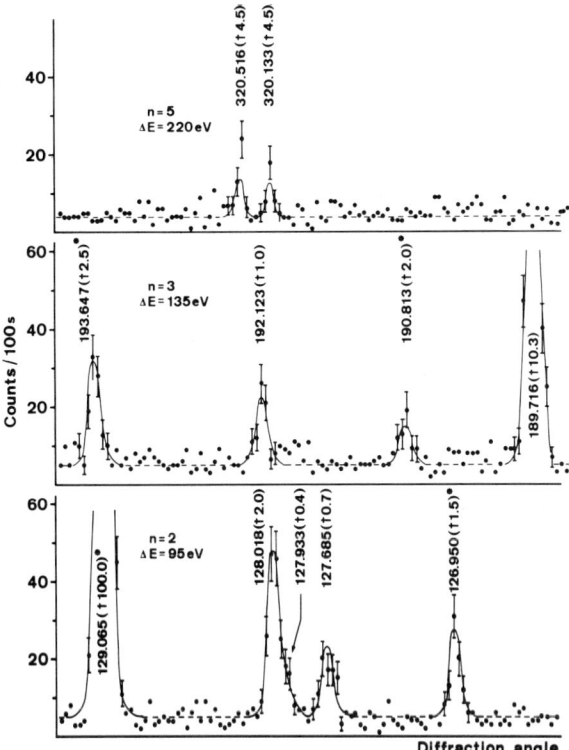

Fig. 1
Fraction of the prompt γ-ray spectrum following the ^{176}Yb(p,3n)^{174}Lu reaction observed in a small angular region in 2nd, 3rd and 5th order. The transitions are labeled by their energies in keV. In parentheses the relative intensity is given: an intensity of 100 corresponds approximately to 30 photons/100 reactions. An asterisk indicates a delayed transition, so that the total intensity does not appear in the prompt spectrum shown here.

These values can also be predicted from the properties of the neighbouring odd-A and even nuclei. The results we obtain for the ground state band are in good agreement with the predictions. The determination of a branching ratio within the isomeric band was made difficult

Table I Observed and predicted g-factors for the 6⁻ isomeric state in ^{174}Lu

Quantity	observed value	predicted value[a]
$\frac{7}{2} g_{\Omega_p} + \frac{5}{2} g_{\Omega_n}$	2.30(34)	1.32(3)
g_R	0.44(6)	0.29(1)

[a] From odd-A and even neighbours

since both the cascade (171.1 keV) and the cross-over (320.5 keV) transitions depopulating the $I^\pi K = 8^-6$ level are members of narrow multiplets which cannot be resolved with semiconductor detectors. This is however possible with the crystal spectrometer (see e.g. Fig. 2). From the branching ratio and the magnetic moment, the anomalous values listed in Table I are obtained [3].

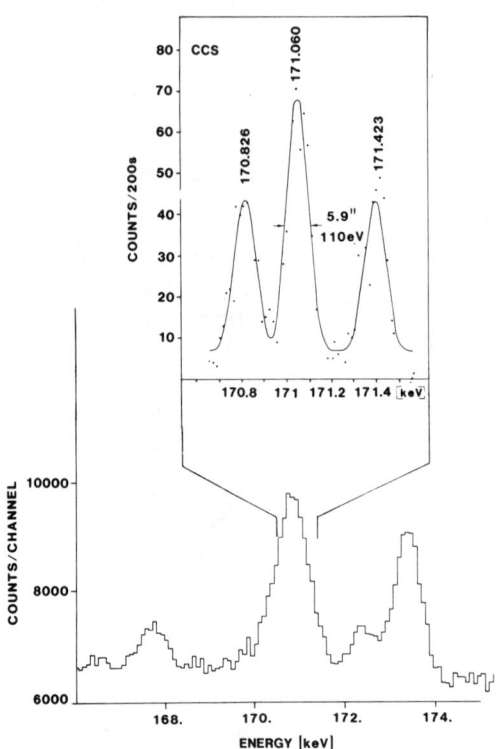

Fig. 2
Portion of the ^{176}Yb(p,3n)^{174}Lu gamma ray spectrum observed at $\Theta = 55°$ with a Ge-detector having 700 eV FWHM resolution at 170 keV. The region of interest (insert) has been observed with a curved crystal spectrometer in DuMond geometry.

Anomalously large g_R have also been observed for a few other high K two-quasiparticle configurations in odd-odd and even nuclei, so that the effect has to be taken seriously. We are still investigating its origin.

REFERENCES

1) K.S. Krane, C.E. Olsen, S.S. Rosenblum and W.A. Steyert, Phys. Rev. C12 (1975) 1999.

2) J. Kern and G.L. Struble, Nucl. Phys. A286 (1977) 371.

3) J. Kern, A. Bruder, J.-Cl. Dousse, M. Gasser, V.A. Ionescu, R. Lanners, B. Perny, B. Piller, Ch. Rhême and B. Schaller (to be published).

NUCLEAR SPECTROSCOPY WITH FEW-NUCLEON TRANSFER REACTIONS
ON LIGHT NUCLEI+

H.J.Hauser,T.Rohwer,F.Hoyler*,G.Staudt,S.Abd el-Kariem**,
P.Grasshoff,H.V.Klapdor***,A.Körber,W.Leitner**,V.Rapp****,
M.Walz,D.Weinmann
Physikal.Institut der Universität Tübingen,
Fed.Rep.of Germany

ABSTRACT

Nuclei in the 1p- and 2s1d-shell have been investigated with (p,α)-, (n,α)-, $(p,^3He)$- and (d,α)-reactions. The experiments were performed with projectile energies between 14 MeV and 45 MeV. From a systematic analysis of the reaction data two- and three-nucleon spectroscopic factors have been obtained, which are compared to predictions of shell model calculations. Spectroscopic results for target nuclei between ^9Be and ^{27}Al are discussed.
The shape of the differential cross sections can be described fairly well by DWBA calculations in zero range as well as in finite range approximation. Both microscopic and cluster form factors have been used. As a general result the microscopic calculations strongly underestimate the absolute cross sections.
The influence of energy dependent optical potentials, collective excitations and sequential transfer processes on the calculated cross sections is discussed. The optical potentials and the deformation parameters used in the analysis of the (p,α) reactions have been determined from elastic and inelastic scattering data which were measured at the appropriate energies in additional experiments.

INTRODUCTION

The analysis of transfer reactions has attractive features for nuclear spectroscopy. Especially few-nucleon transfer reactions are useful for extracting information on two-, three-, or four-nucleon correlations in nuclei and therefore they offer a valuable tool for the study of nuclear wave functions.
In the last years we have studied some (p,α)-, (n,α)-, $(p,^3He)$- and (d,α) reactions on light nuclei. In order to calculate the spectroscopic amplitudes we have taken the shell-model wave functions given by Kurath et al.[1]

+Supported by the BMFT
*ILL, Grenoble
**DAAD fellowship
***MPI f. Kernphysik, Heidelberg
****Department of Physics, University of Edinburgh

(1p-shell) and Chung and Wildenthal[2] (sd-shell). It is the main point of this paper to test the shell-model predictions with regard to two- and three-nucleon transfer reactions, and therefore we present a comparison between our experimental results and spectroscopic strengths for a number of studied transitions.

EXPERIMENTS

The experiments have been carried out at the cyclotron of the KFA Jülich [(p,α),(p,³He), E_p=22-45MeV] and at the MP tandem accelerator of the MPI für Kernphysik Heidelberg [(p,α),(d,α), E =18-24MeV]. The (n,α) experiments (E_n=14MeV) have been performed using the 3 MV Van de Graaff accelerator of the University of Tübingen. The outgoing particles were detected either by means of counter telescopes or using the magnetic spectrographs BIG KARL in Jülich and the Heidelberg multigap.

In order to deduce optical potentials in additional experiments proton and α-particle elastic and inelastic scattering data were measured using the proton beam of the Jülich cyclotron and the α-beams of the cyclotron of the University of Bonn (E_α=48 and 54MeV) and of the MP tandem together with the post accelerator of the MPI Heidelberg (E_α=63MeV).

DWBA ANALYSIS, ABSOLUTE CROSS SECTION VALUES

In order to analyse the experimental data DWBA calculations have been done. Concluding the results it can be stated that the shape of the angular distributions can be well fitted in most cases using both a microscopic or a cluster form-factor in a zero range as well as in a finite range calculation. Furthermore the relative transition strengths are fairly well reproduced by both the microscopic and the semi-microscopic calculations (see fig. 1).

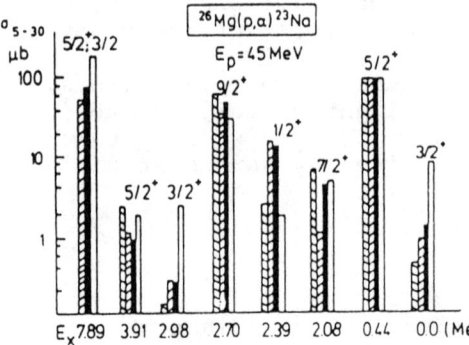

Fig.1: Comparison between angle-integrated cross sections (⊓), bare spectroscopic factors (■), DWBA with cluster form-factor (▧), and DWBA with microscopic form-factor (▨). Normalisation at the first excited state at E_x = 0.44 MeV.

However, in the microscopic calculations the absolute cross section values are about one or two orders of magnitude too small. This result has also been obtained in DWBA analyses of other (p,α) reactions[3] and a similar situation is found in (p,t) reactions[4]. Several attempts have been made to explain this discrepancy: 1) Choice of other optical α-potentials, 2) Consideration of two-step processes via inelastic channels in deformed nuclei, 3) Contributions of sequential transfer processes[5] e.g. (p,α)=(p,t)(t,α), 4) Consideration of knock-out processes[6]. But all these efforts only give rise to weak modifications of the calculated data. However, an enhancement of the form factor at the nuclear surface tends to increase the absolute cross section values significantly[7]. In the microscopic model such an enhancement may originate from two-body correlations and centre-of-mass corrections in the bound-state wave functions as already discussed elsewhere[4].

SPECTROSCOPIC RESULTS

Although further work is necessary to obtain a real understanding of few-nucleon transfer processes, fig.1 shows that the <u>relative</u> intensities can be well described by the bare SU(3) spectroscopic factors indicating the minor importance of the dynamics of the processes for the relative values. This statement is confirmed by the results given in figs.2-4.

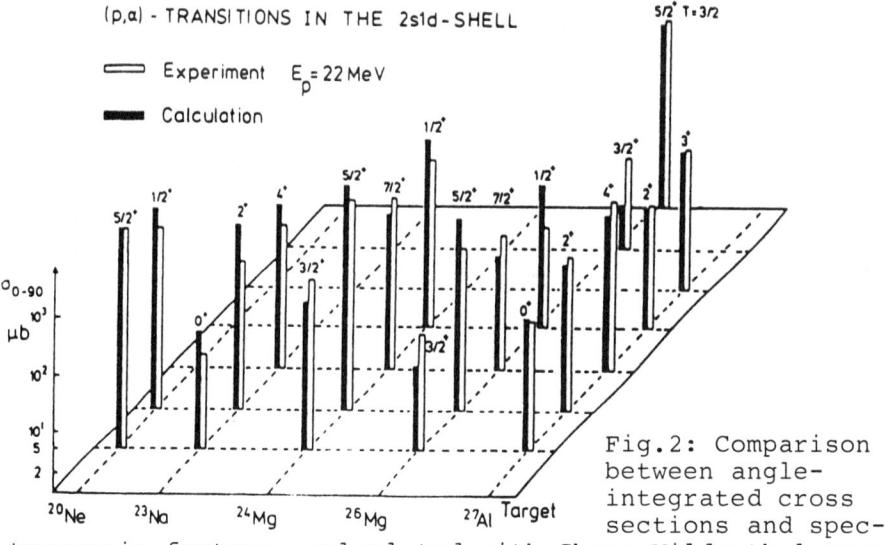

Fig.2: Comparison between angle-integrated cross sections and spectroscopic factors, calculated with Chung-Wildenthal wave functions. The sum of the spectr. factors is normalized to the sum of the experimental data.

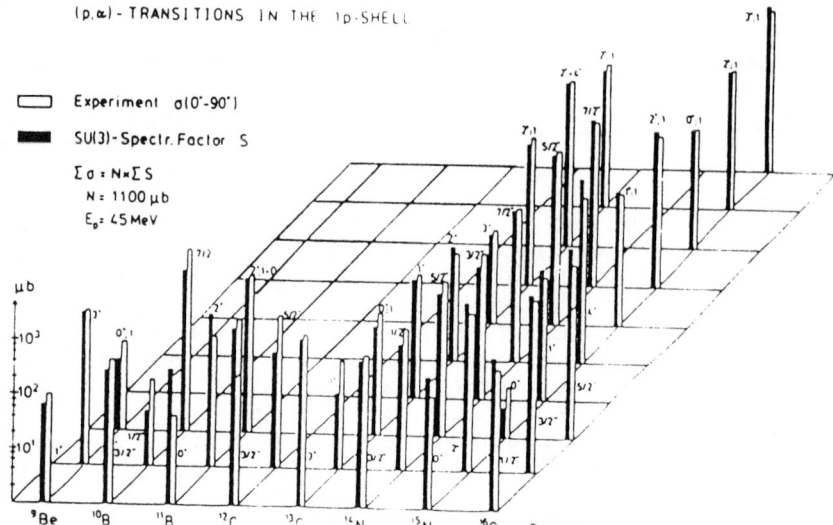

Fig.3: Comparison between angle-integrated cross sections and spectroscopic factors, calculated with Cohen-Kurath wave functions. Normalisation as indicated above.

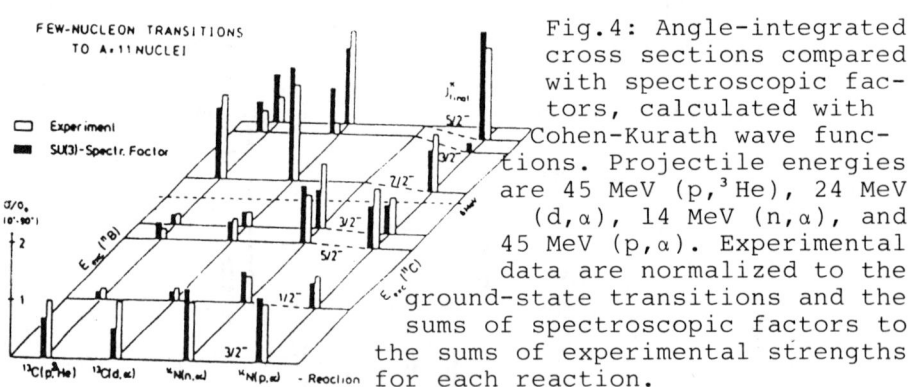

Fig.4: Angle-integrated cross sections compared with spectroscopic factors, calculated with Cohen-Kurath wave functions. Projectile energies are 45 MeV (p,^{3}He), 24 MeV (d,α), 14 MeV (n,α), and 45 MeV (p,α). Experimental data are normalized to the ground-state transitions and the sums of spectroscopic factors to the sums of experimental strengths for each reaction.

REFERENCES

1. S. Cohen, D. Kurath, Nucl. Phys. 73, 1 (1965)
2. W. Chung, B. Wildenthal, priv. communication
3. F. Brunner et al., Nucl. Phys. A398, 84 (1983)
4. W. T. Pinkston, Comm.Nucl.Part.Phys. 12, 133 (1984)
5. F. Hoyler et al., Proc. 4.Int.Conf. on Cluster Aspects of Nucl.Struct. and Nucl.React., Chester, 23.-27.7.84, to be publ.
6. E. Gadioli et al., Z. f. Physik, to be publ.
7. F. Hoyler et al., Phys. Rev., to be publ.

HEAVY-ION RADIATIVE CAPTURE OF $^{12}C + ^{14}C$ INTO ^{26}Mg

L. Ricken
State University of New York, Stony Brook, NY 11794
and Ruhr-Universität Bochum, 4630 Bochum, West Germany[+]

A.M. Sandorfi, D.H. Dowell
Brookhaven National Laboratory, Upton, NY 11973

P. Paul
State University of New York, Stony Brook, NY 11794

ABSTRACT

The radiative capture of ^{12}C by ^{14}C into the fused nucleus ^{26}Mg has been investigated for $E_{CM} = 6.2 - 8.1$ MeV. Structures are observed in the excitation functions for γ-decay to the bound states up to $E_x = 5.7$ MeV in ^{26}Mg. The lack of any structure in elastic scattering and fusion suggests that these might reflect the electromagnetic strength functions built upon the low-lying levels of ^{26}Mg. However, the experiment indicates considerable differences between these strength functions for the different final states.

I. INTRODUCTION

Recent studies of heavy-ion radiative capture have revealed a rather varied picture of this rare type of reaction. The reaction $^{12}C(^{12}C,\gamma)^{24}Mg$ exhibits unusual resonances whose structure appears closely linked to the ground-state band of the fused nucleus ^{24}Mg.[1] In $^{12}C(^{16}O,\gamma)^{28}Si$, similar resonances were found in the γ_1 transitions, but not in γ_0,[2] while unusually large cross sections for capture into the excited prolate band of ^{28}Si were observed, their strong enhancement probably resulting from the large overlap with the inherently prolate heavy-ion entrance channel.[3]

To investigate further the extent to which heavy-ion capture occurs in other systems, we have studied the $^{12}C + ^{14}C$ reaction. It could reveal features quite different from the $^{12}C + ^{12}C$ and $^{12}C + ^{16}O$ cases since no structures in the carbon width have been observed near the Coulomb barrier.

II. EXPERIMENTAL PROCEDURE

Two separate experiments were performed at the Brookhaven MP7 Tandem Van de Graaff, using in one a radioactive beam, and in the second a radioactive target, to study the $^{12}C + ^{14}C$ capture reaction. In the first experiment, a ^{14}C beam of up to 0.5 particle-μA bombarded natural carbon foils, 0.5 - 1.0 MeV (CM) in thickness. The second experiment was done with ^{14}C targets of typically 70 μg/cm^2 thickness,[4] corresponding to an energy loss of the incident ^{12}C beam of at most 200 keV (CM) in the energy range $E_{CM} = 6.2 - 8.1$ MeV investigated.

High-energy γ-rays were detected in the BNL MKIII NaI(Tl) spectrometer.[5] Fig. 1 shows two spectra from the second experiment with a ^{14}C target. Transitions to the ground and first two excited states of ^{26}Mg are clearly resolved. All spectra are dominated, however, by

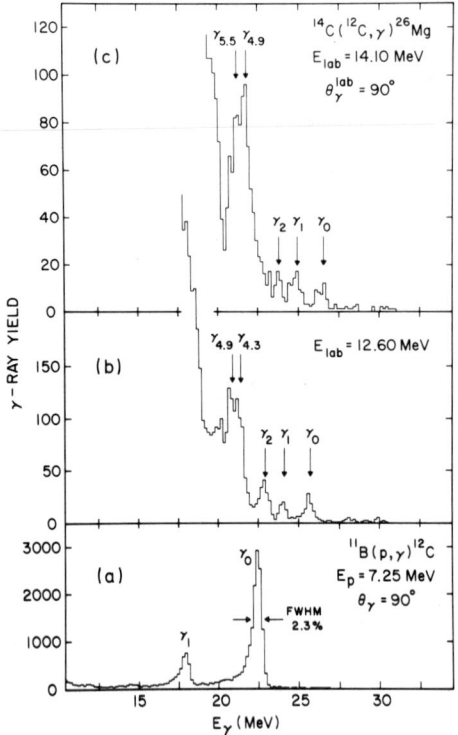

FIG. 1. High-energy portions of γ-ray spectra from the (a) $^{11}B(p,\gamma)^{12}C$, and (b,c) $^{14}C(^{12}C,\gamma)^{26}Mg$ reactions. Transitions to final states are indicated.

a large peak due to unresolved transitions to groups of excited states around $E_x = 4.3$, 4.9 and 5.5 MeV. Transitions to the third and fourth excited states ($E_x = 3.59$ and 3.94 MeV) are comparatively weak throughout the energy region investigated. A line-shape decomposition was used to extract cross sections from the capture spectra. Except for the worse resolution in the first run with thicker targets, the results from both runs are consistent. We concentrate here on the cross sections obtained with good resolution in the $^{14}C(^{12}C,\gamma)^{26}Mg$ experiment.

III. RESULTS AND DISCUSSION

The excitation functions for the six observed transitions are shown in Fig. 2. The size of the cross sections is comparable to those of the $^{12}C(^{12}C,\gamma)^{24}Mg$ [1] and $^{12}C(^{16}O,\gamma_1)^{28}Si$ [2] reactions. Electric dipole radiation is isospin-allowed for the $^{12}C + ^{14}C$ system, and is thus expected to contribute strongest. The measured angular anisotropy indeed indicates a dominant dipole component, most clearly in the case of γ_0. In addition, since the low-lying levels of ^{26}Mg are prolate, radiative capture in $^{12}C + ^{14}C$ is not inhibited by deformation, as is the case for $^{12}C(^{16}O,\gamma_0)^{28}Si$.[2] All of the excitation functions exhibit a certain amount of structure. In marked contrast to the resonances observed in $^{12}C(^{12}C,\gamma)^{24}Mg$,[1] all of the structures occurring in the $^{12}C + ^{14}C$ energy range presently investigated are uncorrelated in center-of-mass energy. However, the $^{12}C + ^{12}C$ system is exceptional in that it exhibits a large number of resonances in many decay channels. Furthermore, the spins of the $^{12}C(^{12}C,\gamma_0)$ resonances were limited to $J^\pi = 2^+$, as opposed to the dipole transitions in $^{12}C + ^{14}C$.

The capture cross sections are proportional to the product of the elastic and radiative partial widths, $\Gamma_{HI}\Gamma_\gamma$. Since the elastic scattering[6] and particle decay cross sections[7] are both smooth functions of energy, we have investigated the possibility of structures in Γ_γ. According to the Brink hypothesis, Giant Dipole Resonances are built upon every state of a nucleus. In order to convert our heavy-ion capture yields into their corresponding total photon ab-

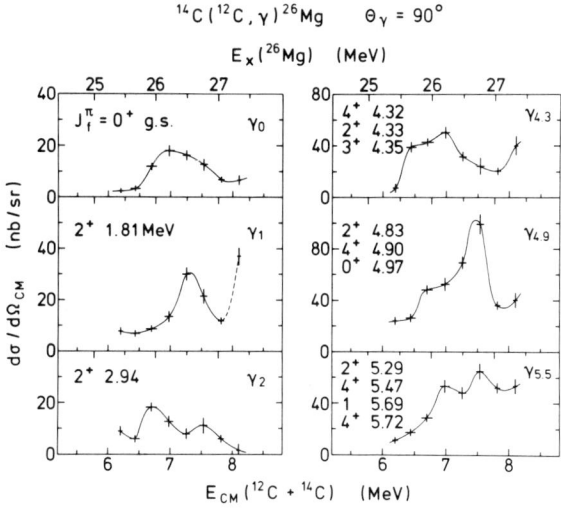

FIG. 2. Excitation functions for ^{14}C(^{12}C,γ)^{26}Mg. Cross section errors give statistical uncertainties from the fitting procedure only, horizontal bars represent the energy loss of the beam in the target. Lines are drawn merely to guide the eye.

sorption cross sections, we approximate angular distributions by pure E1 multipole radiation patterns, taking into account all possible partial waves in the heavy-ion entrance channel. Via detailed balance, we then deduce total absorption cross sections, assuming that the dipole states excited by photon absorption fission statistically into the ^{12}C + ^{14}C channel.

Fig. 3 shows the deduced total absorption cross sections for the various final states. There is little correlation evident with the ground-state GDR of ^{26}Mg from photoneutron data.[8] The structures in γ_0 and γ_1 occur at excitation energies where the GDR appears to be flat, while

FIG. 3. Total absorption cross sections deduced from the ^{12}C + ^{14}C capture data, plotted versus the excitation energy above the final states indicated. The photoneutron data for ^{26}Mg [8] are shown for comparison (bottom).

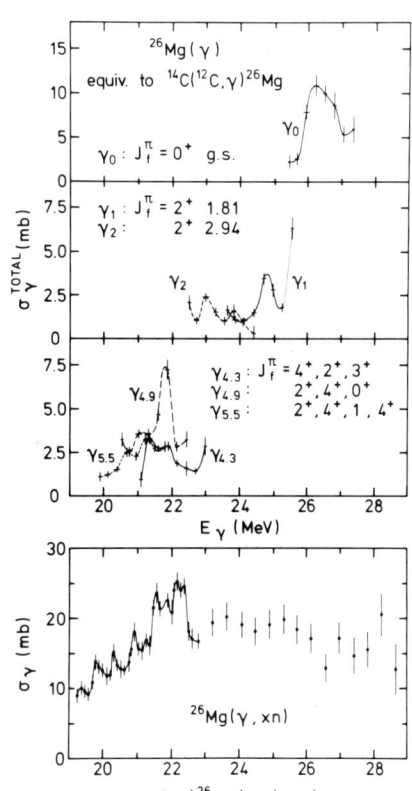

those in $\gamma_{4.3}$, $\gamma_{4.9}$ and $\gamma_{5.5}$ fall into a region of increased structure in the GDR. However, the significance of correlations is difficult to interpret because of the lack of definite isospin identifications in the GDR. In the absence of isospin mixing, the $^{12}C + ^{14}C$ capture reaction is probing the $T = 1$ component of the electromagnetic strength functions only. The prominent bump in the GDR at 22 MeV is supposed to be a $T = 2$ part of the isospin-split GDR in ^{26}Mg. An unknown amount of $T = 1$ strength is located underneath this bump and at still higher energies.[9] This might explain the fact that our deduced total absorption cross sections are smaller than the photo-neutron yields[8] by more than the factor of 2 from schematic isospin coupling considerations alone.

In conclusion, we have observed high-energy γ-ray transitions from the radiative capture of ^{12}C by ^{14}C into the bound states of ^{26}Mg up to $E_x = 5.7$ MeV. If the structures observed in capture are indeed due to structures in the radiative widths, then the experiment indicates considerable differences between Giant Dipole Resonances built upon different final states. This is in contrast to the results of the $^{12}C + ^{16}O$ capture reaction populating highly-excited states in ^{28}Si.[3] The Giant Dipole excitation of the prolate intrinsic shape of ^{28}Si resembles the GDR upon the oblate ground state, although in fact it appears narrower.

One of us (L.R.) would like to thank the members of the Nuclear Structure Laboratory at Stony Brook for their kind hospitality, and the Heinrich Hertz-Stiftung (Düsseldorf, FRG) for financial support. We would like to acknowledge the fine beam preparation by the BNL Tandem staff. Work supported by the U.S. Department of Energy, contract No. DE-AC02-76CH00016.

[+]Present address

1. A.M. Nathan, A.M. Sandorfi, T.J. Bowles, Phys. Rev. C24, 932 (1981)
2. A.M. Sandorfi and M.T. Collins, in Resonances in Heavy Ion Reactions, Proc. Bad Honnef 1981, ed. K.A. Eberhard, Lecture Notes in Physics Vol. 156 (Springer-Verlag, Berlin 1982), p. 264
3. M.T. Collins, A.M. Sandorfi, D.H. Hoffmann, M.K. Salomaa, Phys. Rev. Lett. 49, 1553 (1982)
4. Targets fabricated by J.L. Gallant, Chalk River (Canada)
5. A.M. Sandorfi and M.T. Collins, Nucl. Instr. and Meth. 222, 479 (1984)
6. H.G. Bohlen et al., Phys. Lett. 41B, 425 (1972); D. Konnerth et al., to be published
7. W. Galster et al., Nucl. Phys. A277, 126 (1977)
8. B.L. Berman, At. Data Nucl. Data Tables 15, 319 (1975)
9. M. Sugawara et al., Nucl. Phys. A248, 477 (1975), and references therein

A STUDY OF CERIUM AND NEODYMIUM NUCLEI CLOSE TO THE PROTON DRIP LINE

B J Varley, R Moscrop, S Babkair, C J Lister, W Gelletly
University of Manchester, Manchester M13 9PL, U.K.

and H G Price
Daresbury Laboratory, Warrington WA4 4AD, U.K.

ABSTRACT

The γ-rays from the ^{40}Ca + ^{92}Mo and ^{40}Ca + ^{96}Ru reactions have been studied at 180 MeV bombarding energy in coincidence with neutrons and charged particles. The yrast γ-ray sequences in a number of even-even Ce and Nd nuclei were observed. Preliminary analysis indicates that the light Ce and Nd nuclei are deformed ($\epsilon_2 \sim 0.3$). The first band crossings in these nuclei were observed.

INTRODUCTION

Nuclei with large numbers of neutrons and protons outside closed shells are expected to be deformed in their ground states and exhibit collective properties in their excited states as a result of the polarization of the nuclear core by the large number of valence nucleons. Consequently one would anticipate[1] that very light rare-earth nuclei are deformed in their ground states. A number of calculations [2-4] have now quantified these expectations and one of them [4] predicts a promontary of deformed nuclei ($\epsilon_2 > 0.3$) jutting out towards the line of stability.

Until recently the experimental investigation of this proposed new region of deformation has been prohibited by the lack of suitable beam-target combinations. We have now begun studies at Daresbury Laboratory of nuclei in this region with the ^{40}Ca + ^{92}Mo (180 MeV) and ^{40}Ca + ^{96}Ru (180 MeV) reactions using the neutron wall. We report here some of our preliminary findings.

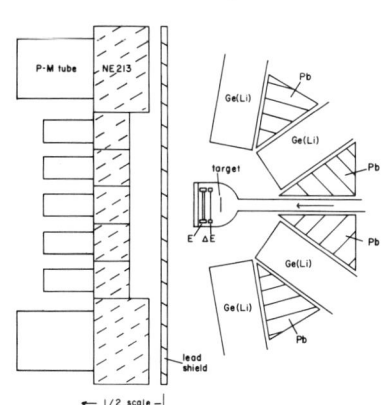

Fig. 1. A schematic view of the neutron multiplicity array.

MEASUREMENTS

At 180 MeV bombarding energy many reaction channels are open and the most neutron-deficient nuclei are produced with small cross-sections. To observe the γ-rays from the most neutron-deficient nuclei in-beam we have constructed the neutron wall [5], an array of neutron detectors and an E-ΔE telescope for charged particles all operating in coincidence with Ge detectors in the backward hemisphere. There are 37 neutron detectors in the array, 25 central detectors each 10 x 10 x 10 cm^3, and 12 outer detectors each 25.4 x 25.4 x 15 cm^3. Thus the neutron wall forms a 1m^2 vertical 'wall' typically mounted 20 cm behind the target with a 2.5 cm thick Pb sheet in front of it.

In order to determine the isotopic origins of the observed in-beam γ-rays, we recorded neutron-charged particle-γ-ray coincidences. Particle-γ-γ and neutron-γ-γ coincidences were also recorded.

RESULTS

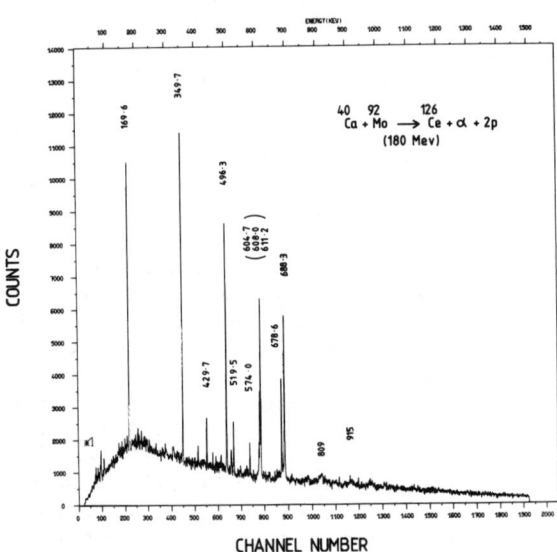

γ-rays from a large number of nuclei were observed in these reactions including the even-even nuclei ^{126}Ce, ^{128}Ce, ^{128}Nd, ^{130}Nd and ^{132}Nd. An example of the particle-gated γ-γ coincidence spectra, showing the yrast sequence in ^{126}Ce, is shown in fig. 2. Our measurements and analysis are at an early stage but if we make the assumption that these yrast transitions represent the decay of the ground state band then we find the values of E(2+), E(4)/E(2) and ϵ_2 listed in Table 1.

Fig. 2. particle-γ-γ coincidence spectrum for yrast transitions in ^{126}Ce.

The values of ϵ_2 were deduced from Grodzins' formula [6];

$$\epsilon_2 = 0.95 \left[\frac{1228}{A^{7/3} E(2^+)} \right]^{1/2} .$$

Fig. 3. shows ϵ_2 vs N for the Nd (Z=60) isotopes. The calculated values of Aberg [3] and Möller and Nix [2] are also shown.

We conclude from these results that the light rare-earth nuclei are deformed with $\epsilon_2 \sim 0.3$, and that the transition to deformation is a smooth one. Ignoring the nuclei near the N=82 closed shell where vibrations are important Aberg's calculations reproduce the

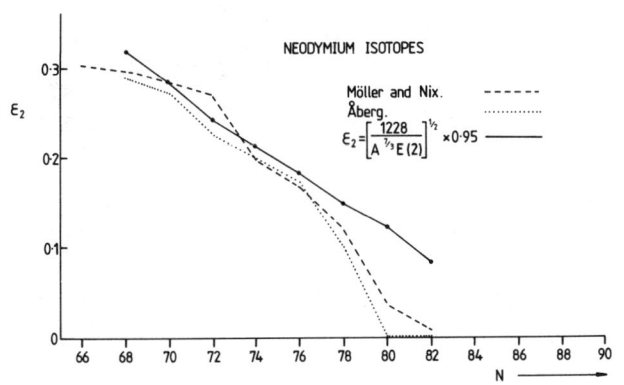

Fig. 3 ϵ_2 vs N for the Nd isotopes.

transition rather better than those of Möller and Nix. In particular there appears to be no sharp change in ϵ_2 between N=72 and N=74. ^{126}Ce exhibits a strong backbend at a frequency close to that of the yrast sequence in ^{128}Ce. The interaction between bands is much stronger in the Nd isotopes, which is not evident a priori in cranked shell model calculations.

Observation of the neighbouring odd nuclei indicates that the band crossings are due to the $(h^{11}/_2)^2$ protons.

Table 1

Nucleus	E(2+)	E(4)/E(2)	ϵ_2^*
$^{126}_{58}$Ce	169.6	3.06	0.28
$^{128}_{58}$Ce	207.0	2.93	0.26
$^{128}_{60}$Nd	134.5	3.17	0.31
$^{130}_{60}$Nd	157.8	3.06	0.28
$^{132}_{60}$Nd	212.6	2.84	0.24

*The quadrupole deformation was derived from Grodzins formula[6]

REFERENCES

1. E Marshalek, L Persson and R K Sheline, Rev. Mod. Phys. 35 (1963).
2. P Möller and J R Nix, At. Data and Nucl. Data Tables 26 (1981) 165.
3. S Aberg, Physica Scripta 25 (1982) 23.
4. G Leander and P Möller, Phys. Letters 110B (1982) 17.
5. B J Varley, C J Lister, R Moscrop, W Gelletly and H G Price, Proceedings of the Conference on Nuclear Instrumentation for Heavy Ion Research, Oak Ridge, Tennessee (1984), to be published.
6. L Grodzins, Phys. Letters 2 (1962) 88.

VI. NUCLEOSYNTHESIS

NEUTRON CAPTURE REACTIONS IN ASTROPHYSICS

F. Käppeler
Kernforschungszentrum Karlsruhe, Institut für Kernphysik III,
Postfach 3640, 7500-Karlsruhe, Federal Republic of Germany

ABSTRACT

About 2/3 of the chemical elements in nature were formed in neutron capture reactions. During the life of a star there are certain evolutionary stages where neutrons are available to build up the elements beyond iron which cannot be synthesized by charged particle reactions.

The observed abundance pattern allows to distinguish a rapid and a slow neutron capture process (r- und s-process). The r-process taking place far from the valley of stability is difficult to investigate because of the required extrapolation of nuclear properties to extreme neutron rich nulcei. The s-process, on the other hand, proceeds along the valley of stability. Therefore, the involved isotopes are accessible to laboratory measurements. This information allows for quantitative calculation of s-process abundances and other parameters which represent constraints for stellar models.

Two examples are outlined: (i) the s-process branching at A = 147, 148 yields a rather accurate value for the neutron density. (ii) Comparison of s-process abundances with observations of stellar atmospheres are particularly interesting for the unstable isotopes ^{93}Zr, ^{99}Tc and ^{147}Pm. Their deficiency with respect to stable neighbors may yield estimates for the transport time from the stellar interior to the surface.

INTRODUCTION

Neutron capture reactions play a role in various fields of astrophysics. In the third minute of the big bang they led to the primordial helium abundance[1] and spallation neutrons from cosmic rays may have modified the isotopic abundances in meteorites[2]. Most important, however, are neutron capture reactions for the synthesis of the majority of the chemical elements[3]. This contribution deals with some aspects of this last topic.

It is commonly accepted that all but the very lightest chemical elements were formed in stars. As under stellar conditions fusion reactions are limited to the mass region $Z \lesssim 26$ because of the increasing coulomb barriers and the decreasing binding energies at higher mass numbers, all elements beyond iron are the result of neutron capture reactions.

Neutrons appear in significant concentrations only in late evolutionary stages of the stars. During most of their life time the stars burn quiescently, fusing in their cores hydrogen into helium, as does the sun. But when the hydrogen is exhausted in the centre, the core heats up by gravitational contraction until the 3α-reaction ignites by which 3α particles are combined to ^{12}C. At this point the star becomes a Red Giant. As time proceeds the burning zones more outward towards the surface as is shown in Fig. 1 for the example of a 7 $M_\odot$ star ($M_\odot$ = mass of sun).[4,5]

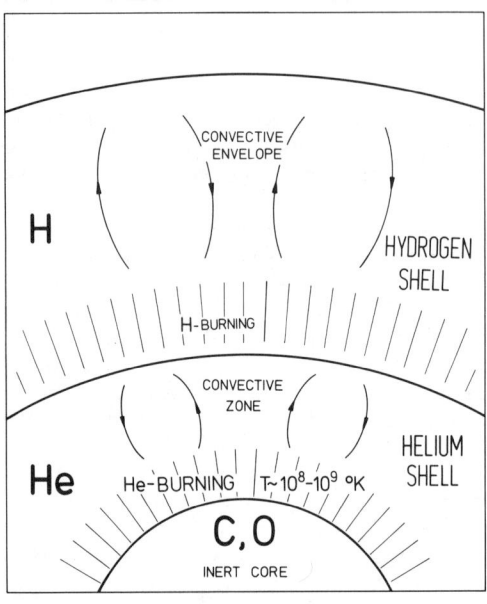

Fig. 1
Schematic structure of a 7 $M_\odot$ pulsating Red Giant.

An inert core of ^{12}C and ^{16}O (the ashes of He-burning) is surrounded by a helium burning shell and a hydrogen burning envelope. It was first found by Weigert[6] and by Schwarzschild and Härm[7] that helium shell burning does not operate continuously but that shell flashes occur. These give rise to large scale convective motions as was shown by Iben[4], and this exchange of material is essential for the production of neutrons as well as for the transport of neutron processed matter to the stellar surface. For a 7 $M_\odot$ star a flash lasts for about 3 y followed by an interpulse period of ~ 500 y, but these times can be much longer for lower mass stars.

In the helium burning zone the temperature is high enough (~3·10^8K) that neutrons are released in (α,n) reactions on ^{13}C and/or ^{22}Ne. If some hydrogen is mixed from the envelope into the convective shell after a flash it is captured by carbon to form ^{13}C by the sequence

$$^{12}C(p,\gamma)\ ^{13}N(\beta^+\nu)\ ^{13}C.$$

^{22}Ne is produced in another reaction chain. During hydrogen burning in the envelope virtually all CNO is transformed into ^{14}N by the CNO cycle. This ^{14}N, when convected down into the hotter He shell becomes ^{22}Ne through the reactions

$$^{14}N\ (\alpha,\gamma)\ ^{18}F\ (\beta^+\nu)\ ^{18}O\ (\alpha,\gamma)\ ^{22}Ne.$$

Which of the two neutron sources dominates depends on the core mass of the star. For core masses $\gtrsim 0.9$ $M_\odot$ ^{22}Ne(α,n) is the main neutron source whereas ^{13}C(α,n) prevails for cores with masses $\lesssim 0.8$ $M_\odot$. A detailed discussion of stellar models for giant stars is given in the comprehensive review of Iben and Renzini[8].

An important feature of both sources is the comparably small neutron density of $\sim 10^8$ cm^{-3}. This results in neutron capture times of ~ 10 to 100 y, much longer than typical beta decay times. For this reason the neutron capture process in Red Giants is called the s-process (s = slow).

Strong evidence for stellar s-process nucleosynthesis was the observation of lines from the (astronomically short lived) radioactive element T_c in S-stars by Merill[9] in 1952. Since then the observational techniques were greatly improved allowing now for quantitative determinations of element abundances in stellar atmospheres[10,11,12]. With these observations it is principally possible to test the s-process models for a given stellar mass, but so far the models are not capable to provide the parameters for a reliable abundance calculation. Therefore, s-process abundances are still be determined with the classical model (see below).

The final fate of a massive star is a Supernova explosion. When the fuel is exhausetd in the core and the core mass approaches the Chandrasekhar limit of ~ 1.5 $M_\odot$ the degenerate electron pressure can no longer counterbalance gravitation. The core collapses within less than a second to nuclear matter density forming a neutron star or even a black hole. The outer layers are driven off in the explosion by a shock wave which forms at the rebound core. A survey of present Supernova models might be found in Ref. 13. It has since long been proposed that a rapid neutron capture process (r-process) is associated with Supernovae[3], which should account for approximately half of the heavy element abundances in the solar system. Originally, it was believed that the r-process occurs close to the rebound core in highly neutronized matter, starting from iron as seed. In his review of this model Hillebrandt[14] noted two major problems: transport of the r-processed material from the deep interior to the surface without changing the composition and overproduction of r-elements and iron in the galaxy. A promising alternative site for the r-process which avoids these difficulties is the explosive He burning as the outgoing Supernova shock wave heats the convective helium shell[15,16].

The quantitative calculation of r-process abundances is very difficult because the short time scale of the explosion implies that the physical conditions (temperature, pressure, density etc.) are rapidly changing. Moreover, the isotopes involved in the process are extremely neutron rich, with neutron excesses of 10 - 20 compared to the stable isotopes. The required extrapolation of nuclear properties (beta decay times, nuclear masses and binding energies, level densities, fission barriers etc.) bear additional uncertainties. Although the beta decay rates have been improved recently[17] a complete calculation of r-abundances with full hydrodynamics and a reaction network (including reliable data on beta decay rates and (n,γ), (γ,n), (nf) reactions, fission feedback, α decay and beta delayed n-emission) remains still to be done.

This short sketch of stellar nucleosynthesis by neutron capture reactions illustrates the progress of the "classical" approach[3] to more realistic models. While such an advanced treatment is certainly necessary for the understanding of the r-process, many aspects of the s-process can be investigated in the classical way, as will be discussed in section II. Examples for the interplay between the stellar models and constraints from the classical approach are given in sections III and IV.

II THE CLASSICAL s-PROCESS

The neutron capture chain for the s-process synthesis starts, of course, at ^{22}Ne or ^{13}C but the abundance contribution to the elements with Z < 26 is not significant because these are copiously produced by charged particle reactions. Nevertheless, these elements act as a filter, absorbing approximately 75 % of all neutrons[18]. The remaining fraction is captured by the elements in the abundance maximum around iron to build-up the heavier elements. Among the iron group, ^{56}Fe is so much more abundant than all the other isotopes that it can be considered as the only seed for the s-process to a good approximation.

Starting from ^{56}Fe as seed the s-process neutron capture path can be worked out easily because usually the beta decay rates are much faster than the neutron capture rates. Fig. 2 shows this path in the mass region 145<A<150 through Nd, Pm and Sm. This figure illustrates two important points: (i) The branchings of the path at A=147 and 148 indicate that in some cases the beta half lives are long enough that neutron capture on these unstable isotopes (dashed boxes)

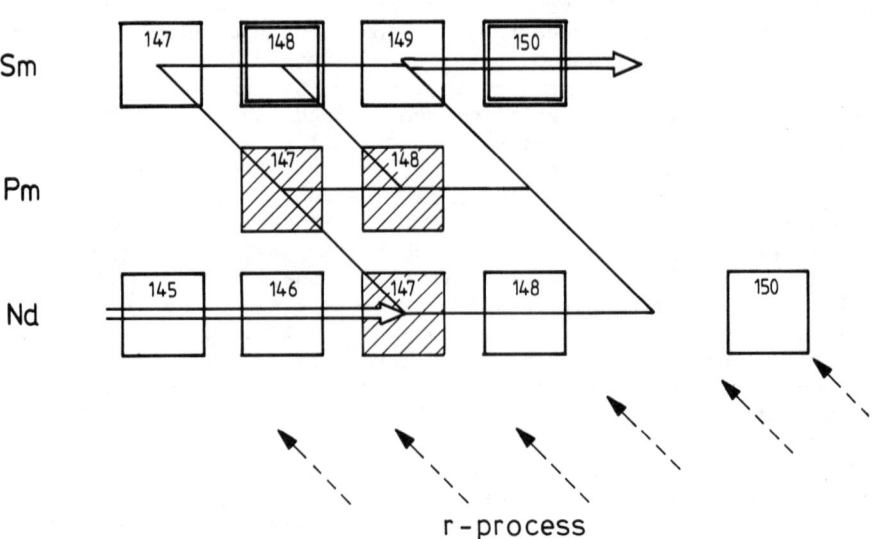

Fig. 2 The s-process neutron capture path through the mass region 145<A<150

can compete with ß-decay. (ii) Most isotopes of the heavy elements
are produced in the s- and r-process as well. The r-process decay
chains as indicated by dashed arrows end at the first stable isobar.
If a second stable isobar occurs on the s-process path it is shielded
against the r-process and, therefore, represents the true s-process
abundance. On Fig. 2, these examples, ^{148}Sm and ^{150}Sm, are marked by
double boxes. In turn, ^{150}Nd is of pure r-process origin.

The classical s-process models are phenomenological. A seed of
^{56}Fe is irradiated by a steady neutron flux and the resulting abundances are compared with the observed abundances of the 26 s-only isotopes which are found along the s-process path. Clayton et al.[19]
showed that a single irradiation of iron group elements could not generate the s-process abundances but that a distribution of neutron
fluences was called for, with smaller amounts of seed exposed to larger fluences. Details of this fluence distribution did emerge when
the nuclear physics input for the model, the neutron capture cross
sections, where improved with time. Seeger, Fowler and Clayton[20]
found that the duration of the s-process was not long enough to establish equilibrium in the s-process mass flow which means that the neutron capture rate was not equal for all isotopes along the chain.
However, the equilibrium condition which can be expressed by the capture cross sections σ and the s-process abundances N as

$$\sigma N(A-1) = \sigma N(A)$$

holds for the mass regions between magic neutron numbers. At shell
closure the cross sections become very small, acting as bottle necks
for the mass flow. As a result the $\sigma N(A)$-curve exhibits a pronounced
ledge-precipice structure. In 1974 Clayton and Ward[21] showed that an
exponential distribution for the neutron fluence τ allows the system
of coupled differential equation describing the mass flow to be solved analytically. At the same time Ulrich[22] found that such an exponential exposure distribution follows naturally from the pulsed s-process in the helium burning shell of Red Giants if always a certain
fraction of s-processed matter is removed from the He shell and mixed
into the outer envelope.

This rather simple model was able to describe at least the overall features of the s-process abundance pattern in the solar system
surprisingly well. It turned out, that the more accurate the neutron
capture cross sections became over the years, the better agreement
was obtained between calculated and observed abundances of the s-only
nuclei. This result strongly supports the attitude to persue the
classical s-process model further.

Fortunately, the solar system abundances were improved simultaneously and it is impressive to see that at present an accuracy of
5 to 10 % is achieved for most elements[23]. The situation with the capture cross section measurements has been discussed by Käppeler
et al.[24] showing a 5 % accuracy to be the limit of established techniques. Recently a new method has been suggested[25] which is expected
to yield a precision of $\sim$ 1 - 2 % thus allowing for an investigation
of more s-process details[26].

However, not all isotopes on the s-process path are accessible to measurements either because samples are not available (e.g. for the important xenon isotopes) or because these isotopes are unstable. In these cases measurements have to be replaced by theory. Extensive calculations with the Hauser-Feshbach model and a global parametrization have been carried out by Holmes et al.[27] and Woosley et al.[28] for isotopes all over the periodic table. In most cases these data agree with experimental values to within $\pm$ 50 %. Harris[29] published improved calculations using local parameter systematics for the level densities according to Reffo[30] thus reducing the uncertainties to $\pm$ 30 % on average. By elaborate treatment of local systematics for all relevant parameters Reffo et al.[31] and Beer et al.[32] found that $\pm$ 20 % seem to be a general limit for Hauser Feshbach calculations. Fig. 3 presents the parameter systematics for the mass region 145<A<150 from Winters et al.[33] which was established to calculate the capture cross sections of unstable ^{147}Nd, ^{147,148}Pm (see section III).

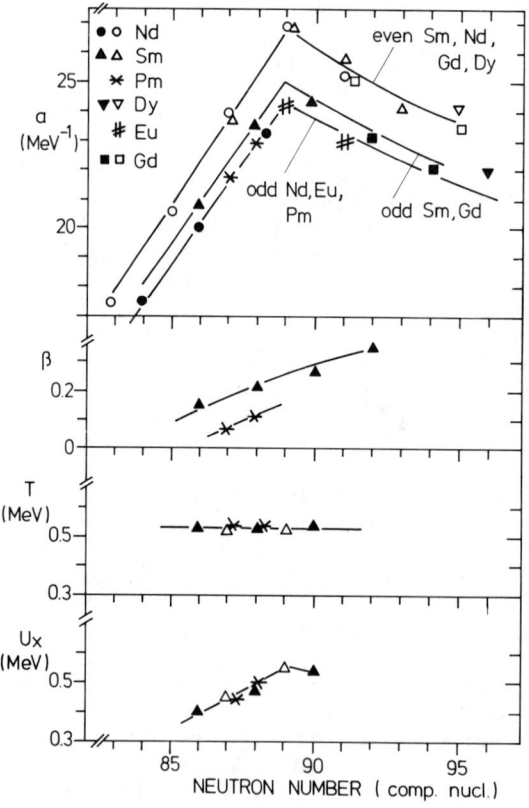

Fig. 3 Local systematics of the relevant parameters for Hauser-Feshbach calculations of neutron capture cross sections of the radioactive isotopes ^{147}Nd, ^{147,148}Pm.

But even for stable isotopes the relevant neutron capture cross sections for s-process studies cannot always be determined without the help of theory. This is due to the high temperature of $\sim 3.5 \cdot 10^8$K at the s-process site. At these temperatures the thermal energy is kT = 30 keV which means that the neutron energy spectrum is Maxwellian ranging from 1 to ∿ 150 keV. Accordingly, neutron capture cross sections to be used in s-process calculations need to be averaged over the Maxwell spectrum, e.g. see Ref. 24. In a plasma of kT = 30 keV low lying excited states are in thermal equilibrium with the ground state with population probabilities

$$P_i = (2J_i+1) \exp(-E_i/kT) / \Sigma_m (J_m+1) \exp(-E_m/kT)$$

where J_i, E_i are spin and energy of level i. The sum in the denominator is the nuclear partition function. An excited state with large spin may be more strongly populated than the ground state due to the larger statistical weight. Then it is not possible to evaluate the stellar neutron capture rate from laboratory measurements alone but capture in excited states needs also to be considered. It is interesting to note that in this situation a competing reaction channel appears: inelastic neutron scattering from an excited state to a lower state by which the neutron gains energy. The most famous example of this type is ^{187}Os where this effect is crucial for the evaluation of the chronometric pair ^{187}Os/^{187}Re [34,35]. (See also the contribution at G. Reffo[36]).

With a complete set of Maxwellian averaged capture cross sections and with the empirical σN products of the s-only isotopes the classical s-process model yields the following informations[24]:

- the σN(A)-curve and consequently the s-process abundances N_s,
- the r-process yields by decomposition of solar system abundances according to $N_r = N_\odot - N_s$,
- the seed abundances,
- the mean neutron exposures,
- the number of neutrons captures per ^{56}Fe seed.

The last three items are important for comparison with the stellar s-process site. As was shown by Almeida and Käppeler[18] these quantities as derived from the classical model for solar system material are consistent with the pulsating He-burning shell in Red Giants with the ^{22}Ne(α,n) reaction as a neutron source.

The first two results impressively justify the classical model. With only 4 inherent parameters the s-process abundances can be determined within the 5 - 10 % uncertainty of the empirical σN-products. Fig. 4 shows an updated σN(A)-curve compared to Ref. 24. The black squars are s-only isotopes with well determined abundances and cross sections, which served as normalization points for the fit. Open symbols denote those s-nuclei for which abundances or cross sections are uncertain or which are involved in s-process branchings (squares) and isotopes with less than 10 % abundance contribution from the r-process (circles). From this σN(A)-curve the s-process-abundances N_s can easily be worked out, and it is remarkable that no overproduction compared to solar abundances is obtained.

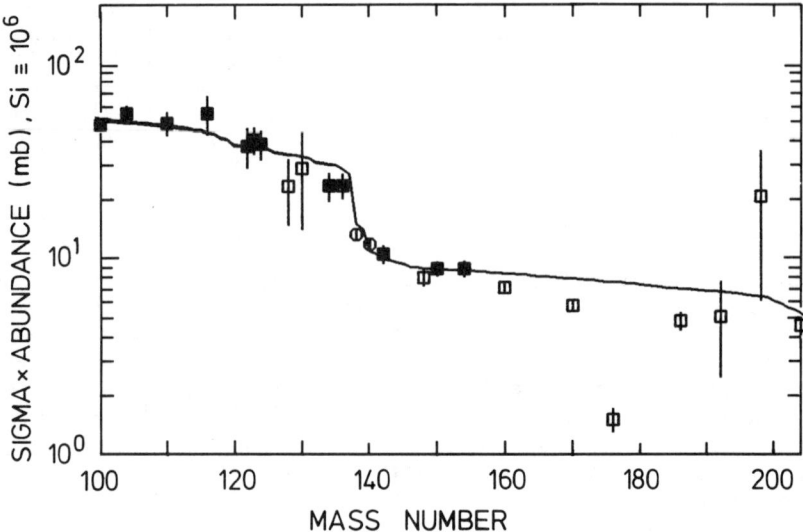

Fig. 4 Updated $\sigma N(A)$-curve compared to Käppeler et al.[24]. The solid line is a fit to the empirical σN-values of the pure s-isotopes indicated by black squares. Open symbols denote pure s-isotopes for which cross sections or abundances are uncertain or which are involved in branchings (squares) and isotopes with less than 10 % r-process contribution (circles).

In turn, the set of s-process abundances can be used to derive the r-process yields via

$$N_r = N_\odot - N_s.$$

Here and throughout this contribution the p-process abundances are neglected which appears to be justified at the present 5-10 % uncertainty level[24]. The resulting r-process yields are plotted in Fig. 5. Most important in this figure is the perfect agreement between the distribution of the pure r-isotopes (black squares) and the calculated yields. The error bars indicate the uncertainties originating from those of the solar and s-process abundances. The so determined r-process abundances are the main testing ground for r-process model calculations[14-17] because they are the only reliable information on the r-process to compare with.

The results of the classical s-process model sketched so far provide a satisfactory description of global or average quantities. But what about the dynamics of the s-process in the He shell flashes depicted by stellar models? This strongly time dependent s-process should exhibit
    (i)   temperature effects and
    (ii)   freeze-out effects.

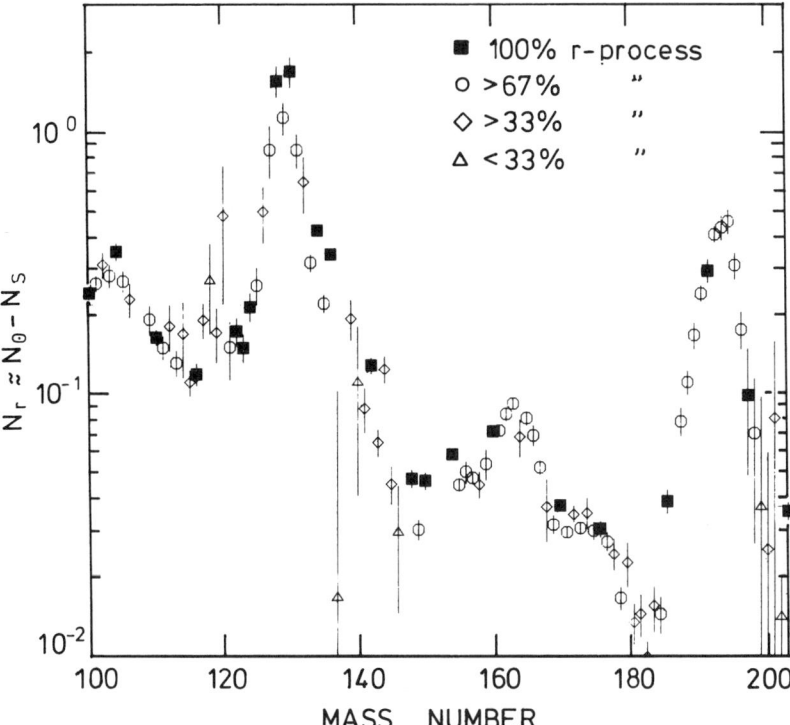

Fig. 5 r-process yields derived by subtraction of s-process abundances from solar abundances. Pure r-isotopes are plotted as black squares, the various other symbols indicate the relative r-contributions to the solar values.

Temperature effects have already been mentioned in connection with neutron capture in excited states. Another, more important consequence of temperature concerns the beta decay rates, because ß-decay from excited states may be much less forbidden than from the ground state[37]. In addition, atomic effects become important due to the high degree of ionization in the stellar plasma. Thus electron capture rates are depressed because of the absence of K shell electrons. Instead, bound state ß⁻-decay (where the electron is emitted into an atomic orbit gaining the electron binding energy) may accelerate those transitions where the beta endpoint energy is comparable to the electron binding energies. These plasma effects have been studied in detail by Takahashi and Yokoi[38] who find in the extreme case of ^{163}Dy that this terrestrially stable isotope becomes unstable at the s-process site. (For this problem see also the contribution by Beer, Walter and Macklin[39]). Other examples for temperature effects will be discussed below (sections III, IV) and by Norman[40] in his contribution to this conference. In summary, temperature effects can be included into the classical s-process model as far as only modifications of neutron capture cross sections and beta decay rates are concerned. The variation of temperature with time during the shell flashes and in the interpulse period is roughly a factor 2-3 and can be neglected in most cases at least to first order.

Contrary to temperature, the neutron density during a shell flash varies by several orders of magnitude and goes to zero in the interpulse period. As a consequence, freeze-out effects become significant as was pointed out by Cosner, Iben and Truran[41]. The abundance pattern in s-process branchings are particularly sensitive to freeze-out effects. Therefore, the neutron densities derived from different branchings by means of the classical steady s-process model are expected to exhibit characteristic discrepancies according to the respective, cross section dependent freeze-out limits. Recent studies of this type yielded, however, rather consistent values for the s-process neutron density[33,42,43] over the entire capture path. So far, it is not clear whether the freeze-out effects on the s-process abundances are small or whether the present analyses are still too uncertain to reveal their influence. This problem of a dynamic s-process is also discussed by Mathews, Howard and Ward[44] at this conference.

III THE s-PROCESS BRANCHINGS AT A=147,148

The s-process neutron capture path as sketched in Fig. 2 indicates three branching points, ^{147}Nd and ^{147,148}Pm. These branchings are the most reasonable explanation for the result of Winters et al.[33], who found from a recent accurate capture cross section measurement that the σN products of the s-only isotopes ^{148}Sm and ^{150}Sm were not equal:

$$\sigma N(^{148}Sm)/\sigma N(^{150}Sm) = 0.91 \pm 0.03.$$

The fact that a small but significant fraction of the s-process flow bypasses ^{148}Sm can then be used to evaluate the neutron density. Most easily this can be done by means of the classical steady flow model keeping in mind that neglection of any time dependence may introduce some uncertainty (see above). At any branching point the split of the s-process flow can be expressed by the branching ratio

$$B_{\beta^-} = \lambda_{\beta^-}/(\lambda_{\beta^-} + \lambda_n)$$

in terms of the beta decay rate $\lambda_{\beta^-} = \ln 2/t_{1/2}$ and the neutron capture rate $\lambda_n = n_n \sigma v_T$. The branching ratio is determined by the σN-values of the relevant isotopes in the branching, in this example by ^{148}Sm and ^{150}Sm. To obtain the neutron density, n_n, it is essential to know the neutron capture cross sections σ and the half lives $t_{1/2}$ of the branching point isotopes under s-process conditions. In the present case, this means that these data are required for ^{147}Nd, and ^{147,148}Pm.

The capture cross sections were calculated by Winters et al.[33] using carefully evaluated parameter systematics (Fig. 3) which were established from as much experimental information on the investigated isotopes and their neighbors as was available from literature. An uncertainty of ± 20 % was estimated for the calculated cross sections from the propagation of parameter uncertainties (for details see Walter et al.[45]). The situation is somewhat complicated by an isomeric state in ^{148}Pm and by the effect of temperature. To illustrate this, the lowest levels in ^{147}Pm and ^{148}Pm are plotted in Fig.6 (from Lederer and Shirley[46]).

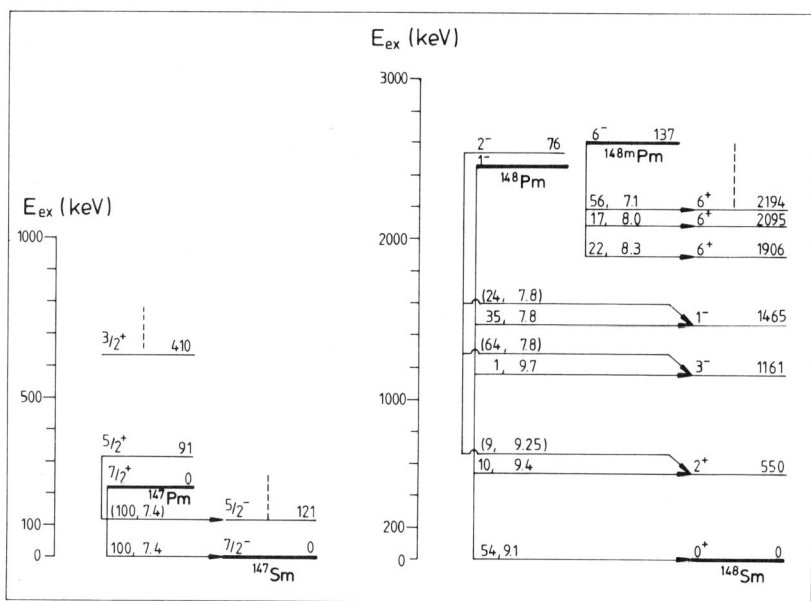

Fig. 6 Low lying levels in ^{147}Pm and ^{148}Pm which contribute to the stellar neutron capture and beta decay rates.

^{147}Pm: Of the excited states in ^{147}Pm the first level at 91 keV is populated to 3.5 % whereas the second level at 410 keV is practically not affected. The influence of the first excited level on the neutron capture rate is negligible because of its small population probability.

Beta decays from the first exited state are first non-unique forbidden. By analogy to similar transitions in this mass region, one may expect that only the 5/2$^+$→5/2$^-$ transition needs to be considered, and for this the Q-value is 30 keV lower than for the ground state 7/2$^+$→7/2$^-$ decay. Though this qualitative agreement would indicate that the ^{147}Pm half life is not affected by temperature, the calculations of Takahashi and Yokoi[38,47] yield a reduction in half life by a factor 2 at kT = 30 keV. In order to derive a lower limit for the neutron density the branching analysis was carried out with the laboratory half life. The influence of the reduced half life is later illustrated in Table I.

^{148}Pm: For this isotope thermal effects are difficult to estimate because its level scheme is poorly known: so far only the three levels shown in Fig. 6 have been determined. As ^{147}Pm targets are easily available it would be possible to remove this shortcoming by standard (n,γ) studies. Given this difficulty, the problem is whether or not the isomeric state at 137 keV and the ground state are in thermal equilibrium during the s-process. Both states are about equally populated by neutron capture in ^{147}Pm but have rather different half lives. The probability for the electromagnetic E4 transition to the first excited state is only 5 % so that it can hardly provide an efficient link for thermal equilibration of isomer and ground state on the bety decay time scale. However, such a link could excist through higher lying states above the isomer. At present, it is not possible to decide whether or not thermal equilibrium is achieved during the s-process, and, therefore, both possibilities have to be considered.

If equilibrium is attained, the isomer is quickly depopulated to 3.8 % whereas the first excited state is populated to 11 %. This is too small to influence the total neutron capture rate which is determined by the ground state (σ=1542 mb). The beta decay is slightly enhanced because the decay from the first excited state is favored by the higher Q-value. Hence, the stellar beta decay rate becomes $\lambda_\beta^* = 1.7 \cdot 10^{-6}$ s^{-1} = 1.15 λ_β^{lab} according to Takahashi and Yokoi^{47}.

If equilibrium is not attained, the isomer has to be treated independently from the ground and first excited state and does not contribute to the nuclear partition function of ^{148}Pm. Then one needs the fractional population of the isomer at keV neutron energies (to date only known at thermal) and its neutron capture cross section (σ = 2453 mb) in order to include the isomer as an additional species into the branching analysis.

^{147}Nd: The first excited state in ^{147}Nd at 50 keV is populated to 31 % at kT = 30 keV whereas only 2 % of all nuclei are in the second excited state at 128 keV. But even the rather large population of the first state is not sufficient to alter the neutron capture rate signifincantly, at least, if the cross section of this state agrees to that of the ground state within a factor of two. For the stellar beta decay rate Takahashi and Yokoi^{47} find only a marginal enhancement of 1.5 % compared to the laboratory rate.

The result which one ultimately obtains for the mean s-process neutron density is summarized in Table I. The question whether isomer and ground state are in thermal equilibrium or not is obviously most important, as the respective values for the neutron density differ by a factor 3. For the case that equilibrium is not achieved, the relevant parameters in the branching analysis have been changed by their estimated uncertainties. All resulting variations in the neutron density are small compared to the effect of equilibration. Nevertheless, the value of $(1 \pm 0.4) \cdot 10^8$ cm^{-3} represents a firm lower limit which is an important result in itself. Further improvements, however, require more information on the level scheme of ^{148}Pm to decide about thermalization of the isomer and better cross sections for ^{148,150}Sm to reduce the uncertainty in the ratio R.

Table I Mean s-process neutron density, n_n, from branchings at A = 147,148

Isomer in ^{148}Pm equilibrated	Isomer in ^{148}Pm not equilibrated
$3.1 \cdot 10^8$ cm^{-3}	$(1.0 \pm 0.4) \cdot 10^8$ cm^{-3}
	variation of R$^{a)}$ by ± 3.5 % → (+30%)
	variation of $\sigma(^{147}$Nd) by ± 20 % → ($\pm$ 2%)
	variation of $\sigma(^{147}$Pm) by ± 20 % → ($\pm$ 4%)
	variation of $\sigma(^{148}$Pm) by ± 20 % → ($\pm$ 4%)
	variation of $\sigma(^{148}$Pm^m) by ± 20 % → ($\pm$12%)
	increase of $\lambda_\beta(^{147}$Pm) by factor 2 (calculated stellar value^{47}). → (+25%)

a) $R = \sigma N(^{148}Sm)/\sigma N(^{150}Sm)$

IV OBSERVATIONS OF UNSTABLE ISOTOPES IN STARS

The study of the elemental composition of stellar atmospheres started in 1814 when Fraunhofer discovered the dark lines in the spectrum of the sun, but it took more than a century before Russell and Adams[48] succeeded to perform quantitative analyses of the sun and other stars in 1928. Similarly, there was a long time between the first discovery of the unstable element technetium by Merrill[9] and its quantitative determination by Smith and Wallerstein[12] in 1983, indicating the difficulty of the problem. The radioactive elements or isotopes with half lives much smaller than the age of the stars on which they are observed can be ascribed to the s-process going on in these particular stars. The surface abundances of unstable species are important because (if compared to the original s-abundances) they provide clues to the mechanisms and time scales of the convective motions which brought them from the production site to the surface[8,49]. This so obtained information can be compared directly with stellar models.

At present, two unstable isotopes, ^{93}Zr and ^{99}Tc, have been quantitatively determined in several Red Giant Stars whereas the discovery of the radioactive element promethium[50] in only one star is still in question. To begin with the safely observed cases, Fig. 7 shows the s-process chain through the mass region 90<A<102. The relevant ß-unstable isotopes are marked by their stellar half lives according to Takahashi and Yokoi[47]. Note, that the ^{99}Tc half life is drastically reduced. It is almost constant for temperatures below 10^8K but drops then from the terrestrial value of $2.1 \cdot 10^5$ y to only 2.2 y at kT = 30 keV, a result which was also independently derived by Cosner and Truran[37] and by Schatz[51]. The consequence of the short stellar half life of ^{99}Tc is that it was either quickly removed from the production site and cooled down to temperatures below 10^8K where it lives sufficiently long to be convected up to the surface, or that

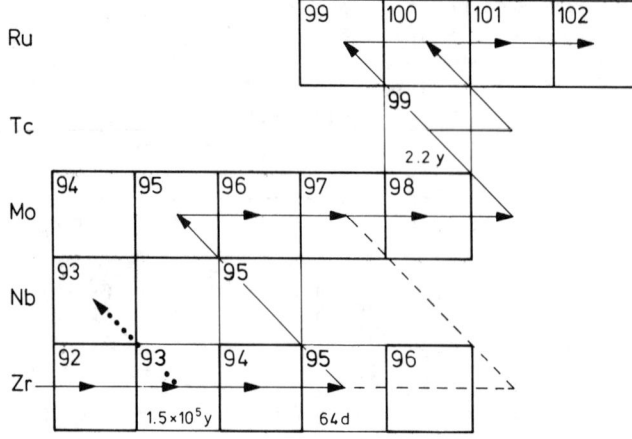

Fig.7 The neutron capture path between A = 90 and 102. For unstable isotopes the calculated stellar half lives[47] are given for kT = 30 keV.

it was synthesized at lower temperatures. This second possibility was suggested by Smith and Wallerstein[12] with the $^{13}C(\alpha,n)$ reaction as the s-process neutron source. If this were correct, the s-process branching at ^{99}Tc as indicated in Fig. 7 would be negligible because of the much longer half life at $T = 10^8 K$ ($t_{1/2} = 1.2 \cdot 10^5$ y) at which the $^{13}C(\alpha,n)$ reaction is supposed to operate.

It was pointed out by Anders[52] that ^{93}Nb is bypassed during the s-process but is afterwards increasing in abundance on the time scale of the ^{93}Zr decay of 1.5 million years. At the same ^{99}Tc decays with a half life of $2.1 \cdot 10^5$ y (if it is at temperatures $<10^8 K$) thus increasing the molybdenum abundance. Hence, the relative abundances of the elements Zr to Mo depend on the time passed since the termination of the s-process.

This feature was investigated by Smith and Wallerstein[12] who measured the abundances of V, Zr, Nb, Mo and Ru relative to Ti in six stars, two of which also showed significant Tc lines. They interpreted their results using $\sigma N = $ const through the mass region from Zr to Ru, and found evidence for multiple neutron capture events, probably shell flashes. The abundance uncertainties in this work of a factor three did not allow for a quantitative discussion.

If the stellar atmosphere is cool enough this picture can be considerably refined by the identification of the zirconium isotope ratios [53]. This rather accurate technique makes use of the isotopic splitting of the band heads in ZrO and avoids the problems in the determination of stellar element abundances due to stratification effects and different excitation potentials. How important these results are was outlined by Beer and Walter[54] who showed that the isotopic ratios of the Zr isotopes are sufficiently sensitive to the mean s-process neutron exposure. In turn, with the observed abundance ratios of Zook[53], they were able to derive the associated mean neutron exposure $\tau_o = 0.075$ mb^{-1}. This value and the results of Tomkin and Lambert[10] ($\tau_o = 0.6$ mb^{-1}) and of Holweger and Kovács[11] ($\tau_o = 0.06$ mb^{-1})

demonstrate that stellar neutron exposures may differ widely from the exposure characteristic for solar system material[55] (τ_o = 0.3 mb^{-1}). Especially for small neutron exposures the $\sigma N(A)$-curve is rapidly decreasing around A = 100 and therefore the assumption σN = const is often too crude for the interpretation of stellar abundances.

To summarize the discussion of ^{93}Zr and ^{99}Tc it should be noted that the observation of these isotopes alone is not sufficient for a penetrating analysis. Instead, simultaneous observations of as many characteristic s-process elements (especially around closed neutron shells) are called for, complemented by identification of isotopic patterns. On this data basis the s-process abundances of the particular investigated stars can be established with the classical s-process model and then be compared to the abundances of unstable isotopes. The first efforts along these lines by Smith and Wallerstein[12] and by Zook[53] have produced promising results but need to be extended further to obtain quantitative data on the mixing mechanisms in Red Giant stars.

In 1970, Aller and Cowley[50] announced the possible identification of promethium in the extremely unusual Ap star HR465. As can be seen from Fig. 2 any reasonable neutron capture synthesis will produce only ^{147}Pm in significant amounts, all other isotopes being either bypassed by the capture chain or too short lived. But also the half life of ^{147}Pm (2.6 y on earth and 1.1 y at kT = 30 keV[47]) is too short that detectable amounts of this isotope could ever reach the stellar atmosphere. Therefore, nuclear reactions on the surface of the star were suggested as an alternative explanation[56]. If the existence of promethium can be confirmed during the present rare earth maximum of this spectrum variable star, it would be a unique probe for its possible atmospheric processes.

V SUMMARY

Neutron capture reactions in astrophysics are most important for the synthesis of the chemical elements heavier than iron. The s-process abundances are directly determined by the respective neutron capture rates while the r-process yields are at least modified by neutron captures during freeze-out. As the s-process follows the valley of stability and is believed to operate at fairly constant temperature, the capture rates for the involved isotopes can be measured in the laboratory.

During the last decade, the classical models for neutron capture nucleosynthesis were complemented by more detailed models related to stellar sites and evolutionary stages. However, as these models are necessarily extremely complex, the classical s-process approach is still important. It has been shown to reproduce the s-process abundances in solar system material within their uncertainties and to yield many important constraints for stellar models, provided that the effect of temperature on certain beta decay rates and (in some cases) on neutron capture rates is accounted for. The success of the classical s-process calls for further improvements of the basic input data.

A field of increasing interest are elemental abundances in Red Giant stars, many of which exhibit clear s-process enhancements. At present, observations are often incomplete and of limited accuracy but some results are detailed enough to allow for classical s-process analyses. In this context, the unstable isotopes ^{93}Zr and ^{99}Tc may provide clues to the convective motions between the s-process site and the surface.

REFERENCES

1. S. Weinberg, The first three minutes (Basic Books, New York 1977)
2. R.L. Macklin, J.H. Gibbons, T. Inada, Nature 197, 369 (1963)
3. E.M. Burbidge, G.R. Burbidge, W.A. Fowler, F. Hoyle, Rev. Mod. Phys. 29, 547 (1957)
4. I. Iben, Jr., Ap. J. 196, 525 (1975)
5. I. Iben, Jr., Ap. J. 208, 165 (1976)
6. A. Weigert, Z. Ap. 64, 395 (1966)
7. M. Schwarzschild, R. Härm, Ap. J. 150, 961 (1967)
8. I. Iben, Jr., A. Renzini, Ann. Rev. Astron. Astrophys. 21, 271 (1983)
9. P.W. Merrill, Science 115, 484 (1952)
10. J. Tomkin, D.L. Lambert, Ap. J. 273, 722 (1983)
11. H. Holweger, N. Kovács, Astr. Ap. 132, L5 (1984)
12. V.V. Smith, G. Wallerstein, Ap. J. 273, 742 (1983)
13. M.J. Rees, R.J. Stoneham, eds., Supernovae: A Survey of Current Research (Reidel, Dordrecht 1981)
14. W. Hillebrandt, Space Science Reviews 21, 639 (1978)
15. W. Hillebrandt, F.-K. Thielemann, Mitt. Astron. Ges. 43, 234 (1977)
16. J.W. Truran, J.J. Cowan, A.G.W. Cameron, Ap. J. 222, L63 (1978)
17. H.V. Klapdor, T. Oda, J. Metzinger, W. Hillebrandt, F.-K. Thielemann, Z. Phys. A 299, 213 (1981)
18. J. Almeida, F. Käppeler, Ap. J. 265, 417 (1983)
19. D.D. Clayton, W.A. Fowler, T.E. Hull, B.A. Zimmerman, Ann. Phys. 12, 331 (1961)
20. P.A. Seeger, W.A. Fowler, D.D. Clayton, Ap. J. Suppl. 11, 121 (1965)
21. D.D. Clayton, R.A. Ward, Ap. J. 193, 397 (1974)
22. R.K. Ulrich, in Explosive Nucleosynthesis, ed. D.N. Schramm and W.D. Arnett (Austin, University of Texas Press) p. 139
23. E. Anders, M. Ebihara, Geochim. Cosmochim. Acta 46, 2363 (1982)
24. F. Käppeler, H. Beer, K. Wisshak, D.D. Clayton, R.L. Macklin, R.A. Ward, Ap. J. 257, 821 (1982)
25. F. Käppeler, G. Schatz, K. Wisshak, KfK-report 3472 (1983)
26. F. Käppeler, Workshop on Challenges and New Developments in Nucleosynthesis, Yerkes Observatory, Williams Bay, Wisconsin, Oct. 19-20 (1983)
27. J.A. Holmes, S.E. Woosley, W.A. Fowler, B.A. Zimmerman, Atomic Data and Nuclear Tables 18, 305 (1976)
28. S.E. Woosley, W.A. Fowler, J.A. Holmes, B.A. Zimmerman, Atomic Data and Nuclear Tables 22, 371 (1978)
29. M.J. Harris, Ap. Space Sci. 77, 357 (1981)
30. G. Reffo, in Nuclear Theory for Applications: Proc. of a Course held at ICTP, Trieste, 17.Jan.-1o.Feb.1978 (IAEA, Vienna) p. 205

31. G. Reffo, F. Fabbri, K. Wisshak, F. Käppeler, Nucl. Sci. Eng. 80, 630 (1982)
32. H. Beer, F. Käppeler, G. Reffo, G. Venturini, Ap. Space Sci. 97, 95 (1983)
33. R.R. Winters, F. Käppeler, K. Wisshak, A. Mengoni, G. Reffo, in preparation
34. S.E. Woosley, W.A. Fowler, Ap. J. 233, 411 (1979)
35. K. Yokoi, K. Takahashi, M. Arnould, Astr. Ap. 117, 65 (1983)
36. G. Reffo, contribution to these proceedings
37. K. Cosner, J.W. Truran, Ap. Space Sci. 78, 85 (1981)
38. K. Takahashi, K. Yokoi, Nucl. Phys. A 404, 578 (1983)
39. H. Beer, G. Walter, R.L. Macklin, contribution to these proceedings
40. E.B. Norman, contribution to these proceedings
41. K. Cosner, I. Iben, Jr., J.W. Truran, Ap. J. (Letters) 238, L91, (1980)
42. H. Beer, G. Walter, R.L. Macklin, J. Patchett, Phys. Rev. C 30, 464 (1984)
43. G. Walter, KfK-Report 3706 (1984) and PhD thesis, Univ. of Heidelberg
44. G.J. Mathews, W.M. Howard, R.A. Ward, contribution to these proceedings
45. G. Walter, B. Leugers, F. Käppeler, Z.Y. Bao, D. Erbe, G. Rupp, G. Reffo, F. Fabbri, KfK-Report 3652 (1984)
46. C.M. Lederer, V.S. Shirley, Table of Isotopes (Wiley, New York, 1978)
47. K. Takahashi, K. Yokoi, private communication (1984)
48. H.N. Russell, W.S. Adams, Ap. J. 68, 9 (1928)
49. J.M. Scalo, G.E. Miller, Ap. J., 246, 251 (1981)
50. M.F. Aller, C.R. Cowley, Ap. J. (Letters), 162, L145 (1970)
51. G. Schatz, Astr. Ap. 122, 327 (1983)
52. E. Anders, Ap. J. 127, 355 (1958)
53. A.C. Zook, Ap. J. (Letters) 221, L113 (1978)
54. H. Beer, G. Walter, Astr. Ap. 133, 317 (1984)
55. G.J. Mathews, F. Käppeler, Ap. J. (in press)
56. H.R.E. Tjin A Djie, R.J. Takens, E.P.J. van den Heuvel, Astr. Letters 13, 215 (1973)

Capture Processes and Element Synthesis in the Universe

Hans Volker Klapdor
Max-Planck-Institut für Kernphysik
Heidelberg, Germany

Contents:

1. Introduction
2. Capture processes and beta decay
3. The new description of β-decay far from stability
3.1 Method of calculation
3.2 Results
3.2.1. Half-lives, rates of β-delayed neutrons and fission
3.2.2. Electron and antineutrino spectra from nuclear reactors and reactor decay heat
3.2.3. Double beta decay
4. The synthesis of heavy elements by the r-process
5. The age of the universe and the neutrino mass

1. Introduction

The understanding of the element synthesis in the universe requires as one prerequisite information about nuclear properties and reaction cross sections for several thousands of nuclei. For example, for calculating the production of heavy elements by the r-process, in principle neutron capture cross sections and beta decay properties of ~6000 nuclides between β-stability line and neutron drip line are needed: The r-process path indicating the element distribution built up for less than a second in supernova explosions runs close to this neutron drip line (Fig. 1).

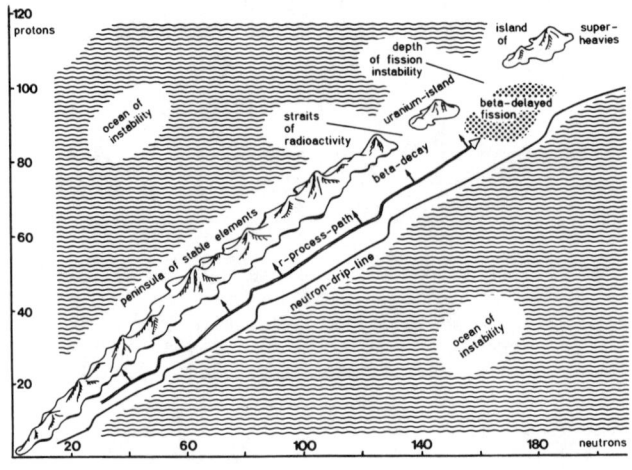

Fig. 1:
Schematic representation [1] of the synthesis of heavy elements by the r-process (see text).

Most of the nuclei of interest are at present and for the foreseeable future not accessible in terrestrial laboratories. So one has to rely on extrapolations. Capture processes have played and play an important role here and far beyond the narrow field of capture cross sections themselves.

Capture processes enter in different ways into element synthesis in the universe:

A) In a 'direct' way:

n,p,γ capture cross sections determine **directly** specific element abundances in synthesizing processes (r(n), s- or p-process). Examples are discussed in the papers of Norman [2] and Käppeler [3]. Another example is given by the neutron capture cross sections in ^{46,48}Ca [4,5] and its implications for the understanding of the Ca-Ti-Cr anomalies in an Allende inclusion [6,7].

B) In two 'indirect' ways:

e^-,ν capture (or scattering) cross sections determine the dynamics of the gravitational collapse of heavy stars (e.g. by neutrino cooling) and of supernova explosions, ν-scattering restricts the possible scenarios of the astrophysical site of the r-process.

p,n,e^-,HI capture, in a wider sense, was **the** tool leading to an understanding of β-decay far from stability in recent years (see Fig. 2) - and we shall restrict ourselves on this type of capture in this paper.

The new information on β-decay [8-12] derived from capture has many consequences for various fields in particle physics and nuclear physics all merging into the problem of element synthesis in the universe or astrophysical questions related with this problem (Fig. 2).

- the new beta decay half-lives fix the astrophysical site of the r-process and thus give a solution for a 25 years old astrophysical problem [1,8]

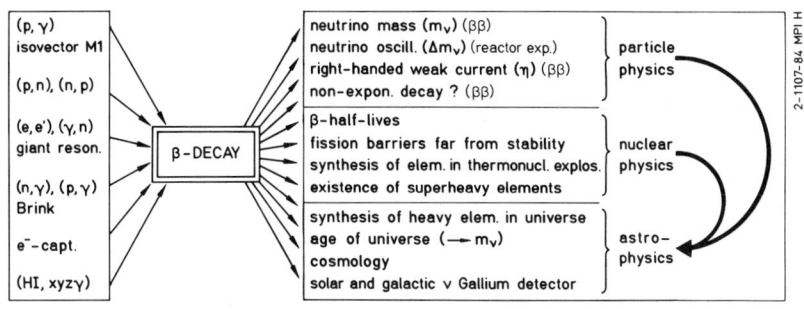

Fig. 2: Capture and its connection to some fundamental problems in particle, nuclear and astrophysics.

- the new rates of beta-delayed neutron emission and fission put the method of cosmochronology for determining the age of our galaxy on a more reliable basis [1,8,9]

- β-delayed fission rates affect the question and decide finally whether superheavy elements are built up in nature or not

- β-decay determines the properties of detectors for solar and galactic neutrinos [8,13], expected to give a solution of the solar neutrino puzzle and the first **direct** information on the fusion process in the sun and the gravitational collapse of heavy stars. Existing phenomenological estimates of the β-decay in this context [76] are insufficient and experimental results [77] (from the Ga (p,n) reaction) are questionable and incomplete (see Refs. 13,78). For a detailed discussion the reader is referred to Ref. 13).

- the nuclear physics experiments, such as reactor ν oscillation experiments and double beta (ββ) decay experiments, aiming at the solution of fundamental problems of particle physics such as a non-vanishing neutrino mass and a right-handed weak current need [1,8,9,11,14-16] for their interpretation the better understanding of 'classical' beta decay derived from capture reactions. On the other hand the results of these experiments will have strong influence on astrophysics: It is well known that neutrino mass and structure and age of the universe are intimately related. Admitting weak interaction theories with non-vanishing Majorana neutrino masses would allow for example flipping the neutrino helicity by Majoron-mediated reactions, $\bar{\nu}_L^e$ produced in e$^-$-capture will - interacting via Majorons - produce final states with all families of left- and right-handed neutrinos, and this would finally lead to models of gravitational collapse of massive stars drastically different from the low-entropy collapse models following from the usual assumption of lepton number conservation [17,18].

In this paper in section 2 we shall briefly remind ourselves of the connection between the various capture reactions and β-decay, in section 3 we discuss the progress achieved in the understanding of β-decay, particularly far from the β-stability line, by capture reactions and in sections 4,5 we discuss the consequences for element synthesis and related problems.

2. Capture processes and beta decay

What do we mean by understanding of β-decay: Predictability of the energy positions of the coherent particle-hole excitations excited by the beta-operators and of the matrix elements of the transitions to them, in other words, knowledge of the β strength distribution in the daughter nuclei which then determines all other quantities like β-decay half-lives, rates of β-delayed neutron emission and fission, shapes of e$^-$ and ν spectra.

Different capture processes have shed light on different aspects of the β strength function (see Figs. 2,3)

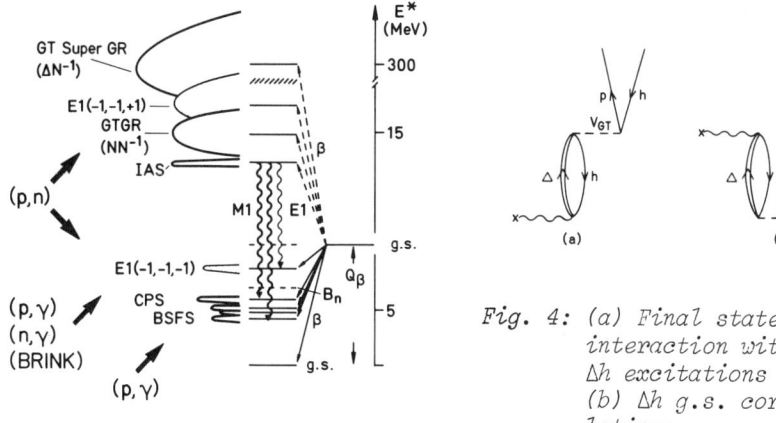

Fig. 4: (a) Final state interaction with Δh excitations
(b) Δh g.s. correlations

Fig. 3: Beta strength distribution in neutron-rich nuclei (schematic, see Ref. 8).

- (p,γ) reactions investigating the isovector M1 γ-decay of isobaric analogue states [8,19,20] as performed since more than ten years ago, gave the first information on the distribution of low-lying Gamow-Teller (GT) strength - by the proportionality of the GT and the spin part of the M1 operator (see, e.g., Ref. 8).

- the (p,n) reaction confirmed [21] the existence of the GT giant resonance (GR) predicted in the early sixties by Ikeda et al. [22] The GTGR, as spin-isospin vibration of the nucleus familiar to the M1 GR [8,23], can be investigated directly in β-decay only in a few cases of β^+ decay of $T_3 < 0$ nuclei. The (p,n) reaction also is the best tool for testing the GT sum rule and together with calculations in a quark model language [1,10] gave the best insight into the quenching of GT strength by Δ excitations (and other mechanisms). The systematics of investigations of the GTGR has been extended recently out to ^{238}U by experiments we performed [24] at the Indiana cyclotron. Recent progress in the experimental techniques by Madey and coworkers made also low-lying GT and E1 spin-flip strength in heavy nuclei visible in the (p,n) reaction [25].

- the inverse (n,p) reaction is a powerful tool to investigate $T_<$ GT strength at low excitation energy and free of admixtures of $T_>$ states

- proton- [26], neutron- [27] and heavy-ion (HI) [28] capture reactions followed by γ-emission have been used to investigate the E1 giant resonance built on excited states, in other words to test the Brink hypothesis (see also Ref. 29). By the strict proportionality between the E1 γ and the first-forbidden β-operators (see, e.g., Ref. 8), such investigations are important for an understanding of the systematics of the positions of low-lying collective E1 charge exchange states and the corresponding first-forbidden β-strength in nuclei with large neutron excess [8,23].

All this information together led to a break-through [8] in the following sense: It led to the understanding of the systematics of the GT strength distribution, particularly at low excitation energies with the consequence that schematic descriptions of β-decay far from stability (like that of the socalled 'gross-theory' [30] which actually was for the last ten years the only theory predicting β-decay properties far from stability) can be replaced now by microscopic calculations which put the achievable accuracy on a new basis. This has been discussed in detail in Refs. 8,9. We, therefore, enter directly into the representation of some typical results obtained by the new calculations.

3. The new description of beta decay far from stability

3.1. Method of calculation

The main features of our calculations are (for details the reader is referred to Refs. 1,8-16): Starting from BCS wave functions of the parent states the spectra of GT states in the daughter nuclei are calculated in the Tamm-Dancoff approximation (TDA) with a residual interaction

$$V_{GT} = 2\chi \Sigma\ \sigma(1)\sigma(2)\ (\tau^+(1)\tau^-(2) + \tau^+(2)\tau^-(1))$$

whose strength χ is taken from experiment, namely from measured positions of GTGR (the strength we extract in this way corresponds to a Landau-Migdal parameter of $g_0' \simeq 0.6$, in good agreement with recent microscopic derivations [31]. To incorporate ΔN^{-1} excitations in the model space the operators $\sigma_i \tau_j$ are replaced by

$$F_{ij} = \frac{3}{5} \sum_{q=1} (\sigma_q)_i (\tau_q)_j$$

acting on the three quarks in the nucleon.

We take further into account the following types of ground state (g.s.) correlations:

(a)   spin-isospin correlations

(b)   quadrupole-quadrupole correlations, i.e. the admixture of two-phonon excitations into the g.s.

(c) ΔN^{-1} correlations (the ΔN^{-1} excitations do not only enter into the GT strength distribution by the final state interaction but also by ΔN^{-1} g.s. correlations (Fig. 4), which leads to a strong energy dependence of the quenching of the GT strength [10]).

Basing on such an approach we have calculated (with suitable approximations where appropriate) the β^- decay (β strength functions $S_\beta(E)$) of all nuclei between β-stability line and neutron drip line and the β^+ decay of various nuclides on the left side of the β-stability line. Ref. 12 shows examples of strength functions and the typical difference in the description of its structure by the present microscopic and the 'gross theory' of Ref. 30, respectively. The results for various quantities derived from the calculated $S_\beta(E)$ are presented in the following sections

3.2. Results

3.2.1. Beta decay half-lives, rates of β-delayed neutron emission and fission

The calculated half lives are published in At. Data Nucl. Data Tables [12]. The accuracy of the predictions is considerably improved over that of the only existing earlier prediction of Ref. 30 (see the discussion in Ref. 12). The microscopic half-lives are, for neutron-rich nuclei, systematically shorter than those predicted earlier. Table A in [12] gives a feeling for the predictive power of the present approach.

Calculated rates of β-delayed fission and neutron emission are given in Refs. 8,9,32, complete Tables of P_n and $P_{\beta df}$ values are under preparation [33]. In these calculations the neutron tranmission coefficients were calculated from an optical model potential obtained [34] from infinite nuclear matter calculations in the Brückner-Hartree-Fock-approximation with a Reid's hard core nucleon-nucleon interaction by applying the local density approximation to finite nuclei and the transmission coefficients for fission were calculated assuming a 'complete damping mechanism' and taking the parameters of the double humped barriers from Howard and Møller [35]. The main result concerning the $P_{\beta df}$ values is that we find [8,9,32] a region of very strong β-delayed fission around $Z \simeq 94$, $N \simeq 167$, (see Fig. 1), coinciding with the long known region of strong spontaneous fission which occurs as a consequence of the 'bay of shells' in the Lund nomenclature of fission barrier height systematics. Of importance for the astrophysical application is, that the range of β-delayed fission extends beyond (versus lower Z) the range of spontaneous fission – as a result of an effective lowering of fission barriers by the β decay to excited states. It should be noted, that the calculated $P_{\beta df}$ values allow a natural explanation of the long-known 'odd-even reversal' in the element distribution produced in thermonuclear explosions (see Refs. 8,9).

3.2.2. Electron and antineutrino spectra from nuclear reactors and reactor decay heat

The calculation of the spectra of electrons and antineutrinos produced in nuclear reactors by the β^- decay of the ~1000 fission products from fission of the various fuel components can be regarded as a kind of global test of our calculated β strength functions. Refining our earlier calculations [11] we used [36] the more recent tabulation of fission product yields of Rider [37] (ENDF/B-VI) and

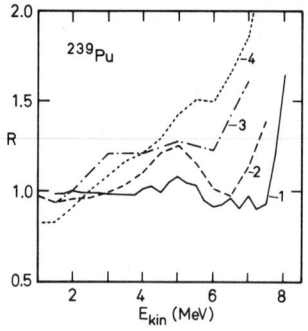

Fig. 5: The electron spectrum from thermal fission of ^{239}Pu. R denotes the ratio between the values of Refs. 41-43,45 (labelled 1,2,4,3, respectively) and our calculation [36] (curve 4 uses the $\bar{\nu}$ spectrum of Ref. [43]

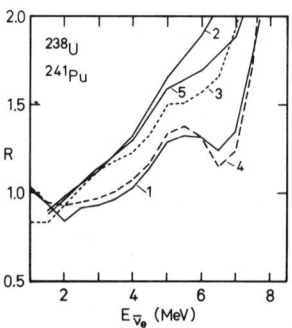

Fig. 6: The predicted $\bar{\nu}$ spectra from fast fission of ^{238}U and thermal fission of ^{241}Pu. R denotes the ratio of the values of Refs. 42,44,43 (labelled 1,2, 3) for ^{238}U, and Refs. 42,44 (labelled 4,5) for ^{241}Pu and our calculation [36].

included radiative corrections following the theory of Ref. 38 (see Ref. 39). The latter do not affect the corresponding $\bar{\nu}$ spectra calculated earlier [11], they affect, however, the conversion procedure used by Schreckenbach et al. [40] and v. Feilitzsch et al. [41] to deduce $\bar{\nu}$ spectra from measured e⁻ spectra to the order of up to 5%. Fig. 5 shows the large progress in the precision of the calculated electron spectrum compared to earlier calculations [42-45] which suffer from a poor description of the beta strength function (see the discussion in Refs. 1,8,9,11,46).

Fig. 6 shows our calculated [36] $\bar{\nu}$ spectra from thermal fission of ^{241}Pu and fast fission of ^{238}U in comparison with the rough estimates of Refs. 42-44 still used in the analysis of the most recent $\bar{\nu}$ oscillation experiments [47,48]. Because of the large sensitivity of the analysis of reactor $\bar{\nu}$ experiments and of the deduced values of the neutrino mass and the mixing angle to the choice of the $\bar{\nu}$ core spectrum [1,8,9,11] the use of microscopically calculated $\bar{\nu}$ core spectra is indispensable for those reactor $\bar{\nu}$ oscillation experiments presently under way at power reactors such as Gösgen [47,48] and Bugey [49], where the fuel composition is complex and varies as function of time. This is true also for 'two-distance' measurements such as that of Ref. 48 (see the discussion in Ref. 1).

Fig. 7 shows by an example that the same calculations giving the correct e⁻ and ν spectra, rather perfectly reproduce the decay heat

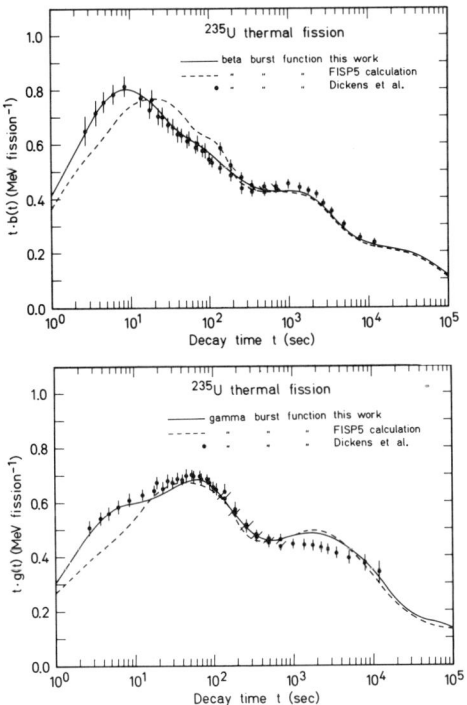

Fig. 7:
Microscopically calculated [36] beta and gamma burst functions as function of time after shut-down of a nuclear reactor for thermal fission of ^{235}U (solid line) compared to a recent experiment [50] and a typical earlier calculation (dashed line, [51]).

of nuclear reactors [36] (shown is the burst function for ^{235}U) – total as well as the parts from beta and γ-transitions separately, to the limited extent where there exist measurements. It should be mentioned here, that calculations of the burst functions using the gross theory of β-decay by Yoshida and Nakasima [52], unfortunately have to be considered as a kind of multiparameter fit (see the discussion in Ref. 8).

3.2.3. Double beta decay

The more realistic description of the beta strength function is reflected also in a better description of double beta decay – particularly of 2ν transition matrix elements $M_{2\nu}$ [1,9,14-16]. Previous calculations of two-neutrino ββ-decay rates by Haxton et al. [53] cannot describe the distribution of the GT strength in the intermediate nucleus properly, since they do not take into account collective effects produced by the spin-isospin force, i.e. the existence of the GTGR, and also not quadrupole-quadrupole and ΔN^{-1} correlations. A consequence is that these authors overestimate the total decay rates of ^{128}Te and ^{130}Te by almost two orders of magnitude. Table 1 shows that this discrepancy is markedly reduced by the new calculations – mainly as an effect of the quadrupole-quadrupole correlations [15] (for the remaining discrepancy see the discussion in Refs. 1,16).

Table 1: 2ν and 0ν ββ-decay half lives and deduced neutrino masses (see text)

Nucleus	$T^{(2\nu)}_{1/2(calc)}$ (in yr) Ref. 15	$R_0 M^{(0\nu)}_{(calc)}$ Ref. 53	ξ Ref. 16	$T^{(total\ or\ 2\nu)a)}_{1/2(exp)}$	$T^{0\nu}_{1/2(exp)}$	m_ν(eV) Ref. 16
^{76}Ge	$2.2*10^{20}$	$4.15*10^{20}$	10.40	5.4	$>2.8*10^{19}$	$>5*10^{22}$ [56]	<2.3
^{82}Se	$1.5*10^{19}$	$2.62*10^{19}$	8.20	4.5	$(1.45\pm0.15)10^{20}$	$>1.5*10^{20}$ [55]	<26
					$(1.0\pm0.4)10^{19}$	$>3.1*10^{21}$ [57]	<5.5
^{128}Te	$5.7*10^{23}$	$8.79*10^{22}$	10.03	19.3	$>8*10^{24}$ (2σ)	$>8.10^{24}$ [54]	<0.35
^{130}Te	$1.2*10^{20}$	$1.71*10^{19}$	9.44	19.7	$(2.55\pm0.20)*10^{21}$	$>2.6*10^{21}$ [55]	<4.2

a) total ββ half lives except for the value for ^{76}Ge and the second value for ^{82}Se.

This observation has a large effect on the upper limits for the neutrino mass deduced from present (mainly geochemical [54]) experiments. In previous analyses [53,54] of experimental data the calculated 0ν ββ decay matrix element $M_{0\nu}^{calc.}$ was 'scaled down' to take care of the (previous) discrepancy between calculation and experiment concerning the 2ν rates: The value of $M_{0\nu}$ used to deduce an upper limit for the neutrino mass from a measured upper limit for the 0ν decay

$$\omega_{0\nu} = |M_{0\nu}|^2 (Am_\nu^2 + Bm_\nu\eta + C\eta^2)$$

was $M_{0\nu} = M_{0\nu}^{calc.} \cdot |M_{2\nu}^{exp}/M_{2\nu}^{calc.}|$

This procedure, however, is found to be no longer correct, since the collective effects arising from spin-isospin forces and phonon correlations can affect the 2ν matrix elements much more than the $M_{0\nu}$ values [16] - as an effect of the different radial dependence of the involved operators (long range correlations such as phonon corelations are unimportant for 0ν ββ decay due to the $1/r_{ij}$ factor in $M_{0\nu}$).

This different behaviour of the two kinds of matrix elements expresses itself in ξ values much larger than 1 ($\xi = M_{0\nu}R_0/M_{2\nu}$, with R_0 denoting the nuclear radius). It means further that the scaling procedure commonly used underestimates $M_{0\nu}$. Table 1 shows in addition to our calculated [15,16] $M_{0\nu}$ and $M_{2\nu}$ matrix elements and ξ values the new limits for the neutrino mass deduced from the experimental values in the following most conservative way: We assume a) the righthanded current admixture η to be zero b) $\omega_{0\nu} \leq \omega_{geochem}$. The limit from the ^{128}Te data, therefore, does not depend on the often used ratio argument for the 2ν rates of ^{128}Te and ^{130}Te (see Refs. 15,54). The new obtained limit for ^{128}Te is by more than an order of magnitude smaller than the previous limit of 8.8 eV (5.6 eV using the 2ν ratio argument [54]). ββ decay thus seems to be at present the most sensitive method to test the neutrino mass. The new lower limit is consistent with the recent results from reactor $\bar{\nu}$ oscillations [48,49].

Concluding this section, from the fact that we reproduce consistently, with the same microscopically calculated β-strength functions, known β decay half lives, rates for β-delayed fission and neutron emission, the electron spectra from thermal fission of ^{235}U, ^{239}Pu, the decay heat (burst functions) from thermal fission of ^{235}U, ^{239}Pu, ^{241}Pu, double beta decay rates — all with an accuracy unknown up to now — it seems that there exists for the first time a reliable description of β-decay far from stability. Some of the consequences will be discussed in the next sections.

4. The synthesis of heavy elements by the r-process

The first element synthesis in the universe occured 226 sec after the big bang. At that time, after the eras of hadrons and leptons were passed, the temperature was low enough that formed deuterons remained stable with the consequence that at this time all neutrons were boiled to He which since this time forms ~25% of the mass of the universe. Only 700 000 years later with the decoupling of matter and radiation the formation of galaxies and stars set in and with this the formation of all other elements — except Li, Be, B which are produced by spallation in cosmic rays. While the elements up to iron can be built up by fusion in hydrostatic burning phases of heavy stars, there are essentially two processes responsible for the synthesis of the heavy elements in the nowadays observed cosmic element distribution: Slow and fast neutron capture (s- and r-process). The fingerprints of these two processes are the peaks in the element distribution at and below the neutron magic numbers. We shall restrict ourselves here to the r-process. Its principle is known since 25 years. An explosive event produces on a time scale of a second a high neutron flux. By successive neutron capture and β decay from some starting element distribution (a CNO and s-processed solar element distribution) there is built up an element distribution ~20 - 30 mass units away from the β stability line (denoted by the r-process path in Fig. 1). After the neutron flux drops this distribution decays back to the β stability line. The resulting final distribution of stable elements depends obviously sensitively on the β decay properties of the several 1000 nuclei between r-process path and β stability line. The β half lives set sharp restrictions on the possible time scales of the dynamics and thus possible scenarios and they determine (together with the n-capture rates) the position of the r-process path in the N-Z plane and thus the position of peaks in the final element distribution [8,9,32].

The question of the site of the r-process in the universe was an open question for more than two decades, particularly, since the old site proposed by Fowler and Seeger to occur close to the neutron star forming in a supernova explosion (Fig. 8) has been ruled out by

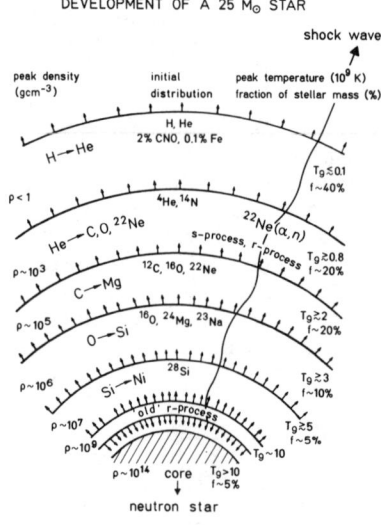
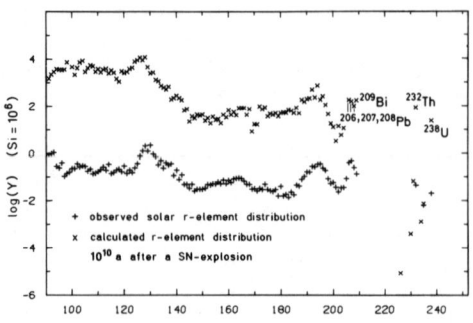

Fig. 9: The calculated [32] final element distribution resulting from β decay of the element distribution built up in the r-process path, and subsequent α decay of transbismuth nuclei, 10^{10} years after a supernova explosion.

Fig. 8: To the site of the r-process: Structure and development of a heavy star (~25 $M_\odot$) undergoing a supernova explosion (from Refs. 1,9).

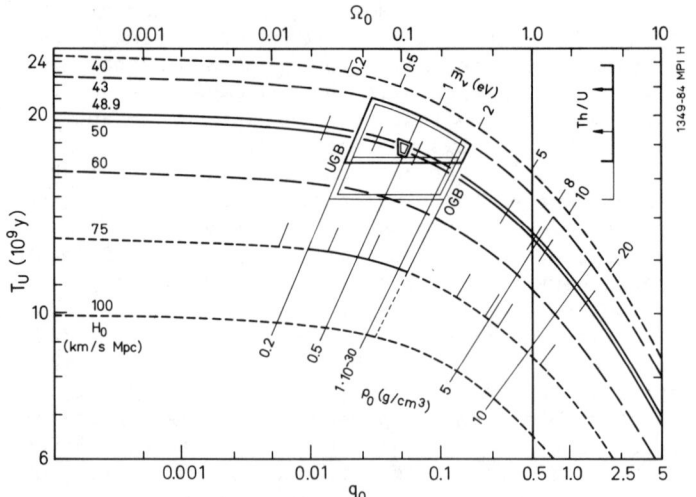

Fig. 10: (following [74]): Connection between age of the universe T_U (in $10^9 y$), deceleration parameter q_0 ($\Omega = 2q_0 = \rho_0/\rho_{crit}$.) and Hubble constant H_0 for Friedmann models ($\Lambda = 0$). UGB and OGB denote the lower and upper limit of the observed baryonic matter density, respectively. The new meteoritic age of the universe is given to the right.

the discovery of weak neutral currents - another example of the
close interplay of weak interaction and 'macroscopic' astrophysics.
The weak neutral currents lead by coherent ν scattering on Fe to a
decrease of the neutron density (by inverse β decay) below the critical value necessary to drive an r-process in that scenario [58,
59]. Other serious inconsistencies of the 'old' r-process were
discussed already in [60].

As an alternative process explosive He-burning in supernovae was
investigated since 1977 [61]: The detonation shock front running
through the outer shells in the supernova explosion leads for
~0.5 sec to an increase of the neutron density in the He-burning
shell by 10 orders of magnitude - up to ~10^{20} cm^{-3}. These neutrons
come from the reaction ^{22}Ne(α,n) which during hydrostatic burning
drives the s-process. The question was: Can this neutron-density
available for 0.5 sec drive an r-process leading to the observed
solar r-element distribution? The answer from hydrodynamical star
explosion calculations was no - as long as the β decay half lives of
the 'gross theory' were used. We have shown, however, that the
situation changes when the shorter microscopic β decay half lives
are used [8,9,62]. The results are as follows (for a detailed
discussion see Refs. 8,9):

- a solar r-element distribution can be obtained in this process
(Fig. 9) when using the microscopic β decay half lives. The calculated absolute enrichment of r-material in the He-burning shell
(more precisely in the 1/10 of the shell participating in the
process) of several 10^3 corresponds to an average enrichment factor
of 10 - 20 per supernova, which is comparable to the enrichment
factor predicted [63] for the production of lighter elements in
explosive burning of inner shells. The new r-process site also
facilitates the explanation of observed r-element overabundances in
meteorites [9,64]

- Beta-delayed fission leads to a cut-off of the calculated r-process path at already small Z (around Z = 92) (see Ref. 32 and
Fig. 1). Consequently, no superheavy elements will be produced in
the r-process and so probably in nature.

- the calculation of the production ratios of the cosmochronometers
^{232}Th, ^{235,238}U, ^{244}Pu (see Fig. 9) puts the method of cosmochronology for the first time on a reliable basis.

The obtained solution describes the production of heavy elements in
the universe with a consistency never reached before and is independent from still unsolved problems (see Ref. 63) of the triggering
of the supernova explosion by the gravitational collapse.

5. The age of the universe and the neutrino mass

Starting with the element distribution in the r-process path as it
results in the scenario of explosive He-burning, using the microscopically calculated rates for β-delayed fission in the r-process
path and during decay back to the stability line and taking into

account β-delayed n-emission we find [8,9,32] production ratios of the chronometric pairs which are remarkably lower than the adopted 'standard' values (see Refs. 1,8,9,32). The influence of these changes in the production ratios on the deduced age of the galaxy is considerable. In a simple exponential model of galactic nucleosynthesis in which its four parameters are deduced from the calculated production ratios of the chronometric pairs $^{232}Th/^{238}U$, $^{235}U/^{238}U$, $^{244}Pu/^{238}U$, $^{129}I/^{127}I$, and their observed ratios (in meteorites) at the time of condensation of the solar system, we find [8,9,32] an age of the galaxy of $T_G = (20.8 (+2,-5))10^9$ a, i.e., almost a factor of 2 larger than found in previous cosmochronological studies of the actinide chronometers. Allowing for $1*10^9$ years for the time between the big bang and the formation of the first galaxies this corresponds to an age of the universe of $(21.8 (+2,-5))10^9$ a. (Use of the meteoritic ratio for $^{232}Th/^{238}U$ of 2.32 of [66] instead of the standard value of 2.50 ± 0.2 given by [65] would reduce this value to $18.9*10^9$ a. Since an improved treatment of level densities [67], however, seems to systematically increase these numbers, we tend more to the first of the above ages.) The new values are consistent with most recent determinations of the ages of globular clusters [68-71] yielding values between 16 and $25*10^9$ a and of the Hubble-time of the universe yielding [72] $T_U = H_0^{-1} = (19.5 (+3.2,-2.3))$ (corresponding to a global value of $H_0 = (50 \pm 7)$ km s^{-1} Mpc^{-1} and an extreme Friedman model with deceleration parameter $q_0 = 0$). (For a detailed discussion, also of the other chronometers, see Refs. 8,9,32).

An interesting conclusion can be drawn [1,9,71,73,74] about an average mass of the neutrino from a comparison of the new actinide age with values of the Hubble constant. In Friedmann models with cosmological constant $\Lambda = 0$ the age of the universe can be represented, for a given Hubble constant H_0, as function of the mass density of the universe $\Omega = \rho_0/\rho_c$ (ρ_c denoting the critical density, beyond which we would have a closed universe). In such a representation the new actinide age of the universe of $T_U = (21.8 (+2,-5))10^9$ a would, with the global value of $H_0 = (50 \pm 7$ km s^{-1} Mpc^{-1} from [7], limit the parameter Ω to ≤ 0.5 (it would not completely exclude, however, $\Omega = 1$ for the $^{232}Th/^{238}U$ meteoritic ratio of 2.32 of [66]) (Fig. 10). Comparing the corresponding mass density ρ_0 with the lower limit of the observed baryonic mass density $\rho_{obs} = (0.5(+0,7,-0.3))10^{-30}$ gcm^{-3} [71,74] and interpreting the difference as totally due to a finite neutrino mass would lead to an upper limit of the latter of $m_\nu \lesssim 3.5$ eV (6 eV for the $^{232}Th/^{238}U$ ratio of 2.32). The values of m_ν would decrease when assuming larger values of H_0. So, the actinide age clearly excludes, within Friedman models of the universe, neutrino masses $\lesssim 10$ eV. This is consistent with a recent estimate of an upper limit of $m_{\nu e}$ from gauge theories [75]. The lowest limit of the actinide age, on the other hand, is not inconsistent with the value of $\Omega = 1$ preferred [18] as a 'natural' value in inflationary theories of the universe.

References

[1] H.V. Klapdor, Invit. talk given at Internat. Symp. on Nuclear Spectroscopy and Nuclear Interactions, Osaka, Japan, 21-24 March 1984
[2] E.B. Norman, contrib. to this conf.
[3] F. Käppeler, contrib. to this conf.
[4] F. Käppeler et al, to be publ.
[5] A.W. Mursin, H.V. Klapdor et al., to be publ.
[6] A.G.W. Cameron, Astrophys. J. 230, L53 (1979)
[7] W.A. Fowler et al., preprint
[8] H.V. Klapdor, Progr. Part. Nucl. Phys. 10, 131 (1983), Phys. Bl. 38, 182 (1982)
[9] H.V. Klapdor, Fortschritte der Physik, in press
[10] K. Grotz, H.V. Klapdor, J. Metzinger, Phys. Lett. 132B, 22 (1983)
[11] H.V. Klapdor, J. Metzinger, Phys. Rev. Lett. 48, 127 (1982), Phys. Lett. 112B, 22 (1982)
[12] H.V. Klapdor, J. Metzinger, T. Oda, At. Data Nucl. Data Tables 31, 81 (1984), Z. Phys. A309, 91 (1982)
[13] K. Grotz, H.V. Klapdor, J. Metzinger, contrib. to this conf.
[14] K. Grotz, H.V. Klapdor, J. Metzinger, J. Phys. G9, L169 (1983)
[15] H.V. Klapdor, K. Grotz, Phys. Lett. 142B, 323 (1984)
[16] K. Grotz, H.V. Klapdor, subm. for publ. and Phys. Rev. C, in press (1984)
[17] E.W. Kolb, D.L. Tubbs, D.A. Dicus, Astrophys. J. 255, L57 (1982)
[18] E.W. Kolb, Proceed. XI Internat. Conf. Neutrino Phys. and Astrophys. (Neutrino 84), Nordkirchen, June 11-16, 1984
[19] H.V. Klapdor, Phys. Lett. 35B, 405 (1971)
[20] H.V. Klapdor, M. Schrader, G. Bergdolt, A.M. Bergdolt, Yu.W. Namow, Isv. Akad. Nauk. 42, 64 (1978)
[21] R.R. Doering, A. Galonsky, D.M. Patterson, G.F. Bertsch, Phys. Rev. Lett. 35, 1961 (1975)
[22] K. Ikeda, S. Fujii, J.I. Fujita, Phys. Lett. 3, 271 (1963)
[23] A. Bohr, B. Mottelson, Nuclear Structure, Vol. II, Benjamin, N.Y., 1975
[24] R. Madey et al, H.V. Klapdor et al, to be publ.
[25] K. Grotz, H.V. Klapdor et al., Phys. Lett. 126B, 417 (1983)
[26] Z. Szeflinski, G. Szeflinska, Z. Wilhelmi, T. Rzaca, H.V. Klapdor, E. Andersen, K. Grotz, J. Metzinger, Phys. Lett. 126B, 159 (1983)
[27] S. Raman et al., Phys. Rev. C23, 2794 (1981)
[28] K.A. Snover, contrib. to this conf.
[29] W. Kitipowa, Yad. Fiz. 36, 597 (1982)
[30] K. Takahashi, M. Yamada, T. Kondoh, At. Data Nucl. Data Tables 12, 101 (1973)
[31] K. Nakayama, S. Krewald, J. Speth, Phys. Lett. 145B, 310 (1984)
[32] F. Thielemann, J. Metzinger, H.V. Klapdor, Z. Phys. A309, 301 (1983), Astron. Astrophys. 123, 162 (1983)
[33] J. Metzinger, H.V. Klapdor, to be publ.

[34] J.P. Jeukenne, A. Lejeune, D. Mahaux, Phys. Rev. C182, 1190 (1977)
[35] W.M. Howard, P. Møller, At. Data Nucl. Data Tables 25, 219 (1980)
[36] J. Metzinger, H.V. Klapdor, to be publ.
[37] B.F. Rider, Compil. of Fiss. Prod. ENDF/B-VI (1981)
[38] A. Sirlin, Phys. Rev. 164, 1767 (1967)
[39] H. Behrens, W. Bühring, Electron Radial Wave Functions and Nuclear Beta Decay, Clarendon Press, Oxford (1982)
[40] K. Schreckenbach et al., Phys. Lett. 99B, 251 (1981)
[41] F.v. Feilitzsch et al., Phys Lett. 118B, 162 (1982)
[42] P. Vogel, G.K. Schenter, F.M. Mann, R.E. Schenter, Phys. Rev. C24, 1543 (1981)
[43] F.T. Avignone, Z.D. Greenwood, Phys. Rev. C22, 594 (1980)
[44] W.I. Kopeikin, Yad. Fiz. 32, 1507 (1980)
[45] A.A. Borowoi, Yu.W. Klimow, W.I. Kopeikin, Yad. Fiz. 37, 1345 (1983)
[46] H.V. Klapdor, Phys. Rev. C23, 126 (1981)
[47] J. Vuilleumier et al., Phys. Lett. 114B, 298 (1982)
[48] K. Gabathuler et al., Phys. Lett. 138B, 449 (1984)
[49] J.F. Cavaignac et al., preprint LAPP-EXP-84-03, May 1984
[50] J.K. Dickens, T.A. Love, J.W. McConnell, R.W. Peelle, Nucl. Sci. Eng. 74, 106 (1980)
[51] A. Tobias, Progr. Nucl. En. 5, 1 (1980)
[52] T. Yoshida, R. Nakasima, J. Nucl. Sci. Technol. 18, 393 (1981)
[53] W.C. Haxton et al., Phys. Rev. Lett. 47, 153 (1981), Phys. Rev. D25, 2360 (1982), Comm. Nucl. Part. Phys. 11, 41 (1983), Progr. Part. Nucl. Phys., in press
[54] T. Kirsten, H. Richter, E. Jessberger, Phys. Rev. Lett. 50, 474(1983), Z. Phys. C16, 189 (1983)
[55] T. Kirsten in Proc. Workshop Science Underground. Los Alamos (1982), Proc. Telemark Mini Conf. on Neutrino Mass and Gauge Structure of Weak Interact., Am. Inst. Phys. (ed.), (1982)
[56] E. Bellotti et al., contrib. to HEP 83, Brighton, July (1983)
[57] B.T. Cleveland et al., Phys. Rev. Lett. 35, 757 (1975)
[58] D.Z. Freedmann, Phys. Rev. D9, 1389 (1974)
[59] W. Hillebrandt, Phys. Bl. 38, 189 (1982)
[60] W. Hillebrandt, Proc. 4th EPS Gen. Conf., 255 (1979)
[61] W. Hillebrandt, F.K. Thielemann, Mitt. Astron. Ges. 43, 234 (1977), J.W. Truran, J.J. Cowan, A.G.W. Cameron, Astrophys. J. 222, L63 (1978)
[62] H.V. Klapdor et al., Z. Phys. A199, 312 (1981), CERN-Report 81-09, 341 (1981)
[63] S.E. Woosley, T.A. Weaver, Proc. NATO Adv. Study Int., Cambridge 1981, p. 79 and Bull. Astr. Soc. 14, 957 (1982)
[64] H.V. Klapdor, Sterne und Weltraum, in press
[65] E.M.D. Symbalisty, D.N. Schramm, Rep. Progr. Phys. 44, 293 (1981)
[66] E. Anders, M. Ebihara, Geochim. Cosmochim. Acta 46, 2363 (1982)
[67] J. Metzinger, H.V. Klapdor, to be publ.
[68] A. Sandage, Astrophys. J. 252, 553 (1982)

[69] P.E. Nissen, ESO Messenger 28, 4 (1982)
[70] A. Ardeberg, H. Lindgren, P.W. Nissen, Astron. Astrophys. 128, 194 (1983)
[71] W. Priester, Vortrag N333 der Rheinisch-Westfälischen Akademie der Wissensch., Westdeutscher Verlag 1984
[72] A. Sandage, G.A. Tammann, Astrophys. J. 256, 339 (1982)
[73] H.V. Klapdor, Proceed. Internat. Sympos. Electromagn. Properties of Atomic Nuclei, Tokyo, November 9-12 , 1983
[74] H.J. Blome, W. Priester, Naturwissenschaften 9, 456 (1984)
[75] P. Cea, Phys. Lett. 146B, 75 (1984)
[76] J.N. Bahcall et al., Rev. Mod. Phys. 50, 881 (1978) and 54, 767 (1982)
[77] H. Orihara et al., Phys. Rev. Lett. 51, 1328 (1983)
[78] A. J. Baltz, J. Weneser, B.A. Brown, J. Rapaport, preprint (1984)

STATUS OF HELIUM BURNING OF ^{12}C

H. P. Trautvetter, A. Redder and C. Rolfs
Institut für Kernphysik, Universität Münster,
Münster, W. Germany

ABSTRACT

The $^{12}C(\alpha,\gamma)^{16}O$ reaction is of crucial importance for the understandingg of stellar evolution and nucleosynthesis. The present state of knowledge for this reaction is presented as well as new experimental approaches.

In spite of enormous experimental efforts that have gone into studies of the $^{12}C(\alpha,\gamma)^{16}O$ capture reaction, there are still considerable uncertainties in the cross section (reaction rate) at the relevant stellar energy region of E_0 = 0.3 MeV (Refs. 1-3). As a consequence, neither the relative amounts of ^{12}C and ^{16}O produced by Red Giant stars nor the stars' subsequent evolution and the nucleosynthesis of the metallic elements can be determined with great confidence. The formidable problems encountered in the direct studies of this reaction arise from the combination of a low γ-ray capture yield ($\sigma \leq 50$ nb) and a high neutron-induced γ-ray background mainly due to the $^{13}C(\alpha,n)^{16}O$ reaction, which is a prolific source of neutrons. In the work of Dyer and Barnes[1] an excitation function was obtained at θ_γ = 90° and $E_{c.m.}$ = 1.4-3.0 MeV, which is sensitive only to the E1-amplitude in the capture mechanism. From γ-ray angular distributions obtained at higher beam energies ($E_{c.m.}$ = 2.2-2.8 MeV) it was suggested that the observed E2-amplitude is negligible at stellar energies. For this suggestion, the E1-data yielded an extrapolated value for the astrophysical S-factor of $S(0.3$ MeV$)$ = 0.08 ± 0.04 MeV-b (Ref. 4). This recommended value has been questioned recently by the data of Kettner et al.[2] because the angle-integrated excitation function obtained at $E_{c.m.}$ = 1.2-3.5 MeV could not be explained alone by an E1-amplitude but required a non-negligible E2-amplitude. Using a simple Breit-Wigner formalism and incorporating data from elastic scattering, which was obtained concurrently,[2] the capture data led to an extrapolated S-factor of $S(0.3$ MeV$)$ = 0.3-0.5 MeV-b (Ref. 2). Recent microscopic model calculations[5] have confirmed the existence of a non-negligible E2-amplitude at stellar energies and led to a value of $S(0.3$ MeV$)$ ~ 0.35 MeV-b for the angle-integrated data,[2] and of $S(0.3$ MeV$)$ ~ 0.25 MeV-b for the earlier 90°-data,[1] to which the theoretically calculated E2-contribution in the capture mechanism was added. An improved extrapolation of both data sets requires additional experimental data, i.e., γ-ray angular distributions obtained over a wide range of beam energies and as low in energy as possible. These distributions should be characterized by E1-E2 interference effects, from which the energy dependence of both multipole transitions can be extracted, thus improving their individual extrapolations.

As a first step in such an experimental program, ^{12}C solid targets depleted in ^{13}C have been produced by an implantation technique. The target was a 1 mm thick Cu sheet (2x2 cm^2 area) with 12 canals of 0.9 mm diameter, which were used to cool the target at a water pressure of 30 bar. A gold layer was evaporated on this Cu sheet and its thickness was chosen to be larger than the range of 4 MeV α-particles. This precaution was necessary in order to avoid a neutron-induced background from the $^{13}C(\alpha,n)^{16}O$ reaction taking place in the "durty" Cu sheet. The ^{12}C beam of 110 keV from the 350 kV Münster accelerator was implanted in the Au layer. The beam was scanned over nearly the full area of the target sheet and extensive LN$_2$ shrouds were used to prevent normal carbon deposition on the target during beam bombardment. The content and distribution of the implanted ^{12}C zone was subsequently determined using the $^{12}C(p,\gamma)^{13}N$ non-resonant reaction.[6] It was found that the ^{12}C

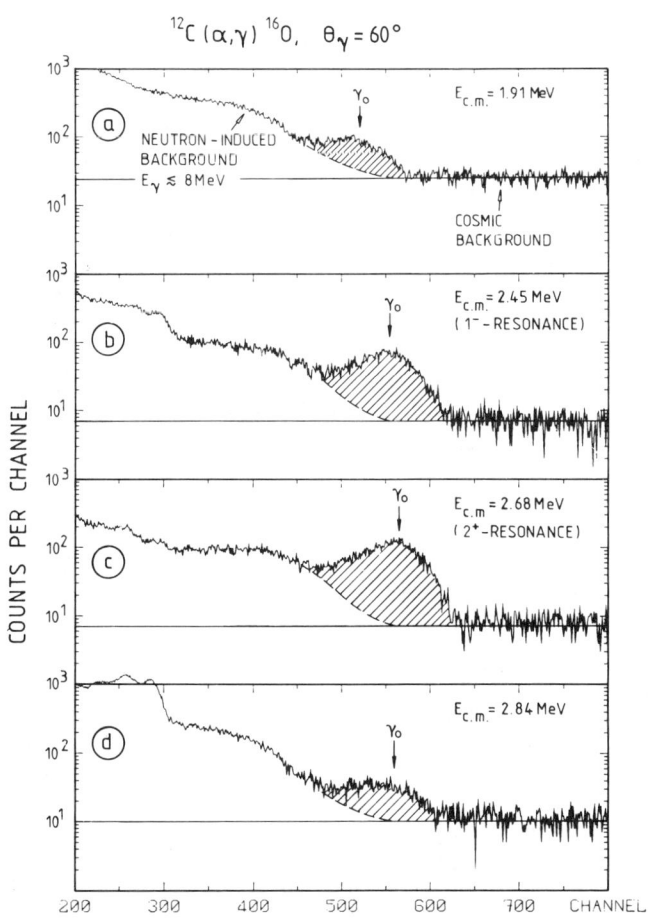

Fig. 1. Sample γ-ray spectra obtained at 60° with a 4x4" NaI(Tl) detector using ^{12}C-implanted targets.

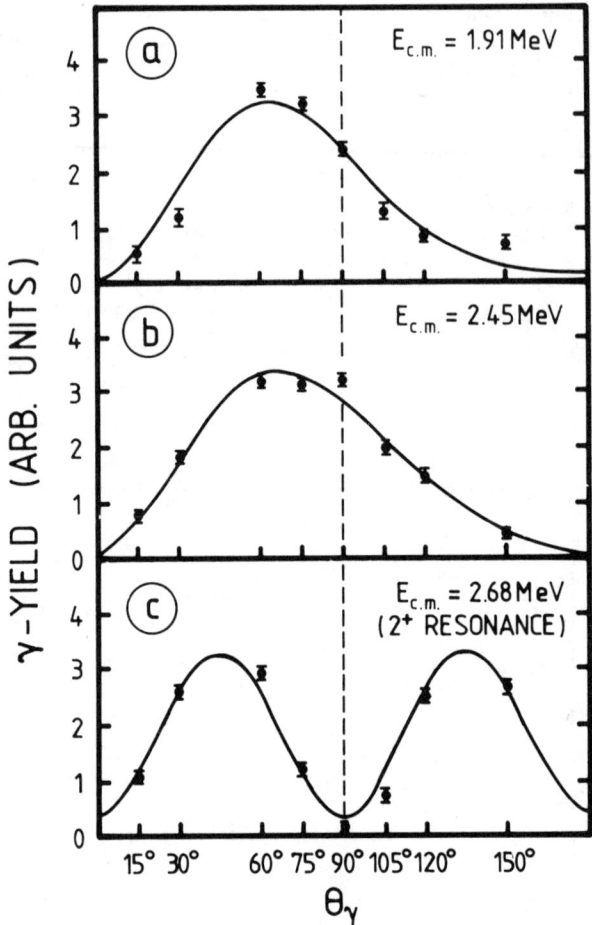

Fig. 2. Angular distributions of the groundstate transition in $^{12}C(\alpha,\gamma)^{16}O$. The lines through the data points are results of χ^2-fits.

targets started at the surface of the Au layer and had a nearly homogeneous distribution with a thickness consistent with the range of the ^{12}C ion beams. Targets with abundance ratios of ^{12}C: Au = 5:1 in the implantation zone were produced. The targets were also found to be fairly stable under extremely heavy beam loads (sputtering of the target material was observed, but no blistering, after bombardment over one week; see below).

In preliminary runs (March 1984) at the 4 MV Dynamitron accelerator in Stuttgart the targets were bombarded with an α-beam of 300-500 μA in the energy range of E_α = 1.6-3.9 MeV. The capture transitions were observed in eight 4x4" NaI(Tl) detectors covering the angular range of θ_γ = 15°-150°. Sample γ-ray spectra obtained in this set-up are shown in Fig. 1. As can be seen in the figure the groundstate transition γ_0 is well separated from the remaining neutron-induced background at $E_{c.m.} \gtrsim 2.1$ MeV. There are indications that this remaining background is created predominantly on the beam defining slits and apertures in the beam line. The angular distribution obtained at the well-known $J^\pi = 2^+$ resonance ($E_{c.m.} = 2.68$ MeV) exhibits the familiar pattern (Fig. 2c), while the distributions obtained at energies outside this resonance are characterized by asymmetries around 90°, indicating clearly the presence

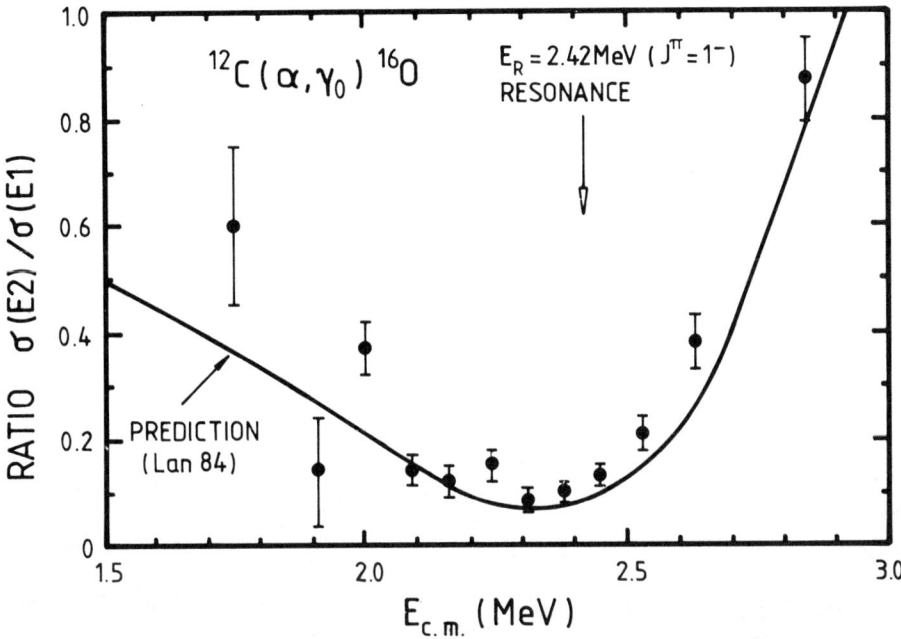

Fig. 3. Energy dependence of the $\sigma(E2)/\sigma(E1)$ ratio as obtained from the analyses of γ-ray angular distributions and as predicted (Ref. 5). The data points at $E_{c.m.} \lesssim 2.0$ MeV are subject to considerable uncertainties due to problems of analyses of NaI(Tl)-spectra.

of both E1 and E2 amplitudes in the capture mechanism. The analysis of these distributions involves two free parameters, the ratio of capture cross sections $\sigma(E2)/\sigma(E1)$ and the relative phase between both multipole transitions. The deduced ratio $\sigma(E2)/\sigma(E1)$ as a function of beam energy is shown in Fig. 3. Shown also in the figure is the ratio predicted by Langanke.[5] As expected, the ratio exhibits a minimum in the neighborhood of the $J^\pi = 1^-$ resonance ($E_{c.m.} = 2.42$ MeV) and their absolute values are not in contradiction to the prediction. At the low energy region, the γ_0-ray line shifts more and more toward the region of the neutron-induced background (Fig. 1a) hampering a quantitative analysis of its intensity. As a consequence, the deduced ratio $\sigma(E2)/\sigma(E1)$ in this energy region (Fig. 3) is subject to significant uncertainties. However, as tested experimentally, the extremely reduced neutron background due to the ^{12}C implanted targets allows the use of high-resolution Ge(Li) detectors in future investigations of this reaction and thus the above problem in spectra analyses should be reduced significantly. The results of such investigations, in particular at $E_{c.m.} \leq 2.0$ MeV (Fig. 3), have to be awaited before safe conclusions can be drawn on the capture mechanisms involved in this astrophysically important ^{12}C$(\alpha,\gamma)^{16}$O reaction.

REFERENCES

1. P. Dyer and C. A. Barnes, Nucl. Phys. A233, 495 (1974).
2. K. U. Kettner, H. W. Becker, L. Buchmann, J. Görres, H. Kräwinkel, C. Rolfs, P. Schmalbrock, H. P. Trautvetter, and A. Vlieks, Zeit. Phys. A308, 73 (1982).
3. C. Rolfs, Proc. of 1983 workshop on "Challenges and New Developments in Nucleosynthesis," Yerkes Observatory (Chicago Press, 1984).
4. S. E. Koonin, T. A. Tombrello, and G. Fox, Nucl. Phys. A220, 221 (1974).
5. K. Langanke, private communication.
6. T. Freye, H. Lorenz-Wirzba, B. Cleff, H. P. Trautvetter, and C. Rolfs, Zeit. Phys. A281, 211 (1977).

SOME EFFECTS OF HIGH TEMPERATURE AND
DENSITY ON NEUTRON-CAPTURE NUCLEOSYNTHESIS

Eric B. Norman* and Stephen E. Kellogg
Nuclear Physics Laboratory, University of Washington
Seattle, WA 98195

ABSTRACT

Examples of nuclear reactions between nuclei in excited states, beta decays of nuclear excited states, and bound-state beta decays are shown. The effects of these processes on selected problems in heavy-element nucleosynthesis are discussed.

INTRODUCTION

In recent years, it has become increasingly apparent that under the conditions of high temperature and density required in the s- and r- neutron-capture processes, the effects of several "exotic" nuclear phenomena should be considered in calculations of heavy-element nucleosynthesis. Examples include nuclear reactions between nuclei in excited states, beta decays of nuclear excited states, and bound-state beta decays. While the effects of such processes in the laboratory are, in general, completely negligible, their influence in stellar environments can be quite large. In this talk the effects of these processes on the nucleosynthesis of selected nuclei of current astrophysical interest are discussed.

^{26}Al

Observations in certain meteorites of ^{26}Mg isotopic abundance excesses that are correlated with the Al/Mg elemental abundance ratios have established that ^{26}Al was present in the early solar system.[1] The astrophysically short half-life of ^{26}Al [7.2×10^5 year, Ref. 2] implies that the nucleosynthetic event which produced this ^{26}Al must have occurred no more than a few million years prior to the formation of these meteorites. An intriguing possibility is that this event was a nearby supernova explosion which not only injected freshly synthesized material into the proto-solar nebula, but more importantly, actually triggered the gravitational collapse of this nebula to form our solar system.[3] As a result of these observations and speculations, there is a great deal of current interest in the nucleosynthesis of ^{26}Al.

The level structure of the nucleus introduces some interesting complications to any study of ^{26}Al. The ^{26}Al ground state, ^{26}Al^g, has $J^\pi = 5^+$ and a half-life of 7.2×10^5 year. The first excited state, ^{26}Al^m, at 228 keV has $J^\pi = 0^+$ and β^+ decays to ^{26}Mg with a

---

* Present address: Nuclear Science Division, Lawrence Berkeley Laboratory, Berkeley, CA 94720.

half-life of 6.3 sec but does not decay to $^{26}Al^g$ (Ref. 2). Thus any ^{26}Al formed in this isomeric level will not survive long enough to become incorporated into a meteorite as ^{26}Al.

Once formed, the ^{26}Al ground state can be destroyed by several processes. At sufficiently high temperatures, the following sequence of reactions may occur:

$$^{26}Al^g + \gamma \rightarrow \, ^{26}Al^* \qquad [1]$$

$$^{26}Al^* \rightarrow \, ^{26}Al^m + \gamma' \qquad [2]$$

$$^{26}Al^m + n \rightarrow \, ^{26}Mg + p \qquad [3]$$

or

$$^{26}Al^m \rightarrow \, ^{26}Mg + \beta^+ + \nu_e \qquad [4]$$

Even though there is no electromagnetic transition that directly connects $^{26}Al^g$ and $^{26}Al^m$, these two states can "communicate" through higher-lying excited states. The rates of the photoexcitation reactions can be calculated for those states whose lifetimes and γ-decay branching ratios are known by using the principle of detailed balance. We have performed these calculations[4] using published experimental values[2] for the required level properties. The results of these calculations indicate that for temperatures $\geq 7.5 \times 10^8$ °K, $^{26}Al^g$ and $^{26}Al^m$ reach their thermal equilibrium populations on time scales <1 sec. Thus $^{26}Al^g$ can be thermally excited to $^{26}Al^m$ and then destroyed via the (n,p) reaction or by β^+ decay. For instance, at a temperature of 1×10^9 °K, even in the absence of any free neutrons, the equilibration between $^{26}Al^g$ and $^{26}Al^m$ reduces the effective half-life of the nucleus from its laboratory value of 7.2×10^5 y to only 24 minutes! Similar results regarding $^{26}Al^{g,m}$ equilibration have been obtained by Ward and Fowler.[5]

In environments where in addition to high temperatures there also exist free neutrons, the (n,p) reactions can rapidly destroy ^{26}Al. While there are not (as yet) any direct measurements of the cross sections for this reaction, measurements of the $^{26}Mg(p,n_0)^{26}Al^g$ cross sections have been used together with the principle of detailed balance to obtain cross sections for the $^{26}Al^g(n,p_0)^{26}Mg_{g.s.}$ reaction.[6] These results are shown as the data points in Fig. 1. The solid curve represents the $^{26}Al^g(n,p_0)^{26}Mg_{g.s.}$ cross sections found by applying the principle of detailed balance to the low energy $^{26}Mg(p,n_0)$ cross sections calculated by Woosley et al.[7]

Woosley et al.[8] have also calculated the "stellar reaction rate," $N_A \langle \sigma v \rangle$, for the $^{26}Al(n,p)^{26}Mg$ reaction using a Hauser-Feshbach technique. From this rate, a Maxwellian averaged (n,p) cross section $\langle \sigma v \rangle / \langle v \rangle$, can be calculated. The dashed curve in Fig. 1 represents the results of these calculations which include the effects of two mechanisms that have not as yet been studied experimentally. The (n,p) reaction on ^{26}Al will populated excited states as well as the ground state of ^{26}Mg. This effect is included in the Hauser-Feshbach calculations but can only be studied experimentally with an ^{26}Al target. Secondly, at temperatures >1×10^9 °K, nuclear excited states are populated and can serve as targets for nuclear reactions. This effect is also included in the stellar rate calculations but cannot be deduced from existing experimental results. However, in an experimental study of the ^{26}Mg(p,n) reaction to excited states in ^{26}Al (Ref. 9), it was shown that the cross sections for the (n,p$_o$) reaction on the first two excited states of ^{26}Al are much larger than those for the (ground state)→(ground state) reaction. It is apparent from Fig. 1 that the inclusion of these two effects results in large cross sections. The implications of these results are that at least at the high temperatures where nuclear excited states are populated and where free neutron densities are reasonably high, the (n,p) reaction will be a very effective ^{26}Al destruction mechanism.

One is therefore led to the conclusion that in order to produce the amounts of ^{26}Al needed to account for the observed meteoritic anomalies, lower temperature environments are required. In hydrogen-rich regions at temperatures on the order of 1×10^8 °K, equilibration between ^{26}Al^{g,m} will not occur and furthermore, free neutron densities are typically quite low. In such locations, the destruction rate of ^{26}Al is determined by the relatively slow ^{26}Al(p,γ) reaction,[10] and as a result, sufficiently large quantities of ^{26}Al may be produced.[5,11]

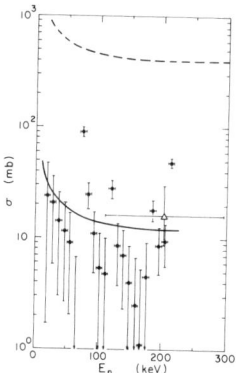

Fig. 1. Data points and solid curve are ^{26}Al$_{g.s.}$(n,p$_o$)^{26}Mg$_{g.s.}$ cross sections calculated from measured (Ref. 6) and theoretically estimated (Ref. 7) ^{26}Mg$_{g.s.}$(p,n$_o$)^{26}Al$_{g.s.}$ cross sections, respectively. Dashed curve represents the Maxwellian averaged cross sections for the ^{26}Al(n,p)^{26}Mg reaction as determined from calculations (Ref. 8) of the stellar reaction rate.

^{99}Tc

Remarkably perhaps, it is still true that one of the strongest pieces of evidence that nuclear reactions actually occur in stellar interiors is the observed presence of technetium on the surfaces of some stars.[12] While the presence of technetium has been established in many stars, it was not until recently that any quantitative Tc abundances had been determined.[13] From the known path of the s-process, it is expected that the only technetium isotope that could be observed in stars is ^{99}Tc, which has a laboratory half-life of

2.1 × 10^5 years (Ref. 14). However, under the conditions of high temperature appropriate for the s-process, the half-life of this nucleus can be dramatically reduced.

As can be seen in Fig. 2, the long half-life of the ^{99}Tc ground state is due to the fact that its beta decay to ^{99}Ru is a second-forbidden transition. However, as discussed by Ulrich,[15] the beta decay of the 141-keV level of ^{99}Tc to the ^{99}Ru ground state is an allowed transition with an estimated β-decay partial half-life of approximately 32 days. Thus if this level could come into thermal equilibrium with the ^{99}Tc ground state, then the effective half-life of this nucleus could be quite short. Using the principle of detailed balance, we find that for $T_8=1$ (where T_8=temperature in units of 10^8 °K) these two levels will reach their thermal equilibrium abundance ratios in less than 10^{-2} second. The resulting effective half-life of the nucleus is essentially the laboratory value for $T_8=1$, but decreases rapidly to 6000 years for $T_8=1.5$, 380 years for $T_8=2$, 25 years for $T_8=3$, and to 6.5 years for $T_8=4$.

As discussed by Smith and Wallerstein,[13] for technetium to be observed at all on stellar surfaces, it cannot have experienced temperatures ≥3 × 10^8 °K for very long. These and future measurements of technetium abundances in stars will help to place constraints on the temperatures at which the s-process occurs and thus should allow the source of the s-process neutrons to be determined.

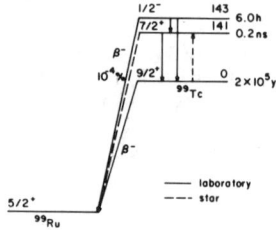

Fig. 2. Partial level scheme of ^{99}Tc illustrating the possibility of photoexcitation of the ^{99}Tc ground state to the $J^\pi = 7/2^+$ level at 141 keV and the subsequent β-decay.

^{176}Lu

^{176}Lu is one of the few naturally-occurring radioactive species that has survived from the era of nucleosynthesis. As can be seen in Fig. 3, it is produced only in the s-process and thus was suggested by Audouze, Fowler, and Schramm[16] and independently by Arnould[17] as a potential s-process chronometer. However, the possible usefulness of ^{176}Lu as such a clock is complicated by its nuclear structure. The $J^\pi = 7^-$ ground state, ^{176}Lu^g, β$^-$ decays to ^{176}Hf with a half-life of 3.6 × 10^{10} years (Ref. 14). The first excited state of the nucleus is a $J^\pi = 1^-$ level at 127 keV, ^{176}Lu^m, which β^- decays to ^{176}Hf with a half-life of 3.7 hours but does not decay to ^{176}Lu^g (Ref. 14). Thus any ^{176}Lu formed in this isomeric state will not contribute to the presently observed abundance of ^{176}Lu. Furthermore, if it were possible for ^{176}Lu^g to be excited to ^{176}Lu^m under the conditions of high temperature and density appropriate for the s-process, the net yield of ^{176}Lu would be further reduced.

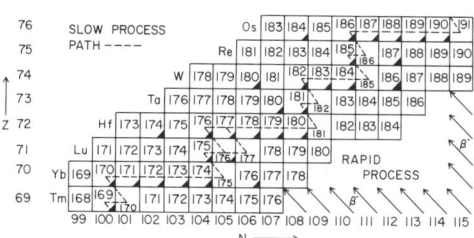

Fig. 3. Expected paths of the slow (s) and rapid (r) neutron-capture processes near A=180. Stable (or very long-lived) nuclei are indicated by shaded triangles.

Among the possible mechanisms for "communication" between ^{176}Lu^g and ^{176}Lu^m are photoexcitation, inelastic neutron scattering, Coulomb excitation, and positron-annihilation excitation. All of these processes undoubtedly involve transitions from ^{176}Lu^{g,m} to some higher-lying state(s) of intermediate spin which subsequently decay(s) to ^{176}Lu^{m,g}. The possibility of photoexcitation of ^{176}Lu^g to ^{176}Lu^m via such a level is shown schematically in Fig. 4.

Photoactivation of ^{176}Lu has been previously studied by Yoshihara[18] and by Veres and Pavlicsek.[19] Using ^{60}Co (E_γ = 1173 and 1332 keV) γ-ray sources, these authors observed increased activities following bombardment but deduced photoactivation cross sections of 1000-2400 nb and 23-47 nb, respectively. To resolve this discrepancy and to investigate the possibility of photoactivation of ^{176}Lu with lower energy photons, Norman et al.[20] irradiated a natural Lu target with large photon fluxes from both ^{137}Cs (E_γ = 662 keV) and ^{60}Co sources and used a plastic scintillator detector to search for the beta particles emitted in the decay of ^{176}Lu^m.

Following the ^{137}Cs bombardment, no excess of counts above the ^{176}Lu^g background was observed. From these measurements, however, an upper limit on the photoactivation cross section for 662-keV photons of 10 nb was established. After the ^{60}Co irradiation, an activity was observed which decayed away with the 3.7 hour. ^{176}Lu^m half-life. Assuming that both the 1173- and 1332-keV photons contribute to the observed yield, infer a photoactivation cross section of 38±10 nb for ^{60}Co photons. This result is in good

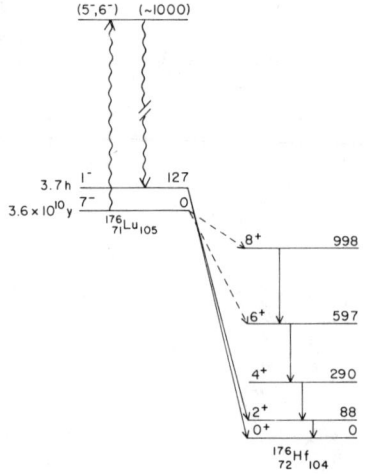

Fig. 4. Partial level scheme of ^{176}Lu and the decay schemes of ^{176}Lu^g and ^{176}Lu^m. The possibility of photoexcitation of the long-lived ^{176}Lu^g to a higher lying state of intermediate spin which subsequently decays to ^{176}Lu^m is shown schematically.

agreement with that of Veres and Pavlicsek[19] but disagrees strongly with that of Yoshihara.[18]

Perhaps one way to interpret these experimental results is to say that the lowest level mediating the transition from ^{176}Lu^g to ^{176}Lu^m is located above 662 keV but below 1332 keV. In the discussion which follows, it will be assumed that any possible transition from ^{176}Lu^g to ^{176}Lu^m proceeds through a level at 1.00 MeV excitation energy.

As has been recently pointed out, non-resonant excitation can be the dominant mechanism for photoactivation[21,22] Thus in calculating the excitation rate from ^{176}Lu^g to the 1.00-MeV level, we will assume that all photons with $E_\gamma \geq 1.00$ MeV contribute. Furthermore, since the energy dependence of non-resonant photoexcitation has not yet been established, it will be assumed in what follows that the photoexcitation cross section is independent of energy above threshold.

The photoexcitation rate from ^{176}Lu^g to ^{176}Lu^m via the assumed 1.00-MeV level is

$$\lambda_\gamma = N_\gamma \sigma_\gamma c, \qquad [5]$$

where N_γ is the number of photons per cm^3 with energy greater than 1.00 MeV, σ_γ is the photoactivation cross section, and c is the speed of light. N_γ is calculated as a function of temperature by integrating the Planck distribution. Inserting the measured value of σ_γ into equation [5] yields the quantity $t^\gamma_{1/2} = (\ln 2)/\lambda_\gamma$. This quantity represents the time required to convert 50% of the initial ^{176}Lu^g to ^{176}Lu^m via photoactivation. If this timescale is short compared to the β-decay half-life of ^{176}Lu^m, then equilibration will occur. Using the measured value for σ_γ, we have calculated the half-life against photoexcitation as a function of temperature. The results of these calculations are shown in Fig. 5. We find that if a 1.00-MeV level were the mediator of the transition from ^{176}Lu^g to

$^{176}Lu^m$, then equilibrium would be guaranteed for $T_8 \geq 4.0$. Below this temperature, $^{176}Lu^g$ will still be destroyed by this process but the rate will be determined by the relatively slow photoexcitation to the 1.00-MeV level.

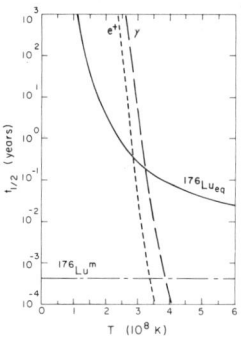

Fig. 5. Calculated half-lives against photoexcitation (— — —), and positron annihilation excitation (- - - - - -) as a function of temperature. The positron annihilation-excitation half-life was calculated for He matter at a density of 1000 g cm^{-3}. Also shown are the effective half-life of $^{176}Lu^g$ as a function of temperature assuming equilibration between $^{176}Lu^g$ and $^{176}Lu^m$ (————). The half-life of $^{176}Lu^m$ which is unaffected by changes in temperature, is also shown for comparison.

Another mechanism by which $^{176}Lu^g$ may be converted to $^{176}Lu^m$ that has recently been observed experimentally is positron annihilation-excitation.[23] In this process in the laboratory, a positron annihilates with a bound atomic electron (most probably a K-shell electron). Then, rather than the usual two-photon annihilation radiation being emitted, it is likely that a single real annihilation photon is emitted and a virtual photon is absorbed by the atomic nucleus producing a nuclear excited state.[24] A possible Feynman diagram of this process for the conversion of $^{176}Lu^g$ to $^{176}Lu^m$ is shown in Fig. 6. The laboratory cross section for this phenomenon is 26±9 μb (Ref. 23) which is nearly a thousand times larger than the photoactivation cross section for ^{60}Co γ-rays. Thus, given a sufficiently large number of positrons, this process could be a very efficient $^{176}Lu^g$ destruction mechanism.

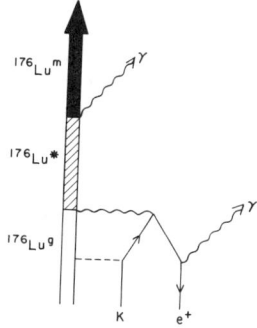

Fig. 6. A possible Feynman diagram for the process of positron annihilation-excitation of $^{176}Lu^g$ to a higher-lying state of intermediate spin which subsequently decays to $^{176}Lu^m$.

The number density of positrons as a function of temperature and density can be easily calculated.[25] In the laboratory, the positron annihilation-excitation process involves bound atomic electrons. In hot stellar environments, annihilations on both bound and free electrons contribute to this reaction rate. However, in order to obtain a conservative estimate for the rate of this reaction, we will consider only annihilation-excitations on bound K-shell electrons. The number of such bound electrons was calculated as a function of temperature and density using the Stromgren[26] formula for Lu atoms embedded in pure He matter. For example, at reasonable s-process conditions of $T_8 = 3.5$ and $\rho = 10^3$ g cm^{-3}, we find that the mean number of bound K electrons, $\bar{n}_1$, for Lu is 0.14 as compared to the normal laboratory value of 2. The rate for positron-annihilation excitation of $^{176}Lu^g$ to $^{176}Lu^m$ is

$$\lambda_+ = \langle N_{e+} \sigma_+^{eff} v_+ \rangle, \qquad [6]$$

where N_{e+} is the number of positrons per cm^3, $\sigma_+^{eff} = (\bar{n}_1/2)\sigma_+^{lab} = (\bar{n}_1)$ 13 μb, and v_+ is the thermal velocity of the positrons. We have calculated the half-life against positron annihilation-excitation as a function of temperature and density. Our results for a density of 1000 g cm^{-3} are also shown in Fig. 5. We conclude that for densities in the range of 10^2-10^4 g cm^{-3}, that this process alone guarantees thermal equilibrium will be established between $^{176}Lu^g$ and $^{176}Lu^m$ for $T_8 \geq 3.5$.

From the results of our experiments and calculations, we find that the processes of photoexcitation and positron annihilation-excitation alone guarantee that $^{176}Lu^g$ and $^{176}Lu^m$ are in thermal equilibrium for $T_8 \geq 3.5$. In our calculations of the photoexcitation rate, we assumed that the lowest-lying level which mediates the conversion of $^{176}Lu^g$ to $^{176}Lu^m$ is located at 1.00 MeV excitation. If subsequent experiments demonstrate the existence of such levels at lower excitation energy, then equilibration of $^{176}Lu^g$ and $^{176}Lu^m$ will occur at even lower temperatures. Other processes such as Coulomb excitation and inelastic neutron scattering will at some level also induce transitions from $^{176}Lu^g$ to $^{176}Lu^m$. Their influence will only strengthen the argument that $^{176}Lu^g$ and $^{176}Lu^m$ are in thermal equilibrium during an s-process that occurs near $T_8 = 3.5$. The conclusion that can be drawn from this work is that the final abundance of $^{176}Lu^g$ that emerges from an s-process environment is a very sensitive function of the thermal history that is has experienced. The fact that a recent estimate for the age of the s-process elements based on the ^{176}Lu chronometer (Ref. 27) yielded an unreasonable age of (39±10) × 10^9 yr provides additional evidence that the abundance of ^{176}Lu has been altered by thermal effects in stellar environments. Thus, ^{176}Lu is not a reliable s-process chronometer.

^{180}Ta

From looking at Fig. 3, it at first appears that the naturally occuring isomer $^{180}Ta^m$ is bypassed in the s-, r-, and p-processes. However, an attractive idea is the suggestion by Beer and Ward[28] that $^{180}Ta^m$ may be produced in the standard s- and r-processes through previously undetected β-decay branches of $^{180}Hf^m$ and ^{180}Lu.

These branches are labelled f_β, f_β', and f_m in Fig. 7. The 5.5 hour, $J^\pi=8^-$ ^{180}Hf isomer lies 204 keV above the long-lived $J^\pi=9^-$ isomer in ^{180}Ta; therefore an allowed Gamow-Teller β-transition should compete with the K-inhibited electromagnetic decays to the ground state rotational band of ^{180}Hf. The ^{180}Hf^m branching ratio to ^{180}Ta^m, f_β, need only by 3.1% to account for the ^{180}Ta abundance by the s-process alone, while log-ft considerations predict a range of from 0.14% to 22%.

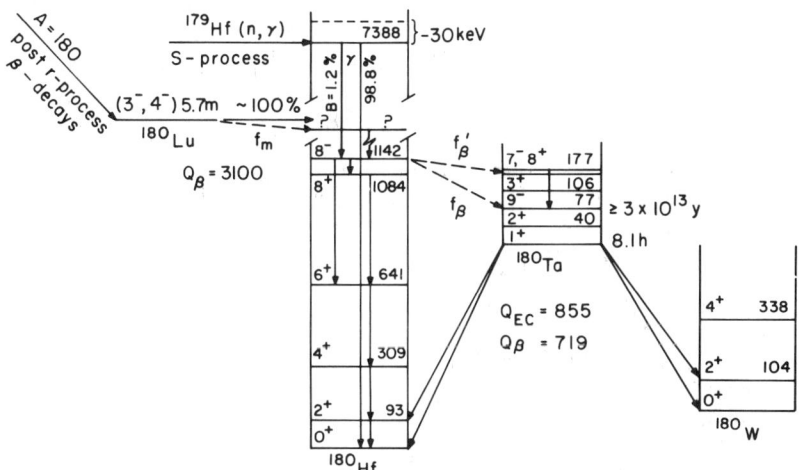

Fig. 7. Partial level and decay schemes of A=180 nuclei. The possible β-decay of ^{180}Lu to a level(s) in ^{180}Hf which subsequently decay(s) to ^{180}Hf^m is labelled f_m. Possible β-decay branches of ^{180}Hf^m to ^{180}Ta are labelled f_β and f_β'.

In 1961, Gallagher et al.[29] established an upper limit of 3.8% on f_β by using a high resolution beta spectrometer to look between the conversion electron lines produced in the electromagnetic decay mode. Our experiment[30,31] attempts to veto the intense conversion peaks by taking advantage of the fact that f_β is a direct transition, while each conversion electron is accompanied by 2-3 γ rays. We first activate enriched ^{179}Hf in the University of Washington reactor to create the ^{180}Hf isomer. The source is then mounted in close geometry to a 500 μm thick surface barrier detector to stop the β's and conversion electrons. This assembly is then sandwiched inside a 4π NaI detector to veto any coincident γ events. Surface barrier spectra are obtained in singles, routing each event according to whether there is a NaI veto or not. After our most recent activation, the conversion lines were still present in the NOVETO route though suppressed by a factor of 100. Most of the counts in the valleys between the peaks are due to backscatter electrons from the conversion line peaks. By taking the ratio of the number of counts in the 120 keV valley of the NOVETO route to the intensity in the K215 conversion line peak in the VETO route, we have obtained a conservative limit of $f_\beta \leq 1\%$.

In addition to the direct branch from $^{180}Hf^m$ to $^{180}Ta^m$, it is energetically possible for $^{180}Hf^m$ to β-decay to the $J^\pi=7^-$ and/or $J^\pi=8^+$ levels known to be located[32,33] near 180-keV excitation energy in ^{180}Ta. Such a possible decay mode is labelled f_β' in Fig. 7. These levels would subsequently decay to the $J^\pi=9^-$ $^{180}Ta^m$ with the emission of a ~100-kev γ-ray. To search for such a decay mode, we replaced the surface barrier detector with a Ge(Li) detector and searched for single γ-rays emitted following the β-decay of $^{180}Hf^m$. In addition to the well-known multiple-γ EM decay mode, we have observed a previously unknown γ ray at 100.8±0.2 keV that decays with the $^{180}Hf^m$ 5.5 hour half-life, but has no other γ rays in coincidence with it. We have most likely observed the decay sequence $^{180}Hf^m(8^-) \xrightarrow{\beta} {}^{180}Ta(8^+) \xrightarrow{\gamma} {}^{180}Ta^m(9^-)$ and deduce a log-ft of 6.56±0.15 for this 103-keV β-transition. The branching ratio, f_β', relative to the EM decay is $(2.6\pm0.3) \times 10^{-4}$ (Ref. 34).

Our limits on f_β and f_β' allow for no more than a ~30% s-process contribution to the solar system abundance of ^{180}Ta. The r-process could account for the difference if $^{180}Hf^m$ is also fed by a small fraction, f_m, of the β-decay of the 5.7 minute ^{180}Lu. To search for such a decay, we bombarded ~1 gram samples of natural Hf metal with the U.W. cyclotron neutron beam (generated by $^9Be(d,n)$) to produce the 5.7 minute ^{180}Lu by the $^{180}Hf(n,p)$ reaction. Also produced was $^{180}Hf^m$ by the $^{180}Hf(n,n')$ reaction with 100 times the cross-section. We therefore extracted the rare earth activities by dissolving the Hf target in HF acid and precipitating out LuF_3 with the use Y as a carrier. A radiochemical Hf reduction factor of 10000 was achieved in less than five minutes. The solution and precipitate were counted with a Ge(Li) detector and searches were made for evidence of ingrowth of $^{180}Hf^m$ decay γ rays as a result of ^{180}Lu decay. No indication of such decays was observed, and by using ^{181}Hf produced by the $^{180}Hf(n,\gamma)$ reaction as a tracer, we measured $f_m \leq 0.06\%$ (Ref. 35). Combining this result with our upper limit on f_β and f_β' implies no more than a 15% r-process contribution to the solar-system abundance of ^{180}Ta.

Our limit on the β-decay branch of ^{180}Lu to $^{180}Hf^m$ seems to stand in contradiction to the recent result of $f_m = (0.46\pm0.15)\%$ of Eschner et al.[36] A possible way to reconcile these two results is the proposal of a short-lived high-spin isomer in ^{180}Lu which decays directly to $^{180}Hf^m$. Such an isomer would have almost completely decayed before our[35] chemical separation was completed, while the technique of Eschner et al.[36] would have been more sensitive to it.

Even if the β branches of the Beer and Ward model[28] or the more recently proposed s-process path of Yokoi and Takahashi[37] prove to be effective $^{180}Ta^m$ production mechanisms, one has to consider the question of the stability of $^{180}Ta^m$ in stellar environments. Norman et al.[38] have shown that photo-deexcitation of $^{180}Ta^m$ to the $t_{1/2} = 8.1$ hour $^{180}Ta^g$ could possibly reduce the effective half-life of the nucleus to the order of one day! If this were the case, it would be impossible to build up any appreciable ^{180}Ta abundance. The possibility of a s- and/or r-process origin of ^{180}Ta would thus be eliminated and one would be forced to reconsider the possibility of a low-temperature p-process or a non-stellar production mechanism to account for the abundance of nature's rarest isotope.

^{187}Re

As can be seen in Fig. 3, the long-lived isotope ^{187}Re ($t_{1/2}$ = 4.3 × 10^{10} year, Ref. 14) is produced only in the r-process. It was therefore suggested[39] as a potential r-process chronometer. Again, however, there are nuclear (and atomic) physics complications which compromise the usefulness of this clock.

In the laboratory, a neutral ^{187}Re atom decays to a neutral ^{187}Os atom with a β-decay endpoint energy of 2.6 keV (Ref. 14). This energy is actually less than the ~14 keV difference in atomic electron binding energy between ^{187}Re and ^{187}Os. Thus, while a neutral ^{187}Re atom is more massive than a neutral ^{187}Os atom, the situation is reversed when one considers the bare nuclei. As discussed previously, at the high temperatures required for both the s- and r-processes, most substances are nearly fully ionized. Thus one must consider the possibilities of the electron-capture decay of ^{187}Os to ^{187}Re, and the bound-state β-decay of ^{187}Re into previously unoccupied atomic orbitals.

The effects of these processes have been studied theoretically by Yokoi et al.[40] The results of their calculations of the rates for the β-decay of ^{187}Re to ^{187}Os and for the electron-capture decay of ^{187}Os to ^{187}Re as a function of temperature (mass coordinate) within a 1 solar-mass star are shown in Fig. 8. It can be clearly seen that there will be competition between these two possible decay modes, and that as a result the final abundances of ^{187}Re and ^{187}Os will be very sensitive to the thermal history they have experienced. Thus ^{187}Re is not a reliable r-process chronometer.

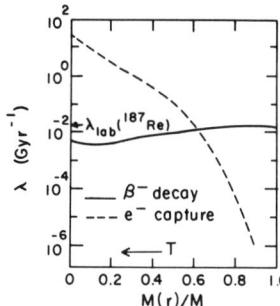

Fig. 8. ^{187}Re β-decay and ^{187}Os electron-capture rates in a main sequence star of 1 solar mass as a function of the mass fraction coordinate. The β-decay rate, $\lambda_{lab}(^{187}Re)$ is indicated for comparison. These calculations are taken from Ref. 40.

CONCLUSIONS

It has been demonstrated that several "exotic" nuclear processes can play significant roles in problems of heavy element nucleosynthesis. The nuclear physics which actually occurs in stellar interiors is considerably more complicated than that which we study in the laboratory but perhaps,
"The fault, dear Brutus, is not in our stars but in ourselves..."

Wm. Shakespeare
Julius Caesar

ACKNOWLEDGEMENTS

This work was supported in part by the U.S. Department of Energy, Division of Nuclear Physics.

REFERENCES

1. G.J. Wasserburg and D.A. Papanastassiou, in Essays in Nuclear Astrophysics, ed. C.A. Barnes, D.D. Clayton, and D.N. Schramm (Cambridge Univ. Press, New York, 1982) p. 77.
2. P.M. Endt and C. van der Leun, Nucl. Phys. A 310, 1 (1978).
3. A.G.W. Cameron and J.W. Truran, Icarus 30, 447 (1977).
4. E.B. Norman, Bull. Am. Phys. Soc. 24, 63 (1979) and Prog. Particle. and Nucl. Phys., Vol. 6, ed. D. Wilkinson (Pergamon, Oxford, 1981) p. 285.
5. R.A. Ward and W.A. Fowler, Astrophys. J. 238, 266 (1980).
6. E.B. Norman, T.E. Chupp, K.T. Lesko, and P.J. Grant, Astrophys. J. 251, 834 (1981).
7. S.E. Woosley, W.A. Fowler, J.A. Holmes, and B.A. Zimmerman, Caltech preprint OAP-422 (1975) p. A20.
8. S.E. Woosley, W.A. Fowler, J.A. Holmes, and B.A. Zimmerman, Atomic Data Nucl. Data Tables 22, 371 (1978).
9. E.B. Norman, K.T. Lesko, T.E. Chupp, and P. Schwalbach, Nucl. Phys. A 357, 228 (1981).
10. L. Buchmann et al., Nucl. Phys. A 415, 93 (1984).
11. M. Arnould, H. Nørgaard, F.K. Thielemann, and W. Hillebrandt, Astrophys. J. 237, 931 (1980).
12. P.W. Merrill, Astrophys. J. 116, 21 (1952) and Pub. Astron. Soc. Pacific 68, 356 (1956).
13. V.V. Smith and G. Wallerstein, Astrophys. J. 273, 742 (1983).
14. C.M. Lederer and V.S. Shirley, Table of Isotopes, 7th ed. (Wiley, New York, 1978).
15. R.K. Ulrich, in Essays in Nuclear Astrophysics, ed. C.A. Barnes, D.D. Clayton, and D.N. Schramm (Cambridge Univ. Press, New York, 1982) p. 301.
16. J. Audouze, W.A. Fowler, and D.N. Schramm, Nature 238, 8 (1972).
17. M. Arnould, Astron. Astrophys. 22, 311 (1973).
18. K. Yoshihara, Isotopes and Radiation 3, 472 (1960).
19. A. Veres and I. Pavlicsek, Acta. Phys. Acad. Sci. Hung. 28, 419 (1970).
20. E.B. Norman, T. Bertram, S.E. Kellogg, S. Gil, and P. Wong, Astrophys. J. (submitted).
21. A. Ljubicic, K. Pisk, and B.A. Logan, Phys. Rev. C 23, 2238 (1981).
22. M. Crcmar, A. Ljubicic, K. Pisk, B.A. Logan, and M. Vrtar, Phys. Rev. C 25, 2097 (1982).
23. Y. Watanabe, T. Mukoyama, and R. Katano, Phys. Rev. C 23, 695 (1981).
24. R.S. Raghavan and A.P. Mills, Phys. Rev. C 24, 1814 (1981).

25. D.D. Clayton, Principles of Stellar Evolution and Nucleosynthesis (McGraw-Hill, New York, 1968) p. 274.
26. B. Strömgren, Zs. Astrophys. 4, 118 (1932).
27. G. Schatz, Proc. 2nd Workshop on Nucl. Astrophys., Ringberg Castle, MPA Report 90 (1983) p. 71.
28. H. Beer and R.A. Ward, Nature 291, 308 (1981).
29. C. Gallagher, M. Jorgensen, and O. Skilbreid, Nucl. Phys. 33, 285 (1962).
30. S.E. Kellogg and E.B. Norman, Bull. Am. Phys. Soc. 27, 705 (1982).
31. E.B. Norman and S.E. Kellogg, Proc. 2nd Workshop on Nucl. Astrophys., Ringberg Castle, MPA Report 90 (1983) p. 92.
32. E. Warde et al., Phys. Rev. C 27, 98 (1983).
33. R.A. Dewberry and R.A. Naumann, Phys. Rev. C 28, 2259 (1983).
34. S.E. Kellogg and E.B. Norman, Bull. Am. Phys. Soc. (submitted for 1984 Nucl. Div. Meeting).
35. S.E. Kellogg and E.B. Norman, Bull. Am. Phys. Soc. 28, 989 (1983).
36. W. Eschner et al., Zs. Phys. A 317, 281 (1984).
37. K. Yokoi and K. Takahashi, Nature 305, 198 (1983).
38. E.B. Norman, S.E. Kellogg, T. Bertram, S. Gil, and P. Wong, Astrophys. J. 281, 360 (1984).
39. D.D. Clayton, Astrophys. J. 139, 637 (1964).
40. K. Yokoi, K. Takahashi, and M. Arnould, Astron. Astrophys. 117, 65 (1983).

DYNAMIC STELLAR NEUTRON CAPTURE NUCLEOSYNTHESIS: THE NEED FOR MORE NUCLEAR DATA FOR THE s-PROCESS

G. J. Mathews, W. M. Howard, K. Takahashi, and R. A. Ward
Lawrence Livermore National Laboratory, Livermore, CA 94550

ABSTRACT

We summarize results from a detailed parameter study of the s-process in models which produce an exponential distribution of exposures by sequential irradiations and dredge up in the stellar environment. The calculations are based on a complete network of measured and calculated neutron capture and beta-decay rates as well as estimates for their temperature dependence. In the framework of these models we identify and systematically vary the astrophysical variables which affect the observed solar-system σN (cross section times abundance) curve. Constraints are placed on the s-process neutron exposure and flux as well as the temperatures, densities, neutron pulse shape and inter-pulse period. The results also highlight important needs for better nuclear data in various mass regions.

INTRODUCTION

Historically the s-process has been associated with red giant stars[1]. From the begining, however, it has been clear[1-3] that a single neutron exposure could not produce the observed behavior of the solar-system s-process σN curve (neutron capture cross section times abundance) as a function of atomic mass. An exponential distribution of exposures was found to give a good fit[3] and was at first presumed to be the result of reprocessing during galactic chemical evolution.

Ulrich[4,5] showed that an exponential distribution of exposures could be achieved in a single star which subjected initial seed material to periodic neutron exposures followed by dredge up of some fraction of the irradiated material to the stellar surface. This s-process scenario has been explored in a series of papers[6-11] based on a $^{22}Ne(\alpha,n)^{25}Mg$ neutron-source s-process during thermal pulses of asymptotic giant branch stars. There is sufficient uncertainty in the stellar models, however, that a different approach is warranted, i.e. to utilize the observed solar-system s-process σN curve to define the constraints on any stellar model for the s-process. This is the subject of the present work. This study reveals that the s-process neutron-capture flow is probably considerably more complicated than the continuous chain assumed in the classical s-process. Deviations of the best fit from the observed solar-system values highlight important needs for improved nuclear data.

S-PROCESS PARAMETERS

In the classical s-process (without beta-decay branching) the abundance of an isotope is given simply by the solution to the set of coupled differential equations,

$$dN_A/d\tau = \sigma(A-1)N_{A-1} - \sigma(A)N_A \quad , \tag{1}$$

where τ is the time integrated neutron flux (i.e. neutron exposure) and σ the Maxwellian averaged neutron capture cross section.

The implication of Eq. (1) is that at equilibrium, a single exposure would lead to a constant σN value for all isotopes in the s-process. This is not observed in the solar-system σN curve (see Fig. 1). The data require a distribution of exposures[2]. A good fit can be obtained[3] with an exponential distribution of neutron exposures, i.e. the probability, $\rho(\tau)$ for a given exposure is taken to be,

$$\rho(\tau) \propto \exp(-\tau/\tau_0) \quad . \tag{2}$$

This exponential distribution of exposures leads to a particularly simple solution[12] for Eq. (1),

$$\sigma(A)N(A) = \frac{\sigma(A-1)N(A-1)}{[1 + 1/(\tau_0\sigma(A))]} \quad , \tag{3}$$

where, τ_0 is the mean neutron exposure from Eq. (2).

In large part the identification of the astrophysical site for the s-process has been the search for an environment which would produce the value τ_0 (or equivalently the number of neutron captures per seed) necessary to reproduce the solar-system σN curve.

In a periodic s-process operating in a single star[4,5], the mean exposure, τ_0, is simply related to the average exposure per pulse, $\Delta\tau$, and the fraction of material, r, which remains after each dredge up. That is, after n exposures, the fraction of material, $\rho(\tau)$, which will have experienced an exposure of $\tau = n\Delta\tau$ will be $r^n = r^{\tau/\Delta\tau}$. Thus, the mean exposure is simply, $\tau_0 = -\Delta\tau/\ln(r)$. Note, that the mixing fraction and exposure per pulse are not independent parameters. For any fraction, r, an average exposure per pulse can be choosen such that the σN curve can be fit. In our calculations, therefore, only the mean exposure is varied in the fit. We have set $r = 0.285$ which corresponds to the fraction remaining after dredge up for the thermal-pulse model with a $1.16 M_\odot$ CO core[8].

In realistic models for the s-process, unstable nuclei have the opportunity to capture a neutron before beta decay, and hence, lead to a branch point along the s-process chain. It has long been appreciated[1] that such branch points are valuable as a constraint on the average neutron flux for the s-process since the neutron-capture flow in the s-process (and therefore the σN values) will depend upon the competition between neutron capture and beta decay[5,13,14].

Once a given exposure is determined, the dynamics of the variation of neutron density with time will affect the σN curve, i. e., a rapidly decreasing flux may produce significantly different nucleosynthesis than a constant-flux for the same neutron exposure. For the purposes of the present study we parameterize the flux history with the analytical form suggested in the thermal pulse models[8], i. e. we write for the flux, Φ_n, as a function of time, t,

$$\ln(\Phi_n) = (\Delta\tau/t_0)\exp(-t/t_0) \qquad (4)$$

where $\Delta\tau$ is the average exposure per pulse. Variations of the decay constant, t_0, characterize the intensity of the peak neutron exposure.

In addition to the flux, beta decay branch points will also depend on the temperature since thermal population of excited states can considerably alter beta decay rates[13,15,16] and even induce beta decay in nuclei which are stable at T=0. Since the temperature does not vary much (after an initial rapid increase) during an individual pulse in the stellar models[8], we assume various constant temperatures during the pulse in this parameter study. We assume an interpulse temperature[8] of kT = 14 keV.

In principle, the s-process abundances could also depend upon whether long-lived unstable nuclei produced at the end of a pulse have time to decay before the next pulse. In the limit of a short interpulse period, this could make a difference in the final abundances. In this parameter study we investigate the extreme limits of zero and infinite time between pulses to study the effects of the interpulse period on the final abundances.

CALCULATION

In dynamic environments such as these the simplicity of the classical s-process (Eq. 1) is lost due to a break down of the assumption that neutron captures are slow compared to beta decay. The only way to compute the nucleosynthesis is with a network:

$$\frac{dN(Z,A)}{dt} = N(Z,A-1)\Phi_n\sigma_{n,\gamma}(Z,A-1) + N(Z-1,A)\lambda_\beta(Z-1,A)$$
$$- N(Z,A)[\Phi_n\sigma_{n,\gamma}(Z,A) + \lambda_\beta(Z,A)] . \qquad (5)$$

The time-dependence of the flux and temperature dependence of the cross sections and beta rates are also included. In a few cases, positron, electron-capture, or alpha decay must also be added to Eq. (5).

Experimental neutron-capture cross sections were used when available. These were extracted from several recent reviews[5,14,16,17] as well as more recent measurements[18-23].

Cross sections for unstable or unmeasured nuclei were taken from Hauser-Feshbach estimates[24]. The temperature dependence of all cross sections was taken from ref. 24 and normalized to the experimental kT = 30keV Maxwellian averaged cross sections where known. The Hauser-Feshbach calculations should give a good representation of the temperature dependence of the cross sections (including capture to excited states) over the range of temperatures of interest in this work[23].

Ground-state beta decay rates were taken from experiment. Decay rates from thermally populated excited states were calculated assuming thermal equilibrium and appropriately choosen ft values[25]. The initial seed abundances were taken from solar-system values[26].

RESULTS

We have adjusted each of the parameters to minimize χ^2 for the fit to 23 s-only nuclei with $Z \geq 40$. The value for ^{187}Os was omitted from the list of s-only nuclei because of the unknown contribution from ^{187}Re decay. Lighter nuclei were also omitted because of possible contribution from other sources[15].

In Figure 1 we determine the mean exposure for the s-process in the classical model, i.e. a constant ($t_0 = \infty$) low flux ($\Phi_n = 2 \times 10^{14} cm^{-2} sec^{-1}$) switched off after an average exposure, $\Delta\tau$, such that the neutron-capture time scale is long ($\sim 10^3$ yrs.) at a standard s-process temperature of kT = 30 keV.

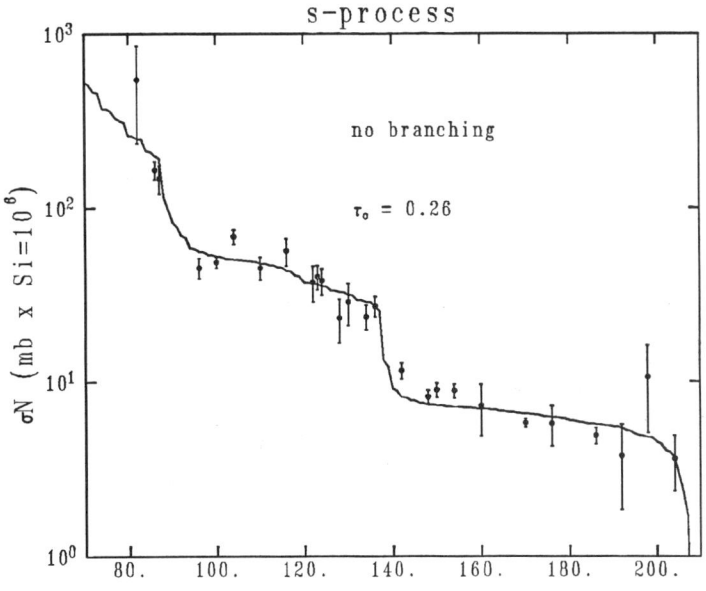

A

Fig. 1. Classical s-process fit to the observed σN curve.

The interpulse period is taken as long so that unstable nuclei decay between pulses. A good fit is obtained with $\tau_0 = 0.26\pm0.01$ mb^{-1}. In the next fit (Fig. 2) we have varied Φ_n and τ_0. Note that the the character of the curve is considerably changed and the quality of the fit is improved. The χ^2 for the fit gradually decreases as the flux is increased until an optimum flux, $\Phi_n = 3.2\times10^{16}$ cm^{-2}sec^{-1}. The quality of the fit then quickly deteriorates for higher fluxes. From this we conclude that the constant fluxes consistent with the observed σN curve for s-only nuclei must be in the range of $\Phi < 4\times10^{16}$ cm^{-2}sec^{-1}. The best fit value of 3.2×10^{16} cm^{-2}sec^{-1} implies a neutron density of $n_n = 1.3\times10^8$ cm^{-3} for kT = 30 keV. This is consistent with previous estimates[5,21].

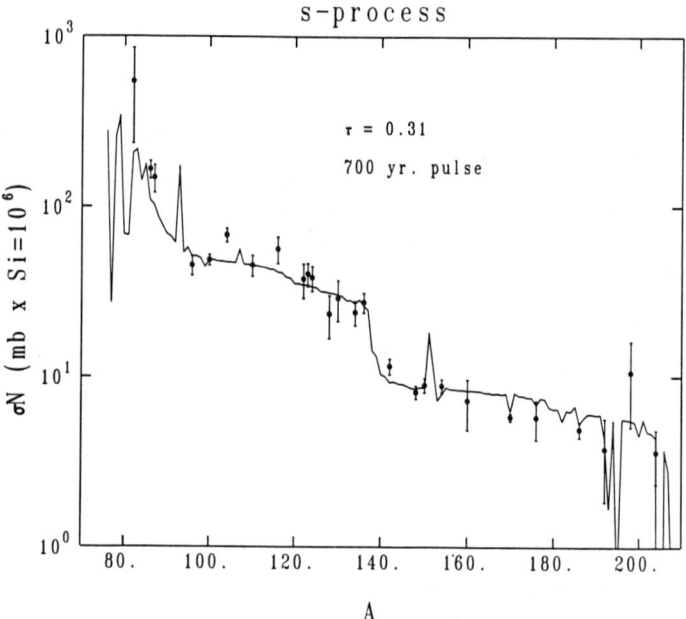

Fig. 2. Fit to σN curve when both flux and mean exposure are varied.

At the present time, the optimization of the fit with respect to the temperature and pulse shape is still in progress. The basic features of Fig. 2 are not significantly affected by these parameters however. The best fit for the solar-system σN curve always exhibits pronounced deviations from the smooth monotonically decreasing classical s-process curve. The reason for this is that many stable nuclei are actually produced as beta-unstable progenitors due to neutron captures on beta-unstable nuclei. If the progenitor has a larger cross section there will be a dip in the σN curve when the cross section for the stable daughter is used. Similarly, if the cross section for the unstable nucleus is smaller there will be a peak. This is the reason for the pronounced peaks on Fig. 2 at ^{93}Nb (produced as ^{93}Zr) and ^{151}Eu (significantly produced as ^{151}Sm).

One check on the importance of beta branching is the effect of these branches on the r-process abundances. A number of isotopes are produced which are normally considered to be produced only in the r-process. It is an additional constraint on the s-process, therefore, not to indroduce unrealistic structure into the r-process abundance curve, and in particular, not to overproduce the r-process isotopes. The new r-process abundance curve is shown in figure 3. The gaps in the abundances correspond to overproduction of isotopes by the s-process. The gap at A = 200, however, is probably due to the large uncertainty in the elemental abundance of Hg and may not be significant. The minimum at A = 182 arises to some extent from the production of ^{182}W as ^{182}Ta and ^{182}Hf and is probably a real minimum in the r-process curve. The gap at A = 137,138 is due to the subtraction of large numbers at the ^{138}Ba s-process peak and is probably not significant. The gaps at A = 92 and 94 are due to an s-process overproduction of Zr. It does not appear that this is due to a problem with the s-process flow through ^{93}Zr since ^{93}Nb is not overproduced. This descrepancy may indicate that the Zr cross sections or meteoritic solar-system abundance are too low.

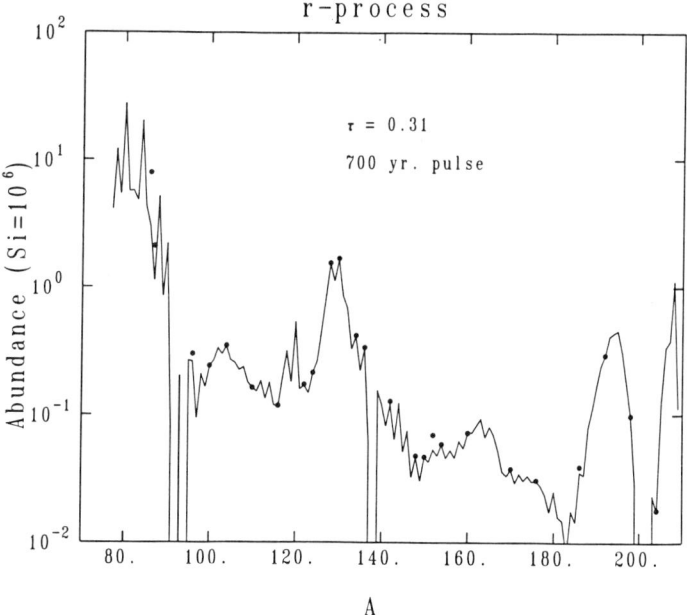

Fig. 3 The r-process abundances from a subtraction of the s-process abundances from the solar-system abundances. The points are abundances of nuclei normally regarded as r-only.

For the purposes of this conference it is important to point out the nuclear data needs which are evident from this study. For one, there are still a number of stable nuclei along the s-process path for which neutron-capture cross sections are not measured. These include, ^{66}Zn, ^{72,73}Ge, ^{77}Se, ^{83,84}Kr, ^{99}Ru, $^{128-131}$Xe, $^{192-195}$Pt, and $^{198-201}$Hg. Amoung these isotopes ^{192}Pt and ^{198}Hg are particularly important s-only nuclei. The σN value for ^{192}Pt is predicted to be substantially less than the σN value for ^{198}Hg (see Fig. 2) due to branching at ^{192}Ir and ^{191}Os. The effect on the σN value of ^{204}Pb due to branching at ^{204}Tl and ^{203}Tl is also important, and would be much clearer if the uncertainties in the ^{204}Pb cross section could be reduced.

A second point is that there are several s-process only isotopes with measured neutron capture cross sections which can not be brought into agreement (within one standard deviation) with the solar-system s-process σN curve for any variation of parameters. The most notable descrepancies are for ^{104}Pd, and ^{116}Sn. The uncertainty in the abundances for these nuclides is probably not very large[25]. Therefore, the nuclear data for these nuclei and neighboring isobars ought to be reexamined. The descrepancies in the light nuclei, ^{76}Se and ^{82}Kr, are probably due to a contribution from a lower mean exposure[15]. There are also several s-only nuclei which may exhibit important structure but the cross sections uncertainties are too large, particularly for ^{160}Dy and ^{176}Hf.

CONCLUSION

We have identified the model parameters which characterize the solar-system s-process abundances. These parameters have been varied to find optimum values which should be taken as constraints any stellar model for the s-process. We find that the s-process is mostly sensitive to the mean exposure, τ_0, and the neutron flux. The best fit σN curve suggests an s-process environment in which beta-branching induces significant changes in the s-process curve. The deviations of this best fit from the solar-system σN curve indicate the needs for improved nuclear data.

Work performed under the auspices of the U.S. Department of Energy by the Lawrence Livermore National Laboratory under contract number W-7405-ENG-48.

REFERENCES

1. G. R. Burbidge, E. M. Burbidge, W. A. Fowler, and F. Hoyle, Rev. Mod. Phys., 29, 547 (1957).
2. D. D.Clayton, W. A. Fowler, T. E. Hull, and B. A. Zimmerman, Ann. Phys., 12, 331 (1961).
3. P. A. Seeger, W. A. Fowler, D. D. Clayton, Astrophys. J. Suppl., 97, 121 (1965).

4. R. K. Ulrich, in "Explosive Nucleosynthesis", eds. D. N. Schramm and W. D. Arnett (Univ. Texas Press, Austin, 1973), p. 139.
5. R. K. Ulrich, in "Essays in Nuclear Astrophysics", eds. C. A. Barnes, D. D. Clayton, and D. N. Schramm (Cambridge Univ. Press, N. Y., 1982) p. 301.
6. I. Iben, Jr., Astrophys. J., 196, 525 (1975).
7. I. Iben, Jr., Astrophys. J., 196, 549 (1975).
8. I. Iben, Jr., Astrophys. J., 217, 788 (1977).
9. J. W. Truran, and I. Iben, Jr., Astrophys. J., 216, 197 (1977).
10. I. Iben, Jr., and J. W. Truran, Astrophys. J., 220, 980 (1978).
11. K. R. Cosner, I. Iben, and J. R. Truran, Astrophys. J. Lett., 238, L91 (1980).
12. D. D. Clayton, R. A. Ward, Astrophys. J., 193, 397 (1974).
13. R. A. Ward, M. J. Newman, and D. D. Clayton, Astrophys. J. Suppl., 31, 33 (1976).
14. B. J. Allen, J. H. Gibbons, and R. L. Macklin, Adv. Nucl. Phy., 4, 205 (1971).
15. F. Käppeler, H. Beer, K. Wisshak, D. D. Clayton, R. L., Macklin, and R. A. Ward, Astrophys. J., 257, 821 (1982).
16. A. G. W. Cameron, Astrophys. J., 210, 489 (1955).
17. M. J. Newman, Astrophys. J., 219, 676 (1978).
18. H. Beer and R. L. Macklin, Phys. Rev., C26, 1404 (1982).
19. H. Beer, F. Käppeler, G. Reffo, and G. Ventorini, Astrophys. Space Sci., 97, 95 (1983).
20. R. R. Winters, F. Käppeler, K. Wisshak, A. Mengoni, and G. Reffo, (submitted to Astrophys. J., 1984).
21. H. Beer, F. Käppeler, K. Yokoi, and K. Takahashi, Astrophys. J., 278, 388 (1984).
22. H. Beer and G. Walter, Astron. Astrophys. (1984 in press).
23. G. J. Mathews, and F. Käppeler, Astrophys. J., 286, (1984 in press).
24. J. A. Holmes, S. E., Woosley, W. A. Fowler, and B. A. Zimmerman, Atom. Nucl. Data Tables, 18, 306 (1976).
25. K. Takahashi and K. Yokoi (to be published).
26. E. Anders and M. Ebihara, Geochim. Cosmochim. Acta, 46, 2263 (1982).

NEUTRON CAPTURE AND TOTAL CROSS SECTIONS FOR ^{48}Ca: ASTROPHYSICAL IMPLICATIONS*

R. F. Carlton
Middle Tennessee State University, Murfreesboro, TN 37132

J. A. Harvey, N. W. Hill and R. L. Macklin
Oak Ridge National Laboratory, Oak Ridge, TN 37831

ABSTRACT

Attempts to understand abundance anomalies of the Ca isotopes in the Allende meteorite via the nβ-process require 30-keV Maxwellian averaged capture cross sections. Experimental data on ^{48}Ca in this energy region of astrophysical significance is important since a single resonance in this vicinity could dominate the capture cross section. Neutron capture and total cross section measurements have been performed at ORELA on a 9.97-g sample of $CaCO_3$, enriched to 96% ^{48}Ca, over the energy ranges 10 eV-500 keV (σ_γ) and 10 keV-4 MeV (σ_T). Only two small resonances were found for ^{48}Ca in the 30-keV energy region (at 19.3 and 106.9 keV) and those only in capture. Their contribution to the 30-keV-averaged cross section is only 50 μb compared to 1.0 mb calculated from direct capture. The p-wave strength and large $d_{5/2}$ strength observed above 150 keV do not contribute significantly to the 30 keV capture.

INTRODUCTION

Allende inclusion (EK-1-4-1) manifests correlated Ca and Ti anomalies in isotopic abundance ratios when compared to solar system values. The most successful attempt[1] at understanding these anomalies is the generalized neutron capture/beta decay (nβ) process wherein both mechanisms are effective. The β-decay rates are not assumed to be either fast or slow compared to the neutron capture rate. This process, however, is not successful in explaining anomalies of other inclusions in the Allende such as C-1. Furthermore, substantial modification (factor of 10) of the global Hauser-Feshbach-based capture cross section[2] for ^{49}Ca and ^{46}K are required to explain the Ca-Ti anomalies. In this regard the capture cross section of ^{48}Ca has been measured[3] at thermal energies in a reactor and the contribution at stellar temperatures (kT=30-keV) is in excellent agreement with (HF) results. A single resonance, however, in this vicinity could dominate the direct capture contribution and render arguments[1] with regard to ^{49}Ca(n,γ) unjustifiable on the basis of nuclear structure in this region. (See Ref. 1 for a complete description of the nβ process applied to Allende (EK-1-4-1) and for arguments justifying the increase in capture cross section required to explain the observed anomalies.)

*Research sponsored by the Division of Basic Energy Sciences, U.S. Department of Energy, under Contract Nos. DE-AC05-84OR21400 with the Martin Marietta Energy Systems, Inc. and DE-AS05-80ER10710 with Middle Tennessee State University.

EXPERIMENT AND DATA ANALYSIS

Neutron capture cross sections for ^{48}Ca were measured at ORELA using a sample of $CaCO_3$, enriched to 96% ^{48}Ca (3.83% ^{40}Ca impurity). The sample thickness was 126 b/atom. Neutron energies were inferred by time of flight. Experimental parameters for both capture and transmission measurements are summarized in Table I. Capture measurements were performed at 40-m from the photoneutron source. Gamma rays were detected with non-hydrogeneous liquid scintillators (C_6F_6) and their energies were calculated by pulse-height weighting.[4] Resonances were fitted to a Breit-Wigner form including Doppler broadening, resonance self protection, multiple scattering, and both Gaussian and exponential resolution functions.

Table I. Experimental Parameters

Quantity	Capture	Transmission
Rep. Rate	400-800 Hz	800 Hz
Pulse Width	7 ns	9 ns
Flight Path	40 m	200 m

Transmission measurements were performed at 200 m. Neutrons were detected by proton recoil detectors. The neutron energy resolution over the energy range 100-4000 keV varied from 0.02% to 0.09%. Resonance parameters were determined by least squares adjustment using a multilevel R-matrix code including Doppler and resolution broadening. Results are shown in Fig. 1.

RESULTS

A capture run below 1 keV neutron energy revealed no resonances. All but two of the resonances observed in the 2-200 keV range were due to the 25 times less abundant ^{40}Ca! The two attributed to ^{48}Ca are shown in Fig. 2. Since they were not observed in transmission, only their capture kernels were determined. Least square fits at the positions of four resonances observed in transmission in the 300-500 keV range indicate reasonable radiative widths for three and an upper limit for the fourth. The total Maxwellian averaged ^{48}Ca(n,γ) cross section (a), together with the resonance-only contribution (b), is shown in Fig. 3. The latter amounts to only 4.5% at 30 keV.

No resonances were observed in transmission below 100 keV, the first resonance appearing at 162 keV. R-matrix fits to the carbon- and oxygen-subtracted total cross section of ^{48}CaCO$_3$ from 100-500 keV are shown in Fig. 1.

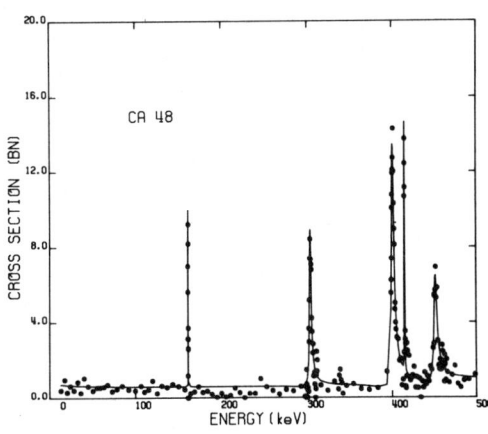

Fig. 1. σ_T with R-matrix fit below 500 keV.

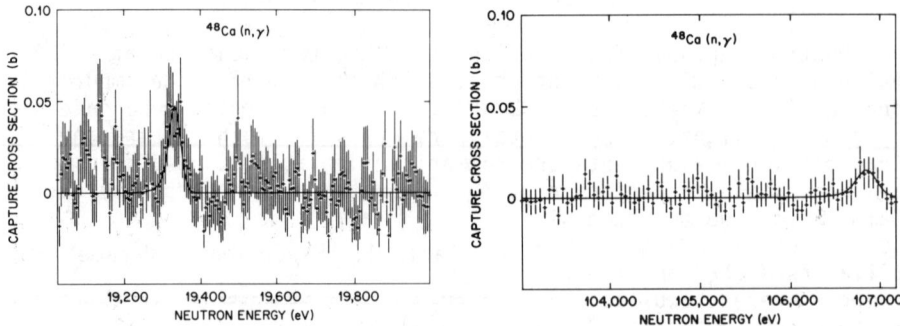

Fig. 2. ^{48}Ca capture resonances below 200 keV.

Resonance parameters based on the combined analyses are presented in Table II. The neutron widths were deduced for an assumed radiation width of 1.0 eV. Transmission data were analyzed up to 4 MeV and no evidence of s-wave strength which might contribute to the 30 keV cross section was found.

Table II. ^{48}Ca(n,γ) Resonance Parameters.

Neutron Energy (Lab-keV)	$g\Gamma_n\Gamma_\gamma/\Gamma$ (eV)	Γ_n (keV)	Γ_γ (eV)	J^π
19.3	0.0095 (14)a			
106.9	0.124 (24)			
161.4	0.15 (10)	0.059 (6)	0.15 (10)	(1/2)
303.6	0.30 (30)	2.50 (16)	0.3 (3)	1/2$^-$
401.2	1.6 (6)	3.33 (13)	0.8 (3)	3/2$^-$
415.5	1.5 (4)	0.38 (4)	0.5 (1)	5/2$^+$
450.4	2.5 (10)	4.30 (28)	2.5 (10)	1/2$^-$

aThe uncertainty in the last digit(s) indicated in parentheses are derived from the statistical standard deviations of the capture data or transmission. Systematic uncertainties are much smaller, not exceeding 4% for capture and 1% for transmission.

DISCUSSION

From Fig. 3 we obtain a value of (1.05±0.13) mb for the Maxwellian averaged capture cross section of ^{48}Ca at 30 keV. This result differs little from the value deduced solely from thermal measurements and is in substantial agreement with the value (1.1 mb) used in calculations for nβ processing.[2]

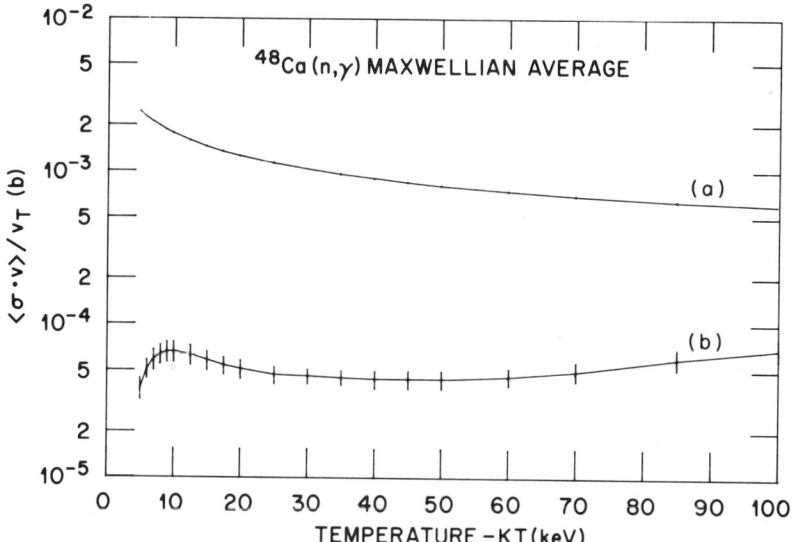

Fig. 3. Total capture cross section (a) and resonance-only contribution (b).

1. D. G. Sandler, S. E. Koonin, and W. A. Fowler, Astrophys. J. 259, 908 (1982).
2. S. E. Woosley, W. A. Fowler, J. A. Holmes, and B. A. Zimmerman, Atomic Data and Nuclear Data Tables 22, 371 (1978).
3. F. P. Cranston and D. H. White, Nucl. Phys. A169, 95 (1971).
4. J. H. Gibbons and R. L. Macklin, Phys. Rev. 159, 1007 (1967).

THE Dy163-Ho163 BRANCHING: AN s-PROCESS BAROMETER

H. Beer and G. Walter
Kernforschungszentrum Karlsruhe, IK III,
P.O.B. 3640, D-7500 Karlsruhe 1, FRG

R.L. Macklin
Oak-Ridge National Laboratory, P.O.B. X, Oak-Ridge, TN 37831

ABSTRACT

The neutron capture cross sections of Dy163 and Er164 have been measured to analyse the s-process branching at Dy163 - Ho163. The reproduction of the s-process abundance of Er164 via this branching is sensitive to temperature kT, neutron density, and electron density n_e. The calculations using information from other branchings on kT and the neutron density n_n give constraints for n_e at the site of the s-process.

INTRODUCTION

In order to obtain information about the site of the slow neutron capture nucleosynthesis (s-process) which means knowledge of the astrophysical parameters of temperature, neutron flux and mass density branchings of the synthesis path at beta unstable nuclei have to be examined. The Dy163 - Ho163 branching is unique as it provides information on all three important astrophysical parameters: s-process temperature, and neutron and electron density. The sensitivity to the electron density is especially interesting as it allows for a calculation of the mass density in the He-shell, the presumed site of the s-process.

MEASUREMENTS

The neutron capture cross section measurements on Dy163 and Er164 were carried out at the Oak-Ridge Linear Accelerator (ORELA) and the Karlsruhe 3.75 MV pulsed Van de Graaff (VDG). The Dy163 cross section measurement was determined using the time-of-flight technique. The measurements were performed both at ORELA and the VDG. The excellent agreement of the two data sets is a secure check of systematic uncertainties (Fig.1). The Er164 cross section was determined with the activation technique at the VDG by counting of the K x-ray activity of Er165 (10.3h) relative to the activity of the 412 keV gamma line of Au198 (2.69d). The gamma activity was recorded with a gamma-x Ge-detector (efficiency: 7.2% ; resolution at 1.33 MeV: 1.6 keV) with good response at the very low x-ray energy of Ho as well as at the energy of the Au198 line. More details about the time-of-flight facilities and the activation technique can be found elsewhere[1,2]. The sample characteristics and the results are summarized in Tables I-II.

ANALYSIS

Under the stellar s-process conditions Dy163, a stable isotope on the synthesis path becomes radioactive so that competion between beta decay and neutron capture allows for an s-process synthesis of Er164 via neutron capture of Ho163 and subsequent beta decay of Ho164 to Er164. The surprising fact that Dy163 in stellar interiors might be radioactive is due to the small Q-value between Ho163 and Dy163 of only 2.3 keV[3,4] which makes possible bound state beta decay and facilitates excited state beta decay. The dependency of Dy163, Ho163, and Ho164 half-lives on temperature kT and electron density has been treated in ref.[5,6] In order to reproduce the Er164 s-process abundance (solar minus p-process abundance estimated via Er162) the half-lives of Dy163, Ho163, and Ho164 and the capture cross sections of Dy163, Ho163 and Er164 are the indispensable input parameters for our branching calculation. As the astrophysical parameters, temperature and neutron density, are known from other branchings[1] the electron density could be adjusted to yield the s-process abundance of Er164. Fig.2 shows the σ-times-abundance Ns curve at the site of the Dy163 branching to illustrate our calculations. Fig.3 gives the range of allowed values for kT, n_n and n_e.

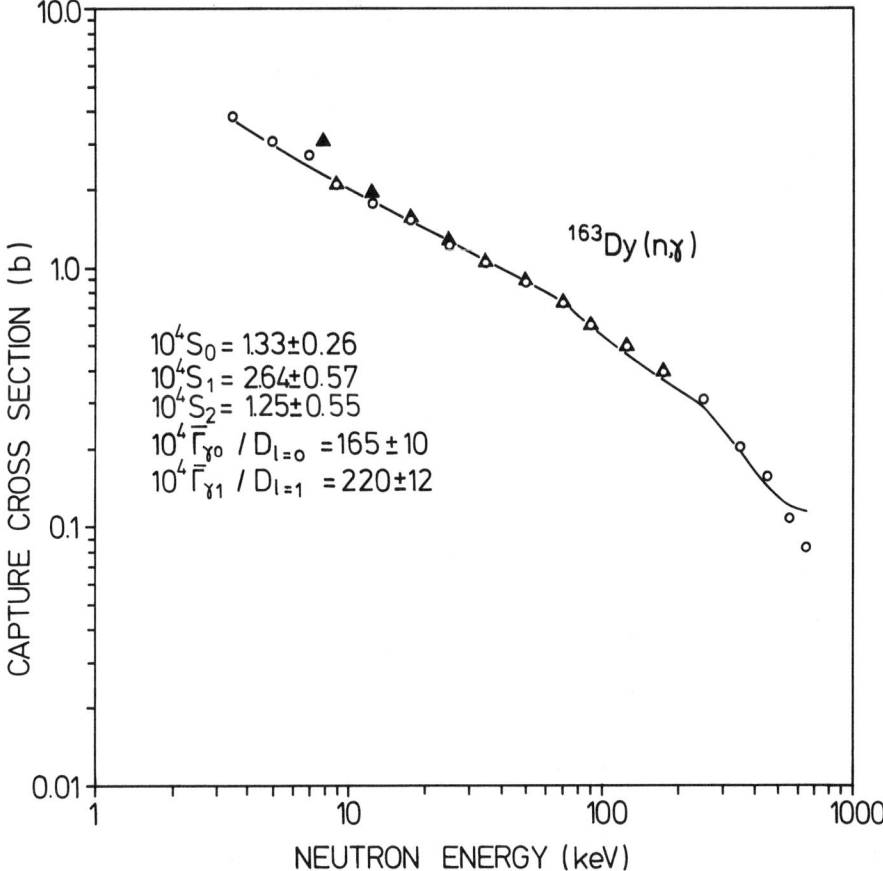

Fig.1 Effective cross section for Dy163(n,γ). The curve is a statistical model fit to the data[8] with the indicated values for s-,p-,and

d-wave strength functions, radiation widths and level spacings. Circles are data from ORELA, triangles from VDG.

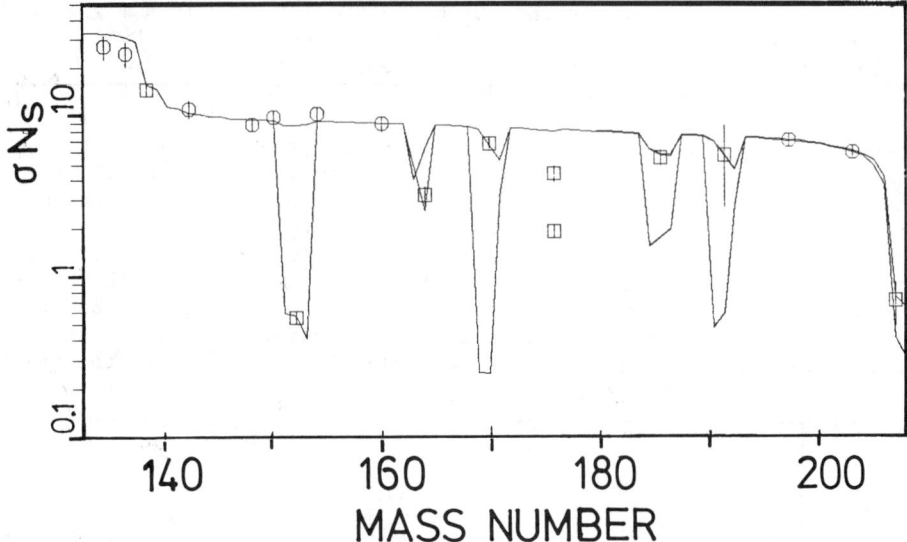

Fig.2 The product of capture cross section σ and s-process abundance Ns as a function of mass number. The symbols are empirical data. The solid line is the s-process calculation with the exponential distribution of neutron exposures. At Pb and Bi an additional exposure component has been added.[9] At the branchings kT, n_n, and n_e are adjusted to reproduce Gd152, Er164, Os186, and Pt192. Only in the case of Er164 is the branching sensitive to the electron density n_e.

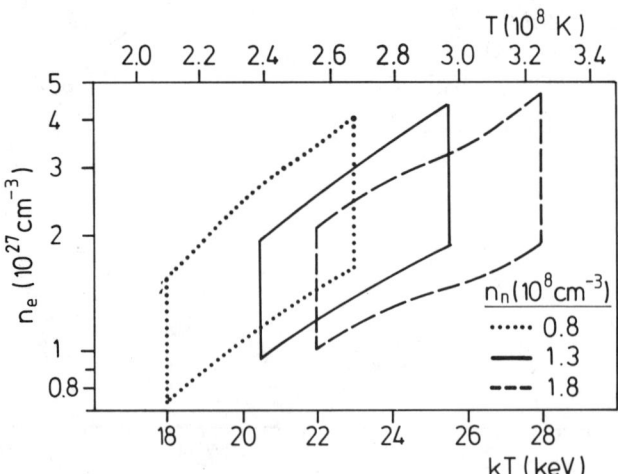

Fig.3 The curves circumscribe allowed ranges of n_e and kT for three different neutron densities. The variation of n_e is due to the various uncertainties in the Dy163-Ho163 branching and the range of kT= 18-28 keV and n_n=(0.8-1.8) 10^8/cm^3 from the other branchings.[1]

CONCLUSIONS

There is a simple relation between electron density and the mass density ρ in the He-shell of a red giant star which is the presumed site of s-process nucleosynthesis. We can assume that He is totally ionized and the electrons are still not degenerate. Therefore $n_e = \frac{1}{2}\rho N_o$ (No: Avogadro's number) is a measure of the mass density in the He-shell. The mass densities found (2600-13000 g/cm^3) are higher than currently assumed (2000 g/cm^3), but still compatible for example with Ulrich's s-process model of a 5$M_\odot$ star[7] ($\rho \leq 3500$ g/cm^3).

Table I Sample characteristics and Maxwellian averaged capture cross section σ at the thermal energy kT=30 keV.

Sample	Diameter (mm)	Weight (mg)	% Isotopic composition	σ(mb) at kT=30 keV
Dy$_2$O$_3$	15	907.39	96.85 Dy163	1153±44 (VDG)
				1130±45 (ORELA)
Er-metal	6	8.30	1.61 Er164	714±61 (VDG)

Table II Dy163(n,γ) resonance capture areas. The stated uncertainty is statistical only.

Eo (eV)	gΓn$\Gamma\gamma/\Gamma$ (meV)	Eo (eV)	gΓn$\Gamma\gamma/\Gamma$ (meV)
2651	19.4±1.9	2732	28.6±1.7
2657	66.6±0.8	2743	36.9±1.9
2665	12.2±1.7	2758	29.4±1.8
2675	35.2±2.1	2768	35.8±2.1
2681	34.9±2.1	2776	31.3±2.3
2686	17.6±2.0	2781	30.4±2.9
2700	24.1±1.7	2785	70.3±0.9
2713	65.7±0.7		

REFERENCES

1. H. Beer, G. Walter, R.L. Macklin, P.J. Patchett, Phys. Rev. C (in press)
2. H. Beer, F. Käppeler, Phys. Rev. C21,534(1980)
3. J.U. Andersen et al. Phys. Lett. 113B,72(1982)
4. P.A. Baisden et al. Phys. Rev. C28,337(1983)
5. K. Takahashi, K. Yokoi, Nucl. Phys. A404,578(1983)
6. K. Takahashi, K. Yokoi, private communication
7. R.K. Ulrich, in Explosive Nucleosynthesis, ed. D.N. Schramm, W.D. Arnett (Univ. of Texas Press) Austin 1973, p.139
8. F. Fröhner, private communication
9. H. Beer, R.L. Macklin, in preparation

TARGET THERMALIZATION EFFECT IN ^{187}Os NEUTRON CAPTURE

G. Reffo

ENEA - C.R.E. "E. Clementel", Bologna, Italy

PROBLEM

The thermalization process in the stellar environment may lead to the excitation of certain s-process targets in their lower lying levels.
This fact implies the possibility of neutron captures from excited levels.
In order to draw safe nucleosynthesis conclusions the impact of this effect has to be investigated.
Laboratory measurements, mostly, can only be made for targets in their ground state.
Only reliable model calculations can be used to investigate this effect.
Here we present the results of a study on ^{187}Os neutron capture, which is important for its implications on the Re/Os cosmochronology.

CALCULATIONS

Capture cross sections were calculated with Hauser-Feshbach theory with inclusion of the width fluctuation correction factor according to Moldauer theory.
Neutron transmission coefficients were determined with a spherical optical model.
Gamma transmission coefficients were determined from Brink-Axel hypothesis.
Capture from excited levels of ^{187}Os was calculated including the competition of neutron inelastic channels feeding all lower levels down to the ground state.
Coupled channel calculations were also performed according to a partition of the experimental levels into 3 rotational bands as from Nilsson deformed model: (0, 1/2$^-$; 74, 3/2$^-$; 187, 5/2$^-$)(9.8, 3/2$^-$; 75, 5/2$^-$; 181, 7/2$^-$)(100, 7/2$^-$; 263, 9/2$^-$).

Model parameters were determined from careful model systematics, based on all available experimental information. This leads to a calculated cross section accuracy which can be estimated to be 10%.

RESULTS

As one can see in Table 1, capture cross sections

TABLE 1 - Average Maxwellian Cross Sections for Different Capturing States and Amxwellian Characteristic Temperatures in ^{187}Os

Maxwellian Energies KT=KeV and population probabilities(%) / Energy levels KeV		10	(%)	20	(%)	30	(%)	50	(%)	70	(%)
0	EXP	1988 ± 100°		1171 ± 39°		874 ± 28°		614 ± 12°			
	CALC1	1878	58	1176	43	904	33	684	23	598	19
10		2563	42	1619	51	1238	48	921	38	796	32
74		1328	<.1	880	2	699	6	548	11	488	13
75		2418	<.1	1521	3	1165	8	870	16	755	19
101		2957	<.1	1910	1	1486	5	1133	12	994	17
$<\sigma>^*$		2168		1401		1102		842		744	
$<\sigma>^*/<\sigma_{LAB}>^{CALC}$		.87		.84		.82		.81		.80	
$<\sigma>^*/<\sigma_{LAB}>^{EXP}$		.92		.84		.79		.73			

$<\sigma>^*$ are average maxwellian cross sections weighted on the capturing state population probabilities.

$<\sigma_{LAB}>^{CALC}$ and $<\sigma_{LAB}>^{EXP}$ are the calculated and experimental average maxwellian cross section for ground state capture respectively.

° Winters,R.R. and Macklin,R.L.,1982,Phys.Rev.C25,208.

from excited states can be very different from each other and from the ground state one, due to different neutron inelastic channels competitions as well as to different spin of capturing levels.

Target thermalization effects on maxwellian average cross sections depend on the energy and spin distribution of target levels involved as well as on their population probabilities.

From the last two rows of Table 1, one observes that this effect increases with the characteristic maxwellian energies and exceeds experimental errors as well as the estimated calculation uncertainties.

CONCLUSIONS

Target thermalization results in an even-odd effect of the σN curve, because only even-odd or odd-odd targets exhibit excited levels low enough.

Competition of those neutron inelastic scattering channels cooling the target below the energy of the capturing level is very important.

Coupled channel calculations did not give final results appreciably different from the spherical approximation. In fact at the involved energies direct collective scattering does not affect appreciably the partition of the reaction cross section into its direct and compound nucleus components.

From this study one can deduce that in certain cases spin effects, population probabilities and neutron inelastic competition may combine to give even larger effects than observed in ^{187}Os, therefore for the purpose of nucleosynthesis considerations the effect investigated here cannot be neglected a priori.

INDIRECT INVESTIGATION OF PROTON CAPTURE REACTIONS AT STELLAR ENERGIES*

P.Schmalbrock,T.R.Donoghue,H.J.Hausman,M.Wiescher,V.Wijekumar
Ohio State University, Columbus, Ohio 43210
C.P.Browne
University of Notre Dame, Notre Dame, Indiana 46556
A.A.Rollefson
University of Arkansas at Little Rock,Little Rock, Arkansas 72204
C.Rolfs
WWU Munster, 44 Munster, FRG

ABSTRACT

The levels in the vicinity of the proton thresholds in ^{23}Mg, ^{26}Al and ^{27}Si were investigated via (τ,α), (τ,p) and (τ,d) reactions to provide spectroscopic information for (p,γ)-stellar reaction rate calculations for ^{22}Na, ^{25}Mg and ^{26}Al. A number of new levels are reported which could influence the stellar burning processes.

Knowledge about the nucleosynthesis of Ne, Na, Mg and Al isotopes is important since it is a key to the investigation of several astrophysical problems,[1][3] including the question of the origin of the observed anomalous high abundances of ^{22}Ne and ^{26}Mg in meteorites. One widely discussed hypothesis (with possible consequences for the formation of the solar system) is the live capture of freshly produced ^{22}Na and ^{26}Al during the condensation of the solar matter and the subsequent in situ decay to ^{22}Ne and ^{26}Mg. On the other hand, recent data from the γ-ray satellite observatory suggest continuous ongoing production[3] of ^{26}Al; no comparable evidence for ^{22}Na has been reported. It is generally assumed that the above isotopes are produced predominantly during hydrogen-burning in stars with $T_8 > 0.5$ ($T > 0.5 \times 10^8$ K), e.g. in massive second generation stars, novae or supernovae envelopes. The main reaction sequence for the production of neon and sodium is the NeNa-cycle

^{20}Ne$(p,\gamma)^{21}$Na$(p,\gamma)^{22}$Mg$(\beta^+\nu)^{22}$Na$(p,\gamma)^{23}$Mg$(\beta^+\nu)^{23}$Na$(p,\alpha)^{20}$Ne

and of magnesium and aluminium the MgAl-cycle

$$^{24}\text{Mg}(p,\gamma)^{25}\text{Al}(\beta^+\nu)^{25}\text{Mg}(p,\gamma) \begin{Bmatrix} ^{26}\text{Al}^o(p,\gamma)^{27}\text{Si}(\beta^+\nu) \\ ^{26}\text{Al}^m(\beta^+\nu)^{26}\text{Mg}(p,\gamma) \end{Bmatrix} ^{27}\text{Al}(p,\alpha)^{24}\text{Mg}$$

Due to the different lifetimes of the ^{26}Al ground state ($T_{\frac{1}{2}} = 7.2 \times 10^5$ y) and the first excited isomeric state ($T_{\frac{1}{2}} = 6.3$ s), two different branches of MgAl-cycle have to be considered in reaction-network calulations. These calculations require the stellar reaction rates of all involved reactions as input data.

In general, the direct investigation of these reactions at stellar energies is difficult due to the very low cross sections typical far below the Coulomb barrier. Special difficulties occur for

*Work supported in part by the NSF, NATO and DFG.

Table I: Preliminary spectroscopic information for ^{23}Mg near the proton threshold (Q = 7576 keV)

E_x keV	ℓ	(τ,α) $\sigma_{exp}/\sigma_{calc}$	J^π	E_p^{cm} keV	Γ_p eV a)	$\omega\gamma$(calc) eV	$\omega\gamma$(exp) eV
7587	2	1.1	$3/2^+,5/2^+$	(11)	$5.4*10^{-36}_{-11}$	$2.3*10^{-36}_{-12}$	
7635	3	1.8	$5/2^-,7/2^-$	(59)	$1.1*10_{-3}$	$5.5*10_{-3}$	
7777	3	3.1	$5/2^-,7/2^-$	(201)	$6.3*10$	$3.1*10_{-2}$	
7793	2	0.6	$5/2^+$	(217)	0.09	$3.7*10$	
7852	4	2.5	$7/2^+,9/2^+$	(275)	1.1	0.3	
8016				(444)	80.	0.5	b)
8054				(478)	142.	0.5	b)
8076				(500)	198.	0.5	b)
8164			$5/2^+$	(588)	608.	0.5	b)

a) using smallest possible ℓ and $\theta_\ell^2 = 0.1$
b) in this mass region typically $\Gamma_\gamma \approx 1 eV \ll \Gamma_p$, therfore $\omega\gamma = \omega\Gamma_\gamma$

Table II: Preliminary spectroscopic information for ^{26}Al near the proton threshold (Q = 6306 keV)

E_x keV	ℓ	(τ,α) $\sigma_{exp}/\sigma_{calc}$	ℓ	(τ,d) $\sigma_{exp}/\sigma_{calc}$	E_p^{cm} keV	Γ_p eV a)	$\omega\gamma$(calc) eV	$\omega\gamma$(exp) eV a)
6346	1+3	0.034,0.082	1+3	0.7,0.2	(41)	$< 8*10^{-16}$ c)	$< 4*10^{-16}_{-11}$	
6363	0+2	0.013,0.054	1+2	2.3,1.1	(58)	$7.0*10^{-11}$ b)	$3.5*10^{-9}$	
6398	1+3	0.021,0.017	1+3	0.4,0.4	(93)	$< 6*10^{-9}$ c)	$< 4*10^{-9}_{-6}$	$< 4*10^{-8}_{-7}$
6410	0+2	0.003,0.016	0+2	0.4,0.1	(105)	$3.3*10^{-6}$ b)	$1.7*10^{-6}$	$< 1*10^{-8}$
6436	1+3	0.010,0.030	1	0.25	(130)			$< 4*10^{-8}_{-7}$
6463			1	0.60	(157) a)			$< 1*10^{-7}$
6498	0+2	0.002,0.600	2	0.98	198 a)			$6.7*10^{-7}$

a) Ref. 6   b) using $\theta_\ell^2(E_x)/\theta_\ell^2(6678) = C^2S(E_x)/C^2S(6678)$ c) using $\theta_\ell^2 = 0.1$

Table III: Preliminary spectroscopic information for ^{27}Si near the proton threshold (Q = 7465 keV)

E_x keV	ℓ	(τ,α) $\sigma_{exp}/\sigma_{calc}$	E_p^{cm} keV	Γ_p eV a)	$\omega\gamma$(calc) eV	$\omega\gamma$(exp) eV b)
7465			(0)			
7530			(65)	$< 6*10^{-12}$	$< 3*10^{-12}$	
7563 c)						
7596			(131)	$<1.3*10^{-5}_{-3}$	$< 6*10^{-6}_{-3}$	
7654			(189)	$<4.4*10$	$< 2*10$	$< 4*10^{-5}$
7681 c)						
7703			(238) b)			$< 2*10^{-4}_{-3}$
7742			276 b)			$3.8*10$
7765 c)						
7796			(331) b)			$< 2*10^{-3}_{-3}$
7837			362 b)			$6.5*10$

a) using $\theta_\ell^2 = 0.1$ and $\ell = 0$ b) Ref. 5   c) probable state

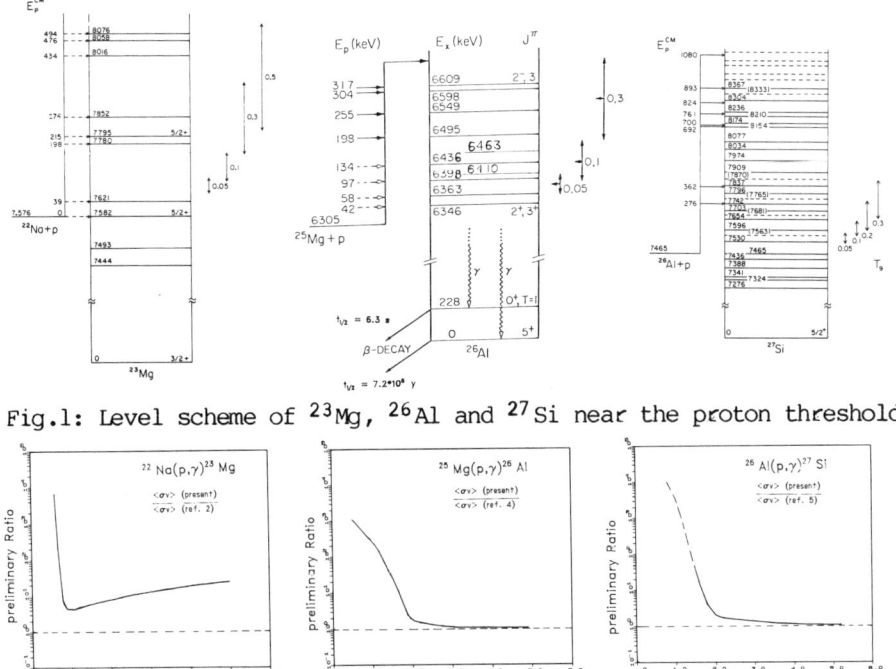

Fig.1: Level scheme of ^{23}Mg, ^{26}Al and ^{27}Si near the proton threshold

Fig.2: Ratios of present to previous stellar rates v/s T_8

the reactions ^{22}Na(p,γ)^{23}Mg and ^{26}Al(p,γ)^{27}Si which require radioactive targets, and for the ^{25}Mg(p,γ)^{26}Al reaction, since the γ-branching ratios to the ground state and the first excited state have to be determined. However, calculations for stellar reaction rates can be performed, if complete spectroscopic information for the proton threshold states is available instead of measured cross sections and/or resonance strengths. To circumvent the difficulties associated with the radioactive target production, we obtained spectroscopic information for the above three reactions by populating the levels of interest with (τ,α),(τ,d) and (τ,p) reactions. The experiments were performed using the Notre Dame FN tandem at $E_τ$ = 12, 15 or 18 MeV and the broad range magnetic spectrograph in conjunction with a position sensitive proportional counter as well as photographic plates from 7.5°-60°. In many cases, angular distributions were analysed with the DWBA-code CHUCK to yield ℓ-values for the transfer.

The results of these investigations are summarized in Tables I-III together with information obtained in direct (p,γ)-measurements. The new information provides an improved basis for calculation of upper limits for the resonance strength and reaction rates assuming typical values of $\Gamma_γ$=1 eV and $\Gamma_p = \theta_\ell^2 P_\ell$ using θ_ℓ^2=0.1, if no further information is available. The influence of the various resonances at different stellar temperatures is shown in

Fig.1. Preliminary ratios of the present and previous[2,4,5] stellar reaction rates are displayed in Fig.2. For example, the inclusion states at higher excitation energy increases the rate for ^{22}Na(p,γ)^{23}Mg as much as tenfold for $T_8>1$. Further the state at E_x=7587 keV with possible $J^π$=5/2$^+$ may yield an increase of several orders of magnitude for $T_8<1$. Discrepancies in the ^{26}Al level scheme lead to as much as three orders of magnitude larger ^{25}Mg(p,γ)^{26}Al rates at $T_8<1.5$ compared to previous rates[4]. For $T_8>1.5$ the rate is well determined by the direct (p,γ)-measurement[3]. Further investigations of the γ-branching ratio of the E_x=6363 and 6410 keV states via α-γ coincidences from ^{27}Al(τ,αγ)^{26}Al are in progress at OSU. We have identified a number of states that could contribute to the stellar reaction rate for ^{26}Al(p,γ)^{27}Si at $T_8<2$. The influence of these states was calculated assuming ℓ=0 proton capture, since further information for the ℓ-values is not yet available due to the very high level density in this excitation energy region.

We are planning to continue these indirect investigations for (p,γ)-reactions which are not amenable to direct measurements and incorporate the results in stellar reaction network calculations.

REFERENCES

1. W.S.Rodney, C.Rolfs, Essays in Nuclear Astrophysics
  ed. by C.A. Barnes et al.(Cambridge Press,1982) p.171
2. R.K.Wallace, S.E.Woosley, Astrophys.J.Suppl. 45 389 (1981)
3. W.A.Fowler, Rev.Mod.Phys. 56 149 (1984)
4. A.E.Champagne et al., Nucl.Phys. A402, 159 (1983)
5. L.Buchmann et al., Nucl.Phys. A415, 93 (1984)
6. K.Elix et al., Z.Phys. A293, 261 (1979)

THERMONUCLEAR REACTION RATES FROM (p,n) REACTION

S. Kailas
Nuclear Physics Division, Bhabha Atomic
Research Centre, Bombay-400 085
India

ABSTRACT

Thermonuclear reaction rates (TNRR) in the temperature range $1 \leq T_9 \leq 5$ have been extracted from experimentally measure (p,n) cross section for ^{92,94}Zr, ^{93}Nb, ^{95,96,98}Mo, ^{103}Rh, ^{110}Pd, ^{107,109}Ag, ^{115}In, ^{117,122}Sn nuclides and for proton energies below 7 MeV. To enhance the usefulness of these reaction rates in astrophysical calculations, they have been fitted to an empirical expression of the from $(A+B\,T_9^C)\exp(-11.605|Q|/T_9)$ where A, B and C are constants. (Q = Q value of the reaction).

INTRODUCTION

Astrophysical studies involving the synthesis of heavier elements require a large number of thermonuclear reaction rates (TNRR) in the temperature range 10^9°K to 10^{10}°K. In the present work, the TNRR in the temperature range $1 \leq T_9 \leq 5$, have been calculated, starting from experimentally measured (p,n) reaction cross section[1,2] for nuclides with A in the range 92 to 122 and for proton energies below 7 MeV.

PROCEDURE AND RESULTS

Following the procedure as discussed by Vlieks et al[3], Kailas and Mehta[4], the averaged (p,n) reaction excitation function from threshold to the highest energy measured, has been fitted with an expression

$$\sigma_{p,n} = \frac{1}{E} \sum_{i=1}^{5} a_i \exp(-b/\sqrt{E})\, E^{i-1} \quad \text{----- 1}$$

where $b = Z - 0.5$ and a_i are adjustable parameters (Z = Target atomic number). Substituting this in the expression for thermo-nuclear reaction rate (TNRR)

$$N_A \langle \sigma v \rangle = N_A \left(\frac{8}{\pi \mu}\right)^{1/2} (kT)^{-3/2} \int_{|Q|}^{\infty} E\, \sigma_{p,n}(E) \exp(-E/kT)\, dE$$

$$\text{----- 2}$$

Table I. Thermonuclear reaction rates

REACTION	\multicolumn{5}{c}{TNRR ($cm^3 g^{-1} sec^{-1}$) at T_9 ($10^9 °K$)}
	1	2	3	4	5
$^{92}Zr(p,n)^{92}Nb$	0.13(-7)	3.42(-1)	1.46(+2)	3.64(+3)	2.82(+4)
$^{94}Zr(p,n)^{94}Nb$	0.49(-4)	3.99(+0)	4.36(+2)	6.92(+3)	4.36(+4)
$^{93}Nb(p,n)^{93}Mo$	0.80(-4)	2.11(+0)	2.32(+2)	3.88(+3)	2.47(+4)
$^{95}Mo(p,n)^{95}Tc$	0.74(-7)	0.36(+0)	1.13(+2)	2.76(+3)	2.20(+4)
$^{96}Mo(p,n)^{96}Tc$	< (-11)	0.70(-2)	1.74(+1)	1.03(+3)	1.27(+4)
$^{98}Mo(p,n)^{98}Tc$	0.72(-7)	0.32(+0)	1.20(+2)	3.36(+3)	2.88(+4)
$^{103}Rh(p,n)^{103}Pd$	0.16(-4)	0.77(+0)	1.09(+2)	2.17(+3)	1.62(+4)
$^{110}Pd(p,n)^{110}Ag$	0.17(-4)	0.86(+0)	9.66(+1)	1.91(+3)	1.44(+4)
$^{107}Ag(p,n)^{107}Cd$	0.18(-6)	0.20(+0)	5.09(+1)	1.25(+3)	1.04(+4)
$^{109}Ag(p,n)^{109}Cd$	0.99(-5)	0.44(+0)	6.69(+1)	1.43(+3)	1.11(+4)
$^{115}In(p,n)^{115}Sn$	0.24(-5)	0.17(+0)	3.41(+1)	8.60(+2)	7.84(+3)
$^{117}Sn(p,n)^{117}Sb$	0.33(-8)	0.39(-1)	1.92(+1)	6.37(+2)	6.34(+3)
$^{122}Sn(p,n)^{122}Sb$	0.24(-7)	0.93(-1)	3.22(+1)	9.02(+2)	8.29(+3)

The number in the parenthesis is the power to which ten has to be raised e.g. (-1) means 10^{-1}.

where N_A = Avogadro number, k = Boltzman Constant, T = temperature, μ = reduced mass, the TNRR value is the temperature range T_9 = 1 to 5 have been computed. The results are tabulated in Table I. In order to enhance the usefulness of these reaction rates in astrophysical calculations, the TNRR values have been fitted to an empirical expression of the from

$$TNRR = (A + B T_9^C) \exp(-11.605 |Q|/T_9) \quad \text{----- 3}$$

where A, B and C are constants for given nucleus. The results of this analysis for the cases where this empirical expression has been successful in predicting the TNRR data, are listed in Table II.

Table II
Parameter fitting the TNRR data

REACTION	A	B	C
$^{92}Zr(p,n)$	7.609(+5)	7.394(+5)	1.963
$^{94}Zr(p,n)$	1.402(+4)	4.120(+3)	3.914
$^{95}Mo(p,n)$	9.610(+4)	5.100(+4)	2.980
$^{98}Mo(p,n)$	4.232(+4)	2.107(+4)	3.613
$^{107}Ag(p,n)$	1.855(+4)	3.548(+3)	3.831
$^{117}Sn(p,n)$	3.317(+4)	8.949(+3)	3.518
$^{122}Sn(p,n)$	2.720(+4)	6.605(+3)	3.596

CONCLUSION

The TNRR values extracted in the present work are broadly consistent with those from thick target activation measurements determined by Roughton et. al for the cases where the nuclei in the two works are common. However, these are noticeable differences in some cases in the temperature range below $T_9 = 3$. The most discrepant are in the TNRR for the $^{117}Sn(p,n)$ reaction. In this case the TNRR's from the two works differ by a factor of atleast two in the entire T_9 range.

REFERENCES

1. C.H. Johnson et al., Phys Rev. C 15 196(1977); Phys Rev. C 20 2052 (1979)
2. R.L. Hershberger et al., Phys. Rev. C 21, 896 (1980), D.S. Flynn et al., Phys Rev. C 20, 1700 (1979) R. Schrills et al. Phys. Rev. C 20 1706 (1979)
3. A.E. Vlieks et al. Nucl. Phys. A 224, 492 (1974)
4. S. Kailas and M.K. Mehta, Pramāna 7, 6(1976)
5. N.A. Roughton et al., Atomic Data and Nuclear Data Tables 23, 177 (1979).

NEUTRINO CAPTURE AND RELATED PROBLEMS:
THE CAPTURE CROSS SECTION IN ^{69,71}Ga

K. Grotz, H.V. Klapdor and J. Metzinger
MPI für Kernphysik, Heidelberg, Germany

ABSTRACT: Calculations of the ν-capture cross sections of ^{69,71}Ga based on microscopic treatment of the Gamow Teller matrix elements are presented. A strong enhancement of the cross section for high energetic neutrinos is found compared to previous phenomenological estimates. The important implications for a gallium neutrino detector are discussed.

1. INTRODUCTION: It is generally believed, that neutrinos play a key part in resolving the unification problem. Therefore a large number of efforts have been undertaken to answer the fundamental questions: Are neutrinos Dirac or Majorana particles? Do they have a finite mass and if so, does mixing occur between neutrinos of different families?

The experimental problems arising from the smallness of all quantities connected with neutrinos can be overcome by either searching for very delicate effects as is done in ν-oscillation, beta decay and double beta decay experiments, or by doing experiments on the very large scale provided by astrophysics. Besides the relic neutrinos from the big bang, which should be trapped in the gravitational field of galaxies in the case that they are massive (~30 eV), there are expected to exist a large number of discrete ν-sources with energies up to >10 TeV.

One powerful technique of extraterrestrial ν-spectroscopy is the capture of neutrinos by atomic nuclei. However, the capture cross sections depend on nuclear structure and have to be treated carefully. We discuss in this paper the ν-capture cross section of a gallium detector in respect to solar neutrino and collapsing star neutrino detection.

2. SOLAR NEUTRINO DETECTION: The so-called solar neutrino problem[1] might be caused by the long sought ν-oscillations. But the result from the famous ^{37}Cl solar ν-experiment is not unambiguous, because solar model uncertainties could be responsible for the observed neutrino deficit, too. ^{37}Cl, due to the high Q-value is sensitive mainly for neutrinos from the decay of ^{8}B produced in a very model dependent rare side branch.

It is believed[2] that measuring the model independent solar pp-neutrinos with a ^{71}Ga detector could give a unique answer. An observed deficit for these neutrinos too, would indicate ν-oscillations. For 2ν mixing with mixing lengths smaller than the solar radius corresponding to $\Delta m^2 \geq 2 \cdot 10^{-8}$ eV^2 a maximum reduction by 50% is expected. The reduction might be less for 10^{-12} eV$^2 \leq \Delta m^2 \leq 10^{-8}$ eV^2. Therefore, a comparison of the measured rate with theoretical predictions requires accurate knowledge of the neutrino capture cross section $\sigma_\nu(E)$ of ^{71}Ga. We shall show that the phenomenological estimate of Bahcall[3] is insufficient.

$\sigma_\nu(E)$ is dominated by the Fermi transition to the isobaric analog state (IAS) and the Gamow Teller (GT) matrix elements. It will

Table 1: Calculated GT-strength
a) for ^{71}Ga   b) for ^{69}Ga

E* (MeV)	B(E*)	E* (MeV)	B(E*)
0	0.071	0$^\alpha$	0.089
0.164	0.0001	0.03$^\beta$	0
0.522	0.061	0.32	0.090
1- 2	0.006	1- 2	0.006
2- 3	0.730	2- 3	0.816
3- 4	0.717	3- 4	2.777
4- 5	1.412	4- 5	0.403
5- 6	4.359	5- 6	2.092
6- 7	0.023	6- 7	0.496
7- 9	0	7- 9	0
9-10	1.676	9-10	6.996
10-11	5.739	10-11	7.022
11-12	17.447	11-12	8.069
12-13	0.233	12-13	0.029

α lowest 1/2$^-$ state
β lowest 5/2$^-$ state

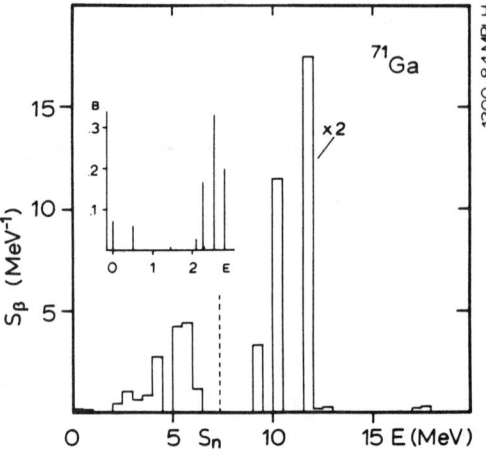

Fig. 1: GT-strength distribution in ^{71}Ge. $S_\beta(E) = \sum_{\Delta E} B/\Delta E$.

(ΔE = 0.5 MeV in this figure.
x2 = to be multiplied by 2.)

turn out that details of the low-lying GT strength of ^{71}Ga, calculated microscopically in this paper, are decisive for an analysis of a gallium solar neutrino experiment. Refining earlier studies[4], the BCS pairing model is used to diagonalize a spin-isospin-force of Yukawa type (range = m_π^{-1}) coming from the π-meson exchange contribution to the nucleon-nucleon force. A similar treatment has recently lead to a better understanding of double beta decay rates[5].

We show the calculated distribution of GT-strength for ^{71}Ga and ^{69}Ga in table 1a) and b) and for ^{71}Ga also in fig. 1. G.s. capture of the low-energetic pp neutrinos (E_{max} = 0.42 MeV) dominates the total capture rate of solar neutrinos in ^{71}Ga. However, to decide the question whether the solar neutrino problem is caused by solar model uncertainties (these should not affect the pp neutrinos) or by ν-oscillations, other processes have to be considered, too. From the recent (p,n) measurement Orihara et al.[6] claimed a GT-strength for the first excited 5/2$^-$ state nearly equal to the g.s., enhancing the capture of low-energetic pp neutrinos. In contrast, our calculation gives only very litte strength for this state. This result is supported by the log ft values of more than 6 for 5/2$^-$ ↔ 3/2$^-$ transitions between low lying states in the neighbouring nuclei (^{67}Cu → ^{67}Zn, ^{69}Ge → ^{69}Ga, ^{73}As → ^{73}Ge, ^{73}Ga → ^{73}Ge). From fig. 2 it is seen, that $\sigma_\nu(E)$ for E < 2 MeV is reasonably approximated by taking only the g.s. transition into account, but for higher energies this approximation and also the Bahcall estimate which is only about a factor of 2 bigger fail seriously. This is already true, if only transitions to states below the neutron separation energy S_n leading not to the stable ^{70}Ge isotope are included (dashed line in fig. 2). The reason is a considerable amount of GT strength between 4 and 6 MeV excitation energy.

In contrast to the general assumption that the ^{8}B neutrinos are of minor importance for the ^{71}Ga detector, we predict them to contri-

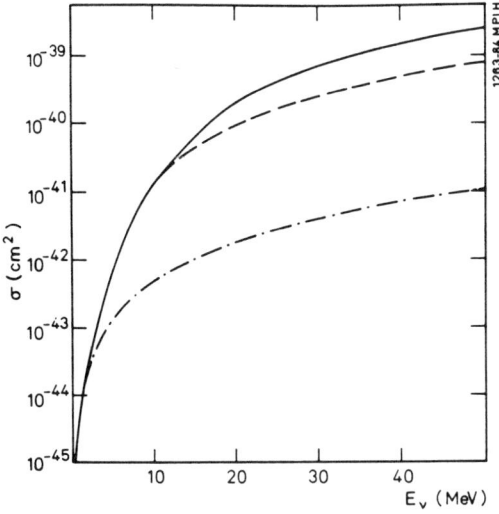

Fig. 2: $\sigma_\nu(E)$ for ^{71}Ga. The full line results from all GT strength shown in fig. 1. The dashed line includes only strength below S_n and the dashed-dotted line represents the contribution from the g.s. transition only.

bute ~20 SNU to the capture rate, which is about 30% the rate from the pp neutrinos. In this value we have already included a reduction factor of 0.7 for quenching effects produced by the Δ_{33} resonance, relying on previous studies[7]. It must, however, be recognized, that according to the ^{37}Cl experiment the ^{8}B ν-flux should be smaller than the standard solar model prediction, on which the 20 SNU are based. Probing the capture cross section with a ^{51}Cr-source is of relevance only for low-energetic neutrinos. It is therefore highly desirable to compare this calculation with the results from a high resolution, low-background (p,n) measurement of the GT-strength for energies up to the GT-giant resonance.

3. NEUTRINOS FROM A COLLAPSING STAR: In hydrodynamical models of the gravitational collapse of a massive star ν-interactions are an essential ingredient. The trapping of neutrinos in the core and also the ν-pair production rate depends not only on hydrodynamical model parameters but also on the Weinberg angle θ_W, the number of families of quarks and leptons[8] and on whether neutrinos have majorana masses or not[9]. The observation of the ν-burst emitted by a collapsing star is therefore of great interest.

Hampel[10] has estimated the feasibility of measuring with a gallium detector a ν-burst of 1.8×10^{53} erg emitted at 1 kpc distance, however, also based on oversimplified assumptions for the ν-capture cross section. Three different mechanisms have to be considered: (i) Capture by ^{71}Ga leading to states in ^{71}Ge below the neutron separation energy S_n. These events can only be distinguished statistically from solar neutrino events. (ii) Capture by ^{69}Ga to states below S_n in ^{69}Ge. (iii) Capture by ^{69}Ga to ^{69}Ge states above S_n, decaying to the ^{69}Ge g.s.. - It has been argued, that process (i) is ineffective because the largest fraction of the GT strength and also the IAS lies above S_n in ^{71}Ge, so that capture by those states will be not detectable. However, fig. 3 shows, that the GT-strength above S_n is <u>not</u> dominating the ν-cross section for ν-energies lower than 20 MeV. For a 30 t Ga detector the number of ^{71}Ge nuclei not undergoing particle emission produced by such a ν-burst would be 120 (255) for the ν-spectra taken from Wilson[11] (Roberts[12]). Such a signal should be easily measurable if it is compared to the 18 nuclei saturation value expected from the solar neutrinos. Process (ii), which, because the

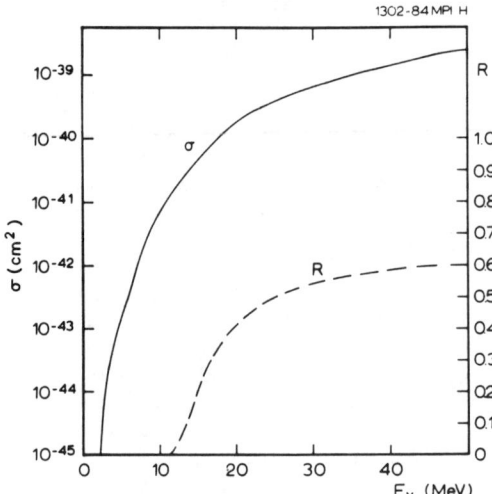

Fig. 3: $\sigma_\nu(E)$ for ^{69}Ga including all strength. Shown is also R, the fraction of the total cross section resulting from population of states above S_n.

IAS in ^{69}Ge is <u>below</u> S_n, has the largest cross section (147 (370) events), but has the disadvantage of the short half live of ^{69}Ge, therefore needing a trigger signal[4]. Although the expectation of 67 (275) ^{68}Ge nuclei by process (iii) is larger than the one for the production of ^{71}Ge nuclei by process (i), at least for the Roberts spectrum, due to the large half life of ^{68}Ge of 288 d the decay rate would not exceed 0.7 per day. In this respect detection by processes (i) and (ii) are more favourable.

4. CONCLUSION: Correct treatment of the Gamow Teller strength is decisive for reliable predictions of the capture cross section of solar and galactic ν-detectors. Considering only the ground state and some lower excited states, as done up to now[3], is by far not sufficient. For the gallium detector the result of a refined treatment is an increased solar model dependent background in the signal from solar pp neutrinos and at the same time larger sensitivity for galactic neutrinos.

REFERENCES
1. R. Davis Jr., D.S. Harmer and K.C. Hoffmann, Phys. Rev. Lett. <u>20</u> 1205 (1968)
2. W. Hampel, Int. Conf. Neutrino Phys. and Astrophys., Nordkirchen 1984
3. J.N. Bahcall et al., Rev. Mod. Phys. <u>50</u>, 881 (1978) and <u>54</u>, 767 (1982)
4. H.V. Klapdor, Prog. Part. Nucl. Phys. <u>10</u>, 131 (1983)
5. H.V. Klapdor and K. Grotz, Phys. Lett. <u>142B</u>, 323 (1984)
6. H. Orihara et al., Phys. Rev. Lett. <u>51</u>, 1328 (1983)
7. K. Grotz, H.V. Klapdor and J. Metzinger, Phys. Lett. <u>132B</u>, 22 (1983)
8. S.W. Bruenn, W.D. Arnett and D.N. Schramm, Astrophys. J. <u>213</u>, 213 (1977)
9. E.W. Kolb, D.L. Tubbs and D.A. Dicus, Astrophys. J. <u>255</u>, L57 (1982)
10. W. Hampel, Proc. Int. Conf. Neutrino Phys. and Astrophys., Erice, Italy 1980, p. 61
11. J.R. Wilson, Astrophys. J. <u>163</u>, 209 (1971)
12. A. Roberts, H. Blood, J. Learned and F. Reines, Proc. Int. Neutrino Conf., Aachen 688 (1977)

VII. APPLICATIONS

IN SITU NEUTRON CAPTURE SPECTROSCOPY OF GEOLOGICAL FORMATIONS

J. A. Grau, S. Antkiw, R. C. Hertzog[*], R. A. Manente,
and J. S. Schweitzer

Schlumberger-Doll Research, Ridgefield, CT. 06877

ABSTRACT

Neutron-capture-induced gamma-ray spectroscopy has been useful for determining the geological properties of porosity, lithology, and water salinity of the rock and fluid outside the iron casing which is cemented into most oil wells. This paper discusses such measurements using a pulsed source of 14-MeV neutrons and a NaI(Tl) gamma-ray detector. The most difficult task in interpreting these measurements is to separate signals generated by elements in the formation rock and fluid from those due to elements inside the well bore. The physical properties at work here include fast-neutron thermalization times, thermal-neutron diffusion, competing neutron-capture cross sections, and gamma-ray transport to the detector. Results from studying the geophysical parameters which most affect these physical properties, porosity and water salinity, are presented here. Methods for enhancing the spectral contribution from the earth formation relative to that from the borehole are discussed. These include understanding the time dependence of the spectral shape, optimizing the source-to-detector spacing, and employing absorptive borehole fluid displacers.

INTRODUCTION

Due to the relative ease with which both neutrons and gamma rays can pass through a quarter-inch iron casing, the (n,γ) reaction has for many years provided a useful means to probe the formation rock and pore fluid surrounding oil wells. Using a pulsed source of 14-MeV neutrons and a NaI(Tl) gamma-ray detector, the original tools[1] determined the total capture cross section of the rock and fluid by measuring the time decay of the total gamma ray yield. This measurement is still in demand, based on its ability to distinguish salty water from oil and shale from sand. Current generation tools[2,3], using much the same hardware, perform an elemental analysis of the earth formation by measuring the energy spectrum of the gamma-ray yield. This paper will examine the downhole spectroscopy measurement in general and in particular the effect that the unique borehole geometry has on the results. We will show that an understanding of these borehole effects requires an understanding of the four main physical processes which govern the measurement: fast-neutron thermalization times, gamma-ray transport, thermal-neutron diffusion, and total capture cross sections.

DESCRIPTION OF THE MEASUREMENT

Since more complete descriptions of this tool have been published elsewhere[2,3], only a brief description will be given here. The neutron source, NaI(Tl) detector, amplifier, ADC, spectral memory, and telemetry electronics are all enclosed in a 9.2-cm-diameter iron pressure housing which is lowered into the well at the end of thousands of feet of cable. At the top end of the cable, a

---

[*] Current address: Schlumberger Well Services, Houston, TX 77001

computer controls tool operation, accepts spectral data from the down-hole memory, and computes and displays results. The tool is operated with a burst of 14-MeV neutrons from a (D,T) accelerator, followed by a time delay to allow for thermalization and for decay of the borehole flux. After this delay, which may be from 20 to several hundred micro seconds, an energy spectrum of gamma rays produced by the capture of thermal neutrons is accumulated for a fixed time period, which is usually comparable to the delay time. This cycle is repeated thousands of times each second.

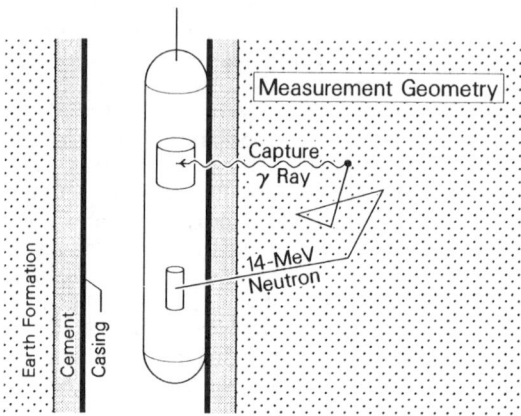

Fig. 1 - Measurement geometry for a sonde eccentered in a cased borehole. The fluid inside the casing is mostly either water or oil, which for our purposes means it contains a large amount of hydrogen.

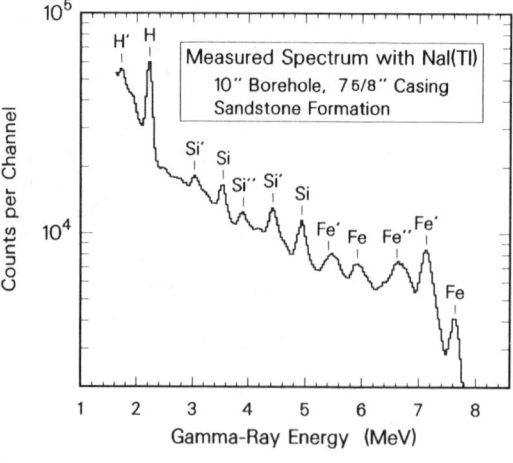

Fig. 2 - Spectrum measured in the geometry of Fig. 1. The formation is oil-filled sandstone. Prominent peaks have been identified, with primes indicating one or two escapes. Our quantitative analysis, however, is not based upon peak areas.

Figure 1 portrays a typical measurement geometry with the sonde eccentered in a cased borehole. The fluid inside the casing is mostly either water or oil with some heavy compounds added to provide sufficient pressure to balance the pressure from the earth formation. Figure 2 is a sample spectrum measured in this geometry, the formation being sandstone with oil filling the pores and a column of fresh water filling the casing. Easily identified are gamma-ray peaks from hydrogen in the oil and water, silicon in the sandstone, and iron in the casing and tool pressure housing. A small amount of Ca, which is contained in the cement around the casing, shows up in the weighted-least-squares analysis[4] of the spectrum, but it is not visually obvious. One goal of this measurement is to do an elemental analysis of the formation rock and pore fluid. Using a NaI(Tl) detector, an element is generally considered detectable only if it can contribute at least one percent to the total measured spectrum. Based upon elemental concentrations typically found in geologic formations and their respective (n,γ) cross sections, Table I lists the elements considered detectable along with the geologically important minerals and fluids in which they are found.

Table I Geologically significant elements detectable from (n,γ) reactions using NaI(Tl)

Element	Found in these Compounds
H	H_2O, Hydrocarbons, Clays
Si	SiO_2, Clays
Ca	$CaCO_3$, $CaMg(CO_3)_2$, $CaSO_4$, CaO
S	$CaSO_4$, FeS_2, S
Fe	Fe, Clays, FeS_2
Cl	NaCl

The ultimate goal of the measurement is to extract a mineral and fluid-type analysis from the elemental analysis. Before this can be attempted, however, a serious problem must be solved: Much of the hydrogen in Figure 2 comes from the oil in the formation fluid, however a large fraction of it comes from the water inside the casing and in the cement. A similar problem can exist for other elements, thus it is necessary to separate the borehole contribution from the formation contribution for each elemental yield. We will show below how this borehole contribution can be measured in the laboratory, how it varies with environmental parameters, and how these variations can be understood in terms of simple physical models of the measurement.

PHYSICS OF THE MEASUREMENT

An elemental yield from the (n,γ) reaction can be expressed as the following time, space and energy integral:

$$Y_i = \int_{E_n} \int_t \int_V \Phi_n(\vec{r},t,E_n)\, \sigma_i(E_n)\, \eta_i(\vec{r}) \sum_j b_{j,i}\, \Gamma(\vec{r},E_{\gamma_{j,i}})\, d^3\vec{r}\, dt\, dE_n , \qquad (1)$$

where:

Y_i = Total detected γ-ray yield from element i
Φ_n = Neutron flux per unit energy
σ_i = Microscopic cross section for element i
η_i = Atomic density of element i
Γ = Transmission and detection probability for $\gamma_{j,i}$
$b_{j,i}$ = Branching fraction of γ ray j for element i

An accurate evaluation of this integral has been accomplished thus far only using Monte-Carlo simulation. To gain physical insight, therefore, it is useful to simplify Equation 1 by doing some averaging and by identifying homogeneous regions:

$$Y_{i,k} = \eta_{i,k}\, \overline{\sigma}_i\, m_i\, g_{i,k} , \qquad (2)$$

where:

$Y_{i,k}$ = Detected γ-ray yield of element i from homogeneous region k
$\eta_{i,k}$ = Atomic density of element i in region k
$\overline{\sigma}_i$ = Energy-averaged cross section for element i
m_i = γ-ray multiplicity for element i
$g_{i,k}$ = Sensitivity for detecting element i in region k:
= $\int_{V_k} \int_t \overline{\Phi}_n(\vec{r},t)\, \Gamma(\vec{r},\overline{E}_{\gamma_i})\, dt\, d^3\vec{r}$
$\overline{\Phi}_n$ = Energy-averaged neutron flux
$\overline{E}_{\gamma_i}$ = Average γ-ray energy for element i

While not rigorously correct, the advantage of this formulation is that it separates all of the elusive physics into a single parameter, $g_{i,k}$, which can be measured in many instances. The parameter which is most easily measured is the relative borehole sensitivity for environments where the well contains no casing or cement:

$$G_{i,B} = g_{i,B}/g_{i,F} , \qquad (3)$$

where B and F refer to the borehole and formation, respectively. We will next show various examples in which this parameter has been measured and will explain the systematics in terms of the physics of the measurement.

POROSITY AND BOREHOLE SIZE DEPENDENCE

A good way to study relative borehole sensitivities is to examine the porosity indicator ratio, PIR, which is used to measure formation porosity. PIR is the ratio of the hydrogen to silicon plus calcium yields, where each yield has been normalized to an appropriate mineral or fluid volume sensitivity. Laboratory measurements[5], from which this ratio can be determined, have been made over a wide range of borehole and formation environments. In the absence of borehole contribution, PIR would equal $\phi/(1-\phi)$, ϕ being the porosity fraction.

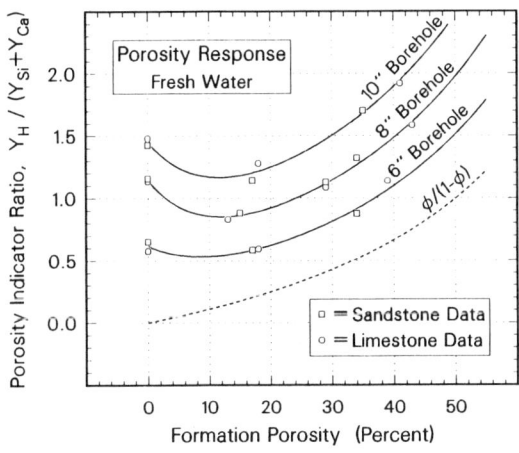

Fig. 3 - Response of the porosity indicator ratio at various borehole diameters. The dashed line would be measured if there were no borehole contribution to the hydrogen yield.

As Figure 3 shows, however, the measurement is normally far from this ideal, the difference attributable to borehole contributions to the hydrogen yield. From these data we can calculate the relative borehole sensitivity for hydrogen, the result of which has been plotted in Figure 4. There is an obvious borehole size dependence, which for high porosities can be approximated by:

$$G_{H,B} = 1.4(1-e^{-A/A_0}) \quad \text{(at high porosity)}, \qquad (4)$$

where A is the cross-sectional area of the borehole fluid. $G_{H,B}$ does not become infinite at infinite borehole sizes because the tool is eccentered in the borehole. There is also a porosity dependence to $G_{H,B}$ which can be explained only by examining the physics of the measurement. The increase from 40 to 15 percent porosity is predicted well by the increased attenuation of the formation hydrogen gamma rays by the increasingly dense formation. This argument, however, predicts a nearly linear change with porosity and thus does not explain the sharp upturn at zero percent porosity. For a complete understanding we must examine the thermalization process. The 14-MeV source neutrons are most effectively thermalized by elastic collisions with hydrogen nuclei, losing on balance half their remaining energy per collision. In the absence of hydrogen, the formation neutrons will penetrate more deeply into the formation before they thermalize. The formation signal will then suffer both solid angle losses and increased attenuation, thereby increasing the relative borehole sensitivity to that which is observed in Figure 4.

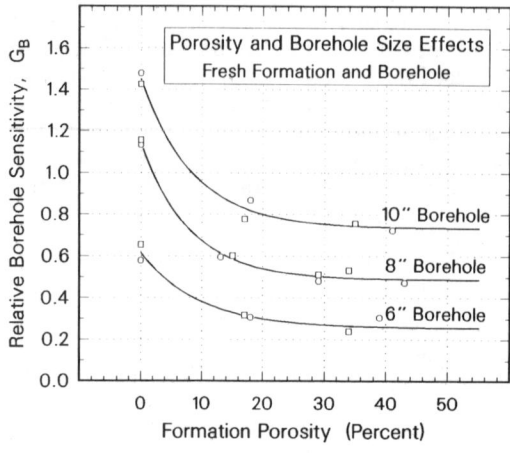

Fig. 4 - Relative borehole sensitivity for hydrogen in freshwater environments. The borehole size dependence is purely geometrical, while the porosity dependence reveals some physical principles governing the measurement.

WATER SALINITY DEPENDENCE

The largest observed changes in borehole sensitivity are due to changes in water salinity, as is shown in Figure 5. In this case both the porosity and borehole size are fixed while the salinity of the borehole fluid and formation fluid are independently changed from fresh water to saturated salt water. There is a substantial increase in $G_{H,B}$ with formation salinity and an equally large decrease with borehole salinity. These results were obtained as before from the porosity indicator ratio. A similar result, however, can be obtained by looking at

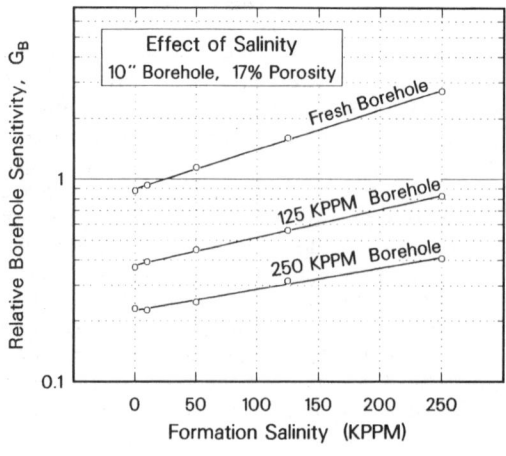

Fig. 5 - Relative borehole sensitivity for hydrogen as a function of borehole and formation salinity, as determined from the porosity indicator ratio. The borehole sensitivity increases with formation salinity and decreases with borehole salinity.

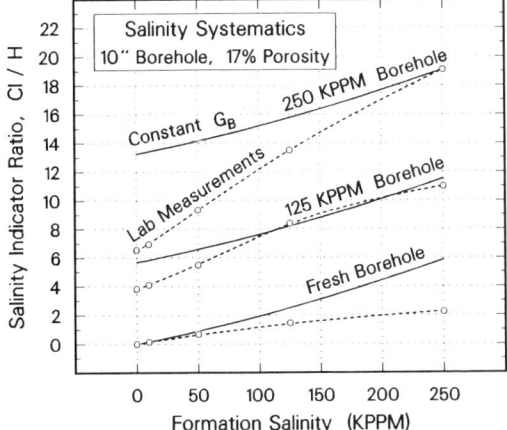

Fig. 6 - Salinity indicator ratio as a function of formation and borehole salinity. The solid lines are calculations assuming a constant borehole sensitivity. The dashed lines are the measured values of this ratio. The non-constant $G_{Cl,B}$ needed to reproduce the lab measurements can readily by calculated.

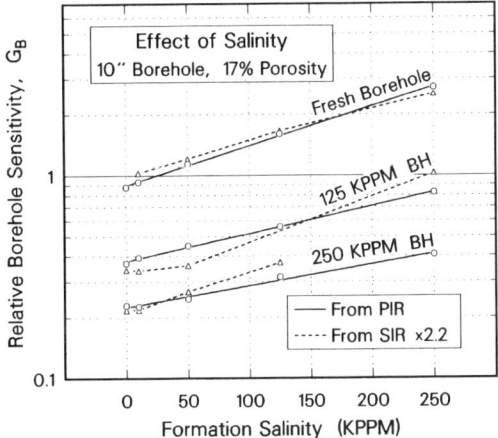

Fig. 7 - Comparison of the relative borehole sensitivities for hydrogen (solid lines) and chlorine (dashed lines). The trends are the same, save for an overall factor of 2.2 which can be explained by reduced attenuation of the chlorine gamma rays.

the salinity measurement itself. The solid lines in Figure 6 are plots of the ratio of the chlorine to hydrogen yields one would expect with changes in borehole and formation salinity, assuming a constant relative borehole sensitivity. The dashed lines in Figure 6 are the actual measurements of that ratio. The difference between these curves can be used to calculate the actual, non-constant relative borehole sensitivity for chlorine. This determination is plotted in Figure 7, along with the original sensitivity for hydrogen determined from the porosity ratio. The agreement is quite good provided the salinity-determined sensitivity is multiplied by 2.2. Since the average energy of the chlorine gamma rays is larger than that for hydrogen, there is less attenuation of the formation chlorine signal. The factor of 2.2 would be predicted from a fairly reasonable average depth of penetration into the formation of 20 centimeters.

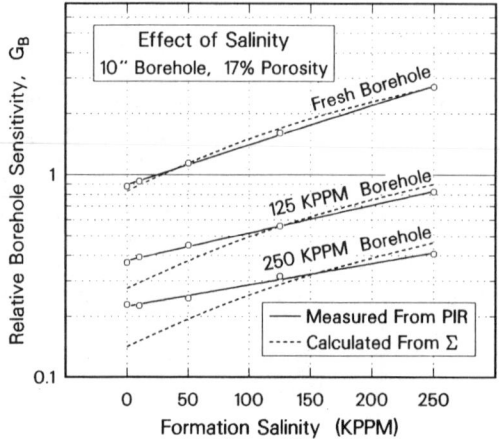

Fig. 8 - Comparison of the measured relative borehole sensitivity for hydrogen with the ratio of the time-integrated neutron fluxes, calculated from the macroscopic capture cross sections in the borehole and the formation assuming no diffusion effects.

The explanation of the salinity dependence of these borehole sensitivities is actually rather simple. The presence of chlorine substantially increases the total capture cross section, increasing the rate at which the neutron flux decays. This decay rate can easily be calculated for both formation and borehole from the macroscopic cross sections of each region, ignoring the effects of neutrons diffusing back and forth between the regions. Integrating these calculated fluxes over the measurement time intervals, as is called for in Equation 2, produces a salinity dependence closely resembling the measured borehole sensitivity, as is shown in Figure 8. Small discrepancies remain because diffusion was ignored in the flux calculation. Diffusion would moderate the differences between borehole and formation, changing the calculation in the needed direction. For the salty borehole, fresh formation case, for example, there will be a net diffusion of neutrons into the borehole, thereby increasing the borehole sensitivity toward the measured value.

REDUCTION OF BOREHOLE SENSITIVITY

There are three simple methods of reducing borehole sensitivity, although each produces its own penalty which must be weighed against the benefits. The first method is to increase the delay time between the end of the neutron burst and the start of data accumulation. Why this helps can be seen clearly in Figure 9, which plots the decay of the neutron flux in both formation and borehole for a typically encountered environment. Ignoring diffusion effects, the relative borehole sensitivity would satisfy the following proportionality:

$$G_{i,B} \sim e^{-T_D(\Sigma_B-\Sigma_F)/4.55}, \tag{5}$$

where Σ_B and Σ_F are the macroscopic capture cross sections of the borehole and formation in cm^{-1} and T_D is the delay time in micro seconds. Since Σ_B is frequently larger than Σ_F, especially if casing is present, the longer you wait the less borehole relative to formation signal will be measured. Figure 10 is a comparison of measured borehole sensitivities for a short and long delay. In practice one must weigh the enhanced formation signal against the overall reduction in count rate to reach an optimum delay, being mindful also that with

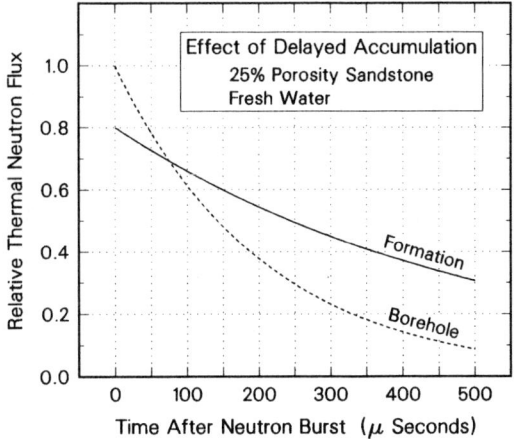

Fig. 9 - Time decay of the thermal-neutron flux in the formation and borehole for a typically encountered environment. The borehole flux frequently decays more rapidly than the formation flux.

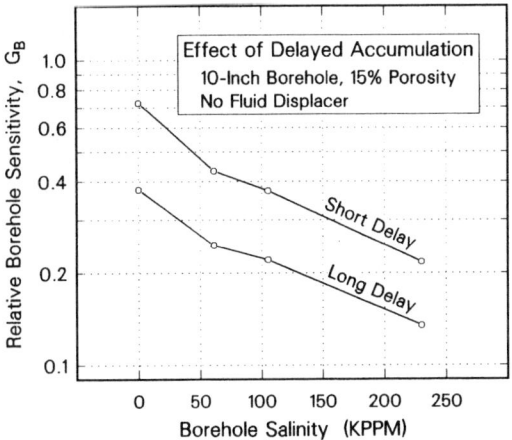

Fig. 10 - Reduction of borehole sensitivity by delayed accumulation. The short delay is 20 μ seconds, while the long delay is about 500 μ seconds.

too much delay the measurement will eventually be dominated by activation, predominantly of oxygen and iodine.

The second method is to encase the outside of the tool with an absorptive borehole-fluid displacer, containing for example a suitable concentration of ^{10}B. Since this increases the total capture cross section of the borehole, the neutron flux in the borehole will decay more quickly, thereby reducing the borehole contribution during a suitably delayed accumulation time. Figure 11 shows the effect of such a displacer when used with both the long and short delays shown previously. The reduction in borehole sensitivity is quite significant, even though for this case only about 20 percent of the borehole fluid had been displaced. The disadvantage to this method is that restrictions near the top of the well sometimes preclude using a displacer whose diameter is large enough to be helpful.

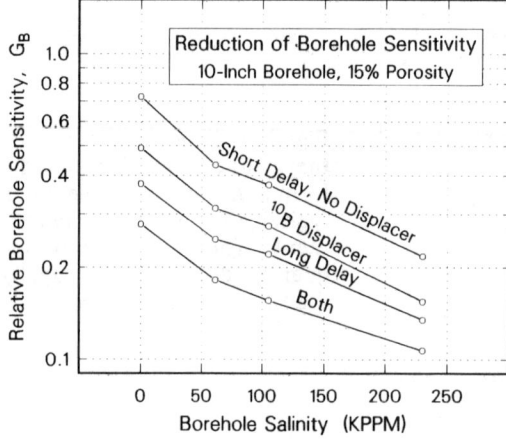

Fig. 11 - Reduction of borehole sensitivity due to an absorptive fluid displacer (rubber impregnated with B_4C). For this case only about 20 percent of the borehole fluid was displaced.

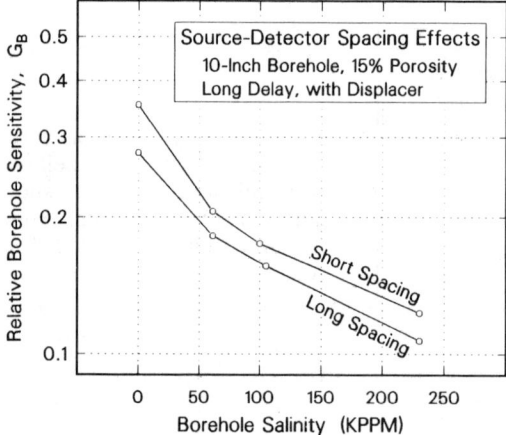

Fig. 12 - Reduction of borehole sensitivity by increasing the source-to-detector spacing. The long spacing is about 1.5 times the short spacing.

The third reduction method is simply to increase the spacing between the neutron source and the gamma-ray detector. The reason for this is purely a solid-angle argument having little to do with nuclear physics. Figure 12 shows the effect of increasing the spacing by about 50 percent while also employing the previous two reduction techniques. The tradeoff here is that the total count rate will drop quickly, so that in practice the optimum spacing will still have some borehole contribution.

All three of these techniques to reduce borehole signal also reduce formation signal, albeit to a lesser degree. Since time comes at a premium during oil-well logging, increased statistical uncertainties due to count-rate losses can never be fully recouped. Rather, the optimum source-to-detector spacings and time delays are those which minimize the ultimate uncertainties in the geologically significant measurement results.

SUMMARY

Neutron-capture-induced gamma-ray spectroscopy is a useful means to study earth formations, although the necessary borehole environment complicates interpretation of the measurement. While it is not possible to fully model the measurement without resorting to computer-intensive Monte Carlo techniques, a simplified treatment, accounting for the macroscopic effects of fast-neutron thermalization times, thermal-neutron diffusion, total capture cross sections, and gamma-ray attenuation, adequately describes the observed laboratory data.

ACKNOWLEDGEMENT

The authors wish to acknowledge R. Plasek of Schlumberger Well Services, Houston, TX for much of the early engineering development and B. Roscoe, also of Schlumberger Well Services, for intellectual support and for providing much of the data discussed here.

REFERENCES

1. J. S. Wahl, W. B. Nelligan, A. H. Frentrop, C. W. Johnstone, and R. J. Schwartz, Soc. Pet. Eng. Jour. **10**, 365 (1970).
2. R. Hertzog and R. Plasek, IEEE Trans. Nucl. Sci. **NS-26**, 1558 (1979).
3. R. C. Hertzog, Soc. Pet. Eng. Jour. **20**, 327 (1980).
4. J. A. Grau and J. S. Schweitzer, Trans. Am. Nucl. Soc. **43**, 260 (1982).
5. J. S. Schweitzer, R. A. Manente, and R. C. Hertzog, Jour. Pet. Tech. **36**, (1984).

Analytical Neutron-Capture Gamma-ray Spectroscopy:
Status and Prospects

Richard M. Lindstrom
Center for Analytical Chemistry
National Bureau of Standards
Gaithersburg, Md. 20899
and
David L. Anderson
Department of Chemistry
University of Maryland
College Park, Md. 20742

ABSTRACT

The use of neutron-capture gamma rays for elemental analysis has become an established technique, applicable for the measurement of a list of elements which complements conventional delayed-gamma neutron activation analysis. Three distinct areas of application of the prompt method have been laboratory-based analysis using reactor neutrons, field measurements (especially borehole logging for mineral exploration), and industrial process stream analysis, the latter two using neutron generators or isotopic neutron sources. Continued improvements in detector systems and the increasing availability of clean, high-intensity beams from cold-neutron guide tubes are opening quantitatively and qualitatively new analytical possibilities. Second-generation instruments using cold neutrons promise an increase in reaction rate of up to two orders of magnitude. This will increase the productivity of these facilities by a similar factor, and will make the use of capture gamma rays more nearly comparable in sensitivity to conventional neutron activation analysis whenever the sample may be brought to the neutrons.

HISTORY

The use of neutron-capture gamma-ray spectroscopy as a tool in chemical analysis is an old idea made practical by new technology. Bibliographies of publications on the method have been compiled by Gladney [1] and by Glascock [2]; the latter lists 522 papers through 1983. Unlike most recently-developed instrumental methods of chemical analysis, this one has not acquired a universally accepted name or acronym. We will use "PGAA" throughout this discussion: others that have been used are listed in Table 1. The explosion of papers on PGAA since the mid-60s was made possible by advances in detectors, sources, and electronics, particularly the development of Ge gamma detectors, sealed-tube neutron generators, Californium-252 sources, and intelligent pulse-height analyzers.

Table 1
Synonyms and Acronyms for Analytical Neutron-capture Gamma-ray Spectroscopy

NCGA	Neutron Capture Gamma-ray Analysis
PGAA	Prompt Gamma-ray Activation Analysis
PGNAA	Prompt Gamma-ray Neutron Activation Analysis
PNCAA	Prompt Neutron Capture Activation Analysis
PNAA	Prompt Neutron Activation Analysis
RNC	Radiative Neutron Capture
TCGS	Thermal-neutron Capture Gamma-ray Spectroscopy

Field and process applications have been thoroughly reviewed in recent years [2,3,4,5]. Both are well established, with a strong component of commercial interest. These applications will be treated in these Proceedings by Grau, by Schweitzer and Manente, and by Gozani. The present review will focus on measurements by PGAA centered at several research reactors.

The groundwork was laid for high-resolution reactor-based PGAA eighteen years ago by Isenhour and Morrison [6,7], who showed the applicability of the technique, in steady state and with a modulated neutron beam, using a small Ge(Li) detector; sensitivities under their experimental conditions were not sufficient for the method to become widely used. Other early work was reported by Greenwood [8] and Duffey et al [9,10]. Henkelmann and Born [11] used capture gamma rays analytically in a high-intensity cold neutron beam, but the work did not lead to the establishment of a permanent PGAA facility at the reactor. The current surge of interest is partly due to the construction and analytical application of the internal-sample facility at Los Alamos [12] and external-beam facilities at the NBS Reactor [13] and at the University of Missouri[14], all of which have roughly comparable analytical capabilities. These three installations feature intense neutron sources and large, Compton-shielded Ge gamma detectors. Characteristics of these and other analytical PGAA facilities have been tabulated by Anderson et al [15]. While others are under construction or in the planning stage, we know of no new facilities built in the past few years.

There is a synergism between PGAA and conventional instrumental neutron activation analysis (INAA), in that the two techniques employ the same tools and technology to determine complementary sets of elements. INAA is best known as a method of trace analysis, insensitive in general to the major and minor elements that make up many materials. On the other hand some of the strongest lines in PGAA spectra are often from these very elements, especially hydrogen, silicon, and sulfur. Carbon, nitrogen, and phosphorus may also be prominent in some samples. If PGAA is applied to an unknown material before INAA, the difficult question of neutron self-shielding by strongly absorbing elements in the sample is directly answerable, using the same sample if necessary. The two methods together are capable of measuring nondestructively, in a wide variety of materials of scientific, economic, or regulatory interest, all the major elements except oxygen, all minor elements, and many trace elements, as many as fifty in all. In oxidized matrices such as rocks, a partial check is possible on the accuracy and completeness of the analysis by adding the concentrations of the elements as oxides and comparing the sum with 100 percent. This synergism of prompt and delayed analysis is a major justification for PGAA with the neutron intensity in presently available reactor neutron beams, for an analytical method which may require many hours to irradiate each sample is not ideal for sample-intensive work.

PHYSICS AND CHEMISTRY

An old calumny says that chemists make bad measurements on good material, while physicists make good measurements on bad material [16]. Since the goal of analytical chemists is to characterize "things as they are" [17], the materials we deal with are often bad, and we must take the physicist's approach. Our materials are often very bad indeed, giving spectra with hundreds of peaks of all sizes. A major part of the early work at the Maryland-NBS facility and elsewhere to make the PGAA method reliable for chemical analysis was to separate full-energy from escape peaks and from contaminants. The only way to make certain of the identity of a particular gamma ray in a spectrum is to show that the intensity per unit mass of the capturing element is constant or that the ratio of intensity of two peaks is constant for several samples of the element obtained from different sources. We found, for instance, that in one of the standard tables available in 1976, 38 of the 81 stable elements showed a gamma ray within 5 keV of the capture line of cadmium. It is likely that many of these tabulated lines were in fact due to traces of cadmium in the samples irradiated or in the apparatus.

An essential task in making the PGAA method reliable has been the enumeration of interfering lines which are nearly at the same energy as the line used for analysis. Tables of the most sensitive gamma rays and their major interferences have been published by several analytical laboratories [12,14,18]; this effort is continuing. In addition to peaks from neutron capture in the sample and in its packaging material, a PGAA spectrum contains both peaked and continuum background arising from the environment and apparatus, chiefly from the scattering of neutrons and gamma rays in the probe beam by the sample holder, the collimation, and the sample itself. This residual background limits the sensitivity of the analysis of common low-Z elements, especially C, H (in the organic materials of the shielding) and N (in air and the liquid nitrogen-filled Dewar for the Ge gamma detector).

APPLICATIONS OF REACTOR-BASED PGAA

The PGAA facilities which have been established at several research reactors are capable of simultaneously determining the concentrations of up to twenty major, minor, and trace elements in a variety of matrices. In Table 2 are given the sensitivities and minimum detection limits (C(mdl), defined according to Jaclevic and Walter [19] and Currie [20]) of a number of elements in three widely different sample types. It can be seen that the range of C(mdl) values for the "crustal" elements is close to ideal for multielement analysis of most geological materials. Graham et al [21] and Anderson et al [22] have used PGAA to characterize NBS, USGS, and IAEA geochemical standard materials for up to twenty-one elements. In biological matrices, Failey et al [18] showed that the concentrations of about a dozen elements can be routinely measured.

Many applications of PGAA involve the measurement of only one or a few elements which exhibit especially good sensitivity. For example, boron is a uniquely interesting element, its broad Doppler-shifted gamma-ray peak being easily measurable at concentrations ranging from percent to ppb levels. Few other methods are available for the determination of boron, and volatile boron compounds may be lost during dissolution of refractory materials. The use of boron-doped silicon in the semiconductor industry is

Table 2
Sensitivity and Limits of Detectability of PGAA
for Various Elements in Coal, Basalt, and Bovine Liver

Element	Energy (kev)	Sensit. cps/mg	Detection Limit C(mdl) Coal	Basalt	Liver
H	2223	0.86	10	20	30
B	478	530	0.05	0.02	0.07
C	1262	0.00039	27000	13000	37000
N	1885	0.0030	3400	1400	3500
F	1634(d)	0.0011	6000	3000	8000
Na	472	0.15	300	60	100
Mg	585	0.0085	1200	700	1900
Al	1779(d)	0.024	500	200	500
Si	3539	0.0066	600	300	800
P	637	0.0062	1300	1000	2600
S	841	0.054	200	100	300
Cl	516	1.4	7	4	12
K	770	0.13	70	50	120
Ca	1942	0.022	500	200	1000
Ti	1382	0.38	30	10	30
V	1434(d)	0.33	25	10	35
Cr	835	0.11	100	50	150
Mn	847(d)	0.46	9	4	10
Fe	352	0.046	300	300	400
Co	277	1.2	20	14	20
Ni	465	0.12	1200	700	1700
Cu	278	0.12	200	130	200
Zn	1077	0.019	400	300	600
As	165	0.17	200	200	200
Se	239	0.31	80	80	80
Br	244	0.18	120	150	190
Sr	898	0.031	400	200	600
Mo	778	0.11	90	60	90
Cd	558	170	0.06	0.04	0.09
In	162	2.9	14	15	14
Ba	627	0.55	20	10	20
Nd	697	1.3	7	4	8
Sm	334	640	0.03	0.02	0.03
Gd	182	680	0.05	0.05	0.05
Pb	7368	0.0004	5000	5000	5000

$C(mdl) = 3.29 (R/t)^{1/2}/S$, where R = background counting rate under the peak; t = duration of count, taken as 12 h for basalt and 20 h for bovine liver. The notation (d) indicates that the gamma ray is emitted as a result of the subsequent decay of the nucleus.

too familiar to require comment. Depth profiling of B
by thermal neutron reactions [23] can be made
quantitative by comparison with PGAA analysis of the
same sample, if a single-sided thin B film is measured
by both techniques.

Geochemical applications of boron have been
studied by PGAA at several laboratories, including
measurements of the cosmic abundance of B [24],
measurement of B in geochemical standard materials
[22,25], and the use of gas-phase B for tracing air
masses from coal-fired power plants [26]. In the
latter study boron (which occurs primarily as H_3BO_3 in
the atmosphere) was measured by PGAA both on particles
and in the gas phase, along with S, Cl, and other
important natural and anthropogenic elements by INAA.
It is hoped that the addition of gas-phase species
(measured by PGAA) to the chemical element balance
studies of the atmosphere will help identify a number
of emission sources.

A recent paper [27] shows the use of Sm measured
by PGAA as a non-absorbable biological tracer in
studies of equine digestion.

Hydrogen will be an important focus of future PGAA
measurements with higher quality neutron beams. The
present detection limit (set chiefly by the background)
is barely too high to be of use in measuring hydrogen
in metals at physically interesting concentrations.

COLD SOURCES AND ADVANCED METHODS IN THE FUTURE

Just as the successful application in the recent
few years has been due to the exploitation of
pioneering work by the employment of modern equipment,
we may expect a similar flourishing as cold neutron
beams become established at research reactors around
the world and provide the event rates necessary for
real trace analysis and high throughput of analytical
samples. If a factor of 100 in capture rate can be
gained, then the sensitivities in Table 3 are
attainable.

In addition to greater flux and longer wavelength
than simple neutron beams, a guided beam of cold
neutrons [28] can be virtually free of gamma rays and
epithermal and fast neutrons. However, a great deal can
be accomplished in this regard by incorporating filters
into conventional neutron beams [29]. A radial beam
used for neutron depth profiling at NBS, for instance,
uses single-crystal sapphire and silicon filters to

Table 3

Detection Limits Expected for Cold-Source PGAA Facility

Detection limits in micrograms

H	0.1	K	1	Se	0.1
B	0.0001	Ca	5	Br	1
C	200	Ti	0.2	Mo	1
N	200	V	0.2(d)	Cd	0.0005
F	50(d)*	Cr	0.5	In	0.1
Na	1	Mn	0.1(d)	Nd	0.05
Mg	10	Fe	2	Sm	0.0002
Al	2(d)	Co	0.05	Gd	0.0001
Si	10	Ni	0.5	Pb	50
P	1	Cu	0.5		
S	1	Zn	5		
Cl	0.1	As	2		

*(d) signifies a decay line, not prompt.
Assumptions: Irradiation time 20 hr
Flux 1×10^9 n/cm^{2}s
One 40% Ge detector at 40 cm

deliver a thermal flux of 3×10^8/cm^{2}s at a cadmium ratio of 10,000, with a gamma-ray component of only 200 mrad/hr [23]. Non-thermal components in analytical PGAA beams make necessary the use of large masses of thermalizing and absorbing shielding in the immediate vicinity of the sample and detector, which by scattering contribute to the background. With purely thermal neutrons, hydrogenous materials can be eliminated, absorbers can be more selectively located, and the target-detector assembly can be made more compact and thus more efficient.

Substantial elimination of the low-energy scattered gamma-ray component will make many low-energy capture lines useful for quantitation. Elements which will be measurable at lower levels include Sc, Mn, Co, Cu, Zn, Rh, several rare earths, Ir, Au, and Hg.

Additional sensitivity will be a boon for the analysis of much smaller samples, such as those obtained from new atmospheric sampling techniques which collect particulates from smaller volumes of air.

A capture rate one or two orders of magnitude greater than at present, even with no low-energy scattered gamma rays present, generally implies high-rate gamma spectroscopy. Fortunately, this can be handled by the new generation of transistor-reset preamplifiers, fast amplifiers, loss-correction circuits [30], fixed-dead-time ADCs, high-rate coincidence and anticoincidence electronics, and inexpensive computers now becoming widely available on the commercial market (i. e., to people without electronic design and construction facilities, which includes most analytical chemists). Reactor-based PGAA, as always, will borrow from other branches of gamma spectroscopy and neutron physics, and has especially much to learn from the borehole logging pulsed-neutron-generator technology, which now routinely measures rates of 200,000 counts per second above 0.2 MeV [5].

"...The use of a two Ge(Li) detector coincidence system can provide fruitful information of such high quality that NaI-Ge(Li) coincidence studies are obsolete." So concluded Bolotin at the first meeeting in this series at Studsvik in 1968 [31] -- and this with 15-30 cc detectors! A number of new technologies are ready for transfer to analytical PGAA [29]. Since capture rates may be increased by two orders of magnitude with new sources, and detection efficiencies of Ge detectors are several times higher than just a decade ago, Ge coincidence rates of a factor of several thousand higher than hitherto may be expected, an improvement in counting rate that makes the selective exploitation of individual decay schemes of interest for analytical purposes. A further improvement by increasing solid angle with the use of compact bismuth germanate [32] or cadmium tungstate scintillators for gating detectors may improve coincidence rates still more, the limiting factors probably being neutron scattering by the sample and the ability of laboratory computers to acquire and process the data as fast as they are generated.

SUMMARY

Gamma rays resulting from the capture of reactor neutrons have proven to be of great utility in chemical analysis. The advent of cleaner, more intense neutron beams will make this branch of spectroscopy an equal partner with conventional neutron activation analysis.

REFERENCES

1. E. S. Gladney, A Literature Survey of Chemical Analysis by Thermal Neutron-Induced Capture Gamma-Ray Spectroscopy (Report LA-8028-MS) (Los Alamos Scientific Laboratory, Los Alamos, N.M., 1979).

2. M. D. Glascock, A Literature Survey of Elemental Analysis by Neutron-Induced Prompt Gamma-Ray Spectroscopy and Related Topics (U of Mo (Unpublished), Columbia, Mo, 1984).
3. R. C. Greenwood, In Proceedings of the Third International Symposium on Neutron-Capture Gamma-Ray Spectroscopy and Related Topics, R. E. Chrien, and W. R. Kane, Eds., (Plenum, New York, 1979), pp. 441-460.
4. M. D. Glascock, In Neutron-Capture Gamma-Ray Spectroscopy and Related Topics, T. von Egidy, F. Gonnenwein, and B. Maier, Eds., (Institute of Physics, London, 1982), pp. 641-654.
5. C. L. Herzenberg, Application Potential of Advanced Instrumental Methods for On-Line Automated Composition Analysis of Solid/Liquid Fossil Energy Process Materials, Vol 1: Nuclear Methods (USDOE: ANL/FE-83-21, Argonne, Illinois, 1983).
6. T. L. Isenhour, and G. H. Morrison, Anal. Chem. 38, 162- 167 (1966).
7. T. L. Isenhour, and G. H. Morrison, Anal. Chem. 38, 167- 169 (1966).
8. R. C. Greenwood, Trans. ANS 10, 28-29 (1967).
9. D. Duffey, A. El-Kady, and F. E. Senftle, Nucl. Inst. Methods 80, 149-171 (1970).
10. F. E. Senftle, H. D. Moore, D. B. Leep, A. A. El-Kady, and D. Duffey, Nucl. Inst. Methods 93, 425-459 (1971).
11. R. Henkelmann, and H. J. Born, J. Radioanal. Chem. 16, 473-481 (1973).
12. E. S. Gladney, D. B. Curtis, and E. T. Jurney, J. Radioanal. Chem. 46, 299-308 (1978).
13. D. L. Anderson, M. P. Failey, W. H. Zoller, W. B. Walters, G. E. Gordon, and R. M. Lindstrom, J. Radioanal. Chem. 63, 97-119 (1981).
14. A. G. Hanna, R. M. Brugger, and M. D. Glascock, Nucl. Inst. Methods 188, 619-627 (1981).
15. D. L. Anderson, W. H. Zoller, G. E. Gordon, W. B. Walters, and R. M. Lindstrom, In Neutron-Capture Gamma- Ray Spectroscopy and Related Topics, T. von Egidy, F. Gonnenwein, and B. Maier, Eds., (Institute of Physics, London, 1982), pp. 655-668.
16. E. B. Wilson, Jr, An Introduction to Scientific Research (McGraw-Hill, New York, 1952).
17. G. E. F. Lundell, Ind. Eng. Chem. (Anal. Ed.) 5, 1-15 (1933).
18. M. P. Failey, D. L. Anderson, W. H. Zoller, G. E. Gordon, and R. M. Lindstrom, Anal. Chem. 51, 2209-2221 (1979).
19. J. M. Jaklevic, and R. L. Walter, In X-Ray Fluorescence Analysis of Environmental Samples, T. G. Dzubay, Ed., (Ann Arbor Science, Ann Arbor, 1977), pp. 95-106.

20. L. A. Currie, In X-Ray Fluorescence Analysis of Environmental Samples, T. G. Dzubay, Ed., (Ann Arbor Science, Ann Arbor, 1977), pp. 289-306.
21. C. C. Graham, M. D. Glascock, J. J. Carni, J. R. Vogt, and T. G. Spalding, Anal. Chem. 54, 1623-1627 (1982).
22. D. L. Anderson, Y. Sun, M. P. Failey, and W. H. Zoller, Geostand. Newslett. , (1985).
23. R. G. Downing, R. F. Fleming, J. K. Langland, and D. H. Vincent, Nucl. Inst. Meth. Phys. Res. 218, 47-51 (1983).
24. D. B. Curtis, E. S. Gladney, and E. T. Jurney, Geochim. Cosmochim. Acta 44, 1945-1953 (1980).
25. M. D. Higgins, Geostand. Newslett. 8, 31-34 (1984). and D. Duffey, Nucl. Inst. Methods 93, 425-459 (1971).
26. M. E. Kitto, D. L. Anderson, G. E. Gordon, and J. M. Ondov, ACS Annual Meeting, Washington, D.C. , (1983).
27. W. D. James, F. F. Arnold, K. R. Pond, M. D. Glascock, and T. G. Spalding, 83, 209-214 (1984).
28. H. Maier-Leibnitz, In Neutron Capture Gamma-Ray Spectroscopy, N. Ryde, Ed., (IAEA, Vienna (STI/PUB/235), 1969), pp. 105-112.
29. W. R. Kane, In Proceedings of the Third International Symposium on Neutron-Capture Gamma-ray Spectroscopy and Related Topics, R. E. Chrien, and W. R. Kane, Eds., (Plenum, New York, 1979), pp. 485-502.
30. G. P. Westphal, J. Radioanal. Chem. 70, 387-410 (1982).
31. H. H. Bolotin, In Neutron-Capture Gamma-Ray Spectroscopy, N. Ryde, Ed., (IAEA, Vienna (STI/PUB/235), 1969), pp. 15- 34.
32. D. M. Drake, L. R. Nilsson, and J. Faucett, Nucl. Inst. Methods 188, 313-317 (1981).

IN VIVO DETERMINATION OF PROTEIN BY PROMPT
NEUTRON CAPTURE IN FIBROCYSTIC DISEASE

B.J. Allen, N. Blagojevic
Australian Atomic Energy Commission Research Establishment
Lucas Heights Research Laboratories
Private Mail Bag, Sutherland, NSW 2232, Australia

K. Gaskin, V. Soutter, R. Howman-Giles
Royal Alexandra Hospital for Children
Camperdown, NSW 2050, Australia

ABSTRACT

Measurement of total body nitrogen (TBN) and muscle mass is a useful tool in assessing nutritional repletion of malnourished subjects. The present study examines a prototype facility for measuring TBN by prompt neutron activation with a Pu-Be or ^{252}Cf fast neutron source and observing the 10.8 MeV ground state γ-ray from the ^{14}N(n,γ) reaction with NaI detectors. Dose distributions, background sensitivity and nitrogen yields are reported for phantoms and a maximum radiation dose of 20 mrem per exposure is determined. We propose to measure TBN and muscle mass in a group of malnourished patients before and during nutritional repletion.

INTRODUCTION

Cystic fibrosis (CF) is the commonest autosomal recessive disease in Caucasian children with an incidence of one in 2500 live births. Chronic suppurative lung disease with death due to respiratory failure and pancreatic insufficiency are the major manifestations of CF. With improved therapy for chronic suppurative lung disease, medium survival rates have increased from 5 years in the 1950s to 20 years in the late 1970s. In addition, nutritional factors are considered to influence survival, as malnutrition is associated with a poor prognosis, and the maintenance of near normal nutrition and growth in one clinic has been associated with an even higher medium survival age of 35 years. However, it has yet to be proven that nutritional repletion of a malnourished CF subject will improve prognosis, and we wish to assess the effect of nutritional supplementation in such patients using elemental or polymeric formulae delivered at night by infusion via nasogastric or gastrostomy tubes. To optimise the use of these expensive formulae and to ensure their efficacy in fat and protein repletion, careful monitoring of body composition is essential before and during nutritional therapy.

Lean body mass can be determined by measuring ^{40}K in a whole body monitor because potassium is nearly absent in adipose tissue. For normal subjects, total body potassium (TBK) and nitrogen correlate well, but in malnutrition, a tissue potassium deficit can occur and lean body mass may be underestimated.

It is therefore important to determine both potassium and nitrogen and, to this end, we have evaluated a prototype TBN facility with Pu-Be and ^{252}Cf neutron sources.

TOTAL BODY NITROGEN MEASUREMENT

The technique of TBN measurement was first developed by Biggin et al.[1] at Birmingham using 2.6 MeV neutrons from a cyclotron. Later isotropic neutron sources were used to provide the neutron flux by Mermagh et al.[2] at Toronto, Vartsky et al.[3] at Brookhaven National Laboratory and, most recently, Beddoe et al.[4] at Auckland. While the $^{14}N(n,2n)^{13}N$ reaction can be used to produce the positron emitter ^{13}C, the prompt neutron capture reaction $^{14}N(n,\gamma)^{15}N$ with a neutron source allows a TBN facility to be readily installed in a hospital. About 15% of nitrogen capture γ-ray decays go to the ground state, emitting a 10.83 MeV γ-ray which can be detected by a NaI spectrometer because the energy is well above the intense background of γ-rays from the source and from neutron capture in the detector shielding and the detector itself. However, it is important to minimise background count rates and the presence of pile-up events in the 10 to 11 MeV region using a pile-up inspector and inhibit gate.

NEUTRON SOURCES

We have made measurements with a 200 x 150 mm NaI detector and ^{239}Pu-Be, ^{238}Pu-Be and ^{252}Cf fission sources (∿2 x 10 neutrons s^{-1}) using a 6 L water phantom with 4.7 mol/L NH₄OH, about twice the tissue equivalent nitrogen concentration, and find the ^{252}Cf source gives a fivefold improvement in nitrogen to background ratio compared with Pu-Be, the spectrum being shown in Fig. 1. The 2.65 y half-life of ^{252}Cf counts against use of this source, and the fission neutron spectrum is somewhat softer than the Pu-Be sources. However, the neutron distribution in the 12 L phantom is comparable to that for Pu-Be.

SHIELDING GEOMETRY

A suitable exposure time for a TBN determination is nominally 1000 s for which an uncertainty of 4% can be obtained. We found the optimum source-skin distance (SSD) of 300 mm for our Cf source. Lead shielding in the collimator is very effective in reducing the detector background and count rate, as well as 100 mm lead on top of the shielding block. The skin-detector distance is ∿150 mm with 50 mm of 70% boron carbide in paraffin as a fast neutron shield for the NaI detector. Having available a 200 x 150 mm NaI detector in a massive Pb and borated paraffin shield, we have chosen the supine/prone geometry with source below and bilateral NaI detectors. The second detector is 150 x 100 mm NaI. The patient will traverse a 230 x 230 mm neutron beam on a moving table, first supine, then prone, to remove asymmetry effects and average out the neutron dose to the body.

LINEARITY

Linearity of the system was determined with 6 L phantoms with different concentrations of NH₄OH ranging from 1 to 4.7 mol L^{-1} (i.e. 13.8 to 66.9 g N kg^{-1}) (Fig. 2). The system is, of course, non-linear

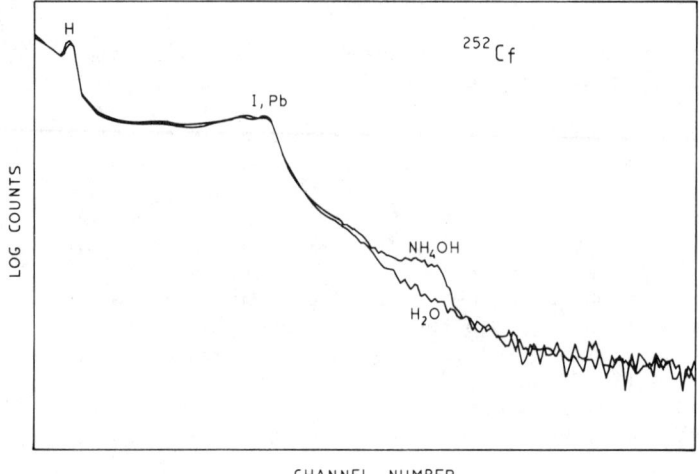

Fig. 1 Gamma-ray spectrum for 3 mCi ^{252}Cf neutron source & 6 L of 4.7 mol L^{-1} NH$_4$OH at SSD = 305 mm

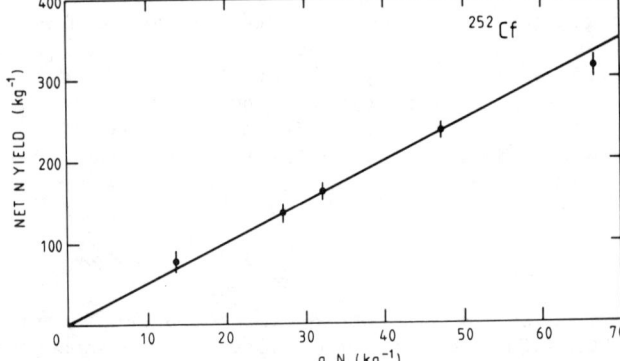

Fig. 2 Net nitrogen yield as a function of nitrogen content of 6 L NH$_4$OH phantoms

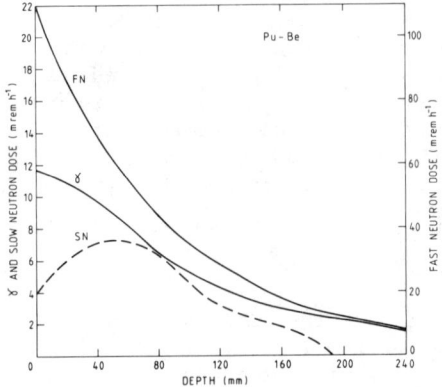

Fig. 3 Dose-equivalent distributions (mrem h^{-1}) for 11.5 Ci ^{238}Pu-Be source for 2-12 L phantoms at SSD = 305 mm. Gamma (γ), slow neutron (SN) and fast neutron (FN) distributions are given

in response to different thicknesses (or volumes) of phantom at the same NH$_4$OH concentration. This is because of inadequate thermalisation for small volumes (<4 L) and neutron attenuation at large volumes (>10 L) for the same phantom geometry. In the region of interest for CF children (4 to 8 L), the ratio of N and H yields was found to be linear, in agreement with Vartsky et al.[5]

BODY MASS BACKGROUND

The volume of the neutron scattering phantom influences the overall background observed by the NaI detector. By using three containers with different areas and varying the fluid volume, the standard deviation of the nitrogen peak background in the region of interest was found to be ∿3% for the Pu-Be source. A volume correction further reduces this uncertainty.

DOSE-EQUIVALENT DISTRIBUTIONS

Thermal neutron, γ-ray and fast neutron films were placed at 40 mm intervals between 2 L bags of 4.7 mol L^{-1} NH$_4$OH solution in a 330 x 245 mm container and measurements were made with a calibrated fast neutron rem counter. The dose-equivalent distributions as a function of phantom thickness are shown in Fig. 3 for SSD = 305 mm and the Pu-Be fission source. The fast neutron dose equivalent peaks at the lower surface and decreases exponentially through the thickness of the phantom. The thermal neutron dose is relatively low at the surface and builds up to a maximum at about 50 mm thickness.

The surface dose is the maximum dose which is ∿120 mrem h^{-1}. By changing the patient from supine to prone position half way through the exposure and slowly moving the patient across the beam from shoulder to lower thigh, the maximum dose is reduced to about 70 mrem h^{-1} or about 20 mrem for 1000 s exposure. The average dose equivalent to the patient is ∿12 mrem (120 μSv) in 1000 s.

It is planned to measure the TBN of each patient in the pilot study three times over a six month period (at 0, 3 and 6 months). The expected average dose to the patient would be less than one half of the annual X-ray dose for exposed patients in the USA in 1980.

ACKNOWLEDGEMENT

The assistance of T. Nantawisarakul is gratefully acknowledged.

REFERENCES

1. H.C. Biggin et al., Nature New Biol., 236, 187 (1972).
2. J.R. Mermagh et al., Phys. Med. Biol. 22, 831 (1977).
3. D. Vartsky et al., J. Nucl. Med. 20, 1158 (1979).
4. A.H. Beddoe et al., Phys. Med. Biol. 29, 371 (1984).
5. D. Vartsky et al., J. Radioanal. Chem. 48, 243 (1979).

In Situ Neutron-Induced Spectroscopy of Geological Formations With Germanium Detectors

J. S. Schweitzer and R. A. Manente

Schlumberger-Doll Research, Ridgefield, CT. 06877

Abstract

Neutron-induced gamma-ray spectroscopy, when performed with a germanium detector, can provide quantitative elemental analysis of geological formations from measurements performed in a well. The most difficult source of background to deal with is that due to the spectral yield from the elements present in the material within the well that are also of interest in the elemental analysis of the rock. The use of different types of spectroscopy and an illustration of the spectral constraints for performing a geologically significant elemental analysis, rather than ore body or major element analysis, are presented from borehole and laboratory studies.

Introduction

The possibility of using germanium detectors under field conditions has resulted in a number of applications where the improved energy resolution compared with scintillation detectors has produced spectroscopic analyses that were not previously practical.[1] In general, these applications have been focused on very narrow spectroscopic problems such as the evaluation of ore-quality minerals and coal or the analysis of a few major elements. Now that the technology for performing various types of gamma-ray spectroscopy has become routinely established under field conditions, the desire to provide a complete elemental analysis to describe generally the environment of the measurement[2] leads to the need of a basic understanding of the complexities introduced by performing a measurement, from inside a borehole, of the surrounding formation which is infinite in extent and heterogeneous. In particular, not only do the geometrical effects introduced by the borehole in a subsurface measurement have to be accounted for, but the contents of the hole and the hardware used for the measurement, often containing elements that are also present in the rock and fluids, become a part of the measurement. To obtain elemental concentrations that are indeed representative of only the rock and fluid requires both an understanding of the effects on the measurement of the borehole environment and, where practical, the use of multiple measurements to unravel the elemental concentration of a particular measurement from competing sources of production.

Many of the physical effects induced in an elemental measurement are discussed in another paper at this conference.[3] We focus here on the additional considerations which are necessary when using germanium detectors for multielement analysis. The first obvious consideration is that speed of measurement is critical. Under field conditions, the luxury of spending a great deal of time to overcome the effects of poor peak-to-background ratios does not exist. Thus, it is necessary to ensure the best possible resolution and counting rates. One aspect involves the quality of the detector and the electronics with an optimized source strength.[4] However, when attempting to perform multielement analysis, the ambiguity in the identification of particular gamma rays to an initial element may require a compromise on the optimum count rate. For example, the highest neutron flux can be obtained with d,t neutron generators which produce 14-MeV neutrons. However, when using 14-MeV neutrons to produce delayed activity, the same final product can be created from more than one initial element. Thus, an attempt to identify the concentration of sodium through the ^{24}Na decay from the ^{23}Na(n,γ)^{24}Na reaction with thermalized neutrons would be severely complicated by the ^{24}Na activity created by the ^{24}Mg(n,p)^{24}Na and ^{27}Al(n,α)^{24}Na reactions. To obtain the most statistically significant result, it may be advantageous to use a lower energy chemical source of neutrons such as ^{252}Cf which would produce ^{24}Na activity only from the neutron capture on ^{23}Na and not from the high-energy-neutron-induced reactions, even at the expense of a smaller neutron flux. In other cases, however, the concentrations can be obtained from other reaction branches using the high-energy-neutron-induced activities where the higher flux attainable with an accelerator produced source of 14-MeV neutrons overcomes the generally lower cross sections for high energy reactions compared to thermal capture cross sections.

Results

For complete elemental analysis, the spectrometer must be designed to perform all possible types of gamma ray spectroscopy induced by neutrons: thermal capture, inelastic scattering, high energy particle producing, high energy reaction- and thermal capture-induced delayed activation, and the detection of natural activity. This is necessary as no single type of spectroscopy is sufficient for detecting the broadest group of elements. The prompt gamma rays produced by inelastic scattering or high energy neutron reactions can most efficiently be produced by an accelerator source of 14-MeV neutrons and requires the detector to be located near the target of the accelerator (typically less than a meter). Neutron capture spectroscopy can be performed with d,t neutrons, lower energy d,d neutrons produced by accelerators, or chemical sources such as ^{252}Cf and also requires that the neutron source be located close to the detector. However, when delayed activation is performed, it is important that the neutron source be located sufficiently far from the detector (typically 3-10 m) to ensure that activity is not generated in the cryostat. This is particularly important when trace element detection is desired.

This latter point is illustrated in Fig. 1 which shows a comparison of downhole measurements of thermal neutron Al and V activation produced with a ^{252}Cf source to the results obtained with laboratory neutron activation analyses, NAA, performed on cores from the well. The solid bars on the figure which extend from the axis show the magnitude of the NAA results obtained from the cores. The solid curve on the Al portion of the figure shows the downhole results when the measurement was performed continuously. The error bar symbols for both Al and V data indicate the mean value $\pm$ 1 σ from the results of 300 sec stationary measurements. The downhole measurements are scaled in counts/sec and the laboratory results are scaled at 20% maximum for Al and 200 ppm maximum for V. When comparing the results, it is important to note that absolute depth adjustment is difficult and apparent disagreements may well be due to improper depth matching or to the difference between the small size of the core sample used for laboratory NAA ($\sim$ 1 gm) and the heterogeneous volume measured in the borehole of many tens of kilograms. Al concentration is related to the clay minerals found in rock, while the V concentration is related to the asphaltine content in oil.

An example of the difference between spectroscopic logging for an element in an ore body, as opposed to when that element is desired as part of an elemental analysis when the concentration is more typical of the average crustal abundance, is shown in Fig. 2. These spectra are of thermal capture gamma rays obtained with the ^{252}Cf source and the samples located near the detector. The upper spectrum was obtained from an artificial sample that contained 10% by weight of TiO_2. This concentration is typical of the concentration in an ore body[5] and the Ti capture lines are easily detected. The lower spectrum is obtained from a core sample which contains ~ 0.2% by weight of TiO_2. Two important features can be seen. The first is that the spectrum contains many gamma ray peaks from components of the cryostat such as Fe, Ni, and Cr which would make the quantitative analysis of these elements in the formation difficult. In addition, the primary capture lines from Ti at 6759.7 keV and 6418.0 keV are difficult to separate from common contaminant peaks when there is a small Ti concentration. A primary interference is provided by the single and double escape peaks of the 7278.9 keV and the full energy and single escape peaks of the 5920.5 keV capture peaks from the Fe contained in the cryostat and other mechanical parts. Cl is also commonly present in the logging environment and contributes interfering lines from the double escape of the 7790.0 keV peak, the double escape from the 7413.8 keV peak, and the full-energy 5715.2 keV peak. In this situation, the sensitivity for determining the concentration of Ti is critically dependent on maintaining optimum resolution to separate the Ti lines from the contaminant lines from Fe and Cl to achieve a significant quantitative analysis.

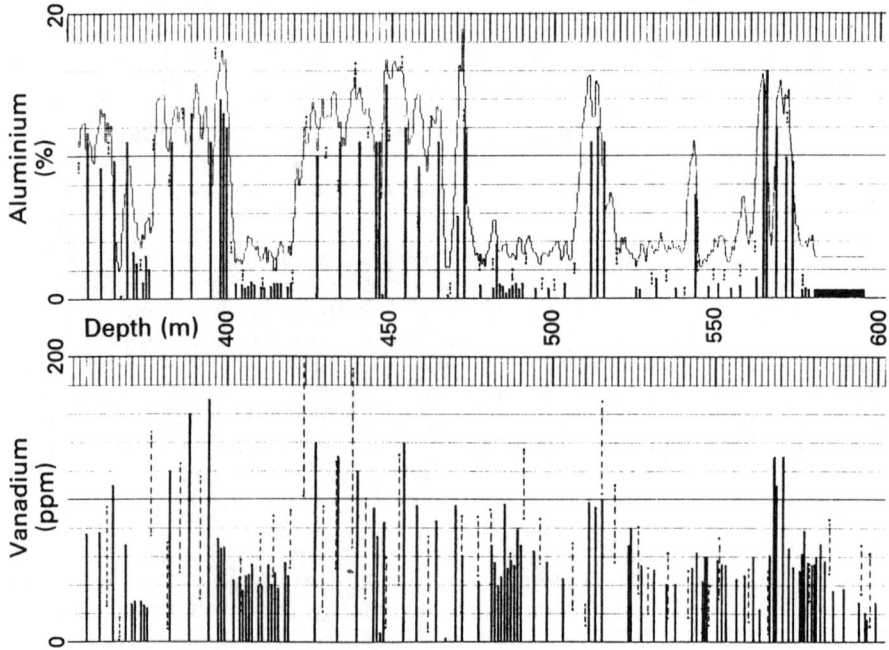

Figure 1. Comparison of downhole measurements of Al and V concentrations with results of laboratory NAA on core samples.

References

[1] R. C. Greenwood, Third International Symposium on Neutron Capture Gamma-Ray Spectroscopy and Related Topics, (BNL, New York, 1978) p. 441, and references therein.

[2] M. M. Herron, The Twenty-First Annual Meeting of the Clay Minerals Society, Baton Rouge, Oct. 1-4, 1984.

[3] J. A. Grau, S. Antkiw, R. C. Hertzog, R. A. Manente, and J. S. Schweitzer, This conf.

[4] J. L. Mikesell, F. E. Senftle, and R. J. Macy, Third International Symposium on Neutron Capture Gamma-Ray Spectroscopy and Related Topics, (BNL, New York, 1978) p. 693.

[5] P. F. Wiggins, D. Duffey, and F. E. Senftle, Trans. Am. Nucl. Soc., **13**, 490 (1970).

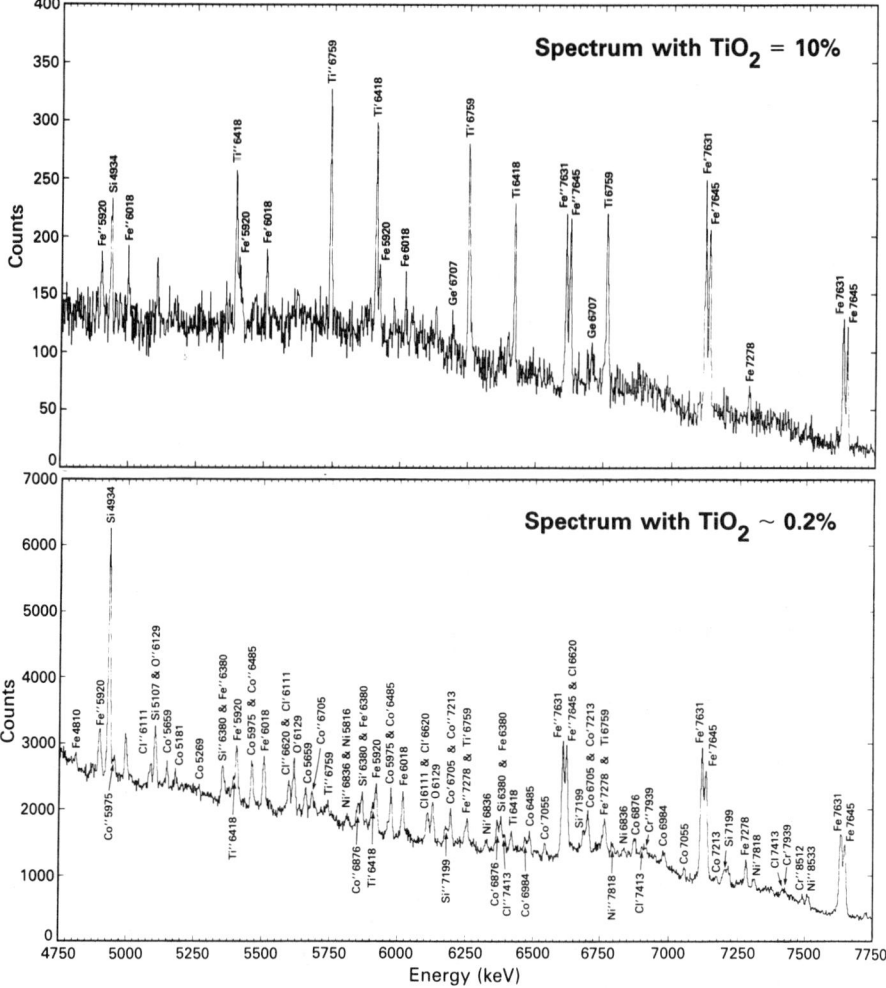

Figure 2. Capture spectra from samples with 10% and 0.2% TiO_2 by weight.

PHYSICS OF RECENT APPLICATIONS OF PGNAA FOR ON-LINE ANALYSIS OF BULK MINERALS

Tsahi Gozani
Science Applications International Corporation
1257 Tasman Drive, Sunnyvale, CA 94089

ABSTRACT

In the last few years the coal, cement and other mineral industries are slowly but steadily coming to realize the potential economical, technological and environmental benefits from using nuclear and especially PGNAA techniques. This realization along with the great efforts and the technological developments of the last decade culminated in a successful production and operation of a family of on-line analyzers under the name "Nucoalyzer."

The main technical factors that contributed to this success are:

- The development of a very stable high count rate gamma spectroscopy for high (Ge) and medium (NaI or BGO) resolution detectors

- the emergence of reliable large radiation-hardened Ge detectors

- careful study and understanding of the transport processes of neutron and the associated gamma rays in various bulk media

- the incorporation, from the outset, of realistic industrial constraints along with those determined by the physics and electronics

- the availability of small but reliable and powerful multichannel analyzers and microprocessors

- incorporation of non-nuclear technique, such as micro-wave transmission, to supplement the nuclear techniques

The physics and engineering principles of the Nucoalyzer are reviewed. The successful system optimization, that led to a major reduction in the effect of extraneous effects, such as bulk density, moisture variation, geometrical instabilities, radiation damage, etc., are discussed and some highlights of results are presented.

INTRODUCTION

Prompt capture gamma rays emitted as a result of neutron (mostly thermal) radiative capture provide an invaluable tool for near instantaneous elemental analysis of material streams.

Many industrial processes use large amounts of valuable but highly heterogeneous materials. They are difficult to accurately and quickly characterize (e.g., to determine composition) thus prohibiting process control. Examples of such situations are provided by the coal industry, other fossil based industries, like oil-shale, the mineral industries like copper, phosphate and cement, and the metal industry. Efficient use of such material requires some kind of process control, e.g., mineral beneficiation, coal washing and blending to meet quality requirements. Conventional measurement methods require extensive sampling and chemical- analytical techniques which virtually rule out the possibility of process control. The prompt gamma neutron activation analysis (PGNAA), where neutrons from an external isotopic source, e.g., ^{252}Cf, ^{241}Am-Be, are the interrogating radiation and the capture gamma rays provide the elemental signatures, provides the means for such process control with quality equals or surpassing conventional sampling-chemical methods. Considerable effort over the last 8 years was expended to research, develop, design, build and field test fully industrial on-line nuclear analyzers of coal based on the PGNAA technique.

Several technological developments over that period along with a thorough neutron-gamma physics study of the problem greatly contributed to the successful completion of the project. The main contributing factors for the accomplishments are enumerated below:

- the development of a very stable high count rate gamma spectroscopy for high (Ge) and medium (NaI or BGO) resolution detectors

- the emergence of reliable large radiation-hardened Ge detectors

- careful study and understanding of the transport processes of neutron and the associated gamma rays in various bulk media

- the incorporation from the outset of realistic industrial constraints along with those determiend by the physics and electronics

- the availability of small but reliable and powerful multichannel analyzers and microprocessors

- incorporation of non-nuclear technique, such as micro-wave transmission, to supplement the nuclear techniques

- continuous interactions with the potential industrial users

The general theoretical usefulness of the PGNAA method can be expressed by the relative detectability of various elements by

Table I Relative Detectability of Various Elements

Element	E (KeV)	D
H	2223	2.0×10^{-1}
Be	6809	3.9×10^{-4}
B	7005	1.3×10^{-4}
C	4945	1.1×10^{-4}
N	10829	4.5×10^{-4}
O	3271	1.8×10^{-6}
F	3589	3.3×10^{-5}
Na	3982	1.9×10^{-3}
Mg	2828	6.6×10^{-4}
Al	7724	1.4×10^{-3}
Si	3539	2.3×10^{-3}
S	5421	5.8×10^{-3}
Cl	6111	1.1×10^{-1}
K	5381	2.9×10^{-3}
Ti	6760	1.8×10^{-2}
V	7163	7.8×10^{-3}
CR	8884	9.6×10^{-3}
MN	7244	1.8×10^{-2}
Fe	7631	1.4×10^{-2}
CO	6876	3.1×10^{-2}
Ni	8999	3.4×10^{-2}
Cu	7915	1.1×10^{-2}
Zn	7863	1.1×10^{-3}
Ge	3028	7.4×10^{-3}
Zr	6294	2.0×10^{-4}
Mo	6919	5.4×10^{-4}
Ag	5698	4.1×10^{-3}
Cd	5824	2.8×10^{-1}
Gd	4843	$1.6 \times 10^{+1}$
W	6190	2.9×10^{-3}
Au	6251	1.7×10^{-2}
Pb	7368	4.6×10^{-4}
Bi	4171	4.8×10^{-5}

thermal neutron capture gamma rays. This detectability[1] (see Table 1) is defined by

$$D = \frac{0.6}{A} \sigma_o I_\gamma \text{ (max)} \qquad (1)$$

where A is the atomic weight, σ_o is the standard thermal (n,γ) capture cross sections (in barns) and I is the highest branching ratio for that element for gamma ray energies above 2.223 MeV, with generally the least amount of interference in bulk mineral. This table and the fact that in PGNAA all elements are measured simultaneously explains why the method is generally not suitable for trace elements, but very appropriate for major and minor elements with concentrations of roughly above 0.1 wt.%.

Some of the nuclear physics considerations in the system design and performance will be discussed briefly. The schematic of the CONAC (Continuous On-line Analyzer of Coal) is shown in Figure 1. This figure shows the key features of the system that

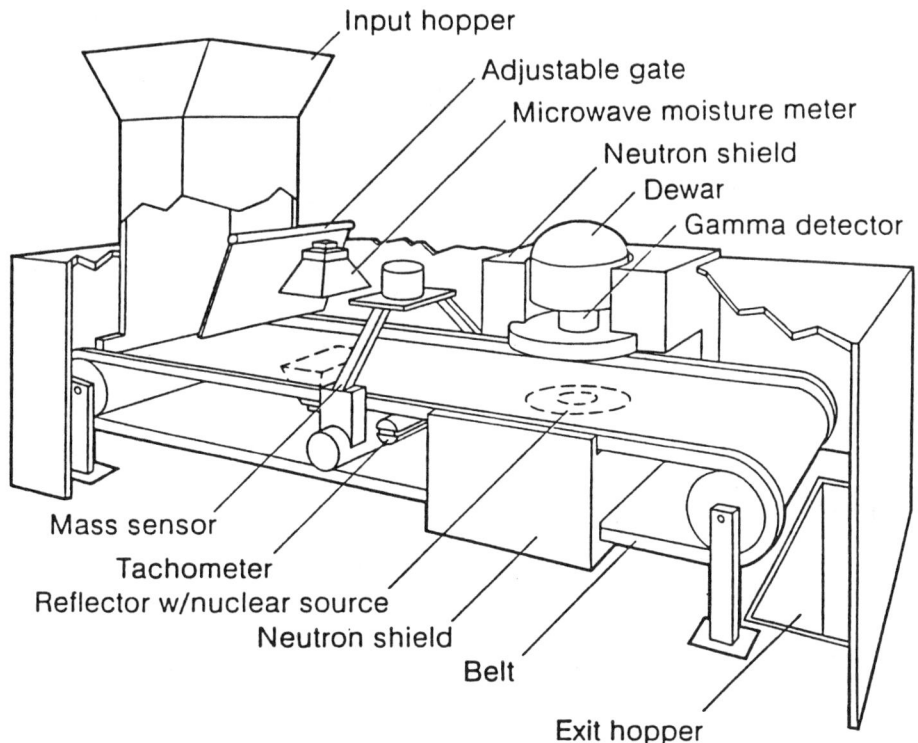

Fig. 1. Schematic Impression of CONAC

are tied closely to the physical processes and optimization which are discussed in the ensuing chapters. The coal is fed into a hopper then directed onto a wide belt, maintaining a constant height. The belt carried the coal through various measurement stations: microwave attenuation for moisture determination, gamma-ray densitometer for density (and mass flow rate) measurements and finally through neutron irradiation zone where the prompt gamma rays are generated and detected by an appropriate detector system.

PHYSICAL PRINCIPLES

The signal measured in a PGNAA system for a bulk material sample is the capture gamma rays from a specific element (or isotope) that escape absorption in the sample and interact within the detector volume. The relationship between this signal and the desired quantity, namely the fractional weight of the specific element, measured usually in units of weight percent (wt.%), is a complicated one. It involves the neutron (mostly thermal) flux level and spatial distribution, as a source for the capture gamma rays, the bulk density of the sample, which affects both the neutron flux and the gamma ray attenuation, the geometry (e.g., sample thickness) of the system and finally the elemental composition of the sample. The presence and abundance of elements other than the measured one can affect the latter in two distinct ways: one effect is direct via spectral interference, namely closely packed or overlapping gamma lines or/and contribution to Compton background. Chlorine provides an excellent example of an element that causes serious spectral interference requiring great care in extracting correct net peak area independent of interferences. The second and indirect way the composition influences the signal is through neutron and gamma moderation and absorption. Hydrogen plays a dominant role in this effect. For example, in coal it is the main neutron absorber and virtually the only neutron moderator. Trace concentration (0-100 ppm) of natural boron, chlorine (5000 ppm and above) and nitrogen (1% and above) are the other important absorbers. A graphic summary of these effects is shown in Figure 2.

The functional relationship between the desired signal, R_{ij}, namely the net peak area count rate of the j-th gamma line emitted by the i-th element, and the fractional weight, W_i, of that element in the sample is given below for the case of thermal neutron capture:

$$R_{ij} = W_i \left\{ 0.6 \, v_0 \, \frac{I_{ij} \sigma_{oi}}{A_i} \, \epsilon_j \int_{D_s} \vec{dr} [N(r) \int \frac{1}{4\pi r_d^2} e^{-\rho \mu_j r_d} \, \vec{dr_d}] \right\} \quad (2)$$

MACROSCOPIC INTERACTION

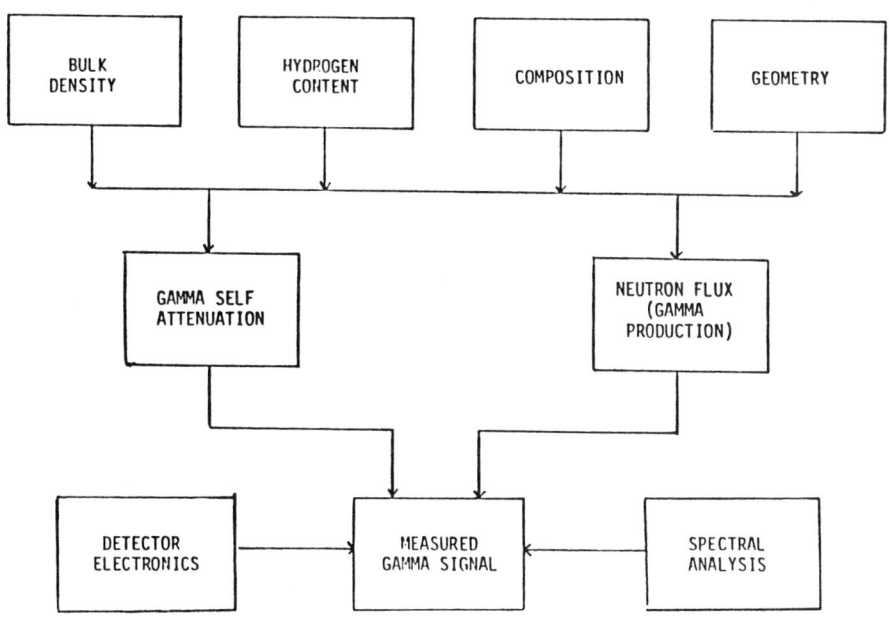

Fig. 2. Schematic illustration of the effect of various parameters on the measured gamma signal.

where

R_{ij} = the net-interference free-peak area count rate from the j-th line emitted by the i-th element, W_i = the concentration of the i-th element, v_o = 2200 m/s, ρ = sample bulk density, I_{ij} = branching ratio for the j-th gamma line, σ_{oi} = the microscopic thermal neutron capture cross-section at v_o, A_i = the atomic weight, ϵ_j = the intrinsic detector efficiency, $N(r)$ = the thermal neutron density in units of neutron/cm^3/sec., μ_j = the total attenuation coefficient of the specific gamma ray in units of cm$_2$/g, D_s = detector surface, $\vec{r}$ and $\vec{r}_d$ are defined in Figure 3.

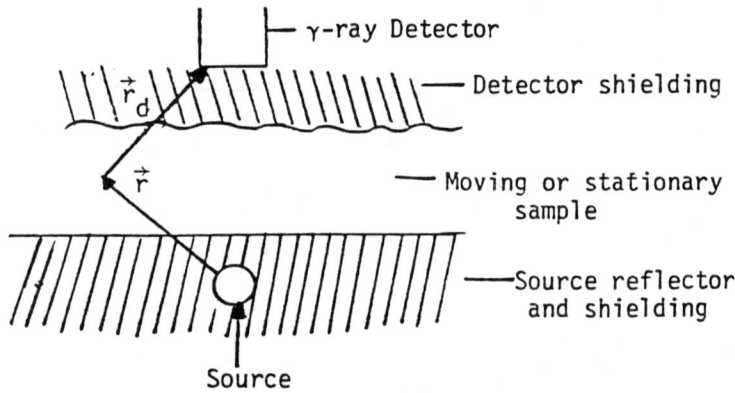

Fig. 3. Schematic of a basic PGNAA system.

The purpose of a good system design and optimization is to make the expression, excluding W_i (or W_j), a constant independent of those parameters that are difficult or impossible to control, namely and other elemental compositions. This constant is then determined during calibration. The possibility, in principle, to optimize the system is provided by the causal relationship between a neutron (mainly thermal) and a gamma-ray. Increasing bulk and/or hydrogen density usually will increase the neutron flux especially in the vicinity of the fast neutron source. This increase hence results in higher capture gamma ray production. But the same change in densities will increase the attenuation of the gamma rays on their way to the detector. These two counteracting processes can result in, under the right condition (e.g., proper geometry), a full or partial cancellation of these effects making the signal nearly independent of the above mentioned parameters.

The presence of such an optimum geometry was demonstrated calculationally using neutron-gamma ray coupled transport codes solving the Boltzman equation and then confirmed experimentally. Some of the results for coals of different density in a spherical shell geometry with a point source at the center are shown in Figure 4. Similar results are shown for variable hydrogen concentration with a fixed density (0.8 g/cm^3) in Figure 5. Similar calculations were done for a more realistic slab geometry with a plane ^{252}Cf source on one side and a detector on the other side of the sample. The results are shown in Figures 6 and 7.

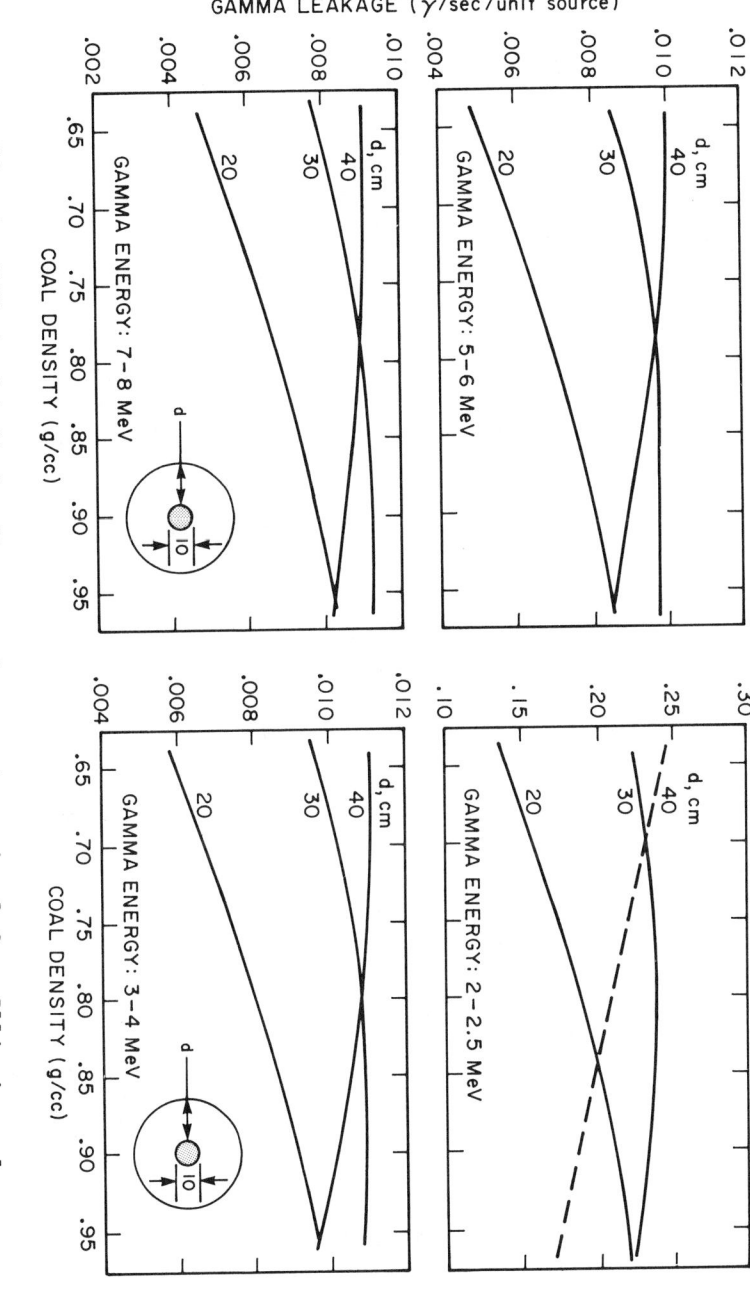

Fig. 4. Effect of bulk density on neutron capture signal from Illinois coal
Note: At 30 cm. thickness of spherical shell, the density effects are minimal for all gamma ray energies.

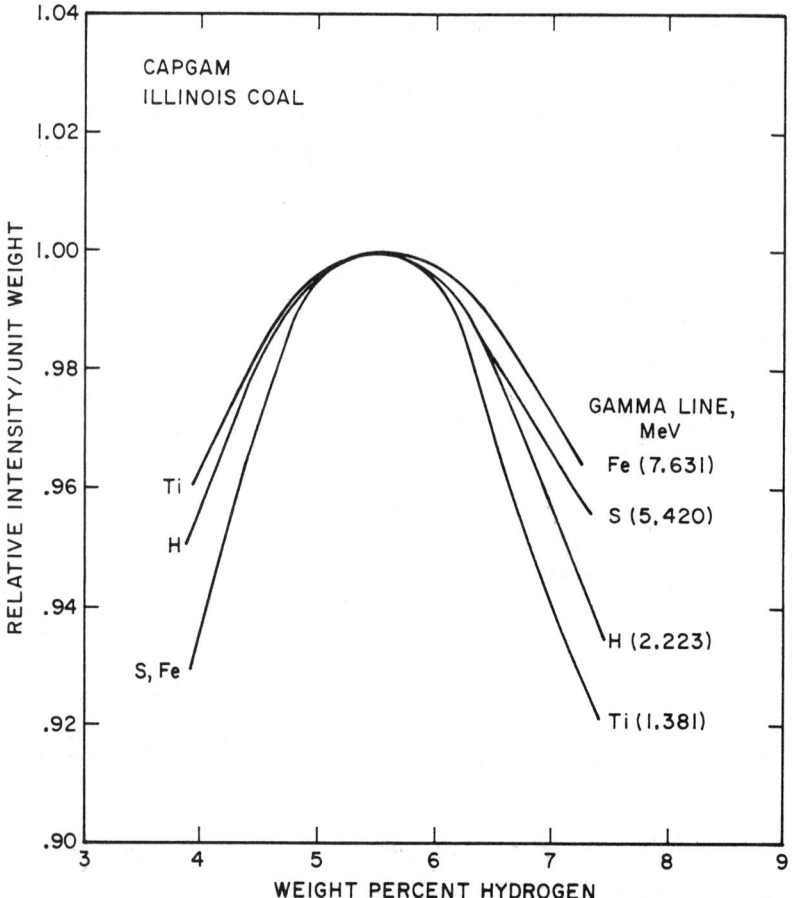

Figure 5. Effect of hydrogen weight percent on the capture gamma intensity.

Thus the proper choice of geometry that must, however, be compatible with the industrial requirments, reduces considerably the effect of ρ and W_H. The residual effects and those of the other absorbing elements can be further reduced via theoretical correction and/or experimental normalization.

DETECTORS AND ELECTRONICS

The need to obtain timely measurements requires high count rates, thus strong sources and relatively large detectors. This usually compromises the high energy resolution of the germanium detector and also limits its life-span because of damage from fast neutrons. The high resolution is a fundamental requirement in order to be able to determine 12 or so different major and minor

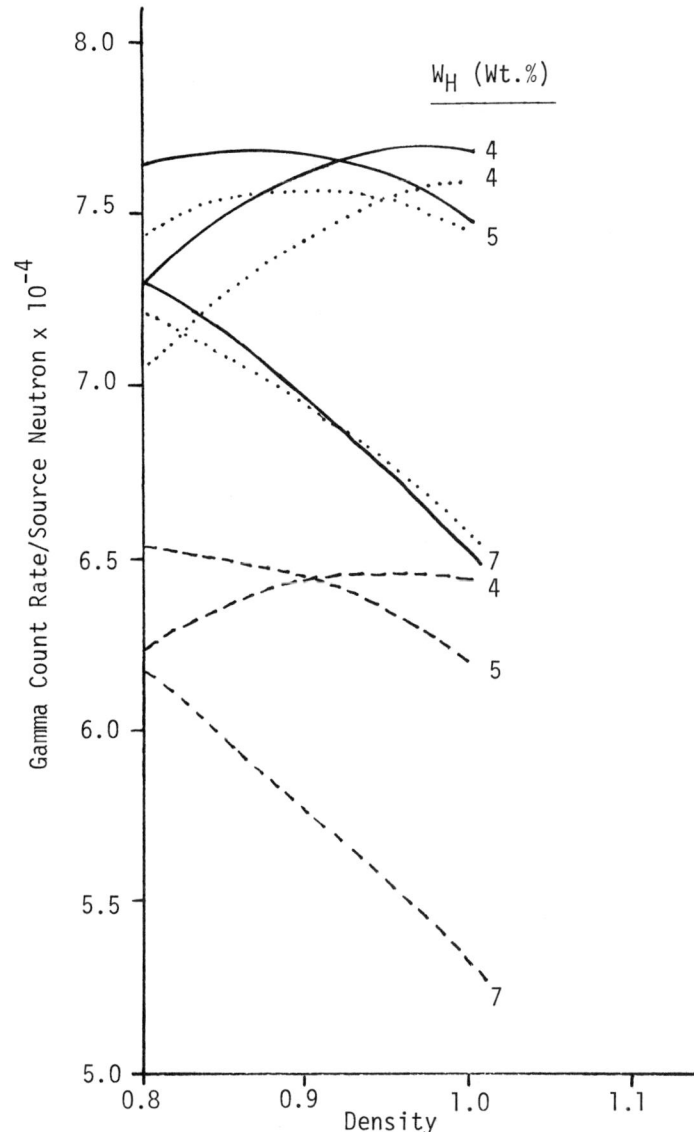

Figure 6. Effect of bulk density on neutron capture signal from Illinois coal for different hydrogen concentrations (slab geometry, 25 cm. thick).

Note the relative flat maxima around $\rho = 0.9$ g/cm^3 for W_H around 5 ± 5 wt.%

———— = 5-6 MeV
········ = 3-4 MeV
— — — = 7-8 MeV

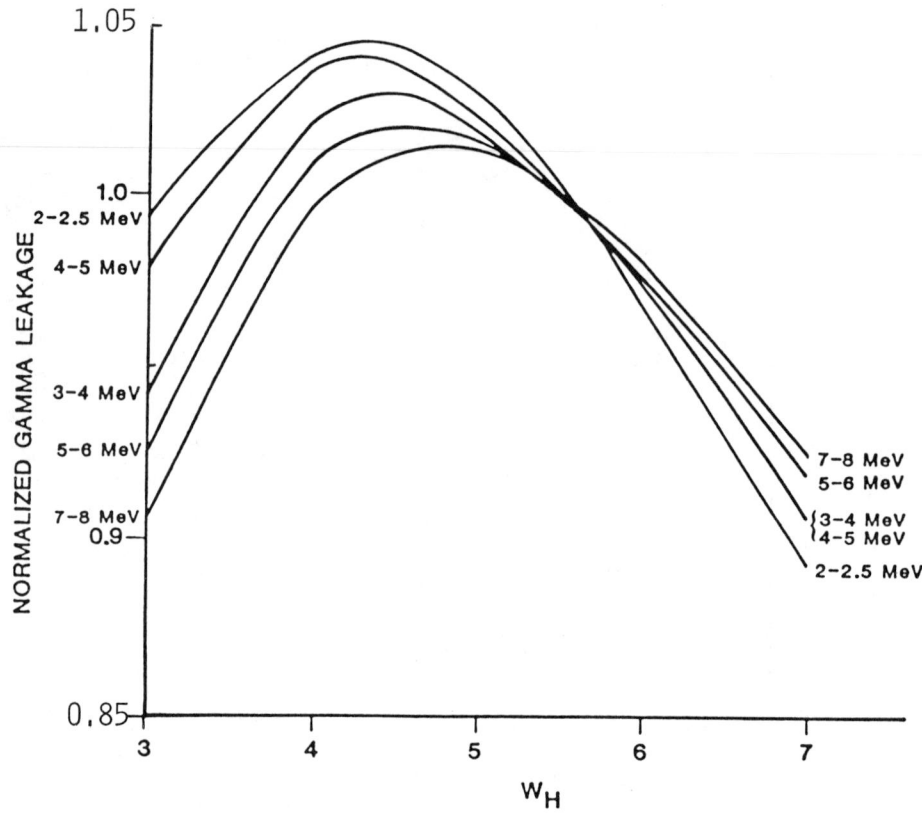

Fig. 7. Gamma ray leakage normalized to its value at W_H = 5.58%, as a function of W_H for ρ = 0.9 g/cm^3 (Illinois coal)

elements. Using the radiation resistant N-type hyper-pure Ge detector with a carefully designed shielding that maximizes the attenuation of the fast neutrons with only a small impact on the neutron capture gamma rays greatly extends the useful life of the detector. Under this condition little or no resolution degradation (which is constantly monitored) is expected over a period in excess of a year. At that time the detector may be sent in for restoration.

The high count rate entails pulse pile up that causes spectral degradation. In order to reduce this effect, time variant gated amplifiers[1,2] are being used, allowing reasonable throughput and very good resolutions even with total countrates in excess of 10^5 cps. When substantially higher count rates are necessary in order to measure important elements such as carbon, sulfur, nitrogen, chlorine, silicon, and iron, in a short time (e.g., within minutes), a highly efficienct (15cm diameter x 15cm thick) a medium resolution NaI(Tl) scintillator is used in conjunction with a Ge detector. Much higher count rates are attained with the former detector. However, the specificity of the elemental determination by that detector is achieved and maintained by a continuous calibration with the Ge detector, as both are viewing virtually the same incident gamma ray spectrum. As an example, multi-element spectra of medium complexity, (of a calcium bearing mineral), measured by the two detectors is shown in Figure 8. A more complex spectra, that of capture gamma ray in coal are shown in Figures 9 and 10 as measured with a Ge and NaI detector, respectively. These spectra show that for some elements with dominant lines, one can get the line intensity from the NaI spectrum with relative ease using some simple unfolding techniques. Extracting intensity from NaI spectra of other lines, which are either weak and/or crowded together, is impossible or impractical. These lines can be determined by the high resolution Ge detector. At the end of the Ge counting cycle, which is longer than that of the NaI's, to achieve satisfactory precision, the NaI detector is calibrated by the former detector, e.g., the ratio between the interfering lines is transferred from the Ge to NaI. This allows the NaI to furnish the elemental concentration in a considerable shorter time than otherwise. The ratio of the interfering lines is used until it is updated at the end of the next Ge cycle. Through this means one achieves both the required high elemental specificity (via Ge) and short response time down to a few minutes (via NaI).

APPLICATIONS

Industrial need, industrial requirements, physics calculations and experiments culminated into the design and construction of a line of nuclear based non-intrusive on-line analyzers for mineral composition. The most sophisticated of these, called CONAC (Continuous On-line Analyzer of Coal) alluded to earlier will be briefly described here. This analyzer encompasses all the nuclear-neutron physics attributes discussed in the previous chapters, along with all important industrial prerequisites.

The components of the CONAC system are schematically shown in Figure 11. This figure highlights all the key components of the CONAC, including the hybrid (Ge and NaI) detector concept discussed earlier.

The accuracy of CONAC in on-line operation is demonstrated in Figures 12 and 13. These figures show a comparison between the given composition of coals from all over the world as determined by careful sampling and chemical analysis (according to established ASTM standards) and PGNAA measurment (calibrated already in wt.% units). These figures and many others show excellent agreement with the conventional analytical techniques.

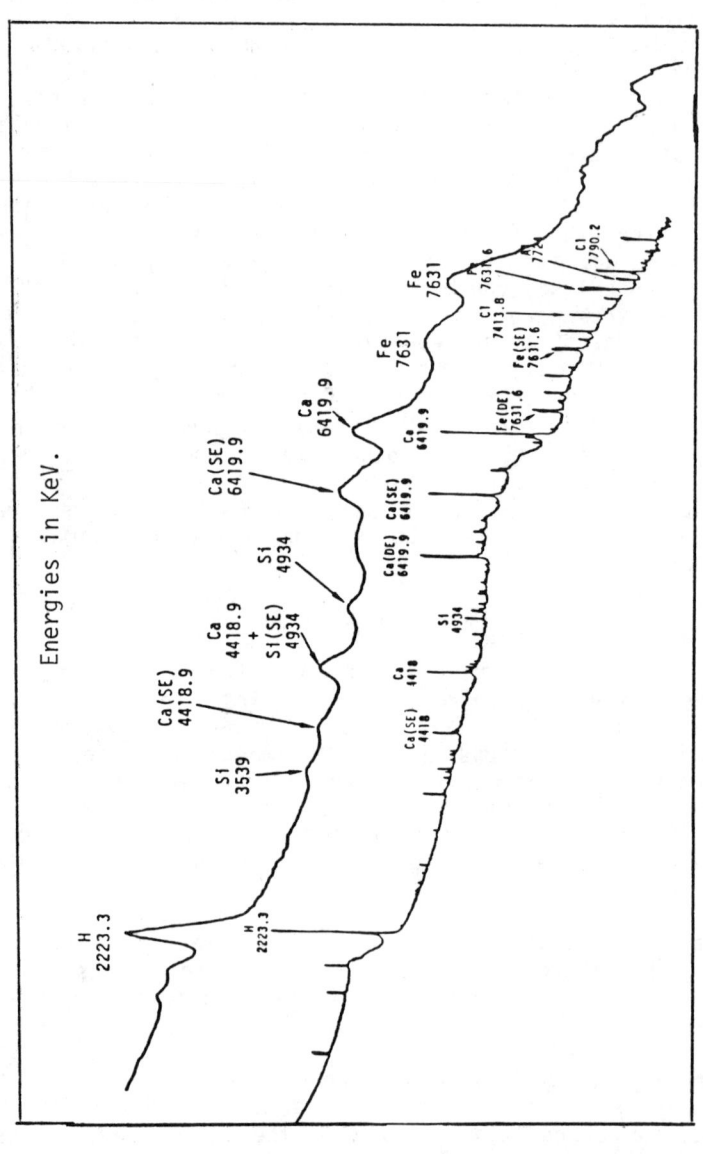

Fig. 8. Comparison of Ge and NaI spectra for calcium bearing minerals. Ge - 27% efficient, 40000 sec. measurement. NaI - 12.5 cm. diameter x 12.5 cm. thick, 1000 sec. measurement.

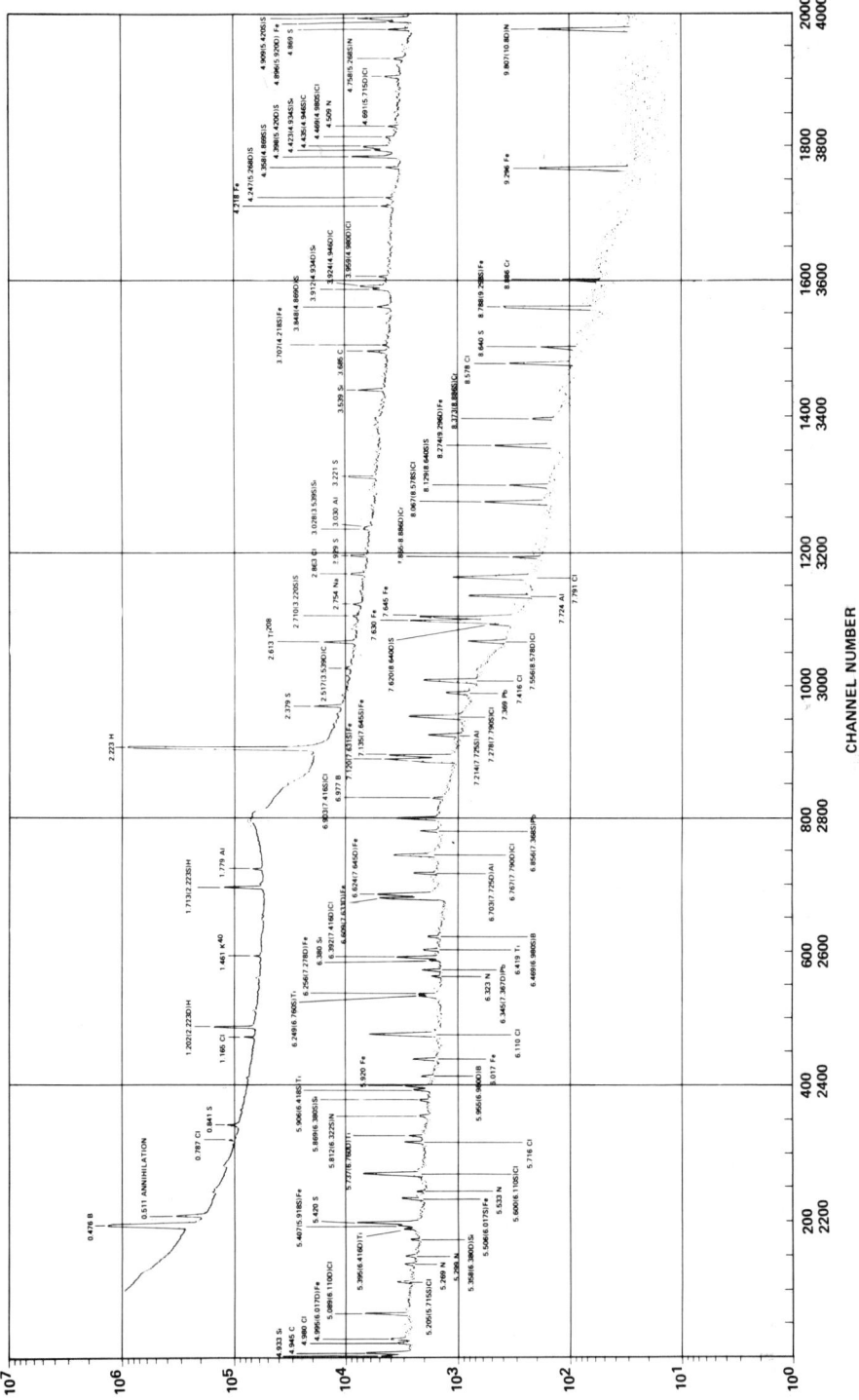

Figure 9. Ge Spectrum of Pittsburgh #8 coal. Note the wealth of information (about 200 usable peaks) and its complexity.

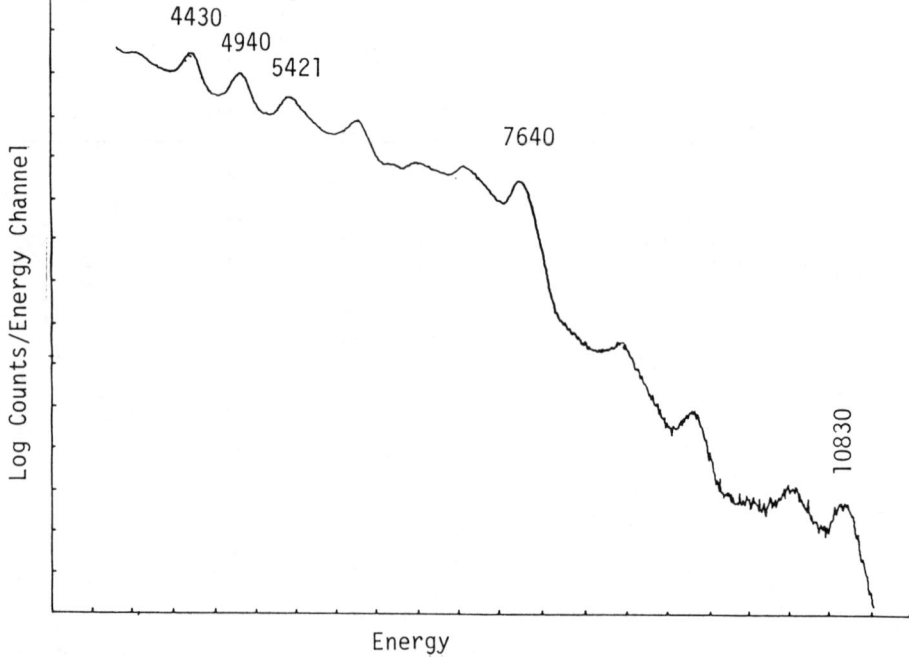

Fig. 10. High energy spectrum of Pitts. #8 coal as measured by 15 cm. diameter x 15 cm. thick NaI detector.

Indeed in most instances the limitation to the CONAC accuracy was traced to inherent errors in the chemical (including sampling) analysis. Thus the practical accuracies of the PGNAA instrument were shown to equal or exceed those of the conventional techniques.

CONAC has also achieved another of its key objectives--high speed, namely yielding good precision (i.e., low statistical error) in a relatively short time. Relative precision of 0.5 to 5%, depending on the specific elements, are obtained in an accumulation time of 3-5 and 20 minutes with NaI and Ge detectors, respectively.

Based on the same or similar principles other instruments were developed in order to respond to different industries and requirements. These instruments under the label "Nucoalyzer," are shown surrounding the CONAC in Figure 14.

There is a host of other nuclear techniques (see Table 2) that are applicable to some facets of compositional analyses. These techniques and especially PGNAA are a testimony to the great utility and service nuclear methods can provide the industry.

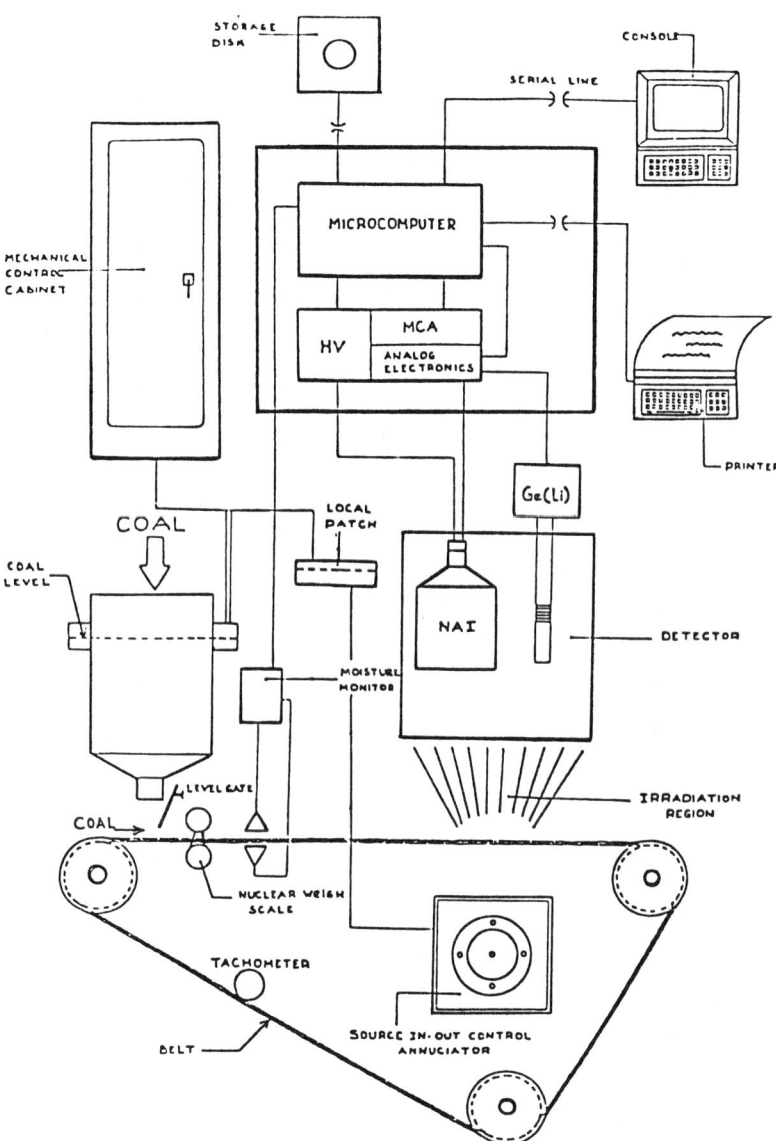

Fig. 11. CONAC system components.

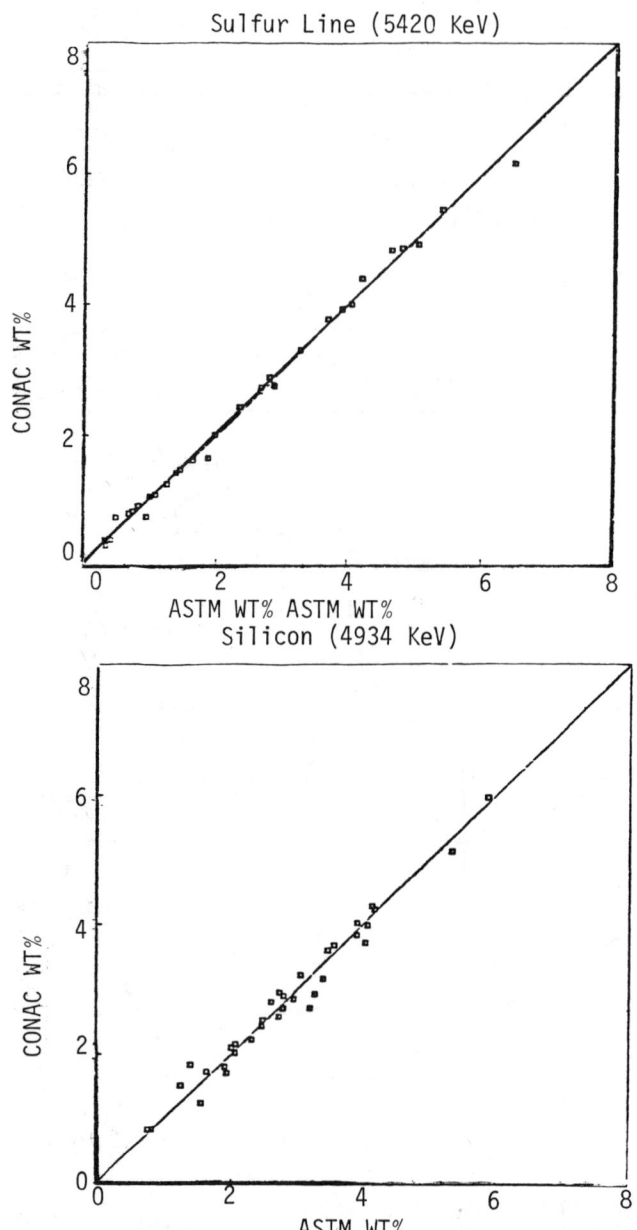

Figure 12 and 13. Comparison of calibrated CONAC weight percent with ASTM weight percent for the Germanium detector.

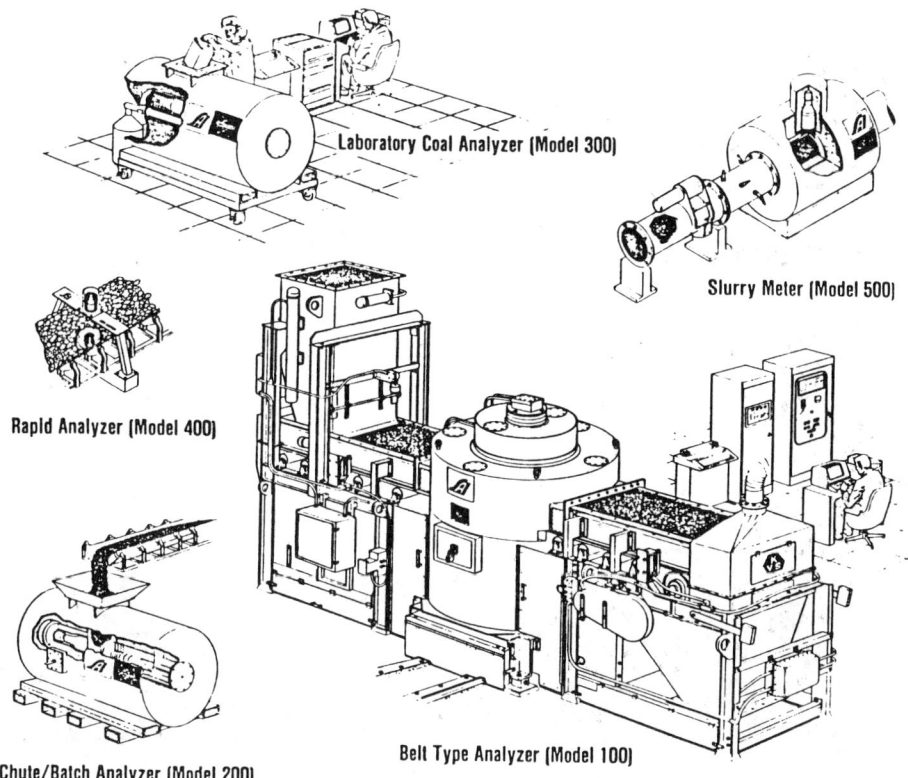

Fig. 14. Nucoalyzer family of instruments.

Table 2.
ACTIVE NUCLEAR TECHNIQUE FOR ON-LINE MEASUREMENT OF BULK MEDIA

RADIATION SOURCE	MAIN INTERACTION	DETECTOR	INFERRED QUANTITY	COMMENTS (Status, major problems, etc.)
Cf-252 neutron source	(n,γ)	NaI(Tℓ)	Abundance of a few prominent elements (e.g. S in coal)	2 systems operational; (requires unfolding and great stability)
(α,n), Cf-252 neutron sources	(n,γ)	Ge or Ge(Li)	Abundance of all major and minor elements (12 elements in coal)	One system operational (highest accuracy available to-date)
(α,n), Cf-252 neutron sources	Inelastic scattering $(n,n'\gamma)$	NaI(Tℓ)	Abundance of C in coal; S, Zn, and other elements in some mineral ores	Experimental, (large interferences)
14 MeV neutron generator	$(n,n'\gamma)$ (n,p) (n,α)	NaI(Tℓ)	Abundance of H,C,O,Cl,Si,Aℓ. Velocity & mass flow measurements of multi-phase streams	Commonly used in oil bore-hole logging; prototype built for solid-liquid slurry mass flow rate measurements

REFERENCES

1. J. H. McQuaid, D. R. Brown, T. Gozani and H. Bozorgmanesh, "High Rate Spectroscopy for On-Line Nuclear Coal Analyzer," IEEE Trans. Nucl. Sci. NS-28 (1) (Feb., 1981).

2. C. Britton, T. Becker, T. J. Paulus and R. Trammell, "Characteristics of High Rate Energy Spectrosocpy System Using HPGe Coaxial Detectors and Time Variant Filters," IEEE Trans. Nucl. Sci., NS-31 (1) (Feb., 1983).

Reports describing specific aspects of the work discussed in this paper are available as Electric Power Research Institute Reports, Research Project RP983-1 and RP983-4, Nuclear Assay of Coal, Vol. 1 through 10.

NUCLEAR RESONANCE-FLUORESCENCE ANALYSER
OF ORES AND SURFACES

J.C. Palathingal
University of Puerto Rico, Mayaguez, Puerto Rico 00708

ABSTRACT

Nuclear resonance fluorescence of the continuous spectrum of gamma rays generated by Compton scattering of radiation from ^{60}Co or, alternately, ^{124}Sb is investigated as a possible technique for analysis of materials. The sensitivities for various elements are estimated. With appropriate variations in the design of equipment, the method can be employed for nondestructive analysis of hard surfaces, ocean floors and special rock samples; and also for analysis of ores in quantities of the order of hundreds of kilograms and bore holes. General geometries of plausible designs of equipment are given.

INTRODUCTION

Nuclear gamma-ray fluorescence was first put in practice as a technique for ore analysis in bore holes[1] by B.D. Sowerby in 1971. The method was highly restricted, requiring very special radiation sources for each element, which could not be available in most cases. The possibility of using ^{60}Co or ^{124}Sb as a common radiation source for a simultaneous investigation of several elements has been investigated and results are presented pointing out the feasibility of the technique in analysing ores, rock samples and hard surfaces.

METHOD

The gamma radiation from ^{60}Co or, alternately, ^{124}Sb is passed through the material to be analysed. A small section, ~ 10^{-6}, of the photons of the continuous spectrum generated by Compton scattering has energy appropriate for nuclear resonance scattering. The radiation scattered through a large angle is recorded by a HPGe detector-analyser system. The relative high energy of the nuclear resonance photons ($\simeq$ 1 MeV in most cases) makes it possible to distinguish them from the far more numerous Compton photons. The intensity of the latter can however be overwhelming. An absorber is hence used in the radiation path which attenuates the intensity selectively.

APPARATUS

The details of the equipment are expected to be dependent on the physical form of the material that is analysed; ores, surfaces, bore holes or rock samples. In each case, however, an apparatus of cylindrical geometry seems to be appropriate. General outlines of convenient designs are given in Fig. 1. Sources of strength around 100 Ci are found typically adequate, although strengths up to 10 kCi may be employed in field models for speedy analysis.

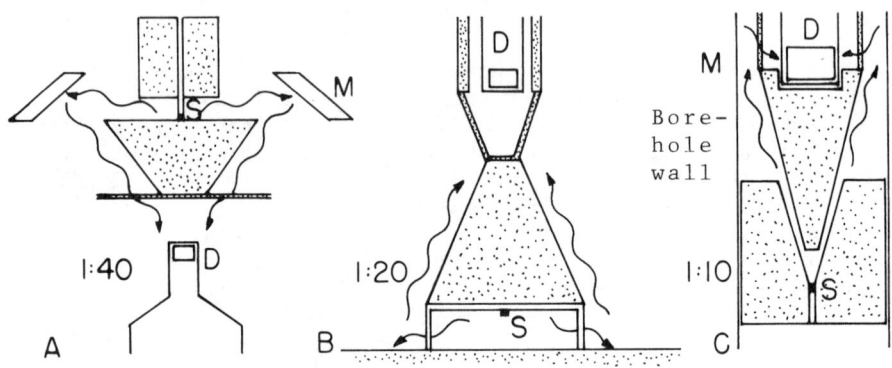

Fig.1

A: Ore Analyser
B: Surface Analyser
C: Bore-Hole Analyser

S: Source
D: Detector
M: Material Analysed
▩ Lead

SPECTRUM DETAILS AND SENSITIVITY

The nuclear resonance-fluorescence-count rates for an incident radiation intensity, I_o is derived, assuming a thick scatterer, in an approximation, as

$$C_n = I_o \omega_{sc} \frac{\sigma_e f(E_o)}{\sigma_e + a\sigma'_e} \int_{E_{min}}^{E_{max}} \frac{\sigma_n(E)dE}{\sigma'_e(1+a) + \sigma'_n(E)} \cdot P(\theta_o) t \omega_D \varepsilon_D(E_o) \qquad (1)$$

where the symbols ω, $P(\theta_o)$ and a represent, respectively, the geometrical efficiency, angular distribution factor and the ratio of path lengths of the ongoing and backscattered radiation inside the scatterer. Also, t is the transmission through the preferential absorber of Compton photons and $\varepsilon_D(E_o)$ the detection efficiency for the incident resonance photons (energy, E_o). The electronic and nuclear cross sections are denoted by σ, with appropriate subscripts. The term, $f(E_o)$ represents the fraction of photons per unit energy interval[2], at energy, E_o.

The major competition to the resonance-count rates is expected to arise from the Compton tail of the Rayleigh and Thomson-scattered high-energy photons. Contributions to the resonance-energy channels from natural background radiation and pile up of Compton-photon pulses can be negligible. Since the Rayleigh and Thomson photons of high energy cannot be cut off preferentially, the ultimate limits to employable source strengths are dictated by such coherent scattering. With a 100-Ci source, a pure aluminum scatterer (thick) yields a resonance-peak-count rate, 2.0 per sec, of which about a third is from background, predominantly from Rayleigh and Thomson processes. The sensitivity of the method estimated for a list of elements (not comprehensive) is seen in the last column of the Table, which represents the percentage-minimum content of each element detectable at a 90% confidence level from a one-hour data, assuming that the ore has an average atomic number, Z_{av} = 10. The columns 4 and 5 indicate the photopeak-count rates of resonance-fluorescence events and the background (natural + source-induced), with a scatterer solely of the element listed.

TABLE I SENSITIVITY FOR VARIOUS ELEMENTS

Element Probed	Probe Nucleus	Resonance Level (keV)	Photopeak Count Rate (cps)		Sensitivity (%)
			Nucl. Reson.	Back-Ground	
(1)	(2)	(3)	(4)	(5)	(6)
With ^{60}Co as source:					
Li	^{7}Li	478	13	0.09	0.4
Al	^{27}Al	1,014	1.3	0.7	2.5
P	^{31}P	1,266*	2.7	0.1	0.4
Cl	^{35}Cl	1,219*	1.9	0.2	0.65
Sc	^{45}Sc	720	6.2	1.4	0.35
Ti	^{48}Ti	984	1.5	1.7	1.5
Fe	^{56}Fe	847	1.1	2.2	1.8
Co	^{59}Co	1,190*	5.9	0.7	0.2
Cu	^{63}Cu	670	2.5	2.3	0.4
		962	2.2	3.3	
		1,327*	0.6	0.3	
	^{65}Cu	771	2.6	2.6	
		1,116	1.7	2.2	

(1)	(2)	(3)	(4)	(5)	(6)
Zn	^{64}Zn	992	1.7	2.9	0.9
	^{66}Zn	1,039	1.0	2.7	
	^{68}Zn	1,077	0.7	2.5	
Ga	^{69}Ga	872	2.8	3.4	0.4
		1,106	3.3	2.7	
	^{71}Ga	910	2.3	3.4	
Sn	^{116}Sn	1,294*	0.45	0.7	1.0
	^{118}Sn	1,230*	0.6	1.1	

With ^{124}Sb as source: (Additional Elements)

F	^{19}F	1,459	3.5	0.01	0.15
		1,554	3.2	0.01	
Mg	^{24}Mg	1,369	0.5	0.06	0.9
	^{25}Mg	1,612	0.85	0.02	
Si	^{28}Si	1,779*	0.06	0.005	4.5
Cr	^{52}Cr	1,434	0.4	0.09	1.2
Mn	^{55}Mn	1,528	0.7	0.06	0.8
Ni	^{58}Ni	1,454	0.36	0.10	1.5
	^{60}Ni	1,332	0.18	0.16	

* Only the high-energy component of the incident radiation produces resonance fluorescence.

REFERENCES

1. B.D. Sowerby, Nucl. Instr. & Meth. 94, 45 (1971); 108, 317 (1973).

2. F.R. Metzger, Progr. Nucl. Phys. 7, 53 (1959).

AMBIGUITIES IN PIGE CAUSED BY DIFFERENT REACTIONS

A.Z. Kiss, E. Koltay, B. Nyakó, E. Somorjai
Institute of Nuclear Research, Hungarian Academy of
Sciences, H-4001 Debrecen, P.O.Box 51, Hungary

A. Anttila, J. Räisänen
Department of Physics, University of Helsinki,
SF-00170 Helsinki 17, Finland

ABSTRACT

On the basis of relative thick target PIGE yield curves determined experimentally the role of different (p,x) reactions exciting a given γ-transition in a nucleus is treated. It is shown, that the proper selection of bombarding energies can solve the ambiguity caused by possible reactions in the analysis of a matrix.

INTRODUCTION

Particle induced prompt gamma-ray emission (PIGE) is normally considered as a tool for sensitive microanalytical determination of elemental concentrations complementary to proton induced X-ray emission (PIXE) in detecting lighter elements not seen in PIXE measurements. As it was shown[1] in special cases PIGE may compete with PIXE for heavier elements, too. Furthermore, contrary to the elemental character of PIXE, PIGE distinguishes among different isotopes of a given element. Consequently, the field of possible applications of PIGE is broader than considered earlier.

For an absolute determination of the constituents of a sample, a complete set of absolute (p,γ) and (p,bγ) reaction cross sections would be needed as the function of the bombarding energy, on a high level of accuracy. In the case of measurements relative to standard samples relative thick target yields are to be known for the selection of experimental parameters best fitted to a given matrix. General surveys of low Z element data were given for low energy (1-2.4 MeV) proton beams by Anttila et al.[2] and for higher energies (2.4-4.2 MeV) in our paper[3] while for higher Z values data are available in ref.[1]. In the work of Gihwala and Peisach[4,5] yield data are given for 4.5 MeV proton energy and practical applications are treated for a broader Z interval.

The unambiguity of the analysis is sometimes questionable because of the possibility of exciting a given γ-transition on two or three different sample constituents of neighbouring Z numbers in different simultaneous reactions. The aim of the present paper is to comment on such cases and to show how ambiguities can be avoided, on the basis of experimental data presented in ref.[3].

EXPERIMENTS

Relative thick target PIGE yields were determined for elements Z = 3-9, 11-17, 19-21 in the energy range E_p = 2.4-4.2 MeV with a 25 cc. Ge(Li) detector at the Institute of Nuclear Research, Debrecen. The data were normalised to those from earlier work of Anttila at al.[2] for 1-2.4 MeV. The sets of spectra and yield data were used in searching for reactions exciting the same transitions.

Details of the experimental conditions are given elsewhere[2,3].

RESULTS

The γ-transitions observed in the interval of Z treated are excited in (p,γ), (p,p'γ), (p,αγ) and (p,nγ) reactions. It means, that the appearance of a given γ-line of nucleus X_Z^A may indicate the presence of target nuclei Y_{Z-1}^{A-1}, X_Z^A, V_{Z+1}^{A+3} and W_{Z-1}^A, respectively. The cases where a given gamma line may be assigned to different target nuclei are as follows:

^{7}Li(p,nγ)^{7}Be and ^{10}B(p,αγ)^{7}Be	429 keV
^{9}Be(p,γ)^{10}B and ^{10}B(p,p'γ)^{10}B	718, 1022 keV
^{13}C(p,γ)^{14}N and ^{14}N(p,p'γ)^{14}N	2313 keV
^{24}Mg(p,p'γ)^{24}Mg and ^{27}Al(p,αγ)^{24}Mg	1369 keV
^{27}Al(p,γ)^{28}Si, ^{28}Si(p,p'γ)^{28}Si and ^{31}P(p,αγ)^{28}Si	1779 keV
^{30}Si(p,γ)^{31}P, and ^{31}P(p,p'γ)^{31}P	2233 keV
^{31}P(p,γ)^{32}S and ^{32}S(p,p'γ)^{32}S	2230 keV
^{34}S(p,γ)^{35}Cl and ^{35}Cl(p,p'γ)^{35}Cl	1219 keV
^{34}S(p,p'γ)^{34}S and ^{37}Cl(p,αγ)^{34}S	2127 keV
^{37}Cl(p,γ)^{38}Ar and ^{41}K(p,αγ)^{38}Ar	2168 keV
^{41}K(p,γ)^{42}Ca and ^{42}Ca(p,p'γ)^{42}Ca	1525 keV.

As it is known for nuclear reactions, at low bombarding energies gamma transitions from radiative capture processes will be predominant, while at higher energies particle channels (p,p'γ), (p,αγ) and (p,nγ) will be opened and the gamma rays emitted in the decay of excited low lying states of the end nuclei will compete more and more with gamma transitions from (p,γ) process. The yield curves for different particle channels will be of different shape because of the difference in Coulomb penetrations.

As examples, yield curves are compared for three cases in Fig.1. From here it is clear that the origin of a gamma ray or the proportion of the contribution from different reactions can be determined when measurements are made at different proton energies

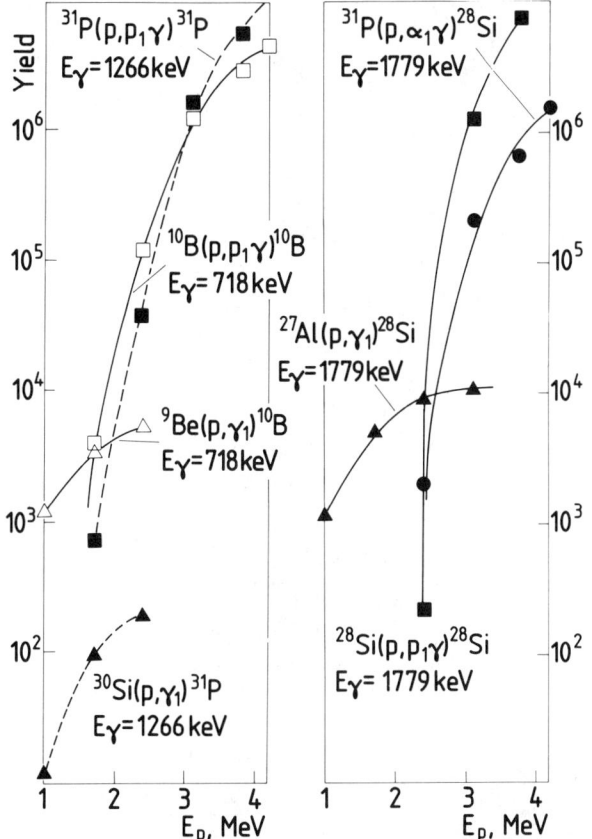

Fig. 1. Gamma yields from (p,γ) and (p,bγ) reactions.

properly selected.

As a general conclusion we can formulate the statement that for each matrix the optimum bombarding energies should be properly selected aiming at the highest selectivity, unambiguity and speed in analysis. The spectra and yield values presented in our earlier paper referred to could serve as a basis for the planning and performing actual PIGE measurements.

REFERENCES

1. J. Räisänen, R. Hänninen, Nucl. Instr. Meth. 205, 259 (1983).
2. A. Anttila, R. Hänninen, J. Räisänen, J. Radioanal. Chem. 62, 441 (1981).
3. Á.Z. Kiss, E. Koltay, B. Nyakó, E. Somorjai, A. Anttila, J. Räisänen, to be published.
4. D. Gihwala, M. Peisach, J. Radioanal. Chem. 70, 287 (1982).
5. M. Peisach, D. Gihwala, J. Radioanal. Chem. 61, 347 (1981).

VIII. RELATED TOPICS

COINCIDENCE ELECTRON SCATTERING (e,e'f) AND MULTIPOLE STRENGTH FUNCTIONS IN ^{238}U*

K. A. Griffioen, [a]P. J. Countryman, [b]K. T. Knöpfle,
[c]K. Van Bibber, and M. R. Yearian
Department of Physics and High Energy Physics Laboratory
Stanford University, Stanford, CA 94305

J. G. Woodworth and D. Rowley
Lawrence Livermore National Laboratory, Livermore, CA 94550

J. R. Calarco
Department of Physics, University of New Hampshire, Durham, NH 03824

ABSTRACT

The inherent power of inelastic electron scattering to probe the nuclear response function in the nuclear continuum is completely exploited only in the coincident detection of decay products. The coincidence requirement effectively eliminates the strong elastic radiative tail which hitherto plagued the analysis of single arm (e,e') experiments. Large physical backgrounds (e.g. due to multistep processes as encountered in inelastic hadron scattering) are negligible due to the dominance of single photon exchange. Thus, coincidence electron scattering is a unique tool for quantitatively testing modern calculations of the nuclear response function. We have measured ^{238}U(e,e'f) cross sections at three values of the momentum transfer ($q \approx 0.26$, 0.40, and 0.55 fm^{-1}, which correspond closely to the first maxima of the E1, E2/E0, and E3 form factors respectively) from below fission threshold to 23 MeV excitation energy. The angular correlations were measured at six or seven angles in each run. The data permit a clear multipole decomposition and a rigorous comparison with recent theoretical results from the Quasiparticle Random Phase Approximation (QRPA). The extracted E1 strength function agrees well with photofission results. The E2/E0 distribution displays prominent resonances at 10 and 13 MeV which suggest an identification with collective isoscalar E2 and E0 strength respectively, as predicted within the QRPA. However, the observed E2 strength is anomalously low, corroborating some earlier $(\alpha, \alpha'f)$ reports. E3 and higher order multipoles contribute a sizable amount of strength in the region from 5 to 17 MeV.

*Work supported in part by the Lawrence Livermore National Laboratory under Department of Energy Contract No. W-7405-ENG-48.

[a]NSF Graduate Fellow.

[b]Supported in part by Alexander von Humboldt Stiftung; permanent address: Max Planck Institut für Kernphysik, D6900 Heidelberg 1, FRG.

[c]Alfred P. Sloan Research Fellow.

INTRODUCTION

In spite of the fact that many reactions and experimental techniques have been used to study the new giant resonances in ^{238}U—$(\alpha, \alpha')^{1,2}$, $(\alpha, \alpha'f)^{3-6}$, $(^6Li, ^6Li'f)^7$, $(e,f)^{8,9}$, $(e^+,f)/(e^-,f)^{10}$, $(e,e')^{11}$, and $(e,e'f)^{12}$—a clear picture of the location and strength of these resonances in the fission channel is still lacking, most notably for the isoscalar giant quadrupole resonance (GQR). Although it is impossible to review all of these measurements, one should recall that early inclusive electrofission measurements implied a large fraction of the E2 isoscalar energy-weighted sum rule in the fission channel[8,9], located in a broad structure from 6 to 20 MeV excitation energy[9]. These determinations required very accurate (e,f) and (γ,f) cross section measurements in conjunction with calculated virtual photon spectra for both E1 and E2. The comparison of electron and positron induced fission, which removes the sensitivity to the absolute normalization of the (γ,f) cross sections, however, produced little evidence, if any, for the fission decay of the GQR. Extensive $(\alpha, \alpha'f)^3$ measurements also implied a small quadrupole fission decay, but this interpretation was contradicted by later $(\alpha, \alpha'f)^4$ work in which the angular correlation was mapped out less completely. Although ^{6}Li appeared to be a promising probe for accentuating the resonance contribution above the physical background associated with hadron scattering, a firm conclusion did not result[7].

Coincidence electron scattering, which is just now beginning to be exploited at a few laboratories around the world, will perhaps prove to be the single most valuable tool for deducing the nuclear response function. The coincidence requirement eliminates the strong elastic radiative tail and leaves only the contribution from the giant resonances. The connection between cross sections and nuclear structure information is rigorous and straightforward[13]. Indeed, if the nuclear response function is mapped out over a sufficiently broad range of momentum transfer q, detailed transition charge, current, and magnetization densities will result in a model-independent way; this is presently the situation in the discrete-state region[14]. If the measurements are confined to the low-q region (up through the first maximum for a given multipolarity) recourse to some model must be made. The first measurements of ^{238}U(e,e'f) at the University of Illinois[12] and the present work are in this category, although the present data significantly extend the excitation energy and improve the statistics of Reference 12.

EXPERIMENT

The experimental program was carried out at the Stanford Superconducting Recyclotron, utilizing beams of $\approx 20\mu$A instantaneous current and $\approx 50\%$ duty factor. Spectra were recorded at 80.3, 118.4, and 163.8 MeV in each of three different runs in order to minimize any problems of relative normalization between the three energies. Given the electron scattering angle of $\theta_{e'} = 40°$, the bombarding energies of 80.3, 118.4, and 163.8 MeV were chosen such that the mean momentum transfers (q = 0.26, 0.40, and 0.55 fm^{-1}) correspond roughly to the first maxima of the electric dipole (E1), quadrupole and monopole (E2/E0), and octupole (E3) form factors respectively. Scattered electrons were detected in a 36-inch, 180° double-focussing spectrometer with a momentum acceptance of 4%. The spectrometer was instrumented with 24 overlapping plastic scintillator counters which provided 47 channels of data with resolution of 0.1%. The spectrometer was positioned at 40° throughout the measurements; this angle was chosen to allow reasonable coincidence counting rates and to yield appropriate values of momentum transfer for the available beam energies. The solid angle of the spectrometer was determined geometrically to be 3.6 msr. Measurements of elastic scattering on carbon and uranium verified that the spectrometer was

100% efficient to within 5% error. The relative efficiencies of the individual channels were determined using the smooth portions of the radiative tail from elastic scattering on carbon. Coincidence data were collected in such a way as to insure a 30% overlap between spectra of different spectrometer settings. The data in the regions of overlap agreed with each other within statistical errors in each case.

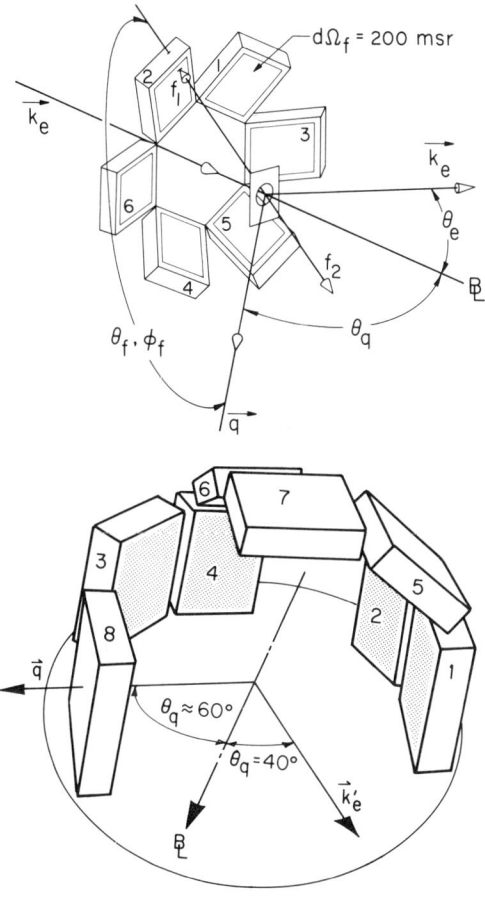

Fig. 1. PPAC arrays for the first and third runs (top), and the second run (bottom).

The fission fragments were detected in an array of parallel plate avalanche counters (PPAC's)[15]. These detectors proved to be virtually insensitive to the large electron backgrounds present and clearly distinguished the fission fragments from the electrons and noise. Figure 1 shows the configurations of PPAC's used. In the first and third runs, six PPAC's were arranged azimuthally about the beam axis in the backward hemisphere; they were not, however, symmetric with respect to the momentum transfer axis **q** about which angular correlations are measured. In the second run, the array was configured to cover as much of a full octant as possible. Solid angles (≈ 200 msr) of the PPAC's were determined geometrically by specifying the coordinates of the detector and integrating numerically over the detector face. These values were checked against a measurement using a californium-252 source and a solid state detector of known solid angle. The two measurements agreed within the accuracy of each (5%), thus confirming the 100% efficiency of these devices.

The targets consisted of $\approx 450\,\mu g/cm^2$ of uranium enclosed by a $40\,\mu g/cm^2$ backing of carbon on each side. Since the carbon produces no fission, its presence on the target is of no concern. The thicknesses of the targets were determined to better than 4% accuracy by Rutherford scattering of 12 MeV alpha particles at the Lawrence Livermore National Laboratory Cyclograph facility.

The beam current was integrated from the output of a non-interactive toroid beam current monitor which sat directly upstream from the scattering chamber. The accuracy of this device has been established to better than 2% by drawing a known current through a wire which passes axially through the toroid.

The pulse-height signals from the PPAC's, the time differences between electron and fission signals, and the electron channel number were all recorded on magnetic tape via CAMAC for each coincidence event. The coincidence event was determined by the logical AND of a 100 ns NIM pulse from the spectrometer and a 5 ns pulse from one of the fission detectors. The recorded time differences, therefore, show a spectrum with a large peak corresponding to true coincidences and a flat background, 100 ns in duration, resulting from accidental coincidences. Choice of instantaneous beam current dictated the trues to accidentals ratio; this was kept at values greater than 1 to 1. A coincidence spectrum was generated from an electron spectrum obtained from events within the true time window, less an appropriate fraction of the spectrum generated from events outside of the true time window.

The resulting counts were redistributed into uniform bins of 200 keV width and then normalized with respect to target thickness, solid angles, and integrated beam current. The individual spectrometer settings were then merged together by averaging the data at each point of overlap. The full coincidence spectrum was corrected for radiative losses. This amounted to a 20% increase in the cross section near fission threshold and less than an 8% increase above 7 MeV.

CROSS SECTIONS AND ANGULAR DISTRIBUTIONS

The differential coincidence electrofission cross sections, $d^3\sigma/d\Omega_{e'} d\Omega_f d\omega$, were fitted as a function of fission angle to the general formalism of Kleppinger[16]. No azimuthal asymmetry within the 10% statistical accuracy of the measurements was observed. As a consequence the data were fitted to the remaining three important terms of the expansion, $a_0 + a_2 P_2(\cos\theta) + a_4 P_4(\cos\theta)$, in which θ is the angle between the momentum transfer direction and the fission axis and the $P_n(\cos\theta)$ are the regular Legendre polynomials averaged over the finite extent of the PPAC's. Odd order polynomials do not contribute because fission fragments are emitted back-to-back.

Figures 2-4 show the results of the least squares fit for each of the three incident beam energies. The upper two panels show the angular correlation coefficients a_2/a_0 and a_4/a_0. These indicate large anisotropies near fission threshold and isotropy for excitation energies $E_x > 10$ MeV. The bottom panel displays $4\pi a_0$ normalized to the Mott cross section; this is the coincidence cross section integrated over fission angle. The prominent peak at $E_x = 6$ MeV results from the fact that the fission barrier ($B_f = 5.9$ MeV) lies lower than the neutron separation energy ($S_n = 6.14$ MeV). The solid and dashed curves in Figures 2-4 show the E1 contribution to the electron scattering spectra resulting

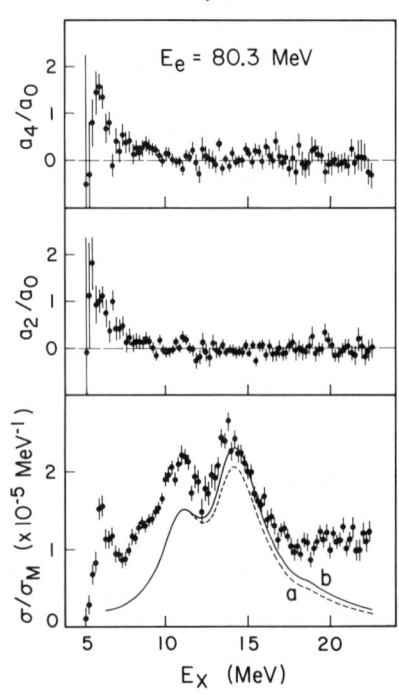

Fig. 2. Coincidence data at 80.3 MeV.

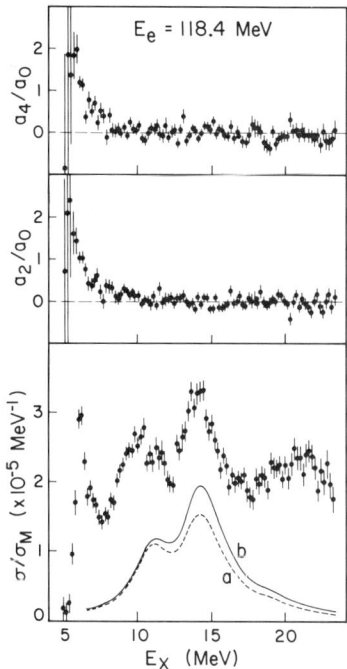
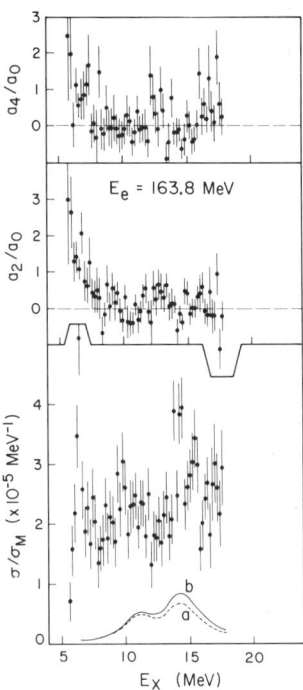

Fig. 3. Coincidence data at 118.4 MeV.

Fig. 4. Coincidence data at 163.8 MeV.

from the E1 strength function deduced from the photofission data[17] measured at Saclay and Quasiparticle Random Phase Approximation (QRPA)[18] (solid line) or Tassie model[19] (dashed line) form factors. Indeed, the dipole absorption accounts for most of the cross section at 80.3 MeV bombarding energy, but only for a small fraction at 163.8 MeV. The bimodal character of the spectra results largely from the increase of fission probability at $E_x = 12$ MeV due to the opening of the second chance fission (nf) channel.

According to the Bohr fission hypothesis[20], the angular distribution of fission fragments will depend on the quantum numbers J, K, and M (the total angular momentum, the projection along the body-fixed axis, and the projection along the space-fixed axis respectively) of the transition states at the saddle point. With **q** as the axis of quantization, the plane wave Born approximation (PWBA) condition $M = 0$ should be satisfied even at our lowest bombarding energy. Though J and M are rigorously conserved, K probably is not. If K is not conserved, then substantial K mixing will occur as the nucleus proceeds to the saddle point and the angular distribution will become isotropic. Near the barrier, however, the number of available states is small and some anisotropy should be observed. A simple Fermi gas model establishes that these low-lying states at the saddle point are predominantly $K = 0$[21]. Although our angular distributions are not accurate enough to establish the values of J represented, the forward-peaking of the data indicates the predominance of $K = 0$ states. The isotropy above 7 MeV is consistent with K-mixing as the nucleus proceeds from giant resonance to fission.

THE QRPA

Zawischa and Speth[18] have calculated the nuclear strength functions and transition charge densities for ^{238}U within the QRPA for multipolarities $L = 0, 1, 2$. Figure 9a (located in the next section) exhibits the $\Delta T = 1$, E1 strength function. The prolate deformation ($\beta \approx 0.23$) splits the giant dipole resonance (GDR) into $K = 0, 1$ modes at 11.3 and 12.7 MeV respectively in qualitative agreement with photofission results, although the magnitude of the splitting is too small. The E2 and E0 strength distributions are shown as solid and dashed lines respectively in Figure 9f. The isoscalar GQR is evident at 10 MeV and the isoscalar giant monopole (GMR) is seen at 15 MeV. The peaks at 17 and 20 MeV are isovector in nature; no isovector E2 or E0 states are expected below 17 MeV. Deformation is expected to mix the $J = 0, 2$ states with $K = 0$, but the QRPA does not predict a large effect. The strength functions of Figures 9a and 9f have been binned in 1 MeV intervals and multiplied by the E1 fission probability for comparison with the data.

Proton and neutron contributions to the transition charge density are calculated independently for each state. For the very collective states the densities tend to be surface-peaked as expected, but do exhibit some oscillatory behavior in the nuclear interior not present in the simple hydrodynamic models. From these densities, form factors have been calculated.

Since the QRPA states contain no intrinsic width arising from coupling to multiparticle-multihole configurations, a phenomenological spreading width of

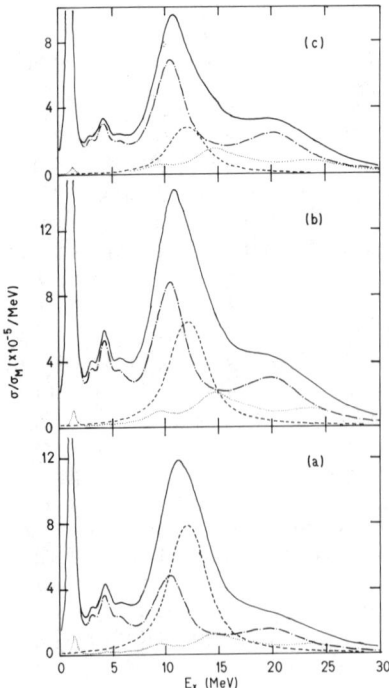

Fig. 5. Predicted QRPA cross sections by multipole.

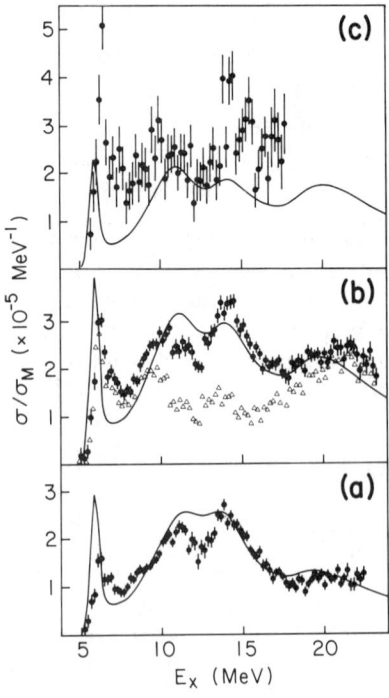

Fig. 6. Coincidence data integrated over fission angle (points) and the QRPA predictions (lines).

$\Gamma = 0.28 E_x$ as suggested by References 1 and 11 has been applied to the strength functions in the form of a Breit-Wigner line shape. For comparison to the fission data, this must be multiplied by a fission probability for each multipole. For lack of better knowledge, the E1 fission probability $P_f(E1, E_x)$ deduced from the real photon work has been used in each case. The E1 fission probability is well-determined up to 18 MeV and can be extrapolated with confidence to 23 MeV.

With QRPA form factors, then, absolutely normalized spectra may be calculated and compared with the experimental data. Figure 5 shows the total QRPA cross sections at 80.3 (a), 118.4 (b), and 163.8 (c) MeV for E0 (dotted line), E1 (dashed line), E2 (dot-dashed line), and the sum (solid line). Figure 6 presents the solid curves of Figure 5 scaled down by $P_f(E1, E_x)$ and overlayed on the coincidence data. The overall magnitude and shape of the calculations match the data at 80.3 (a) and 118.4 (b) MeV, but fall short at 163.8 (c) MeV. This is not surprising since the E3 form factor is expected to reach its maximum at our highest incident energy, and octupole vibrations have not been included in the QRPA calculations, except as rotational states built upon the GDR. These 3^- states, however, contribute only 2% to the E3 isovector energy-weighted sum rule (EWS$_1$R).

MULTIPOLE DECOMPOSITION

It should be evident that a satisfactory comparison of the QRPA calculations to the data, in which all multipoles are summed, may simply be fortuitous. To be sure, departures from prediction in any one multipole could easily be masked in the summation. Consequently, we have performed a multipole decomposition of our data by a least squares fit to the functional forms of the multipole form factors. In order to explore the model-dependence of these results, we have chosen to employ Tassie form factors, which are directly related to the ground state nuclear charge density determined by elastic scattering and parameterized in terms of a half-density radius, $c_0 = 6.805$ fm, and surface thickness, $t = 0.605$ fm[22]. It is well-known[11] that up through the region of the first maximum, the form factors are characterized by the transition charge radius $c_{tr} = [(\int r^\lambda \rho_{tr} d\tau)/(\int \rho_{tr} d\tau)]^{1/\lambda}$ and are otherwise insensitive to details of ρ_{tr}. Thus, following Pitthan[11], the Tassie form factors may be varied easily by altering the ground state radius. Normalizing the form factor to 1.0 e^2fm$^{2\lambda}$ implies that as the transition charge radius increases, the form factor decreases and its first maximum moves to lower q (Figure 7).

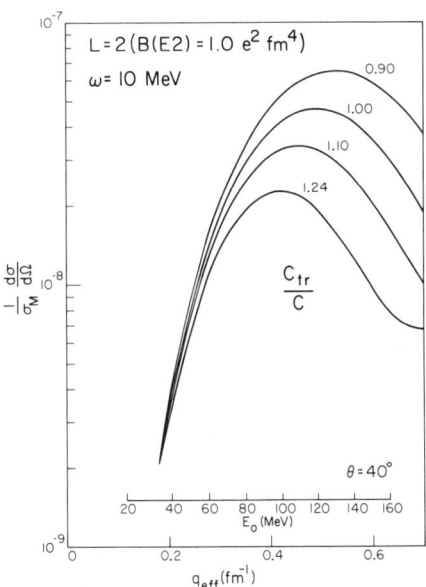

Fig. 7. The dependence of the E2 Tassie model form factor on the transition radius.

To gauge what should be considered realistic or reasonable variations of the transition charge densities, we have derived transition charge radii from the calculated densities of the collective QRPA states. For the dipole states $(J,K) = (1,0)$ and $(1,1)$, $c_{tr}/c_0 = 1.03$ and 0.96 respectively; for the highly collective quadrupole states $(J,K) = (2,1)$ and $(2,2)$, $c_{tr}/c_0 = 0.98$ and 0.95 respectively. Thus the Tassie form factors derived from the unmodified ground state charge density seem justified for collective states, and the variations of $\pm 10\%$ which we will explore are extreme limits.

The functional form of the monopole form factor in the low-q region is identical to that of the quadrupole, but reduced in magnitude by a factor of two. Therefore, the E0 and E2 strength functions cannot be separated without measurements which permit a full Rosenbluth separation.

Form factors for both the QRPA and the Tassie model were calculated using the code FOUBESFIT[14]; small corrections were included for the proton and neutron form factors, and for the small transverse component to the scattering.

The multipole decomposition is complicated by the fact that the resonances of differing multipolarity overlap in excitation energy. In fact, if all multipolarities lay directly on top of each other, the experimental electrofission form factor σ/σ_{Mott} would increase indefinitely as a function of q. Figure 8 shows the result of placing one sum rule of strength at 15 MeV for the first four electric multipoles. The full form factor only begins to turn over beyond the maximum of the E4 form factor. Perhaps if E5 and higher order multipolarities are included, the total form factor may never decrease. As a consequence of this, the E4 and E5 strength may significantly effect our coincidence data, even though our largest q extends only to the maximum of the E3 form factor.

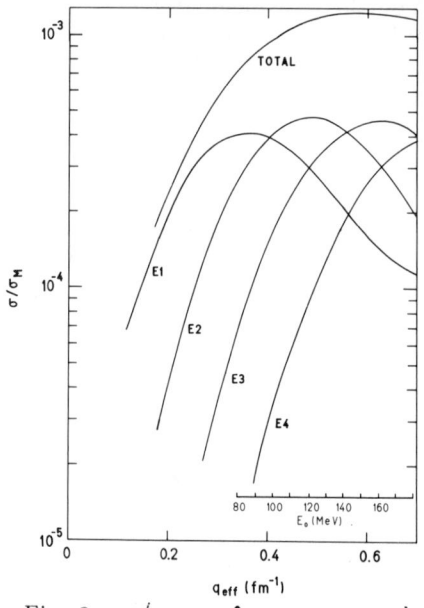

Fig. 8. σ/σ_{Mott} for one sum rule of E1–E4 strength at 15 MeV using Tassie model form factors.

Least squares fits were made for E1, E2/E0, and E3 to the three (e,e'f) spectra using the q-dependence of the Tassie model form factors. Transition charge radii were allowed to take on all possible combinations of the values $c_{tr}/c_0 = 0.9, 1.0,$ and 1.1 for each multipole. Some of these fits are shown in Figure 9c for E1, Figure 9e for E2/E0, and Figure 11a for E3. Representative error bars reflect the propagation of the statistical errors of the data. Although no χ^2 results, several criteria are invoked to determine goodness of fit. First of all, the strength functions must be non-negative at each value of excitation energy. Second, the dipole strength must agree in magnitude and shape with the photofission results, e.g. the Saclay (γ,f) work (Figure 9b). Third, a substantial fraction of the isoscalar monopole energy weighted sum rule (EWS$_0$R) must be found in agreement with the $0°$ (α,α'f) measurements—80% assuming the same fission probability for E0 as for E1. (Although differing in location and width, the E0 strength extracted from (α,α') was also consistent

865

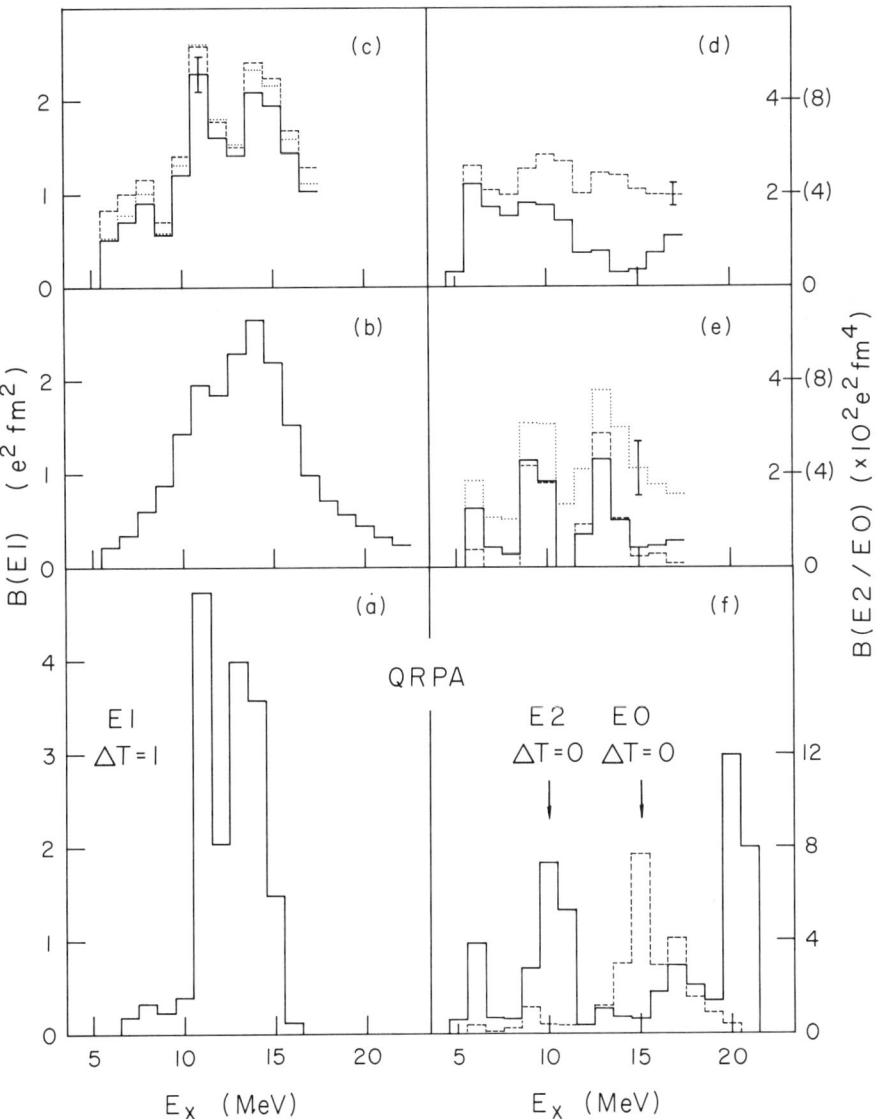

Figure 9. Experimental and theoretical strength functions in the fission channel. Left panels: E1; right panels: E2/E0. (c) and (e) result from the fits unconstrained by $\sigma_{\gamma,f}$; c_{tr}/c_0 (E1,E2/E0,E3) = (1.1,1.0,0.9), dotted; (1.0,1.0,1.0), dashed; (0.9,0.9,0.9), solid. (d) results from the constrained fits; (1.0,1.0,1.0), solid; (1.1,1.0,0.9), dashed. (b) is the experimental B(E1) distribution from the Saclay photofission data.

with a full sum rule[1].)

A comparison of the extracted E2/E0 strength and the QRPA predictions (Figures 9e and 9f respectively), strongly suggests that the peaks observed at 9.5 MeV and 13.0 MeV should be identified with the isoscalar quadrupole and monopole resonances respectively. These features are plain with even the most extreme choices of form factors. Hereafter in comparison with other data and sum rules, we shall assume that the strength below 11 MeV excitation is entirely quadrupole in character, and that above 11 MeV it is entirely monopole.

Table I summarizes these findings. The two most satisfactory fits are for $(c_{tr}^{E1}/c_0, c_{tr}^{E2}/c_0, c_{tr}^{E3}/c_0) = (1.0, 0.9, 0.9)$ and $(1.0, 1.0, 0.9)$. These radii bracket those predicted within the QRPA. The value of $c_{tr}/c_0 = 0.9$ for E3 can be explained by the presence of E4 and higher order strength which already contributes substantially at 163.8 MeV (see Figure 8). In fact, the E4 form factor is already at half of its maximum value at 163.8 MeV. Therefore, the best E3 form factor which mimics the combined E3-E4 behavior should peak higher out in q. Decreasing the transition radius in the Tassie model form factor has precisely this effect. Regardless of choice of transition radii, the isoscalar GQR is reduced in strength relative to the QRPA predictions by at least a factor of two, assuming that the E2 fission probability is the same as the E1. Thus a diminished E2 strength in the fission channel could imply corroboration of the findings of Reference 3, namely that the fission probability for the quadrupole is significantly smaller than for the dipole. More extreme assumptions concerning the form factors (e.g. inflating the transition radii by 10%) move the E2 strength upward but erode the good agreement for E1 and E0.

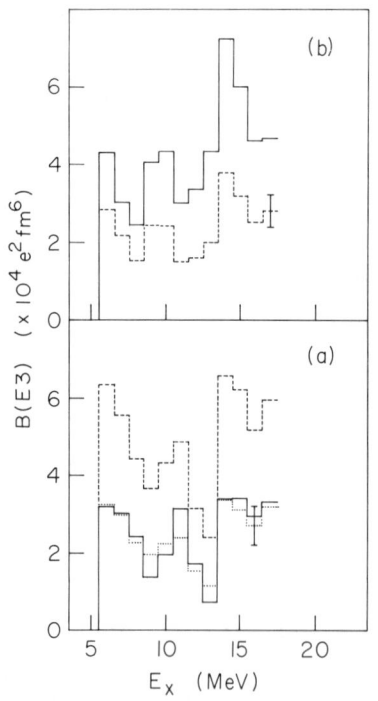

Fig. 10. Extracted E3 strength functions for the conditions listed in Fig. 9.

In addition to the previous analysis, multipole decompositions have been carried out in which the fits[12] were constrained by the real photon data of Saclay[17]. In this case, the goodness of fit is indicated by χ^2 with a single degree of freedom. Again we conclude that the fits are best with a reduced E3 transition radius. These constrained fits, however, are problematic. The best fit employing the Saclay data requires $c_{tr}/c_0 = 1.1$ for the dipole oscillations with a minimal average χ^2 per degree of freedom of 2.3. The E2/E0 and E3 distributions are more sensitive to excursions in transition radii in this case, and the E2/E0 strength function (Figure 9d) barely shows the GQR and GMR resonances which are plain enough even in the 118.4 MeV spectrum with the dipole contribution subtracted off (open triangles in Figure 6). The problem stems from the fact that the photofission data dominate in the determination of the B(E1) distribution. Although it would seem that the real-photon-point constraint would greatly improve knowledge of the higher multipoles, inaccuracies in the photofission data can make large errors in the extracted E2 strength. This can be seen by considering how much of the low-q (e,e'f) spectrum may be accounted for by the giant dipole resonance (see Figure 2). Relatively small perturbations may exert a large lever arm on the experimental form factor σ/σ_{Mott} for the remainder. Against this backdrop we note that whereas the newer (γ,f) data from Giessen[23] agree well with the older Saclay[17] data above 16 MeV, the cross sections below that suggest differences in shape and magnitude comparable to the differences between the Saclay and Livermore[24] measurements (Figure 11). Even though certain choices of form factor lead to large fractions of the E2 sum rule, we view the results of the constrained fits with reservation.

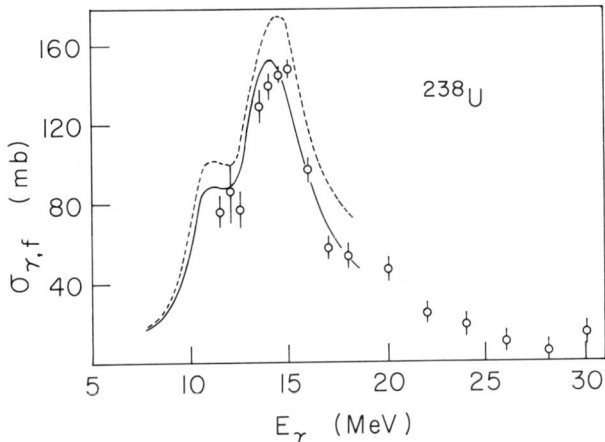

Fig. 11. Photofission cross sections. Dashed: Ref. 24; solid: Ref. 17; circles: Ref. 23.

Figure 10 shows the constrained (a) and unconstrained (b) E3 strength functions accompanying Figures 9e and 9d respectively. Under all assumptions concerning the form factors, peaks at 6, 10, and 14 MeV are observed. The features at 6 and 14 MeV may reflect nothing more than the fission barrier and the second chance fission threshold, although the low energy octupole resonance which from systematics should lie at ≈ 5.5 MeV and exhaust 30% of the E3 EWS_0R may contribute to the lower peak. The feature at 10 MeV however has not been observed before. We reiterate that the B(E3) cannot be directly compared to sum rules since it represents only an upper limit due to the presence of E4 excitations.

Table I. Observed Strengths as Percentages of the EWSR's for $E_x < 17.5$ MeV.

	E1[c]	E0	E2	E3
QRPA[18]	(87)	(66)	(84)	
(γ,f)[17]	(88)			
(e,e')[11]	(117–122)			
(e,e'f)[12]			3.7(8)[d]	10(45)[e]
(e,e'f)[a]	(82 ± 4)	28 ± 4 (88)	0.9 ± 0.4 (1.3)	6.7 ± 1.0 (30)	24 ± 2 (76)
	(80 ± 4)	32 ± 5 (102)	1.0 ± 0.4 (1.5)	7.9 ± 1.2 (33)	29 ± 2 (94)
	(88 ± 4)	42 ± 4 (134)	1.1 ± 0.4 (1.7)	9.3 ± 1.0 (39)	27 ± 2 (86)
(e,e'f)[b]	(87 ± 0.4)	18 ± 2 (52)	1.5 ± 0.2 (2.2)	10.7 ± 0.4 (44)	31 ± 2 (101)
	(87 ± 0.4)	35 ± 2 (103)	1.5 ± 0.2 (2.3)	14.1 ± 0.4 (59)	26 ± 2 (84)

Values in parentheses indicate total strengths and all other values refer to strengths in the fission channel alone. For the present data, the values in parentheses result from assuming that each multipole fissions with the E1 fission probability.

(a)Fit unconstrained by the photofission data. The three rows correspond to $c_{tr}/c_0 = (1.0, 0.9, 0.9)$, $(1.0, 1.0, 0.9)$, and $(1.1, 1.0, 0.9)$ respectively. The columns correspond to ranges of excitation energy: $E_x \leq 17.5$, $11 \leq E_x \leq 17.5$, $E_x \leq 6.5$, $E_x \leq 11$, and $E_x \leq 17.5$ MeV respectively.

(b)Fit constrained by the photofission data. The two rows correspond to $c_{tr}/c_0 = (1.0, 1.0, 0.9)$ and $(1.1, 1.0, 0.9)$ respectively. The columns correspond to ranges of excitation energy: $E_x \leq 17.5$, $12 \leq E_x \leq 17.5$, $E_x \leq 6.5$, $E_x \leq 12$, and $E_x \leq 17.5$ MeV respectively.

(c)E1 sums are with respect to the isovector E1 sum rule.

$(d)$$5.7 \leq E_x \leq 7$ MeV.

$(e)$$7 \leq E_x \leq 11$ MeV.

DISCUSSION

This work presents perhaps the first detailed comparison of a modern theoretical calculation of the nuclear response in the continuum region. The QRPA accounts for the overall shape and magnitude of the spectra at the two lower values of q. Although not free from model dependence and guidance from theory, a multipole decomposition of the data indirectly corroborates the long standing report of an anomalously small E2 fission probability. Our best estimate is that, if the fission probabilities were identical, the strength observed in the fission channel below 11 MeV would be no more than a third of the E2 EWS$_0$R. Such a result might imply either an anomalously large non-statistical component to the neutron decay channel, or some dynamic inhibition of fission several MeV above the barrier. Neither would be easy to understand. A direct measurement of the neutron decay spectrum might at least help us to decide which mechanism is responsible.

The 2$^+$ strength near the fission barrier is less well determined. The model dependent uncertainty here may be somewhat larger yet, since the non-collective states tend to display more complicated radial dependencies in the nuclear interior. Nevertheless, it is fair to say that for $5.5 \leq E_x \leq 6.5$ MeV, we observe 2% or less of the EWS$_0$R in the fission channel for reasonable variations of c_{tr}/c_0 in the unconstrained fits. This is probably consistent with the value of 3.7% (for $5.7 \leq E_x \leq 7$ MeV) from previous (e,e'f) work[12], for which a larger value is expected since their fits did not include E3. It is hard to reconcile with the value of 6% derived from inclusive electrofission angular distributions[27]. Beyond that, little quantitative evidence exists concerning near-barrier 2$^+$ states. An early $(\alpha, \alpha'f)$ study of fission fragment angular correlations in ^{238}U gave results consistent with the transition spectrum below 6 MeV containing either $(J, K) = (2, 0)$ and $(2, 2)$ states or $(2, 0)$ and $(2, 1)$ states[28], but reported no B(E2).

The amount of cross section above the high energy tail of the giant dipole resonance, at 80.3 and 118.4 MeV, is large. Certainly a major component of the broad yield around $E_x = 21$ MeV is the $\Delta T = 1$ quadrupole. However, the data do not extend far enough in excitation energy to permit a strength extraction.

APPENDIX: SUM RULES

The classical energy-weighted sums for electric multipolarities are given by

Bohr and Mottelson[25] as follows:

$$S(E0) = \frac{2\hbar^2}{M} Ze^2 <r^2>_{prot}$$

$$S(E1) = \frac{9}{4\pi} \frac{\hbar^2}{2M} \frac{NZ}{A} e^2$$

$$S(E\lambda) = \frac{\lambda(2\lambda+1)^2}{4\pi} \frac{\hbar^2}{2M} Ze^2 <r^{2\lambda-2}>_{prot} \quad \text{for} \quad \lambda \geq 2$$

in which M is the proton mass and $<r^{2\lambda-2}>_{prot}$ is the $(2\lambda-2)$th moment of the nuclear ground state charge distribution. The isoscalar contribution to these sums is generally taken to be the fraction Z/A of the total[26]. We have calculated these sum rules using the deformed ground state charge distribution measured by Cooper[22]. Table II lists these results which are used in the present analysis.

Table II. Energy-Weighted Sum Rules

λ	$<r^{2\lambda-2}>_{prot}$	$S(E\lambda)$	$S(E\lambda, \Delta T = 0)$
0	—	2.60×10^5 MeV-fm^4	1.01×10^5 MeV-fm^4
1	—	839 MeV-fm$^2 (\Delta T = 1)$	—
2	34.1 fm^2	2.59×10^5 MeV-fm^4	1.00×10^5 MeV-fm^4
3	1547 fm^4	3.45×10^7 MeV-fm^6	1.33×10^7 MeV-fm^6
4	1.181×10^5 fm^6	4.21×10^9 MeV-fm^8	1.63×10^9 MeV-fm^8
5	7.686×10^6 fm^8	5.11×10^{11} MeV-fm^{10}	1.98×10^{11} MeV-fm^{10}

ACKNOWLEDGMENTS

We wish to thank the accelerator staff at the High Energy Physics Laboratory, which includes H. A. Schwettman, T. I. Smith, M. S. McAshan, C. M. Lyneis, R. E. Rand, and S. Brain, for providing us with excellent quality electron beams in the course of this experiment. We also wish to thank J. Garibaldi and A. Combs III of Lawrence Livermore Laboratory for their technical support, and J. T. Hoeksema for helping with the computer analysis.

REFERENCES

[1] H. P. Morsch et al., Phys. Rev. **C25**, 2939(1982).
[2] H. P. Morsch et al., Phys. Lett. **119B**, 311(1982).
[3] J. van der Plicht et al., Nucl. Phys. **A346**, 349(1980).
[4] F. E. Bertrand et al., Phys. Lett. **99B**, 213(1981).
[5] H. P. Morsch et al., Phys. Lett. **119B**, 315(1982).
[6] S. Brandenburg et al., Phys. Rev. Lett. **49**, 1687(1982).
[7] A. C. Shotter et al., Phys. Rev. Lett. **43**, 569(1979).
[8] A. C. Shotter et al., Nucl. Phys. **A290**, 55(1977).

[9] J. D. T. Arruda-Neto et al., Phys. Rev. **C18**, 863(1978).
[10] H. Ströher et al., Phys. Rev. Lett. **47**, 318(1981).
[11] R. Pitthan et al., Phys. Rev. **C21**, 28(1980).
[12] D. H. Dowell et al., Phys. Rev. Lett. **49**, 113(1982).
[13] T. W. Donnelly, "Coincidence and Polarization Measurements with High Energy Electrons", CEBAF Workshop, June 1984 (MIT CTP Preprint 1184).
[14] J. Heisenberg, Adv. Nucl. Phys. **12**, 61(1981).
[15] J. D. T. Arruda-Neto et al., Nucl. Instr. Meth. **190**, 203(1981).
[16] W. E. Kleppinger, Thesis, Stanford University, 1982.
[17] A. Veyssière et al., Nucl. Phys. **A199**, 45(1973).
[18] D. Zawischa, J. Speth, in "Dynamics of Nuclear Fission and Related Collective Phenomena", from *Lecture Notes in Physics*, Vol. 158, Proceedings Bad Honnef, Germany 1981, Ed. P. David, T. Mayer-Kuckuk, A. van der Woude, Springer Verlag, Berlin, 1982, pg. 231; and private communication.
[19] L. J. Tassie, Aust. J. Phys. **9**, 407(1956).
[20] A. Bohr, *Conf. on Peaceful Uses of Atomic Energy*, Geneva, Vol. 2, (UN, New York, 1956) pg. 151.
[21] R. Vandenbosch and J. R. Huizenga, *Nuclear Fission*, (Academic Press, NY, 1973).
[22] T. Cooper et al., Phys. Rev. **C13**, 1083(1976).
[23] H. Ries et al., Phys. Rev. **C29**, 2346(1984).
[24] J. T. Caldwell et al., Phys. Rev. **C21**, 1215(1980).
[25] A. Bohr and B. R. Mottelson, *Nuclear Structure*, Vol. II, (Benjamin, Reading, Mass., 1975) pg. 399.
[26] J. Weneser and E. K. Warburton, in *The Role of Isospin in Nuclear Physics*, ed. D. H. Wilkinson, (North Holland, Amsterdam, 1969).
[27] J. D. T. Arruda-Neto et al., Rev. Brasil. d. Fis. **13**, 117(1983).
[28] H. C. Britt et al., Phys. Rev. **144**, 1046(1966).

SENSITIVE SEARCH FOR NEUTRON-ANTINEUTRON TRANSITIONS AT THE ILL REACTOR

M. Baldo-Ceolin
Dipartimento di Fisica "G. Galilei", Università di Padova, Padova-Italy
Istituto Nazionale di Fisica Nucleare, Sezione di Padova, Padova-Italy

ABSTRACT

A sensitive search for neutron-antineutron transitions in a neutron beam at the ILL reactor in Grenoble is presented.

Neutrons will be transported under vacuum in a magnetic field free region. Antineutrons present in the beam, as they come out from the field free region, will annihilate on a thin Carbon target. The annihilation events will be identified by their characteristic signature of ~ 2 GeV energy release in the form of several pions.

Background events are expected to be negligible. In a run of ~ 300 days we will observe the neutron-antineutron transitions if the oscillation time is less or equal to 10^8 sec.

MOTIVATIONS

According to the grand unified theories the baryon (B) and lepton (L) number conservation laws are only approximate [1]: thus a search for breakdown of B conservation appears as one of their most effective experimental tests [2]. If B is not conserved every proton, as well as every neutron bound in nuclei, must eventually decay into lighter particles. Conservation of energy, angular momentum and electric charge require that any change in the sum of baryon and lepton numbers $\Delta(B+L)$ be an even number, in any possible decay mode.

The first generation of proton decay experiments measured the limit $\Gamma < 0.5^{-32}$ years at 90% C.L. [3] for the decay rate of the process $p \to e^+ + \pi^0$, in disagreement with the SU(5) predictions ($\Gamma \sim 10^{-(29\pm 2)}$ years).

Consequently the theoretical predictions for nucleon decay processes are undetermined and all the decay channels are in principle possible.

A conceivable interesting process is the $\Delta B=2$, $\Delta L=0$ transition [4], which in addition to the decay process

$$NN \to \text{pions}$$

may give rise to neutron-antineutron oscillations, characterized by an oscillation time τ_{osc}, which depends at first order on the corresponding interaction Hamiltonian H' through

$$(\tau_{osc})^{-1} = \delta m = \langle \bar{n}|H'|n\rangle \sim \sqrt{\Gamma_{\Delta B=2} \cdot M} \tag{1}$$

where $\Gamma_{\Delta B=2}$ is the $\Delta B=2$, $\Delta L=0$ nucleus decay rate, and M is the nucleon mass.

Actually the theoretical models leave a rather large incertitude in the value expected for τ_{osc}. The so-called "left-right symmetric" models, however, suggest mainly for the neutron oscillation time a value $\tau_{osc} \simeq 10^8$ sec [5].

Moreover in the quark-lepton picture the simplest term in the effective Lagrangian that can induce $\Delta B=2$ processes is [6]

$$L_{eff} \sim M_x^{-5}(qqq \cdot qqq) + h.c.$$

and therefore, if neutron-oscillation processes are observed in the experimentally accessible region up to $\tau_{osc} \sim 10^8$ sec or more, this will open a new physics in the mass range $M_x \sim 10^4 \div 10^6$ GeV.

NEUTRON OSCILLATION PHENOMENOLOGY AND EXPERIMENTAL METHOD

The $\Delta B=2$ interaction produces a mixing among neutron and antineutron states, so that n and $\bar{n}$ are no longer "two distinct states", but belong to "one two state system" whose time evolution (in the c.m.s.), assuming CP invariance and neglecting decays, is governed by the equation

$$\frac{d\psi}{dt} = -i\,M\,\psi\,; \qquad \psi = \binom{n}{\bar{n}} \qquad (2)$$

with

$$M = \begin{pmatrix} M & \delta m \\ \delta m & \bar{M} \end{pmatrix} \qquad (3)$$

where the non diagonal terms represent the $n \rightleftarrows \bar{n}$ transition energy, while the diagonal terms represent the effective neutron and antineutron masses: $M = m_n + V_n$, $\bar{M} = m_n + V_{\bar{n}}$, m_n is the neutron(antineutron) mass, and V_n, $V_{\bar{n}}$, the nuclear and/or electromagnetic potential for neutrons and antineutrons in the surrounding medium.

To the eigenstates of eq.(2)

$$n_1 = \cos\Theta\, n + \sin\Theta\, \bar{n} \qquad\qquad n_2 = -\sin\Theta\, n + \cos\Theta\, \bar{n}$$

belong the masses

$$m_{1,2} = \frac{1}{2}(M+\bar{M}) \pm \sqrt{\delta m^2 + \Delta E^2}$$

where

$$\text{tg}2\Theta = \frac{\delta m}{\Delta E} \qquad \text{and} \qquad \Delta E = \frac{1}{2}(V_n - V_{\bar{n}})$$

The time evolution of a neutron state

$$n = \cos\Theta \; n_1 + \text{sen}\Theta \; n_2$$

is described by the equation (neglecting β decay)

$$n(t) = \cos\Theta \; n_1 e^{-im_1 t} + \text{sen}\Theta \; n_2 e^{-im_1 t},$$

and the probability of finding an antineutron component at time t, in a state that at t = 0 was a pure neutron state, is given by

$$P(\bar{n},t) = \left(\frac{\delta m}{\Delta M}\right)^2 \text{sen}^2 (\Delta M \cdot t) \tag{4}$$

Eq.(4) shows that the probability $P(\bar{n},t)$ oscillates with amplitude $A = (\delta m/\Delta M)^2$ and angular frequency $\omega = \Delta M$, where

$$\Delta M = \frac{1}{2}(m_1 - m_2) = \sqrt{\delta m^2 + \Delta E^2} \; .$$

Since in practice, neutrons are never free, ΔM is much larger than δm, and neutron oscillations are strongly suppressed, the probability $P(\bar{n},t)$ going to zero as $(\delta m/\Delta M)^2$.

However, real neutrons behave as free neutrons for values of t and ΔM such that [7]

$$\Delta M \cdot t \ll 1 \tag{5}$$

Eq. (5) represents the "quasi free neutron condition": by it $P(\bar{n},t)$ grows as t^2, while for $t > (\Delta M)^{-1}$, $P(\bar{n},t)$ oscillates between $(\delta m/\Delta M)^2$ and zero with the angular frequency $\omega = \Delta M$.

Thus in order to reach an appreciable value for $P(\bar{n},t)$ it is imperative to make ΔE as small as possible.

This condition can be fulfilled primarily using neutron beams propagating in a region properly evacuated (in order to avoid nuclear interactions) and shielded against any external magnetic field.

Summarizing the above discussion we may conclude that, first, an experimental search on neutron oscillations requires:

i) a very intense neutron source;
ii) a propagation region free from any kind of interactions.

It has to be noted that, since condition i) is largely satisfied in the deep underground experiments searching for ΔB=1 nucleon decays, the possibility that the same experiments might also detect n-$\bar{n}$ oscillations in nuclei has been explored [8], thus allowing a mesurement of the interesting parameter $\delta m = (\tau_{osc})^{-1}$.

From proton-decay type experiments one gets $\Gamma_{ann} < 2 \cdot 10^{-31}$ years [9], and this result due to unavoidable background effects appears to be the experimental limit of the present generation experiments.

Moreover, as is well known, there are substantial uncertainties in relating free and bound n-n̄ mixing arising from uncertainty in the n̄ optical potential.

Furthermore the strength of n ⇄ n̄ transitions need not be the same for neutrons isolated or contained in nuclear matter, and consequently Γ_{ann} does not constrain the value of the free neutron oscillation time [10].

Following the neutron oscillations assumption, a state initially composed of pure neutrons should become, within a finite time, a mixture of neutrons and antineutrons. In order to test this hypothesis, one has to detect the antineutrons bringing them to annihilate in nuclear matter.

The antineutron signature will be the typical release of ~2 GeV, shared by several pions (5 in average) with total momentum $\vec{p} \cong 0$.

Once the "quasi free condition" is satisfied, the constraint which defines the quality of a neutron-oscillation experiment may be deduced from eq.(5) as

$$\tau_{osc} \propto \sqrt{N \cdot \varepsilon} \; t = \sqrt{I \cdot T \cdot \varepsilon} \; \frac{L}{v} \propto \left\{ \frac{I \cdot \varepsilon \cdot A}{E} \right\}^{\frac{1}{2}} \tag{6}$$

where: I, the neutron current in n sec^{-1}, depends on the power of the neutron source; ε, the fraction of annihilation events which can be unambiguously identified, depends on the properties ("quality") of the detector; t = L/v, is the time in sec of the "quasi free propagation"; v, the neutron velocity in m sec^{-1}; Nt^2, for a given source, depends upon neutron energy and annihilation target area.

An important peculiarity of n-n̄ oscillation experiment carried out with neutron beams lies in the fact that the background can be measured directly. The background is mainly due to neutral cosmic ray interactions that the detector would not be able to discriminate from annihilation processes. However, the antineutron yield may indeed be suppressed at will by a magnetic field B applied in the propagation region (see eq.(4)) leaving the cosmic ray background to be measured separately.

In this type of experiment both
a) gamma rays and fast neutrons from the neutron source, coming along the moderated neutron beam, and
b) neutrons scattered from the beam at the annihilation target and neutron capture gamma rays

can give a large radiation flux on the detector.

Although the effects of a) and b) will in general not be such as to mimick an annihilation interaction, they may strongly reduce the detection efficiency.

Then, in order to achieve a high antineutron detection efficiency

minimizing at the same time the neutral cosmic ray background, the annihilation target and the detector should satisfy the following requirements: first of all both must be protected by a most effective shielding and anticoincidence system against cosmic rays, and further
a) the target should be characterized by
   i) large annihilation probability for the antineutron component;
   ii) low Z material to preserve the characteristic features of the annihilation events as much as possible;
   iii) high transparency to the neutron beam to minimize neutron scattering and capture;
   iv) a total mass as small as possible in order to reduce residual cosmic ray interactions;
b) the detector should:
   i) be large and massive enough to accept and stop as many annihilation products as possible;
   ii) allow good energy resolution, particle identification and track pattern recognition to ensure vertex reconstruction and measurement of the energy emitted from the annihilation event.

EXPERIMENTAL SET-UP FOR THE PROPOSED EXPERIMENT

The proposed experiment is the second part of the long term experiment initiated in 1980 at the ILL reactor in Grenoble: it aims at measuring τ_{osc} up to $(1-5) \cdot 10^8$ sec.

The experimental set-up sketched in Fig. 1 has been designed taking into account the results of the first part of the experiment [11], which was planned as an exploratory search with the definite purpose of singling out the main factors to emphasize in planning an experiment capable of reaching a significant sensitivity in τ_{osc}.

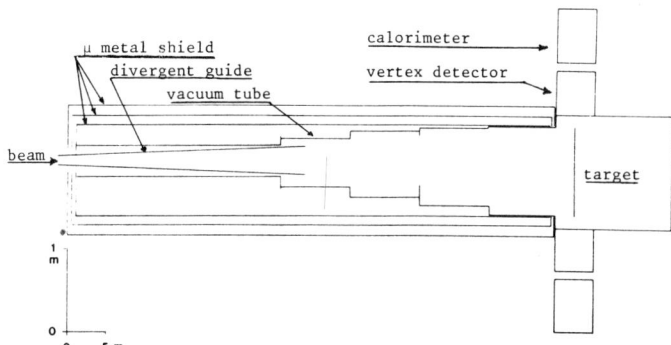

Fig. 1 - Quasi free propagation region and magnetic shielding. $B = 10^{-4}$ gauss will provide 100% $n \to \bar{n}$ effect; $B \sim 10^{-3}$ gauss will suppress it by 20%.

The proposed experimental set-up consists of
a) a cold neutron beam transported to the experimental area by a curved guide;
b) a long channel for the quasi free beam propagation;
c) a thin isolated annihilation target;
d) a fine grain detector with high spatial and energy resolution;
e) an efficient cosmic ray shield.

The relevant quantity $I \cdot A \cdot \epsilon / E$ has been optimized taking into account:
i) the neutron beam properties;
ii) the target which should be small enough to keep the final detector dimensions reasonable.

Cold neutrons will be transported to the experimental area by the new guide H52 ∼60 m long, made of several elements with 6×12 cm² cross-section, arranged on a curve of 5000 m radius. The direct radiations from neutron source will thus be switched off and, at the same time, the lower energy neutrons selected: the effective neutron temperature will be reduced to ∼15° K, corresponding to an average velocity v ∼460 m sec^{-1}. The beam intensity will be $I = 3.3 \cdot 10^{11}$ n sec^{-1}.

In order to keep the dimensions of the annihilation target and of the experimental apparatus within reasonable values, a simple device is envisaged based on neutron reflection properties within a guide [12]. It consists of a guide with slightly divergent walls. With this simple device, if Θ is the angle of the neutron trajectory with the axis of the guide, a reflection at the walls reduces Θ by 2δ, where δ is the divergence of the walls. So, if Θ_{in} is the divergence of a neutron entering the guide, its divergence at the exit after n reflection, will be

$$\Theta_{out} = \Theta_{in} - 2n\delta.$$

The required target area A needed to contain the beam will be correspondingly decreased by a considerable factor.

The quasi free propagation region will consist of:
i) a (30-70) m long straight guide, with slightly divergent walls (3 mrad);
ii) a ∼30 m long conical vessel (Θ = 15 mrad)
where a gas pressure $P < 10^{-6}$ torr and a residual magnetic field $B < 10^{-4}$ gauss warrant the quasi free condition.

The average time interval from the last neutron reflection in the divergent guide to the end of the quasi free propagation region will be

$t \cong (0.1 - 0.15)$ sec

The target will be placed 2.5 m down-stream of the magnetically shielded region where the beam cross-section will be ~70×100 cm² (see Fig. 2).

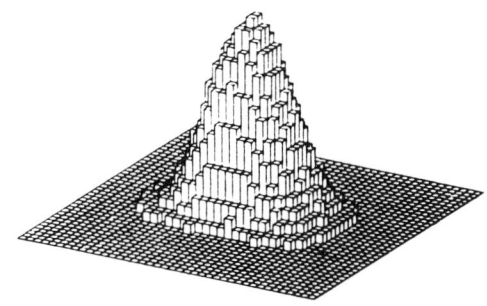

Fig. 2 - Lego-plot of the neutron beam at the annihilation target.

The target, a Carbon foil ~100 μ thick, ϕ = 110 cm, ~0.1 kg weight, is enough to annihilate the antineutron component with more than 99% probability and it will scatter only 0.5 per cent of the crossing neutrons. It has been evaluated through a Monte Carlo calculation, based on experimental data, obtained during a test run with an antiproton beam from the CERN PS, that ~38 per cent of the annihilation processes remain undisturbed in a Carbon nucleus, while for the rest the momentum unbalance will be in the average $\Delta p \cong$ 300 MeV/c and the total π visible energy E $\cong$ 1600 MeV.

Consequently the signature for an annihilation event will be:
i) the interaction vertex inside the target;
ii) a visible energy E $\simeq$ (1500±500) MeV;
iii) a total momentum p < (400±400) MeV/c.

The detector has been designed in order to have high resolution for the vertex reconstruction and high probability for detecting and measuring the energy emitted in the annihilation process.

In order to recognize and measure the annihilation products:
a) the detector is as near as possible to the target;
b) there is as little material as possible between the target and the detector;
c) the vertex detector is practically immaterial.

A cross sectional view of the detector is shown in Fig. 3.

It is shaped as a box surrounding the target with a solid angle $\Delta\Omega/4\pi \simeq$ 1. The walls of the box will consist of Limited Streamer Tube planes [13] and scintillation counter plates, immediately outside the propagation vessel. The ensemble works as:
i) a charged particle vertex detector: 10 L.S.T. planes and 2 scintillator counter planes in each detector side;
ii) a neutral particle vertex detector: 10 L.S.T. planes in each side

interleaved with .5 cm Aℓ plates and with a ∿.5 cm Pb plate placed
in front;

iii) a calorimeter able to absorb the whole energy emitted in annihilation processes, consisting of 16 limited streamer tube planes
interleaved with Fe plates of increasing thickness.

A Monte Carlo calculation shows that ∿80% of the annihilation processes will be fully contained in this ensemble and the vertex reconstructed to within a few centimeters.

Fig. 3 - Cross sectional view of the detector.

The experiment will be protected by proper material against cosmic
ray neutrals and by a veto system against charged penetrating particles.

The main characteristics of the planned experiment are summarized
in Table I.

On the basis of the analysis of the cosmic ray background performed in the previous experiment, less than 1 background event in 10^8 sec

TABLE I

Neutron Beam

$I = 3.3 \cdot 10^{11} n \; sec^{-1}$
$R = 5000 \; m$
$\sqrt{<\lambda>^2} \simeq 8.5 \; \text{Å}$

Propagation Region

1) (30-70)m long straight divergent guide
2) 30 m drift vessel
   $t = (0.10 \div 0.15) \; sec$
   $Nt^2 = (3.3 \div 7.4) \cdot 10^9 n \; sec^2$

Annihilation Target

Carbon foil ∼100 μm thick
Neutron beam cross-section
at the target 70×100 cm^2
Vessel $\phi = 140$ cm

Detector

Limited streamer tube planes and
scintillation counter plates
(5×4 m^2) × 144 tube planes
Good vertex reconstruction
Good energy measurement
$\frac{\Delta\Omega}{4\pi} \simeq 1$; $\varepsilon \simeq 0.8$

Background

$N_B < 1/10^8 \; sec$
$T \geq \pi \cdot 10^7 \; sec$

Sensitivity

$\tau_{osc} \; (2 \div 5) \cdot 10^8 \; sec$

would be expected. Therefore, the experiment can be run for a year effective time with negligible background, so as to reach a sensitivity in τ_{osc} up to $2 \cdot 10^8$ sec or more.

In this experiment in the minimal configuration one expects ∼3 annihilation events per day if $\tau_{osc} = 10^7$ sec, and still 10 events if $\tau_{osc} = 10^8$ sec.

REFERENCES

1. -J.C. Pati and A. Salam, Phys. Rev. D8, 1240 (1973).
    -H. Georgi and S.L. Glashow, Phys. Rev. Lett. 32, 438 (1974).
2. For a review see: H. Primakoff and S.P. Rosen, Ann. Rev. Nucl. Part. Sci. Vol. 31, 145 (1981).
3. M. Koshiba, "XXII International Conference on High Energy Physics", Leipzig, July 1984.
4. -S.L. Glashow, Proceedings of "Neutrino '79", Bergen, Vol. I, p.518.
    -V.A. Kuzmin, Pisma JEPT 13, 335 (1970).
    -R.E. Marshak and R. Mohapatra, Phys. Rev. Lett. 44, 1316 (1980).
    -L.N. Chang and N.P. Chang, Phys. Lett. 92B, 103 (1980).
5. R.E. Marshak and R.N. Mohapatra, VPI-HEP-84/10.
6. -S. Weinberg, Phys. Rev. D22, 1694 (1980).
    -F. Wilczek and A. Zee, Phys. Rev. 43, 1571 (1979).
7. -M. Baldo-Ceolin, Proceedings of the "Conference on Astrophysics and Elementary Particles: Common Problems", Roma (1980), p. 251.
    -R.E. Marshak and R.N. Mohapatra, Phys. Lett. 94B, 183 (1980).
8. C.B. Dover, A. Gal and M. Richard, Phys. Rev. D27, 1090 (1983), ref. therein.
9. -T.W. Jones et al., Phys. Rev. Lett. 52, 720 (1984).
    -G. Battistoni et al., Phys. Lett. 133B, 454 (1983).
10. -P.K. Kabir, Phys. Rev. Lett. 51, 231 (1983).
    -J. Basecq and L. Wolfenstein, Nucl. Phys. B224, 21 (1983).
11. M. Baldo-Ceolin, "Workshop on Reactor Based Fundamental Physics", Grenoble, Nov. 1983, Journal de Physique, Colloque C3, Suppl. n.3, Tome 45, C3 - 173-183 (1984).
12. H. Maier-Leibnitz and T. Springer, Ann. Rev. Nucl. Sci. 16, 207 (1962).
13. G. Battistoni et al., Nucl. Instr. and Methods 176, 297 (1980).

THE BETA DECAY OF POLARIZED NEUTRONS

P. Bopp, D. Dubbers, L. Hornig, E. Klemt, J. Last, H. Schütze
Physikalisches Institut der Universität Heidelberg, Germany

S.J. Freedman
Argonne National Laboratory, Argonne, USA

Otto Schärpf
Institut Laue-Langevin, Grenoble, France

ABSTRACT

Beta decay of polarized neutrons has been studied with the superconductive spectrometer PERKEO at the Institut Laue-Langevin. The energy spectrum of the ß-decay asymmetry has been measured for the first time; from the absolute value of the asymmetry we obtain a new value for the ratio of weak coupling constants g_A/g_V, which is compared to similar data from hyperon decays. The measurement of further weak interaction parameters from neutron decay is in progress.

INTRODUCTION

The ß-decay of the free neutron is the simplest and most fundamental of nuclear ß-decays. In neutron decay the weak interaction of quarks can be studied without the disturbing effects due to nuclear structure. For example, neutron decay is by now the only source to measure the size of the weak axial coupling constant g_A, which determines the strength of the charged axial weak current which couples an up quark to a down quark*. In nuclei, on the other hand, g_A is quenched to a value $g_A \lesssim g_V$, where g_V is the weak vector coupling constant.

The actual size of the ratio g_A/g_V is predicted by quark theory only with low accuracy. The precise determination of g_A/g_V from neutron decay is nonetheless of great theoretical importance, because it can be compared to similar quantities obtained from the semileptonic decays of the heavier baryons, like the strange hyperons Λ, Σ, and Ξ; this allows to test current ideas on the weak mixing of different quark generations.

Equally important, a good knowledge of the weak coupling constants from neutron decay is desirable because they determine the weak cross sections of other processes with similar Feynman diagrams, like neutrino-nucleon and weak nucleon-nucleon interactions, which are important in cosmological models, in solar cycle calculations, and in neutrino physics in general, and which so far cannot be directly measured.

---

*W-boson production in the CERN p-p̄ collider may become another source for g_A.

Finally, in the long run, it is desirable to measure other weak interaction parameters directly from neutron decay, like the weak magnetism form factor, whose value, in the frame-work of electroweak theory, is determined solely by the known electromagnetic properties of the nucleons, and further to set new limits on the scalar, tensor, and pseudo-scalar contributions to the weak interaction.

POLARIZED NEUTRON DECAY

The weak interaction form factors in neutron decay can be obtained[1] from measurements of: i. the neutron half-life; ii. the various correlation coefficients between electron momentum, antineutrino momentum, and neutron polarization; and iii., for the weak magnetism form factor, the precise shapes of the β-energy spectra.

When, for instance, the β-decay of polarized neutrons is studied, the angular and energy distribution of the electrons emitted is given by

$$W(E,\vartheta)dEd\Omega = F(E)\{1 + b(E) + \frac{v}{c} PA[1+c(E)]\cos\vartheta\}dEd\Omega \quad (1)$$

The statistical shape function, $F(E)$, is a function of β kinetic energy E, approximately given by the expression

$$F(E) = (E_o-E)^2 (E + mc^2) \sqrt{E(E + 2mc^2)} . \quad (2)$$

Here E_o is the β-endpoint energy, v/c is the electron speed in units of the speed of light, ϑ is the emission angle relative to the neutron polarization P, and A is the electron momentum-neutron polarization-correlation coefficient, also called β-asymmetry parameter, which is nonzero because of parity violation. The terms $b(E)$ and $c(E)$ account for small effects (~1-2% at $E=E_o$) both from weak magnetism and radiative corrections[2].

A is predicted in V-A theory to depend solely on the ratio g_A/g_V. With $\lambda = |g_A/g_V|$, we have

$$A = -2\lambda \frac{\lambda - 1}{1 + 3\lambda^2} \quad (3)$$

Since λ turns out to be near unity, measuring A is in general a good way to determine λ.

EARLIER EXPERIMENTS

While the β-decay of free neutrons is important for weak interaction theory, it is, on the other hand, very difficult to study experimentally. In fact, neutron decay experiments had been going on for the past thirty five years, and progress was rather slow. The reasons for this are, first, the small energy release in neutron decay, with rather low β-endpoint energy of some 780 keV, and second, as a consequence, the rather long neutron half-life of approximately 10 minutes.

Due to the long half-life, only one out of 10^7 neutrons decay within the active volume of a typical neutron decay apparatus,

installed at a thermal neutron beam from a nuclear reactor. The
neutron decay experiments are usually performed in a hostile environment, with many gamma rays penetrating the apparatus, both prompt
and ß-delayed, giving rise to signals in the same energy range as
the true ß-events. To suppress background most earlier experiments
used sophisticated proton-electron coincidence techniques; they
therefore suffered from low statistics, and, in addition, had to
rely on Monte Carlo simulations to define the effective beam volume,
the effective solid angles, and the detector efficiencies. Furthermore,
the backscattering probability for low energy ß-particles is rather
high and difficult to correct for. For these reasons, the two ß-energy
spectra from neutron decay published so far[3,4] were useful only
in the upper half of the ß-energy interval, and energy spectra from
polarized neutron decay did not exist at all. - Nevertheless, the
very careful measurements of neutron correlation coefficients[5,6,7]
gave mutually consistent results, whereas recent direct measurements[4,8,9] of the neutron half-life scatter as much as 7%, with
about 1% individual error bars.

THE SUPERCONDUCTING NEUTRON DECAY SPECTROMETER PERKEO

We have developped a neutron decay spectrometer, called PERKEO,
which is installed at the High Flux Reactor of the Institut Laue-
Langevin at Grenoble. The main features of PERKEO are its long decay
length, its large decay volume, the effectively loss-free detection
of the decay electrons over a 4π solid angle, and the complete suppression of electron backscattering effects. The apparatus is installed outside the far end of the long I.L.L. neutron guide hall, where
reactor associated background is very low.

In the first experiment with PERKEO we studied the ß-decay
of polarized neutrons. The count rates achieved are several orders
of magnitude higher than in previous experiments, see table 1. At
the same time, systematic errors are largely suppressed and for
the first time we obtain very clean ß-energy spectra from neutron
decay, both polarized and unpolarized.

PERKEO is shown schematically in Fig.1. The principal component
is a 1.7 m long 20 cm diameter superconducting selonoid which produces
a ≈ 1.5 Tesla field along the direction of the neutron beam. Electrons
from neutrons decaying inside the spectrometer move in helical paths
< 1 cm in diameter with axis along magnetic field lines. Four super-
conducting trim coils at either end distort field lines causing

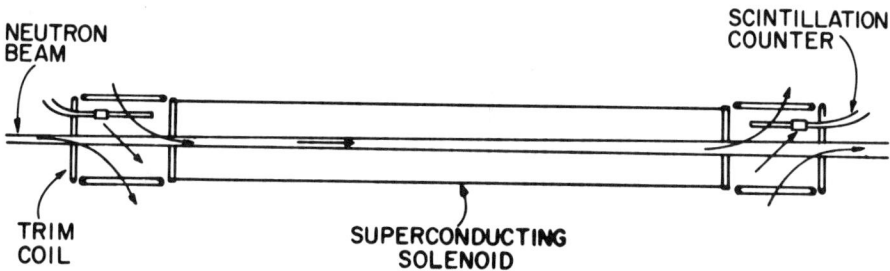

Fig. 1. Schematic layout of PERKEO.

electron trajectories to bend away from the neutron beam. The electron trajectories are finally intercepted by plastic scintillation counters located at each end of the spectrometer. The magnetic field strength decreases continuously towards both ends of the spectrometer, so no electrons can be magnetically trapped. The electron flight times between the detectors is effectively within the range 7 ns $\lesssim$ t $\lesssim$ 300 ns.

Each scintillator is coupled to two photomultipliers and coincidence signals with thresholds below the single photoelectron level are required to reduce noise. The detectors are calibrated with various conversion line sources. Fig. 2 shows the typical response. The observed resolution corresponds to $\approx$160 photoelectrons/MeV.

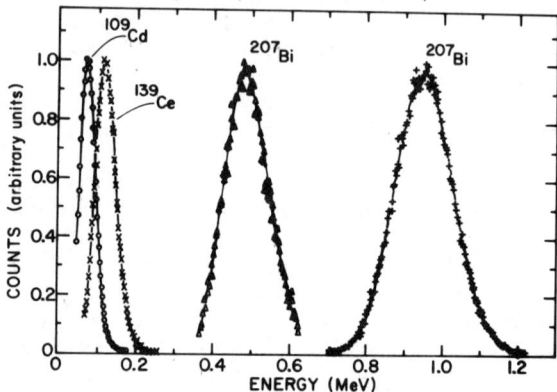

Fig. 2. Detector response to various conversion line sources.

The scintillators are large enough so that the total detection solid angle is 4π-str within the spectrometer. An electron backscattered from a detector is either reflected back by the magnetic mirror effect or it moves along a field line to the other detector. Since both detector signals are summed the total electron energy is recorded. About 1% of the neutron decay events have the measured electron energy shared in the two detectors. The detector hit first and thus the initial electron velocity component along the spectrometer axis is identified by the timing of the detected signals.

PERKEO is located at the polarized neutron experimental station PN7 at ILL. The total capture flux into the spectrometer from the PN7 supermirror polarizer is (8 ±3) x $10^9 sec^{-1}$. The beam is collimated both horizontally and vertically by a 1.8 m long 3 x 3 channel collimator made entirely from sintered ^{6}LiF. The beam cross section thus defined is 3.8 x 5.6 cm^2 at the exit of the spectrometer. No evidence for a halo on the neutron beam is observed at the 10^{-5} level. During measurements of the β-asymmetry neutrons are longitudinally polarized and the polarization is switched every 5 sec with a current sheet spin flipper.

MEASUREMENTS

Figure 3 is the background subtracted unpolarized β-spectrum obtained by summing the spectra taken for equal times from each detector and both neutron polarizations. The solid curve through the data is a fit to the expected shape including detector resolution. The total neutron decay count rate is 165 sec^{-1}. The relative number of electrons below threshold is less than 4%.

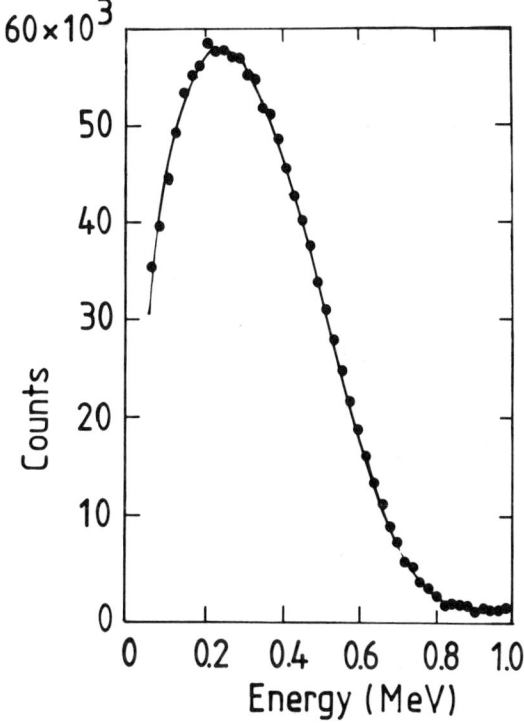

Figure 3. The β-decay energy spectrum from one run. The solid curve is a fit to the resolution corrected Fermi shape.

We define the experimental asymmetry as

$$A_\beta = \frac{N_1^\uparrow - N_1^\downarrow}{N_1^\uparrow + N_1^\downarrow} - \frac{N_2^\uparrow - N_2^\downarrow}{N_2^\uparrow + N_2^\downarrow} \qquad (4)$$

where the N's are energy spectra for equal times: 1 and 2 refer to the detector, ↑ and ↓ to the polarization.

From (1) we have

$$A_\beta = \frac{1}{2}(1 + f) \frac{v}{c} \text{PSA} \frac{1 + c(E)}{1 + b(E)} . \qquad (5)$$

The factor of 1/2 is from averaging over the detector solid angle; f is the spin flip reversal efficiency; and S corrects for magnetic mirror reflections in the inhomogeneous fields at the end of the spectrometer. S is reliably calculated from the known field distribution, we obtain S = 0.890(15). The energy dependence of A_β is dominated by the factor v/c but the additional factor $(1 + c(E))/(1 + b(E))$ accounts for the small but not negligible effect from weak magnetism and radiative corrections.

Figure 4 shows the β-decay asymmetry A_β, whose dependence on the β-particle energy has been measured for the first time. The data are in accord with the energy dependence expected from the v/c factor. The theoretical values for b(E) and c(E) are included in the fit; statistics is not yet good enough to extract them from the data.

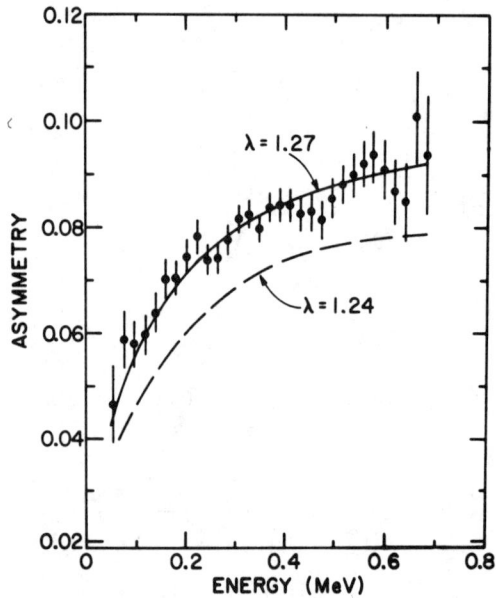

Fig. 4. Experimental β-ray asymmetry as a function of β-energy. The curves are expectations for two values of λ.

The neutron polarization and spin flip efficiency are determined to be P = 96.7(7)% and f = 99.8(3)% from measurements with a second supermirror and spin flipper.

From the absolute size of the β-decay asymmetry we obtain the value A = -0.118(3) which is corrected for weak magnetism and radiative effects. The error is half from statistics, half from systematics, both subject to further improvements. From Eq.3 we obtain the preliminary result λ = 1.27(1). The solid curve in Fig.4 is the prediction for the experimental A_β with λ = 1.27; the dashed curve is the corresponding prediction for λ = 1.24 which is the value suggested in Ref.12.

Table 1. Count rates and polarizations in various polarized neutron decay experiments.

Year	Ref.	β-counts/hour N	n-polarization P	Relative Quality Factor $Q \propto P^2 N$
1970	10	100	0.870(30)	1
1971	11	200	0.770(20)	1.6
1975	6	420	0.790(15)	3.5
1979	7	420	0.733(20)	3.0
1983	This work	600,000	0.967(7)	7,500

Table 2. Summary of recent λ-measurements ($\lambda = |g_A/g_V|$)

Year	Ref.	Quantity Measured	λ
1972	4	$T_{1/2}$ = 636.6(9.6)	1.245(11)
1978	9	$T_{1/2}$ = 607.8(5.4)	1.280 (7)
1980	8	$T_{1/2}$ = 649(12)	1.230(14)
1982	12	PDG recomm. value $T_{1/2}$ = 641(8)	1.239 (9)
1978	5	a = -0.1017(51)	1.259(17)
1975	6	A = -0.113(6)	1.254(15)*
1979	7	A = -0.114(5)	1.257(12)*
1983	This work (preliminary)	A = -0.118(3)	1.270 (9)
1983	13	Various hyperon decay parameters + Cabbibo hypothesis	1.182(22)**

* the values of λ given in refs.6 and 7 include the statistics of refs. 10 and 11.

**the overall value of λ = 1.233(16) given in the abstract of ref.13 includes neutron decay λ-values, but only those from the PDG-recommended half-life measurements, refs.4 and 8. Here we quote the λ-value of ref.13 derived solely from the strange hyperon decay data.

Table 2 lists given values of λ determined by other recent experiments. Our value of λ is somewhat higher than but consistent with the value of λ derived from earlier correlation coefficient measurements. It is several standard deviations higher than the Particle Data Group recommended value derived from two recent half-life measurements. There is a clear discrepancy with the λ-value derived from hyperon decays at CERN using Cabbibo's theory. Inclusion of the third quark generation through the Kobayashi Maskawa matrix makes things even worse[14].

CONCLUSION

We have built a new type of spectrometer in which neutron decay can be studied with high statistics. At the same time, sources of systematic errors are largely suppressed. We have measured clean β-spectra for both polarized and unpolarized neutron decay, and derived a new value for the ratio of weak coupling constants g_A/g_V. At present, we continue our measurements on the β-decay asymmetry. The measurement of other neutron decay parameters is in preparation.

We thank the ILL management and staff for their hospitality. This work has been supported by the Bundesministerium für Forschung und Technologie and the United States Department of Energy.

REFERENCES

1. E.D. Commins, P.H. Bucksbaum, Weak Interaction of Leptons and Quarks, Cambridge University Press, 1983.
2. D.H. Wilkinson, Nuclear Physics A377, 424 (1982).
3. J.M. Robson, Phys. Rev. 83, 349 (1951)
4. C.J. Christensen et al., Phys. Rev. D5, 1628 (1972)
5. Chr. Stratowa et al., Phys.Rev. D18, 3970 (1978)
6. V.E. Krohn, G.R. Ringo, Phys. Lett. 55B, 175 (1975)
7. B.G. Erozolimskii et al., Sov. J. Nucl. Phys. 30, 356 (1979)
8. J. Byrne et al., Phys. Lett. 92B, 274 (1980).
9. L.N. Bondarenko et al., JETP Lett. 28, 303 (1978)
10. C.J. Christensen, V.E. Krohn, G.R. Ringo, Phys.Rev. C1, 1693 (1970)
11. B.F. Erozolimskii et al., JETP Lett. 13, 252 (1971)
12. Particle Data Group, Phys. Lett. 111B, 91 (1982)
13. M. Bourguin et al., Z. Phys. C21, 27 (1983)
14. W. Siebert, private communication

STATISTICAL METHODS OF SPIN ASSIGNMENT
IN COMPOUND NUCLEAR REACTIONS

H. Mach

Brookhaven National Laboratory, Upton, New York, 11973, USA
and
McMaster University, Hamilton, Ontario, Canada, L8S 4K1

and

M. W. Johns
McMaster University, Hamilton, Ontario, Canada, L8S 4K1

ABSTRACT

Spin assignment to nuclear levels can be obtained from standard in-beam gamma-ray spectroscopy techniques and in the case of compound nuclear reactions can be complemented by statistical methods. These are based on a correlation pattern between level spin and gamma-ray intensities feeding low-lying levels. Three types of intensity and level spin correlations are found suitable for spin assignment: shapes of the excitation functions, ratio of intensity at two beam energies or populated in two different reactions, and feeding distributions. Various empirical attempts are examined and the range of applicability of these methods as well as the limitations associated with them are given.

INTRODUCTION

Spin assignments to the low-lying levels in medium and heavy nuclei are generally based on indirect techniques sensitive to transition multipolarities and the relative spin difference between nuclear levels. For example, measurements of angular distributions and correlations, γ-ray linear polarization and electron internal conversion coefficients, independently or in any combination, seldom provide an unique spin assignment. In the domain of compound nuclear reactions, statistical methods can complement these standard in-beam γ-ray spectroscopy techniques and increase the potential for reliable spin assignments.

So far no systematic effort has been made to explore the broad application of the statistical methods of spin assignment or to provide a "common code of practice". These methods have been applied to numerous specific cases and were empirical in character, occasionally supported by systematic tests of the method or detailed model calculations. Furthermore, these methods are not uniformly accepted throughout the research community. A short review of these methods may clarify some of the problems listed above.

*Research was performed in part under contract DE-AC02-76CH00016 with the U.S. Department of Energy and was also supported by a NSERC Canada grant.

The statistical methods are based on: a) the ability to detect the change in the intensity feeding low-lying levels of interest due to changes in the average spin of the formation states (the latter caused by modification of the reaction process), and b) the statistical process governing the decay of the formation states to the states of interest. After an extensive literature search, three basic correlation patterns of level-spin and γ-ray intensities feeding these levels were found to be reliable spin indicators.

EXCITATION FUNCTIONS

The shape of the excitation functions has two features, which are directly spin dependent. The first such feature is that the excitation function characteristic of a higher spin reaches a maximum at a higher beam energy. This correlation can be expressed quantitatively in a functional form. The second feature is that, when normalized to a weighted average of a group of curves of a given spin, the excitation functions of higher spin show higher slope. Although this relation is a qualitative one, the information contained in it can be transformed into a quantitative functional dependence using the ratio of intensities method.

The excitation function can be given for the following quantities associated with the level: "total feeding intensity" usually measured by summing over all observed processes deexciting a level, and "side-feeding intensity" measured by subtracting from the total feeding intensity the intensity due to all observed discrete transitions populating this level from above. Although the side-feeding is considered to be the closest approximation to the statistical feeding, the total feeding may also represent an adequate approximation if a level is fed by several discrete transitions and if none of them carries a significant amount of the total intensity.

In (n,n'γ) studies the γ-ray excitation functions of the level of unknown spin are compared to the empirically determined standards[1]. In (p,nγ) reactions the shape of the function is compared to the predictions derived from simple Hauser-Feshbach calculations[2]. In deuteron, alpha and heavy-ion induced reactions the slope of these functions is used in a qualitative way and is usually combined with angular distribution results to provide spin assignments[3,4]. Furthermore, the maxima of these functions can be used in a quantitative way as spin indicators[5] (see Fig. 1a).

RATIO OF INTENSITIES

The ratio of level intensity obtained at higher beam energy to that at the lower one shows a smooth functional dependence, which is characterized by a higher ratio for higher spin value. A similar function can be constructed using level intensities populated in two different reactions and leading to the same nucleus of interest. Individual γ rays deexciting the same level must have identical intensity ratios which can be used to test the

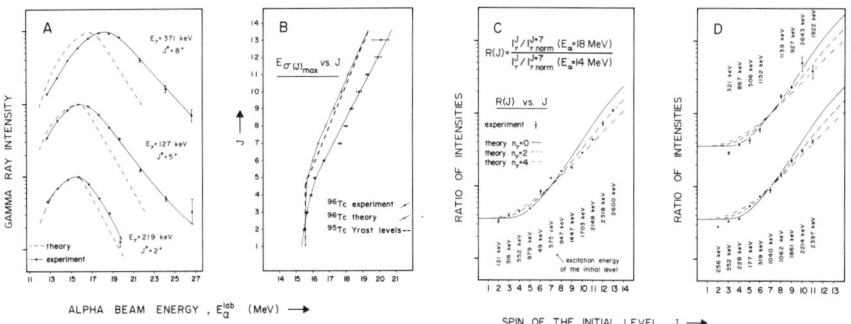

Fig. 1. A) A comparison of the experimental and theoretical excitation functions for states in ^{96}Tc populated via $(\alpha,n\gamma)$ reaction (from Ref. 5), B) The dependence of the position of the maximum in the excitation function on the spin of the initial state. C) and D) Ratios of photon intensity measured at an α energy of 18 MeV to the one at 14 MeV plotted against the spin value of the state in which the transition originated.

validity of the decay scheme or to extract the individual intensities of an unresolved doublet.

The ratio of intensity method has been researched carefully and is used in (n,γ) resonance capture reactions to study the spins of the initial as well as final levels[6]. In reactions induced by p, d, α and Li the ratio of intensities provide quantitative information on the spin of the level[3,5,7] (see also Fig. 1B, C, D), while in the heavy-ion induced reactions it is used only in a qualitative way.

FEEDING DISTRIBUTION

A fast decrease of the side-feeding intensity with increasing excitation energy is observed for levels of the same spin. It can be represented by a smooth function, which in general is separate for each spin and distinct up to some fixed excitation energy. The same data can be used in a different way; one can plot level side-feeding against level spin for a sequence of states. Although this representation may provide support for the spin assignment already made, its spin predictive power is limited.

One should note that the feeding distribution provides spin information independent of the information already enclosed in the excitation functions and intensity ratios. The first one is derived from absolute cross sections under particular experimental conditions, while the other two give a relative change in feeding intensity caused by modification of the experimental conditions.

This method is used extensively in reactions induced by light ions, frequently in conjunction with simple Hauser-Feshbach calculations[8,9]. As a method, it is particularly sensitive to the interference of nuclear structure effects in the statistical decay process[10]. In the domain of α-induced reactions, this method is less powerful than excitation functions or the ratio of intensities[3]. In heavy-ion reactions the pattern of side-feeding is used to study the decay process and the reaction mechanism itself.

LIMITATIONS OF THE METHODS

These methods are limited to levels populated in compound nuclear reactions through statistical decay processes. Consequently, one may exclude yrast levels, first few excited states or those predominantly fed by one or two transitions (due to nuclear structure selection rules). Meaningful corrections can be applied in cases when contribution from nonstatistical feeding can be clearly identified and extracted. It is recommended to use empirically determined functions for quantitative analysis.

Statistical models based on the Hauser-Feshbach formalism provide an accurate picture even in the case of complex reactions[3,5,11]. One can use simple models to predict the range of spins populated and choose optimum conditions for the experiment.

Recently an innovative method of spin assignment has been proposed for (p,γ) capture reactions[12]. It is beyond the scope of a short paper to discuss this rather elaborate technique, which has opened new opportunities for statistical methods.

REFERENCES

1. A. J. Filo, S. W. Yates, D. F. Coope, J. L. Weil, and M. T. McEllistrem, Phys. Rev. C23, 1938 (1981).
2. I. D. Fedorets, V. M. Mishchenko, I. I. Zalyubovskii, A. I. Popov, and V. E. Storizhko, Izv. AN SSSR Ser. Fiz. 44, 937 (1980).
3. D. G. Sarantites and E. J. Hoffman, Nucl. Phys. A180, 177 (1972), also Nucl. Phys. A173, 177 (1971).
4. P. Taras and B. Haas, Nucl. Instr. Meth. 123, 73 (1975).
5. H. Mach and M. W. Johns, to be published.
6. C. Coceva and P. Giacobbe, Nucl. Phys. A385, 301 (1982), also Nucl. Phys. A218, 61 (1974).
7. M. R. Macphail and R. G. Summers-Gill, Nucl. Phys. A263, 12 (1976).
8. G. Doukellis, C. McKenna, R. Finlay, J. Rapaport, and H. J. Kim, Nucl. Phys. A229, 47 (1974).
9. V. A. Bondarenko and P. T. Prokofiev, Izv. AN SSSR Ser. Fiz. 43, 2131 (1979).
10. M. R. Macphail, R. F. Casten, and W. R. Kane, Phys. Lett. 58B, 39 (1975).
11. P. Herges, H. V. Klapdor, and T. Oda, Nucl. Phys. A372, 253 (1981).
12. J. A. Cameron, Can. J. Phys. 62, 115 (1984).

TWO PHOTONS CORRELATED PRODUCTION AT THE 25MWTh REACTOR

H.Abramowicz,K.Doroba and R.Walczak
Institute of Experimental Physics,
University of Warsaw,00-681 Warsaw,Poland.

M.Górski,A.Jasiński,T.Kozłowski,W.Ratyński,
M.Szeptycka,M.Szymczak and A.Tucholski
Institute for Nuclear Studies,05-400 Świerk,Poland.

ABSTRACT

We have performed an extensive search for correlated production of two or more photons at the 25MWTh reactor at Świerk/Poland/. We were looking for a prompt signal originating from decays of axions. Our experimental set-up consisted of an array of nine 3"x3" NaI/Tl/detectors,surrounded by active/plastic scintillators/ and passive shieldings,thus increasing experimental sensivity as compared to other experiments in which two detectors were used. Data were collected during 26 weeks,with four days of reactor"ON", and three days of reactor"OFF" each week.
Preliminary results are presented.

INTRODUCTION

Weinberg /1/ and Wilczek /2/ have proposed a light,neutral and long-lived elementary particle,called the axion.Since its production should compete with electromagnetic transitions,nuclear reactors may produce an axion flux due to de-excitation of nuclei after interactions with neutrons.Experiments performed at Julich /3/,Villigen /4/,Grenoble/5/ and Dubna/6/ which searched for two coincident gamma rays originating from the expected decay of light penetrating particle-used two NaI/Tl/ counters.

Such an experiment was repeated at the Świerk reactor with an array of nine NaI/Tl/ crystals.

EXPERIMENTAL SET - UP AND DATA ACQUISITION

Measurements were performed at the Świerk "Maria" reactor. The array of nine NaI/Tl/ detectors of size 3"x3" each was situated at the axis of closed horizontal beam-hole at the distance of 4.8 m from the edge of the reactor core / see Fig.1 /.
The thermal neutron flux inside the core was $\sim 2 \times 10^{13}$ neutrons/cm^2 and the water content 245 liters.The source strength of 2.220MeV gamma transition in the $np \to d\gamma$ reaction was no less than 1.6×10^7 Ci.

The shielding from the reactor side includes a standard beam shutter,a 10 cm layer of borated polyethylene and five layers of lead with 50 cm of total thickness.

The available decay volume is 1.2 m long, 0.6 m wide and 0.6 m high.It is surrounded by an extra shielding against gammas and neutrons scattered in the reactor hall and coming from other experiments.

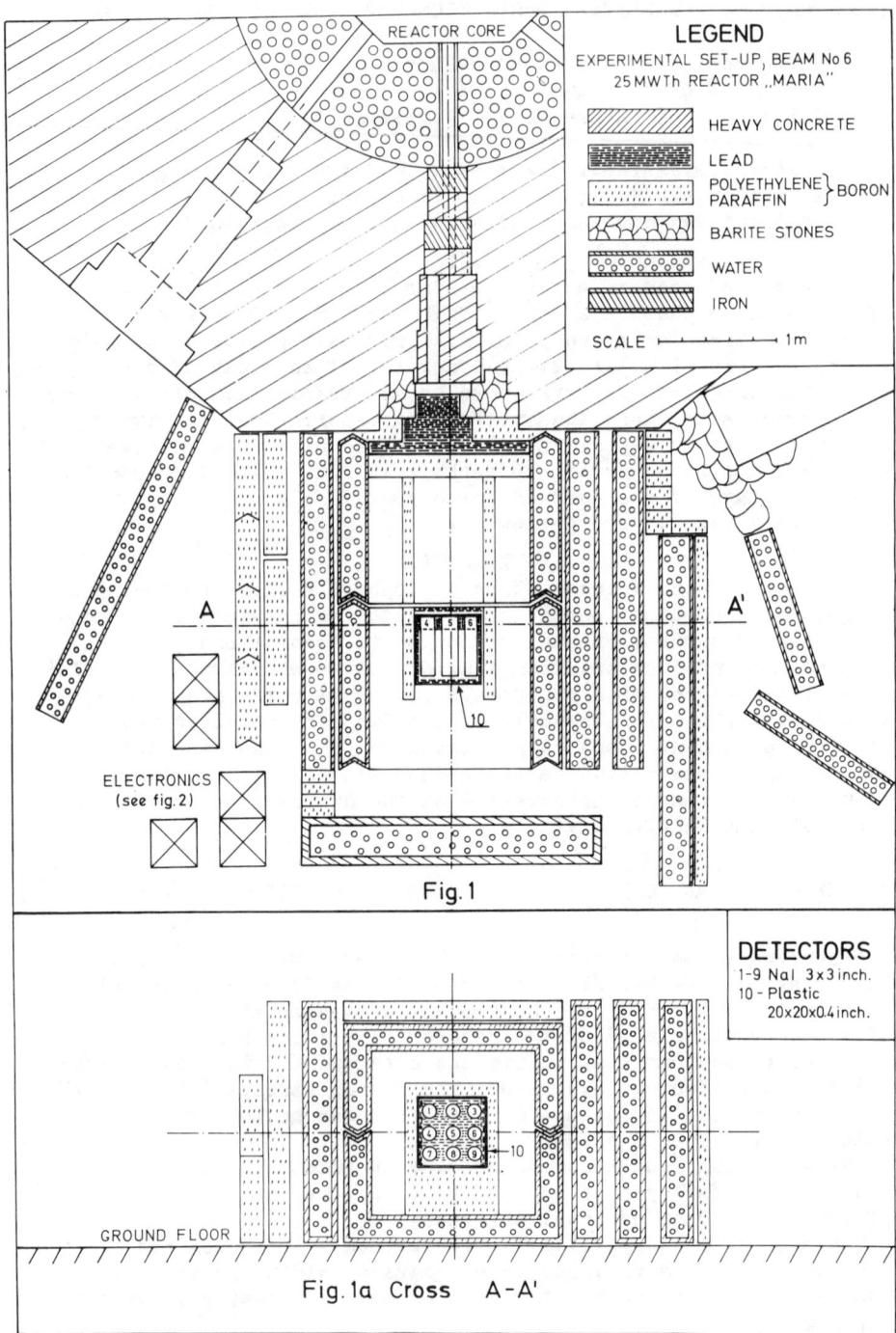

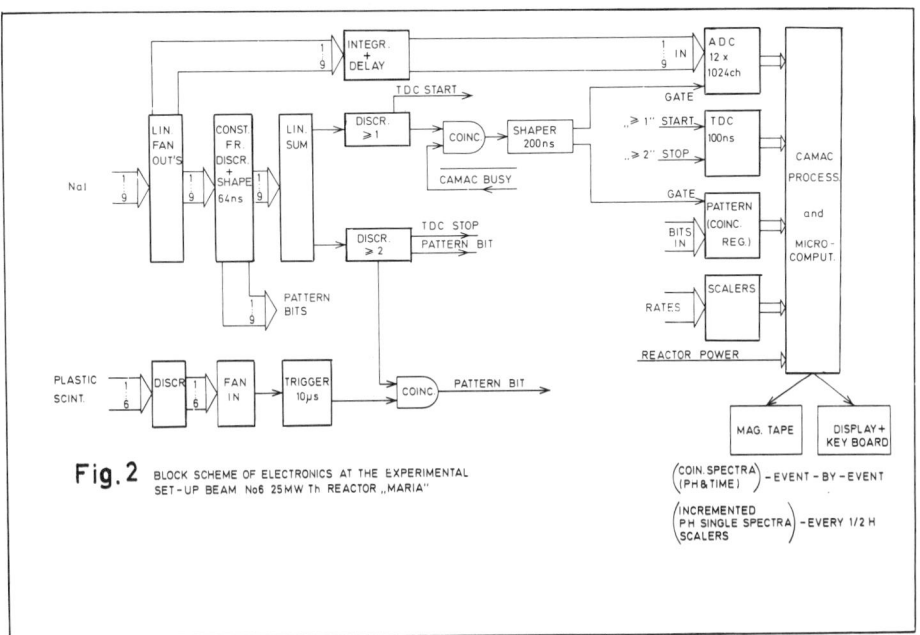

Fig. 2 BLOCK SCHEME OF ELECTRONICS AT THE EXPERIMENTAL SET-UP BEAM No6 25MW Th REACTOR „MARIA"

The block-diagram of electronics is shown in Fig.2. A veto detector surrounds the array of NaI crystals / placed in a lead-copper shielding / to reject events induced by cosmic rays. Each crystal is viewed by an XP2020 photomultiplier connected to the Le Croy ADC 2249A and constant fraction discriminator Polon 1503. The overall resolving time of the electronics is below 50 ns/FWHM/ NaI signals within time window of 200 ns were recorded on tape event-by-event. The single rate of such counter was $\sim$13/s and $\sim$12/s for reactor "ON" and "OFF", respectively.

DATA ANALYSIS AND RESULTS

For each run from 20 to 60 hours long the two-gamma events were selected off-line, inside a time window of 50 ns, and with gamma energies in the range 350 KeV $< E_\gamma <$ 5000 KeV. The single spectra for each NaI crystal and the spectrum of energy sum for six pairs /1-3, 4-6, 7-9, 1-7, 2-8 and 3-9/ of detectors were reconstructed. Spectra for all the runs have been added after the energy calibration. The coincidence rate above 350 KeV was 0.0018/s with reactor "ON" and 0.0017/s with reactor "OFF".

The reactor "OFF" sum spectrum after the correction for measuring time was subtracted from the reactor "ON" spectrum and the difference of these spectra is shown in Fig.3. Experimental data have been analyzed for 469.5 hours of live time measurement with

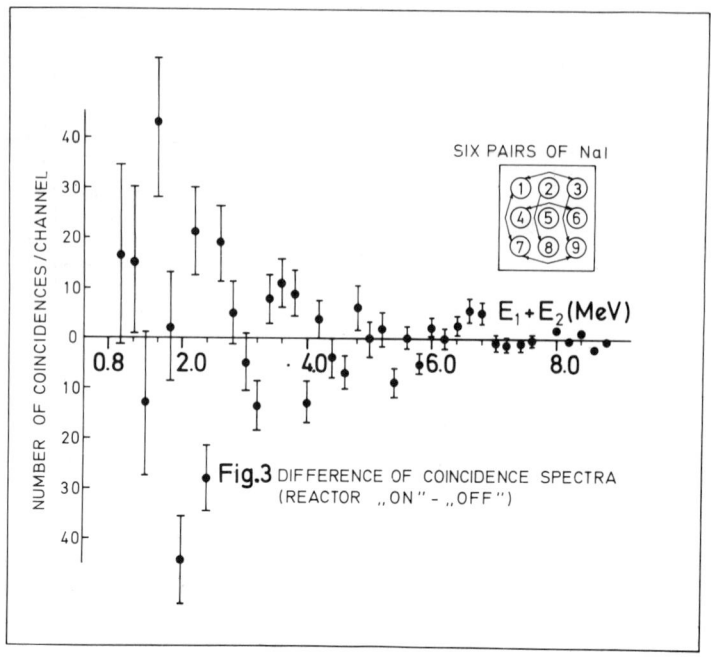

Fig.3 DIFFERENCE OF COINCIDENCE SPECTRA (REACTOR „ON" - „OFF")

the reactor "ON" and 346.6 hours with the reactor "OFF". The total number of coincidences for the reactor "ON" - "OFF" and six pairs of NaI crystals in the energy range 800 KeV $< E_1 + E_2 <$ 2500 KeV is found to be 33 + 67.

This means that an expected prompt signal $/E_1 + E_2 \sim 2.2$ MeV/ originating from axion decays has not been observed in this experiment. The upper limit for the experimental $R_{\gamma\gamma} = 8.5 \times 10^{-5}$/s at 95% C.L.

After calculations of $R_{\gamma\gamma}$ versus the mass of the axion m according to formulas taken from refs./4,5/ and taking into account the acceptance of our experiment we exclude axion masses smaller than 269 KeV and larger than 285 KeV. The analysis of remaining experimental data will be continued.

We would like to thank the management of Reactor Maria at Świerk for a permission to run the experiment charge free and the technical staff of the reactor for the supply of shielding materials and many important informations about the reactor. We are also grateful to dr J.Białkowski from Electronics Dept. of the Institute of Nuclear Research for his help.

R E F E R E N C E S

1/ S.Weinberg, Phys.Rev.Lett. 40/1978/ 223.
2/ F.Wilczek, Phys.Rev.Lett. 40/1978/ 279.
3/ H.Faissner et al., KFA Annual Report 1981 p.69 /1982/
4/ A.Zehnder et.al., SIN Report PR-82-01 /1982/
5/ A.Cavaignac et al., Phys.Lett 121B/1983/ 193.
6/ G.D.Alekseev et al., Dubna Report E1-82-387 /1982/.

AN INVESTIGATION OF PARITY-NON-CONSERVATION IN THE $^{10}B(n,\alpha)$ REACTION

F. Stecher-Rasmussen and P.J. Kok
Netherlands Energy Research Foundation
P.O. Box 1, 1755 ZG Petten, The Netherlands

O.N. Ermakov, J.L. Karpikhin, P.A. Krupchitsky, G.A. Lobov
and V.F. Perepelitsa
Institute for Theoretical and Experimental Physics,
Moscow, USSR.

INTRODUCTION

The weak nucleon-nucleon interaction has been studied by an investigation of the forward-backward asymmetry of the α-particle emission after capture of longitudinally polarized thermal neutrons by ^{10}B nuclei:

$\vec{n} + {}^{10}B \rightarrow {}^{11}B^* \rightarrow {}^{7}Li + \alpha$ (α_0-transition, intensity 6,7%)

$\vec{n} + {}^{10}B \rightarrow {}^{11}B^* \rightarrow {}^{7}Li^* + \alpha$ (α_1-transition, intensity 93,3%)

For this reaction, the emission asymmetry is proportional to F, the parity-non-conservation factor. F is expected to be of the order of 10^{-7} but enhancement effects might yield an emission asymmetry for the studied reaction of 10^{-5}-10^{-6} [1]).

EXPERIMENTAL

To detect low energy α-particles (E_{α_0}=1.79 MeV, E_{α_1}=1.48 MeV) the α-detector was split up into 12 modules each one having a thin (170 μg/cm^2) boron film as internal target.
One detector module is shown in fig. 1. On each side of the target the ionization chamber had two sensitive gaps, B1 and B2 close to the target and D1 and D2 far from the target. By adjustment of the pressure of the ionizing gas (argon) a selection of the proper α-transition was made. The pressure was adjusted in such a way that the ionization in the B-gaps was caused by both α- and Li-particles while the D-gaps only were ionized by the α-particles. In this way the ionization currents in the B-gaps gave an internal check of the instrumental asymmetry, since the p.n.c. asymmetry for α- and Li-particles emitted on the same side of the target would have opposite signs. For an investigation of very small asymmetries a continuous symmetry check is a very important feature.

In fig. 2 the positioning of the stack of detector modules in the polarized beam of thermal neutrons[2]) from the HFR in Petten is shown. The neutron polarization was 90% and the neutron flux $4\times10^8 s^{-1}$. Every T seconds the direction of neutron spin was reversed. The value of T (0.85 s for the α_1-experiment and 3.28 s for α_0) was determined on the basis of reactor noise analysis in order to avoid accidental coincidence between T and resonances in the neutron noise spectrum. Furthermore the longitudinal field was

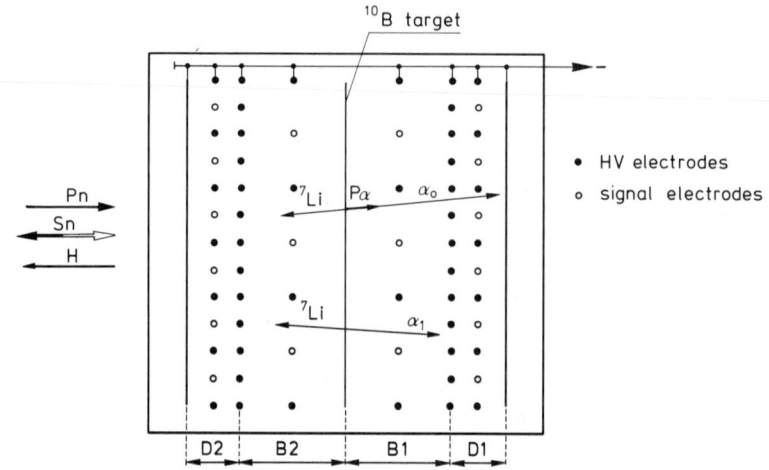

Fig.1. A module of (n, α) multiwire ionization chamber.

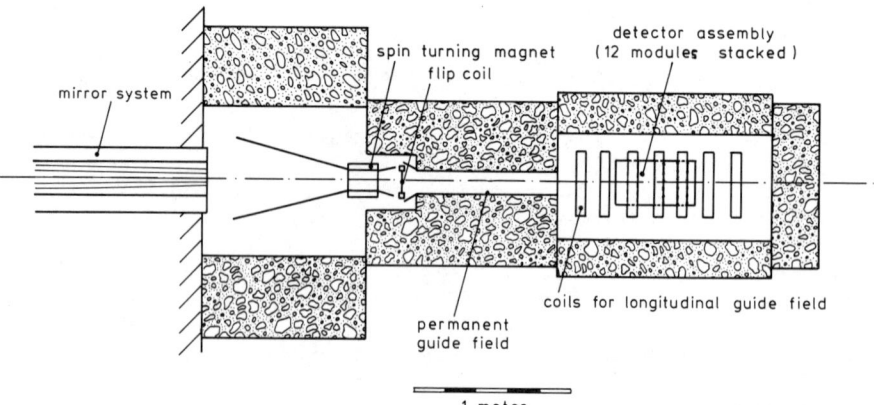

Fig. 2. The detector assembly placed at the polarized neutron beam

reversed three times per day for the α_0-measurement and once per day for α_1. As a final control the experiment was repeated with depolarized neutrons.

RESULTS

The results of the measurements are presented in the form

$$a = \frac{(J_1^+ - J_1^-) - (J_2^+ - J_2^-)}{\sum J}$$

where J_1^+, J_2^+ are the ionization currents in corresponding gaps on opposite sides of the target for one spin direction and J_1^-, J_2^- the same currents for the reverse spin.

Table of the results ($a \times 10^6$)

	α_0-transition		α_1-transition	
	D-gaps	B-gaps	D-gaps	B-gaps
Polarized neutrons	−0.50±0.54	0.01±0.06	0.20±1.50	0.20±0.80
Depolarized neutrons	−0.61±0.42	0.04±0.05	0.70±3.60	0.50±1.30

It follows from the table that no asymmetry effects were observed neither in the α_0- nor in the α_1-transition. The upper limits for the asymmetries at a 90% confidence level are as follows:
- α_0-transition: 3.3×10^{-6}
- α_1-transition: 3.3×10^{-6}

REFERENCES

1. G.A. Lobov, G.V. Danilyan, Izv. Akad. Nauk SSSR, vol. 41, nr. 8, p. 1548 (1977).
2. F. Stecher-Rasmussen, K. Abrahams and J. Kopecky, Nucl. Phys. A181 (1972) 535.

PARITY NON-CONSERVATION IN RESONANCE INTERACTION OF POLARIZED NUCLEONS WITH NUCLEI

G. A. Lobov
Institute for Theoretical and Experimental Physics,
Moscow, USSR

Recently a series of experimental results connected with the effects of the parity non-conservation in the interaction of resonant neutrons and protons with the nuclei was obtained. P-odd asymmetries into inelastic scattering cross section of longitudinally polarized neutrons by the nuclei ^{81}Br, ^{117}Sn and ^{139}La were experimentally investigated in Ref. 1. P-odd asymmetry in the differential cross section of the reaction ^{19}F$(\vec{p},\alpha)^{16}$O for longitudinally polarized protons with the energy E_p = 670 keV, corresponding to the resonant states 1^+ and 1^- of compound nucleus ^{20}Ne was studied in Ref. 2. These results are shown in Table I. In the third column of the table $\sigma^\pm$ are the cross sections corresponding to the helicities of nucleons ±1.

Table I

Target nucleus	Nucleon energy	P-odd asymmetry of cross sections (10^{-3})
^{81}Br	0.88 ± 0.1 eV	$A_n = \dfrac{\sigma^+-\sigma^-}{\sigma^+-\sigma^-} = 24 \pm 4$
^{117}Sn	1.33 ± 0.01 eV	4.5 ± 1.3
^{139}La	0.75 ± 0.01 eV	73 ± 5
^{19}F	670 keV	$A_p = \dfrac{d\sigma^+-d\sigma^-}{d\sigma^+-d\sigma^-} = 6.6 \pm 2.4$

The considered processes occur through the high excited resonant states of compound nuclei. Assumption about the resonance mechanism of P-odd effects due to the interaction of nucleons with nuclei uniquely defines their energy dependence as well as dependence on nuclear characteristics.[3] Let us suppose that at the energy neighboring to incident particle energy, compound nucleus has resonance P-state. Then in the absence of any prohibition on spins, parities or other quantum numbers the nucleon due to strong interaction will be captured in the state of compound nucleus corresponding to this P-resonance. Moreover, let us assume in the absence of weak interaction of the nucleons this process is the main one. However, the weak interaction among the nucleons of compound nucleus result in the fact that the intermediate state is the superposition of P and S states with the opposite parities. P-odd asymmetry into inelastic scattering of longitudinally polarized neutrons by spinless (for simplicity) nuclei may be written as

$$A(k_n) = 2\left(\frac{\Gamma_s^n(k)}{\Gamma_p^n(k)}\right)^{1/2} \text{Re}\langle P | H_W | S \rangle (E-E_s + \frac{i}{2}\Gamma_s)^{-1} \quad (1)$$

where $\langle P | H_W | S \rangle$ is the matrix element of weak interaction Hamiltonian which mixes up resonance P and S states of compound nucleus, E, E_s are the energies of the neutrons and the P-resonance respectively, Γ_s is the total width of S-resonance ($\Gamma_s \ll E_s$). In formula (1) it is assumed that neutron width of S and P-resonances for a neutron wave number K are reduced by the following equations:

$$\Gamma_s^n(k) = \Gamma_s^n \left(\frac{k}{k_s}\right), \quad \Gamma_p^n(k) = \Gamma_p^n \cdot \left(\frac{k}{k_p}\right)^3 \quad (2)$$

where k_s and k_p are the neutron wave numbers corresponding to S and P-resonances. If it is known which S-resonances mix with the P-resonances, then from $A_n(k_p)$ value and Eqs. (1) and (2) one may derive the magnitudes of the mixing coefficients

$$\alpha = \langle P | H_W | S \rangle / (E_p - E_s + \frac{i}{2}\Gamma_s) \quad (3)$$

and the matrix elements $\langle P | H_W | S \rangle$. The last columns in the Table II give the results of calculations of $|\alpha|$ and $\langle P | H_W | S \rangle$ under the assumption that the S-resonances with maximum $\Gamma_s^n/(E_p-E_s)^2$ value mix with the P-resonance. The parameters $g\Gamma_s^n$ and E_s of these resonances from Ref. 1 are also listed here.

Table II

Nucleus	E_s(eV)	$g\Gamma_s^n$ (10^{-3}eV)	$g\Gamma_p^n$ (10^{-8}eV)	$\|\alpha\|$ (10^{-5})	$\|\langle P\|H_W\|S\rangle\|$ (10^{-3}eV)
^{81}Br	101	10	5	3	3
^{117}Sn	-29	5	20	1.5	0.4
^{139}La	-48.6	84	4	2.7	1.4

It is known that the matrix elements of weak interaction between the many-particle state of the compound nucleus are hindered[4]:

$$\langle P | H_W | S \rangle = \langle | H_W | \rangle N^{-1/2}$$

where $<|H_w|> = 0.1$ eV is the single-particle matrix element, $N = \omega/D = 10^4 \div 10^6$, $\omega \simeq 1$ MeV is the spacing of the single-particle levels, $D = 10 \div 100$ eV is the spacing between the real levels of the heavy compound nucleus. P-odd asymmetry A_p of the differential cross section for the process $^{19}F(\vec{p},\alpha)^{16}O$ is defined by the relation:

$$A_p = \left(\frac{8}{3}\right)^{1/2} \left(\frac{\Gamma_s^p}{\Gamma_p^p}\right)^{1/2} \text{Re } e^{i\delta} <P|H_w|S>(E-E_s + \frac{i}{2}\Gamma_s)^{-1} \quad (4)$$

where δ is scattering phase difference in S and P-states of proton, the meaning of the other values is evident.

Using the experimental value A_p from Table I and characteristics of S and P resonance states of compound nucleus ^{20}Ne from Table III (Ref. 2), one can find on the basis of (4) the values of matrix element $<P|H_w|S>$ and the mixing coefficient α (last column of Table III). In that calculation the pure Coulomb's phase shift δ was used.

Table III

(JP;T) of compound nucleus	E_{res}(keV)	Γ_s^p (keV)	Γ_s (keV)	$|\alpha|$	$\|<P\|H_w\|S>\|$ (eV)
(1⁻;0)	650 ± 20	0.02	200	2.7 × 10⁻⁴	0.8
(1⁺;1)	671 ± 1	6	6 ± 0.7		

From the results of Tables II and III it follows that the value of matrix elements $<P|H_w|S>$ for the process $^{19}F(\vec{p},\alpha)^{16}O$ is greater by approximately three orders than for neutron scattering processes. The reasons of it are the following. In the case of the process $^{19}F(\vec{p},\alpha)^{16}O$ the resonances of compound nucleus ^{20}Ne are strongly overlapped, whereas in other ones these resonances are considerably separated. Moreover, mixing of states (1⁺;1) and (1⁻;0) in ^{20}Ne is carried out by isovector part of H_w, which is enhanced.[5] The latter one arises as a result of exchange among nucleons by charged pions. It can be noted that intermediate state (1⁺;1) with account of the parity and isospin selection rules, in the absence of weak interaction between the nucleons, doesn't give any contribution in proton absorption process by nuclei ^{19}F with the successive emission of α-particles. The present work has shown that the P-odd effects in interaction of polarized nucleons with nuclei has the clear-cut resonance behavior previously predicted in Ref. 3. The effects are explainable in the frame of the presently used ideas of the weak interaction of nucleons, and there is no need to introduce any new parity-violating forces (as in Ref. 6). The investigations of resonance P-odd effects give important information about the nuclear structure as well as the structure of weak interaction of nucleons in nuclei.

REFERENCES

1. V. P. Alfimenkov et al., Nucl. Phys. A398, 93 (1983).
2. J. Ohlert et al., Phys. Rev. Lett. 47, 475 (1981).
3. V. A. Karmanov and G. A. Lobov, Pisma v ZhETP 10, 332 (1969); G. A. Lobov, Izv. Acad. Nauk USSR (Ser. Phys.) 34, 1141 (1970); G. A. Lobov, Preprint ITEP-45 (1981).
4. G. A. Lobov, Preprint ITEP-20 (1982).
5. G. A. Lobov, Yad. Phys. 30, 1353 (1979); Izv. Akad. Nauk USSR (Ser. Phys.) 44, 2364 (1980).
6. L. Stodolsky, Phys. Lett. 96B, 127 (1980).

INVESTIGATION OF P-WAVE NEUTRON RESONANCES NEAR NEUTRON BINDING ENERGY WITH LASER RADIATION

Yu Kun Ho

Physics Department, Zhengzhou University, Honan Province, China

F.C. Khanna† and M.A. Lone

Chalk River Nuclear Laboratories, Chalk River, Ontario, Canada

ABSTRACT

Laser-stimulated enhancement of p-wave radiative capture of low energy neutrons is investigated. The cross sections for such a process are calculated in second-order perturbation theory and expressed in terms of the intensity of the laser radiation and the nuclear matrix elements. Four different intermediate states are assumed in the calculations. These are: continuous plane wave, continuous distorted scattering wave, discrete resonances with large p-wave neutron reduced widths and compound nucleus wave functions. Numerical estimates show that an appreciable enhancement of the radiative capture will not be observed until the laser electric field strength reaches a magnitude of 10^5 to 10^8 V/cm, depending whether an s-wave resonance exists simultaneously in the entrance channel.

LASER INDUCED RADIATIVE CAPTURE CROSS SECTIONS

In the presence of a strong laser radiation field, capture of an incident low-energy s-wave neutron and a low-energy photon can excite a nearby p-wave resonance. This process can be identified by comparing the intensities of a primary γ-ray transition, from a nuclear capturing state in a low-lying s-wave bound state, in the presence and absence of the laser field.

We use second-order perturbation theory to calculate the cross sections. The interaction Hamiltonian consists of two terms: the first, $H^{(1)}$, for the electromagnetic field of the nuclear system, the second, $H^{(2)}$, for the laser field interacting with nucleus. Their matrix elements, for E1 transition, are[4]

$$\langle \phi_f | H^{(1)} | \phi_t \rangle = \frac{1}{3} k_\gamma^{3/2} \left(\frac{8\pi\hbar c}{R_o} \right)^{1/2} \langle \phi_f | e_f r\, Y_{1\mu} | \phi_t \rangle \qquad (1)$$

and

$$\langle \phi_t | H^{(2)} | \phi_i \rangle = - \langle \phi_t | e_f\, \underline{\varepsilon}\cdot\underline{r} | \phi_i \rangle, \qquad (2)$$

where k_γ is the wave number of the γ-photon, $e_f = -\frac{Z}{A} e$ is the neutron effective charge, R_o is a normalization radius for the photon wave function, and $\underline{\varepsilon}$ is the electric field vector of the photons.

The cross section can be written as

$$\sigma_{if}^{(L)} = \frac{1}{v} \frac{1}{2(2I+1)} \sum_{m_i m_I M_f(\mu)} \frac{2\pi}{\hbar} \left| \frac{\langle \phi_f | H^{(1)} | \phi_t \rangle \langle \phi_t | H^{(2)} | \phi_i \rangle}{E_t - E \pm \hbar\omega} \right|^2 \frac{R_o}{2\hbar c} \quad (3)$$

where the indices i, t and f designate respectively the initial, intermediate and final states. The laser photon energy is $\hbar\omega$. The reader is referred to ref. (2) for nomenclature and explicit expressions for ϕ_i and ϕ_f. In the calculation of the cross section we have investigated four types of intermediate-state wave functions.

PLANE WAVE STATES

The wave function of the intermediate state can be written as

$$\phi_t = e^{i\underline{k}_t \cdot \underline{r}} \chi_{im_i'} \chi_{Im_I'} \quad (4)$$

where $\chi_{i'm_i'}$ is the spin wave function; $E_t = \frac{1}{2m} \hbar^2 k_t^2$ is the energy of the intermediate state. For simplicity we make the reasonable assumptions that

$$k_f^2 \gg k_\gamma^2 \gg k^2 \approx k_t^2 \gg k_\omega^2$$

where k_f, k and k_ω are respectively the wave numbers of final state, incident neutron and laser photon. Then we get the cross section

$$\sigma_{if}^{(LVB)} = \frac{8}{3} \pi \frac{1}{\hbar v} \frac{(2J_f+1)}{2(2I+1)} e_f^4 \varepsilon^2 S_f \frac{k_\gamma^3}{k_f^7} \frac{1}{(E\mp\hbar\omega)^2} \left| 1 + \frac{2i}{3}(1 \mp \frac{\hbar\omega}{E})^{1/2} \right|^2. \quad (6)$$

DISTORTED SCATTERING WAVES STATES

The intermediate state wave function has the same form as the initial state wave function[2]. The results contain four terms corresponding to the combinations of background and resonances in the initial and intermediate states[5]. The results for the background contribution from the intermediate state are similar to the previous section. The case of resonances in the intermediate state is discussed in the next section.

DISCRETE RESONANCES

When discrete p-wave resonances are chosen for the intermediate states, their wave function can be written as

$$\phi_{\lambda'}^{J'M'} = \sum_{\ell'j'} \left[\frac{M}{\hbar^2} \frac{\sqrt{E_{\lambda'}(eV)}}{k_{\lambda'}} \Gamma_{\lambda'n}^{\ell'}\right]^{1/2} \frac{\phi_{\ell'j'I}}{r} \frac{v_{\ell'j'}(r)}{v_{\ell'j'}(R)}, \qquad (7)$$

where $\Gamma_{\lambda'n}^{\ell'}$ is the reduced neutron width and $v_{\ell'j'}(r)$ is the radial wave function defined by an optical model potential between neutron and the nucleus. We obtain the cross section for the laser induced valence component through resonant intermediate states (LVR) as

$$\sigma_{if}^{(LVR)} = \frac{\pi}{k^2} \sum_{\lambda'J} \frac{(2J+1)}{2(2I+1)} \frac{\Gamma_{\lambda'n}^{(L)} \Gamma_{\lambda'\gamma f}^{(V)}}{(E_{\lambda'} - E \pm \hbar\omega)^2 + \frac{\Gamma_{\lambda'}^2}{4}} \qquad (8)$$

where $\Gamma_{\lambda'\gamma f}^{(V)}$ is the valence radiative capture width[3] for resonance λ'.

The partial neutron width given by

$$\Gamma_{\lambda'n}^{(L)} = \frac{4}{9} \frac{1}{\hbar v} e_f^2 \varepsilon^2 \left|A_{\ell'j'}^{\lambda'}\right|^2 \frac{1}{\left|1 - i K_{\ell J}^{J}\right|^2} (2j'+1)(2J'+1)$$

$$\times W^2(j'J'j J, I1) \left|Q_{\ell'j'\ell j}^{(B)} + \frac{1}{2} \sum_{\lambda(J)} \frac{\Gamma_{\lambda n}}{E_\lambda - E - \frac{i}{2}\Gamma_\lambda} Q_{\ell'j'\ell j}^{(R)}\right|^2 \qquad (9)$$

can be regarded as the laser induced effective p-wave neutron width for the resonance λ'.

COMPOUND NUCLEUS PROCESS

Instead of $\Gamma_{\lambda'\gamma f}^{(V)}$ in Eq. (8), we substitute $\Gamma_{\lambda'\gamma f}$, the experimentally determined partial radiative width. Then we get the laser induced partial radiative capture cross section including compound nucleus process.

ENHANCEMENT INDUCED BY LASER RADIATION FIELD

We define the laser induced enhancement factor as the ratio of probability of a p-wave intermediate state excitation via two step process (capture of low energy s-wave neutron plus capture of a electric dipole photon from laser) to the probability of a p-wave capture of the incident neutron itself.

From Eqs. (8) and (9), the enhancement factor induced by the laser radiation field for the γ-ray transition from the p-wave resonance λ' is

$$F = \frac{\Gamma^{(L)}_{\lambda'n}}{\Gamma_{\lambda'n}}$$

where $\Gamma_{\lambda'n}$ is the p-wave neutron width.

Now we estimate the order of magnitude of the quantities given above. By using the following typical data $E_f = 5$ MeV, $E = 0.024$ eV, $\hbar\omega = 1$ eV, $A = 100$, $Z = 42$, $I = 0$, $S_f = 1$, we obtain

$$\sigma_{if}^{(LVB)} \approx 1.0 \times 10^{-27} [\varepsilon \, (V/cm)]^2 \text{ (in barns)},$$

$$\sigma_{if}^{(LVR)} \approx 3.6 \times 10^{-16} [\varepsilon \, (V/cm)]^2 \text{ (in barns)},$$

and

$F = 5.2 \times 10^{-16} (\varepsilon(V/cm))^2$, when we ignore the second term in Eq. (9),

$\quad = 10^{-10} \quad (\varepsilon(V/cm))^2$, when the second term in Eq. (9) plays a major role.

These results show that:

1. As the energy $E \pm \hbar\omega$ approaches a p-wave resonance energy, the cross section will be about 10^{11} times larger than that when $E \pm \hbar\omega$ is far away from any p-wave resonance energy (background).

2. In order to obtain an enhancement for the γ-ray transition from a shallow-bound or low-energy p-wave resonance level to a low-lying s-wave neutron final state, the required laser electric field should be greater than 10^5 to 10^8 V/cm. The lower value 10^5 V/cm corresponds to a rarely observed case where a p-wave and an s-wave resonance energies differ by the laser photon energy. Since the breakdown threshold for ionization of an ordinary gas target is of the order of 10^5 V/cm, this implies that an experimental observation of this phenomena may be very difficult.

REFERENCES

† Present address: Physics Department, University of Alberta, Edmonton, Canada
1) D.F. Zaretskii and V.V. Lomonosov, Sov. Phys. JETP 54(1981)229; JETP Letter 30(1979)508.
2) Y.K. Ho and M.A. Lone, Nucl. Phys. A406(1983)1.
3) A.M. Lane and J.E. Lynn, Nucl. Phys. 17(1960)563;586.
4) R.R. Roy and B.P. Nigam, Nuclear physics theory and experiment, John Wiley & Sons Inc. New York, 1867.
5) Y.K. Ho, F.C. Khanna and M.A. Lone, to be published.

INVESTIGATION OF CAPTURE REACTIONS FAR OFF STABILITY BY ß-DELAYED NEUTRON EMMISSION

M. Wiescher, B. Leist, W. Ziegert, H. Gabelmann, B. Steinmüller,
H. Ohm, K.-L. Kratz
Universität Mainz, D-6500 Mainz, FRG

F.-K. Thielemann, W. Hillebrandt
MPI für Physik und Astrophysik, D-8046 Garching, FRG

Abstract

Beta-delayed neutron spectroscopy is applied to determine reaction rates of neutron capture on several neutron rich nuclei. The results of these experiments are presented and discussed in the light of their astrophysical implications. Furthermore, the experimental possibilities and limits of planned measurements are advertised.

Introduction

In stellar burning processes at high temperatures $T_9 \geq 0.4$ and high neutron densities $n \geq 10^{18}$ cm^{-3} a dynamical r-process (nß-, n-process [1,2]) may contribute significantly to the formation of heavy elements. In this process the reaction path is expected to be far outside the line of stability and is determined by the ß-decay rates and the rates of neutron capture on radioactive nuclei. A theoretical model of this process requires therefore a detailed knowledge of the neutron capture rates involved. These rates, however, cannot be measured directly yet.

The high level density of most of the medium and high mass compound nuclei of such reactions allow statistical Hauser-Feshbach calculations for determining the particular cross sections and reaction rates [3]. The results, however, depend significantly on the assumed input parameters, such as level densities, neutron binding energies, spin and parity of the target and of the compound nuclei. Without such experimental information on nuclei far outside the line of stability, theoretical predictions may not be reliable. Therefore effort has to be made in determining these parameters.

For nuclei in the r-process path near closed neutron shells the level densities drop considerably due to the small neutron binding energies.[4] For such nuclides Hauser-Feshbach calculations are no longer applicable. Single resonances, besides direct neutron capture, will determine the capture reaction rates on these nuclei. In these cases, detailed nuclear structure information about possible neutron unbound levels are important to calculate the resonant contribution of these states to the reaction rate.

High resolution ß-delayed neutron spectroscopy seems to be a tool to investigate the nuclear structure of neutron rich radioactive nuclei. In the case of ^{87}Kr it was shown recently [5] that the neutron peaks in the ß-delayed neutron spectrum of ^{87}Br correspond to the same neutron unbound states as were found as resonances in ^{86}Kr(n,γ)^{87}Kr reaction. This proofs the applicability of ß-delayed neutron spectroscopy for yielding indirectly information on level parameters relevant for neutron capture on far unstable nuclei.

Experiments

The aim of the experiments was first to investigate the nuclear structure of high-level density neutron rich nuclei, to determine experimentally the level density for calculating the reaction rate by the Hauser-Feshbach method and to proof also its validity far off the valley of stability.

Secondly, it was searched for single neutron unbound states in low-level density neutron rich nuclei in the r-process path. The level parameters of those excited states can be determined by measuring the different decay channels (n, γ). From such data one is able to extract the resonant contribution of these levels to the reaction rates of the particular neutron capture reaction.

The experiments were performed at different isotope separators (Isolde, CERN; Ostis, ILL Grenoble; Mafia, Kernchemie Mainz). The neutron spectra were measured using standard time-of-flight techniques or applying high resolution ^{3}He-ionisation chambers. In addition γ-single and γγ-coincidence spectra were measured with Ge(Li)-detectors.

Results

As example for determining capture rates to high level density compound nuclei the ß-delayed neutron decay of ^{95}Rb was discussed.[6] By the allowed Gamow-Teller decay of ^{95}Rb (J^π=5/2$^-$) to ^{95}Sr states with J^π = 3/2$^-$, 5/2$^-$, 7/2$^-$ are populated. Neutron unbound states will decay predominantly by neutron emmission to the ground state (J^π=0$^+$) and the first excited state (J^π=2$^+$) in ^{94}Sr.[7] The states observed in the ground state neutron spectrum (Fig.1.) occur mainly as p-wave resonances in the neutron capture on ^{94}Sr. The single resonance strengths, calculated on the basis of the Breit-Wigner formalism,[8] yields in an experimental lower limit (∿50%) for

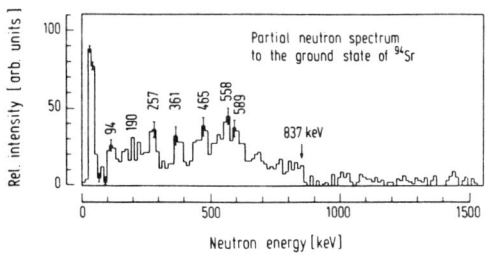

Fig.1. Partial energy spectrum of ß-delayed neutron emmission to the ground state of ^{94}Sr

From the neutron spectra the level density for the 3/2$^-$-levels in ^{95}Sr can be determined to a = (13.0 ± 0.3) MeV^{-1}.[7] From this experimental number, a reliable value for the level density of the states not populated can be evaluated. Figure 2 shows the resulting contributions of synthetic resonances with particular J^π to the neutron capture rate. A comparison of the total rate with the results of a Hauser-Feshbach calculation shows good agreement (Tab.I). Theoretically derived partial widths [8], the experimental level density as well as the known spins and parities for the first excited states of the compound nucleus are the input parameters for this calculation [6].

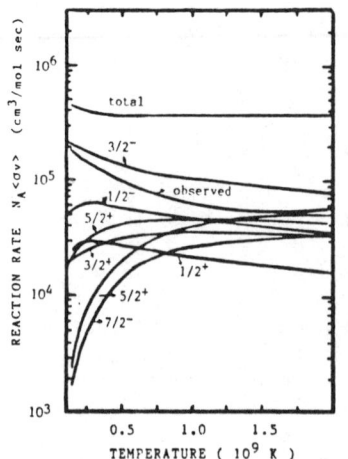

Fig.2. Resonant contributions to the rate of ^{94}Sr(n,γ)^{95}Sr

This method is also applied for determining capture rates on other neutron rich nuclei such as ^{96}Sr(n,γ)^{97}Sr and ^{136}Xe(n,γ)^{137}Xe. Sufficiently high level density is expected for the respective compound nuclei. Therefore, using correct input data, the Hauser-Feshbach rates should be reliable. Table I shows the comparison between the particular reaction rates at kT = 30 keV determined in the different ways described above.

Table I Evaluated neutron capture reaction rates (mbarn)

Reaction	Hauser-Feshbach	Breit-Wigner exp.	tot.	Ref.5
^{86}Kr(n,γ)^{87}Kr	4.12	1.30	3.43	4.8 ± 1.2
^{94}Sr(n,γ)^{95}Sr	2.71	1.06	2.33	
^{96}Sr(n,γ)^{97}Sr	2.51	0.37	2.73	
^{136}Xe(n,γ)^{137}Xe	1.40	0.10	1.39	

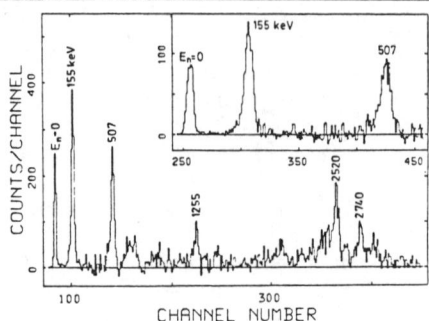

Fig.3. The ß-delayed neutron spectrum of ^{50}K

As a second example for a low level density nucleus the ß-delayed neutron decay of ^{50}K will be considered to derive the neutron capture rate for ^{49}Ca(n,γ)^{50}Ca. In an attempt to explain Ti-isotopic anomalies in meteoritic material [9], this rate was suggested to be considerably higher, compared to Hauser-Feshbach predictions. The ground state of ^{50}K is suggested [10] to have $J^\pi = 0^-$.

Therefore Gamow-Teller decay will populate $J^\pi = 1^-$ states in ^{50}Ca.

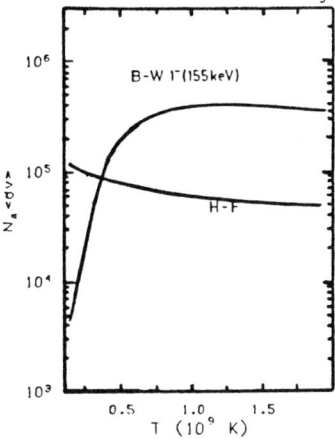

Fig.4. Predicted reaction rates for ^{49}Ca(n,γ)^{50}Ca

The neutron unbound 1^- states are expected to determine the neutron capture rate on ^{49}Ca ($J^\pi = 3/2^-$) as s-wave resonances. The ß-delayed neutron spectrum (Fig.3) reveals the existence of a strong state at E_x = 6.515 MeV, equivalent to a (n,γ)-resonance at 155 keV.[8]

From the spectroscopic results the neutron width $\Gamma_n \leq 7.4$ keV and an upper limit for the γ-width $\Gamma_\gamma \leq 45$ eV were derived. This results in an upper limit for the reaction rate, considerably higher compared to the Hauser-Feshbach predictions at high temperatures $T > 0.5 \cdot 10^9$ as shown in figure 4.

Future experiments

It is planned to determine by this technique the rates of different (n,γ)-reactions on high and low level density nuclei. In particular, in capture reactions on ^{46}K, ^{47,48,50}Ca and ^{51}Sc single resonances may determine the reaction rates. The investigation of these raections may help to understand the isotopic anomalies in the Ca-Ti region.[9]

1. J.B. Blake, D.N. Schramm, Ap.J. 209, 846 (1976)
2. G.J. Mathews, LLNL-preprint UCRL 89914 (1983)
3. S.E. Woosley, W.A. Fowler, J.A. Holmes, B.A. Zimmermann, Nucl. Dat. Tab. 22, 371 (1978)
4. G.J. Mathews, A. Mengoni, F.-K. Thielemann, W.A. Fowler, Ap.J. 270, 740 (1983)
5. S. Raman, B. Fogelberg, J.A. Harvey, R.L. Macklin, P.Stelson, A. Schröder, K.-L. Kratz, Phys. Rev. C28, 602 (1983)
6. B. Leist, Diploma thesis, Universität Mainz (1984)
7. K.-L. Kratz, H. Ohm, A. Schröder, H. Gabelmann, W. Ziegert, B. Pfeiffer, G. Jung, E. Monnand, J.A. Pinston, F. Schussler, G.I. Crawford, S.G. Prussin, Z.M. de Oliveira, Z. Phys. A312, 43 (1983)
8. K.-L. Kratz, W. Ziegert, W. Hillebrandt, F.-K. Thielemann, Astron. Astrophys. 125, 381 (1983)
9. D.E. Sandler, S.E. Koonin, W.A. Fowler, ApJ. 259, 908 (1982)
10. A. Dobado, A. Povès, priv. comm. (1982)

DIRECT MEASUREMENT OF NATURAL LINE WIDTHS IN DELAYED-NEUTRON ENERGY SPECTRA*

Robert D. McElroy and David D. Clark
Ward Laboratory, Cornell University, Ithaca, NY 14853

R. L. Gill and A. Piotrowski
Brookhaven National Laboratory, Upton, NY 11973

ABSTRACT

Delayed-neutron spectra of mass-separated fission products from the TRISTAN facility at Brookhaven are being investigated to develop systematics of neutron resonances in the 1-100 keV range in neutron-rich nuclides as a unique experimental basis for testing theoretical extrapolations of level parameters and neutron cross sections from stable to very unstable uclides. Aims, methods, and early results have been described elsewhere. The time-of-flight system yields resolved peaks, >10 between 3 and 100 keV in Rb-95 and Rb-97. Natural line widths have been deduced to date for two Rb-95 peaks: 414 ± 33 eV for the 14.44 ± 0.09 keV resonance and 670 ± 270 eV for the 26.70 ± 0.13 keV resonance, found by fitting the experimental data to a Lorentzian convoluted with a calculated system response function. Similar analysis of further peaks and deduction of level densities and other parameters await additional data from newly improved apparatus and planned experimental checks of the resolution calculation by measuring the Br-87 spectrum and comparing it with known resonance parameters of the Kr-86 + n cross section.

INTRODUCTION

The phenomenon of beta-delayed neutron emission affords unique empirical clues to neutron cross sections of far-unstable nuclides. The key point is to exploit the inverse relationship between neutron emission and absorption by compound-nuclear levels above the neutron binding energy. Data of sufficient quality and quantity to allow construction of empirically based cross sections for very neutron-rich nuclides are of interest for astrophysics and for nuclear theory. These cross sections may in some cases be dominated by individual isolated resonances and in others may be viewed as averages over a number of closely spaced resonances. In either case, delayed neutron spectra are a unique source of data for nuclides that are completely inaccessible to direct measurement. The validity of this approach has been demonstrated recently in an unpublished report by Kratz et al.[1] who applied it in the case of two precursors: Br-87 and K-50.

---

*Supported in part by U.S. Department of Energy

METHOD

The basic experimental method has been described elsewhere.[2] Briefly, the time-of-flight (TOF) apparatus consists of a plastic scintillator telescope to detect betas that feed neutron-emitting levels and a Li-6 glass scintillator to detect the emitted neutrons. The TOF signal and slow signals from the neutron and thick beta detectors are digitized and recorded in three-parameter event mode. The neutron detector pulse height is used to eliminate many of the pulses due to gammas, the beta pulse height to measure the beta energy. In a recent modification, not yet used, two additional neutron detectors have been added and the electronics modified so that data can be taken at three times previous rates.

The event mode data are analyzed in straightforward fashion to eliminate unwanted events such as randoms and energy spectra are derived. The system response function has been extensively studied in Monte Carlo and other calculations. With typical flight paths of 50 to 70 cm the FWHM energy resolution is less than 4% below 50 keV. The calculated response to monoenergetic neutrons is a very clean, nearly symmetric peak. For a few lines in the spectra the resolution is sufficient to permit deduction of the natural level width. The experimental data in the TOF spectrum are fitted to a Lorentzian of width Γ convoluted with a system response function composed of a Gaussian timing curve (2.97-ns FWHM), a trapezoidal contribution from detector thickness and non-axial path lengths, and an exponential tail (calculated by Monte Carlo) from multiple scattering in the neutron scintillator.

The calculated system response function can be quantitatively checked experimentally by measuring the spectrum of Br-87 and comparing the analyzed results with an expected spectrum derived from the precisely known resonance energies and widths found by Raman et al.[3] in cross section measurements. (This is the only case in which it is feasible to compare a delayed neutron spectrum with the cross section of a stable target nuclide.) We plan such a calibration of our system when the appropriate ion source is available to us at TRISTAN.

RESULTS

The principal isotopes that have been suitable for initial experiments with the TOF system have been the rubidiums, and two have been studied so far: Rb-95 and Rb-97. Tables of energies and intensities are attached. In the table for Rb-95 results by Ohm[4] are shown; they are taken with a Cuttler-Shalev He-3 gridded ionization chamber. The TOF system has better energy resolution in the range below 100 keV; hence more peaks are apparent. In studies with a proton recoil spectrometer, Greenwood[5] reports peaks at 14.1, 26.4, 47, and 93.3 keV. His results and those of Ohm above 100 keV, where the TOF system resolution is worsening, are not shown here.

The two cases in which the natural line width has been found are stated above in the abstract.

Table I Energies and Intensities of Rb-95 Delayed Neutrons

this work		Ohm[a]	
E_n(keV)	I_n(rel)	E_n(keV)	I_n(rel)
9.8 ± 0.1[b]	7.6 ± 3.9		
12.0 ± 0.2	13.7 ± 5.1		
14.44± 0.09	(100)	13.6 ± 1.1	(100)
19.7 ± 0.2[b]	7.8 ± 2.4		
24.0 ± 0.23	8.7 ± 4.4		
26.70± 0.13	30.4 ± 5.8	25.5 ± 1.1	28.2 ± 1.6
31.2 ± 0.2	8.0 ± 2.0		
36.3 ± 0.2	7.5 ± 2.3		
38.9 ± 0.24	12.5 ± 3.3	41.4 ± 1.3	6.4 ± 0.9
41.0 ± 0.35	7.2 ± 3.2		
45.52± 0.25	9.6 ± 2.6		
47.68± 0.25	12.9 ± 3.1		
50.75± 0.40	8.7 ± 3.8		
57.0 ± 0.35	7.6 ± 2.6		
60.5 ± 0.4	5.9 ± 2.8		
64.72± 0.32	13.1 ± 3.2		
72.9 ± 0.5[b]	11.6 ± 3.2	72.6 ± 1.5	2.7 ± 0.3
78.3 ± 0.4	12.3 ± 3.6		
81.7 ± 0.4	10.2 ± 2.6	83.5 ±1.2	10.3 ± 0.7
86.6 ± 0.5	7.0 ± 3.7		
94.6 ± 0.7	∼20	94.2 ± 1.2	15.5 ± 0.9

[a]Ref. 4   [b]Possibly doublet

Table II Energies and Intensities of Rb-97 Delayed Neutrons

E_n(keV)	I_n(arb)
4.3 ± 0.15	2.3 ± 0.7
8.9 ± 0.2	5.5 ± 1.6
11.4 ± 0.2	3.4 ± 1.4
14.2 ± 0.2	3.0 ± 1.1
16.8 ± 0.3	2.4 ± 1.0
19.9 ± 0.3	6.5 ± 2.1
27.6 ± 0.3	5.2 ± 1.4
31.5 ± 0.4	3.7 ± 1.1
38.0 ± 0.4	2.5 ± 1.0
49. ± 0.5	
53. ± 0.5	
70.0 ± 0.5	13.3 ± 2.4
82.2 ± 0.5	7.8 ± 1.3
107. ± 1.	

SUMMARY

Experimental studies of delayed neutron spectra are a unique source of information on level parameters of nuclides far from the valley of stability which are of astrophysical and theoretical interest. TOF experiments to date have revealed a number of new peaks and have permitted deduction of natural line widths for two strong lines. Future work will exploit the high resolution, clean response function, and beta-energy sensitivity of the TOF method to derive level densities and positions of a series of neighboring nuclides in the fission product region. Level widths, though of interest, cannot be determined for a large number of case; however, the level densities expected for neutron-rich nuclides that lie between the stable valley and the neutron drip line should be readily determined from experiments of the type presented here.

ACKNOWLEDGMENTS

The continuing assistance of the TRISTAN staff is gratefully acknowledged. We thank R. C. Greenwood for permission to quote his Rb-95 results and K.-L. Kratz for sending us numerous informative preprints and reprints. Special thanks are due H. E. Jackson of Argonne National Laboratory for providing the Li-6 detectors.

REFERENCES

1. K.-L. Kratz, W. Zeigert, W. Hillebrandt, F.-K. Thielemann, report MPA-65. Max-Planck-Institut für Astrophysik, Garching (April 1983)
2. D. D. Clark, R. D. McElroy, T.-R. Yeh, R. E. Chrien, R. L. Gill, in Proceedings of Specialists Meeting on Yields and Decay Data of Fission Product Nuclides, Brookhaven, October 1983 (in press)
3. S. Raman, B. Fogelberg, J. A. Harvey, R. L. Macklin, P. H. Stelson, A. Schröder, K.-L. Kratz, Phys. Rev. C $\underline{28}$, 602 (1983)
4. H. Ohm, unpublished thesis, Mainz, 1981
5. R. C. Greenwood, private communication

ON THE LIMIT RESOLUTION OF A CURVED-CRYSTAL GAMMA-RAY SPECTROMETER

V.L.Alexeev, E.K.Leushkin, L.I.Molkanov, V.L.Rumiantsev
Leningrad Nuclear Physics Institute, Gatchina, USSR 188350

ABSTRACT

Superhigh resolution with use of a bent natural quartz crystal has been obtained. The instrumental line-width at $E_\gamma = 176.8889 \pm 0.0002$ keV was 0.19 s of arc (5.9 eV) in the 2nd of reflection and 0.20 s of arc (2.5 eV) in the 5th order.

The highest resolution in curved-crystal gamma-ray spectrometers is obtained by diffracting from the (110) planes of quartz, which does not cause line broadening because of elastic quasi-mosaic[1]. J.W.Knowles states in his review[2] that the effective mosaic width ω_m is $\approx$ 1 s of arc for nearly perfect crystals of natural quartz. Since the resolution of the best curved-crystal spectrometers is close to this value, it is concluded that a further improvement of the angular (energy) resolution with quartz crystals is impossible. This assertion was checked experimentally in the course of this study.

Measurements were carried out with a modified 4 m Cauchoistype curved-crystal gamma-ray spectrometer ГСК-2[3,4]. The 3.78 mm thick crystal of natural quartz reflecting from the (110) planes was prepared using the crystal cutting and adjusting technique described in the preprint [4]. Steps were taken to reduce the contribution to the angular width of a line from the geometry and from the residual aberration of the crystal area used (50–100 mm^2) down to $\leq$ 0.2 s of arc.

At $E_\gamma = 176.8889 \pm 0.0002$ keV (from the Pr(n, γ)^{142}Pr reaction) the instrumental line width was 0.19 s of arc (5.9 eV) in the 2nd order of reflection and 0.20 s of arc (2.5 eV) in the 5th order, which corresponds to the resolution $\Delta E/E$ of 0.0033% and 0.0014%, respectively. The line shape is shown in the figure.

The findings were analyzed and the instrument parameters optimized for the superhigh resolution using a program developed for the computer BESM-6. The calculated line shape agrees well with the experimental one for an effective mosaic width $\omega_m \leq 0.1$ s of arc.

Thus it has been proved experimentally that a substantial increase (by an order of magnitude) of curved-crystal gamma-ray spectrometers resolving power is possible using natural quartz crystals.

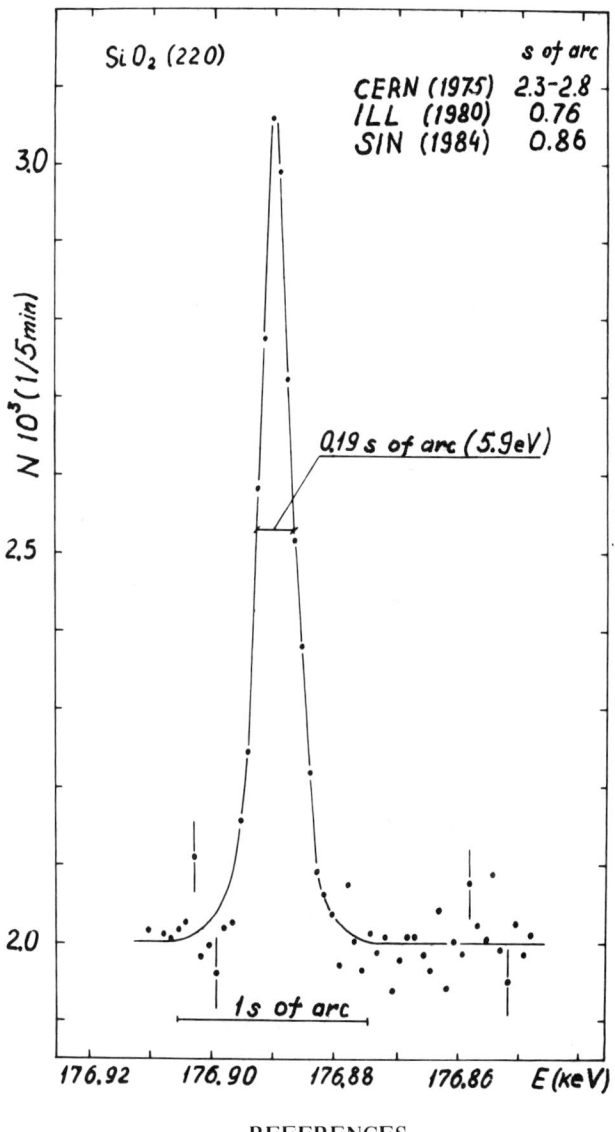

REFERENCES

1. O.I.Sumbaev, Sov. Phys. JETP, 5, 170 (1957); ibid 5, 61 (1957); Zh. Eksp. Teor. Fiz., 54, 1362 (1968).
2. J.M.Knowles, Nucl. Instr. Meth., 162, 677 (1979).
3. O.I.Sumbaev, A.I.Smirnov, Nucl. Instr. Meth., 22, 125 (1963).
4. V.L.Alexeev, E.K.Leushkin, V.L.Rumiantsev, V.M.Samsonov, LNPI Preprint No. 621, Leningrad, 1980.

CRYSTAL REFLECTIVITY FOR GAMMA RAYS

E. Kaerts and P.H.M. Van Assche
University of Leuven, 3030 Leuven and SCK/CEN, B-2400 Mol Belgium

ABSTRACT

The integral reflectivity of a 3.5 mm thick, bent Si(220) crystal was measured. Its energy dependence was found to be $E^{-0.96\pm0.04}$ in the energy range 344-1408 KeV. This result is discussed and it is shown that the use of these highly perfect silicon crystals can substantially extend the high energy range of crystal diffraction spectrometers.

INTRODUCTION

The utility - and the limitations - of a Bent-Crystal Diffraction (BCD) spectrometer depend strongly on the behaviour of the analyzing crystal. A study of large crystals for a bent-crystal gamma diffractometer by L. Jacobs (1), L. Jacobs and M. Hart (2) already showed that the use of highly perfect silicon crystals instead of the traditionally used quartz crystals can substantially improve both spectrometer efficiency and energy resolution. Through its reflectivity behaviour, the crystal also determines the high energy limit of the spectrometer. With regard to the favourable energy resolution and precision of highly perfect silicon crystals over quartz crystals, it is interesting to study their influence on the high energy range of BCD spectrometers. In practice, the standard can be set by the two bent quartz crystals as used in the spectrometers at the I.L.L. high flux reactor in Grenoble (3). For γ-energies above 150 KeV (350 KeV) in the 4 mm (13 mm) case the crystal reflectivity shows an E^{-2} energy dependence.

EXPERIMENTAL

The energy dependence of the integral reflectivity of a 3.5 mm thick bent Si(220) crystal was measured using γ-transitions following the ε/β^+ and β^- ground state decay of both ^{152}Eu ($T_{1/2}$ = 13.3 y) and ^{154}Eu ($T_{1/2}$ = 8.8 y). Reflectivities of only those transitions that could be measured in both the diffracted and the transmitted beam were taken into account. By doing so, no afterwards corrections for absorption in the crystal had to be made. A total of ten measuring points between 100 KeV and 1500 KeV was selected.

RESULTS

A relative value for the integral reflectivity was obtained as

$$R = \frac{1}{I_0} \sum_j I_j$$

where I_0 is the intensity in the transmitted beam and the summation goes over the diffraction profile. Because we only wanted to

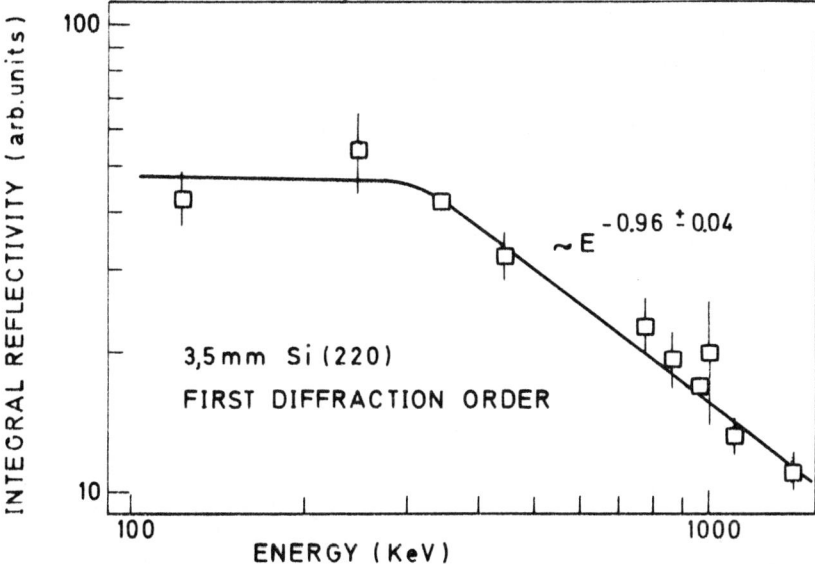

Fig. 1 Integral reflectivity curve for a 3.5 mm thick bent Si(220) crystal.

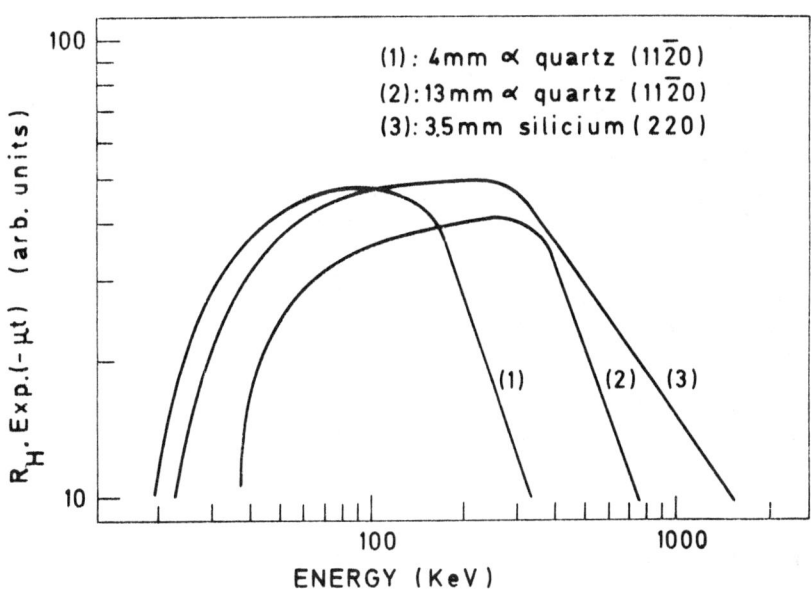

Fig. 2 Comparison of the total γ-diffraction efficiency for bent quartz- and silicon crystals.

study the energy dependence of the reflectivity, no attempt was made to reach absolute values. The energy dependence was measured to be $E^{-0.96\pm0.04}$ in the energy range 344-1408 KeV. The saturation point of the reflectivity is found to lay around 300 KeV. The obtained integral reflectivity curve is shown in Fig. 1. The observed behaviour of the integral reflectivity in this artificial grown crystal brings new insight in theories of diffraction and of crystal structure, both influencing it. Indeed, no simple explanation of the observed linearity in the high energy range can be found in standard diffraction theories (4). In collaboration with the National Bureau of Standards in Washington and the Institut Laue Langevin in Grenoble a theoretical explanation will be given later.

DISCUSSION

The saturation point of the crystal reflectivity depends on the crystal thickness. By thickening the crystal, the saturation point can be shifted towards higher energies. However, the total γ-diffraction efficiency of a crystal is not only determined by the crystal reflectivity but also by the absorption of γ-radiation passing through the crystal. A comparison between the total diffraction efficiencies of the two quartz crystals that are used in the spectrometers at the I.L.L. and our silicon crystal is shown in Fig. 2. For energies beneath the saturation point, the theoretical saturation value of 50 % is assumed for the reflectivity. The absorption of γ-radiation in the crystal is taken into account through the factor exp $(-\mu T/\cos\Theta)$, with μ the lineair absorption coefficient for the concerned crystal and T the crystal thickness. Because of the low reflectivity for high energetic (E ⩾ 200 KeV) γ-rays diffracted by the 4 mm quartz crystal and the high absorption of low energetic (E ⩽ 200 KeV) γ-rays in the 13 mm quartz crystal, it is obvious that a combination of the two crystals is necessary for a meaningful spectroscopic study of γ-transitions up to an energy of one MeV. The reflectivity behaviour of the 3.5 mm silicon crystal - i.e. lineair instead of quatratic wavelength dependence - offers the possibility of scanning the whole energy range from a few tens KeV up to a few MeV with only one crystal without disturbance due to the absorption of low energy γ-rays. Beside improving the energy resolution, the use of these nearly perfect silicon crystals will also substantially extend the high energy range of crystal diffraction spectrometers.

REFERENCES

(1) L. Jacobs, Thesis (Katholieke Universiteit Leuven, 1977)
(2) L. Jacobs and M. Hart, Nucl. Instr. and Meth. 143 (1977) 319
(3) H.R. Koch, H.G. Börner, J.A. Pinston, W.F. Davidson, J. Faudou, R. Roussille and O.W.B. Schult, Nucl. Instr. and Meth. 175 (1980) 401
(4) W.H. Zachariasen, Theory of X-ray diffraction in crystals (Dover, New York)

ACCURATE DETERMINATION OF GAMMA-RAY
ENERGIES FOR E ≤ 2 MeV

E.G. Kessler, Jr., G.L. Greene, and R.D. Deslattes
Center for Basic Standards, National Bureau of Standards
Gaithersburg, Maryland 20878

H.G. Börner
Institut Laue-Langevin, Grenoble, France

ABSTRACT

A flat crystal spectrometer facility has been established at the 57 MW reactor at the Institut Laue-Langevin (ILL). The high flux reactor with the associated source changing facilities produces the intense capture gamma-rays needed for the high-resolution low-efficiency spectrometer. Initial measurements of gamma-ray energies up to 2 MeV from the reaction $^{35}Cl(n,\gamma)$ have clearly demonstrated that sub-ppm measurements of intense sources are possible in the 2 to 4 MeV region. Energy values for the 517, 786, 788, 1165, 1951, 1959 keV lines are available with uncertainties of ∼ 1 ppm. Three of these lines (786 + 1165 = 1951) satisfy the sum rule to better than 1 ppm. Future prospects for high energy capture γ-ray measurements which impact on the neutron mass and the fundamental constants are briefly discussed.

INTRODUCTION

The goal of the experiment described here is to extend gamma-ray energy measurements to as high an energy and accuracy as possible. To achieve this goal a gamma-ray spectrometer developed at NBS has been coupled to the high flux reactor at the ILL.[1] This new flat crystal facility has been operational for about nine months and has produced ∼ 1 ppm energy measurements up to E ≃ 2 MeV.

The new facility incorporates the technology which was used to establish low energy (E < 500 keV) gamma-ray standards.[2,3] The measurement scheme is designed to relate in a rigorous way, γ-ray energies to the optical wavelength scale and the Rydberg constant, R_∞. Only intense gamma-ray lines can be measured because of the high resolution and low efficiency of the gamma-ray spectrometer.

In the region E > 1 MeV, intense gamma-ray lines are only available from prompt (n, γ) reactions. Thus, the gamma-ray source must reside within the reactor during measurement. The high flux (∼ 5 x 10^{14} $ncm^{-2}s^{-1}$) and existing tangential port with source changing apparatus at the high flux reactor of the Institut Laue-Langevin provide a unique facility for this work.[4]

MEASUREMENT FACILITY

The measurement scheme which links the gamma-ray and optical

0094-243X/85/1250921-04 $3.00 Copyright 1985 American Institute of Physics

regions consists of three distinct steps.[2] Only the third step takes place at the ILL facility. In the first step the lattice spacing of a particular nearly perfect silicon crystal is compared to an optical wavelength using simultaneous optical and x-ray interferometry. In the second step the lattice spacings of other crystals (Si and Ge) are compared to the interferometrically measured crystal. In the third step, crystals whose lattice spacings have been determined in step two are used to diffract gamma-rays. The spectrometer used in this last step is a precision absolute angle measuring device. Gamma-ray wavelengths are obtained by combining the lattice spacing and diffraction angle measurements via the Bragg relation.

The gamma-ray spectrometer is a two axis flat crystal spectrometer whose crystals are used in transmission. The support plate for the spectrometer rides on rails which permit the instrument to be removed from the beam when measurements are being made with GAMS 2/3. The spectrometer is approximately 17 m from the source and the two crystals are separated by 53 cm. Slit collimators are used before, between, and after the crystals in order to prevent the direct beam from entering the detector and to reduce background.

Each axis is equipped with a polarization sensitive Michelson interferometer having an angular sensitivity of 10^{-9} radians. The interferometers are absolutely calibrated by requiring the sum of the external angles of an optical polygon to equal 360 degrees. The calibration procedure is accurate to better than a 0.1 ppm, but thermal and temporal drifts require calibration of the interferometers before and after a measurement.

Three sets of crystals are available for diffracting γ-rays -Ge 400, Si 220, and Si 111. The crystals are 5 cm wide x 2.5 cm high x a few mm thick and are mounted in a manner which minimizes strains. A study of crystal widths as a function of energy and order clearly indicates that the perfection of the Ge crystals is superior to that of the silicon crystals. The Ge crystals have been used for all of the measurements at ILL to date. The ILL source manipulating apparatus limits the source area to approximately 0.2 x 2.5 cm^2. Thus, only the central few mm of the crystals are used for diffraction.

^{36}Cl DATA

The source for the ^{36}Cl capture gamma-rays was 1.8 gm of NaCl contained in a graphite holder 0.25 x 1.3 x 2.5 cm^3. The intense gamma-rays which were measured are the 517, 786, 783, 1165, 1951 and 1959 keV lines. The first order Bragg angles ranged from 0.49 degrees for the 517 keV line to 0.13 degrees for the 1959 keV line. Gamma-ray intensity profiles are recorded by step scanning the second crystal in angle while the first crystal is held fixed.

Preliminary data analysis indicates that all energies are

accurate to approximately 1 ppm except the 1959 keV line for which
only very limited data was recorded. Three of the lines are constrained by the level scheme (786 + 1165 = 1951). After correction
of the measured energies for recoil, the Ritz sum rule appears to be
satisfied within the uncertainty. Thus, the ability to accurately
measure energies which differ in magnitude by a factor of three appears to be demonstrated. Final energy values for these intense
lines are expected to be available in the near future.[5]

FUTURE DIRECTIONS

The availability of high accuracy, high energy gamma-ray energies opens a number of interesting experimental possibilities.
These include an improved value of the neutron mass, an alternate
route to the fundamental constants combination $N_A h/c$ and the fine
structure constant α, and the establishment of consistency between
gamma-ray energies based on the wavelength scale and those based
on the mass difference scale.

All of these possibilities involve the deuteron binding energy
which is determined by measuring the (n,p) 2.2 MeV capture γ-ray.
This energy currently has an uncertainty of 3 ppm.[6] A significant
collection of data in this gamma-ray line has been recorded with the
new flat crystal facility. A value accurate to near 1 ppm is anticipated from these measurements.

The measurement of this 2.2 MeV capture gamma-ray presents an
opportunity for an improved determination of the neutron mass. The
relevent relation for the neutron-hydrogen atomic mass difference is
$M(n) - M(H) = E_\gamma(2.2 \text{ MeV}) - M(H_2 - D)$. Because this mass difference
depends on the capture gamma-ray energy and the $H_2 - D$ mass doublet
splitting, significant improvements in the neutron mass will only
result if both the capture gamma-ray and the mass doublet splitting
are improved.

To obtain a value for $N_A h/c$, the same precision energy and mass
spectroscopic techniques used in the neutron mass measurement need
to be re-applied to larger energy intervals. An example of such a
measurement is discussed in detail in Ref. 1. For our purpose here
it is sufficient to remark that measurement of an energy interval
in terms of a wavelength, λ, and an atomic mass difference, ΔM,
leads via the mass-energy conversion relation to the equation
$N_A h/c = \Delta M \lambda$. Both the wavelength and the mass measurements need to
be improved since the current value of $N_A h/c$ has an uncertainty of
~ 0.2 ppm. However, the scientific interest in such a measurement
is high because it provides an independent route to the fine structure constant.

Energy and mass measurements which contribute to $N_A h/c$ also
provide an important link between the wavelength and mass difference
gamma-ray energy scales. The high energy gamma-ray energy scale has
traditionally been based on mass spectroscopic measurements of iso-

topic mass differences.[7,8] The extension of the wavelength measurements into the high energy region and the improvement in the mass difference measurements should provide gamma-ray energies up to 5 MeV which are significantly more consistent and accurate.

REFERENCES

1. R.D. Deslattes, G.L. Greene, and E.G. Kessler, Jr., Journal de Physique 45, C341 (1984).

2. R.D. Deslattes, E.G. Kessler, W.C. Sauder, and A. Henins, Annals of Physics 129, 378 (1980).

3. E.G. Kessler, Jr., R.D. Deslattes, A. Henins, and W.C. Sauder, Phys. Rev. Lett. 40, 171 (1978).

4. H.R. Koch, H.G. Börner, J.A. Pinston, W.F. Davidson, J. Faudou, R. Roussille, and O.W.B. Schult, Nucl. Instr. and Meth. 175, 401 (1980).

5. E.G. Kessler, Jr., G.L. Greene, R.D. Deslattes, H. G. Börner, to be published.

6. C. van der Leun and C. Alderliesten, Nucl. Phys. A 380, 261 (1982).

7. R.G. Helmer, P.H.M. Van Assche, and C. van der Leun, Atomic Data and Nucl. Data Tables 24, 39 (1979).

8. L.G. Smith and A.H. Wapstra, Phys. Rev. C 11, 1392 (1975).

STUDY OF RADIATIVE CAPTURE GAMMA-RAYS ARISING FROM NEUTRON INTERACTIONS WITH IRON BARRIERS

R.M.A. Maayouf and A.S. Makarious
Reactor & Neutron Physics Department, NRC,
Atomic Energy Authority, Cairo, Egypt.

ABSTRACT

This work represents the measured data of the spatial and angular energy distributions of radiative capture γ-rays produced from the interactions of reactor neutrons with iron barriers of different thicknesses. The measurements were performed, at different angles of neutron incidence and emission of γ-rays, using a γ-spectrometer with stilbene scintillator. The measurements show that for barrier's thickness ≥ 5 cm, the increase of the angle of neutron incidence (to $70°$) decreases the value of the capture γ-ray flux by $1.2-3$ times the value obtained when the angle of incidence is $0°$. The data also show that the increase in the angle of emission, for barriers ≥ 5 cm, by $50°-70°$ decrease the capture γ-ray flux respectively by $\sim 2-3$ times. The emperical formula describes the present measurements with adequate accuracy (10-15%), is also given.

INTRODUCTION

The knowledge of the secondary gamma radiation characteristics allows, where nuclear facility shields are concerned, an optimal choice of thickness and alternation of layers. This is from the point of view of both the minimum radiation heat generation and the minimum dose of mixed radiation behind the shield. The information yielded, from studying gamma rays emitted from different thicknesses of materials widely used for radiation shielding, is important both for studying the production and attenuation of gamma rays through matter and for calculation and design of reactor shields[1]. Regardless the existing measurements[2-8] carried out using reactor neutron beam, are few and mainly concerned with integral values.

In this work are presented the results of measurements carried out with iron barriers, of different thicknesses, exposed to a primary neutron beam from one of the horizontal channels of the ET-RR-1 reactor and to a beam filtered through boron carbide.

EXPERIMENTAL DETAILS

Neutrons emitted from one of the horizontal channels of the ET-RR-1 reactor, see Fig. 1, are first collimated and then incident on the sample with an angle Θ_0 to the normal. The emitted gamma rays were detected, using a stilbene scintillator. A detailed description of the spectrometer and the technique used for discrimination against neutrons are given elsewhere[8]. The measurements were carried out for iron samples of infinite diameters and of thicknesses 2,5 and 10 cm. Impurities in the Fe, as determined by chemical analysis, were C(3.0% by weight), Si(0.05%), Mn(1.0%), P(2.0%), and S(0.04%). The space energy distributions behind iron samples of different thicknesses, were measured at different angles of incidence Θ_0 and emission Θ; both angles were varied. The sum of the angles Θ_0 and Θ was varied for each sample between 30 and 120 degrees. All the experimental runs were made with the ET-RR-1 reactor operating at full power (2MW). The measurements were performed with a direct neutron beam and with the beam filtered through a 1-cm thick B_4C filter positioned at the beam exit. Also a background measurement was carried out for each run, by blocking the beam with a lead filter (45 cm length) placed on the symmetry axis of the neutron beam between the beam collimator and the sample.

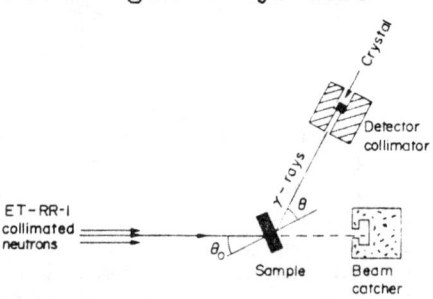

Fig. 1. The experimental arrangement.

ANALYSIS OF THE EXPERIMENTAL DATA

The measured pulse amplitude distributions were transformed to γ-energy distributions exactly as described in reference [8]. The contribution due to primary γ-rays, emitted from the ET-RR-1 reactor, in the measured γ-spectra was eliminated by subtracting estimated values of primary γ-rays spectra from the measured ones. The estimation of the contribution due to primary γ-rays was performed as described in reference[7].

RESULTS AND DISCUSSION

Some of the radiative capture γ-ray spectra, obtained in the present measurements, emitted from the iron barriers are represented in Fig. 2. The spectra are for different thicknesses, angles of neutron incidence Θ_0, and angles of γ-emission Θ. The characteristic capture γ-peaks, at 6.02 and 7.64 MeV, are observed in all the displayed spectra.

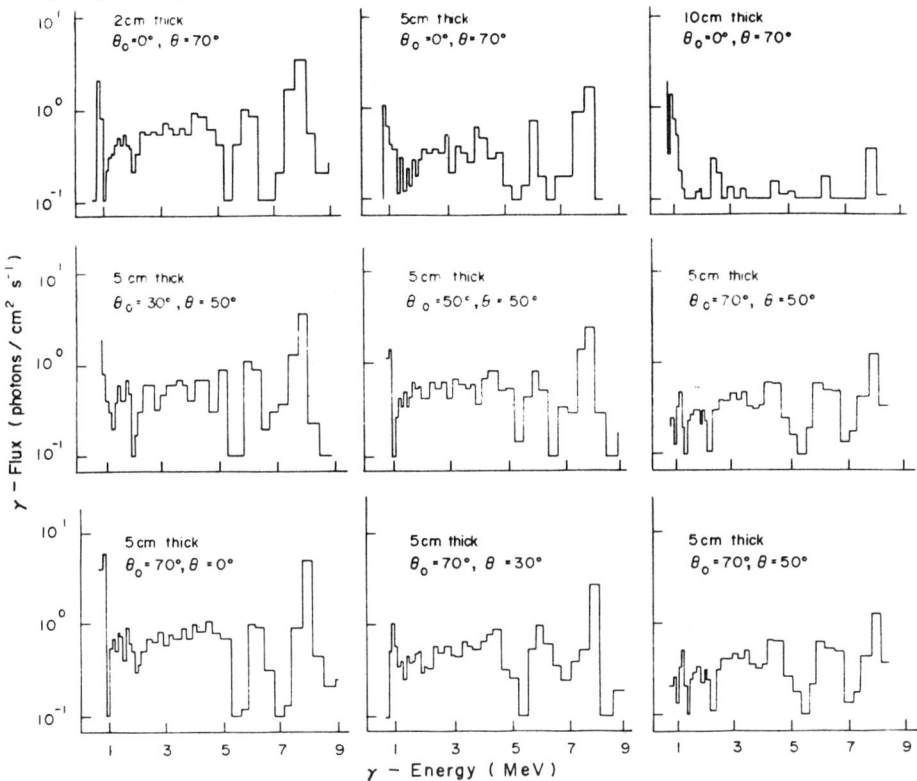

Fig. 2. Radiative capture γ-ray spectra measured at different conditions.

The dependencies of the intensity of the observed γ-peaks, deduced from the spectra displayed in Fig.2, on sample thickness, Θ_0 and Θ are represented in Fig.3. It is noticeable that the increase of the barrier's thickness from 2-10 cm (at $\Theta_0 = 0°$, $\Theta = 70°$) decreases the intensity of the peaks observed at 6.02 and 7.64 MeV respectively by $\sim$ 5 and 10 times the intensities observed for a barrier 2 cm thick. The increase of the angle of neutron incidence Θ_0 from 30°-70°, while the angle Θ is fixed at 50°, decreases the intensity of the peaks observed for 5 cm thick barrier at 6.02 and 7.64 MeV respectively by $\sim$ 1.6 and 2.6 times the values observed when $\Theta_0 = 30°$. At an angle $\Theta_0 = 70°$, the increase of Θ, up to 50°, decreases the intensity of the peaks emitted, from the 5 cm thick barrier, at 6.02 and 7.64 MeV respectively by $\sim$ 1.4 and 3.8 times the values observed

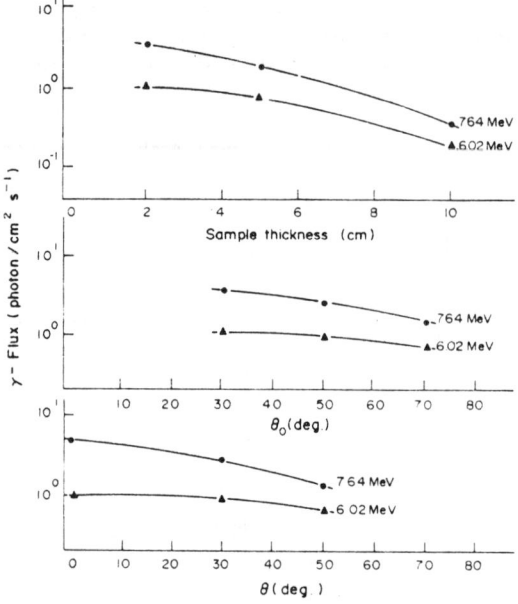

Fig. 3. The dependencies of the intensity of the observed γ-peaks.

at θ=0°. Thus, the oblique incidence of neutrons on the shielding barrier influences the γ-ray flux in a way similar to an effective increase of the barrier's thickness. The angular distributions of all the radiative capture γ-ray fluxes, obtained from the present measurements, are displayed in Fig. 4 for different thicknesses of the iron barrier and angles θ_0. The display show the dependency between the radiative capture γ-ray flux, at different angles of neutron incidence θ_0, on the angle of emission θ.

It was found in the present work, that the angular distribution of the radiative capture γ-ray flux $\emptyset(\theta_0, \theta, t)$ can be described by the following empirical formula:

$$\emptyset = \frac{A\left[e^{-at/\cos\theta_0} - e^{-bt/\cos\theta}\right]}{bt/\cos\theta - at/\cos\theta_0}$$

Where a and b depend on the samples composition while A depends on both the incident neutron beam and barriers thickness t. The solid curves, represented in Fig. 4, are calculated using this formula with the parameters a =0.0878, b=0.0809 and A=17.42, 8.5 and 3.67 respectively for barrier's thicknesses 2,5 and 10 cm. The agreement between the trends, calculated from the empirical formula, and the experimental points is reasonable within the accuracy of measurements (10-15%). It is noticeable that for barrier's thickness ≥5 cm the increase of the angle of neutron incidence θ_0 (up to 70°) decreases the value of the capture γ-flux by 1.2-3 times the value when the angle of incidence is 0°. This seems to be consistent with the results of calculations reported by Barsov et al.[4], where they found that the increase of angle of incidence (to 60°), on the iron barrier, decreases the density of the

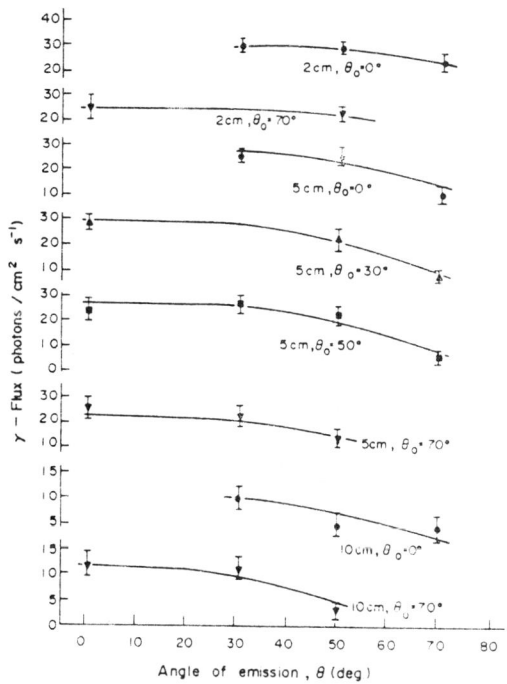

Fig. 4. The angular distributions of the capture γ-ray fluxes measured at different barrier thicknesses and angles of neutron incidence θ_0.

capture γ-radiation sources by 1.5-2 times the value obtained when the angle of incidence is 0°. Also one can see from Fig. 4, that the measured distributions are close to the isotropic at least for angles $\theta < 50°$. This well agrees with the conclusions given by Abagjan et al.[6], where they stated that the angular distribution of the secondary γ-radiation flux, behind the shield of heavy materials, is slightly anisotropic (close to the isotropic at least for angles $\theta < 60°$).

REFERENCES

1. J.A. Kazansky, etal., Physical Investigations in Reactor Physics, Mosatomizdat, Moscow, (1968).
2. R.E. Maerker and F.J. Muckenthaler, ORNL-4475 (1970).
3. E.V. Goryachev, etal., Atomnaya Energia, 31,3,209(1971).
4. P.A. Barsov, etal., Voprosi Fiziki Zachiti Reactorov, 5, Atomizdat, Moscow, (1972).
5. S.F. Degtyarev, et al., Atomnaya Energia, 33,1,575(1972).
6. A.A. Abagjan etal., Reactor Shielding, Princeton N.J. Science Press, (1977).
7. R. Megahid, A.S. Makarious and R.M.A. Maayouf, Int. Journ. Appl. Rad. & Isotopes, 32, 513 (1981).
8. A.S. Makarious, etal., ORNL-5678 (1977).

REPLACEMENT OF THE VESSEL AND BEAMTUBES OF THE HIGH FLUX REACTOR IN PETTEN

K. Abrahams and F. Stecher-Rasmussen
Netherlands Energy Research Foundation
P.O. Box 1, 1755 ZG PETTEN, The Netherlands
M. J. W. Weel, Video Team JRC, Petten

ABSTRACT

In Petten the Nuclear Structure Group uses neutron radiative capture as a tool for the study of bound and unbound states. Neutron rich nuclei provide a testing ground for physical theories and are relevant for nuclear applications. Moreover, neutron beams do not heat the target and therefore work with polarized target nuclei is feasible. In this paper -a slightly modified text of a video presentation - it will be shown how replacement of the reactor vessel was involving our group.

INTRODUCTION

The reactor vessel, which is located in the centre of a pool filled with water, had become brittle in course of 23 years of operation and had to be replaced. This replacement of the vessel could only be performed if the experimental hall was cleared and the beamtubes were removed. We had a unique possibility to upgrade the neutron beam facilities. In the new configuration for example the large facility at the previous thermal collumn will be converted into a twin beam arrangement, where one channel will provide a beam of filtered neutrons (24 keV) and the other channel a beam of thermal neutrons. Both beams are optimized to yield very high neutron flux densities. Further to this new installation the present beam channels for polarized thermal neutrons will be upgraded.

It is clear that this project took quite some time spent for preparation. Many people have been involved in designing, drawing and planning. As we are not the only users of the reactor we have to discuss the refitting procedure extensively and you will understand that the time schedule was a point of concern.

The new reactor vessel was still not completely mounted while an identical copy of the vessel was already available. In case that one of the vessels is not fulfilling the requirements it is always possible to take the duplicate. This was done because hundreds of people are waiting for the reactor to start up again and it will cost much less to make a second vessel than to risk delay.

REMOVAL AND STORAGE

In october 1983 the old reactor was scrammed for a year and in the meantime the electronics were removed and the concrete shielding blocks had to be taken away.

Also in the operation of removal and storage many people were involved because this was rather delicate equipment as for example the nuclear orientation setup (which is unique in the world). With this setup nuclei in the target can be oriented with their magnetic moment in interaction with the field generated by a super conductive magnet. The target is cooled in the large cryostat. The temperature which can be reached there is of the order of a few millikelvin in fields up to eight tesla.

This setup was rather compact and easily transportable and mounted in a special handle to secure it. Also the pumping equipment is rather large, but it is so compactly designed that it does not have to be demounted before it is removed, which is convenient because we intend to restart the system in a few months. A big truck was used to carry out the most bulky parts of the shielding. Main problem during the removal procedure was the weight and delicacy of the instruments and the radiation background. While the shielding blocks are removed health physicists controlled the contamination and the radiation level.

Also quite some time was devoted to the removal of the system which used to polarize neutrons. It was cleared after fifteen years of intensive use. During the month November 1984 there were still many loose ends and these were sad moments for the physicists who saw their ideas and dreams reduced to concrete and steel again.

The large facility, which replaced the earlier thermal collumn was most difficult to remove. The removal of the external blocks of concrete had to be followed by collimator extraction and by extraction of a large nickel mirror system, which focussed a thermal neutron beam into an intense beam spot. A lead coffin housed the mirror system during the shut down time of the reactor. This large lead container is needed because this system is higly radioactive, and it will not be disposed off but stored because it has to be re-installed again. First this coffin was carefully adjusted to allow extraction of the mirror system and then we pulled the system into the coffin.

In this way we could remove the system easily although there was a dose level of hundreds of rems per hour inside the coffin.

Large shielding walls were removed by riding on rolls just as used by pyramide builders to remove large blocks.

After removal of the external shielding we entered into the other beamtubes and with a special envelope (a cylinder) the collimators were extracted in the same method as was done earlier for the nickel mirror system.

First the envelope was adjusted and then the operator pulled the beam collimator into the envelope, while health physicists monitored the radiation levels continuously. At the tip of the collimators the dose levels are of the order of a few rem per hour but at the place where people worked of course the dose level was a thousand times less.

Inside the reactor pool the reactor had been disconnected from the cooling systems in the meantime so that the next step in the procedure was the removal of the beamtubes. These operations have been performed under water because the radiation level should be reduced. The top flange of the reactor was carried out to be replaced by a wooden dummy flange, because no bolts or loose ends should trouble the removal procedure. The next step was the removal of all bolts by means of which the vessel was connected to the bottom of the pool. After this was done it was possible to lift the vessel in a rather smooth procedure. Two operators on the top of the vessel were lifting the vessel and moved it under water to the other side of the pool where it was sliced with a pneumatically driven saw. It turned out that the aluminium was not at all radioactive and there even were plans to convert the slices into sundials to present it to some of our relations. It is very interesting to note that in some places quite a fraction of the aluminium was converted into silicium by the neutron radiation.

The most difficult part of the whole operation was the removal of the thermal collumn where the nickel mirror system was placed. It is much bigger than the reactor vessel itself and, what was more troublesome, it was much more radiating (up to thousands of rems per hour). A special envelope was made for this procedure and adjusted to extract the thermal collumn. All the work from then on should be done very carefully planned on a large distance with a good shielding. Some force had to be applied but not to much because it turned out that everything moved almost as easily as more than twenty years ago. After sealing off the whole thing was rolled out and moved to the waste disposal area.

START OF REFITTING PROCEDURE

Inside the pool, which had been dried, special lead blocks were placed to protect against radiation from the activated shielding around the beam ports.
Next the system, which will be replacing the large facility at the place of the previous thermal collumn, was mounted. This was an aluminium twin beam construction, which allows a much better shielding with water, concrete, and steel plates. Large solid angles are possible as both beams can see one quarter of the whole reactor vessel and with a specially designed collimator this will yield a rather background free set of beams with extremely high intensity. The new reactor vessel came in from the factory in July 1984 and it was placed into the reactor main shielding (the dried pool) in order to see wether all measures are within tolerances.

UPGRADING THE BEAM FACILITIES OF THE HIGH FLUX REACTOR IN PETTEN

F. Stecher-Rasmussen
Netherlands Energy Research Foundation ECN,
P.O. Box 1, 1755 ZG Petten, The Netherlands.

1. INTRODUCTION

The High Flux Reactor in Petten, which is of the ORR type, is a light water cooled, Be moderated research reactor operating at 45 MW. For neutron beam experiments the reactor is provided with radially positioned horizontal beam tubes, among which 4 was used for nuclear physics. After 20 years of operation it was decided to upgrade the reactor by renewing the reactor vessel. The vessel replacement, which goes on during this conference, offers a unique possibility for important improvements of the beam facilities.

2. THE TWIN-BEAM LARGE FACILITY

In the new beam configuration the former thermal column will be converted into a twin-beam arrangement (fig. 1). Each of the beams will face the entire area of one side of the reactor core. In this way a flux density increase of about a factor 15 with respect to a standard beam tube can be expected. One channel will be provided with an iron filter producing a beam of 24 keV neutrons. In the other channel a focussing system of Ni mirrors [1] will be installed yielding a clean beam of thermal neutrons. Both facilities are optimized to give high neutron flux densities. In table 1 the expected beam parameters can be found.

3. FACILITIES FOR POLARIZED THERMAL NEUTRONS

The Petten group is provided with two facilities for polarized thermal neutrons. One is used for capture of polarized thermal neutrons by polarized nuclei. Here the present neutron polarizer, a Heussler crystal, will be replaced by a polarizing mirror system [2]. In this way the flux density will be improved considerably. Furthermore, the magnetic field of the nuclear polarization installation will be increased to 8 tesla, thus improving the degree of nuclear polarization.
In the other facility, which provides polarized thermal neutrons for capture experiments with unpolarized nuclei, the present polarizing mirror system will be replaced by a similar system with better parameters [3], thus improving the beam characteristics.

4. OTHER FILTERS

In addition to the iron filter the fifth neutron beam facility for nuclear physics will be provided with a scandium filter. In this way a neutron beam of 2 keV with an estimated flux density of about 10^7 $s^{-1}.cm^{-2}$ will be obtained.

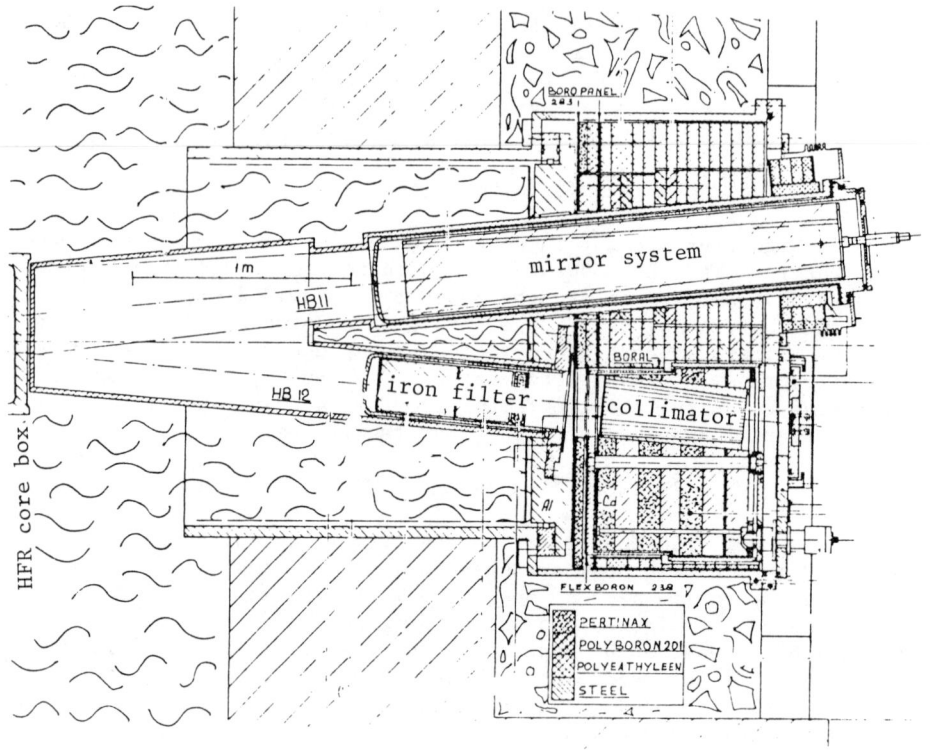

Fig. 1. The twin-beam facility.

Table I. The new parameters of the neutron beams for nuclear physics at the HFR.

beam nr.	flux density (in $cm^{-2}.s^{-1}$)	energy	special equipment
HB2	2×10^7	thermal (polarized)	nuclear polarization (T=5 mK, H=8 tesla)
HB4	10^7	2 keV	
HB7	4×10^7	thermal (polarized)	analyser for γ-ray circular polarization
HB11	3×10^7	thermal	
HB12	10^7	24 keV	nuclear polarization (proton scatterer for neutron polarization)

REFERENCES

1. F. Stecher-Rasmussen, Inst. Phys. Conf. Ser. No. 62 (1982), p. 638 (Proc. of 4th Int. Symp. on Neutron Capture Gamma-Ray Spectroscopy, Grenoble, Sept. 1981).
2. F. Stecher-Rasmussen, K. Abrahams and J. Kopecky, Nucl. Phys. A181 (1972) 535.
3. K. Dörr, H. Ackermann, B. Bader, H.-J. Stöckmann and P. von Blanckenhagen, Nucl. Inst. Methods 190 (1981) 211.

IX. AFTER-DINNER SPEECH

(n,γ) RESEARCH IN THE FIFTIES

Bernard B. Kinsey
102 Skyline Drive, Austin, Texas 78746

ABSTRACT

Neutron research in the fifties, prompted by atomic energy development, and with the stimulus of the coming of the national laboratories, created a huge new body of knowledge, new experimental techniques, and theoretical insight. The capture γ-ray work confirmed much of what had been forseen in the thirties by Rasetti, Kikuchi and his associates, Fleischmann, and others. It started with the use of Compton spectrometers in Russia, and a pair spectrometer in Canada, and later adaptions of magnetic spectrometers with coincidence counting methods by Motz, and others. Probably the most interesting result, not forseen in the thirties, was the identification and measurement of electric dipole radiation. The surge of experimental effort which commenced with the fifties, carried on into the 60's. An attempt will be made to trace its influence over the years on the development of research establishments and universities.

This occasion reminds me of the last time I attended a meeting at Oak Ridge, when Art Snell, well primed for the occasion, talked about his experiences in a train out of Paddington Station, in the company of three British brigadiers. I am not blessed with such good fortune, but looking back over the years a lot of very peculiar things have taken place. But let me first try to recapture the days of capture gamma rays, which is what I am supposed to talk about.

I can't quite remember how I got into capture gamma rays. Sent to Canada during the war along with many others, we had a year or two in Montreal to prepare for what we were to do at the new Atomic Energy laboratory at Chalk River. At Montreal we had the use of a part of the incomplete University of Montreal. The chemists had a wonderful time there - if you had any radioactivity you wanted to dispose of, you just poured it down the sink. So, after we moved out of it, a good part of the building had to be torn apart to get it out. Just imagine the fuss that that would have caused if done now! I had had some interest in RaD at the time, and acquired an instant reputation as the Great Contaminator, after I had got a whole room contaminated with polonium.

Although the technical methods for the study of gamma rays were more advanced than they were before the war, what we discovered after was, to some extent, foreshadowed by the earlier work of Rasetti, of Fleischmann, and of Kikuchi and his associates in the 1930's. They found (as, with hindsight, one might expect) high energy gamma rays comparable with the neutron binding energy, specially in the iron region, and many low energy gamma rays of high intensity. Subsequently, in the 50's, measurements verified these earlier conclusions, clarified, and extended them.

Following the war, there was considerable development of magnetic spectrometers, NaI detectors appeared somewhere around 1950; and at the end of the decade, there were further developments of magnetic spectrometers in conjunction with coincidence methods, all very useful for studies of neutron capture gamma rays in particular energy ranges.

What made all these things possible was the development of nuclear reactors producing gamma ray sources of strengths unheard of before the war. One could either use a small source outside the reactor, activated by a slow neutron beam, or a very much larger and stronger source located at the surface of the reactor or even within it, with a long clear path to the detector. Bartholomew and I chose to use the latter with a pair spectrometer for reasons which are not now very clear to me, possibly because it seemed easier to make a magnet suitable for a pair spectrometer, than the more sophisticated developments that were then going on in Russia, possibly also because we had not bothered sufficiently to estimate what sort of counting rates we might have expected to get. In fact, had we done so, we would probably have been shocked into doing something quite different. Or maybe it was because the method had only just been invented by McDaniel and Walker. Anyway we had four and a half years together, when everything we exposed to the neutrons in the reactor produced something interesting. Our method, depending on pair creation, was capable of surprisingly good resolution, accurate energy measurements, and even absolute measurements of intensity.

The Compton spectrometer was developed mainly in Russia. In the first half of the decade, considerable experience in the use of magnetic spectrometers had been usefully applied to the study of radioactivities, the determination of conversion coefficients, and the like. The Compton method had distinct advantages over the pair method, in that the interpretation of intensities was much easier, the sensitivity of the instrument being much less energy dependent. Some very high quality work was done by Groshev and his associates.

There were also further interesting applications of magnetic spectrometers which were employed mostly for low energy studies, with registration of photoelectrons combined with a cleaning up of the spectrum by elimination of Compton electrons by coincidence methods (e.g., Motz, Alburger, about 1955). At the end of the decade, methods were found to improve the resolution of NaI crystals. However, I doubt if anyone in that period had much idea of the vast improvements which were to come from the use of germanium detectors, which, with their great sensitivity, made possible detailed studies of resonance emission both for neutrons and for charged particles.

To me the most interesting thing was the identification of electric dipole gamma rays, made possible by the shell model which predicted the parities of the ground state and that of the capturing state. If you put your mind back to the 30's, with the exception of a few high energy gamma rays produced by proton reactions, of which little was known, the only gamma rays of which we had detailed information were the few produced in radioactive decay, for which the internal conversion coefficients were known, and the lifetimes could be inferred from the competition with α-emission. The few available were all exceedingly slow and gave the impression that all nuclear transitions would prove to be of second order. Now for the first time we had positive identification of multipole order, and in some cases could definitely identify transitions as proceeding from one specific state to another. And what is more, their transition probabilities were not so very far off Weisskopf's predictions.

Details of the cascade of gamma rays between excited states has evaded investigation because of its complexity. Quite possible Coulomb excitation and resonance studies of charged particles have given more insight into the nuclear processes than capture gamma rays.

The 50's saw, if not the establishment of, possibly the most fruitful period in the lives of the government laboratories. The driving force behind the neutron work in this period, was the need for nuclear data for reactor development. It came to its full fruition, I think, in the 60's. Hence the enormous effort that went into such things as thermal neutron capture cross sections, cross sections in the keV region, fast neutron cross sections, inelastic scattering, and neutron resonances. Much of this effort didn't even involve nuclear reactors. Would this have happened in the absence of the drive furnished by the requirements of nuclear power? Almost certainly not. Remember the neutron-proton scattering and related subjects like the hydrogen capture cross section? Many of my generation had a hand in it and learned a lot from it. When we started during the war, no one had a clear notion how to measure a neutron flux, thermal or fast (and many in universities haven't a clue even now). Another example: the keV neutron capture cross sections and the huge efforts to determine the cross section of gold - an element (at that time in good supply), very pure, and with the most convenient activation and resonance properties. A lot of people now in this room had a hand in that. It took 20 years to establish the cross section as a function of energy to an accuracy of even 10 percent.

Undoubtedly none of this essential knowledge would have come into existence were it not for the government laboratories. In the early 50's they were something new. It wasn't at all clear how long the national laboratory would last or whether it would be desirable for them to continue. Remember Kowarski's Principle? It predated that marvel of dismal science, the Laffer curve. Kowarski's point was that a government operated laboratory starts with zero production, will rise to a peak, and then fall off as bureaucracy raises its ugly head, finally falling to zero and becoming negative. The discussions I remember of it at Chalk River, in the early 50's, suggested that the period between the zeros would be 20 years. At Chalk River, as elsewhere, this brave prognostication came to nothing, partly because government run laboratories are not necessarily inefficient, but, but like others, depend very much on who runs them. Most got off to a good start, Chalk River thanks first to Crockroft, and then later became a model of how such a place should be run under Ben Lewis' wise direction.

Since those days the national laboratories have expanded, some to attain enormous dimensions. Not all are as interesting places as they used to be for many have had to yield to pressure to become self sufficient by diversification. Some came under pressure to stop things they were doing. Even Blackett in the 50's - bless his name - tried to keep the study of air showers out of Harwell, on the grounds, I suppose, that public money would be better spent in universities. But it never occurred to him that if he had had his way, the money spent that way at Harwell very probably would not have been available to universities. There is much prejudice against the national laboratories in universities, most of it without foundation. On occasion, when talking about national laboratories, I have rubbed up ex-colleagues the wrong way. They reacted with an explosive outburst against them.

In those early days before Chalk River became an independent entity, it was run from Ottawa by C. J. McKenzie, president of the National Research Council of Canada, who didn't hesitate to put us in our place. One very wet morning, before the roads had been paved, I left home to catch the bus and fell in a pothole up to my knees. Arriving at the laboratory a little later in a bad temper. I found a circular on my desk with McKenzie's comments. He said that he now wanted to remind research workers that they could be had for two a penny, and what was really needed were experienced engineers. Everyone read this memo; it shattered their morale until it was realized that every word of it was true. Some 25 years later, Ted Hincks and I had lunch with McKenzie in the basement of the NRC building in Ottawa, long after he had retired. After we left him, I mentioned this incident to Ted, who was at Chalk River at the time. He didn't remember it, but said that on another occasion he had put a question to CJ whose reply was so devastating, and yet so obviously true, that it took his ego five years to recuperate. (He didn't say what it was!)

About the same time, the spring of 1952 or 53, the American Scientist published those articles written by a New York psychologist, in all seriousness, about why people became scientists. In short, he said that it was usually something going wrong in early youth. Arguments with father, for example, might result in a determination to go one better. The articles, of which there were two in succession, had a profound effect in Chalk River. Years later, at a conference somewhere, I discussed this with Martin Johns, who was then chairman of the physics department at McMaster University. He remembered it well and said that his department made it required reading for all new graduate students. Such matters weren't important, but they were stimulating, and helped to keep in check the innate tendencies to lunacy common to academics.

In those days there was a more productive and free and easy atmosphere both in universities and in the national laboratories. Since then it seems to me that academics have become more conscious of money and much less willing to attempt to do anything without it. One must admit, though, that without it, it is indeed difficult to get student help; in this there has been little change over the years. What is at fault, if anything, is the reluctance of academics to make measurements themselves without such assistance.

I put it to you that the national laboratories have a success story to relate not entirely due to the lavish resources available to their research workers; but more, I think, to the fact that doing research was what they were hired to do. In universities academics are supposed primarily to teach, and secondarily to do research. In fact the priorities are the other way round. And with this confusion are sown the seeds of their decay.

In the 50's none of this was yet visible. Visiting American students discussing their work always seemed to me to be more coherent than Europeans; and likewise visiting professors would be much better acquainted with their subjects. I have the impression that university professors, since, have gone the way of their European counterparts, most of whom have ceased to do any experimental work themselves, preferring to work through students, thereby following a path of least resistance. So what do these characters do, for you have to do something. Well, there is teaching, of course, but it gets you nowhere. However, professors are inventive: some get away with doing just nothing, by looking distinguished and rise in the heirarchy on bogus credentials. Others devise techniques of their own. You can be very highly regarded by your colleagues if you pester the dean once a week; if you don't he will never hear of you and you will be conveniently forgotten next time salary increases come around, even if they are mandated by state law. Then you can improve your prospects by holding press conferences, or get your name in the paper. This is called "visibility," another word for personal publicity, a quality now equated with excellence. Or you might land a defense contract, for which you may draw a salary in addition to that which you draw from the university, put a "management fee" in your pocket and become a millionaire in five years. And for this you don't even have to succeed in your project, for neither the defense agency nor the university cares a damn and the universities always work on the assumption that there is always more public money where the first lot came from. If you become wealthy in the process, you will be greatly admired by your colleagues, for nothing is more admired in academia than money.

Then there is the technique of fast talk, which, like fast food, is very popular these days. For this you don't have to be understood, you throw in a technical word here and there, and your audience will be much impressed even if they have not understood a word of it. An ex-colleague was a specialist in this art. He embellished it with the pretense that he was expert in getting federal grants, for which he was very much admired, though in fact, while he was at the university, he never raised even one. But, sad to relate, we lost him partly because he had wangled a leave of absence under false pretenses, in order to carry on his private business, and also because someone disclosed that one of the companies of which he was president was involved in manipulations running afoul of the Securities and Exchange Commission.

There are many other kinds of making do. Some combine a genuine capacity for research with image making. Why not?

This is not to say that diddling the taxpaper is exclusively American. A friend in England told me recently about an ingenious system worked out in a university in southern England, which I have not seen emulated here. It can be described as an initial preparation of the ground followed by infiltration. My friend was in the position of a member of a board of trustees - they call it something else over there - and at one meeting of the board, a professor came forward to say that there was a wonderful man in his field, who would be willing to come, if he were given a top salary. I won't say what the field was, apart from suggesting that it was one of the less exacting of the least exact sciences, or, shall we say, one of the more bullworthy sciences. The board agreed; then three months later, the same guy appeared before them to say that this wonderful man had agreed to come, and that they need staff to work with him. This was all in a period of great financial stringency, so the board found this suggestion unwelcome, but agreed to it nevertheless. Three months later they came to him again, this time to say that now that they have this wonderful man coming and having engaged the staff to work with him, they need space to put them in, and could they have an extension to the building? My friend asked me: do you have this sort of thing in the U.S.? Indeed we do, I said, though nothing quite like this one. When public money is squandered, as it so often is, it is usually done on ten times the scale.

All this makes one wonder where we are going, if it is downhill, it is, for statistical reasons, already much too late to put in reverse. Partly, no doubt, it is the lust for money, which, for academics, is no more pathological than it is for other professions. And, after all, allowance must be made for the fact that most academics have only just come down from the trees. However, real scientific and technical jobs are less well paid than administrative jobs. This is becoming clear here in the popularity of the softer sciences. In this country we may already be at or near the end of the road, determined by the large number of universities and the paucity of people really capable of doing research. Other countries are up against similar problems. In Italy, for example, the end of the road seems to be where all members of a faculty are related to each other by marriage.

This reminds me of a story told me by a friend who went to Lima, Peru, on a junket for university presidents. In conversation with the head of the University of Buenos Ayres, he learned that that year out of some 40 students in electrical engineering, four were native Argentines, and the rest were foreigners. Of the total, a couple were killed in riots, and the rest left the country after getting their degrees. Returning to England, my friend discovered that at his university there had been only fourteen students graduating in electrical engineering that year, all but one being foreigners. Shocked to the core, he invited me over to a meeting on the subject, my job being to talk about the state universities in the U.S. What I had to say just wasn't believed!

This raises further questions, but let us go back once more to earlier times. Put your mind back to the compound theory of nuclear reactions. When Paul and Clarke drew attention to convincing evidence of direct reactions in 1953, theoreticians leaped aboard the new band wagon and it wasn't until Allen (from Harwell) went on a trip to the U.S. and called in at MIT that the compound theory was taken out of the cupboard where it had got lost. It had been demonstrated beyond doubt in Ghoshal's experiment (1950). Allen's and Ghoshal's work must probably be reckoned as landmarks almost in the class of the experiments of Rutherford and Marsden. Note that they were performed in national laboratories. Could they, would they have been done if left to the enterprise only of universities? Or consider the Dubcek effect. This was the results of years of painstaking photographic plate work at a government Laboratory. There are now very few institutions which possess such facilities, let alone know how to use them. At my institution, anyone foolhardy enough to start such a project would long since have been fired for lack of productivity!

Looking back on all this, I wonder to what extent many advances in physics owe their existence to chance and to what extent many of our ideas have been corrected by chance occurrences. Most people of my age must have come across things in the course of their work which they didn't have time or equipment to examine further and which didn't fit into the expected pattern. There is usually the means available somewhere, for few machines become obsolete. They are scrapped more often as a result of the drying up of funds. At McGill, Stewart Foster once remarked to me that after the Berkeley people had stormed over a field, like a forest fire, there would always be something to be gleaned by those who follow after. It was the moderate energy cyclotron he built there which was used by Bell to discover proton radioactivities. The similar machine at Harwell, much vilified because it didn't have the energy to produce mesons, was used to discover the Taylor-Woods effect, and the Schwinger scattering of neutrons. With the vastly improved methods of detection and registration now available, how many of the accepted principles of physics couldn't be reexamined?

I have to finish somewhere, and I am tempted to mention a proposal recently put forward in Canada for a club for provincial prime ministers, past, present, failed, flawed or just moribund. During the discussions, someone quoted that sage tyrant, Dionysius the Elder, who said "If your speech is no more substantive than silence, then let there be silence."

X. CONFERENCE SUMMARY

CONCLUDING REMARKS

J. E. Lynn

Atomic Energy Research Establishment
Harwell, England

1. INTRODUCTION

This has been the Fifth International Symposium on Capture Gamma-Ray Spectroscopy and Related Topics. It has differed in some ways from its predecessors. Firstly, by dropping "neutron" as the first word in the title it has achieved a generalization of its subject matter. Thus, gamma-induced reactions appear as a spectroscopic tool in their own right rather than as an inverse reaction, and heavy-ion and proton reactions move into the central arena rather than appearing as related topics. Secondly, "democratic" principles have been employed in the paper selection, all the members of the International Advisory Committee, the Program Advisory Committee and the Program Selection Committee having voted on the choice of proposed papers to be read at the Symposium.

Yet there has been evidence of some submerged non-democratic features and I have a feeling that a strong man has been manipulating this democratic process. For example, I never saw a proposal for the very last item of the program, and I certainly did not have a vote on the matter. If I had, the choice might have been different.

Thus I have found myself with the task of attempting to summarize this conference, and I have found that there are some acute difficulties that face the would-be conference summarizer. One is the matter of nuances and ambiguities in the spoken English language. Has one really picked up the correct interpretation of a speaker's remarks? I am reminded of a story relating to Sir Thomas Beecham, the noted English orchestral conductor. In the days before the Second World War, when British railways were the property of private companies, Beecham was travelling in a non-smoking compartment when a lady entered and sat opposite him. Presently she pulled a cigarette and lighter from her handbag saying "You don't mind if I smoke, do you?" Beecham replied "Madam, if you smoke I shall be violently sick." At this the lady said imperiously, "My dear man, you don't seem to understand; I am one of the directors' wives." Beecham responded "Madam, that makes no difference. Even if you were the Director's only wife, I should still be violently sick."*

Another difficulty is the sheer amount of information that is transmitted (but not necessarily received) in a week of symposium activity. There have been 45 invited papers at this meeting, and well over 80 posters. It is my impression that during the last several years the almost universal adoption of the overhead projector has led to a much increased density of information per presented paper, which

―――――――
*This story becomes a little less ambiguous in its written rather than its spoken version, but perhaps it makes the point more clearly.

0094-243X/85/1250951-12 $3.00 Copyright 1985 American Institute of Physics

only exacerbates the problem of summarising a meeting. Selection of material to include in the summary is an acute problem. Even when it is obvious which item should be selected in a given body of material, irrational motives allied to shortage of time can lead one to make wrong decisions. I can illustrate this with another railway story.

At the early period I mentioned in my last anecdote, many British railway carriages, especially on the branch lines, were arranged as a series of compartments with no connecting corridor, each compartment having eight seats. One day a man jumped aboard such a train just as it was pulling away from the platform and found that in the compartment he had entered seven seats were occupied, and on the eighth, by the window, obviously enjoying the experience, was a medium-sized female dog. Its owner was clearly the lady sitting opposite, so the man addressed her, asking if she would remove the dog so that he could occupy the seat. "Certainly not," she replied tartly, "I have paid for the dog's ticket, so she has a right to a seat." At this the man grew angry, opened the window, picked up the dog and threw it out. The lady was speechless with rage; but another man in the opposite corner lowered the newspaper he was reading, looked over his spectacles and mildly rebuked the dog ejector. "My dear fellow," he said, "you have just thrown the wrong bitch out of the window."

One further difficulty in my case* is that I have been unable to listen to all the invited papers, nor spend much time at the poster sessions. I have also been unable to spend much time in discussion with the other delegates so as to distill some kind of synthesis of views of the most important achievements of this meeting. I must emphasize therefore that the scheduled heading of this talk in the program is "Concluding Remarks" and what I am now going to present is very much a personal impression of this symposium rather than a balanced summary. But I hope as I proceed that not too many of you will have cause to mutter, "That fellow has just thrown the wrong bitch out of the window."

2. CLASSICAL SPECTROSCOPY

At the heart of the conference subject matter is the topic that might be called "classical spectroscopy," the detailed exploration of energies, transition rates and structure of nuclear state.

I suppose it could be said that this subject had its "classical age" in the 1950's, and we heard the history of that period delightfully described by Bernard Kinsey on Tuesday evening. I would just like to add one anecdote to that account. Bernard mentioned that he was known as the "Great Contaminator" when he was at Chalk River. In his subsequent period at Harwell he could fairly be described as the "Great Administrator". His major contribution to administration at that period was the introduction of the round file, to which he consigned all communications from the bureaucracy. It was never far from his feet, under his office desk.

---

*How I acquired these temporary crutches is a story that I do not wish to share.

In this topic of classical spectroscopy there has been renewed
momentum, starting about a decade ago, which might be termed the
"neo-classical" era, resulting from a combination of highly refined
experimental techniques, based on high intensity of sources and extremely high energy resolution, and new theoretical models for
interpreting the wealth of new spectroscopic data. It is impressive
to see at this symposium how this momentum has been maintained.

On the experimental side, there has been a tremendous production
of extensive and precise data. It has been encouraging to see the
ways that experimental groups have combined, pooling together complementary data. The experimental techniques in (n,γ) and (n,electron)
spectroscopy, especially on the Grenoble crystal spectrometers and the
electron spectrometer BILL could hardly have been improved in this
period, as the complexity of raw experimental spectra shown at this
conference have emphasized (especially when irradiation times in the
ILL reactor have allowed daughter nuclides to grow in), but there
seems to me to have been a steady improvement in resolution of (d,p)
and (d,t) meaurements, so that a few keV resolution is now commonplace. New methods of using and analyzing the data, as exemplified
by the papers on E0 transitions and average resurance capture, have
been developed, and quite new reactions for this field, such as $(\alpha,n\gamma)$
and $(^{13}C,n\gamma)$ have been drawn in to give fundamentally new information.

The foremost theoretical development has undoubtedly been the
Interacting Boson Approximation model. As it was first set up this
can best be described as a phenomenological model, rather different
from normal phenomenology, which usually rests on an intuitive physical
idea, but, instead, a sophisticated mathematical phenomenology arising
from a very general view-point. By switching on different kinds of
coupling amongst the bosons the overall $U(6)$ symmetry of the system is
broken in different ways, giving rise to the well-known limits
described by the lower symmetry groups $U(5)$, $SU(3)$ and $O(6)$.

By exploring special details of this rich mathematical formalism
new facets and relationships of the spectroscopic data have been
revealed. In the meantime the basic model has been developed. By the
time of the Grenoble conference, IBM-2, in which separate neutron and
proton boson-pairs are considered, was being rather fully discussed,
and so was the interpretation of the spectra of odd-A nuclides by
means of the IBFM in which the symmetry of the fermion is included.

The new developments reported at this conference include IBM-3,
and supersymmetry (SUSY) which was barely hinted about at Grenoble.
Supersymmetry again originates from abstract and general theortical
concepts. As Feng explained, supersymmetry is a theory in search of
experimental manifestation, and in nuclear spectra it appears to have
found its clearest evidence. The experimental evidence was discussed
in some detail for two cases, $^{194,195}Pt$, corresponding to the $O(6)$
reduction of $U(6)$ in the IBM model and $^{184,185}W$, corresponding to the
$SU(3)$ reduction. The success of SUSY in explaining these pairs of
spectra is indicated by the consistency of the parameters used in
fitting the spectra and transition rates of the two partners in each
case. In the case of the platinum isotopes the paramater sets agree,
but there is a problem in the magnitude of the Q.Q term in the
tungsten case, which may be resolved by demanding consistency in the

definition of Q in the Hamiltonian and E2 operator. A more stringent test of SUSY would require the incorporation of the study of seniority 2 states in the next higher isotone, but these will be too mixed with a dense pattern of other states not describable within the limited sub-space covered by IBM to allow this extension to be practicable. Thus within the context of nuclear structure theory, the concept is perhaps of limited usefulness. Nevertheless, the enthusism among theorists is high and Paar presented a detailed set of calculations, culminating in a complete spectrum prediction for a hypernucleus. This resulted in some pessimism, not to say outright scepticism among experimenters about the prospects of verifying such a prediction.

Elliott, in his fine review talk, which clearly delineated the significance of IBM within the overall picture of nuclear structure dynamics, also stated that, from a philosophical point of view, the interpretation of IBM as a collective theory is not interesting. But from the point of view of interpretation and unification of data embracing a wide range of collective models the impact of IBM has been enormous.

Of concern therefore to the experimental physicist, are the relationships of IBM to physically based pictures of the nucleus. Since IBM aims to explain the spectra of nuclei traditionally considered to have motion of collective character, intense interest has focussed on parallels and differences between these and IBM predictions. Comparison with the so-called geometrical models was well developed by the time of the Grenoble conference. For example, in the comparison of the SU(3) limit with well deformed axially symmetric rotators, the spectra showed features of direct correspondence but transition strengths showed some remarkable differences, which further provoked the question of their detailed relationships. Work on this was well advanced by the time of the Grenoble symposium, but it is interesting to see here at Knoxville that there are still valuable insights appearing. In the work reported by Casten, in which the Q-term of the Hamiltonian is expressed consistently with the E2 operator of the IBM as the change from the SU(3) limit to the O(6) limit is made, thus allowing the free parameter χ in the E2 operator to act as measure of this change, various transition strength rates and energy eigenvalue ratios have been calculated both in the Davydov triaxial rotator model and the SU(3) to O(6) model. From this comparison an effective γ-angle of axial distortion in the Davydov equivalent of the IBM can be obtained, and interpreted as a kind of mean axial asymmetry due to axial softness in the potential energy of the IBA model. It is interesting to see that even in the SU(3) limit this mean γ-softness approaches 10°, for a realistic finite number of bosons.

A new region of nuclides to which the O(6) limit is applicable was also reported at this conference, as a result of application of the $(\alpha,n\gamma)$ and $^{13}C,n\gamma)$ reactions which I mentioned earlier. These comprise some 10 to 20 nuclei around mass 130. A significant outcome of the 3-parameter analysis of these spectra is that two of the parameters are found very nearly to satisfy a constraint that would be required by the application of the consistent Q-formalism for the Hamiltonian and E2-operators, suggesting a deeper physical interpretation of the O(6) limit.

However, the most fundamental interest in the IBM lies in its microscopic basis and work on this topic has been pursued now for a number of years following the introduction of IBM-2 with explicit treatment of 2 kinds of s- and d-bosons, identified as neutron and proton. An excellent example of the success of IBM-2 was to be found among the poster contribuions to this conference. Hamilton and his colleagues have studied the N=84 isotones of barium, cerium and neodymium, the spectra of which can be quite well fitted by the SU(5) vibrational limit of IBM-1 apart from a 2+ level at about 2 MeV in each nuclide for which spin assignment, branching ratios and multipole-mixing ratios of their radiative decay were measured. These levels can be described as a mixed symmetry state within the vibrational description, and as such can only be encompassed by IBM-2. Their properties can be well described within this framework.

The remote ancestry of the IBM can probably be traced back to the classic work of Elliott that demonstrated rotational level schemes inherent in the shell-model, so it was particularly fascinating at this conference to hear Elliott's exposition of this topic, revealing clearly the approximate identification of bosons with nucleon pairs, and the maximum number of bosons as half the valence nucleon number, and also some of the limitations of the model, such as the small subset of shell-model space it encloses, and the lack of a Pauli principle mechanism for shell closure. The pursuit of this microscopic understanding has led to the consideration of neutron-proton pairs and hence the introduction of IBM-3, new, I think, to most of us at this conference. This allows the application of the IBM to light and medium nuclei, where isospin is required to give full shell-model descriptions of the states. The quantitative illustration of mapping from the $f_{7/2}$ shell onto IBM-3 was illuminating. The failure of spin 4 states to approximate to pure boson states especially when the maximum boson number in the shell is small could be attributed to the lack of a g-boson in the IBM description. This topic was pursued in the paper by Otsuka, who showed how a unitary transformation between the d- and g-boson to minimize their coupling could result in an effective s-d description close to that of the full s, g, d, description.

While the scheme outlined by Elliott for deducing the parameters of the Hamiltonian of the IBM from shell-model calculations with only a small number of valence nucleons, and then using the IBM thus found to calculate the collective state properties of nuclei closer to midshell seems a sensible and useful one, I can hardly forebear to comment after listening to Wildenthal's paper that it seems unnecessary to apply it to the sd-shell. The success of modern shell-model computations illustrated in his paper in describing the properties of all the levels up to 6 MeV or so of a wide range of nuclei using a limited set of universal parameters can only be described as impressive. New data on the sd-shell continue to accumulate, as classical methods, such as Doppler-shift attenuation following proton capture yielding half-lives and isoscalar E2 transition strengths, continue to be improved.

At the opposite end of the mass table we have heard about more rigorous methods to improve models of the higher vibrational levels that are found in these fully collective spectra, and specifically to

explain the apparent anharmonicities found in them. While this
multiphonon treatment seemed to have success in explaining the relative positions of the low-lying K=0, J^π= 1⁻ octupole state and much
higher-lying 0^+ state (assumed to be the 2 octupole-phonon state) of
nuclei like ^{222}Ra in terms of phonons built up from pairs of quasiparticles interacting through a monopole pairing force and octupole-
octupole force, an interesting old question was raised from the floor;
can these low-lying K^π=0⁻ states be explained alternatively by stable
octupole asymmetry of such nuclei? This is an example of one of those
problems that hangs around, more or less in the subconscious, for
perhaps 20 years or more, never finding a solution, yet posing a
challenge to our sense of fully understanding the subject. In the
meantime our concepts of the possibilities in this area have been
enlarged by work in a different area, namely that of fission, where
the data imply a wide variety of different nuclear shapes, including
meta-stable octupole asymmetric shapes at highly extended elongations,
which appear to account for the strong vibrational resonances found in
fission cross-sections.

In many ways the development of nuclear structure theories is
still being outstripped by the amassment of experimental data. As von
Egidy explained complete level schemes are now commonplace for heavy
even nuclides up to 2.5 MeV, for even-odd to 1-3 MeV and odd-odd to
0.8 MeV. In many ways there are advantages in treating such large
bodies of data statistically. The analyses reported by von Egidy
showed some interesting results. For example, analysis of the level
spacing distribution of several even-even nuclides show evidence for
extra quantum numbers other than angular momentum and parity,
suggesting the persistence of perhaps the K quantum number to quite
significant excitation energies.

3. HIGHER-EXCITATION ENERGIES

These statistical aspects of the classical spectroscopic region
form a natural link to gamma-ray properties at higher excitation
energies. But conversely there are some detailed spectroscopic
aspects of the higher excitation energy regime that have been particularly in evidence at this conference because of its broader remit
than in the past. It has been particularly fascinating to hear about
discrete γ-ray emission from very high spin states excited by heavy
ions from the new generation of very high energy tandem accelerators
such as those at Daresbury and Oak Ridge. Although at comparatively
high excitation energies these states are quite close to the yrast
line and hence may be capable of detailed spectroscopic interpretation. The important developments at Daresbury in very large total
energy γ-ray detectors incorporating high resolution elements for
single γ-rays are giving us a remarkably complete picture of the
structure of states with angular momentum as high as 40-60 ℏ. These
very high spins have a major effect on both single particle motion in
the nucleus and the collective motion, which can be followed by
observing band-crossings and changes in moments of inertia. Strong
evidence has been presented that major shape changes occur as the

angular momentum passes through critical values. In ^{158}Er for example the nucleus changes from a prolate rotor to a very marked triaxial shape, while in ^{152}Dy the most deformed nuclear shape of all (apart from the fissioning actinides) has been discovered, the high moment of inertia for states in the 35-55$\hbar$ range indicating a prolate deformation parameter $\beta \sim 0.5$ (to be compared with normal rare-earth values of $\beta \sim 0.3$).

In few-nucleon systems of course it is also possible to give detailed descriptions of the nuclear structure at normal spin values up to moderately high excitation, and to have a reasonable hope of theoretical interpretation. Thus, relatively fast nucleon and light-ion capture has its part to play in classical spectroscopy, and when carried out with polarized systems as we have heard described this week can give valuable information, which does not seem to be completely accounted for within our current theoretical understanding of these "simple" systems.

For heavy nuclei the study of primary gamma-rays following slow and intermediate energy neutron capture has also made an important contribution to the detailed information on low energy level structure, especially through the resonance-averaged capture (RAC) technique, which exploits the statistical nature of low spin, highly excited nuclear states. The detailed understanding of statistical properties of highly excited levels that was accumulated in the classical period of neutron resonance spectroscopy has enabled the RAC technique to be refined and sharpened. But this method is also giving information on the properties of the highly excited initial resonance states. It was a surprise to many of us, several years ago, when the first measurements of M1 transitions from resonances indicated that they had very appreciable strength relative to E1 transitions (greater than 1/10). Kopecky showed us graphs of how the picture has been filled in across the mass table, showing that this high relative strength is quite universal. But the new surprise is how the E2 transitions are not very weak compared to M1. Again the ratio is only of the order of 5 or so, but apparently this high E2 strength can be explained by application of a Brink type of hypothesis to the E2 giant resonance.

The RAC technique depends on the non-selectivity, on the average, of resonances for transitions to final states irrespective of the detailed structure of the final state. Two presented papers at this conference have shown results that throw some doubt on that assumption in the important rare earth region. Of course in lighter nuclei it is now well-established that selective non-statistical mechanisms operate. The best-known is the valence-nucleon transition, which for s-wave neutrons, incorporates important channel and direct effects, a new presentation of which was given this week. Valance neutron capture has been known to be significant in the 3s and 3p giant resonance strength function peaks (that is, for mass number regions 50-60 and 90-100), but in recent work carried out at Oak Ridge and Los Alamos it has been found to be important for very distant light nuclids, i.e., the sulphur-isotopes. The observations are of direct capture at thermal neutron energy, which shows a strong correlation with (d,p) single-particle p-wave strength, but they also indicate that valence capture similarly correlated must predominate over statistical compound nucleus mechanisms in the resonances, and this has been corrobrated in a poster contribution by Allen.

This proven existence of the highly selective valence capture mechanism suggests that there must be mass bounds for the applicability of the RAC technique for the determination of J^π values of low-lying states reached by the primary transitions, and some effort ought to be made to probe these boundaries. It also brings up the question of the general nature of the primary radiation from slow neutron capture. The statistical description now generally relies on the Brink hypothesis that it can be described in terms of the wing of the Lorentzian form of the giant dipole resonance, but the existence of valence transitions suggest that, in some mass regions at least, a more general microscopic description, partially decoupled from the giant dipole resonance may be applicable. In this connection it was interesting to hear Dietrich's and Iachello's theoretical expositions of the E1 giant resonance. Von Egidy's statistical analysis of transitions amongst extensive discrete level series in lighter nuclei, indicates the power law of the gamma-ray energy dependence to be ~3, rather than 4 to 5 expected on the Brink hypothesis. Doubts about use of the smooth Lorentz form in this excitation energy region also arise from some of the detailed gamma-ray elastic scattering work we have seen presented this week, in which, incidentally, the gamma-ray spectroscopy theme has been inverted to give a technique that provide precise, monoenergetic gamma-rays from strong neutron capture sources. It is not surprising that structure in the 6-10 MeV range occurs for nuclides in the lead region, as reported by Shahal, but the apparent strong structure at about 5.5 MeV in ^{238}U reported by Schumacher is certainly remarkable. One desperately wishes that these gamma-rays were tunable in energy, so that this phenomenon could be explored in greater detail.

One source of monoenergetic gamma-rays that is tunable to some extent is fast-proton capture, and this technique was described by Smith, with presentation of results on lead that raised the old question: where is the strength of the expected M1 giant resonance? Another very interesting example of the use of this technique was to be found in the poster session. Zhang and Lancman have measured photofission cross-sections of ^{232}Th in sampled limited energy regions between 6 and 7 MeV. At lower energies considerable structure with spacing ~1 keV is revealed; this is probably class-II level structure associated with the secondary well of the fission barrier and could give important information on barrier parameters. Side by side with this poster was one on the photofission cross-section of ^{232}Th measured with bremsstrahlung and unfolded with great care. This showed much grosser resonance-structure believed to be associated with a tertiary minimum in the fission barrier. While on this topic I would like to mention a problem that I hope might be answered at a future symposium: the observation of the gamma-ray spectrum for de-excitation of class-II resonances through the states associated with the highly deformed secondary well.

4. GIANT RESONANCES

At much higher effective excitation energies detailed spectroscopic descriptions of medium to heavy nuclei are not feasible. The aim is to get a statistical description, usually formulated as a dissolution of basic, simple modes of motion amongst the dense continuum of compound nuclear states. For gamma-spectroscopy the most significant of these basic modes are the electromagnetic giant resonances.

At this conference much has been heard and learned about these giant resonances, not so much, interestingly enough, from the traditional photo-and electron excitation methods as from proton and heavy-ion capture. It is impressive to see how experimental techniques, such as the 72-segment NaI spectrometer on the heavy-ion facility at ORNL, and sophisticated data analysis methods have been developed to isolate and enhance effects that are comparatively weak against the intense background of competing particle emission processes. It was also encouraging to learn that these incident projectiles will be joined in the not-too-distant future by powerful fast-neutron excitation methods. A foretaste of the quality of time-of-flight fast neutron capture measurements expected from the LAMPF/WNR facility at Los Alamos was demonstrated by the $(n,n'\gamma)$ data taken on the "Mark 1" version of this source. Since some evidence for the signature of the isovector E2 giant resonance utilizing much weaker neutron sources was already presented the auguries for the future of this work seem good.

At the same time as we are seeing these important developments in hadron excitation of the electromagnetic giant resonances, it is obvious that electron excitation methods are not going to be left behind. The use of fission as a signature for excitation of the compond nucleus by electron inelastic scattering in order to avoid the background from the elastic scattering radiative tail marks a significant step forward in these techniques.

So, what have we learned in this area of work at this symposium? First of all, a major topic of interest has been the revival of Brink's old idea that an E1 giant resonance can be built on every excited state of the nucleus, and this is similar in energy, strength, and width to the classical E1 photo-resonance built on the ground state. This idea was tested a little on a few states of very light nuclei by means of (p,γ) reactions around 1960, but it has taken another twenty years or so for it to be subject to much more extensive investigation both in mass range and range of excitation energy, despite the fact that the idea has been incorporated into neutron resonance capture theory for very many years now. I was particularly intrigued by the results presented by Snover on ^{12}C bombardment of ^{154}Sm which indicated that the giant dipole resonance profile of radiative transitions down to excited states even at some 30 MeV excitation was close to that of the strongly prolate ground state. That this conclusion can be extracted at all is so startling that I felt it right and proper to be questioned on the conference floor, and I hope that considerable effort will go into independent confirmation of this important result.

By contrast, the fusion of nearly symmetric very heavy projectile-target systems is reported as giving yields for gamma-ray de-excitation of the compound nucleus that are unexpectedly high on the statistical model picture and show little or no sign of a giant dipole resonance profile in the gamma-ray strength function. A possible explanation is that such symmetric fusion leads to very highly deformed compound systems. In this connection there is some evidence from the collision of much lighter nearly symmetric nuclei. The interpretation of the data on $^{12}C + ^{12}C$ radiative capture to the ground state and its rotational band suggests that resonances in the cross-section of this reaction correspond to unusually highly-deformed states of the ^{24}Mg nucleus. Also, definitive evidence was reported for strong radiative transitions to shape-isomeric states. For example, bombardment of ^{12}C by ^{16}O leads to strong excitation of the 0_3^+ state at 6.69 MeV in ^{28}Si, a prolate shape isomer, the ground state of ^{28}Si being oblate.

Some modifications to Brink's original idea have resulted from all this experimental work. The most significant is that in lighter nuclei at least, for the (p,γ) reaction, the integrated strength of the excited state giant dipole resonance is not independent of the structure of the state but is proportional to its single-particle proton strength, indicating a direct coupling of this reaction to the GDR. A simular effect is observed in (He^3,γ) reactions. Also in light systems there is strong evidence that isospin remains a rather pure quantum number for states de-exciting through the GDR mechanism. Another modification of the Brink hypothesis, but an expected one, is that the width of the dipole resonance increases significantly with excitation energy.

Quite clear effects of the isoscalar E2 giant resonance at 9 MeV or so have been demonstrated in the (e,e'f) and ^{17}O inlastic scattering reactions, but this has in turn raised the old problem of why it has not been observed in $(\alpha,\alpha'f)$. Clearly very significant selection rules operate from this resonance, as witness, in the ^{17}O inelastic scattering experiment, the absence of a branch to the 3⁻ collective state at 2.6 MeV in ^{208}Pb.

I have already mentioned the signature of the isovector E2 giant resonance observed in fast neutron capture; this component is so weak as to be scarcely observable in the total n,γ cross-section, or indeed in certain charge-exchange reactions. This is also true in the photon scattering cross-section, but again its signature has been observed in the angular distribution.

Finally, we have heard of the fascinating data described by Evans Hayward, taking us into a new regime of excitation altogether, on the excitation by photons of the Δ nuclear resonance, showing a universal excitation curve for nucleons in nuclei over the entire mass region, differing significantly from that of the free nucleon itself.

5. APPLICATIONS

In these times when scientific activity is increasingly intensive and increasingly specialized it is particularly rewarding to seek the connections of one's own field with other branches of science and technology. Indeed I would go further and say that without these connections, work in a given speciality can become rather sterile. I would also add that in many countries where many expanding scientific disciplines are competing for funds from non-expanding or even decreasing financial resources it is almost imperative to have interest and encouragement for one's endeavours from colleagues in other scientific and technological fields. For these reasons, as well as the sheer interest of the work reported, I was particularly glad to listen to some of the papers in the sessions on nuclear astrophysics, nuclear applications, and related topics that we have had in the last two days.

In astrophysical theory, some of the charged particle cross-sections of light elements are still of key importance, and we have heard of new experimental initiatives to measure these extremely small cross-sections at low energies. The importance of neutron capture cross-sections for establishing the astrophysical mechanisms for creating the heavier elements, especially by the s-process, is well-known. Accurate measurements of these cross-sections have played a crucial role in the development of this subject. What has been particularly well brought-out at this symposium is the importance of our theoretical understanding of capture reaction and nuclear decay mechanisms, thus allowing reasonably confident estimates of capture cross-sections and effective half-lives of excited nuclear states that can be strongly populated at stellar temperatures. These theoretical estimates are making a particularly significant impact in stellar chronometry.

The far reaching connections of β-decay properties of nuclides with other sciences are particularly fascinating. We have heard how observations and theoretical understanding of nuclear reactions, resting especially on (p,γ) studies of isovector M1 transitions and (p,n) excitation of the Gamow-Teller giant resonance, including its low energy fragmentation have led to reliable calculations of the beta-strength distribution in nuclei far from the stability line. These in turn, often in alliance with data and estimates of nuclear capture cross-sections have enabled calculations to be made on effects ranging from the shut-down decay heat of power-reactors, through the electron spectrum for the neutrino oscillation experiment at the ILL reactor, to estimates of the age of the universe and hence to upper limits on the neutrino mass. This connection with the very fundamental properties of particles was neatly rounded out this morning with the two interesting papers on very difficult experiments to detect neutron-anti-neutron oscillations and to measure the beta-dacay of polarised neutrons.

Nuclear power applications of our subject have not been explicitly mentioned at this conference, but some of the work described here has certainly been motivated by the nuclear data requirement of this major technology. Thus the work described by Hoff on the level schemes of the odd-odd actinides can be viewed as a direct application of the theoretical understanding of nuclei gained by basic studies, largely in the gamma-spectroscopy field, to a major improvement of the nuclear data base for power reactor design. These data can not only be used for Hauser-Feshbach type calculations of neutron capture cross-sections, but have also been used according to a poster contribution by Gardner et. al. for calculations of meta-stable states arising from nuclear transmutations. The value of such calculations is emphasized by the fact that the experimental measurement of these important isomer ratios as a function of neutron energy is an extremely difficult experimental task.

It seems to me that few people outside of neutron research circles such as ours are aware of the extent to which neutrons are now becoming used in non-nuclear industrial applications. The last two talks of the symposium gave a survey of how neutrons are capable of great sensitivity in a host of analytical applications, or alternatively, of how their penetrating power makes them capable of bulk analysis, in increasingly flexible ways, in quite remarkable industrial situations such as the oil-well bore-hole logging application described today. One general comment I would like to make is that as industries of quite different kinds become gradually aware of the capabilities of nuclear techniques, they take them up apparently unworried in general about the nuclear apprehensions borne by the general public. I believe therefore that it is vital for all of us to give maximum support and aid to these industrial developments, because they may well prove to be the back-door route to gaining public acceptability to nuclear power, and hence vindicate much of our own claims for public support in carrying out research in this still fascinating field of nuclear structure physics.

Finally, in closing my talk, I would like to pay my personal tribute to Dr. Raman, the Symposium Chairman, and his team for the excellent organization of this meeting. Its success could not have been achieved without their great devotion and effort.

LIST OF PARTICIPANTS

Kees Abrahams
Energieonderzoek Centrum Nederland
Postbus 1
1755 ZG Petten
THE NETHERLANDS

Joerge Aichelin
University of Tennessee
Physics Department
Knoxville, TN 37996-1200
UNITED STATES OF AMERICA

Barry J. Allen
Lucas Heights Research Lab
Private Mail Bag
Sutherland, NSW 2232
AUSTRALIA

Robert A. August, Jr.
Duke University
Department of Physics
Durham, NC 27706
UNITED STATES OF AMERICA

Cyrus Baktash
Oak Ridge National Laboratory
Building 6000
Oak Ridge, TN 37831
UNITED STATES OF AMERICA

A. Baha Balantekin
Oak Ridge National Laboratory
Buiding 6003
Oak Ridge, TN 37831
UNITED STATES OF AMERICA

Milla Baldo-Ceolin
Instituto Di Fisica G. Galilei
via Marzolo 8
Padova 35100
ITALY

Ron Baldry
EG&G ORTEC
100 Midland Road
Oak Ridge, TN 37830
UNITED STATES OF AMERICA

James Ball
Oak Ridge National Laboratory
Building 6000
Oak Ridge, TN 37831
UNITED STATES OF AMERICA

Frantisek Becvár
Charles University
Faculty of Mathematics and Physics
V Holesovická ch 2
180 00 Prague 8
CZECHOSLOVAKIA

James R. Beene
Oak Ridge National Laboratory
Physics Division
Building 6000
Oak Ridge, TN 37831
UNITED STATES OF AMERICA

Fred E. Bertrand
Oak Ridge National Laboratory
Building 6000
Oak Ridge, TN 37831
UNITED STATES OF AMERICA

George Bertsch
University of Tennessee
Physics Department
Knoxville, TN 37996-1200
UNITED STATES OF AMERICA

Carrol R. Bingham
University of Tennessee
Department of Physics
Knoxville, TN 37996-1200
UNITED STATES OF AMERICA

S. Leslie Blatt
Ohio State University
1302 Kinnear Road
Columbus, OH 43212
UNITED STATES OF AMERICA

Mirjana Bogdanovic
Boris Kidric Institute
P. O. Box 522, Lab. 010
11001 Beograd
YUGOSLAVIA

Giovanni Bonsignori
Ist. di Fisica A. Righi
Dipartimento de Fisica
via Irnerio 46
40 127 Bologna
ITALY

Hans Borner
Institut Laue-Langevin
Avenue des Martyrs
156 X - 38042 Grenoble Cedex
FRANCE

Joe Bradley
TENNELEC, Inc.
P.O. Box 2560
Oak Ridge, TN 37830-2560
UNITED STATES OF AMERICA

Alison M. Bruce
Brookhaven National Laboratory
Physics Department
Upton, NY 11973
UNITED STATES OF AMERICA

Thomas W. Burrows
Brookhaven National Laboratory
National Nuclear Data Center
Building 197D
Upton, NY 11973
UNITED STATES OF AMERICA

Robert F. Carlton
Middle Tennessee State University
Department of Chemistry and Physics
Murfreesboro, TN 37132
UNITED STATES OF AMERICA

Boris Castel
Queen's University
Physics Department
Kingston, Ontario
CANADA K7L 3N6

Richard F. Casten
Brookhaven National Laboratory
Department of Physics
Upton, Long Island, NY 11973
UNITED STATES OF AMERICA

Robert E. Chrien
Brookhaven National Laboratory
Department of Physics
Upton, Long Island
NY 11973
UNITED STATES OF AMERICA

Jolie A. Cizewski
Yale University
Wright Nuclear Structure Lab
272 Whitney Avenue
P.O. Box 6666
New Haven, CT 06511
UNITED STATES OF AMERICA

David D. Clark
Cornell University
Ward Laboratory
Ithaca, NY 14853
UNITED STATES OF AMERICA

Hans-Georg Clerc
Technische Hochschule Darmstadt
6100 Darmstadt
FEDERAL REPUBLIC OF GERMANY

Claudio Coceva
CRE "E. Clementel"
Via G. Mazzini, 2
40138 Bologna
ITALY

Glenn G. Colvin
Institut Laue-Langevin
38042 Grenoble Cedex
FRANCE

John W. T. Dabbs
Oak Ridge National Laboratory
Building 6010
Oak Ridge, TN 37831
UNITED STATES OF AMERICA

Rod Dayton
BICRON
12345 Kingsman Road
Newbury, OH 44065
UNITED STATES OF AMERICA

J. Kirk Dickens
Oak Ridge National Laboratory
Building 6010
Oak Ridge, TN 37831
UNITED STATES OF AMERICA

Frank S. Dietrich
Lawrence Livermore National Lab
P.O. Box 808
Livermore, CA 94550
UNITED STATES OF AMERICA

Timothy R. Donoghue
Ohio State University
Physics Department
Columbus, OH 43210
UNITED STATES OF AMERICA

David H. Dowell
Brookhaven National Laboratory
Building 901A
Upton, NY 11973
UNITED STATES OF AMERICA

Gerard J. Dreiss
American Physical Society
Editorial Office Building
1 Research Road, Box 1000
Ridge, NY 11961
UNITED STATES OF AMERICA

Stamatina Dritsa
N.R.C. "Demokritos"
Physics Division
Aghia Paraskevi
Athens
GREECE

Dirk Dubbers
Universitat Heidelberg
Physikalisches Institut
Philosophenweg 12
D-6900 Heidelberg 1
FEDERAL REPUBLIC OF GERMANY

Joe D. Eddlemon
PULCIR, Inc.
9209 Oak Ridge Highway
Oak Ridge, TN 37830
UNITED STATES OF AMERICA

Till von Egidy
Technische Universitat Munchen
Physik Department
D-8046 Garching bei Muchen
FEDERAL REPUBLIC OF GERMANY

Hiroyasu Ejiri
Osaka University
Faculty of Science
Toyonaka, Osaka, 560
JAPAN

James P. Elliott
University of Sussex
Physics Department
Brighton BN1 9QH
UNITED KINGDOM

Yurdanur A. Ellis-Akovali
Oak Ridge National Laboratory
Building 4500-N
Oak Ridge, TN 37831
UNITED STATES OF AMERICA

Henning Esbensen
University of Tennessee
Physics Department
Knoxville, TN 37996-1200
UNITED STATES OF AMERICA

J. M. Feagin
California State University
Department of Physics
Fullerton, CA 92634
UNITED STATES OF AMERICA

Da Hsuan Feng
National Science Foundation
Physics Division
1800 G. Street N.W.
Washington, DC 20550
UNITED STATES OF AMERICA

David J. S. Findlay
AERE Harwell
Nuclear Physics Division
Building 418
Oxon OX11 ORA
UNITED KINGDOM

Birger Fogelberg
Studsvik Science Research Lab
S-61182 Nykoping
SWEDEN

Bob Fortner
NUCLEAR DATA, Inc.
Golf and Meacham Roads
Schaumburg, IL 60196
UNITED STATES OF AMERICA

Donald G. Gardner
Lawrence Livermore National Lab
L-234
Livermore, CA 94550
UNITED STATES OF AMERICA

Maureen A. Gardner
Lawrence Livermore National Lab
L-234
Livermore, CA 94550
UNITED STATES OF AMERICA

William Gelletly
University of Manchester
Schuster Laboratory
Manchester M13 9PL
UNITED KINGDOM

C. W. Glover
Oak Ridge National Laboratory
Building 6000
Oak Ridge, TN 37831
UNITED STATES OF AMERICA

Cynthia A. Gossett
University of Washington
Nuclear Physics Lab., GL-10
Seattle, WA 98195
UNITED STATES OF AMERICA

James A. Grau
Schlumberger-Doll Research
P.O. Box 307
Ridgefield, CT 06877
UNITED STATES OF AMERICA

Wayne P. Graves
TENNELEC, Inc.
601 Oak Ridge Turnpike
Oak Ridge, TN 37831-2560
UNITED STATES OF AMERICA

David Greaves
Nuclear Physics
c/o Nordita
Blegdamsvej 17
2100 Copenhagen
DENMARK

Geoffrey L. Greene
National Bureau of Standards
Room A141, Bldg. 221
Gaithersburg, MD 20878
UNITED STATES OF AMERICA

Keith A. Griffioen
Stanford University
High Energy Physics Lab.
W. W. Hansen Laboratories of Phys
Stanford, CA 94305
UNITED STATES OF AMERICA

Klaus Grotz
Max Planck Institut fur Kernphysik
Postfach 10 39 80
D-6900 Heidelberg 1
FEDERAL REPUBLIC OF GERMANY

Melvyn L. Halbert
Oak Ridge National Laboratory
Building 6000
Oak Ridge, TN 37831
UNITED STATES OF AMERICA

Dennis Hamilton
University of Sussex
Physics Department
Brighton BN1 9QH
UNITED KINGDOM

J. A. Harvey
Oak Ridge National Laboratory
P.O. Box X, Bldg. 6010
Oak Ridge, TN 37831
UNITED STATES OF AMERICA

Hans J. Hauser
Universitat Tubingen
Physikalisches Institut
D-7400 Tubingen
FEDERAL REPUBLIC OF GERMANY

Hershel J. Hausman
Ohio State University
Van de Graaff Laboratory
1302 Kinnear Road
Columbus, OH 43212
UNITED STATES OF AMERICA

Evans Hayward
National Bureau of Standards
Physics Division
Gaithersburg, MD 20899
UNITED STATES OF AMERICA

Robert L. Hershberger
University of Kentucky
Department of Physics
Lexington, KY 40506
UNITED STATES OF AMERICA

Nat W. Hill
Oak Ridge National Laboratory
Building 6010
Oak Ridge, TN 37831
UNITED STATES OF AMERICA

Yu-Kun Ho
Zhengzhou University
Physics Department
Honan Province
PEOPLE'S REPUBLIC OF CHINA

Richard W. Hoff
Lawrence Livermore Laboratory
P.O. Box 808
Livermore, CA 94550
UNITED STATES OF AMERICA

Christoph Hofmeyr
NUCOR
Private Bag X256
Pretoria 0001
SOUTH AFRICA

Daniel J. Horen
Oak Ridge National Laboratory
Building 6010
Oak Ridge, TN 37831
UNITED STATES OF AMERICA

Friedrich Hoyler
Institute Laue-Langevin
BP 156X
F-38042 Grenoble Cedex
FRANCE

Francesco Iachello
Yale University
A. W. Wright Nuclear Structure Lab
P.O. Box 6666
272 Whitney Avenue
New Haven, CT 06511
UNITED STATES OF AMERICA

Masayuki Igashira
Tokyo Institute of Technology
Research Lab. for Nuclear Reactors
2-12-1 O-Okayama, Meguro-ku,
Tokyo 152
JAPAN

Ashok K. Jain
University of Roorkee
Department of Physics
Roorkee- 247667
INDIA

Cleland H. Johnson
Oak Ridge National Laboratory
P.O. Box X, Bldg. 6010
Oak Ridge, TN 37831
UNITED STATES OF AMERICA

Noah R. Johnson
Oak Ridge National Laboratory
P.O. Box X, Bldg. 6000
Oak Ridge, TN 37831
UNITED STATES OF AMERICA

Serge C. Joly
Centre D'Etudes de Bruyeres le
   Chatel
BP n°12
91680 Bruyeres-Le-Chatel
FRANCE

Erik Kaerts
SCK-CEN
Boeretang 200
B-2400 Mol
BELGIUM

Sylvian Kahane
Nuclear Research Centre-Negev
Physics Department
P. O. Box 9001
Beer-Sheva 84-190
ISRAEL

Franz Kappeler
Kernforschungszentrum Karlsruhe
Postfach 3640
D 7500 Karlsruhe 1
FEDERAL REPUBLIC OF GERMANY

Juhani Keinonen
University of Helsinki
Department of Physics
Siltavuorenpenger 20 D
SF-00170 Helsinki 17
FINLAND

Jean Kern
University of Fribourg
Physics Department
CH-1700 Fribourg
SWITZERLAND

Bernard B. Kinsey
102 Skyline Drive
Austin, TX 78746
UNITED STATES OF AMERICA

Hideo Kitazawa
Tokyo Institute of Technology
Research Lab. for Nuclear Reactors
2-12-1 O-Okayama, Megro-ku,
Tokyo 152
JAPAN

Hans V. Klapdor
Max Planck Institut fur Kernphysik
Postfach 10 39 80
D-6900 Heidelberg 1
FEDERAL REPUBLIC OF GERMANY

Rainer W. Kohler
Central Bureau for Nucl. Meas
Steenweg naar Retie
B-2440 Geel
BELGIUM

Jiri Kopecky
Energy Research Foundation
P. O. Box 1
1755 ZG Petten
THE NETHERLANDS

Raymond L. Kozub
Tennessee Tech. University
Department of Physics
Cookeville, TN 38505
UNITED STATES OF AMERICA

Chris E. Laird
Eastern Kentucky University
351 Moore Building
Richmond, KY 40475
UNITED STATES OF AMERICA

Henry Lancman
Brooklyn College
Physics Dept.
Brooklyn, NY 11210
UNITED STATES OF AMERICA

James L. Langenbrunner
TUNL Duke University
Physics Department
Durhams, NC 27707
UNITED STATES OF AMERICA

Arlene J. Larabee
University of Tennessee
Physics Department
Knoxville, TN 37996-1200
UNITED STATES OF AMERICA

Duane C. Larson
Oak Ridge National Laboratory
Building 6010
Oak Ridge, TN 37831
UNITED STATES OF AMERICA

Georg A. Leander
ORNL/UNISOR
Building 6003
Oak Ridge, TN 37831
UNITED STATES OF AMERICA

Ronald A. Lewandowski
HARSHAW/FILTROL
6801 Cochran Road
Solon, OH 44139 USA

Anund Lindholm
Tandem Accelerator Lab.
Box 533
75121 Uppsala
SWEDEN

Richard M. Lindstrom
National Bureau of Standards
Analytical Chemistry Division
Gaithersburg, MD 20899
UNITED STATES OF AMERICA

M. Aslam Lone
Chalk River Nuclear Laboratories
Chalk River
Ontario
CANADA K0J 1J0

J. E. Lynn
AERE Harwell
Oxfordshire
OX11 ORA
UNITED KINGDOM

Refaat M. A. Maayouf
Atomic Energy Authority
Nuclear Research Centre
Reactor & Neutron Physics
Cairo
EGYPT

Henryk A. Mach
Brookhaven National Laboratory
510A, Physics Dept.
Upton, NY 11973
UNITED STATES OF AMERICA

Richard L. Macklin
Oak Ridge National Laboratory
Building 6010
Oak Ridge, TN 37831
UNITED STATES OF AMERICA

Grant J. Mathews
Lawrence Livermore Laboratory
Mail Code L-405
P.O. Box 808
Livermore, CA 94550
UNITED STATES OF AMERICA

Robert D. McElroy
Cornell University
Ward Laboratory
Ithaca, NY 14853
UNITED STATES OF AMERICA

Francis K. McGowan
Oak Ridge National Laboratory
P. O. Box X
Oak Ridge, TN 37831
UNITED STATES OF AMERICA

Gary E. Mitchell
North Carolina State University
Physics Department
Raleigh, NC 27695-8202
UNITED STATES OF AMERICA

Ronald L. Mlekodaj
Oak Ridge National Laboratory
Building 6008
Oak Ridge, TN 37831
UNITED STATES OF AMERICA

Gabor Molnar
Institute of Isotopes
P.O. Box 77
H-1525 Budapest
HUNGARY

Terry C. Moore
HARSHAW/FILTROL
6801 Cochran Road
Solon, OH 44139
UNITED STATES OF AMERICA

Brian R. More
Institut Laue-Langevin
BP 156X
I38042 Grenoble-Cedex
FRANCE

Henry T. Motz
Los Alamos National Lab.
MS A103
Los Alamos, NM 87545
UNITED STATES OF AMERICA

Said F. Mughabghab
Brookhaven National Laboratory
National Nuclear Data Center
Upton, Long Island, NY 11973
UNITED STATES OF AMERICA

Alan. M. Nathan
University of Illinois
23 Stadium Drive
Champaign, ILL 61820
UNITED STATES OF AMERICA

Leif R. Nilsson
Tandem Accelerator Laboratory
Box 533
S-751 21 UPPSALA
SWEDEN

Paul J. Nolan
University of Liverpool
Physics Department
Liverpool L69 3BX
UNITED KINGDOM

Eric B. Norman
Building 88
Lawrence Berkeley Laboratory
1 Cyclotron Road
Berkeley, CA 94720
UNITED STATES OF AMERICA

Marvin Nushan
PULCIR, Inc.
9209 Oak Ridge Highway
Oak Ridge, TN 37830
UNITED STATES OF AMERICA

Masumi Oshima
Oak Ridge National Laboratory
Building 6000
Oak Ridge, TN 37831
UNITED STATES OF AMERICA

Takaharu Otsuka
Los Alamos National Laboratory
P. O. Box 1663
T-5, MS B283
Los Alamos, NM 87545
UNITED STATES OF AMERICA

Vladimir Paar
University of Zagreb
Physics Department
Marulicev TRG 19/I
41000 Zagreb
YUGOSLAVIA

Jose C. Palathingal
University of Puerto Rico
Physics Department
Mayaguez, P.R. 00708
PUERTO RICO

Themis Paradellis
NRC Demokritos
Tandem Accel. Lab.
15341 Aghia Paraskevi
GREECE

Philip A. Parkhurst
Bicron Corporation
12345 Kinsman Rd.
Newbury, OH 44065
UNITED STATES OF AMERICA

Claire Perey
Oak Ridge National Laboratory
Building 6010
Oak Ridge, TN 37831
UNITED STATES OF AMERICA

Francis G. Perey
Oak Ridge National Laboratory
Building 6010
Oak Ridge, TN 37831
UNITED STATES OF AMERICA

Fred Petrovich
Florida State University
Department of Physics
Tallahassee, FL 32306
UNITED STATES OF AMERICA

Robert Piepenbring
Institut des Sciences Nucleaires
53 Avenue des Martyrs
38026 Grenoble
FRANCE

Judy Potter
EG&G ORTEC
100 Midland Road
Oak Ridge, TN 37830
UNITED STATES OF AMERICA

S. (Ram) Raman
Oak Ridge National Laboratory
Building 6010
Oak Ridge, TN 37831
UNITED STATES OF AMERICA

Wojciech Ratynski
Institute for Nuclear Studies
Nuclear Spectroscopy Department
05-400 Otwock - Swierk
POLAND

Gianni Reffo
ENEA "E-Clementel"
via Mazzini 2
Bologna
ITALY

Ken Renner
EG&G ORTEC
100 Midland Road
Oak Ridge, TN 37830
UNITED STATES OF AMERICA

Bill Richards
BICRON
12345 Kinsman Road
Newbury, OH 44065
UNITED STATES OF AMERICA

Ludger Ricken
Ruhr-Universitat Bochum
Institut fur Experimentalphysik I
Postfach 10 21 48
D-4630 Bochum
FEDERAL REPUBLIC OF GERMANY

John C. Riley
Duke University
Physics Department
Durham, NC 27706
UNITED STATES OF AMERICA

N. Russell Roberson
Duke University
Physics Department
Durham, NC 27706
UNITED STATES OF AMERICA

Russell L. Robinson
Oak Ridge National Laboratory
Building 6000
Oak Ridge, TN 37831
UNITED STATES OF AMERICA

William S. Rodney
National Science Foundation
Physics Division Room 341
1800 G. Street N.W.
Washington, DC 20550
UNITED STATES OF AMERICA

Claus Rolfs
University Munster
Institute fur Kernphysik
Domagkstrasse 71
4400 Munster
FEDERAL REPUBLIC OF GERMANY

Bradley A. Roscoe
Schlumberger Well Services
P. O. Box 4594
Houston, TX 77210-4594
UNITED STATES OF AMERICA

Csaba Rozsa
HARSHAW-FILTROL
6801 Cochran Road
Solon, OH 44139
UNITED STATES OF AMERICA

Peter Ryge
Science Applications, Inc.
1257 Tasman Drive
Sunnyvale, CA 94089
UNITED STATES OF AMERICA

Andrew M. Sandorfi
Brookhaven National Laboratory
Building 901A
Upton, NY 11973
UNITED STATES OF AMERICA

Karl-Peter Schelhaas
Max-Planck-Institute for Chemistry
Saarstrasse-23
D-6500 Mainz
FEDERAL REPUBLIC OF GERMANY

Gregory K. Schenter
Cornell University
Ward Laboratory
Ithaca, NY 14853
UNITED STATES OF AMERICA

Petra Schmalbrock
Ohio State University
Van de Graaf Laboratory
1302 Kinnear Road
Columbus, OH 43212
UNITED STATES OF AMERICA

Hans H. Schmidt
Technischen Universitat Munchen
Fakultat fur Physik
James-Franck-Strabe
D-8046 Garching bei Munchen
FEDERAL REPUBLIC OF GERMANY

Marcel R. Schmorak
Oak Ridge National Laboratory
P. O. Box X
Building 6000
Oak Ridge, TN 37831
UNITED STATES OF AMERICA

Olaf Scholten
Michigan State University
Cyclotron Laboratory
East Lansing, MI 48824
UNITED STATES OF AMERICA

Otto W. B. Schult
Kernforschungsanlage Julich
Institut fur Kernphysik
Postfach 1913
D-5170 Julich
FEDERAL REPUBLIC OF GERMANY

Martin H. W. Schumacher
Universitaet Goettingen
II. Physikalisches Inst.
Bunsenstrasse 7-9, D-3400
  Goettingen
FEDERAL REPUBLIC OF GERMANY

Jeffrey S. Schweitzer
Schlumberger-Doll Research
P.O. Box 307
Ridgefield, CT 06877
UNITED STATES OF AMERICA

Paul B. Semmes
Georgia Institute of Technology
School of Chemistry
Atlanta, GA 30332
UNITED STATES OF AMERICA

Hellmut Seyfarth
Kernforschungsanlage Julich
Institut fur Kernphysik
P.O.B. 1913
D-5170 Julich
FEDERAL REPUBLIC OF GERMANY

Richard G. Seyler
Ohio State University
Physics Department
174 W. 18th Ave.
Columbus, OH 43210
UNITED STATES OF AMERICA

Oded Shahal
Nuclear Research Centre-Negev
P.O. Box 9001
Beer-sheva
ISRAEL

Norman K. Sherman
National Research Council of Canada
X-25, M-35
Division of Physics (XNR)
Ottawa, Ontario
CANADA K1A0R6

Zong Ren Shi
Institute of Atomic Energy
P.O. Box 275, Beijing
PEOPLE'S REPUBLIC OF CHINA

Michio Shimizu
Tokyo Institute of Technology
Research Lab. for Nuclear Reactors
2-12-1 O-okayama, Megro-ku,
Tokyo 152
JAPAN

Philip B. Smith
Lab. voor Algemene Natuurkunde
University of Groningen
Westersingel 34
9718 DB Groningen
THE NETHERLANDS

Arthur H. Snell
Oak Ridge National Laboratory
Building 6003
Oak Ridge, TN 37831
UNITED STATES OF AMERICA

Kurt A. Snover
University of Washington
Department of Physics
Seattle, WA 98195
UNITED STATES OF AMERICA

Prakash C. Sood
Banaras Hindu University
Physics Department
Varanasi 221005
INDIA

Eugene H. Spejewski
Oak Ridge National Laboratory
Building 6008
Oak Ridge, TN 37831
UNITED STATES OF AMERICA

Robert R. Spencer
Oak Ridge National Laboratory
Building 6010
Oak Ridge, TN 37831
UNITED STATES OF AMERICA

Adriaan M. J. Spits
SCK/CEN Mol BR1
Boeretang 200
B-2400 MOL
BELGIUM

Finn Stecher-Rasmussen
Energy Research Foundation
E.C.N. Postbus 1
NL-1755 ZG Petten
THE NETHERLANDS

Mario Stefanon
ENEA, C.R.E. "E.Clementel"
via Mazzini-2
40138 Bologna
ITALY

Paul H. Stelson
Oak Ridge National Laboratory
P.O. Box X, Bldg. 6000
Oak Ridge, TN 37831
UNITED STATES OF AMERICA

Roger K. Stevens
THE NUCLEUS
461 Laboratory Road
Oak Ridge, TN 37830
UNITED STATES OF AMERICA

Marija P. Stojanovic
Boris Kidric Institute
Laboratory of Physics
P. O. Box 522
11001 Beogard
YUGOSLAVIA

Hongzhou Sun
Drexel University
Department of Physics
Philadelphia, PA 19104
UNITED STATES OF AMERICA

Janos Sziklai
Central Research Institute
P.O. Box 49
Budapest 114
HUNGARY H-1525

D. Ronald Tilley
North Carolina State University
Department of Physics
Raleigh, NC 27695
UNITED STATES OF AMERICA

Kenneth S. Toth
Oak Ridge National Laboratory
Physics Division
Building 6000
Oak Ridge, TN 37831
UNITED STATES OF AMERICA

James S. Tsang
Schlumberger-Doll Research
P.O. Box 307
Ridgefield, CT 06877
UNITED STATES OF AMERICA

Worasit Uchai
Oak Ridge National Laboratory
Building 6010
Oak Ridge, TN 37831
UNITED STATES OF AMERICA

Pieter H. M. Van Assche
Nuclear Energy Centre
SCK/CEN
B-2400 MOL
BELGIUM

Karl A. Van Bibber
Stanford University
Department of Physics
Stanford, CA 94305
UNITED STATES OF AMERICA

Cor van der Leun
Rijksuniversiteit Utrecht
Fysisch Laboratorium
Princetonplein 5
Postbus 80.000
3508 TA Utrecht
THE NETHERLANDS

C. Randy Vane
Oak Ridge National Laboratory
Building 6010
Oak Ridge, Tennessee 37831
UNITED STATES OF AMERICA

Fons Van Hees
Duke University
Physics Department
Durham, NC 27706
UNITED STATES OF AMERICA

George Vourvopoulos
Department of Physics
Western Kentucky University
Bowling Green, KY 42101
UNITED STATES OF AMERICA

Douglas J. Wagenaar
TUNL Duke University
Physics Department
Durham, NC 27706
UNITED STATES OF AMERICA

Bob Ward
HARSHAW-FILTROL
6801 Cochran Road
Solon, OH 44139
UNITED STATES OF AMERICA

David D. Warner
Brookhaven National Laboratory
Physics Department
Upton, NY 11973
UNITED STATES OF AMERICA

Oren A. Wasson
National Bureau of Standards
Physics Division
Gaithersburg, MD 20899
UNITED STATES OF AMERICA

Bernard W. Wehring
North Carolina State University
Box 7909
Nuclear Engineering
Raleigh, NC 27695
UNITED STATES OF AMERICA

Jesse L. Weil
University of Kentucky
Department of Physics
Lexington, KY 40506
UNITED STATES OF AMERICA

Henry R. Weller
Duke University
Physics Department
Durham, NC 27706
UNITED STATES OF AMERICA

Steve A. Wender
Los Alamos National Laboratory
Physics Division
Los Alamos, NM 87545
UNITED STATES OF AMERICA

Robert M. Whitton
Duke University
Department of Physics
Durham, NC 27706
UNITED STATES OF AMERICA

Michael C. F. Wiescher
Universitat of Mainz
Institut fur Kernchemie
Postfach 3980
D-6500 Mainz
FEDERAL REPUBLIC OF GERMANY

Bryan H. Wildenthal
Drexel University
Department of Physics
Philadelphia, PA 19104
UNITED STATES OF AMERICA

Harold Winslett
EG&G ORTEC
100 Midland Road
Oak Ridge, TN 37830
UNITED STATES OF AMERICA

Ron R. Winters
Denison University
Department of Physics
Granville, OH 43023
UNITED STATES OF AMERICA

Alexander Wolf
Nucl. Res. Centre Negev
Physics Dept.
P. O. Box 9001
Beer-Sheva 84-190
ISRAEL

Cheuk Yin Wong
Oak Ridge National Laboratory
Building 6003
Oak Ridge, TN 37831
UNITED STATES OF AMERICA

John L. Wood
Georgia Institute of Technology
School of Physics
Atlanta, GA 30332
UNITED STATES OF AMERICA

Steven W. Yates
University of Kentucky
Chemistry Department
Lexington, KY 40506
UNITED STATES OF AMERICA

Glenn R. Young
Oak Ridge National Laboratory
Building 6003
Oak Ridge, TN 37831
UNITED STATES OF AMERICA

Phillip G. Young
Los Alamos National Laboratory
Mail Stop B243
Los Alamos, NM 87545
UNITED STATES OF AMERICA

Hong-Xin Zhang
Brooklyn College of CUNY
Physics Deparament
Brooklyn, NY 11210
UNITED STATES OF AMERICA

Bill Zimmer
EG&G ORTEC
100 Midland Road
Oak Ridge, TN 37830
UNITED STATES OF AMERICA

Martin Zirnbauer
California Institute of Technology
Physics Department
Pasadena, CA 91109
UNITED STATES OF AMERICA

AUTHOR INDEX

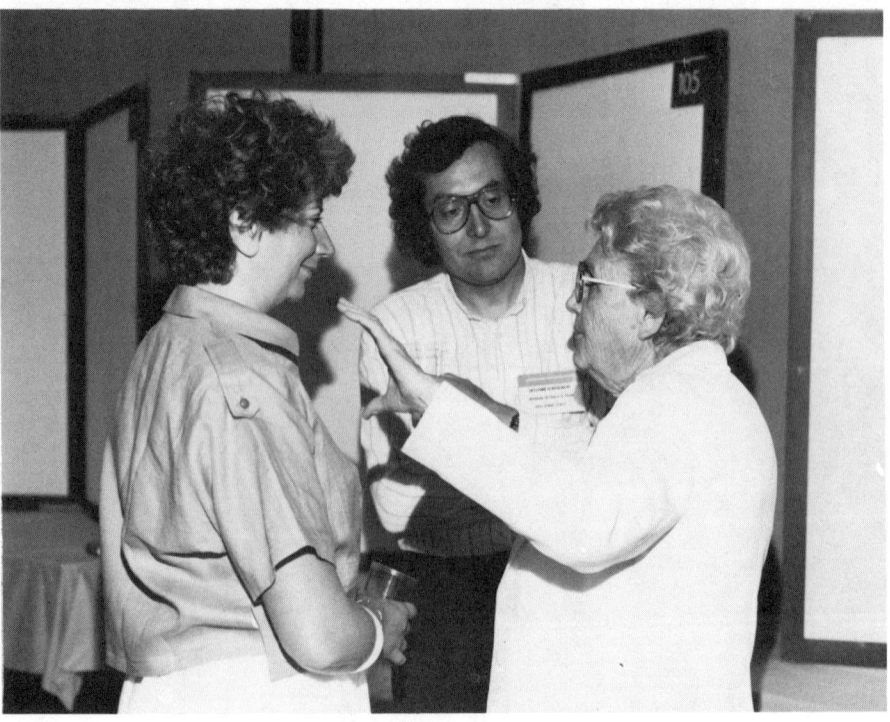

AUTHOR INDEX

(Speakers are identified through bold type)

A
Abd el-Kariem, S. 701
Abid, H. E. 545
Abrahams, K. 930
Abramowicz, H. 893
Ahmed, S. T. 414
Al-Amili, M. A. 545
Alexeev, V. L. 916
Al-Janabi, T. J. 414
Allaart, K. 113
Allen, B. J. 501, 505, 509, 820
Al-Obeidi, S. R. 545
Anderson, D. L. 810
Antilla, A. 851
Antkiw, S. 799
Arthur, E. D. 530
Auchampaugh, G. F. 483

B
Babkair, S. 709
Baldo-Ceolin, M. 871
Barreau, G. 378, 386, 423
Becvár, F. 345
Beene, Jr. 623
Beer, H. 778
Behkami, A. N. 305
Belgya, T. 534
Berant, Z. 213, 232
Bertrand, F. 623
Birenbaum, Y. 208, 213, 232
Blagojevic, N. 820
Blatt, S. L. 570
Bogdanović, M. 382, 386
Boisson, J. 274
Bonsignori, G. 113
Bopp, P. 881
Börner, H. G. 262, 305, 378, 382, 386, 403, 406, 410, 416, 423, 427, 921
von Brentano, P. 423, 427
Brissot, R. 378, 386, 423
Browne, C. P. 785
Bruce, A. M. 431, 435

C
Calarco, J. R. 857
Cameron, J. A. 688
Carlton, R. F. 774
Casten, R. F. 24, 406, 423, 435
Chalupka, A. 386
Clark, D. D. 912
Clerc, H.-G. 636
Colvin, G. G. 290, 382, 390, 403, 406, 410, 416, 420, 427
Company, F. Z. 501, 505, 509
Countryman, P. J. 857

D
Davidson, W. F. 221, 403
Decowski, P. 684
Delbianco, W. 221
Dennis, L. C. 495
Deslattes, R. D. 921
Dietrich, F. S. 445
Ding, Dazhao 376
Diószegi, I. 542
Donoghue, T. R. 785
Doroba, K. 893
Dousse, J.-Cl. 698
Dowell, D. H. 597, 705
Dubbers, D. 881

E
Edwards, G. 237
von Egidy, T. 305, 386, 406
Ejiri, H. 582
Elliott, J. P. 3, 106
Ermakov, O. N. 897

F
Faust, H. 378
Fazekas, B. 534
Feng, Da Hsuan 48
Findlay, D. J. S. 237
Flynn, D. 695
Fogelberg, B. 493
Förster, I. 423, 427
Franklyn, C. 378
Freedman, S. J. 881

G
Gabbard, F. 692, 695
Gabelmann, H. 908
Gácsi, Z. 539
Gardner, D. G. 513, 547

Gardner, M. A. 513, 547
Gaskin, K. 820
Gasser, M. 698
Gelberg, A. 423
Gelletly, W. 420, 709
Gill, R. L. 912
Górski, M. 893
Gozani, T. 828
Grasshoff, P. 701
Grau, J. A. 799
Greene, G. L. 921
Greenwood, R. C. 403
Griffoen, K. A. 857
Grotz, K. 793
Guidotti, R. 225
Guo, Taichang 376

H

Halbert, M. L. 623
Hamilton, W. D. 106
Hanna, S. S. 217
Haque, A. M. I. 423, 427
Harvey, J. A. 493, 774
Hauser, H. J. 701
Hausman, H. J. 785
Hawkes, N. P. 237
Hayward, Evans 131
Hessberger, F. 636
Hershberger, R. L. 692, 695
Hertzog, R. C. 799
Hill, N. W. 774
Hillebrandt, W. 908
Hlawtsch, G. 305
Ho, Yu-kun 362, 551, 904
Hoff, R. W. 274, 513, 547
Hofmeyr, C. 378
Hornig, L. 881
Howard, W. M. 766
Howman-Giles, R. 820
Hoyler, F. 382, 406, 410, 416, 701
Hungerford, P. 305, 386, 406

I

Iachello, F. 153
Igashira, M. 517, 520, 523
Irbäck, A. 106

J

Jain, A. K. 117
Jain, K. 117
Jasinski, A. 893
Johns, M. W. 889

Joly, S. 526
Jurney, E. T. 539

K

Kaerts, E. 416, 918
Kahane, S. 208, 213, 232
Kailas, S. 789
Kajrys, G. 221
Kamikubota, N. 582
Kamoon, S. S. 414
Kane, W. 386, 406
Käppeler, F. 715
Karpikhin, J. L. 897
Keinonen, J. 557
Keller, J. -G. 636
Kellogg, S. E. 753
Kern, J. 274, 698
Kerr, S. A. 305, 382, 386, 406, 416, 423, 427
Kessler, E. G., Jr. 921
Khanna, F. C. 904
Khitrov, V. A. 396, 399
Kicińska-Habior, M. 680, 684
Kikstra, S. W. 217
Kinsey, B. B. 939
Kishimoto, T. 582
Kiss, A. Z. 851
Kitazawa, H. 517, 520, 523
Klapdor, H. V., 701, **732**, 793
Klemt, E. 881
Knat'ko, V. A. 439
Knöpfle, K. T. 857
Kok, P. J. 897
Koltay, E. 851
Komano, H. 517, 520, 523
Kondurov, I. A. 386
Kopecky, J. 318
Körber, A. 701
Kozlowski, T. 893
Kratz, K.-L. 908
Krupchitsky, P. A. 897
Krusche, B. 305

L

Laird, C. E. 692, 695
Lancman, H. 241
Last, J. 881
Leander, G. A. 125
Leist, B. 908
Leitner, W. 701
van der Leun, C. 217
Leushkin, E. K. 916
Li, Guohua 376

Lieb, K. P. 305
Lindstrom, R. M. 810
Lister, C. J. 709
Liu, Jing-Feng 551
Lobov, G. A. 897, 900
Loginov, Yu. E. 386
Lone, M. A. 362, 551, 904
Lynn, J. E. 951

M
Maayouf, R. M. A. 925
Mach, H. 889
Macklin, R. L. 774, 778
Maino, G. 228
Makarious, A. S. 925
Manente, R. A. 799, 824
Manfredi, V. R. 103
Maráczy, Cs. 542
Martynov, V. V. 386
Mathews, G. J. 766
Matulewicz, T. 680, 684
McElroy, R. D. 912
Metzinger, J. 793
Mheemeed, A. K. 414
Mizumoto, M. 493
Molkanov, L. I. 916
Molnár, G. 534
Moreh, R. 179, 208, 213, 232
Moscrop, R. 709
Münzenberg, G. 636

N
Nagai, Y. 582
Nathan, A. M. 142
Nilsson, L. 458
Nolan, P. J. 612
Norman, E. B. 753
Nyakó, B. 851

O
Ohm, H. 908
Ohsumi, H. 582
Otsuka, T. 10

P
Paar, V. 70
Palathingal, J. C. 847
Paradellis, T. 677
Paul, P. 705
Perepelitsa, V. F. 897
Perny, B. 698
Piepenbring, R. 38, 274

Piotrowski, A. 912
Popov, Yu P. 396, 399
Price, H. G. 709

Q
Quint, B. 636

R
Räisänen, J. 851
Raman, S. 221, 493, 495, 539
Rapp, V. 701
Rascher, R. 423, 427
Ratyński, W. 893
Redder, A. 748
Reffo, G. 782
Reich, C. W. 403
Reisdorf, W. 636
Rhême, Ch. 698
Richter, R. 423, 427
Ricken, L. 705
Rohwer, T. 701
Rolfs, C. 748, 785
Rollefson, A. A. 785
Rowley, D. 857
Rumiantsev, V. L. 916

S
Sa, J. 539
Sahm, C.-C. 636
Sandorfi, A. M. 647, 705
Saporetti, F. 225
Satoh, T. 582
Savoia, M. 113
Schärpf, Otto 881
Scheerer, H. J. 305, 386, 406
Schmalbrock, P. 785
Schmidt, H. H. 305, 386, 406
Schmidt, K.-H. 636
Scholten, O. 109
Schreckenbach, K. 290, 305, 378, 382, 386, 390, 403, 406, 410, 416, 420, 423, 427
Schumacher, M. 166
Schütze, H. 881
Schwab, W. 636
Schweitzer, J. S. 799, 824
Sené, M. R. 237
Seyfarth, H. 382, 386
Shahal, O. 179
Sherman, N. K. 221
Shi, Zongren 376, 435
Shibata, T. 582

Shimizu, M. 517, 520, 523
Sikora, B. 684
Simić, J. 390
Simon, R. S. 636
Smith, Ph. B. 192
Snover, K. A. 660
Soloviev, V. G. 499
Somorjai, E. 851
Sood, P. C. 121
Soutter, V. 820
Spits, A. M. J. 403
Stachel, J. 435
Staudt, G. 701
Stecher-Rasmussen, F. 897, 930, 933
Stefanon, M. 335
Steinmüller, B. 908
Stojanović, M. P. 390
Stoyanov, Ch. 499
Sukhovoj, A. M. 396, 399
Sushkov, P. A. 386
Szeptycka, M. 893
Sziklai, J. 688
Szöghy, I. M. 688
Szymczak, M. 893

T
Takahashi, K. 766
Thielemann, F.-K. 908
Töke, J. 684
Trautvetter, H. P. 748
Tschöp, E. 636
Tucholski, A. 893

V
Van Assche, P. H. M. 403, 416, 918
Van Bibber, K. 857
Van der Leun, C. 217

Varley, B. J. 709
Ventura, A. 228
Veres, A. 534, 542
von Brentano, P. 423, 427
von Egidy, T. 305, 386, 406
Voronov, V. V. 499

W
Walczak, R. 893
Walter, G. 778
Walz, M. 701
Ward, R. A. 766
Warner, D. D. 247, 403, 406, 420, 423, 431
Weel, M. J. W. 930
Weil, J. L. 539
Weinmann, D. 701
Weller, H. R. 470
Wender, S. A. 483
Wiescher, M. 785, 908
Wijekumar, V. 785
Wildenthal, B. H. 89
Wolf, A. 213, 232
Woodworth, J. G. 857

Y
Yamamuro, N. 523
Yazvitsky, Yu. S. 396, 399
Yearian, M. R. 857
Youhana, H. M. 545
Young, P. G. 530

Z
Zeng. Xiantang 376
Zhang, H. X. 241
Ziegert, W. 908
Zijderhand, F. 217
Zuffi, L. 228

AIP Conference Proceedings

No. 29	Magnetism and Magnetic Materials - 1975 (21st Annual Conference, Philadelphia)	76-10931	0-88318-128-2
No. 30	Particle Searches and Discoveries - 1976 (Vanderbilt Conference)	76-19949	0-88318-129-0
No. 31	Structure and Excitations of Amorphous Solids (Williamsburg, VA., 1976)	76-22279	0-88318-130-4
No. 32	Materials Technology - 1976 (APS New York Meeting)	76-27967	0-88318-131-2
No. 33	Meson-Nuclear Physics - 1976 (Carnegie-Mellon Conference)	76-26811	0-88318-132-0
No. 34	Magnetism and Magnetic Materials - 1976 (Joint MMM-Intermag Conference, Pittsburgh)	76-47106	0-88318-133-9
No. 35	High Energy Physics with Polarized Beams and Targets (Argonne, 1976)	76-50181	0-88318-134-7
No. 36	Momentum Wave Functions - 1976 (Indiana University)	77-82145	0-88318-135-5
No. 37	Weak Interaction Physics - 1977 (Indiana University)	77-83344	0-88318-136-3
No. 38	Workshop on New Directions in Mossbauer Spectroscopy (Argonne, 1977)	77-90635	0-88318-137-1
No. 39	Physics Careers, Employment and Education (Penn State, 1977)	77-94053	0-88318-138-X
No. 40	Electrical Transport and Optical Properties of Inhomogeneous Media (Ohio State University, 1977)	78-54319	0-88318-139-8
No. 41	Nucleon-Nucleon Interactions - 1977 (Vancouver)	78-54249	0-88318-140-1
No. 42	Higher Energy Polarized Proton Beams (Ann Arbor, 1977)	78-55682	0-88318-141-X
No. 43	Particles and Fields - 1977 (APS/DPF, Argonne)	78-55683	0-88318-142-8
No. 44	Future Trends in Superconductive Electronics (Charlottesville, 1978)	77-9240	0-88318-143-6
No. 45	New Results in High Energy Physics - 1978 (Vanderbilt Conference)	78-67196	0-88318-144-4
No. 46	Topics in Nonlinear Dynamics (La Jolla Institute)	78-057870	0-88318-145-2
No. 47	Clustering Aspects of Nuclear Structure and Nuclear Reactions (Winnepeg, 1978)	78-64942	0-88318-146-0
No. 48	Current Trends in the Theory of Fields (Tallahassee, 1978)	78-72948	0-88318-147-9
No. 49	Cosmic Rays and Particle Physics - 1978 (Bartol Conference)	79-50489	0-88318-148-7
No. 50	Laser-Solid Interactions and Laser Processing - 1978 (Boston)	79-51564	0-88318-149-5
No. 51	High Energy Physics with Polarized Beams and Polarized Targets (Argonne, 1978)	79-64565	0-88318-150-9
No. 52	Long-Distance Neutrino Detection - 1978 (C.L. Cowan Memorial Symposium)	79-52078	0-88318-151-7
No. 53	Modulated Structures - 1979 (Kailua Kona, Hawaii)	79-53846	0-88318-152-5

AIP Conference Proceedings

No.	Title		
No. 54	Meson-Nuclear Physics - 1979 (Houston)	79-53978	0-88318-153-3
No. 55	Quantum Chromodynamics (La Jolla, 1978)	79-54969	0-88318-154-1
No. 56	Particle Acceleration Mechanisms in Astrophysics (La Jolla, 1979)	79-55844	0-88318-155-X
No. 57	Nonlinear Dynamics and the Beam-Beam Interaction (Brookhaven, 1979)	79-57341	0-88318-156-8
No. 58	Inhomogeneous Superconductors - 1979 (Berkeley Springs, W.V.)	79-57620	0-88318-157-6
No. 59	Particles and Fields - 1979 (APS/DPF Montreal)	80-66631	0-88318-158-4
No. 60	History of the ZGS (Argonne, 1979)	80-67694	0-88318-159-2
No. 61	Aspects of the Kinetics and Dynamics of Surface Reactions (La Jolla Institute, 1979)	80-68004	0-88318-160-6
No. 62	High Energy e^+e^- Interactions (Vanderbilt, 1980)	80-53377	0-88318-161-4
No. 63	Supernovae Spectra (La Jolla, 1980)	80-70019	0-88318-162-2
No. 64	Laboratory EXAFS Facilities - 1980 (Univ. of Washington)	80-70579	0-88318-163-0
No. 65	Optics in Four Dimensions - 1980 (ICO, Ensenada)	80-70771	0-88318-164-9
No. 66	Physics in the Automotive Industry - 1980 (APS/AAPT Topical Conference)	80-70987	0-88318-165-7
No. 67	Experimental Meson Spectroscopy - 1980 (Sixth International Conference, Brookhaven)	80-71123	0-88318-166-5
No. 68	High Energy Physics - 1980 (XX International Conference, Madison)	81-65032	0-88318-167-3
No. 69	Polarization Phenomena in Nuclear Physics - 1980 (Fifth International Symposium, Santa Fe)	81-65107	0-88318-168-1
No. 70	Chemistry and Physics of Coal Utilization - 1980 (APS, Morgantown)	81-65106	0-88318-169-X
No. 71	Group Theory and its Applications in Physics - 1980 (Latin American School of Physics, Mexico City)	81-66132	0-88318-170-3
No. 72	Weak Interactions as a Probe of Unification (Virginia Polytechnic Institute - 1980)	81-67184	0-88318-171-1
No. 73	Tetrahedrally Bonded Amorphous Semiconductors (Carefree, Arizona, 1981)	81-67419	0-88318-172-X
No. 74	Perturbative Quantum Chromodynamics (Tallahassee, 1981)	81-70372	0-88318-173-8
No. 75	Low Energy X-ray Diagnostics - 1981 (Monterey)	81-69841	0-88318-174-6
No. 76	Nonlinear Properties of Internal Waves (La Jolla Institute, 1981)	81-71062	0-88318-175-4
No. 77	Gamma Ray Transients and Related Astrophysical Phenomena (La Jolla Institute, 1981)	81-71543	0-88318-176-2
No. 78	Shock Waves in Condensed Matter - 1981 (Menlo Park)	82-70014	0-88318-177-0
No. 79	Pion Production and Absorption in Nuclei - 1981 (Indiana University Cyclotron Facility)	82-70678	0-88318-178-9
No. 80	Polarized Proton Ion Sources (Ann Arbor, 1981)	82-71025	0-88318-179-7
No. 81	Particles and Fields - 1981: Testing the Standard Model (APS/DPF, Santa Cruz)	82-71156	0-88318-180-0

AIP Conference Proceedings

No. 82 Interpretation of Climate and Photochemical
Models, Ozone and Temperature Measurements
(La Jolla Institute, 1981)   82-071345   0-88318-181-9

No. 83 The Galactic Center
(Cal. Inst. of Tech., 1982)   82-071635   0-88318-182-7

No. 84 Physics in the Steel Industry
(APS.AISI, Lehigh University, 1981)   82-072033   0-88318-183-5

No. 85 Proton-Antiproton Collider Physics - 1981
(Madison, Wisconsin)   82-072241   0-88318-184-3

No. 86 Momentum Wave Functions - 1982
(Adelaide, Australia)   82-072375   0-88318-185-1

No. 87 Physics of High Energy Particle Accelerators
(Fermilab Summer School, 1981)   82-072421   0-88318-186-X

No. 88 Mathematical Methods in Hydrodynamics and
Integrability in Dynamical Systems
(La Jolla Institute, 1981)   82-072462   0-88318-187-8

No. 89 Neutron Scattering - 1981
(Argonne National Laboratory)   82-073094   0-88318-188-6

No. 90 Laser Techniques for Extreme Ultraviolet
Spectroscopy (Boulder, 1982)   82-073205   0-88318-189-4

No. 91 Laser Acceleration of Particles
(Los Alamos, 1982)   82-073361   0-88318-190-8

No. 92 The State of Particle Accelerators and
High Energy Physics (Fermilab, 1981)   82-073861   0-88318-191-6

No. 93 Novel Results in Particle Physics
(Vanderbilt, 1982)   82-073954   0-88318-192-4

No. 94 X-Ray and Atomic Inner-Shell Physics-1982
(International Conference, U. of Oregon)   82-074075   0-88318-193-2

No. 95 High Energy Spin Physics - 1982
(Brookhaven National Laboratory)   83-70154   0-88318-194-0

No. 96 Science Underground
(Los Alamos, 1982)   83-70377   0-88318-195-9

No. 97 The Interaction Between Medium Energy
Nucleons in Nuclei - 1982 (Indiana University)   83-70649   0-88318-196-7

No. 98 Particles and Fields - 1982
(APS/DPF University of Maryland)   83-70807   0-88318-197-5

No. 99 Neutrino Mass and Gauge Structure
of Weak Interactions (Telemark, 1982)   83-71072   0-88318-198-3

No. 100 Excimer Lasers - 1983
(OSA, Lake Tahoe, Nevada)   83-71437   0-88318-199-1

No. 101 Positron-Electron Pairs in Astrophysics
(Goddard Space Flight Center, 1983)   83-71926   0-88318-200-9

No. 102 Intense Medium Energy Sources
of Strangeness (UC-Santa Cruz, 1983)   83-72261   0-88318-201-7

No. 103 Quantum Fluids and Solids - 1983
(Sanibel Island, Florida)   83-72440   0-88318-202-5

No. 104 Physics, Technology and the Nuclear Arms Race
(APS Baltimore - 1983)   83-72533   0-88318-203-3

No. 105 Physics of High Energy Particle Accelerators
(SLAC Summer School, 1982)   83-72986   0-88318-304-8

AIP Conference Proceedings

No. 106	Predictability of Fluid Motions (La Jolla Institute, 1983)	83-73641	0-88318-305-6
No. 107	Physics and Chemistry of Porous Media (Schlumberger-Doll Research, 1983)	83-73640	0-88318-306-4
No. 108	The Time Projection Chamber (TRIUMF, Vancouver, 1983)	83-83445	0-88318-307-2
No. 109	Random Walks and Their Applications in the Physical and Biological Sciences (NBS/La Jolla Institute, 1982)	84-70208	0-88318-308-0
No. 110	Hadron Substructure in Nuclear Physics (Indiana University, 1983)	84-70165	0-88318-309-9
No. 111	Production and Neutralization of Negative Ions and Beams (3rd Int'l Symposium, Brookhaven, 1983)	84-70379	0-88318-310-2
No. 112	Particles and Fields - 1983 (APS/DPF, Blacksburg, VA)	84-70378	0-88318-311-0
No. 113	Experimental Meson Spectroscopy – 1983 (Seventh International Conference, Brookhaven)	84-70910	0-88318-312-9
No. 114	Low Energy Tests of Conservation Laws in Particle Physics (Blacksburg, VA, 1983)	84-71157	0-88318-313-7
No. 115	High Energy Transients in Astrophysics (Santa Cruz, CA, 1983)	84-71205	0-88318-314-5
No. 116	Problems in Unification and Supergravity (La Jolla Institute, 1983)	84-71246	0-88318-315-3
No. 117	Polarized Proton Ion Sources (TRIUMF, Vancouver, 1983)	84-71235	0-88318-316-1
No. 118	Free Electron Generation of Extreme Ultraviolet Coherent Radiation (Brookhaven/OSA, 1983)	84-71539	0-88318-317-X
No. 119	Laser Techniques in the Extreme Ultraviolet (OSA, Boulder, Colorado, 1984)	84-72228	0-88318-318-8
No. 120	Optical Effects in Amorphous Semiconductors (Snowbird, Utah, 1984)	84-72419	0-88318-319-6
No. 121	High Energy e^+e^- Interactions (Vanderbilt, 1984)	84-72632	0-88318-320-X
No. 122	The Physics of VLSI (Xerox, Palo Alto, 1984)	84-72729	0-88318-321-8
No. 123	Intersections Between Particle and Nuclear Physics (Steamboat Springs, 1984)	84-72790	0-88318-322-6
No. 124	Neutron-Nucleus Collisions - A Probe of Nuclear Structure (Burr Oak State Park-1984)	84-73216	0-88318-323-4
No. 125	Capture Gamma-Ray Spectroscopy and Related Topics-1984 (Internat. Symposium, Knoxville)	84-73303	0-88318-324-2